AF517377

MORPHOLOGY AND TAXONOMY OF FUNGI

MORPHOLOGY AND TAXONOMY OF FUNGI

Ernst Athearn Bessey, Ph.D.

Distinguished Professor of Botany, Retired, and Dean Emeritus of the School of Graduate Studies, Michigen State College

1st Reprint in India

Published by:

United Book Prints
4806/24, Ansari Road,
Daryaganj,
New Delhi - 110 002
Tel.: 011-41511055
E-mail: unitedbookprints@gmail.com

Print: 2015

ISBN: 978-93-83692-26-2

Printed in India

PREFACE

In the over fifteen years that have elapsed since the publication of the author's "A Text-book of Mycology,"[1] the subject has experienced many changes. The subject of Medical Mycology has gained greatly in interest and there has arisen a great realization of the importance of those fungi whose antibiotic products have become indispensable in the treatment of many maladies of Man and other animals. It has therefore become apparent that an entirely new work is necessary, not merely as a textbook for students but also to assist the many persons now undertaking research in the various fields of study in which the knowledge of fungi is fundamental to their work. Although in some features the present work follows the framework of the older Text-book it has been entirely rewritten and greatly enlarged.

Aside from the above-mentioned special fields of study, investigations of the microscopic soil and water fungi, both parasitic and saprophytic, which comprise the simplest forms, in structure at least, have revealed hundreds of species and many new genera, families, and even orders. Thus great numbers of fungi are now known where but few were recognized less than two decades ago. This has compelled radical revisions in the classification of the lower fungi.

In the higher fungi, especially the Basidiomyceteae, many new species and genera have been described. Above all, studies in the anatomy and chemical reactions of these new as well as of the old, long known species have shown the artificiality of the old Friesian system of classification. This has led to radical revisions which beyond doubt indicate far more correctly the true relationships and phylogeny of the members of this class. Yet here, there still remains so much to be done that mycologists are not at all in agreement as to the ultimate system of classification to be used.

With so many points still in disagreement, it is impossible for a book of this type to settle these matters beyond dispute. The author, therefore, while taking a middle ground in many disputed areas, has attempted to show those using this book the various points of conflict and opposing

[1] Bessey, Ernst Athearn: A Text-book of Mycology, xv + 495 pp., 139 figs., Philadelphia, Blakiston, 1935.

ideas so that they may judge for themselves what seems to be the more logical. Accordingly numerous references are given at the close of each chapter from which these conflicting ideas may be sought. It is the hope of the author that the recognition of these conflicts may stimulate further study until the points at issue may be settled by a fuller knowledge of the facts.

In this work an attempt has been made to give the user the opportunity to learn the more important characters of almost all the generally recognized orders of fungi, and to have a foundation upon which the more intensive study of any of these groups may be based. So many of these organisms are still known but scantily, so that their true relationships are very doubtful. As examples, the Protomycetales and Eccrinales may be mentioned. Many species of each order have been recognized, but the gaps in knowledge still make their relationships to other fungi uncertain.

Little more than mere mention is made of those fungi that make up the subject of Medical Mycology, as the author does not feel competent to enter far into that field. Furthermore, he feels that a rather broad foundation in the knowledge of fungi in general, such as this book should give the student, is absolutely necessary before reasonable progress can be made in the study of such a specialized subject. Similarly, but little discussion is made of the physiology of fungi, as that subject, too, requires a good knowledge of the fungi themselves as well as of the physiology of green plants and an adequate training in chemistry. The genetics of fungi is scarcely touched upon except where it is necessary to understand somewhat the relationships of certain fungus groups. In other words, this book must not be expected to be a complete encyclopedia of fungi, but rather the foundation upon which to lay the various superstructures of the edifice of Mycology.

So much has been published in the last fifteen years or more in the way of manuals and monographs that the final chapter, "Guide to the Literature for the Identification of Fungi," is already large in size, in spite of the fact that many older references have been omitted where newer ones are readily available. Here, especially, the effects of World War II have been deplorable. Contacts with mycologists in many portions of the world have been broken; continuity of exchange of literature has been interrupted. The mycologic publications in many of the countries where fighting was intense have been destroyed, so that much that was of value cannot even be located. As a consequence, in this chapter some glaring omissions doubtless will become apparent. It is hoped that these will not lessen the value of the book.

As before, the author owes a great deal to the generous coöperation of many botanists in the New World and in portions of Europe. To begin to name them would result in a list too long to be printed here. Furthermore, he expresses his gratitude to the many mycologic friends who have assisted him by granting permission to use illustrations from their publications. To the management of *Mycologia* especial thanks are due for their blanket permission to use illustrations from that indispensable periodical. To M. B. Walton, of Cleveland, Ohio, the author's thanks are due for the generous gift of prints of some of his wonderfully fine photographs of fungi.

To his wife, especially, the author owes thanks for the continued stimulus to persist in the arduous work of preparing this new book, for writing the manuscript, for assistance in reading proof, and for attending to the many other tiring details that go into the making of a book.

ERNST A. BESSEY
Distinguished Professor of Botany, retired

Michigan State College
East Lansing, Michigan
June 1950.

CONTENTS

Preface vii

1. Introduction 1
2. Mycetozoa and Related Organisms 22
3. Phycomyceteae: Chytridiales and Hyphochytriales 42
4. Phycomyceteae: Blastocladiales and Monoblepharidales 78
5. Phycomyceteae: Lagenidiales and Saprolegniales 94
6. Phycomyceteae: Peronosporales and Protomycetales 126
7. Phycomyceteae: Mucorales, Entomophthorales, Zoopagales, Eccrinales 150
8. The Higher Fungi: Carpomyceteae 192
9. Class Ascomyceteae: Laboulbeniales and Discomycetes 200
10. Class Ascomyceteae: The "Pyrenomycetes" 262
11. Class Ascomyceteae: Erysiphales, Aspergillales, Myriangiales, Saccharomycetales 307
12. Class Basidiomyceteae: Subclass Teliosporeae 366
13. Class Basidiomyceteae: Subclass Heterobasidiae 436
14. Class Basidiomyceteae: Subclass Eubasidiae, "Hymenomyceteae" 464
15. Class Basidiomyceteae: Subclass Eubasidiae, "Gasteromyceteae" 530
16. Fungi Imperfecti: The Imperfect Fungi 572
17. The Phylogeny of the Fungi 628

18. GUIDE TO THE LITERATURE FOR THE IDENTIFICATION OF FUNGI.......... 660
List 1. General Works Covering the Whole Field of Systematic Mycology.......... 662
2. Host Indexes, Local Fungus Lists with Host Indexes, Lists of Fungi on Special Hosts or Substrata, Bibliographies, Fungi of Man and Other Animals.......... 668
3. Mycetozoa, Including Myxogastrales, Plasmodiophorales, Acrasiales, Labyrinthulales.......... 672
4. General Works on Phycomyceteae.......... 673
5. Chytridiales.......... 673
6. Hyphochytriales.......... 675
7. Blastocladiales and Monoblepharidales.......... 675
8. Lagenidiales.......... 676
9. Saprolegniales (Including Leptomitales).......... 676
10. Peronosporales, Also Protomycetales.......... 677
11. Mucorales, Entomophthorales.......... 678
12. Zoopagales, Eccrinales.......... 680
13. Ascomyceteae, Miscellaneous.......... 682
14. Laboulbeniales.......... 682
15. Lecanorales and Pyrenulales.......... 683
16. Pezizales Operculati and Inoperculati, Including "Phacidiales".......... 688
17. Tuberales.......... 693
18. Taphrinales.......... 694
19. Hysteriales.......... 695
20. Sphaeriales.......... 695
21. Hypocreales.......... 699
22. Dothideales.......... 700
23. Pseudosphaeriales.......... 701
24. Hemisphaeriales.......... 702
25. Erysiphaceae.......... 703
26. Meliolaceae (Perisporiaceae).......... 704
27. Remainder of Erysiphales: Capnodiaceae, Englerulaceae, Trichothyriaceae, Atichiaceae.......... 705
28. Myriangiaceae.......... 706
29. Aspergillales (Plectascales).......... 707
30. Saccharomycetales and Asporogenous Yeasts.......... 707
31. Ustilaginales (Including Graphiolaceae).......... 709
32. Uredinales.......... 712
33. Heterobasidiae.......... 718
34. "Hymenomyceteae": General Works.......... 720
35. Thelephoraceae and Exobasidiaceae.......... 721

36. Clavariaceae 723
37. Hydnaceae 724
38. Meruliaceae 725
39. Polyporaceae 726
40. Boletaceae 729
41. Agaricaceae (in the Broader Sense) 730
42. Gasteromyceteae—General Works 742
43. Hymenogastrales, Sclerodermatales, etc 743
44. Lycoperdales, including Tulostomataceae and Podaxaceae 744
45. Nidulariales, including Arachniaceae 746
46. Phallales 746
47. Sphaeropsidales 747
48. Melanconiales 750
49. Moniliales: Moniliaceae 750
50. Moniliales: Dematiaceae 753
51. Moniliales: Tuberculariaceae, Stilbellaceae, and Mycelia Sterilia 755

INDEX 757

1

INTRODUCTION

There are many kinds of living organisms whose interactions serve to maintain or to break the balance of Nature. Some of them, green plants, build up organic substances from the inorganic matter at their disposal. They are basic to all life on the earth, for without them almost all other organisms would eventually perish. The animals reach probably the greatest degrees of complexity but they must take their basic organic foods directly or indirectly from the chlorophyll-containing plants. Besides these two great groups of organisms there are the numerous bacteria —minute breakers-down of the complex organic substances built by the green plants and by the animals and also builders-up of some of the inorganic substances that the green plants need. Aside from these there is the great horde of organisms called fungi. Perhaps there are 100,000 species of them, according to Bisby and Ainsworth (1943), but of the 100,000 named fungi probably not over 40,000 are, according to these authors, valid species, leaving 60,000 or more to be recognized and described in the future. It is with the fungi that mycology has to do. Ordinarily they are considered by most botanists to be plants, but Martin (1932) and others who followed him, have suggested that fungi may be neither animals nor plants but a third kingdom, of common origin with them, which has undergone a parallel development to the animal and plant kingdoms.

Fungi

There is no general agreement as to the limits of the forms that should be called fungi. Some of the earlier mycologists included the bacteria with them, but that is rarely done nowadays. The majority of botanists, or to be more specific, of mycologists, include the Slime Molds (Mycetozoa or Myxomycetes) among the fungi. Yet the great German mycologist and plant pathologist Anton de Bary (1831–1888) said (1887), "I have since the year 1858 placed the Myxomycetes under the name Mycetozoa outside the limits of the vegetable kingdom, and I still consider this to be their true position." With this viewpoint the author is in agreement, but since most mycologists accept these organisms as fungi they are included in this textbook.

From the foregoing it is apparent that the fungi do not possess chlorophyll and also that the chlorophyll-less bacteria and Mycetozoa are excluded from their ranks. There are many other organisms which lack chlorophyll and still are not fungi. Thus in the diatomaceous genus *Nitzschia* some species are known that possess no chloroplasts while the majority of species possess them. Otherwise they are so similar that they are retained in the same genus. In the Red Seaweeds (Florideae) there are several species that lack chlorophyll and are parasitic upon other Florideae—e.g., *Harveyella mirabilis* (Reinsch) Schmitz & Reinke (see Sturch, 1899). Yet in their modes of sexual and asexual reproduction they can be assigned definite positions among these algae. Among the higher plants (Anthophyta or Angiospermae) many widely separated chlorophyll-less species are found, e.g., in the Orchidaceae and Burmanniaceae among the Monocotyledoneae, and *Cuscuta, Cassytha, Monotropa, Rafflesia, Orobanche,* and many others in the Dicotyledoneae. Thus it is necessary to delimit the fungi by further characters than merely the lack of chlorophyll. Such a definition is in the main negative.

Definition. As a group the fungi may be defined as chlorophyll-less nonvascular plants whose reproductive or vegetative structures do not permit them to be assigned to positions among recognized groups of algae or higher plants, and as excluding the Bacteria (which are typically one-celled and lack a typical nucleus) and the Mycetozoa (which have an animal type of structure and reproduction).

Whether the fungi represent a single phylum of organisms with a common origin or have arisen in independent lines from several ancestral types of plants is still a matter of debate among students of their phylogeny and classification. They range from very simple short-lived, one-celled structures whose single cell becomes the organ of reproduction (e.g., *Olpidiopsis,* which at maturity becomes the zoosporangium from which escape the zoospores) to massive perennial mycelia giving rise to great spore fruits as in some of the puffballs, pore fungi, etc. Except for the lack of chlorophyll and the saprophytic or parasitic mode of life thereby necessitated, these two extremes have no single character in common: manner of reproduction, structure of the vegetative body, chemical composition of the cell wall, etc. The extreme simplicity of the one type of fungus might be considered to indicate a low position in evolution, i.e., great primitiveness, but on the other hand it might be the result of a great degree of simplification from a much more complex fungus. The lack of any good fossil record of these lower fungi prevents us from obtaining direct evidence in this matter.

Structure. The vast majority of fungi consist vegetatively of more or less elongated, septate or nonseptate filaments. These are called individually hyphae (singular, hypha) and collectively the mycelium. They may be uniform in thickness or tapering from broad to slender portions in the same hypha or in different portions of the same mycelium. They may be branched or unbranched (simple). In thickness they may be less than 0.5 μ up to over 100 μ (in some Saprolegniales). In size the whole mycelium may be only a few microns in length or it may produce great sheets or strands that extend many meters.

The composition of the cell wall is very variable among the different fungi and sometimes in the same individual at different stages of maturity. Basically the chief components appear to be various types of carbohydrates or mixtures of these: cellulose, pectose, callose, etc. Mixed with these and probably often in chemical combination with them there may be other substances. Cellulose predominates in many of the Phycomyceteae so that the characteristic cellulose reaction is shown upon treatment with chloriodide of zinc, but sometimes where it makes up the bulk of the wall it does not respond to this reagent until certain fatty deposits in the outer portion of the cell wall are first dissolved away, as in *Monoblepharis*. In a great many fungi, especially the Ascomyceteae, Basidiomyceteae and the higher Phycomyceteae, either cellulose is entirely lacking, being replaced by some other carbohydrate, or the considerable amount of chitin with which the wall is impregnated prevents the cellulose from showing its presence. Chitin is never alone in the wall but it may form a considerable portion of the component substances. Aside from the foregoing substances calcium carbonate or other salts may be deposited upon or within the wall. Although von Wettstein (1921) and others have identified the chief component of the cell wall of many fungi, apart from the carbohydrates, as chitin, identical with the chitin of the Arthropods, Dous and Ziegenspeck (1926) after a careful comparative study of the animal chitin with the fungus chitin conclude that these are parallel compounds derived from different basic substances, but with much the same general characters. Throughout this book, wherever the word chitin is used it should be understood as referring to this fungus chitin, not to the true animal chitin.

There are two main types of mycelium; in one the hyphae are cellular and in the other, coenocytic. A cellular hypha usually contains either one or two nuclei per cell and the division of the cell is initiated by the division of the nucleus or by the simultaneous division of both nuclei, respectively. In a coenocyte there are many nuclei and the formation of septa occurs without immediate reference to any preceding nuclear division. A coenocytic hypha may be "tubular," i.e., lacking septa, or septate. In the latter case each segment is multinuclear. A tubular coenocyte, such as is characteristic of the majority of the Phycomyceteae, produces septa to set apart the reproductive organs (sporangia or gametangia) from the hypha, or to fence off an injured region, or to separate an empty portion of the hypha from those portions still containing protoplasm. A cellular mycelium may have multinucleate cells in the younger portions but by intervention of septa the older portion is transformed to cells with one or two nuclei, depending upon the phase of the mycelium. An old mycelium, no longer able to continue normal growth, may undergo nuclear division without septum formation so that the old cells may have several nuclei. In the Higher Fungi that generation of development in which the cells are uninuclear is sometimes called the monocaryon phase and that in which the cells are binuclear the dicaryon phase, or primary and secondary mycelium, respectively. In general the growth in length and the formation of new cells takes place in the terminal portion of the hypha or of its branches.

Septum formation occurs, as in most algae, by the production of a circular shelf which gradually grows inward until (in the Phycomyceteae) it makes a complete septum or (in the Higher Fungi) leaves a small central perforation through which there is a continuous protoplasmic connection from cell to cell. In *Allomyces*, one of the Phycomyceteae, septa may be formed but these are imperfect with large openings and are called pseudosepta.

The mycelia of the majority of fungi are hyaline, especially the hyphae that are embedded in the substratum and functioning as organs for obtaining nutriment. These are sometimes colored but the color is much more frequent in those hyphae that are external. The pigment causing this dark color is related to melanin and is largely confined to the cell walls. The hyphae that bear the conidia or that protect the other types of reproductive structures are especially apt to be dark-colored.

Aside from their elongated thread-like form the hyphae may be packed together tightly so that they adhere in elongated strands which sometimes have a hard black external layer and creep long distances. These are called rhizomorphs. In *Armillariella mellea* (Vahl) Karst., these black, shoestring-like strands creep under the bark of roots and trunks of trees and through the soil. Storage organs called sclerotia are frequently produced by fungi. Dense masses of hyphae arise and the short cells enlarge laterally until a compact pseudoparenchymatous tissue is formed whose cells become polyhedral by mutual pressure. These are filled with food materials and the walls may remain thin or become somewhat thickened, especially the outermost layers of cells which thus form a protective cortex which may be light in color or more often brown or black. These sclerotia if small, as in *Sclerotium rolfsii* Sacc., may be transported easily by surface water during heavy rains. Larger sclerotia may remain near where they are produced and then develop typical reproductive organs for the fungus when favorable conditions arrive. This is the case with sclerotia of *Claviceps, Sclerotinia*, and some species of *Polyporus*. From the sclerotia themselves under certain conditions new mycelium may grow out instead of reproductive organs. This is the case with the overwintering stage of a species of *Pellicularia*, the common black scurf ("*Rhizoctonia*") of potatoes and other plants.

Reproduction. Asexual reproduction in the true fungi may occur by the formation within a zoosporangium of naked cells, zoospores, which upon their release swim away by means of anteriorly, laterally, or posteriorly attached flagella, either one or two in number depending upon the order of fungi concerned. This production of zoospores is confined to some orders of the Phycomyceteae. The zoospores eventually settle down and encyst, and the encysted cell becomes the start of the new plant. In the majority of fungi, including many of the Phycomyceteae and all of the Higher Fungi in which asexual reproduction occurs, no motile spores are produced but the spores are provided with a wall and are distributed by air currents, by water, by insects, etc. These spores are of several types of origin. In the Mucoraceae they are produced internally in a sporangium; upon the rupture or dissolution of its walls, they are set free and distributed by air or water currents. Such spores are sometimes called aplanospores in contradistinction to the motile naked zoospores which may be called planospores. Conidia arise as single separable cells of the mycelium. They may arise by the fragmentation of the whole mycelium or of special hyphae into cylindrical, ovoid, or spherical cells (oïdial mode of conidium formation) or by the cutting off of terminal or lateral cells from special hyphae or conidiophores. In a number of genera the conidia are pushed out one by one from the neck of a flask-like cell (or phialid). The conidia, by whatever means they arise, may be hyaline or colored, and may remain one-celled or by formation of transverse or longitudinal septa may become two-celled to many-celled. They may be released singly or remain attached in a chain. In some cases, instead of producing conidia, a tangled mass of hyphae may form a rounded ball as in the genus *Papulospora*. For the purpose of enabling the fungus to survive unfavorable conditions such as cold, lack of water, etc., chlamydospores are produced by many fungi. These are terminal or intercalary cells of a hypha (or even single cells in a conidium made up of a row of cells) which enlarge and round up, store supplies of food, and form a thick wall. Such cells may live for years until favorable conditions arrive. They have nothing to do with the sexual stage of the fungus so that the use of the term chlamydospore for the reproductive cells (teliospores) of the Ustilaginales is unwarranted.

Sexual reproduction, or substitutes for it, may be found in most of the groups of fungi except the artificial class called Fungi Imperfecti which was established to include those fungi in which sexual reproduction has not been discovered. The simplest type of sexual reproduction is the union of two cells of equal size and to all appearances alike in all characteristics. These two gametes may both be nonmotile (aplanogametes) or motile (planogametes), as in the Yeasts and Chytridiales, respectively. The zygote formed by their union may become a new one-celled plant or may produce mycelium of various types. The gametes are more often unequal in size and the motility may be limited only to the smaller one, then designated as the male gamete or sperm. Still more often flagella are lacking in both gametes and fertilization is brought about by the union of the cells by dissolution of a portion of the intervening walls. Sometimes the male nucleus is introduced into the female gamete, or egg, through a tube extending into the latter. When the zygote is the product of the union of clearly dissimilar gametes, forming a definite zygote cell which often serves as a resting spore, it may be called an oospore. When a similar resting spore results from the union of similar gametes it is called a zygospore. It must be noted that the gradation between isogametes (i.e., equal gametes) and anisogametes (i.e., gametes that are dissimilar) is gradual.

In some fungi, instead of producing definite gametes, any cell of one mycelium may unite with any cell of another compatible mycelium, so that no definite part of the fungus can be distinguished as a male or female reproductive organ. This is especially true in the Ustilaginaceae. However, many fungi do produce clearly distinguishable male and female organs of reproduction. Very often the gametes may develop into new plants without union, i.e., by parthenogenesis. This phenomenon is observed from some of the simplest fungi up to many of the Higher Fungi. Sometimes a vegetative cell may be substituted for the normal male gametangium (antherid), the nucleus of this substituting cell functioning in place of a normal sperm nucleus. So it comes about that, with parthenogenesis occurring in many fungi and substitution of vegetative cells for gametes in others, it is difficult to follow the evolutionary sequence of the development of sexual reproduction in these organisms.

The occurrence of the union of the sexual nuclei naturally leads to the production of a diploid nucleus. In very few fungi does this nucleus multiply in the diploid state (perhaps in a few Yeasts). Usually its first divisions are meiotic, so that throughout the life history of the fungus the nuclei are always haploid except immediately after the union of the gamete nuclei. Yet among the fungi, particularly in the Ascomyceteae and Basidiomyceteae, we find a contrast of mycelia that are diploid in nature and haploid in nature although all the nuclei are haploid. These are respectively the dicaryon and monocaryon phases of mycelia. Whether the two nuclei (of separate sexual origin) are present in a cell within one nuclear membrane (i.e., a diploid nucleus) or each in its own separate membrane (i.e., two haploid nuclei in the cell), the effect on the cytoplasm is practically the same, so that a dicaryon cell is to all intents and purposes a diploid cell. Thus when Buller (1930) speaks of the "diploidization" of a monocaryon mycelium by the introduction of a compatible haploid nucleus which multiplies and spreads from cell to cell this term is essentially correct although the nuclei, now in pairs in each cell, are still haploid. He was speaking of diploidization of the cell as a whole not of the contained nucleus.

In the simplest fungi the whole plant, consisting of but one cell, becomes the reproductive unit that produces the asexual or sexual cells. As we study fungi of greater complexity we find that the vegetative and reproductive portions of the organism are more and more segregated. In many fungi the process of sexual reproduction becomes increasingly complex and leads to the formation of, not a single zygote, but a very complex structure, the spore fruit, many of whose cells become the ultimate reproductive spores. Some of the puffballs, e.g., *Calvatia gigantea* (Batsch ex Pers.) Lloyd, attain a diameter of over one meter. This is the spore fruit. The vegetative mycelium is subterranean and not noticed except when hunted for.

Parasitism. Because of their lack of chlorophyll all fungi must obtain their organic food from sources external to themselves. The whole mycelium may have the power to absorb these foods or this task may be relegated to special portions such as rhizoids or to haustoria, the knob-like or finger-like processes that enter the cells of the host plant. In many of the Phycomyceteae there is a great difference between the much branched mycelium within the substratum and that portion outside which bears the reproductive organs.

Probably the majority of fungi are saprophytic, i.e., feed upon the organic products or remains of plants or animals but not upon the living organisms themselves. The substances utilized by the fungus are often very varied in nature. Simple sugars, starches, cellulose, or organic acids may satisfy its needs provided the necessary mineral nutrients are present. On the other hand some fungi require the presence of various growth factors, such as biotin and thiamin, and some proteins or their building blocks, the amino acids. Some saprophytic fungi are much more limited as to their organic foodstuffs. In general, however, their range of foods is far wider than that of strict parasites. Some facultative parasites are able to grow apparently indefinitely as saprophytes. Thus a species of *Fusarium* capable of causing the death of *Sesamum indicum* L. was studied in culture by the author (1904). When provided with the necessary mineral nutrients, it developed upon cellulose, simple and complex sugars, various organic acids, asparagin, peptone, gelatin, glycerine, etc., and under both aerobic and anaerobic conditions.

On the other hand many fungi are obligatory parasites and methods have not yet been devised by which they can be grown except upon suitable living host tissues. This is true of the Rusts (Uredinales), the White Rusts (Albuginaceae), most of the Downy Mildews (Peronosporaceae), the minute insect parasites belonging to the Laboulbeniales, etc. This would seem to indicate that for such parasites the choice of food is very strictly limited. Among some of the commoner parasites, e.g., stem rusts of wheat, *Puccinia graminis tritici* Erikss. & Henn., and powdery mildew of various grasses, *Erysiphe graminis* DC., there have been found what are called biologic or physiologic forms. These are races of the fungus that are indistinguishable except for the fact that one will grow only on certain species or varieties while the other races grow only on other varieties.

Parasitic fungi may be destructive or balanced parasites. The former may kill the host cells or tissues by means of some poisonous substances which may even diffuse out in advance of the fungus hyphae with the consequence that the latter actually enter dead tissues. Some species of *Botrytis*, *Sclerotinia*, *Pythium*, etc., are of this type. On the other hand a balanced parasite is so well adapted in its demands upon the host to the ability of the latter to supply these needs and to continue to live and grow that both fungus and host develop together until the time comes when the fungus is ready to produce its spores. Then it too destroys the surrounding tissues. Many of the Smuts (Ustilaginales) are balanced parasites. Actually all gradations between these two extremes may be found. Mostly the destructive parasites are less strictly confined to definite hosts than are the balanced parasites.

Some fungi belonging to the Polyporaceae attack and destroy only the dead cells (wood fibers, tracheary tissues, etc.) of the wood of living trees and from that viewpoint are saprophytes, yet their growth ceases when the death of the tree occurs. The conditions within the host tissues that favor the growth of the fungus are evidently sufficiently changed when the tree dies so that the fungus no longer finds the conditions of environment that are requisite for its continued growth. Many fungi are parasitic in certain stages of their growth and saprophytic later on. Thus the fungus of apple scab (*Venturia inaequalis* (Cke.) Wint.) grows and produces its conidia upon the living leaves and fruits of the apple but overwinters saprophytically and produces its sexual stage of reproduction within the dead leaves on the ground. This is true of very many of the leaf spot fungi of economic as well as wild plants.

Just how a parasitic fungus obtains its food from its host is not clear in all cases. Apparently it may be by the action of some secretion from the fungus hyphae or haustoria upon the plasma membrane of the host cells, making this more permeable to the contained solutes so that they diffuse out and are absorbed by the fungus.

The whole field of fungus physiology offers many interesting lines of study and cannot be entered upon within the limits of this book. Nor will space permit the extensive discussion of medical mycology (see C. W. Dodge, 1935) or technical mycology (Lafar, 1903, 1910). Only brief mention is made in Chapter 16 of the production of antibiotics. The genetics of fungi is noted where necessary for the understanding of the development of various groups of these organisms.

History of Mycology

A very brief sketch of a few steps in the history of mycology should not be omitted. The larger fungi, or rather their conspicuous fruiting bodies, were well known to the ancients, but knowledge of their true nature and their manner of growth had to await the invention of the microscope. The Romans knew and distinguished various edible and poisonous mushrooms. The Emperor Nero is reported to have been very fond of *Amanita caesarea* (Schaeff.) Fr., which owes its specific epithet to this association. The word fungus (related to the verb *fungor*, to flourish) was applied to mushrooms and to excrescences from the ground or from trees. The Greek word mykes (μύκης) was applied to some types of fungi. From this comes the characteristic part of the word mycology. For untold centuries the Chinese have known and used certain fungi for food and others for medicine but, as in the Occident, with little real knowledge as to the true nature of these organisms.

After the invention of printing in Europe there began to appear various "herbals," describing and, in many cases, illustrating more or less elaborately the plants of southern and western Europe. In some of these the larger fungi are illustrated. Thus Clusius (Charles de la Cluse, 1529–1609) in 1601 devoted many illustrations and many pages of text to the discussion of edible and poisonous fungi. No attempt was made to classify these into genera or families as these terms are now used. In 1623 in his "Pinax Theatri Botanici," Gaspard Bauhin (1560–1624) attempted to bring together all plants known to him or to his predecessors. He divided the approximately 100 species of fungi and lichens into groups to which he gave names. The idea of the genus as a definite category for the purposes of classification had not yet become firmly established, so that some of his group names include directly, as the next subordinate rank, the species, while in other cases there are intermediate categories. All lichens he included in the group *Muscus Saxatilis vel Lichen* (9 species). Under the name *Fungus* he included 81 species which are distributed now among the Agaricaceae, Boletaceae, Polyporaceae, Clavariaceae, Auriculariaceae, Lycoperdaceae, Phallaceae, Clathraceae, Pezizaceae, and perhaps other families. *Agaricum Fungus* corresponds practically to the laterally attached Polyporaceae, especially *Fomes. Tubera*, with 2 species, was applied to truffles (*Tuber*) and other subterranean firm fungi.

Tournefort (1656–1708) is the botanist who, more than any other, brought to general acceptance the concept of the genus as the classificatory category next above the species. He still maintained the cumbersome method of naming a species with the genus name followed by a descriptive phrase, now universally abandoned for the binomial manner of writing a name which was popularized by Linnaeus. Six genera of fungi and one of lichens were recognized by Tournefort in his "Elemens de Botanique" in 1694. The generic names used were adopted from his predecessors. *Fungus* corresponds to all centrally stipitate Agaricaceae, Boletaceae, and Polyporaceae. *Boletus* includes *Morchella*, *Clathrus*, and *Phallus*. *Agaricus* was applied to fungi attached laterally to trees, logs, etc., such as various Polyporaceae, *Auricularia*, etc. *Lycoperdon* included the Lycoperdaceae and also the larger rounded Mycetozoa. *Coralloides* included various branched fungi, among others the branching species of *Clavaria*. *Tubera* was used as by Bauhin.

Dillenius (1687–1747) added a good many species and some new genera. He also changed the names of some groups from those used by Tournefort or, retaining the name, changed its application. Thus all centrally stipitate Agaricaceae were placed in the genus *Amanita*. *Boletus* was entirely changed and made to include the present Boletaceae and centrally stipitate Polyporaceae. *Morchella* and *Phallus* were introduced to take up the species included in Tournefort's *Boletus*. *Bovista* was substituted for *Lycoperdon* and *Fungoides* for *Coralloides*. For cup- or saucer-shaped fungi Dillenius used the name *Peziza*. Mention must be made of Sebastien Vaillant (1669–1722) whose book "Botanicon Parisiense" in 1727 gave illustrations of fungi and other plants whose accuracy and beauty were scarcely equalled for over a century. He listed all genera alphabetically, regardless of their real relationship, hence the fungi are scattered throughout the work. *Agaricus* and *Boletus* are used as they were by Tournefort. Most of the Agaricaceae are included in the genus *Fungus* which is however a very heterogeneous assemblage of organisms. *Fungoides* is in part *Peziza*, and *Corallofungus* includes some of the species of *Clavaria*. Aside from the beautiful illustrations and the descriptions, Vaillant added little to mycology.

The foremost student of fungi before the time of Linnaeus was the Italian botanist Pier' Antonio Micheli (1679–1737). He was apparently the first student of these organisms to use the microscope on them, crude as was his instrument. His great work "Nova Plantarum Genera" was completed by 1719 but, for lack of funds, the first part only was published, after a delay of ten years, in 1729. His biographer, Targioni-Tozzetti (1858), reports that the second part was completed but never published for lack of means. There is no doubt that Micheli knew his fungi far better than any of his forerunners or contemporaries. He gave usable keys by which genera could be identified and, for the larger genera, keys to the species. Many of his figures and descriptions were so excellent that there is no difficulty now in identifying them. He distinguished *Fungi lamellati* (Agaricaceae), *Fungi porosi* (Polyporaceae and Boletaceae), *Fungi pulverentes* (Lycoperdaceae and some others), *Fungi Ramosi* (the branching Clavariaceae), etc. Among the generic names used by him and still recognized are *Phallus, Clavaria, Clathrus, Lycoperdon, Geaster,* and *Tuber.* He used *Agaricum* as did his forerunners for laterally attached *Fomes, Trametes, Fistulina,* and *Stereum. Polyporus* was confined to the stipitate polypores; *Suillus* to the present *Boletus* and its allies; *Erinaceus* to the stipitate Hydnaceae; *Fungus* to the stipitate Agaricaceae. *Boletus* as used by him is now known as *Morchella;* his *Puccinia* is now called *Gymnosporangium. Coralloides* was equivalent to the branched species of *Clavaria,* etc. In addition to collecting and studying the larger fungi, Micheli was perhaps the first botanist to attempt cultures of molds. He sowed spores of "*Mucor*" (evidently *Rhizopus nigricans* Ehr.) on one side of pieces of squash and "*Aspergillus*" on the other. Each produced its own kind of fungus. He inoculated two pieces of squash with "*Botrytis,*" covering one with a bell jar and leaving the other exposed. The covered piece developed only *Botrytis* while the uncovered piece developed *Mucor* as well, thus showing, as Micheli pointed out, that the spores of these various molds were distributed through the air.

Linnaeus (Carl von Linné, 1707–1778), who is often called the "Father of Botany," advanced the knowledge of fungi little if at all. In his great work "Speecis Plantarum" (1753) he attempted to bring together descriptions of all of the known species of plants. His adoption of the two-word form of name for species (which we call generic and specific epithets) marked a very great advance in convenience and simplicity. In his twenty-fourth class, "Cryptogamia," the fungi are to be found chiefly under the heading Cryptogamia Fungi but a few are located among the Cryptogamia Algae. His treatment of the fungi, which he mostly knew only from the study of botanical literature, and only superficially at first hand, is far less scientific than that of Micheli or of Dillenius. The lichens were included in the genus *Lichen,* among the algae, as was the genus *Tremella.* This genus includes the alga *Nostoc* as well as the rust *Gymnosporangium,* the Basidiomycete *Auricularia,* several lichens, and probably one or more species now included in *Tremella.* All of the Agaricaceae, as we know the family, were included in the genus *Agaricus* and all the pore fungi in the genus *Boletus,* differing from Micheli and his predecessors.

Erinaceus of Micheli and earlier botanists became *Hydnum*. *Phallus* was made to include both *Phallus* and *Morchella*, being equivalent to Tournefort's *Boletus*. *Lycoperdon* included Lycoperdaceae and some Mycetozoa. *Mucor* included all molds such as Mucorales, Fungi Imperfecti, and Erysiphaceae. Other genera recognized by Linnaeus were *Elvela*, *Peziza*, and *Clavaria*.

The most significant advance in the classification of fungi after Linnaeus is to be found in the works of Christiaan Hendrik Persoon (1755–1837). The number of recognized species had become greatly increased and the great improvements in the microscope made it possible to study the manner by which the spores were borne, so that the major groups as now recognized began to appear.

Probably the greatest contribution to the knowledge of the larger fungi, particularly the "Hymenomycetes," was made by Elias Magnus Fries (1794–1878), whose active mycological work extended over a period of more than half a century. The impetus given to mycology by these two great botanists was felt over the whole world and fungi unknown to science were discovered by the thousands.

In the first third of the nineteenth century the smaller Ascomyceteae, especially Sphaeriales, and the pycnidial Fungi Imperfecti, were mostly described superficially, often being thrown together in the same genus. Little was known of the Rusts, Smuts, molds of all sorts, various Moniliales, Melanconiales, etc. Only when these were studied carefully with the compound microscope did order begin to arise out of chaos. Of the many workers in that period, mention may be made of August Carl Joseph Corda (1809–1849) whose "Icones Fungorum," a six-volume work, published from 1837–1854, showed the detailed structure of many of the larger fungi but also threw light on hundreds of the microscopic forms. Soon following this came the beautifully illustrated three-volume work of the Tulasne brothers, "Selecta Fungorum Carpologia," 1861–1865 (Louis René Tulasne, 1815–1885, being the chief author).

In the United States the first extensive study of fungi was undertaken by Lewis David von Schweinitz (1780–1834), a minister in the United Brethren Church, who collected extensively in North Carolina and Pennsylvania. His publications on American fungi (1822 and 1832) were the first noteworthy ones that appeared. A third of a century later Charles Horton Peck (1833–1917) began his work at Albany, New York, as state botanist, a position which he held from 1867–1915. His chief interest was in the fungi of which he described about 2500 species previously unrecognized. His collections and descriptions formed the foundation for many monographic studies of various genera by later students, especially of the Agaricaceae. Other nations had similar lovers of fungi who added greatly to the knowledge of these organisms, but space does not permit the mention of their names, except Carlos Spegazzini (1858–1926), of Argentina, who explored the mycologically almost unknown territory of southern South America and gave the first descriptions of several thousand species.

As new species of fungi were recognized in all parts of the world their descriptions appeared in all sorts of scientific journals, reports of learned societies, and even in textbooks, so that it became increasingly difficult for a student of fungi to know whether a fungus under study by him was new to science or already described. Many fungi were unavoidably named several times by different investigators. In Germany and other European countries floras were published in which were described all species of fungi known to occur in those regions, such, for example, as Rabenhorst's "Kryptogamen-Flora" (1844, 1845). Valuable as were such works, they did not include fungi from other regions, and since fungi are much more cosmopolitan in their distribution than higher plants the probability always existed that many fungi described from elsewhere would be found to occur within the area covered by the work. This uncertainty deterred many mycologists from describing supposedly new species for fear of duplication. This was especially true of students of fungi who lived away from the great European centers of mycological activity. This condition became so bad that the great Italian mycologist Pier' Andrea Saccardo (1845–1920) decided, before 1880, to bring together in one work the descriptions of all fungi hitherto recognized. Thus began the monumental "Sylloge Fungorum," the first volume of which appeared in 1882 and the twenty-fifth in 1931. With the appearance of the first volume of this work systematic mycology again took a great leap forward. In the meantime various periodicals were established which were devoted partly or entirely to fungi. These are found in many countries and in many languages, for the science of mycology is bounded by no political or linguistic boundaries.

With the greatly increased knowledge of the structures of fungi, for which the Tulasne brothers were in great degree responsible, there began a new phase of mycological work about the middle of the nineteenth century. This was the study of the life histories of fungi. The earlier mycologists had been, in the main, satisfied to describe the different forms as they found them, often not dreaming that in many cases they were treating of different stages of the same organism. To be sure some of the earlier describers of fungi suggested that different forms found in close association might be different stages of the same fungus, suggestions which in many cases were found by later investigators to be correct. The work of Anton de Bary began at about this period and was carried on with such enthusiasm and skill that a great series of life history investigations followed. His first outstanding work was his investigation of the life history of the Mycetezoa (1859). He maintained his interest in this group of organisms for many years and inspired the work of many outstanding students in this field. Probably the most outstanding of de Bary's earlier studies was that in which he determined the life cycle of the rusts (Uredinales) and proved the heteroecious nature of black stem rust of small grains (*Puccinia graminis* Pers.). In the two decades after the appearance of this work (1865) we find many of de Bary's pupils following and extending life history studies in all groups of fungi. Preeminent among these was Oscar Brefeld, who employed, not for the first time but probably most extensively up to then, the method of growing the fungi under study in pure culture on various types of culture media and under various external conditions. In this way he was enabled to study the developmental stages of many fungi and to learn much of their physiology as well. His chief series of contributions began in 1872 and the last volume appeared in 1912. One of de Bary's early students and collaborators was the Russian, M. S. Woronin (1838–1903), who returned to Russia after several years of association with his great teacher, there becoming in his turn the center of a group of very able mycologists and plant pathologists. The life history studies thus stimulated by de Bary and his students have continued up to the present with so great a number of investigators that mention of their names, even, must be omitted.

The introduction of the recently developed cytological methods to the study of fungus life histories began on a large scale with the investigations by P. A. Dangeard (1894) in France, and of R. A. Harper (1896–1897). Their first publications along these lines appeared between 1894 and 1897 and were soon followed by the contributions of a host of other eager students in all parts of the world. The correction or confirmation of previously held ideas—particularly with reference to the nature of the sexual act in fungi—thus made possible, has proved to be of the utmost value in assisting the determination of relationships among the fungi.

In 1904 and succeeding years A. F. Blakeslee made known the occurrence of those sexual phenomena in the Mucorales to which he gave the names heterothallism and homothallism. The study of these types of sexual reaction has been extended to other groups of fungi: In the Basidiomyceteae by Mlle. Bensaude (1918), Hans Kniep (1913–1917), Miss Mounce (1922), Vandendries (1923), Hanna (1925), and many others from 1915 to the present; in the Ustilaginales by Bauch (1922), Hanna (1929), Kniep, Stakman (1927), and others in the last twenty years; in the Uredinales by Craigie (1927, 1931), Andrus (1931), Miss Allen (1930), etc., since 1927; in the Ascomyceteae by B. O. Dodge (1927), Ames (1932), Drayton (1932), and several others, mainly since 1927. Dodge (1928), Lindegren (1933), and others have made intensive studies on the genetics of fungi in the last fifteen or more years, particularly in various species of *Neurospora*, while many others have studied the smuts from the genetic standpoint. Hybrids have been produced in both these groups as well as in the rusts and their structure and their behavior studied. So much has been published in these fields in the last fifteen to twenty years that the names of the investigators cannot be listed here. Of those mentioned above only a few of their earlier papers are noted, though in many cases they produced many later contributions.

The present day finds systematic mycologists active all over the world. Life histories are being studied in all groups. The sexual relations are being scrutinized from the lowest to the highest fungi and genetic studies are revealing results somewhat parallel, but on a smaller scale as yet, to those attained by the study of *Zea mays* and *Drosophila*. Now, as never before, a knowledge of the fungi themselves is necessary.

The discovery of the production of antibiotic substances by various fungi has encouraged intense research in that field, the results of which have only recently become of great importance in the medical field.

Rules for Nomenclature

The rules for botanical nomenclature, especially as applicable to fungi, have been given an extended discussion by Bisby (1945). Only the more fundamental points will be taken up here. These rules have been formulated and added to and modified at a series of International Botanical Congresses in 1867, 1905, 1910, 1930, 1935, and 1950. The aims are expressed in Articles 2 and 4 of the Rules, from which the following sentences are quoted:

> The object of the rules is to put the nomenclature of the past into order and to provide for that of the future. They are always retroactive: names and forms of nomenclature contrary to a rule (*illegitimate names* or *forms*) cannot be maintained. . . . The essential points in nomenclature are: (1) to aim at fixity of names; (2) to avoid or to reject the use of forms and names which may cause error or ambiguity or throw science into confusion. . . . Next in importance is the avoidance of all useless creation of names.
>
> Art. 7. Scientific names of all groups are usually taken from Latin or Greek. When taken from any language other than Latin, or formed in an arbitrary manner, they are treated as if they were Latin. Latin terminations should be used so far as possible for new names.

The earliest name properly applied to a plant shall be retained, provided it is a binomial, i.e., consists of the generic name and specific epithet. Since, however, it is impractical to go back to the classical authors or those of the Middle Ages it has been agreed that names applied before the appearance of Linnaeus' "Species Plantarum" (1753) shall not be considered, nor those subsequent to that date which do not use the binomial nomenclature. Because many of the fungi were not well known to Linnaeus the basic dates for the earliest authoritative names have been assigned variously in accordance with the groups of fungi concerned. Thus for the Mycetozoa and Lichens the date remains 1753; for the Uredinales, Ustilaginales, and Gasteromycetes it is 1801, based on Persoon's "Synopsis Methodica Fungorum"; and for all other fungi 1821–1832, based on the appearance of the various volumes of Fries' "Systema Mycologicum." So the beginning date for the Hymenomycetes is based upon Volume I of this work, part of which is said to have appeared late in 1820 and the remainder in 1821. Any names given by later authors are valid only if the species concerned were not included in that volume. The first section of Volume II appeared in 1822 and included, and therefore is authoritative for their nomenclature, the Discomycetes and many of the larger Heterobasidiae, and the sclerotioid Fungi Imperfecti. Part II of this volume appeared in 1823 and contained some of the Gasteromycetes, whose basic date is Persoon's work in 1801, and the Pyrenomycetes (and Sphaeropsidales). Volume III, Section I (1829) contains (in addition to the remainder of the Gasteromycetes and the Mycetozoa) the Erysiphales. For this last group it is the basic work. Section II (1832) consists mainly of the Fungi Imperfecti, but also Hypodermii (i.e., Uredinales and Ustilaginales) for which the work of Persoon is basic.

A generic name is always a noun in the singular number. It is always written with an initial capital. The specific epithet is mostly an adjective in Latin form, which must agree in gender with the generic name (e.g., *Lepiota procera*), or it may be the genitive case of some noun (e.g., *Mycosphaerella fragariae*), or it may be a noun in the nominative case (e.g., *Xylaria hypoxylon* or *Fomes pinicola*). Mostly the specific epithet is written without an initial capital, except where it is a noun in the nominative case that was an old generic name, or where it is based upon a personal name or a generic name. Many authors prefer to decapitalize all specific epithets, even when based upon personal or generic names. By the rules adopted at the 1950 International Botanical Congress at Stockholm, all specific epithets must be decapitalized.

Valid publication of a hitherto undescribed plant consists of the assignment of a name and the description of the organism, in proper manner. After 1935 all descriptions to be valid must be in Latin aside from any descriptive text in any other language. The description of a species must be based upon a definite collection, specimen, or culture, which is designated as the "type specimen." This should be preserved and the place where it is deposited should be indicated. A genus must be based upon some particular species, the "type species." A family must be based upon a "type genus" and an order upon a "type family." Ordinarily the familial and ordinal names are based upon the stem of the name of the type genus, with the appropriate endings, *-aceae* or *-ales*, respectively. If a genus, family, or order is divided into two or more parts the original name must be retained for the part in which the type species, genus, or family remains.

When a species is transferred to another genus than that under which it was first described, or when for any other valid reason the generic name is changed, the specific epithet must be retained, subject to change of gender in case it is an adjective, if the new genus name is of different gender. Exceptions are as follows: If the new genus name is the same as the specific epithet or if in the new genus a similar combination already exists (e.g., when *Leontodon taraxacum* was split off from the genus *Leontodon* and placed in the new genus *Taraxacum*, the epithet had to be replaced; *Uredo sorghi* (1897) was transferred to the genus *Puccinia*, but as there was already a *Puccinia sorghi* (1832), another epithet had to be chosen). In this case the oldest available epithet of a synonym must be taken or in lack of such a name a new epithet must be provided by the author making the transfer.

In describing a species *de novo* the describer's (abbreviated) name follows the specific epithet, but if an author changes the genus to which the species must be assigned the name of the person who first gave the epithet must be placed in parentheses followed by that of the author of the new combination. For example, Bessey and Thompson described a new species of fungus under the name *Genea cubispora* Bessey & Thompson. When Miss Gilkey determined that this belonged to the genus *Hydnotria*, the name became *Hydnotria cubispora* (Bessey & Thompson) Gilkey. If the epithet was given to a plant accompanied by an adequate description before the basic date for that group of plants it is not necessary to refer to the author of that epithet but only to the person who first used it on or after the basic date. However, this later author could indicate that he took this epithet on the authority of the earlier author by placing, as authority for the combination, the abbreviated name of the earlier author followed by "ex" and the name of the later author. Thus Persoon in 1801 described a puffball under the name *Lycoperdon giganteum*, ascribing this combination to Batsch who published the name before 1790. It is permissible simply to give "Pers." as the authority, but it is preferable to write "Batsch ex Pers."

To avoid confusion a genus name once applied to any plant may not ever thereafter be used for another genus, even though the name is invalid for the first genus. Thus, unless an exception is made at some future Botanical Congress, the generic name *Empusa*, given about 100 years ago to an orchid, is not available for a genus of the Entomophthorales. A specific epithet once applied in a given genus may never be used for another species in the same genus.

In fungi with several stages of development to which different names have been given, the species epithet that is to be retained is the one applied to the "perfect" stage of the fungus, regardless of the fact that names may have been given earlier to some of the other stages. Thus Persoon (1801) recognized the following three species of rusts which are now known to be different stages of the same rust: *Aecidium berberidis* for the aecial stage occurring on *Berberis vulgaris* L.; *Uredo linearis* for the uredial stage on small grain; and *Puccinia graminis*, for the telial stage on the same host. Only the last name may be used for the species, for it is the name of the "perfect" stage. The name *Lycoperdon poculiforme* Jacq., given 10 or 12 years earlier may not be used for two reasons: it was given *before* the basic date 1801 and was applied to the aecial stage of the rust.

The fungi, including the Mycetozoa, which probably do not belong at all in this group of organisms, may be divided as follows:

Key to the Major Groups of Fungi

Vegetative stage permanently naked and either flagellate or amoeboid. Encysted spores produced to serve as organs of distribution or to carry the organisms over unfavorable conditions.
Subclass Mycetozoa (Chap. 2)

Vegetative stage for all or part of its course of development with cell walls.
True Fungi

Plant one-celled and then giving rise to planocytes, or producing a coenocytic mycelium. Sexual reproduction resulting in the formation of a zygospore or an oospore. Class Phycomyceteae (Chaps. 3–7)

Plant one-celled, not producing planocytes, or producing a cellular mycelium. Sexual reproduction resulting in the formation of a spore fruit.
Higher Fungi (Phylum Carpomyceteae, Chaps. 8–16)

Ultimate reproductive spores of the spore fruit produced internally in an ascus. Class Ascomyceteae (Chaps. 9–11)

Ultimate reproductive spores of the spore fruit produced externally upon a basidium or its equivalent.
Class Basidiomyceteae (Chaps. 12–15)

Sexual reproductive stage not known.
Class Fungi Imperfecti (Chap. 16)

At the close of each chapter is a key to the families and more important genera of the orders considered there.

Literature Cited

Allen, Ruth F.: A cytological study of heterothallism in Puccinia graminis, *J. Agr. Research*, **40**(7):585–614. *Pls.* 1–17. 1930. (This paper was followed by many others in the same field in the next six years.)

Ames, Lawrence M.: An hermaphroditic self-sterile but cross fertile condition in Pleurage anserina, *Bull. Torrey Botan. Club*, **59**(6):341–345. *Fig.* 1. 1932.

Andrus, C. F.: The mechanism of sex in Uromyces appendiculatus and U. vignae, *J. Agr. Research*, **42**(9):559–587. *Figs.* 1–11. 1931.

de Bary, Anton: Die Mycetozoen. Ein Beitrag zur Kenntnis der niedersten Thiere, *Z. wiss. Zool.*, **10**(1):88–175. *Pls.* 6–10. 1859.

———: Neue Untersuchungen über die Uredineen, insbesondere die Entwickelung der Puccinia graminis und den Zusammenhang derselben mit Aecidium berberidis, *Monatsber. kgl. preuss. Acad. Wiss. Berlin*, **1865**, pp. 15–49. 1 *pl.* 1865.

———: Comparative Morphology and Biology of the Fungi, Mycetozoa and Bacteria. Authorized English translation by Henry E. F. Garnsey, revised by Isaac Bayley Balfour, xix + 525 pp. 198 *figs*. Oxford, Clarendon Press, 1887.

BAUCH, ROBERT: Kopulationsbedingungen und sekundäre Geschlechtsmerkmale bei Ustilago violacea, *Biol. Zentr.*, **42**:9–38. 1922.

BAUHIN, GASPARD: Pinax theatri botanici, xviii + 522 pp. + index, 22 pp. Basel, Ludovicus Rex, 1623.

BENSAUDE, MATHILDE: Recherches sur le cycle évolutif et la sexualité chez les Basidiomycètes, 156 pp. 13 *pls*. 30 *figs*. Nemours, 1918.

BESSEY, ERNST A.: Die Bedingungen der Farbbildung bei Fusarium, *Flora*, **93**(4):301–334. 1904.

BISBY, G. R.: An Introduction to the taxonomy and nomenclature of fungi, 117 pp. Kew, Surrey, Imperial Mycological Institute, 1945.

——— AND G. C. AINSWORTH: The numbers of fungi, *Brit. Mycol. Soc. Trans.*, **26**(1–2):16–19. 1943.

BLAKESLEE, ALBERT F.: Sexual reproduction in the Mucorineae, *Proc. Am. Acad. Arts Sci.*, **40**:205–319. *Pl.* 14. 1904.

BREFELD, OSCAR: Botanische Untersuchungen über Schimmelpilze. Hefte 1–4. 1872–1881. Title changed to Botanische Untersuchungen über Hefenpilze, Fortsetzung der Schimmelpilze, for Heft 5, 1883, thereafter Untersuchungen aus dem Gesammtgebiet der Mykologie, Hefte 6–15. 1884–1912. In all over 2500 pp., 110 *pls*. Hefte 1–8, Leipzig, Arthur Felix; Hefte 9–15, Münster i. W., Heinrich Schöningh.

BULLER, A. H. REGINALD: The biological significance of conjugate nuclei in Coprinus lagopus and other Hymenomycetes, *Nature*, **126**(3183):686–689. *Figs*. 1–7. 1930.

CLUSIUS, CAROLUS: Fungorum in Pannoniis observatorum brevia historia, in Rariorum plantarum Historia, pp. 211–295. *Many figs*. Antwerp, 1601.

CORDA, A. C. I.: Icones fungorum hucusque cognitorum, 6 vols. 64 *large pls*. Prag, vols. 1–4, J. G. Calve; vols. 5–6, Fr. Ehrlich, 1837–1854.

CRAIGIE, J. H.: Experiments on sex in rust fungi, *Nature*, **120**(3012):116–117. *Fig*. 1. 1927.

———: An experimental investigation of sex in the rust fungi, *Phytopathology*, **21**(11):1001–1041. *Figs*. 1–14. 1931.

DANGEARD, P. A.: La reproduction sexuelle des Ascomycètes, *Le Botaniste*, **4**(1–2):21–58. *Figs*. 1–10. 1894.

DILLENIUS, JOHANN JAKOB: Catalogus plantarum sponte circa Gissam nascentium. Cum appendice, etc., 314 pp. 16 *pls*. Frankfort a. M., 1719.

DODGE, B. O.: Nuclear phenomena associated with heterothallism and homothallism in the Ascomycete Neurospora, *J. Agr. Research*, **35**(4):289–305. *Pls*. 1–3. *Figs*. 1–5. 1927.

———: The production of fertile hybrids in the Ascomycete Neurospora, *ibid.*, **36**(1):1–14. *Pls*. 1–4. 1928.

DODGE, CARROLL W.: Medical Mycology. Fungous Diseases of Men and Other Mammals, 900 pp. 142 *figs*. St. Louis, C. V. Mosby Co., 1935.

DOUS UND ZIEGENSPECK: Das "Chitin" der Pilze, *Z. Pilzkunde*, N.F., **5**(18):292–296. 1926.

Drayton, F. L.: The sexual function of the microconidia in certain Discomycetes, *Mycologia*, **24**(3):345–348. 1932.

Fries, Elias Magnus: Systema mycologicum, sistens fungorum ordines, genera et species huc usque cognitas, 3 vols., 1866 pp. Greifswald, Ernest Mauritius, 1821–1832.

Hanna, W. F.: The problems of sex in Coprinus lagopus, *Ann. Botany*, **30**(154): 431–457. 1925.

———: Studies in the physiology and cytology of Ustilago zeae and Sorosporium reilianum, *Phytopathology*, **19**(5):415–442. *Pl.* 7. *Figs.* 1–3. 1929.

Harper, R. A.: Die Entwickelung des Peritheciums bei Sphaerotheca Castagnei, *Ber. deut. botan. Ges.*, **13**(10):475–481. *Pl.* 39. 1896.

———: Beitrag zur Kenntnis der Kerntheilung und Sporenbildung im Ascus, *ibid.* (Generalversammlungsheft 1) **13** (67)–(78). *Pl.* 27. 1896.

———: Ueber das Verhalten der Kerne bei der Fruchtentwickelung einiger Ascomyceten, *Jahrb. wiss. Botan.*, **29**:655–685. *Pls.* 11–12. 1896.

———: Kerntheilung und freie Zellbildung im Ascus, *ibid.*, **30**:249–284. *Pls.* 11–12. 1897.

Killermann, S.: Elias Fries (1794–1878), *Z. Pilzkunde*, N.F., **6**(3):33–38, (4):49–56, (5):65–68. *Portrait.* 1927.

Kniep, Hans: Beiträge zur Kenntnis der Hymenomyceten, I–V, *Z. Botan.*, **5**:593–637, *Pls.* 2–5, 1913; **7**:369–398, *Pl.* 2, *Figs.* 1–20, 1915; **8**:353–359, *Pl.* 3, 1916; **9**:81–118, *Pls.* 1–3, *Figs.* 1–14, 1917.

———: Vererbungserscheinungen bei Pilzen, *Bibliographia Genetica*, **5**:371–478. *Figs.* 1–9. 1929.

Lafar, Franz: Technical Mycology. The Utilization of Microorganisms in the Arts and Manufactures. Translated (from the German) by Charles T. C. Salter, vol. 1, 312 pp., *Pl.* 1, *Figs.* 1–90, 1910; vol. 2, 748 pp., *Figs.* 91–208, 1903; London, Charles Griffin and Co.

Lindegren, Carl C.: The genetics of Neurospora, *Bull. Torrey Botan. Club*, **59**(2):85–102, *Figs.* 1–4; (3):119–138, *Figs.* 1–5, 1932; **60**(3):133–154, *Pl.* 9, *Figs.* 1–6, 1933.

Linnaeus, Carolus: Species plantarum, exhibentes plantas rite cognitas, ad genera relatas, cum differentiis specificis, nominibus trivialibus, synonymis selectis, locis natalis, secundum systema sexuale digestas, 2 vols., x + 1200 pp. and index. Stockholm, Laurentius Salvius, 1753.

Martin, G. W.: Systematic position of the slime molds and its bearing on the classification of the fungi, *Botan. Gaz.*, **93**(4):421–435. 1932.

Micheli, Pier' Antonio: Nova plantarum genera juxta Tournefortii methodum disposita, xxii + 234 pp. 108 *pls.* Florence, 1729.

Mounce, Irene: Homothallism and heterothallism in the genus Coprinus, *Brit. Mycol. Soc., Trans.* **8**:256–269. 1922.

Persoon, Christiaan Hendrik: Synopsis methodica fungorum. Pars prima et secunda, xxx + 706 pp. and index. *Pls.* 1–5. Göttingen, Heinrich Dieterich, 1801.

———: Mycologia Europaea seu completa omnium fungorum in variis Europaeae regionibus detectorum enumeratio, methodo naturali disposita, descriptione succincta, synonymia selecta et observationibus criticis additis, 3 sections, 9 vols. 852 pp. 30 *colored pls.* Erlangen, Johann Jacob Palm, 1822–1828.

Rabenhorst, Gottlob Ludwig: Deutschlands Kryptogamen-Flora oder Handbuch zur Bestimmung der kryptogamischen Gewächse Deutschlands, der Schweiz, der Lombardisch-Venetianischen Königreichs und Istriens. Band 1

Pilze, xxii + 614 pp. Band 2, Teil 1, Lichenen, xii + 129 pp. Leipzig, E. Kummer, 1844–1845. (For citation of ed. 2, see Chapter 18 of this book.)

SACCARDO, P. A.: Sylloge Fungorum omnium hucusque cognitorum, 25 vols., Pavia, Italy, 1882–1931. (For citation of the contents of each volume, see Chapter 18 of this book.)

SCHWEINITZ, LUDOVICUS DAVIDUS DE: Synopsis fungorum Carolinae superioris. Edited by D. F. Schwaegrichen, *Schriften der Naturforschenden Gesellschaft zu Leipzig*, **1**(4):20–131. *Pls.* 1–2. 1822.

———: Synopsis fungorum in America Boreali media degentium, *Trans. Am. Phil. Soc.*, N.S., **4**(8):141–318. *Pl.* 19. 1832.

STAKMAN, E. C., and J. J. CHRISTENSEN: Heterothallism in Ustilago zeae, *Phytopathology*, **17**(12):827–834. 1927.

STURCH, H. H.: Harveyella mirabilis (Schmitz & Reinke), *Ann. Botany*, **13**(49): 83–102. *Pls.* 3–4. 1899.

TARGIONI-TOZZETTI, GIOVANNI: Notizie della vita e opere di Pier' Antonio Micheli, vi + 446 pp. Florence, 1858.

TOURNEFORT, PITTON: Elemens de botanique ou methode pour connoître les plantes. Tome I, text, xviii + 562 pp. + index, 18 pp. Tomes II and III, 451 *pls.* Paris, L'Imprimerie Royale, 1694.

TULASNE, L. R. ET C.: Selecta fungorum carpologia, 3 vols., 782 pp. 61 *large pls.* Paris, Typographie Impériale, 1861–1865.

VAILLANT, SEBASTIAN: Botanicon Parisiense ou dénombrement par ordre alphabétique des plantes qui se trouvent aux environs de Paris, preface + 205 pp. 33 *pls.* Leyden and Amsterdam, Jean & Herman Verbeck, 1727.

VANDENDRIES, RENÉ: Recherches sur déterminisme sexuel des Basidiomycètes, *Mém. Acad. roy. Belg.*, 2me sér., **5**(1):1–98. 1923.

VON WETTSTEIN, FRITZ: Das Vorkommen von Chitin und seine Verwertung als systematisch-phylogenetisches Merkmal im Pflanzenreich, *Sitz. ber. Akad. Wiss. Wien, Math. naturw. Klasse, Abt. I*, **130**(1):3–20. 1921.

2

MYCETOZOA AND RELATED ORGANISMS

UNDER this title are brought together several groups of organisms which are probably interrelated but whose relationship to the true fungi is very doubtful. As mentioned in the preceding chapter the author follows de Bary (1887) in considering them to be more nearly related to members of the Animal Kingdom. They are discussed in this chapter because, for several hundred years before their life histories were known, botanists looked upon their mature fruiting bodies as fungi and gave them names, many of which are still maintained. Because they are more usually considered, even at this date, as fungi, they are given a place in this text.

These organisms all agree in the fact that they are naked during all stages of development except the culminating spore stage. Mostly for a portion or for the whole of this naked stage they are more or less amoeboid and ingest particles of food, rejecting the portions remaining after digestion is accomplished. At germination of the spore the escaping naked protoplast may be a simple uninucleate amoeboid cell, or myxamoeba, or it may be provided with one or two anterior flagella which eventually are retracted, leaving the cell a myxamoeba. To a greater or less degree both the planocytes and the myxamoebae are capable of multiplication by fission. Usually after a while the myxamoebae, through the division of the protoplast, become multinuclear plasmodia, or these may arise by the fusion of separate myxamoebae as well as by growth and nuclear division. In the better known groups there is usually a sexual stage, and eventually, just before spore formation, a meiotic division of the nuclei of the plasmodium so that the encysted spores contain haploid nuclei.

These organisms are usually classified by the zoologists as belonging to Phylum Protozoa, Class Sarcodina (or Rhizopoda). This classification is quite different from that mostly used by botanists. In the following pages an attempt will be made to reconcile these two viewpoints, yet maintaining so far as possible the nomenclature with which botanists, in particular mycologists, are more or less familiar. Four orders will be considered. The ordinal limits as here treated are far wider than those found in the monographs of the Slime Molds by Lister (1925), or Macbride and Martin (1934), and others. These four orders may be included in the Subclass Mycetozoa of Class Sarcodina, Phylum Protozoa. The four orders are Myxogastrales, the true Slime Molds, Acrasiales, Plasmodiophorales and Labyrinthulales. There are other organisms probably closely related to these that have been studied by zoologists but which have usually been neglected by botanists and are not considered here.

Order Myxogastrales or Slime Molds. Often called Myxomycetes or Myxogastres, the Myxogastrales or Slime Molds compose the most numerous group of this subclass. They are terrestial organisms or inhabitants of manure, decaying wood, bark, fungi, etc. Their spores are produced upon or within aerial sporangia and are more or less dependent upon wind for their distribution. In general the life history (with some modifications) is as follows: The spore, upon absorbing water, cracks open its cell wall and escapes as a single (in some species by division of the nucleus before germination, two) naked uninucleate swarm spore or planocyte, rounded posteriorly and tapering to the anterior end from which arises a single flagellum or in many cases two flagella. De Bary (1859) described and figured occasional biflagellate zoospores of *Fuligo*

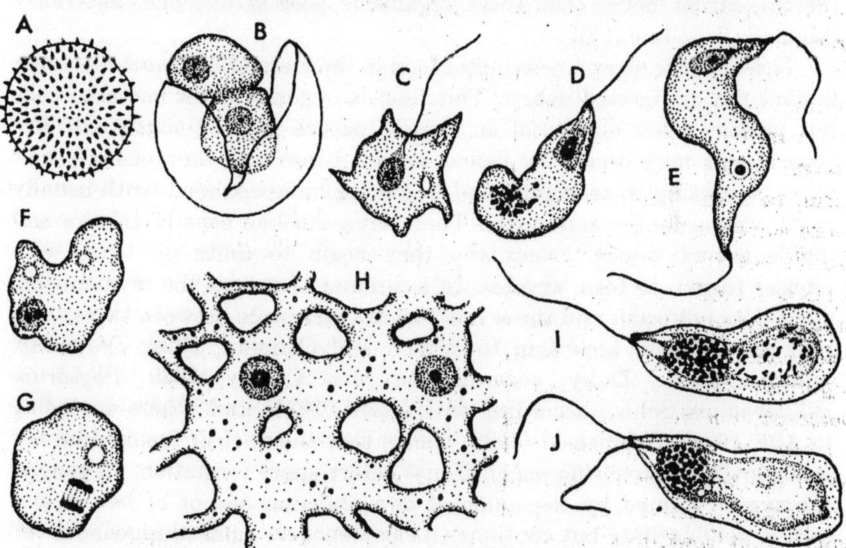

FIG. 1. Myxogastrales. (A–H) *Physarum polycephalum* Schw. (A) Spore. (B) Germinating spore. (C, D) Swarm spores. (E) Uniting swarm spores. (F, G) Amoeboid zygotes. (H) Portion of plasmodium. (I–J) *Physarella oblonga* (B. & C.) Morg. (I) Uniflagellate swarm spore. (J) Biflagellate swarm spore. (All figures much magnified.) (A–H, after Howard: *Am. J. Botany*, 18(2). I–J, after Sinoto and Yuasa: *The Botanical Magazine* [Tokyo], 48(574):722.)

septica (L.) Gmel. and *Trichia varia* Pers., as well as the normal uniflagellate zoospores. Gilbert (1927) reports that in one collection of *Stemonitis fusca* Roth he studied about one-fourth of the swarm spores bore two flagella and three-fourths bore only one. Sinoto and Yuasa (1934) and Yuasa (1935) showed that whether one or two flagella are produced the cell is uninucleate and possesses two blepharoplasts usually connected by a slender rod and joined to the nucleus by means of a single rhizoplast. In the case of two flagella each blepharoplast bears one, in the case of one flagellum one of the blepharoplasts remains without a flagellum. Ellison (1945) confirmed Yuasa's findings and demonstrated that the flagella whether single or two in number were all of the whiplash type, i.e., have a firmer, outer tubular portion beyond which the interior portion projects more or less or may be gathered into a ball. No flagellum of the tinsel type is produced, i.e., with numerous minute cilium-like lateral outgrowths. Elliott (1948) made intensive studies of the germinating spores of 11 species of this order and found that biflagellate swarm spores were present in all these species, up to nearly 100 per cent in *Fuligo septica* (L.) Gmel. In all cases except in the genus *Stemonitis* one flagellum is long and the other very short and often somewhat recurved. Unless seen in profile the shorter flagellum is difficult of demonstration. This probably accounts for the earlier belief that these organisms possess but one, anteriorly directed, flagellum. (Fig. 1.)

These motile swarm cells ingest food in the manner of *Amoeba,* leaving behind the undigested debris. This food is in many cases bacterial cells but various other objects of an organic nature may be consumed. The swarm cells may divide by fission several times and then change their form, retracting their flagella and becoming more rounded, with usually more conspicuous pseudopodia. These myxamoebae usually enlarge and divide several times. Eventually they begin to unite by twos, with nuclear fusion, to form zygotes. In a number of genera the myxamoebal stage does not occur and the sexual fusion takes place between two swarm cells (*Reticularia,* according to Wilson and Cadman, 1928; *Didymium difforme* (Pers.) Duby, according to Miss Cayley, 1929; *Physarum polycephalum* Schw., according to Howard, 1931; and others according to Abe, 1934). Skupienski (1928) claims that the nuclear fusion does not occur in *Didymium difforme* until just before spore formation. In general the zygote formed by the union of two myxamoebae or of two swarm cells is nonflagellate but continues its existence as a naked amoeboid cell which ingests its food and grows in size, with accompanying mitotic division of the nucleus. This multinucleate structure is called a plasmodium. Zygotes or small plasmodia may fuse with other zygotes and plasmodia so that growth is both internal and by accretion. Plasmodia may ingest and feed upon swarm spores and myxamoebae. The plas-

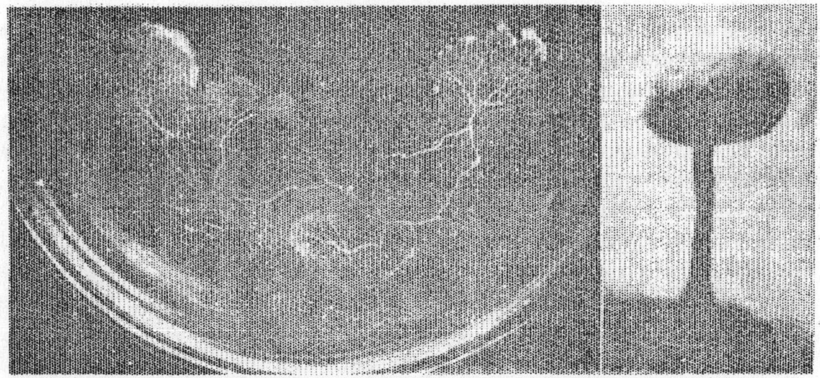

FIG. 2. Myxogastrales. *Didymium squamulosum* (A. & S.) Fr. (A) Plasmodium. (B) Sporangium. (Courtesy, D. M. Cayley.)

modium creeps through the soil or rotten wood or decaying vegetable matter, fruiting bodies of fungi, etc., digesting the food suitable for it and increasing in size and in the number of contained nuclei. Howard and Currie (1932) showed that the plasmodia of various species of slime molds destroy the mycelium and sporophores of many types of Hymenomyceteae with great rapidity. On the other hand some species can be grown very successfully upon various standard culture media. Eventually it emerges as a fine or coarse network upon the surface of the substratum. It may be lobed and exhibit a continual creeping motion in various directions. In diameter it may vary from a few millimeters up to 15 or 20 centimeters, and in color from white to yellow, orange, red, brown, violet, and other colors. Camp (1937) showed that it contains true contractile vacuoles. Food is obtained by invagination, the lining of the food vacuole being a portion of the original external surface of the plasmodium.

Eventually the plasmodium heaps itself up somewhat, on the exterior of its substratum, or even creeps up adjacent objects, and there undergoes the changes which lead to the production of the fructifications. These may be separate from one another or may be crowded together into a compound structure. In the Suborder Endosporeae there appears externally, a noncellular peridium secreted by the plasmodium. This may be of various thicknesses according to the species, and may or may not be encrusted with lime. Within the fructification are secreted numerous noncellular threads and beams which form a sort of framework. This is the capillitium. Its structure and arrangement are of great value in classifying this difficult group of organisms. Between the threads of the capillitium, usually following simultaneous meiotic divisions of all nuclei of the plasmodium, the protoplasm rounds up into innumerable small uninucleate cells which secrete cell walls. In *Physarum polycephalum* it has been shown by Howard (1931) that the cell walls are secreted before the spores round up, so that at first a continuous mass of polyhedral cells is produced. The cells then begin to round up and the cell walls split, thus forming the spores. The walls of some species are said to contain cellulose, but some investigators deny this. C. van Wisselingh (1898) claimed the presence of cellulose in *Didymium squamulosum* (A. & S.) Fr., but denied its presence in *Fuligo septica* (L.) Gmel. F. von Wettstein (1921) studied the composition of the spore wall in seven other genera and could not demonstrate the presence of cellulose except, doubtfully, in *Stemonitis* and *Reticularia*. Neither author found chitin in any of these. The spores are most often violet, purple, or brown. (Fig. 2.)

By the rupture or dissolution of the peridium the spores are permitted to escape, the capillitial threads preventing all the spores from escaping at once. The fructifications may be sessile or stalked, round or elongated, scattered or crowded, almost microscopic or, in the case of some of the compound fructifications, up to 10 cm. in length and 4–5 cm. in width and thickness.

In the genus *Ceratiomyxa*, the only genus of the Suborder Exosporeae, the spores are extruded externally from the fructification instead of being produced internally. The studies of H. C. Gilbert (1935) seem to indicate that this genus may be looked upon as an extreme modification of the condition in Suborder Endosporeae. As in that suborder the sexual fusion occurs by the union of two or more swarm spores followed after some time by the union of pairs of nuclei. The zygotes then enter decaying wood within which they feed, perhaps upon the mycelium of the fungi causing the decay. Eventually the plasmodium creeps out of the wood in which it has been growing and secretes a massive central core of mucilaginous nature, on the outside of which the protoplasm creeps as a thin, somewhat reticulate, sheet. In this sheet a further division of the nuclei is followed by a cleavage of the protoplasmic layer into uninucleate naked cells, flattened at the contact surfaces. Each of these secretes a stalk of material similar to that of the main central core, on whose tip the thin-walled spore takes its position. Within this the nucleus divides meiotically to form four haploid nuclei, the spore falling off before or after this stage. It cleaves into four cells in each of which mitotic nuclear division and cleavage occur, so that 8 naked flagellate cells arise from each spore. Gilbert homologizes the stalk upon which the spore sits and the spore with the stalk and sporangium respectively in the Endosporeae. Olive (1907) disagrees with Gilbert, claiming that the nuclear fusions do not occur until just before the formation of the stalks for the external spores. Jahn (1936) agrees with Gilbert that sexual union occurs by the fusion of the swarm cells and their nuclei but locates the meiotic division of the nuclei, as Olive did, in the plasmodium before the stalked spores are pro-

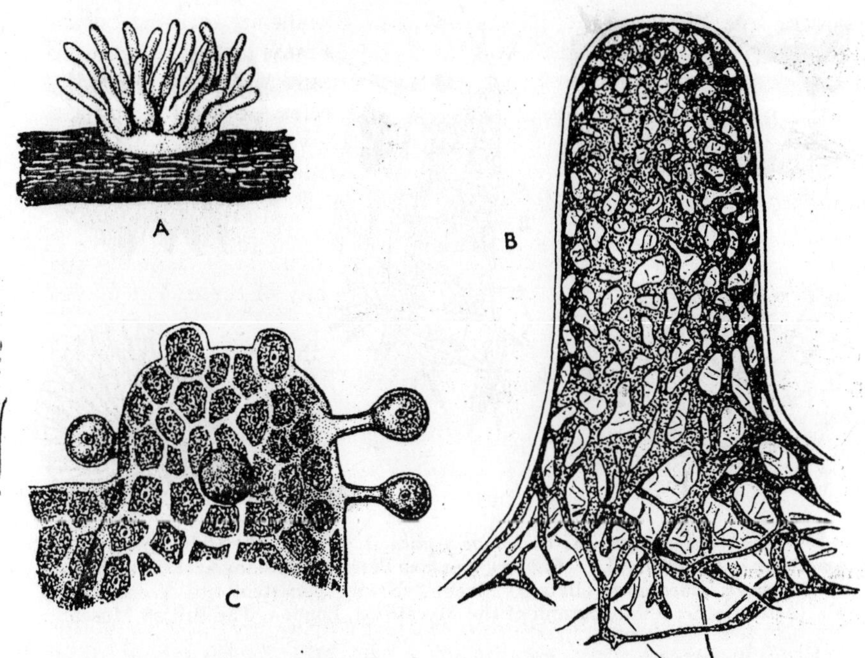

FIG. 3. Myxogastrales. *Ceratiomyxa fruticulosa* (Muell.) Macbr. (A) Habit sketch. (B) Development of young fruiting body from plasmodium. (C) Portion of mature sporophore, showing production of external spores. (After Famintzin and Woronin. From Engler and Prantl: Die natürlichen Pflanzenfamilien, Leipzig, W. Engelmann.)

duced. In the Endosporeae, Skupienski (1928) claims that the nuclear fusion does not occur in *Didymium difforme* until the time for spore formation is at hand. Then the hundreds of nuclei in the plasmodium unite simultaneously in pairs, this being followed closely by the meiotic divisions which give rise to the nuclei of the spores. His interpretation of the nuclear cycle in *Didymium* thus agrees with Olive's for *Ceratiomyxa.*

The Suborder Exosporeae contains but one genus, *Ceratiomyxa*, with two or three species. Their fructifications appear as simple or branched columns or even as poroid structures 2–6 mm. tall, light in color, growing on decayed logs, stumps, etc. (Fig. 3.)

The Suborder Endosporeae is recognized by Macbride and Martin (1934) as possessing 59 genera and 380 species. The classification of the numerous genera and species is based upon the type of the peridium and the presence or absence of lime in it and in the elements of the capillitium, the microscopic structure of the latter, the formation of simple or compound fructifications and their size, shape, and color. Further characters are the color, shape, size, and markings of the spores as well as the characters of the plasmodium.

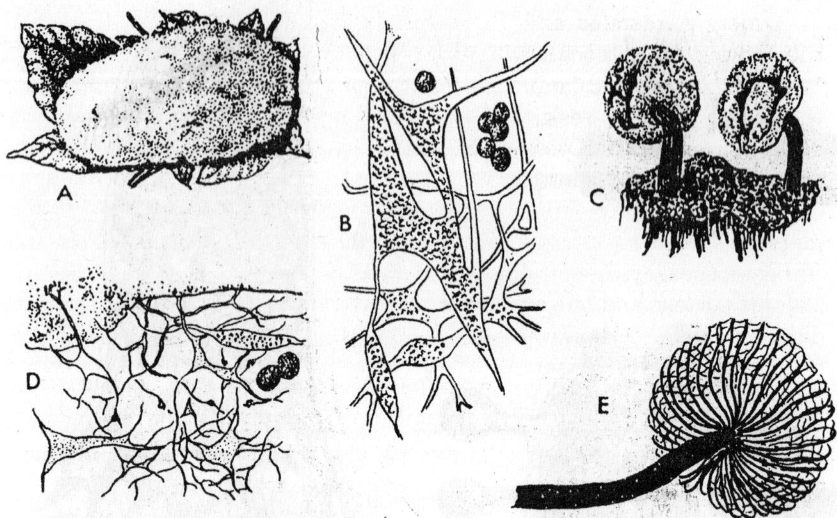

Fig. 4. Myxogastrales. (A, B) *Fuligo septica* (L.) Gmel. (A) Habit drawing. (B) Capillitium and spores. (C, D) *Physarum nutans* Pers. (C) Sporangia. (D) Capillitium. (E) *Dictydium cancellatum* (Batsch) Macbr., showing persistent ribs of sporangium wall. (After Lister: A Monograph of the Mycetozoa, London, The British Museum.)

Mention should be made of *Stemonitis*, with purple or rust-brown spores, clustered, stalked fruit bodies with the stalk extending upward as an axial strand (columella) from which branch off the capillitial threads which are combined into a loose network that is surrounded by an evanescent peridium. *Physarum* has clustered, separate or compacted, sessile or stalked, lime-encrusted sporangia whose capillitial threads are expanded here and there and filled with lime granules. *Fuligo* has a similar capillitium but the spore fruits are united into a single large convolute "aethalium." This is one of the largest of the Slime Molds. In *Dictydium* the stalked sporangium has no internal capillitium but when the peridium disappears it leaves numerous longitudinal ribs that run from base to apex like the lines of the meridians on the globe. *Arcyria* and *Trichia* have sessile or stalked fruits. The capillitium lacks a columella and consists of a tangle of tubular threads with characteristic thickenings such as spines, warts, rings, etc. (Fig. 4.)

The fructifications of Slime Molds may be found on rotten logs or stumps, on sawdust, leaves, beams in moist cellars, and frequently on blades of grass or other vegetation in lawns. Very dry habitats do not favor their occurrence. In moist weather the plasmodia may be found creeping about on the surface of, or emerging from, the various substrata within which they develop. It is possible to grow some species in pure cultures, from the spores to the maturity of the fructifications.

Orders Acrasiales and Labyrinthulales. Sometimes associated with the Subclass Mycetozoa are the two orders Acrasiales and Labyrinthulales. Whether they should be included in this subclass or in distinct subclasses the author will not seek to decide. In the first of these two orders the swarm spores do not possess flagella although they are amoeboid upon emerging from the spore wall. They consist of saprophytic or parasitic organisms occurring on dung, decaying wood, leaf mold or other organic matter. The spores upon germination give rise to naked amoeboid cells (myxamoebae) with or without conspicuous pseudopodia and containing a single nucleus and one or more food vacuoles. Within these vacuoles are digested the bacteria and other bits of organic matter that serve as food for the organism. Raper (1937) has shown that the myxamoebae of *Dictyostelium discoideum* Raper are strictly parasitic upon bacteria, many kinds of which may serve as their food. There is no indication of a symbiotic relation between the bacteria and the myxamoebae. The latter enlarge and divide several times and in *D. mucoroides* Bref., according to Skupienski (1920), then unite by twos. The resulting zygotes seem now to be mutually attracted to one another and draw together into heaps of naked cells that maintain their individuality. These heaps of separate cells are called pseudoplasmodia in contrast to the true plasmodia that are found in the Myxogastrales. At this stage pressure or, in some cases, exposure to bright light will cause the pseudoplasmodium to separate into its individual cells which reassemble again elsewhere. Skupienski claims that in *Dictyostelium* the cells eventually unite into a true plasmodium within which the nuclei undergo two more divisions (probably meiotic).

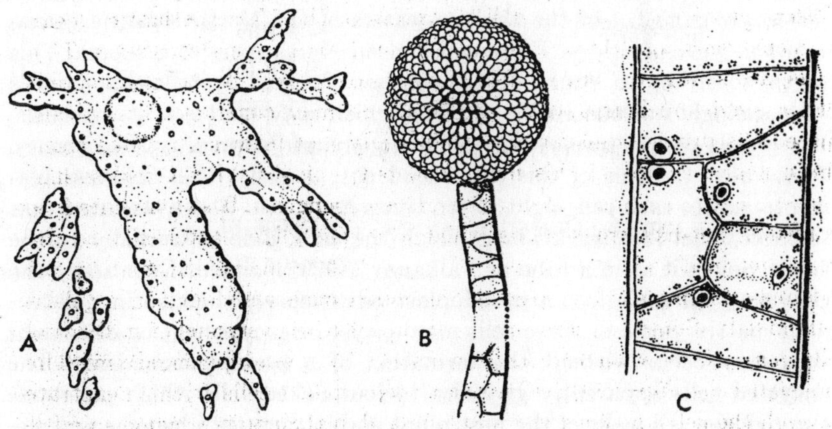

FIG. 5. Acrasiales. (A) Pseudoplasmodium of *Polysphondylium violaceum* Bref (B, C) *Dictyostelium mucoroides* Bref. (B) Stalk and terminal ball of spores. (C Details of cells of stalk. (After Olive: *Proc. Boston Soc. Natural History*, **30**:451–513.

In this genus the plasmodium heaps itself up, at first, into a conical structure in the basal part of which cellulose walls are produced, separating it into the more or less hexagonal cells of the stalk. In the upper, head-like part it breaks into separate, rounded cells, which also develop cellulose walls and become the spores. These are embedded in a slimy drop. Sexual reproduction and the eventual formation of a true plasmodium have not been reported but possibly occur in the other genera of the order. The myxamoebae of *Copromyxa* and *Guttulina* lack conspicuous pseudopodia. The fruiting bodies of the former are sessile, of the latter short-stalked. *Dictyostelium, Acrasis*, and *Polysphondylium* produce myxamoebae with well-developed pseudopodia. All produce stalked fruiting bodies (branched in the last named) with ovoid or spherical heads of spores. Olive (1902) gives a monograph of this order, describing all known species. (Fig. 5.)

The Acrasiales differ from the Myxogastrales in the absence of a flagellate swarm-spore stage preceding the formation of the myxamoebae, and in the fact that the naked cells remain distinct for a long time in the pseudoplasmodium, forming a true plasmodium only a short time before the fruiting body is produced, if at all. Furthermore no capillitium is formed nor a peridium, although the slime that lies between and around the spores may represent these structures.

The Labyrinthulales include parasitic forms attacking algae and other aquatic plants in both marine and fresh-water habitats, and possibly one genus of dung-inhabiting saprophytes. There are several species in the genus *Labyrinthula*, of which *L. macrocystis* Cien. was shown by Renn (1935) to be the cause of the destruction of most of the eel grass (*Zostera marina* L.) of the shallow waters of the North Atlantic Ocean in recent years, both on European and on North American shores. This genus, according to Young (1943) consists of naked, spindle-shaped cells with a single nucleus and a vacuole which may contract at intervals. These cells divide transversely or obliquely and form a rope-like mass. Those near the tips of this mass send out, apically, thin colorless filaments, one to each cell, eight to ten times its length. These filaments fuse to form a net-like track along which the cells glide—externally to the track, not in it as in a tube as Valkanov (1929) claimed. Eventually the cells may assemble into a pseudoplasmodial mass embedded in a gelatinous matrix. Some of these cells may encyst or encystment of individual cells may occur without the formation of a pseudoplasmodium. The encysted cells apparently give rise to four naked cells which penetrate through the cell walls of the host plant and in their turn become spindle-shaped and start the development of new "net-plasmodia." Dangeard (1932) suggested that possibly there is a sexual stage somewhere in the life cycle but did not demonstrate where it occurred. The naked cells emerging from the cysts are nonflagellate in *Labyrinthula* and anteriorly uniflagellate in the possibly related *Labyrinthomyxa.* (Fig. 6.)

Order Plasmodiophorales. This order to which eight or more genera have been ascribed is not definite as to its limits or relationships. The type genus *Plasmodiophora* differs in so many points from many of the genera assigned to the order that it may be necessary to limit the order

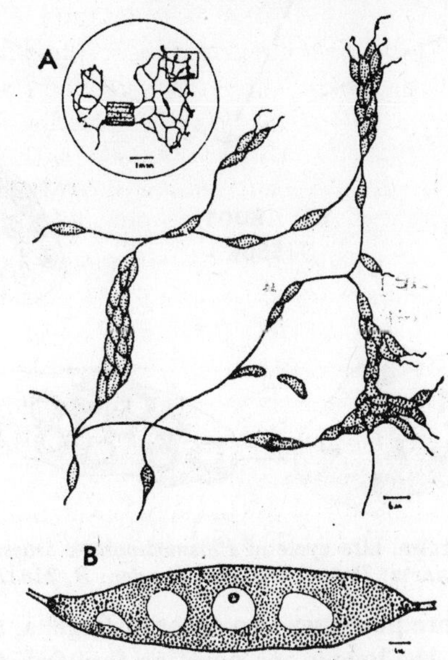

FIG. 6. Labyrinthulales. *Labyrinthula macrocystis* Cien. (A) "Net plasmodium," showing clumps of cells on the "tracks" and cells sending out threads which are the beginnings of new tracks. (B) Highly magnified single cell lying on track. (Courtesy, Young: *Am. J. Botany,* **30**(8):586–593.)

to one genus and to place the others in another order, perhaps not at all closely related. The type species *P. brassicae* Wor., causing swellings and malformations of the roots of Brassicaceae (Cruciferae), will be discussed first and then some of the others with remarks as to their differences and possible relationship. *Woronina,* placed here by Sparrow (1943), is returned to a position in the Order Lagenidiales because of its cellulose walls and the possession of flagella of two types. Possibly other genera may have to be relegated to that position also. In the following account many points still are disputed or assumed without adequate confirmation.

The spores of *Plasmodiophora brassicae* possess a dark wall which, according to van Wissenlingh (1898), contains chitin but not cellulose. Upon germination, usually a single zoospore emerges, although Honig (1931) and Rochlin (1933) claim that only a nonflagellate amoeba is

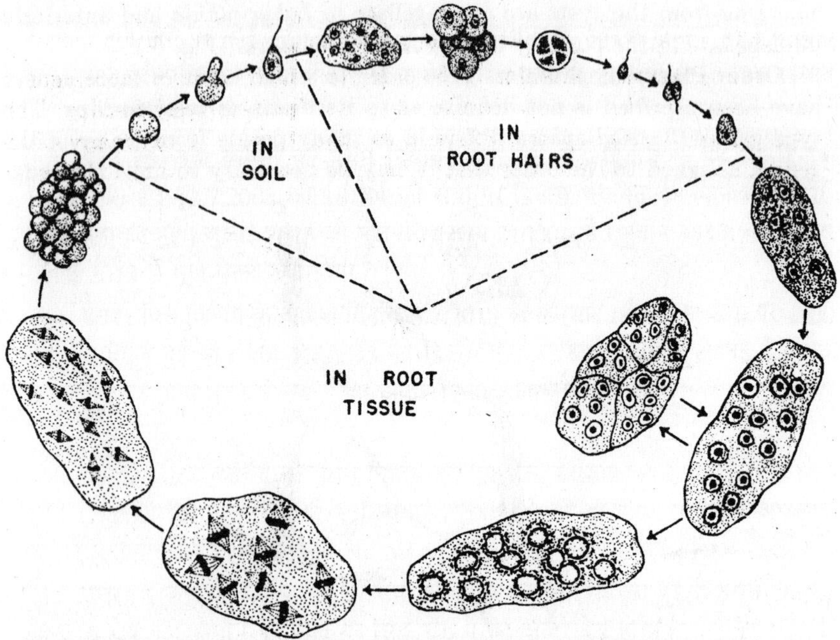

FIG. 7. Plasmodiophorales. Life cycle of *Plasmodiophora brassicae* Wor. (After Cook and Schwartz: *Trans. Roy. Soc. London, B*, **218**:283–314.)

formed. The zoospore possesses two anterior flagella, as demonstrated by Ledingham (1934), the longer one pointing forward, the shorter directed almost at right angles. Ellison (1945) showed that both of these flagella are of the truncated whiplash type, neither showing any tinsel structures. This is a character that indicates that *Plasmodiophora* is not related to the Olpidiopsidaceae, whose anteriorly biflagellate zoospores possess one flagellum of each of these two types. The zoospore is more or less amoeboid. When it comes into contact with a root hair or epidermal cell of the root of a suitable host, the flagella disappear and a hole is dissolved in the host cell wall through which the amoeba enters, then this hole is closed, presumably by host action. Cook and Schwartz (1930) showed that within the root hair this amoeba enlarges and mitotic divisions of the nucleus occur until a small plasmodium is formed, containing from a few up to one hundred nuclei according to Fedorintschik (1935). This plasmodium cleaves into uninucleate cells around each of which a thin wall is formed. Whether this wall contains cellulose or chitin has not been determined. They give rise to structures called sporangia or gametangia. The nucleus divides mitotically two or three times and thus there are formed four to eight uninucleate, anteriorly flagellate swarm cells, smaller than those that emerge from the resting spores. The number and character of the

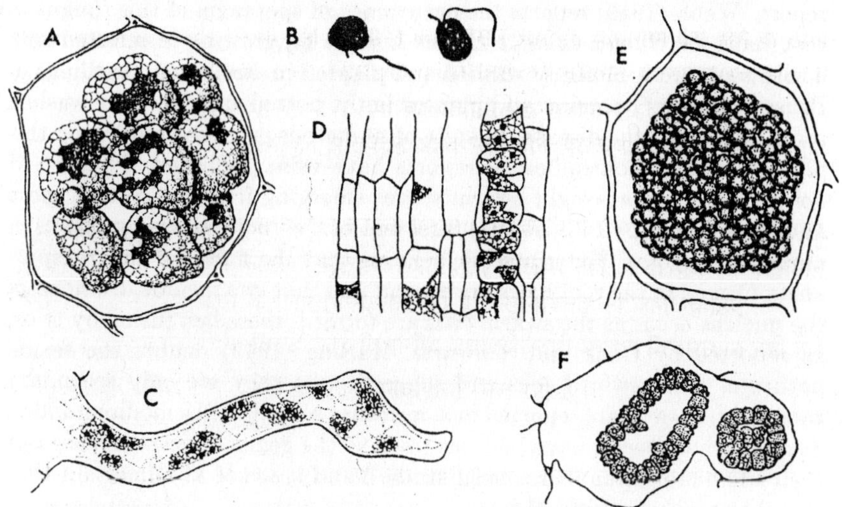

FIG. 8. Plasmodiophorales. (A) Spore balls of *Spongospora subterranea* (Wall.) Lagerheim. (B–E) *Plasmodiophora brassicae* Wor. (B) Bifiagellate zoospores. (C) Amoebae in root hairs. (D) Young plasmodia in root cortex. (E) Host cell filled with spores. (F) *Sorosphaera veronicae* Schroet. (A, after Osborn: *Ann. Botany*, **25**(98):327–341. B, after Ledingham: *Nature*, **133**(3362):534. C–E, after Chupp: *Cornell Univ. Agr. Sta. Bull.*, **387**:421–452. F, after Palm and Burk: *Arch. Protistenk.*, **79**(3):263–276.)

flagella have not been determined in this species. The foregoing authors believe that the swarm cells unite by twos and that the resulting amoeboid zygotes are the origins of the large plasmodia which give rise to the resting spores. The zygotes and young plasmodia arising from them may possibly unite to form larger plasmodia. These are slowly amoeboid and at least in their younger stages seem to be able to pass from cell to cell of the host. The young plasmodia apparently may undergo division. As growth of the parasite progresses, the host cells multiply hyperplastically and the infected cells undergo hypertrophy. Eventually the cell contents of the invaded cells are almost completely exhausted and the cell is practically filled by the plasmodium. The nuclei of the latter then undergo two rapidly succeeding divisions which are believed to be meiotic, although cytologic studies have not definitely proved this. Then the protoplasm rounds up into uninucleate spores around each of which a dark chitin wall is secreted. There is no enclosing membrane around the mass of spores which lie free in the host cell. Upon decay of the root the spores are set free. In the laboratory they are brought to germination only with difficulty. The wall cracks open and the zoospore emerges with the two flagella in advance of the cell body. At this stage, the author has observed the two flagella on the living zoospore before it has escaped from the spore wall, so that there is no doubt as to the correctness of Ledingham's report. Webb (1949) reports the occurrence of sporangia of this fungus in root hairs of *Rumex* sp. and *Holcus lanatus* L., growing in infested soil. These roots were carefully washed and planted in sterile soil. Seedlings of Brussels sprouts (*Brassica* sp.) planted in the pots along with these washed roots developed the disease. In spite of numerous extended studies on this parasite, many contradictory reports have been published. Thus P. M Jones (1928) reports eight swarm spores emerging from the resting spore and their fusion by twos so that infection of the root hairs is by means of amoeboid zygotes. Fedorintschik believes that the nuclei of the zoosporangia formed in the root hairs are diploid and that two meiotic divisions of the nucleus occur as the swarm cells are formed, these last fusing by twos, as reported by Cook and Schwartz. Karling (1942) doubts the sexual nature of these swarm cells and suggests that they are only secondary zoospores, such as are reported in some other so-called Plasmodiophorales. Several other species have been ascribed to the genus *Plasmodiophora* but their life histories have not been studied and it is not at all certain that they belong here. (Figs. 7, 8.)

The other genera which Karling admits to this order, which has but one family, are as follows:

Tetramyxa: causing pronounced hypertrophy of host tissues, developing spores in twos or fours, and with sporangia and sexual stage unknown, and zoospores not seen.

Two or possibly three species in stems or roots of seed-plants.

Octomyxa: spores usually develop in eights, sporangia numerous, zoospores anteriorly biflagellate, one flagellum directed forward, a longer one posteriorly, when swimming. No sexual stage known. Cell walls not of cellulose. Causes gall-like enlargements in *Achlya glomerata* Coker.

Sorosphaera: spores compacted into a hollow sphere, zoosporangia are formed (Ledingham, 1939) and their zoospores anteriorly biflagellate and heterocont. Spore walls do not contain cellulose.

Two species: *S. veronicae* Schroeter forms galls in the stems of *Veronica; S. radicalis* Cook & Schwartz forms galls in the roots of grasses. In the former zoosporangia are unknown, in the latter thin-walled zoosporangia are formed. Sexual reproduction not demonstrated.

Sorodiscus: spores formed in a disk-shaped sorus usually in two layers, much like a flattened spherical sorus of *Sorosphaera.* Zoosporangia and character and number of flagella unknown. Sexual reproduction not observed.

Two species: one in stems of *Callitriche* and one in *Chara.*

Spongospora: spores in a hollow sphere with several openings, zoosporangia are formed according to Ledingham (1935), zoospores anteriorly biflagellate and heterocont, similar in size whether from sporangia or from resting spores. Sexual fusion of myxamoebae reported by Cook (1933).

Best known species is *S. subterranea* (Wall.) Lagerheim causing the powdery scab of the tubers of potato (*Solanum tuberosum* L.). It also attacks the stems and roots of this host and of related plants.

Ligniera: spores in irregular clusters, usually not causing hypertrophy of tissues except of root hairs in one species. This, according to Palm and

Burk (1933), is a very doubtful genus. Zoosporangia are known but further study is needed to determine whether the zoospores are truly uniflagellate anteriorly as reported or biflagellate.

Polymyxa: spores in irregular clusters in the roots of grasses, zoosporangia with long necks are abundant, zoospores from the resting spores as well as those from the zoosporangia are anteriorly biflagellate and heterocont (Ledingham, 1939). Sexual reproduction has not been observed. Hypertrophy of the host cells is not brought about.

Several other genera have been ascribed to this family but Karling (1942) is doubtful as to the correctness of their assignment here (*Sorolpidium, Anisomyxa, Trematophlyctis, Sporomyxa, Peltomyces, Cystospora*).

The relationship of the four orders—Myxogastrales, Acrasiales, Labyrinthulales, and Plasmodiophorales—described in the foregoing, is not admitted by all. Sparrow (1943) places the last named order in his group Biflagellatae, in which he also includes the Saprolegniales, Leptomitales, Lagenidiales, and Peronosporales. The author admits the similarity of *Octomyxa* and *Polymyxa* to *Woronina*, which he includes in the Lagenidiales. It is possible that with further study of these and some other genera now assigned to the Plasmodiophorales they may be found to have cell-wall composition and flagellar structure that will compel their removal from their present position. Be that as it may, the order stands or falls by its type species *Plasmodiophora brassicae*. Therefore it is with this species that comparisons must be made.

In these four orders the resting spores possess a cell wall whose chief component is chitin in *Plasmodiophora* but is not definitely determined in the other groups. Upon germination a naked amoeboid uninucleate cell is set free, which may lack flagella (Acrasiales, *Labyrinthula* of the Labyrinthulales) or which may have one or two anteriorly attached flagella (Myxogastrales, *Labyrinthomyxa* of the Labyrinthulales, and *Plasmodiophora*). Where two flagella are present they are both of the whiplash type, neither being of the tinsel type. Sexual reproduction where reported (Myxogastrales and *Plasmodiophora*) is by the union of two myxamoebae or two flagellate swarm spores. The vegetative body of the organism is a plasmodium or a more or less loose aggregation of myxamoebae. Eventually this separates into naked uninucleate spores around which a spore wall is produced. Because of the naked vegetative body of plasmodial nature the zoologists have included these four groups among the Protozoa. The protozoologist Kudo (1946) places these four groups in Phylum Protozoa, Class Sarcodina, including the Labyrinthulales in his Order Proteomyxa, and the other three in his Order Mycetozoa.

The genus *Reticulomyxa* has recently been described by Miss Nauss (1949). The organism consists of a central multinucleate plasmodium with radiating, forking, and anastomosing branches which appear to be the chief organs for the capture of the food particles upon which it lives.

In some regards this seems to lie in a position intermediate between the Myxogastrales and the Proteomyxa, while in some structural features it shows similarity to the Labyrinthulales. No definite formation of sporangial structures has been observed.

In the author's opinion the primitive ancestors of these four groups were more or less colonial amoeboid organisms consisting of encysted cells at one stage of their life history. From these emerged biflagellate swarm cells, both flagella being of the whiplash type and nearly or quite equal. This biflagellate condition has persisted in *Plasmodiophora* and is found in a considerable number of the swarm cells of the Myxogastrales, the presence of two blepharoplasts remaining even when one flagellum is missing. In the Labyrinthulales in one genus no flagella are known and in one genus the swarm spores are described as anteriorly uniflagellate. The questions as to blepharoplast number and type of flagellum are not solved. In the Acrasiales the flagella are entirely lacking. The Myxogastrales are undoubtedly the furthest developed from the evolutionary standpoint, in the development of sporangia and capillitium and adaptation to aerial dispersal of the encysted spores. The Acrasiales are probably closely related. The Labyrinthulales and *Plasmodiophora* are water or soil organisms and lack the complicated structure of the Myxogastrales.

Key to the More Important Orders of Mycetozoa

Saprophytes or surrounding and ingesting fungi, bacteria, etc. Sporangia aerial.
  Sporangia with thin or thick peridium, and mostly with a capillitium.
    Spores upon germination producing an anteriorly uni- or biflagellate swarm cell (rarely nonflagellate myxamoeba). — Order Myxogastrales
  Sporangia without peridium and capillitium, spores embedded in a mass of slime.
    Spores producing nonflagellate myxamoebae. — Order Acrasiales
Parasites in the cells of algae and of submerged aquatic plants, forming net-plasmodia.
  No aerial sporangia. Spores upon germination producing myxamoebae or anteriorly uniflagellate zoospores. — Order Labyrinthulales
Parasites producing plasmodia within the cells of roots and stems of higher plants, a few in algae and aquatic fungi. Swarm spores anteriorly biflagellate. — Order Plasmodiophorales

Key to the More Important Families and Genera of Myxogastrales

(*Based in Part Upon Macbride and Martin*)

Spores produced externally. — Family Ceratiomyxaceae
  Only genus. — *Ceratiomyxa*
Spores produced internally.
  Spores violet, brown, or purplish gray (rarely ferruginous or colorless). Capillitium always present.
    Sporangia with lime granules (calcium carbonate) in capillitium and often in peridium also. — Family Physaraceae

Fructification aethalioid, capillitium with lime knots. *Fuligo*
Fructification of plasmodiocarps or of separate sporangia.
Capillitium of nearly uniform anastomosing tubules, containing lime granules throughout. *Badhamia*
Capillitium of threads containing lime knots.
Peridium incrusted with lime.
Sporangium dehiscent circumscissilely. *Craterium*
Sporangium by introversion thimble-like or vase-like, dehiscence by petal-like lobes. *Physarella*
Sporangium not introverted, dehiscence irregular. *Physarum*
Peridium smooth and shining. *Leocarpus*
Sporangia with lime on or in peridium, none in the capillitium. Family Didymiaceae
Calcareous deposits in form of stellate crystals.
Aethalioid. *Mucilago*
Plasmodiocarpous or of separate sporangia. *Didymium*
Calcareous deposits not stellate.
Lime in form of closely adjacent peg-like processes. *Physarina*
Lime in scattered flattened scales. *Lepidoderma*
Lime forming a continuous shell, peridium mostly double. *Diderma*
Sporangium without lime in capillitium and peridium (in stipe and columella in some species of *Diachea*). Capillitium of more or less reticulately anastomosing threads.
Columella usually well developed, capillitium arising along its whole length. Family Stemonitaceae
Stipe and columella calcareous or waxy. *Diachea*
Stipe and columella never calcareous or waxy.
Capillitial branches forming definite outer network. *Stemonitis*
Capillitial branches not forming surface network. *Comatricha*
Columella short or well developed, capillitium arising at apex or in apical portion. Family Lamprodermaceae
Columella reaching apex of sporangium, capillitium arising from a disk at its top. *Enerthenema*
Columella one-third to half the height of sporangium, peridium iridescent, capillitium dense, bushy, branches tapering. *Lamproderma*
Columella short, capillitium bushy, the tips of the branches with disk-like fragments of peridium. *Clastoderma*
Spores violet to ochraceous or pale, columella and true capillitium lacking.
Outer layer of peridium flaking off leaving the inner layer of reticulate thickenings which surround the spore mass. Family Cribrariaceae
Meshes of the reticulum more or less isodiametric in the upper portion of the sporangium. *Cribraria*
Thickenings of the peridium wall like the meridians on a globe with very delicate cross connections. *Dictydium*
Peridium wall not reticulately thickened.
Sporangia separate, sometimes plasmodiocarpous. Family Liceaceae
Sporangia mostly sessile, not dehiscing by a lid. *Licea*
Sporangia mostly stalked, opening by a lid. *Orcadella*

Sporangia closely appressed, retaining their walls, dehiscent at apex. Family Tubiferaceae
Sporangia cylindrical, densely clustered. *Tubifera*
Sporangia ovate in a loose cluster on a common stalk. *Alwisia*
Sporangia sessile, flattened, closely clustered into a pseudoaethalium, rarely scattered. *Liceopsis*
Forming an aethalium, no sporangial walls remaining at maturity.
Pseudocapillitium thread-like or of perforate or frayed sheets, spores ochraceous. Family Reticulariaceae
Surface alveolar, made up of the caps of the sporangial units. *Dictydiaethalium*
Surface not alveolar.
Pseudocapillitium of flat irregular plates fraying out into threads. *Reticularia*
Pseudocapillitium of broad perforated plates. *Enteridium*
Pseudocapillitium of colorless branched tubes, spores pale. Family Lycogalaceae
Only genus. *Lycogala*
Spores yellow to ochraceous, capillitial threads with characteristic markings: spirals, rings, spines, cogs, etc., sometimes faint or wanting.
Capillitium a network or of separate threads, marked with spiral bands, threads coarse. Family Trichiaceae
Capillitial threads separate, spirals irregular or faint. *Oligonema*
Capillitial threads separate, spirals distinct, regular. *Trichia*
Capillitial threads forming a network.
Spirals regular. *Hemitrichia*
Spirals irregular or obscured by reticulations. *Calonema*
Capillitium a network of coarse threads attached to lower part of peridium, markings various, never of spirals. Family Arcyriaceae
Capillitium elastic, pushing out far beyond the cup-like base of the peridium. *Arcyria*
Capillitium not elastic. *Lachnobolus*
Capillitial threads slender, warted or spinulose or smooth.
Capillitial threads solid, peridium usually single. Family Dianemaceae
Capillitial threads hair-like, coiled. *Margarita*
Capillitial threads nearly straight. *Dianema*
Capillitial threads hollow, peridium double. Family Perichaenaceae
Capillitium warty or spiny, dehiscence irregular. *Ophiotheca*
Capillitium as above, dehiscence circumscissile. *Perichaena*

Key to the Genera of Order Acrasiales

Myxamoebae with inconspicuous, rounded pseudopodia.
Fruiting bodies sessile. *Copromyxa*
Fruiting bodies short-stalked. *Guttulina*
Myxamoebae with well-developed, more or less acute, pseudopodia. Fruiting bodies stalked.
Spores in rounded slime-covered heads.
Stalks not branched. *Dictyostelium*
Stalks branched. *Polysphondylium*
Spores in chains, stalks not branched. *Acrasis*

Key to the Genera of Order Labyrinthulales

Germinating resting spores producing nonflagellate cells. ***Labyrinthula***
Germinating resting spores producing anteriorly uniflagellate cells. ***Labyrinthomyxa***

Key to the Genera of Plasmodiophorales

(*Based upon Karling, 1942*)

Resting spores not united, mostly nearly filling the host cell. Zoosporangia small, producing few zoospores. ***Plasmodiophora***
Resting spores in small clusters or united in more or less compact cystosori. Except *Octomyxa* parasitic in tissues of Higher Plants.
- Spores mostly in 4's or 2's. Zoosporangia unknown. ***Tetramyxa***
- Spores mostly in 8's. Zoosporangia small. In *Achlya.* ***Octomyxa***
- Spores united to form a hollow sphere. Zoosporangia small. ***Sorosphaera***
- Spores forming a two-layered flattened disk. Zoosporangia unknown. ***Sorodiscus***
- Spores in a rounded sponge-like mass perforated by large canals. Zoosporangia mostly small. ***Spongospora***
- Spores in variable-sized masses, sometimes loose in the cell which they usually do not fill. Zoosporangia small. ***Ligniera***
- Spore masses variable in size. Zoosporangia large, elongated, with prominent exit tubes. ***Polymyxa***

Literature Cited

ABE, SEIJI: On the syngamy of some Myxomycetes, *Science Repts. Tokyo Bunrika Daigaku, B, Zoöl.* **2**(18):193–202. *Illustrated.* 1934.

DE BARY, ANTON: Die Mycetozoen. Ein Beitrag zur Kenntniss der niedersten Thiere, *Z. wiss. Zool.*, **10**(1):88–175. *Pls.* 6–10. 1859.

———: Comparative Morphology and Biology of the Fungi, Mycetozoa and Bacteria. Authorized English translation by Henry E. F. Garnsey, revised by Isaac Bayley Balfour, xix + 525 pp. 198 *figs.* Oxford, Clarendon Press, 1887.

CAMP, W. G.: The structure and activities of Myxomycete plasmodia, *Bull. Torrey Botan. Club*, **64**(5):307–335. *Figs.* 1–10. 1937.

CAYLEY, DOROTHY M.: Some observations of Mycetozoa of the genus Didymium, *Brit. Mycol. Soc. Trans.*, **14**:227–248. *Pls.* 5–6. 1929.

COOK, W. R. IVIMEY: A monograph of the Plasmodiophorales, *Arch. Protistenk.*, **80**(2):179–254. *Pls.* 5–11. *Figs.* 1–14. 1933.

———, AND E. J. SCHWARTZ: The life-history, cytology and method of infection of Plasmodiophora brassicae Woron., the cause of finger-and-toe disease of cabbage and other crucifers, *Trans. Roy. Soc. London, B*, **218**:283–314. 3 *pls.* 1930.

DANGEARD, P. A.: Observations sur la famille des Labyrinthulées et sur quelques autres parasites des Cladophora, *Le Botaniste*, **24**(3/4):217–258. *Pls.* 22–24. *Fig.* 1. 1932.

DUBOSCQ, O.: Labyrinthomyxa sauvageaui n.g.n.sp., Protéomyxée parasite de Laminaria lejolisii Sauvageau, *Compt. rend. soc. biol.*, **84**:27–30. *Fig.* 1. 1921.

———: Les plasmodes de Labyrinthomyxa sauvageaui, *ibid.*, **84**:30–33. *Figs.* 2–3. 1921.

ELLIOTT, EUGENE W.: The swarm-cells of Myxomycetes, *J. Wash. Acad. Sci.*, **38**(4):133–137. *Figs.* 1–8. 1948.

ELLISON, BERNARD R.: Flagellar studies on zoospores of some members of the Mycetozoa, Plasmodiophorales and Chytridiales, *Mycologia*, **37**(4):444–459. *Figs.* 1–4. 1945.

FEODORINTSCHIK, N. S.: Osnobyne zakonomerosti biologii kapustnoi kily (Plasmodiophora brassicae Wor.). Russian with English summary, *Summ. Sci. Res. Wk. Inst. Pl. Prot. Leningrad*, 1935, pp. 69–70. 1936. (Reviewed in *Rev. Applied Mycol.*, **16**(1):10. 1937.)

GILBERT, FRANK A.: On the occurrence of biflagellate swarm cells in certain Myxomycetes, *Mycologia*, **19**:277–283. *Figs.* 1–2. 1927.

GILBERT, HENRY C.: Critical events in the life history of Ceratiomyxa, *Am. J. Botany*, **22**(1):52–74. *Pls.* 1–3. *Fig.* 1. 1935.

HONIG, F.: Der Kohlkropferreger (Plasmodiophora brassicae Wor.). Eine Monographie, *Gartenbauwiss.*, **5**:116–225. *Figs.* 1–11. 1931.

HOWARD, FRANK L.: The life history of *Physarum polycephalum*, *Am. J. Botany*, **18**(2):116–133. *Pls.* 12–19. *Fig.* 1. 1931.

———, AND MARY E. CURRIE: Parasitism of Myxomycete plasmodia on the sporophores of Hymenomycetes, *J. Arnold Arboretum, Harvard Univ.*, **13**(2): 270–284. *Pls.* 48–49. *Figs.* 1–2. 1932.

———, AND ———: Parasitism of Myxomycete plasmodia on fungus mycelia, *ibid.*, **13**(4):438–447. *Pl.* 54. 1932.

JAHN, E.: Myxomycetenstudien: 16. Die Kernphasen und die Zahl der Chromosomen, *Ber. deut. botan. Ges.*, **54**(8):517–528. *Pl.* 36. 1936.

JONES, PHILIP M.: Morphology and cultural history of Plasmodiophora Brassicae, *Arch. Protistenk.*, **62**:313–327. *Pls.* 15–21. *Fig.* 1. 1928.

KARLING, JOHN S.: The Plasmodiophorales, ix + 144 pp. 17 *pls.* 17 *figs.* New York, published by the author, 1942.

KUDO, RICHARD ROKSABRO: Protozoology, ed. 3, xiii + 778 pp. 336 *figs.* Springfield, Ill., Charles C Thomas, Publisher, 1946.

LEDINGHAM, G. A.: Zoospore ciliation in the Plasmodiophorales, *Nature*, **133** (3362):534. 1 *fig.* 1934.

———: Occurrence of zoosporangia in Spongospora subterranea (Wallroth) Lagerheim, *ibid.*, **135**:394. *Figs.* 1–4. 1935.

———: Studies on Polymyxa graminis, n. gen. n. sp., a plasmodiophoraceous root parasite of wheat, *Can. J. Research, C*, **17**:38–51. *Pls.* 1–5. *Figs.* 1–3. 1939.

LISTER, ARTHUR: A Monograph of the Mycetozoa. A descriptive catalogue of the species in the herbarium of the British Museum. Ed. 3, revised by Gulielma Lister, xxxiii + 296 pp. 223 *pls.* 56 *figs.* London, Trustees of the British Museum, 1925.

MACBRIDE, THOMAS H., AND G. W. MARTIN: The Myxomycetes. A descriptive list of the known species with special reference to those occurring in North America, 339 pp. 21 *pls.* New York, Macmillan Co., 1934.

NAUSS, RUTH W.: Reticulomyxa filosa Gen. et Spec. nov., a new primitive plasmodium, *Bull. Torrey Botan. Club*, **76**(3):161–173. *Figs.* 1–13. 1949.

OLIVE, EDGAR W.: Monograph of the Acrasiae, *Proc. Boston Soc. Natural History*, **30**:451–513. *Pls.* 5–8. 1902.

———: Cytological studies on Ceratiomyxa, *Trans. Wisconsin Acad. Sci.*, **15**(2): 753–774. *Pl.* 47. 1907.

PALM, B. T., AND MYRLE BURK: The taxonomy of the Plasmodiophoraceae, *Arch. Protistenk.*, **79**(2):263–276. *Figs.* 1–15. 1933.

RAPER, KENNETH B.: Growth and development of Dictyostelium discoideum with different bacterial associates, *J. Agr. Research*, **55**(4):289–316. *Figs.* 1–4. 1937.

RENN, CHARLES E.: Wasting disease of Zostera in American waters, *Nature*, **134**(3385):416. *Fig.* 1. 1934.

———: A Mycetozoan parasite of Zostera marina, *ibid.*, **135**(3414):544–545. 1935.

———: The wasting disease of Zostera marina, *Biol. Bull.*, **70**(1):148–158. *Figs.* 1–5. 1936.

ROCHLIN, EMILIA: Zur Frage der Widerstandsfähigkeit der Cruciferen gegen die Kohlhernie (Plasmodiophora brassicae Wor.), *Phytopath. Z.*, **5**(4):381–406. *Figs.* 1–7. 1933.

SINOTO, Y., AND A. YUASA: Studies in the cytology of reproductive cells: I. On the planocytes in five forms of Myxomycetes, *The Botanical Magazine* (*Tokyo*), **48**(574):720–729. *Figs.* 1–19. 1934.

SKUPIENSKI, F. X.: Recherches sur le cycle évolutif de certains Myxomycètes, 83 pp. 2 *pls.* 2 *figs.* Paris, Imprimerie M. Flinikowski, 1920.

———: Badania bio-cytologiczne nad Didymium difforme. Czesc pierwsza (Biocytological study of D. difforme. First part. Polish with French summary), *Acta Soc. Botan. Polon.*, **5**(3):255–336. 7 *pls.* (2 *colored*). 13 *figs.* 1928.

SPARROW JR., FREDERICK K.: Aquatic Phycomycetes, Exclusive of the Saprolegniaceae and Pythium, xix + 785 pp. 634 *figs.* Ann Arbor, Univ. Michigan Press, 1943.

VALKANOV, ALEXANDER: Protistenstudien: 4. Die Natur und die systematische Stellung der Labyrinthuleen, *Arch. Protistenk.*, **67**:110–121. *Figs.* 1–10. 1929.

WEBB, P. C. R.: Zoosporangia believed to be those of Plasmodiophora brassicae in the root hairs of non-cruciferous plants, *Nature*, **163**(4146):608. 1949.

VON WETTSTEIN, FRITZ: Das Vorkommen von Chitin und seine Verwertung als systematisch-phylogenetisches Merkmal im Pflanzenreich, *Sitz. ber. Akad. Wiss. Wien, Math. naturw. Klasse, Abt. I*, **130**(1):3–20. 1921.

WILSON, MALCOLM, AND ELSIE J. CADMAN: The life history and cytology of Reticularia Lycoperdon Bull., *Trans. Roy. Soc. Edinburgh*, **55**:555–608. *Pls.* 1–6. *Figs.* 1–4. 1928.

VAN WISSELINGH, C.: Mikrochemische Untersuchungen über die Zellwände der Fungi, *Jahrb. wiss. Botan.*, **31**(4):619–687. *Pls.* 17–18. 1898.

WORONIN, M.: Plasmodiophora brassicae, Urheber der Kohlpflanzen-Hernie, *Jahrb. wiss. Botan.*, **11**:548–574. *Pls.* 29–34. 1878.

YOUNG III, EDWARD LORRAINE: Studies on Labyrinthula. The etiologic agent of the wasting disease of eel-grass, *Am. J. Botany*, **30**(8):586–593. *Figs.* 1–2. 1943.

YUASA, AKIRA: Studies in the cytology of reproductive cells: III. The genesis of the flagellum in the planocytes of Fuligo septica Gmelin, *The Botanical Magazine* (*Tokyo*), **49**(584):538–545. *Figs.* 1–27. 1935.

3

PHYCOMYCETEAE: CHYTRIDIALES AND HYPHOCHYTRIALES

The first class of true fungi, for the Mycetozoa discussed in Chapter 2 are not considered to be fungi or even plants, is that of the Phycomyceteae. In this class the fungus body may consist of

1. One cell—with a cell wall and with or without non-nucleate rhizoids—which either becomes directly a zoosporangium or gametangium or else empties its contents into an external zoosporangium or gametangium.

2. Instead of a single cell with non-nucleate rhizoids, nuclei may pass from the first cell through modified rhizoids and take their positions in swellings in the latter, there forming secondary centers. Such fungi are polycentric in contrast to the monocentric forms with but a single cell. When the fungus is entirely contained in the cell of its host and completely lacks rhizoids or haustoria, it is called holocarpic; when it is external to its substratum, or internal and gains its nourishment by means of rhizoids or haustoria, it is called eucarpic.

3. A still further advance is the development of a true coenocytic mycelium containing many nuclei, usually with rhizoids or nucleate trophic hyphae within the substratum. In general this type of mycelium is not provided with true septa (although greatly perforate pseudosepta occur in *Allomyces*) except to fence off sporangia and gametangia, injured regions, or portions of the mycelium that have exhausted their protoplasmic contents.

In seven of the twelve orders treated here as belonging to the Phycomyceteae asexual reproduction and, to a much lesser degree, sexual reproduction is by means of planocytes—i.e., flagellate, naked cells. These are of three types which appear to characterize three series of orders. The types are as follows: flagellum single, posterior; flagellum single, anterior; flagella two, anterior or lateral. In the other five orders (the relationship of two of which is greatly in doubt), asexual reproduction is by means of nonflagellate cells and sexual reproduction is by the union of two gametes or two gametangia. The Keys given below will serve to distinguish the orders of this class.

Key to the Orders of Posteriorly Uniflagellate Fungi

Without true mycelium, at most a rhizomycelium. Alternation of sporophytic and gametophytic generations wanting. Zoospores mostly without large nuclear cap or "side body." Order Chytridiales (Chap. 3)

Rarely without mycelium, mostly with multinucleate, coenocytic mycelium. Sexual reproduction, where known, by the union of isoplanogametes or anisoplanogametes. Alternation of sporophytic and gametophytic generations often present. Zoospores with large nuclear cap and "side body." Order Blastocladiales (Chap. 4)

With multinucleate, coenocytic mycelium. Sexual reproduction by the union of motile sperm with non-motile egg, to form thick-walled oospore. Alternation of sporophytic and gametophytic generations wanting. Zoospores with numerous small granules at the apex. Order Monoblepharidales (Chap. 4)

Key to the Order of Anteriorly Uniflagellate Fungi

Only one order recognized. Order Hyphochytriales (Chap. 3)

Key to the Orders of Anteriorly or Laterally Biflagellate Fungi

Holocarpic fungi, parasitic within animal or plant hosts (rarely saprophytic), the whole organism being converted into a single reproductive organ (gametangium or zoosporangium) or a series of such organs, separated by septa. Zoospores of primary type or much oftener of secondary type, preformed in the zoosporangium or becoming organized in a vesicle formed at the apex of the exit papilla or tube. Sexual reproduction by the union, sometimes through a short conjugation tube, of two gametangia of equal size or more often of an antherid and oogone. Oospore usually not surrounded by periplasm. Order Lagenidiales (Chap. 5)

Eucarpic fungi usually (but not always) with well-developed branched and tapering holdfast system. The external fungus body a system of usually stout cylindrical, branched coenocytic hyphae, nearly uniform in thickness or with large basal segments and slender branches. Reproductive organs borne on the branches and making up only a small part of the whole fungus. Asexual reproduction by biflagellate zoospores, often dimorphic, rarely of primary type only, often only of secondary type. Oogones with one or more eggs, with or without periplasm. Fertilization by antherids to form thick-walled oospores. Mostly saprophytic in soil and fresh water, in some cases parasitic in roots, algae, fungi, or fresh-water animals. Order Saprolegniales (Chap. 5)

Eucarpic fungi consisting of slender, cylindrical hyphae with or without haustoria, mostly growing within the tissues of living hosts (mostly plants), or in dead organic matter in soil or water. Zoospores of secondary type only, produced under water in slender or rounded zoosporangia and organized in a vesicle at the mouth of the exit tube, or the zoosporangia (conidia) are borne aerially and distributed by air currents, the formation of the zoospores occurring in these conidia after they fall into water. In some genera zoospore formation is omitted and the conidia germinate by germ tubes. Sexual reproduction by oogones with usually one egg and periplasm, fertilized by an antherid through a conjugation tube. Order Peronosporales (Chap. 6)

Key to the Orders of Nonflagellate Phycomyceteae

Walls of the mycelium readily responding to the test for cellulose by chloriodide of zinc.

Parasites entirely within the tissues of the aerial parts of higher plants and forming on their branched, occasionally septate, coenocytic hyphae large, thick-walled intercalary or terminal resting spores within which, upon germination, are produced numerous small spores which upon escaping may unite by twos and then infect other host plants. Relationship very doubtful.
Order Protomycetales (Chap. 6)

Parasites in the alimentary canals of Arthropoda, forming slender, at first nonseptate, coenocytic hyphae which very rarely branch. The contents of the hyphae are transformed by formation of septa into endospores which escape through basal or apical openings. Larger endospores may be formed in which by a sexual (?) process there is a union of nuclei. Relationship very doubtful.
Order Eccrinales (Chap. 7)

Walls of the mycelium not readily responding to tests for cellulose. Chitin-like substances present.

Saprophytes or more rarely parasites (mainly in fungi). Mycelium relatively large and abundant, more often nonseptate, at least when young. Asexual reproduction by aplanospores produced in sporangia or the latter reduced to conidium-like sporangioles. Sexual reproduction by the union of unequal or almost equal gametangia to form "zygospores."
Order Mucorales (Chap. 7)

Parasites in insects or in desmids or fern-prothallia or fungi or saprophytes on dung of lizards and frogs. Mycelium at first coenocytic but early becoming septate and very often breaking up into short "hyphal bodies." Asexual reproduction mostly by formation on external conidiophores of terminal conidia which are shot off with violence. Sexual reproduction by union of apparently equal gametangia to form thick-walled zygospores.
Order Entomophthorales (Chap. 7)

Parasites in soil-inhabiting amoebae and nematodes or in the alimentary canal of aquatic insects. Mycelium (except where thickened haustoria are produced) slender, at first coenocytic. Asexual reproduction by the formation of terminal, ellipsoidal or fusiform conidia, single or in chains, or of similar lateral conidia. Sexual reproduction by the union of apparently equal gametangia, adjacent in the same hypha or in separate hyphae lying in close proximity, or at the ends of long, slender filaments.
Order Zoopagales (Chap. 7)

Order Chytridiales. The Chytridiales are comparatively simple in their structure and for that reason are considered first. Whether this simplicity is due to their primitive nature or whether it is due to simplification from more complex forms is a matter of dispute and will be discussed at more length in Chapter 17 on the Phylogeny of the Fungi.

This order includes a number of families of fungi that are either largely aquatic or depend upon the presence of water for their dispersal. They are either parasitic in the roots, stems, or leaves of higher plants; in algae;

in fungi; in the eggs or larvae of worms, Arthropoda, or simpler animals; or, perhaps more often, saprophytic in dead plant or animal material. They are abundant in some types of soils and in fresh and salt water. Their life histories are poorly known in many cases but have been worked out very well in some species. In general the life history is as follows: The posteriorly uniflagellate zoospore after a period of swimming settles down on the substratum and penetrates it or encysts on the outside and then enters it wholly or merely as a more or less extensive haustorium or system of rhizoids. Unlike the Mycetozoa it produces a cell wall of its own, usually at an early stage. The single nucleus divides repeatedly as the cell enlarges, and eventually the cytoplasm fragments into very many uninucleate zoospores or naked gametes or into uninucleate cells which produce cell walls and sooner or later divide internally into zoospores or naked gametes. Sexual reproduction where known may be accomplished by the union of two motile gametes before entry into the substratum, by the union of two cells attached to or residing within the substratum, or by the union of two cells by means of a rhizoid through which a gamete nucleus passes.

In this order the zoospores or motile gametes escape through an exit papilla or tube whose apex softens and permits the motile cells to push out (the inoperculate series), or they escape through a sort of cap that opens like a trap door, the so-called operculum (the operculate series). This operculum is more generally at the apex of the papilla or tube, but in *Karlingia* (Johanson, 1944) and *Catenomyces* (Hanson, 1945) it may be formed within the tube some distance from the apex, which softens and deliquesces much as in the inoperculate forms. The single posterior flagellum is of the "whiplash" type, i.e., the basal portion consists of a firmer outer tube whose more fluid contents extend beyond its tip as a more slender, very flexible structure resembling the lash of an old-fashioned buggy whip. This lash is sometimes very short or scarcely distinguishable.

The cell wall of the Chytridiales does not ordinarily give the characteristic blue or violet coloration of cellulose when chloriodide of zinc is applied, but there are a number of exceptions in which this reaction is distinct. Whether chitin is actually present in all cell walls that do not show the cellulose reaction is not certain. Harder (1937) reported that in one species of the Family Rhizidiaceae different individuals possess cellulose or chitin, perhaps in cells of different ages. Nabel (1939) demonstrated the presence of chitin but not of cellulose in some species of *Rozella* and *Synchytrium*. Schwartz and Cook (1928) report the presence of cellulose in the cell walls of *Olpidium radicale* while Scherffel (1925) states that in certain structures of *Micromycopsis cristata* Scherffel and *Synchytrium mercurialis* (Lib.) Fuckel cellulose is present. Tischler (1927) reports the haploid chromosome number in the Chytridiales to be usually 4 or 5 but 10 to 12 in *Polyphagus euglenae* Now.

A plant body which consists of but a single rounded cell without any rhizoids or haustoria is said to be holocarpic, but if such structures are present it is eucarpic.

In the majority of the eucarpic genera (the monocentric series) the zoospores produce but one enlargement from which, directly or indirectly, the zoosporangium arises. In several genera (the polycentric series) the germinating zoospore produces an enlarged cell from which arise rhizoids which here and there in their course produce other enlargements into which enter nuclei derived from the original cell. These secondary enlargements may become zoosporangia themselves or may give rise to zoosporangia as well as to other rhizoids with further swellings.

Until recently it has been customary (Fitzpatrick, 1930) to include in the Order Chytridiales all those organisms with structures and life histories similar to those outlined above. The differences in the types of flagellation of the motile cells and of the composition of the cell wall have led mycologists recently to exclude from this order those genera with a single anterior flagellum of the tinsel type (Order Hyphochytriales) and those with two anterior or lateral flagella, one of the tinsel and one of the whiplash type (Family Olpidiopsidaceae).

Sparrow (1942) makes the primary division of this order, on the basis of mode of escape of the zoospores, into two series: Inoperculatae and Operculatae. Miss Whiffen (1944) on the contrary considers the monocentric or polycentric structure to be of prime importance, using the method of zoospore escape as a subordinate character.

In the treatment of this order the author follows mainly the arrangement of Miss Whiffen, with constant reference to the work of Sparrow. First, the holocarpic forms are considered. Whether these are more primitive than the eucarpic monocentric forms, or have arisen from some of these by loss of the rhizoidal or haustorial apparatus in connection with their habitat entirely within the host cell, must remain undecided at present. Sparrow recognizes three families, the relationship of the second of which is somewhat doubtful. They are distinguished as follows:

Olpidiaceae: vegetative cell enlarging to form a single sporangium.

Achlyogetonaceae: vegetative cell elongated and by septation forming a linear series of sporangia.

Synchytriaceae: vegetative cell dividing internally into numerous sporangia, or the latter formed within an outgrowth from the vegetative cell.

Family Olpidiaceae. This family contains six or more genera, totaling forty or more well-substantiated species and many more described species whose status, according to Sparrow, is uncertain. They are entirely endobiotic, i.e., living entirely within the host cells or tissues.

They are more frequently parasitic but may occur as saprophytes within the dead host cells. The posteriorly uniflagellate zoospore settles on the exterior of the host and the flagellum disappears, probably in most cases being withdrawn into the body of the cell which then becomes covered with a thin wall. A slender infection tube grows through the host cell wall and the contents of the encysted zoospore pass into the host. The empty cyst soon disappears. Within the host cell the uninucleate fungus enlarges and usually early produces a cell wall. In some genera, e.g., *Olpidium*, the fungus cell does not completely fill that of the host, but in *Rozella*, as it expands, the fungus eventually completely fills the host cell, the walls of the fungus and host coming into close apposition. The most important genus of the family is *Olpidium* which is parasitic in algae in fresh or salt water, in pollen grains or fungus spores that have fallen into the water, in small aquatic animals or their eggs, as well as in the roots of various land plants. *O. brassicae* (Wor.) Dang. is parasitic in the roots of cabbage and other plants of the genus *Brassica*, and *O. viciae* Kus. inhabits the roots of vetch (*Vicia*). In these the posteriorly uniflagellate zoospore becomes attached to the root of the host plant and encysts there. The contents of the cyst dissolve a small hole through the host cell wall and enter the cell. It forms at first a naked uninucleate mass near the center of the host cell. It may remain there or may dissolve its way into the next underlying cell or even further. Eventually the fungus cell encysts and begins to grow. More than one fungus cell may infect the same host cell, up to 15 or 20 in *O. viciae*. In this case they do not attain so great a size as when single and become flattened where they press against one another. As the fungus cell enlarges the nucleus divides repeatedly. Eventually the organism consists at maturity of a smooth, thin-walled, more or less round or ellipsoid zoosporangium, within which cleavage into uniflagellate zoospores occurs. These escape through one (rarely more) exit tube which grows to the outside of the host and then permits the zoospores to escape by the softening and giving way of the apex of the tube. Sexual reproduction in *O. viciae* and in *O. trifolii* Schroet. occurs, according to Kusano (1912, 1929), by the union of two zoospores (in this case functioning as gametes) outside the host, the resulting biflagellate zygote encysting and infecting the host in the manner described above. The resultant cell becomes thick-walled and is often more or less angular. It is usually smaller than the asexually produced sporangia. These thick-walled resting spores may live for a period of several months until favorable conditions occur, when they swell, cracking open the outer thick wall and permitting the emergence of an exit tube from which the zoospores escape. The union of the two nuclei does not occur until the following spring shortly before the germination of the resting spores. It is probable that meiosis occurs early in the series of nuclear divisions within the resting spore.

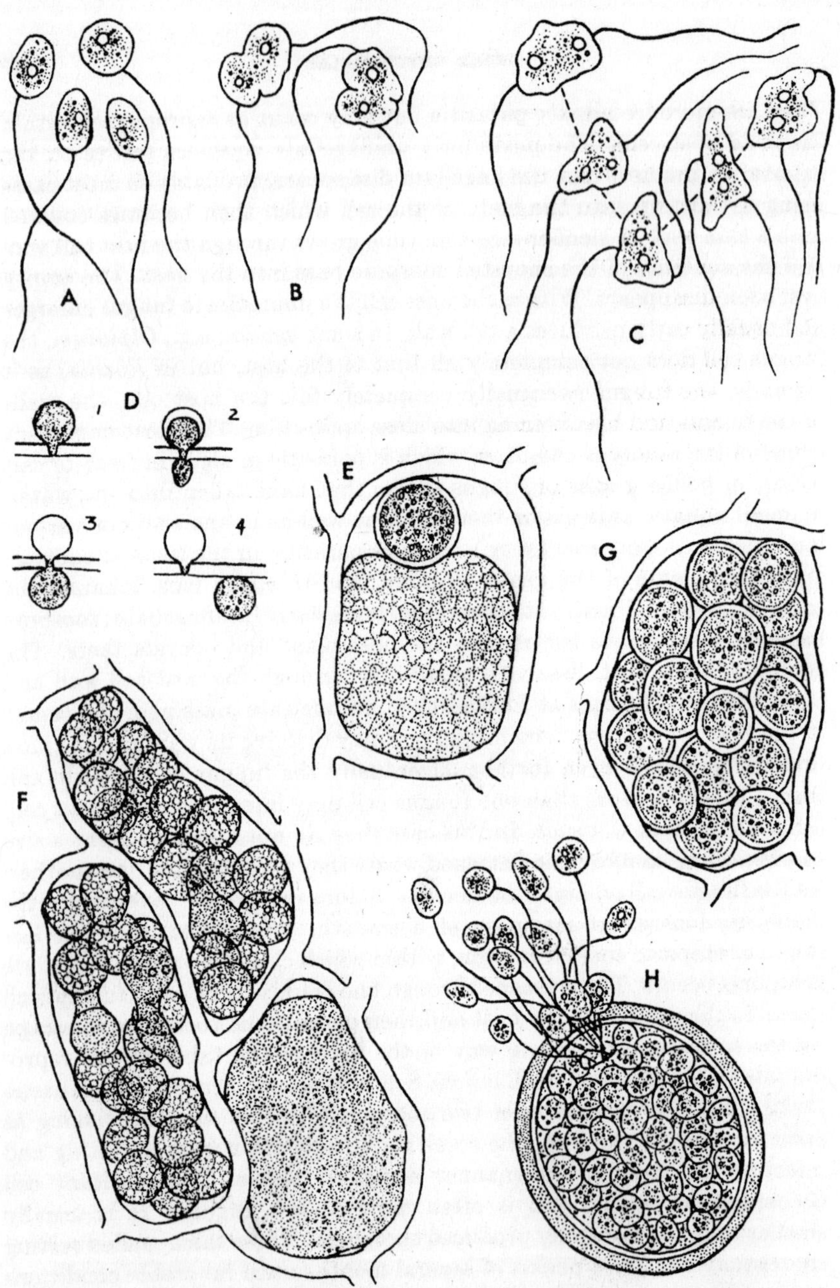

Fig. 9. Chytridiales, Family Olpidiaceae. *Olpidium viciae* Kus. (A) Zoospores. (B) Zoospores uniting. (C) Biflagellate zygotes. (D) Infection of host cell by zygote. (E) A thin-walled zoosporangium and a resting zoosporangium in the same host cell. (F) Thin-walled zoosporangia in host cells. (G) Host cell with many resting zoosporangia. (H) Resting zoosporangium discharging zoospores. (After Kusano: *J. Coll. Agr., Imp. Univ. Tokyo*, **4**(3):141–199.)

It is likely that all zoospores of these species are potentially either gametes or zoospores, depending upon the environment and the opportunity for union of the cells. The earlier emerging zoospores usually do not fuse but those emerging later, especially if from separate zoosporangia, do. For most species of the genus, sexuality has not been demonstrated. Treatment with chloriodide of zinc does not reveal any cellulose reaction in those species of the genus that have been tested except in *O. radicale* Schwartz and Cook (1928). (Fig. 9.)

The genus *Pseudolpidiopsis* was founded by Minden (1915) to include an organism studied by Zopf (1884), parasitic in algae of the family Zygnematacea, which was similar to *Olpidium* in asexual reproduction but different in sexual reproduction. He named it *Olpidiopsis schenkiana* Zopf. Sexual reproduction occurs within the host cell by the union of two adjacent cells, the contents of one passing into the other which forms a thick wall to which the empty smooth cell wall of the male cell adheres. Since this is the type of sexual reproduction characteristic of the genus *Olpidiopsis* in the Order Lagenidiales doubt has been thrown on the accuracy of Zopf's description and figures (1884) of the zoospore as posteriorly uniflagellate. Sparrow (1942) and others are inclined to unite the two genera and place them in the Olpidiopsidaceae. *Pleotrachelus*, with many exit tubes from the sporangium is otherwise similar to *Olpidium*. Nothing is known as to the sexual reproduction although resting spores are produced in one of the three definitely known species. These species occur in molds of the genus *Pilobolus* and in the fresh-water alga *Oedogonium*. (Fig. 10.)

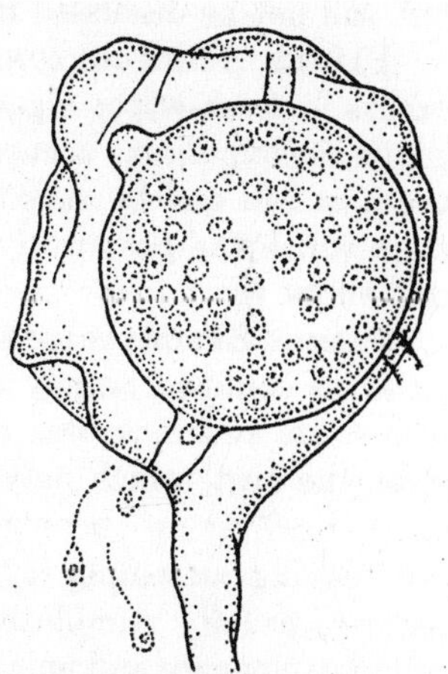

FIG. 10. Chytridiales, Family Olpidiaceae. *Pleotrachelus fulgens* Zopf. Zoosporangium in *Pilobolus* sp. (After Zopf. Courtesy, Sparrow: Aquatic Phycomycetes, Exclusive of the Saprolegniaceae and Pythium, Ann Arbor, University of Michigan Press.)

Rozella includes eleven species recognized by Sparrow. They are all parasitic in fungi of the Blastocladiales, Monoblepharidales, Saprolegniales, Peronosporales, Chytridiales, and a few other groups, mostly in fresh water but one species in salt water. After entering the host cell or filament, which frequently becomes considerably hypertrophied, the organism remains naked for some time, finally producing a thin wall which fuses with the host wall. In some species the organism becomes septate into a series of several sporangia. The posteriorly uniflagellate zoospores escape through one or more inoperculate exit papillae. Resting spores are formed in several species and are smaller than the host cells, lying free from the host wall. They may be smooth or spiny. Sexual reproduction is unknown. *Sphaerita* is an incompletely known genus of which *S. dangeardii* Chatton and Brodsky is parasitic in species of *Euglena*. The description of the single flagellum attached anteriorly but trailing posteriorly casts doubt on the correctness of its position here since in the genus *Pseudosphaerita* of the Olpidiopsidaceae the two flagella, attached near the anterior end are unequal in length, the shorter one anterior and the longer one trailing posteriorly. Several other genera are recognized but will not be discussed here.

FAMILY ACHLYOGETONACEAE. This family containing three genera, occurs in fresh-water algae, and reportedly in some Nematodes. In the genus *Septolpidium*, occurring in Diatoms, the fungus elongates within the host cell and by successive septation forms a series of zoosporangia from which the posteriorly uniflagellate zoospores escape through single exit tubes.

FAMILY SYNCHYTRIACEAE. In this family we find fungi whose parasitism is confined largely to the higher plants. *Synchytrium* is the chief genus. Its swarm spores possess one posterior flagellum. After settling upon the host, which only occurs in a drop or film of water, the swarm spore dissolves an opening into an epidermal cell and enters as a whole, not leaving an empty cyst attached externally. The presence of the parasite usually stimulates the host cell to hypertrophic growth. This cell may elongate and enlarge to a balloon-shaped external structure as in *Erodium cicutarium* (L.) L'Her. infected by *S. papillatum* Farlow, the other cells of the host remaining normal. The enlargement of the epidermal cell may be chiefly lateral and inward in some plants so that it does not project from the surface but more or less crushes the adjacent host cells. The enlargement may be both outward and inward, some of the adjacent cells developing hyperplastically and even undergoing hypertrophy, the result being a wart-like structure, as in *S. vaccinii* Thom. on the cranberry, *Oxycoccus macrocarpus* (Ait.) Pursh. When great hyperplasia and hypertrophy both occur in the host tissues, the fungi may remain in the inwardly enlarged epidermal cells or by the division of the latter may come to lie more deeply in the massive hyperplastic tissues. This is the case in *S. endobioticum* (Schilb.) Perc., the cause of the wart disease of the potato, *Solanum tuberosum* L. M. T. Cook (1945) has made a study of the different types of response by the host to infection by different species of *Synchytrium*. (Fig. 11.)

Within the host cell the parasite enlarges and soon develops a cell wall. This is reported by von Wettstein (1921) and von Gutenburg (1909) to contain chitin, a component of the cell walls of most of the higher fungi. At first the parasites remain uninucleate. By division of the nucleus

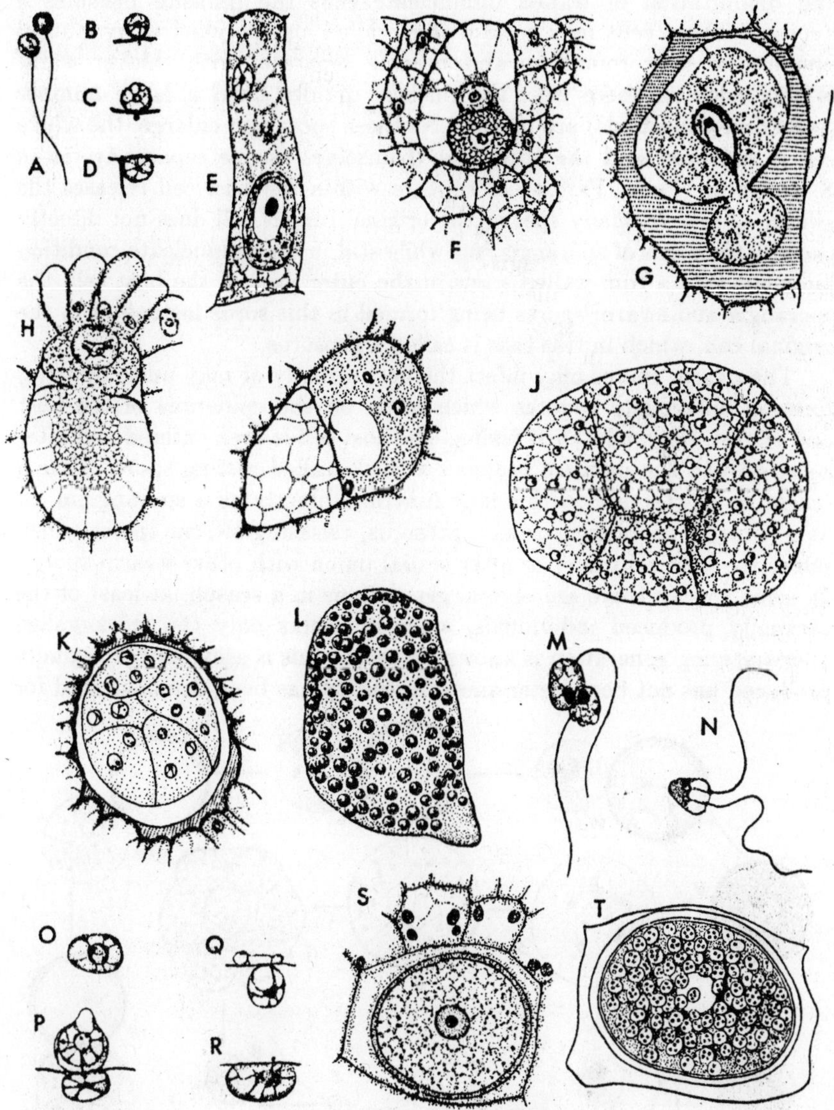

FIG. 11. Chytridiales, Family Synchytriaceae. *Synchytrium endobioticum* (Schilb.) Perc. (A) Zoospores. (B–D) Stages in the infection of the host. (E) Parasite in the upper part of host cell. (F) Parasite enlarged. (G, H) Sorus pushing out from prosorus. (I) Beginning of nuclear multiplication in sorus. (J) Beginning of segmentation in sorus. (K) Sorus containing five zoosporangia. (L) Single zoosporangium containing swarm cells. (M–O) Stages in union of swarm cells. (P–R) Infection of host by zygote. (S) Resting sporangium in host cell. (T) Resting sporangium producing swarm cells. (After Curtis: *Trans. Roy. Soc. London, B*, **210**:409–478.)

and delimitation of walled uninucleate cells the parasite becomes a "sorus." These cells may remain thin-walled and polyhedral by mutual pressure or may round up and become separate, with thicker walls. Within each of these cells the nucleus divides until a large number (sometimes hundreds) are present. As these sporangia enlarge, the whole cell is ruptured and the sporangia themselves escape separately, as in *S. decipiens* (Peck) Farlow, or rupture within the host cell releases the swarm spores. In many species the original fungus cell does not directly become the sorus of sporangia but while still in the uninucleate condition buds out into a thin-walled sorus in the outer part of the host cell, the sporangia and swarm spores being formed in this sorus instead of in the original cell, which in this case is called a prosorus.

The swarm spores may infect the host directly or may unite by twos, forming biflagellate zygotes which settle on the epidermis of the host and encyst, the nuclei then fusing. The host cell is then entered and after some growth the parasite becomes a thick-walled resting spore. After a longer or shorter period this may function directly as a sporangium, or as a sorus of sporangia, or as a prosorus, releasing swarm spores which infect the host asexually or after sexual union with other swarm spores. In some species there are several generations in a season, at least of the asexually produced individuals, while in others only the thick-walled overwintering generation is known. Whether this is asexually or sexually produced has not been determined. Sexuality has been demonstrated for

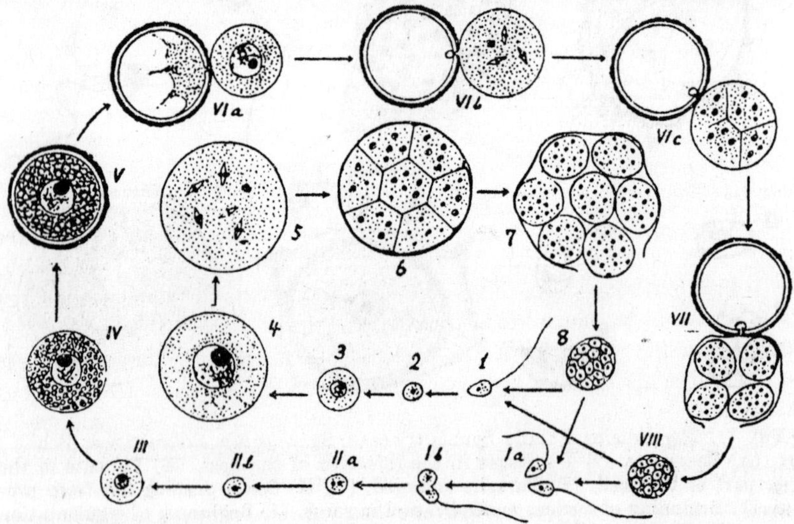

FIG. 12. Chytridiales, Family Synchytriaceae. *Synchytrium fulgens* Schroet. (1–8) Summer cycle. (I–VIII) Winter cycle. (After Kusano: *Japanese J. Botany*, 5(1):35–132.)

only a few species as yet, e.g., *S. endobioticum* by Miss Curtis (1921) and *S. fulgens* Schroet., on *Oenothera*, by Kusano (1930). In the former the overwintering sexually produced resting cells become deep-seated in the tissues of the host by the hyperplastic division of the cells. These resting cells become sporangia directly, not sori of sporangia. In *S. fulgens* both the summer and winter generations become prosori. (Figs. 11, 12.)

Two species of *Synchytrium* are of economic interest, *S. vaccinii*, causing small galls on the leaves and fruits of the cranberry (*Oxycoccus macrocarpus*), and *S. endobioticum*, the cause of the very destructive wart disease of the potato.

Three other genera occur in this family as parasites in algae. One of these is *Micromyces* of which the species *M. zygogonii* Dang. and *M. longispinosus* Couch have been observed in America by Couch (1937) in *Mougeotia* and *Spirogyra*, respectively. The infected cells of the host are usually somewhat enlarged. The plant body of the fungus is spherical with numerous long spines. Through a small opening the contents emerge and form a thin-walled sorus which divides into several sporangia. Within these are produced numerous uniflagellate swarm spores, which may infect the host directly or may first unite by twos. The smaller, thick-walled resting spores, which are likewise spiny, are probably the product of infection by the zygotes, but this has not been proved. The resting spore germinates by the extrusion of a sorus within which sporangia are produced. The zoospores escape when the host cell, which has usually become much swollen, bursts at one side. Scherffel (1925) sets apart as the genus *Micromycopsis* some species in which there is an exit tube from the fungus body through the wall of the algal host so that the sorus, containing two

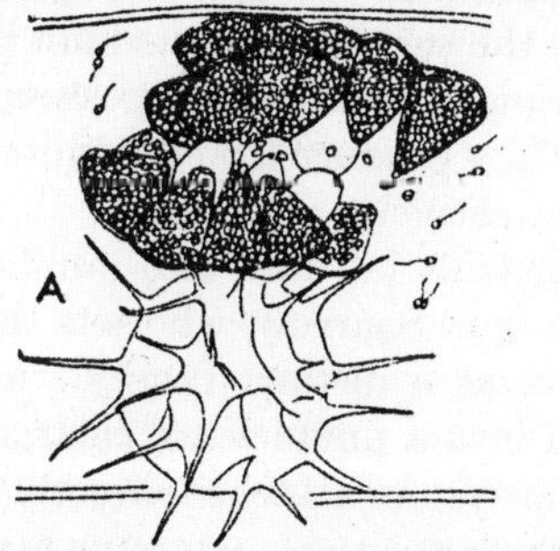

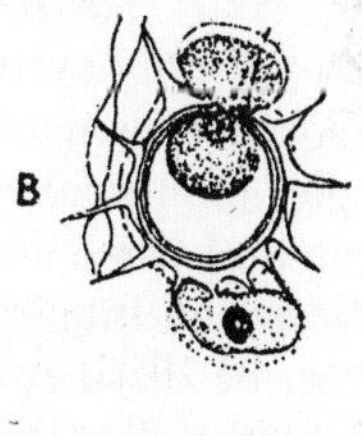

FIG. 13. Chytridiales, Family Synchytriaceae. *Micromyces longispinosus* Couch. (A) Sorus, previously extruded from the spiny prosorus, breaking up into numerous zoosporangia, some of which are discharging swarm cells. (B) Spiny resting spore discharging a sorus, only one nucleus present. (Courtesy, Couch: *Mycologia*, **29**(5):592–596.)

or three sporangia, is external to the host cell instead of internal as in *Micromyces*. He points out that though in general the cell walls of the Synchytriaceae do not show a cellulose reaction, yet the wall of the sporangial sorus and of the tube on which this is borne becomes red-violet in color with iodine in potassium iodide solution, in *Micromycopsis cristata* Scherffel and also in *Synchytrium mercurialis* (Lib.) Fuckel. Miss Canter (1949) has described a third genus in this family parasitic in fresh-water algae, the genus *Endodesmidium*, parasitic in Desmids. Like the two foregoing genera this is endobiotic and holocarpic. The prosorus of the only recognized species is smooth-walled, not spiny as in *Micromyces* and most species of *Micromycopsis*. The relatively large sorus grows out of one end of the prosorus within the cavity of the host cell. The contents divide into numerous, mostly spherical sporangia which escape through papillae into the cavity of the host wall or into the surrounding medium if the papilla pierces the cell wall. These sporangia occasionally possess a posterior flagellum which is feebly active. They produce 2 to 5 minute, posteriorly uniflagellate zoospores which swim actively. (Fig. 13.)

The eucarpic, monocentric Chytrids fall into two families according to Miss Whiffen and three according to Sparrow, who sets the operculate forms apart from the inoperculate as a distinct family. In general the fungus body can be distinguished into a uninucleate enlargement and a non-nucleate haustorial or rhizoidal system attached directly or indirectly to the former. This varies from a short peg to an extensive mass of usually tapering and more or less branched rhizoids. These may be entirely intramatrical or, except for the tips of the branches, may be entirely extramatrical. The body of the encysted zoospore may enlarge and become the sporangium or it may become a prosporangium from whose apex the sporangium arises. On the other hand an enlargement of the germ tube may be formed, the contents of which eventually pass into the sporangium which arises by enlargement of the original zoospore cyst. In Miss Whiffen's family Entophlyctaceae the zoospore cyst empties itself completely into the subjacent portion of the germ tube, this enlargement becoming the sporangium or a prosporangium out of which the sporangium buds. The old zoospore cyst falls away or remains only as an empty fragment.

The eucarpic monocentric families may be distinguished as follows according to Miss Whiffen:

Rhizidiaceae: zoospore cyst enlarging into a sporangium or prosporangium.
Entophlyctaceae: zoospore cyst not further functional; the upper part of the germ tube enlarging into a zoosporangium or prosporangium.

Both of these families as delimited by Miss Whiffen contain operculate as well as inoperculate genera. Sparrow segregates the operculate forms into the family Chytridiaceae. He divides the inoperculate genera on the basis of their position with reference to the substratum into family Phlyctidiaceae, epibiotic and endobiotic, and family Rhizidiaceae, interbiotic, i.e., only the tips of the rhizoids penetrating into the substratum.

FAMILY RHIZIDIACEAE. As in the Olpidiaceae and Synchytriaceae, the swarm spores in this family are provided with but one, posterior, flagellum. Upon reaching the host they send each a haustorial process into the matrix, the main body of the encysted swarm spore remaining outside. The haustorium may be a short undivided peg-like structure or slender and more or less branched, sometimes penetrating other host cells as well. The external portion may enlarge directly to form the zoosporangium or the latter may be formed above it, or as a swelling of the subjacent part of the haustorium. The organism remains uninucleate until the sporangium begins to develop when the nucleus divides many times to form the nuclei of the swarm spores. The haustorium does not at any time contain any nuclei and can hardly be considered as homologous to a mycelium. Resting cells are produced in some species but their mode of origin is unknown in most cases. Sexual reproduction by the union of two well-developed cells has been observed in a few forms. Fusion of swarm spores has been reported but whether it is a true sexual process or merely a sort of rejuvenescence of weak or exhausted cells has not been determined. Petersen (1903), Sparrow (1936), and Karling (1945) have demonstrated sexual reproduction in *Siphonaria*, and Sparrow (1937) also in *Rhizoclosmatium* and in *Asterophlyctis*. A special rhizoid-like outgrowth proceeds from one cell, presumably to be considered the male cell, to another cell. This tube may be long or short or in some cases the two cells are in direct contact. When this process is completed the female cell enlarges and forms a thick-walled resting spore. After a variable length of time this zygote serves as a prosporangium, a small pore being produced and the contents emerging, enclosed by a thin cell wall, and forming an external zoosporangium. The host organisms of the parasitic forms of this family are mostly algae, pollen grains, or small aquatic animals, but *Rhizophydium graminis* Ledingham is parasitic in the roots of *Triticum* and *Panicum*. Many species are saprophytic.

Phlyctochytrium grows on algae into which its much-branched haustorium penetrates. This arises from an apophysis just within the host wall while the portion external to the wall enlarges to become the zoosporangium. Only the presence of the apophysis distinguishes it from *Rhizophydium*. (Fig. 14.)

Rhizophydium occurs mostly in water on various substrata. It consists, when mature, of an enlarged, more or less spherical, thin-walled external sporangium, with a usually tufted haustorium within the host cell. Upon the maturity of the numerous swarm spores they escape through one or more inoperculate pores in the sporangium wall. Sometimes the

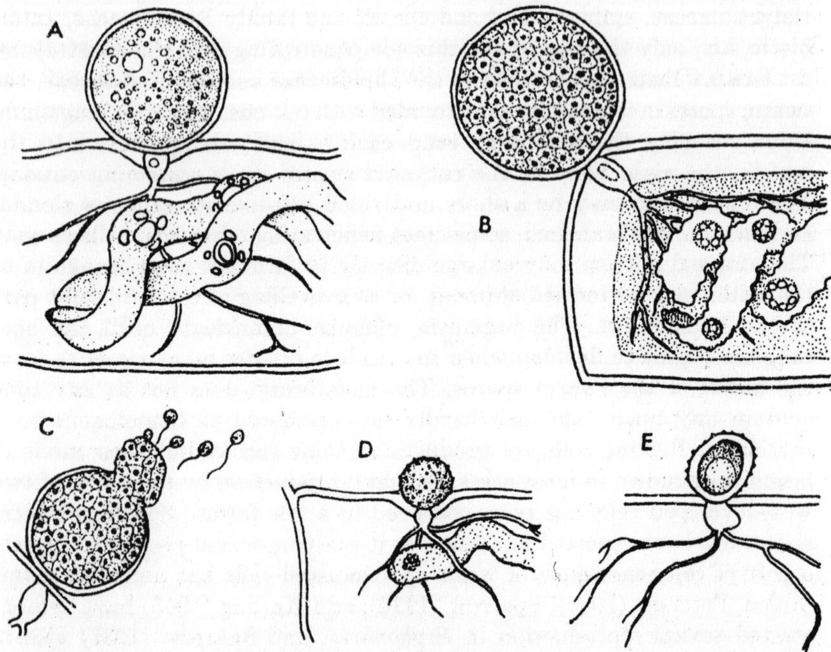

FIG. 14. Chytridiales, Family Rhizidiaceae. *Phlyctochytrium hallii* Couch. (A) Plant nearly mature. (B) Zoospores differentiated. (C) Discharge of zoospores. (D, E) Resting sporangia. (After Couch: *J. Elisha Mitchell Sci. Soc.*, **47**(2):245-260.)

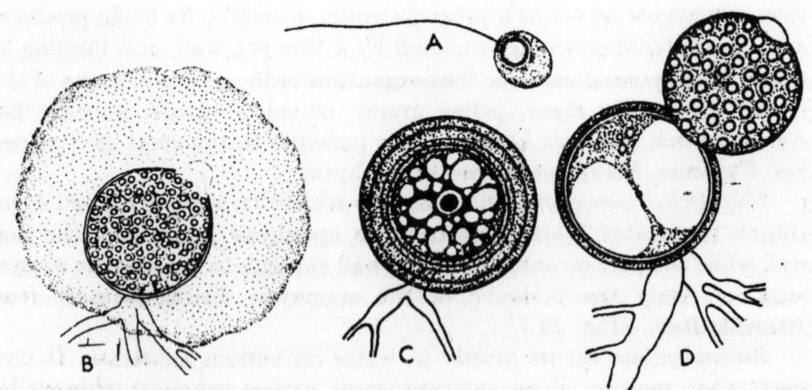

FIG. 15. Chytridiales, Family Rhizidiaceae. *Rhizophydium coronum* Hanson. (A) Zoospore. (B) Zoosporangium with swollen exit papilla. (C) Mature resting spore (stained). (D) Germinating resting spore. (Courtesy, Hanson: *Am. J. Botany*, **32**(8):480-483.)

external cell becomes a thick-walled resting spore which produces its swarm spores only after some time. Couch (1932) has shown for *R. couchii* Sparrow that these resting spores arise as follows: A zoospore "comes to rest on the host, penetrates the wall, and develops apparently just like a sporangium. Later another spore comes and attaches itself to the larger body. . . . The smaller cell discharges its entire contents into the larger. . . . This now secretes around itself a thick wall and goes into the resting state." In *R. ovatum* Couch the same author (1935) reports that the male cell encysts on the algal host (*Stigeoclonium*) and produces its rhizoids. The female cell comes to rest upon the male cell and both cells enlarge, the wall between becoming perforated. The male nucleus passes into the larger female cell and unites with its nucleus. This cell becomes a thick-walled resting spore. (Fig. 15.)

Rhizophlyctis differs from *Rhizophydium* mainly in that the rhizoids are numerous and arise from more than one point on the sporangial wall. The four or five species are parasitic on fresh-water algae or saprophytic on insect exuviae and vegetable debris. *R. petersenii* Sparrow can be cultivated on cellophane or filter paper in pure water cultures. *Karlingia rosea* (de Bary and Wor.) Johanson which differs in the production of an operculum near the base of each of the several exit papillae was formerly assigned to this genus under the name *R. rosea*. This species grows saprophytically on organic matter in the soil, the large sporangia sometimes exceeding 0.1 mm. in diameter. These become rose-colored and give a rosy tinge to the soil. This species also may be cultivated on cellophane and various kinds of vegetable matter such as sterilized onion skin, grass leaves, etc. The position of the operculum at the base of the exit papilla instead of at its apex is of interest.

Obelidium grows upon the exuviae of aquatic insects. Its external body bears a spine and a cup or funnel-like base, and the branching rhizoids arise from a small apophysis. The zoospores escape from a lateral, inoperculate opening below the spine. No resting stage is known. It occurs in Europe and Asia and Sparrow (1938) reports *O. mucronatum* Now. also from the United States. *Siphonaria* forms a round or ellipsoid zoosporangium and strongly developed rhizoids. One species has spines laterally and apically but does not possess the cup or funnel-like base of the preceding genus. Sexual reproduction occurs by the union of two thalli through a short or long slender tube, the female thallus becoming a thick-walled resting spore. (Fig. 16.)

The genus *Polyphagus* is usually placed in this family but its relationship to the other genera is doubtful. *P. euglenae* Now. is parasitic on species of *Euglena* and other one-celled green organisms. Instead of being an internal parasite it lives externally. According to Wager (1913), the germinating zoospore sends out in various directions slender processes

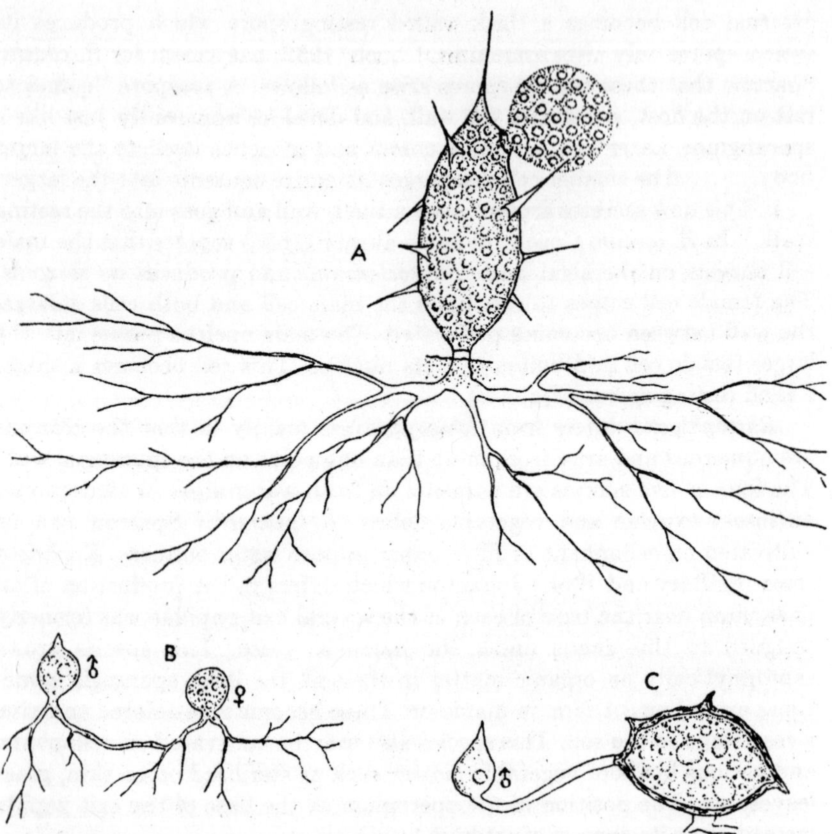

FIG. 16. Chytridiales, Family Rhizidiaceae. *Siphonaria petersenii* Karl. (A) Mature plant with zoospores escaping. (B) Union of male and female plants through a narrow tube. (C) Sexually produced thick-walled zygote with empty male plant still attached. (Courtesy, Karling: *Am. J. Botany*, **32**(9):580–587.)

which enter the host cells encountered. Sometimes, where the latter are crowded in considerable numbers, as many as fifty may be attacked by the haustoria from one parasite. The latter remains uninucleate and is invested by a firm thin wall. Within the swelling (prosporangium) representing the original zoospore, the nucleus divides to form the nuclei of the new zoospores, and the whole contents bud out into a somewhat elongated thin-walled sporangium within which the division into the uninucleate zoospores takes place. As many as several hundred zoospores may be produced. Upon the occurrence of conditions unfavorable for further asexual reproductions there may occur the conjugation of two cells. A somewhat smaller cell sends out a slender process (perhaps a modified haustorium), the tip of which enlarges when it comes into contact with a larger cell. Into this enlarged tip the nucleus and contents of the original cell pass. Then the nucleus of the other cell with which it is in contact enters through a small opening and the wall thickens to form a thick-walled resting spore. After several months a zoosporangial sac is formed in which the two nuclei fuse and then divide to form the nuclei of the zoospores. Three species of *Polyphagus* are recognized by Sparrow.

The genus *Harpochytrium* is sometimes placed in this family. The plant consists of a slender or stout tubular, straight or curved, and sessile or stalked zoosporangium with blunt or pointed apex. It adheres to the surface of the host (mostly fresh-water filamentous algae) by a foot, rarely penetrating the algal wall. When young it is uninucleate but becomes multinucleate as it grows. The upper half to three-quarters of the protoplasm becomes divided transversely into four or five up to many posteriorly uniflagellate zoospores which escape out of an inoperculate apical opening, attaching themselves to a host cell by the tip of the flagellum. The protoplasm, containing one or more nuclei, which remains in the zoosporangium grows and produces a new sporangium by proliferation. Very much resembling this genus is the alga *Chytridiochloris*, based on *H. viride* Scherf., which has a chloroplast and reproduces in the same manner. For this reason the correctness of the assignment of *Harpochytrium* to this family is doubted by some authors, especially by Jane (1946) who has made a monographic revision of the genus.

Because of its operculate manner of dehiscence Sparrow (1943) places the genus *Chytridium* in a separate family, Chytridiaceae, while Miss Whiffen includes it and other operculate forms of the same general structure in the Rhizidiaceae. The original encysted zoospore becomes the sporangium. At its base is the endobiotic haustorium or rhizoidal system which may produce an apophysis in some species. This haustorial system may be a simple unbranched peg or a typical branched system of rhizoids may arise from the tip of the peg, from the base of the sporangium, or from the apophysis. The operculum is apical. Resting spores are endobiotic, apparently asexually produced. They act as prosporangia, giving rise to extramatrical operculate sporangia. Sparrow recognizes twenty-seven species besides a number of doubtful ones. They are mostly parasitic in fresh-water algae but a few occur in marine algae. They are reported from Europe, Asia, and North America.

Family Entophlyctaceae. In this family the zoospore cyst becomes emptied and usually soon disappears or remains as an empty cap. The fungus body enlarges within the host and sends out rhizoids which in some species are very extensive, over 0.5 mm. long and up to 10 μ thick. Some forms are strictly parasitic, mostly upon algae, but many are saprophytes on various types of material. In *Entophlyctis* the germ tube enlarges to become the sporangium within the cell of the algal host. The usually single exit papilla is inoperculate. Resting spores are known in some species. In *E. vaucheriae* (Fisch) Fischer the latter acts as a prosporangium, giving rise to a thin-walled spherical sporangium. Sexual reproduction is not known. In *Endochytrium* the exit tubes are operculate, otherwise the fungus is like *Entophlyctis*. In *Diplophlyctis* the germ tube enlarges as in the two foregoing genera but this enlargement is a prosporangium from which grows out an external inoperculate sporangium. The resting spores in *D. intestina* (Schenk) Schroet. are formed, according to Sparrow (1936), by the anastomosis of the rhizoidal systems of two plants, the contents of the smaller passing into the larger which becomes thick-walled and covered externally by minute short sharp spines. They may function directly as zoosporangia or as prosporangia. In *Phlyctorhiza* (Hanson, 1946) the rhizoids radiate from the germ tube, branching freely and anastomosing to form a reticulum. The proximal portions of these rhizoid branches form, at first, a thin, angular sporangium which eventually becomes a round, shallow, somewhat lenticular, thin-walled zoosporangium lying underneath the radiating branches. Through a low, inoperculate papilla the zoospores escape into a vesicle where they soon become active and swim away. Thick-walled somewhat tuberculate resting spores may be formed, apparently asexually, in place of the zoosporangia. In germination they act as prosporangia. Rarely some of the rhizoid branches produce secondary zoosporangia so that such individuals are polycentric. The only known species, *P. endogena* Hanson, grows in the basement membrane of insect integuments.

Nephrochytrium, saprophytic mostly on algae or grass leaves in water, corresponds to *Diplophlyctis* except that the exit tubes are operculate. Whether its resting spores are produced by a sexual process or not is unknown.

Macrochytrium may possibly belong in the Entophlyctaceae but needs further study. The zoospore gives rise to a stout germ tube which produces a cluster of coarse rhizoids. Laterally from the germ tube, near its point of origin, arises the subspherical sporangium which may attain a diameter of 0.5 mm. It opens at the apex by a large operculum allowing the escape of up to a thousand zoospores. This has been found in Europe and in the United States, saprophytic on submerged fruits or twigs. Whether this is a typical member of the Chytridiales or not is uncertain. The coarse rhizoidal system suggests the possibility of a coenocytic mycelial structure with many nuclei instead of an enucleate system.

The polycentric, eucarpic Chytridiales are placed by some students in one, two, or three families. Until their life histories and cytology are known more completely any classification of these genera will have to be tentative. The author recognizes two families as well distinguished, with the probability that more may have to be recognized when the life histories of the members of the group are better known.

The dozen or so genera of Chytridiales that are polycentric differ from the eucarpic, monocentric genera in the production of many nucleate expansions on the rhizoidal system, each of which may become a center for the production of zoosporangia, resting spores or other rhizoidal branches. As a result the fully developed individual consists of many centers of varying size and shape, connected by (usually) slender, non-nucleate rhizoidal strands or tubes. This system of tubes and rhizoids was given the name "rhizomycelium" by Karling (1932). One of the crucial points in the recognition of the affinity of these organisms is to determine whether they possess a true mycelium, with nuclei along its course, or a rhizomycelium. Furthermore, does this distinction necessarily separate these fungi into different orders or do we have here a gradual gradation within the same order from organisms scarcely distinguishable from the Rhizidiaceae or Entophlyctaceae to those in which the rhizomycelium has become truly mycelial?

The two families tentatively accepted by the author are as follows:

Cladochytriaceae: intra- or extramatrical, eucarpic, with many centers connected by slender, branching, rhizoidal or tubular threads, with swellings here and there. Sporangia thin-walled, terminal or intercalary, operculate or inoperculate. Resting spores thick-walled, apparently not sexually produced.

Physodermataceae: epibiotic, monocentric, and eucarpic at first, then by independent infection endobiotic and strongly polycentric, producing resting spores only. The latter eventually produce internally one to several thin-walled, inoperculate zoosporangia. Probably the epibiotic stage produces gametes while the resting spore sporangia produce zoospores but this has not been proved.

Family Cladochytriaceae. In the Cladochytriaceae the chief genus with inoperculate sporangia is *Cladochytrium,* which is saprophytic and forms a system of fine branching rhizoids within the tissues of the substratum. Here and there arise swellings which may become the round or pyriform sporangia or which may become spindle-shaped, once septate "turbinate organs," one of these cells becoming a sporangium and perhaps later the other. Septa are formed to separate the sporangia and resting cells from the remainder of the rhizomycelium. The zoosporangia empty through an inoperculate exit tube which may be quite long. Some of the swellings become colorless resting spores with a thickened wall. These, according to Karling (1934), germinate by becoming sporangia in which numerous zoospores are produced by cleavage and escape through a short sporangial neck, in *C. replicatum* Karl. The cytology of this fungus has been studied by Karling (1937). (Fig. 17A–C.)

Physocladia (Sparrow, 1932) differs in habit from the foregoing in that it is external to the matrix (pine pollen in the case of *P. obscura* Sparrow) which is penetrated only by the fine tips of the rhizomycelium. It consists of a slender, branched thread with rhizoidal branches and

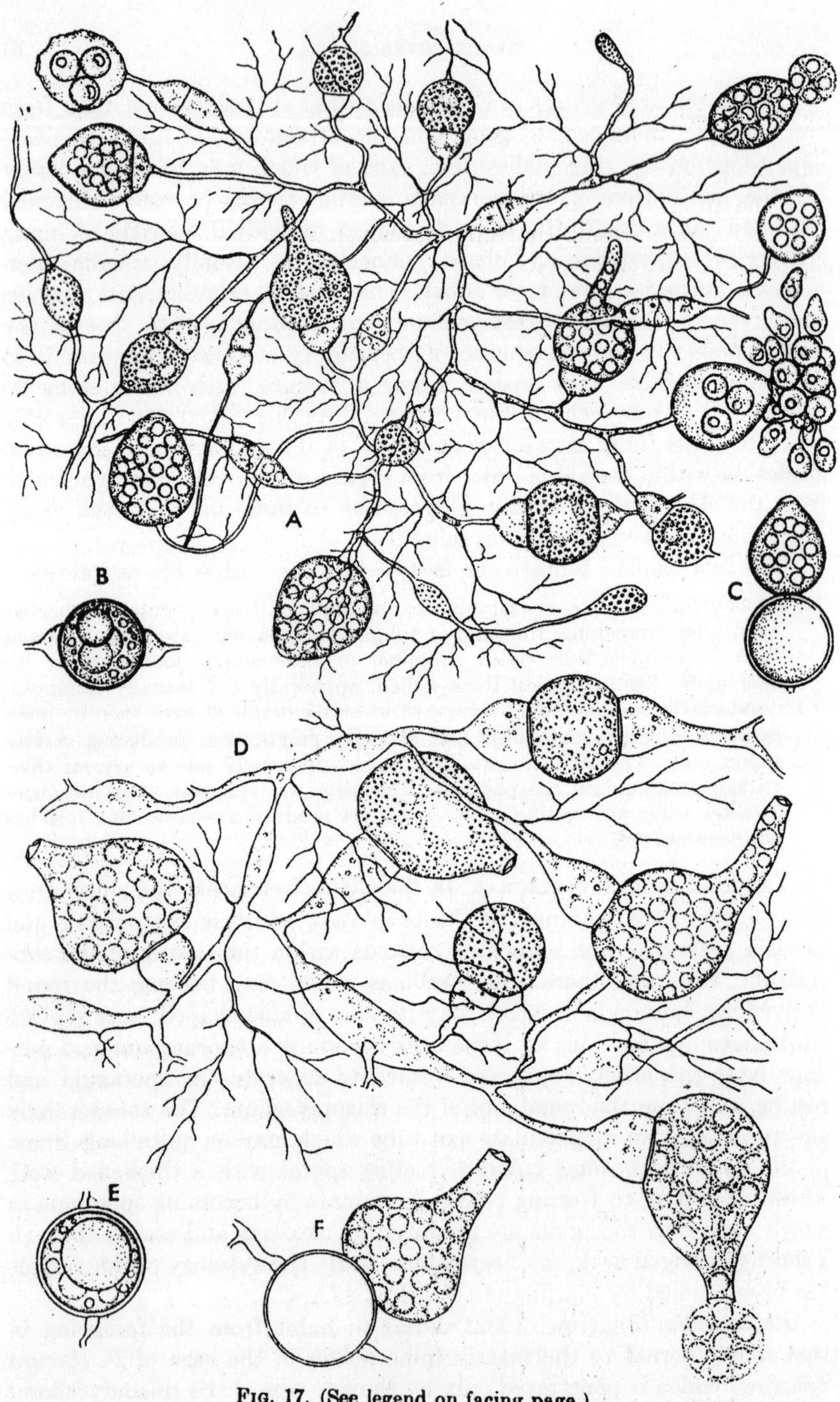

FIG. 17. (See legend on facing page.)

intercalary swellings here and there and, rarely, with septate turbinate cells, and with large terminal spherical zoosporangia, each with an apophysis. These produce large numbers of zoospores which escape through a nonoperculate opening and swim actively for some time in a large vesicle before its membrane is ruptured. Large spherical, thick-walled resting spores are formed terminally. Their further fate has not been determined.

Polychytrium (Ajello, 1942) differs from both the foregoing genera in its coarse branched rhizomycelium which has occasional rhizoids but no turbinate cells or swellings. At the ends of the branches and sometimes in intercalary positions the zoosporangia arise, usually in pairs, sometimes several close together. They are pyriform. In them the zoospores are developed and push out through the apical exit papilla or tube surrounded by a slimy matrix in which at first they lie motionless. Soon they become active and break away and swim off. They have a long posterior flagellum and a conspicuous lunate nuclear cap. After the discharge of the zoospores new zoosporangia are formed by proliferation. In addition to these typical zoosporangia, yellowish-brown very tuberculate zoosporangia may occur, in pairs or in clusters. From these the zoospores escape in the same manner. No resting spores have been observed.

The principal operculate genera of this family are *Nowakowskiella*, *Septochytrium*, *Megachytrium*, and *Catenomyces*. In *Nowakowskiella* the slender rhizomycelium is nonseptate, except where the enlargements become zoosporangia, with usually a large apophysis. Zoospores escape in a mass before separating or those few remaining behind creep out one by one. The zoosporangia sometimes proliferate. Resting spores are formed in some species of the genus. Roberts (1948) shows that in some species of *Nowakowskiella* there is a marked morphological differentiation between the vegetative (or trophic) and the reproductive portions of the rhizomycelium. The former is mostly confined to the interior of the substratum and produces no reproductive organs from the enlargements. The nuclei are found only in the swellings but not in the isthmuses or rhizoids. In the reproductive portion, which is mostly external to the substratum, the nuclei occur not only in the enlargements, which may eventually become the zoosporangia, but also in the intervening filamentous structures. In *Septochytrium* the primary swelling becomes a

---

FIG. 17. Chytridiales, Family Cladochytriaceae. (A–C) *Cladochytrium tenue* Now. (A) General view of part of the thallus, showing spindle organs, zoosporangia in various stages of development, and rhizomycelium. (B) Resting spore which developed in a spindle organ. (C) Resting spore germinating to produce external zoosporangium. (D–F) *Nowakowskiella macrospora* Karl. (D) General view of part of the thallus, showing rhizomycelium and swellings and operculate zoosporangia. (E) Resting spore. (F) Resting spore germinating to produce external zoosporangium. (Courtesy, Karling: *Am. J. Botany*, **32**(1):29–35.)

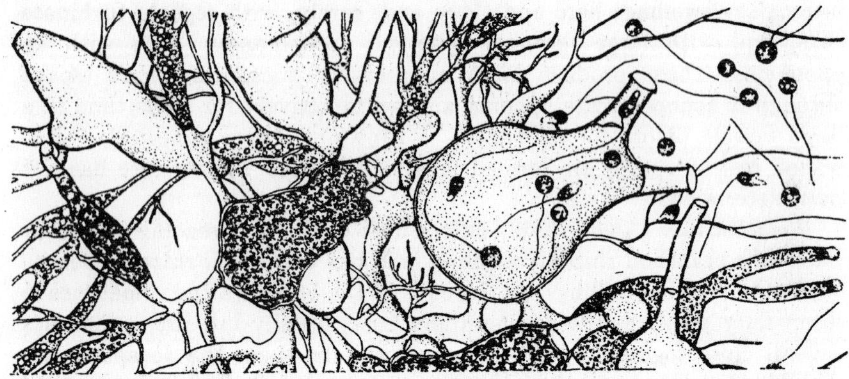

FIG. 18. Chytridiales, Family Cladochytriaceae. *Catenomyces persicinus* Hanson. Portion of plant showing coarse rhizomycelium and zoosporangia with endo-operculate exit tubes. (Courtesy, Hanson: *Am. J. Botany*, **32**(7):431–438.)

large sporangium from whose sides arise many moderately stout rhizoidal tubes with occasional constrictions and septa, and fusiform swellings, some of which may become large or small zoosporangia or resting spores. The latter act as prosporangia when they germinate. In *S. variabile* Berdan, Miss Berdan (1939) demonstrated that the sporangial walls and intercalary swellings give a pronounced violet color on treatment with chloriodide of zinc, indicating the presence of cellulose. This reaction is absent in most of the members of the family. *Megachytrium* differs from the two foregoing genera in the possession of a very coarse, tubular, branched rhizomycelium instead of fine tapering threads. The swellings of this rhizomycelium become operculate sporangia or resting spores. The latter act as prosporangia when they germinate. The genus *Catenomyces* (Hanson, 1945) is quite similar to the three foregoing genera but has an endo-operculum deep down in the exit tube, not at its apex. Resting spores have been doubtfully observed. The zoosporangia may be terminal or intercalary, with one to several exit tubes. The rhizomycelium is stout but tapers to slender rhizoid-like tufts. (Fig. 17D–F, Fig. 18.)

The inoperculate genus *Catenaria*, assigned by Sparrow (1943) to the subfamily Catenarioideae, has been shown by Couch (1945) to be probably more closely related to Order Blastocladiales and will accordingly be discussed in the next chapter.

FAMILY PHYSODERMATACEAE. Two genera, *Physoderma* and *Urophlyctis* are usually included in this family, but since the only constant difference is the effect upon the host the validity of this distinction is doubtful. Both are parasitic on and in higher plants. *Physoderma* may discolor and eventually kill the infected tissues without causing marked hypertrophy, while *Urophlyctis* induces strong gall formation. Tentatively

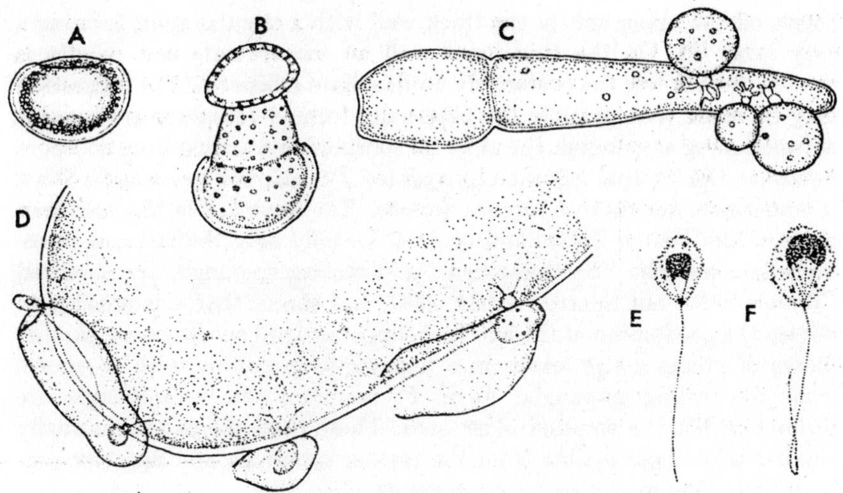

FIG. 19. Chytridiales, Family Physodermataceae. *Physoderma zeae-maydis* Shaw. (A) Resting sporangium, side view. (B) Resting sporangium germinating to form zoosporangium. (C) Young ephemeral zoosporangia. (D) Emptied ephemeral zoosporangia, one with beginning of proliferation. (E) Swarm spore from ephemeral zoosporangium. (F) Swarm spore from resting spore sporangium. (Courtesy, Sparrow: *Am. J. Botany*, **34**(2):94–97.)

the genera are kept distinct, especially since phytopathological literature maintains this separation. They differ from the foregoing polycentric forms in that the primary infection produces an external sporangium with rhizoids penetrating the epidermal cells of the host. This rather flattened, or even slipper-shaped, zoosporangium opens without an operculum and sets free numerous swarm cells. Successive sporangia may arise by proliferation within the empty walls. The polycentric stage is originated in a manner characteristic of the Cladochytriaceae, probably by infection from these swarm cells

In *Physoderma maculare* Wallr., Clinton (1902) has shown that the germinating zoospore sends rhizoids into the host cell (submerged leaf of *Alisma*) and enlarges externally to become a zoosporangium much in the manner of *Rhizophydium*. After the zoospores escape through an exit papilla a second zoosporangium may be formed within the empty wall of the first, and so on two to four times. Other zoospores settling on older leaves, or perhaps upon leaves not permanently emersed, send in a fine filament which enlarges to become a storage cell (or "Sammelzelle") and produces haustorial processes or fine filaments which may penetrate to other cells and there in their turn produce similar cells, and so on. From each such cell there may arise a bud which eventually becomes larger than the original cell. This is a resting sporangium and is thick-walled and somewhat flattened on one side. After a period of rest it enlarges and bursts off the upper side of the thick wall with a circular split, forming a very large lid. On the thin inner wall an inoperculate exit papilla is formed, setting free the posteriorly uniflagellate zoospores. The suggestion may be made that possibly the externally formed zoospores are capable of functioning as gametes, the external zoosporangia arising from zoospore infection, the internal infection by zygotes. *Physoderma zeae-maydis* Shaw is sometimes destructive to corn (maize, *Zea mays* L.) in the southern parts of the United States and in Asia. Usually only the internal rhizomycelium and the "Sammelzellen" and resting sporangia are observed (Tisdale 1919) but Sparrow (1934, 1947) has shown that it is possible to obtain the production of the external slipper-shaped sporangia by placing pieces of young maize leaves in a hanging-drop culture with zoospores from the resting sporangia. As in *P. maculare* new zoosporangia are formed within the emptied older ones. These planospores are markedly smaller than those arising from the resting sporangia and Sparrow suggests that they may possibly be gametes. (Fig. 19.)

Jones and Drechsler (1920) and also Bartlett (1926) have shown that *Urophlyctis* has much the same life history as *Physoderma* but causes extensive gall production by the host, this being practically the only distinction between the two genera. The external zoosporangia have been reported in several species of *Urophlyctis*. Within the infected epidermal cell there develops a "Sammelzelle" or "turbinate cell," at first uninucleate but soon multinucleate. At its distal end is formed a terminal tuft of haustoria. At several places on the cell, buds are formed into each of which a nucleus passes, following which the bud grows out as a somewhat enlarged end of a very slender non-nucleate filament. This enlargement in turn becomes a turbinate cell and may give rise to other similar cells, usually three to five cells from each. In the center of the distal tuft of haustoria there soon buds out a thin-walled cell which grows rapidly and becomes much larger than the cell from which it originates. This becomes thick-walled and bears a crown of haustoria. During its growth the cytoplasm and nuclei from the turbinate cell pass into it. This is not an act of fertilization, as was believed by earlier mycologists. The connecting rhizomycelium soon disappears and finally only the resting sporangia are to be found in the gall tissue. After some time these sporangia are capable of germination. Scott (1920) has studied this process in *U. alfalfae* Magn. It produces one to fifteen or more zoosporangia varying in diameter from 10 to 40 microns, which push out through irregular fissures in the brown wall. Zoospores escape through short exit papillae, there being several such papillae on the larger sporangia. The zoospores are 4 to 8 microns long with a posterior flagellum 30 to 50 microns in length. Scott observed no conjugation of these zoospores. On the other hand, O. T. Wilson (1920) reports that the zoospores are

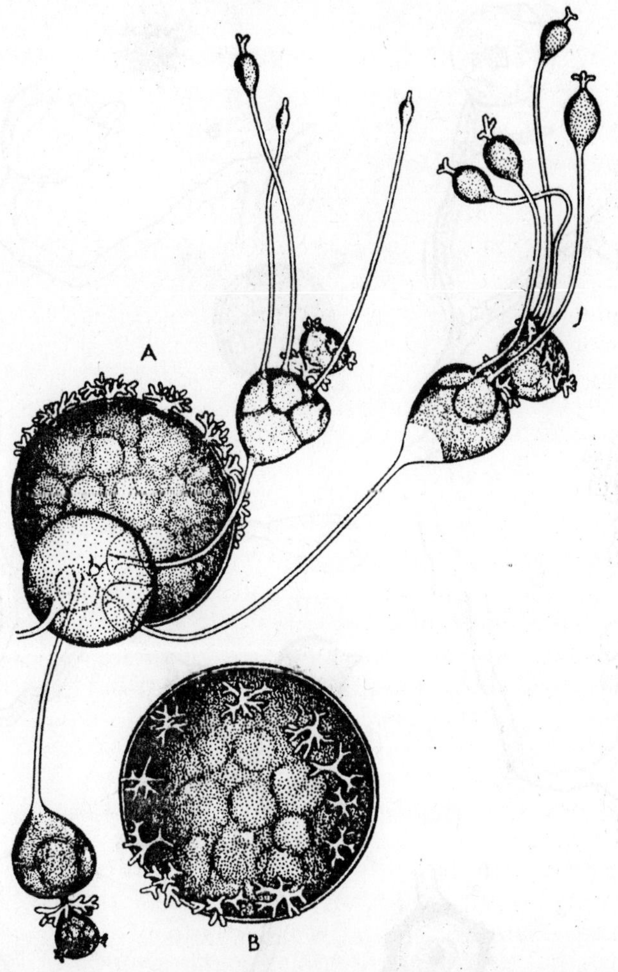

FIG. 20. Chytridiales, Family Physodermataceae. *Urophlyctis alfalfae* Magn. (A) Rhizomycelium, "Sammelzellen," and resting sporangia. (B) Top view of resting sporangium. (After Jones and Drechsler: *J. Agr. Research*, **20**(4):295–324.)

biflagellate, one flagellum being very short, and that they are of two sizes. These conjugate before infection takes place. This needs confirmation as there is a possibility that these supposed zoospores or gametes may have been organisms parasitic within the resting sporangia of the *Urophlyctis*. Several species of *Urophlyctis* are known. (Fig. 20.)

As a tentative addition to the operculate Chytridiaceae, *Tetrachytrium* and *Zygochytrium*, reported by Sorokin (1874) from Russia may be considered here for want of further information. They are soil or water fungi,

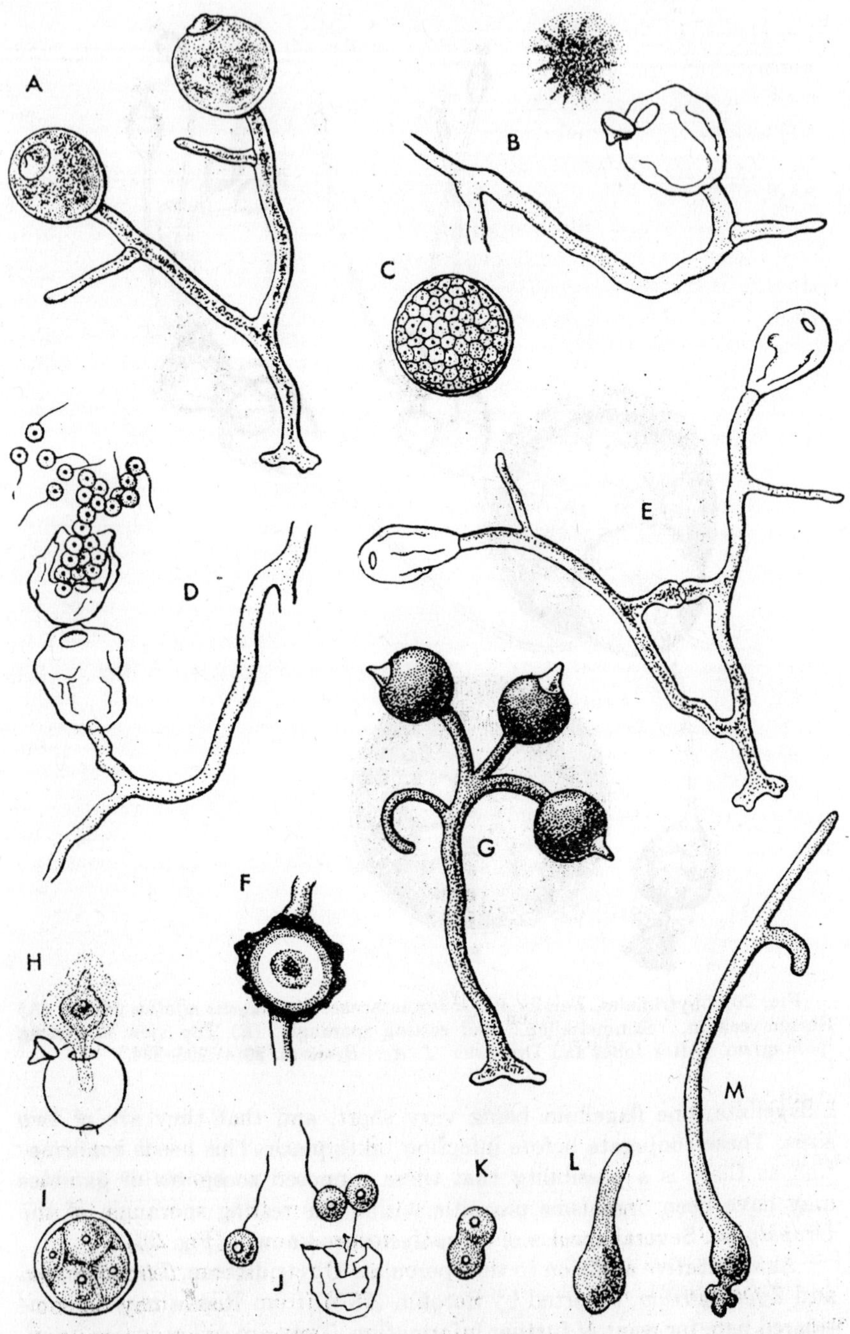

FIG. 21. (See legend on facing page.)

less than 0.1 mm. tall, and consist of an upright, stout and few-branched, nonseptate hypha with a basal holdfast. Whether this is a true hypha or a rhizomycelium is not known, since neither species has been found again since their original description.

In *Zygochytrium* the main stalk is forked, each branch terminating in a spherical, operculate sporangium. The contents escape as a naked mass which soon forms a thin membrane. In the interior the protoplasm divides into many posteriorly uniflagellate zoospores which escape by the irregular rupture of the enclosing membrane, and soon germinate. Between the main forks of the fungus short branches grow which meet and unite, forming a zygospore in a manner very suggestive of zygospore formation by the Mucorales. This thick-walled spore germinates soon by a slender hypha and produces a new plant resembling the parent plant. This fungus is yellow and grows on dead, submerged insects. (Fig. 21A–F.)

In *Tetrachytrium*, which is blue-green in color, the nonseptate main stalk bears three or four somewhat recurved branches on some of which develop terminally spherical, operculate gametangia. Their contents escape in the same manner as from the zoosporangia of *Zygochytrium* and in a similar manner become invested in a thin membrane. Within this are formed four posteriorly uniflagellate gametes which unite by twos to form thin-walled, nonflagellate zygospores. After a little while these germinate to form new plants. The gametes which do not unite do not germinate. Occurring on dead plant material and even on dead beetles in water. (Fig. 21G–M.)

Until these fungi can be found again it will be impossible to decide whether they really belong to the Chytridiales or to a group of fungi that have a true mycelium. Their relationship may be with the operculate Cladochytriaceae.

Order Hyphochytriales. The organisms making up this order show a close parallelism in body structure to the Chytridiales with which they have been usually associated. They differ, however, in the type of zoospore which is anteriorly uniflagellate. In *Rhizidiomyces apophysatus* Zopf, Couch (1941) demonstrated that this single anterior flagellum is of the tinsel type in contrast to the posterior flagellum of the Chytridiales which is of the whiplash type. The cell walls may show positive cellulose reaction in a few cases but in some species chloriodide of zinc calls forth no characteristic cellulose coloring. Karling (1943) has grouped these forms with a

Fig. 21. Chytridiales. (A–F) *Zygochytrium aurantiacum* Sor. (A) Mature plant. (B) Emptied zoosporangium, showing operculum. (C) Extruded protoplasm has formed zoospores. (D) Zoospores escaping. (E) Plant with beginning of conjugation. (F) Mature zygospore. (G–M) *Tetrachytrium triceps* Sor. (G) Mature plant. (H) Protoplasm escaping from gametangium. (I) Four planogametes formed in the extruded protoplasm. (J) Gametes escaping. (K) Gametes uniting. (L) Zygote beginning to germinate. (M) Young plant. (After Sorokin: *Botan. Ztg.*, **32**(20):305–315.)

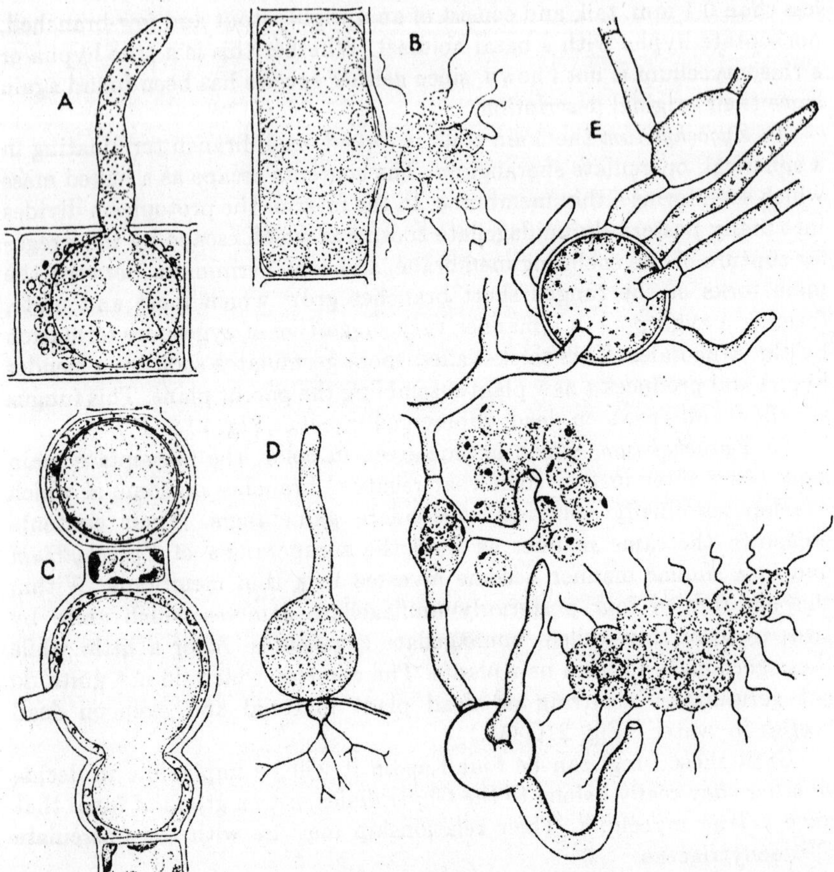

FIG. 22. Hyphochytriales. (A–C) Family Anisolpidiaceae. *Anisolpidium ectocarpii* Karling. (A) Zoosporangium with unopened exit tube. (B) Zoosporangium discharging the anteriorly uniflagellate zoospores. (C) Portion of filament of *Ectocarpus* containing one empty zoosporangium and one resting spore. (D) Family Rhizidiomycetaceae. *Rhizidiomyces apophysatus* Zopf. (E) Family Hyphochytriaceae. *Hyphochytrium catenoides* Karling. (A–C, courtesy, Karling: *Am. J. Botany*, 30(8):637–648. D–E, courtesy, Sparrow: Aquatic Phycomycetes, Exclusive of the Saprolegniaceae and Pythium, Ann Arbor, Univ. Michigan Press.)

single anterior flagellum into three families, employing the ordinal name Anisochytridiales. They may be distinguished as follows:

Anisolpidiaceae: holocarpic, intramatrical, monocentric, the thallus becoming a sporangium or a resting spore.
Three genera: *Anisolpidium*, *Reesia*, and *Cystochytrium*.

Rhizidiomycetaceae: eucarpic, monocentric, sporangia and resting spores extramatrical, rhizoidal system intramatrical.
Two genera: *Rhizidiomyces* and *Latrostium*.

Hyphochytriaceae[1]: polycentric, intramatrical, forming a more or less branched mycelium-like structure which here and there, terminally as well as in intercalary positions, enlarges to form the sporangia or resting spores.

Two genera: *Hyphochytrium* and *Catenariopsis*.

FAMILY ANISOLPIDIACEAE. *Anisolpidium* resembles *Olpidium* very closely in its life history except that sexuality has not been observed. The pyriform zoospore possesses a long, anteriorly attached flagellum. Two or three species parasitic in marine Phaeophyceae. In *A. ectocarpii* Karling, the chromosome number is five or six in the dividing nuclei in the sporangium. *Reesia* differs from the foregoing genus in the persistence of the naked, amoeboid stage until the organism has attained almost full size when a cell wall is produced. In *R. amoeboides* Fisch, the resting spores arise from infection by biflagellate zygotes formed by the union of two swarm spores. Two species, parasitic in *Lemna*. (Fig. 22A–C.)

FAMILY RHIZIDIOMYCETACEAE. *Rhizidiomyces* resembles *Rhizophydium* (or *Phlyctochytrium*, in one species, because of the presence of an apophysis), but the zoospore is anteriorly uniflagellate, with the tinsel type of flagellum. Four species are known, two saprophytic in soil, one parasitic in algae, and one parasitic in the oogonia of Saprolegniaceae. *Latrostium*, parasitic in the oogonia of *Vaucheria*, possesses only one recognized species. The zoospores escape not through an exit tube as in *Rhizidiomyces* but by the deliquescence of a large exit papilla. (Fig. 22D.)

FAMILY HYPHOCHYTRIACEAE. *Hyphochytrium infestans* Zopf was first described by its author in 1884 and has not been recognized since. It was parasitic in the apothecium of a species of *Helotium*. It consists of an extensive, coarse, tubular, branched, occasionally septate rhizomycelium (or true mycelium?). The zoosporangia arise as terminal and intercalary swellings on the mycelium. They open by a subapical orifice releasing the anteriorly uniflagellate zoospores. No resting spores have been observed. Two other species are known: *H. hydrodictii* Valkanov on *Hydrodictyon* and *H. catenoides* Karling, mostly saprophytic. The former produces resting spores. *Catenariopsis* possibly belongs in the foregoing genus. Its difference appears to be a greater contrast between the enlarged zoosporangia and the short connecting isthmuses. (Fig. 22E.)

Key to the More Important Families and Genera of Order Chytridiales

Holocarpic, i.e., endobiotic, without rhizoids, discharge tubes inoperculate.
  Vegetative cell enlarging to form a single zoosporangium.
    Family Olpidiaceae
    Zoosporangium not completely filling the host cell.
      Flagellum anteriorly attached but trailing posteriorly. (Doubtfully belonging in this family.) *Sphaerita*

[1] This family as here delimited differs greatly from the treatment accorded in the first edition of this textbook, both as to its suggested relationship and contents.

Flagellum posterior, discharge tubes rarely more than one. *Olpidium*

Flagellum posterior, discharge tubes usually numerous. *Pleotrachelus*

Zoosporangium completely filling the host cell, the cell walls of the fungus and host in close contact. *Rozella*

Vegetative cell elongating and by septation forming a series of zoosporangia. Family Achlyogetonaceae

More than two zoosporangia formed in a series.

Zoospores encysting at mouth of exit tube. *Achlyogeton*

Zoospores swimming away from exit tube without encysting. *Septolpidium*

Only two zoosporangia formed, separated by an isthmus. *Bicricium*

Vegetative cell dividing internally into numerous zoosporangia or serving as a prosorus, the zoosporangia arising in a cell developing from the prosorus. Family Synchytriaceae

Parasitic in higher plants; vegetative cell rather large, becoming a sorus or prosorus or resting spore. *Synchytrium*

Parasitic in algae; vegetative cell small, becoming a prosorus or a resting spore.

Sorus of zoosporangia formed within the host cell.

Sorus dividing into a few sporangia which show no signs of flagella, each producing a number of zoospores. *Micromyces*

Sorus dividing into many zoosporangia which are set free, part within the host cell, part through a papilla into the surrounding water. Occasionally a zoosporangium bears a feebly active posterior flagellum. Zoospores minute, 2–5. *Endodesmidium*

Sorus of zoosporangia formed outside the host cell at apex of a discharge tube which pierces the host cell wall and produces only a few nonmotile zoosporangia. *Micromycopsis*

Eucarpic, i.e., endobiotic, epibiotic, or interbiotic, with rhizoids or haustoria. Monocentric. Including both operculate and inoperculate genera.

Zoospore cyst enlarging into a zoosporangium or prosporangium. Family Rhizidiaceae

Zoospore cyst enlarging into a zoosporangium.

Exit papilla or tube inoperculate.

Subsporangial swelling (apophysis) lacking.

Interbiotic (i.e., not closely attached to host cell, which is entered only by the tips of the rhizoidal branches).

Rhizoidal system arising from the main axis. *Rhizidium*

Rhizoidal system with several branches from the body of the sporangium. *Rhizophlyctis*

Epibiotic (i.e., closely associated with the host cell which is penetrated by the rhizoidal system). *Rhizophydium*

Subsporangial swelling present (scarcely differentiated in *Obelidium*).

Epibiotic. *Phlyctochytrium*

Interbiotic.

Apex of zoosporangium with mucro; a cup- or funnel-like base. *Obelidium*

No pronounced mucro; no cup- or funnel-like base.

Zoosporangium somewhat stellate. *Asterophlyctis*

Zoosporangium spherical or ellipsoidal.
Rhizoids delicate. *Rhizoclosmatium*
Rhizoids coarse. *Siphonaria*
Exit tube operculate near base, interbiotic. *Karlingia*
Exit tube operculate at apex, epibiotic. *Chytridium*
Zoospore cyst enlarging into a prosporangium.
Epibiotic, operculate, small, rhizoids not massive. *Chytridium*
Epibiotic, operculate, rhizoidal system massive. *Macrochytrium*
Interbiotic, inoperculate. *Polyphagus*
Zoospore cyst not enlarging into zoosporangium or prosporangium, but upper part of germ tube so enlarging.
Family Entophlyctaceae
Exit tube or papilla inoperculate. Endobiotic.
Upper part of germ tube becoming a zoosporangium. *Entophlyctis*
Zoosporangia arising by enlargement of the proximal portions of the rhizoidal branches. *Phlyctorhiza*
Upper part of germ tube becoming a prosporangium. *Diplophlyctis*
Exit tube operculate, germ tube enlarging into a zoosporangium.
Epibiotic. Rhizoids constricted into catenulate segments. *Catenochytrium*
Endobiotic. Rhizoids not constricted. *Endochytrium*
Exit tube operculate. Endobiotic. Germ tube enlarging into a prosporangium. *Nephrochytrium*
Eucarpic and polycentric.
Mostly saprophytic, intra- and extramatrical, with many centers connected by slender (rarely stout) tubes. Sporangia thin-walled, terminal and often intercalary. Resting spores thick-walled, apparently not sexually produced.
Family Cladochytriaceae
Sporangial openings inoperculate.
Rhizomycelium slender, branching, intramatrical, with turbinate cells. Zoosporangia intercalary and terminal. Resting spores thick-walled, acting as prosporangia when germinating. *Cladochytrium*
Rhizomycelium slender, branching, mostly extramatrical, rarely with turbinate cells. Zoosporangia and resting spores single at the apex of hyphae or branches. On pine pollen in water. *Physocladia*
Rhizomycelium coarse, branched, without turbinate cells or intercalary swellings. Zoosporangia usually in clusters of two or more at tips of hyphae or intercalary. No resting spores although thicker-walled, tuberculate zoosporangia are formed which set free their zoospores promptly. *Polychytrium*
Sporangia operculate at the mouth of the opening.
Intra- and extramatrical rhizomycelium slender and branched, with various more or less fusoid swellings. Zoosporangia formed mostly on the extramatrical rhizomycelium, terminal or intercalary, apophysate. Thicker-walled resting spores sometimes formed from a mass of parenchyma-like cells. Zoosporangia sometimes proliferating. *Nowakowskiella*
Mainly intramatrical, rhizomycelium moderately stout, constricted occasionally and septate or partially so at the constrictions.

branching and tapering to fine points. Numerous fusiform swellings which may become large or small zoosporangia or resting spores, the latter acting as prosporangia when they germinate. Zoosporangia not proliferating. *Septochytrium*

Extra- and intramatrical, rhizomycelium rather thick and undulate, branched, not tapering into fine points. Zoosporangia terminal and intercalary, rarely proliferating. Resting spores intercalary, on germination acting as prosporangia. *Megachytrium*

Exit tubes of zoosporangia operculate near base (endo-operculate). Rhizomycelium moderately stout. *Catenomyces*

Parasitic epibiotically and endobiotically on higher plants, the epibiotic stage monocentric, consisting of a proliferating, slipper-like zoosporangium with a tuft of rhizoids in the epidermal cell, the endobiotic stage being a slender branching rhizomycelium with numerous centers which give rise to further rhizomycelium and to thick-walled resting sporangia, mostly flattened on one side. Family Physodermataceae

Endobiotic stage not causing warts or galls on the host. *Physoderma*

Endobiotic stage causing warts or galls. *Urophlyctis*

Appendix to the Operculate Cladochytriaceae, but with relationship doubtful.

Stalk forked and bearing terminally the operculate zoosporangia. Sexual reproduction by formation of thick-walled zygospore by a mucoroid process of conjugation. *Zygochytrium*

Stalk branched and bearing at its tips operculate gametangia producing 4 gametes in each. No thick-walled resting spores known. *Tetrachytrium*

Key to the Families and Genera of Order Hyphochytriales

(*Based upon Karling, 1943*)

Holocarpic, monocentric, endobiotic. Family Anisolpidiaceae

Vegetative stage provided early with a cell wall and becoming a round or ellipsoidal zoosporangium with 1–3 exit tubes. Sexual reproduction unknown. Parasitic in Phaeophyceae. *Anisolpidium*

Vegetative stage naked and amoeboid almost until maturity of the rounded zoosporangium. Exit tube single. Sexual reproduction by the union of uniflagellate isogametes. Parasitic in Lemnaceae. *Reesia*

Vegetative structure becoming elongated, sometimes septate, to form one elongated or several rounded zoosporangia with thick walls. Zoospores escaping through a median aperture. Sexual reproduction unknown. Parasitic in roots. *Cystochytrium*

Eucarpic, monocentric, epibiotic. Family Rhizidiomycetaceae

Zoosporangia with or without apophysis, zoospores undergoing cleavage within a vesicle at the apex of the exit tube. Parasites or saprophytes. *Rhizidiomyces*

Zoosporangia without apophysis, zoospores fully developed within the zoosporangium. Parasites in oospores of *Vaucheria*. *Latrostium*

Eucarpic, polycentric, hypha-like, without rhizoids, with terminal and intercalary zoosporangia. Family Hyphochytriaceae

Only one well established genus. Parasitic in various hosts.

Hyphochytrium

The genus *Catenariopsis* may be distinct from *Hyphochytrium*. In it the zoosporangia are separated from one another by short isthmuses.

Literature Cited

AJELLO, LIBERO: Polychytrium: a new cladochytriaceous genus, *Mycologia*, **34**(4):442–451. *Figs*. 1–16. 1942.

BARTLETT, A. W.: On a new species of Urophlyctis producing galls on Lotus corniculatus Linn., *Brit. Mycol. Soc. Trans.*, **11**:266–281. *Pls*. 11–14. 1926.

BERDAN, HELEN: Two new genera of operculate chytrids, *Am. J. Botany*, **26**(7): 459–463. *Figs*. 1–2. 1939.

CANTER, HILDA M.: Studies on British chytrids: VI. Aquatic Synchytriaceae, *Brit. Mycol. Soc. Trans.*, **32**(1):69–94. *Pls*. 7–11. *Figs*. 1–13. 1949.

CLINTON, G. P.: Cladochytrium alismatis, *Botan. Gaz.*, **33**(1):49–61. *Pls*. 2–4. 1902. This is really *Physoderma maculare* Wallr.

COOK, MELVILLE T.: Species of Synchytrium in Louisiana: III. The development and structure of the galls, *Mycologia*, **37**(6):715–740. *Figs*. 1–12. 1945.

COUCH, JOHN N.: Rhizophidium, Phlyctochytrium and Phlyctidium in the United States, *J. Elisha Mitchell Sci. Soc.*, **47**(2):245–260. *Pls*. 14–17. 1932.

———: New or little known Chytridiales, *Mycologia*, **27**(2):160–175. *Figs*. 1–64. 1935.

———: Notes on the genus Micromyces, *Mycologia*, **29**(5):592–596. *Figs*. 1–4. 1937.

———: The structure and the action of the cilia in some aquatic Phycomycetes, *Am. J. Botany* **28**(8):704–713. *Figs*. 1–58. 1941.

———: Observations on the genus Catenaria, *Mycologia*, **37**(2):163–198. *Figs*. 1–78. 1945.

CURTIS, K. M.: The life-history and cytology of Synchytrium endobioticum (Schilb.) Perc., the cause of the wart disease in potato, *Trans. Roy. Soc. London, B*, **210**:409–478. *Pls*. 12–16. 1921.

FITZPATRICK, HARRY MORTON: The Lower Fungi—Phycomycetes, xi + 331 pp. 112 *figs*. New York, McGraw-Hill Book Co., 1930.

VON GUTENBURG, H. L.: Cytologische Studien an Synchytrium Gallen, *Jahrb. wiss. Botan.*, **46**:453–477. 1909.

HANSON, ANNE MARIE: A morphological, developmental, and cytological study of four saprophytic chytrids: I. Catenomyces persicinus Hanson, *Am. J. Botany*, **32**(7):431–438. *Pls*. 1–3. 1945; IV. Phlyctorhiza endogena gen. nov. sp. nov., *ibid.*, **33**(9):732–739. *Figs*. 1–49. 1946.

HARDER, RICHARD: Über das Vorkommen von Chitin und Zellulose und seine Bedeutung für die phylogenetische und systematische Beurteilung der Pilze, *Nachr. Ges. Wiss. Göttingen. Math. physik. Klasse, Fachgruppe VI*, N.F., **3**(1):1–7. 1937.

JANE, FRANK W.: A revision of the genus Harpochytrium, *J. Linnean Soc. London*, **53**(348):28–40. *Figs*. 1–28. 1946.

JOHANSON, ALFHILD E.: An endo-operculate chytridiaceous fungus: Karlingia rosea gen. nov., *Am. J. Botany*, **31**(7):397–404. *Figs*. 1–37. 1944.

JONES, FRED REUEL, AND CHARLES DRECHSLER: Crownwart of alfalfa caused by Urophlyctis alfalfae, *J. Agr. Research*, **20**(4):295–324. *Pls*. 47–56. 1920.

KARLING, JOHN S.: Studies in the Chytridiales: VII. The organization of the chytrid thallus, *Am. J. Botany*, **19**(1):41–74. *Figs*. 1–138. 1932.

KARLING: Resting sporangia in Cladochytrium, *Science*, **79**(2052):390. 1934.

———: The cytology of the Chytridiales with special reference to Cladochytrium replicatum, *Mem. Torrey Botan. Club*, **19**:1–22. *Pls.* 1–6. *Figs.* 1–2. 1937.

———: The life history of Anisolpidium ectocarpii gen. nov. et sp. nov., and a synopsis and classification of other fungi with anteriorly uniflagellate zoospores, *Am. J. Botany*, **30**(8):637–648. *Figs.* 1–21. 1943.

———: Brazilian chytrids: V. Nowakowskiella m..crospora n. sp. and other polycentric species, *ibid.*, **32**(1):29–35. *Figs.* 1–51. 1945a.

———: Brazilian chytrids: VII. Observations relative to sexuality in two new species of Siphonaria, *ibid.*, **32**(9):580–587. *Figs.* 1–53. 1945b.

KUSANO, SHUNSUKE: On the life-history and cytology of a new species of Olpidium, with special reference to the copulation of motile isogametes, *J. Coll. Agr., Tokyo Imp. Univ.*, **4**:144–199. *Pls.* 15–17. *Fig.* 1. 1912.

———: Observations on Olpidium trifolii Schroet., *ibid.*, **10**(2):83–99. 7 *figs.* 1929.

———: The life-history and physiology of Synchytrium fulgens Schroet., with special reference to its sexuality, *Japanese J. Botany*, **5**(1):35–132. *Figs.* 1–19. 1930.

MARTIN, G. W.: Systematic position of the slime molds and its bearing upon the classification of the fungi, *Botan. Gaz.*, **93**(4):421–435. 1932.

VON MINDEN, M.: Chytridiineae, Ancylistineae, Monoblepharidineae, Saprolegniineae, in Kryptogamenflora der Mark Brandenburg, vol. 5, pts. 2–3, pp. 209–630. Leipzig, Gebrüder Borntraeger, 1911–1915.

NABEL, KURT: Über die Membran niederer Pilze, besonders von Rhizidiomyces bivellatus nov. spez., *Arch. Mikrobiol.*, **10**:515–541. *Figs.* 1–7. 1939.

PETERSEN, HENNING EILER: Note sur les Phycomycètes observées dans les téguments vides des nymphes de Phryganées avec description de trois espèces nouvelles de Chytridinées, *J. Botan.*, **17**(6–7):214–222. *Figs.* 1–11. 1903.

ROBERTS, JOHN MAURICE: Developmental studies of two species of Nowakowskiella Schroeter: N. ramosa Butler and N. profusa Karling, *Mycologia*, **40**(2): 127–157. *Figs.* 1–2. 1948.

SCHERFFEL, A.: Endophytische Phycomyceten. Parasiten der Bacillariaceen und einige neue Monadinen. Ein Beitrag zur Phylogenie der Oomyceten (Schroeter), *Arch. Protistenk.*, **52**(1):1–141. *Pls.* 1–5. 1925.

SCHWARTZ, E. J., and W. R. IVIMEY COOK: The life-history and cytology of a new species of Olpidium: Olpidium radicale sp. nov., *Brit. Mycol. Soc. Trans.*, **13**:205–221. *Pls.* 13–15. 1928.

SCOTT, C. EMLEN: A preliminary note on the germination of Urophlyctis alfalfae, *Science*, N.S., **52**(1340):225–226. 1920.

SOROKIN, N.: Einige neue Wasserpilze, *Botan. Ztg.*, **32**:305–315. *Pl.* 6. 1874.

SPARROW, JR., FREDERICK K.: Observations on the aquatic fungi of Cold Spring Harbor, *Mycologia*, **24**(3):268–303. *Pls.* 7–8. *Figs.* 1–4. 1932.

———: The occurrence of true sporangia in the Physoderma disease of corn, *Science*, N.S., **79**(2060):563–564. 1934.

———: Evidences for the possible occurrence of sexuality in Diplophlyctis, *Mycologia*, **28**(4):321–323. *Figs.* 1–2. 1936.

———: Some chytridiaceous inhabitants of submerged insect exuviae, *Proc. Am. Phil. Soc.*, **78**(1):23–53. *Pls.* 1–4. *Figs.* 1–5. 1937.

———: The morphology and development of Obelidium mucronatum, *Mycologia*, **30**(1):1–14. *Figs.* 1–44. 1938.

———: A classification of aquatic Phycomycetes, *ibid.*, **34**(1):113–116. 1942.

SPARROW: Aquatic Phycomycetes, exclusive of the Saprolegniaceae and Pythium, xix + 785 pp. 634 *figs.* Ann Arbor, Univ. Michigan Press, 1943.

———: Observations on chytridiaceous parasites of Phanerogams: II. A preliminary study of the occurrence of ephemeral sporangia in the Physoderma disease of maize, *Am. J. Botany*, **34**(2):94–97. *Figs.* 1–17. 1947.

TISCHLER, G.: Pflanzliche Chromosomen-Zahlen, *Tabulae Biologicae*, **4**:1–83. 1937.

TISDALE, W. H.: Physoderma disease of corn, *J. Agr. Research*, **16**(5):137–154. *Pls. A and B (colored) and* 10–17. *Fig.* 1. 1919.

WAGER, H.: The life-history and cytology of Polyphagus Euglenae, *Ann. Botany*, **27**(106):173–202. *Pls.* 16–19. 1913.

VON WETTSTEIN, FRITZ: Das Vorkommen von Chitin und seine Verwertung als systematisch-phylogenetisches Merkmal im Pflanzenreich, *Sitz. ber. Akad. Wiss. Wien, Math. naturw. Klasse, Abt. I*, **130**(1):3–20. 1921.

WHIFFEN, ALMA J.: A discussion of taxonomic criteria in the Chytridiales, *Farlowia*, **1**(4):583–597. 1944.

WILSON, O. T.: Crown-gall of alfalfa, *Botan. Gaz.*, **70**(1):51–68. *Pls.* 7–10. 1920.

ZOPF, WILHELM: Zur Kenntnis der Phycomyceten: I. Zur Morphologie und Biologie der Ancylisten und Chytridium, zugleich ein Beitrag zur Phytopathologie, *Nova Acta Leopoldina*, **47**:141–236. *Pls.* 12–21. 1884.

4

PHYCOMYCETEAE: BLASTOCLADIALES AND MONOBLEPHARIDALES

THE organisms treated in this chapter show in very many points a close relationship to the Rhizidiaceae in the Chytridiales, but exhibit a greater development of the vegetative structure and a greater complexity of sexual reproduction in most of the cases where this is known. Instead of being a single uninucleate cell with a non-nucleate haustorial system which immediately becomes a sporangium by enlargement and internal division into zoospores, the vegetative body is a multinucleate clavate, cylindrical or spherical, or hypha-like structure which may be simple or branched. True septa are wanting except to delimit injured regions or sporangia or gametangia, but coarsely perforated pseudosepta occur in *Allomyces* and "cellulin" plugs may occur at the constrictions in *Gonapodya.* From this vegetative body arise one to many sporangia or gametangia. These organisms are aquatic or perhaps more often inhabit the soil. Most are saprophytic on vegetable material but the genus *Catenaria* contains species which may grow parasitically, in nematodes, in fluke eggs, and in the case of one species (Couch, 1945a) in the hyphae of the aquatic fungi *Allomyces* and *Blastocladiella. Coelomomyces* is parasitic in the larvae of mosquitoes.

The zoospores and gametes, where motile, are as in the Chytridiales posteriorly uniflagellate, with the whiplash type of flagellum. In a few species in the genera *Blastocladia* and *Monoblepharis* biflagellate zoospores have been observed occasionally. Cotner (1930a,b) demonstrated that these are abnormal structures, being binucleate as well as biflagellate, and that their occurrence is due to unfavorable conditions of development, usually unfavorable temperature, so that the typical zoospore of this whole group should be looked upon as being uniflagellate. The cell walls do not give the cellulose reaction with chloriodide of zinc except after treatment for some time with warm KOH or NaOH solutions which apparently saponify fatty deposits in the cell walls. After such treatment and thorough washing the wall in *Monoblepharis*, according to von Wettstein (1921) and to Harder (1937), shows a violet coloration with chloriodide of zinc. The author obtained the same reaction with an unidentified species of *Blastocladia*. Harder claims that in *Allomyces* and certain other members of the family the cell wall consists fundamentally of chitin. Nabel (1939) makes the same claim for *Allomyces*, *Blastocladia*, and *Blastocladiella*. In *Coelomomyces* a cell wall has not been demonstrated on the vegetative mycelium.

Order Blastocladiales. In the first edition of this textbook Blastocladiales and the following order, Monoblepharidales, were included under the latter name. Because of the slight differences in the structure of the zoospores and of the type of sexual reproduction, where it is known, and the production of resting sporangia the author now follows Sparrow (1943) in recognizing two orders. In the Blastocladiales the posteriorly uniflagellate swarm cells (zoospores or gametes) usually possess a well-marked "nuclear cap" attached mostly to the anterior surface of the nucleus. In the Monoblepharidales the nuclear cap is not noticeable but in the swimming cells numerous small granules are gathered at the apex sometimes forming a sort of papilla.

In the Blastocladiales sexual reproduction, where known, is by the union of two equal or unequal motile gametes to form a biflagellate zygote which germinates without becoming a resting spore. In the Monoblepharidales the egg is nonflagellate and after fertilization forms a thick wall to become a resting spore. In the Blastocladiales, in addition to the thin-walled zoosporangia, there are found also thick-walled resting sporangia which, in *Blastocladiella* and *Allomyces*, have a close connection with the alternation of sporophytic and gametophytic generations—an alternation that is unknown in the Monoblepharidales. These resting sporangia are spherical, ovoid, or pyriform, with a thick outer wall often perforated by numerous pits, and a thin inner wall. On germination the outer wall cracks open and the expanding inner wall produces one or more exit papillae from which the swarm spores emerge. These latter are called "planonts" by Sparrow (1943).

Three families are tentatively recognized, Catenariaceae (Couch 1945a), Coelomomycetaceae (Couch, 1945b) and Blastocladiaceae. They are distinguished as follows:

Family Catenariaceae. This family is parasitic in worms or fungi, or saprophytic in various other plant or animal substrata. The plant body at first is tubular, mostly unbranched, coenocytic, swelling at more or less regular intervals to form reproductive organs which are connected by short, narrow isthmuses, septate at each end and sometimes in the middle. These reproductive bodies are either thin-walled zoosporangia discharging by exit tubes or, on more exhausted media, thick-walled

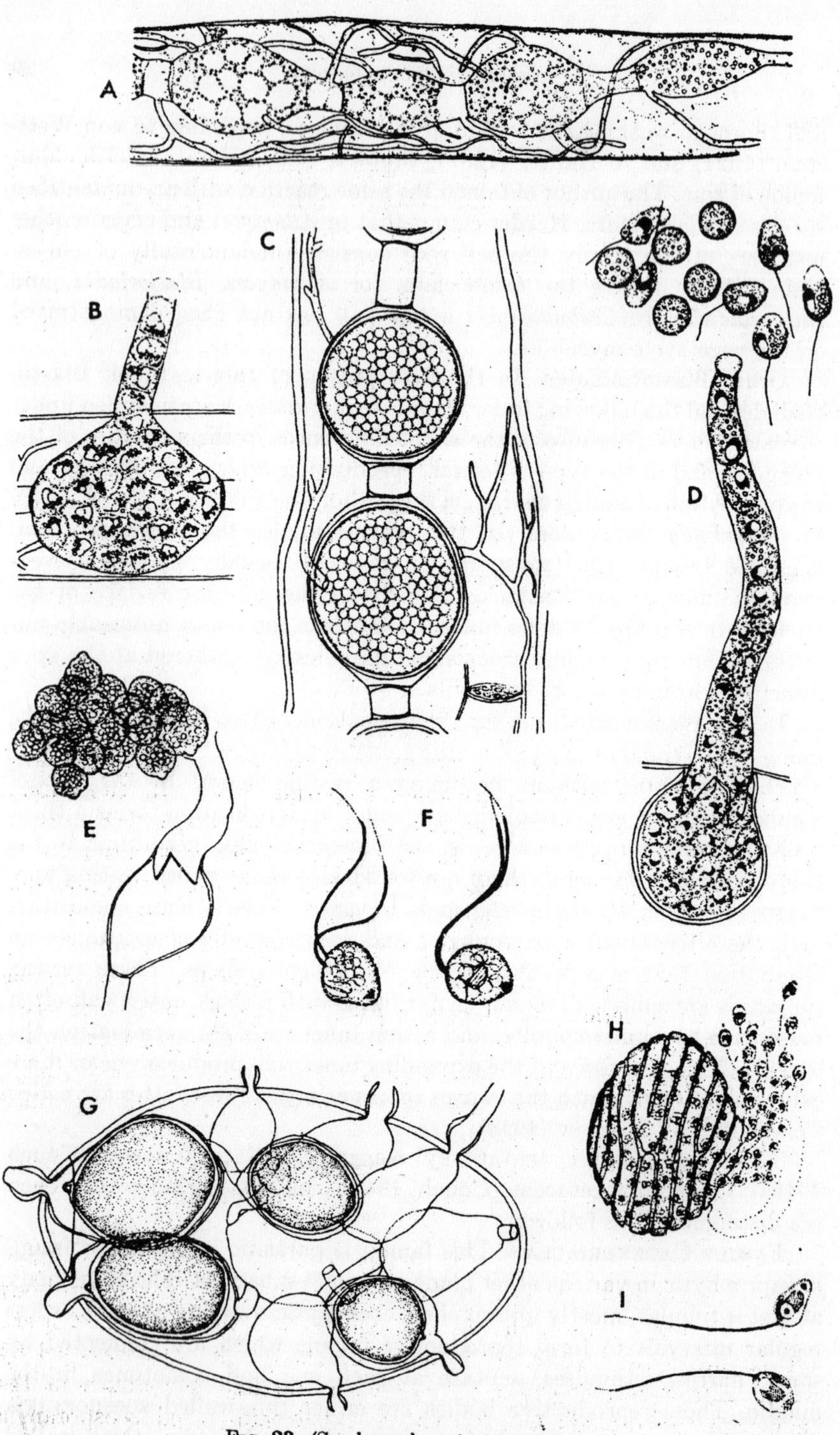

FIG. 23. (See legend on facing page.)

resting spores usually free from the hyphal membrane. These resting sporangia are smooth or minutely granular. In germination the outer wall cracks open and a tube emerges through which the zoospores are discharged. Rhizoids are produced at almost any point on the thallus, from the isthmuses as well as the reproductive swellings. Sexual reproduction occurs by the union of equal, motile gametes. One genus only is known, *Catenaria*, which was formerly placed in the family Cladochytriaceae of the order Chytridiales until Couch (1945a) demonstrated that the mode of reproduction is typical of some forms of the Blastocladiales.

Of the two well-studied species of the genus, sexual reproduction is unknown in *C. anguillulae* Sorokin. It produces both thin-walled and resting zoosporangia which appear to produce exactly similar zoospores. In *C. allomycis* Couch, the zoospores from the thin-walled zoosporangia produce plants with either or both types of sporangia. Those from the resting sporangia produce less active zoospores which encyst at or near the mouth of the exit tube. After about two hours these cysts produce four gametes, each, which escape through a short papilla and sooner or later unite by pairs to form posteriorly biflagellate zygotes. These swim for a time, the two flagella lying close together and being synchronous in their motions. The zygotes eventually encyst and penetrate the host tissue by means of a slender germ tube. Apparently, pairing of the gametes is rare between those from any one cyst. This type of reproduction is practically identical with that of *Blastocladiella cystogena* Couch and Whiffen (1942) and comparable to that of *Allomyces cystogenus* Emerson as interpreted by McCranie (1942). (Fig. 23A–G.)

FAMILY COELOMOMYCETACEAE. Coelomomycetaceae are parasitic in the larvae of insects, chiefly mosquitoes. Mycelium coenocytic, nonseptate, lacking rhizoids and without demonstrable cell walls, branched and somewhat anastomosing. Terminal portions of the short branches enlarge and break free and develop resting spores surrounded by the plasma membrane of the mycelium. These spores possess a thick outer, colored wall, usually pitted or striate, and a thinner inner wall. Dehiscence by a longitudinal slit. Zoospores posteriorly uniflagellate with an, at most, imperfectly formed nuclear cap. No thin-walled sporangia nor sexual stage

FIG. 23. Blastocladiales. (A–G) Family Catenariaceae. (A–F) *Catenaria allomycis* Couch. (A) Chain of zoosporangia within filament of *Allomyces*. (B) Zoosporangium with zoospores about to escape. (C) Resting sporangia within host filament. (D) Zoospores emerging from germinating resting sporangium and encysting. (E) Cysts at mouth of exit tube of resting sporangium. (F) Biflagellate zygotes formed by union of gametes emerging from the cysts. (G) *Catenaria anguillulae* Sorokin, fluke egg with four resting sporangia and several emptied thin-walled zoosporangia. (H, I) Family Coelomomycetaceae. *Coelomomyces lativittatus* Couch and Dodge. (H) Germinating resting sporangium. (I) Zoospores. (A–G, courtesy, Couch: *Mycologia*, **37**(2):163–193. H–I, courtesy, Couch and Dodge: *J. Elisha Mitchell Sci. Soc.*, **63**(1):69–79.)

observed. Twelve or more species. Only genus known *Coelomomyces* (Keilin, 1921; Couch, 1945b; Couch and Dodge, 1947). (Fig. 23H, I.)

FAMILY BLASTOCLADIACEAE. The vegetative portion of the plants of this family consists of a more or less extensive system of tapering and branching rhizoids and of a globular or clavate external portion from which may arise directly the sporangia and gametangia or a system of branches on which these organs are borne. In *Allomyces* the hyphae are more nearly uniform, without a broad basal portion, and are frequently constricted here and there. At these points usually occur coarsely perforate "pseudosepta." Three or more genera are recognized in this family, which may be distinguished as follows:

Blastocladia: plant body with a simple, lobed or branched basal portion, often with secondary axes. Zoosporangia with a single apical exit papilla. Alternation of sporophytes and gametophytes unknown.

Blastocladiella (including according to Couch and Whiffen (1942) *Clavochytrium* and *Sphaerocladia*): plant body a spherical vegetative structure bearing rhizoids at the base or on all sides, or more or less clavate, becoming directly a reproductive organ or bearing at its apex the single reproductive organ. Alternation of generations demonstrated in some species. Discharge papillae one or several.

Allomyces: plant body a cylindrical basal segment giving rise, usually dichotomously, to cylindrical, often constricted, branches with pseudosepta. Discharge papillae usually more than one. Alternation of sporophytes and gametophytes known in some species.

Blastocladiella. In *Blastocladiella stübenii* Couch and Whiffen (1942) the plant body of the sporophyte is spherical, with tapering, much-branched rhizoids emerging at all sides. It bears a thin-walled zoosporangium with one to several discharge tubes or a dark, thick-walled, resting sporangium which produces swarm cells that give rise to indistinguishable male or female gametophytes similar to, but smaller than, the sporophytes. The emerging gametes are equal in size and cannot be distinguished by color. They fuse in pairs and at once produce the sporophytes. In *B. variabilis* Harder and Sörgel (1938) the sporophyte is cylindrical, bearing at its base much-branched rhizoids and at its apex either a thin-walled zoosporangium or a thick-walled resting sporangium. The zoospores from the former give rise to sporophytes again but those from the latter (planonts) give rise to similar gametophytes, each with a single clavate gametangium, orange-colored in the male and colorless in the female gametophyte. The biflagellate zygote gives rise at once to a sporophyte. Five other species are distinguished. In *B. stomophila* (Couch and Cox) Couch and Whiffen only thin-walled zoosporangia have been observed. In *B. simplex* Matthews and two other species both zoosporangia and resting sporangia are produced, but the zoospores from the latter reproduce sporophytes, no gametophytes having been recognized.

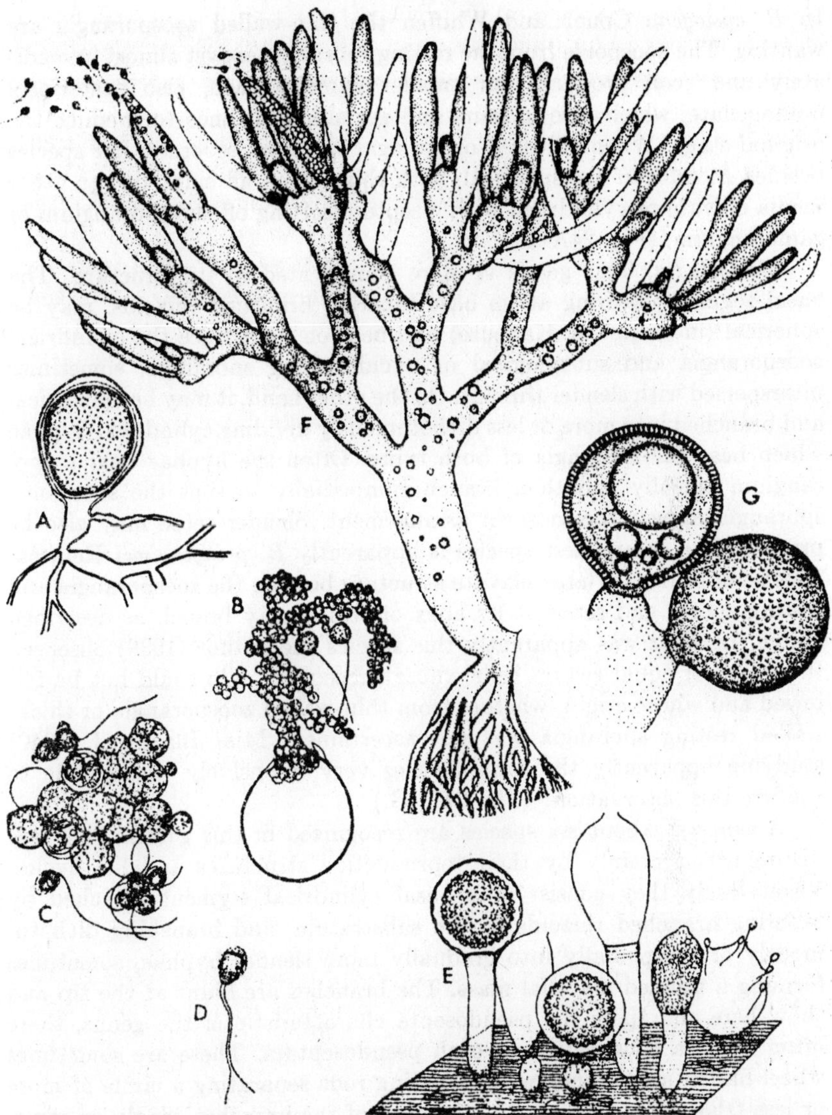

Fig. 24. Blastocladiales, Family Blastocladiaceae. (A–D) *Blastocladiella cystogena* Couch and Whiffen. (A) Plant with mature resting sporangium. (B) Uniflagellate zoospores, two greatly magnified, just after discharge from resting sporangium. (C) Cysts, some still containing and some discharging gametes. (D) Biflagellate zygote. (E) *Blastocladiella asperosperma* Couch and Whiffen, cluster of plants, some with resting sporangia and some with thin-walled sporangia. (F, G) *Blastocladia pringsheimii* Reinsch. (F) Plant with thin-walled sporangia. (G) Resting sporangia. (A–E, courtesy, Couch and Whiffen: *Am. J. Botany*, **29**(7):163–193. F–G, courtesy, Thaxter: *Botan. Gaz.*, **21**(2):45–52, Univ. Chicago Press.)

In *B. cystogena* Couch and Whiffen the thin-walled zoosporangia are wanting. The zoospores from the resting sporangia encyst almost immediately and soon produce four smaller gametes each, also posteriorly uniflagellate, which fuse in pairs and germinate at once to produce the original stage. When cultured on an agar medium, several other species besides *B. stübenii* are spherical, with rhizoids on all sides, but on other media clavate or cylindrical with a septum setting off the sporangium or gametangium. (Fig. 24A–E.)

Blastocladia. This genus is more complicated in its structure. The basal segment, bearing at its base tapering branched rhizoids, may be spherical (in *B. globosa* Kanouse) and bear on its surface the cylindrical zoosporangia and subspherical or ovoid resting sporangia, sometimes interspersed with slender threads. On the other hand, it may be cylindrical and branched into more or less dichotomously dividing cylindrical hyphae which bear the sporangia of both types. Often the hyphae bear a sporangium apically and then branch sympodially so that the successive sporangia appear racemose in arrangement. Slender setae may also be present. The commonest species is apparently *B. pringsheimii* Reinsch, which varies from a large clavate structure bearing the zoosporangia and resting spores and setae at its apex or which may branch as described above. In what was apparently this species the author (1939) observed the union of equal swarm cells whose subsequent fate could not be followed and whose origin, whether from thin-walled zoosporangia or thick-walled resting sporangia was not ascertained. Miss Blackwell (1940) studying apparently the same species very intensively was unable to confirm this observation. (Fig. 24F, G.)

Allomyces. About six species are recognized in this genus. They are distinguished mainly by their reproductive structures and life cycles. Vegetatively they consist of a basal cylindrical segment attached by tapering branched rhizoids to the substratum, and branching dichotomously or sympodially into gradually more slender hyphae, sometimes forming a tangled mycelial mass. The branches are blunt at the tip and show here and there the pseudosepta characteristic of the genus, there often being a constriction at each pseudoseptum. These are sometimes wheel-like in appearance with radiating rods separating a circle of more or less triangular openings. At the ends of the branches, singly or sometimes in chains, are produced the cylindrical or somewhat oval, thin-walled zoosporangia. By sympodial branching these may come to have the appearance of a racemose arrangement. The zoospores escape by one or several inoperculate openings. The same plants may also bear the thick-walled resting sporangia which vary in shape from spherical to ovoid or even lemon-shaped and with the brown outer wall deeply punctate. The zoospores from the thin-walled zoosporangia produce the

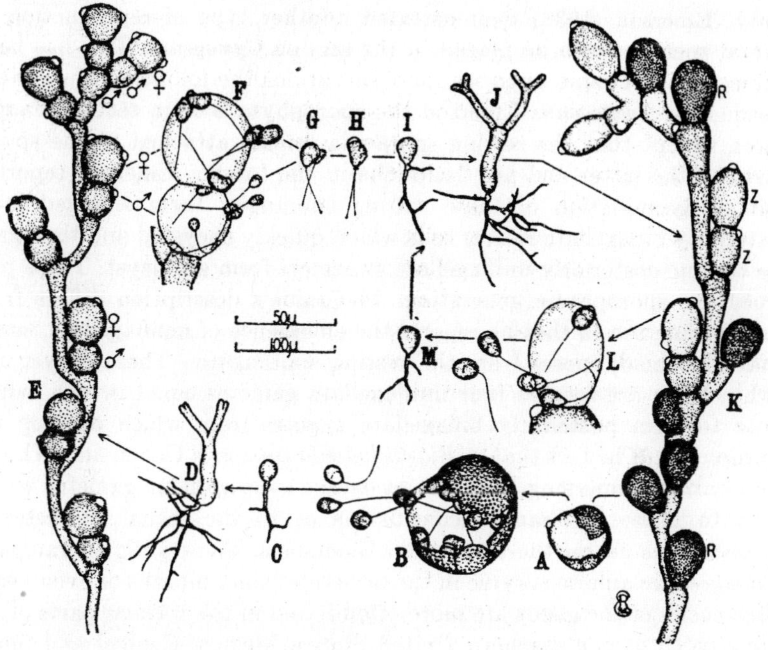

FIG. 25. Blastocladiales, Family Blastocladiaceae. *Allomyces arbuscula* Butler. Life cycle. (After Emerson. Courtesy, Sparrow: Aquatic Phycomycetes, Exclusive of the Saprolegniaceae and Pythium, Ann Arbor, Univ. Michigan Press.)

same stage of development. Those from the thick-walled resting sporangia behave differently according to the species. In the main the reproduction is of two distinct types. In the section *Euallomyces*, represented typically by *A. javanicus* Kniep and *A. arbuscula* Butler, the zoospores from the thick-walled resting spores produce gametophytes of approximately the same structure as the sporophytes described above. In place of the sporangia, chains of gametangia are produced terminally on the branches. There are usually two, rarely more, in each chain. These consist of a smaller and a larger gametangium, the former being terminal in *A. javanicus*, the latter terminal in *A. arbusculus*. The smaller, male gametangium is salmon-pink to orange in color, the larger, female gametangium is colorless. The gametes emerge from one to three exit papillae. The male gametes are smaller and faintly colored and about twice as numerous as the colorless female gametes. They fuse promptly and the biflagellate zygote germinates almost at once to produce the sporophyte. This type of sexual reproduction and alternation of generations in this genus was first reported by Kniep (1929, 1930) in *A. javanicus*. Hatch (1933) demonstrated it in *A. arbuscula*. Since then Hatch (1935, 1938), Emerson (1941), McCranie (1942), and Wolf (1941) have given the genus intensive study. Emerson (1938) demonstrated another type of reproduction in several species which he placed in the section *Cystogenes.* This has been studied by McCranie upon whose observations the following life cycle is based. In *A. cystogenus* Emerson the sporophyte is as in section *Euallomyces*, except that the resting sporangia remain attached to the sporophyte in the latter and are deciduous in the former. Emerson reported that on germination of these resting sporangia, there emerged large posteriorly biflagellate swarm cells which quickly encysted and then gave rise to four posteriorly uniflagellate swarmers from each cyst. These produced the sporophytic generation. McCranie's description differs from that of Emerson in that he reports the emergence of nonflagellate, somewhat amoeboid, spores from the resting sporangium. These encyst and within them are formed four uniflagellate gametes equal in size, which unite to form posteriorly biflagellate zygotes from which develop the sporophytes. The fact that in *Blastocladiella cystogena* Couch and Whiffen the swarmers emerging from the cysts act as isogamous gametes would seem to make McCranie's conclusions as to the sexual processes in *A. cystogenus* nearer the truth than Emerson's. The resting sporangia of *Allomyces* are able to survive in the dry state many months or even years. The species of the genus are more often found in the warmer parts of the world, such as the southern United States, Mexico, Central and South America, West Indies, Southern Asia, East Indies, Africa, and southern Europe. (Fig. 25.)

Order Monoblepharidales. The chief differences between this order and the Blastocladiales are the fertilization of large, nonmotile eggs by posteriorly uniflagellate sperms; the hypha-like mycelium, like that of *Allomyces* but lacking the perforated pseudosepta of that genus; the absence of thick-walled resting sporangia and of alternation of generations, with minor differences in the structure of the swarm cells. Of the three genera here included in this order sexual reproduction is unknown in *Gonapodya* and the correctness of the inclusion of that genus in the order is doubtful. Possibly it should be placed in the foregoing order. The two other genera, *Monoblepharis* and *Monoblepharella*, are closely related. With the doubtful inclusion of *Gonapodya* but one family is recognized. The genus *Myrioblepharis* has been placed in this order by some investigators, but it is probable that this is not an independent organism but some other aquatic fungus parasitized by a ciliate Protozoan. The only species, *M. paradoxa*, was described by Thaxter in 1895, and found once by von Minden in 1915. The papers of these authors should be referred to.

Family Monoblepharidaceae. *Gonapodya* is a genus of two species, growing saprophytically on vegetable matter in fresh water. When rose fruits are placed in water for several weeks, this fungus frequently develops in little tufts, usually covered by bacterial slime, on the surface of the fruits. The mycelium is variously branched, arising from a tuft of branching, tapering rhizoids. The external mycelium is constricted at

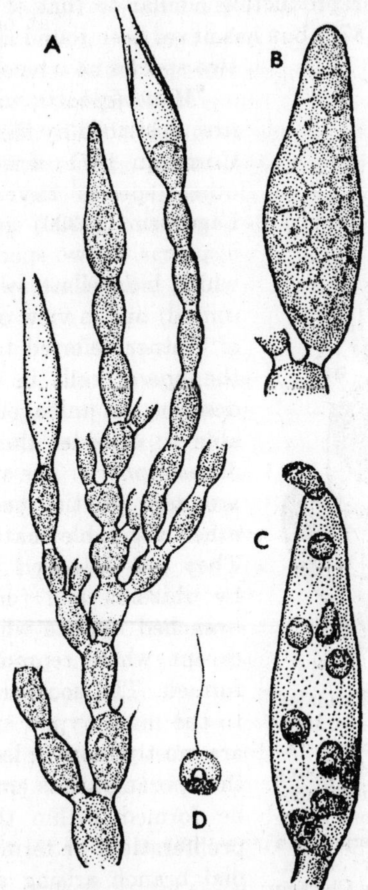

Fig. 26. Monoblepharidales, Family Monoblepharidaceae. *Gonapodya siliquaeformis* (Reinsch) Thaxt. (A) Portion of fungus showing typical habit. (B) Unopened zoosporangium with zoospores outlined. (C) Zoosporangium with escaping zoospores. (D) Zoospore. (After Thaxter: *Botan. Gaz.*, 20(11):477–485, Univ. Chicago Press.)

frequent intervals producing cylindrical or moniliform coenocytic segments. At the constrictions there are usually plugs of a carbohydrate substance that has been called cellulin. At the ends of the branches the terminal segment becomes the zoosporangium, ovate or lanceolate in outline, and producing numerous zoospores of the same type as those of *Monoblepharis*. After these have escaped through the apical pore successive zoosporangia may be formed in the empty sporangial wall by proliferation. Sexual reproduction similar to that of *Monoblepharis* was reported by Cornu (1871) but has never been found again in either of the two species now recognized. (Fig. 26.)

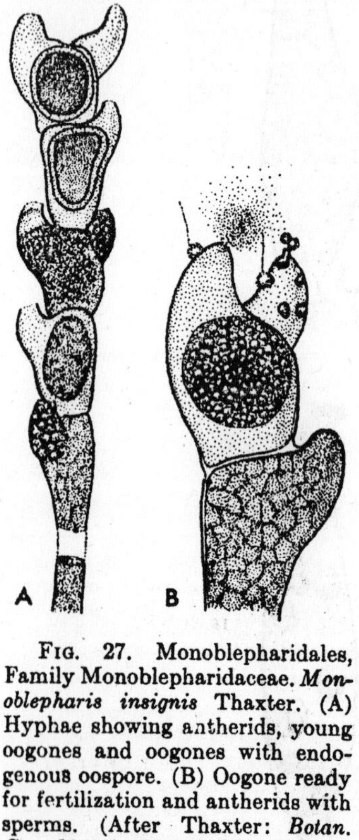

A B

FIG. 27. Monoblepharidales, Family Monoblepharidaceae. *Monoblepharis insignis* Thaxter. (A) Hyphae showing antherids, young oogones and oogones with endogenous oospore. (B) Oogone ready for fertilization and antherids with sperms. (After Thaxter: *Botan. Gaz.*, **20**(10):433–440, Univ. Chicago Press.)

Monoblepharis, with two species, was first described by the French mycologist Cornu in 1871; since then four or five other species have been recognized. Lagerheim (1900) gave the name *Diblepharis* to two species of this genus in which biflagellate swarm cells were observed, but in view of the investigations of Cotner referred to earlier, and since the sperm cells in those species were described as uniflagellate, it seems desirable to consider these as belonging to *Monoblepharis*. The species of this genus are saprophytic, usually on twigs or other vegetable matter in fresh water. They are attached to the substratum by rhizoids and form unbranched or branched hyphae which are nonseptate except where reproductive organs are formed. The zoosporangia are terminal to the main hypha or its branches, and are mostly more or less cylindrical. After the sporangium is emptied another may be formed within the empty wall by proliferation, or terminally on a sympodial branch arising at the base of the old sporangium. The zoospores escape, fully formed, from an opening dissolved in the apex of the sporangium and are typically posteriorly uniflagellate. As mentioned above the frequent biflagellate condition of the zoospores is probably abnormal.

The oogone in *Monoblepharis* may be formed terminally or by the enlargement of a subterminal segment. In the former case a second segment immediately below the oogone becomes the antherid; in the latter case the terminal segment becomes the antherid. In *M. macrandra* (Lagerheim) Woronin the antherids may be terminal on slender branches distant from the oogones. The number of sperms produced in an

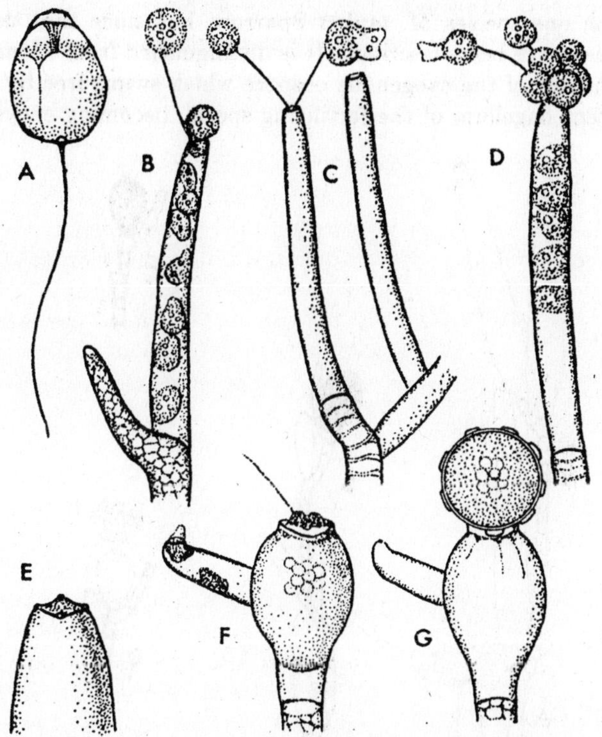

FIG. 28. Monoblepharidales, Family Monoblepharidaceae, *Monoblepharis polymorpha* Cornu. (A) Zoospore. (B–E) Zoosporangia. (F) Fertilization of oogone. (G) Exogenous oospore. (After Sparrow: *Ann. Botany*, **47**(187):517–542.)

antherid is usually 4–7, but in *M. insignis* Thaxter there may be up to 24–32. They are posteriorly uniflagellate and swim to the oogone or creep along its surface in a jerky manner as the flagellum waves. The oogones become uninucleate according to Laibach (1927) before the basal septum is formed. They are pyriform or ovoid and when ready for fertilization become open at or near the apex, sometimes at the top of a short broad neck. The protoplasmic contents round up into a more or less spherical egg near the base of the oogone or close to its opening. Upon entry of the sperm into the egg the flagellum of the former disappears and the fertilized egg remains in the oogone and forms a thick wall (endogenous oospore) or creeps out through the oogonial opening and encysts externally (exogenous oospore), in some cases then falling off. Rarely both modes of oospore formation may occur in the same species, e.g., *M. macrandra*. The two sex nuclei do not unite in the oospore until shortly before its germination by a slender germ tube. (Figs. 27, 28.)

Monoblepharella was distinguished as a separate genus by Sparrow

(1940) with one species *M. taylori* Sparrow, but since that date a few other species have been described. It is distinguished from *Monoblepharis* by the behavior of the exogenous oospore which swims free by means of the persistent flagellum of the fertilizing sperm, becoming encysted after

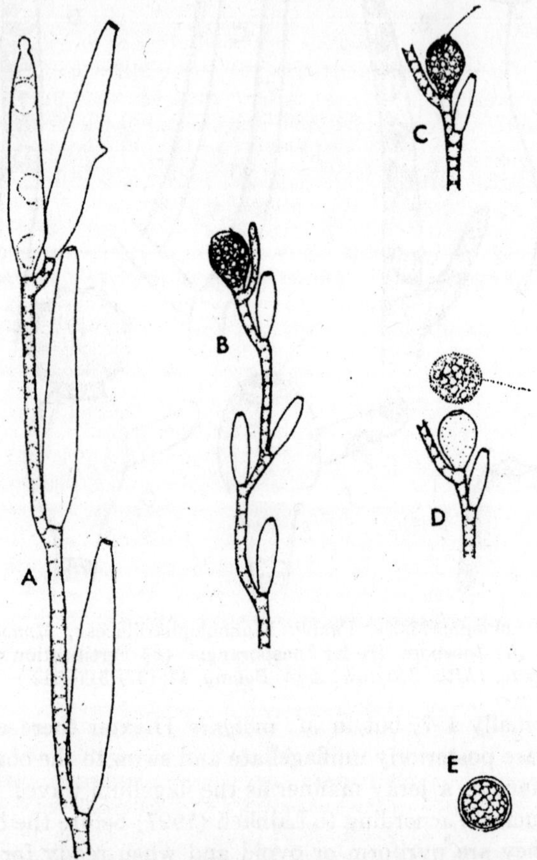

Fig. 29. Monoblepharidales, Family Monoblepharidaceae. *Monoblepharella mexicana* Shanor. (A) Hypha with zoosporangia. (B) Hypha with oogones and antherids. (C) Oogone just after entry of sperm cell. (D) Zygote swimming away by means of the flagellum of the sperm cell. (E) Encysted zygote. (Courtesy, Shanor: *Mycologia*, **34**(3):241–247.)

having progressed some distance. Occasionally in *M. taylori* the oogone may contain up to four eggs, though one is the more usual number. In *M. elongata* Springer, in about half of the oogones more than one egg is formed, sometimes as many as eight. (Fig. 29.)

Looking back from *Monoblepharis* we can see a fairly unbroken series from the eucarpic, monocentric, inoperculate Rhizidiaceae, through *Blastocladiella*, *Allomyces*, and *Monoblepharella* to *Monoblepharis*, with the uninucleate vegetative body in the more primitive forms becoming transformed directly into a zoosporangium or gametangium, as in *Rhizophydium*, while in *Blastocladiella* the vegetative structure is ovoid or elongated with many nuclei, bearing the reproductive organs terminally. Here the motile gametes are isogamous. In *Allomyces* the plant body is more hypha-like and branched, and the reproductive organs are numerous. In section *Euallomyces* the motile gametes are heterogamous. In *Monoblepharella* the female gametes are one to eight in the oögone and are motile, not by their own flagella, but by means of the flagellum of the sperm. In *Monoblepharis* the egg is single in the oogone and is not at all motile except in the species with exogenous oospores, where it creeps out of the oogone after fertilization but has no power of further locomotion.

Key to the Families and Genera of Order Blastocladiales

Plant body at first tubular, coenocytic, then forming alternately zoosporangia and narrow isthmuses. Rhizoids arising at isthmuses or at all points. Resting spores formed sometimes in place of zoosporangia.
Family Catenariaceae

Only genus. *Catenaria*

Plant body coenocytic, branching and anastomosing, apparently without a cell wall in its vegetative stage, multinucleate. Tips of the branches enlarging to form thick-walled resting spores with the outer wall marked punctately and with longitudinal lines. These produce posteriorly uniflagellate zoospores soon, or after a resting period in the dry state. Parasitic in the larvae of mosquitoes. Family Coelomomycetaceae

Only genus. *Coelomomyces*

Plant body with basally attached tuft of branching rhizoids and a globular or clavate external basal segment upon which directly, or upon branches of which, arise the reproductive organs (zoosporangia, resting spores, gametangia). Sometimes the distinction between the basal piece and the branches arising from it is only slight. Family Blastocladiaceae

Plant body with a simple or lobed or branched basal portion, often with secondary axes. Zoosporangia with single apical exit papilla. Alternation of sporophytic and gametophytic generations unknown. Sexual reproduction mostly unknown. *Blastocladia*

Plant body a more or less spherical structure which is directly transformed into a reproductive organ, or more or less clavate, bearing at its top a single reproductive organ. Discharge papilla one or several. Alternation of generations demonstrated in some species. *Blastocladiella*

Plant body consisting of a cylindrical basal segment giving rise—usually dichotomously—to cylindrical, often constricted, branches with pseudosepta at the constrictions. Usually more than one discharge papilla. Sexual reproduction by anisoplanogametes or isoplanogametes. Alternation of sporophytic and gametophytic generations known in some species.
Allomyces

Key to the Family and Genera of Order Monoblepharidales

Single family. Family Monoblepharidaceae

Branching mycelium constricted into cylindrical or rounded coenocytic segments, terminal ones becoming zoosporangia. Sexual reproduction not certainly demonstrated. *Gonapodya*

Mycelium often not extensively branched, not constricted. Zoosporangia terminal. Antherids and oogones terminal or subterminal. Egg cells not flagellate.

Egg cells not motile by means of the flagellum of the sperm. *Monoblepharis*

Egg cells motile by means of the flagellum of the fertilizing sperm. *Monoblepharella*

Literature Cited

BESSEY, ERNST A.: Isoplanogametes in Blastocladia, *Mycologia*, **3**(3):308–309. 1939.

BLACKWELL, ELIZABETH: A life cycle of Blastocladia Pringsheimii Reinsch, *Brit. Mycol. Soc. Trans.*, **24**(1):68–86. *Figs.* 1–9. 1940.

CORNU, MAXIME: Note sur deux genres nouveaux de la famille des Saprolégniées, *Bull. soc. botan. France*, **18**:58–59. 1871.

COTNER, FRANK B.: The development of the zoospores in the Oomycetes at optimum temperatures and the cytology of their active stages, *Am. J. Botany*, **17**(6):511–546. *Pls.* 30–32. *Fig.* 1. 1930a.

———: Cytological study of the zoospores of Blastocladia, *Botan. Gaz.*, **89**(3): 295–309. *Figs.* 1–10. 1930b.

COUCH, JOHN N.: Observations on the genus Catenaria, *Mycologia*, **37**(2):163–193. *Figs.* 1–78. 1945a.

———: Revision of the genus Coelomomyces, parasitic in insect larvae, *J. Elisha Mitchell Sci. Soc.*, **61**(1–2):124–136. *Pls.* 1–2. 1945b.

———, AND H. R. DODGE: Further observations on Coelomomyces, parasitic on mosquito larvae, *ibid.*, **63**(1):69–79. *Pls.* 15–20. 1947.

———, AND ALMA J. WHIFFEN: Observations on the genus Blastocladiella, *Am. J. Botany*, **29**(7):582–591. *Figs.* 1–66. 1942.

EMERSON, RALPH: A new life cycle involving cyst-formation in Allomyces, *Mycologia*, **30**(2):120–132. *Figs.* 1–11. 1938.

———: An experimental study of the life cycles and taxonomy of Allomyces, *Lloydia*, **4**(2):77–144. *Figs.* 1–16. 1941.

HARDER, RICHARD: Über das Vorkommen vom Chitin und Zellulose und seine Bedeutung für die phylogenetische und systematische Beurteilung der Pilze, *Nach. Ges. Wiss. Göttingen. Math. physik. Klasse, Fachgruppe VI*, N.F., **3**(1):1–7. 1937.

———, UND GEORG SÖRGEL: Über einen neuen planoisogamen Phycomyceten mit Generationswechsel und seine phylogenetische Bedeutung, *ibid.*, **3**(5): 119–127. *Figs.* 1–4. 1938.

HATCH, WINSLOW R.: Sexuality in Allomyces arbuscula Butler, *J. Elisha Mitchell Sci. Soc.*, **49**(1):163–170. *Pl.* 12. 1933.

———: Gametogenesis in Allomyces, *Ann. Botany*, **49**(196):623–650. *Figs.* 1–33. 1935.

———: Conjugation and zygote germination in Allomyces arbuscula, *ibid.*, N.S.. **2**(7):583–614. *Pls.* 18–22. *Figs.* 1–13. 1938.

KEILIN, D.: On a new type of fungus: Coelomomyces stegomyiae n.g., n.sp., parasitic in the body cavity of the larva of Stegomyia scutellaria Walker (Diptera, Nematocera, Culicidae), *Parasitology*, **13**(3):225–234. *Figs.* 1–7. 1921.

KNIEP, HANS: Allomyces javanicus n. sp. ein anisogamer Phycomycet mit Planogameten, *Ber. deut. botan. Ges.*, **47**:199–212. *Figs.* 1–7. 1929.

———: Über den Generationswechsel von Allomyces, *Z. Botan.*, **22**(9):433–441. *Figs.* 1–2. 1930.

LAGERHEIM, G.: Mykologische Studien: II. Untersuchungen über die Monoblepharideen, *Kgl. Svenska Vetenskapsakad. Handl.*, **25**, Afd. 3, No. 8: 1–42. *Pls.* 1–2. 1900.

LAIBACH, FRIEDRICH: Zytologische Untersuchungen über die Monoblepharideen, *Jahrb. wiss. Botan.*, **66**: 596–630. *Pls.* 12–13. *Figs.* 1–12. 1927.

MCCRANIE, JAMES: Sexuality in Allomyces cystogenus, *Mycologia*, **34**(2):209–213. *Fig.* 1. 1942.

VON MINDEN, M.: Chytridiineae, Ancylistineae, Monoblepharidineae, Saprolegniineae, in Kryptogamenflora der Mark Brandenburg, vol. 5, pt. 3, pp. 353–496. Leipzig, Gebrüder Borntraeger, 1911.

NABEL, KURT: Über die Membran niederer Pilze, besonders von Rhizidiomyces bivellatus nov. spez., *Arch. Mikrobiol.*, **10**:515–541. *Figs.* 1–7. 1939.

SPARROW JR., FREDERICK K.: Phycomycetes recovered from soil samples collected by W. R. Taylor on the Allan Hancock 1939 expedition, *Allan Hancock Pacific Expeditions*, **3**(6):101–112. *Pls.* 16–17. 1940.

———: The Monoblepharidales, *Ann. Botany*, **47**(187):517–542. *Pl.* 20. *Figs.* 1–2. 1933.

———: Aquatic Phycomycetes, exclusive of the Saprolegniaceae and Pythium, xix + 785 pp. 634 *figs.* Ann Arbor, Univ. Michigan Press, 1943.

THAXTER, ROLAND: New or peculiar aquatic fungi: 1. Monoblepharis, *Botan. Gaz.*, **20**:433–440. *Pl.* 29. 1895a; 2. Gonapodya Fischer and Myrioblepharis nov. gen., *ibid.*, **20**:477–485. *Pl.* 31. 1895b; 3. Blastocladia, *ibid.*, **21**:45–52. *Pl.* 3. 1896.

VON WETTSTEIN, FRITZ: Das Vorkommen von Chitin und seine Verwertung als systematisch-phylogenetisches Merkmal im Pflanzenreich, *Sitz. ber. Akad. Wiss. Wien, Math. naturw. Klasse, Abt. I*, **130**(1):3–20. 1921.

WOLF, FRED T.: A contribution to the life history and geographic distribution of the genus Allomyces, *Mycologia*, **33**(2):158–173. *Figs.* 1–2. 1941.

5

PHYCOMYCETEAE: LAGENIDIALES AND SAPROLEGNIALES

In the two foregoing chapters have been discussed organisms, chiefly aquatic or soil inhabiting, with posteriorly uniflagellate swarm cells and whose cell walls are deficient in cellulose or do not show the cellulose reaction without preliminary treatment to remove some masking substances. Also organisms with a single anterior flagellum of the tinsel type were discussed. In this and the next following chapter are treated forms, many of them aquatic or soil inhabiting, in which the swarm cells are anteriorly or laterally biflagellate and in which the cellulose reaction is normally shown immediately upon application of chloriodide of zinc without other preliminary treatment. The anterior flagellum is of the tinsel type, the posterior one of the whiplash type. Here, too, a rather well connected series can be followed from holocarpic, endobiotic, monocentric forms up to fungi with extensive mycelium and complicated modes of reproduction, both asexual and sexual. This series, of course, may be read in the reverse direction, leading from the complex to simplified forms. The author does not follow Sparrow (1943) in considering that the Plasmodiophoraceae are related to the other groups included in the chapter since *Plasmodiophora* is apparently more closely related to the Mycetozoa (see Chapter 2).

Order Lagenidiales. The limits of this order are far from definite and it may be that, as in the first edition of this textbook, the genera and families here included should be placed in the order Saprolegniales. Sparrow recognized the close affinity of some of these to that order by placing the Ectrogellaceae and Thraustochytriaceae there. At present the author is not convinced that enough is known of the structures and life histories of all of these rather simple or reduced forms to warrant their distribution in separate orders. It therefore seems preferable to use the order Lagenidiales as a temporary catchall for a number of genera which perhaps are not too closely related, but which have in common the characters given below. The members of this order are (except Thraustochytriaceae) endobiotic and holocarpic in that they live within their host cells, mostly as parasites, and lack rhizoids, the whole fungus body becoming one or several reproductive units. Asexual reproduction is accomplished by the transformation of the plant body into a zoosporangium which empties its anteriorly or laterally biflagellate zoospores into the water surrounding the host cell through an inoperculate exit tube or tubes in the manner of *Olpidium* or *Achlyogeton* in the Chytridiales. The zoospores may escape singly and swim away independently or may encyst at the mouth of the exit tube and later escape and swim away. In some cases the zoospores are not distinguishable in the sporangium within the host cell but emerge as a mass of protoplasm to form an external vesicle (as in *Pythium*) and there become organized and break through the plasma membrane of the vesicle and swim away. In almost all cases the zoospores are of the secondary type, i.e., more or less kidney-shaped with one of the two lateral flagella directed anteriorly and the other posteriorly.

Sexual reproduction where known is by the contact of whole plants or of special segments and passage of the contents of one into the other, which immediately becomes a zoosporangium or which may become a thick-walled resting spore (oospore). In the latter case there is usually no periplasm. The uniting organs (gametangia) may be approximately equal and similar in appearance or the antherid may be much smaller than the oogone and fertilization may be accomplished by a conjugation tube. The thick-walled oospore has been observed in some cases to germinate as a zoosporangium.

In the order as here defined it is exceedingly uncertain whether we have a series progressing from a simple monocarpic fungus to a short hypha of several reproductive units on the way toward the evolution of a well-developed hypha such as we find in the Saprolegniales or Peronosporales or, what may be equally likely, the fungi of this order represent different degrees of reduction from various genera of those orders. In conformity with the procedure in the Chytridiales-Monoblepharidales series and in the Hyphochytriales the organisms in the series with biflagellate zoospores (the Biflagellatae of Sparrow) are arranged with those of simple structure first.

Karling (1942) recognizes five families in the group that he calls "Simple Holocarpic Biflagellate Phycomycetes." He does not wish to convey the idea that these are necessarily a single phylogenetic series, especially in view of the insufficient knowledge of the life histories of the majority of the described species and of the uncertainty of their relationship, either as primitive or as reduced forms. If the latter, some of them may represent reductions from Saprolegniaceae, Pythiaceae, or Leptomitaceae that have a greater or less similarity because in reduction to simpler vegetative structure they perforce become very similar. Sparrow (1943) includes these simple or simplified forms in five families in two orders (not coinciding entirely with the families recognized by Karling).

The author follows Sparrow in part in his distinction of three families to which he adds doubtfully the Woroninaceae and Thraustochytriaceae.

Woroninaceae: fungus remaining naked and amoeboid for a considerable time, forming a "plasmodium"(?). Eventually separating into several segments which produce cell walls and become zoosporangia or the whole "plasmodium" enlarging to form a single zoosporangium with its wall in close contact with the host wall. Zoospores preformed in the zoosporangia. Clusters of angular or single round resting spores may be formed. Parasitic in algae and fungi.

Olpidiopsidaceae: one-celled, free in the host cell and early producing a cellulose wall. Zoospores preformed in the zoosporangium. Resting spores formed sexually or by parthenogenesis. Parasitic in algae or fungi.

Sirolpidiaceae: forming a linear series of zoosporangia, occasionally a single, elongated zoosporangium. Zoospores preformed in the zoosporangium. Sexual stage unknown. Parasites in marine algae.

Lagenidiaceae: one-celled or more often a short, constricted or unconstricted, unbranched or branched row of cells, each of which becomes a zoosporangium or gametangium. Zoospores completing their development in a vesicle at the opening of the exit tube. Oospores formed within the female gametangium. Parasitic in algae (mainly fresh-water forms) or microscopic animals or roots of grasses.

Thraustochytriaceae: resembling *Rhizophydium*. Parasitic upon marine algae. Zoosporangium epibiotic, obpyriform, attached to a branched endobiotic rhizoidal system. Zoospores formed in the zoosporangium but not motile, and set free by the dissolution of the apical portion of the wall, as angular cells, from which after some time the biflagellate motile stage emerges. Zoosporangia proliferating after the discharge of the zoospores. A family of very doubtful relationship.

Family Woroninaceae. In this family of possibly three genera the zoospores encyst on the external surface of the host and empty their contents into the host cell through a slender tube, the empty spore walls remaining attached for some time. Within the host cell the uninucleate fungus remains naked and amoeboid and grows, accompanied by multiplication of the nuclei, until it largely fills the cell. Eventually this plasmodium-like structure separates into a group of several segments around each of which a cellulose wall is produced. The zoospores produced in these sporangia escape through exit tubes, those further inside the sorus emptying through those nearer the surface. The zoospores are biflagellate anteriorly. Instead of producing zoosporangia the naked fungus mass may divide into very numerous small angular resting spores, also with cellulose walls, which are clustered together in more or less definite compact cystosori of a few to many spores. Each resting spore upon germination gives rise to one zoospore or to several zoospores. The foregoing description applies to *Woronina*, the type genus of the family, which is parasitic in the hyphae of Saprolegniaceae and in *Vaucheria*. Two other genera, *Rozellopsis*, parasitic in Pythiaceae, and *Pyrrhosorus*, parasitic in Red Seaweeds, are tentatively also placed in this family but differ in not forming clusters of zoosporangia and in their failure to pro-

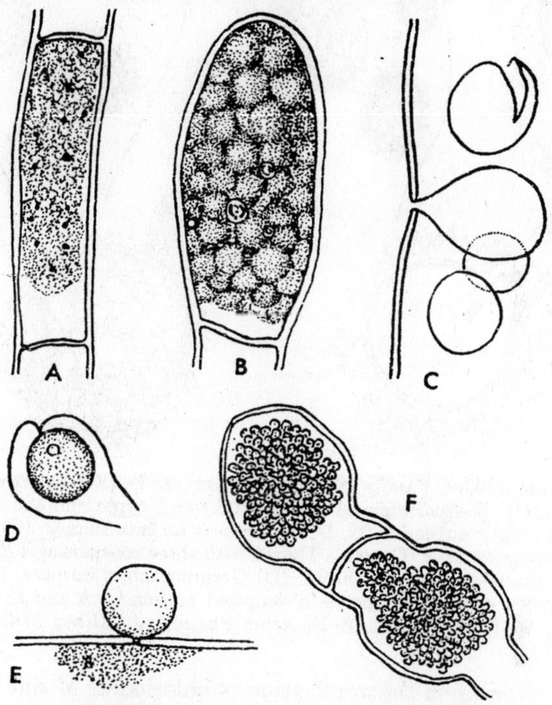

FIG. 30. Lagenidiales, Family Woroninaceae. *Woronina polycystis* Cornu. (A) "Plasmodium" nearly filling host cell. (B) Early stage in the formation of zoosporangia. (C) Empty zoosporangia. (D) Zoospore. (E) Infection of host by zoospore. (F) Cystosori within host. (After Cook and Nicholson: *Ann. Botany*, **47**(188):851–859.)

duce cystosori. The vegetative stage within the host is, as in *Woronina*, more or less plasmodial in nature. (Fig. 30.)

FAMILY OLPIDIOPSIDACEAE. With the removal of *Woronina* to a distinct family the remainder corresponds in the main to the Woroninaceae of the previous edition. Vegetatively and in their asexual reproduction the members of this family show great similarity to those of the Olpidiaceae in the Chytridiales. They differ strongly, however, in producing zoospores with two anteriorly or laterally attached flagella, one of the tinsel type, the other of the whiplash type, and in possessing walls which give the

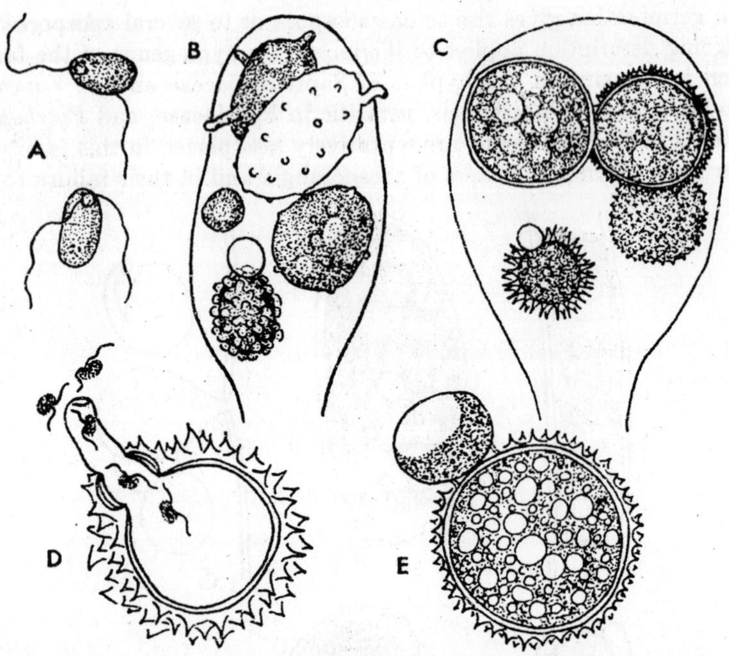

FIG. 31. Lagenidiales, Family Olpidiopsidaceae. (A, B) *Olpidiopsis vexans* Barrett. (A) Zoospores. (B) Zoosporangia of smooth and rough types and one mature oogone with attached empty antherid. (C, D) *Olpidiopsis varians* Shanor. (C) Swollen tip of hypha of *Saprolegnia ferax* (Gruith.) Thuret with three zoosporangia and one mature oogone with attached, empty antherid. (D) Germination of oospore. (E) *Olpidiopsis luxurians* Barrett, oogone and partially emptied antherid. (A and E, after Barrett: *Ann. Botany*, **26**(101):209–238. B–D, after Shanor: *J. Elisha Mitchell Sci. Soc.*, **55**(1):167–195.)

cellulose reaction upon the application of chloriodide of zinc without any preliminary treatment. It is not at all certain that all the genera assigned to this family really belong here. The principal genus, of about 20 species, is *Olpidiopsis*. The zoospores are anteriorly biflagellate or the flagella may be somewhat laterally attached in a groove, one directed forward (tinsel type) and one posteriorly (whiplash type), as demonstrated by Couch (1941). The species of this genus are strictly parasitic in the hyphae of Saprolegniaceae, Pythiaceae, and various algae. One species is reported in *Riccia* and a doubtful species in an insect. The zoospore settles on the outside of the host cell and encysts with a cellulose wall. It produces a slender infection tube through which the uninucleate naked protoplasm enters the host. It remains naked for a while but eventually forms a cellulose wall and when fully grown becomes a spherical or ellipsoidal zoosporangium. An exit tube pierces the host cell wall and the zoospores escape through the softened inoperculate tip. Before the exit tube opens the zoospores become visible in the zoosporangium and after a few moments the contents again appear homogeneous, the zoospores again becoming visible and actively motile just before the tube opens. Depending upon the number of zoospores infecting the host cell there may be from one to many sporangia formed. The host cell may not show much enlargement or may be quite strongly hypertrophied. In the species parasitic in fungi the zoosporangium may be ornamented by few or many, short or elongated, slender or stout spine-like processes which are produced by the host protoplasm, not by the parasite. Sexual reproduction where known is by the union of two adjacent equal or unequal cells of the parasite which at the point of contact produce a perforation of the walls through which the multinucleate contents of one cell pass into the other, also multinucleate, cell. Barrett (1912) reported the apparent union of these numerous nuclei in pairs. The zygote cell enlarges somewhat and then forms a thick, angular, smooth, or coarsely spiny or knobby, wall. After a little while this resting spore becomes a zoosporangium and the biflagellate zoospores emerge through an inoperculate exit tube. The empty male cell, sometimes more than one, remains attached to the oospore and is sometimes partly covered by the thickened wall of the latter. Over 20 species have been described but careful culture work with pure cultures is necessary before the validity of all these species can be established. The genus was first set up by Cornu in 1872 for several species occurring in the hyphae of Saprolegniaceae. He described the zoospores as posteriorly uniflagellate but Fischer (1882) studying what he believed to be the same species found that the zoospores were anteriorly biflagellate. In 1884 the great German mycologist Wilhelm Zopf described *O. schenkiana* in the filaments of *Spirogyra*, but described and figured the zoospores as posteriorly uniflagellate. On this account von Minden (1915) transferred the species to a new genus which he named *Pseudolpidiopsis*. Scherffel and others studying what they believed to be the same species found that the zoospores were biflagellate. If the usually accurate observer Zopf was incorrect the genus *Pseudolpidiopsis* (tentatively included in this book in the Olpidiaceae) must be rejected but if he was correct the genus must be maintained. It must be noted that several cases are known where apparently almost identical species have been found on careful study to differ in their flagellation (e.g., *Sphaerita dangeardii* Chat. and Brod., in the Olpidiaceae and *Pseudosphaerita euglenae* Dangeard, in the Olpidiopsidaceae, both parasitic in *Euglena*). (Fig. 36.)

In some species of *Olpidiopsis* the resting spores may arise parthenogenetically under certain conditions and sexually under other conditions, while in other species the sexual stage is unknown as yet. These were set apart by Fischer (1892) as a separate genus *Pseudolpidium*. Probably this does not deserve this generic distinction but Karling (1942) uses this as a temporary resting place for the species in which resting spores, either sexually or parthogenetically produced, are unknown while Sparrow (1943) uses this name for the forms with parthogenetically produced resting spores. They occur in algae or fungi.

Petersenia was described by Sparrow (1934) for two species parasitic in marine Florideae and one species in *Saprolegnia*, which differ from *Olpidiopsis* in the production of lobed or tubular zoosporangia, mostly with more than one discharge tube, and for which, so far, no resting spores are known. In *Pythiella* (Couch, 1935) the primary zoospores are without flagella and creep out through the mouth of the exit tube where they encyst and after about an hour emerge as zoospores of the secondary type, with the two flagella laterally attached. The female cell produces a distinct egg with periplasm and the male cell fertilizes it through a short conjugation tube. It is possible that this genus represents a very much reduced form of *Pythium*. The only known species is parasitic in the hyphae of species of *Pythium*. *Pseudosphaerita* resembles *Sphaerita* of the Olpidiaceae. It is imperfectly known and needs further study.

FAMILY SIROLPIDIACEAE. In this as in the foregoing family the fungus lies free within the host cell and produces a cellulose wall. Usually, however, this cell elongates and becomes divided by cross walls into several zoosporangia which disarticulate, in *Sirolpidium*, or remain tubular, sometimes shortly branched, with cross walls at maturity but not disarticulating, in *Pontisma*. Resting spores are not known in either genus. Both are parasitic in marine algae, the former in *Bryopsis* and *Cladophora*, the latter in the red alga *Ceramium*. In both genera the laterally biflagellate zoospores are produced in the zoosporangia and escape fully formed through an inoperculate exit tube. Karling (1942) places *Petersenia* in this family because of its elongated, tubular, or lobed cell body. Its type species was originally described as a species of *Pleotrachelus*, of the Olpidiaceae, but differs from that in the biflagellate zoospores.

FAMILY LAGENIDIACEAE. This family was originally named Ancylistidaceae, but with the discovery by Miss Berdan (1938) that the genus *Ancylistes* really belongs in the Entomophthorales, the present designation was given by Karling (1939). In this family the zoospores are mostly of the secondary type, the primary type being rare. The contents of the zoosporangium are set free through an exit tube into a vesicle in which the zoospores attain their final form and then escape by bursting the vesicle membrane. They encyst and germinate on the outside of the host cell and form a slender infection tube at whose apex a short or long, simple or branched, coenocytic plant body is formed. This may remain nonseptate in *Lagena* or septate with cylindrical segments in *Lagenidium* or constricted at the septa into bead-like segments as in *Myzocytium*. Each segment may become a zoosporangium or a gametangium. Adjacent seg-

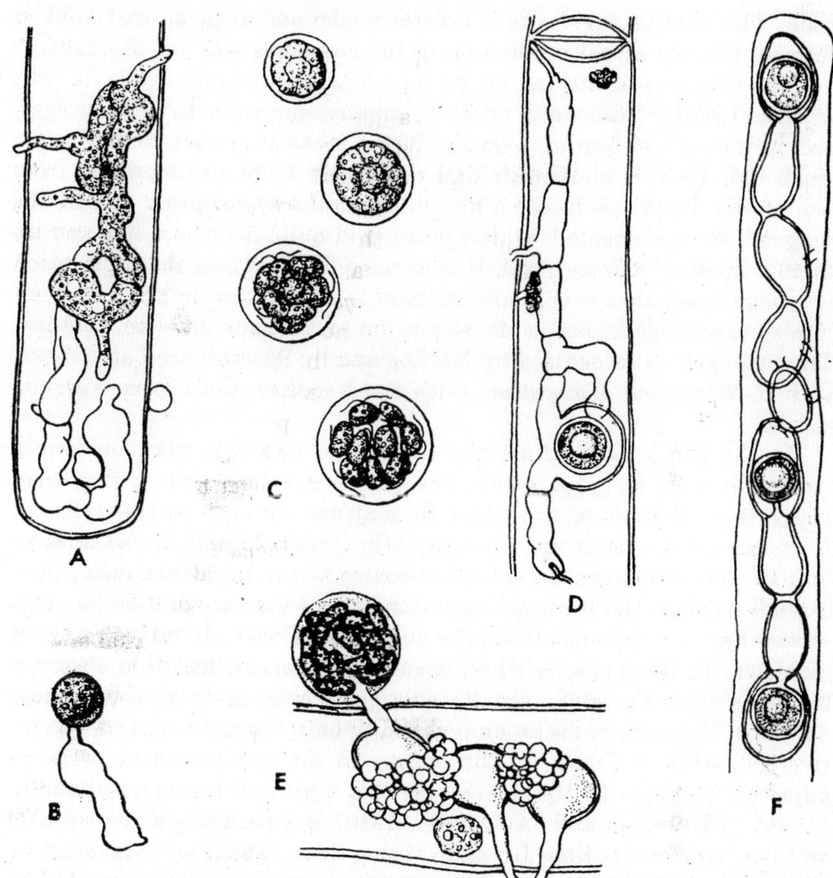

Fig. 32. Lagenidiales, Family Lagenidiaceae. (A–D) *Lagenidium rabenhorstii* Zopf. (A) The coenocyte has divided into several zoosporangia. (B) Forty-five minutes later, the cytoplasm is escaping from one of these zoosporangia into a vesicle. (C) Four successive stages in the development of the same vesicle. (D) Plant with four empty zoosporangia and an empty antherid and an oogone containing a mature oospore. (E, F) *Myzocytium proliferum* Schenk. (E) Plant with empty zoosporangia, one with a vesicle containing zoospores. (F) Plant with three antherids and three oogones containing mature oospores. (After Zopf: *Nova Acta Leopoldina*, **47**(4):141–236.)

ments in the same hypha or in separate hyphae that are in contact may function as antherid and oogone, respectively. They may be alike in size and other appearances or slightly different. By means of a conjugation tube the contents of the antherid pass into the oogone and then a thick-walled, rounded oospore is formed. There is no distinct periplasm. Germination of the oospore was described by Dangeard (1903) in *Myzocytium vermicola* (Zopf) Fischer, which is parasitic in a free-living nematode. The nucleus divides into several nuclei and then an exit tube is formed, but the actual production of the zoospores was not observed.

The members of this family are parasitic, rarely saprophytic, in cells of algae (mostly fresh-water species), microscopic animals, or their eggs, and in the case of *Lagena*, parasitic in the roots of grasses. Sometimes a small individual is nonseptate and is scarcely to be distinguished from one of the Olpidiopsidaceae while in *Lagenidium giganteum* Couch the fungus body is elongated, multiseptate, and quite mycelium-like and up to 40 μ thick in extreme cases. It is in cases like this that the supposition is strengthened that this family represents a reduction in size and complexity from higher, better developed fungi, perhaps close to *Pythium*. The three genera recognized by Karling and by Sparrow are *Lagenidium*, with 15–20 species, *Myzocytium*, with 4 or 5 species, and *Lagena* with one species.

In the genus *Lagenidium* the coenocytic segments may be few in number and all included within one cell of the algal host or they may pierce the cell walls of the latter to continue through several cells. In *L. giganteum* the hosts are small aquatic Crustacea and larvae of mosquitoes. Not only does the extensive coarse mycelium fill the host but it extends out into the surrounding water up to a distance of 0.1 mm. This species has been brought into pure culture by Couch. Its sexual stage is unknown. In those species where sexual reproduction has been observed the female gametangium usually enlarges considerably as the oospore develops. The antherid is an unmodified or only slightly modified vegetative cell, either adjacent to the oogone in the same filament or in an adjacent filament. In the latter case this may indicate heterothallism. *Lagena* (Vanterpool and Ledingham, 1930) is a parasite in the roots of wheat (*Triticum aestivum* L.) and other grasses, which are weakened by the presence of this fungus. The organism may be much elongated or even more or less coiled in the host cell but not septate. It produces a single exit tube through which the contents of the zoosporangium emerge into a vesicle in which they undergo their final transformation and swim away after rupturing its membrane. Near-by thalli may unite by a conjugation tube through which the contents of one cell pass into the other, rounding up there to form a thick-walled oospore. The mode of germination of the latter is unknown. (Figs. 32, 33.)

Possibly belonging to this family is *Aphanomycopsis* (Scherffel, 1925) forming a more or less elongated and coiled, often somewhat branched, coenocytic thallus in the cell of a Diatom. This thallus is transformed into a zoosporangium with one or more exit tubes at whose mouths the primary (nonflagellate) zoospores encyst, later emerging and swimming away as zoospores of the secondary type. Sexual reproduction has not been demonstrated although thick-walled resting spores are formed in enlarged

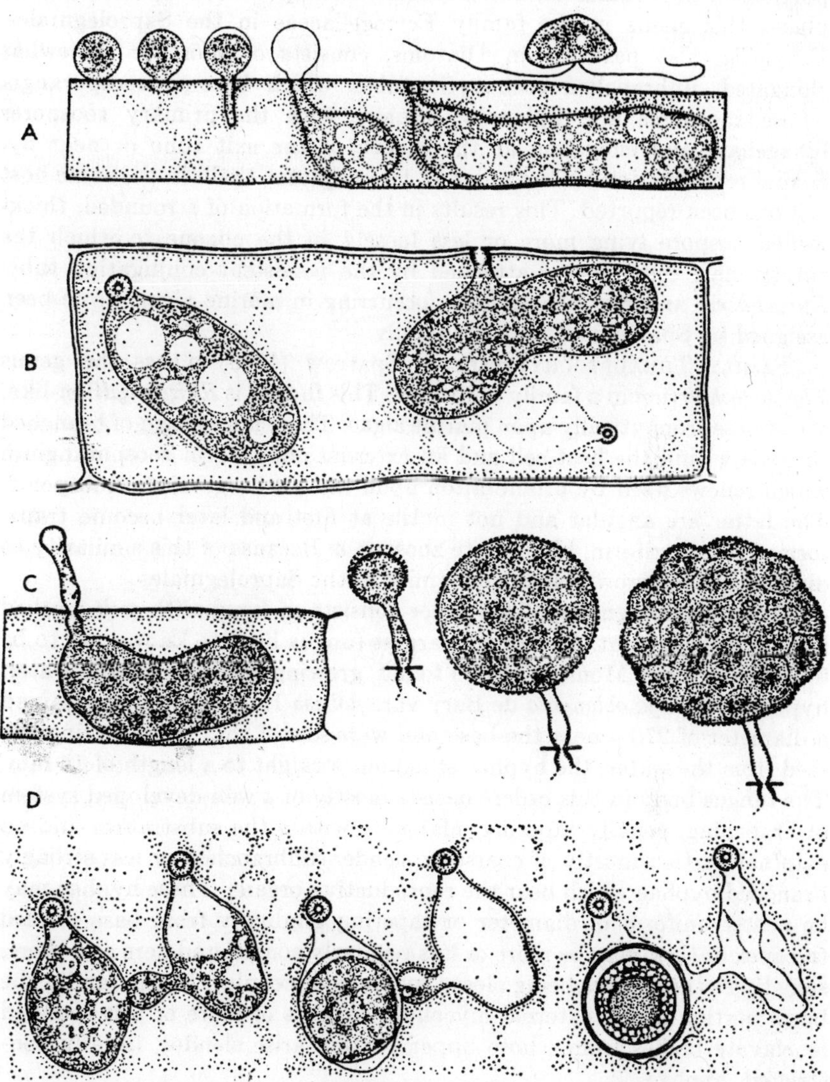

FIG. 33. Lagenidiales, Family Lagenidiaceae. *Lagena radicicola* Vanterpool & Ledingham. (A) Zoospore and semidiagrammatic representation of infection of root hair of wheat. (B) Zoosporangia in top and side view and an empty zoosporangium. (C) Formation of vesicle at apex of discharge tube of zoosporangium. (D) Semidiagrammatic representation of the conjugation of two thalli. (Courtesy, Vanterpool and Ledingham: *Can. J. Research*, **2**(3):171–194.)

portions of the tubular thallus. Because of its diplanetism Sparrow (1943) places this genus in the family Ectrogellaceae in the Saprolegniales. *Ectrogella*, also parasitic in Diatoms, consists of short or somewhat elongated, unbranched coenocytic tubes which become zoosporangia whose zoospores are diplanetic, usually with the primary zoospores biflagellate and encysting at the mouth of the exit tube or near by. Sexual reproduction by the union of two adjacent thalli in the same host cell has been reported. This results in the formation of a rounded, thick-walled oospore lying more or less loosely in the oogone to which the empty male cell remains attached by the persistent conjugation tube. *Eurychasma* and *Eurychasmidium*, occurring in marine algae, have been assigned by Sparrow to the same family.

FAMILY THRAUSTOCHYTRIACEAE. Sparrow (1943) places the genus *Thraustochytrium* in a family of its own. This fungus is *Rhizophydium*-like, growing saprophytically upon marine algae. There is a system of branched rhizoids within the host cell and an external obpyriform zoosporangium which renews itself by proliferation upon the discharge of the zoospores. The latter are angular and not motile at first and later become transformed into pyriform, biflagellate zoospores. Because of this similarity to diplanetism Sparrow places this family in the Saprolegniales.

Order Saprolegniales. This order consists of fungi with well-marked hyphal development. In fact the largest fungus hyphae known are to be found here. Thus Monsma (1937) found, growing on hemp seed in water, hyphae of *Achlya oblongata* de Bary var. *globosa* Humphrey that attained a diameter of 270 μ near the base and were so stiff that, on removing the seed from the water, the hyphae stood out straight to a length of 15 mm. The fungus body in this order consists mostly of a well-developed system of branching, usually tapering, filaments within the substratum and an external portion mostly of coarse or slender, unbranched or less strongly branched hyphae which bear the reproductive organs. These hyphae may be nearly uniform in diameter or tapering gradually from base toward the apex or they may be more or less regularly constricted here and there, sometimes the orifices being closed by plugs of cellulin. In one family the basal portion of the external mycelium consists of more or less rounded or clavate bodies from whose upper portion arise slender, usually constricted, hyphae.

The mycelium of this order is coenocytic with a multinucleate layer of cytoplasm surrounding a large central vacuole in the larger hyphae. The cell walls show the cellulose reaction immediately upon treatment with chloriodide of zinc. Usually septa are lacking in the mycelium except: (1) to delimit zoosporangia, (2) to set apart the gametangia from the remainder of the hyphae, and (3) to delimit injured portions. In *Achlya polyandra* de Bary, Horn (1904) observed and showed to the author the extensive and rapid formation of transverse and oblique walls until the hypha was cut up into innumerable angular segments, under the influence of exceedingly minute amounts of copper in the water. Growth of the external hyphae is terminal and by the limited formation of branches. Unlike the Lagenidiales the mycelium does not become all converted into reproductive organs but these are mostly terminal or subterminal on the external hyphae and their branches.

The Saprolegniales are saprophytic on dead plant or animal matter in the soil or in water (fresh water, rarely brackish) or parasitic in algae or small animals or even fish, and in some cases in the roots of plants in the soil. Although spoken of usually as water molds perhaps the majority are inhabitants of moist soil.

The zoosporangia are of the same diameter as the hyphae or somewhat enlarged, cylindrical or ovoid. At their inception the protoplasm from the lower portions of the hyphae crowds into the terminal portions destined to become the zoosporangia until these are filled with dense, multinucleate protoplasm with a much reduced central vacuole. At the base of each a septum is formed setting it apart from the supporting hypha. Within this zoosporangium cleavage of the protoplasm begins next to the wall and progresses inwardly to the central vacuole, cutting out uninucleate naked portions of protoplasm which round up somewhat and eventually become the zoospores. These are usually pyriform, with two anterior flagella, one of the tinsel type and the other of the whiplash type. Such zoospores are called primary zoospores. Escape of the zoospores occurs through the softened tip and sometimes through a lateral papilla. In all cases the opening is inoperculate. After the zoosporangium is emptied the basal septum may arch up into the empty space, filling it and becoming a second zoosporangium. This formation of zoosporangia by proliferation may occur repeatedly. Instead of proliferating, the hypha just below the emptied zoosporangium may branch out laterally, the branch quickly turning upward to form a new zoosporangium parallel to the empty one. This sympodial formation of zoosporangia is characteristic of some genera.

The primary zoospores may escape and swim away some distance before encysting. In some species of *Pythiopsis* this encysted primary zoospore eventually germinates by a slender tube and forms a new plant, or the encysted zoospore may escape from its cell wall as a primary zoospore again. More often in other genera the encysted primary zoospore escapes from its cyst after a few minutes or hours in a different form, the secondary zoospore. It is kidney-shaped or resembles a grape seed with the two flagella arising in the groove, usually nearer the more pointed end than the base. One flagellum, of the tinsel type, is directed forward while the whiplash type flagellum is directed posteriorly. This zoospore encysts after a while and may germinate by a slender hypha or re-escape and re-encyst several times. In some cases the primary zoospore apparently produces no flagella but creeps to the opening of the zoosporangium and there encysts, escaping as a secondary zoospore some time later. In still other genera the primary, nonflagellate, zoospores encyst within the zoosporangium as separate round cells or become compacted into polyhedral cells by mutual pressure. In the latter case they form short exit tubes that pierce the zoosporangial wall so that their contents escape as secondary zoospores. In a few species the zoospores that encyst within the zoosporangium do not escape at all as swimming cells but germinate directly by germ tubes.

It is customary to refer to the characteristic of forming primary zoospores only, as monoplanetic and of the formation of two successive types of zoospores as diplanetic. More correctly these two terms should be monomorphic and dimorphic. Properly speaking monoplanetic means wandering once, or with only one swimming stage, while diplanetic means with two swimming stages. Since in *Pythiopsis* the primary type of zoospore may swim and encyst several times and in *Achlya*, *Dictyuchus*, and other genera, the secondary type of zoospore may also do the same thing the customary terms are not used in their correct etymological sense.

Sexual reproduction occurs by the formation of male and female gametangia, antherids and oogones, respectively. They may be terminal on the main hyphae or on short lateral branches, rarely intercalary. The oogones round up and become filled with multinuclear protoplasm and then a cross wall is formed. Most of the nuclei degenerate leaving a much reduced number. A single egg or several eggs may be formed, containing a few nuclei each, of which only one remains sexually functional, the others moving toward the outer wall and usually disintegrating. The whole of the original protoplasm may be used up in the formation of the egg or eggs or a portion of it may be left surrounding a single egg. This is called the periplasm and may contain numerous nuclei which eventually disintegrate. In the genus *Araiospora* the periplasm becomes divided by radial sheets of protoplasm in which these nuclei lie and then radial cell walls are formed so that the egg is surrounded by a layer of cells. Probably in most cases the periplasm produces much of the exterior ornamentation of the oospore wall.

The antherids, depending upon the species of the fungus, may be formed on separate plants from those forming the oogone (heterothallic species), or both kinds of sexual organs may be produced on the same mycelium (homothallic species). This distinction is not in all cases sharp as there may be varying degrees of maleness or femaleness, as shown by Bishop (1940). The antherids are terminal on long slender branches or on short branches or may arise as enlargements of the oogonial stalk, just below the oogone. When arising at a distance these antheridial hyphae are attracted to the oogones apparently by a secretion from the latter. J. R. Raper (1939, 1940) shows evidence that these are in the nature of hormones. In heterothallic species of *Achlya* he demonstrated that a secretion from the male plant stimulates the formation of oogones on the female plant and that in turn the substances given off by the developing oogones lead to the production of antheridial branches in the male plants and, probably, guide chemotropically the direction of their growth.

Upon reaching the oogone the usually somewhat enlarged tip of the antheridial branch flattens against it and, if it has not already occurred, a septum is produced, separating the antherid from the supporting hypha. The antherid is usually plurinucleate and the nuclei may divide again. Eventually most of them degenerate. At or near the center of the contact surface a tube grows from the antherid through the oogone wall to the egg, or if there are several eggs this conjugation tube may become branched so that one antherid may fertilize several of them. Couch (1924) showed that in *Leptolegnia caudata* de Bary an opening is dissolved between the antherid and oogone permitting fertilization without the formation of a conjugation tube. According to Kevorkian (1925) this is true also of *Apodachlya brachynema* (Hild.) Pringsh. while Cooper (1929) also demonstrated this for *Brevilegnia diclina* Harvey. After the entry of a single sperm nucleus into the egg the latter secretes a definite wall which may become thick, with a smooth or rough exterior. The union of nuclei does not occur until much time has elapsed. Although it has not been demonstrated it is assumed that meiosis occurs at the germination of the oospore. This usually occurs by the formation of a hypha which may develop to form a new plant or which may produce a zoosporangium. In many species of this order the oospores develop parthenogenetically. The formation or nonformation of antherids depends upon the conditions of nutrition, temperature, etc. It has been asserted that in some cases antherids, though present, may not function.

It is difficult to make a decision as to whether the pluriovulate condition should be considered the more primitive or a derived condition in this order. If the ancestral forms were fungi whose female gametangia contained several large motile eggs (as in *Allomyces*), we would expect the more primitive Saprolegniales to be pluriovulate, but if we look to the Lagenidiales for the stock whose evolution led to the Saprolegniales, we find that they have but a single egg. If we look to the Siphonales, we find that in these green algae in the genus *Vaucheria* the oogone contains but a single egg while there are other members of that group in which the oogone contains several eggs. The author is inclined to favor the hypothesis that the pluriovulate condition is derived from the uniovulate condition. Of the more than 20 genera included in this order the egg is single in all but 6 or 8 genera and also in some of the species of each of these genera that are normally pluriovulate. The species of this order have been monographed by Coker (1923) and the North American species by Coker and Matthews (1937).

The author recognizes three families in this order: Saprolegniaceae, Leptomitaceae and Rhipidiaceae. The two latter are considered by Miss Kanouse (1927) and by Sparrow (1943) to deserve segregation into a separate order, the Leptomitales, with which the author does not agree. Until the various forms included in this book in the Lagenidiales are better known, it is uncertain whether some of them should be more closely associated with the Saprolegniales. There is undeniably a close relationship between these two orders and also with the next order, the Peronosporales.

The three families of Saprolegniales may be distinguished as follows:

Saprolegniaceae: mycelium not definitely constricted at intervals nor with cellulin plugs, mostly cylindrical or gradually narrowing toward the extremities but not consisting of an enlarged basal portion and slender branches. Zoospores mostly dimorphic or with modifications of dimorphism, only primary zoospores produced in one genus. Oogone with one or more eggs and lacking periplasm.

Leptomitaceae: mycelium definitely constricted at more or less regular intervals, frequently with cellulin plugs. Zoospores dimorphic. Oogones usually with one egg, in one species with more, with no periplasm.

Rhipidiaceae: mycelium more or less well differentiated into an enlarged basal portion with slender, mostly constricted, branches which bear the zoosporangia and sexual organs. (In *Mindeniella* these arise directly on short stalks from the basal segment.) Oogones with single eggs and with periplasm.

Family Saprolegniaceae. The members of this family are in some cases strictly aquatic, but a considerable number of species are soil inhabitants. Contrary to the belief prevailing earlier they are mostly saprophytic on vegetable matter, less often on animal matter. Only a few species of *Achlya* and *Saprolegnia* are sometimes destructive to young fish and fish eggs in fish hatcheries. Several species of *Aphanomyces* and one or two other genera are parasitic in algae, in the roots of higher plants, or in aquatic animals.

In about half of the *genera* the oogone contains but one egg, but the pluriovulate *species* are far in the majority. The number of eggs per oogone in these may vary from 2 or 3 up to 50. Fertilization of the eggs is accomplished by the passage of sperm nuclei, one to each egg, usually through conjugation tubes that penetrate the oogone wall from the adhering antherids. In some cases although antherids are present there seems to be no opening for the passage of the sperm nucleus so that the egg develops parthenogenetically. The oospore may germinate by a germ

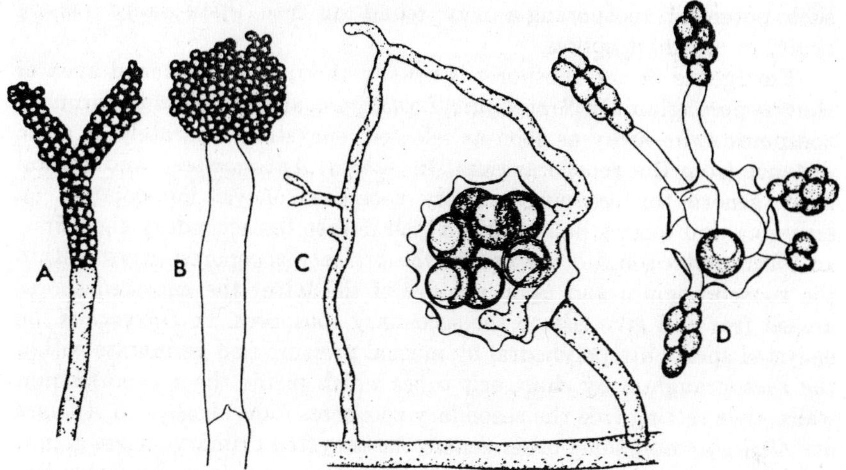

FIG. 34. Saprolegniales, Family Saprolegniaceae. *Thraustotheca primoachlya* Coker & Couch. (A) *Thraustotheca* type of zoosporangium. (B) *Achlya* type of zoosporangium. (C) Oogone and antherid. (D) Oospore germinating and producing several small zoosporangia. (Courtesy, Coker and Couch: *J. Elisha Mitchell Sci. Soc.*, **40(3–4)**: 197–202.)

tube which produces a new mycelium. In *Thraustotheca primoachlya* Coker & Couch the germinating oospore may divide into several internal spores or these may be formed in short germ tubes growing out through the pits of the oogone wall (Coker and Couch, 1924). Ziegler (1948) studied the germination of 26 species of this family representing 6 genera and found the following 4 types. (1) "A long or short germ tube is formed, with an apical sporangium"; (2) "the germ tube produces a sparsely branched mycelium with a sporangium at the apex of the main hypha or a branch"; (3) "the primary germ tube forms a branched mycelium"; (4) "the primary germ tube forms a long unbranched hypha." (Fig. 34 D.)

Apparently the primitive form of zoospore in this family is pear-shaped with two equal anterior flagella. Only such primary zoospores are formed in the two known species of *Pythiopsis*. Far more often the species of this order are dimorphic. Other forms show various modifications of the dimorphic plan.

The zoosporangia are typically terminal segments of hyphae, but sometimes several are formed one behind the other. When the zoospores are discharged a new zoosporangium may arise by proliferation, sometimes five or six times. In other cases the new zoosporangia are formed on short or long sympodially produced branches. Usually they are slender, like the supporting hypha, or clavate or ovoid. Under certain cultural conditions the hypha may form ovoid or clavate zoosporangia in chains, each opening by an exit pore near its apical end. Under some conditions such potential zoosporangia may round up into thick-walled resting spores or chlamydospores.

Emergence of the zoospores is usually through the softened apex of the zoosporangium. In *Saprolegnia, Leptolegnia,* and *Isoachlya* the primary zoospores swim away as soon as released, encysting separately at some distance from the zoosporangium. In *Achlya, Aphanomyces,* and several other genera the escaping primary zoospores encyst immediately on emerging and form a ball of cells which release the secondary zoospores. In *Thraustotheca* and other genera the primary zoospores encyst within the zoosporangium and upon rupture of the latter the encysted spores are set free and give rise to the secondary zoospores. In *Dictyuchus* the encysted spores are polyhedral by mutual pressure and germinate within the zoosporangium by short exit tubes which pierce the zoosporangium walls, thus setting free the secondary zoospores individually. In *Aplanes* and *Geolegnia* and some other genera the encysted primary spores germinate by germ tubes within the zoosporangium or after the latter has disintegrated. Under varying conditions of culture the same species of *Saprolegnia* or *Achlya* may be induced to produce its zoospores in the manner typical of *Saprolegnia, Achlya, Thraustotheca,* or *Aplanes,* showing that these modifications are not of very deep fundamental importance. This is corroborated by the fact that Salvin (1942) was able to succeed in attempts at mating *Thraustotheca clavata* (de Bary) Humphrey with *Achlya flagellata* Coker, the former producing the antherids and the latter the oogones. The oospores so produced could not be brought to germination by the methods attempted. (Figs. 34–38.)

Pythiopsis. In this rarely studied genus the sympodially produced sporangia may be ovoid (*P. cymosa* de Bary) or slender. The mycelium is rather stout, as is typical for the family. The zoospores which escape are of the primary type. After encysting they may germinate by a germ tube or may produce zoospores again, but these are still of the primary type in the two species. The oogone usually has but one egg. The antherids may be up to three in number and may arise from immediately below the oogone. The species of this genus occur in soil or in fresh water. A third species assigned to this genus by Harvey (1925) should be transferred, according to Coker and Matthews (1937) to the genus *Isoachlya.*

Saprolegnia. This is the most commonly studied genus of the family. It contains about 20 species, mostly saprophytic, rarely parasitic, on animal or vegetable matter in water or soil. Dead insects in water or even larger animals become surrounded by a fringe of the long external hyphae, the much-branched trophic hyphae being within the animal tissues. These external hyphae are fairly stout (up to 50 to 100 microns in diameter in extreme cases) and more or less straight and but little branched. They terminate in club-shaped zoosporangia within which numerous zoospores

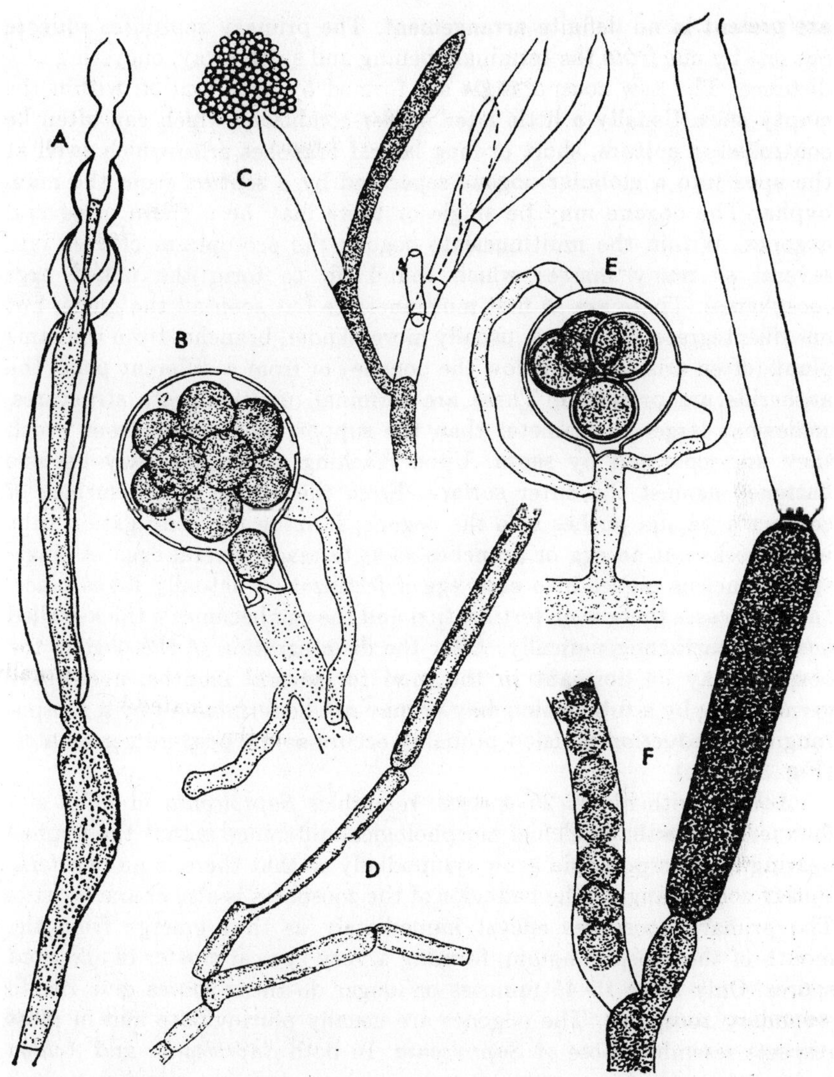

FIG. 35. Saprolegniales. (A–E) Family Saprolegniaceae. (A, B) *Saprolegnia monoica* Pringsh. var. *glomerata* Tiesenh. (A) Proliferated zoosporangia. (B) Oogone and antherid. (C–E) *Achlya racemosa* F. Hildeb. (C) Cluster of zoosporangia. (D) Gemmae. (E) Oogone and antherids. (F) Family Leptomitaceae, *Leptomitus lacteus* (Roth) Agardh. Portion of filament showing one empty sporangium, one containing zoospores, and the other not yet mature. (Courtesy, Coker: The Saprolegniaceae with Notes on Other Water Molds, Chapel Hill, Univ. North Carolina Press.)

are present in no definite arrangement. The primary zoospores squeeze out one by one from the terminal opening and swim away, encysting at a distance. The new zoosporangia are formed by proliferation within the empty ones. Usually a little later, under conditions which can often be controlled in culture, short or long lateral branches arise which swell at the apex into a globular oogone separated by a septum from the main hypha. The oogone may be single or there may be a chain of several oogones. Within the multinucleate oogone the protoplasm cleaves into several or many masses which round up to form the naked eggs (oospheres). These are at first multinucleate but soon all the nuclei but one disintegrate. On longer, usually more slender, branches from the same plant (often arising just below the oogone) or from a different plant the antherids are produced. These are terminal, multinucleate structures, somewhat larger in diameter than the supporting hyphae, from which they are separated by septa. Upon reaching an oogone they become flattened against the outer surface. From the center of the surface of contact a papilla pushes into the oogone, forming the conjugation tube which seeks out an egg or branches so as to reach several eggs. A single sperm nucleus passes into each egg if fertilization actually takes place. In many cases there is no fertilization and the egg becomes a thick-walled oospore, parthenogenetically. After the disintegration of the oogone the oospore may lie dormant in the mud for several months, eventually germinating by a tube which may or may not be terminated by a zoosporangium. Reduction division probably occurs as the oospore germinates. (Fig. 35A, B.)

Achlya, with about 25 species, resembles *Saprolegnia* in habit and manner of growth. The chief morphological difference is that the hyphae bearing the zoosporangia grow sympodially so that there is no proliferation of zoosporangia. The behavior of the zoospores is also characteristic. The primary zoospores encyst immediately as they emerge from the mouth of the zoosporangium, forming a very typical cluster of encysted spores. Only after 15–45 minutes or longer do these spores give rise to secondary zoospores. The oogones are usually pluriovulate and in most respects resemble those of *Saprolegnia*. In both *Saprolegnia* and *Achlya* and in a few other genera the presence or absence of thin places or "pits" in the oogone walls is of diagnostic importance. Many species of *Achlya* are parthenogenetic. Both of these genera are usually exceedingly sensitive to minute traces of copper salts as discovered by Horn (1904), yet Gäumann (1919) found *S. monoica* Pringsh. growing in the shaft of an abandoned copper mine in Lapland where the dissolved salts gave the water a green color. (Fig. 35C–E.)

Aphanomyces has slender zoosporangia with but a single row of zoospores which behave on emerging as do those of *Achlya*. The oogone

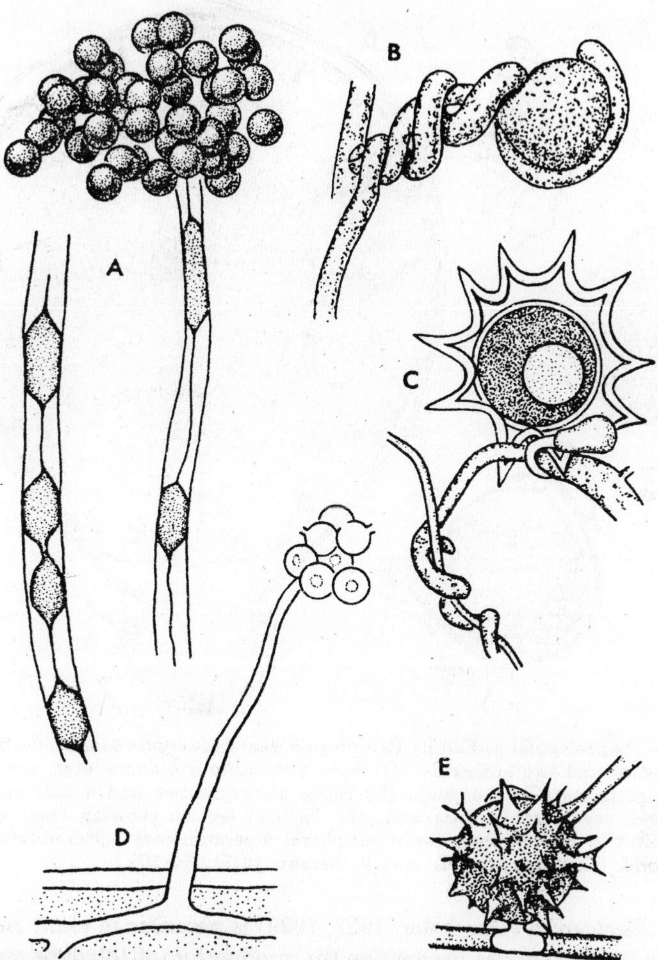

Fig. 36. Saprolegniales, Family Saprolegniaceae. (A–C) *Aphanomyces exoparasiticus* Coker & Couch. (A) Encysted zoospores at mouth of zoosporangium. (B) Young oogone and antherid. (C) Mature oogone containing oospore. (D, E) *Aphanomyces phycophilus* de Bary. (D) Discharged zoospores. (E) Oogone and antherid. (A–C, after Couch: *J. Elisha Mitchell Sci. Soc.*, **41**(3–4):213–227. D–E, after Sparrow: *Mycologia*, **22**(3):118–121.)

has but a single egg. The species of this genus are parasitic in algae and in the roots of higher plants, where they may cause root rots, as well as upon some aquatic animals, mostly Crustaceans. *A. acinetophagus* Bartsch and Wolf (1938) has been described from a fresh-water protozoan. Scarcely distinguishable from *Aphanomyces* is *Hydatinophagus* parasitic upon Rotifers. Another closely related genus, also parasitic upon Rotifers, is *Sommerstorffia*, with special spike-like branches which catch the host

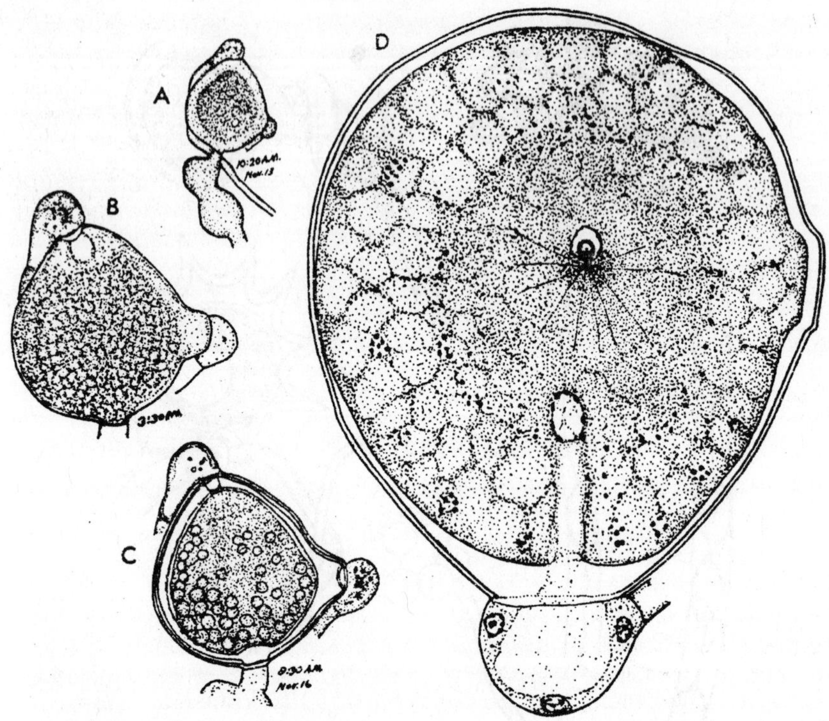

Fig. 37. Saprolegniales, Family Saprolegniaceae. *Leptolegnia caudata* de Bary. (A) Young oogone and two antherids. (B) Same structures five hours later, one antherid discharging contents into oogone. (C) Same structure two and a half days later, oospore wall considerably thickened. (D) Stained section showing entry of sperm nucleus into the egg, with remains of peripheral supernumerary nuclei in both oogone and antherid. (Courtesy, Couch: *Am. J. Botany*, **19**(7):584–599.)

victims. *Plectospira* (Drechsler, 1927, 1929) is parasitic in plant roots and resembles *Aphanomyces* except for the production of lobulate masses of hyphae which apparently serve as auxiliary storage parts of the zoosporangia, much as occurs in some species of the genus *Pythium* (see p. 127). The oogones have a single egg but no periplasm and become surrounded by many antherids, up to over 50, of which only a few reach full development. *Leptolegnia* resembles *Aphanomyces* in its slender hyphae and its slender zoosporangium with a single row of zoospores, as well as in the production of only a single egg in the oogone. It differs in that the primary zoospores swim away immediately and encyst at a distance as in *Saprolegnia*. (Figs. 36, 37.)

Dictyuchus resembles *Achlya* and like it produces clavate or cylindric zoosporangia which may be single or in chains or more or less sympodially clustered. They may separate from the supporting hyphae and float around carrying within them the encysted primary zoospores. Normally these zoospores encyst within the zoosporangium and by mutual pressure become more or less polyhedral in shape, depending upon their number. The zoosporangial wall may be persistent, in which case the secondary type zoospores escape through short exit papillae which pierce the surrounding wall, so that eventually a net-like structure remains. In other species the zoosporangial wall is evanescent early so that the encysted primary zoospores are rounded on their outer face and somewhat flattened where they are in contact. They may become separated under pressure. The emerging zoospores in both types are of the secondary type and may encyst and emerge again several times before germinating by germ tubes. In the species with persistent zoosporangium walls this is the only type of zoosporangium known. In those with evanescent walls under certain conditions of culture medium and temperature the first zoosporangia (according to Couch, 1931) empty their zoospores through an apical pore where they encyst as in *Achlya.* The later zoosporangia are as in *Dictyuchus.* The oogone contains a single egg. The one species with several eggs, *D. polysporus* Linds., assigned here has been determined by Apinis (1930) to belong to the genus *Protoachlya.* Most of the known species have been isolated from the soil but some occur also in fresh water. Both homothallic and heterothallic species are known (Couch, 1926b; Coker and Braxton, 1926). In *Thraustotheca* a similar preliminary production of *Achlya*-like zoosporangia has been reported (Coker and Couch, 1924). (Fig. 38.)

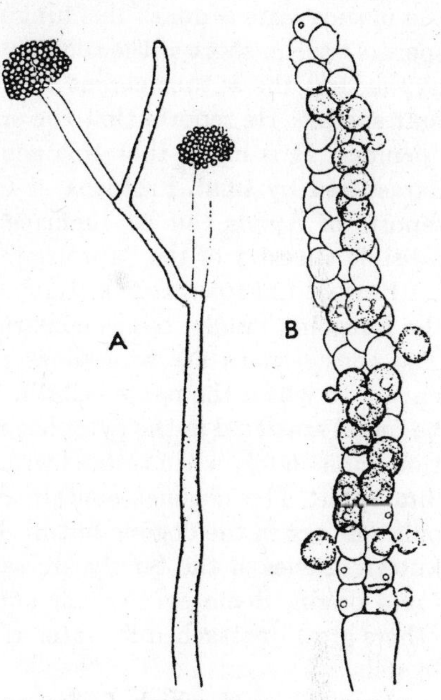

FIG. 38. Saprolegniales, Family Saprolegniaceae. *Dictyuchus missouriensis* Couch. (A) Early zoosporangia of *Achlya* type. (B) More enlarged zoosporangium of *Dictyuchus* type. (Courtesy, Couch: *J. Elisha Mitchell Sci. Soc.*, **46**(2):225–230.)

Miss Huneycutt (1948) has described a new genus, *Aphanodictyon*, with the single species *A. papillatum.* This has the vegetative structure of *Aphanomyces*, with very slender, branching mycelium. The zoosporangia are globose or subglobose and the primary zoospores encyst within them as in *Dictyuchus.* The secondary type zoospores escape through short exit tubes. The oogones contain 1–8 eggs (mostly 3–6), which are fertilized by antherids of androgynous or diclinous origin. The oogones have few to many papillae about 2 μ thick and of variable length up to 20 μ. The fungus occurs on keratinized material in the soil and may be cultured on thin slivers of horse hoof.

Apinis (1935) described a genus which he named *Archilegnia* in which fertilization is claimed to be by motile uniflagellate sperm cells. In other respects such as the mode of asexual reproduction and the structure of the pluriovulate oogones this fungus is like *Saprolegnia*. Not only do the sperms arise in short antherids growing out at right angles from the main hypha, but the author claims that encysted zoospores may give rise to four sperms. He reports that the sperms enter the oogone through small openings. It is likely that this genus was described from a *Saprolegnia* parasitized by small Protozoa or Chytridiales. If the correctness of the reports of Apinis can be confirmed it will throw great light upon the possible ancestry of the Saprolegniaceae.

FAMILY LEPTOMITACEAE. In this family the mycelium is slender as in the foregoing family, but is constricted at more or less regular intervals. The constrictions are sometimes plugged by granules of carbohydrate nature to which the name cellulin has been given. Similar granules may be found scattered in the cytoplasm. The cell walls give the cellulose reaction immediately when treated with chloriodide of zinc. The zoospores are dimorphic. The oogones contain no periplasm. In *Apodachlya* there is only one egg in the oogone but in *Apodachlyella* there are several. All the known species of the family are saprophytic, chiefly on matter of vegetable origin. Some species are aquatic, growing on sticks, fruits, etc.; others grow unattached in water rich in organic matter; and some grow in soil.

Leptomitus, of which *L. lacteus* (Roth) Agardh is perhaps the only species, consists of branching, cylindrical hyphae rather uniform in size, found in water containing large amounts of organic matter, such as drainage water from sugar factories, near the mouths of sewers, etc. It is easily distinguished by its constrictions which may remain open or become plugged by granules of cellulin. The terminal segments first and then successively the segments behind them become converted into zoosporangia in which are produced numerous pear-shaped primary zoospores all of which may escape successively through the terminal zoosporangium or from separate openings from each zoosporangium. The zoospores scatter as do those of *Saprolegnia*, eventually giving rise to zoospores of the secondary type. Sexual reproduction is unknown. (Fig. 35 F.)

Apodachlya is also much branched and has no specialized holdfast hyphae. The zoosporangia are distinct from the main hyphae, and are mostly ovoid or pyriform, tapering below to a short pedicel. The oogone contains a single egg without periplasm and may be terminal or lateral.

The antherid may subtend the oogone, directly, or arise on a short stalk from immediately below it. No conjugation tube seems to be produced. Germination of the oospore is by means of one or two germ tubes.

In *Apodachlyella* zoosporangia have not been observed. The spherical or pyriform oogones contain 2 to 12, more often 4 to 7 eggs, which become thick-walled oospores without surrounding periplasm. The slender constricted antheridial branches, 2 or 3 in number, arise from the segment below the oogone. A conjugation tube is produced by each functioning antherid.

FAMILY RHIPIDIACEAE. In this family the mycelium is provided with well-developed holdfast hyphae bearing a more or less thickened, sometimes even spherical, basal segment from whose upper portion arise slender, often constricted, hyphae terminated by zoosporangia or by the sexual organs, or both may arise directly from the basal segment (*Mindeniella*). Only zoospores of the secondary type are produced. The oogone contains a single egg surrounded by periplasm which in one genus (*Araiospora*) forms a cellular layer closely investing the oospore.

Sapromyces, with two or more species, grows on sticks, etc., in water, being attached by its rather few rhizoids. The upright main axis is not much thickened in comparison to the several slender, constricted branches that arise at its apex. These bear at their apices one to several obovoid, clavate, or cylindrical zoosporangia which give rise to numerous biflagellate, kidney-shaped zoospores which escape directly or into an evanescent vesicle. Possibly due to sympodial growth some of the zoosporangia may appear to be lateral. On the same branches with the zoosporangia or on separate branches the oogones and antherids arise. These also may be in terminal clusters of two or more, or single, or may appear to be lateral. The obovoid oogones contain each a single egg with abundant periplasm. The clavate antherid is borne on a slender, sometimes coiled hypha arising from just below the point of attachment of the oogone. It becomes attached to the oogone at the apex.

Rhipidium, with four species, consists of a thick, more or less cylindrical body with numerous rhizoids, growing on fruits, twigs, etc., in water. At its top it gives off slender branches which are constricted here and there. Terminally on these slender branches arise the ovoid zoosporangia which may later appear lateral on account of the sympodial mode of growth of the hypha. The protoplasmic contents of the zoosporangium divide into numerous zoospores which push out into a cylindrical vesicle. Upon the rupture of the latter the biflagellate, kidney-shaped zoospores escape. The oogones are also terminal on slender branches and contain each a single egg surrounded by a layer of periplasm. The antherid may arise on a slender branch just below the oogone or on a longer branch from another plant. It attaches itself to the basal portion of the oogone.

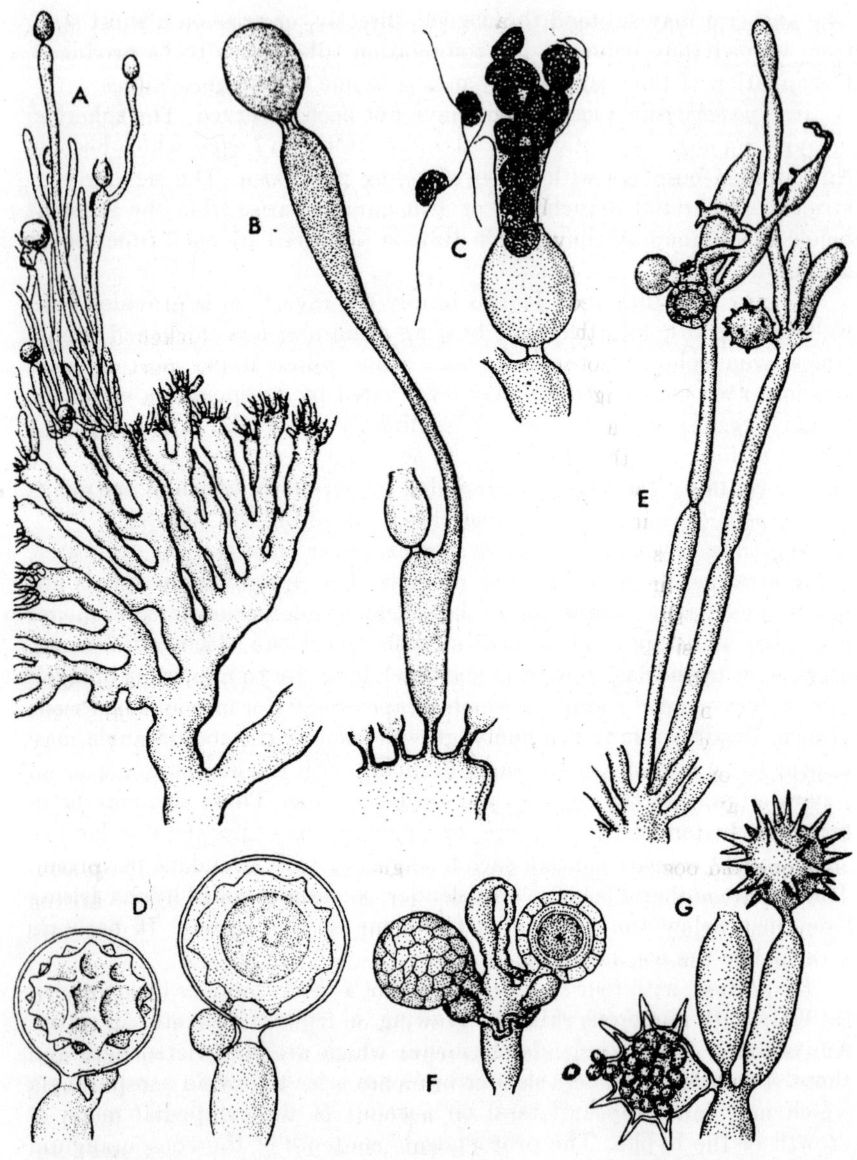

FIG. 39. Saprolegniales, Family Rhipidiaceae. (A–D) *Rhipidium americanum* Thaxt. (A) Habit sketch. (B) Detail of branch bearing zoosporangia. (C) Discharge of zoospores. (D) Two oogones, each with a mature oospore. (E–G) *Araiospora pulchra* Thaxt. (E) Portion of plant bearing an oogone and zoosporangia of both types. (F) Oogones and antherids. (G) Ordinary and spiny zoosporangia. (After Thaxter: *Botan. Gaz.*, **21**(6):317–331. Univ. Chicago Press.)

The oospore is thick-walled and roughened areolately. Its manner of germination has not been reported. (Fig. 39 A–D.)

Araiospora, with four species, grows on vegetable matter in water. It has a thick supporting or storage body from which arise the numerous slender constricted branches bearing the reproductive organs. The zoosporangia are of two kinds, ovoid or club-shaped with smooth walls, and subspherical and covered with stout spines. The zoospores are similar in each type, being biflagellate and kidney-shaped. The spherical oogones are borne on similar branches, sometimes on distinct plants. They are characterized by a cellular periplasm layer around the oospore. The oogone is fertilized by a basally applied antherid which may arise near by. (Fig. 39 E–G.)

Mindeniella also has zoosporangia of two kinds. The thin-walled clavate or ovoid sporangia are pedicellate on or near the apex of the stout cylindrical or clavate basal segment. They may be spiny at the upper end or without spines. The resting zoosporangia are ovoid or almost spherical and pedicellate. They are spiny and thick-walled. Their germination has not been observed. No sexual organs are known.

It is apparent that the Rhipidiaceae form a group of more highly specialized genera which have probably arisen from the Leptomitaceae. Whether the latter arose from the Saprolegniaceae or vice versa is uncertain. In view of the somewhat more specialized sexual organs in the genus *Apodachlya* it is possible that this family arose from the Saprolegniaceae or that both arose from the Lagenidiales. However it is also possible that the latter represent ends of series of reductions from various ancestral forms in the Saprolegniales and possibly in the Peronosporales. Because of the difference in zoospore structure and in the composition of the cell wall it seems unlikely that these orders have any close relationship with the Chytridiales-Blastocladiales-Monoblepharidales series.

Keys to the Families and Genera of Lagenidiales

Key to the Genera of Family Woroninaceae

"Plasmodium" forming a cluster ("sporangiosorus") of numerous cellulose-walled zoosporangia or a cluster ("cystosorus") of thick-walled, angular resting spores. Parasitic in Saprolegniaceae and green algae. *Woronina*

"Plasmodium" becoming surrounded by cellulose walls forming an elongated and sometimes branched thin-walled tube. The contents separate into naked "spore mother cells" each of which produces 8 zoospores. Resting spores unknown. Saprophytic in marine Florideae. *Pyrrhosorus*

"Plasmodium" at maturity filling the infected portion of the host and forming its wall pressed closely against that of the host. Forming a single more or less rounded zoosporangium or a row of cylindrical zoosporangia. Resting spores where known round and spiny and free from the cell wall of the host. Parasitic in Pythiaceae and Saprolegniaceae. *Rozellopsis*

Key to the Genera of Family Olpidiopsidaceae

Zoospores escaping through discharge tubes.
Zoospores monomorphic (monoplanetic).
Zoosporangia spherical to ellipsoid, mostly with one discharge tube.
Resting spores produced, usually by a sexual process. *Olpidiopsis*
Resting spores unknown. *Pseudolpidium*
Zoosporangia irregularly lobed or tubular, usually with more than one discharge tube. Resting spores unknown. *Petersenia*
Zoospores dimorphic (diplanetic), encysting at the mouth of the discharge tube and then escaping singly. Zoosporangia spherical or nearly so. Fertilization of oogone by a conjugation tube. Oogone with one egg and periplasm. *Pythiella*
Zoospores escaping by a large irregular break in the wall of the spherical or ellipsoidal zoosporangium. Resting spores not certainly known. *Pseudosphaerita*

Key to the Genera of Family Sirolpidiaceae

Thallus elongate, narrowly tubular, becoming septate and then disarticulating into separate zoosporangia. *Sirolpidium*
Thallus elongate, broadly tubular, becoming septate into distinct zoosporangia which remain attached. *Pontisma*

Key to the More Important Genera of Family Lagenidiaceae

Thallus elongated, not becoming septate, sometimes coiled, parasitic in roots of grasses and other plants. Zoospores formed in a vesicle at the tip of the single exit tube. Sexual reproduction by union of contents of adjacent thalli through a conjugation tube to form a thick-walled resting spore. *Lagena*
Thallus short or elongated, not septate, sometimes branched, in the cells of diatoms. Primary zoospores encysting at mouth of exit tube and emerging as zoospores of the secondary type.
Thallus short, not branched, primary zoospores flagellate. Sexual reproduction by union of adjacent thalli through a permanent conjugation tube, the empty male thallus remaining attached. *Ectrogella*
Thallus elongated and coiled, sometimes branched. Primary zoospores not flagellate. Resting spores formed here and there within the elongated thallus but no sexual process observed. *Aphanomycopsis*
Thallus more or less rounded, parasitic in the exterior cells of Phaeophyceae or Rhodophyceae. Resting spores not observed.
Sporangia becoming extramatrical at maturity. Primary zoospores encysted at the mouth of the exit tube or more often against the inner surface of the zoosporangium. Upon emergence they are of the secondary type. *Eurychasma*
Sporangia remaining intramatrical with many discharge tubes. Primary zoospores encysting near the mouths of these tubes and emerging as secondary zoospores. *Eurychasmidium*
Thallus elongated and coenocytic, mostly later divided into multinucleate segments by cross septa. If the latter are wanting perhaps the species should be transferred to the genus *Lagena*. Each segment becomes a zoosporangium

or a gametangium. Zoospores forming in a vesicle at the tip of the exit tube. In fresh-water algae, pollen grains, and microscopic aquatic animals, etc.

Thallus strongly constricted at each septum. Antheridial cell poorly differentiated. *Myzocytium*

Thallus not, or only slightly, constricted at the septa. Antherids sharply distinguished from the oogones, with well-developed conjugation tube. Sometimes septa fail to be formed. *Lagenidium*

Thallus and sexual reproduction much as in *Lagenidium*, but no zoospores formed. Aerial conidiophores discharge the single conidia violently from the apex. See *Ancylistes* in Order Entomophthorales (Chap. 7)

Key to Family Thraustochytriaceae

Single genus. Epibiotic zoosporangium with rhizoids. *Thraustochytrium*

Keys to the Families and Genera of Saprolegniales

Key to the More Important Genera of Family Saprolegniaceae

(Based in part upon Coker and Matthews, 1937)

Fertilization of oogones reportedly by uniflagellate sperm cells, otherwise as in *Saprolegnia*. Probably a species of the latter with Chytridiaceous or Protozoan parasites. *Archilegnia*

Oogone fertilized by male nuclei introduced from adhering antherids; development sometimes parthenogenetic.

Sporangia rare or wanting, the spores encysting in the sporangium without a swimming stage and germinating by germ tubes, very rarely by swimming cells. *Aplanes*

Sporangia abundant, the spores encysting within them.

Oogones usually with more than one egg.

Encysted spores liberated by the irregular rupture of the sporangial wall, then germinating by germ tubes or by emergence of zoospores of the secondary type. *Thraustotheca*

Encysted spores liberated in successive groups by the breaking off of the apical portion of the sporangium, later germinating by emergence of zoospores of the secondary type. *Calyptralegnia*

Encysted spores escaping as zoospores of the secondary type through exit papillae which pierce the zoosporangium wall. Mycelium very slender, growing on keratinized media. *Aphanodictyon*

Oogones with only one egg.

Mycelium of vigorous and extensive growth; encysted spores in several rows producing a net-like appearance, germinating by germ tubes or more often by emergence of zoospores of the secondary type. *Dictyuchus*

Mycelium of very limited growth, dense and opaque.

Encysted spores in one to several rows, thin-walled, escaping by dissolution or rupture of sporangial wall, usually germinating by germ tubes, more rarely by emergence of zoospores of the secondary type. *Brevilegnia*

Encysted spores in one row, thick-walled, multinucleate, never forming a swimming stage. *Geolegnia*

Sporangia abundant, the spores emerging from an apical mouth before encysting.
Zoospores of primary type only, after escaping from sporangium germinating by germ tube or by repetition forming zoospores again of the primary type. *Pythiopsis*
Zoospores of primary type swimming some distance before encysting, then emerging as secondary type zoospores.
Zoospores in a single row in the zoosporangium. *Leptolegnia*
Zoospores in more than one row.
New sporangia formed by proliferation. *Saprolegnia*
New sporangia formed by cymose branching. *Isoachlya*
Zoospores of primary type with or without flagella and encysting at the mouth of the zoosporangium, later emerging as secondary type zoospores.
Zoospores in more than one row in the sporangium.
Zoospores all encysting at the mouth of the zoosporangium. New sporangia formed by cymose branching. *Achlya*
Zoospores encysting in part at the mouth of the sporangium and some swimming away before encysting. New sporangia formed by cymose branching and sometimes also by proliferation. *Protoachlya*
Zoospores in one row in the slender zoosporangium.
Masses of lobulate inflated segments auxiliary to the zoosporangia are produced. *Plectospira*
No lobulate auxiliary masses.
Branches of mycelium of usual type. *Aphanomyces*
Spike-like branches with sticky tips which catch and parasitize rotifers. *Sommerstorffia*

Key to the Genera of Family Leptomitaceae

(Based upon Sparrow, 1943)

Thallus coarse, branched, constricted into numerous cylindrical segments. Beginning at the apex the segments become zoosporangia in basipetal succession without change of size or shape. No sexual reproduction known. Growing saprophytically, in heavily polluted water. *Leptomitus*
Thallus more slender, constricted, the zoosporangia with definite pedicels and mostly ovoid or pyriform. Usually producing sex organs.
Oogones with one oospore. *Apodachlya*
Oogones with more than one oospore; zoosporangia not reported. *Apodachlyella*

Key to the Genera of Family Rhipidiaceae

Basal cells giving rise to branches which bear the reproductive organs.
Basal cells slender, sporangia smooth walled, oospores with undulate outer wall. *Sapromyces*
Basal cells usually stout, sporangia smooth walled or spiny or both in the same species. Oospores with reticulate or cellular outer wall.
Oospore wall cellular, both smooth and spiny sporangia present. *Araiospora*
Oospore wall reticulate, sporangia with smooth walls. *Rhipidium*
Reproductive organs arising directly from the basal cell and of varying degrees of spininess. *Mindeniella*

Literature Cited

Apinis, Arv.: Untersuchungen über die in Lettland gefundenen Saprolegniaceen nebst Bemerkungen über andere Wasserpilze, *Acta Horti Botan. Univ. Latviensis*, **4**:201–246. *Pls.* 1–4. *Figs.* 1–4. 1929 (1930).

———: Fertilization of oospheres by planogametes in Saprolegniaceae. Preliminary note, *ibid.*, **8**(1–3):103–110. 1 *pl.* 1933 (1935).

Barrett, J. T.: Development and sexuality of some species of Olpidiopsis (Cornu) Fischer, *Ann. Botany*, **26**(101):209–238. *Pls.* 23–26. 1912.

Bartsch, Alfred F., and Fred T. Wolf: Two new Saprolegniaceous fungi, *Am. J. Botany*, **25**(6):392–395. *Figs.* 1–11. 1938.

Berdan, Helen: Revision of the genus Ancylistes; *Mycologia*, **30**(4):396–415. *Figs.* 1–22. 1938.

Bishop, Harlow: A study of sexuality in Sapromyces Reinschii (Schröt.) Fritsch, *Mycologia*, **32**(4):505–529. *Figs.* 1–6. 1940.

Coker, William Chambers: The Saprolegniaceae with Notes on Other Water Molds, pp. 1–201. 63 *pls.* Chapel Hill, Univ. North Carolina Press, 1923.

———, and H. H. Braxton: New water molds from the soil, *J. Elisha Mitchell Sci. Soc.*, **42**:139–147. *Pls.* 10–15. 1926.

———, and J. N. Couch: Revision of the genus Thraustotheca, with a description of a new species, *ibid.*, **40**(3–4):197–202. *Pls.* 38–40. 1924.

———, and Velma Dare Matthews: Saprolegniales. Saprolegniaceae, Ectrogellaceae, Leptomitaceae, *North American Flora*, **2**(1):15–67. 1937.

Cook, W. R. Ivimey, and W. H. Nicholson: A contribution to our knowledge of Woronina polycystis Cornu, *Ann. Botany*, **47**(188):851–859. *Figs.* 1–16. 1933.

Cooper, George Olds: Cytological studies on the sporange development and gametogenesis in Brevilegnia diclina Harvey, *Trans. Wisconsin Acad. Sci.*, **24**:309–322. *Pls.* 3–5. 1929.

Cornu, Maxime: Monographie des Saprolégniales. Étude physiologique et systématique, *Ann. sci. nat. Botan.*, 5me sér., **15**:1–198. *Pls.* 1–7. 1872.

Couch, John N.: Some observations on spore formation and discharge in Leptolegnia, Achlya, and Aphanomyces, *J. Elisha Mitchell Sci. Soc.*, **40**:27–42. *Pls.* 4–5. 1924.

———: Notes on the genus Aphanomyces, with a description of a new semiparasitic species, *ibid.*, **41**(3–4):213–227. *Pls.* 26–33. 1926a.

———: Heterothallism in Dictyuchus, a genus of the water molds, *Ann. Botany*, **40**(160):849–881. *Pls.* 35–38. *Figs.* 1–3. 1926b.

———: Observations on some species of water molds connecting Achlya and Dictyuchus, *J. Elisha Mitchell Sci. Soc.*, **46**(2):225–230. *Pl.* 15. 1931.

———: The development of the sexual organs in Leptolegnia caudata, *Am. J. Botany*, **19**(7):584–599. *Pls.* 42–44. 1932.

———: New or little known Chytridiales, *Mycologia*, **27**(2):160–175. *Figs.* 1–64. 1935.

———: The structure and actions of the cilia in some aquatic Phycomycetes, *Am. J. Botany*, **28**:704–713. *Figs.* 1–58. 1941.

Dangeard, P. A.: Les ancêtres des champignons supérieurs, *Le Botaniste*, **9**:1–263. *Pls.* 1–18. 2 *figs.* 1903–1906.

Drechsler, Charles: Two water molds causing tomato rootlet injury, *J. Agr. Research*, **34**(3):287–296. *Figs.* 1–2. 1927.

———: The beet water mold and several related root parasites, *ibid.*, **38**(6):309–361. *Figs.* 1–17. 1929.

Fischer, Alfred: Untersuchungen über die Parasiten der Saprolegnieen, *Jahrb. wiss. Botan.*, **13**:286–371. *Pls.* 13–15. 1882.

———: Phycomyceten, in L. Rabenhorst: Kryptogamen-Flora von Deutschland, Oesterreich und der Schweiz, Zweite Auflage, Band 1, Abt. 4, pp. 1–505. *Figs.* 1–74. Leipzig, Verlag von Eduard Kummer, 1892.

Gäumann, Ernst: Ein Beitrag zur Kenntnis der lappländischen Saprolegnieen, *Botan. Notiser*, **1918**:151–159. 1919.

Harvey, James Vernon: A study of the water molds and Pythiums occurring in the soils of Chapel Hill, *J. Elisha Mitchell Sci. Soc.*, **41**:151–164. *Pls.* 12–19. 1925.

Huneycutt, Maeburn Bruce: Keratinophilic Phycomycetes: I. A new genus of the Saprolegniaceae, *J. Elisha Mitchell Sci. Soc.*, **64**(2):277–285. *Pls.* 35–36. 1948.

Horn, L.: Experimentelle Entwickelungsänderungen bei Achlya polyandra de Bary, *Ann. Mycol.*, **2**(3):207–241. *Figs.* 1–21. 1904.

Kanouse, Bessie B.: A monographic study of special groups of the water molds: I. Blastocladiaceae; II. Leptomitaceae and Pythiomorphaceae, *Am. J. Botany*, **14**:287–306, 335–357. *Pls.* 32–34, 48. 1927.

Karling, John S.: A new fungus with anteriorly uniciliate zoospores, Hyphochytrium catenoides, *Am. J. Botany*, **26**:512–519. *Figs.* 1–18. 1939.

———: The Simple Holocarpic Biflagellate Phycomycetes, x + 123 pp. 25 *pls.* New York, published by the author, 1942.

Kevorkian, Arthur G.: Studies in the Leptomitaceae: II. Cytology of Apodachlya brachynema and Sapromyces reinschii, *Mycologia*, **27**(3):274–285. *Pls.* 19–20. 1935.

von Minden, M.: Chytridiineae, Ancylistineae, Monoblepharidineae, Saprolegniineae, in Kryptogamenflora der Mark Brandenburg, vol. 5, pt. 2, pp. 193–352. Leipzig, Gebrüder Borntraeger, 1911.

Monsma, Edwin Y.: A study of the water molds of the Lydell State Fish Hatchery at Comstock Park, Michigan, *Papers Mich. Acad. Sci.*, **22**:165–182. 1936 (1937).

Raper, J. R.: Heterothallism and sterility in Achlya and observations on the cytology of Achlya ambisexualis, *J. Elisha Mitchell Sci. Soc.*, **52**:274–289. *Pls.* 22–24. 1936.

———: Rôle of hormones in the sexual reaction of heterothallic Achlyas, *Science, N.S.*, **89**(2310):321–322. *Fig.* 1. 1939.

———: Sexuality in Achlya ambisexualis, *Mycologia*, **32**(6):710–727. *Figs.* 1–4. 1940.

Salvin, S. B.: Preliminary report on the intergeneric mating of Thraustotheca clavata and Achlya flagellata, *Am. J. Botany*, **29**(8):674–676. *Figs.* 1–2. 1942.

Scherffel, A.: Endophytische Phycomyceten-Parasiten der Bacillariaceen und einige neue Monadinen. Ein Beitrag zur Phylogenie der Oomyceten (Schröter), *Arch. Protistenk.*, **52**(1):1–141. *Pls.* 1–5. 1925.

Shanor, Leland: Studies in the genus Olpidiopsis: I. Resting spore germination in a new species; II. The relationship of Pseudolpidium Fischer and Olpidiopsis (Cornu) Fischer, *J. Elisha Mitchell Sci. Soc.*, **55**(1):167–177, 179–195. *Pls.* 24–26. 4 *figs.* 1939.

Sparrow Jr., Frederick K.: The non-sexual stage of Aphanomyces phycophilus, *Mycologia*, **22**(3):118–121. *Fig.* 1. 1930.

———: Observations on marine Phycomycetes collected in Denmark, *Dansk Botanisk Arkiv*, **8**(6):1–24. 4 *pls.* 1934.

SPARROW JR., FREDERICK K.: Aquatic Phycomycetes, Exclusive of the Saprolegniaceae and Pythium, xix + 785 pp. 634 *figs.* Ann Arbor, Univ. Mich. Press, 1943.

THAXTER, ROLAND: New or peculiar aquatic fungi: 4. Rhipidium, Sapromyces, and Araiospora, nov. gen., *Botan. Gaz.*, **21**(6):317–331. *Pls.* 21–23. 1896.

VANTERPOOL, T. C., and G. A. LEDINGHAM: Studies in "browning" root rot of cereals: I. The association of Lagena radicicola n. gen., n. sp., with root injury of wheat, *Can. J. Research*, **2**(3):171–194. *Pls.* 1–2. *Figs.* 1–7. 1930.

ZIEGLER, ARTHUR WILLIAM: A comparative study of zygote germination in the Saprolegniaceae, *J. Elisha Mitchell Sci. Soc.*, **64**(1):13–40. *Pls.* 1–6. 1948.

ZOPF, WILHELM: Zur Kenntnis der Phycomyceten: I. Zur Morphologie und Biologie der Ancylisteen und Chytridiaceen, zugleich ein Beitrag zur Phytopathologie, *Nova Acta Leopoldina*, **47**(4):141–236. *Pls.* 12–21. 1884.

6

PHYCOMYCETEAE: PERONOSPORALES AND PROTOMYCETALES

Order Peronosporales. The fungi discussed in the previous chapters were mainly inhabitants of soil and water, and largely saprophytic or parasitic in habit. This order contains many genera and species that are very strict parasites on Flowering Plants (Anthophyta or Angiosperms) and which have hitherto resisted all attempts to grow them on artificial culture media. However, most if not all of the members of the family Pythiaceae respond more or less readily to efforts to bring them into culture upon nonliving media. Some genera of this family were placed in the past among the Saprolegniales because of characters intermediate between the two orders.

In the main the asexual reproduction of the Peronosporales exhibits a step in evolution above that in the Saprolegniales. In the latter the zoospores are mostly produced from zoosporangia that remain attached to the main mycelium when the zoospores are set free. However, it must be noted that the hyphae may break up into separate rounded segments called "gemmae" which eventually are capable of functioning as zoosporangia. In the majority of Peronosporales the tips of aerial hyphae enlarge and are separated from the main hypha by a cross wall and are then set free and distributed, usually by air currents. These so-called "conidia" show their true nature as potential zoosporangia when they fall into water, as then their contents may divide internally into zoospores. The zoospores produced in this order are always of the secondary type. These zoospores may in some cases encyst and become free-swimming several times.

Sexual reproduction occurs by the fertilization, through a conjugation tube from an antherid, of the normally single egg in an oogone to form a thick-walled oospore. With a few possible exceptions the egg is surrounded by periplasm as in the Rhipidiaceae of the order Saprolegniales and *Pythiella* in the order Lagenidiales.

There is no sharp distinction of root-like holdfast hyphae and external hyphae of quite different nature as occurs in many of the Saprolegniaceae and in the Rhipidiaceae. The mycelium usually shows a cellulose reaction upon application of a solution of chloriodide of zinc. The hyphae are mostly much more slender than in the above-mentioned families but, like them, are normally coenocytic and nonseptate except to set off reproductive organs or injured portions or sometimes empty portions of the mycelium.

Following Fitzpatrick (1930) the 400–500 species are divided into three families as follows:

Pythiaceae: saprophytic or parasitic, mycelium intracellular, less often intercellular with haustoria. Zoosporangia filamentous to ovoid or spherical, in some cases the ovoid forms becoming separable "conidia," if aerial. These conidia are borne singly, rarely in chains, at the tips of unbranched conidiophores or at the tips of the branches of a sympodially branched conidiophore. They germinate by the formation of zoospores or by a germ tube. The oospore germinates by becoming a zoosporangium or by forming a short hypha terminated by a conidium or by producing mycelium.

Albuginaceae: strictly parasitic in herbaceous Anthophyta (Angiospermae). Mycelium intercellular with globular haustoria. Conidiophores clavate, clustered in extensive sori under the epidermis of the host which is ruptured by the pressure of the conidia which are produced successively in chains at the apex of each conidiophore. Conidia germinating by the formation of zoospores or of germ tubes. Oospore germinating by the formation of zoospores.

Peronosporaceae: strictly parasitic in herbaceous, and, more rarely, woody Anthophyta (Angiospermae). Mycelium intercellular with globular or filamentous haustoria. Conidiophores emerging through the stomata singly or two or three together, unbranched or monopodially much branched, bearing the conidia singly at the tips of the branches or on short sterigmata on the top of the unbranched conidiophore. Conidia germinating by the formation of zoospores or by a stout germ tube. Oospore germinating by a germ tube or by a short unbranched conidiophore.

Family Pythiaceae. The Pythiaceae have been variously divided into from 4 to 14 genera. Of these, *Pythium* (in its wider delimitation) and *Phytophthora* contain the greater number of species and are best known.

The species of *Pythium* are mostly soil or water inhabitants, probably living most of the time as saprophytes. Some of these soil species, however, are capable of becoming destructive parasites upon plants, causing rotting of the tissues or damping-off. A few are parasitic on other species of the genus. Some species (e.g., *P. proliferum* de Bary, which can be caught by hanging fruits in bodies of water) are saprophytic upon various vegetable objects submerged in water. Of the 66 species recognized by Middleton (1943) three are saprophytic and one parasitic on animal matter, nine parasitic in fresh-water algae, and one saprophytic in a marine red seaweed, about nine in soil and vegetable debris and not known to be parasites, and forty or so capable of parasitism in higher

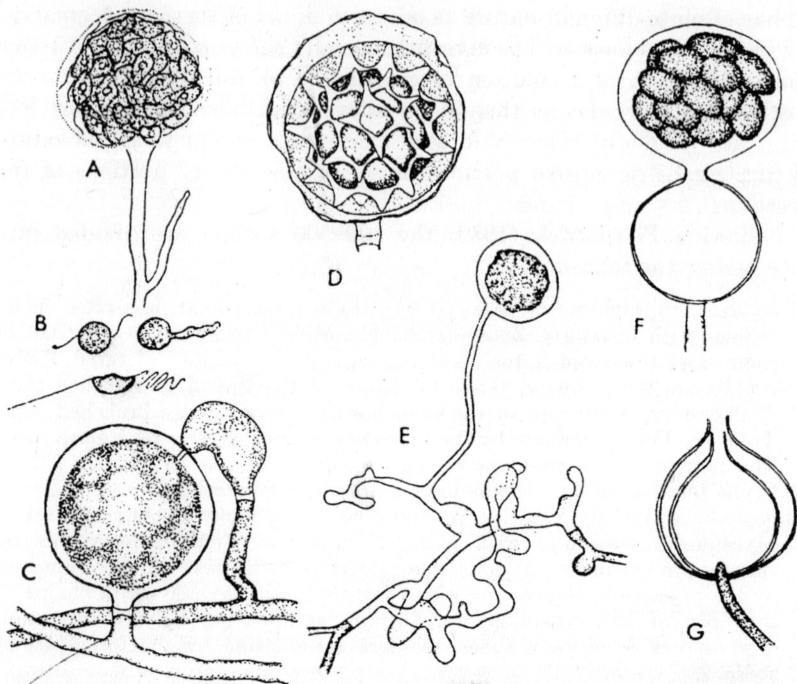

Fig. 40. Peronosporales, Family Pythiaceae. (A–D) *Pythium dictyosporum* Racib. (A) Formation of zoospores within vesicle at the open tip of the discharge tube of the filamentous sporangium. (B) Zoospores. (C) Young oogone and antherid. (D) Mature oogone and oospore. (E) *Pythium torulosum* Coker and Patterson, showing the toruloid zoosporangium and vesicle. (F, G) *Pythium proliferum* de Bary. (F) Spherical zoosporangium with vesicle. (G) Proliferated zoosporangium. (A–D, courtesy, Sparrow: *Mycologia*, **23**(3):191–203. E–G, courtesy, Matthews: Studies on the Genus Pythium, Chapel Hill, Univ. North Carolina Press.)

plants, mainly roots and stems and in one species in the leaves. In the forms parasitic in land plants the mycelium is mainly intracellular, usually killing the cells rapidly, in some cases appearing to kill them in advance of the arrival of the hyphae.

The mycelium is slender and, as in the Saprolegniales, is a branching coenocyte. Septa are formed to set off the antherids and oogones and, usually, the zoosporangia, and to separate the empty older portions of the mycelium from those containing protoplasm. Sexual reproduction has been studied by Trow (1901), Patterson (1927), Edson (1915), and others, and has been found to resemble most closely that in the Rhipidiaceae. The oogone arises terminally, less often intercalarly on longer or shorter unmodified hyphae. It is ovoid or spherical and smooth or papillate or even echinulate. Sometimes it is tubular-elongate at one or both ends. Of the several nuclei in the oogone, all but one migrate to the periphery into the periplasm which surrounds the centrally placed egg except in the rare cases in which two to six eggs are produced. In *P. polysporum* Poitras (1949) nearly one fourth of the oogones produce from two to six oospores. The antherid is also plurinucleate at first but only one nucleus functions in fertilization of each egg. The filament bearing the terminal antherid may arise at a great distance from the oogone ("diclinous") or from near by on the hypha bearing it ("monoclinous"). In some cases the antherid is a small segment of the hypha immediately under the oogone, the conjugation tube entering the latter through the base ("hypogynous"). In a few species the antheridial filament wraps in a spiral around the hypha bearing the oogone. Often the oospore develops parthenogenetically. Sometimes many antherids may become attached to the same oogone although only one antherid functions in its fertilization. Through a conjugation tube one male nucleus is introduced into the egg which then forms a thick wall and becomes an oospore. The oospore may completely fill the oogone ("plerotic") or may leave a space between it and the oogone wall ("aplerotic"). It is usually thick-walled and smooth but may be reticulately thickened externally. Germination may be delayed for a long while. It is effected by the production of a long germ tube or of a short germ tube terminating in a zoosporangium, or in some cases the oospore may produce the zoospores internally, emptying them through a short beak or exit tube into a vesicle as in the germination of a conidium (Drechsler, 1947). (Fig. 40.)

Asexual reproduction is mostly by means of zoospores. Typically the zoosporangium forms a beak (usually short but sometimes several times as long as the diameter of the zoosporangium). The tip of the beak softens and out of it flows the protoplasm of the zoosporangium, to form a spherical mass, the so-called vesicle. Within this the differentiation into zoospores is completed and the kidney-shaped biflagellate zoospores rupture the plasma membrane of the vesicle and escape singly, or several in a clump which separate subsequently. After swimming for a while the zoospore encysts and germinates by a germ tube or, as apparently first reported by Cornu (1872), sends out a short papilla from whose apical opening emerges a single zoospore, with or without a vesicle, or sometimes a vesicle in which are produced several zoospores. The process may be repeated three or four times, the zoospores of the successive crops being smaller each time.

Three more or less distinctive types of zoosporangium may be distinguished as follows:

1. Slender filaments, simple or branched, of the same size and appearance as the vegetative mycelium, and opening at the tip or tips of the hyphae to form a short or long emission tube at whose apex a vesicle is formed. This filamentous zoosporangium may be separated from the re-

mainder of the mycelium by a septum (subgenus *Nematosporangium* of Fischer, 1892) or the septum may be lacking (subgenus *Aphragmium* of Fischer). (Fig. 40 A, B.)

2. The more or less filamentous zoosporangium is somewhat thickened and lobed, often only near the base, to form the so-called toruloid structure, the plasmatoönkosis of Sideris (1931). This is a storage organ enabling the zoosporangium to produce larger numbers of zoospores in the vesicles at the tips of the emission tubes. (Fig. 40 E.)

3. The zoosporangium is spherical, ovoid or limoniform, terminal or intercalary, single or in short chains (*Sphaerosporangium* of Fischer). (Fig. 40 F.)

Schroeter (1897) and Sideris (1931) included types 1 and 2 in a genus which they called *Nematosporangium*, applying the name *Pythium* to Fischer's subgenus *Sphaerosporangium*. Inasmuch as the type species of the genus, *P. monospermum* Pringsheim (1858), forms zoosporangia of type 1, the name *Pythium* must be retained for Schroeter's *Nematosporangium*, if the genus is divided. The later monographers of the genus (Butler, 1907; Miss Matthews, 1931; and Middleton, 1943) do not make this division but include all three types of sporangia in one genus.

In some submerged species with spherical or ovoid sporangia new zoosporangia arise within the emptied ones by proliferation as in *Saprolegnia*, or a branch arises just below the point of attachment of the zoosporangium and is terminated by a new one, sympodial development proceeding as in *Achlya*. In some species the zoosporangia may be produced on aerial hyphae in which case they may become detached and distributed by currents of air, germinating when they fall into water either in the usual manner or by a germ tube. These wind-carried detached zoosporangia are called conidia. Usually the submerged zoosporangia and the aerial conidia are alike in the same species. *Pythium debaryanum* Hesse, with its zoosporangia representing potential conidia, is common in soil as a damping-off parasite of seedlings when the soil is too moist or the seedlings too much crowded. It can often be obtained by placing a little soil in a dish with some cooled boiled water and placing in the latter a few boiled hemp seeds (not too many or the bacteria will become numerous). On these seeds *Pythium* and various Saprolegniales will appear in a few days. *Pythium aphanidermatum* (Edson) Fitzpatrick, the cause of black-root disease of radishes (especially noticeable in the White Icicle variety) and rotting of sugar-beet seedlings and roots of many other plants, is a species with the second type of sporangia. It was originally described as the type of a new genus, *Rheosporangium*, by Edson (1915).

The genus *Phytophthora* contains 15 or 20 species some of which live as saprophytes in the soil, developing as parasites in the presence of suit-

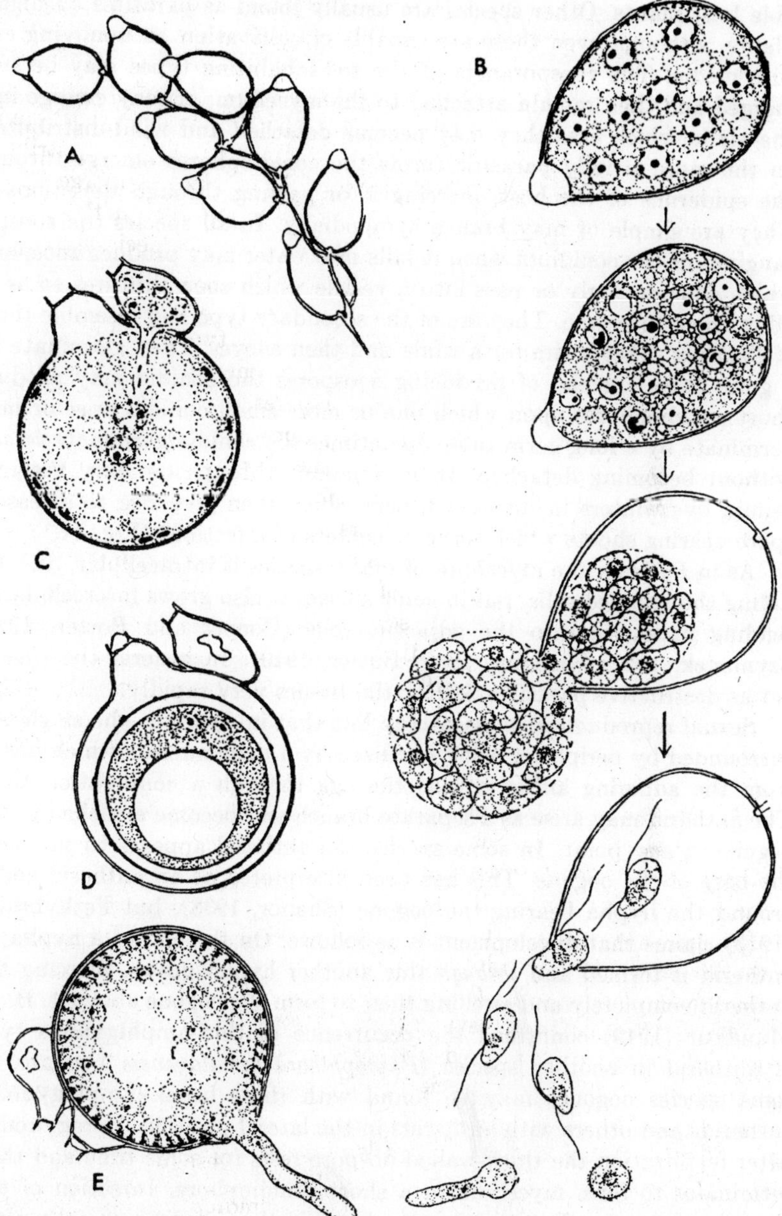

FIG. 41. Peronosporales, Family Pythiaceae. *Phytophthora cactorum* (L. & C.) Schröt. (A) Aerial filament in moist air, producing sympodially a succession of sporangia ("conidia"). (B) Formation of zoospores within the sporangium and emission into a vesicle. (C) Fertilization of oogone. (D) Mature oospore. (E) Germinating oospore. (Courtesy, Blackwell: *Brit. Mycol. Soc. Trans.*, **26**(1–2):71–89.)

able host plants. Other species are usually found as parasites on higher plants, although even these are capable of cultivation on nonliving culture media. The zoosporangia of the soil-inhabiting types may be submerged and then remain attached to the mycelium, or may emerge into the air in which case they may become detached and wind-distributed. In the more strictly parasitic forms the conidiophores emerge through the epidermis of the host, piercing it or passing through the stomata. They are simple or may branch sympodially. In all species the zoosporangium or the conidium when it falls into water may produce zoospores which emerge singly or pass into a vesicle which soon ruptures so as to allow them to escape. They are of the secondary type and resemble those of *Pythium*. They swim for a while and then encyst. They germinate by a germ tube. Instead of producing zoospores the conidia may produce short conidiophores upon which one or more small conidia arise or may germinate by a long germ tube. Sometimes the conidia germinate *in situ* without becoming detached. In *P. infestans* (Mont.) de Bary the mycelium overwinters in infected tubers which then give rise to diseased, spore-bearing shoots which serve as centers of infection. (Fig. 41.)

As in *Pythium* the mycelium of many species is intracellular, directly killing the invaded cells, but in some species it also grows intercellularly, sending haustoria into the adjacent cells (Cooper and Porter, 1928; Szymanek, 1927; Klebahn, 1909; Butler, 1910). In general the species act as destructive parasites, killing the tissues very rapidly.

Sexual reproduction is essentially like that in *Pythium*, the single egg surrounded by periplasm being fertilized by a male nucleus which passes from the adhering antherid into the egg through a conjugation tube. The antherid may arise as a separate branch and become attached to the oogone at any point. In some species the antherid appears to surround the base of the oogone. This has been interpreted as an antherid coiled around the hypha bearing the oogone (Shanor, 1938), but Pethybridge (1913) claims that development is as follows: On the end of a hypha an antherid is formed and *through* this another hypha grows, piercing the antherid completely and swelling then to form the oogone above it. B. D. Mundkur (1949) confirmed the occurrence of this amphigynous type of antherid in another species (*Phytophthora himalayensis* Dastur). In some species oogones may be found with these basal (amphigynous) antherids and others with antherids in the lateral position (paragynous). After fertilization the thick-walled oospore rests for some time and then germinates to form mycelium or a short conidiophore. Infection of the host may occur by the zoospores produced in the conidia or by the germ tubes from the oospores. With the soil-inhabiting species the mycelium may directly penetrate the subterranean portions of the host. Narasinhan (1930) reports that *P. arecae* (Colem.) Pethyb. is heterothallic but Tucker (1931) believes that this needs further study. (Fig. 42.)

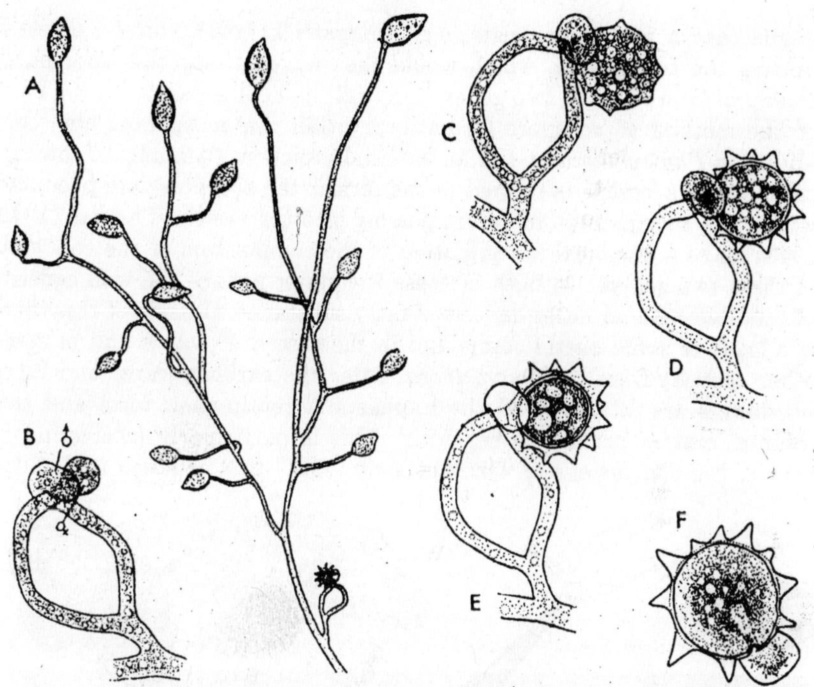

Fig. 42. Peronosporales, Family Pythiaceae. *Phytophthora stellata* Shanor. (A) Habit sketch of hypha with numerous sporangia and a sexual reproductive branch. (B–E) Successive stages in the development of the same antherid and oogone, showing possible explanation of the so-called amphigynous development. (F) Fertilization of oogone. (Courtesy, Shanor: *J. Elisha Mitchell Sci. Soc.*, **54**(1):154–162.)

Phytophthora infestans (Mont.) de Bary, the cause of the late blight of potato (*Solanum tuberosum* L.), tomato (*Lycopersicon esculentum* Mill.), and rot of potato tubers, was first observed as a serious enemy of the potato about 1845. After being studied by various investigators it was first fully described by de Bary in 1876. For many decades oogone formation was unknown in this species until Clinton in 1911 reported their production in culture on oat agar. In 1927 Murphy reported finding them on the surface of tubers and in the surrounding soil. As in some other species these oospores were mostly parthenogenetic in origin although a basal antherid was observed in one case. In *P. phaseoli* Thaxter, the oogones were shown by Clinton (1906) to be produced in the seeds while the conidiophores covered the surface of the pods of the lima bean (*Phaseolus limensis* Macf.). *P. cactorum* (L. & C.) Schröt. and some other species are troublesome rot-producing and damping-off fungi of many kinds of cultivated plants.

The border line between some of the root-inhabiting species of *Phytophthora* and some of the conidium-producing species of *Pythium* is so vague that a few mycologists, e.g., Fitzpatrick (1923), have suggested uniting the two genera, which would then have to take the older name *Pythium*. However the two genera may still be kept separate on the basis of the method of zoospore formation: within the zoosporangium (conidium) in *Phytophthora* and in an external vesicle in *Pythium*. In the rare cases where a vesicle is formed in the former the zoospores are produced within the zoosporangium before passing into the vesicle. Thomas (1942, 1943) made a chemical investigation of the composition of the cell walls of these two genera. In both of these the major part of the wall consists of cellulose with an outer deposit of fatty substance. Outside of this there is a layer of some pectic compound in the case of *Pythium* and of some other carbohydrate in *Phytophthora*. After the carbohydrates and fatty substances are all dissolved the hyphae still retain their form and this residual matter proved to be chitin. This is particularly interesting in view of the conclusions of von Wettstein (1921) that cellulose and chitin

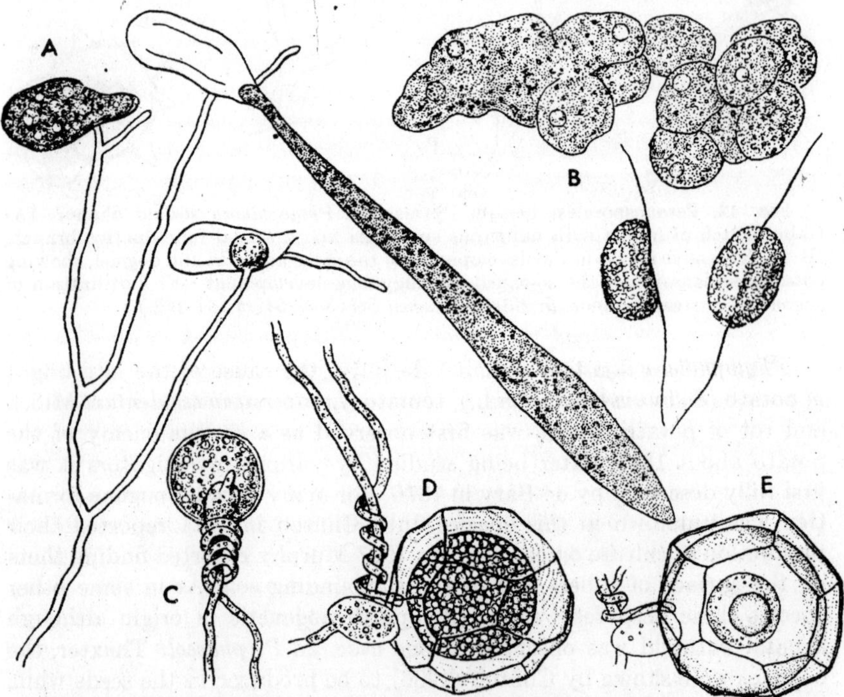

FIG. 43. Peronosporales, Family Pythiaceae. *Pythiogeton transversum* Minden. (A) Portion of plant with a zoosporangium not quite mature, a zoosporangium discharging its contents, and an empty zoosporangium. (B) Portion of the discharged contents separating into zoospores. (C) Young oogone and antherid. (D) Oogone and antherid after fertilization. (E) Mature oospore. (After von Minden: *Mykologische Untersuchungen und Berichte*, 1(2):146–255.)

are mutually exclusive. Schröter (1897), recognizing the similarity to the Saprolegniales, placed *Pythium* in that group, but retained *Phytophthora* in the Peronosporales. This cannot be upheld in view of the closeness of the two genera, but merely shows the difficulty of drawing sharp delimiting lines in some cases. A number of other genera have been described in this family, some being soil-inhabiting or aquatic saprophytes and others being serious parasites of various economic plants. By some mycologists they are merged with the genus *Phytophthora* and by others are maintained as distinct genera.

Pythiogeton, described by von Minden (1916), is an aquatic saprophyte whose elongated or sac-like zoosporangium has its axis more or less transverse to the hypha on which it is borne terminally or in an intercalary position. The zoospores are expelled in a mass which seems to lack the tough plasma membrane characteristic of the vesicle of *Pythium* and soon breaks up into the individual zoospores. Where known the oospores have very thick walls. (Fig. 43.)

Other genera assigned to this family are *Diasporangium* (Höhnk, 1936), a soil-inhabiting parasite, *Trachysphaera* (Tabor and Bunting, 1923), etc. *Pythiomorpha* was described by Petersen (1910) and considered by him to be worthy of the establishment of a separate family. Several species have been ascribed to this genus but recent studies by Blackwell, Waterhouse, and Thompson (1941) seem to indicate that these all represent various species of *Phytophthora* growing in water.

Family Albuginaceae. This family consists of the single genus *Albugo* or, as it is often called, *Cystopus*. In view of the fact that the former name was given in 1821 and the latter in 1847 the latter must be abandoned. The species number about twenty-five. In contrast with the members of the foregoing family the species are strictly parasitic, never occurring as saprophytes. They do not lend themselves to cultivation on culture media. Within some species—e.g., *A. candida* (Pers.) Kuntze—there are numerous specialized races that are adapted only to certain host species or groups of species. The mycelium is strictly intercellular except for the small globular haustoria which are borne on the ends of short, very slender processes which pierce the host cell wall. Melhus (1915) has shown that the mycelium of *A. candida* may overwinter in the tissues at or below the crown of such host plants as are winter annuals or biennials, growing out into the new shoots in the spring. The conidiophores are formed on ends of short sympodially branching hyphae which arise from a mass of mycelium gathered in a limited area underneath the epidermis of the host (the so-called sorus). They are club-shaped and stand, closely packed together, perpendicular to the surface of the epidermis, between it and the subepidermal cells. From the apex of each conidiophore are abstricted successively the spherical or ovoid pluri-

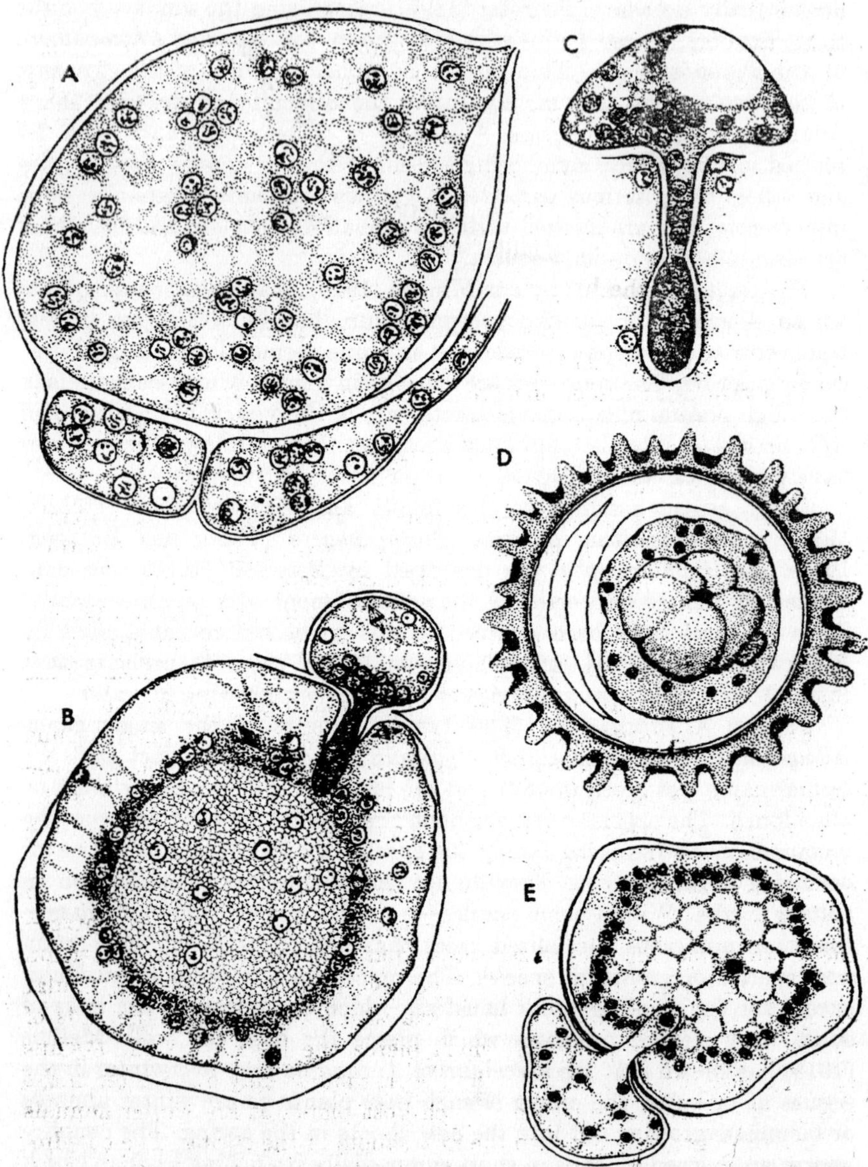

FIG. 44. Peronosporales, Family Albuginaceae. (A–D) *Albugo bliti* (Biv.-Bern.) Kuntze. (A) Oogone and antherid before delimitation of the egg (oösphere). (B) Multinucleate egg about to be fertilized, surrounded by periplasm containing supernumerary nuclei. (C) Antherid and multinucleate conjugation tube. (D) Section of mature oospore, showing numerous nuclei. (E) *Albugo portulacae* (DC.) Kuntze, egg and conjugation tube each with a single functional nucleus. Supernumerary nuclei remaining in antherid and in the periplasm. (A–D, after Stevens: *Botan. Gaz.*, **28**(3):149–176, Univ. Chicago Press. E, after Berlese: Icones Fungorum, Padua.)

nucleate conidia which are separated from each other by slender connections, the disjunctors, whose dissolution permits the conidia to fall apart. The chains of conidia thus formed raise and eventually rupture the overlying epidermis, permitting the conidia to escape and to be distributed by air currents. The similarity of these sori, except for the color, to those of Rusts led to the name "White Rust" often applied to fungi of this genus. Upon falling into water the conidia divide internally into several uninucleate, biflagellate, kidney-shaped zoospores which escape by dissolution of a special spot in the conidial wall. After swimming for a short time these zoospores encyst and germinate by a germ tube. In some species Palm (1932) has shown that the conidia germinate usually by the production of a stout germ tube which infects the host without producing zoospores. Sexual reproduction takes place in the tissues of the host. Often the portions of the host plant in which this occurs are much hypertrophied. This is especially the case with *Albugo candida* in which the inflorescence and individual flowers of the host may be much thickened and enlarged. The distorted flowers remain green and are sometimes several times as large as the normal flowers. On the ends of hyphal branches the almost spherical oogones are separated from the hyphae by septa. Stevens (1899, 1901) studied the process of fertilization in several species. The oogone is at first multinucleate, the number of nuclei being as high as 300. These may all pass to the periplasm leaving but a single egg nucleus in the egg or after passing to the periplasm they may divide, half of the daughter nuclei remaining in the periplasm and the other half in the egg so that eventually the latter may contain 100 or more nuclei. The multinucleate antherid on the end of a hyphal branch attaches itself to the oogone and eventually sends into the egg a conjugation tube through which one male nucleus passes, in the first case mentioned above, or 100 or more in the second case. These male and female nuclei fuse by pairs. The fertilized egg produces a thick wall, consisting of a thin endospore and a thick roughened epispore. In the first type of fertilization the zygote nucleus divides repeatedly so that the oospore overwinters as a multinucleate structure. In the spring zoospores are formed and the epispore is ruptured, the endospore pushing out through the break as a bladder which in its turn ruptures and permits the zoospores to escape. Just where meiosis occurs is not yet certain. It has been suggested that the nuclear divisions occuring in the antherid and oogone before fertilization represent this process, or it may occur in the first nuclear division in the fertilized egg. In North America the common species are *A. candida* (Pers.) Kuntze, on various crucifers (Brassicaceae); *A. portulacae* (DC.) Kuntze, on purslane (*Portulaca oleracea* L.); *A. bliti* (Biv.-Bern.) Kuntze, on various species of *Amaranthus*; *A. tragopogonis* (DC.) S. F. Gray, on salsify (*Tragopogon porrifolius* L.) and

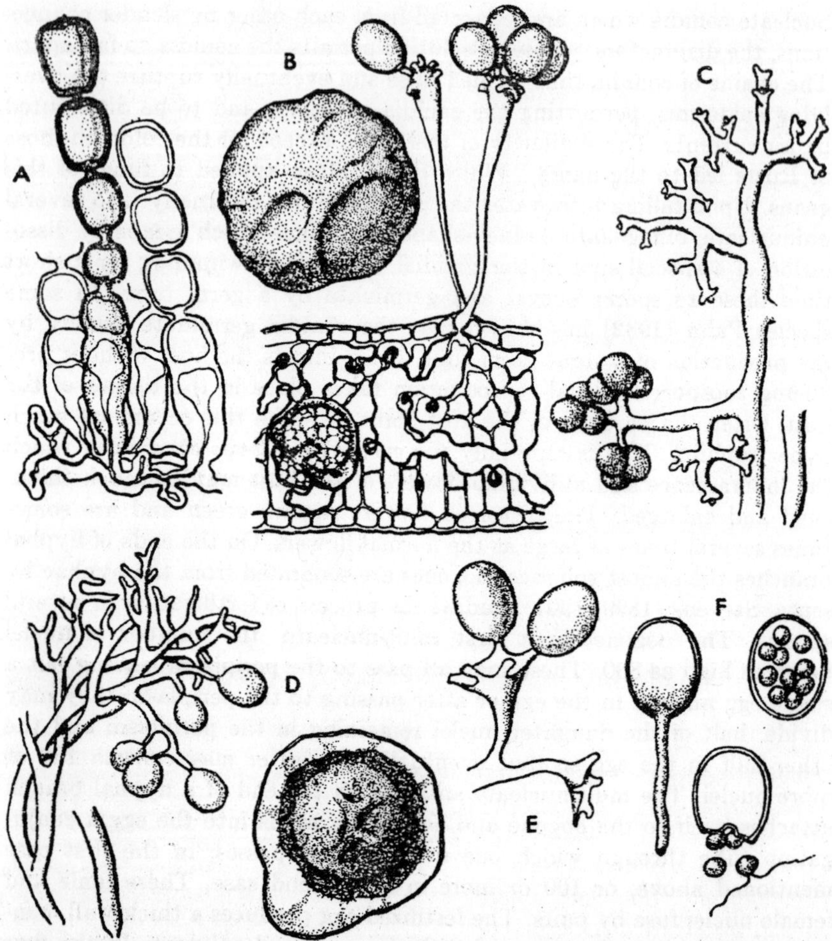

FIG. 45. Peronosporales. (A) Family Albuginaceae. *Albugo portulacae* (DC.) Kuntze. Conidiophores and conidia. (B–F) Family Peronosporaceae. (B) *Basidiophora entospora* Roze & Cornu, conidiophores and mature oospore. (C) *Rhysotheca australis* (Speg.) Wilson, conidiophore. (D) *Peronospora ficariae* Tul., conidiophore and oogone with oospore. (E–F) *Bremia lactucae* Regel. (E) Tip of conidiophore branch. (F) Conidia showing germination by germ tube and by zoospores. (A–E, after Berlese: Icones Fungorum, Padua. F, after Milbrath: *J. Agr. Research,* **23**(12):989–994.)

other composites; and *A. ipomoeae-panduranae* (Schwein.) Swing., on the sweet potato (*Ipomoea batatas* (L.) Lam.) and related plants. In most cases the disease caused is of minor importance. (Figs. 44 and 45 A.)

FAMILY PERONOSPORACEAE. The Peronosporaceae, like the Albuginaceae, differ from the Pythiaceae in their strictly parasitic habit, the mycelium always being intercellular with haustoria penetrating the ad-

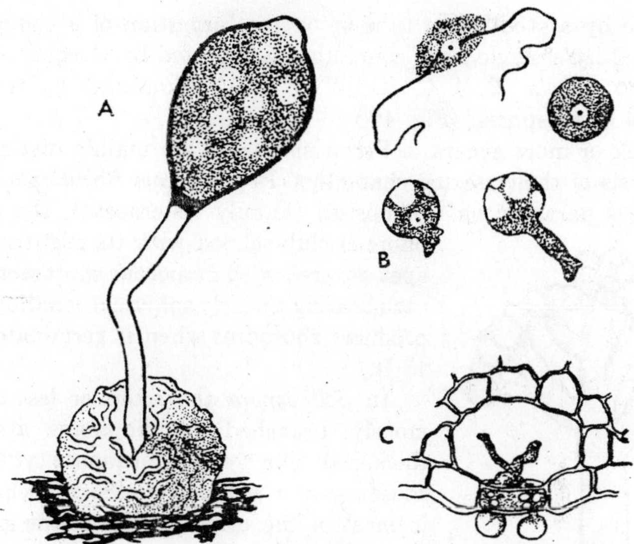

FIG. 46. Peronosporales, Family Peronosporaceae. *Rhysotheca viticola* (B. & C.) G. W. Wilson. (A) Germination of oospore to form one large conidium. (B) Zoospores with flagella, and encysted and germinating. (C) Infection of host tissue through a stoma. (After Gregory: *Phytopathology*, **2**(6):235–249.)

jacent host cells. Like those of the Albuginaceae the haustoria of this family may be knob-like but in many species they are filamentous or finger-like. The conidiophores are external to the host and produce the conidia singly on the ends of the branches. These conidia are plurinucleate and germinate in most cases by the formation of zoospores, as in *Albugo*. In *Peronospora* and *Bremia* the typical mode of germination is by means of a stout germ tube without the formation of zoospores. In *Rhysotheca* (included in *Plasmopara* by many authors) the conidia germinate by the formation of zoospores but in *Plasmopara* (in the narrower sense) the whole protoplasmic contents of the conidium escape as a naked plurinucleate but nonflagellate mass which quickly rounds up and encysts and then germinates by a germ tube. This difference in the mode of germination is the basis for Wilson's (1907) division of *Plasmopara* into the two genera. In nearly all genera in which germination by means of zoospores is typical the conidia may, under special conditions, germinate directly by germ tubes. A. de Bary (1876), Melhus (1915), Jones and Torrie (1946), have shown that as in *Albugo candida*, so also in this family, particularly in the genus *Peronospora*, the mycelium can live over winter in the tissues of a biennial, winter annual, or perennial host and thus infect the new plants in the spring without the aid of conidia or oospores. Sexual reproduction is like that in those species of *Albugo* in which the mature oogone contains a uninucleate egg. Mostly the oospores germinate by a stout germ tube or by the formation of a conidiophore terminated by a single large conidium (as reported by Gregory, 1912, in *Rhysotheca viticola* (B. & C.) G. W. Wilson), sometimes by the direct formation of zoospores. (Fig. 46.)

The six or more genera of Peronosporaceae are mainly distinguished on the basis of their asexual characters. In the genus *Basidiophora*, with two species parasitic on Composites (Family Asteraceae), the conidiophore is club-shaped with its slightly swollen apex covered with numerous short sterigmata, each bearing a nearly spherical conidium which produces zoospores when it germinates. (Fig. 45 B.)

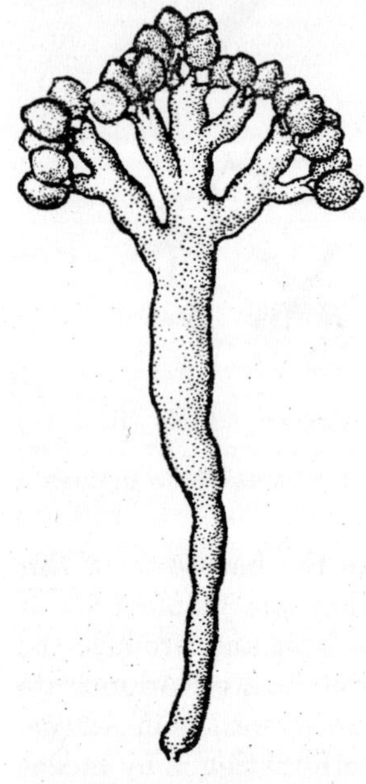

Fig. 47. Peronosporales, Family Peronosporaceae. *Sclerospora graminicola* (Sacc.) Schröt., conidiophore. (After Weston: *J. Agr. Research*, **27**(10):771–784.)

In *Sclerospora* the more or less dichotomously branched conidiophores are much thickened. The conidia germinate typically by zoospores in *S. graminicola* (Sacc.) Schröt., but in most of the other species of the genus by germ tubes. The oogone wall may remain thin and lie closely against, but not grown fast to, the thick-walled oospore (in *S. graminicola* (Sacc.) Schröt., according to McDonough, 1937) or it may become thickened and somewhat folded, separate from the oospore wall which has a thin wrinkled outer layer and a thick inner layer (in *S. macrospora* (Sacc.) McDonough, according to McDonough, 1947). The thirteen or more species are chiefly parasites of grasses (Family Poaceae). In the East Indies they cause serious injury to sugar cane (*Saccharum officinarum* L.) and to Indian corn or maize (*Zea mays* L.). *Sclerospora graminicola* is found frequently throughout the north temperate regions on foxtail grasses (*Setaria*). The conidiophores form a downy layer on the under side of the infected leaves early in the morning but quickly dry down as the air becomes warmer and drier. After the oospores develop the leaves die and shred longitudinally into thread-like strips on which the oospores may be seen readily by the aid of a hand lens. Weston (1920, 1921, 1923, 1924) has given various species of this genus very careful study. (Fig. 47.)

Plasmopara and *Rhysotheca*, usually united under the former name, produce slender, much branched conidiophores whose branches arise nearly at right angles. The tips of the branches are truncate. The two genera differ, as mentioned above, by the mode of germination of the conidia: by zoospores in *Rhysotheca* and by a single naked mass in *Plasmopara*. The most important species from the economic standpoint is *R. viticola* (B. & C.) G. W. Wilson, which causes the downy mildew and brown rot of the foliage and fruit, respectively, of grape (various species of *Vitis*).

Pseudoperonospora (called *Peronoplasmopara* by some although the former name has priority) has slender conidiophores branching at acute angles and with pointed tips. The usually violet-tinged conidia germinate by zoospores. *Ps. cubensis* (B. & C.) Rostow., first described from Cuba, is probably native to Russia where its destructive effects have been known for many years although the fungus was first recognized there in 1903 by Rostowzew. It is a very serious enemy of the cucumber (*Cucumis sativus* L.) and muskmelon (*C. melo* L.). Another species, *Ps. celtidis* (Waite) G. W. Wilson, attacks the hackberry (*Celtis*) while other species are found on hemp (*Cannabis*), on hops (*Humulus*), and on nettle (*Urtica*). Hoerner (1940) has shown that the species on hops (*Ps. humuli* Miyabe) is capable of infecting *Celtis*, *Cannabis*, and *Urtica* and suggests that there is one species infecting Urticaceae, perhaps with various physiological strains.

Bremia and *Peronospora* have slender conidiophores, branching at acute angles in a more or less dichotomous manner. Their conidia germinate typically by germ tubes, although zoospore production also has been reported for *Bremia* by Milbrath (1923) and for *P. spinaciae* (Grew.) Laub. by Eriksson (1919). Schultz (1937) could not observe zoospore production in the strains of *B. lactucae* Regel growing in Germany. In *Bremia* the tips of the branches enlarge into disk-like structures bearing sterigmata on their edges. *B. lactucae* is sometimes destructive to lettuce grown under glass. In *Peronospora* the tips of the branches taper to a point. Gäumann (1923) recognizes 268 species of this genus. They are of economic importance in but few cases. *P. spinaciae* is sometimes destructive in plantings of spinach (*Spinacia oleracea* L.) and *P. parasitica* (Pers.) de Bary on various crucifers (Brassicaceae). (Fig. 45 D–E.)

Gäumann (1918a, b), Wartenweiler (1918) and others have shown that the earlier recognized species of this family are separable by biometric and cultural means into large numbers of closely related species confined to very limited numbers of host species and differing constantly, but only slightly, in the size and shape of the conidia and conidiophores. When such studies have been extended to all parts of the world and to all the forms occurring on different host species the number of species of Peronosporaceae will doubtless be very greatly increased.

Several other genera have been described which may be found to be justified. The physiology of conidial germination must be studied carefully in these as well as in the older genera. Only after such studies can we be certain that some of the generic distinctions now maintained, or recently proposed, are really valuable. Should zoospore formation in *Peronospora* be confirmed it would seriously weaken the distinction between that genus and *Pseudoperonospora*.

The evolutionary tendencies within the Peronosporales are of interest to students of phylogeny. The species of *Pythium* with long, narrow, hypha-like zoosporangia would undoubtedly be included in the Saprolegniales were it not for those other species of *Pythium* with ovoid or spherical zoosporangia which may even function as separable conidia, thus forming a transition to *Phytophthora*. This genus still shows in some species a close relationship to *Pythium*, in the production of submerged zoosporangia and facultative saprophytic habits, while in other species with well-developed conidiophores and strictly parasitic habits the genus approaches closely the Peronosporaceae. The fact that the single egg in the oogone is surrounded by periplasm, as in the Rhipidiaceae in the Saprolegniales, would suggest that the relationship of the Pythiaceae is closer to this family than to the Saprolegniaceae in which there is no periplasm and the majority of species have numerous eggs in the oogone. It has been suggested by some mycologists that from the Pythiaceae have been derived on the one hand the Saprolegniales (through the Rhipidiaceae) and on the other hand the remainder of the Peronosporales. As obligate parasitism became prevalent in the Peronosporales evolution appears to have proceeded in several lines. The catenulate conidia of *Albugo* call to mind the proliferating zoosporangia of some species of *Pythium* as well as of *Saprolegnia*, while the sympodial conidiophores of the more advanced species of *Phytophthora* remind one of the sympodial branching in *Achlya* and some species of *Pythium*. The monopodial conidiophores of the Peronosporaceae do not resemble so closely any structures in *Pythium*. As parasitism has progressed we also find the transition from attached zoosporangia to separable zoosporangia (conidia) leading finally to the conidium as found in *Peronospora*, in which zoospore formation has been lost, although the plurinucleate condition persists.

The fungi that have been considered in the preceding chapters of this book have been largely aquatic in habit or reveal their aquatic ancestry by producing naked flagellate cells in some stages of their development, although in some genera these have been suppressed. This has been true even for the majority of the strictly parasitic species of the Peronosporales which have abandoned the aquatic habit to assume that of parasitism in land plants. Besides the production of zoospores the great majority of the foregoing organisms, except most of the Chytridiales series, show the cellulose reaction promptly upon the application of chloriodide of zinc solution. Many of those which fail to show this reaction promptly do so when certain masking substances are removed. The presence of true chitin is demonstrated for only a minority of the species. The forms with well-developed hyphae show pronounced anisogamy in most cases. For the latter reason the Monoblepharidales, Lagenidiales, Saprolegniales, and Peronosporales were included in the subclass Oomyceteae of the Class Phycomyceteae in the first edition of this book, following the practice of Fitzpatrick (1930) and other mycologists. These were set apart from the subclass Zygomyceteae, in which approximate isogamy was held to warrant such a distinction, accompanied as it was by loss of the power to produce motile cells and by the greater predominance of chitin in the cell walls. The fact that the simpler forms included in the Oomyceteae, such as Chytridiales, and some of the Blastocladiales and Lagenidiales, show isogamy and that many forms whose closest kinship seems to lie with the Mucorales and Entomophthorales have pronounced anisogamy makes it doubtful whether this character should be used to distinguish subclasses. Hence these two group names have been abandoned in this edition.

Order Protomycetales. As a very doubtful appendix to the Phycomyceteae so far considered must be added the Order Protomycetales, with a single family Protomycetaceae. The true position of this group among the fungi has long been the subject of speculation. The author follows Fitzpatrick (1930) in placing these fungi among the Phycomyceteae but with very little idea as to what groups of that class may have given rise to them. The most recent and extensive investigations on the group are those of Sappin-Trouffy (1897) and of von Büren (1915, 1922). The family seems to have no affinity to the Ascomyceteae or to the other Higher Fungi, though it has been assigned to various positions among these. It consists of one well-established genus, *Protomyces*, with 12 or more species to which 3 other genera have been added, *Taphridium*, *Volkartia*, and *Protomycopsis*. Until life history studies are more complete, the validity of their segregation is uncertain. All the well-known species of these 4 genera are parasitic in the stems, leaves, or fruits of Ammiaceae (Umbelliferae), Asteraceae (Compositae), and Cichoriaceae.

A number of species have been described from various other families but need further study before their validity is assured.

The mycelium is subepidermal or intercellular in the underlying tissues. It is septate at occasional intervals, each segment being plurinucleate. The cell walls give a strong cellulose reaction with chloriodide of zinc. Some of the segments of the mycelium enlarge and become multinucleate (30–40 nuclei, in *Protomyces inundatus* Dang.). Within the original wall the cell enlarges and becomes surrounded by a thick, three-layered cellulose wall, the nuclei dividing several times to become 100–200 in number. These resting sporangia (or "chlamydospores" as some authors call them) may remain in the tissues over winter or may germinate the same season. In some species they are subepidermal and in

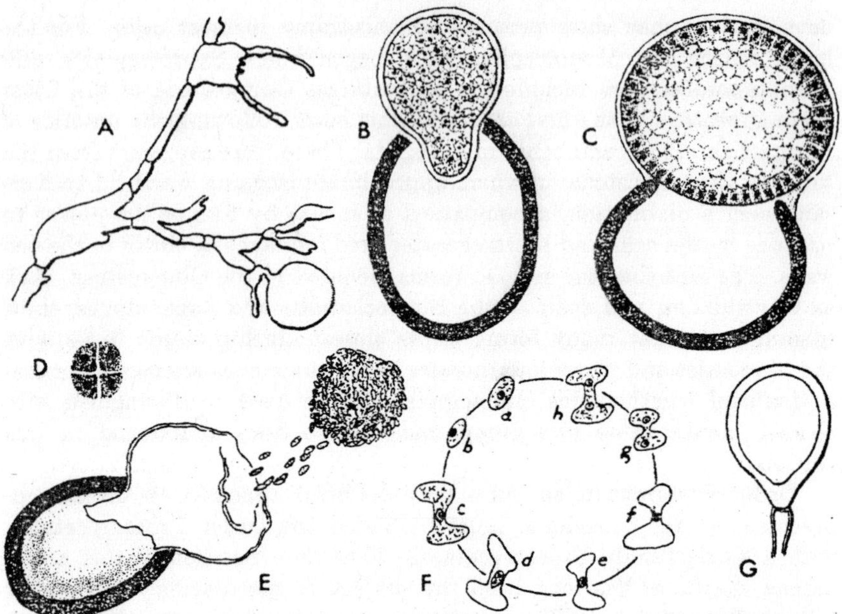

Fig. 48. Protomycetales, Family Protomycetaceae. (A–F) *Protomyces macrosporus* Unger. (A) Young mycelium with intercalary young resting sporangia. (B) Germination of resting sporangium. (C) Protoplasm forming parietal layer of spore mother cells. (D) Spore mother cell divided into spores. (E) Discharge of ball of spores. (F) Successive stages in fusion of spores, showing union of nuclei in the conjugation tube and division of the zygote nucleus. (G) *Protomycopsis leucanthemi* Magn., terminally formed resting sporangium. (A–E, G, after von Büren: *Beitr. Kryptogamenflora Schweiz,* **5**(1):1–95. F, *ibid.,* **5**(3):1–94.)

others scattered at various depths. In germination the sporangia of *Taphridium* produce their spores without breaking the exospore, while those of the other genera burst the exospore on one side and the contents bulge out like a balloon, still surrounded by the inner wall. The nuclei in *Protomyces* all migrate to the periphery and a large central vacuole is formed. The thin layer of peripheral cytoplasm is divided by cleavage planes, starting at the outside, into little uninucleate cells which divide twice (meiosis?) forming four ellipsoidal spores out of each cell. These spores then mass at the center or apex of the sporangium and by breaking of the latter are thrown out along with the slimy contents of the vacuole. In some species, apparently not in others, the spores fuse by twos after being set free, sometimes before. The nuclei unite in the conjugation tube. The spores in culture media germinate to form yeast-like cells but when inoculated on the proper host produce endophytic mycelium. (Fig. 48.)

The affinities of this family are exceedingly uncertain. The final division of the nuclei in the sporangium may be a reduction division. The cells that divide into four spores each have been called asci and the whole sporangium a "synascus" but the absence of nuclear fusion in these "asci" before the formation of the spores seems to exclude that possibility. The cellulose nature of the cell wall, it has been suggested, speaks for the phycomycetous relationship. If related to the Lagenidiaceae we must assume the loss of motility of zoospores or gametes. Relationship to the Chytridiales is less likely because of the presence of well-developed mycelium with cellulose walls. Baker, Mrak, and Smith (1943) suggest that *Coccidioides immitis* Rix. and Gil., the fungus causing the disease of man called coccidioidomycosis may possibly belong in the Protomycetales.

Keys to the Families and More Important Genera of Peronosporales

Key to the More Important Genera of Family Pythiaceae

Zoospores not preformed in the zoosporangium but developing in an extruded mass which may or may not be enclosed in a plasma membrane.
  Extruded mass of protoplasm not in a definite vesicle, zoosporangia elongated transversely to the supporting hypha. *Pythiogeton*
  Zoospores developed in a definite vesicle.
    Hyphae bearing short lateral branches adapted to the capture of the rotifers upon which the fungus feeds. Zoosporangia hyphal. *Zoophagus*
    No special hyphal branches for capturing prey. Zoosporangia hyphal or toruloid or more or less spherical. *Pythium*
Zoospores preformed in the mostly more or less spherical zoosporangium (conidium) and escaping individually or sometimes in a temporary vesicle. *Phytophthora*
No zoospores. External warted spherical conidia. Oogones with amphigynous type of antherid. Of doubtful relationship. *Trachysphaera*

Key to the Genus of Family Albuginaceae

  Only genus. Conidia catenulate on subepidermal conidiophores. *Albugo*

Key to the More Important Genera of Family Peronosporaceae

Conidiophores clavate or somewhat cylindrical, somewhat swollen above with numerous short sterigma-like branches. Conidia without appendages, germinating by formation of zoospores. Parasitic on Composites. *Basidiophora*
Conidiophores not much branched. Conidia with prominent beak and a basal appendage consisting of the adhering upper end of the conidiophore. Conidia germinating by production of zoospores. Possibly belonging to the Pythiaceae. *Kawakamia*
Conidiophores with prominent branches.
  Conidiophores stout, with heavy branches clustered near the apex, quickly fugacious. Conidia germinating by germ tubes or by the formation of zoospores. Parasitic on grasses. *Sclerospora*
  Conidiophores more slender, branching monopodially, usually nearly at right angles. Tips of branches obtuse.

Conidia germinating by the formation of a single large naked nonflagellate plasmatic mass which encysts and then germinates by a single germ tube. *Plasmopora*

Conidia germinating by the production of biflagellate zoospores. *Rhysotheca*

Conidiophores slender, more or less dichotomously branched, mostly at acute angles. Conidia usually germinating by single germ tube, or by zoospores.

Tips of branches enlarged into disks with sterigmata at the margin. Parasitic upon Cichoriaceae. Zoospores rare. *Bremia*

Tips of branches swollen, but not disk-like. Parasitic on *Viola*. Zoospores rare or none produced. *Bremiella*

Tips of branches acute, conidia germinating by a germ tube. *Peronospora*

Tips of branches acute, conidia germinating by zoospores. *Pseudoperonospora*

Literature Cited

BAKER, E. E., E. M. MRAK, AND C. E. SMITH: The morphology, taxonomy, and distribution of Coccidioides immitis Rixford and Gilchrist 1896, *Farlowia*, **1**(2):199–244. *Pls*. 1–8. 1943.

DE BARY, ANTON: Researches into the nature of the potato fungus, Phytophthora infestans, *J. Roy. Agr. Soc. Engl.*, 2d ser., **12**(1):239–269. 8 *figs*. 1876.

BERLESE, AUGUSTO NAPOLEONE: Icones fungorum ad usum Sylloges Saccardianae accomodatae Phycomycetes: I. Peronosporacee, pp. 1–40. *Pls*. 1–67. Padua, published by the author, 1898.

BLACKWELL, ELIZABETH M.: The life history of Phytophthora cactorum (Leb. & Cohn) Schroet., *Brit. Mycol. Soc. Trans.*, **26**(1–2):71–89. *Figs*. 1–7. 1943.

———, G. M. WATERHOUSE, AND M. V. THOMPSON: The invalidity of the genus Pythiomorpha, *ibid.*, **25**(2):148–165. 2 *figs*. 1941.

VON BÜREN, GÜNTHER: Die schweizerischen Protomycetaceen mit besonderer Berücksichtigung ihrer Entwicklungsgeschichte und Biologie, ***Beiträge zur Kryptogamenflora der Schweiz***, **5**(1):1–95. *Pls*. 1–7. *Figs*. 1–28. 1915.

———: Weitere Untersuchungen über die Entwicklungsgeschichte und Biologie der Protomycetaceen, *ibid.*, **5**(3):1–94. *Pls*. 1–2. *Figs*. 1–27. 1922.

BUTLER, E. J.: An account of the genus Pythium and some Chytridiaceae, *Mem. Dept. Agr. India, Botan. Ser.*, **1**(5):1–160. *Pls*. 1–10. 1907.

———: The bud-rot of palms, *ibid.*, **3**(5):221–280. *Pls*. 1–5. 1910.

CLINTON, G. P.: Downy mildew, Phytophthora phaseoli Thaxt. of lima bean, *Connecticut Agr. Expt. Sta. Ann. Rept.*, **1905**:278–303. *Pls*. 20–22. 1906.

———: Oospores of potato blight, *Connecticut Agr. Expt. Sta. Bienn. Rept.*, **1909–1910**:753–774. *Pls*. 38–40. 1911.

COOPER, D. C., AND C. L. PORTER: Phytophthora blight of peony, *Phytopathology*, **18**(11):881–899. *Pl*. 29. *Figs*. 1–5. 1928.

CORNU, MAXIME: Monographie des Saprolégniales. Étude physiologique et systématique, *Ann. sci. nat. Botan.*, 5me sér., **15**:1–198. *Pls*. 1–7. 1872.

DRECHSLER, CHARLES: Germination of oospores of Pythium Butleri and Pythium tardicrescens, *Phytopathology*, **37**(6):438–439. 1947.

EDSON, H. A.: Rheosporangium aphanidermatus, a new genus and species of fungus parasitic on sugar beets and radishes, *J. Agr. Research*, **4**(4):279–292. *Pls*. 44–48. 1915.

ERIKSSON, JAKOB: Zur Entwickelungsgeschichte des Spinatschimmels (Peronospora Spinaciae (Grew.) Laub.), *Arkiv för Botanik*, **15**(15):1–25. *Pls.* 1–4. *Figs.* 1–3. 1919.

Fischer, Alfred: Phycomyceten: Pythium, in L. Rabenhorst: Kryptogamen-Flora von Deutschland, Oesterreich und der Schweiz, Zweite Auflage, Band 1, Abt. 4, pp. 393–410. *Figs.* 64–65. Leipzig, Verlag von Eduard Kummer, 1892.

Fitzpatrick, Harry Morton: Generic concepts in the Pythiaceae and Blastocladiaceae, *Mycologia*, **15**:166–173. 1923.

———: The Lower Fungi—Phycomycetes, xi + 331 pp. 112 *figs.* New York, McGraw-Hill Book Co., 1930.

Gäumann, Ernst: Über die Spezialisation der Peronospora auf einigen Scrophulariaceen, *Ann. Mycol.*, **16**:189–199. *Figs.* 1–6. 1918a.

———: Ueber die Formen der Peronospora parasitica (Pers.) Fries. Ein Beitrag zur Speziesfrage bei den parasitischen Pilzen, *Botan. Centr. Beihefte, Erste Abt.*, **35**:1–143, 395–533. *Figs.* 1–47. 1918b.

———: Beiträge zu einer Monographie der Gattung Peronospora Corda, *Beiträge zur Kryptogamenflora der Schweiz*, **5**(4):1–360. *Figs.* 1–166. 1923.

Gregory, C. T.: Spore germination and infection with Plasmopara viticola, *Phytopathology*, **2**(6):235–249. *Figs.* 1–7. 1912.

Hiura, M.: A simple method for the germination of oospores of Sclerospora graminicola, *Science*, N.S., **72**:95. 1930.

Hoerner, G. R.: The infection capabilities of hop downy mildew, *J. Agr. Research*, **61**(5):331–334. *Fig.* 1. 1940.

Höhnk, Willy: On three Pythiaceous Oomycetes, *Botan. Centr. Beihefte, Abt. A*, **55**(1):89–99. *Figs.* 1–4. 1936.

Jones, Fred R., and J. H. Torrie: Systemic infection of downy mildew in soybean and alfalfa, *Phytopathology*, **36**(12):1057–1059. *Fig.* 1. 1946.

Klebahn, Heinrich: Krankheiten des Flieders, 75 pp. 45 *figs.* Berlin, Gebrüder Borntraeger, 1909.

Matthews, Velma Dare: Studies on the Genus Pythium, 136 pp. 29 *pls.* Chapel Hill, Univ. North Carolina Press, 1931.

McDonough, E. S.: The nuclear history of Sclerospora graminicola, *Mycologia*, **29**(2):151–172. *Figs.* 1–2. 1937.

———: A cytological study of the development of the oospore of Sclerospora macrospora (Sacc.), *Trans. Wisconsin Acad. Sci.*, **38**:211–218. *Figs.* 1–8. 1946 (1947).

Melhus, I. E.: Perennial mycelium in species of Peronosporaceae related to Phytophthora infestans, *J. Agr. Research*, **5**(2):59–70. *Pl.* 3. *Fig.* 1. 1915.

Middleton, John T.: The taxonomy, host range and geographic distribution of the genus Pythium, *Mem. Torrey Botan. Club*, **20**(1):1–171. *Figs.* 1–17. 1943.

Milbrath, D. G.: Downy mildew on lettuce in California, *J. Agr. Research*, **23**(12):989–994. *Pls.* 1–3. 1923.

von Minden, M.: Beiträge zur Biologie und Systematik einheimischer submerser Phycomyceten, *Mykologische Untersuchungen und Berichte*, **1**(2):146–255. *Pls.* 1–8. *Figs.* 1–26. 1916.

Mundkur, Belaji D.: Morphology and cytology of development of the sex organs of Phytophthora himalayensis Dastur, *Botan. Gaz.*, **110**(3):475–486. *Figs.* 1–28. 1949.

Murphy, Paul A.: The production of the resting spore of Phytophthora infestans on potato tubers, *Sci. Proc. Roy. Soc. Dublin*, **18**(34):407–412. *Pl.* 18. 1927.

Narasinhan, M. J.: Studies in the genus Phytophthora in Mysore: I. Heterothallic strains of Phytophthora, *Phytopathology*, **20**(3):201–214. *Figs.* 1–5. 1930.

PALM, B. T.: Biological notes on Albugo, *Ann. Mycol.*, **30**(5–6):421–426. *Figs.* 1–3. 1932.

PATTERSON, P. M.: Oogenesis in Pythium torulosum, *J. Elisha Mitchell Sci. Soc.*, **43**:124–128. *Pl.* 11. 1927.

PETERSEN, HENNING E.: An account of Danish freshwater Phycomycetes, with biological and systematical remarks, *Ann. Mycol.*, **8**(5):494–560. *Figs.* 1–27. 1910.

PETHYBRIDGE, G. H.: On the rotting of potato tubers by a new species of Phytophthora having a mode of sexual reproduction hitherto undescribed, *Sci. Proc. Roy. Soc. Dublin*, **13**:529–565. *Pls.* 42–44. 1913.

POITRAS, ADRIAN W.: A new aquatic species of Pythium, *Mycologia*, **42**(2):171–176. *Figs.* 1–17. 1949.

PRINGSHEIM, N.: Beiträge zur Morphologie und Systematik der Algen: II. Die Saprolegnieen, *Jahrb. wiss. Botan.*, **1**:284–306. *Pls.* 19–21. 1858.

ROSTOWZEW, S. J.: Beiträge zur Kenntnis der Peronosporeen, *Flora*, **92**:405–430 *Pls.* 11–13. 1903.

SAPPIN-TROUFFY, P.: Note sur la place du Protomyces macrosporus Unger dans la classification, *Le Botaniste*, **5**:285–288. *Fig.* 1. 1897.

SCHRÖTER, J.: Saprolegniineae, in A. ENGLER UND K. PRANTL: Die Natürlichen Pflanzenfamilien, Teil 1, Abt. 1, pp. 93–105, *Figs.* 76–89, 1897a; Peronosporineae, *ibid.*, pp. 108–119, *Figs.* 92–102, 1897b. Leipzig, Wilhelm Engelmann.

SCHULTZ, HELMUT: Zur Biologie der Bremia lactucae Regel, des Erregers des falschen Mehltaus des Salats, *Phytopath. Z.*, **10**(5):490–503. *Figs.* 1–10. 1937.

SHANOR, LELAND: Observations on the development of a new species of Phytophthora, *J. Elisha Mitchell Sci. Soc.*, **54**(1):154–162. *Pls.* 15–16. 1938.

SIDERIS, C. P.: Taxonomic studies in the Pythiaceae: I. Nematosporangium, *Mycologia*, **23**(4):252–295. *Figs.* 1–12. 1931; II. Pythium, *ibid.*, **24**(1):14–61. *Figs.* 1–21. 1932.

SPARROW JR., FREDERICK K.: Observations on Pythium dictyosporum, *Mycologia*, **23**(3):191–203. *Pl.* 20. *Text Fig.* 1. 1931.

STEVENS, F. L.: The compound oosphere of Albugo bliti, *Botan. Gaz.*, **28**(3):149–176; (4):224–245. *Pls.* 11–15. 1899.

———: Gametogenesis and fertilization in Albugo, *ibid.*, **32**(2):77–98; (3):159–169; (4):238–261. *Pls.* 1–4. 1 *text figure.* 1901.

SZYMANEK, JOSEPH: Contribution à l'étude du Phytophthora infestans, parasite de la pomme de terre, *Ann. épiphyt.*, **13**(4):213–282. 4 *pls.* 31 *figs.* 1927.

TABOR, R. J., AND R. H. BUNTING: On a disease of cocoa and coffee fruits caused by a fungus hitherto undescribed, *Ann. Botany*, **37**(145):153–157. *Figs.* 1–3. 1923.

THOMAS, R. C.: Composition of fungus hyphae: III. The Pythiaceae, *Ohio J. Sci.*, **42**:60–62. 1942; IV. Phytophthora, *ibid.*, **43**:135–138. 1943.

TROW, A. H.: Observations on the biology and cytology of Pythium ultimum n.sp., *Ann. Botany*, **15**(58):269–312. *Pls.* 15–16. 1901.

TUCKER, C. M.: Taxonomy of the genus Phytophthora de Bary, *Univ. Missouri Agr. Expt. Sta. Research Bull.* **153**:1–208. *Figs.* 1–30. 1931.

WARTENWEILER, ALFRED: Beiträge zur Systematik und Biologie einiger Plasmopara-Arten, *Ann. Mycol.*, **16**:249–299. *Pls.* 1–3. *Figs.* 1–12. 1918.

WESTON JR., W. H.: Philippine downy mildew of maize, *J. Agr. Research*, **19**(3): 97–122. *Pls. A and B (colored) and* 16–25. 1920.

———: Another conidial Sclerospora of Philippine maize, *ibid.*, **20**(9):669–684. *Pls.* 76–79. 1921.

———: Production and dispersal of conidia in the Philippine Sclerosporas of maize, *ibid.*, **23**(4):239–278. *Pls.* 1–10. 1923.

———: Nocturnal production of conidia by Sclerospora graminicola, *ibid.*, **27**(10):771–784. *Pls.* 1–2. 1924.

VON WETTSTEIN, FRITZ: Das Vorkommen von Chitin und seine Verwertung als systematisch-phylogenetisches Merkmal im Pflanzenreich, *Sitz. ber. Akad. Wiss. Wien, Math. naturw. Klasse, Abt. I*, **130**(1):3–20. 1921.

WILSON, GUY WEST: Studies in North American Peronosporales: II. Phytophthoreae and Rhysotheceae, *Bull. Torrey Botan. Club*, **34**:387–416. 1907.

7

PHYCOMYCETEAE: MUCORALES, ENTOMOPHTHORALES, ZOOPAGALES, ECCRINALES

The first three orders that form the subjects of this chapter agree in the more or less coenocytic structure of the mycelium, in the union of equal or unequal gametangia to form "zygospores," and in the total absence of flagellate sexual or asexual cells. Also, chitin is present in most of these forms and the cell walls do not show the cellulose reaction upon the application of a solution of chloriodide of zinc.

These three orders may be distinguished as follows:

Mucorales: mycelium very extensive, nonseptate or septate in older aerial hyphae. Asexual reproduction typically by aplanospores formed in terminal sporangia. In a few genera these are reduced to indehiscent sporangioles which function as conidia. Sexual reproduction usually present.

Entomophthorales: mycelium not very extensive, at first coenocytic but sooner or later becoming septate and often falling apart into hyphal bodies. Asexual reproduction typically by conidia which are usually shot off with violence. Sexual reproduction frequently replaced by the parthenogenetic development of "azygospores."

Zoopagales: mycelia very slender, nonseptate or septate when older, attached to their hosts (aquatic or soil-inhabiting amoebae, nematodes, or insect larvae) by more or less complicated haustoria. Asexual reproduction by various types of conidia. Sexual reproduction by the union of gametangia to form zygospores of various shapes.

In addition to the foregoing are the Eccrinales which form a group of fungi whose relationship is very doubtful. They are placed here as a sort of appendix to the Phycomyceteae so that they may not escape attention.

Order Mucorales. The Mucorales are widely distributed fungi with a stout, well-developed, much branched coenocytic mycelium, very similar to that of some representatives of the Saprolegniales. In the older mycelium, especially in the aerial portions, septa may divide it into plurinucleate segments, but the young mycelium and that submerged in the substratum usually remain nonseptate. Ayers (1935) has shown that septation in *Dispira cornuta* van Tiegh begins in the germ tube and that numerous septa are formed in both young and old mycelium. The cell wall is reported by von Wettstein (1921) to contain chitin and pectose compounds and no cellulose, but Mangin (1899) reports true cellulose in young sporangia of some species and Hopkins (1929) found both cellulose and chitin in *Mucor rouxianus* (Calmette) Wehmer. It must be noted that Nabel (1929) could find no cellulose in this species nor in any other members of the order.

Asexual reproduction is typically by the formation of nonmotile, encysted spores (aplanospores) in sporangia terminal to the hyphae. These sporangia are formed in the same manner as in the Saprolegniales and Peronosporales, by the passage of a portion of the contents of the hypha into a terminal enlargement which is then cut off from the hypha by a septum. The multinuclear contents of the sporangium are divided by cleavage planes into naked, at first polyhedral, cells containing one or more nuclei each. These then round up and encyst and escape by the rupture or dissolution of the sporangium wall. They germinate by a stout germ tube. Swingle (1903) has shown that all of the protoplasm in the sporangium is used up in the formation of the spores. There are very interesting evolutionary modifications of this typical sporangium, leading to the development of structures comparable to conidia.

The greater part of the mycelium may be represented by that within the substratum, the aerial portion consisting of scarcely more than enough to give rise to the sporangia. On the other hand in many species the aerial mycelium may be very extensive, forming a large cottony mass from which arise, here and there, the sporangiophores. It is usually white but often the sporangiophores are dark colored. As it grows older this aerial mycelium may become septate but not truly cellular for the segments formed are plurinucleate coenocytes. A number of species of *Mucor* when growing in a medium rich in nutrients and of rather high osmotic pressure (e.g., a rather concentrated sugary medium) form a yeast-like growth instead of the normal filamentous mycelium.

The Mucorales are mainly saprophytic on vegetable matter, more rarely on animal matter, and are abundant in the soil and in plant debris. Many are coprophilous. Some are weak parasites on living plant tissues which are rich in stored food but not active, such as the roots of the sweet potato (*Ipomoea batatas* (L.) Lam.). A number of species are parasitic upon other fungi, even upon other Mucorales. A few have been described as parasites on animal tissues.

The probable course of sporangial evolution within the Mucorales can best be followed by studying the sporangia of a selected list of genera. The genera chosen may not represent direct lines of descent, since the

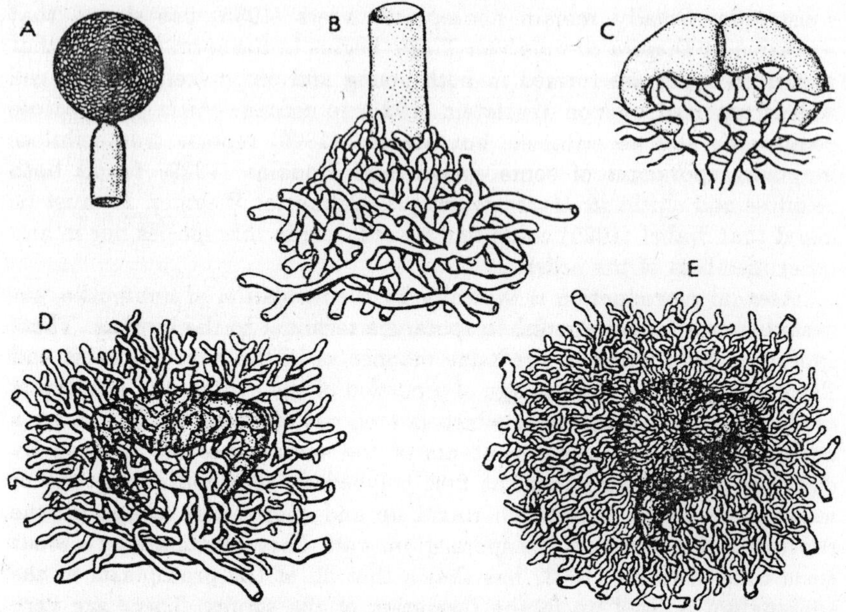

FIG. 49. Mucorales, Family Mortierellaceae. *Mortierella rostafinskii* Bref. (A) Sporangium. (B) Hyphal cluster at base of sporangiophore. (C–E) Stages in production of zygospore. (A–B, after Brefeld und von Tavel: Untersuchungen aus dem Gesammtgebiete der Mykologie, Heft 9, pp. 1–156, Münster i. W., Heinrich Schöningh. C–E, after Brefeld: Botanische Untersuchungen über Schimmelpilze, Heft 4, pt. 5, pp. 81–96, Leipzig, Arthur Felix.)

sexual reproduction is left unconsidered, but they probably indicate in general the directions that the modifications followed.

The simplest and probably the most primitive type of sporangium is that found in the genus *Mortierella*. It must not be understood that this genus is considered to be the most primitive of the Mucorales, for in its sexual reproduction it is so much modified from the more typical representatives of the order that it is clearly to be recognized as well advanced in evolution. However, in its sporangial development it seems to have retained a very primitive structure. This is merely one of the very many cases where evolution has advanced far along certain lines of development (in this case the manner of sexual reproduction) while remaining about at a standstill in its mode of asexual reproduction. The sporangium is a spherical enlargement of the apex of the sporangiophore, set off from the latter by a cross wall at the point where the enlargement begins. In sporangial evolution the next step appears to have been the development, as in *Mucor*, of a "columella." This in reality represents a displacement of the septum separating the sporangium from the sporangiophore so that it arches up into the former. The columella is laid down in the position it is to occupy and does not represent a cross wall that subsequently bulged up into the sporangium. This gives a much larger surface to the septum and permits a much freer transfer of food into the sporangium. Accompanying this more efficient food supply we find the sporangia to be larger, almost in proportion as the columella increases in size. In these two types of sporangium the numerous spores escape by the dissolution or breaking up of the relatively thin sporangial wall. The columella often remains firm and unchanged after the sporangium has ruptured and the spores have been set free. Schostakowitsch (1896) reported that in *Mucor proliferus* Schost. the columella forms a new sporangium by proliferation, reminding one of the condition in *Saprolegnia*. (Figs. 49A, 50A–C.)

The sporangium of *Pilobolus* represents a special modification of the foregoing type. In it the apical wall of the many-spored sporangium is very much thickened. The columella is rather small Below the sporangium the sporangiophore is enlarged into a subsporangial vesicle that may be two or three times the diameter of the somewhat flattened sporangium. The sporangiophore tip is sensitive to light and this leads to the curvature of its lower part so that the sporangium is directed toward the source of the light. As the vesicle enlarges the turgor finally becomes so great that the apex ruptures and the sporangium is blown off, along with the watery contents of the vesicle, sometimes to a distance of over a meter. The sporangia adhere to vegetation and are eaten by herbivorous animals through whose digestive tracts the spores pass unharmed. In the dung of these animals the mold grows and produces its conspicuous fructifications. (Figs. 51A, 52.)

Returning to the genus *Mucor* we find that some species have unbranched sporangiophores while in others the sporangiophores may branch sympodially or monopodially, each branch terminating in a sporangium. Usually these sporangia are approximately equal in size although frequently the terminal one is slightly larger. In *M. proliferus* and some other species, the spores produced in the lateral sporangia are smaller than those produced in the larger, terminal sporangium. In *Thamnidium* the terminal sporangium, which is often the first one formed, is larger and possesses a well-developed columella; somewhat below it there grow out from the sporangiophore short branches (often much forked dichotomously), all terminating in small sporangia or some in pointed spines. These small sporangia (sporangioles) are few-spored, sometimes with not over two or three spores, and possess no columella. The whole sporangiole becomes detached and distributed by air currents. As water is absorbed the spores swell and burst the sporangial wall and escape. In some species of *Thamnidium,* under certain conditions only the sporangioles are formed. In the genus *Dicranophora* the sporangioles are only one- to two-spored and their spores are much larger than those

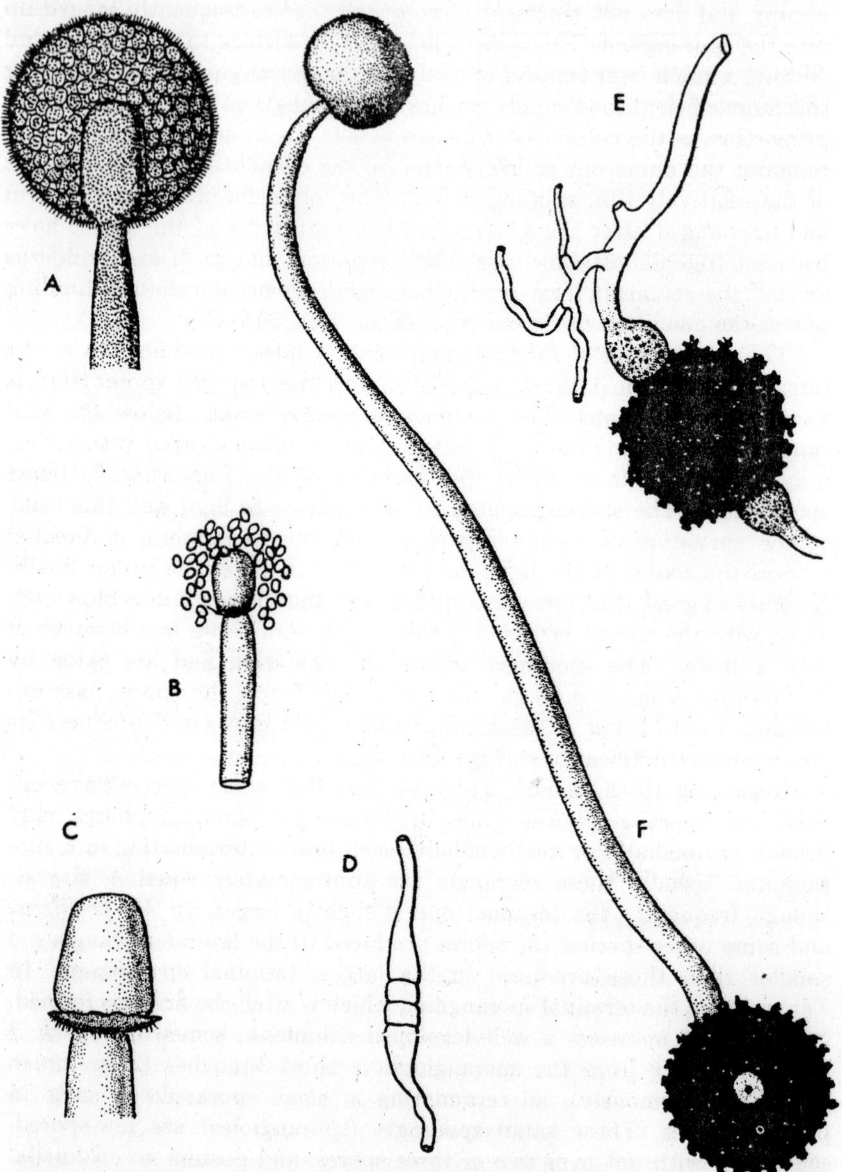

Fig. 50. Mucorales, Family Mucoraceae. *Mucor mucedo* (L.) Fres. (A) Mature sporangium just before escape of spores. (B) Escape of spores upon dissolution of sporanginm wall. (C) Columella after escape of spores. (D) Sexual reproduction in early stage, with thin-walled gametangia in contact. (E) Mature zygospore. (F) Germinating zygospore forming sporangiophore. (After Brefeld: Botanische Untersuchungen über Schimmelpilze, Heft 1, pp. 1–64, Leipzig, Arthur Felix.)

in the terminal sporangium, which is often lacking. In *Chaetocladium* the large terminal sporangium is entirely lacking. The sporangioles are one-spored and may be indehiscent in some species or in other species permit the spore to escape on germination. These monosporous sporangioles are often called conidia. (Fig. 53C, K.)

In *Blakeslea* the sporangium has a large columella when the fungus is well nourished and is smaller and few-spored without the columella

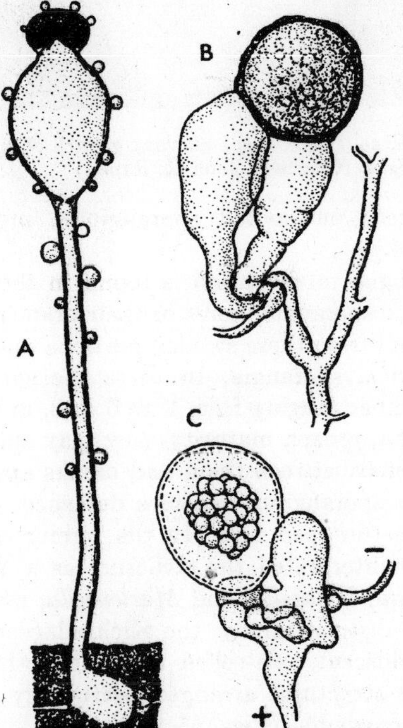

FIG. 51. Mucorales, Family Pilobolaceae. (A, B) *Pilobolus kleinii* van Tiegh. (A) Diagram of optical longitudinal section of sporangiophore and sporangium. (B) Zygospore. (C) *Pilobolus crystallinus* Tode ex van Tiegh. Heterothallic formation of zygospore. (A, after Buller: Researches on Fungi, vol. 6, pt. 1, pp. 1–224, London, Longmans, Green and Co. B, after Zopf: *Nova Acta Leopoldina*, **52**(7):352–358. C, after Krafczyk: *Ber. deut. botan. Ges.*, **49**(3):141–146.)

when poorly nourished. In addition numerous small two- to four-spored sporangioles are formed on short sterigmata from the surface of large rounded heads clustered at the apex of a sporangiophore. These spores differ markedly in size and shape from those produced in the sporangia. In *Choanephora* the sporangium possesses a columella and the sporangioles are borne as in *Blakeslea* on round heads at the apex of the sporangiophore of the other type. They are monosporous and indehiscent and function, as in *Chaetocladium*, as conidia. In *Cunninghamella* the sporangia are never found and only the heads of indehiscent sporangioles ("conidia") are developed. In *Mycotypha* the stalk of the sporangiophore is septate and the head is a little enlarged and much elongated. It is covered with in-

↘ Source of light

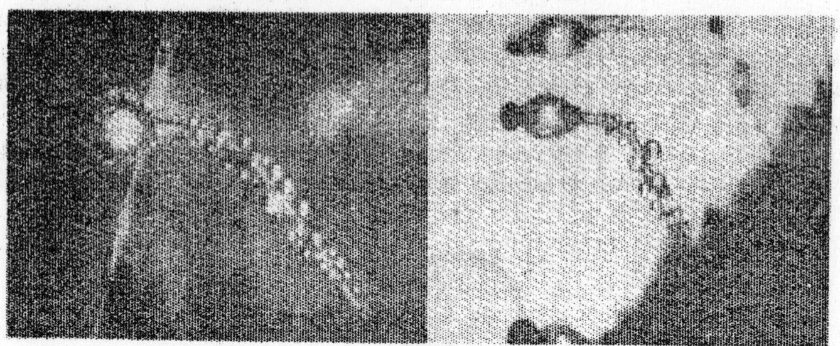

FIG. 52. Mucorales, Family Pilobolaceae. *Pilobolus* sp., showing growth of sporangiophores toward the light. (Courtesy, John R. Raper.)

numerable ellipsoidal indehiscent one-spored sporangioles, mostly in closely crowded whorls. (Fig. 54.)

Another direction of sporangial modification is found in the Piptocephalidaceae. In this family the sporangiophores or their branchlets are somewhat swollen terminally and from these swollen portions there grow out radially numerous cylindrical sporangia. In these the spores are formed in a single row, the number ranging from 2 to 6 or 8, or more—rarely, up to 30. As the spores approach maturity they may enlarge so that the sporangium is constricted between them and breaks apart into one-spored pieces or the whole sporangium becomes detached and the spores escape one by one from the open base. In the former case the earlier students of these forms interpreted the structure as a chain of conidia. In the genera *Coemansia*, *Kickxella*, and *Martensella*, which are included by Linder (1943) in a distinct family, the Kickxellaceae, these sporangia are reduced to indehiscent, one-celled structures which are borne on distinct sterigma-like structures arranged pectinately on one side of lateral, several-celled sporocladia. (Figs. 55, 56.)

Sexual reproduction is typically by the union of two approximately equal gametangia to form a so-called zygospore which usually occupies the cavities of the two gametangia and develops thick walls. The gametangia are multinucleate. The zygospores germinate, usually after considerable time, by the formation of a germ tube which may branch and start new mycelium or remain unbranched and terminate in a sporangium.

The study of the sexual reproduction of the Mucorales reveals an interesting state of affairs. In perhaps the majority of species tested in this respect a culture started from a single spore as well as all cultures derived from sporangia produced on this culture will, when grown separately or in contact with each other produce no zygospores. On bringing such cultures into contact with other cultures of the same species, origi-

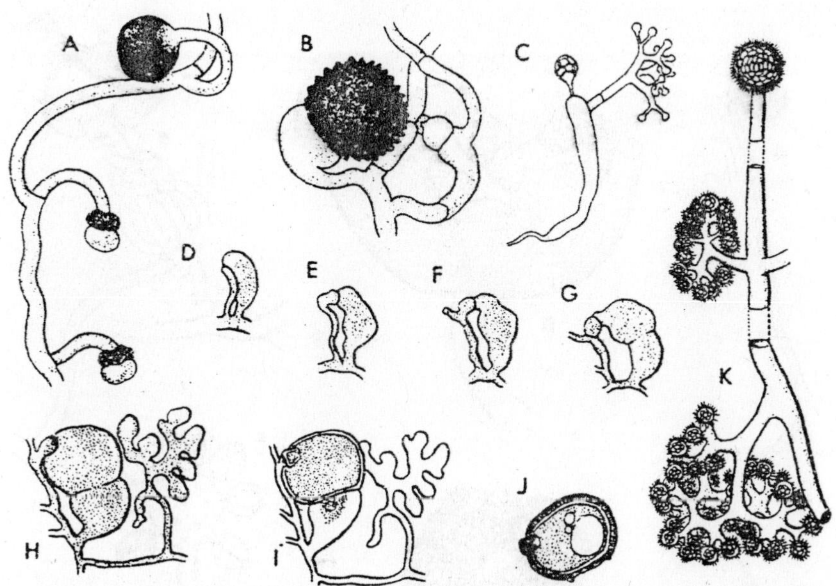

FIG. 53. Mucorales. (A–B) Family Mucoraceae. *Zygorhynchus macrosporus* Ling-Young. (A) Sporangiophores and sporangia. (B) Anisogamous formation of zygospores. (C–K) Family Thamnidiaceae. (C–J) *Dicranophora fulva* Schroet. (C) Primary sporangium and clusters of sporangioles. (D–J) Successive stages in sexual reproduction. (K) *Thamnidium elegans* Link, sporangiophore showing large terminal sporangium and lateral clusters of sporangioles. (A–B, after Ling-Young: *Rev. gén. botan.*, **42**(495):152. C–J, after Dobbs, *Brit. Mycol. Soc. Trans.*, **21**(1–2):172, 183. K, after Brefeld from Comparative Morphology of Fungi, by Gäumann and Dodge. McGraw-Hill Book Company.)

nated from plants obtained in various places, it is found that sometimes at the line of contact between two cultures very abundant zygospore formation occurs. On their part these other plants are self-sterile or sterile when grown in contact with each other so far as sexual reproduction is concerned. Thus A. F. Blakeslee (1904) determined that for many species of Mucorales there are two sexes, each capable of almost indefinite perpetuation by means of the asexual spores, but producing zygospores only when the mycelium of one sex comes into contact with that of the other sex. He named this phenomenon heterothallism and called such molds heterothallic. In contrast to these he found many species in which, when the proper conditions of environment and nutrition were met, zygospore production would occur within the mycelium, originating from a single spore. Such molds he called homothallic. The common bread mold, usually called *Rhizopus nigricans* Ehr., is a good example of a heterothallic mold while *Sporodinia grandis* Link, a mold frequently found on decaying mushrooms, is homothallic. Ling-Young (1930) showed that

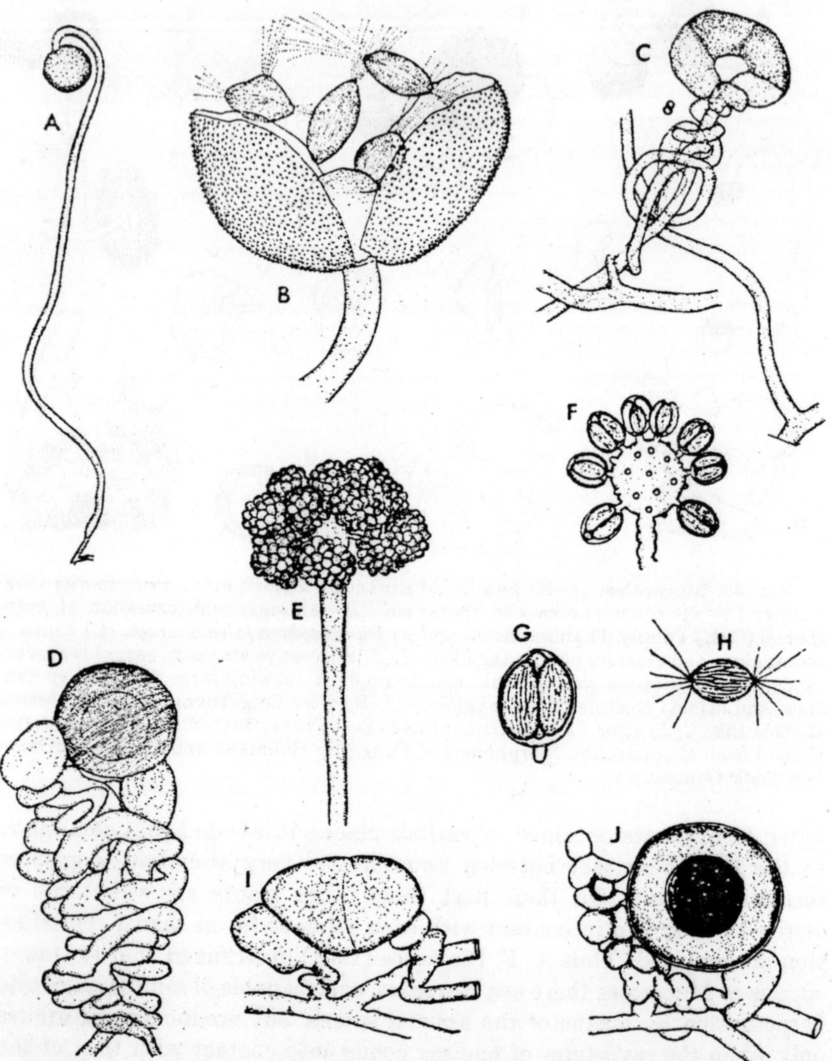

FIG. 54. Mucorales, Family Choanephoraceae. (A–E) *Choanephora conjuncta* Couch. (A) Sporangium and sporangiophore. (B) Opening sporangium showing spores and columella. (C) Young stage of sexual reproduction. (D) Sexual reproduction with mature zygospore. (E) Sporangiophore with sporangioliferous heads. (F–J) *Blakeslea trispora* Thaxter. (F) Single sporangioliferous head. (G) Single sporangiole. (H) Spore from sporangiole. (I) Young zygospore. (J) Mature zygospore. (A–E, courtesy, Couch: *J. Elisha Mitchell Sci. Soc.*, **41**(1–2):141–150. F–J, courtesy, Weber and Wolf: *Mycologia*, **19**(6):302–307.)

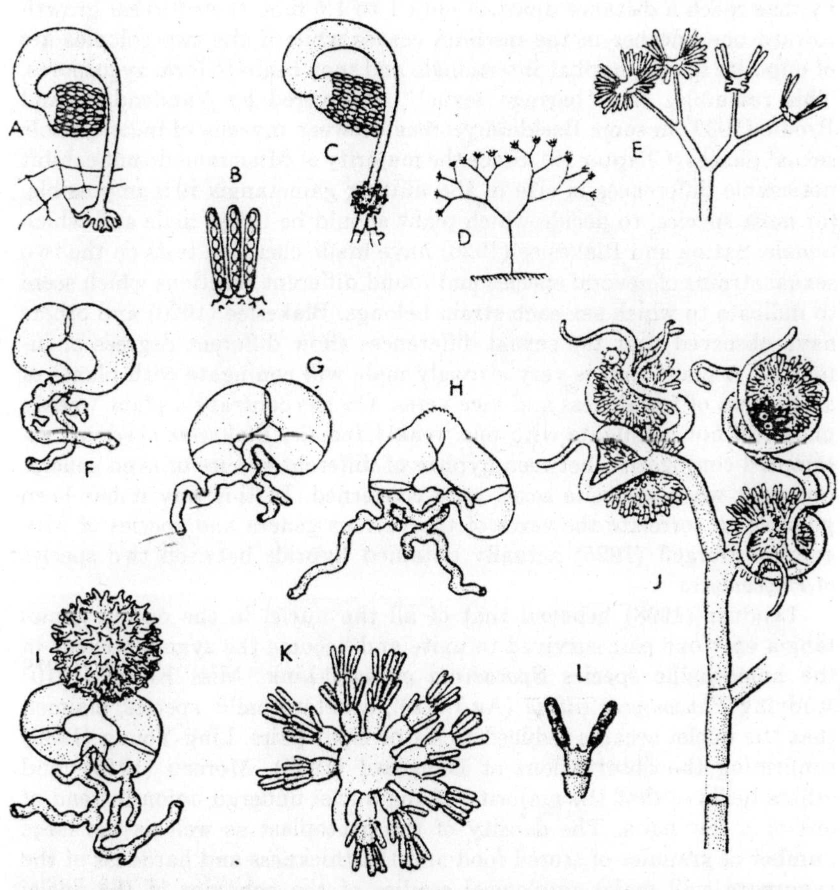

FIG. 55. Mucorales, Family Piptocephalidaceae. (A–C) *Syncephalis cornu* van Tiegh. (A) Sporangiophore. (B) Sporangia containing spores. (C) Germination of zygospore. (D, E). *Piptocephalis cruciata* van Tiegh. (D) Branched sporangiophore. (E) Terminal branches with sporangia borne on cruciate sterigmata. (F–I) *Piptocephalis freseniana* de Bary, stages in zygospore formation. (J–L) *Dispira americana* Thaxt. (J) General habit of fungus. (K) Head with not quite mature sporangia. (L) Two mature sporangia on sterigma. (A–E, after van Tieghem: *Ann. sci. nat. Botan.*, 6me sér., **1**:5–175. F–I, after Brefeld: Botanische Untersuchungen über Schimmelpilze, Heft 1, 1–64, Leipzig, Arthur Felix. J–L, after Thaxter: *Botan. Gaz.*, **20**(12):513–518, Univ. of Chicago Press.)

when two colonies of the same sex of *Phycomyces nitens* (Ag.) Kze. are grown in a culture medium rather poor in nutritive substances, as the hyphae reach a distance apart of only 1 to 1.5 mm. their further growth toward one another in the medium ceases while if the two colonies are of opposite sex the hyphae intermingle and soon begin to form zygospores. This resembles the "barrage sexuel" discovered by Vandendries and Brodie (1933) in some Basidiomyceteae between mycelia of incompatible sexual phases (Chapter 12). Since the majority of Mucorales do not exhibit noticeable differences in size in the uniting gametangia it is impossible, for most species, to decide which plant should be called male and which female. Satina and Blakeslee (1926) have made chemical tests on the two sexual strains of several species and found different reactions which seem to indicate to which sex each strain belongs. Blakeslee (1920) and others have observed that the sexual differences show different degrees of intensity. A plant that is very strongly male will conjugate with plants of all degrees of femaleness and vice versa. On the contrary a plant weakly male will not conjugate with one weakly female. Blakeslee observed attempted conjugation between hyphae of different species or even genera, but only when opposite sexes were concerned. In this way it has been possible to correlate the sexes of the various genera and species of Mucorales. Burgeff (1925) actually obtained hybrids between two species of *Phycomyces*.

Lendner (1908) believed that of all the nuclei in the united gametangia only one pair survived to unite and become the zygote nucleus in the homothallic species *Sporodinia grandis* Link. Miss Keene (1919) studying *Phycomyces nitens* (Ag.) Kze., a heterothallic species, claimed that the nuclei became reduced to six or eight pairs. Ling-Young (1930) confirming the observations ot Dangeard (1906), Moreau (1913), and others believes that the majority of the nuclei undergo union instead of one or a few pairs. The density of the protoplast as well as the large number of granules of stored food and the thickness and hardness of the zygospore wall make cytological studies of the behavior of the nuclei very difficult. Cutter (1942a, b) reports the results of very extensive studies which in some particulars contradict the conclusions of Lendner and of Miss Keene and to a large degree substantiate those of Ling-Young. He finds several types of behavior. In *Mucor hiemalis* Wehm., *Blakeslea trispora* Thaxt., *Absidia spinosa* Lendn., all three heterothallic, and a number of other Mucorales, all of the nuclei entering the zygote unite by twos within a few days and quickly undergo meiosis, the dormant zygospore containing only haploid nuclei. In *Rhizopus nigricans* Ehr., *Absidia glauca* Hagem, and *A. coerulea* Bain. (*Tieghemella glauca* (Hagem) Naum. and *T. coerulea* (Bain.) Naum., respectively), also heterothallic, some of the nuclei entering the zygote degenerate and some unite by twos. The dormant zygospore contains only diploid nuclei which undergo meiosis shortly before their germination. In *Phycomyces blakesleanus* Burg. and *P. microsporus* van Tiegh., of which at least the former is heterothallic, the nuclei increase in number in the zygote and become gathered in groups of several nuclei each, with a few scattered single nuclei. Shortly before germination of the zygospore some of the nuclei unite by twos and in the sporangium arising from it are found some diploid nuclei, some haploid nuclei, the product of meiosis, and probably

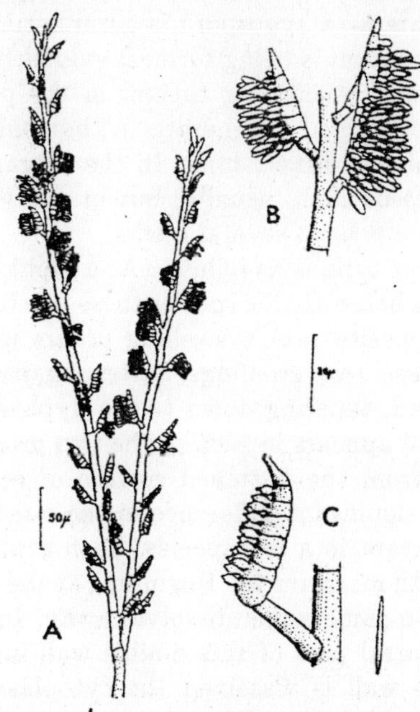

Fig. 56. Mucorales, Family Kickxellaceae. (A, B) *Coemansia erecta* Bainier. (A) Vigorously growing conidiophores. (B) Sporocladium. (C) *Coemansia aciculifera* Linder. Mature sporocladium and a single detached conidium. (Courtesy, Linder: *Farlowia*. 1(1):49 77.)

also some haploid nuclei which represent the scattered single nuclei which never united. In the homothallic *Sporodinia grandis*, contrary to Lendner's report, there is nuclear division in the zygote but no nuclear fusion and no meiosis.

Germination generally does not occur until after a considerable time has elapsed. Usually a stout upright sporangiophore is sent out after the outer wall is cracked open, and numerous spores are formed in a large sporangium. In the plant usually called *Mucor mucedo* (L.) Fres., a heterothallic species, the spores produced in this sporangium are all of one sex, showing that the differentiation of the sexes must have occurred in some nuclear division within the zygospore prior to its germination. On the other hand in *Phycomyces nitens* spores of both sexes are found in the sporangium produced by the germinating zygospore, as well as, occasionally, a few spores that are not yet differentiated sexually and which produce homothallic plants. Even in these plants the spores in the sporangia that they bear become more and more differentiated into the two sexes. Evidently this is connected with the nuclear condition found by Cutter to exist in the zygospores of this species. Sjöwall (1945) claims that in *M. mucedo* and *M. hiemalis* the gamete nuclei unite and undergo meiosis early in the zygospores so that for most of the resting period the nuclei are haploid. These nuclei undergo degeneration until finally only a few, apparently all of the same sexual phase, enter the sporangium. In *Rhizopus* the union of nuclei occurs as a zygospore is germinating and the reduction division as the sporangium is being formed, evidently only one diploid nucleus being concerned. According to him in the process of meiosis three of the four nuclei formed degenerate so that only one haploid nucleus is available for further functioning. In the sporangium the spores start out with one nucleus each, usually, but in many cases this divides so that the mature spores have several nuclei.

The sexual process of the *Mucor* type is as follows: At a point where the hyphae come in contact (if of a heterothallic species these two hyphae must originate from plants of opposite sex) a swelling occurs in each hypha, pushing them apart. These two swellings flatten against one another and become much enlarged, tapering down to the hyphae from which they arose. Soon a cross wall appears in each of the two processes, parallel to and a little distance from the flattened surface of contact, forming the two gametangia. No significant difference in the size of the two gametangia can be noticed except in a few species. Each contains a rather dense mass of cytoplasm with many nuclei. Beginning at the center the double wall separating the two gametangia dissolves away. In some cases there is a bulging of the central part of this double wall into one of the gametangia and when the wall is dissolved the cytoplasm and nuclei pass into this receptive gametangium. Soon, however, the whole intervening wall is dissolved away and the mingled cytoplasm fills out the cavities of the two gametangia. In the very young zygospore the thin wall represents that of the original gametangia. Within this is laid down a layer of more or less distinct dark plates or pyramids below which a two- or three-layered cell wall is secreted. As the zygospore enlarges the original gametangial wall is broken into small pieces that flake off so that finally the exterior is covered by the dark plates which may resemble short broad spines. Ling-Young showed that in many cases of heterothallic species the more vigorous strain (female) usually produces a larger gametangium and a larger suspensor than does the male mycelium. In *Mucor hiemalis* Wehm. not only is this the case but according to this author the contents of the male gametangium pass into the female gametangium through a small opening which then becomes closed so that the zygospore is formed solely from the female gametangium. (Fig. 50D–F.)

In a few species of Mucorales the gametangia and supporting suspensors are very unequal so that the product of the union of the two gametangia might be called an oospore. In general practice, however, they are called zygospores since their mode of origin is similar to that in isogamous species. In *Dicranophora* the one gametangium is many times larger than the other. In *Zygorhynchus* also, the difference between the two gametangia is marked. Both these genera are homothallic. (Fig. 53A–J.)

In *Pilobolus*, *Mortierella*, *Piptocephalis*, *Phycomyces*, and other forms two hyphae coming into contact may become attached by finger-like processes or wind around each other once or twice and then grow parallel and in contact. Very soon they curve away and bend so that only the tips are in contact, much as in a pair of tongs. In *Piptocephalis* and *Endogone* the zygospore does not form in the space occupied by the two gametangia but buds out from these to form an external zygospore. (Figs. 51B–C, 54C–D, 55F–I, 59A–F.)

The zygospore remains naked in the majority of species. In *Phycomyces nitens* branches grow out from the suspensors, surrounding the zygospore loosely with stiff, black, more or less dichotomously branched processes. In *Absidia glauca* these protective structures are curved and hooked. Often they arise more abundantly from the female suspensor in both these species. In *Mortierella* a dense hyphal mass several layers in thickness is formed around and closely appressed to the zygospore. (Figs. 49C–E, 57A–B, H–J.)

The order Mucorales is divided into several families. Mycologists are not in accord as to their number or limits. Probably these differences of opinion will remain until greater agreement is attained as to the probable course of phylogenetic development within the order and as to the relative importance to be ascribed to the evolutionary processes involving the sporangia and those concerning sexual reproduction. If the group is primitively isogametangic then the heterogametangic genera represent a later modification while if heterogamy is considered primitive then the isogametic condition found in *Mucor* is a modification of this ancestral condition. The customarily used classification is based largely upon the asexual reproductive structures and development and is probably artificial in many particulars. It can be replaced by a more natural system only when the above-mentioned questions are settled satisfactorily. In this book eight families are recognized, following Fitzpatrick (1930) in the main. These agree in most points with the six families into which Zycha (1935) divides this order. Naumov (1935, 1939) recognizes eight families. He divides many of the older genera so that he includes 38 genera and

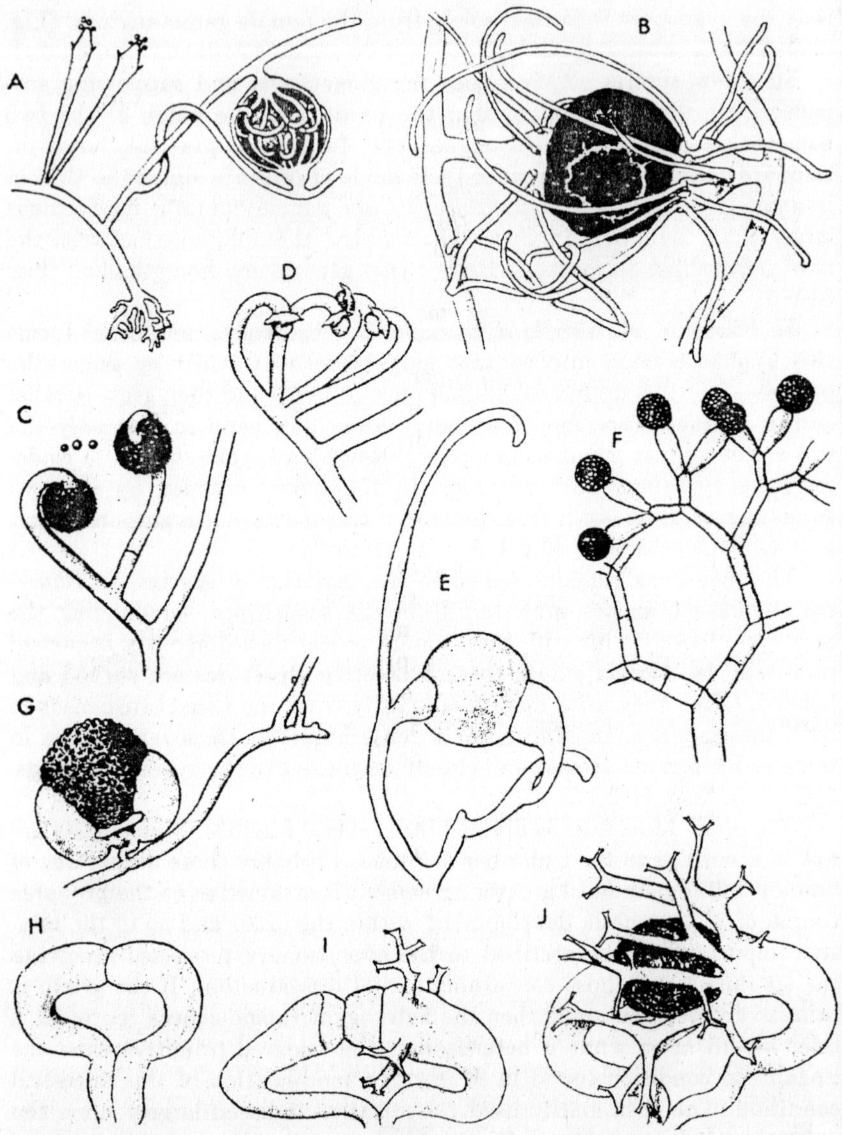

FIG. 57. Mucorales, Family Mucoraceae. (A) *Absidia septata* van Tiegh. & Le Monnier, sporangiophores and zygospore. (B) *Absidia glauca* Hagem, mature zygospore. (C–D) *Circinella minor* Lendner. (C) Unopened sporangia. (D) Opened sporangia showing columella. (E) *Circinella umbellata* van Tiegh. & Le Monnier, zygospore. (F–G) *Sporodinia grandis* Link. (F) Dichotomously branched sporangiophore. (G) Zygospore. (H–J) *Phycomyces microsporus* van Tiegh., stages in formation of zygospore. (A–F, after Lendner: *Beitr. Kryptogamenflora Schweiz,* **3**(1):103, 104, 110, 134, 137. G, after Ling-Young: *Rev. gén. botan.*, **42**(496):206. H–J, after Christenberry, *J. Elisha Mitchell Sci. Soc.*, **56**(2):332–366.)

340 species and subspecies, excluding the Endogonaceae and a few other forms. Including these the number would reach 44 genera and about 370 species.

FAMILY MUCORACEAE. Sporangia relatively large and many spored, oval or spherical, with a well developed columella. The sporangia are all alike and are borne singly, or several may be produced on a racemosely or otherwise branched sporangiophore. The wall of the mature sporangium may dissolve in the presence of moisture or may break into pieces and fall away, thus releasing the contained spores. The most important genera are: *Mucor*, whose unbranched or branched sporangiophores arise from the main mycelium; *Sporodinia*, whose sporangiophores are repeatedly and closely dichotomously branched; *Rhizopus*, in which long stout creeping hyphae (stolons) form tufts of sporangiophores at the points where the stolons attach themselves by rhizoids. *Absidia, Zygorhynchus*, and *Circinella* belong in this family as do several other genera. The zygospores are naked or more or less protected and are produced by both the *Mucor* type and the *Pilobolus* type as described above. It should be noted that some of the names used above, e.g., *Mucor, Rhizopus*, and *Sporodinia*, do not conform to the international rules of nomenclature requiring the earliest generic names to be used. Until some future International Botanical Congress shall settle the status of *Syzigites* vs. *Sporodinia, Rhizopus* vs. *Mucor, Mucor* vs. *Hydrophora*, it will perhaps be well to use the more customary generic names. (Figs. 50, 53A, 57.)

FAMILY PILOBOLACEAE. Sporangia more or less flattened vertically with a thick dark colored apical wall which does not break or dissolve. The sporangiophore is swollen at the base in *Pilobolus* and has a pronounced enlargement or subsporangial vesicle (in *Pilobolus*) or a slight one (in *Pilaira*) subtending the sporangium. The columella is more or less conical or may project almost to the top of the sporangium. In *Pilobolus* when the maximum osmotic pressure has been attained in the subsporangial vesicle it ruptures in a circular slit so that the columella and adhering sporangium are violently expelled by the mass of liquid forced out by the contraction of the ruptured vesicle. This sporangium turns over in flight or on striking some object, such as a blade of grass, and adheres by its gelatinous base. In *Pilaira* the elongating sporangiophore places the sporangium in contact with some near-by object to which it remains attached when the sporangiophore collapses. The dung-inhabiting species of *Pilobolus* are very easily obtained by bringing freshly dropped horse manure into the laboratory and enclosing it in a large dish. After a very few days the spores, which have passed unharmed through the alimentary canal of the horse, will produce an abundant growth of mycelium and the large conspicuous sporangiophores will appear in large numbers. When the dish is placed in a box closed on all sides except for a small window, the sporangiophores can be induced to turn and discharge their sporangia in the direction of the window, without much scattering of aim, if the illumination is good. Miss R. F. Allen and Miss H. D. M. Jolivette (1913) and later Miss Jolivette (1914) studied the relation of the discharge to different colors of light qualitatively. Miss Parr (1918) studied these relations quantitatively. The light at the violet end of the spectrum is the most efficient, grading to the red without intermediate maxima or minima. The presentation time, i.e., the minimum length of exposure necessary to bring about the effect, "varies in inverse ratio to the square root of the wave frequency" (Parr). Flint (1942) demonstrated that an unidentified species of *Pilobolus* gave no curvature but abundant discharge of sporangia in random directions in red light of the range of 6576–7024 Å units, and strong phototropic curvature and discharge with blue light centering at 4600 Å units.

As the sporangiophore is elongating, before the sporangium has begun to form, it is pointed and positively phototropic, so that it points toward the direction of the strongest light of effective wave length. With the formation of the subsporangial vesicle the latter has been shown by Buller (1921, 1934) to form a very effective light-perceiving organ or "eye" which brings about the final aiming so that the sporangium is shot off in the most favorable direction. Buller showed that in the vertical direction the sporangium may be shot over six feet and horizontally, eight feet.

In this family the suspensors of the gametangia are curved like a pair of tongs.

Grove (1934) recognizes 9 species of *Pilobolus* with 7 additional doubtful species, and 5 species of *Pilaira*, while Naumov (1939) recognizes respectively 18 and 5. (Figs. 51, 52.)

Family Thamnidiaceae. The genera of this family show the beginning of the differentiation into two types of sporangium, viz., the large terminal sporangium with a columella and the small lateral separable sporangioles which lack a columella. The latter are borne on short, variously divided branches arising laterally on the main axis of the sporangiophore at whose apex the large sporangium is produced, if present. In *Thamnidium* the spores are alike in sporangium and sporangiole. In some species some of the branches upon which the sporangioles are borne are sterile at the tips and form spine-like processes. The sporangiole-producing branches are dichotomously forked. The zygospores are formed on approximately equal suspensors about as in *Mucor*. In *Helicostylum* the sporangioles are borne on short circinate branchlets from the unforked lateral branches. In *Dicranophora* the sporangioles, according to Dobbs (1938), are one- to few-spored and have a rounded columella and thin sporangial wall. The spores are very variable in size and mostly bean-shaped, while they are mostly ellipsoid and more numerous and on the whole smaller in the terminal sporangium. The gametangia are very unequal in size. The genus *Chaetocladium* is parasitic upon other Mucorales. It shows close relationship to *Thamnidium* both in its sexual reproduction, which resembles that of *Mucor*, and in the production of numerous sporangioles on forking branches some of which terminate in sterile spines. The two genera differ in the complete absence of typical sporangia in *Chaetocladium* as well as in the fact that the sporangioles are monosporous and indehiscent, and hence are called conidia by many authors. Other genera are recognized by some students of the group. (Fig. 53C–K.)

FAMILY CHOANEPHORACEAE. The large sporangia are produced in two genera and possess a columella and resemble those of *Thamnidium*. The sporangioles instead of arising singly at the tips of forked branches are found crowded on the surface of the swollen apical portion of a large sporangiophore or of its branches. They are monosporous and indehiscent in three genera and several spored and dehiscent in one genus, *Blakeslea*. Most of the species are parasitic or saprophytic on flowers or other vegetable matter. The proportion of sporangia to sporangiole-bearing heads in *Blakeslea* has been shown by Goldring (1936) to depend considerably upon the amount of food present in the medium. *Choanephora* produces sporangia and indehiscent monosporous sporangioles, while *Cunninghamella* produces only the latter. Gäumann interprets the swollen apex of the sporangiophore of the latter as a homologue of the sporangium and the sporangioles clustered on its outer surface as in reality spores which have been, as it were, pushed out so as to become external instead of being produced internally. This view is not at all in accord with that of the author, who considers the sporangioles to be homologous to those of *Thamnidium*, i.e., reduced lateral sporangia. Possibly related to *Cunninghamella* is the genus *Mycotypha*, described by Miss Fenner (1932). The mycelium is like that of most Mucorales, coenocytic and much branched and only occasionally septate. The sporangioles are reduced to minute "conidia" closely covering the sides and apex of a cylindrical or clavate enlargement of the upper portion of the eventually septate sporangiophore. The resemblance of the head of sporangioles to the flowering head of *Typha* suggested the name given to the organism. So far the formation of zygospores has not been observed in *Mycotypha*. In *Choanephora* and *Blakeslea* zygospore formation is much like that in *Pilobolus* but in *Cunninghamella* it resembles that of *Mucor*. The genus *Sigmoideomyces* probably belongs to this family. Its heads of spores (sporangioles) are borne laterally on branched septate sporangiophores whose branches are curved more or less like the letter S. Sexual reproduction is unknown. Possibly *Thamnocephalis* also is related. (Fig. 54.)

Note: An additional genus of the Choanephoraceae was published too late to appear in the text of this chapter. *Cokeromyces* Shanor (type species *C. recurvatus* Poitras) produces numerous sporangioles on elongated recurved stalks arising from the terminal capitate swelling of the sporangiophore. Twelve to twenty spores are produced in each sporangiole, lacking the striae and appendages found in other genera of the family. No true sporangia are produced. Zygospores arise in the manner of *Mucor* and *Cunninghamella.*

Shanor, Leland, Adrian W. Poitras, and R. K. Benjamin: A new genus of the Choanephoraceae, *Mycologia*, **42**(2):271–278. *Figs.* 1–12. 1950.

Family Piptocephalidaceae. The species of this family are largely parasitic on other Mucorales although some are saprophytic. The sporangia are narrow and more or less clavate or cylindrical, with the spores usually in one row, often appearing when mature like chains of conidia. The number of spores formed in the sporangium varies from two (rarely one) to 30. Usually the sporangium breaks into monosporous segments, the spore being enclosed in the sporangial wall. In a few cases the whole sporangium may break off and the spores escape individually from the opening. In *Syncephalis* the sporangiophore is upright and unbranched, tapering from the more thickened basal portion up to the enlarged, usually spherical head which may be upright or nodding. The radiating sporangia may arise directly from the surface of this head or singly or several together upon short "sterigmata." These sporangia may be dehiscent at maturity letting the contained two to several spores escape. In *Syncephalastrum* the sporangiophore is mostly branched as in the racemose type of *Mucor.* On the spherical heads terminating the branches the narrow sporangia arise radially without the intervention of sterigmata. These break apart into monosporous segments so that they were long considered to be true conidia and compared to the chains of conidia formed on the swollen heads of the conidiophores of *Aspergillus.* In *Piptocephalis* the sporangiophores are dichotomously branched, bearing at the tips of the branches slightly enlarged segments ("sterigmata") of various shapes, from which arise several sporangia which break up into one-spored pieces. In *Dispira* the terminal heads of the much branched sporangiophore bear two-celled "sterigmata" from each of which arise short sporangia of from two to several spores. The genus *Spinalia* (Vuillemin, 1904) probably belongs in this family. It has a coenocytic mycelium and produces heads covered by "conidia." On the spherical heads the so-called "conidia" occur in chains of two. According to Vuillemin (1922) the zygospores are formed by the *Mucor* type of sexual reproduction in *Syncephalastrum.* In *Piptocephalis* and *Syncephalis* sexual reproduction is more like the *Pilobolus* type except that the zygospore buds out from the united gametangia and matures externally. (Fig. 55.)

FAMILY KICKXELLACEAE. Linder (1943) erected this family to contain *Kickxella*, *Coemansia*, and *Martensella*. In these the sporangia (conidia) are indehiscent and one-celled, but in the first-named genus an incomplete septum indicates that perhaps it is descended from a two-spored ancestral form. In all three genera there are special lateral sporangium-

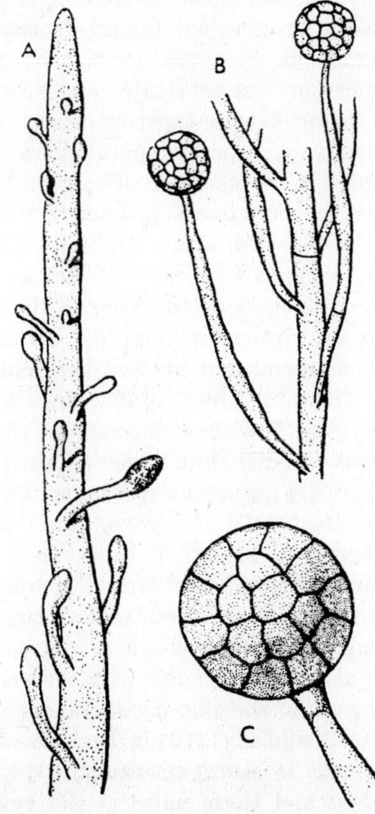

FIG. 58. Mucorales Family Mortierellaceae. *Dissophora decumbens* Thaxt. (A) Young apical portion of fertile hypha. (B) Portion of fertile hyphae bearing sporangiophores. (C) Mature sporangium. (After Thaxter: *Botan. Gaz.*, 58(4): 353–366, Univ. of Chicago Press.)

bearing sporocladia consisting of several cells. From each cell of these except one or more terminal cells arise several sterigmata each bearing a single conidium. The sporocladium with its conidia somewhat resembles a comb. The nearest relationship of this family appears to be with some members of the preceding family. Sexual reproduction has been observed in only one species of *Coemansia* and resembles considerably that of *Pilobolus*. (Fig. 56.)

FAMILY MORTIERELLACEAE. The position of this family is uncertain. In sporangial structure the absence of a columella would indicate a position below that of the Mucoraceae, but the occurrence of sporangioles which are reduced to conidia in some forms would suggest a higher position. This is also indicated by the formation of a dense hyphal protective coat around the base of the sporangiophore in some species and around the zygospores which are formed much as in *Pilobolus*. Of the several genera assigned to this family *Mortierella* is the only one that has been fairly well studied. In addition to the spherical, many-spored sporangia there occur in some species indehiscent, monosporous sporangioles ("conidia" or "stylospores"). The thirty or more species are mostly saprophytic on vegetable matter or are weak parasites. In *Haplosporangium* the main sporangiophores are more or less prostrate, septate and constricted at the septa. From these segments arise tapering lateral branches bearing terminally or also laterally one-spored sporangioles (in one species) or two-spored sporangioles in *H. bisporale* Thaxter. No large sporangia are produced and sexual reproduction is unknown. *Dissophora* resembles a racemosely branched *Mortierella*. (Figs. 49, 58.)

FAMILY ENDOGONACEAE. The twenty-six or more species at present included in this family are grouped together on the basis of their sporocarps. These are loose or moderately firm sclerotium-like bodies a few millimeters to 2–3 centimeters in diameter and produced in humus-rich soil, leaf mold or among mosses. The interwoven hyphae are coenocytic with occasional septa in the old hyphae. Scattered throughout the sporocarp are three types of reproductive bodies, but never all three in the same sporocarp. In two species there are numerous sporangia. They have a transverse basal septum as in *Mortierella*. They contain from 4–12 spores in one species and many spores in the other. In a dozen or more species the reproductive bodies are thick-walled chlamydospores of various shapes and sizes. Some authors consider these to be azygospores. True sporangia are entirely lacking. In a few species only zygospores are formed in much the same manner as in *Piptocephalis*, with the suspensors parallel and the zygospore budding out of the united gametangia or out of the larger of the two. According to Bucholtz (1912) in *Endogone lactiflua* Berkeley most of the numerous nuclei in each gametangium degenerate leaving only one privileged nucleus and these unite in the zygospore. Atkinson (1918) reports that in *E. sphagnophila* Atk. many pairs of nuclei unite. In *E. fasciculata* Thaxter and *E. occidentalis* Kanouse both chlamydospores and zygospores are present in the sporocarp. Miss Walker (1923) obtained *E. malleola* Harkness in pure culture and procured mycelial growth and sporangia of the *Mortierella* type but no sporocarps were produced nor chlamydospores nor zygospores. Miss Kanouse (1936) succeeded in obtaining a culture from some of the mycelium in a not quite mature sporocarp of *E. sphagnophila*. This produced a *Mucor*-like mycelium bearing numerous sporangiophores and very closely re-

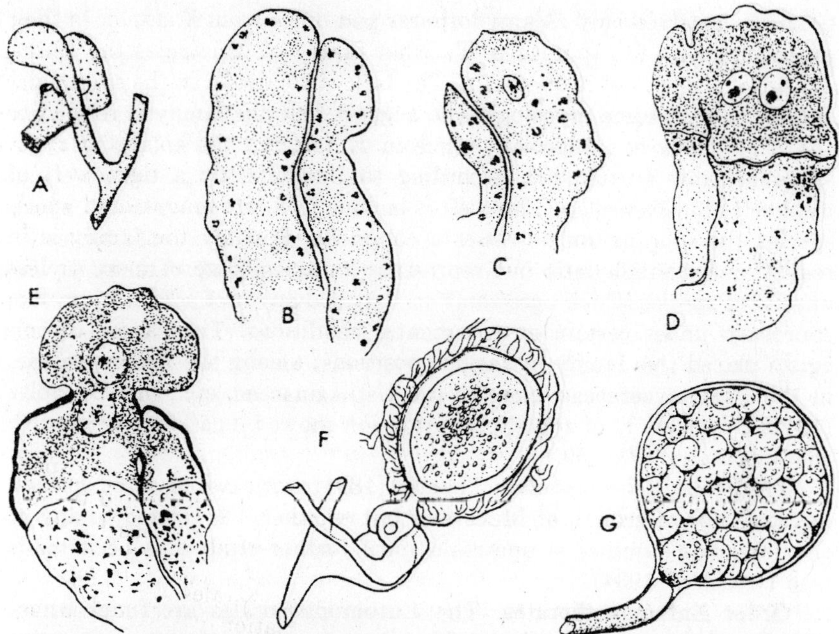

FIG. 59. Mucorales, Family Endogonaceae. (A–F) *Endogone lactiflua* Berkeley. (A) Young gametangia. (B) Gametangia in longitudinal section, showing numerous nuclei. (C) Gametangia in longitudinal section, showing one privileged nucleus in the right-hand gametangium, the others receding basally. (D) Gametangia set off by septa and male nucleus entered into the female gametangium. (E) Zygospore budding out from top of female gametangium. (F) Practically mature zygospore. (G) *Endogone pisiformis* Link, mature sporangium. (After Bucholtz: *Botan. Centr. Beihefte*, II Abt., **29**(2):147–225.)

sembling the mycelium and sporangiophores of *M. ramannianus* Möller, and like that species with a pink to rose color. Very numerous chlamydospores were produced also and eventually, when grown on 1 per cent malt agar, zygospores of the typical *Endogone* type. The sporangial walls quickly break up or, when placed in water, deliquesce, letting the spores escape and revealing a spherical columella. The species is homothallic, for cultures from a single spore produce zygospores. *Endogone occidentalis* contains in its sporocarps both chlamydospores and zygospores, but when brought into culture by Miss Kanouse gave rise to chlamydospores only, producing neither zygospores nor sporangia. In none of the cultures by Miss Walker or Miss Kanouse could typical sporocarps be obtained. Because the sporangia obtained from cultures of *E. sphagnophila* are *Mucor*-like, not like those of *E. malleola* Harkness and *E. reniformis* Bresadola, Miss Kanouse segregates the latter two in a separate genus *Modicella*. The two other genera assigned to this family, *Sclerocystis* and

Glaziella, produce only chlamydospores and differ from *Endogone* in their arrangement in the sporocarp. To what family of Mucorales the Endogonaceae are nearest is uncertain. The lack of columella in the sporangial species has led some mycologists to suggest that the family is related to the Mortierellaceae, especially since in *E. lactiflua* the enlarging zygospore becomes surrounded, according to Bucholtz, by a tight weft of hyphae as in *Mortierella*. This weft is lacking in *E. sphagnophila* in which species the sporangium possesses a columella. Possibly the family is in reality not monophyletic but represents an assemblage of more or less unrelated species which agree only in the production of a sclerotium-like sporocarp under certain environmental conditions. The earlier mycologists placed this family in various positions; among the Ascomyceteae, in the Protomycetaceae, close to the Ustilaginaceae, etc., but Bucholtz (1912) by his study of zygospore formation showed its affinity to be with the Mucorales. (Fig. 59.)

The genus *Helicocephalum* Thaxter (1891), with two or three species, has the type of growth of Mucorales but whether it is related to any of the foregoing families is uncertain until further study has been made (see Drechsler, 1934).

Order Entomophthorales. The Entomophthorales are fungi whose mycelium is often much reduced. Upon germination of the spore the germ tube is usually coenocytic but sooner or later septa appear which divide it into plurinucleate or even uninucleate segments, forming a septate mycelium in some genera or in other genera falling apart into the so-called "hyphal bodies." These latter may multiply by fission. Of the five or more genera two are saprophytic or rarely parasitic, one is parasitic in the gametophytes of ferns, one in the Desmidiaceae, and the remainder in the bodies of insects. The latter group includes the vast majority of the known species of the order. Asexual reproduction is by means of reduced sporangia (probably more properly sporangioles) which may be uninucleate or plurinucleate and are shot off singly from the apex of a somewhat club-shaped sporangiophore, except in the genus *Massospora* in which they are produced internally in the body of the insect host. These sporangia are usually called conidia, as in the homologous structures in the Peronosporaceae and Mucorales. In many cases such a "conidium" may produce another "conidiophore" and shoot off a secondary conidium, and that may produce a tertiary conidium, and so on. The ultimate germination is by a germ tube except in the genus *Basidiobolus* in which the conidium shows its true sporangial nature by producing internal spores. Sexual reproduction is by the union of mycelial segments (or of hyphal bodies) to form zygospores which may lie *in* one of the uniting gametangia (*Basidiobolus*), *between* the suspensors (*Conidiobolus*), or laterally to the fusion cell or to one of the conjugating gametangia (*Ento-*

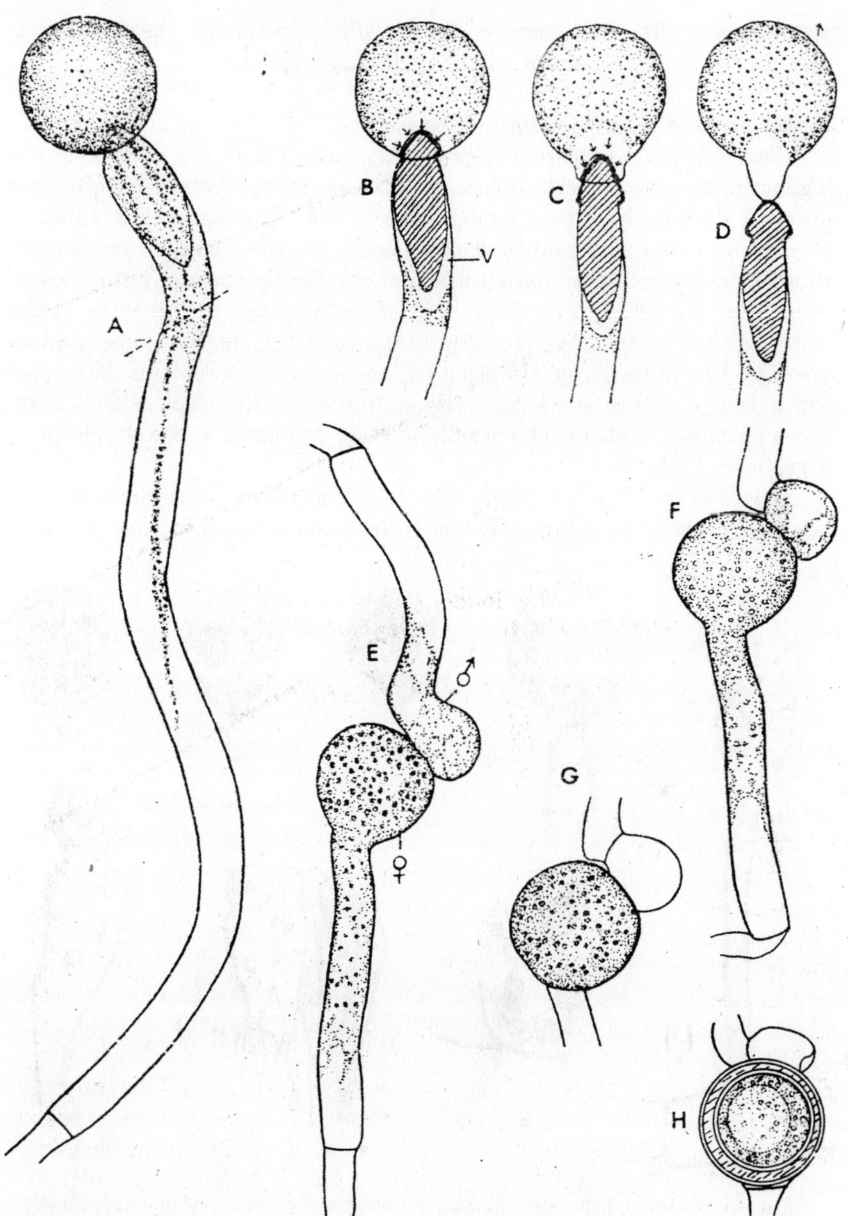

FIG. 60. Entomophthorales, Family Entomophthoraceae. *Conidiobolus brefeldianus* Couch. (A) Conidiophore and almost mature conidium. (B–D) Successive stages in discharge of conidium. (E–H) Stages in sexual reproduction. (Courtesy, Couch, *Am. J. Botany*, **26**(3): 119–130.)

mophthora). The zygospore is thick-walled. In many species similar spores are formed parthogenetically, the azygospores.

FAMILY ENTOMOPHTHORACEAE. Usually only one family is recognized in this order: the Entomophthoraceae.

Conidiobolus is saprophytic or weakly parasitic on the fruit bodies of higher fungi and on certain insects. Its asexual spores are conidia, not dividing internally into sporangiospores. The branching mycelium is divided by septa into multinucleate segments. The spherical conidia are shot off in the manner characteristic of the family. Sexual reproduction occurs by the union of adjacent cells of unequal size in the same hypha or by contact of two hyphae. The contents of the smaller gametangium are cut off from the supporting cell (or suspensor) by a septum. They pass through a pore into the female cell within which the thick-walled zygospore becomes mature. On germination it produces a conidiophore or mycelium. (Fig. 60.)

Basidiobolus is most frequently found growing saprophytically on the excrement of frogs, lizards, etc. It has been isolated by van Overeem

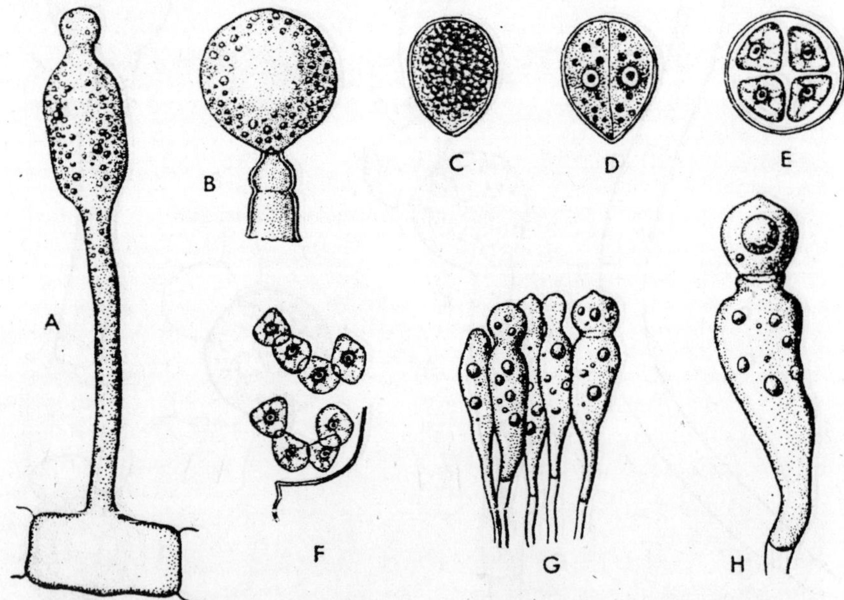

FIG. 61. Entomophthorales, Family Entomophthoraceae. Asexual reproduction. (A–F) *Basidiobolus ranarum* Eidam. (A) Developing conidiophore. (B) Apex of mature conidiophore. (C–E) Development of conidium into sporangium in stomach of frog. (F) Spores set free from sporangium. (G, H) *Entomophthora muscae* (Cohn) Fresenius. (G) Group of conidiophores. (H) Single conidiophore. (A, B, G, H, after Thaxter: *Mem. Boston Soc. Natural History*, **4**(6):133–201. C–F, after Levisohn: *Jahrb. wiss. Botan.*, **66**(3):513–555.)

(1925) in pure culture from a fungus mass deep in a fistular abscess in the leg of a horse in Java. On the excrement it forms sporangiophores which shoot off the not yet divided sporangia, much as occurs in *Pilobolus*. These sporangia are eaten by beetles which in turn are devoured by frogs or lizards in whose stomachs the beetles are digested, setting free the sporangia which only then divide internally to produce the spores. These escape from the sporangium and multiply in the alimentary canal by fission or by budding. They are set free in the excrement and then germinate to form mycelium on which are produced other sporangia. This mycelium is at first coenocytic but soon septa appear and divide it into a multiseptate branched mycelium. Zygospore formation is as follows: Two adjacent segments of a hypha send up parallel beaks in contact, into each of which a nucleus migrates and divides. One nucleus of each pair remains in the apex at its beak, being cut off by a septum. One of the two original segments enlarges considerably, an opening is dissolved through the separating septum, and the nucleus and part of the cytoplasm of the smaller cell pass into the larger cell. There the nuclei fuse and the thick-walled "zygospore" is formed. Couch (1939) reported that the cell wall responds to the chloriodide of zinc test by the cellulose reaction. Because of this and of the uninucleate condition of the cells he advocated placing *Basidiobolus* in a separate family, Basidiobolaceae. (Figs. 61A–F, 62A–E.)

Ancylistes was formerly included in the Ancylistidaceae (now called Lagenidiaceae) because of the similarity of its vegetative structure and sexual reproduction to *Lagenidium*. No zoospore production is known. With the discovery by Miss Berdan (1938) that "conidia" similar to those in the Entomophthorales were produced and that these were shot off in the same manner the genus had to be transferred to the latter order. The three species are parasitic in Desmidiaceae of the genus *Closterium*. Two adjacent cells in the same filament or cells in contact in parallel filaments may conjugate. One cell is usually smaller than the other and the thick-walled zygospore is formed in a protuberance from the larger cell. Its germination is unknown.

Completoria is parasitic in the cells of the gametophytes of ferns. Infection spreads from cell to cell of the host by means of hyphae penetrating the cell walls. Azygospores are formed in the host cells. The conidiophores emerge into the air and bear conidia which are violently discharged. The mycelial masses in the host cells are much branched and lobed and not conspicuously septate.

Entomophthora is the largest genus of the family, containing, according to Fitzpatrick (1930), about 40 species, all parasitic within insects. Some species of this genus are called *Empusa* by some writers, but since this name was used earlier for a genus of Orchids the next later name, *Ento-*

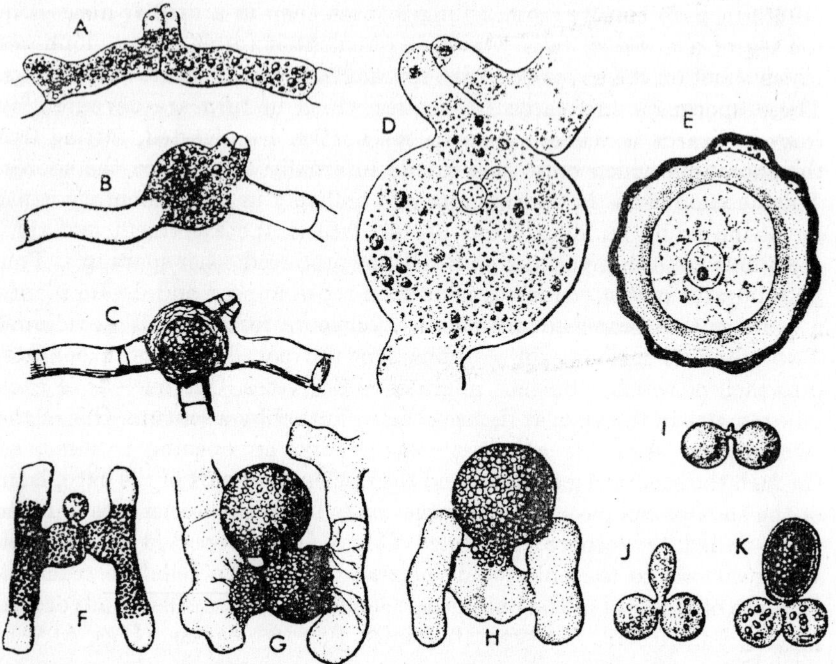

FIG. 62. Entomophthorales, Family Entomophthoraceae. Sexual reproduction. (A–E) *Basidiobolus ranarum* Eidam. (F–H) *Entomophthora sepulchralis* Thaxt. (I–K) *Entomophthora fresenii* Now. (A–C, after Eidam: *Beitr. Biol. Pflanz.*, **4**(2):181–251. D–E, after Fairchild: *Jahrb. wiss. Botan.*, **30**:285–296. F–K, after Thaxter: *Mem. Boston Soc. Natural History*, **4**(6):133–201.)

mophthora, must be used. The internal organs of the host are dissolved, presumably by enzymes secreted by the fungus. The mycelium may be quite extensive and filamentous with only occasional septa or may become septate at frequent intervals or break apart into numerous hyphal bodies. The clavate conidiophores emerge through the thinner parts of the body wall of the insect and shoot off their conidia with great violence. The conidiophores may be simple and packed together in a palisade layer or they may be branched at the base. In the latter case the conidia are uninucleate, in the former they are plurinucleate. Both azygospores and zygospores are known in this genus. It is not settled yet whether only one pair or several pairs of nuclei are functional, nor is it known at what stage of development of the zygospore the nuclei fuse. *E. muscae* (Cohn) Fresenius is the frequent cause of the death of house flies (*Musca domestica*) in the autumn months. The affected flies cling to window panes and other objects and die there, and the glass immediately surrounding them becomes whitened by a halo of discharged conidia. True zygospores are unknown but azygospores are produced abundantly in Europe but very rarely in America. *E. grylli* Fresenius is very abundant in some seasons in the plains states of the United States, causing the death of immense numbers of grasshoppers which climb up on stalks of grasses and other plants and there die. In moist weather their abdomens are first covered by belts of conidiophores emerging from between the segments, but in dry weather these are not conspicuous. Azygospores are produced in great numbers in the body cavity of the host. (Figs. 61G–H, 62F–K.)

Several other genera have been described, including *Massospora*, parasitic in the 17-year cicada (*Tibicina septendecim*). In this genus the conidia are produced within the host, not on extruded conidiophores, and are distributed by the gradual sloughing off of the affected parts while the insect is still capable of creeping around. Sexual reproduction is unknown.

Speare (1912) and Sawyer (1929) devised means of growing species of *Entomophthora* in culture and thus have been able to follow out the life histories of some species more fully than formerly.

Order Zoopagales. The fungi tentatively brought together in this order are parasitic on soil-inhabiting and aquatic animals: amoebae, nematodes, and insect larvae. The hyphae are coenocytic and at first nonseptate and slender. Later they develop occasional cross walls. Sexual reproduction is by conjugation of short or long filaments and the production of zygospores of various forms, spherical to boat-shaped. Asexual reproduction is by the formation of conidia, either laterally and singly or apically, in the latter case sometimes in chains. Whether these conidia

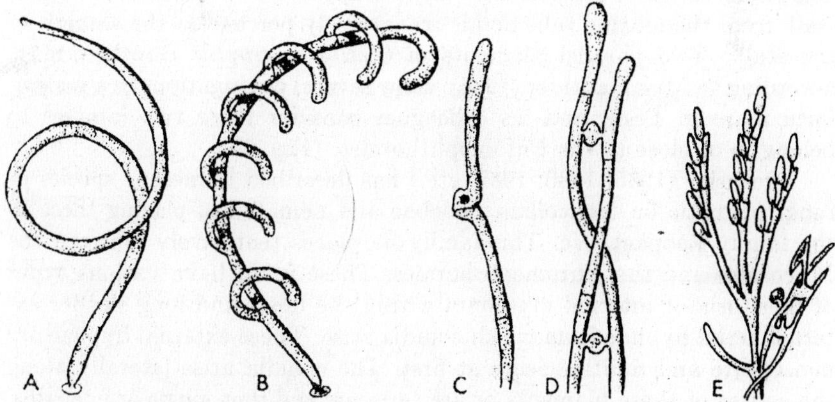

FIG. 63. Zoopagales. (A–D) Family Harpellaceae. *Harpella melusinae* Leger & Duboscq. (A) Vegetative hypha not yet septate. (B) Septate hypha bearing curved conidia. (C, D) Formation of zygospore. (E) Family Genistellaceae. *Genistella ramosa* Leger & Gauthier. Whole fungus with spikes of conidia and formation of young and mature boat-shaped zygospores. (A–D, after Leger and Duboscq: *Compt. rend.*, **188**(14):951–954. E, after Leger and Gauthier: *Compt. rend.*, **194**(26):2262–2265.)

are comparable to the sporangioles of some Mucorales or to the sporangia of Piptocephalidaceae needs further study.

Two groups of fungi placed by their discoverers (Leger and Duboscq, 1929b and Leger and Gauthier, 1932) in the families Harpellaceae and Genistellaceae, occur as parasites (possibly only as commensals) in the alimentary canals of various aquatic insects. They consist of slender unbranched flexuous coenocytic hyphae (Harpellaceae) or branched coenocytic hyphae (Genistellaceae). They are attached to the host tissues by disk-like or lobed expansions, even finger-like structures in some cases, but do not penetrate through the epidermal layer. The hyphae eventually become septate and bear slender or stout, straight or curved, usually uninucleate conidia, mostly one from each cell of the hypha, often only along one side or in a crown near the apex. Leger and Gauthier (1935) showed that these conidia bear at their point of attachment one (*Stachylina* and *Typhella*), two (*Genistella*), or four (*Harpella*) fine filaments which are coiled in the supporting cell and pulled out of the latter as the conidium becomes detached. These appendages may be 3 to 6 times the length of the conidium. Their function is not known but possibly they have something to do with the manner of infection of the host. Sexual reproduction occurs through the union of adjacent cells in the same filament, the thick-walled zygospore being formed in one of these cells as in *Basidiobolus* or (often in the same species) by the union of conjugation tubes from filaments lying side by side, the zygospores being formed in the center of the enlarged connecting tube or on a stalk growing from this point. In the Harpellaceae the zygospore is spherical while in the Genistellaceae it is biconical or boat-shaped and surmounts a lateral stalk from the uniting tube and is transversely perched at the summit of the stalk. Two parallel filaments of *Glotzia centroptili* Gauthier may, according to Miss Gauthier (1936), show several conjugations in a scalariform manner. Leger and his colleagues consider these two families to belong in or close to the Entomophthorales. (Fig. 63.)

Drechsler (1935, 1936, 1937, etc.) has described numerous species of fungi parasitic on terricolous amoebae and nematodes, placing these in the family Zoopagaceae. This family he places tentatively between the Mucorales and the Entomophthorales. These fungi have various types of haustoria or internal mycelium within the hosts and long slender external aerial hyphae from which conidia arise. These external hyphae are nonseptate and multinucleate at first. The conidia arise laterally along the course of these filaments or are terminal and then single or in chains. They are more or less spindle-shaped. The haustorium of *Endocochlus*, parasitic in *Amoeba*, is a stout two or more times helicoidally wound structure formed at the end of the infection tube which penetrated the host from a conidium that adhered to the external surface of its victim by means of a sticky substance. More rarely infection results from ingestion of the conidia. From this coil, after the death of the host, one or more slender aerial unbranched or slightly branched hyphae which are at first nonseptate emerge to the exterior. The terminal portion becomes septate and from a few of the cells thus formed arise laterally sessile, spindle-shaped conidia. Other short slender hyphae arising from the haustorial coil grow parallel for a short distance and unite at the apex. Just beyond the point of union a spherical thin-walled enlargement (called by Drechsler, 1935a, the zygosporangium) is formed, equally upon the apices of the two uniting hyphae or on an extension from near the apex of one of them. Within this zygosporangium is produced an angularly lobed, thick-walled zygospore. The zygosporangial wall may remain intact and does not break up into small pieces and disappear as does the corresponding structure in Mucorales. (Fig. 64.)

In *Cochlonema* the thick spiral haustorial body much resembles that of *Endocochlus* but the spindle-shaped conidia are in chains on the aerial conidiophores. They are apparently ingested by the host and thus infect it. Sexual reproduction is much as in that genus but in some cases the zygosporangial wall is lobed or warty to correspond to the lobes or warts of the zygospore. In *Bdellospora* the spindle-shaped conidia are also produced in chains. Instead of being ingested by the amoeba and forming a coiled haustorium in its interior the conidium adheres to the outer surface of the host cell and sends in a slender infection tube which divides dichotomously into several short lobes. The external conidium enlarges until many times its original size and from it arise the aerial conidiophores and also the slender branches (from separate individuals) that coil around one another many times and then conjugate at the tips to form a zygospore as in the foregoing genera. In *Zoopage* the mycelium is external, somewhat branched and nonseptate. When an amoeba comes in contact with such a hypha it adheres to it and a short-lobed haustorium penetrates the host cell. The external mycelium produces short aerial chains of elongated spindle-formed conidia. Sexual reproduction is much as in the preceding genera. *Acaulopage* and *Stylopage* are similar but their conidia are single, not in chains, practically sessile in the former and on short conidiophores in the latter. The latter genus may attack Amoebae and Nematodes. *Cystopage* produces no distinct conidia but intercalary or lateral chlamydospores in the intramatrical mycelium as well as in the extramatrical mycelium. It attacks Nematodes and Rhizopoda.

Drechsler (1935) points out the similarity of the mycelium and catenulate conidia of some species of *Actinomyces* to those of some of the more delicate forms in the Zoopagaceae and suggests that there may be some

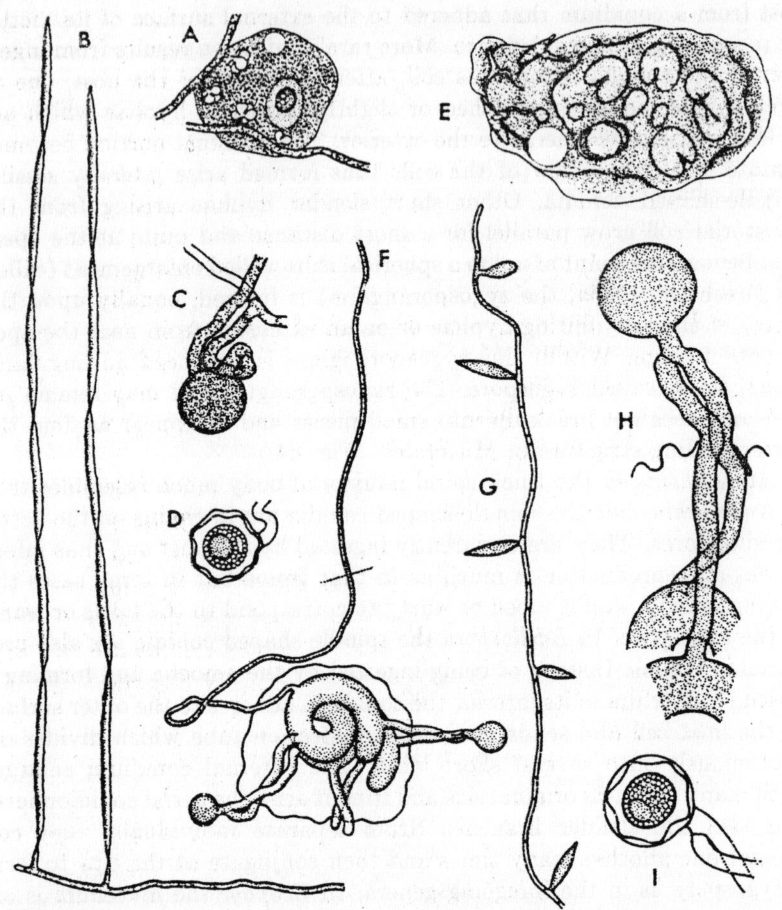

FIG. 64. Zoopagales, Family Zoopagaceae. (A–D) *Zoopage phanera* Drechs. (A) Parasitized amoeba with two haustoria from each of two hyphae. (B) Creeping hypha with upright chains of conidia. (C) Early stage of sexual reproduction. (D) Mature zygospore. (E–I) *Endocochlus asteroides* Drechs. (E) Large amoeba with stages of infection and young thalli of fungus. (F) Coiled thalli of fungus producing zygospores and sending out a long conidiiferous hypha (only basal portion drawn). (G) Portion of fertile region of conidiiferous hypha. (H) Young stage of sexual reproduction. (I) Mature zygospore. (Courtesy, Drechsler: *Mycologia*, **27**(1):6–40.)

relationship here. He believes that the latter as well as the Harpellaceae and Genistellaceae are related and have affinities with the Syncephalideae in the Mucorales as well as with the Entomophthorales.

Order Eccrinales. Like the family Harpellaceae the members of this order also grow as parasites attached to the wall of the alimentary canal, stomach, intestine, or anal plates, of Arthropoda: insects, crustaceans, myriapods, etc., both aquatic and land-inhabiting species. The genera

Eccrina and *Enterobryus* were described by Leidy (1849) but were considered by him to be endobiotic algae. Leger and Duboscq (1905) first assembled these and other genera into a distinct order of fungi. Since that date these authors (1916, 1929a, and other papers), Poisson (1929, 1931), Lichtenstein (1917), and others have recognized twelve or more genera which Leger and Duboscq divide among three families. In general these fungi consist of unbranched or not extensively branched hyphae attached to the chitinous wall of the alimentary canal by a cup-like or disk-like holdfast of callose. The hyphal walls of the fungi are reported to be composed of cellulose. These hyphae are coenocytic at first and may attain the length of 100 μ in some species of *Amoebidium* up to over 10 mm. in *Arundinula capitata* Leg. and Dub. Reproduction occurs by the formation of microspores, macrospores, and resting spores. The microspores arise as follows: In the distal portion of the hypha (this may be only a small portion or nearly the whole hypha) the nuclei arrange themselves axially in close proximity. Cross walls are then formed producing uninucleate microspores that are thin disks or short cylinders rarely attaining an axial length equal to the diameter of the filament. These usually have their walls free eventually from the hyphal wall and

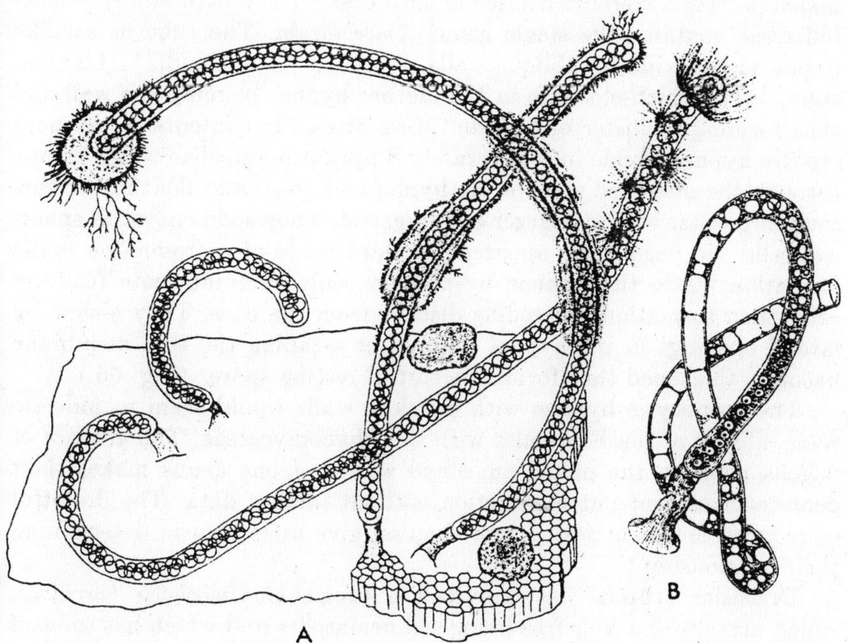

FIG. 65. Eccrinales, Family Eccrinaceae. *Enterobryus elegans* Leidy. (A) Young, not yet septate specimens. (B) Mature specimen, somewhat artificially coiled to save space in drawing. (After Leidy: *Smithsonian Inst. Pubs. Contribs. to Knowledge,* **5**(2):1–67.)

by the breaking of the filament at the apex or through lateral openings the spores escape. The macrospores are larger and may be uninucleate at first, the septation of the hyphae being often oblique. The nuclei divide until usually there are four nuclei to each macrospore. Their walls also become separate from the hyphal walls and from the septa. Holes are dissolved in the lateral wall near the apex or near the base of the containing cell and the macrospores escape. The resting spores are usually formed in the hyphae at about the time the chitinous wall of the posterior portion of the intestine begins to become free at the molting stage of the host. These cells may arise by the union of two naked protoplasts or "gametes" within the segment of the filament, the resulting zygotes becoming thick-walled (*Arundinula capitata*) or the segments of the hyphae may be binucleate and the two nuclei unite. Then a thick wall is formed around each such zygote. The cross septa of the hypha may disappear leaving the numerous resting spores free in the tube. They may escape by a distal opening (*Taeniellopsis orchestiae* Poisson, 1939). In some genera of the family Taeniellaceae the resting spores possess one nucleus at maturity. In the family Arundinulaceae[1] there are two nuclei and in the family Eccrinaceae there are four nuclei in each zygote. The sexual process is merely surmised in most cases. The fourth family, Amoebidiaceae contains the single genus *Amoebidium*. The tube or sac-like hypha may divide by oblique walls into 4–16 "endoconidia" (Lichtenstein, 1917) which elongate in the mother hypha, piercing the wall and thus forming a cluster of hyphae. More often the contents of the nonseptate hyphae divide into 2–4, rarely 8 pyriform amoebae which escape through the dissolved apex of the hypha and creep and float in the surrounding water with the larger end foremost. They soon encyst as spherical cells. No flagella are apparent. A third mode of reproduction is the formation inside the hyphae by oblique walls of uninucleate fusiform cells, their formation proceeding distally from the base. They escape by lateral openings in the hyphal wall. After escaping the thin membrane becomes thickened thus forming a sort of resting spore. (Fig. 65.)

The coenocytic hyphae with cellulose walls would seem to indicate some affinity of the Eccrinales with the Phycomyceteae. The absence of flagella even on the naked amoeboid spores of one genus makes their definite placement out of question without further data. The doubtful or reduced modes of sexual reproduction give little help in determining their relationship.

Drechsler (1946b) has described a fungus *Gonimochaete horridula*, which attacks and kills free-living soil nematodes and which has some of the characteristics of the Eccrinales, but which differs in many respects.

[1] Leger and Duboscq (1929a) used the name *Arundinula* (and the family name Arundinulaceae) instead of *Arundinella*, a preoccupied name.

Only asexual reproduction has been observed so far. In hyphal outgrowths from short one-celled thalli within the host are produced a few to numerous endoconidia with definite thin walls. These are pushed out several at a time at short intervals until all have been discharged. In germination they produce a small sticky knob by which they become attached to the living nematode host, into whose body a germ tube penetrates and then divides repeatedly to form the unicellular thalli. Possibly this fungus may have some relationship to this order.

Keys to the Families and More Important Genera of Mucorales

Key to the Families of Order Mucorales

Asexual sporangia, sporangioles or "conidia" aerial.

All the sporangia many-spored, with a well-developed columella; sporangium wall relatively thin and breaking or deliquescent. — Family Mucoraceae

All the sporangia many-spored, with a moderate-sized columella; sporangium wall thickened above and not breaking up or deliquescent. — Family Pilobolaceae

Terminal primary sporangium of the sporangiophore many-spored, with a well-developed columella; sporangium wall thin and breaking up or deliquescent; secondary sporangia in the form of few-celled or one-celled sporangioles which are usually indehiscent. Primary sporangia lacking under unfavorable conditions, and never formed in a few genera.

Sporangioles on more or less dichotomous branches formed laterally along the main sporangiophore. (Primary sporangium lacking in the genus *Chaetocladium*.) — Family Thamnidiaceae

Sporangioles on the surface of rounded or elongated heads terminating sporangiophores apart from the primary sporangiophore. Primary sporangium lacking in the genera *Cunninghamella, Mycotypha*, etc. — Family Choanephoraceae

Sporangia spherical, many-spored, with a basal septum and no columella. — Family Mortierellaceae

Sporangia narrow, one- to several-spored, with no columella, usually more or less capitately borne, often breaking apart into one-spored segments. — Family Piptocephalidaceae

Sporangia reduced to one-celled, indehiscent sporangioles ("conidia") borne singly on sterigmata arranged on one side of a branch ("sporocladium") so as to resemble a comb. — Family Kickxellaceae

Sporangia, zygospores and chlamydospores in the interior of rounded masses of hyphae; often buried in humus or soil. — Family Endogonaceae

Key to the More Important Genera of Family Mucoraceae

Sporangiophores arising from the thin or thick mat of aerial mycelium, not from stolons. Not repeatedly forked dichotomously.

Primary sporangium present at the apex of the sporangiophore.

Sporangiophore tall, dark, unbranched, with metallic appearance. — *Phycomyces*

Sporangiophore slender and uniformly thick its whole length, unbranched, or racemosely or cymosely branched—in the latter case the branches sometimes circinately curved; not dark-colored, nor of a metallic appearance. Sexual reproduction practically isogamous. Not parasitic on fungi (but sometimes saprophytic). *Mucor*

Like *Mucor* but parasitic upon Mucorales. *Parasitella*

Much like *Mucor* but sporangiophore much broader below, tapering upward, somewhat metallic in appearance. Parasitic upon Basidiomycetes. Mycelium often with short lateral spine-like branches. *Spinellus*

Much like *Mucor* but sexual reproduction very strongly heterogamous; homothallic. *Zygorhynchus*

Primary sporangium missing at the apex of sporangiophore which is cymosely branched with circinately recurved branches.

Sporangia spherical, without apophysis, spores spherical. *Circinella*

Sporangia pyriform, with apophysis, spores ellipsoid. *Pirella*

Sporangiophores repeatedly forked dichotomously; homothallic, isogamous. *Sporodinia*

Sporangiophores arising from stolons.

Sporangia mostly large, mostly spherical, spores dark and striate; sporangiophores arise at rooting nodes of the stolon. *Rhizopus*

Sporangia small, pyriform, spores colorless or colored; sporangiophores arising from the summit of the arch of curving stolons which mostly root at the nodes.

Stolons forming a series of regular, rather short arches with the sporangiophores arising from their summits. *Absidia*[2]

Stolons forming longer, less regular curves, the sporangiophores mostly clustered at their summits but some scattered singly. *Tieghemella*

Stolons as in *Tieghemella* but the sporangiophores in a whorl midway along the curve. *Mycocladus*

Key to the Genera of Family Pilobolaceae

Sporangiophore swollen to a subsporangial vesicle which is broader than the sporangium, which is violently discharged. *Pilobolus*

Sporangiophore uniform in thickness, without subsporangial vesicle. Sporangium not discharged. *Pilaira*

Key to the Genera of Family Thamnidiaceae

Sporangiophore terminated by a primary sporangium that is many-spored and with a columella, and bearing lateral variously branched clusters of secondary sporangiophores with dehiscent or indehiscent sporangioles.

Sporangiole-bearing branches dichotomously forked, walls of sporangia not strongly cutinized. *Thamnidium*

Sporangiole-bearing branches dichotomously forked, walls strongly cutinized. Very strongly heterogamous. *Dicranophora*

Sporangiole-bearing branches in whorls, many of the hyphae terminating in awl-like extensions. *Chaetostylum*

Sporangiole-bearing branches circinately curved. *Helicostylum*

No primary sporangium formed, main axes of sporangiophore and of the branches terminating in awl-like extensions. *Chaetocladium*

---

[2] *Tieghemella*, *Mycocladus*, and three other genera, *Proabsidia*, *Protoabsidia*, and *Lichtheimia*, are often all included in *Absidia*.

Key to the Genera of Family Choanephoraceae

Primary sporangia with columella, terminal to an upright sporangiophore. Sporangioles borne on the surface of spherical heads clustered at the upper end of special sporangioliferous hyphae.

Sporangioles containing 3 (to 6) spores, resembling those of the sporangium. *Blakeslea*

Sporangioles one-spored, indehiscent. Sexual reproduction of the *Mucor* type. *Choanephora*

Primary sporangia lacking, only the indehiscent one-spored sporangioles formed.

Sporangioles borne on rounded heads on short lateral branches, mostly rough.

Fertile hyphae in a dense, tangled mass of aerial mycelium and bearing the heads of sporangioles terminally and on lateral branches without regular order, sporangioles smooth or rough, round or ovoid. Sexual reproduction of the *Mucor* type. *Cunninghamella*

Fertile hyphae upright, forking into S-shaped branches, sterile at the tips but bearing heads of spherical rough sporangioles on short lateral branches. *Sigmoideomyces.*

Fertile hyphae upright, supported at the base by mostly four spreading branches and at the top dichotomously dividing several times into branches sterile at their tips but bearing on short lateral branches the heads of spherical rough sporangioles. *Thamnocephalis*

Sporangioles small, ellipsoidal, smooth, borne on an elongated head which becomes septate at maturity and with the sporangioles resembling a head of *Typha.* *Mycotypha*

Key to the Genera of Family Mortierellaceae

Sporangiophores erect, unbranched, or branched. *Mortierella*

Sporangiophores creeping. *Herpocladiella*

Sporangiophores short and unbranched along the sides of a hypha of indeterminate growth. *Dissophora*

(*Haplosporangium* possibly belongs to this family.)

Key to the Genera of Family Piptocephalidaceae

Main sporangiophores with large heads covered by the radiating narrow sporangia.

Main sporangiophore branched, sporangia arising directly, not from sterigmata. *Syncephalastrum*

Main sporangiophores stout, not branched, tapering toward the terminal head. In most cases sporangia borne on sterigmata, one or more on each. *Syncephalis*

Main sporangiophores branching freely above by true or false dichotomy with small heads or only slight enlargements from which the sterigmata arise.

Branches straight, sterigmata one-celled. *Piptocephalis*

Branches more or less spirally curved, sterigmata two-celled, two-celled sporangia arising from the apex of each cell. *Dispira*

Main sporangiophores with small or medium size heads terminal to the hyphae and their branches, bearing on all sides two-celled sporangioles. Origin of these latter not definitely worked out and relationship uncertain. *Spinalia*

Key to the Genera of the Family Kickxellaceae

Sporocladia in a whorl at the apex of the sporangiophore. *Kickxella*
Sporocladia pleurogenous on the sporangiophore.
 Sporocladia producing sporangioles (conidia) on the upper side. *Martensella*
 Sporocladia producing sporangioles on the lower side. *Coemansia*

Key to the Genera of Family Endogonaceae

Fruiting body containing reproductive cells throughout, either sporangia, or chlamydospores or zygospores. *Endogone*
Fruiting body hollow, containing chlamydospores scattered irregularly in the thin wall. *Glaziella*
Fruiting body not hollow, rather firm, chlamydospores arranged in regular layers. *Sclerocystis*

Key to the Genera of Order Entomophthorales

Parasitic in the gametophytes of ferns. *Completoria*
Parasitic in the desmid genus *Closterium*. *Ancylistes*
Saprophytic in the excrement of frogs and lizards. *Basidiobolus*
Saprophytic or weakly parasitic on fungi (rarely on insects); the septate mycelium extensive: conidiophores elongated, gametangia unequal, the zygospore developed in the larger, receptive gametangium. *Conidiobolus*
Parasitic in insects.
 Conidia produced internally in the body of the insect, not discharged with violence. *Massospora*
 Conidia produced on conidiophores extruded through the body wall, discharged with violence. *Entomophthora*[3]

Key to the Families and Genera of Order Zoopagales

Saprophytes or commensal parasites, attached to, but not penetrating through, the membrane of the alimentary canal of aquatic insects (chiefly larvae).
 Fungus filament not branched, zygospores approximately spherical. Family Harpellaceae
  Conidia curved cylindrical, with 4 basal filaments. *Harpella*
  Conidia navicular, stalked, with one basal filament. *Stachylina*
  Conidia tubular, borne side by side on the racket-shaped terminal cell of the filament. *Opuntiella*
 Fungus filament branched, zygospores navicular or biconical, stalked. Family Genistellaceae
  Conidia elongate-ovoid, with 2 basal filaments, borne on one-sided spike. *Genistella*
  Conidia long stalked, curved like a banana, in a crown at apex of plant. *Orphella*
  Conidia terete on a one-sided spike. *Stipella*
  Conidia rod-shaped arising laterally on a somewhat twisted stipe, with one basal filament. *Glotzia*
  Conidia ovoid, with single basal filament, on one-sided spikes. Plant bulbous at base. *Graminella*

---

[3] The name *Empusa* is not valid for this, having been used previously for an orchid.

Parasitic by large or small haustoria or internal hyphae, in soil-inhabiting Rhizopods or Nematodes. Family Zoopagaceae

Main fungus body a thick spiral structure in the cell of an amoeba, conidia borne on slender aerial hyphae.

Conidia borne singly at intervals on long aerial hyphae. *Endocochlus*

Conidia borne in chains from erect aerial hyphae. *Cochlonema*

Main fungus body a very much enlarged conidium on the outside of the body of an amoeba, attached by dichotomously forked internal haustoria. Conidia in long chains on erect aerial hyphae. *Bdellospora*

Main fungus body an effuse mycelium, adhering to Rhizopods or Nematodes and sending lobed haustoria into them.

Conidia in vertical chains on short lateral branches. On Rhizopods. *Zoopage*

Conidia single, sessile or almost so, upright on the creeping mycelium. On Rhizopods. *Acaulopage*

Conidia single on upright branches from creeping mycelium. On Rhizopods and Nematodes. *Stylopage*

No conidia found but asexual reproduction by means of terminal, lateral or intercalary chlamydospores. On Rhizopods and Nematodes. *Cystopage*

Main fungus body an extensive mycelium in the body of Nematodes. External hyphae formed and bearing at right angles a number of short lateral conidiiferous branches which produce first a terminal conidium and then successively in basipetal direction other conidia. *Euryancale*

Key to the Families of Order Eccrinales

No amoeboid spores.

Resting spore at maturity with a single nucleus. Family Taeniellaceae

Resting spores at maturity with two nuclei. Family Arundinulaceae

Resting spore at maturity with four nuclei. Family Eccrinaceae

Amoeboid spores produced, also some resting spores. Family Amoebidiaceae

Literature Cited

ALLEN, RUTH F., AND HALLY D. M. JOLIVETTE: A study of the light reactions of Pilobolus, *Trans. Wisconsin Acad. Sci.*, **17**(1):533–598. *Figs*. 1–16. 1913.

ATKINSON, G. F.: The genus Endogone, *Mem. Brooklyn Botan. Garden*, **1**:1–17. 1918.

AYERS, THEODORE T.: Parasitism of Dispira cornuta, *Mycologia*, **27**(3):235–261. *Figs*. 1–4. 1935.

BERDAN, HELEN: Revision of the genus Ancylistes, *Mycologia*, **30**(4):396–415. *Figs*. 1–22. 1938.

BLAKESLEE, ALBERT F.: Sexual reproduction in the Mucorineae, *Proc. Am. Acad. Arts. Sci.*, **40**:205–319. *Pls*. 1–4. 4 *figs*. 1904.

———: Sexuality in Mucors, *Science*, N.S., **51**:375–409. 1920.

BREFELD, OSCAR: Zygomycetes, in Botanische Untersuchungen über Schimmelpilze, Heft. 1, pp. 1–64. *Pls*. 1–6. Leipzig, Arthur Felix, 1872.

———: Mortierella Rostafinskii, *ibid.*, Heft. 4, pt. 5, pp. 81–96. *Pl*. 5. Leipzig, Arthur Felix, 1881.

———: Die Hemiasci und die Ascomyceten, in Gemeinschaft mit DR. FRANZ VON TAVEL UND GUSTAV LINDAU, in Untersuchungen aus dem Gesammtgebiete der Mykologie, Heft. 9, pp. 1–156. *Pls*. I–III*B*. Münster i. W., Heinrich Schöningh, 1891.

Bucholtz, Fedor: Beiträge zur Kenntnis der Gattung Endogone Link, *Botan. Centr. Beihefte, Zweite Abt.*, **29**:147–225. *Pls.* 3–10. 1912.

Buller, A. H. Reginald: Upon the ocellus function of the subsporangial swelling of Pilobolus, *Brit. Mycol. Soc. Trans.*, **7**:61–64. 1921.

———: The biology and taxonomy of Pilobolus, in Researches on Fungi, vol. 6, pt. 1, pp. 1–224. 111 *figs.* London, Longmans, Green and Co., 1934.

Burgeff, H.: Über Arten und Artkreuzung in der Gattung Phycomyces Kunze, *Flora*, **118–119**:40–46. *Figs.* 1–2. 1925.

Couch, J. N.: A new dioecious species of Choanephora, *J. Elisha Mitchell Sci. Soc.*, **41**(1–2):141–150. *Pls.* 8–11. 1925.

———: A new Conidiobolus with sexual reproduction, *Am. J. Botany*, **26**(3):119–130. *Figs.* 1–53. 1939.

Cutter Jr., Victor M.: Nuclear behavior in the Mucorales; I. The Mucor pattern, *Bull. Torrey Botan. Club*, **69**(7):480–508. *Figs.* 1–66. 1942a; II. The Rhizopus, Phycomyces and Sporodinia patterns, *ibid.*, **69**(8):592–616. *Figs.* 1–21. 1942b.

Dangeard, P. A.: Les Mucorinées, *Le Botaniste*, **9**:227–253. 1906.

Dobbs, C. G.: The life history and morphology of Dicranophora fulva Schröt., *Brit. Mycol. Soc. Trans.*, **21**:167–192. *Pls.* 10–11. *Text figs.* 1–80. 1938.

Drechsler, Charles: A new species of Helicocephalum, *Mycologia*, **26**(1):33–37. *Pl.* 4. 1934.

———: Some conidial Phycomycetes destructive to terricolous Amoebae, *ibid.*, **27**(1):6–40. *Pls.* 1–7. 1935a.

———: Some non-catenulate conidial Phycomycetes preying on terricolous Amoebae, *ibid.*, **27**(2):176–205. *Figs.* 1–5. 1935b.

———: A new species of conidial Phycomycetes preying on nematodes, *ibid.*, **27**(2):206–215. *Fig.* 1. 1935c.

———: A new species of Stylopage preying on nematodes, *ibid.*, **28**(3):241–246. *Fig.* 1. 1936a.

———: New conidial Phycomycetes destructive to terricolous Amoebae, *ibid.*, **28**(4):363–389. *Figs.* 1–7. 1936b.

———: New Zoopagaceae destructive to soil Rhizopods, *ibid.*, **29**(2):229–249. *Figs.* 1–6. 1937.

———: New Zoopagaceae capturing and consuming soil amoebae, *ibid.*, **30**(2): 137–157. *Figs.* 1–4. 1938.

———: A few new Zoopagaceae destructive to large soil rhizopods, *ibid.*, **31**(2): 128–153. *Figs.* 1–7. 1939a.

———: Five new Zoopagaceae destructive to rhizopods and nematodes, *ibid.*, **31**(4):388–415. *Figs.* 1–5. 1939b.

———: Four Phycomycetes destructive to nematodes and rhizopods, *ibid.*, **33**(3):248–269. *Figs.* 1–5. 1941.

———: New species of Acaulopage and Cochlonema destructive to soil amoebae, *ibid.*, **34**(3):274–297. *Figs.* 1–6. 1942.

———: Several additional Phycomycetes subsisting on nematodes and amoebae, *ibid.*, **37**(1):1–31. *Figs.* 1–4. 1945.

———: Three new Zoopagaceae subsisting on soil amoebae, *ibid.*, **38**(2):120–143. *Pls.* 1–6. 1946a.

———: A nematode-destroying Phycomycete forming immotile spores in aerial evacuation tubes, *Bull. Torrey Botan. Club*, **73**(1):1–17. *Figs.* 1–36. 1946b.

———: Three Zoopagaceous fungi that capture and consume soil-inhabiting rhizopods, *Mycologia*, **39**(3):253–281. *Figs.* 1–8. 1947a.

———: Three new species of Zoopage predaceous on terricolous rhizopods, *ibid.*, **39**(4):379–408. *Figs.* 1–7. 1947b.

EIDAM, EDUARD: Basidiobolus, eine neue Gattung der Entomophthoraceen, *Beitr. Biol. Pflanz.*, **4**(2):181–251. *Pls.* 9–12. 1887.

FAIRCHILD, D. G.: Ueber Kerntheilung und Befruchtung bei *Basidiobolus ranarum* Eidam, *Jahrb. wiss. Botan.*, **30**:285–296. *Pls.* 13–14. 1897.

FENNER, E. ALINE: Mycotypha microspora, a new genus of the Mucoraceae, *Mycologia*, **24**(2):187–198. *Pls.* 2–3. *Fig.* 1. 1932.

FITZPATRICK, HARRY MORTON: The Lower Fungi—Phycomycetes, xi + 331 pp. 112 *figs.* New York, McGraw-Hill Book Co., 1930.

FLINT, LEWIS H.: Note on phototropism in Pilobolus, *Am. J. Botany*, **29**(8):672–674. *Fig.* 1. 1942.

GAUTHIER, MARCELLE: Sur un nouvel entophyte du groupe des Harpellacées Leg. et Dub.; parasite des larves d'Éphémérides, *Compt. rend.*, **202**(12):1096–1098. *Figs.* 1–4. 1936.

GOLDRING, DAVID: The effect of environment upon the production of sporangia and sporangioles in Blakeslea trispora Thaxter, *Ann. Missouri Botan. Garden*, **23**:527–542. *Pl.* 25. 1936.

GROVE, W. B.: A systematic account and arrangement of the Pilobolidae; in BULLER, A. H. REGINALD: Researches on Fungi, vol. 6, chap. 4, pp. 190–224. *Figs.* 97–111. London, Longmans, Green and Co., 1934.

HOPKINS, E. W.: Microchemical tests on the cell walls of certain fungi. Cellulose and chitin, *Trans. Wisconsin Acad. Sci.*, **24**:187–196. 1929.

JOLIVETTE, HALLY D. M.: Studies on the reactions of Pilobolus to light stimuli, *Botan. Gaz.*, **57**(2):89–121. *Figs.* 1–12. 1914.

KANOUSE, BESSIE B.: Studies of two species of Endogone in culture, *Mycologia*, **28**(1):47–62. *Figs.* 1–33. 1936.

KEENE, MARY LUCILE: Cytological studies of the zygospores of Sporodinia grandis, *Ann. Botany*, **28**(111):455–470. *Pls.* 35–36. 1914.

———: Studies of the zygospore formation in Phycomyces nitens Kunze, *Trans. Wisconsin Acad. Sci.*, **19**:1195–1220. *Pls.* 16–18. 1919.

KRAFCZYK, HANS: Die Zygosporenbildung bei Pilobolus cristallinus, *Ber. deut. botan. Ges.*, **49**(3):141–146. *Figs.* 1–2. 1931.

LEGER, L., ET O. DUBOSCQ: Les Eccrinides, nouveau groupe de Protophytes parasites, *Compt. rend.*, **141**(9):425–427. 1905.

———, ET ———: Sur les Eccrinides des Hydrophilides, *Arch. zool. exptl. et gén.*, **56**(2):21–31. *Figs.* 1–4. 1916.

———, ET ———: Eccrinoides Henneguyi n.g. et n.sp. et la systématique des Eccrinides, *Archives d'Anatomie Microscopique*, **25**:309–324. *Figs.* 1–6. 1929.

———, ET ———: L'évolution des Paramoebidium, nouveau genre d'Eccrinides, parasite des larves aquatiques d'insectes, *Compt. rend.*, **189**:75–77. *Figs.* 1–15. 1929a.

———, ET ———: Harpella melusinae n.g. et n.sp. Entophyte eccriniforme parasite des larves de Simulie, *ibid.*, **188**(14):951–954. *Figs.* 1–6. 1929b.

———, ET MARCELLE GAUTHIER: Endomycètes nouveaux des larves aquatiques d'insectes, *ibid.*, **194**(26):2262–2265. *Figs.* 1–3. 1932.

———, ET ———: La spore des Harpellacées (Leger et Duboscq), champignons parasites des insectes, *ibid.*, **200**(17):1458–1460. 1935.

LEIDY, JOSEPH: Enterobryus, a new genus of Confervaceae, *Proc. Acad. Nat. Sci. Phila.*, **4**:225–227. 1849.

———: A flora and fauna within living animals, *Smithsonian Inst. Pubs. Contribs. to Knowledge*, **5**(2):1–67. *Pls.* 1–10. 1851 (1853).

LENDNER, A.: Les Mucorinées de la Suisse, *Beiträge zur Kryptogamenflora der Schweiz*, **3**(1):1–180. *Pls.* 1–3. *Figs.* 1–59. 1908.

LEVISOHN, IDA: Beitrag zur Entwicklungsgeschichte und Biologie von Basidiobolus ranarum Eidam, *Jahrb. wiss. Botan.*, **66**(3):513–555. *Figs.* 1–15. 1927.

LICHTENSTEIN, JEAN L.: Sur un Amoebidium à commensalisme interne du rectum des larves d'Anax imperator Leach, Amoebidium fasciculatum n.sp., *Arch. zool. exptl. et gén.*, **56**(3):49–62. *Figs.* 1–7. 1917.

———: Sur un mode nouveau de multiplication chez les Amoebidiacées, *ibid.*, **56**(4):95–99. *Fig.* 1. 1917.

LINDER, DAVID H.: The genera Kickxella, Martensella and Coemansia, *Farlowia*, **1**(1):49–77. *Pls.* 1–4. 1943.

LING-YOUNG: Étude biologique des phenomènes de la sexualité chez les Mucorinées, *Rev. gén. botan.*, **42**:144–158, 205–218, 283–296, 348–365, 409–428, 491–504, 535–552, 618–639, 681–704, 722–752. *Figs.* 1–36. 1930; **43**:30–43. 1931.

MANGIN, L.: Observations sur la membrane des Mucorinées, *J. botan.*, **18**:209–216, 339–348, 371–378. *Pls.* 7–8. *Figs.* 1–7. 1899.

MOREAU, FERNAND: Recherches sur la reproduction des Mucorinées et de quelques autres Thallophytes, *Le Botaniste*, **13**:1–127. *Pls.* 1–14. 1913.

NABEL, KURT: Über die Membran niederer Pilze, besonders von Rhizidiomyces bivellatus nov. spez., *Arch. Mikrobiol.*, **10**:515–541. *Figs.* 1–7. 1939.

NAUMOV, N. A.: Opredelitel Mukorovych (Identification of Mucorales), 2d rev. ed., 140 pp. 49 *figs.* 1 *diagram.* Moscow and Leningrad, Botanical Institute of the Academy of Sciences U.S.S.R., 1935. (In Russian.)

———: Clés de Mucorinées (Mucorales). Translated by S. Buchet and I. Mouraviev, Encyclopédie Mycologique, vol. 9, pp. 1–137, i–xxxiv. 83 *figs.* Paris, Paul Lechevalier, 1939.

VON OVEREEM, C.: Ueber ein merkwuerdiges Vorkommen von Basidiobolus ranarum Eidam. Beiträge zur Pilzflora von Niederländisch Indien No. 10, *Bull. Jardin Botanique de Buitenzorg*, sér. III, **7**:423–431. *Pl.* 1. *Figs.* 1–4. 1925.

PARR, ROSALIE: The response of Pilobolus to light, *Ann. Botany*, **32**(126):177–205. *Figs.* 1–4. 1918.

POISSON, RAYMOND: Recherches sur quelques Eccrinides parasites de Crustacés Amphipodes et Isopodes, *Arch. zool. exptl. et gén.*, **69**(3):179–216. *Figs.* 1–23. 1929.

———: Recherches sur les Eccrinides. Deuxième contribution, *ibid.*, **74**(4):53–68. *Figs.* 1–7. 1931.

SATINA, SOPHIA, AND A. F. BLAKESLEE: Studies on biochemical differences between (+) and (−) sexes in Mucor: 2. A preliminary report on the Manilov reactions and other tests, *Proc. Nat. Acad. Sci. U.S.*, **12**:191–196. 1926.

SAWYER JR., W. H.: Observations on some entomogenous members of the Entomophthoraceae in artificial culture, *Am. J. Botany*, **16**:87–121. *Pls.* 9–12. 1929.

SCHOSTAKOWITSCH, W.: Mucor proliferus n.s. Eine neue sibirische Mucorart, *Ber. deut. botan. Ges.*, **14**:260–263. *Pl.* 18. 1896.

SJÖWALL, MALTE: Studien über Sexualität, Vererbung und Zytologie bei einigen Mucoraceen, 97 pp. 27 *figs.* Lund, 1945.

SPEARE, A. T.: Fungi parasitic upon insects injurious to sugar cane, *Hawaiian Sugar Planters' Assoc., Rept. Expt. Sta., Pathol. Physiol. Ser. Bull.* 12:1–62. *Pls.* 1–6. 1912.

SWINGLE, D. B.: Formation of spores in the sporangia of Rhizopus nigricans and of Phycomyces nitens, *U.S. Dept. Agr. Bur. Plant Industry Bull.* 37:1–40. *Pls.* 1–6. 1903.

THAXTER, ROLAND: The Entomophthoreae of the United States, *Mem. Boston Soc. Natural History*, **4**(6):133–201. *Pls.* 14–21. 1888.

———: On certain new or peculiar North American Hyphomycetes: II. Helicocephalum, Gonatorrhodiella, Desmidiospora nova genera and Everhartia lignatilis n.sp., *Botan. Gaz.*, **16**(6):201–205. *Pls.* 19–20. 1891.

———: New or peculiar American Zygomycetes: I. Dispira, *ibid.*, **20**(12):513–518. *Pl.* 34. 1895.

———: New or peculiar Zygomycetes: 3. Dissophora, and Haplosporangium, nova genera, *ibid.*, **58**(4):353–366. *Pls.* 26–29. 1914.

VAN TIEGHEM, P.: Nouvelles recherches sur les Mucorinées, *Ann. sci. nat. Botan.*, 6me sér., **1**:5–175. *Pls.* 1–4. 1875.

VANDENDRIES, RENÉ, ET H. J. BRODIE: Nouvelle investigations dans la domaine de la sexualité des Basidiomycètes et étude expérimentale des barrages sexuels, *La Cellule*, **42**(2):165–209. 1 *pl.* 27 *figs.* 1933.

VUILLEMIN, PAUL: Le Spinalia radians g. et sp. nov. et la série des Dispirées, *Bull. soc. mycol. France*, **20**:26–33. *Pl.* 2. 1904.

———: Une nouvelle espèce de Syncephalastrum; affinités de ce genre, *Compt. rend.*, **174**(15):986–988. 1922.

WALKER, LEVA B.: Some observations on the development of Endogone malleola Hark., *Mycologia*, **15**(6):245–257. *Pls.* 26–27. *Figs.* 1–3. 1923.

WEBER, GEORGE F., AND FREDERICK A. WOLF: Heterothallism in Blakeslea trispora, *Mycologia*, **19**(6):302–327. *Pls.* 28–30. 1927.

VON WETTSTEIN, FRITZ: Das Vorkommen von Chitin und seine Verwertung als systematisch-phylogenetisches Merkmal im Pflanzenreich, *Sitz. ber. Akad. Wiss. Wien, Math. naturw. Klasse, Abt. I*, **130**(1):3–20. 1921.

ZOPF, WILHELM: Zur Kenntniss der Infections-Krankheiten niederer Thiere und Pflanzen: IV. Einfluss von Parasitismus auf Zygosporenbildung bei Pilobolus crystallinus, *Nova Acta Leopoldina*, **52**(7):352–358. *Pl.* 22(6). 1888.

ZYCHA, H.: Mucorineae, in Kryptogamenflora der Mark Brandenburg, vol. 6a, pp. 1–264. 114 *figs.* Leipzig, Gebrüder Borntraeger, 1935.

8

THE HIGHER FUNGI: CARPOMYCETEAE

THE fungi that have been the subjects of discussion in the foregoing chapters are frequently designated as the Lower Fungi or Phycomyceteae (excluding the Mycetozoa) in contrast to the often much more highly developed forms usually designated as the Higher Fungi or Carpomyceteae. It may be well here to contrast the two groups as to their main points of difference, recognizing that though these distinctions are in general valid yet there are many forms of Higher Fungi in which these differences are not recognizable.

Structure

1. The Phycomyceteae are prevailingly tubular coenocytes, nonseptate except at regions of injury or where reproductive organs are cut off from the rest of the mycelium. (Exceptions may be noted in old aerial mycelium of some Mucorales and in most Entomophthorales.) The simpler members of this class may consist of single cells. The Higher Fungi are cellular, i.e., their hyphae are divided by septa into true cells, usually uninucleate or binucleate. (Exceptions are numerous. Thus old cells sometimes become multinucleate; on the other hand, many young hyphae from germinating spores or young ascogenous hyphae may delay the formation of cross walls for some time. Aside from these, numerous species are scattered throughout the whole group in which almost all the cells are plurinucleate.)

2. In the Lower Fungi sexual reproduction leads to the formation of a single, usually thick-walled, oospore or zygospore. In the Higher Fungi, wherever true sexual organs or processes can be distinguished, the union of sperm and egg, or of cells substituted for these, leads to the production of a many-celled structure, called by Sachs (1874) a "spore-fruit," all the cells of which may become the reproductive cells (as the ascospores in the asci of the order Saccharomycetales) or of which only a part are reproductive cells, the remainder being necessary components of a more or less complicated fruit body (e.g., the sum of the ascogenous hyphae

and asci arising from the oogone of *Pyronema*, along with the various hyphae of vegetative origin, all of which form the apothecium, while only the ascospores are the reproductive cells).

3. In the Phycomyceteae, except the Mucorales and the Entomophthorales and allied forms, uniflagellate or biflagellate zoospores are generally produced as a means of asexual reproduction, while in these orders the asexual reproductive structures are clearly only modifications of zoosporangia. In the Higher Fungi the asexual reproductive cells are conidia, which are separable single cells (sometimes several-celled structures) of the mycelium with no indication of ever having had any homology with zoosporangia.

4. The aquatic Phycomyceteae and the terrestrial order Peronosporales have walls composed of cellulose (or carbohydrates closely related to cellulose). Cellulose is lacking in the Higher Fungi, or when present has been shown by Thomas (1928, 1930) to play only a subordinate part. In these the basis of the wall is chitin surrounded or mixed with carbohydrates and more or less fatty substances. Hopkins (1929) reports that neither chitin nor cellulose is present in several species of the class Ascomyceteae and of the class Basidiomyceteae, although other species of both classes tested by him showed chitin but not cellulose to be present, as is claimed by von Wettstein (1921) to be probably universal in these two classes. Senft (1916) reports that the ascogenous hyphae of *Chrysothrix nolitangere* Mont., a lichen, respond to the standard cellulose tests. The same is true, according to Honey (1936), of the disjunctors in the chains of conidia of *Monilinia*. Malençon (1931) has shown that the epispore of the spores of some Agaricaceae, e.g., species of *Russula* and *Lactarius*, is blued by iodine reagents. He believes the epispore to contain a mixture of cellulose and callose. This was first observed by Rolland in 1887. Kühner (1936) finds a similar reaction in the outer layer of the internal hyphae of the pileus and stipe of some species of *Mycena*. The Mucorales and Entomophthorales seem to occupy an intermediate position in that they have chitin in their cell-walls and rarely have true cellulose, although other carbohydrates are present in the wall.

Parasitism

The Higher Fungi are with few exceptions terrestial or epiphytic, a very few species are truly aquatic, some being parasitic upon marine algae, but they possess none of the characteristics of a primary aquatic habit such as zoospores, etc. One order, the Laboulbeniales, consists exclusively of insect-infesting parasites. The remainder are mostly either saprophytes, or parasites upon plants (exceptionally upon animals). All degrees of parasitism are found, varying from destructive parasitism in which the tissues of the host are killed even before the invading hyphae reach them (e.g., *Sclerotinia sclerotiorum* (Lib.) de By.) to extreme cases of balanced parasitism (e.g., some of the Ustilaginales in which the fungus and host, together grow for a long time with little apparent harm to the latter). One large group has developed a special type of parasitism upon terrestrial or epiphytic fresh-water algae, forming peculiar structures which enclose the algal hosts. These are the Lichens. Some of the most harmful fungi from the economic standpoint, the Rusts and Smuts, are found among the Higher Fungi.

Reproduction

Asexual reproduction by means of conidia is widely distributed throughout the various classes and orders, but seems to be entirely lacking in some of the groups. The conidial production may consist of the abstriction of a single cell from the tip of a short, unbranched conidiophore or the conidiophore may be longer and branched. The conidia may be produced successively at the tip, separating from the conidiophore as soon as formed or clinging together in a mucilaginous drop or remaining attached in a chain. A chain of conidia may produce new conidia acrogenously, i.e., each new conidium arises from the apical conidium of the chain, so that the basal conidium is oldest and the apical one the last formed. This is the type of conidial formation in *Cladosporium*, *Alternaria*, etc. On the other hand the conidial formation may be basigenous, i.e., each new conidium is produced at the apex of the conidiophore just below the last formed conidium (e.g., Erysiphaceae, Aspergillaceae). A third method of formation of chains of conidia is the almost simultaneous rounding up of the cells of a simple or branched hypha into catenulate conidia. These conidia then fall apart almost simultaneously (e.g., *Monilinia*, *Oospora*, etc.). Where the conidia formed in this manner are small they are often called "oïdia."

The conidiophores may be scattered or crowded or enclosed in a hollow structure provided with an apical opening (i.e., produced in a pycnidium). If the conidiophores become laterally crowded and adherent into a column we have a coremium, such as is formed in species of *Penicillium* under certain environmental conditions.

The conidia themselves may vary greatly in size and shape as well as color and number and arrangement of cells. In general we may distinguish conidia destined for distribution by air currents and those that are distributed by other means. The latter are usually surrounded by a sticky substance when wet and cling to anything with which they come in contact, drying down and becoming firmly attached when dry. When wet by rain such conidia may be scattered by the currents of water running on the surface of the substratum or may be splashed about by the falling rain drops and distributed by the wind which carries these droplets about. Insects may become contaminated by the sticky spores and thus carry the fungus from place to place.

Chlamydospores are very frequent in many of the orders of the Carpomyceteae. They arise by the rounding up of mycelial cells, filled with food substances, and the production of a thick wall. They may be intercalary or terminal, single or two to many in a series. In some fungi they germinate promptly, in others they are able to remain dormant a long time.

The mycelium usually consists of long, slender branching hyphae with terminal growth. Usually the cell is several to many times as long as wide. The septa are centrally perforated by a pore which may be small or may approach in size the perforations found in the septa of the Florideae. In the order Laboulbeniales the cells are short and broad and the whole structure does not resemble typical mycelium. In the Phycomyceteae where septa do occur they are formed, as in the Higher Fungi, by an annular growth on the inner surface of the longitudinal wall which progresses inward, but instead of leaving a central pore it continues its development until the opening is obliterated.

Buller (1933) and Brodie (1942) give the diameter of these openings in the Carpomyceteae as of the order of 1.0 to 1.5 μ. The flow of cytoplasm through the septa has been observed by Ternetz (1900) and by Buller, sometimes to carry small vacuoles with the flowing stream. The rate of flow through the hyphae varies but Ternetz observed the speed of 10.5 cm. per hour in *Ascophanus carneus* (Pers.) Boud., and Buller the speed of 6 cm. per hour in *Fimetaria fimicola* (Roberge) Griffiths and Seaver.

Storage Organs

In many species in the class Basidiomyceteae peculiar structures called "clamp connections" occur at each septum. These will be described in connection with that class. The mycelium may be packed together laterally into compact strands with a firm hard outer layer, forming long "rhizomorphs" which may extend many meters. In many species storage organs, sclerotia, are formed in which the hyphae are packed tightly together and the short cells are pressed mutually to become polyhedral in shape. The adjacent cells adhere to form a parenchyma-like tissue, more properly called pseudoparenchyma. True parenchyma which is rare in the fungi consists of moderately thin-walled cells, isodiametric or at least not greatly elongated, which have arisen through the division in various planes of larger cells or of small meristematic cells. The walls separating these cells are the septa between two daughter cells. Pseudoparenchyma arises from closely crowded or interwoven hyphae divided by cross septa into rather short cells which by their enlargement and mutual pressure become similar in appearance to the cells of true parenchyma. Except for the septa separating the cells in the same hypha the cell walls are double, i.e., walls of separate cells pressed together and usually adhering, but not the result of division of one preexisting cell. The interior cells of a sclerotium are filled with stored food and the outer cells are thick-walled, almost sclerenchymatous, and usually dark in color. Pseudoparenchyma may be found also in the fruiting bodies of some of the Higher Fungi, such as the perithecia and apothecia of some of the Ascomyceteae. True parenchyma is apparently present in some fungi of this group. The body of the spore fruit is more often built up of elongated, more or less interwoven hyphae of elongated cells. This is often, though erroneously, as Starbäck (1895) points out, called prosenchyma. In parasitic species the mycelium often sends haustoria of various shapes into the host cells.

Fruiting Structures

Unlike the Lower Fungi whose fruiting structures are microscopic or at most only a few millimeters or centimeters in length (some Mucorales), the fruiting bodies of the Higher Fungi often attain considerable size. Thus *Calvatia gigantea* (Batsch ex Pers.) Lloyd, the giant puffball, was recorded by C. E. Bessey (1884) as producing a spore fruit 1.6 meters long, 1.35 meters wide, and about 24 cm. high. That of *Fomes officinalis* (Vill.) Fr. sometimes (rarely it is true) reaches the height of 60 cm. and a diameter of 15 to 20 cm. Clements (1910) reported that *Polyporus squamosus* (Huds.) Fr. is "said to attain a width of 7 feet and a weight of 40 pounds." Roger Heim (1936) has described three species of Boletaceae from South Africa and Madagascar 40 to 60 cm. in pileus diameter. Ljungh (1804) described a species of cup-fungus from Java, *Peziza cacabus* (now called *Geopyxis cacabus* (Fr.) Sacc.), with a spore fruit 3 ft. tall, the cup being 20 in. tall and 25 in. broad, and the hollow stipe 16 in. tall and 3 in. thick. Specimens of a form of *Agaricus arvensis* Fr. collected by the author had a pileus diameter of 30 cm. Such enormous spore fruits produce almost incredible numbers of spores. Thus Buller (1909) estimated that a puffball 40 × 28 × 20 cm. would produce about 7 trillion (7,000,000,000,000) spores. At the same rate the enormous puffball mentioned above would produce about 160 trillion (160,000,000,000,000) spores. A specimen of *Agaricus campestris* Fr. only 8 cm. in diameter produced over 1,800,000,000 spores at the rate of about 40,000,000 spores per hour. At this rate the *Agaricus arvensis* found by the author would have produced about 27,000,000,000 spores. A spore fruit of *Ganoderma applanatum* (Pers. ex Fr.) Pat. with an area of one square foot lower surface produced, according to White (1920), 30,000,000,000 spores a day for about six months, or a total of over 5,000,000,000,000 spores. On the other hand many of the Higher Fungi have microscopic spore fruits.

Coloring

In the majority the vegetative mycelium is colorless and that is true of the reproductive structures in many cases. In rhizomorphs and the outer layer of cells of sclerotia and sometimes in individual hyphae the color may be dark. This color seems to reside in the cell wall and is probably related chemically to melanin. Some fungi produce pigments in the interior of the hyphae. These may be soluble in various solvents and are sometimes variable in color depending upon the chemical reaction. Thus a *Fusarium* studied by the author (1904) and parasitic on *Sesamum orientale* L. produces a red- or violet-colored pigment that turns blue when the surrounding medium becomes alkaline. It is soluble in acids and the red form is soluble in solutions of their salts. Litmus is the product of the cells of one of the lichen-producing fungi, as is orcein. Some hyphae cause the coloration of the substratum in which they are growing. Thus wood in which *Chlorociboria aeruginosa* (Fr.) Seaver is growing takes on a green color, due to a pigment secreted by the mycelium. The reproductive structures are in the majority of cases colored, the pigments in some being within the cell (e.g., the bright red color of the hymenium of some Pezizales), in others in the cell wall. The latter is mostly the case with the light to dark brown or almost black coloration found in the majority of apothecia and perithecia and in the teliospores of Rusts and Smuts, the spore fruits of Tremellales, Auriculariales, etc. The fungus pigments, especially those occurring in the lichen-producing fungi, have been studied by various authors but no really comprehensive modern study has been made of the subject taking advantage of the more recent investigations in organic chemistry.

Nomenclature

The Higher Fungi were called by Charles E. Bessey (1907) the Phylum Carpomyceteae, i.e., fruit-producing fungi, in reference to the production of spore fruits in this phylum. The name Eumyceteae is often applied to this group but is here discarded in view of the fact that it has also been applied to include all the filamentous fungi in contrast to the nonfilamentous Mycetozoa, Chytridiales, etc. The Higher Fungi are divided into several classes whose distinctions are based on the type of the ultimate reproductive cells of the spore fruits.

Key to the Classes of Higher Fungi

The ultimate reproductive cells of the spore fruit are ascospores, produced mostly eight in number within the cell called an ascus, which starts out as a dicaryon cell in which the nuclei fuse, the resultant diploid nucleus then dividing meiotically until usually eight nuclei are produced, around which the ascospores are developed. Over 42,000 species have been described.
Class Ascomyceteae

The ultimate reproductive cells of the spore fruit are the basidiospores or sporidia, borne externally on a cell called the basidium which was originally a dicaryon cell in which the two nuclei united, or on the outside of a group of four cells produced by the division of, or as an outgrowth from, the cell with the diploid nucleus. About 32,000 species have been described. With considerable reluctance the author follows the practice of most mycologists and includes the Class Teliosporeae of the first edition of this textbook as a distinct subclass within the Class Basidiomyceteae. The teliospore characteristic of this subclass originates as a dicaryon cell whose nuclei unite. From this cell grows out a thin-walled, usually four-celled, filament, the promycelium, upon whose cells are borne the sporidia. The limits of this subclass coincide rather closely with the Hypodermii which form Order IV of Class IV, Coniomycetes, of Elias Fries (1832). Ascus, teliospore, and basidium appear to be homologous structures, originating as binucleate cells in which karyogamy occurs, followed by reduction division of the fusion nucleus, the nuclei thus formed becoming the nuclei of the ascospores, sporidia, or basidiospores, respectively.
Class Basidiomyceteae

In addition to these classes there is another class, the Fungi Imperfecti, often called Deuteromyceteae, consisting of fungi whose vegetative structures or mode of asexual reproduction show their relationship to the Higher Fungi, but which lack any sexual type of reproduction or structures substituted for such sexual reproduction. Perhaps the majority are asexual stages of Ascomyceteae though some are undoubtedly corresponding stages of Basidiomyceteae. Until the perfect (sexual) stages can be found, their true relationship cannot be ascertained. About 32,000 species have been recognized.
Class Fungi Imperfecti

Literature Cited

Bessey, Charles E.: An enormous puff-ball, *Am. Naturalist*, **18**(5):530. 1884.

———: A synopsis of plant phyla, *Univ. (Nebraska) Studies*, **7**(4):1–99. *Pl.* 1. 1907.

Bessey, Ernst A.: Die Bedingungen der Farbbildung bei Fusarium, *Flora*, **93**(4):301–334. 1904.

Brodie, Harold J.: Protoplasmic continuity in the powdery mildew, *Erysiphe graminis* DC., *Can. J. Research*, **C**, **20**:595–601. *Pl.* 1. *and Figs.* 1–9. 1942.

Buller, A. H. Reginald: Researches on Fungi: vol. 1, xi + 287 pp., 5 *pls.* 83 *figs.*, 1909; vol. 2, xii + 492 pp., 157 *figs.*, 1922; vol. 3, xii + 611 pp., 227 *figs.*, 1924; vol. 4, xiii + 329 pp., 4 *pls.*, 149 *figs.*, 1931; vol. 5, xiii + 416 pp., 174 *figs.*, 1933; vol. 6, xii + 513 pp., 231 *figs.*, 1934. London, Longmans, Green and Company.

Clements, Frederick E.: Minnesota Mushrooms, *Minnesota Plant Studies IV*, 169 pp. 124 *figs.* Univ. Minnesota, 1910.

Fries, Elias: Systema mycologicum, sistens fungorum ordines, genera et species huc usque cognitas, vol. 3, viii + 524 pp., index 200 pp. Greifswald, Ernest Mauritius, 1829–32.

Heim, Roger: Observations sur la flora mycologique malgache: III. Trois bolets gigantesques d'Afrique et de Madagascar, *Rev. Mycol.*, N.S., **1**(1):3–18. *Pls.* 1–4. 1936.

Honey, Edwin E.: North American species of Monilinia: I. Occurrence, grouping and life histories, *Am. J. Botany*, **23**(2):100–106. *Figs.* 1–4. 1936.

HOPKINS, E. W.: Microchemical tests on the cell walls of certain fungi. Cellulose and chitin, *Trans. Wisconsin Acad. Sci.*, **24**:187–196. 1929.

KÜHNER, ROBERT: Sur la réaction à l'iode des parois des hyphes des carpophores des Mycena, *Compt. rend.*, **203**(23):1287–1289. 1936.

LJUNGH, SVEN INGEMAR: Peziza Cacabus, en ny och besynnerlig svamp fran Java, *Kgl. Svenska Vetenskapsakad. Handl.*, **25**:39–41. *Pl.* 1. 1804.

MALENÇON, GEORGES: Considérations sur les spores des Russules et des Lactaires, *Bull. trimestr. Soc. mycol. France*, **47**:72–86. *Pl.* 3. *Figs.* 1–3. 1931.

ROLLAND, LEON: De la coloration en bleu developee par l'iode sur divers champignons et notamment sur un agaric, *Bull. Soc. mycol. France*, **3**:134–137. 1887.

SACHS, JULIUS: Lehrbuch der Botanik, vierte umgearbeitete Auflage, xvi + 928 pp. *Figs.* 1–492. Leipzig, Wilhelm Engelmann, 1874.

SENFT, E.: Beitrag zur Anatomie und zum Chemismus der Flechte Chrysothrix Nolitangere Mont., *Ber. deut. botan. Ges.*, **34**(8):592–600. *Pl.* 17. 1916.

STARBÄCK, K.: Discomyceten-Studien, *Bihang til Handl. Kgl. Svenska Vetenskapsakad.*, Band 21, Afd. III, No. 5, 1895.

TERNETZ, CHARLOTTE: Protoplasmabewegung und Fruchtkörperbildung bei Ascophanus carneus Pers., *Jahrb. wiss. Botan.*, **35**:273–312. *Pl.* 7. 1900.

THOMAS, R. C.: Composition of fungus hyphae: I. The Fusaria, *Am. J. Botany*, **15**(9):537–547. 1928; II. Sclerotinia, *ibid.*, **17**(8):779–788. 1930.

VON WETTSTEIN, FRITZ: Das Vorkommen von Chitin und seine Verwertung als systematisch-phylogenetisches Merkmal im Pflanzenreich, *Sitz. ber. Akad. Wiss. Wien, Math. naturw. Klasse, Abt. I*, **130**(1):3–20. 1921.

WHITE, J. H.: On the biology of Fomes applanatus (Pers.) Wallr., *Trans. Roy. Can. Inst.*, **12**(2):133–174. *Pls.* 2–7. 1920.

9

CLASS ASCOMYCETEAE: LABOULBENIALES AND DISCOMYCETES

Introduction

THE members of the two extremes of this class have little in common beyond the production of the ascus. The type of sexual union (which is often absent), the plan of the spore fruit, even the nature and size of the vegetative mycelium vary tremendously. At the one extreme we find the Yeasts (Order Saccharomycetales) in some of which the unicellular plant becomes transformed directly into an ascus, while near the other extreme are the "Discomycetes" (Lecanorales, Pezizales, etc.) in which there is a well-developed mycelium and in some species a sexual union of a nonmotile sperm with a trichogyne, leading to the production of a well-organized apothecium with many asci.

It is therefore essential to study the ascus, as being the one structure common to all Ascomyceteae. Omitting the apogamous forms for the present, we find the young ascus to be a binucleate cell well supplied with food. The two nuclei are usually considerably larger than those of the vegetative mycelium. They fuse, forming a diploid nucleus with double the number of chromosomes found in each of the original pair. This nucleus usually enlarges quite considerably. It divides meiotically to form four nuclei which in the great majority of cases divide again. In a few species the nuclear division may be repeated until 16, 32, 64, or more nuclei are present in the young ascus. In one species of *Schizothecium* (*Pleurage*) the number, according to L. M. Ames,[1] is 512, while in *Thelebolus stercoreus* Tode ex Fr. the number is over 1000. A part of the ascus cytoplasm gathers around each nucleus and is soon set off from the remaining cytoplasm (epiplasm) by a cell wall thus forming the ascospore. The epiplasm may assist in the formation of the outside layer of the ascospore wall (epispore) which is often beautifully sculptured. The

[1] In a letter to the author.

cytoplasm of the ascospore builds the endospore, the inner layer of the spore wall. The spore nucleus may subsequently divide, usually followed by septum formation, so that the ascospore may eventually be two-celled

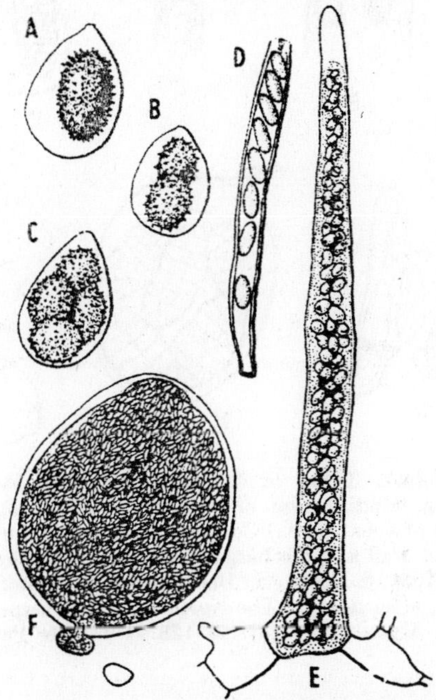

FIG. 66. Ascomyceteae. Variations in ascospore number in asci of various species. (A–C) *Tuber candidum* Hark. 1-spored, 2-spored, and 4-spored asci, respectively. (D) *Septotinia podophyllina* (E. & E.) Whetzel. 8-spored ascus. (E) *Dipodascus uninucleatus* Biggs. Multisporous ascus. (F) *Thelebolus stercoreus* Tode ex Fr. Section through apothecium showing the single ascus with 1000 or more ascospores. (A–C, after Gilkey: *Oregon State Monographs. Studies in Botany*, **1**:1–63. D, after Whetzel: *Mycologia*, **29**(1):128–146. E, after Biggs: *Mycologia*, **29**(1):34–44. F, after Ramlow: *Botan. Ztg.*, **64**(1):85–99.)

or even multicellular, though perhaps the one-celled condition is the more frequent, as it is probably the more primitive. (Fig. 66.)

The ascus varies in shape from cylindrical or clavate in those forms with a well-developed hymenium, to ovoid or subglobose in those in which the asci are scattered or only loosely clustered. The ascospores escape in various manners. In many cases, particularly in the forms with

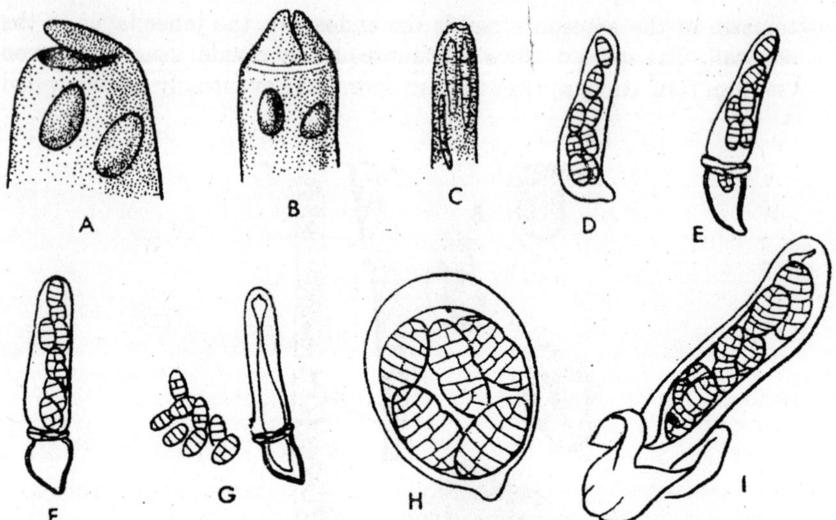

Fig. 67. Ascomyceteae. Types of dehiscence of ascus. (A) Typical operculum. (B) Bilabiate opening, modification of operculum. (C) Inoperculate opening by softening and bursting of apex. (D–G) Contraction of external wall and expansion and lateral rupture of inner wall and discharge of spores in *Pyrenophora*. (H, I) Ascus of *Myriangium duriaei* Mont. & Berk. (H) Unexpanded. (I) Inner wall expanded, outer wall contracted. (A–C, after Seaver: The North American Cup-fungi (Inoperculates). D–G, after Atanasoff: *Mycologia*, **11**(3):125–128. H–I, after Petch: *Brit. Mycol. Soc. Trans.*, **10**:45–80.)

a typical hymenium, the asci absorb water as they reach maturity and become considerably distended. At the apex there is an area which undergoes softening and stretching until the wall suddenly gives way under the pressure from within the ascus, permitting the escape of the epiplasm, ascospores, and vacuolar liquid while the ascus wall contracts. The ascospores may be shot off for a distance of several centimeters, many times the length of the ascus. In the family Pezizaceae the apex of the ascus develops a little lid (operculum) which is forced out, often remaining attached by one edge like a trap door. In many Ascomyceteae the whole ascus undergoes digestion at maturity, thus setting free the ascospores in a mucilaginous liquid. Other methods of ascus dehiscence or rupture have been reported by Atanasoff (1919), Falck (1916, 1923), Ziegenspeck (1926), and other investigators. (Fig. 67.)

The spore fruits of the Ascomyceteae may be classified in general as either apothecia or perithecia or as stromatic structures not referable to either of these forms. In addition there are structures which fit none of these categories such as the naked asci of the order Saccharomycetales and the spore fruits of the order Laboulbeniales.

In the typical apothecium we find a disk or saucer-shaped or even cup-like structure usually from a few millimeters up to several centimeters in diameter. The texture is usually fleshy, fragile to tough, sometimes leathery, and the color from pale brown to black, sometimes red, yellow, or other colors, or even colorless. The upper surface constitutes the hymenium, a layer of elongated cells standing at right angles to the surface like a palisade. It consists of asci intermingled with supporting and protective cells or hyphae, the paraphyses. Immediately below the hymenium is a layer, thin or fairly thick, the hypothecium, consisting mainly of light colored hyphae running parallel to the surface of the hymenium and from which the asci and paraphyses arise. Often sharply contrasted with the hypothecium but sometimes grading into it is the excipulum which makes up the larger part of the basal portion of the apothecium. Its tissue may be pseudoparenchymatous or may be formed by interwoven hyphae. The outer (lower) surface may be filamentous or may resemble an epidermis. Varying from this type we may find apothecia borne on stipes (as in *Sclerotinia*) or the hymenium may be convex (as in *Pyronema*). In some cases the body of the apothecium is bent back along the stipe so as to form a clavate structure with the upper portion covered by an external hymenium (*Geoglossum* and *Morchella*). In other cases the apothecium is subterranean and variously folded internally to form passages and chambers lined by the hymenium (various Tuberales). The apothecium proper is the product of the growth of the hyphae adjacent to the ascogone, when this organ is present. It may develop upon, underneath or within a more or less fleshy stroma or the stroma may be entirely absent (most Pezizaceae). Corner (1929–1931) has made a detailed study of the mode of growth and development of various types of apothecia and concludes that their structure indicates relationship to algal ancestors, possibly a group ancestral to the present Florideae. (Fig. 68.)

The typical perithecium is small, usually less than a millimeter in diameter, and more or less spherical in shape. It is more often dark-colored and somewhat hard and brittle, though not always so. Thus in the forms customarily included in the order Hypocreales, the perithecium may be bright-colored and fleshy or leathery. Usually, but not always, there is an apical opening, the ostiole, through which the ascospores eventually escape. It may be a simple opening or may have a low lip or be drawn out to a long slender neck. As limited by Nannfeldt (1932) the true perithecium is lined over the whole inner surface or only in its basal portion by a hymenium composed of thin-walled asci (sometimes thickened at the apex) intermingled with true paraphyses and with periphyses in the ostiolar region. Julian Miller (1928) has shown that the true wall of the perithecium lies within a stromatic structure which may form simply a thin, darker-colored external layer or may form a massive struc-

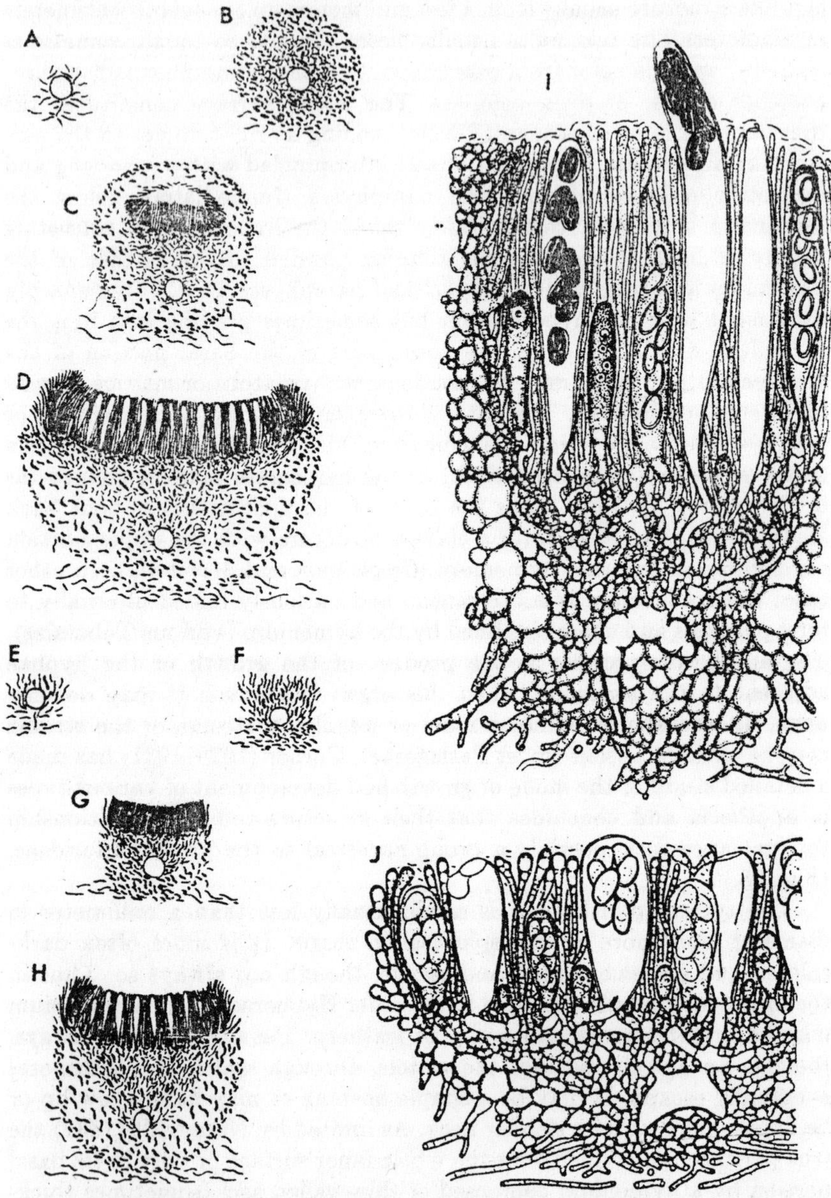

Fig. 68. Pezizales, Family Pezizaceae. (A–D) Diagram of development of angiocarpic apothecium. (E–H) Development of gymnocarpic apothecium. (I) Apothecium of *Ascobolus stercorarius* (Bull.) Schroet. (J) Apothecium of *Ascophanus granuliformis* (Cr.) Boud. (Courtesy, Corner: *Brit. Mycol. Soc. Trans.*, **14**:275–291.)

ture within which numerous perithecia are imbedded or on which the perithecia are seated, each with a thin outer stromatic layer. The true perithecial wall is colorless or light-colored and is formed from one or more layers of hyphae arising from the supporting cell of the oogone or antherid, producing a hollow structure surrounding the ascogonium and the ascogenous hyphae and the asci which arise from it.

In the works of the older mycologists other structures were also called perithecia, which they often resemble very greatly. The studies of von Höhnel (1902–1923), Theissen (1913), Nannfeldt (1932), and others have shown that these structures are entirely stromatic, without any true perithecial wall, and with single asci or tufts of asci without paraphyses, arising in cavities of the stroma. Fungi with this type of spore fruit form the group *Ascoloculares* of Nannfeldt (1932) in contrast to the *Ascohymeniales* which produce apothecia or perithecia in the sense indicated above. A third type of ascocarp with perithecium-like structures, usually without any ostiole, and with the asci scattered throughout the interior, neither in tufts nor forming a hymenium, is considered by some mycologists to be a true perithecium, by others to be of a different nature. Fungi with this type of spore fruit form the *Plectascales* of Nannfeldt and others.

What seems to be, in the author's opinion, a rather primitive but characteristic type of sexual reproduction is that described by Higgins (1936) in *Mycosphaerella tulipiferae* (Schw.) Higgins. In this species, as in other species of the genus studied by the same investigator (1914, 1920, 1929), the male gametes are nonmotile, thin-walled sperm cells produced usually by fours within sperm mother-cells in the interior of more or less spherical, hollow spermogonia. The sperms are imbedded in a mucilaginous mass and escape through an apical opening in the spermogonium as the mass expands with the absorption of moisture. Within a loose mass of hyphae an archicarp is formed consisting of a spherical or ovoid cell, the oogone, with a single large nucleus, and a trichogyne, extending as a slender hypha several times the length of the oogone. The loose hyphae surrounding the archicarp grow and those at the exterior eventually cohere into a dark-colored, firm outer wall, the interior hyphae forming a pseudoparenchymatous mass of thin-walled colorless cells. In the meantime one or more sperm cells have adhered to the trichogyne which forms a papilla at whose tip an opening is formed through which the sperm nucleus enters and passes down into the oogone. The male nucleus gradually enlarges as it progresses and eventually the two nuclei are approximately equal in size and side by side in the oogone. The trichogyne disintegrates, a wall cutting it off from the oogone. The latter enlarges and becomes more or less lobed, the two nuclei in the meantime dividing conjugately many times. From the lobed oogone arise ascogenous hyphae

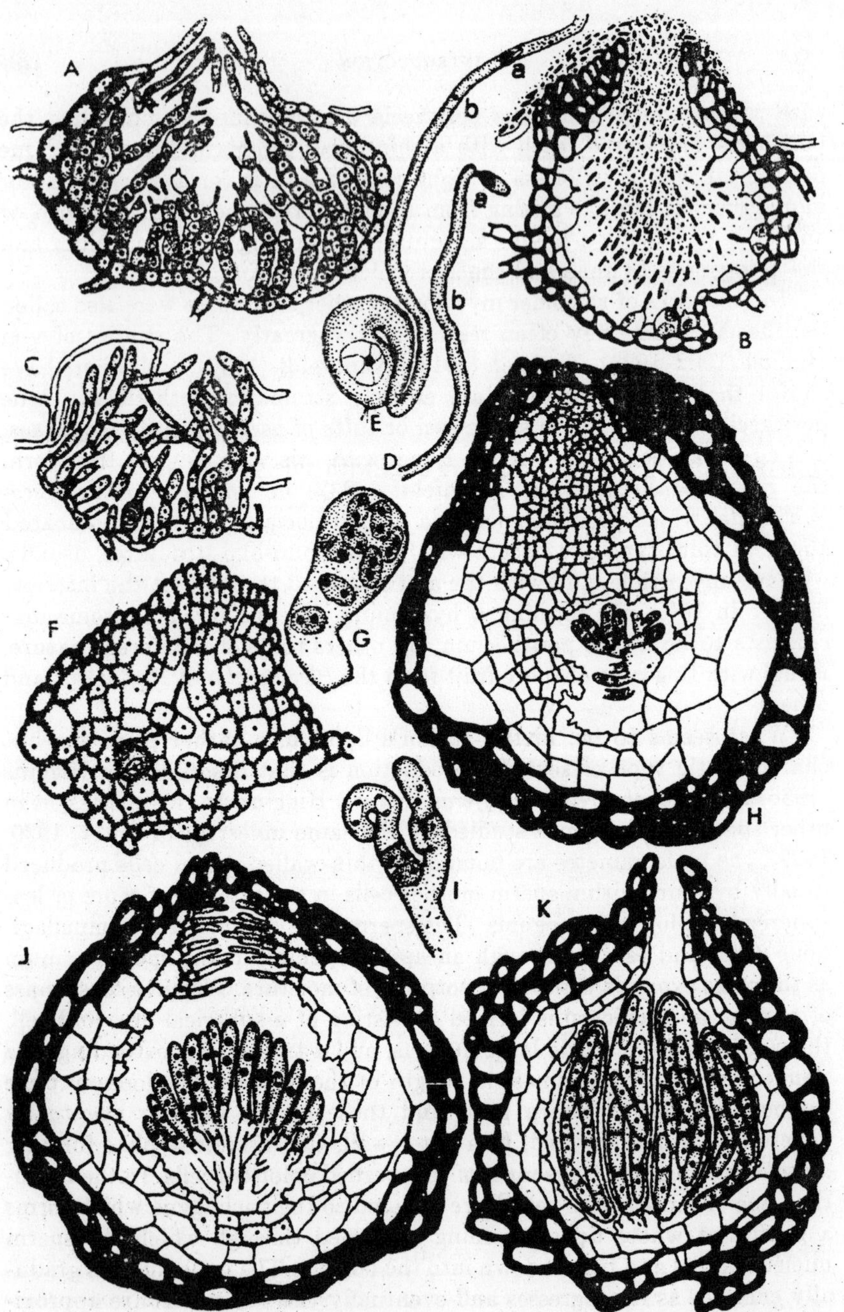

FIG. 69. Ascomyceteae. Supposedly rather primitive type of sexual reproduction in *Mycosphaerella tulipiferae* (Schw.) Higgins. (A) Spermogonium showing some sperm mother-cells with several contained sperms as well as such cells with sperms emerging.

(*Continued on facing page.*)

containing pairs of nuclei and forming asci at their tips by means of hooks or croziers in the manner described below in *Pyronema.* As the asci enlarge, the thin-walled hyaline pseudoparenchymatous cells of the perithecium are destroyed except for a number of slender hyphae at the top (periphyses) which seem to play a part in the formation of an apical opening (osticle) in the spore fruit, through which the mature asci protrude as they discharge their ascospores. (Fig. 69.)

An appreciation of the characteristic sexual reproductive processes in the more complex Ascomyceteae can perhaps best be obtained by a study of the phenomena in *Pyronema omphalodes* (Bull. ex Fr.) Fckl. This is by no means a very primitive form nor is it a simple structure. It illustrates, however, most of the features that occur in this class. This fungus is found in nature most frequently on patches of soil where there has recently been a fire, such as the site of a camp fire. It also appears frequently in greenhouses on flower pots that have been steamed to sterilize the soil. At first there appears a thin whitish, moldy growth on which arise groups of orange-colored apothecia which give the whole surface of the soil an orange color lasting only a few days, after which the fungus disappears, to be followed by other fungi. The mycelium is colorless and septate, with its cells mostly multinucleate. Tufted branches are produced, each bearing terminally a more or less spherical, multinucleate oogone, from whose apex there grows out a curved hypha, also multinucleate, the trichogyne. From one of the basal cells supporting the oogone arises an obovoid or clavate multinucleate antherid. The trichogyne grows to the antherid and coils upon or around its apex. An opening is then formed from one to the other. The majority of the hundred or more antheridial nuclei pass into the trichogyne, whose nuclei have already begun to degenerate, and then through an opening in the septum at the base of the trichogyne into the oogone which itself contains 100 to 200 nuclei. Here they pair with the oogone nuclei. According to Harper (1900) and to Gwynne-Vaughan and Williamson (1931) the paired nuclei fuse, forming about half as many diploid zygote nuclei. According to Claussen (1912) they do not fuse but merely pair closely. According to Dangeard (1907) and his followers no opening is formed between the

---

Fig. 69—(*Continued*)

(B) Spermogonium with sperm cells mostly discharged. (C) Very young perithecium showing oogone with one large nucleus and short trichogyne. (D) Apical portion of fully developed trichogyne with attached sperm (a) and the small trichogyne nucleus (b). (E) Oogone and lower half of trichogyne showing sperm nucleus (a) and trichogyne nucleus (b). (F) Oogone containing two nuclei (male and female). (G) Oogone with eight pairs of nuclei, each in a mass of denser cytoplasm. (H) Ascogenous hyphae, branching out from oogone. (I) Early stages of ascus formation on branching ascogenous hypha. (J, K) Nearly mature and mature perithecia. (Courtesy, Higgins: *Am. J. Botany*, **23**(9):598–602.)

trichogyne and the antherid (which he considers to be a degenerate structure and calls the trophogone), and the nuclei in the antherid and trichogyne disintegrate *in situ,* the oogone nuclei then arranging themselves in pairs. Following this stage all agree that soon 10 to 20 buds appear on the surface of the oogone and elongate to become ascogenous hyphae into which the diploid zygote nuclei or the pairs of haploid nuclei pass until many nuclei are present in each hypha. The nuclei probably divide in the oogone as some of them do in the ascogenous hyphae. The latter elongate, forking somewhat. Eventually septa are formed, producing cells that are plurinucleate toward the base of the hypha and fewer nucleate toward the apex where the last few cells are binucleate. In the meantime from the cells supporting the tufts of oogones and antherids there have been growing outward and upward numerous hyphae which intermingle with the ascogenous hyphae derived from the oogone and also form a mass of external hyphae.

From the terminal binucleate cell of each ascogenous hypha a lateral branch forms just beneath the apex and the two nuclei divide simultaneously (conjugate division) so that the lateral cell also becomes a dicaryon (binucleate) cell. This may be repeated. Eventually the numerous terminal dicaryon cells thus produced proceed to the production of the asci. The cell curves back upon itself like a hook, with a pair of nuclei in the curve. The nuclei divide conjugately and cross walls are formed, leaving two of the nuclei (a daughter nucleus of each of the two original nuclei) at the curve and one daughter nucleus in the cell at the tip of the hook and another daughter nucleus in the cell cut off at the base of the hook. The two nuclei in the curve of the hook fuse while the cell elongates. This is the young ascus. The basal and apical cells of the hook may fuse and then elongate and form a new hook and a new ascus, etc. The fusion nucleus of the young ascus is diploid or tetraploid according to the interpretation as to the presence or absence of nuclear fusions in the oogone. This nucleus undergoes three successive divisions to produce eight nuclei. The first two divisions are reduction divisions (meiosis) according to either theory, the third division being considered the final division of a second meiosis by the advocates of the tetraploid nature of the young ascus nucleus. By both theories the eight resulting nuclei are haploid. From the centrosome remaining in close proximity to each of the eight nuclei fibrillae appear to radiate and certain of these rays curve downward around the nucleus, at some little distance from it, apparently delimiting a mass of cytoplasm surrounding the nucleus from the remainder of the cytoplasm of the ascus, the epiplasm. Along this delimiting surface the ascospore wall is laid down. In the meantime the asci have been elongating as have the vegetative hyphae surrounding and between them. The latter become the paraphyses while the former

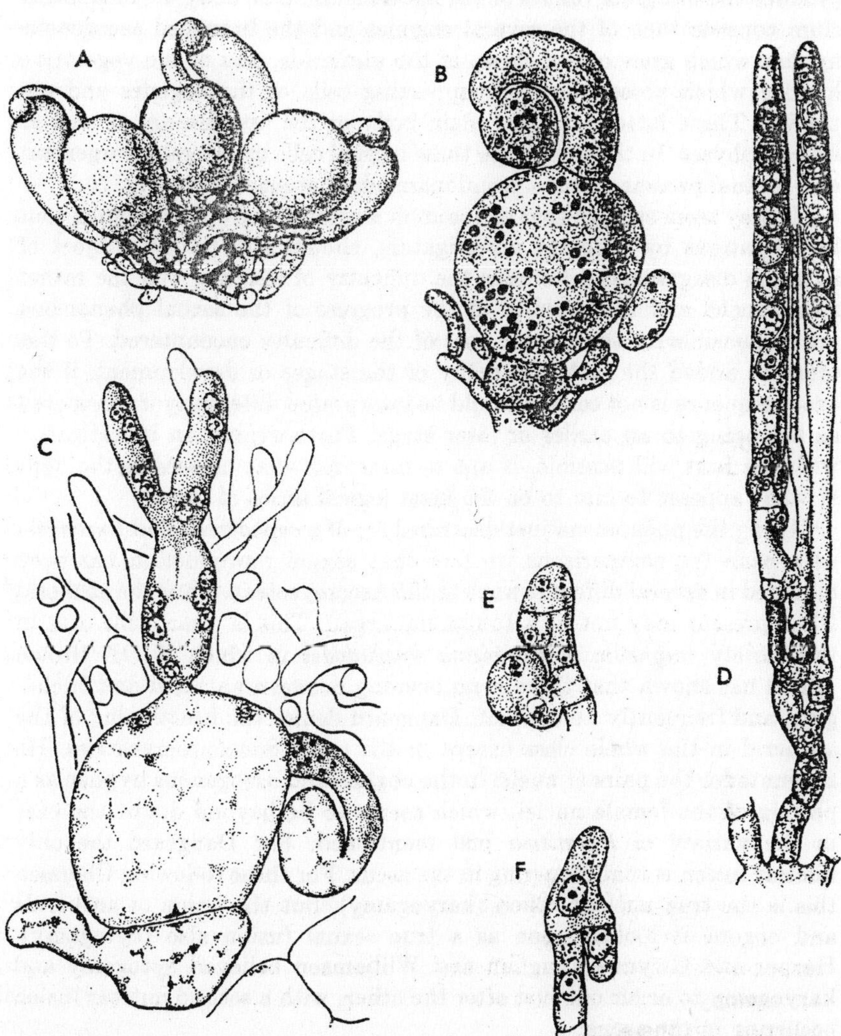

FIG. 70. Pezizales, Family Pezizaceae. *Pyronema omphalodes* (Bull. ex Fr.) Fckl. Sexual reproduction. (A) Group of antherids and oogones. (B) Section through oogone, trichogyne, and apex of antherid, showing opening of latter into the trichogyne. (C) Similar section, later stage, showing a young, forked, ascogenous hypha. (D) Ascogenous hypha from oogone to ascus. (E) Ascogenous hypha showing two hooks. (F) Young ascus with its single diploid nucleus. (A, B, E, F, after Harper: *Ann. Botany*, **14**(55):321–400. C–D, after Claussen: *Z. Botan.*, **4**(1):1–63.)

produce the marginal tissues of the apothecium. The body of the apothecium consists then of the several oogones and the branched ascogenous hyphae which grew out of these, of the antherids, and of the vegetative hyphae which arose from the supporting cells of the oogones and antherids. These latter form the main body of the apothecium as well as its paraphyses. In the excipulum these hyphal cells by lateral enlargement and mutual pressure form a pseudoparenchymatous tissue. (Fig. 70.)

It may seem strange that so common a species, the object of numerous investigations by different investigators, should still be the subject of so much disagreement. Perhaps the difficulty of staining well the rather small nuclei and the rapidity of the progress of the sexual phenomena are responsible for the greater part of the difficulty encountered. To this must be added the fact that many of the stages of development, if the exact sequence is not certain, could be interpreted differently if considered as belonging to an earlier or later stage. Furthermore, an investigator, with the best will possible, is apt to interpret what he sees in the light of what appears to him to be the most logical series of events.

Using the phenomena just described for *Mycosphaerella* and *Pyronema* as a basis for comparisons we find that sexual reproduction has been modified in several different ways in the Ascomyceteae. Thus the antherid when present may not be a functional organ. This is clearly the case in the variety *inigneum* of *Pyronema omphalodes* in which W. H. Brown (1915) has shown that there is no opening between antherid and trichogyne and frequently no contact. Dangeard denies the functioning of the antherid in the whole class except in the order Saccharomycetales. He accounts for the pairs of nuclei in the oogone and ascogenous hyphae as a pairing of the female nuclei, which seems to be beyond doubt the case in the variety of *Pyronema* just mentioned. For Dangeard the only nuclear fusion is that occurring in the ascus. For those following Claussen this is the true nuclear fusion (karyogamy), but the union of antherids and oogone is looked upon as a true sexual fusion also (cytogamy). Harper and Gwynne-Vaughan and Williamson believed cytogamy and karyogamy to occur one just after the other, with a second nuclear fusion occurring in the ascus.

There is a marked tendency toward the production of a more or less coiled series of cells, usually considerably greater in diameter than the cells of the vegetative mycelium and often tapering to a long slender, multicellular trichogyne. Such a structure is called an ascogonium and the cell out of which the ascogenous hyphae bud may properly be considered the true oogone. This may be multinucleate or uninucleate. In the Laboulbeniales, a few genera of the Lecanorales, and apparently also in a few of the Pezizales and Sphaeriales, minute nonmotile sperms are produced internally or externally on short antheridial branches and upon reaching the trichogyne fuse with it, the sperm nucleus entering and passing from cell to cell until it enters the oogone. In other cases the trichogyne fuses with an antherid without the formation of separate sperm cells and the nucleus or nuclei from the antherid enters the trichogyne and eventually reaches the oogone. All degrees of reduction may be found from a multicellular ascogonium with a long trichogyne to a one-celled oogone and one-celled trichogyne as it occurs in *Pyronema*. In many cases the trichogyne is lacking, so that the antherid comes into direct connection with the oogone. Sometimes no antherid at all is formed. In such a case pairing of the nuclei often does not occur in the oogone but the female nucleus divides and the nuclei pass out in pairs into the ascogenous hyphae. In *Ascophanus granulatus* (Bull.) Speg. (*Humaria granulata* Quel.), which does not possess an antherid, Gwynne-Vaughan and Williamson (1930) report that the oogone nuclei unite by pairs and the resultant zygote nuclei enter the ascogenous hyphae. Apparently a cell of the series of ascogonial cells may in some cases be substituted in function for an antherid, its nucleus taking the place of the antherid nucleus. Sometimes no recognizable oogone or ascogonial cells can be found. Some of the vegetative hyphae of the spore fruit become converted in a manner not known into ascogenous hyphae with dicaryon cells. In many families of the Ascomyceteae the terminal cell or cells of the ascogenous hypha become asci without the formation of a hook as described above in *Pyronema*. The two nuclei of the cell fuse and the cell enlarges terminally or laterally, with successive nuclear divisions and ascospore development. In the sexual species of the order Saccharomycetales two cells fuse to form a single ascus and no ascogenous hyphae are produced. Kharbush (1927) reports[2] a similar origin of the asci in the highly developed apothecium of *Botryotinia fuckeliana* (de Bary) Whetzel. In this apothecium, according to him, there are no distinguishable ascogenous hyphae. At the base of the hymenium the apices of adjacent hyphae unite and the nuclei fuse, thus giving rise to the young asci, one fusion of paired hyphae for each ascus. Greis (1940) has found a quite similar origin of asci in two species of *Morchella* studied by him. In the subhymenial layer, or shortly below it, terminal multinucleate cells of adjacent vegetative hyphae unite and grow out to form a stouter binucleate cell, with one nucleus contributed by each hypha. This cell may enlarge and become the ascus or the nuclei may divide conjugately with the production of a short ascogenous hypha, whose terminal cell becomes the ascus, without the formation of the usual hook. The opposite extreme is found in some species of *Taphrina* (*Exoascus*) in which Miss Wieben (1927) has shown that the ascospores

[2] In view of the entirely different mode of sexual reproduction reported by Drayton (1934) for *Stromatinia gladioli* (Drayton) Whetzel it is evident that the process in *Botryotinia fuckeliana* needs re-examination.

are of two opposite sexual tendencies, four of each in each ascus. These ascospores, or the spores that bud off from them, give rise to slender germ tubes which fuse with those from spores of opposite sexual tendency, producing a dicaryon mycelium which becomes the vegetative mycelium within the host. Eventually some of the cells of this mycelium enlarge, the nuclei fuse and the asci are formed. Thus in *Botryotinia fuckeliana*, if Kharbush's report is correct, the mycelium and the apothecium lack entirely the dicaryon phase except as the cells fuse to initiate the asci, while in *Taphrina* the whole vegetative mycelium is of dicaryon nature.

Some botanists suggest that the production of nonmotile sperms which fuse with trichogynes and the budding out of ascogenous hyphae from the oogone are indications that the Ascomyceteae may have descended from some algae related to the Red Seaweeds (Florideae). Other botanists (e.g., Gäumann, 1926; Atkinson, 1915; Nannfeldt, 1932) considered the nonmotile cells that fuse with the trichogyne to be merely modified conidia which have been substituted for antherids, just as certain fusions of vegetative cells have taken the place of the union of sexual organs in the Class Basidiomyceteae. Dangeard (1907) and the Moreaus (1926, 1928) deny any fusion of sperms and trichogyne, at least so far as any transference of nuclei occurs. Since the author follows those that consider the production of a trichogyne and nonmotile sperms and of an oogone producing numerous ascogenous hyphae to be primitive characters for the Ascomyceteae the orders will be arranged in a sequence according to that viewpoint. This matter will be discussed in detail in Chapter 17. Those botanists who take the opposite viewpoint and consider the occurrence of sperms and trichogynes as representing no more than accidental convergence of evolutionary development in both Ascomyceteae and Florideae would probably prefer to start with the Order Saccharomycetales.

Order Laboulbeniales. These are minute, almost microscopic parasites upon insects. They develop externally upon the host except for a haustorium or "foot" that is rooted in the chitinous body wall of the host or less often may penetrate it and form a branching hyphal growth in the body cavity. In the more usual form the foot usually enters the body wall at a pore and thus obtains an ample supply of food without penetrating clear into the body cavity. The fungi vary from plants with only a few cells in number and considerably less than 0.1 mm. in height to forms with hundreds of cells and 2 or 3 mm. tall. The cell walls are usually thick and firm, often dark in color. Between adjacent cells which have arisen by the division of a common parent cell a perforation in the septum is distinctly visible as is usual in the Class Florideae. The plant may consist essentially of a row of cells which give off laterally some branched filamentous appendages and a female reproductive branch. On or near the appendages are borne the antherids. This simple type of structure may become more complex by the longitudinal division of the cells to form a body several cells in thickness from whose sides the appendages and the male and female organs may arise. In *Zodiomyces vorticellarius* Thaxter the main plant body is multicellular and widened at the top, bearing on the flattened upper surface many filamentous appendages and the sexual organs. Some of the larger forms lie prostrate on the body of the host, rooting at various points by means of rhizoids. (Fig. 71.)

In most of the genera of the order the antherids are flask-shaped organs. The apex opens and a uninucleate sperm is pushed up into the neck by the division of the nucleus and cytoplasm of the body of the antherid. By successive formation of sperms in this way those previously formed are pushed out of the neck. They are apparently naked cells, entirely devoid of cilia or flagella. Sometimes several flask-shaped antherids open into a common cavity with a single opening to the outside. In a few genera the sperms are exogenous, being produced by the abstriction of a terminal cell of a short, slender branch from an appendage. Such sperms appear to possess a very thin cell wall. (Fig. 71E, H.)

The female reproductive branch ("archicarp" of some authors) usually consists of a row of three cells, from the base to the apex respectively the oogone ("carpogenic cell"), trichophore, and trichogyne. The first two are nearly or completely surrounded by, usually, a single layer of closely adhering protective cells. The trichogyne is usually one-celled but may divide into several cells and is simple or extensively branched. Perhaps in most cases the sperms are brought into contact with the trichogyne by the active movements of the insect host as it brushes against surrounding objects or other insects. Possibly in the case of aquatic insects, or those frequenting wet places, water currents may bring about the transfer of sperms to the trichogyne. In the genus *Zodiomyces* the elongated trichogyne seeks out and unites with the sperm. In some species antherids are unknown and the development of the oogone is probably apogamous. (Fig. 71A–C.)

After fertilization the oogone divides into about three cells, the binucleate middle cell of which now buds out laterally on all sides to form numerous binucleate asci in which the two nuclei fuse and then divide in the usual way to form eight nuclei. All eight or only four of these nuclei serve as the centers of origin of the ascospores, in the latter case the other four nuclei undergoing degeneration. The ascospores usually are elongated and become two-celled. The ascus walls digest and leave the numerous ascospores in a probably somewhat sticky gum in the cavity of the considerably enlarged spore fruit whose walls have increased in thickness. Eventually the ascospores are discharged between the apical cells and because of their sticky walls adhere to objects with which they come in contact, such as the body of another insect. In the latter case

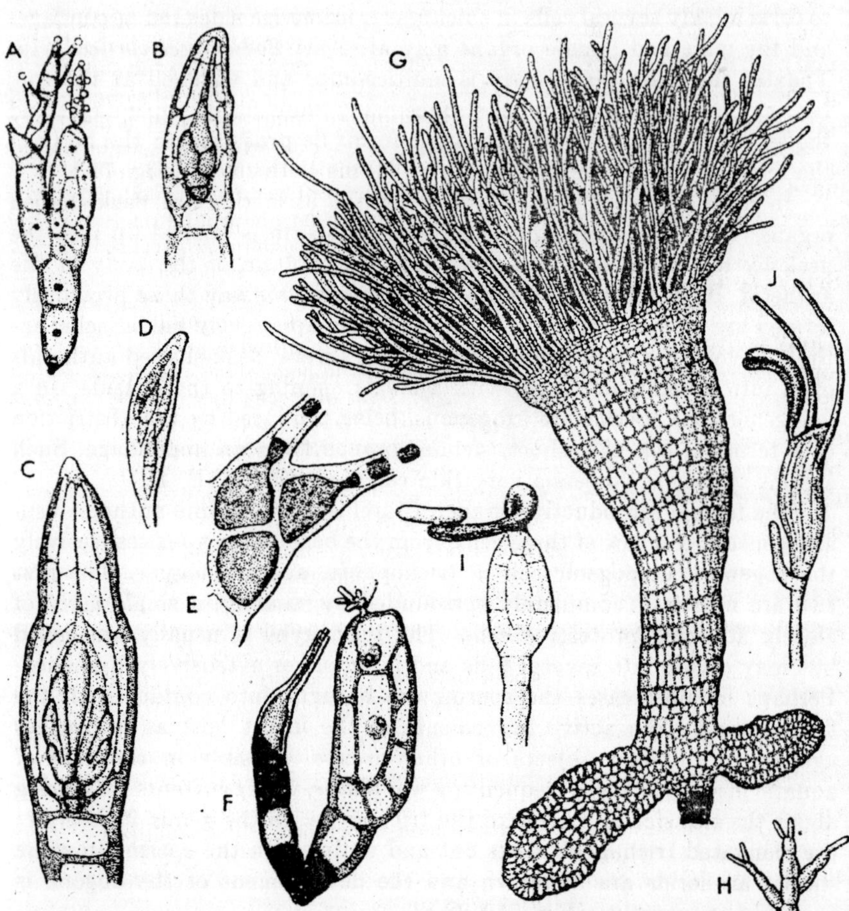

FIG. 71. Laboulbeniales, Family Laboulbeniaceae. *Stigmatomyces baeri* (Knoch) Peyritsch. (A) Mature plant showing at the left the appendage with several antherids and at the right the enclosed oogone with its papillate trichogyne. (B) The fertilized oogone has divided into two basal cells and several upper cells. (C) Asci are budding out of some of the upper cells. (D) An ascus. (E) Two antherids with escaping sperms. (F) *Amorphomyces falagriae* Thaxt., adjacent male and female plants. (G–J) *Zodiomyces vorticellarius* Thaxt. (G) Mature plant. (H) Antheridial branch with exogenous sperms. (I) Oogonial branch with trichogyne and attached sperm. (J) Mature ascocarp. (After Thaxter: *Mem. Am. Acad. Arts Sci.*, **12**:195–429.)

they germinate, one cell becoming the foot and the other developing the remainder of the plant. The full nuclear phenomena of fertilization have not been worked out because of the difficulty of finding the fungi in the proper stages of development and the extremely difficult technique of sectioning and staining.

Only a very few genera and but few species of this order were known until the monumental work of Dr. Roland Thaxter, published in successive parts in 1895, 1908, 1924, 1926, and 1931, revealed the fact that this order contains hundreds of species, dozens of genera, and several families. Since the appearance of the first volume of this marvelous work other mycologists also have added numerous species to those described by Dr. Thaxter. Not only are these fungi found on beetles (Coleoptera) but also on Hymenoptera, Diptera, and various other orders of insects. In view of the fact that of the 600,000 to 1,000,000 described species (and possibly as many or more as yet undescribed species) of insects only a few thousand have been examined for the presence of these parasites it seems reasonable to suppose that the number of species, genera, and even families of this order may be greatly increased in the future.

The relationship of Laboulbeniales to the other orders of Ascomyceteae is not very close. The spore fruit is unlike that of any other members of this class and the vegetative structure finds few analogies. In the sexual reproduction the formation of functional, separable sperm cells is known in *Collema* and other lichens and in Pezizales, Sphaeriales, and other groups. In most of these however the sperms are not produced endogenously as naked cells as is true of most of the Laboulbeniales but seem to have thin walls even in those cases where endogenous formation does occur. The "archicarp" of the Laboulbeniales reminds one remarkably of the condition in some of the Florideae where the archicarp is surrounded by protective cell layers with merely the trichogyne exposed.

The following orders: Lecanorales, Pezizales, Tuberales, Hysteriales, and Taphrinales all produce spore fruits that may be considered as typical or modified apothecia. They are often or in part included under the group name Discomycetes.

Order Lecanorales (The Disk Lichens). These constitute a large group of organisms which have in common the production of apothecia and which show a specialized form of parasitism on land species of Chlorophyceae and Myxophyceae. The validity of the maintenance of this group apart from the Pezizales is, to say the least, very doubtful, but until the reproductive processes, especially the behavior of the sexual nuclei, are better known in both groups it is perhaps better to follow custom and consider the two orders separately. Possibly when such studies have been carried out in all the more important genera that produce apothecia the system of classification of both orders will have to be entirely revised.

As long ago as 1887 Alfred Möller reported that he had been able to grow various lichens in culture media in the absence of algae, and that these produced small thalli, up to 1.5 cm. in diameter in some cases, with typical cortex and medulla. In some tests the development of typical spermogonia was observed. His experiments had to be cut short after four months so that the further development of the still growing fungi could not be followed.

Vegetatively the Lecanorales vary from a loose branching mycelium penetrating in all directions the gummy colony of a species of *Nostoc*, the form of that colony setting the limits to the size and shape of the lichen body (*Collema*), to a very complexly branched, firm thallus with a cortical outer layer of hyphae enclosing the algal hosts and thus making possible unlimited growth, without reference to the natural shape of the unparasitized algal colony.

The mycelium is in general slender, light-colored, septate, and branching. The septa are centrally pierced by a rather minute perforation. In the forms with a cortex (the majority of lichens) the mycelial cells composing it are short, broad, rather thick-walled and compacted together into a pseudoparenchyma one to several cells in thickness. In the interior of the lichen the mycelium is filamentous and loose. Usually the algal hosts are found in definite layers. In some cases definite penetration of the algal cell by the mycelium can be observed. Such cells are eventually killed. Geitler (1933) reported that more often no such penetration can be seen but that the mycelium is applied to the host cell in the manner of an appressorium. Just how the fungus draws its nourishment from the alga in the absence of direct penetration is a matter of conjecture. Possibly some substance secreted by the fungus increases the permeability of the plasma membrane of the algal cells, thus permitting sugars and other soluble food stuffs to diffuse out from the cell to be picked up by the fungus. The fungus furnishes a certain amount of protection to the alga and probably gives it a more equable habitat, protecting it from the rapid extremes of drying and moisture, sun and shade, heat and cold. In so far as this is true it is to the advantage of the alga. On the other hand, however, the constant tribute levied in the form of food substances diffusing out from the cells must reduce their vigor somewhat. It is worthy of note that algae whose free-living development includes zoospore formation as a normal mode of reproduction usually have this entirely suppressed, reproduction being limited mainly to fission.

The earlier students of lichens consider the enclosed algal hosts to be a part of the lichen organism, possibly reproductive in nature, hence the term "gonidia" applied to them. By some they were considered to be the photosynthetic organs of the lichen. Their similarity to algae was early noted and Schwendener (1867, 1868) supplied the evidence that they were truly algae. He and others following him were able to synthesize lichens from cultures of algae and lichen ascospores (Bonnier, 1889). In spite of this evidence some lichenologists were loath to give up the old idea and as late as 1913 Elfving, observing green bodies within the hyphae of some lichens, maintained that they were integral parts of the lichen. Liro (1914) confirmed the occasional presence of these green bodies within the hyphae but believed them to be portions of cells or whole small cells of the host algae that had entered the hyphae through openings. Tobler (1925) summarized the results of his own work and that of others on the biology of the lichens, confirming the belief that they consist of fungi parasitizing upon algae with a mutual interaction of the two organisms to produce the characteristic structures.

The algal hosts are usually Blue-green Algae (Myxophyceae) or Green Algae (Chlorophyceae) whose habitat consists of moist situations on land, such as on the ground, rocks, trees, etc. More often the one-celled algae are preferred (*Chroococcus, Chlorococcum, Protococcus*, etc.), but some of the filamentous forms are also captured and made prisoners (*Nostoc, Trentepohlia*, etc.). Most genera of lichens are confined to a single species or to closely related species of algae but a few lichens are able to use as algal hosts forms from widely diverse groups. It is noteworthy that the thalli formed by the same species of fungus with these widely separated algal hosts are very different, according to the host present.

Lichens may be grouped morphologically in accordance with the type of thallus into crustose, foliose, fruticose, and pendent lichens. The first form closely adhering crusts on the substratum to which they are held fast by hyphal strands; the foliose lichens are flat and thin ("leaf-like") but adhere to the substratum only at definite points; the fruticose lichens are upright in habit and more or less branched, resembling little shrubs (*frutex* means shrub). The pendent forms are attached only at localized spots and are long and slender and branching, hanging from the twigs or branches of the trees to which they are attached. They are often considered only a special type of fruticose lichen. It must be noted that these distinctions are not absolute for there are gradations between these types. Most species of *Cladonia* are at first crustose or foliose (but with small "leaves"), later forming upright podetia upon which the apothecia are developed, at which time the prostrate foliar growth may disappear. The distinction between crustose and foliose lichens is not always clear. (Fig. 72.)

In tropical and subtropical regions many lichens grow on the algae attached to the surface of leaves. The fungus hyphae may then enter through the stomatal openings into the interior of the leaf where it seems probable that they are somewhat parasitic upon the leaf tissues. McWhorter (1921) showed that some lichens of the genera *Cladonia* and

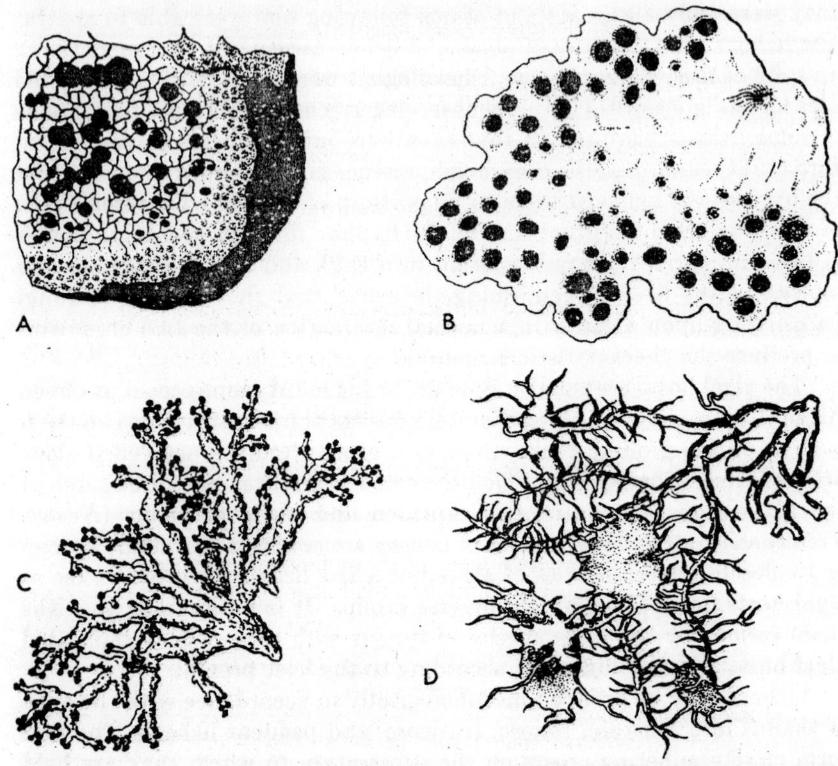

FIG. 72. Morphologic types of lichens. (A) Crustose type, *Lecidea platycarpa* Ach. (B) Foliose type, *Gyrophora muhlenbergii* Ach. (C) Fruticose type, *Stereocaulon coralloides* E. Fr., fruticose podetium ("secondary thallus"). (D) Pendent type, *Usnea barbata* (L.) Wigg., portion of thallus with apothecia. (A, after Reinke: *Jahrb. wiss. Botan.*, **28**:70–150. B–D, after Schneider: A Text-book of General Lichenology, Binghamton, N.Y., Willard N. Clute & Co.)

Amphiloma growing in colonies of mosses (Musci) are capable of destroying the latter by direct parasitic attack. Rock-inhabiting lichens often penetrate the rock itself for some distance by means of their holdfast hyphae. Miss Mellor (1922) called attention to serious damage done to glass windows in some old churches in France where lichens attached themselves to the glass and gradually corroded it. Probably the requisite mineral nutrients of the lichen and enclosed algae are obtained by the hyphae that attach the lichen to its substratum. A very few species of lichens develop in aquatic habitats.

A great many organic acids, the so-called lichen acids, have been found in the lichens. They were given intensive study by Zopf (1907). Following a suggestion of his, one of his students, F. Tobler (1909) conducted experiments which demonstrated that this production of lichen acids is the result of mutual physiological interaction of alga and fungus, as they do not occur in either organism when grown alone. From some of these acids may be obtained brilliant pigments such as orcein, litmus, etc. (Miss Smith, 1926).

Many lichens have no known asexual mode of reproducing themselves. The pendent forms are frequently torn to pieces by the wind and carried considerable distances, thus achieving distribution. A good many lichens produce pycnidia containing conidia. For some species these conidia have been germinated to produce a mycelium. On a great many lichens special asexual reproductive structures, the soredia, are produced. In barest details a soredium arises as an outgrowth of the interior mycelium of the thallus, carrying with it some of the algal cells. Having grown out through the surface of the thallus this mycelial mass rounds up into a ball with a sort of cortex, containing in its interior loose hyphae and a few host cells. This structure breaks loose and is distributed by wind or rain to other locations where the hyphae grow fast, thus starting a new lichen thallus.

In their sexual reproduction the Lecanorales are rather uniform in their end product, the apothecium, which differs in detail but not in fundamental plan in the various families and genera. These variations have to do with shape (concave, flat, convex); color; structure of the paraphyses; number, color, structure, and shape of the ascospores (colorless or brown, one-celled or divided into two or more cells, ellipsoidal, fusoid, needle-like, etc.); structure of the hypothecium and excipulum; size; location on the thallus, etc. Two different structures called excipula or exciples are of importance in systematic arrangement of the families and genera of lichens. The "thalloid exciple" is a marginal wall around the apothecium consisting of an upgrowth of the thalloid hyphae, often with enclosed algal cells. It does not arise from the developing apothecium. The "proper exciple" is the cup-like margin at the edge of the apothecium which is formed by the outgrowth and upgrowth of the apothecial tissues. Sometimes both types are present but more usually only one type, and in many lichens neither type of exciple is observable. In contrast with most of the Pezizales the apothecia of the Lecanorales are usually slow in development and persist for a long time, maturing a few asci at intervals. The asci are inoperculate and usually thickened at the apex. The details of the sexual process, particularly the behavior of the sexual nuclei, are sadly in need of further study in almost all genera of the order. It may be safely said that no lichen has been satisfactorily studied from all these standpoints. The conditions in *Collema* and *Collemodes* will illustrate the main features of the sexual process in this order.

In its interior *Collema* consists of a slender, branched mycelium, loosely penetrating and limited in outline to the shape of the *Nostoc* colony it inhabits. There is no cortical layer. On the interior mycelium there arise here and there somewhat thicker hyphae which are noticeable because of their dense contents and more or less loosely coiled structure. These are the ascogonia. The cells like those of the vegetative mycelium are uninucleate. Each ascogonium consists of from one to three coils of cells terminated by a filament (the trichogyne) which turns toward and projects just through the surface of the colony. Its exposed tip is slightly enlarged and covered with a somewhat thickened wall which is sticky when wet. In its multicellular structure it differs greatly from the trichogyne of the Florideae. The male organs are branched hyphae projecting into a conceptacle-like spermogonium which opens at the surface of the colony. From these branches there bud off minute, uninucleate, nonmotile sperms possessing a delicate cell wall. When wet by rain the gummy mass filling the spermogonium swells and oozes from the opening, where the gum dissolves and the sperm cells are floated off by the film of rain water. Such a sperm coming in contact with the sticky tip of the trichogyne adheres to it. Stahl (1877) and Baur (1899) have demonstrated that an opening is dissolved quickly through which the sperm nucleus enters into the apical trichogyne cell. Successive swellings and disappearance of the septa of the trichogyne seem to indicate the passage of the sperm nucleus down to the coils of the ascogonium. The nuclear behavior has not been followed in detail however. From one of the ascogonial cells, which therefore corresponds in function to the oogone of *Pyronema*, ascogenous hyphae begin to grow outward and upward. The surrounding vegetative hyphae also become actively involved in growth and produce the vegetative part, including the paraphyses, of the apothecium. The ascogenous hyphae produce their asci by the hook method as described for *Pyronema*. It should be noted that in some genera of lichens the sperm cells, or cells resembling them and produced in similar conceptacles, are capable of growing in pure culture in nutrient media until normal thalli are produced bearing similar conceptacles. This was reported by Möller in 1887. He obtained thalli by culturing such cells from *Buellia punctiformis* Hoffm., *Opegrapha subsiderella* Nyl., *O. atra* Pers., *Arthronia* sp., *Calicium parietinum* Ach., and other species of *Calicium*. He also cultured the conidia from the pycnidia which in some cases are present in the same thallus. He therefore drew the conclusion that the spermogonia as well as pycnidia were both asexual reproductive structures and that the supposed sexual function of the spermatia was erroneous. (Fig. 73A, B, D–G.)

Miss Bachmann (1912) erroneously identified the very similar lichen *Collemodes bachmannianum* Fink with *Collema pulposum* (Bernh.) Ach. She found that in *Collemodes* the branches which produce the sperm cells are not produced together in spermogonia but are scattered here and

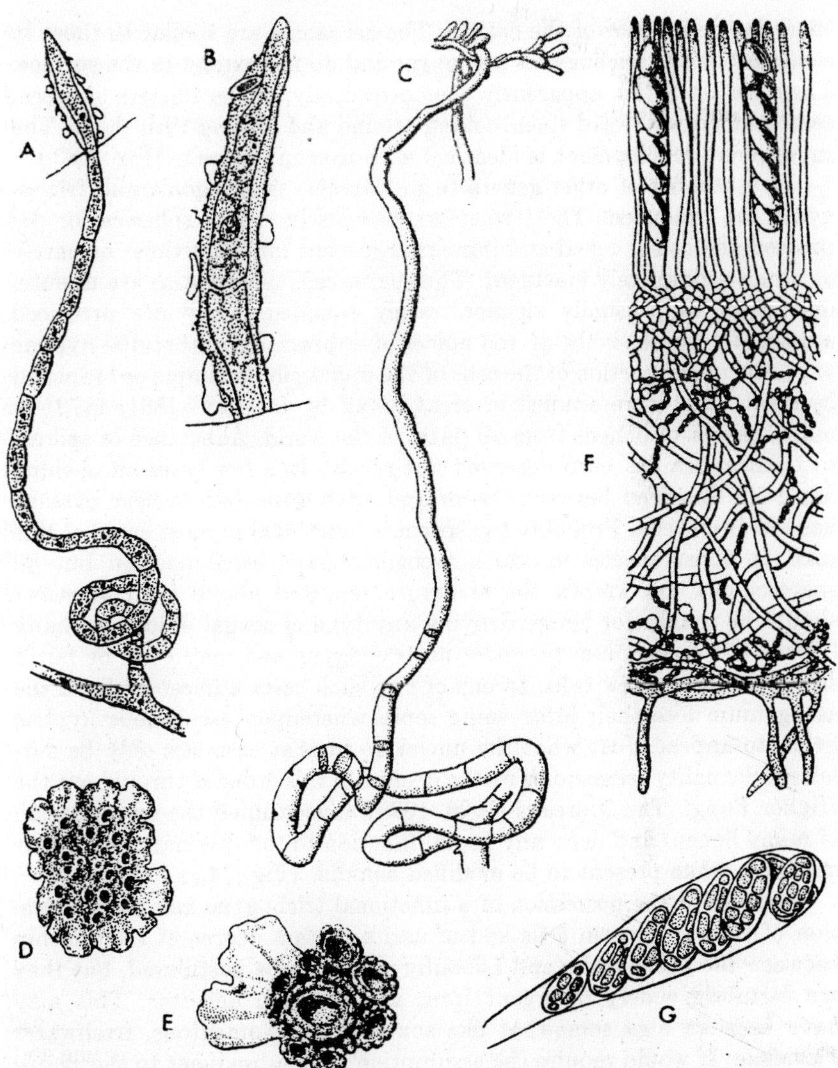

FIG. 73. Lecanorales, Family Collemaceae. (A, B) *Collema crispum* Ach. (A) Ascogonium and trichogyne. (B) Apical cell of trichogyne, showing sticky surface. (C) *Collemodes bachmannianum* Fink. Ascogonium and trichogyne growing to clusters of sperm cells produced within the thallus. (D–G) *Collema pulposum* (Bernh.) Ach. (D) Habit sketch. (E) Cluster of apothecia. (F) Section through apothecium. (G) Ascus. (A–B, after Baur: *Ber. deut. botan. Ges.*, **16**(10): 363–367. C, after Bachmann: *Ann. Botany*, **26**(103):747–760. D–G, after Schneider: A Text-book of General Lichenology, Binghamton, N.Y., Willard N. Clute & Co.)

there in the interior of the colony. The ascogonia are similar to those in *Collema* but the trichogynes are longer and do not extend to the surface. They are attracted, apparently chemotropically, to the clusters of sperm cells and grow toward them, coiling around and uniting with them. The subsequent development is identical with that in *Collema*. (Fig. 73C.)

In a number of other genera (e.g., *Physcia*) spermogonia and trichogynes are produced. The former are various in shape, spherical or depressed globose or lobed and immersed (except for the ostiole) or partly or almost completely emergent. The sperm cells (spermatia) are minute, one-celled, and usually slender, rarely rounded. They are produced apparently successively at the apices of unbranched antheridial hyphae or on the upper portion of the cells of the multicellular simple or branched hyphae. They were studied in great detail by Lindsay (1861, 1872) in many genera of lichens from all parts of the world. Adherence of sperms to trichogynes has been observed but rarely. In a few cases an opening has been observed between sperm and trichogyne but nuclear passage has not been seen. Probably the sperm is functional in most cases of this sort. In those species in which ascogonia have been reported but no spermogonia are known the structure reported above for *Collemodes* should be sought for before denying any type of sexual union. In many lichens the ascogonium produces no trichogyne and may even be but a straight row of a few cells. In one or two such cases adjacent cells of the ascogonium lose their intervening septa whereupon ascogenous hyphae begin to appear. Just what the nuclei do in that case can only be surmised. Sexuality seems to be on the decline in this order as throughout the Higher Fungi. The Moreaus (1926, 1928) have studied the reproduction of many lichens and deny any sexual function to the spermatia, considering them when present to be modified conidia. (Fig. 74.)

Because of the possession of a functional trichogyne and the production of separate sperm cells and of asci a certain degree of relationship between the Lecanorales and Laboulbeniales can be postulated, but they are certainly widely divergent from any common ancestor. This may have been an alga somewhat like some of the filamentous, freshwater Florideae. It would require the assumption that subsequent to the acquisition of the ascus-producing habit the one series developed as parasites on insects with little modification of the protected procarp while the other series developed as parasites on algae, presumably at first submerged forms, later land algae. At the same time the spore fruit deviated far from the simpler procarp type shown in the Laboulbeniales. Here again it must be noted that many mycologists hold that the sperm cells are nothing but modified conidia which have taken up secondarily the sexual function in place of an antherid. The external similarity between these groups and the Florideae would be looked upon from this viewpoint as a case of convergence, not as an indication of true phylogenetic relationship.

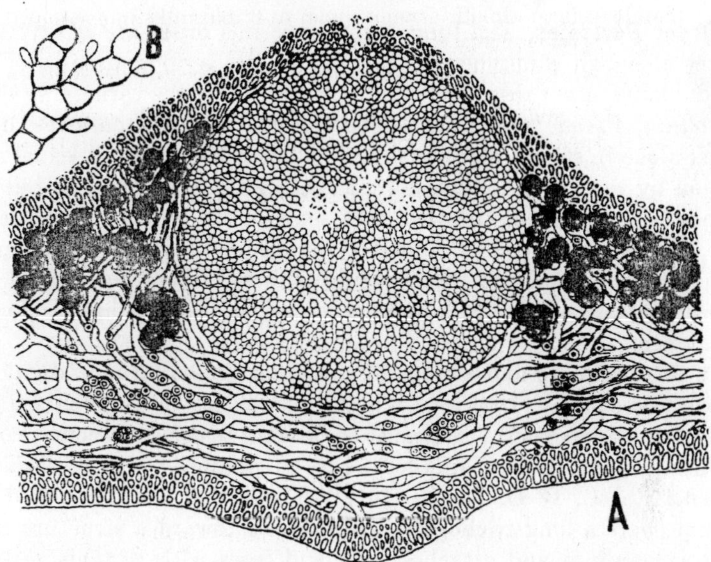

FIG. 74. Lecanorales, Family Physciaceae. *Physcia* sp. (A) Spermogonium. (B) Antheridial filament with sperm cells. (After Kny: Botanische Wandtafeln.)

Zahlbruckner (1926) believed that the group treated here as a single order, the Lecanorales, is really polyphyletic, i.e., derived from nonlichen forming fungi at many different points. Nannfeldt (1932) discussed this problem and showed that some of the perithecial lichens (Order Pyrenulales) are in reality more closely related to the Order Pseudosphaeriales and that of the lichens with true apothecia many have their closest relationship not with other lichens but with Pezizales that are not lichen producing. Perhaps the ultimate logical disposition of the disk lichens will be to distribute them among the other apothecium-forming fungi at the points where their apothecial structure and manner of sexual reproduction seem to fit best. Until the enormous mass of study needed to acquire this information has been carried out it may be best to treat them as a single order.

The forms here included in the Lecanorales are divided by Zahlbruckner into about 37 families, about 275 genera, and over 7400 species. They occur from the tropics to the Antarctic and Arctic zones and from sea level to the tops of the highest mountains where rock is exposed. They furnish the chief food of the caribou, the reindeer, and the musk ox. Some species are used for human food in famine times. Iceland moss, *Cetraria islandica* (L.) Ach., is sometimes used for medicine. Litmus and orcein are derived from lichens. Miss Annie L. Smith's (1921, 1926) publications on these fungi should be studied for a more complete understanding of this extremely variable group of organisms.

Order Pezizales. The fungi comprising this order are mostly saprophytes although a number of the more or less serious diseases of cultivated plants are caused by parasitic species (e.g., various species of *Sclerotinia*, *Pseudopeziza*, etc.). Those forms that are parasitic do not attack algae in such a manner as to produce lichen thalli. Asexual reproduction by means of conidia is found in a good many species but is by no means as widely distributed as in some of the orders to be discussed later. Conidia are formed singly on simple or branched conidiophores or the individual cells of whole segments of mycelium may round up to form chains of conidia. In a number of species sclerotia are produced in abundance.

In this order we find almost all gradations in sexual reproduction. In *Stromatinia gladioli* (Drayton) Whetzel, it was shown by Drayton (1932, 1934) that minute sperm cells must be brought to certain receptive organs of the ascogonia before apothecia can be produced. In *Ascobolus carbonarius* Karst., B. O. Dodge (1912) showed that a much coiled ascogonium bears a long trichogyne which grows toward a structure resembling a conidium and attaches itself and fuses with it. This resembles greatly what Miss Bachmann described for *Collemodes*. In the genus *Ascobolus* other species have an ascogonium which coils directly around and fuses with an upright antherid (Dodge, 1920). In still other species both Schweizer (1923) and Ramlow (1914) have indicated that antherid and trichogyne are both absent. In *Scutellinia stercorea* (Fr.) Kunze, according to Miss Fraser (1907), the oogone is rounded and multinucleate, as in *Pyronema*, but the trichogyne is several-celled. S. G. Jones (1930) found in *Pseudopeziza trifolii* (Biv.-Bernh.) Fckl. the production within the leaf of the host (*Trifolium pratense* L.) of numerous ascogonial coils made of heavily staining uninucleate cells. Around these but without any visible cell fusion the vegetative mycelium develops into actively growing hyphae, some of which emerge from the stomata. Jones called these "trichogynes" but denied any reproductive function, considering them to be "respiratory hyphae." Other hyphae become true ascogenous hyphae with binucleate cells and still others produce the paraphyses and other portions of the apothecium while the original ascogonium degenerates without having served any other function than as a center of attraction for the surrounding hyphae. There is a nuclear fusion in each young ascus and only the first of the following three nuclear divisions is reductional. In some Pezizales even the oogone or ascogonium is not to be found. The eventual product is an apothecium which in its general plan is like that of the Lecanorales but usually larger and more fleshy. (Fig. 75.)

In this order two series of forms may be distinguished, depending upon the mode of dehiscence of the ascus at maturity, viz., the Operculatae

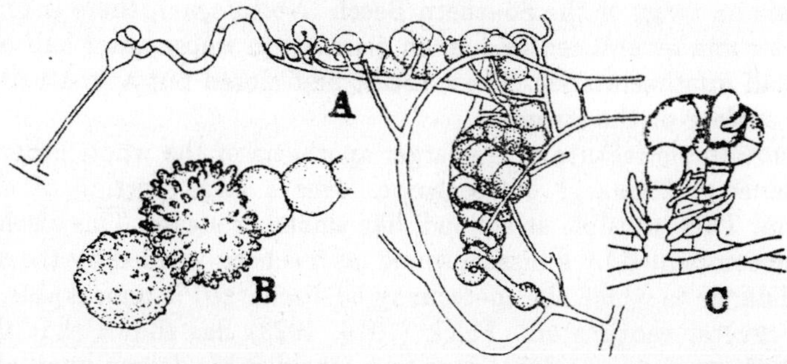

FIG. 75. Pezizales, Family Pezizaceae. (A, B) *Ascobolus carbonarius* Karst. (A) Ascogonium with trichogyne reaching out to a distant antherid. (B) Ascogenous hyphae beginning to bud out from ascogonial cells. (C) *Ascobolus magnificus* Dodge, ascogonium coiling around antherid. (A–B, after Dodge: *Bull. Torrey Botan. Club*, **39**(4):139–197. C, after Dodge, *Mycologia*, **12**(3):115–134.)

and the Inoperculatae. In the former a little lid (operculum) is formed at the apex of each ascus. This gives way when the turgor pressure reaches a certain degree, thus allowing the escape of the ascospores and the surrounding liquid. The operculum may be shot off entirely but more often remains attached at one edge like a trap door. A modification of the typical operculate type is apparently the bilabiate type. In the Inoperculatae the thickened apex of the ascus gradually softens and suddenly yields to the internal pressure forming a pore through which the ascus contents escape.

In the Pezizales the ascospores are one-celled and ellipsoidal to subspherical to a much greater extent than in the Lecanorales where many-celled ascospores are common. The apothecia in this order vary greatly in size. In a few species of *Ascobolus* and some other genera the apothecium is less than a millimeter in diameter; in the larger number of genera and species it is from 5 to 20 mm. in diameter. As mentioned in the preceding chapter a specimen of *Geopyxis cacabus* (Fr.) Sacc. was collected in Java that was nearly a meter high and about 50 cm. across. Seaver (1942) reported that Dr. Helen M. Gilkey and Dr. S. M. Zeller found a specimen of *Daleomyces phillipsii* (Massee) Seaver in Oregon that had a diameter of 40 in. (about 1 meter). The shape is also subject to great variation. In *Pyronema* and some other genera it is convex and naked from the beginning. In more forms it is flat or cup-shaped and in most of them the hymenial surface is at first covered with a more or less evanescent layer. Some species have a subspherical apothecium, at first closed and then opening at the apex by an enlarging pore or ostiole. The apothecium is more often sessile but yet is stalked in many genera. A few genera produce their apothecia just under the surface of the soil, opening by a small pore at the surface when the apothecium is mature. In *Cyttaria*, parasitic on twigs of the Southern Beech (*Nothofagus*), there is formed a fleshy stroma several centimeters in diameter in whose outer half or more the small apothecia arise. These are at first closed but at maturity open at the surface of the stroma.

A noticeable feature of the larger apothecia of the whole order is the simultaneous discharge of ascospores over a large portion of the hymenium. This is visible as a cloud, like smoke or steam. This discharge is often accompanied by a hissing sound, as has been verified by the author. The distance to which the spores may be discharged is remarkable, sometimes several centimeters. Falck (1916, 1923) has shown that the discharge of ascospores is dependent to a considerable degree upon changes in temperature or illumination or upon contact of some other object with the apothecia. Even the stimulus of a gentle current of air is sufficient to cause spore discharge in some species. Buller (1934) gave an extended discussion of the conditions that induce this simultaneous discharge of spores and that affect its direction. In species of *Ascobolus* and some other genera the ascus at maturity becomes greatly elongated and distended laterally by the absorption of a large amount of water. When the operculum gives way the greatly enlarged ascus contracts with much violence and the contained liquid and ascospores are ejected to an amazing distance. In *Saccobolus* the ascospores are massed together in a ball which is expelled further than would be possible for separate spores.

It has been shown that differentiation into two distinct sexual strains occurs among some species of this order. Thus Miss Green (1931) showed for *Ascobolus furfuraceus* Fr. and Betts (1926) for *A. carbonarius* Karst. that they will not produce apothecia when grown in culture from a single ascospore but require the meeting of mycelia developed from different ascospores and then not from any two but from two of opposite sexual strains. On the other hand some species of *Ascobolus* are fertile when grown from but a single ascospore. This is a phase of investigation that has attracted the attention of students but much still remains to be learned. Drayton demonstrated that in *Stromatinia gladioli* there are formed on the mycelium arising from one ascospore minute cells (microconidia or sperms) and certain receptive structures within which are developed ascogonia with long trichogynes. Such a mycelium remains without producing apothecia. The mycelia produced by the eight ascospores of the ascus represent two phases, four of each. The sperms of any mycelium of one phase can fertilize the receptive bodies of any mycelium of the other phase and vice versa. Apparently this is not true heterothallism or condition of maleness and femaleness of the different strains such as occurs among the Mucorales. It is comparable to the self-sterility of many flowering plants to their own pollen. For example the Bartlett pear pistil rarely develops to a fruit when pollenized by pollen from the same variety and the same is true for the Kieffer pear, but these two varieties are usually fertile to each other's pollen. It is perhaps more comparable still to the dimorphic species of *Primula* studied by Darwin (1889), in which the seeds of a capsule will produce about equal numbers of the two types of primrose plants, those with flowers possessing a long style and low-placed stamens and those whose flowers have short styles and stamens high in the corolla tube. Each strain is relatively sterile with pollen from plants of its own type but fertile with pollen from plants of the other type. Whether the condition in *Ascobolus magnificus* Dodge is like the foregoing, i.e., a case of self-sterility, or is true heterothallism (a real difference in sex) remains to be discovered by further study.

The "Discomycetes" were classified by the earlier investigators Persoon (1801) and Fries (1822), largely on the basis of external characters. Later the ascus and ascospore characters were also taken into consideration. The internal structure of the apothecium proved to be of great importance. Durand (1900) used this as a basis for a tentative classification. Nannfeldt (1932) has used these features in his extensive writings on this group. Boudier (1907) pointed out that the mode of dehiscence of the ascus, whether by a lid or by a pore, i.e., operculate or inoperculate, is of great diagnostic value. Seaver (1928) in his volume on the Operculate Cup-fungi recognized only two families in this suborder in place of a larger number recognized by Schroeter and Lindau (1896) in Engler and Prantl.

Order Pezizales: Suborder Operculatae. FAMILY PEZIZACEAE. Apothecia flat, convex, or concave or cup-like, sessile, or short stalked, rarely long stalked, and then the hymenium concave or at most flat. Apothecium pseudoparenchymatous throughout, with few exceptions. Typical representative genera in this family are: *Ascobolus*, growing on animal excrement or on soil, with mature asci much protruding and with mature spores violet in color. The apothecia vary, according to species and environment, from less than 1 mm. up to nearly 3 cm. in diameter. In *A. immersus* Fr. the ascospores may attain a size of 50 to 75 μ in length by 20 to 35 μ in thickness, almost the largest ascospores known. *Pyronema*, grows on soil, especially after a fire or after steaming. The apothecia are 1 to 2 mm. in diameter and the convex hymenium is practically naked from the first. The ascospores are hyaline. *Humarina* (*Humaria* Sacc.) is less than 1 mm. to 1 cm. in diameter, growing on the ground, forming white or bright-colored, mostly disk-shaped apothecia, with hyaline ascospores, differing from *Ascophanus* only in that the latter grows on dung. *Patelia* forms disk-shaped apothecia up to 1 cm. wide, and with the outside clothed with hairs, at least at the edge. *P. scutellata* (L.) Morgan (*Lachnea scutellata* (L.) Gill.) forms its brilliant red disks with a fringe of dark hairs, on rotten wood and is strikingly beautiful. *Plectania*

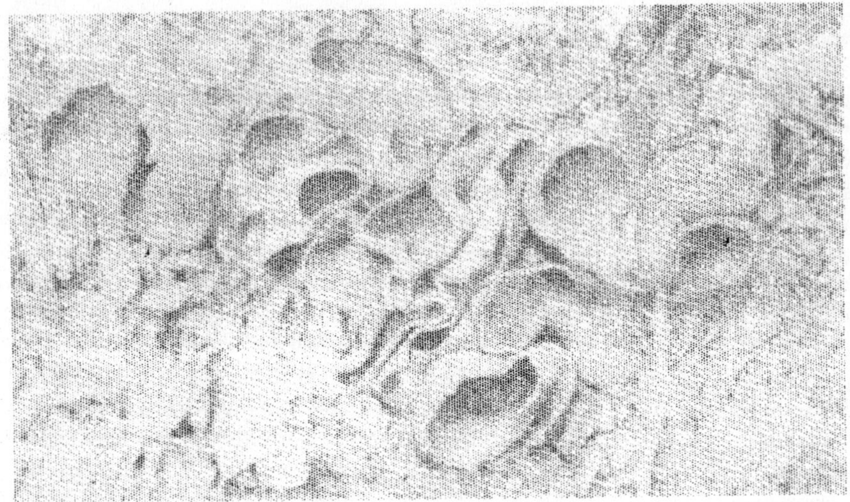

Fig. 76. Pezizales, Family Pezizaceae. *Peziza repanda* (Pers.) Fr. (Courtesy, F. C. Strong.)

has stipitate apothecia growing on sticks lying on or buried in the ground. Its hymenial surface is brilliantly colored. *P. coccinea* (Scop.) Fckl. is very abundant in some regions in early spring forming scarlet apothecia sometimes 3 or more cm. in diameter. The stipe is buried in the soil. *Peziza* has saucer- to cup-like apothecia, usually of considerable size, 2 cm. up to 30 or 40 cm. *P. badia* Fr. and *P. repanda* (Pers.) Fr. are quite frequent in woods and *P. vesiculosa* Fr. in greenhouse soil that has been heavily manured. *Geopyxis* is quite similar to *Peziza* except that the hemispherical or acorn-cup-like apothecium is supported by a relatively slender stalk. (Fig. 76.)

Family Helvellaceae. Apothecia stalked, convex, attached at the apex of the stalk or grown fast to its upper portion. To be mentioned are *Helvella*, with the apothecium more or less saddle-shaped, attached by the center of the under side of the saddle. The hymenial surface may be smooth or gyrosely folded. In the latter case a separate genus *Gyromitra* is often recognized. *Morchella*, with the apothecium grown fast down the side of the upper part of the stalk and with its surface thrown into strong longitudinal and transverse folds so as to be coarsely pitted, thus greatly increases the hymenial surface. The species of this genus are known as Morels or Sponge Mushrooms and are among the most delicious edible fungi known. They grow mostly in deciduous woods, fruiting in the spring. *Verpa*, the Bell Morel, has the cap shaped like a bell and free from the stalk except where attached at the top. It may be smooth or longitudinally ribbed. It is edible, coming a little earlier than the true morels. (Figs. 77, 78.)

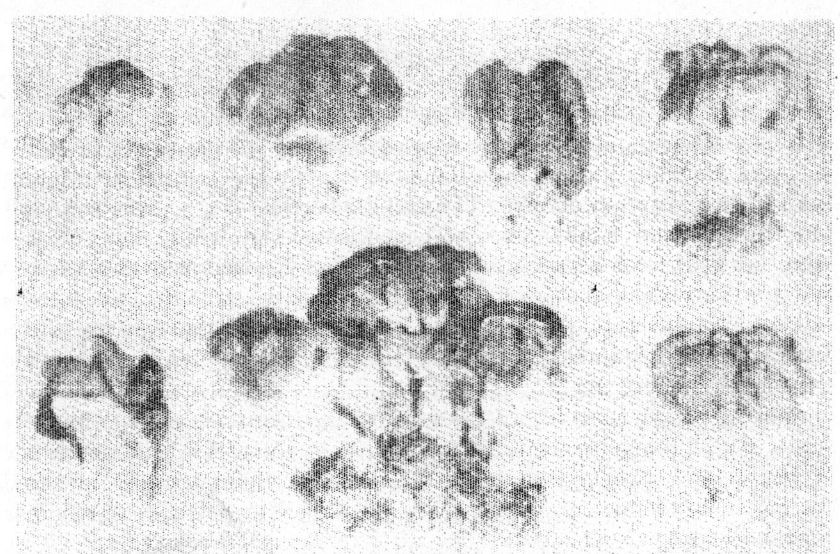

FIG. 77. Pezizales, Family Helvellaceae. *Gyromitra esculenta* (Pers.) Fr. (Courtesy, F. C. Strong.)

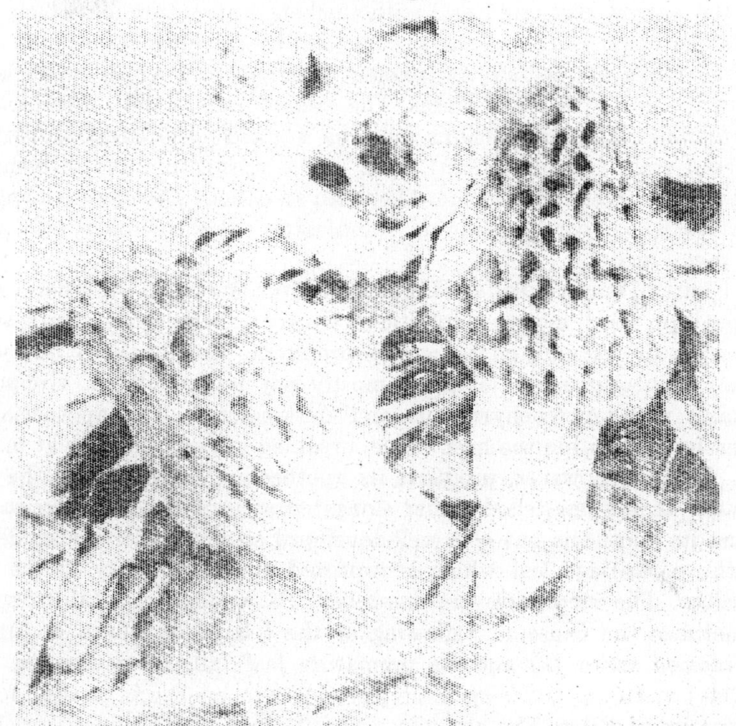

FIG. 78. Pezizales, Family Helvellaceae *Morchella conica* Pers. (Courtesy, F. C. Strong.)

Order Pezizales: Suborder Inoperculatae. This suborder differs from the Operculatae in the absence of an operculum. The apex of the ascus is usually thicker and sometimes depressed. When the spores are mature the ascus elongates and often becomes thicker by the increase in turgor due to the absorption of water. The apical portion of the ascus softens and thickens and finally gives way suddenly, permitting the violent expulsion of the contained liquid and ascospores as the distended ascus wall contracts. The apothecia vary greatly in size, structure, and consistency. Nannfeldt (1932) divided the many genera and species into three orders: Lecanorales (already discussed above), Ostropales, and Helotiales. Because the life histories and inner apothecial structures are thoroughly known in so few of the described species it seems best to the author not to recognize the latter two orders of Nannfeldt until detailed structural and developmental studies in the Lecanorales and in the operculate and inoperculate Pezizales shall enable mycologists to set up a more scientific classification of the whole group of Discomycetes.

Family Ostropaceae. The members of this family are characterized by the narrow elongated asci with thickened apex through which runs a slender canal almost to the surface, and by the long, thread-like ascospores which are septate at frequent intervals and which break up at maturity into cylindrical pieces. The excipulum is mainly pseudoparenchymatous. The apothecium may be stalked (*Vibrissea*), superficial without stalk (*Apostemidium*), sunk in the substratum and disk-shaped (*Stictis*), or rarely perithecium-like, with an ostiole (*Ostropa*). Among the families which would make up Nannfeldt's order Helotiales only a few are mentioned:

Family Dermateaceae (Including Mollisiaceae). Apothecia small or medium sized, mostly epiphytic on woody or herbaceous plants, sometimes on the ground, parasitic or saprophytic, usually fleshy but sometimes cartilaginous or leathery, mostly not bright-colored. Excipulum usually pseudoparenchymatous and dark-colored. The apothecia are often formed within the host tissue, breaking out and opening at maturity. *Mollisia*, however, produces its apothecium externally on the host tissues. It has one-celled rather elongated ascospores. *Pseudopeziza* produces its apothecia out of a well-developed stroma under the epidermis of the parasitized leaf which is ruptured at the maturity of the apothecium. The ascospores are one-celled. *Ps. medicaginis* (Lib.) Sacc. is sometimes the cause of yellowing of the foliage of alfalfa or lucerne (*Medicago sativa* L.) and its premature leaf fall. *Drepanopeziza ribis* (Kleb.) v. Höhn. (*Ps. ribis* Kleb.) produces its apothecia on dead leaves of species of *Ribes*. The actively parasitic stage of this fungus produces only the conidial type of reproduction formerly known as *Gloeosporium ribis* (Lib.) Mont. and Desm. *Diplocarpon* likewise produces its apothecia

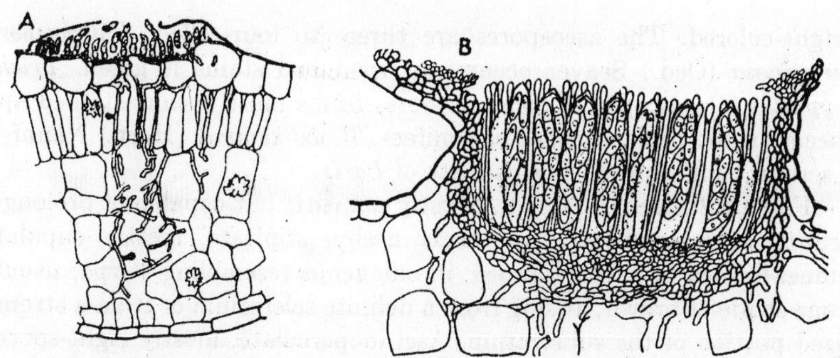

Fig. 79. Pezizales, Family Mollisiaceae. *Diplocarpon earlianum* (E. & E.) Wolf. (A) Acervulus of *Marssonina* stage. (B) Section through apothecium. (Courtesy, Wolf: *J. Elisha Mitchell Sci. Soc.*, **39**(3–4):141–163.)

upon dead leaves, sometimes under a superficial radially arranged shield-like stroma, sometimes not. The parasitic stage produces the conidial form known as *Actinonema* when the radiating dark hyphae are present, or *Marssonina* in their absence, or *Entomosporium* when the conidia have hair-like appendages. On various species of *Prunus* there occur several species of fungi whose conidial stage forms the genus *Cylindrosporium.* It is this stage which is parasitic and causes great damage to the leaves of plums and cherries producing the disease called "yellows" or "shot-hole." The elongated conidia are formed subepidermally in an acervulus. Later in the summer in the same or other acervuli small almost spherical cells are formed, frequently called microconidia. In the stroma developing below the acervulus Higgins (1914a) and Backus (1934) observed the formation of numerous elongated coiled ascogonia which extend up to the microconidial layer. Backus showed that these microconidia grew fast to the terminal cells of the ascogonium, i.e., to the trichogyne, and should therefore be considered as sperm cells. Subsequently ascogenous hyphae are produced and the apothecium develops. Higgins identified this fungus with the genus *Coccomyces* but Nannfeldt indicated that this is incorrect. He therefore gave to these forms the name *Higginsia.* Unfortunately this name is preoccupied and until a valid name is proposed the name *Coccomyces* will probably continue to be used. (Fig. 79.)

Family Helotiaceae. Apothecia mostly fleshy, disk- or cup-shaped, at first closed, often stalked, the excipulum consisting of filamentous hyphae, sometimes grading into an outer layer of shorter, thicker cells. Mostly saprophytic or parasitic upon plant tissues. Apothecia not originating in sclerotia. Conidial stages of reproduction usually not present or at least rarely conspicuous. *Helotium* produces sessile or almost sessile, disk- or cup-shaped apothecia, small to several millimeters in diameter, on plant parts, probably mostly as saprophytes. The hymenium is often bright-colored. The ascospores are three- to four-celled. *Chlorociboria aeruginosa* (Oed.) Seaver occurs on wood and stains it green. *Trichoscyphella* (*Dasyscypha* of some authors) forms mostly short-stalked apothecia on the bark or cones of conifers. *T. willkommii* (Hart.) Nannf. is a serious canker-producing parasite of *Larix*

FAMILY SCLEROTINIACEAE. Mostly parasitic but capable of prolonged growth as saprophytes. Apothecia fleshy, stipitate, mostly cupulate, funnel-formed or saucer-shaped, in one genus resembling *Verpa;* usually some shade of brown; arising from a definite sclerotium or from a stromatized portion of the substratum. Asci inoperculate, mostly eight-spored, ascospores ellipsoidal, often flattened on one side, usually hyaline, unicellular and smooth. Spermatia globose to slightly oval, conidial forms various, in most genera lacking (Whetzel, 1945). As in the preceding family the main portion of the excipulum consists of entangled hyphae, only at the outer surface being reduced to short, somewhat pseudoparenchymatous cells. The stroma may be a free tuberoid sclerotium as in the genus *Sclerotinia* or may be formed in the tissues of the host which are digested and replaced by the fungus hyphae, e.g., *Ciborinia, Ciboria, Monilinia,* or may be plano-convex and attached to the host, *Botryotinia, Septotinia,* etc., or may be indeterminate or of the substratal type, *Lambertella, Rutstroemia,* etc. In all, Whetzel recognized fifteen genera including over ninety species. Many of them were formerly included in the genus *Sclerotinia.* Among the serious pathogens may be mentioned *Sclerotinia sclerotiorum* (Lib.) de Bary, *Monilinia fructicola* (Winter) Honey (*Sclerotinia fructicola* Rehm), the cause of the brown rot of stone fruits in America. Its conidial stage is of the *Monilia* type. *Botryotinia fuckeliana* (de Bary) Whetzel is the cause of a serious disease of the grape vine in Europe. Its conidial stage is a *Botrytis* of the *B.*

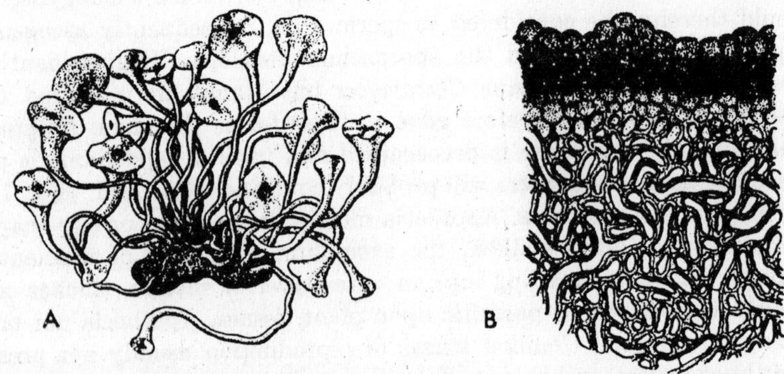

FIG. 80. Pezizales, Family Sclerotiniaceae. *Sclerotinia sclerotiorum* (Lib.) de Bary. (A) Apothecia growing from sclerotium. (B) Magnified section of sclerotium. (After Brefeld: *Unter. Gesammt. Mykol.*, 4(7):112–121.)

cinerea type. In *Stromatinia gladioli* (Drayton) Whetzel, F. L. Drayton (1932, 1934) demonstrated that the fungus has two sexual strains, the sperm cells of one strain fertilizing the receptive female organs of the other strain and vice versa. The conidial stage of this genus also belongs to the genus *Botrytis*. It is probable that all the genera of this family in which "microconidia" are known are of this type of sexual reproduction. (Fig. 80.)

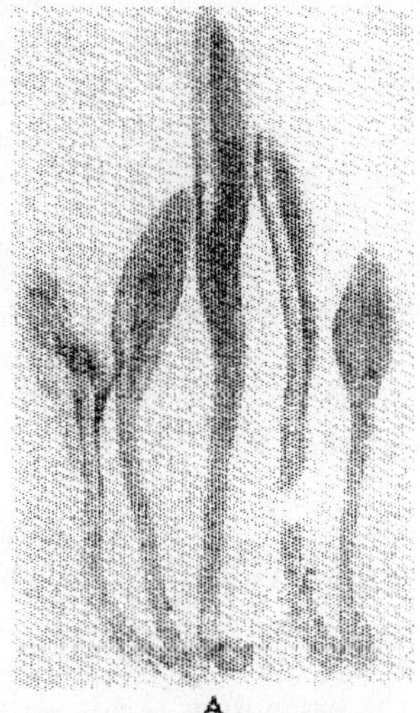

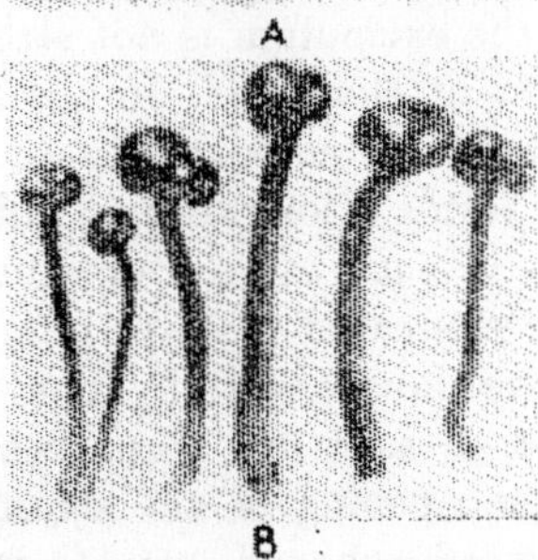

FIG. 81. Pezizales, Family Geoglossaceae. (A) *Geoglossum glabrum* Pers. ex Fr. (B) *Leotia chlorocephala* Schw. (Courtesy, Durand: *Ann. Mycol.*, **6**(5):387–477.)

FAMILY GEOGLOSSACEAE. This family was formerly associated with the Helvellaceae to form the order Helvellales, before the importance of the type of dehiscence of the asci became apparent. The apothecium is stipitate and the hymenium either forms a closely adherent layer to the somewhat thickened upper part of the stipe or forms the upper surface of a head borne at the apex of the stipe. The spore fruit may be entirely made up of hyphae or the cortex may be pseudoparenchymatous. In spite of the external position of the mature hymenium it is really formed endogenously, being covered when young as in most genera of the Pezizales with a convex veil. The ascospores are ellipsoid and one- or two-celled (*Mitrula*) or long elliptical to filiform and several to many septate. They are hyaline to smoky to dark brown in color. The fruiting bodies vary from a few millimeters to five or more centimeters in height. They are mostly found on rotten wood, decaying leaves, moss or soil, usually where plenty of moisture is available. Several genera have bright-colored spore fruits (e.g., *Mitrula*, *Spathularia*, etc.) while those of others are black (e.g., *Geoglossum*). The following genera are worth mentioning: *Mitrula*, spore fruit clavate, bright-colored, ascospores hyaline, ellipsoid; *Microglossum*, similar but with ascospores elongated and many septate; *Geoglossum*, clavate, black, smooth, and dry, ascospores dark, many septate; *Gloeoglossum*, similar but viscid and gelatinous; *Trichoglossum*, similar to *Geoglossum* but beset with spines or setae; *Spathularia*, fan-shaped, bright-colored, ascospores hyaline, many septate; *Leotia*, spore fruit capitate, gelatinous, spores narrowly ellipsoid; *Cudonia*, capitate, leathery, ascospores filiform, multiseptate. Eleven genera and about forty-one species were recognized by Durand (1908) in his excellent monograph of the family. Nannfeldt (1932) believed that this family is related to the stipitate Helotiaceae. (Fig. 81.)

Family Phacidiaceae. The apothecia of this family arise in a well-developed stroma which encloses it below and above, and which is often lenticular in vertical section. These stromata may be superficial or buried in the tissues of the host plant. Possibly as a result of the protection afforded by the stromatic envelope the excipulum is not strongly developed. The stromata may be rounded or elongated. In the former case the stromatic cover often splits stellately at maturity to reveal the apothecium, while in the elongated forms a longitudinal slit is formed. There may be but one apothecium in each stroma or several. In the latter case each apothecium may be elongated more or less, even when the stroma is isodiametric. The asci are clavate, with hyaline, filamentous paraphyses. The ascospores are shot off as in the Pezizaceae when the mature hymenium is exposed by the opening of the stromatic cover. They are elongated and sometimes needle-shaped, one- to many-celled, hyaline or colored. They often have a gummy outer layer. Asexual reproduction is known for many species. These conidial forms usually belong to the form family Leptostromataceae of the Fungi Imperfecti. Some species are saprophytic but perhaps the majority are parasitic upon leaves or twigs. The genus *Hypoderma* is sometimes placed in a separate family, the Hypodermataceae, but the author follows Nannfeldt in uniting the two families. Some authors unite these with other families to form the order Phacidiales.

Phacidium possesses a circular, stellately dehiscing stroma with a single apothecium. Some species are found on the needles of conifers.

FIG. 82. Pezizales, Family Phacidiaceae. *Rhytisma acerinum* (Pers.) Fr. (Courtesy, F. C. Strong.)

Several species of *Lophodermium* also attack the needles of conifers, causing serious leaf fall. In this genus the stroma contains only a single apothecium, but both stroma and apothecium are elongated and narrow, and the dehiscence is by means of an elongated slit. *Rhytisma*, the cause of the tar spot of leaves of maple (*Acer*) and other plants, produces a large more or less isodiametric subcuticular stroma on the upper side of the leaf and usually a smaller sterile stroma on the lower side. After leaf fall the apothecia begin to develop slowly but do not reach maturity until the following spring. In a single stroma are produced numerous elongated apothecia. These do not lie strictly parallel but are more or less sinuately curved or sometimes radiately arranged. At maturity the stroma forms a slit over each apothecium, under proper moisture conditions pulling back at the sides so that the hymenium is fully exposed. At least three species occur on various species of *Acer*, each showing a narrow specialization to only one host or to a group of host species. The Red Maple (*Acer rubrum* L.) is very subject to the disease in some parts of North America. S. G. Jones (1925) has shown that in the stroma there arise ascogonia, with at first one or two cells, which become three- to five-celled. The cross walls become perforated and almost completely absorbed and the nuclei pass into one of the central cells which we must conclude is the oogone. From this arise the ascogenous hyphae with numerous pairs of nuclei. No nuclear fusion occurs until in the young asci which are formed by the hook method. (Fig. 82.)

FAMILY CYTTARIACEAE. The position of this family is not certain, though it probably should be included in the Pezizales. It is not closely related to any of the foregoing families. Apothecia numerous, imbedded in a fleshy stroma produced externally on the twig of the host. Ascospores one-celled, hyaline. One genus, *Cyttaria*, confined with its hosts (species of *Nothofagus*, the Southern Beech) to the South Temperate Zone in South America and Australasia. The fleshy stromata serve the natives for food. The basal portion of the stroma of one or more species produces organs resembling spermogonia with sperm cells. (Fig. 83.)

FIG. 83. Pezizales, Family Cyttariaceae. *Cyttaria gunnii* Berk. (After Lindau, in Engler and Prantl: Die Natürlichen Pflanzenfamilien, Leipzig, W. Engelmann.)

Two genera of inoperculate Discomycetes from Sumatra described by Boedijn (1934) perhaps indicate a transition to the Cyttariaceae. They are *Jacobsonia* on wood and *Myriodiscus* on bamboo stems. From a plectenchymatic stroma there radiate in all directions closely packed, branched stalks of apothecia forming a loose or dense ball respectively. In *Myriodiscus* there are perhaps 800 to 1000 or more of these apothecia which are rather gelatinous. The asci are cylindrical or obovoid and multisporous. These spores are formed in this large number from the beginning and are not the result of budding of a few original spores. Judging from the illustration there must be several thousand of these small ellipsoidal ascospores in a single ascus. In *Jacobsonia*, in which the apothecial branches are not so tightly packed nor so numerous, the asci are only eight-spored and the apothecia are subcoriaceous.

Order Tuberales. The fungi of this order are all terrestrial and produce subterranean spore fruits (ascocarps). Some are probably saprophytic but it seems possible that certain species are perhaps parasites upon the roots of higher plants. The ascocarps vary in diameter from a few millimeters up to three or more centimeters and may be found close to the surface of the soil or at a considerable depth. With no direct means of bringing the ascospores into the air for distribution by air currents as is the case in the Pezizales, the Tuberales depend for distribution largely upon the activities of mycophagous animals, probably to a considerable extent insects, but, for some species at least, ground-inhabiting rodents. Thus in California the so-called ground squirrels dig out the fruiting bodies, doubtless attracted by the odor diffusing up through the soil, and eat them on the spot or carry them away to their burrows or other hiding places. In these processes pieces of the ascocarps are scattered and the ascospores find their way into the soil.

By germination of the ascospores of one species of *Tuber* a conidium-bearing mycelium has been obtained but this has never been grown to the stage where normal ascocarps were produced. Chaze and Mestas (1939) made tissue cultures of *Tuber melanosporum* Vitt. on various culture media and obtained in pure culture an extensive mycelium. Here and there in this mycelium there developed dense dark-colored masses in which were produced the two-spored asci and ascospores typical of the species. However, typical fruiting bodies such as occur naturally in the soil were not produced.

In 1863 Anton de Bary observed typical clamp connections, such as occur in many Basidiomycetes, at the base of the asci in two species of *Tuber*. De Ferry de la Bellone (1886) observed clamp connections on the brown hyphae growing out of the ascocarps of *Tuber brumale* Vitt., *T. mesentericum* Vitt., *T. aestivum* Vitt., and *T. panniferum* Tul. and figured this structure for the last species. Mattirolo (1887) found such clamp connections in mycelium external to but connected with the spore fruits of *T. lapideum* Matt. Greis (1936, 1938) showed that the ascus hook of *T. aestivum* Vitt. has the structure of a clamp connection. If it is confirmed that the external mycelium in this order bears clamp connections it must be concluded that the fungus is dicaryon in nature, as is the case with the secondary phase of mycelium in the Basidiomyceteae

The sexual process is unknown in the Tuberales, the hypogeous habitat of the fruiting body making it very unlikely that it would be observed except by rare accident. The structure of the immature spore fruit has been studied in a good many cases but for many species only the mature ascocarps are known. By comparing the mature stages, taking cognizance of the ontogeny of the ascocarp where it has been observed, Fischer (1896, 1938) and Bucholtz (1902) recognized a graded series leading from types scarcely different from some of the Pezizales that have almost completely subterranean apothecia to very complex forms such as those of *Tuber* in which the apothecial nature of the ascocarp is almost entirely concealed.

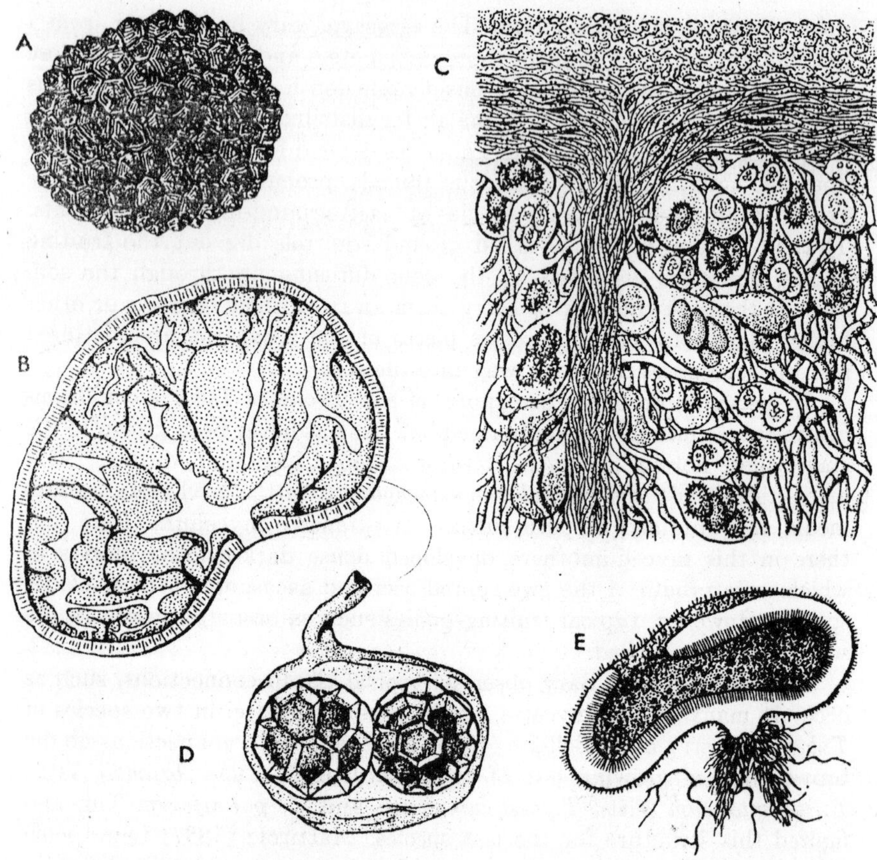

Fig. 84. Tuberales, Family Tuberaceae. (A) *Tuber aestivum* Vitt., external view. (B) *Tuber rufum* Pico, section through ascocarp. (C) Enlarged view of a portion of the ascocarp, the ascospores with spiny surface. (D) *T. magnatum* Pico, ascus showing ascospores with alveolate marking. (E) *Genea hispidula* Berk, vertical section through ascocarp, showing the single cavity and opening. (After Tulasne, in Engler and Prantl: Die Natürlichen Pflanzenfamilien, Leipzig, W. Engelmann.)

Perhaps the simplest structure in the order, according to Miss Gilkey (1939), is that exhibited by some species of *Hydnocystis*. The hypogeous spore fruit is subspherical, with an opening which is usually partly blocked by hairs. The single large cavity is lined by an even hymenium made up of cylindrical, 8-spored asci. The paraphyses are of about the same length as these and do not form an epithecium above them. Except for its permanently hypogeous habit this might well be placed in the Pezizales, a position, in fact, to which Fischer (1938) assigned it. In *Genea* the main cavity may have even walls or may be thrown into folds which much increase the hymenial surface. The paraphyses grow out beyond the asci and unite above the latter to form a thick pseudoparenchymatous epithecium or "secondary cortex" which is generally not so thick as the "primary cortex" (the excipulum and its outer layers). The folding of the surface of the cavity may lead to the formation of canals leading to enlarged chambers. In some of these the hymenium lines both but in *Piersonia* and some other genera the asci arise only in the chambers while the canals are lined only with rudimentary paraphyses. With this increasing complexity the single external opening is lost and several openings develop at points where the canals converge near the surface. In *Tuber* the canals are filled at a very early stage by the ingrowing paraphyses and appear as "veins" in the tissue of the sporocarp. The hymenial chambers become obliterated by the ingrowing asci and paraphyses. In this genus the asci are ovoid or spherical and often but few-spored. They do not form a single hymenial layer at maturity but project into the epithecial tissues at various levels. A number of other genera are recognized with varying degrees of modification of the foregoing structural types. Of all the species of the order a few species of the genus *Tuber* (the truffle) are of economic importance. These species, especially *T. aestivum* Vitt. and *T. melanosporum* Vitt., occur in the rather open forests of Southern Europe (mainly under species of *Quercus*), where they are collected by the use of trained animals (dogs or pigs) which find them by their odor. In recent years many species of Tuberales have been discovered in the Pacific Coast states where the climate is quite similar to the regions of Europe where they are best known. Miss Gilkey (1939) described and figured nineteen genera and fifty-seven species of North American Tuberales. She included in the Tuberales the genera formerly placed by Fischer (1896) in the family Terfeziaceae in the Order Aspergillales. (Fig. 84.)

Order Hysteriales. In this order of plant-inhabiting saprophytes and parasites the apothecia are much reduced in size and compressed laterally, to elongated, often somewhat boat-shaped, structures, opening by a long narrow slit. They are dark-colored, leathery or hard, and show a strong contrast between the dark-colored excipulum and the light-colored hymenial layer. The latter consists of ovoid to elongated cylindrical asci intermingled with mostly septate paraphyses which are frequently enlarged at the apex or branched, even forming at times a well-marked epithecium. The apothecia may be single or in closely packed groups, superficial or emerging from the substratum. The asci open by a pore, not a lid. The ascospores are of very many different forms, as in the Phacidiaceae, to which this order shows many points of relationship. They are ellipsoid and one-celled, or several-celled, or divided both longitudinally and transversely into many cells, or needle-shaped. In color they vary from hyaline to brown. Conidial formation has been reported

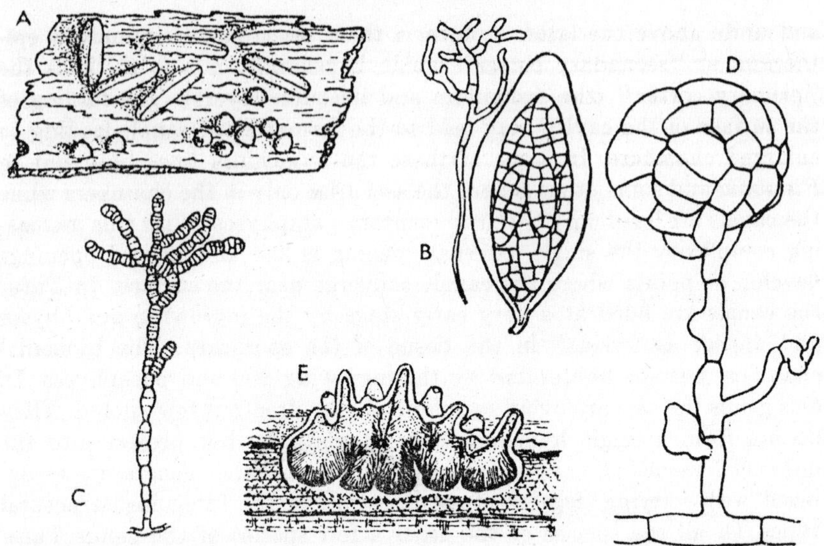

Fig. 85. Hysteriales, Family Hysteriaceae. (A, B) *Hysterographium minutum* Lohman. (A) Apothecia and pycnidia. (B) Ascus and paraphysis. (C) *Hysterium insidens* Schw., asexual stage (*Septonema spilomeum* Berk.). (D, E) *Lophium mytilinum* (Pers.) Fr. (D) Asexual stage, *Papulospora mytilina* (Pers.) Lohman. (E) Pycnidial stage. (After Lohman: *Papers Mich. Acad. Sci.*, **17**:229–288.)

in a few species, usually with the conidiophores included in a pycnidium or somewhat similar structure. Lohman (1933) has shown that some of the conidial forms usually assigned to the genus *Sporodesmium*, of the Fungi Imperfecti, are the conidial stages of several species of Hysteriales. Besides this spore form other conidial stages observed by him represent the genera *Papulospora* and *Septonema* of the Moniliales and various forms of the Sphaeropsidales. The details of sexual reproduction are almost unknown in this group. The relationships within the order and to other orders are still more or less problematic. Usually the more than 400 species are distributed among several families. Of these Nannfeldt (1932) removes the genera assigned to the Hypodermataceae to the Phacidiaceae, where they have been placed in this work. Some of the leaf inhabiting fungi formerly ascribed to this order (e.g., the genus *Parmularia*) are placed by the more recent students of the group in the Pseudosphaeriales. This leaves as the only important family the Hysteriaceae.

Family Hysteriaceae. Apothecia external, black and carbonaceous, single or united in a stroma. Most of the species of this family are saprophytic on bark or decorticated twigs or branches. *Hysterographium fraxini* (Pers. ex Fr.) de Not. is common on various species of ash (*Fraxinus*). Its ascospores are dark-colored and divided by transverse and longitudinal septa. The apothecia are boat-shaped. (Fig. 85.)

The relationship of this order is more or less in dispute. The compact hymenium, with numerous well-developed paraphyses, and the slit-like opening suggest great affinity to some of the Phacidiaceae, as does the great variability of ascospore structure. The suggested relationship to the family Lophiostomataceae in the Order Sphaeriales seems doubtful, in spite of the slit-like ostiole, for in other respects the latter family is typically Sphaeriaceous.

Order Taphrinales (Exoascales). This is a group of approximately 100 recognized species of fungi, all but a very few parasitic. Two or more genera are recognized in two families whose actual relationship to one another is uncertain. Both families are characterized by the production of a superficial hymenium with indeterminate margin and without paraphyses. This may rest upon a thin membranous hypothecium consisting of interwoven nyphae (Ascocorticiaceae) or may burst through the epidermis or cuticle of the host plant without a definite hypothecium (Taphrinaceae). In the author's opinion these represent the ultimate steps in the reduction and simplification of an apothecium. Together with this simplification of apothecial structure there has arisen a marked modification of the sexual process in the Taphrinaceae. According to Miss Wieben (1927) the life history for *Taphrina epiphylla* Sadeb. and *T. klebahni* Wieben is as follows: An ascospore, or one of the smaller spores which the ascospores may produce by budding while still within the ascus, germinates upon the surface of the host plant. Only when the germ tubes from two such spores of opposite sexual phase meet and fuse is active infection of the host possible. As a result of this conjugation a dicaryon hypha is produced which grows intercellularly in the host tissues. R. E. Fitzpatrick (1934) has shown that in *T. deformans* (Berk.) Tul. a single spore can cause infection. The nucleus of this spore divides and thenceforth the two nuclei divide conjugately and the vegetative mycelium is of dicaryon nature as in the case of the species studied by Miss Wieben.

The presence of these hyphae causes hypertrophy and hyperplasia in the affected parts of the host. This may affect the leaves, fruits, and shoots, often causing the formation of "witches' brooms." Eventually dicaryon cells are formed near the surface of the host, often in a subcuticular location, forming a structure resembling pavement epithelium. These cells sometimes become somewhat thick-walled. In them eventually nuclear fusion occurs, followed by elongation of the cell in a vertical direction, rupturing the cuticle or emerging between epidermal cells of the host. The diploid nucleus and most of the cytoplasm pass into the upper part of the cell leaving an empty lower part which is sometimes but not always cut off from the upper part by a cross wall, or the diploid nucleus may divide and one remain in the lower cell while the other passes

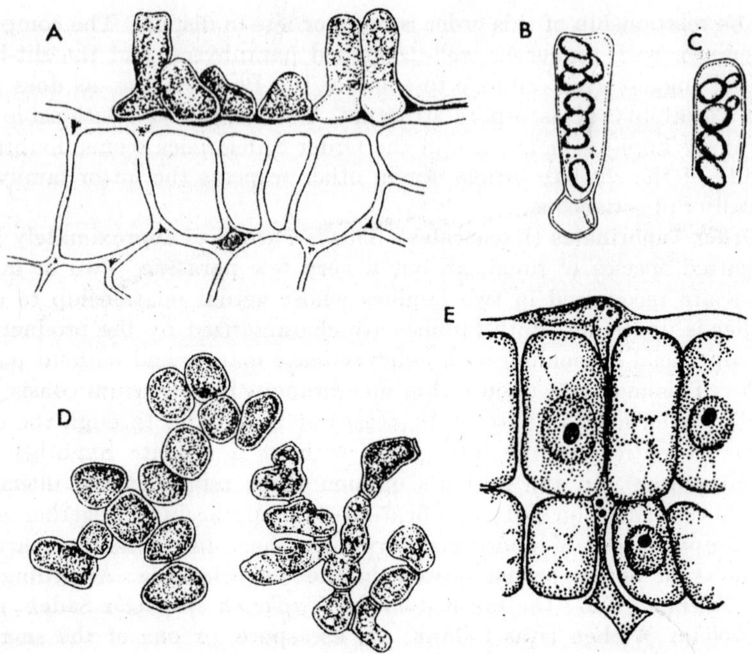

FIG. 86. Taphrinales, Family Taphrinaceae. *Taphrina deformans* (Berk.) Tul. (A) Portion of a leaf section showing asci in various degrees of development. (B) Mature ascus showing empty basal portion. (C) Portion of ascus showing two ascospores beginning to bud. (D) Subcuticular ascogenous cells seen from above. (E) Infection of leaf by dicaryon mycelium. (A–D, courtesy, Pierce: *U.S. Dep. Agr. Vegetable Physiol. Path. Bull.*, **20**:1–204. E, courtesy, Fitzpatrick: *Sci. Agr.*, **14**(6):305–326.)

into the thin-walled upper portion undergoing further divisions there. The diploid nucleus undergoes meiotic and mitotic divisions, the spindle of the first division being transverse according to Juel (1921). Around the eight nuclei thus formed the ascospores develop. In many species each ascospore undergoes budding so that the ascus becomes polysporous. Miss Wieben's germination experiments showed that four of the ascospores (and the spores produced from them by budding) are of one sexual phase and four of the other. From Fitzpatrick's studies it is apparent that in *T. deformans* this distinction does not occur or is not well marked.

Lohwag (1934) advances reasons for believing that *Taphrina* is not a true Ascomycete but a Basidiomycete which has reverted to the ancestral condition. The normal basidium, in his viewpoint, represents an ascus in which the ascospores have pushed out into projections of the ascus wall so as to become apparently external, although reall retained within the wall. In *Taphrina* the transverse position of the nuclear spindle, the habit of budding on the part of the ascospores, and the parietal position assumed by the ascospore nuclei before the spores are delimited are all pointed to as Basidiomycetous characters.

FAMILY ASCOCORTICIACEAE. *Ascocorticium*, the only genus of the Family Ascocorticiaceae, consists of a small number of species whose fructifications produce a gray- or pink-colored thin coating with indeterminate growth, over the surface of bark or bare wood of dead trees. The hypothecium consists of four to six layers of interwoven hyphae running parallel to the surface of the bark. Upon these arise the ovoid eight-spored asci. The asci are formed by the hook or crozier method from dicaryon ascogenous hyphae whose origin is not known. It is not known whether there is any sexual process other than the probable union of two nuclei in each young ascus.

FAMILY TAPHRINACEAE. *Taphrina* (*Exoascus*, *Magnusiella*, or *Taphria*) is the only genus of the Taphrinaceae. Of great economic importance is *T. deformans* (Berk.) Tul., causing the disease of the peach (*Amygdalus persica* L.) known as leaf curl. The ascospores or spores arising from them by budding lie dormant in crevices of the twigs or on the bud scales until the following spring when they germinate and infect the young leaves and even the fruits and young twigs. On the other hand the ascospores may also germinate by budding on the surface of the host and grow in the manner of a yeast, infection taking place in the spring from the yeast-like cells. The affected leaves become much thickened and distorted as well as discolored. The diseased fruits may show irregular bright red patches of thickened tissues. On the diseased areas the asci are formed subcuticularly in the late spring or early summer. In a few cases the spores appear to germinate soon after formation so that Waite (1932) and Poole (1932) report what may be a secondary infection occurring the same season. *Taphrina pruni* Tul. and *T. communis* (Sadeb.) Giesenh. infect the young fruits of *Prunus* and *Padus*, causing them to develop into the hollow hypertrophied structures which, upon the plums, are known as "plum pockets." Here also the asci are subcuticular. *T. cerasi* (Fckl.) Sadeb. causes witches' brooms on various species of cherry (*Prunus cerasus* L., *P. avium* L., etc.). The mycelium lives perennially in the tissues of the host. *T. robinsoniana* Giesenh. causes hypertrophy of the scales of the aments of the alder (*Alnus*). In *T. potentillae* (Farl.) Johans., the mycelium lives beneath the epidermis and sends up, between the epidermal cells, branches which bear separate asci. (Fig. 86.)

In this family the whole vegetative mycelium consists of dicaryon cells and is therefore comparable to the ascogenous hyphae of those forms with definite sexual organs and well-developed apothecia, such as *Pyronema*, etc. Many authors consider this family to be primitive, but it seems more logical to consider it as a much simplified offshoot of the Pezizales, the Ascocorticiaceae being a possible intermediate stage. Miss Catherine

Roberts (1946) describes experiments and observations with *Taphrina deformans* and *Torulopsis pulcherrima* (Lindner) Sacc., which demonstrate many points of similarity in cultural behavior and in structure of the organisms in culture. Both grow in culture in the manner of the budding yeasts (Saccharomycetaceae). *Torulopsis* frequently produces large thick-walled cells which bud out a small thin-walled two- to three-spored ascus. A similar phenomenon was observed in the cultures of *Taphrina* by Miss Roberts and also by Mix (1924, 1935). This would seem to indicate the possibility that *Torulopsis* is a yeast-like derivative of the Taphrinales. It is well known that very many widely unrelated fungi may adopt the yeast manner of growth. It must be noted however that Anton de Bary (1884) believed the yeasts and the Taphrinales to be primitive, closely related Ascomycetes, a viewpoint not shared by the author of this textbook.

Key to the More Important Families of Order Laboulbeniales

(*Modified from Thaxter, 1908*)

Antherids producing naked sperms endogenously.
  Antherids opening by separate tubes. Family Laboulbeniaceae
    12 or more subfamilies. Characteristic genera *Laboulbenia*, *Stigmatomyces*, etc.
  Antherids compound, i.e., opening into a common chamber which has one opening to the outside. Family Peyritschiellaceae
    20 or more genera. *Peyritschiella*, *Dimorphomyces* etc.
Antherids, more or less undifferentiated cells of the appendages or of their branches with thin walls.
  Forming massive multicellular plants. Family Zodiomycetaceae
    3 or more genera. *Zodiomyces*, etc.
  Not forming massive multicellular plants. Family Ceratomycetaceae
    4 or more genera. *Ceratomyces*, etc.
(Many of the groups named by Thaxter should perhaps be made into distinct families.)

Key to the Orders of Apothecial Fungi (Discomycetes)

Parasitic upon fresh-water algae which they enclose in a definite "lichen" thallus.
Order Lecanorales
Not parasitic upon algae or if so not forming a definite "lichen" thallus.
  Apothecia fleshy or leathery, external from the first or emerging more or less from the substratum; round in outline, less often narrowed, sessile or stalked. Usually not in a stroma (except *Cyttaria*). Ascospores discharged into the air. Order Pezizales
    Asci opening by an approximately apical lid (operculum).
Suborder Operculatae
    Asci without an operculum. Suborder Inoperculatae
  Apothecia fleshy, permanently subterranean, with a simple cavity lined with hymenium or the cavity divided by folds and ridges into chambers and passageways. Ascospores not discharged into the air.
Order Tuberales

Apothecia small, laterally compressed, with a narrow, elongated opening; dark-colored and hard. Mostly bark- and wood-inhabiting.
Order Hysteriales

Apothecia with no definite limiting border, the asci forming a hymenium without paraphyses, usually with very limited hypothecial tissues.
Order Taphrinales

Key to the Families and Important Genera of Apothecial Lichens Occurring in the United States (Lecanorales)

(*Based on Fink and Hedrick, 1935*)

Asci disintegrating and the spores forming a powdery coat (mazaedium) on the surface of the hymenium.
- Thallus crustose, without cortex. On trees or decorticated wood.
  - Apothecia borne on stipes. Family Caliciaceae
    - Ascospores one septate; not parasitic on other lichens. *Calicium*
    - Ascospores not septate, not parasitic on other lichens. *Chaenotheca*
    - (Besides the foregoing, five other small genera, two parasitic on other lichens.)
  - Apothecia not borne on stipes. Family Cypheliaceae
    - Ascospores one septate. *Cyphelium*
    - (Besides the foregoing two other small genera.)
- Thallus foliose or bushy, with cortex. Family Sphaerophoraceae
  - One genus in the U.S., mainly on soil or rocks. *Sphaerophorus*

Asci persistent; no mazaedium.
- Apothecia irregular, linear or oblong.
  - Thallus crustose, well-developed "proper exciple" lacking. Family Arthoniaceae
    - Ascospores transversely one to several septate. *Arthonia*
    - Ascospores both transversely and longitudinally septate. *Arthothelium*
  - Thallus crustose, "proper exciple" present.
    - Apothecia not seated in a stroma. Family Graphidaceae
      - Ascospores hyaline or at most light brown.
        - Ascospores not septate; on old wood. *Xylographa*
        - Ascospores septate, paraphyses not branched.
          - Cells of ascospores cylindrical or cubical. *Melaspilea*
          - Cells of ascospores lenticular.
            - Ascospores only transversely septate. *Graphis*
            - Ascospores both transversely and longitudinally septate. *Graphina*
        - Ascospores septate, paraphyses branched and interwoven.
          - Cells of ascospores cylindrical or cubical. Not longitudinally septate. *Opegrapha*
          - Ascospores both transversely and longitudinally septate. *Helminthocarpon*
      - Ascospores brown.
        - Transversely septate. *Phaeographis*
        - Transversely and longitudinally septate. *Phaeographina*
    - Apothecia seated in a stroma. Family Chiodectonaceae
      - Six genera in the U.S.

Thallus fruticose, "proper exciple" present. California. Family Roccellaceae
Cortical hyphae longitudinal. *Dendrographa*
Cortical hyphae transverse.
Spores hyaline. *Roccella*
Spores brown. *Schizopelte*
Apothecia more or less round or cup-like.
Host algae Myxophyceae.
Thallus taking its form from that of the host.
Thallus squamulose to foliose; on Nostoc. Family Collemaceae
Spermatia produced within thallus. *Collemodes*
Spermatia produced in spermogonia opening externally.
Ascospores only transversely septate. *Synechoblastus*
Ascospores both transversely and longitudinally septate.
Definite cortex lacking. *Collema*
Thin definite cortex present. *Leptogium*
Algal host Scytonema or Stigonema. Family Ephebaceae
Seven small genera in the U.S.
Thallus rarely taking its form from that of the algal host.
Thallus large, plainly foliose.
Apothecia without distinct exciple. Family Peltigeraceae
Thallus with cortex above only; apothecium on under surface of the thallus lobes. *Nephroma*
Thallus with cortex above and below; apothecia on upper surface.
Ascospores hyaline. *Peltigera*
Ascospores brownish to brown. *Solorina*
Apothecia with distinct exciple. Family Stictaceae
Single genus in the U.S. *Sticta*
Thallus small, crustose to somewhat foliose or dwarf fruticose.
Several small families, mostly growing on soil or rocks: Pyrenopsidaceae, Lichinaceae, Heppiaceae, Pannariaceae.
Host algae Chlorophyceae.[3]
Apothecia without exciple or this only rudimentary (fairly well developed in *Lecanactis*).
Thallus crustose, ascospores nonseptate or transversely septate. Family Lecanactidaceae
Three small genera, mostly on trees, but some species on rocks or soil. Not widely spread over the U.S.
Thallus crustose, ascospores transversely and longitudinally septate. Family Ectolechiaceae
Only one species, of the genus *Lopadiopsis*, in the U.S.
Apothecia with well-developed exciple.
Both proper and thalloid exciple present, the latter sometimes disappearing.
Apothecia more or less deeply immersed in the thallus. Family Thelotremaceae
Five genera in the U.S., mostly Southern.
Apothecia superficial or not deeply immersed.
Proper exciple dark. Family Diploschistaceae

[3] A few species of the foregoing families that are mostly parasitic on Myxophyceae have Chlorophyceae for their host; see especially Peltigeraceae and Stictaceae.

Two small genera in the U.S. of widely distributed species.
Proper exciple hyaline. Family Gyalectaceae
Five small genera in the U.S., from various parts of the country.
Either proper or thalloid exciple present, very rarely both.
Apothecia with proper exciple.
Thallus taking its form from that of the algal host.
Family Coenogoniaceae
Two small genera in the U.S.
Thallus not taking its form from that of the algal host.
Thallus twofold, a primary crustose, squamulose or foliose portion out of which grow the upright podetia.
Family Cladoniaceae
Podetia short, unbranched, hypothecium hyaline.
Baeomyces
Podetia short to long, more or less branched.
Podetia mostly hollow, spores nonseptate. *Cladonia*
Podetia solid; spores 3 or more septate. *Stereocaulon*
Thallus foliose, apothecia not on podetia.
Family Gyrophoraceae
Spores nonseptate. *Gyrophora*
Spores transversely and longitudinally septate.
Umbilicaria
Spores only transversely septate. *Dermatiscium*
Thallus crustose to squamulose.
Ascospores hyaline, rarely brown, apothecia rarely yellowish, spores nonseptate to septate, with rectangular or cubical cells. Family Lecideaceae
Ascospores nonseptate, very large, thick-walled.
Mycoblastus
Ascospores nonseptate, not large.
Thallus crustose. *Lecidea*
Thallus squamulose. *Psora*
Ascospores one septate, very large, thick-walled.
Megalospora
Ascospores one septate, small. *Catillaria*
Ascospores 1 to 3 to more septate, acicular.
Thallus crustose. *Bacidia*
Thallus squamulose. *Tonina*
Ascospores 3 or more septate, fusiform. *Bilimbia*
Ascospores transversely and longitudinally septate.
Ascospores hyaline. *Lopadium*
Ascospores brown. *Rhizocarpon*
Ascospores hyaline, nonseptate to septate, with lenticular cells.
Apothecia usually yellowish. Thalloid exciple in one genus. Family Caloplacaceae
Apothecium with proper exciple. *Blastenia*
Apothecia with thalloid exciple. *Caloplaca*
Two or more genera present in the U.S.
Ascospores brown. Thalloid exciple in one genus.
Family Buelliaceae
Apothecia with proper exciple, spores one septate.
Buellia

Apothecia with thalloid exciple. *Rinodina*
Three more genera present in the U.S.
Apothecia with thalloid exciple. (See also *Caloplaca* and *Rinodina*, above.)
Thallus foliose or rarely somewhat fruticose.
Thallus more or less yellow in color, ascospores hyaline. Family Teloschistaceae
One genus in the U.S., foliose or some species a little fruticose. *Teloschistes*
Thallus greenish gray to ashy or darker, ascospores brown. Family Physciaceae
Upper cortex plectenchymatous.
Exciple colored like the thallus. *Physcia*
Exciple becoming black. *Pyxine*
Upper cortex not plectenchymatous. *Anaptychia*
Thallus definitely foliose, sometimes also slightly fruticose.
Spores nonseptate, hyaline. Family Parmeliaceae
Thallus yellow, spores 16 to 32 in each ascus. *Candelaria*
Thallus rarely yellow, spores 8 in each ascus.
Thallus flat, apothecia not marginal. *Parmelia*
Thallus more or less upright, apothecia marginal. *Cetraria*
(Besides the foregoing only one other small genus in the U.S.)
Spores septate. Family Stictaceae
One genus in the U.S. *Sticta*
Thallus plainly fruticose or pendent. Family Usneaceae
Ascospores nonseptate.
Thallus dorsiventral. *Evernia*
Thallus upright or pendent.
Greenish gray or straw-colored. *Usnea*
Brownish to brown. *Alectoria*
Ascospores 1 to 3 septate. *Ramalina*
Thallus crustose to squamulose.
Spores minute, many in each ascus. Family Acarosporaceae
Apothecia with thalloid exciple, immersed. *Acarospora*
Apothecium with proper exciple, immersed to sessile. *Biatorella*
Two other small genera in the U.S.
Ascospores usually 8 or less in each ascus. Here may also be sought some genera in Families Caloplacaceae and Buelliaceae (see above).
Ascospores large, with thick wall. Family Pertusariaceae
Single genus in the U.S. *Pertusaria*
Ascospores with thin wall. Family Lecanoraceae
Ascospores nonseptate, small. *Lecanora*
Ascospores 1 to 3 septate. Thallus crustose. *Lecania*
Ascospores transversely and longitudinally septate. *Phlyctis*
(About six other small genera in the U.S.)

Key to the Families and Important Genera of Order Pezizales, Suborder Operculatae

(*Based in part on Seaver, 1928*)

Apothecia cup-shaped or discoid, sometimes convex, sessile or stipitate, never with a pileate structure. Family Pezizaceae

Ascospores globose, hyaline to pale brown. Tribe Sphaerosporeae

Plants growing only on the dung of animals, mature ascospores becoming pale brown to blackish. *Ascodesmis*

Plants not restricted to dung.

Apothecia with well-developed bristles or flexuous hairs.

Apothecia red, brown, or greenish, soft and fleshy or waxy. *Sphaerospora*

Apothecia black, brownish, or orange, tough or cartilaginous, mostly stipitate. *Pseudoplectania*

Apothecia not clothed with well-developed hairs.

Apothecia everted to form subglobose ascocarps clothed externally with the hymenium, somewhat cartilaginous. *Sphaerosoma*

Apothecia discoid to convex, with the hymenium limited to the upper surface and sides; fleshy or waxy.

On the ground or on humus, hymenium strongly convex. *Boudiera*

On the ground or on humus, plane or slightly convex. *Lamprospora*

On living or dead foliage of Conifers, tough. *Pithya*

Ascospores ellipsoid to fusoid, rarely subglobose, becoming violet, later brown or blackish. Tribe Ascoboleae

Spores free in the ascus. *Ascobolus*

Spores united into a ball in the ascus. *Saccobolus*

Ascospores ellipsoid to fusoid, hyaline or pale brown.

Ascospores becoming reticulate at maturity. Tribe Aleurieae

Apothecia clothed with colored hairs. *Melastiza*

Apothecia without colored hairs, hymenium bright orange. *Aleuria*

Apothecia without colored hairs, hymenium dark brown. *Aleurina*

Ascospores smooth or sculptured, but never reticulate.

Apothecia small, up to 1 cm. in diameter, not conspicuously hairy or setose. Tribe Humarieae

Apothecia attached to the substratum clear to their edges, mostly on wood. *Psilopezia*

Apothecia attached to the substratum by the center only, mostly on soil or humus or dung of animals.

Vegetative mycelium superficial, mostly only on burnt places. *Pyronema*

Vegetative mycelium immersed in the substratum.

Asci 8-spored, apothecia fleshy, coprophilous. *Ascophanus*

Asci 8-spored, apothecia fleshy, humicolous. *Humarina*

Asci 8-spored, apothecia tough. *Pseudombrophila*

Asci more than 8-spored, bilabiate. *Streptotheca*

Asci more than 8-spored, with circular operculum.

With small spores. *Ryparobius*

With large spores. *Thecotheus*

Apothecia small- to medium-sized, conspicuously setose or hairy. Tribe Lachneeae

Apothecia partially to entirely buried in the ground. *Sepultaria*

Apothecia superficial on substratum, on soil, wood, humus, or dung.

Densely clothed with brick-red hairs. *Perrotia*

Clothed with hyaline or brown hairs.

Hairs not septate. *Lasiobolus*

Hairs septate, usually erect or bristle-like. *Patella*

Apothecia medium to large, strongly unequal-sided or split on one side. Tribe Otideeae

Springing in clusters from a buried sclerotium. *Wynnea*

Usually isolated, not springing from a sclerotium.

Spores striate with light and dark bands. *Phillipsia*

Spores not striate. *Scodellina*

Apothecia medium to large, symmetrical, stipitate, densely hairy or tomentose, usually tough or leathery. Tribe Sarcoscypheae

Hymenium bright yellow to scarlet, spores striate. *Cookeina*

Hymenium as above, spores not striate. *Plectania*

Hymenium gray or brown, apothecium externally black or brown.

With a thick gelatinous hypothecium. *Bulgaria*

Hypothecium not gelatinous.

Opening more or less stellately, attached to sticks, tough. *Urnula*

Not opening stellately, on the ground, fleshy to tough. *Paxina*

Apothecia medium to large, sessile to short stipitate, not densely hairy or tomentose, fleshy and brittle. Tribe Pezizeae

Apothecium buried in the ground when young. *Sarcosphaera*

Apothecium superficial.

Medium-sized, "shaped like an acorn-cup, with a short slender stem" (but see reference on page 225 to *G. cacabus* (Fr.) Sacc.) *Geopyxis*

Medium-sized to large, sessile or with a stout stem-like base.

Ascospores apiculate, apothecium attached by numerous root-like processes. *Rhizina*

Ascospores apiculate, apothecia centrally attached. *Discina*

Ascospores not apiculate. *Peziza*

Apothecia pileate, subglobose or columnar, pileus always supported by a distinct stem, never cup-shaped or discoid.
Family Helvellaceae

Pileus costate, the ribs anastomosing to form irregular pits.

Sterile stem and fertile head distinct. *Morchella*

Fertile part extending to the base, the ascocarp forming a more or less depressed globose structure. *Daleomyces* (*Durandiomyces*)

Upper side of the pileus smooth or lacunose but not truly costate.

Pileus bell-shaped. *Verpa*

Pileus saddle-shaped or irregularly subglobose. *Helvella*

Pileus gyrosely convolute (perhaps to be included in *Helvella*). *Gyromitra*

Pileus columnar. *Underwoodia*

Key to the Families and Important Genera of Order Pezizales, Suborder Inoperculatae

(*Based upon Nannfeldt, 1932, with additions*)

Apothecia numerous, opening at the surface of a tuberoid fleshy stroma growing on the living twigs of *Nothofagus*. South America and Australasia.
Family Cyttariaceae

Single genus. *Cyttaria*

Apothecia not buried in a tuberoid fleshy stroma.

Asci cylindrical, the apical thickening hemispherical, with a long slender canal. Ascospores hyaline, thread-like, often septate and falling apart into cylindrical cells. Saprophytes. Apothecia sessile, or stalked (*Vibrissea*), or immersed and somewhat perithecium-like. (Order Ostropales of Nannfeldt.)
Family Ostropaceae

Apothecium-like.

Stalked with convex hymenium. *Vibrissea*

Not stalked, superficial. *Apostemidium*

Not stalked, emerging from within the substratum. Paraphyses little if at all branched. *Stictis*

Not stalked, emerging from within the substratum. Paraphyses much branched, forming a strong epithecium. *Schizoxylon*

Perithecium-like, buried in the substratum.

With short conical neck. *Ostropa*

With long, upright or horizontal, then upcurved, neck. *Robergea*

Asci clavate to ovoid, apex not thickened or only slightly so. Ascospores spherical to oblong or needle-shaped, never elongated, thread-like, and falling apart into segments (but see some genera of Family Geoglossaceae).

Hypothecium and underlying tissues pseudoparenchymatous.

Apothecia in an often lens-shaped, externally almost black, often carbonaceous stroma. Excipulum weakly developed. Ascospores oblong to needle-shaped. Family Phacidiaceae

Apothecia single in each stroma, more or less round, opening mostly by stellate lobes.

Ascospores ellipsoid, one-celled, paraphyses not forming an epithecium. *Phacidium*

Ascospores ellipsoid, one-celled, paraphyses forming an epithecium. *Trochila*

Ascospores thread- or needle-like, one- to many-celled. *Coccomyces*

Several apothecia in each stroma, opening by elongated slits.

Ascospores ellipsoid. *Pseudorhytisma*

Ascospores thread- or needle-like. *Rhytisma*

Apothecia single in elongated stromata, opening by elongated slits.

Ascospores fusiform or rod-shaped, eventually two-celled. *Hypoderma*

Ascospores thread-like, one-celled. *Lophodermium*

Apothecia cartilaginous, leathery or fleshy, superficial or emerging from the substratum; sessile or tapering to a central stipe-like base. Stroma if present not dark and hard. Ascospores oblong to needle-shaped, mostly hyaline. Parasites or saprophytes. Family Dermateaceae (including Mollisiaceae)

Apothecia long-lived, cartilaginous or leathery, narrowed at base and formed on a leathery or cartilaginous stroma which breaks out through the bark. Ascospores 8 in number, one-celled, later more-celled. *Dermatea* (*Dermea*)

Similar to *Dermatea* but the ascospores budding in the ascus into innumerable tiny spores. *Tympanis*

Apothecia soft fleshy, no stroma present. Asci small. Ascospores 8 per ascus, ellipsoid, hyaline. Paraphyses slender, branching somewhat near the tips. Conidial stages *Hainesia* and *Pilidium*. Parasitic on herbaceous tissues. *Discohainesia*

Apothecia long-lived, leathery or cartilaginous, not dark-colored, breaking out early from the host tissue. Asci large, ascospores hyaline, simple or septate. Paraphyses thick, often cemented together by their enlarged tips. Parasitic on woody tissues. *Pezicula*

Apothecia waxy or fleshy, breaking out of the host tissue or attached only by the narrow base. Dark- or light-colored, often hairy externally. Asci large. Ascospores hyaline one- to several-celled. Saprophytes or parasites.

Apothecia superficial or attached by the narrow base, rolling together when dry. Ascospores one- to two-celled. *Mollisia*

Apothecia immersed in the host tissues, breaking out at maturity.

Clothed with dark septate hairs. *Pirottaea*

Apothecia hairless except the hyaline marginal hairs. On herbaceous plants except grasses or grass-like species. *Pyrenopeziza*

Apothecia formed on the dead stems or leaves of the host plants, the asexual stages actively parasitic. Exciple not strongly developed.

Apothecia depressed globose, conidial stage *Cylindrosporium*. "*Higginsia*" Nannf.

Apothecia somewhat obconical.

Ascospores one-celled. Conidial stage *Gloeosporium*-like. *Drepanopeziza*

Ascospores two-celled. Conidial stage *Marssonina* or *Entomosporium*. *Diplocarpon*

Apothecia developed on the living host tissues and also on the overwintered dead tissues, soft fleshy, with better developed stroma for the overwintered apothecia. Conidial stage *Sporonema* or *Ramularia* or wanting. Ascospore one-celled. *Pseudopeziza*

Apothecia superficial, attached to the woody substratum by a narrow stalk, flat or saucer-shaped, soft fleshy, mostly bright-colored or hyaline. Asci cylindrical, small; ascospores 8, one-celled, hyaline, small. Paraphyses swollen at the apex with a waxy substance which unites them. Saprophytes. Family Orbiliaceae

(Nannfeldt recognizes three genera, *Orbilia*, *Patinella*, and *Hyalinia*.)

Apothecia superficial or rarely emerging from the substratum, short- or long-stalked or sessile, soft, light-colored, often hairy externally. Ascospores oblong to needle-shaped, hyaline, one- to several-celled. Paraphyses thread-like or lance-like. Saprophytes. Family Hyaloscyphaceae

Apothecia relatively large, almost sessile to long-stalked, mostly clothed with long rough hairs. Paraphyses usually lance-like, longer than the asci. *Lachnum*

Apothecia small, sessile or tapered at base. Paraphyses not lance-like. Hairs with enlarged base and tapering to a fine point. *Hyaloscypha*

(Nannfeldt recognizes some other small genera with small apothecia.)

Hypothecium and most of the epithecial tissues hyphal (not pseudoparenchymatous).

Apothecia superficial or rarely breaking out of substratum, long- or short-stalked, rarely sessile, cartilaginous, leathery or fleshy, mostly medium-sized or large, light-colored or dark, asci clavate, ascospores oblong, hyaline or dark-colored, simple or septate. Not rising from definite stromatic masses. Family Helotiaceae

Apothecia light-colored, fleshy, almost sessile to long-stalked, outer surface of apothecia not hairy. Saprophytes. *Helotium*

Apothecia similar, growing on wood, outer surface clothed with long, cylindrical, and somewhat crisped hairs. Parasites or saprophytes. *Trichoscyphella* (*Dasyscypha* in part)

(Nannfeldt recognizes many other genera, some sessile, most stalked.)

Apothecia arising from a definite sclerotium or a stromatized portion of the substratum, stipitate, asci cylindrical-clavate, 8-spored; ascospores ellipsoid, hyaline (except two genera), smooth, one-celled. Mostly parasitic. (Key to genera modified from Whetzel, 1945.) Family Sclerotiniaceae

Stroma a sclerotium of more or less definite and characteristic form.

Apothecia arising from a tuberoid sclerotium that was formed on the aerial mycelium or in cavities of the host. No conidia known. *Sclerotinia*

Apothecia arising from a thin effuse, subcuticular sclerotium surrounding the affected portion of the host. Small black microsclerotia also formed on the aerial mycelium. No conidia known. *Stromatinia*

Stroma formed in the tissues of the host which it digests and replaces by the dense sclerotial tissues, except for the remnants of the host tissues. No conidia known.

Sclerotium of the discoid type, foliicolous.

Apothecia cupulate or saucer-shaped. *Ciborinia*

Apothecia verpoid (i.e., curved downward at the apex of the stalk). *Verpatinia*

Stroma destroying the male or female flowers and forming a sclerotium. *Ciboria*

Stroma developing in the infected fruits of the hosts, of the hollow spheroid type or pseudosclerotial. Conidial stage of the *Monilia* type. *Monilinia*

Stroma a typical plano-convex sclerotium formed on or just beneath the cuticle or epidermis and firmly attached to it. Conidial stage of the *Botrytis* type, ascospores hyaline.

Conidial branches not twisted. *Botryotinia*

Conidial branches strikingly twisted. *Streptotinia*

Conidial stage unknown, ascospores olive-brown. "*Martinia*" Whetzel

Stroma a more or less angular sclerotium formed in the dead fallen leaves of the host.

Conidia in clusters, elongated, 2 to 3 septate, on host leaf. *Septotinia*

Conidia large, obovoid, produced singly, on host leaf. *Ovulinia*

Stroma of the substratal type, indeterminate, not a definite sclerotium.

Ascospores one-celled, brown. No conidia known. *Lambertella*

Ascospores sometimes septate at maturity, hyaline. No conidial stage known. Stroma sometimes rudimentary or wanting. *Rutstroemia* Rehm, not Karst.

Ascospores hyaline, not septate. Conidial stage of the *Botrytis* type. *Seaverinia*

Spore fruits club-shaped or stalked with pileus. Asci clavate. Ascospores oblong to needle-shaped, one-celled or many times transversely septate, hyaline or brown. Conidial stages not known. Saprophytic. Family Geoglossaceae[4]

Clavate, the ascigerous portion more or less compressed.

Spores small, one-celled. Plants bright-colored. *Mitrula*

---

[4] *Vibrissea* and *Apostemidium*, placed in this family by Durand (1908), are considered by Nannfeldt (1932) to belong to the Ostropaceae (see above).

Spores long ellipsoid, 3 to many septate. Plants bright-colored. *Microglossum*

Spores as in *Microglossum*, hyaline, plants blackish. *Corynetes*

Spores as in *Microglossum*, but fuliginous or brown.

Hymenium without spines or setae, viscid gelatinous. *Gloeoglossum*

No spines or setae, not viscid gelatinous. *Geoglossum*

Hymenium with spines or setae. *Trichoglossum*

Spatulate or fan-shaped, the ascigerous portion decurrent on opposite sides of the stem. *Spathularia*

Pileate, i.e., with stalk and rounded head, not black or blackish.

Spores hyaline, ellipsoid-fusiform, eventually 3 to 5 septate. *Leotia*

Spores hyaline, clavate filiform, multiseptate. *Cudonia*

Key to the More Important Genera of Order Tuberales

(*Based mainly on Miss Gilkey, 1939*)

Asci and paraphyses in a palisade. Ascocarp hollow, lined with hymenium Cavity often reduced by folds or projections to canals or chambers (in some cases filled with hyphae).

Paraphyses not forming a pseudoparenchymatous epithecium beyond the asci.

Cavity simple, closed or with an opening to the surface of the ascocarp.

Ascospores smooth. *Hydnocystis*

Ascospores with knobs or minutely verrucose (some species). *Hydnotrya*

Cavity divided into canals or chambers, or if not so divided, then with several openings.

Ascocarp turbinate, narrowed into a short stipe-like structure, see under Family Helvellaceae. *Daleomyces*

Ascocarp not narrowed to a stipe.

Ascospores smooth.

Canals converging at one conspicuous opening. *Barssia*

Canals not converging; openings absent or at least not conspicuous. *Geopora*

Ascospores knobby or minutely verrucose (some species) *Hydnotrya*

Paraphyses forming a more or less distinct pseudoparenchymatous "secondary cortex" beyond the asci.

Hymenial areas or canals continuous.

Ascospores smooth or minutely granular. *Petchiomyces*

Ascospores papillose, verrucose, or spinose. *Genea*

Hymenial areas nest-like, embedded in pseudoparenchyma. *Genabea*

Hymenial canals or cavities lined by asci but filled with hyphae.

Canals opening at surface of ascocarp.

Converging at apex. *Pachyphloeus*

Not converging at apex, fertile only at the dilated blind ends. *Piersonia*

Canals not opening to surface, fertile their whole length. *Choiromyces*
Asci not in a palisade but irregularly arranged.
Ascocarp containing empty canals or chambers or hypha-filled "venae externae" (i.e., canals reaching toward the surface but filled with hyphae).
Cavities empty, not opening to surface. *Balsamia*
Cavities mostly hypha-filled, opening to surface.
Ascospores smooth. *Pseudobalsamia*
Ascospores variously sculptured.
Venae externae with parallel venae internae. *Tuber*
No venae internae.
Asci 8-spored. *Hydnobolites*
Asci 1- to 4-spored. *Delastreopsis*
Interior of ascocarp divided by sterile veins into nest-like fruiting areas; no venae externae or empty canals or chambers.
Ascospores smooth, very large. *Picoa*
Ascospores sculptured.
Asci 8-spored. *Terfezia*
Asci 2- to 4-spored. *Delastria*
(The genera *Hydnobolites*, *Delastreopsis*, *Picoa*, *Terfezia*, and *Delastria* are placed in some works in Family Terfeziaceae, Order Aspergillales, Fischer, 1896b).

Key to the More Important Genera of Order Hysteriales

(*Modified from von Höhnel, 1918*)

Ascocarp superficial in subiculum or becoming superficial by emerging from the substratum, elongated or linear, not strongly compressed nor with strongly developed keel.
With brown subiculum, ascospores 2-celled, hyaline. *Glonium*
Without subiculum.
Ascospores 2-celled, hyaline. *Psiloglonium*
Ascospores with 2 or more cross septa, hyaline. *Gloniella*
Ascospores similar but colored. *Hysterium*
Ascospores muriform, hyaline. *Gloniopsis*
Ascospores muriform, colored. *Hysterographium*
Ascocarp entirely superficial, strongly compressed laterally and with a well-developed keel, mussel-shaped.
Ascospores brown, 2-celled. *Bulliardiella*
Ascospores brown, with 2 or more cross septa. *Mytilidion*
Ascospores brown, large, fusiform, with many cross septa. *Ostreion*
Ascospores hyaline, thread-formed. *Lophium*
The genera with permanently immersed ascocarps, are to be found in the Phacidiaceae; *Lophodermium*, *Hypoderma*, etc.

Key to the Genera of Order Taphrinales

Asci forming an indeterminate hymenium, on a hyphal hypothecium several cells thick on the surface of the woody substratum.
Family Ascocorticiaceae
Only one genus. *Ascocorticium*

Asci forming an indeterminate subcuticular hymenium or forming tufts of asci emerging from between epidermal cells. Always on living green tissues or fruits. Mycelium sometimes perennial in the host tissues.

Family Taphrinaceae

Only one genus. *Taphrina*

Literature Cited

ATANASOFF, D.: A novel method of ascospore discharge, *Mycologia*, **11**(3):125–128. *Figs.* 1–3. 1919.

ATKINSON, G. F.: Phylogeny and relationships in the Ascomycetes, *Ann. Missouri Botan. Garden*, **2**(1–2):315–376. *Figs.* 1–10. 1915.

BACHMANN, FREDA M.: A new type of spermagonium and fertilization in Collema, *Ann. Botany*, **26**(103):747–760. *Pl.* 69. 1912.

———: The origin and development of the apothecium in Collema pulposum (Bernh.) Ach., *Arch. Zellforsch.*, **10**:369–430. *Pls.* 30–36. 1913.

BACKUS, MYRON P.: Initiation of the ascocarp and associated phenomena in Coccomyces hiemalis, *Contribs. Boyce Thompson Inst.*, **6**(3):339–379. *Figs.* 1–8. 1934.

DE BARY, ANTON: Über die Fruchtentwicklung der Ascomyceten, 38 pp. *Illust.* Leipzig, Wilhelm Engelmann, 1863.

———: Vergleichende Morphologie und Biologie der Pilze, Mycetozoen und Bacterien, xvi + 558 pp. 198 *figs.* Leipzig, Wilhelm Engelmann, 1884.

BAUR, E.: Zur Frage nach der Sexualität der Collemaceen, *Ber. deut. botan. Ges.*, **16**(10):363–367. *Pl.* 23. 1899.

BETTS, EDWIN M.: Heterothallism in Ascobolus carbonarius, *Am. J. Botany*, **13**(7):427–432. 1926.

BIGGS, ROSEMARY: Dipodascus uninucleatus, *Mycologia*, **29**(1):34–44. *Figs.* 1–50. 1937.

BOEDIJN, K. B.: Two new Malaysian genera of Discomycetes, *Bull. Jardin Botanique de Buitenzorg*, sér. III, **13**:478–483. *Figs.* 1–3. 1934.

BONNIER, GASTON: Recherches sur la synthèse des lichens, *Ann. sci. nat. Botan.*, 7me sér., **9**:1–34. *Pls.* 1–5. *Figs.* 1–6. 1889.

BOUDIER, ÉMILE: Histoire et classification des Discomycètes d'Europe, vii + 223 pp. Paris, Paul Klincksieck. Libraire des Sciences Naturelles, 1907.

BROWN, W. H.: The development of Pyronema confluens var. inigneum, *Am. J. Botany*, **2**(6):289–298. 1915.

BUCHOLTZ, F.: Beiträge zur Morphologie und Systematik der Hypogäen nebst Beschreibung aller bis jetzt in Russland angetroffenen Arten, 196 pp. 5 *pls.* *Text figs.* 1902. (Russian with German summary.)

BULLER, A. H. REGINALD: Researches on Fungi, vol. 6, xii + 513 pp. 231 *figs.* London, Longmans, Green and Company, 1934.

CHAZE, JEAN, ET PAUL MESTAS: Essai de culture du mycélium truffier avec production de périthèces, *Compt. rend.*, **208**(26):2099–2101. 1 *fig.* 1939.

CLAUSSEN, P.: Zur Entwicklungsgeschichte der Ascomyceten, Pyronema confluens, *Z. Botan.*, **4**(1):1–63. *Pls.* 1–6. *Figs.* 1–13. 1912.

CORNER, E. J. H.: Studies in the morphology of the Discomycetes: I. The marginal growth of apothecia, *Brit. Mycol. Soc. Trans.*, **14**:263–275. *Figs.* 1–4. 1929; II. The structure and development of the ascocarp, *ibid.*, **14**:275–291. *Figs.* 1–7. 1929; III. The Clavuleae, *ibid.*, **15**:107–120. *Figs.* 1–3. 1930; IV. The evolution of the ascocarp, *ibid.*, **15**:121–134. *Fig.* 1. 1930; V. The evolution of the ascocarp (continued), *ibid.*, **15**:332–350. 1931.

Dangeard, P. A.: Recherches sur le développement du périthèce chez les Ascomycètes, *Le Botaniste*, **9**:59–303. *Pls*. 1–16. 1907a.

———: L'origine du périthèce chez les Ascomycètes, *ibid.*, **10**:1–385. *Pls*. 1–91. 1907b.

Darwin, Charles: The different forms of flowers on plants of the same species, viii + 352 pp. 15 *figs*. New York, D. Appleton & Co., 1889.

de Ferry de la Bellone, C.: Du mycélium des champignons hypogés en général et de celui des Tuberacées en particulier, *Assoc. franç. avance. sci. Compt. rend. de la 14me session, Grenoble, 1885*, pt. 2, pp. 437–443. *Pl*. 11. 1886.

Dodge, Bernard O.: Methods of culture and the morphology of the archicarp in certain species of Ascobolus, *Bull. Torrey Botan. Club*, **39**(4):139–197. *Pls*. 10–15. 1912.

———: The life history of Ascobolus magnificus, *Mycologia*, **12**(3):115–134. *Pls*. 7–8. *Figs*. 1–28. 1920.

Drayton, F. L.: The sexual function of the microconidia in certain Discomycetes, *Mycologia*, **24**(3):345–348. 1932.

———: The sexual mechanism of Sclerotinia gladioli, *ibid.*, **26**(1):46–72. *Pls*. 5–7. *Figs*. 1–4. 1934.

Durand, Elias J.: The classification of the fleshy Pezizineae with reference to the structural characters illustrating the bases of their division into families, *Bull. Torrey Botan. Club*, **27**(9):463–495. *Pls*. 27–32. 1900.

———: The Geoglossaceae of North America, *Ann. Mycol.*, **6**(5):387–477. *Pls*. 5–22. 1908.

Elfving, Fr.: Untersuchungen über die Flechtengonidien, *Acta Soc. Sci. Fennicae*, **4**(2), 71 pp. 8 *pls*. 1913.

Falck, R.: Ueber die Sporenverbreitung bei den Ascomyceten: I. Die radiosensiblen Diskomyceten, *Mykologische Untersuchungen und Berichte*, **1**(2):77–145. *Pls*. 1–2. *Figs*. 1–14. 1916; II. Die taktiosensiblen Diskomyceten, *ibid.*, **1**(3):370–403. *Pl*. 1. *Figs*. 1–5. 1923.

Fink, Bruce: The lichen flora of the United States. Completed for publication by Joyce Hedrick, x + 426 pp. *Pls*. 1–47. *Figs*. 1–4. Ann Arbor, Univ. Mich. Press, 1935.

Fischer, Eduard: Tuberineae., in A. Engler und K. Prantl: Die Natürlichen Pflanzenfamilien, Erster Teil, Abt. **1**:278–290. *Figs*. 202–209. 1896a.

———: Terfeziaceae, in Plectascineae, *ibid.*, Erster Teil, Abt. **1**:312–319. *Figs*. 222–226. 1896b.

———: Tuberineae, *ibid.*, Zweite Auflage, fortgesetzt von H. Harms, Band 5b, 44 pp. 22 *figs*. 1938.

Fitzpatrick, Randal E.: The life history and parasitism of *Taphrina deformans*, *Sci. Agr.*, **14**(6):305–326. *Pls*. 1–2. *Figs*. 1–10. 1934.

Fraser, Helen C. I.: On the sexuality and development of the ascocarp in Lachnea stercorea, *Ann. Botany*, **21**(83):349–360. *Pls*. 29–30. 1907.

Fries, Elias: Systema mycologicum, sistens fungorum ordines, genera et species, huc usque cognitas, vol. 2, Sect. I. pp. 1–274. Lund, Officina Berlingiana, 1822, Sect. II. pp. 275–621. Greifswald, Ernest Mauritius, 1823.

Gäumann, Ernst: Vergleichende Morphologie der Pilze, 626 pp. *Many figs*. Jena, Gustav Fischer, 1926.

Geitler, Lothar: Beiträge zur Kenntnis der Flechtensymbiose, I–III, *Arch. Protistenk.*, **80**(3):378–409. *Figs*. 1–12. 1933.

Gilkey, Helen M.: Tuberales of North America, *Oregon State Monographs. Studies in Botany*, **1**:1–55. *Pls*. 1–5. 1939.

Green, Ethel: Observations on certain Ascobolaceae, *Brit. Mycol. Soc. Trans.*, **15**:321–332. *Figs*. 1–7. 1931.

GREIS, HANS: Entwicklungsgeschichte von Sordaria fimicola (Rob.), *Botan. Arch.*, **38**(2):113–151. 4 *pls.* 5 *text figs.* 1936.

———: Die Entstehung der Basidiomycetenschnallen aus den Ascomycetenhaken, *Jahrb. wiss. Botan.*, **86**(1–2):81–106. *Figs.* 1–3. 1938.

———: Befruchtungsarten bei Morchella, *ibid.*, **89**(2):245–253. *Figs.* 1–3. 1940.

GWYNNE-VAUGHAN, HELEN C. I., AND H. S. WILLIAMSON: Contributions to the study of Humaria granulata Quel., *Ann. Botany*, **44**(173):127–145. *Pls.* 13–14. *Figs.* 1–10. 1930.

———, AND ———: Contributions to the study of Pyronema confluens, *ibid.*, **45**(178):355–371. *Pls.* 12–14. *Figs.* 1–7. 1931.

HARPER, R. A.: Sexual reproduction in Pyronema confluens and the morphology of the ascocarp, *Ann. Botany*, **14**(55):321–400. *Pls.* 19–21. 1900.

HIGGINS, BASCOMBE BRITT: Contribution to the life history and physiology of Cylindrosporium on stone fruits, *Am. J. Botany*, **1**(4):145–173. *Pls.* 13–16. 1914a.

———: Life history of a new species of Sphaerella, *Mycol. Centr.*, **4**(4):187–193. *Figs.* 1–2. 1914b.

———: Morphology and life history of some Ascomycetes with special reference to the presence and function of spermatia: I. *Am. J. Botany*, **7**(10):435–444. *Pl.* 30. *Figs.* 1–2. 1920; II. *ibid.*, **16**(5):287–296. *Pl.* 22. *Fig.* 1. 1929; III. *ibid.*, **23**(9):598–602. *Figs.* 1–13. 1936.

VON HÖHNEL, FRANZ: Mycologische Fragmente: CCLXXII. Über die Hysteriaceen, *Ann. Mycol.*, **16**(1–2):145–154. 1918.

———: Fragmente zur Mykologie, *Sitz. ber. Akad. Wiss. Wien, Math. naturw. Klasse*, 1902–1923. (These contributions, of which there were about 25, began with Band 111 in 1902, continuing to Band 132 in 1923. They include discussions of various groups formerly believed to produce perithecia but shown by the author to have stromatic structures instead.)

JONES, S. G.: Life-history and cytology of Rhytisma acerinum (Pers.) Fries, *Ann. Botany*, **39**(1):41–75. *Pl.* 4. *Figs.* 1–23. 1925.

———: A study of apothecial development in the leaf-spot disease of red clover, *Trans. Roy. Soc. Edinburgh*, **56**(2):507–519. *Figs.* 1–9. 1930.

JUEL, H. O.: Cytologische Pilzstudien: II. Zur Kenntnis einiger Hemiasceen, *Nova Acta Regiae Soc. Scien. Upsaliensis*, Ser. IV, **5**(5):1–43. 2 *pls.* 1921.

KHARBUSH, S.: Évolution nucléaire du Sclerotinia Fuckeliana de Bary, *Bull. Soc. Botan. France*, **74**:257–262. 1 *fig.* 1927.

LINDSAY, W. LAUDER: Memoir on the spermogones and pycnides of flamentous, fruticulose and foliaceous lichens, *Trans. Roy. Soc. Edinburgh*, **22**:101–203. *Pls.* 4–15. 1861.

———: Memoir on the spermogones and pycnides of crustaceous lichens, *Trans. Linnean Soc. London*, **28**:189–318. *Pls.* 8–15. 1872.

LIRO, J. IVAR: Können Algen in den Flechtenhyphen vorkommen? *Acta. Soc. Sci. Fennicae, A*, **6**(1):1–52. *Pl.* 1. *Fig.* 1. 1914.

LOHMAN, MARION LEE: Hysteriaceae: Life-histories of certain species, *Papers Mich. Acad. Sci.*, **17**:229–288. *Pls.* 34–35. *Figs.* 31–41. 1932 (1933).

LOHWAG, H.: Zu Lycoperdellon, *Ann. Mycol.*, **32**(3–4):244–255. *Figs.* 1–2. 1934.

MATTIROLO, ORESTE: Illustrazione di tre nuove specie di Tuberacee italiane, Studio, *Mem. reale accad. sci. Torino*, ser. 2, No. 37, 19 pp. 2 *pls.* 1887.

MCWHORTER, FRANK P.: Destruction of mosses by lichens, *Botan. Gaz.*, **72**(5): 321–325. *Pl.* 13. 1921.

MELLOR, ETHEL: Les lichens vitricoles et la détérioration des vitraux d'église, *Rev. gén. botan.*, **34**:280–286, 336–345. *Pls.* 5–8. *Figs.* 1–4. 1922.

MILLER, JULIAN H.: Biologic studies in the Sphaeriales, *Mycologia*, **20**(4):187–213, (6):305–339. *Pls.* 21–22, 35–38. *Figs.* 1–3. 1928.

MIX, A. J.: Biological and cultural studies of Exoascus deformans, *Phytopathology*, **14**(5):217–233. *Figs.* 1–2. 1924.

———: The life history of Taphrina deformans, *ibid.*, **25**(1):41–66. *Figs.* 1–6. 1935.

MÖLLER, ALFRED: Ueber die Cultur flechtenbildender Ascomyceten ohne Algen, *Untersuchungen aus dem Botanischen Institut der Königlichen Akademie zu Münster i.W.*, pp. 1–52, 1887.

MOREAU, FERNAND, M. ET MME.: Le réproduction sexuelle chez les lichens du genre Collema et la théorie de Stahl, *Compt. rend.*, **182**(12):802–804. 1926.

——— ET ———: Les phénomènes cytologiques de la réproduction chez les champignons des lichens, *Le Botaniste*, **20**(112):1–67. *Figs.* 1–35. 1928.

NANNFELDT, J. A.: Studien über die Morphologie und Systematik der nicht-lichenisierten inoperculaten Discomyceten, *Nova Acta Regiae Soc. Sci. Upsaliensis*, Ser. IV, **8**(2):1–368. *Pls.* 1–19. *Figs.* 1–47. 1932.

PERSOON, CHRISTIAAN HENDRIK: Synopsis methodica fungorum. Pars prima et secunda, xxx + 706 pp. and Index. *Pls.* 1–5. Göttingen, Heinrich Dieterich, 1801.

PETCH, T.: Studies in entomogenous fungi: V. Myriangium, *Brit. Mycol. Soc. Trans.*, **10**:45–80. *Pls.* 2–3. 1 *fig.* 1929.

PIERCE, NEWTON B.: Peach leaf curl: Its nature and treatment, *U.S. Dept. Agr. Vegetable Physiol. Path. Bull. 20*, pp. 1–204. *Pls.* 1–26. 1900.

POOLE, R. F.: Late infection of peach leaf curl in the Carolinas, *Plant Disease Reptr.*, **16**(16):171–172. 1932.

RAMLOW, GUSTAV: Zur Entwicklungsgeschichte von Thelebolus stercoreus Tode, *Botan. Ztg.*, **64**(1):85–99. *Pl.* 4. *Figs.* 1–4. 1906.

———: Beiträge zur Entwicklungsgeschichte der Ascoboleen, *Mycol. Centr.*, 5(4):177–198. *Pls.* 1–2. *Figs.* 1–20. 1914.

REINKE, J.: Abhandlungen über Flechten: I. Das Podetium von Cladonia, *Jahrb. wiss. Botan.*, **26**:495–523. *Figs.* 1–7. 1894; II. Die Stellung der Flechten im Pflanzensystem, *ibid.*, **26**:524–542. 1894; III. Einige Voraussetzungen einer phylogenetischen Morphologie der Flechten, *ibid.*, **28**:39–69. *Fig.* 8. 1895a; IV. Skizzen zu einer vergleichenden Morphologie der Flechtenthallus: Caliciaceen, Lecidiaceen, Graphidaceen, *ibid.*, **28**:70–150. *Figs.* 9–82. 1895b; V. Das natürliche Flechtensystem, *ibid.*, **29**:171–209. 1896.

ROBERTS, CATHERINE: A comparative study of Torulopsis pulcherrima and Taphrina deformans in culture, *Farlowia*, **2**(3):345–383. *Pls.* 1–3. 1946.

SCHNEIDER, ALBERT: A Text-book of General Lichenology, xvii + 230 pp. *Pls.* 1–76. Binghamton, N. Y., Willard N. Clute & Co., 1897.

SCHROETER, J., UND G. LINDAU: Pezizineae, in A. ENGLER und K. PRANTL: Die Natürlichen Pflanzenfamilien, Erster Teil, Abt. 1:173–243. *Figs.* 145–181. 1894.

SCHWEIZER, GEORG: Ein Beitrag zur Entwicklungsgeschichte und Biologie von Ascobolus citrinus nov. spec., *Z. Botan.*, **15**:529–556. *Figs.* 1–20. 1923.

SCHWENDENER, SIMON: Ueber die wahre Natur der Flechten, *Verhandlungen der Schweizerischen Naturforsch. Gesellschaft in Rheinfelden, Aaran.* **1867**:88. 1867.

———: Ueber die Beziehungen zwischen Algen und Flechtengonidien, *Botan. Ztg.*, **26**(18):289–292. 1868.

SEAVER, FRED J.: The North American Cup-fungi (Operculates), 284 pp. *Frontispiece* (*colored*) + 45 *pls.*, 5 *figs.* New York, published by the author, 1928. Supplement, i–viii + 285–377. *Pls.* 46–74. 1942.

SMITH, ANNIE LORRAIN: Lichens, xxviii + 464 pp. 135 *figs.* Cambridge, Cambridge Univ. Press, 1921.

———: Lichen dyes, *Brit. Mycol. Soc. Trans.*, **11**:45–50. 1926.

STAHL, E.: Beiträge zur Entwickelungsgeschichte der Flechten, *Botan. Ztg.*, **32**(12):177–180. 1874.

———: Beiträge zur Entwickelungsgeschichte der Flechten, I and II, Leipzig, 1877. (With many illustrations.)

THAXTER, ROLAND: Contribution toward a monograph of the Laboulbeniaceae, *Mem. Am. Acad. Arts Sci.*, **12**:195–429. *Pls.* 1–26. 1895; **13**:219–460. *Pls.* 28–71. 1908; **14**:309–426. *Pls.* 1–12. 1924; **15**:427–500. *Pls.* 1–24. 1926; **16**:1–435. *Pls.* 1–60. 1 *fig.* 1931.

THEISSEN, F.: Hemisphaeriales, *Ann. Mycol.*, **11**(5):468–469. 1913.

TOBLER, FRIEDRICH: Das physiologische Gleichgewicht von Pilz und Alge in den Flechten, *Ber. deut. botan. Ges.*, **27**(7):421–427. 1 *fig.* 1909.

———: Biologie der Flechten. Entwicklung und Begriff der Symbiose, xiii + 265 pp. 1 *colored plate.* 67 *figs.* Berlin, Gebrüder Borntraeger, 1925.

WAITE, M. B.: Unusual late infections of peach leaf curl, *Plant Disease Reptr.*, **16**(10):109–110. 1932.

WHETZEL, H. H.: Septotinia, a new genus of the Ciborioideae, *Mycologia*, **29**(1): 128–146. *Figs.* 1–18. 1937.

———: A synopsis of the genera and species of the Sclerotiniaceae, a family of stromatic inoperculate Discomycetes, *ibid.*, **37**(6):648–714. *Figs.* 1–36. 1945.

WIEBEN, MAGDALENE: Die Infektion, die Myzelüberwinterung und die Kopulation bei Exoasceen, *Forschungen auf dem Gebiet der Pflanzenkrankheiten und der Immunität im Pflanzenreich*, **3**:139–176. 31 *figs.* 1927.

WOLF, FREDERICK A.: Strawberry leaf scorch, *J. Elisha Mitchell Sci. Soc.*, **39**(3–4):141–163. *Pls.* 9–15. 1924.

ZAHLBRUCKNER, ALEXANDER: Lichenes: B. Spezieller Teil, in A. ENGLER und K. PRANTL: Die Natürlichen Pflanzenfamilien, Zweite Auflage, Abt. **8**:61–270. *Figs.* 32–127. 1926.

ZIEGENSPECK, H.: Schleudermechanismen von Ascomyceten, *Botan. Arch.*, **13**(5–6):341–381. *Figs.* 1–20. 1926.

ZOPF, WILHELM: Die Flechtenstoffe in chemischer, botanischer, pharmakologischer und technischer Beziehung, i–xi, 1–449. 70 *figs.* Jena, Gustav Fischer, 1907.

10

CLASS ASCOMYCETEAE: THE "PYRENOMYCETES"

WITH the exception of the Order Laboulbeniales the Ascomyceteae so far considered have been those whose spore fruits are apothecia or possible modifications of apothecia. In this chapter are discussed those orders which produce true perithecia with ostioles, or stromata containing cavities within which the asci are developed. Because the life histories of all but a few of these thousands of species are entirely unknown the relationships of these orders to one another and of the genera within the orders are admittedly very uncertain. Indeed, with further study it is almost certain that the limits of the orders will have to undergo radical revision. For this reason a rather conservative attitude is taken as to these groups. The author does not believe that the true relationships are shown by the arrangement chosen but this is due to the ignorance of the ontogeny of the species.

Whether the perithecial forms should be treated after or before the apothecial forms is a matter of convenience rather than of expression of phylogenetic history. It seems likely that very early in the development of the Ascomyceteae three tendencies began to be emphasized. In the one the female sexual branch bore an oogone surrounded with a protective layer, as in many Florideae; this group led to the Laboulbeniales. In the two other series the female sexual organs were at first naked. In one of these the number of ascogenous hyphae and their branches was great and a rather wide-spreading spore fruit developed; the apothecial series of fungi. In the third the ascogenous hyphae were shorter and usually few, little branching if at all, and the protective structure formed about them (the perithecium) remained relatively small and was arched over them and closed or with only a comparatively small opening, the ostiole. So far we have followed the first two series of development, we now return to the relatively primitive forms again and start to follow the third series. Because the spore fruits of this series were more often hard-walled and small and like the stones of some small fruits the name "Pyrenomycetes" was applied to include fungi of this nature (from the Greek *pyren*, the stone of a stone fruit or raspberry, etc., and *mykes*, a fungus). It should be noted that the basal groups of each of the three series include forms with free, nonmotile sperms which come in contact with some receptive structure (trichogyne) and thus bring about fertilization. In the apothecial series as well as in the perithecial series this gives place quickly to the direct union of trichogyne and antherid and finally to the entire disappearance of the trichogyne, and in many cases of the antherid also.

Asexual reproduction probably reaches its highest degree of development in this group of orders which may collectively be called the Pyrenomycetes, without necessarily indicating that they are at all closely related. A very large proportion of the species have asexual reproduction by means of conidia. The types of conidial production are most varied. Often sexual reproduction is relatively rare, the conidia serving to maintain the species. In many of the parasitic species the asexual mode of reproduction alone occurs on the living host and the sexual stage is formed only on the dead host tissues. The conidia may be produced on free conidiophores or on conidiophores crowded in an acervulus or on conidiophores enclosed within a pycnidium. In some cases all of these forms occur in the same species of fungus at different stages of growth. Judging by the types of asexual reproduction perhaps the majority of the so-called Imperfect Fungi (Class Fungi Imperfecti) are probably conidial stages of the Pyrenomycetes, although possibly representing in many cases species that have permanently lost their power of sexual reproduction.

Studies have been made of the sexual reproduction in various forms of the group but not only are there great numbers of genera in which studies are lacking, but for many families no such studies have been made. The results are conflicting for some of the supposedly related forms. The stages of sexual reproduction are often difficult to find in a closely connected series of development and it is usually with difficulty that they can be satisfactorily sectioned and stained. This makes the interpretation of the structures observed not at all easy and their interpretation is naturally influenced by previously held views. So in describing the sexual reproduction of various species the author has had to report the findings as described and interpreted by the respective investigators although this course involves some contradictions because of the different interpretations by the various authors. Until many more forms have been investigated, representing all of the more important genera and families, it will be impossible to determine to what extent weight must be given to this process in the classification of these fungi.

A number of cases have been selected that show the various types of structures concerned in sexual reproduction in the Pyrenomycetes, as well as some of the special modifications of sexual behavior. In many of these fungi spermogonium-like structures are known within which are

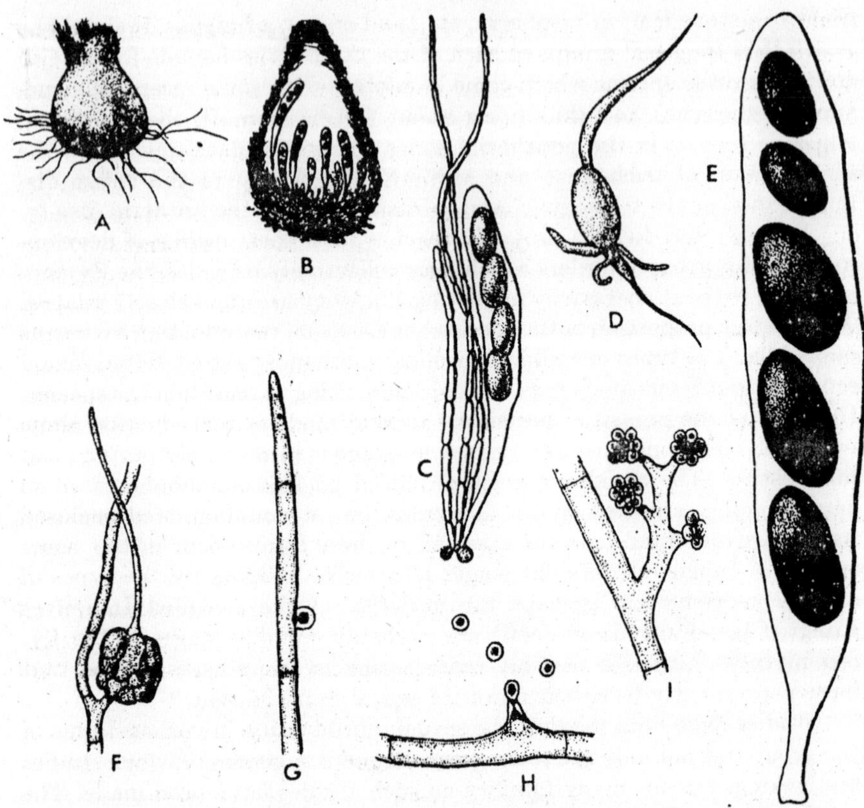

FIG. 87. Sphaeriales, Family Fimetariaceae. *Schizothecium anserinum* (Rabenh.). (A) Mature perithecium. (B) Longitudinal section of perithecium (paraphyses not shown). (C) Ascus and paraphyses. (D) Ascospore showing functional dark cell and empty hyaline cell, and gelatinous appendages. (E) Ascus with three normal ascospores and two smaller ones (only the functional cell shows). (F) Ascogonium with trichogyne. (G) Portion of trichogyne with attached sperm cell. (H) Antherid with emerging sperm. (I) Sperms clustered at tips of antherids. (A, C, D, after Griffiths: *Mem. Torrey Botan. Club*, **11**:1–134. B, E–I, after Ames: *Mycologia*, **26**(5):392–414.)

produced sperm cells, like those of *Collema* and *Physcia* in the Lecanorales. In other forms minute sperm-like bodies (often called microconidia) have been recorded but are not produced in definite spermogonia. In a great many no such structures have been reported.

Ames (1934) found that in *Schizothecium anserinum* (Rabenh.) (*Pleurage anserina* (Rabenh.) Kuntze) small flask-shaped antherids are formed out of whose necks the small sperm cells are successively pushed, much as occurs in most of the Laboulbeniales. The ascogonia, produced from the same mycelium, consist of coiled structures terminating in slender trichogynes. When a suitable sperm comes into contact with a trichogyne, often aided perhaps by insects or mites crawling over the growing mycelium, it adheres to it and soon there occur growth and multiplication of the cells and in about four days the perithecium is mature. Similar structures have been studied in a number of other fungi. Higgins (1936) studied the development of *Mycosphaerella tulipiferae* (Schw.) Higgins which forms its conidial stage (*Cercospora liriodendri* E. and H.) on the living leaves of *Liriodendron tulipifera* L. and its ascigerous stage in the fallen leaves. In the small dark spots formed by the conidial stage ascogonia may be found, each consisting of an ovoid, uninucleate oogone with a slender uninucleate trichogyne several times its length. This trichogyne projects from the surface of the leaf. Ostiolate spermogonia are also produced and within these are many sperm mother cells in each of which are formed four non-motile, thin-walled sperm cells. Under the influence of moisture the gum contained in the spermogonia swells and the sperm cells are extruded and carried passively in the surface film of water to the trichogynes to which they adhere. Usually near the tip of the trichogyne a sperm grows fast to a small papilla and the nucleus enters through the latter and passes down to the oogone, enlarging as it progresses. The sperm and egg nuclei do not unite in the oogone but divide conjugately until eight or more pairs are formed, each pair enclosed in a mass of more deeply staining cytoplasm. Short ascogenous hyphae then arise and the paired nuclei pass into them. Asci are formed by the crozier method and in these asci the nuclear fusion occurs, followed by three divisions which thus give rise to the nuclei of the eight ascospores. When the latter are formed another nuclear division occurs and each ascospore becomes two-celled. The ascocarp walls and internal pseudoparenchymatous contents arise as more or less parallel hyphae growing upward beneath the epidermis before the ascogonium is initiated. By their growth and branching a pseudoparenchymatous stroma is formed with a firm black external wall around the developing ascogonium and ascogenous hyphae. The developing asci press aside or dissolve the colorless thin-walled cells of the interior of the stroma and eventually an ostiole is produced through which the mature asci one by one stretch to the exterior and discharge their ascospores. (Figs. 68, 87.)

In two other species of *Mycosphaerella*, one with *Septoria* as its conidial stage and the other with a *Cercospora* stage Higgins has observed the formation of trichogyne and sperm cells. According to Wolf (1943) only about 12 species out of the 500 species of *Cercospora* listed by Miss Lieneman (1929) as occurring in North America have their perfect stages known, all belonging to the genus *Mycosphaerella*.

The formation of sperm cells is done away with in *Venturia inaequalis* (Cke.) Wint., fertilization being effected by the direct passage of several male nuclei from the antherid through a pore into the tip of the tricho-

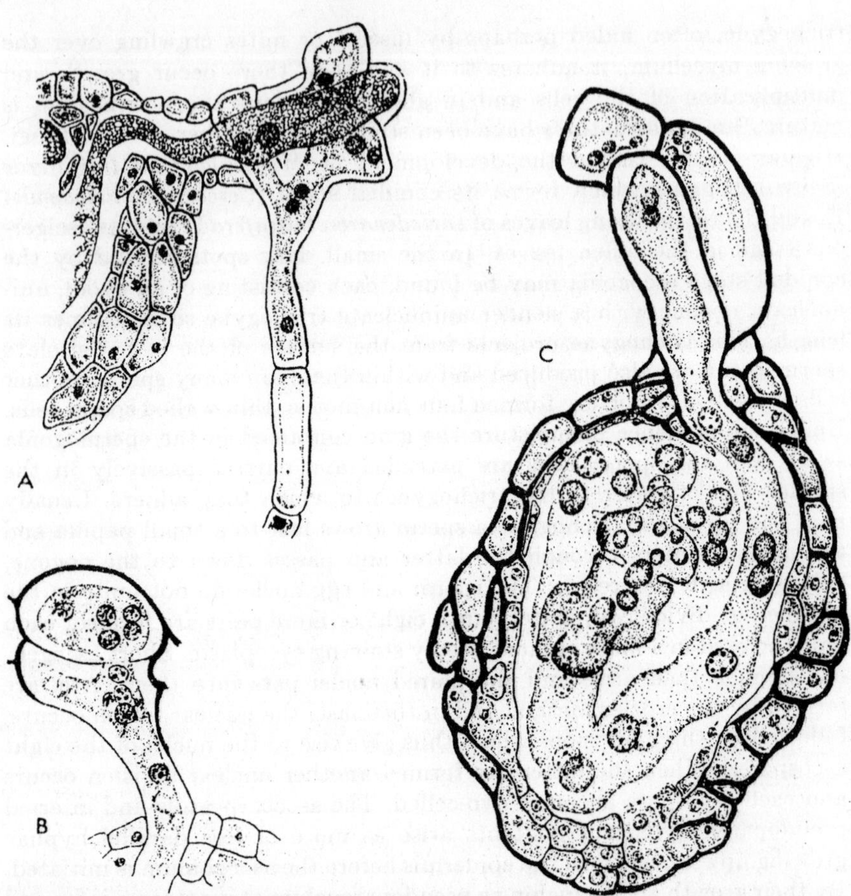

FIG. 88. Pseudosphaeriales, Family Pleosporaceae. *Venturia inaequalis* (Cke.) Wint. Stages in sexual reproduction. (A) The forked antherid is in contact with the exserted trichogyne. (B) The sperm nuclei are passing from the antherid (above) into the trichogyne. (C) The ascogonial cells have become multinucleate and are beginning to form lobes; the cross walls of the trichogyne have dissolved. (After Killian: *Z. Botan.*, 9:353–398.)

gyne, as described by Killian (1917) and by Frey (1924). In *Fimetaria* (*Sordaria*) *fimicola* (Rob.) Griffiths and Seaver it was shown by Greis (1936) that the tip of a slender antherid fuses with the apex of a coiled ascogonium. The antherid may arise from adjacent cells or from separate hyphae. In the lack of an antherid any cell of the ascogonium may send out an extension, a "pseudotrichogyne," to some nearby cell from which a nucleus is obtained. This species is self-fertile, i.e., there are not two sexual phases as in *Schizothecium anserinum*. Elliott (1925) reported that in *Ophiostoma fimbriatum* (E. & H.) Nannf. (*Ceratostomella*) there arise uninucleate oogones on short stalks and drawn out into a trichogyne without nucleus or separating septum. From the same or from a nearby hypha arises an antheridial branch which coils around the oogone and trichogyne. The uninucleate upper cell makes an opening into the trichogyne through which the male nucleus passes and proceeds downward to the body of the oogone, enlarging as it goes. As a result of numerous nuclear divisions occurring conjugately many pairs of nuclei arise which pass out into nonseptate ascogenous hyphae. In the meantime the hyphae below the oogone have grown outward and upward to form a closed perithecium with a thin dark-colored external wall and a mass of hyaline, thin-walled cells to which the ascogenous hyphae attach themselves and from which they grow into the perithecial cavity and produce their asci, apparently without crozier formation, scattered at random through the central hollow. Eight ascospores arise in each ascus and ascus walls, ascogenous hyphae and the remainder of the "nurse cells" digest. In the meantime the long perithecial neck has developed and finally opened at the end permitting the mucilaginous mass of ascospores and digested cell walls to exude as a drop. Gertrud Mittmann (1932) investigating the same fungus, contrary to Elliott's report failed to find any antherid. According to her the ascogonium begins as a single, uninucleate somewhat curved terminal cell of a short lateral branch. The cell elongates and coils into several turns, dividing into three to five uninucleate cells. From the supporting cell, enveloping hyphae begin to grow and eventually give rise to the perithecial wall. Miss Mittmann suggests that the antherid reported by Elliott was one of these enveloping hyphae. When this envelope is several cells in thickness the ascogonial cells enlarge and separate somewhat. One, or sometimes two, of these enlarged cells becomes multinucleate, the number of nuclei usually being eight. This cell (the oogone) divides into several cells which also become eight-nucleate and then divide into binucleate cells which thus form a richly branched system of crowded cells in no definite order. These cells enlarge laterally and their nuclei unite, thus forming the young asci which are nourished by the plasma-rich projecting cells of the inner layer of the perithecial wall. The ascospores are surrounded by a slime layer. *O. coeruleum* (Münch) H. and P. Sydow (*C. coerulea* Münch) and *O. pluriannulatum* (Hedgc.) H. and P. Sydow (*C. pluriannulata* Hedgc.) are reported by Miss Mittmann to be quite similar in development but each falls into two sexual strains, both of which must be present in order that perithecia shall be formed. If she is correct in her statements that no antherid is present a union of hyphae at some other point seems to be indicated. Sartoris (1927) found no antherid near the coiled ascogonium of *O. adiposum* (Butl.) Nannf., studied by him. Varitchak (1931) found in *O. piceae* (Münch) H. and P. Sydow a nonfunctional antherid ("trophogone") around which the young ascogonium coiled without uniting with it. (Fig. 89.)

In *Chaetomium kunzeanum* Zopf and *C. bostrychoides* Zopf a cell of an ascogonial coil is fertilized, according to Greis (1941), by an antherid at the tip of a slender hypha arising near by or at some distance. A coil of thick binucleate cells results. In the second species there grow up around this coil, from below, the hyphae that produce the perithecial wall, while other hyphae grow upward between the ascogenous hyphae arising from the ascogonium, to form the paraphyses. The terminal cell of each ascogenous hypha becomes an ascus, without crozier development. In *C. kunzeanum* the perithecium may develop in the foregoing manner, but usually also there grows out from the thickened ascogonium a stout creeping, branching extension, up to several centimeters in length, from which arise, laterally, coils giving rise to ascogenous hyphae, typical perithecia being formed around them. Thus from one act of fertilization (cytogamy) there may arise, radiating from this point, up to 10 perithecia. This is suggestive of the condition in the genus *Dudresnaya* in the Florideae (Red Seaweeds).

In *Xylaria*, in which numerous perithecia are produced in a stroma, coiled ascogonial hyphae were observed by Brown (1913) in the stroma at points where the future perithecia were due to arise, but their further development has not been reported. Somewhat similar structures were reported by Miss Lupo (1922) in *Hypoxylon*.

Many of the Pyrenomycetes lend themselves readily to cultivation on various media. In some cases cultures from single ascospores will produce perithecia, but very often they have not been produced in such cultures. Edgerton (1914) studied a strain of *Glomerella cingulata* (St.) Spaul. and von Sch. in which a scattering development of perithecia was produced on cultures from a single ascospore but yet when two such cultures were allowed to grow in contact with each other a great mass of perithecia appeared along the meeting line in about half the cases, i.e., when the opposite sexual phases met.

Miss Dowding (1931) and Ames (1932, 1934) have shown for *Schizothecium anserinum*, as B. O. Dodge (1927) had shown previously for *Neurospora tetrasperma* Shear and Dodge, that the ascospores are usually four in number and binucleate. From single spore cultures of such spores perithecia are produced in abundance. Occasionally in place of a binucleate spore two uninucleate spores are produced. These, Ames has shown, produce two mutually compatible and self-sterile mycelia while the binucleate spores produce fertile mycelium, each type of mycelium bearing both male and female organs. By cutting off and transplanting hyphal tips containing one or a very few nuclei Ames determined that in the mycelium from the binucleate spores there are two sorts of nuclei which when separated by the hyphal tip planting are shown to belong to the two different strains. On this mycelium with both kinds of nuclei some branches produce both male and female organs of one phase, self-sterile, and other branches produce similar organs, also self-sterile, but the two sorts are inter-fertile. Thus it is clear that genes for incompatibility are present in the diploid nucleus of the young ascus and during meiosis in

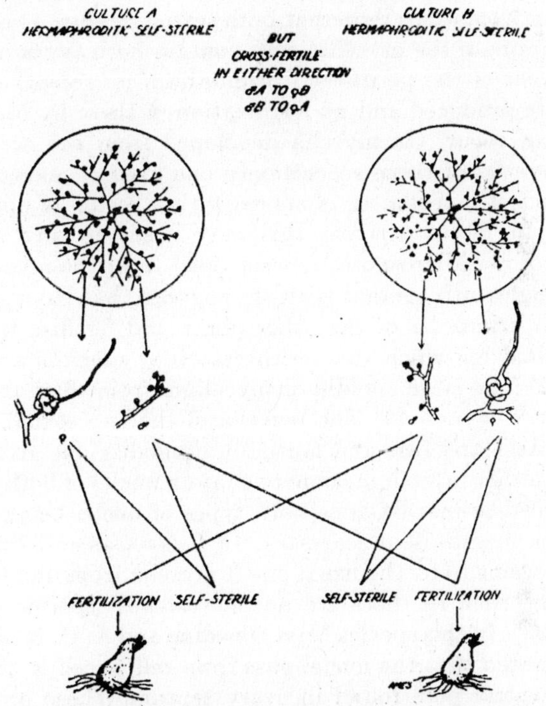

FIG. 89. Sphaeriales, Family Fimetariaceae. *Schizothecium anserinum* (Rabenh). Diagrammatic representation of the sexual condition. (Courtesy, Ames: *Bull. Torrey Botan. Club*, **59**(6):341–345.)

the ascus two sorts of nuclei arise, mutually compatible, but self-incompatible. There is however no segregation of sex as both kinds of nuclei carry the potentiality for the production of male and female organs. (Fig. 89.)

In the genus *Neurospora*, as well as in *Gelasinospora*, Dodge, Shear, Aronescu, Wilcox, and Lindegren (1927–1938) have made an exhaustive study of the sexual reproduction and of the genetics of the segregation of self-incompatibility and mutual compatibility as well as of various other genetic factors. In *N. sitophila* Shear and Dodge and some other species there are formed eight uninucleate spores, each becoming binucleate. Four of these represent one sexual strain and the other four another sexual strain. As with the uninucleate ascospores of *N. tetrasperma* and of *Schizothecium anserinum* the mycelia formed by their germination will form perithecial primordia and "microspores" (sperm cells) which are self incompatible but mutually fertile if both sexual strains are present. In *N. tetrasperma* and *G. tetrasperma* Dowding the ascospores normally contain two nuclei which with rare exceptions represent both sexual strains. As a result when the spores germinate the mycelial cells contain both types of nuclei and at the formation of the perithecial primordium no receptive hyphae or trichogynes are produced and no fertilization of these by microspores or sperm cells can occur. On mycelia developed from the occasional uninucleate ascospores of these species only one kind of nucleus is present and the contact of mycelia, or of spores with mycelia, of opposite phase is necessary. In *N. tetrasperma* this may occur in any of the three following ways: (1) Microspores (sperm cells) from the mycelium of one phase are brought into contact with trichogynes that have grown out of the perithecial primordia of the other phase and fertilize these. This is the same method by which the perithecia of *N. sitophila* are brought to production. (2) The large conidia or mycelium from them may similarly fertilize the trichogynes. (3) The mycelia of the two sexual phases may come into contact and fuse and mutually diploidize one another so that all the cells of either mycelium come to contain nuclei of both types. Then as the perithecial primordia arise both types of nuclei being present fertilization from outside is unnecessary. In *Gelasinospora tetrasperma* this third method seems to be the usual one if mycelia from uninucleate ascospores are concerned as there are no microspores (sperms) nor conidia (Dowding, 1933). In this species Miss Dowding and A. H. R. Buller (1940) have demonstrated that the nuclei pass from cell to cell in the mycelium through the central pore found in every septum. These proceed at the rate of 4 to 5 mm. per hour which is nearly twice to over twice the rate at which the mycelium itself grows. Normally in *Schizothecium anserinum* the first method is the usual one even when the mycelium grew from a binucleate spore for the perithecial primordia and the antherids often arise from hyphal branches that contain only one kind of nucleus. Dodge (1936) showed that in this species there are various races some of which produce no sperm cells. When cultures of the two sexual phases of these are mated perithecia are formed indicating that probably the third method was the one that was effective in perithecial formation.

Keitt and Palmeter (1938) and Keitt and Langford (1941) have shown that also in *Venturia inaequalis* (Cke.) Winter there are two sexual phases, four of the ascospores representing one phase and four the other. Segregation of the genes that determine the two phases may occur in the first meiotic division in the ascus, in which case the four spores of similar phase are placed in succession followed by the four of the other phase. Segregation may occur in the second meiotic division in which case the nuclei are distributed otherwise so that two of the four at either end of the row of eight ascospores will be of one sexual phase and the other two will be of the other phase.

Lindegren (1932) reported the results of experiments with *Neurospora crassa* Shear and Dodge, a species in which each ascus contains eight uninucleate ascospores. He picked out the individual ascospores in order one by one from the ascus and obtained cultures which were then mated to determine their sexual phase. He showed that in about 85 per cent of the asci the two strains are segregated in the first nuclear division and in about 15 per cent in the second. No evidence was obtained to indicate that this segregation ever occurs in the third nuclear division. In *M. sitophila*, with eight ascospores per ascus, and *N. tetrasperma*, with normally four binucleate ascospores, the segregation also may occur in either the first or second division, more often in the first.

From the foregoing it seems possible that the frequent failure to obtain perithecial development in pure cultures from a single ascospore may in some cases be due to the occurrence of two sexual phases in that species.

The groups of families included in the collective term "Pyrenomycetes" differ from those collectively called "Discomycetes" in the nature of the spore fruit. In the latter group, described in the preceding chapter, the spore fruit is an apothecium or a modification of it, in which the asci are in a more or less extensive hymenium which is eventually, in the typical forms, exposed to the air with numerous paraphyses separating and supporting the asci. In the Pyrenomycetes the spore fruits are mostly much smaller, thicker-walled, and rarely opening wide, the ascospores escaping through a small opening, the ostiole, or by the rupture of the whole structure. Formerly these small closed or ostiolate structures were all called perithecia, in contrast to the open apothecia of the Discomycetes. More recent studies in the last forty years, especially by von Höhnel (1918), Theissen (1913), Arnaud (1925), Nannfeldt (1932), and others, have shown that three distinct structural types have been included under the name perithecium. Of the thousands of species with so-called perithecia a great many still remain to be studied carefully to determine to which of these three types they belong.

The first group, called by Nannfeldt the Ascohymeniales, includes the apothecial orders Pezizales, Lecanorales, and others included in Chapter 9, and three orders with true perithecia, Sphaeriales, Hypocreales, and Pyrenulales, discussed in this chapter. The second group, called by Nannfeldt the Ascoloculares, includes what the author considers to constitute the three orders Pseudosphaeriales, Dothideales, and Hemisphaeriales, also included in this chapter. The order Erysiphales, with nonostiolate perithecia, probably contains a mixture of forms with true perithecia, and therefore more closely related to the Sphaeriales, and possibly some Pseudosphaeriales or Hemisphaeriales as well as some Aspergillales. The latter order corresponds to what Nannfeldt and others have called Plectascales. It represents the third group. In it the asci are scattered throughout the tissues of the more or less perithecium-like structure, not forming in any manner what might be considered a hymenium. The Erysiphales, Aspergillales, Myriangiales, and the very simple fungi forming the Saccharomycetales form the subject of Chapter 11.

A typical perithecium consists of a more or less hollow structure whose wall arises from below the sexual organs or at least from below the ascogonium, growing outward and around it and the developing ascogenous hyphae, closing in at the top. Within this there is formed from the ascogenous hyphae a hymenium of asci lining the perithecial wall or forming a cluster at the base. Usually paraphyses arising from the vegetative mycelium which produced the perithecial wall are intermingled with the asci. Nearer the apex of the perithecial cavity there arise the periphyses, also of vegetative origin, which converge and drive upward, helping to produce the ostiole. Apparently in many cases, even in the perithecia developing apart from one another, the vegetative mycelium growing up over the perithecial wall forms a more or less distinct stromatic layer, in which case the true perithecial wall may remain colorless. This stroma may sometimes be massive, enclosing the whole or the lower part of the perithecium.

In general the Sphaeriales, Pyrenulales, and Hypocreales are parallel groups that in all probability will not be maintained separately after life history studies have shown the true kinships of the fungi which are now included in them. They all agree in having (with very few exceptions) ostiolate perithecia. Those of the Sphaeriales are dark-colored and with fairly firm to hard perithecial walls or surrounding stroma. In the Hypocreales the perithecia or enclosing stroma are colorless or bright-colored and are usually softer. The members of both these groups are saprophytic or parasitic, usually on vascular plants, and in the rare cases where they grow upon algae they do not form a lichen thallus. The Pyrenulales are parasitic upon algae which they enclose in a typical lichen thallus, their chief distinction from the Lecanorales being the production of perithecia instead of apothecia.

Julian Miller (1941), for many years a student of the Pyrenomycetes, has had the courage to combine the Sphaeriales and Hypocreales into one order under the former name, treating the latter merely as a family, the Hypocreaceae, in this order.

Many of the species, a number of genera, and several families, that under the older systems of classification were confidently placed in one or another of these three orders, have been found to possess structures that are not true perithecia and so have had to be removed to other orders. Many more have not been studied from this standpoint and are still kept in their customary position but some of these, too, will probably have to be transferred. In the following, where it seems probable that whole families should be transferred, that has been done but in cases of doubt where only a few genera have been carefully studied the old classification (that of Lindau in Engler and Prantl, 1897) is retained.

The term paraphysis has been used differently by various students of the Pyrenomycetes. Theissen, Sydow, and perhaps the majority of mycologists, including the author, define as paraphyses those slender hyphae that develop in the hymenium, growing up from below and ending free above. In between them arise the asci to which they give protection and perhaps nourishment. In some groups of Discomycetes, e.g., various genera of Tuberales and Lecanorales, these paraphyses may enlarge and branch above the asci and grow fast to the enlargement or branches of other paraphyses to form a more or less continuous pseudoparenchymatous layer over the tips of the asci, the epithecium. In the Pseudosphaeriales the sheets of more or less crushed stromatic tissue that remain between the developing asci and which may be reduced to interascal hyphae, attached to the base of the hymenium and the roof of the ascocarp are called by Petrak (1923) true paraphyses while he uses the term metaphyses for the structures called paraphyses by Theissen, etc. The term paraphysoid is used with various meanings by different mycologists. By some it is applied to the filamentous often deliquescent remains of the stromatic tissue between the asci (the true paraphyses of Petrak) while by others it is applied to quickly evanescent true paraphyses.

Order Sphaeriales.[1] This order is described first not from any conviction that it is more primitive than the other two but because it offers a better field in which to point out the various developmental directions in the modification of the perithecia. The perithecial wall is in general dark-colored, at least in its outer stromatic layer, pseudoparenchymatous in structure and free from the enclosed asci which arise from its base or part way up its sides. Sometimes the asci form a dense hymenium but frequently they are more loosely arranged. Paraphyses may be present but are usually delicate and evanescent and not very numerous. When true paraphyses are absent the suspicion is aroused that perhaps these organ-

---

[1] It should be noted that this name and that of the family Sphaeriaceae are not properly valid since the generic name *Sphaeria* has been entirely abandoned because of its former application indiscriminately to fungi with conidia (i.e., without asci) and perithecia. Rather than to propose a new set of names the old familiar ones are retained here.

isms do not rightly belong here but in the Pseudosphaeriales. Periphyses, i.e., paraphysis-like threads at the edge of the hymenium but not intermingled with the asci, are much more frequent. The asci may reach maturity at different ages within the perithecium a phenomenon not rare in the apothecial forms also. The ascospores may be expelled violently through a pore (not an operculum) at the apex of the ascus or the osmotic pressure within may rupture the ascus at its middle so that the upper half is forced off, or the outer layer may break near the apex and contract, allowing the inner layer to expand and finally burst, or the asci, paraphyses, etc., may be digested within the perithecium leaving the ascospores embedded in the gummy mass. This absorbs water during a rain and swells, emerging from the ostiole, where the gum is dissolved away and the spores carried off by the currents in the film of rain water or splashed off in the droplets caused by the striking rain drops. In some genera only the ascus stalk digests, the resultant gummy mass containing the unchanged bodies of the asci. Nannfeldt and some others before him have suggested that in the true Sphaeriales the asci discharge their ascospores through an apical pore and mostly do not undergo autodigestion and that forms in which the latter occurs and no paraphyses are present belong in the Aspergillales or some other order.

The asci vary from cylindrical to club-shaped or obovate, sometimes being drawn out below into a narrow stalk-like portion. The ascospores may lie in one or more rows in the ascus or in a ball-like cluster. The number is mostly eight but in a few cases only four or even fewer spores are formed and in others 16 to 32 or even as many as 256 or 512 are reported. They vary exceedingly in size, shape, structure, and color. They may be colored or hyaline, 1-celled, 2-celled, several-celled in one row (phragmosporous), many-celled by both cross and longitudinal walls (muriform or dictyosporous), long and slender, or even tetrahedral. The recurrence of certain spore and ascus types in what the current classifications consider to be widely distant families has led (and probably rightly) some mycologists (e.g., Vincens, 1918, 1921; Julian Miller, 1928; von Höhnel, 1918; Wehmeyer, 1926; and others) to attempt to amend the classification so as to bring together those forms with similar asci and spores.

The current classifications place first in this order those families in which the perithecia stand separately upon the surface of the substratum or but slightly sunken in it. Two of these families have very thin perithecial walls. These are the Fimetariaceae (Sordariaceae) with naked or almost naked perithecia and the Chaetomiaceae with the perithecia covered with long hairs and with a special tuft of much longer hairs about the ostiole. The fungi in both families grow on dung or decaying plant tissues. They are distinguished further by the fact that the asci of the former expel the ascospores through the ostiole while in the latter the asci are digested and the ascospores escape in a mass of slime. Nannfeldt (1932) therefore places the Chaetomiaceae in the group he calls Plectascales, in this book called Aspergillales. His reason for so doing is that there are no paraphyses or periphyses and that the asci dissolve into slime. Closely related to the Fimetariaceae is the family Melanosporaceae, also with thin perithecial walls, often but not always with a more or less well-developed neck, and with ascospores mainly inclined to be dark-colored and lemon-shaped as in most of the species of the two other families. *Neurospora* and *Gelasinospora* probably belong here or in the Fimetariaceae. The Melanosporaceae are often placed in the order Hypocreales because of the almost colorless or light brown perithecia but it seems best to place them here since the chief difference is only their sometimes lighter color. (Fig. 90 C, D.)

FAMILY FIMETARIACEAE. The chief genera are *Fimetaria* (*Sordaria*) and *Schizothecium* (*Pleurage* or *Podospora*). Both are mostly found on the dung of various animals. Some species have four-spored asci and others asci with eight spores, while in a few species the number of spores is much larger. In *Fimetaria* the spores are surrounded by a layer of water-soluble slime on all sides except a small spot on one end. In *Schizothecium* the spores are two-celled, one cell slender and empty and the other enlarged and dark-colored. There is a long gelatinous appendage at each end, sometimes several at one of the ends. *Hypocopra* is like *Fimetaria* but the perithecia are immersed in a stroma, an exceptional case for this group of families. (Fig. 87.)

FAMILY CHAETOMIACEAE. *Chaetomium* is the characteristic and most frequently found genus of the Chaetomiaceae. Its long ostiolar hairs may be stiff and straight, or wavy, or loosely or tightly coiled, depending upon the species. The spores are mostly lemon-shaped and dark-colored and one-celled. The species are very numerous on damp straw, pasteboard, etc., as well as on manure. Some species are destructive to cloth and other vegetable fabrics, especially in the warmer and more humid portions of the world. (Fig. 90 A, B.)

FAMILY SPHAERIACEAE.[2] Perithecia with firmer wall and with simple ostiole or at most with a low papilla. Perithecia superficial on the substratum or sometimes on a felty mass of mycelium (subiculum). The twenty or more genera in the family are mostly of little economic interest except the genus *Rosellinia*. This deserves mention because of the widespread occurrence of its species (over 200 are known from all parts of the world), mostly on wood and bark. Several species are dangerous parasites, e.g., *R. necatrix* (Hart.) Berl. on the roots and underground portions of the stems of the grape (*Vitis*). The perithecia of *Rosellinia* are nearly

[2] See footnote on page 273.

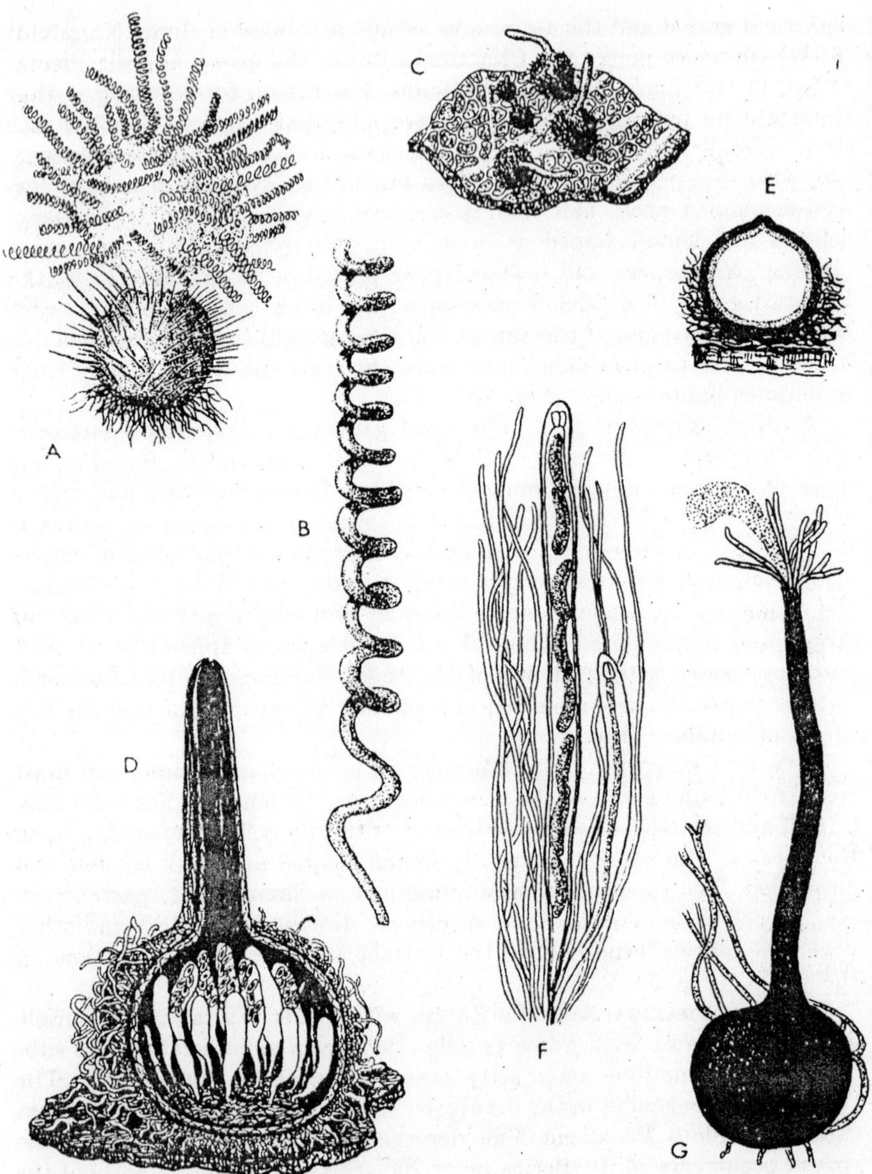

FIG. 90. Sphaeriales. Various types. (A, B) Family Chaetomiaceae. *Chaetomium aterrimum* E. & E. (A) Perithecium showing the straight body hairs, the coiled oral hairs, and the mass of exuded ascospores. (B) Terminal portion of an oral hair. (C, D) Family Melanosporaceae. *Melanospora chionea* (Fr.) Corda. (C) Several perithecia on decaying leaf. (D) Vertical section through perithecium. (E, F) Family Sphaeriaceae. *Rosellinia aquila* (Fr.) de Not. (E) Vertical section through perithecium and

(*Continued on facing page.*)

spherical and have a small ostiolar papilla. They sit externally on the host with their bases sunk in a more or less well-developed subiculum. The ellipsoidal, colored, one-celled ascospores are eight to each ascus. Filamentous paraphyses are present. Because of the ascospore characters, Vincens (1921), Wehmeyer (1926), and Miller (1928) suggest that this genus belongs more properly near the Xylariaceae. Other genera in the family vary as to color and number of cells in the ascospores as well as to the hairiness of the perithecia. In many genera the asci ripen successively and project one or two at a time from the ostiole to discharge their spores, the emptied asci contracting back into the perithecium and giving place to the next maturing asci. (Fig. 90 E, F.)

Family Ceratostomataceae. In this family the perithecia have a distinct, sometimes fairly long neck, otherwise much like the Sphaeriaceae. The perithecial walls are mostly leathery rather than brittle. The asci are mostly accompanied with paraphyses and do not digest as in the next family. The ascospores are one-celled, two-celled, phragmosporous or muriform, and hyaline or brown. The species are mostly saprophytic, growing on wood, bark, or sometimes on stems of herbaceous plants. The genus *Ceratostomella*, more properly called *Ophiostoma*, has been segregated to form the following family, which is discussed here although Nannfeldt (1932) places it in Order Aspergillales.

Family Ophiostomataceae. The perithecia have very long necks. In the genus *Ophiostoma* (*Ceratostomella*) the neck is often several times as long as the diameter of the perithecium. It has a thin perithecial wall and the roundish asci are scattered throughout the perithecial cavity, without paraphyses. The ascus walls undergo autodigestion and with the absorption of water the resultant gummy mass swells and escapes from the apex of the neck as a hyaline drop containing thousands of the one-celled hyaline spores. Because of these characters Nannfeldt placed this family in the Aspergillales. Several species of *Ophiostoma* grow on the wood of various trees whose sapwood takes on a blue color, the so-called "sap stain," owing to the presence of the mycelium in the wood cells. The conidial stages of the various species of this genus have been described under a number of names in accordance with the type of conidia and conidiophores. *Chalara* or *Thielaviopsis* produce their conidia endogenously. In *Cephalosporium* the conidia are produced externally on separate

---

Fig. 90—(*Continued*)

subiculum. (F) Asci, mature and young, and paraphyses. (G) Family Ophiostomataceae. *Ophiostoma ulmi* (Buis.) Nannf. Perithecium with emerging mucilaginous mass of ascospores. (A–B, after Greathouse and Ames: *Mycologia*, **37**(1):138–155. C–D, after Ellis and Everhart: The North American Pyrenomycetes. E–F, after Berlese: ***Riv. patol. vegetale***, **1**(1):5–17; (2):33–46. G, after Buisman: ***Tijdschrift over Plantenziekten***, **38**(1):1–5.)

conidiophores. In *Graphium* similar conidia are produced but the conidiophores are united together in a tall dark-colored stalk (technically a synnema). In most cases the conidia are embedded in slime as are the ascospores and are distributed mainly by insects, which bore in the wood or bark, to whose bodies the spores adhere. The dreaded "Dutch Elm Disease" has for its perfect stage *O. ulmi* (Buis.) Nannf., *Graphium ulmi* Buis. being the conidial stage. This as well as several other species has two sexual strains which must be brought into contact before perithecia will appear. Dade (1928) found that this was also true of *O. paradoxum* (Dade) Nannf. This has for its conidial stage *Thielaviopsis*. (Fig. 90 G.)

Richard and Olga Falck (1947) create another "class" of the "Ascomycetales," which they call "Class Haerangiomycetes." In this class they place those species of *Melanospora* and *Ceratostomella* (*Ophiostoma*) in which the ascus does not possess a definite cell wall but merely a plasma membrane or where the ascus wall is almost immediately dissolved after its formation. In these fungi, therefore, the ascus does not exercise its normal function of ascospore dispersion but these spores are carried out through the ostiole in a mass of "mucus" and rest in a drop in the funnel-like "haerangium" formed by filaments diverging from the edge of the ostiole. In the ascus the spores are formed, eight in number, shaped like the segments of an orange and arranged in a similar manner. This structure the authors call an "octophore." This "class" is considered to be an evolutionary development from Sphaeriales in which definite, functional asci occur.

FAMILY CUCURBITARIACEAE. This family formerly included in this order undoubtedly belongs in the Order Pseudosphaeriales under which it is discussed.

FAMILY LOPHIOSTOMATACEAE. Except for the base which is partly sunk in the substratum and the laterally compressed ostiolar papilla this family differs very little from the Sphaeriaceae. The ostiole compressed into a slit resembles somewhat that of the Hysteriales. This resemblance is only superficial for the lateral compression is confined mainly to the ostiole and ostiolar papilla in this family while in the Hysteriales the whole spore fruit is laterally compressed. *Lophiostoma* is the largest genus of the family. The fungi of this family are mostly saprophytic on bark, wood, or dead herbaceous stems while only a few species are possibly parasitic. Conidial stages are known for only a few forms. (Fig. 91 A.)

FAMILY AMPHISPHAERIACEAE. In general appearance except for its circular ostiole this resembles the preceding family, but the study of the development and the inner structure of the perithecia indicate that it probably may belong to the order Pseudosphaeriales.

Usually placed next are the families with perithecia entirely sunken in

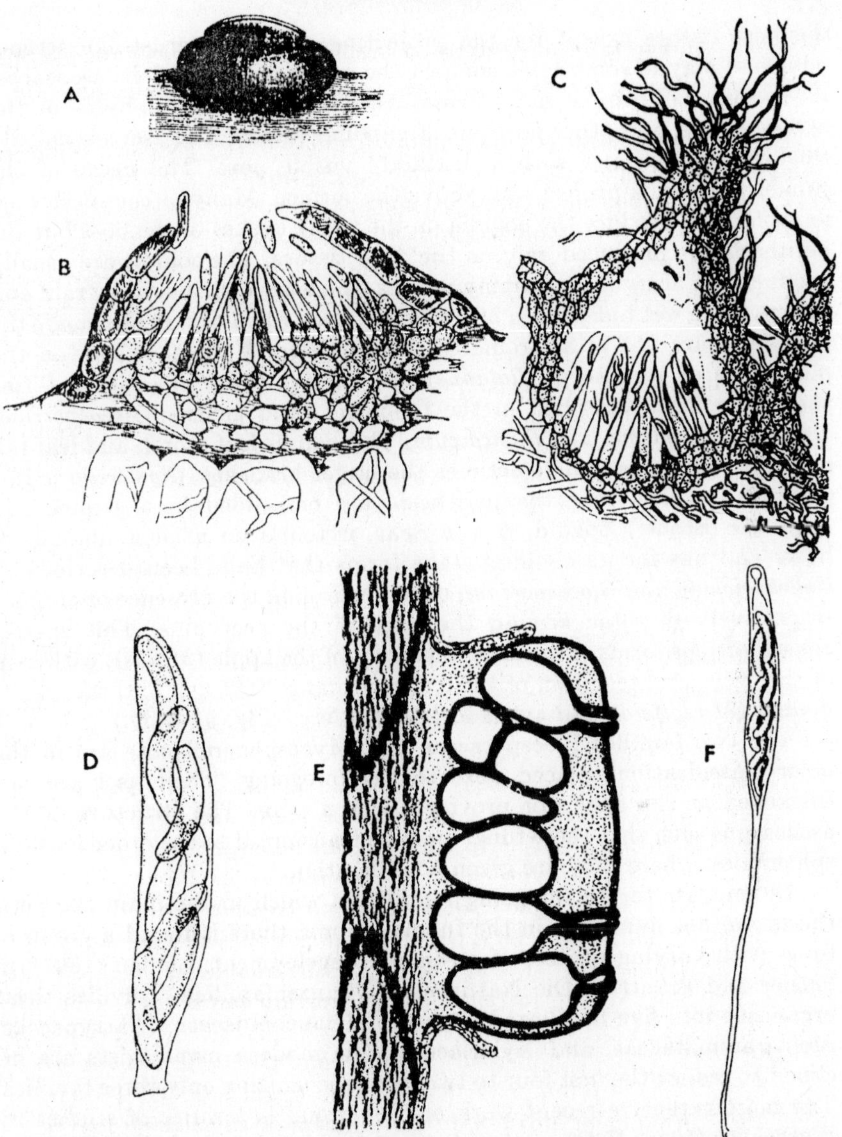

Fig. 91. Sphaeriales. Various types. (A) Family Lophiostomataceae. *Schizostoma montellicum* Sacc. Perithecium. (B–D) Family Gnomoniaceae. *Glomerella cingulata* (Stone.) Spauld. & von Schr. (B) Acervulus with conidia. (C) Perithecia in a stromatic base. (D) Ascus. (E, F) Family Allantosphaeriaceae. *Diatrype virescens* (Schw.) Cke. (E) Vertical section through emergent stroma. (F) Ascus. (A, after Berlese, from Engler and Prantl: Die Natürlichen Pflanzenfamilien, Leipzig, W. Engelmann. B–D, courtesy, Stoneman: *Botan. Gaz.*, **26**(2):69–120, Univ. Chicago Press. E–F, from Berlese: Icones Fungorum, **3**:1–120.)

the host tissues except for the projecting ostiole. A crust-like stroma (clypeus) may in some cases connect the upper portions of the ascocarps.

FAMILY GNOMONIACEAE. These are parasitic or saprophytic in the leaves, stems, and other portions of vascular plants. The asci are usually thickened above and with a distinctly visible pore. The necks of the sunken perithecia project well above the surface. Many of the species are parasitic and produce the conidia on the living tissues of the host but the perithecia are produced only on the dead tissues. The conidia are usually produced in acervuli, in gummy masses which are distributed by rain and insects when wet but harden into a horny mass when dry. In *Gnomonia* the perithecia are not in a stroma. *G. veneta* (Sacc. & Speg.) Kleb. on the plane tree or sycamore (*Platanus*) has various types of conidial forms that have been described in the following form genera: *Gloeosporium*, *Discula*, *Sporonema*, and *Fusicoccum*. This causes leaf scorch and leaf fall and kills the twigs and sometimes the larger branches. *Glomerella* is like *Gnomonia* except that the perithecia are embedded in a stroma. *Gl. cingulata* (Stone.) Spauld. & von Schr. is found on a large number of hosts and has for its conidial stage forms that have been described as *Colletotrichum* and *Gloeosporium*, depending upon the presence or absence respectively of setae around the edge of the acervulus. This species causes various forms of diseases: bitter rot of the apple (*Malus*), withertip of the twigs and tearstain of the fruits of orange (*Citrus*), anthracnose of mango (*Mangifera*) and avocado (*Persea*), etc. (Fig. 91 B–D.)

The two families Pleosporaceae and Mycosphaerellaceae are in the older classifications placed next to the foregoing. Their asci are not thickened at the apex nor provided with a pore. The structure of the ascocarp is such that these fungi must be transferred to the order Pseudosphaeriales, where they are given consideration.

In contrast to the foregoing families in which in the main the perithecia are not immersed in the fungus stroma there is found a group of fungi with varying degrees of stromatic development. Lindau (1897), in Engler and Prantl's "Die Natürlichen Pflanzenfamilien," divides these organisms into five families: Valsaceae, Melanconidaceae, Diatrypaceae, Melogrammataceae, and Xylariaceae. The modern mycologists are inclined to reduce the first four to two, thus recognizing only three families. The most extensive recent work on this group is a series of studies by Wehmeyer (1926, 1933) on the life histories of these fungi. He points out that these stromatic forms exhibit a gradual transition from fungi in which the stroma is vague in outline and not very definite in structure to those with a highly organized stroma. In the simplest type of stroma the surface of the substratum is blackened by the coloring of the mycelium. Wehmeyer states: "The next step in stromatic development comes about by the proliferation of the mycelium within the substratum. As this formation of mycelium increases it usually becomes more or less localized about the forming perithecia." Wehmeyer distinguishes between ectostroma and entostroma thus: "Ectostroma is that portion of the stroma which is formed on the surface of the bark, beneath or within the periderm, and which consists typically of fungous tissue only, except that when it is developed within the periderm it may contain the remnants of the periderm cells, but never of the bark cortex cells. An entostroma is that portion of the stroma which develops within the cortical or woody tissue of the host or substratum and is made up of components of both fungous and host tissues or substratum tissues." Concerning the development of the entostroma in the progressive specialization of the stroma Wehmeyer remarks: "There is usually correlated, very often beneath a differentiated ectostroma, a clustering of the perithecia." This entostroma is often delimited from the surrounding tissues by a thin zone of blackened tissue, forming the black line visible on cutting through the host tissue. Among the variations found in this group of organisms may be noted the following: In a fruiting area (i.e., the region where the perithecia or clusters of perithecia are formed) the perithecia may be scattered or clustered, with or without an entostromatic mycelium about them. If present, according to Wehmeyer, the "entostromatic area may or may not be surrounded by a darkened unorganized zone." It may be lighter in color than the surrounding bark tissue. The ostioles of the perithecia may be separately or collectively erumpent, even clustered perithecia not necessarily being collectively erumpent. A stroma may be *effused*, i.e., containing numerous separately erumpent perithecia or several clusters of perithecia, or *isolated*, when it contains only one cluster of perithecia. The portion of the stroma which is erumpent through the periderm or epidermis is the disk. It may be conical or cushion shaped and well distinguished from the entostroma or grading into it. In contrast to the foregoing are those fungi in which the stroma is compact and composed entirely of fungous tissues and very early becoming external to the substratum.

The three families of the stromatic Sphaeriales may be distinguished as follows:

Allantosphaeriaceae: stroma showing all degrees of development described above, but not entirely of fungal structure. Asci with more or less elongated tapering stalks, forming a persistent hymenial layer. Paraphyses mostly evanescent at maturity. Ascospores mostly allantoid, yellowish hyaline, sometimes inequilaterally ellipsoid and brown. Conidia long cylindrical to filiform.

Diaporthaceae: stroma as in the foregoing family. Asci with short or long evanescent stalks soluble in water so that at maturity the free asci and spores form a loose central mass. Paraphyses present. Ascospores ellipsoid, fusoid, less commonly allantoid, or long cylindrical, hyaline or colored.

Conidia of two types: short cylindrical to filiform and ellipsoid to long cylindrical.

Xylariaceae: stroma well developed, entirely fungal, almost always external, at least eventually, and covered at first by a conidial layer. Asci long, cylindrical, ascospores one-celled, inequilaterally ellipsoid, dark brown, paraphyses filiform.

To these stromatic families probably should be added:

Phyllachoraceae (formerly included in Order Dothideales): the members of this family are leaf parasites with the stroma extending from the upper to the lower surface or between cuticle and epidermis or between epidermis and palisade layer. The perithecial walls are present, and true paraphyses are produced..

FAMILY ALLANTOSPHAERIACEAE. The following genera may be mentioned as they are mostly very frequent: In the Allantosphaeriaceae one of the commonest genera is *Diatrype* with stroma effuse or isolated, ectostroma deciduous, exposing a widely erumpent entostromatic disk. Perithecia parallel, separately erumpent. Eight ascospores, allantoid. Many species. Saprophytes or weak parasites on twigs and branches. *Diatrypella* is similar in many respects but the ascospores are numerous in the ascus. *Eutypella* is much the same as *Diatrype* but the perithecia are clustered and collectively erumpent. *Anthostoma* has stroma effuse or isolated, perithecia separately or collectively erumpent. Asci, in contrast to the foregoing genera, cylindrical, short-stalked, the eight ascospores inequilaterally ellipsoid and dark brown. A transitional form in ascus and spore structure to the Xylariaceae, and perhaps more properly placed in that family. (Fig. 91 E, F.)

FAMILY DIAPORTHACEAE. In this family the genus *Diaporthe* with 600 or more species is the largest or almost so. Stroma effuse or isolated and entostroma light-colored with a dark border zone. Ascospores ellipsoid or fusoid, hyaline, two-celled. Imperfect stage belonging to the form genus *Phomopsis*. In the genus *Valsa* the stromata are isolated and the perithecia clustered in the unaltered bark tissues beneath a distinct conical ectostroma. No marginal zone. Eight ascospores, allantoid, one-celled, hyaline. Imperfect stage belonging to the form genus *Cytospora*. In *Leucostoma* the stromata are isolated or confluent with a dark marginal zone about each perithecial cluster. Asci and spores as in *Valsa*. *Valsella* resembles *Leucostoma* but the asci are polysporous. In *Endothia* the stromata are isolated or confluent with strongly developed, colored entostroma. Eight ascospores, allantoid to ellipsoid, one- or two-celled. *E. parasitica* (Murr.) And. & And. is the fungus which has destroyed nearly all the trees of the American chestnut (*Castanea dentata* (Marsh.) Borkh.) since the fungus was introduced from Eastern Asia on nursery stock about 1900 or a little earlier. (Fig. 92 A–E.)

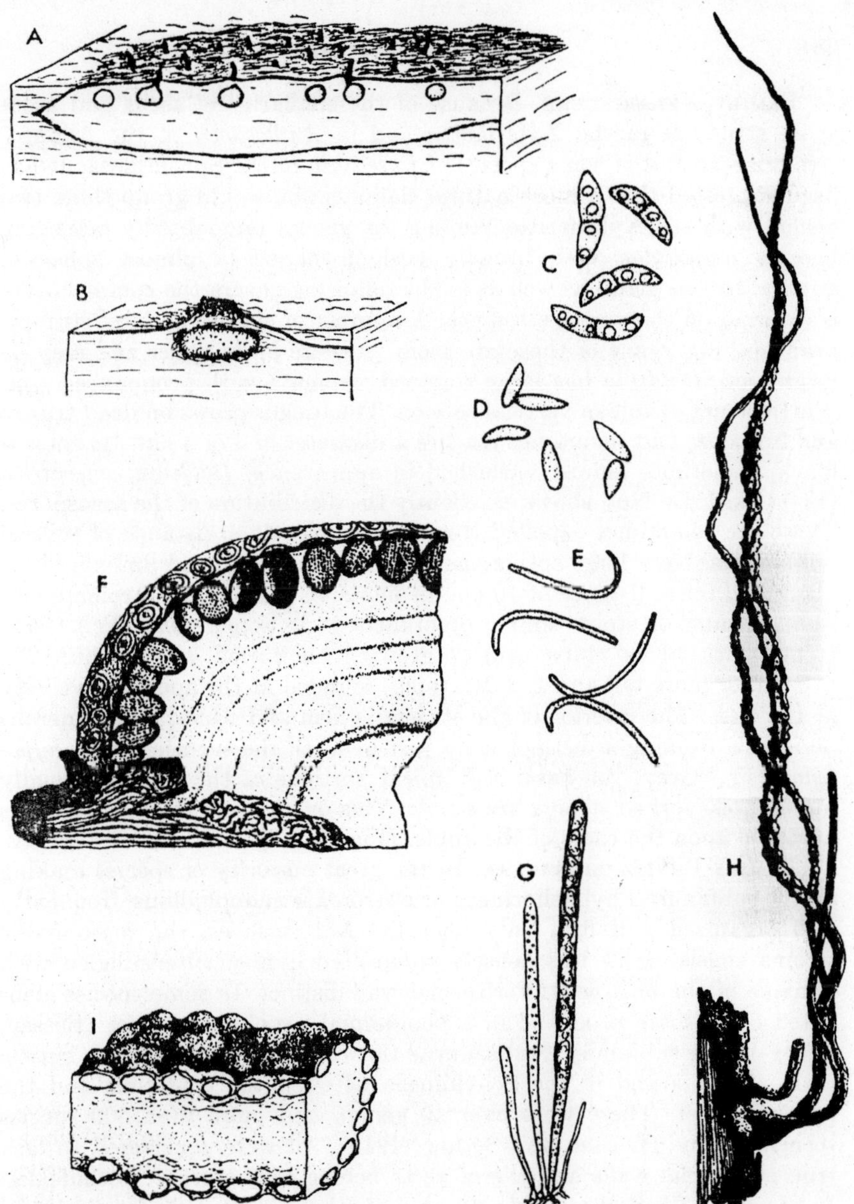

Fig. 92. Sphaeriales. Various types. (A–E) Family Diaporthaceae. *Diaporthe arctii* (Lasch) Nit. (A) Section through stroma with perithecia. (B) Section through stroma with pycnidium. (C) Ascospores. (D) Alpha conidia. (E) Beta conidia. (F–I) Family Xylariaceae. (F, G) *Hypoxylon marginatum* (Schw.) Berk. (F) Vertical section through stroma showing perithecia. (G) Asci of various ages. (H, I) *Xylaria subterranea* (Schw.) Sacc. (H) Stromata growing from piece of wood. (I) Section of stroma enlarged to show perithecia. (A–E, courtesy, Wehmeyer: The Genus Diaporthe Nitschke and Its Segregates, Ann Arbor, Univ. Michigan Press. F–I, from Ellis and Everhart: The North American Pyrenomycetes.)

Family Xylariaceae. Because of the similarity of ascus and ascospore structure in the Xylariaceae and in *Anthostoma* of the Allantosphaeriaceae and of some species of *Rosellinia* in the Sphaeriaceae it has been suggested that a more natural classification would group these two genera with the Xylariaceae. Among the genera undoubtedly belonging here is *Hypoxylon* with broadly cushion-shaped to almost spherical stroma. In this genus as well as in the following genera the conidial layer is external on the young stromata. A number of species on logs, stumps, branches, etc. Some of these are 1 cm. or more in diameter and may be bright red. *Daldinia* has large rounded stromata with pronounced concentric zones visible in vertical section. The fungus grows on dead trunks and branches and sometimes reaches a diameter of 3 or 4 cm. Its color is black, sometimes almost varnished in appearance. *Daldinia concentrica* (Fr.) Ces. & De Not. shows excellently the distribution of the ascospores. These are sometimes expelled from the ostioles to a distance of several millimeters. Near large specimens of this species the dead limbs may be blackened for a distance of 10 cm. or more. In *Xylaria* the stroma is upright, slender or stout, simple or branched. *X. polymorpha* (Fr.) Grev. forms thick black clubs usually growing on buried wood. These are 5–8 cm. or more tall and 1–2 cm. thick, rounded at the apex and velvety at the base. The interior of the stroma is firm and white, the numerous perithecia forming a distinct layer just beneath the surface. In *X. hypoxylon* (Fr.) Grev. the basal and apical portions of the slender, usually more or less forked stroma are sterile. This or a closely related species is parasitic upon the roots of the apple. (Fig. 92 F–I.)

Family Phyllachoraceae. In the great majority of species making up the subfamily Phyllachorineae, the stroma is endophyllous, frequently well organized and firm only near the leaf surfaces, the mesophyllic stroma consisting of less densely compacted hyphae intermingled with remains of the host cells. Perithecial wall distinct. In some species elongated conidia are produced in subepidermal pycnidial cavities (Bessey, 1919). In the subfamily Trabutiineae the stroma lies between the cuticle and epidermis and in the Scirrhiineae between the epidermis and the palisade layer. There were over 40 genera and more than 500 species recognized by Theissen and Sydow (1915). All were considered to lack true perithecial walls and therefore to belong to the order Dothideales. In the genus *Phyllachora* and a number of other genera definite perithecial walls are present (Orton, 1924; Petrak, 1924). This character and the presence of true paraphyses in the vast majority of the species justify placing the family in the Sphaeriales, although some of the smaller genera may still have to be retained in the Dothideales. The greater number of the genera are tropical. *Phyllachora graminis* (Fr.) Fckl. and other species are very frequent parasites of various grasses in the United States and temperate Eurasia. The elongated black stromata resemble unopened rust sori.

Order Pyrenulales. The perithecial lichens make up a group of about 15 families, over 80 genera, and more than 2000 species. They form typical lichen thalli in combination with various algal hosts. Most of the species are crustose or foliose, in only two genera fruticose. The order does not seem to form a compact monophyletic group but its various families seem rather to show relationship to different families of the Sphaeriales and perhaps to some of the Pseudosphaeriales. The perithecia may be sunk singly in the thallus or may be produced in a stroma strongly resembling that of *Diatrype*. Among the common genera may be mentioned *Verrucaria*, with nearly 300 species forming crustose growths on rocks into which the hyphae may penetrate to a considerable distance. The perithecia are black and sunken in the thallus. The eight ascospores are one-celled, ellipsoidal, and hyaline or brown. The algal host is *Protococcus* or *Palmella*. *Pyrenula* includes about 175 species usually on bark, growing on the alga *Trentepohlia*. The perithecia resemble those of *Verrucaria*, but the ascospores are several-celled. Long slender conidia are produced in pycnidia. *Trypethelium* consists of about 75 bark-inhabiting species, mostly tropical and subtropical, whose perithecia are produced in a cushion-like stroma. The 37 or more species of *Astrothelium* are also tropical or subtropical, on bark. Their perithecia are arranged radially in the stroma with their long necks approximated or joining into a common ostiole. Practically nothing is known as to the sexual reproduction of the plants assigned to this order. Spermogonia are known in many species and may function as they are known to do elsewhere.

Order Hypocreales. This order shows a close parallelism with the Sphaeriales as to perithecial form and arrangement. The two orders are customarily distinguished from one another by the consistency and color of the perithecia. In the latter order the perithecia are dark-colored and leathery or brittle while in the former they are bright-colored (rarely dark) and fleshy to leathery. There are border forms such as the genus *Melanospora* which has sometimes been placed in one and sometimes in the other order but which has been treated under the Sphaeriales in this work. Miller (1941) did not recognize the validity of separating these groups of fungi as distinct orders and included the family Hypocreaceae as a distinct family in the Sphaeriales, but in a more recent paper (1949) concludes that until further studies have been made on the structure and development of the perithecia the Order Hypocreales should be retained, except for Family Clavicipitaceae which definitely should be placed in the Sphaeriales. The genera may be arranged with first the forms with scattered superficial perithecia, then those with perithecia crowded on the surface of a stroma, and those with perithecia buried in the substratum or in a stroma. A further group includes forms in which the perithecia buried in the stroma do not have well-developed walls of their own but represent perithecial cavities in the stroma. Only this last type does not have its counterpart in the Sphaeriales. The ascospores, as in the latter order, vary from ellipsoidal and one-celled to two-celled, phragmosporous, muriform, or even thread-like. Some are brown but the majority are hyaline or bright-colored. Conidial fructifications are rather widespread in this order. The conidiophores may be separate and external or they may be packed closely together side by side or may be enclosed in a pycnidium or united into a stalked head (*Stilbella* type). Many of the approximately 1000 species are saprophytic; others are parasitic in the leaves, stems, and roots or other portions of higher plants; still others are parasitic on fungi or upon insects.

The course of sexual reproduction has been worked out completely in a few forms, but, as mentioned for the Sphaeriales, only enough is known to make certain that vastly more must be found out before the knowledge may be used to modify the current system of classification. The latter, as in Sphaeriales, is largely based upon the characters of the mature perithecium.

In a number of Hypocreales a coiled ascogonium and antherid are known. In *Polystigma rubrum* (Fr.) DC., parasitic in the leaves of the plum (*Prunus domestica* L.), Blackman and Welsford (1912) and Nienburg (1914) have shown that the ascogonium is a stout hypha with several coils of mostly plurinucleate cells and tapering into a slender, sometimes branched, trichogyne which may extend through a stoma but apparently more frequently does not do so. Organs exist which have been called spermogonia. Whether they really are properly so called remains in doubt. They usually appear some time after the ascogonia, and their spores are long and slender and curved like some of the conidia of the Diaporthaceae. No connection between one of these spores and a trichogyne has been observed. Eventually, according to Nienburg, the wall breaks down between a multinucleate ascogonial cell and the large uninucleate oogone cell next to it and one nucleus passes into the oogone. Later ascogenous hyphae are sent out from the latter and eventually give rise to asci. Blackman and Welsford disagree with the foregoing and claim that the ascogonium degenerates and that the ascogenous hyphae arise from near-by vegetative hyphae. In *Claviceps purpurea* (Fr.) Tul., ergot, the germinating sclerotia give rise to stalked heads in which arise the perithecial primordia. This consists for each perithecium, according to Killian (1919), of a multinucleate rounded oogone from whose base branch out one or two antherids which also have many nuclei. One of these antherids comes into contact with the oogone at its tip and an opening is formed through which the male nuclei enter. This gives rise, how is not known because certain stages were missed in the investigation, to a series of binucleate cells which develop into ascogenous hyphae and form asci by the hook method.

The 60 or more genera making up the order are variously assigned to one family, to three families, or to still more. The basis of distinction is the presence or absence of stromata, the location of the perithecia, and the type of the ascospores. A number of genera with perithecial cavities in a stroma but without well-developed perithecial walls, and with long slender ascospores and lacking paraphyses, seem to constitute a natural well-defined family, the Clavicipitaceae. Those with ascospores not of this type and with distinct perithecia buried in the stroma are usually called the Hypocreaceae, while those with perithecia external, with or without a stroma, are placed in the Nectriaceae. The following genera should be noted: *Nectria* has perithecia external to the substratum or to an external stroma on the substratum, round with short ostiolar papilla or none, usually light-colored, asci in a tuft at the base of the perithecial cavity, ascospores two-celled, hyaline. Often parasitic on twigs or other plant tissues. In some species, e.g., *N. cinnabarina* Fr., there first arises a cushion-like pseudoparenchymatous stroma which bears on its outer surface a dense layer of slender conidiophores, each bearing a small ellipsoidal spore (*Tubercularia* stage). Later around the base and eventually all over the stroma arise the round, rather thick-walled perithecia from whose ostioles escape the ascospores. Over 250 species of *Nectria* have been described. The presence or absence of the stroma has been used by some mycologists to distinguish two genera. *Hypomyces*, with 50 or so species, produces a felt-like stroma or subiculum over the surface of various species of Agaricaceae, Polyporaceae, etc. The perithecia are essentially like those of *Nectria* except for the presence of the stroma or subiculum. *Gibberella* produces its blue- or violet-colored perithecia on the stems, grains, etc., of various plants. Its ascospores vary from two to several cells. *G. zeae* (Schw.) Petch (*G. saubinetii* Oud.) is the cause of scab and root rot of wheat and other cereal grasses and of the root rot of maize. Its asexual reproduction is by the abundant production of several-celled, sickle-shaped conidia (*Fusarium* stage). Subsequently the perithecia appear. *Sphaerostilbe* has perithecia and ascospores as in *Nectria* but these arise around the base of a stalked conidial head of the *Stilbella* type. It is mostly parasitic on scale insects and other insects infesting the twigs or leaves upon which the fungus occurs. *Polystigma* develops its perithecia in stromata within the host leaf. *Hypocrea* has perithecia much like those of *Nectria*, but buried in the bright-colored stroma which resembles in many ways that of *Hypoxylon* of the Xylariaceae. Over 110 species are known. In the Clavicipitaceae may be mentioned *Epichloe* whose stroma develops as a thick white band around the stems of various grasses. In this white

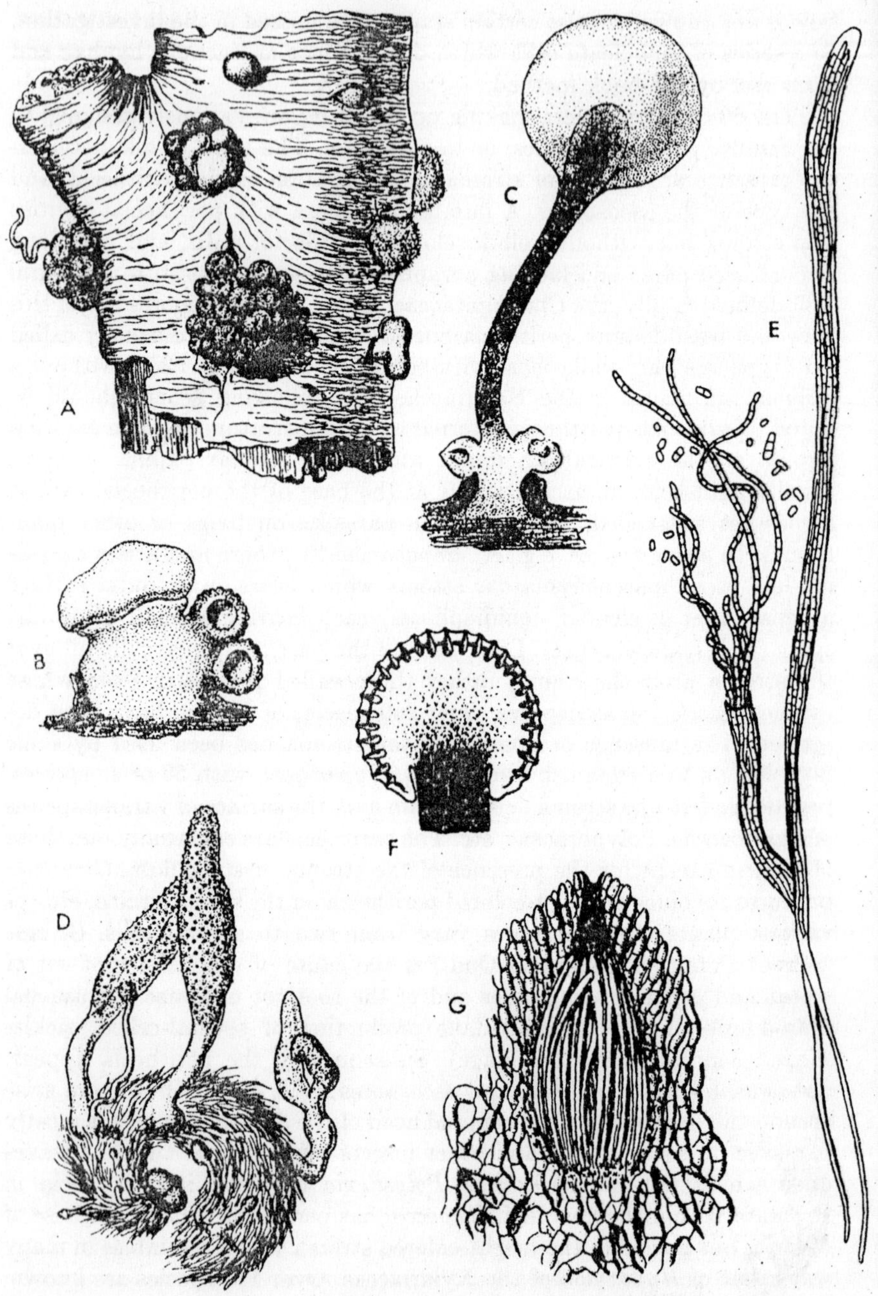

Fig. 93. (See legend on facing page.)

stroma the orange-colored perithecia develop, projecting from the surface by their short ostioles. *Cordyceps* consists of many species mostly parasitic on insects but with one or two species growing on subterranean fungi. They produce stout or slender stalks bearing a round or more often elongated, usually pointed, stromatic head in which arise the numerous perithecia. The latter may be almost completely buried in the stroma or they may project from it so as to be almost free. The well-known "vegetable caterpillar" is a species of this genus. This fungus attacks a caterpillar which has entered the ground to pupate and from its body a stalk several inches high emerges into the air bearing the perithecia in the stroma in its upper part. *Claviceps*, the ergot fungus, with a dozen or more species, produces its purple sclerotia in the spikelets of grasses and related plants. On the ground, usually after overwintering, they send out stalked stromatic heads in which the perithecia arise. The ascospores infect the flower heads of the host species where an external conidia-bearing layer is produced (*Sphacelia* stage). These conidia are borne by insects or rain to other grasses. Eventually the ovaries are completely filled, or all except a small portion of the upper end, with a firm stromatic mass which may be many times as large as the normal ovary. *C. purpurea* (Fr.) Tul. is the commonest species of ergot. It occurs in cultivated rye, less often in wheat, and in many other grasses. The fresh sclerotia have considerable medicinal value. They are poisonous when eaten in large quantity as often happens in time of famine when highly ergotized rye or wheat is consumed by the underfed populace. Pastures in which this fungus is abundant sometimes cause serious diseased conditions to develop in the animals feeding there. (Figs. 93, 94.)

The author has long contended that the color and consistency of the perithecium or stroma are not at all satisfactory as a basis of distinction of the Sphaeriales and Hypocreales. The structure and mode of development of the perithecium are of far greater value in determining the true relationships. Many of the Hypocreales should be placed in the Sphaeriales and perhaps some in the Pseudosphaeriales. The fact that many, perhaps the majority, of the species of the Hypocreales are described as lacking paraphyses suggests the need of more intensive research upon the

---

FIG. 93. Hypocreales. (A–C) Family Nectriaceae. (A, B) *Nectria cinnabarina* Fr. (A) Portion of tree branch with conidial stromata (*Tubercularia* stage) with and without surrounding perithecia. (B) Section through stroma with two perithecia at one side, the remainder still conidiiferous. (C) *Sphaerostilbe gracilipes* Tul. Section through stroma showing two perithecia and a synnema and conidial head of the *Stilbella* stage. (D–G) Family Clavicipitaceae. (D) *Cordyceps militaris* Link. Caterpillar with several stalked stromata. (E) *Cordyceps ophioglossoides* Link. Asci containing ascospores. (F, G) *Claviceps purpurea* (Fr.) Tul. (F) Section of stromatic head. (G) Section through a single perithecium. (A–E, after L. R. and C. T. Tulasne: Selecta fungorum carpologia, vol. 3, pp. 1–221. F–G, after Tulasne from Engler and Prantl: Die Natürlichen Pflanzenfamilien, Leipzig, W. Engelmann.)

FIG. 94. Hypocreales, Family Clavicipitaceae. *Claviceps purpurea* (Fr.) Tul. Stromata growing from sclerotium. (Courtesy, F. C. Strong.)

development of the ascocarp to determine whether it is a true perithecium or stromatic in structure.

Order Dothideales. In contrast to the undoubted Sphaeriales, i.e., those forms possessing true ostiolate perithecia with asci arising from the bottom and sides of the perithecial wall in a common perithecial cavity, and with periphyses near the ostiole and paraphyses among the asci, are a large number of genera formerly more or less closely associated with that order. Of these the Order Dothideales is recognized as a distinct order by most mycologists. It is mostly defined as consisting of fungi parasitic usually on leaves, producing endophyllous or epiphyllous stromata within which arise perithecial cavities which lack definite perithecial walls. Apparently in this order the ascogenous hyphae arise in the center or base of the stroma and, spreading outward and upward through the stromatic tissues, dissolve out cavities within which the numerous asci are produced. Most of the species are tropical or subtropical, but a few forms included in the order reach the north temperate zone. Theissen and Sydow (1915) in their monograph of this order included four families of which all except the Dothideaceae have been found to have their closer relationship with other orders: the Phyllachoraceae with the Sphaeriales, the Polystomellaceae with the Hemisphaeriales and the Montagnellaceae with the Pseudosphaeriales. This leaves but the single family Dothideaceae with about 34 genera and over 100 species.

The two orders Hemisphaeriales and Pseudosphaeriales are stromatic forms, without true perithecia. In these the asci arise from ascogenous hyphae developed in the midst of a more or less pseudoparenchymatous stromatic tissue. These ascogenous hyphae spread through this stroma making their way by pressure and by dissolution of the tissues, eventually forming cavities within which single asci are produced, separated from one another by a thinner or thicker remnant of the original stromatic tissue. In some cases these stromatic remnants entirely disappear so that a cluster of asci arises in the cavity dissolved by their actions in the stroma, as in the Dothideales. The chief difference from the latter is that but one such locule is formed in a perithecium-like stroma instead of many locules in a more massive stroma. The intervening stromatic tissue, where the asci are close together in monascous cavities, was in many cases formerly mistaken for paraphyses but can be distinguished by the fact that it is fastened above as well as below, and often laterally. Those forms in which the "paraphyses" are described as attached reticulately to each other belong in this series as do those where their tips form a continuous pseudoparenchyma above the apices of the asci. Such structures must be sharply distinguished from the epithecium found in many Tuberales and Lecanorales and in some other groups of fungi in which true paraphyses overtop the asci and fuse with one another above the latter. The asci are never operculate. They are usually much thickened, at least upward, and are obovoid or clavate, rarely slender and cylindrical. The ascospores vary from one-celled and hyaline to phragmosporous or muriform, sometimes hyaline and sometimes colored. The stromata may resemble simple perithecia, the central portion of the apex breaking away as a pseudoostiole. The asci may arise parallel in a row at the bottom of the cavity or they may arise in a fan-shaped cluster from a raised "placenta" at the center of the base. In the latter type there are usually no paraphysis-like remnants of the stromatic tissue, this having been dissolved or pushed back as the asci grew, while in the former type these fragments usually persist until the maturity of the asci. Petrak (1923), Gäumann (1928), and others are inclined to consider all perithecium-like structures with a spreading, nonparaphysate basal cluster of asci as pseudosphaeriaceous, even though the spore fruit may appear to possess a true perithecial wall and a typically developed ostiole lined with periphyses. Whether such forms are in reality intermediate, as they believe, between the Pseudosphaeriales and the Sphaeriales or not, will require much further investigation to determine. It seems clear that the typical structure is very different in the two orders. Perhaps Nannfeldt (1932) is right in questioning the validity of the assumption of such intermediate forms, drawing the distinction not so much on the external and structural characters of the wall as on the nature of the hymenium. If true paraphyses (the metaphyses of Petrak) arise from the floor of the perithecium among the slender mostly parallel asci, ending free above, these are to be considered as true Sphaeriales, while the spreading broader asci which entirely crowd back the stromatic tissues, or the rather broad, thick-walled asci which arise in monascous cavities, leaving paraphysis-like threads attached both above and below, indicate their connection to the Pseudosphaeriales, even if the external structures are similar. Unless the latter position is taken we have the anomaly of one genus, *Leptosphaeria*, with some species considered as Pseudosphaeriales, others as intermediate forms, and still others as true Sphaeriales (Petrak, 1923, Gäumann, 1928). The forms with more massive stromata which have at maturity several cavities would find their place better in the Dothideales. Where these perithecium-like stromatic structures arise separately or crowded in clusters we have the Pseudosphaeriales. Where the stromata have been developed superficially on the leaves of the host they are mostly small, hemispherical or disk-like, with the upper surface firmer and the lower portion less firmly developed. They open at the top by a more or less tearing of the tissue. In the interior there may be separate monascous cavities or the central tissues may be dissolved, as in some of the Pseudosphaeriales, leaving a cavity with a cluster of aparaphysate asci. Such fungi form the Order Hemisphaeriales, perhaps the majority of which were formerly included in the old Order Perisporiales. Von Höhnel considers some of the Microthyriaceae to be related to *Meliola* in this latter order.

The more massive stromatic Myriangiales do not show such similarities to the Sphaeriales as either of the foregoing orders. Studies by Miller (1938) demonstrate that this order formerly associated with the Pseudosphaeriales does not belong here but has its closest relationship with the Aspergillales next to which order they are given consideration.

Order Hemisphaeriales. The older interpretation of the fruit bodies was a perithecium with the basal portion poorly developed while a more or less shield-shaped perithecial wall formed the upper half. It is now interpreted as a stroma with a hyphal or pseudoparenchymatous basal portion and a firmer upper part. At the apex, by breaking of the tissues, a pseudo-ostiole is often formed. Another view which once received strong support on the part of a number of mycologists was to consider this structure as a small apothecium with a poorly developed hypothecium and with a more or less permanent and rather late-opening cover, as in the Phacidiales, to which by this theory the order was believed to be related. Theissen and Sydow (1917) held this viewpoint and von Höhnel (1919) believed that certain Microthyriaceae are forms transitional to the Discomyceteae. The order consists of fungi which are entirely superficial or subcuticular or which have a hypodermal stroma connected with an epiphyllous stroma by strands of hyphae emerging through the stomata or other openings. Asexual reproduction is known in a few forms and consists of the formation of conidia from some of the external hyphae or in pycnidial structures. With few exceptions the 111 or more genera and over 300 species are leaf parasites, largely tropical but represented in the temperate zones by a number of genera. Arnaud (1930) has studied the structures of many of these fungi, especially in relation to the host tissues.

Killian (1922) studied the sexual reproduction in *Stigmatea robertiani* Fr., a form previously included in the Family Mycosphaerellaceae of the Sphaeriales. A subcuticular pseudoparenchymatous stromatic layer is produced and in the thicker central portion of this appear several short cells one of which becomes a binucleate oogone with a receptive papilla and another a binucleate antherid. After the fusion of these cells there follow several nuclear divisions and a fusion of male and female nuclei. The resultant diploid nuclei pass out into the ascogenous hyphae which give rise to asci arising from the floor of the ascocarp between the loose stromatic hyphae. This hypothecial floor is pseudoparenchymatous and the top is hemispherical, of radially arranged hyphae which break to leave a central ostiole. Several functional oogones and antherids may be found in each stroma as occurs in formation of the apothecium in *Pyronema*. The aberrant nuclear behavior described by Killian suggests that this process should be reinvestigated.

Luttrell (1940) studied the reproduction in *Morenoella quercina* (Ell. & Mart.) Theissen of the Family Microthyriaceae. This species grows on the leaves of various species of oaks (*Quercus*). The mycelium is superficial and forms a network of fine dark hyphae. By the division and radial growth of several cells of a superficial hypha an elongated shield, one cell in thickness, is produced. On the under side of this shield is developed a plectenchymatous layer, two or three cells thick, of hyaline cells. In this appear ascogenous hyphae of binucleate cells whose terminal cells become the asci. As these enlarge the roof is broken open along a longitudinal slit, and eventually the intervening stromatic tissues are destroyed so that finally the asci stand side by side. While these ascocarps are developing circular spermogonia are produced, with a single top layer of dark cells and a central ostiole and a floor of ovoid hyaline cells which break off successively small rod-shaped spermatia. No union of these spermatia to other cells was observed.

Theissen and Sydow (1917) recognized five families in this order as follows:

Family Stigmateaceae. Upper surface of radially arranged hyphae, arising subcuticularly, vegetative mycelium lacking or almost so. Eleven genera of which *Stigmatea* is the type genus of the family.

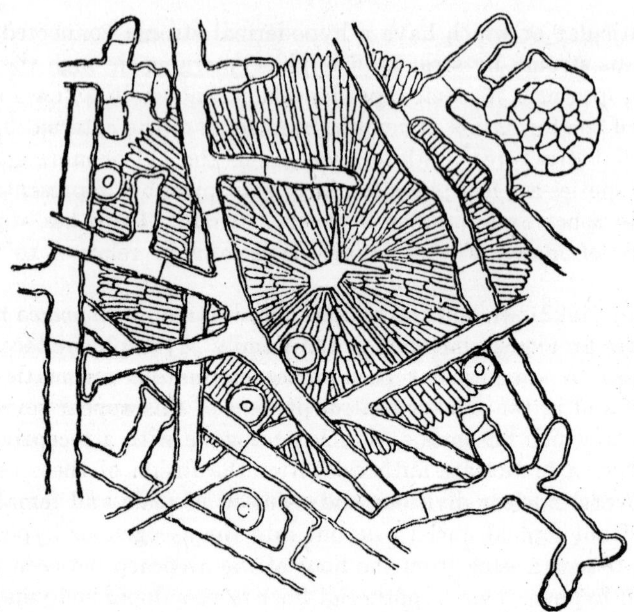

FIG. 95. Hemisphaeriales, Family Microthyriaceae. *Asterina camelliae* Syd. & Butl. Perithecium and mycelium. (After Theissen and Sydow: *Ann. Mycol.*, **15**(6):389–491.)

FAMILY POLYSTOMELLACEAE. Stromata with radial structure as in the preceding, but external to the cuticle and arising from an internal mycelium ("hypostroma") from which emerge strands through the epidermis at various points to give rise to the stromata. There are 39 genera of which 20 form very narrow perithecia which formerly led to their being placed in the Hysteriales. *Parmularia* (perhaps more correctly named *Schneepia*) belongs here. *Polystomella* is the type genus of the family.

FAMILY MICROTHYRIACEAE. Stromata with radial structure, vegetative mycelium and stromata entirely superficial. There are 36 genera with over 150 species; all but a few are leaf parasites. Here and there, on the more or less reticulately arranged coarse brown vegetative mycelium (which is lacking in a few genera), appear the almost lens-shaped stromata. In each, under the radial centrally ostiolate cover, is a hymenium of vertically standing asci intermingled with conspicuous or inconspicuous (rarely lacking) paraphysis-like remnants of the stromatic tissues, which in a few cases form a definite epithecium-like layer. The stromata are mostly round but are in some cases laterally compressed. Among the forms without vegetative mycelium is the genus *Microthyrium* in which the stromata appear as little black superficial dots on the leaves or stems of various plants. *Asterina* forms small round stromata and *Lembosia* linear stromata in the brown vegetative mycelium on the surface of the host. (Fig. 95.)

FAMILY TRICHOPELTACEAE. The conspicuous mycelium is radial in arrangement, or forms sterile parallel hyphae. The cover of the stroma appears to be merely a local thickening of the vegetative mycelium. Paraphysis-like threads are lacking. Six genera are known and from 10 to 15 or more species, all tropical.

FAMILY HEMISPHAERIACEAE. Mycelium lacking or reticulate, superficial. Cover of the stroma not radial in structure. Under the cover there may be a single hymenium with or without paraphysis-like threads or several smaller hymenia may be produced under one cover. In some cases these are reduced to a considerable number of "monascous hymenia," i.e., embedded in the hypothecium are scattered single asci. Nineteen genera of mostly tropical fungi. *Micropeltis* occurs in the Old and New World tropics on leaves. Its stroma contains a single hymenium with many asci and with paraphysis-like structures and with hyaline ascospores of four or more cells.

Order Pseudosphaeriales. The fungi included in this order have been segregated from the Sphaeriales, Perisporiales, and Dothideales, mainly, and even from the Pezizales. Their true relationship is not known; indeed it is doubtful whether all the genera assigned to this order are really related. They are largely tropical, but many occur in temperate regions. They may be parasitic on plants or even on insects, or saprophytic. The fruiting bodies resemble superficially the perithecia or stromata of the Sphaeriales or some Dothideales. They are almost external or at least their tops become external by rupture of the host tissues. They are distinguished from all the preceding orders (except a few species of the Hemisphaeriaceae with "monascous hymenia") by the mode of occurrence of the asci. These arise in separate stromatic cavities, one ascus to each cavity. In the majority of genera they are oblong or nearly spherical and mostly eight-spored. They appear to develop somewhat as follows: Within a pseudoparenchymatous stromatic structure arise branching ascogenous hyphae, probably—in many cases, if not in all—from an ascogonium. These hyphae grow out into the stromatic tissue, dissolving it so that eventually each terminal ascus lies in a cavity of the original sterile tissue. These asci may be separated rather widely or the separating tissue may be but a thin sheet of cells. The developing asci may arise in a fan-shaped cluster destroying the stromatic tissue as they enlarge. The ascospores vary from hyaline to brown and from one-celled to many-celled, in many genera being muriform. The fact that the many-celled type of ascospore is the most typical for the order, the one-celled hyaline spore being found only in two genera of the supposed transitional family Dothioraceae, casts doubt upon the idea that the Pseudosphaeriales are a somewhat primitive order from which have arisen the Sphaeriales. The asci become exposed by the weathering away of the outer part of the stroma or of its apical portion. Rarely the central apical tissues dissolve away to form an ostiole. In the forms with a small perithecium-like stroma the tissues between the ascus locules break or dissolve away leaving shreds that have been taken for paraphyses, so that the numerous asci appear to stand in a true perithecial cavity. Careful investigation has shown that many species formerly assigned to the genera *Pleospora* and *Leptosphaeria* (Family Pleosporaceae, Order Sphaeriales) have the foregoing structure, hence must be transferred to this order. Just how far this may apply to the many remaining species of these and other genera can be determined only by careful study of the development of the young ascocarps. To say, as does Gäumann, that these represent a transition from one order to the other may represent the truth but the author prefers to reserve judgment until further light is thrown on the subject by ontogenetic investigations throughout the various genera of the Sphaeriales. It is possible that these are transitional forms but that evolution has progressed in the contrary direction, from the Sphaeriales to the Pseudosphaeriales. Sexual reproduction has not been investigated in enough undoubted members of the order to enable the information gained to be used in classification. Conidia are produced in a number of families.

Theissen and Sydow recognize several families. The most important are the following:

Family Pseudosphaeriaceae. The stromata resemble separate perithecia of the Sphaeriales and open at the apex either by crumbling of the tissues or by their dissolution to produce an ostiole. The asci are clustered at the base, separated by paraphysis-like hyphae which are attached at the top as well as bottom. Or it may be that the interior tissues and part or all of the interlocular tissues may dissolve, leaving a cluster of asci at the base of a hollow perithecium-like structure so that with mature ascocarps it can not be determined easily whether the fungus belongs to the Pseudosphaeriales or to the Sphaeriales. Among the fungi assigned here with more or less confidence are some species of *Pyrenophora*, of which *P. teres* (Died.) Drechsl. (*P. trichostoma*) is the perfect stage of *Helminthosporium teres* Sacc., the cause of net blotch of barley. In this genus the ascocarps are hairy and the phragmosporous or muriform ascospores colored. Whether *Dothiora* should be included here or in a separate family is doubtful. In its general structure it resembles somewhat the foregoing but the perithecium-like stromata are broader and flatter and the muriform ascospores hyaline. (Fig. 96.)

Family Mycosphaerellaceae. In the Mycosphaerellaceae the genera *Guignardia* and *Mycosphaerella* deserve attention. *G. bidwellii* (Ellis) V. & R. is the cause of the very destructive black rot of the grape.

Native to North America it has been introduced into Europe where it has caused great damage to the more susceptible *Vitis vinifera* L. The fungus attacks the leaves on which it produces brown dead spots and the fruits which become dark-colored and shrunken. In these leaf spots and

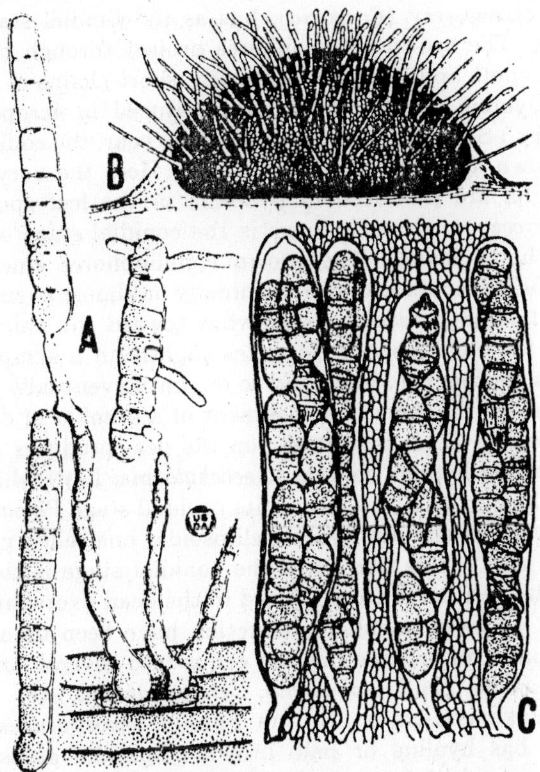

FIG. 96. Pseudosphaeriales, Family Pseudosphaeriaceae. *Pyrenophora teres* (Died.) Drechsl. (A) Conidiophores and conidia (*Helminthosporium teres* Sacc.). (B) Perithecium. (C) Asci in cavities dissolved in the stromatic tissue. (Courtesy, Drechsler: *J. Agr. Research*, **24**(8):641–740.)

shriveled berries are produced numerous pycnidia in which arise hyaline, ellipsoidal spores (*Phoma* stage of the fungus). In the leaves and berries which overwinter on the ground there develop perithecium-like structures in which are produced clusters of eight-spored asci. Paraphyses are not present. Each hyaline ascospore just at maturity forms a septum dividing it into two quite unequal parts. The formation in pycnidium-like structures of "microconidia" which do not appear to be capable of germination leads to the suggestion that we should search here for fertilization of a trichogyne by sperm cells. The genus *Mycosphaerella* contains over 1000 species, many of them parasites of great economic importance. The ascospores are hyaline or pale green with a nearly median septum. The conidial forms are of several types. *M. fragariae* (Schw.) Lind., which causes the leaf spot of strawberry (*Fragaria*), has as its conidial stage *Ramularia tulasnei* Sacc. The brown conidiophores project through the stomata of the diseased spots and produce terminally short chains of cylindrical or rod-shaped hyaline conidia which are produced in acropetal order. In *M. sentina* (Fr.) Schroet., on the leaves of the pear, the conidial stage was formerly known as *Septoria piricola* Desm. Here the very long slender hyaline conidia are produced in pycnidia in the leaf spots. The form known as *Cercospora cerasella* Sacc. is the conidial stage of *M. cerasella* Aderh. on cherry leaves. The brown conidiophores emerge from the stomata in the leaf spots and bear terminally or almost so single elongated several-celled conidia, tapering somewhat toward the apical end. When one falls off the conidiophore elongates slightly in a sympodial manner and produces another conidium, and so on, until eventually an old conidiophore may show the scars of attachment of a number of conidia. *Mycosphaerella pinodes* (B. & Bl.) Stone, on the pea, produces pycnidia containing two-celled hyaline conidia (*Ascochyta pisi* Lib.) while *M. tabifica* (P. & D.) Johns., of the beet, has as its conidial stage *Phoma betae* Fr., in which the pycnidia contain hyaline ellipsoidal one-celled conidia. In all these species and others like them the conidial stage is the destructive stage while the perithecia are produced in the dead overwintering tissues. The different types of conidial production have been used by Klebahn (1918) as a basis for segregating the genus into several subgenera, e.g., *Ramularisphaerella*, *Septorisphaerella*, *Cercosphaerella*, etc.

Family Pleosporaceae. In the Family Pleosporaceae the genus *Physalospora* has hyaline or pale brown ellipsoidal ascospores. In *P. cydoniae* Arnaud, which forms its perithecial stage only on dead twigs, the conidial stage is the destructive *Sphaeropsis malorum* Pk., which causes the black rot of the fruit and the twig blight and canker of the apple and quince. The conidia are produced in pycnidia and are large, ellipsoidal and dark-colored when mature, sometimes becoming uniseptate when old. *Venturia inaequalis* (Cke.) Wint., the cause of the scab on apple leaves and fruits, forms its perithecia in the overwintered leaves infected the previous summer. Its two-celled, slightly colored ascospores are expelled sometimes to a height of 15 mm. The conidial stage (*Fusicladium dendriticum* (Wallr.) Fckl.) develops subcuticularly on the leaves and fruit. In this species both Killian (1917) and Frey (1924) have shown that a well-developed antherid unites with the trichogyne which terminates the coiled ascogonium and several male nuclei pass into it and by successive dissolution of the intervening septa reach the oogone cell. Subsequently ascogenous hyphae are produced. *Pleospora* has dark-colored muriform ascospores. In some species the conidia belong to the form genus *Stemphylium* in which the muriform conidia are produced singly and in other species the conidia are produced in acropetal chains belonging to the form genus *Alternaria*.

FAMILY BOTRYOSPHAERIACEAE. Probably also to be included in the Order Pseudosphaeriales are the families Botryosphaeriaceae and Cucurbitariaceae. These differ from the foregoing families by possessing a more massive stromatic ascocarp. The ascocarps of these two families possess a subcortical hypostroma the upper part of which bursts through the bark and develops as a more or less spherical, perithecium-like body. In the Botryosphaeriaceae the thick wall has a thin blackened and carbonaceous outer layer and the asci are broadly clavate.

FAMILY CUCURBITARIACEAE. In this family the main portion of the stroma, the hypostroma, is well developed or consists of a limited mass of mycelium intermingled with the remnants of the host tissue. From the surface of this basal stroma there bulge out numerous, crowded, more or less spherical stromatal projections or pseudoperithecia in which the development of asci occurs without the production of true paraphyses. These ascocarps have thinner walls than those of the foregoing family, with a thinner blackened outer layer. Most members of the family are saprophytic on wood and bark but possibly some are parasitic. In some species conidia are produced in pycnidia. *Cucurbitaria* is the largest and most frequently observed genus. It is placed by Theissen and Sydow (1916) in this order.

The relationships and phylogeny of the orders considered in this chapter are not at all certain. Theissen and Sydow, Gäumann (1928), and Petrak (1923) would derive the Sphaeriales from the Pseudosphaeriales which in turn they would derive from the Myriangiales and these from near the Aspergillales. The author is inclined to consider that the simpler forms of Sphaeriales and Hypocreales, with hyaline or light-colored, one-celled ascospores and a definite hymenial layer are the more primitive and that from these have arisen the other orders. This will be discussed more fully in Chapter 17.

Key to the Orders of Ostiolate "Pyrenomycetes"

Ascocarps are true perithecia with or without marked stromatic outer layer or surrounding stromatic tissues. Periphyses and true ostioles usually present. True paraphyses present or early disappearing, in some cases apparently entirely lacking.

Perithecia with dark-colored, usually firm walls. Not lichen forming. Order Sphaeriales

Perithecia dark-colored. Lichen forming. Order Pyrenulales

Perithecia and sustaining or enclosing stromata, when present, light-colored or red or blue. Not lichen forming. Order Hypocreales

Ascocarps dothideal, i.e., consisting of cavities dissolved in a fairly massive internal or external stroma, each such perithecial cavity containing a cluster of parallel or diverging asci separated by the paraphysis-like remains of the stromatic tissue, or these disappearing very early. Ostioles formed by the breaking or dissolving away of the apical portion of the stroma above the perithecial cavity. Order Dothideales

Ascocarps pseudosphaeriaceous, i.e., consisting of perithecium-like stromata within which the asci are formed in separate monascous cavities, or the intervening tissues dissolve very early so that the asci form a parallel or diverging cluster in a single large cavity. No true paraphyses. Ostioles formed by the breaking or dissolving of the apical portions of the stromata.
Order Pseudosphaeriales

Ascocarps external, consisting of usually flattened stromata with a thin outer, colored crust, the basal portion pseudoparenchymatous, with thin-walled cells. The asci are formed in monascous cavities. Mycelium mostly external or subcuticular, often very slightly developed and not easily detected.
Order Hemisphaeriales

Ascocarp obviously stromatic with rounded asci single in cavities which are in one or two layers or irregularly scattered.
Order Myriangiales (see Chap. 11)

Because of the large number of genera in the families the following keys are made only for the families. A few of the more important genera are merely mentioned under each family but not keyed out.

Key to the More Important Families of Order Sphaeriales

Perithecial walls thin.

Perithecia light brown, rarely colorless, superficial or sunk in a subiculum, ostiole with papilla or well-developed neck. Paraphyses wanting. Ascospores dark-colored, discharged at maturity.
Family Melanosporaceae

Melanospora, Neurospora, Gelasinospora.

Perithecia dark brown, superficial or sunk in a subiculum, rarely partially sunk in the substratum or in a stroma, not conspicuously hairy. Ascospores dark-colored, with slime coat or appendages, paraphyses present, ascospores discharged through the ostiole.
Family Fimetariaceae (Sordariaceae)

Fimetaria (*Sordaria*), *Schizothecium* (*Pleurage*), *Hypocopra, Sporormia.*

Perithecia dark, hairy all over and with a tuft of periostiolar bristles. Ascospores dark-colored. Paraphyses wanting. Asci dissolving at maturity.
Family Chaetomiaceae

Chaetomium, Ascotricha.

Perithecial walls firm and dark.

Perithecia superficial or sitting in a subiculum, ostiole simple or in a low papilla. Paraphyses mostly present. Ascospores hyaline or colored, one-celled or several-celled. Asci not dissolving at maturity.
Family Sphaeriaceae

Zignoella, Melanomma, Rosellinia.

Perithecia superficial or slightly sunken, with a long neck. Ascospores one-celled or several-celled, hyaline or colored, paraphyses usually present, asci not dissolving at maturity.
Family Ceratostomataceae
Ceratostoma, Ceratosphaeria.

Much like the foregoing but the paraphyses wanting and the asci dissolving at maturity and the hyaline one-celled ascospores exuding in a drop at the tip of a long neck.
Family Ophiostomataceae
One genus *Ophiostoma* (*Ceratostomella*).

Perithecia in a cluster on a more or less well-developed subcortical stroma.
Family Cucurbitariaceae
(see Order Pseudosphaeriales)

Base of perithecium remaining sunk in the substratum at maturity.
Ostiole laterally compressed.
Family Lophiostomataceae[3]
Lophiostoma, Lophiotrema, Schizostoma.
Ostiole circular. Family Amphisphaeriaceae[3]
Amphisphaeria, Caryospora, Winteria, Strickeria.

Perithecia remaining entirely enclosed in the substratum except the projecting ostiole. No stroma present except in *Glomerella*.
Asci thickened at the top, with a distinct pore; ostiole mostly with a distinct neck. Family Gnomoniaceae
Gnomonia, Glomerella.
Asci not apically thickened and without pore; ostiole simple or with a low papilla. Families Mycosphaerellaceae and Pleosporaceae
(see Order Pseudosphaeriales)

Perithecia immersed in a stroma, mostly caulicolous.
Stroma in part consisting of remains of the host tissue.
Asci with long tapering stalks, not deliquescent, ascospores mostly allantoid. Family Allantosphaeriaceae
Diatrype, Eutypella, Anthostoma.
Stalks of the ascus dissolving early, ascospores ellipsoidal, fusoid, long cylindrical, less often allantoid.
Family Diaporthaceae
Diaporthe, Valsa, Leucostoma, Endothia.
Stroma consisting entirely of fungous tissue, eventually external, asci cylindrical, paraphyses abundant, ascospores dark.
Family Xylariaceae
Hypoxylon, Daldinia, Poronia, Xylaria.

Perithecia in an endophyllous stroma, the perithecial walls well developed, especially near the ostiole, where they pierce the clypeus-like portion of the stroma. Leaf parasites. Largely tropical.
Family Phyllachoraceae
Phyllachora, Ophiodothella.

Key to the Families of Order Hypocreales

Perithecia without stroma or external to a stroma; ascospores ellipsoid to cylindrical, one- to many-celled. Family Nectriaceae
Nectria, Gibberella, Sphaerostilbe, Scoleconectria, Thyronectria.

---

[3] Possibly belongs to the Order Pseudosphaeriales.

Perithecia buried in a stroma, ascospores as above. Family Hypocreaceae
Hypocrea, Polystigma, Hypomyces.
Perithecia buried in a stroma with poorly developed perithecial walls, sometimes only slightly developed near the ostiole. Ascospores filiform.
Family Clavicipitaceae
Cordyceps, Claviceps, Balansia, Epichloe.

Key to Some Families of Order Pyrenulales[4]

Perithecia opening through round ostioles; not embedded in a stroma.
Perithecium more or less immersed in the crustose thallus. Paraphyses remaining distinct. Algal host *Trentepohlia*. Mostly on trees, rarely on rocks.
Family Pyrenulaceae
Pyrenula.
Perithecia more or less immersed in the crustose thallus. Paraphyses mostly gelatinizing early. Algal host *Pleurococcus*. Mostly on rocks, rarely on trees.
Family Verrucariaceae
Verrucaria.
Perithecia immersed in the squamulose or foliose thallus. Paraphyses gelatinizing early. Algal host *Pleurococcus*. Mostly on soil or rocks.
Family Dermatocarpaceae
Dermatocarpon.
Perithecia embedded in a stroma. Thallus crustose. Algal host *Trentepohlia.*
Perithecia opening separately through round ostioles. Paraphyses remaining distinct. On trees. Family Trypetheliaceae
Trypethelium.
Perithecia arranged radially in the stroma with the ostiolar necks converging and opening separately or through a common ostiole. Paraphyses remaining distinct. Mostly on trees. Family Astrotheliaceae
Astrothelium.
Perithecia in groups of two (rarely one) or more, the intervening walls sometimes partially missing; the large, irregular ostiole sometimes serving two or more perithecia. Paraphyses poorly developed or disappearing early or persisting. Thallus crustose. Algal host *Palmella* or *Trentepohlia*. On trees.
Family Mycoporaceae
Mycoporum.

Key to the Families of Order Dothideales

Only recognized family.[5] Family Dothideaceae
Dothidea, Dothidella, Systremma.

Key to the Families of Order Pseudosphaeriales

Stromata small, internal or eventually external, perithecium-like. Asci more or less parallel, formed in monascal cavities which long remain separate by the persisting intervening stromatal tissues which connect from the base to the top of the stroma. Family Pseudosphaeriaceae
Pyrenophora, Dothiora.

---

[4] Careful studies are needed to determine whether the ascocarps in this order are true perithecia or pseudoperithecia. The distinction of families as usually made is largely based on the kinds of algal hosts as well as the habit of the thallus. Only part of the 10 or more families are mentioned in this key.

[5] For Phyllachoraceae, see Order Sphaeriales.

Stromata small, perithecium-like, subepidermal, often eventually external. At maturity only a single large cavity with a cluster of spreading asci, the intervening stromatic tissues entirely destroyed or remaining as remnants between the basal portions of the asci. Family Mycosphaerellaceae

Mycosphaerella, Didymellina, Guignardia.

Similar to the foregoing but the asci mostly parallel with more pronounced remnants of stromatic tissues between them. Family Pleosporaceae

Physalospora, Venturia, Pleospora, Leptosphaeria.

Stromata more pronounced, subcortical or subepidermal. Walls surrounding the large central perithecial cavities thick.

Outer stromatal wall with thick blackened external layer. Asci broadly clavate. The pseudoperithecia single or scattered. Family Botryosphaeriaceae

Botryosphaeria.

Outer wall with thin blackened layer. Asci elongated, nearly cylindrical. Pseudoperithecia densely clustered on the small or large basal portion of the stroma. Family Cucurbitariaceae

Cucurbitaria.

Key to the Families of Order Hemisphaeriales

(*This follows Theissen and Sydow*, 1917, *and probably needs radical rearrangement*)

Ascocarps with more or less pronounced radial structure.

Ascocarps arising subcuticularly but emerging, at least at maturity. Vegetative mycelium scanty or lacking. Family Stigmateaceae

Stigmatea.

Ascocarps produced externally on strands of hyphae from the internal mycelium ("hypostroma"). Family Polystomellaceae

Polystomella, Parmularia (*Schneepia*).

Vegetative mycelium and ascocarps entirely superficial. Ascocarps round or laterally compressed. Family Microthyriaceae

Microthyrium, Asterina, Lembosia.

Mycelium conspicuous, external, radial or forming parallel ribbons of closely united hyphae. Cover of the ascocarp arises as a thickening of the vegetative mycelium and is radial at least toward the margin. Family Trichopeltaceae

Trichopeltis.

Ascocarps not showing radial structure. Mycelium reticulate and superficial or almost lacking. Family Hemisphaeriaceae

Micropeltis.

Literature Cited

AMES, L. M.: An hermaphroditic self-sterile but cross-fertile condition in Pleurage anserina, *Bull. Torrey Botan. Club*, **59**(6):341–345. *Fig.* 1. 1932.

———: Hermaphroditism involving self-sterility and cross-fertility in the Ascomycete Pleurage anserina, *Mycologia*, **26**(5):392–414. *Figs.* 1–6. 1934.

ARNAUD, G.: Les Astérinées: IV. Études sur la systématique des champignons pyrénomycètes, *Ann. sci. nat. Botan.*, Xme sér., **7**:643–723. *Pls.* 1–16. *Figs.* 1–25. 1925.

———: Les Astérinées: V. Étude sur les champignons parasites: Caliciacées, Hémisphériacées, etc., *Ann. épiphyt.*, **16**:235–302. *Pls.* 1–14. *Figs.* 1–15. 1930.

Aronescu, Alice: Further studies in Neurospora sitophila, *Mycologia*, **25**(1):43–54. *Pl.* 14. 1933.

Berlese, Augusto Napoleone: Rapporti tra Dematophthora e Rosellinia, *Riv. patol. vegetale*, **1**(1):5–17; (2):33–46. *Pls.* 1–3. 1892.

Bessey, Ernst A.: An undescribed species of Ophiodothella on Ficus, *Mycologia*, **11**(2):55–57. *Pl.* 5. 1919.

Blackman, V. H., and E. J. Welsford: The development of the perithecium of Polystigma rubrum DC., *Ann. Botany*, **26**(106):761–767. *Pls.* 70–71. 1912.

Brown, H. B.: Studies in the development of Xylaria, *Ann. Mycol.*, **1**(1):1–13. *Pls.* 1–2. 1913.

Buisman, Christine: *Ceratostomella ulmi*, de geslachtelijke vorm van *Graphium ulmi* Schwartz, *Tijdschrift over Plantenziekten*, **38**(1):1–5. *Pls.* 1–3. 1932.

Dade, H. A.: Ceratostomella paradoxa, the perfect stage of Thielaviopsis paradoxa (DeSeynes) von Höhnel, *Brit. Mycol. Soc. Trans.*, **13**:184–194. *Pls.* 10–12. 1928.

Dodge, B. O.: Nuclear phenomena associated with heterothallism and homothallism in the Ascomycete Neurospora, *J. Agr. Research*, **35**(4):289–305. *Pls.* 1–3. *Figs.* 1–5. 1927.

———: Spermatia and nuclear migrations in Pleurage anserina, *Mycologia*, 28(3):284–291. *Figs.* 1–2. 1936.

———, and Bernice Seaver: The combined effects of the dominant and recessive lethals for ascus abortion in Neurospora, *Am. J. Botany*, **25**(3):156–166. *Figs.* 1–7. 1938.

Dowding, E. Silver: The sexuality of the normal, giant and dwarf spores of Pleurage anserina (Ces.) Kuntze, *Ann. Botany*, **45**(177):1–14. *Pl.* 1. *Figs.* 1–10. 1931.

———: Gelasinospora, a new genus of Pyrenomycetes with pitted spores, *Can. J. Research*, **9**(3):294–305. *Pls.* 1–3. *Figs.* 1–9. 1933.

———, and A. H. Reginald Buller: Nuclear migration in Gelasinospora, *Mycologia*, **32**(4):471–488. *Figs.* 1–6. 1940.

Drechsler, Charles: Some graminicolous species of Helminthosporium, *J. Agr. Research*, **24**(8):641–740. *Pls.* 1–33. 1923.

Edgerton, C. W.: Plus and minus strains in the genus Glomerella, *Am. J. Botany*, **1**(5):244–254. *Pls.* 22–23. *Fig.* 1. 1914.

Elliott, John A.: A cytological study of Ceratostomella fimbriata (E. & H.) Elliott, *Phytopathology*, **15**(7):417–422. *Pls.* 15–16. 1925.

Ellis, J. B., and B. M. Everhart: The North American Pyrenomycetes, iii + 793 pp. 41 *pls.* Newfield, N. J., published by the authors, 1892.

Falck, Richard and Olga: A new class of Ascomycetales. A contribution to the orbis vitae system of fungi, *Palestine Jour. Bot. Rehovot ser.*, **6**(1–2):89–106. *Figs.* 1–5. 1947.

Frey, Charles N.: The cytology and physiology of Venturia inaequalis (Cooke) Winter, *Trans. Wisconsin Acad. Sci.*, **21**:303–343. *Pls.* 10–11. 1924.

Gäumann, Ernst Albert: Comparative Morphology of Fungi. Translated by Carroll William Dodge, xiv + 701 pp. 406 *figs.* 43 *diagrams.* New York, McGraw-Hill Book Co., 1928.

Greathouse, Glenn A., and L. M. Ames: Fabric deterioration by thirteen described and three new species of Chaetomium, *Mycologia*, **37**(1):138–155. *Figs.* 1–7. 1945.

Greis, Hans: Entwicklungsgeschichte von Sordaria fimicola (Rob.), *Botan. Arch.*, **38**(2):113–151. 4 *pls.* 5 *figs.* 1936.

———: Befruchtungsvorgänge in der Gattung Chaetomium, *Jahrb. wiss. Bot.*, **90**(2):233–254. *Figs.* 1–11. 1941.

Griffiths, David: The North American Sordariaceae, *Mem. Torrey Botan. Club*, **11**:1–134. *Pls.* 1–19. *Figs.* 1–6. 1901.

Higgins, Bascombe Britt: Morphology and life history of some Ascomycetes with special reference to the presence and function of spermatia, III, *Am. J. Botany*, **23**(9):598–602. *Figs.* 1–13. 1936.

von Höhnel, Franz: Mycologische Fragmente: CCLXII. Über die allantoidsporigen Sphaeriaceen, *Ann. Mycol.*, **16**(1–2):127–132. 1918.

———: Über Discomyceten vortäuschende Microthyriaceen, *Ber. deut. botan. Ges.*, **36**(8):465–470. 1919.

Keitt, G. W., and M. H. Langford: Venturia inaequalis (Cke.) Wint.: I. A groundwork for genetic studies, *Am. J. Botany*, **28**(9):805–820. *Figs.* 1–5. 1941.

———, and D. H. Palmeter: Heterothallism and variability in Venturia inaequalis, *Am. J. Botany*, **25**(5):338–345. *Figs.* 1–5. 1938.

Killian, Karl: Über die Sexualität von Venturia inaequalis (Cke.) Ad., *Z. Botan.*, **9**:353–398. *Figs.* 1–22. 1917.

———: Sur la sexualité de l'ergot de seigle, le Claviceps purpurea Tulasne, *Bull. soc. mycologique de France*, **35**:182–197. *Pls.* 10–17. 1919.

———: Le développement du Stigmatea robertiani Fries, *Rev. gén. botan.*, **34**:577–588. *Pls.* 14–17. *Fig.* 1. 1922.

Klebahn, H.: Haupt- und Nebenfruchtformen der Ascomyceten, 395 pp. *Illustrated.* Leipzig, Gebrüder Borntraeger. 1918.

Lieneman, Catherine: Host index of the North American species of the genus Cercospora, *Ann. Missouri Botan. Garden*, **16**(1):1–52. 1929.

Lindau, G.: Pyrenomycetineae, in A. Engler und K. Prantl: Die Natürlichen Pflanzenfamilien, Erster Teil, Abt. **1**:321–491. *Figs.* 228–288. 1897.

Lindegren, Carl C.: The genetics of Neurospora: I. The inheritance of response to heat treatment, *Bull. Torrey Botan. Club*, **59**(2):85–102. *Figs.* 1–4. 1932; II. Segregation of sex factors in the asci of N. crassa, N. sitophila and N. tetrasperma, *ibid.*, **59**(3):119–138. *Figs.* 1–5. 1932; III. Pure bred stocks and crossing over in N. crassa, *ibid.*, **60**(3):133–154. *Pl.* 9. *Figs.* 1–6. 1933; IV. The inheritance of tan versus normal, *Am. J. Botany*, **21**(2):55–65. *Pl.* 1. *Fig.* 1. 1934; V. Self-sterile bisexual heterokaryons, *J. Genetics*, **28**(3):425–435. 1934.

———: A six-point map of the sex-chromosome of Neurospora crassa, *J. Genetics*, **32**(2):243–256. *Pl.* 14. *Figs.* 1–2. 1936a.

———: The structure of the sex-chromosome of Neurospora crassa, *J. Heredity*, **27**(7):251–259. *Figs.* 1–4. 1936b.

Lupo, Patsy: Stroma and formation of perithecia in Hypoxylon, *Botan. Gaz.*, **73**(6):486–495. *Pl.* 18. *Figs.* 1–7. 1922.

Luttrell, E. S.: Morenoella quercina, the cause of leaf spot on oak, *Mycologia*, **32**(5):652–666. *Figs.* 1–13. 1940.

Miller, Julian H.: Biologic Studies in the Sphaeriales, *Mycologia*, **20**(4):187–213, (6):305–339. *Pls.* 21–22, 35–38. *Figs.* 1–3. 1928.

———: Studies in the development of two Myriangium species and the systematic position of the order Myriangiales, *ibid.*, **30**(2):158–181. *Figs.* 1–4. 1938.

———: The Ascomycetes of Georgia, *Plant Disease Reptr., Supplement*, **131**: 31–93. 1941.

———: A revision of the classification of the Ascomycetes with special emphasis on the Pyrenomycetes, *Mycologia*, **41**(2):99–127. *Figs.* 1–37. 1949.

Mittmann, Gertrud: Kulturversuche mit Einsporstämmen und zytologische Untersuchungen in der Gattung Ceratostomella, *Jahrb. wiss. Botan.*, **77**(2): 185–219. *Figs.* 1–45. 1932.

NANNFELDT, J. A.: Studien über die Morphologie und Systematik der nichtlichenisierten inoperculaten Discomyceten, *Nova Acta Regiae Soc. Sci. Upsaliensis*, ser. IV, **8**(2):1–368. *Pls.* 1–20. *Figs.* 1–47. 1932.

NIENBURG, WILHELM: Zur Entwicklungsgeschichte von Polystigma rubrum DC., *Z. Botan.*, **6**(5):369–400. *Figs.* 1–17. 1914.

ORTON, C. R.: Studies in the morphology of the Ascomycetes: I. The stroma and compound fructification of the Dothideaceae and other groups, *Mycologia*, **16**(2):49–95. *Pls.* 7–9. 1924.

PETRAK, F.: Mycologische Notizen V: 200. Über die Pseudosphaeriaceen v. H. und ihre Bedeutung für die spezielle Systematik der Pyrenomyzeten, *Ann. Mycol.*, **21**(1–2):30–69. 1923.

———: Mykologische Notizen VII: 301. Über die phylogenetischen Beziehungen der Gattung Phyllachora und ihre Bedeutung für das System der dothidealen Pilze, *Ann. Mycol.*, **22**(1–2):1–10. 1924.

SARTORIS, GEORGE B.: A cytological study of Ceratostomella adiposum (Butl.) comb. nov., the black-rot fungus of sugar-cane, *J. Agr. Research*, **35**(7):577–585. *Figs.* 1–4. 1927.

SHEAR, C. L., AND B. O. DODGE: Life histories and heterothallism of the red bread-mold fungi of the Monilia sitophila group, *J. Agr. Research*, **34**(11): 1019–1042. *Pls.* 1–4. 1927.

STONEMAN, BERTHA: A comparative study of the development of some anthracnoses, *Botan. Gaz.*, **26**(2):69–120. *Pls.* 7–18. 1898.

THEISSEN, F.: Hemisphaeriales, *Ann. Mycol.*, **11**(5):468–469. 1913.

———, UND H. SYDOW: Die Dothideales, *ibid.*, **13**(3–4):149–746. *Pls.* 1–6. 1915.

———, UND ———: Synoptische Tafeln, *ibid.*, **15**(6):389–491. *Figs.* 1–38. 1917.

———, UND ———: Vorentwürfe zu den Pseudosphaeriales, *ibid.*, **16**(1–2):1–34. *Figs.* 1–5. 1918.

TULASNE, LOUIS RENÉ ET CHARLES: Selecta fungorum carpologia, vol. 3, i–xvi, 1–221. *Pls.* 1–22. Paris, Typographie Impériale, 1865.

VARITCHAK, BOGDAN: Contribution à l'étude du développement des Ascomycètes, *Le Botaniste*, **23**:1–183. *Pls.* 1–20. *Figs.* 1–20. 1931.

VINCENS, F.: Valeur taxinomique d'une particularité de la structure des ascospores chez les Xylariacées, *Bull. soc. mycologique de France*, **34**:101–109. *Figs.* 1–4. 1918.

———: Valeur taxinomique du sillon germinatif des ascospores chez les pyrénomycètes, *ibid.*, **37**:29–33. 1921.

WEHMEYER, LEWIS E.: A biologic and phylogenetic study of the stromatic Sphaeriales, *Am. J. Botany*, **13**(10):575–645. *Fig.* 1. 1926.

———: The genus Diaporthe Nitschke and its segregates, Univ. Mich. Studies, Scientific Ser. 9, i–x, 1–349. *Pls.* 1–18. Ann Arbor, Univ. Mich. Press, 1933.

WILCOX, MARGUERITE S.: The sexuality and arrangement of the spores in the ascus of Neurospora sitophila, *Mycologia*, **20**(1):3–17. *Pl.* 2. *Figs.* 1–2. 1928.

WOLF, FREDERICK A.: The perfect stage of Cercospora sordida, *ibid.*, **35**(5):503–509. 1 *fig.* 1943.

11

CLASS ASCOMYCETEAE: ERYSIPHALES, ASPERGILLALES, MYRIANGIALES, SACCHAROMYCETALES

As was true for several other orders of the Class Ascomyceteae the investigations of the last three decades have necessitated numerous changes in the limits of the orders to be taken up in this chapter. Even as presented in this edition much must be recognized as only tentative, for a great deal more very intensive study will be necessary before a satisfactory classification can be attained.

Order Erysiphales (Perisporiales of most authors[1]). Lindau, in Engler and Prantl (1897), placed in this order almost all of the plant-inhabiting Ascomyceteae with external mycelium and with the perithecia also external. Ostioles are mostly lacking except in the Family Microthyriaceae. The latter family has now been transferred to the Order Hemisphaeriales (see Chapter 10) and the remaining two families have been increased to five or six, mainly by the discovery of new forms and by the division of the Perisporiaceae.

In general the Erysiphales carry on a parasitic, less often saprophytic, existence on the surface of the host plant. Exceptions are found in the species feeding on "honeydew" which occur wherever the latter accumulates in sufficient quantity. As parasites the mycelium may not even penetrate through the epidermis but in many cases the epidermal cells of the host are penetrated by haustoria. The mycelium is septate and branched, mostly with uninucleate cells. Rarely no mycelium is visible except that making up the perithecium.

[1] The Order Perisporiales and Family Perisporiaceae are based upon the genus *Perisporium*. In view of the fact that the type species of this genus, *P. gramineum* Fr., has been shown not to belong to this order as customarily limited, it is necessary to base the order upon a generally recognized genus whose connection with the order is beyond doubt. Following the proposal of Gwynne-Vaughan the name Erysiphales has been selected, based upon the genus *Erysiphe*.

Asexual reproduction is lacking or consists of the formation of single conidia or the successive formation of conidia which may separate as fast as formed or remain adherent in chains, the oldest conidium being the terminal one. In the Capnodiaceae and Meliolaceae conidia may be produced in pycnidia. The perithecia arise on the external mycelium or may be partly surrounded by it except in a few cases where they are subcuticular or subepidermal, becoming external through the rupture of the cuticle or epidermis respectively. They do not possess an ostiole in the Erysiphaceae but in the Capnodiaceae ostiolate and inostiolate species occur, in some cases both in the same genus.

Following Theissen and Sydow (1917) this order may be divided into five families as follows:

Erysiphaceae: external, mycelium white (cinnamon-yellow in *Astomella*), conidia hyaline, falling off singly or more often remaining attached in chains. Perithecia without ostioles, external to the host, free or embedded in cottony mycelium, external layer of peridium dark-colored, brittle at maturity, the cells polygonal in outline, with various types of hyphal appendages. Ascospores hyaline (or yellow in one genus).

Meliolaceae: external mycelium colored, reticulately branched, the individual cells cylindrical (not forming a moniliform hypha), sometimes with spines and hyphopodia, not becoming slimy. Conidia mostly in ostiolate pycnidia. Perithecia external (subepidermal in one genus), usually without true ostioles, external cells polygonal, not becoming slimy. Perithecia without true appendages but often with bristles arising from an overlying layer of spiny mycelium. Ascospores rarely one-celled, mostly two- to several-celled, hyaline or more often brown.

Englerulaceae[1]: external mycelium mostly colored, parasitic on leaves or on fungi on the latter. Chief distinction is the release of the asci from the perithecia by the slimy histolysis of the latter.

Capnodiaceae: external mycelium of dark moniliform hyphae or of straight-sided hyphae united laterally into dark sheets. In many cases saprophytic on "honeydew." Conidia in elongated ostiolate pycnidia. Perithecia dark, with or without ostioles, sometimes in the same genus. Walls of the often stalked perithecia of rounded cells or of longitudinal hyphae united by slime.

Trichothyriaceae: mycelium dark-colored, creeping over the epiphyllous mycelium of *Meliola* and other fungi. Perithecium wall of radial hyphae, originating at the tip of an upright hypha which then turns over so that the morphological base of the perithecium is uppermost with the clusters of asci at the upper end. Ascospores two- to several-celled, hyaline or colored.

Very doubtfully belonging in this order is the following:

Atichiaceae: forming small rounded or stellate cushions on leaves, horny when dry, gelatinous when wet, with no free mycelium. Reproducing asexually by clusters of cells (propagula). Perfect stage consists of asci scattered here and there in one level in thickened areas of the thallus.

Arnaud (1925) does not recognize this order but distributes the families here included among various other groups. Thus he places the

[1] See note on this family on p. 319.

Erysiphaceae in what is called here the Order Hypocreales, *Meliola* and *Amazonia* in the Dothideaceae, the Atichiaceae in the Myriangiales and various Capnodiaceae in the Sphaeriales.

FAMILY ERYSIPHACEAE (THE POWDERY MILDEWS). These fungi are parasitic upon Flowering Plants (Anthophyta) the world over, reaching their greatest development in the temperate zones. They are usually confined to leaves and young tissues of other portions of the plant, such as

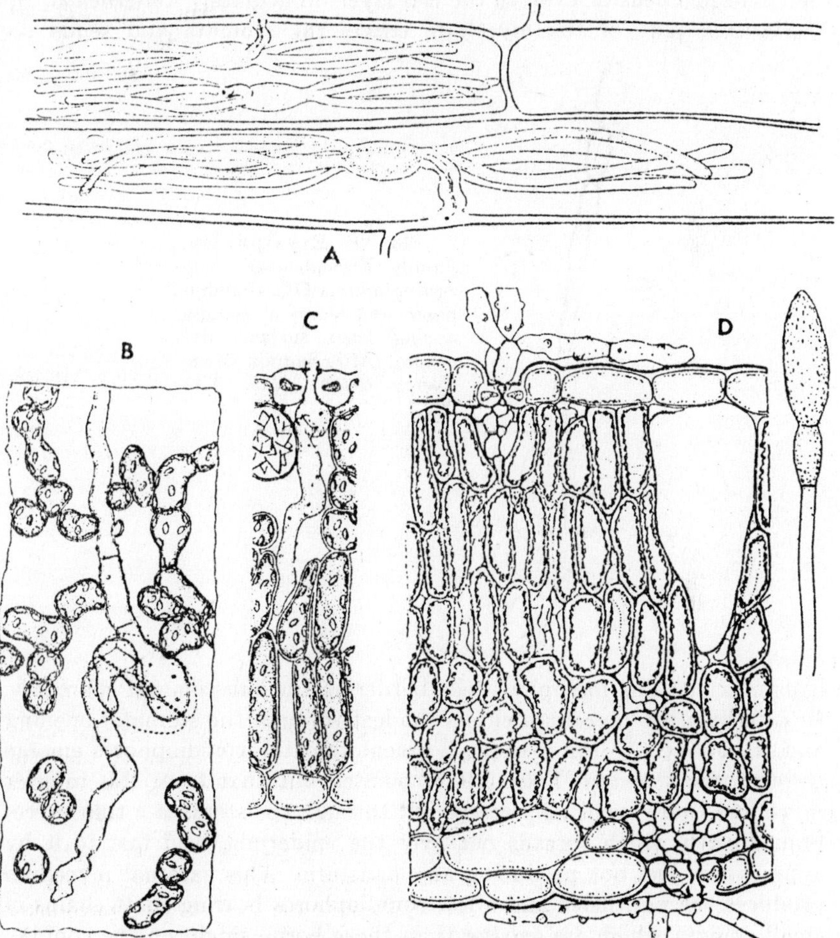

FIG. 97. Erysiphales, Family Erysiphaceae. (A) *Erysiphe graminis* DC. Haustoria in epidermal cells of host, seen from above. (B, C) *Phyllactinia guttata* (Wallr.) Lev. Mycelium in substomatal chamber of host. (D) *Leveillula taurica* (Lev.) Arnaud. External and internal mycelium, and conidiophore. (After Arnaud: *Ann. épiphyt.*, 7:1–115.)

the young shoots, the buds, fruits, etc. *Sphaerotheca phytoptophila* Kell. & Swingle is found only on the lobed galls produced on the hackberry (*Celtis occidentalis* L.) by a species of mite. *Uncinula necator* (Schw.) Burr. attacks not only the leaves and young green shoots but also the immature berries of the grape (*Vitis*) while *Sphaerotheca mors-uvae* (Schw.) B. & C., on the gooseberry (*Grossularia*) is more often on the berries. The mycelium is mainly superficial, obtaining its nourishment by haustoria penetrating the epidermal cells or even to the cell layer immediately underneath. In *Phyllactinia* part of the mycelium enters the stomata and sends its

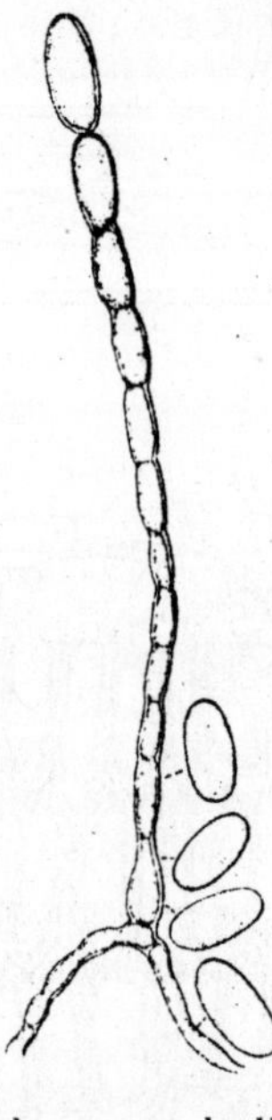

FIG. 98. Erysiphales, Family Erysiphaceae. *Erysiphe graminis* DC. Conidiophore and chain of conidia arising from surface mycelium. (After Salmon: *Mem. Torrey Botan. Club*, 9:1–292.)

haustoria into the mesophyll cells bordering the substomatal chambers. In *Leveillula* the mycelium enters the leaf through the stomatal opening and is confined to the mesophyll except that the conidiophores emerge through the stomata. Foëx (1912) pointed out that from this interior mycelium there eventually creeps out through the stomata a thin mycelium growth which spreads out over the epidermis, held fast to it by appressoria, but not producing any haustoria. This external mycelium produces the perithecia and a few conidiophores bearing short chains of small conidia which are smaller than those borne singly on the conidiophores emerging from the stomata. The mycelial cells are always uninucleate. The conidia arise at the apex of short or elongated conidiophores. In *Leveillula* and *Phyllactinia* the conidium falls off before the next succeeding conidium is formed, but in the remaining genera of the family the conidia remain attached so that a chain of conidia is produced which in

Erysiphe graminis DC. may consist of 20 or even more before the older ones break off. Berlese (1898) reports for this species that after a cell is cut off at the top of the conidiophore it divides into two conidia, the next cell cut off from the conidiophore dividing similarly into two conidia. Thus the terminal two conidia are of equal age and are the oldest pair, followed by younger and younger pairs toward the base of the conidial chain. The conidia are distributed by the wind and germinate on the epidermis of the hosts, producing a short hypha which sends a haustorium into an epidermal cell. Conidia are not described for the genus *Astomella* (Thirumalachar, 1947). (Figs. 97, 98.)

Brodie (1945) summarizes the experiments of himself and others upon the germination of the conidia of Erysiphaceae at different degrees of relative humidity. The conidia of *Sphaerotheca pannosa* (Wallr.) Lev. var. *rosae* Wor. will not germinate at a relative humidity below 95 per cent. On the contrary a considerable percentage of the conidia of *Erysiphe graminis* DC., *E. polygoni* DC., and *Microsphaera alni* (DC.) Wint. germinate in air entirely devoid of moisture, i.e., zero per cent relative humidity. This accounts, at least in part, for the frequently observed phenomenon that some powdery mildews appear abundantly in very dry weather. Possibly also the dryness of the atmosphere keeps down the infection by *Ciccinobolus cesatii* de By. and other fungi that are parasitic upon these mildews. As long as the chains of conidia remain attached to their conidiophores none of the conidia germinate but when detached the terminal conidia at each end of the chain will germinate, but not those between.

After infection has taken place the mycelium grows rapidly, branching in all directions, but gradually spreading radially. Conidial production begins soon and may continue for some time but eventually gives way to the production of perithecia. In some species of powdery mildews conidia only are produced on some hosts, conidia and, later, perithecia developing on other hosts. When perithecial development occurs it is often on only one side of the leaf although conidia may be produced on both sides. This is not universally true. Perithecial production is rare in the tropics.

We owe our first clear knowledge of the sexual processes in this family to Harper (1895 and later) although de Bary and others 20 to 30 years earlier had described the external features. From neighboring hyphae in contact there arise an antheridial and an oogonial branch, each at first uninucleate. The latter is somewhat rounder and the former a little more slender. They become appressed side by side or may even coil about each other a little. The antheridial branch divides into a uninucleate stipe and an apical uninucleate antherid which presses closely against the upper portion of the oogone. The oogonial branch remains unicellular or may divide into a smaller basal and a larger terminal cell (the oogone proper) or the nucleus only may divide, producing a binucleate oogone one of whose nuclei eventually disintegrates. An opening is formed from the antherid into the oogone and through this the single male nucleus passes, or the antherid nucleus divides and one nucleus passes into the oogone, the other remaining in the antherid. According to Harper, Hein (1927), and several other observers the male and female nuclei fuse in the oogone but according to Dangeard (1907) and his students there is no passage of a male nucleus into the oogone hence no nuclear fusion at this stage. Bergman (1941) reinvestigating the development in the species *Sphaerotheca castagnei* Lev. confirms the passage of the antherid nucleus into the oogone, where it passes by the larger oogone nucleus and takes up a basal position. According to him there is no union of the nuclei at this stage. In the meantime from the cell basal to the oogone there grow out branching hyphae which push upward and around the oogone with the attached antherid, thus forming a layer of 16 to 20 cells of which the antherid stalk is one. An additional layer of hyphae arises in a similar manner from the basal cell, within the first layer, and finally a third layer is produced. The innermost layer consists of thin-walled cells, called by Hein "nurse cells," richly filled with food while the two outer layers form the cortex. This eventually becomes dark and the cells become more or less polygonal in outline. The zygote nucleus of the oogone divides rapidly according to Harper and the oogone elongates, becoming an elongated plurinucleate structure which is soon divided by septa into a row of from five to eight cells, all uninucleate except the penultimate cell. Hein finds in his studies that only three cells are produced, the middle one with two nuclei. Bergman describes the development in a different manner. The male and female nuclei in the oogone do not unite immediately but each divides, producing thus four nuclei in the somewhat elongated and curved oogone. Two septa divide this into three cells of which the middle one is binucleate. In *Sphaerotheca* and *Podosphaera* the two nuclei unite and the cell enlarges and becomes the single ascus. The fusion nucleus divides successively until eight nuclei are formed. Around each nucleus part of the cytoplasm of the ascus is cut out by the formation of spore walls, thus producing eight ascospores or, by degeneration of some of the nuclei or partly formed spores, a smaller number. The ascospores are hyaline and broadly ellipsoid. The ascus is obovoid. Miss Allen (1936) reports that in *Erysiphe polygoni* DC. the formation of recognizable antherids is the exception. Usually when any two suitable hyphae come in contact, whether end to end or otherwise, the intervening walls dissolve and nuclear transfer occurs. From the nearby hyphae the perithecium arises containing a maze of interconnecting cells with varying numbers of nuclei. Eventually the outer, colored cortex develops and several of the contained cells enlarge to become the asci, two nuclei uniting in each.

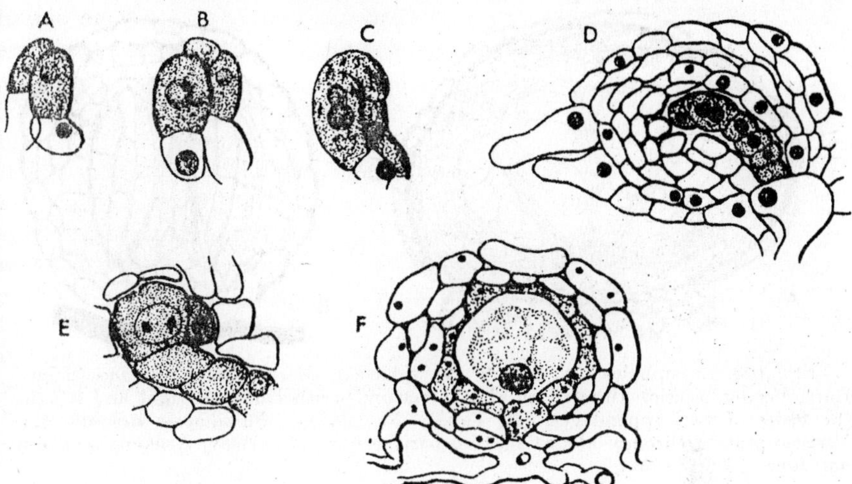

FIG. 99. Erysiphales, Family Erysiphaceae. *Sphaerotheca castagnei* Lev. Stages in sexual reproduction. (A) Oogone and antheridial branch. (B) Antherid set off by a septum. (C) Male nucleus has passed into oogone through an opening. (D) Row of ascogonial cells, the penultimate one binucleate. (E) The nuclei of the penultimate cell have united to form the primordium of the ascus. (F) Young perithecium showing three layers of cells surrounding the young ascus whose nucleus has not yet divided. (A–E, after Harper: *Jahrb. wiss. Botan.*, **29**(4):655–685. F, after Hein: *Bull. Torrey Botan. Club*, **54**(5):383–417.)

Already, therefore, it is apparent in this family as in other cases in the Ascomyceteae that a substitution has occurred of uniting vegetative hyphae for specially formed antherid and oogone. This tendency is one that appears to become more pronounced in the higher groups of fungi where (as in most Basidiomyceteae) no organs can be distinguished as definite oogonial structures or in many cases as antherids. (Fig. 99.)

Harper's accounts indicate two chromosome reductions during the nuclear divisions in the ascus, as would be necessary if there were nuclear fusions both in the oogone and in the ascus. Dangeard, on the contrary, who denies the occurrence of a nuclear fusion in the oogone, admits but one reduction division in the ascus. In the other genera of the family the binucleate penultimate cell of the row arising from the oogone undergoes further division and produces a number of short ascogenous hyphae made up of dicaryon cells. The terminal cell of each ascogenous hypha enlarges and the nuclei fuse and undergo division, thus giving rise to an ascus with ascospores. The latter develop as do those of *Sphaerotheca*. In a few genera the ascospores are two- or four-celled (see Key to the genera of this family on p. 353). The number of asci may vary from 5 to 8 in some species up to 20 to 30 in others. As the asci enlarge the layer of nurse cells is gradually destroyed, leaving only the usually two-layered cortex. (Fig. 100 A.)

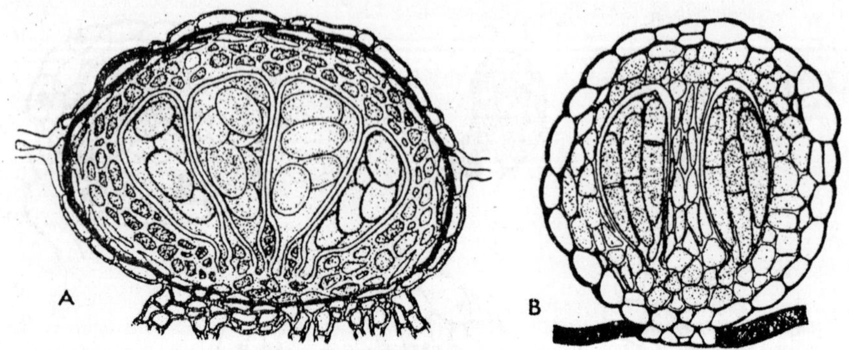

FIG. 100. Erysiphales. (A) Family Erysiphaceae. *Microsphaera quercina* (Schw.) Burr. Vertical section through an almost mature perithecium. At right and left are the bases of two appendages. (B) Family Meliolaceae. *Parodiöpsis stevensii* Arn. Vertical section through almost mature perithecium. (Courtesy, Arnaud: *Ann. sci. nat. Botan.*, **7**:643–723.)

From the outer cells of the cortex, mainly near the equatorial zone, arise the characteristic appendages. These are simple and hypha-like in *Erysiphe*, *Leveillula*, *Sphaerotheca*, *Chilemyces*, and *Leucoconis*, hooked or spirally-coiled at the tip in *Uncinula* and *Uncinulopsis*, straight and once or more dichotomously forked at the apex in *Microsphaera*, *Podosphaera*, and *Schistodes*, or stiff and needle-like, with a bulb-like base, in *Phyllactinia*. In some species the appendages are colorless, but often they are colored basally. They do not seem to have the same function in all cases. They may hold the perithecia fast to the mycelium or may push it up above the surface of the mycelium, or may curve downward and pry the perithecium loose, as in some species of *Uncinula* and in *Phyllactinia*. The hooked or forked appendages would seem fitted for distribution by insects, but that has not yet been demonstrated to be the normal means of distribution. The upper half of the perithecium of *Phyllactinia* and of *Typhulochaeta* bears short, penicillately branched mucilaginous cells. These serve to fasten the perithecium, after its separation from the mycelium, with its top side down, to objects with which it comes into contact. Appendages are lacking in *Astomella*. (Fig. 101.)

When the ascospores are mature, which may not be until the following spring, the asci and the inner cells of the perithecium absorb water and swell until the perithecium is ruptured, at which time the asci also begin to burst, discharging the enclosed ascospores with considerable force, or when the perithecium bursts it may throw out the asci still containing their spores.

Homma (1933) sowed a single conidium of *Sphaerotheca fuliginea* (Schlecht.) Pollacci upon its host plant and on the resultant mycelium were produced conidia and sexual organs. He therefore considered this

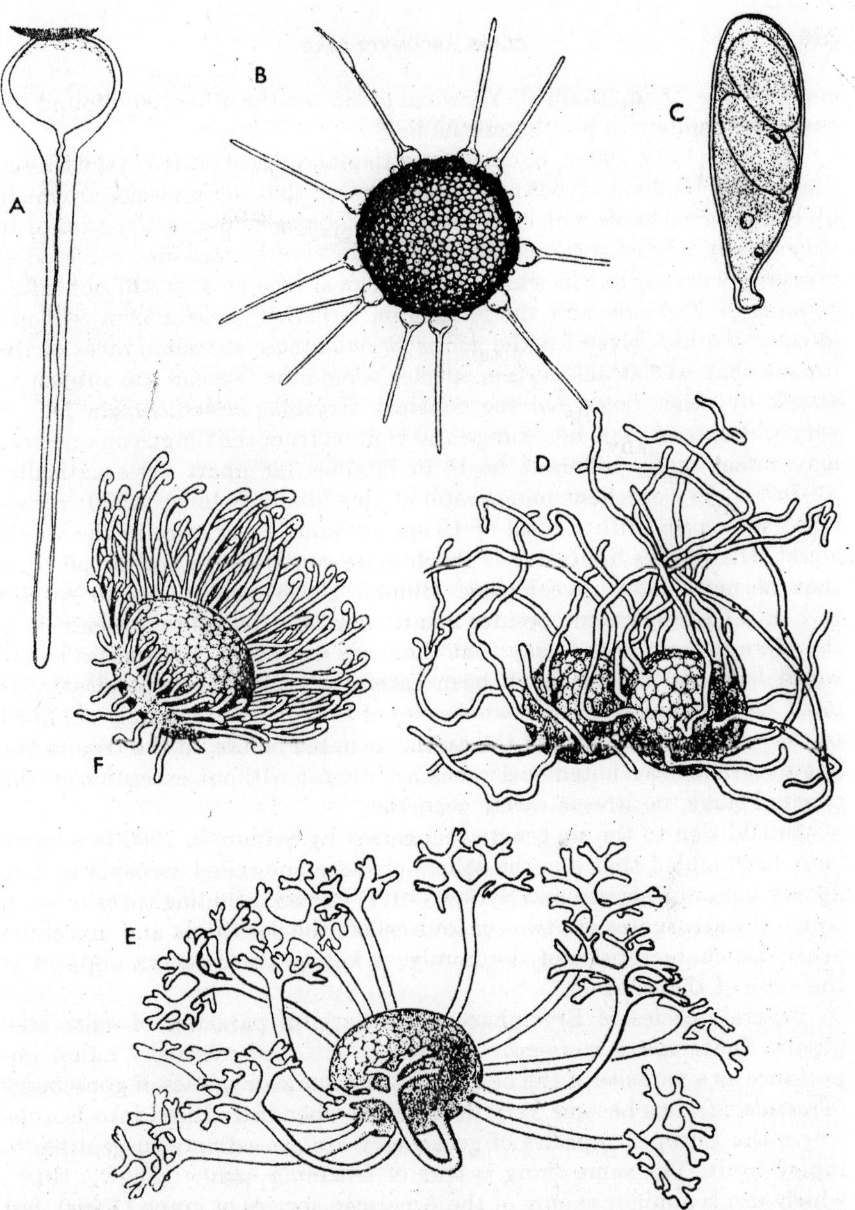

FIG. 101. Erysiphales, Family Erysiphaceae. Perithecia with various types of appendages. (A–C) *Phyllactinia eleagni* Linder. (A) Perithecial appendage, greatly enlarged. (B) Perithecium with appendages. (C) Ascus with ascospores. (D) *Sphaerotheca humuli* (DC.) Burr. Perithecia with appendages. (E) *Microsphaera berberidis* (DC.) Lev. Perithecium with appendages. (F) *Uncinula salicis* (DC.) Wint. Perithecium and appendages. (A–C, courtesy, Linder: *Mycologia*, **35**(4):465–468. D–F, after Tulasne, Selecta Fungorum Carpologia, Paris, Typographie Impériale.)

species to be "homothallic." Yarwood (1935) on the other hand found the sunflower mildew to be "heterothallic."

Salmon (1903, 1904a, b, c, 1905) in England, Reed (1916) in the United States, and Hashioka (1938) in Japan, showed that some species are made up of biological races which are confined to but a single host species or to very closely related species. This is true of *Erysiphe graminis*, confined to grasses (Poaceae) but in which the biological race on *Poa* will not infect *Bromus* or *Triticum* and the strains on either of these genera will not infect the other. Even for the genus *Bromus* there are some races of the fungus that will attack certain species while other species are subject to attack by other races. On the contrary *Erysiphe cichoracearum* DC. is very widespread in its host range and conidia from the fungus on one host may infect many different hosts in families far apart systematically.

In his very excellent monograph of this family Salmon (1900) recognized six genera with a total of 49 species and 11 varieties in the whole world. His species limitation is much more conservative and broad than that of most later mycologists. Blumer (1933) recognized 80 species in Central Europe alone besides mentioning 50 extralimital species. It is therefore perhaps safe to say that there are upward of 150 species in the world when all regions have been carefully explored mycologically. It must be noted that the known species of the family are mainly confined to the extratropical parts of the earth. As noted before, in the tropics the not too frequently noted specimens are almost without exception in the conidial stage, perithecia rarely occurring.

In addition to the six genera recognized by Salmon in 1900 four more have been added that like the others produce one-celled ascospores. The author follows Theissen and Sydow (1917) in also including three more in which the ascospores are two- or four-celled, the perithecia and mycelium otherwise characteristic of the family. A key to these genera appears at the close of this chapter.

Several species of Erysiphaceae are harmful parasites of cultivated plants. *Sphaerotheca mors-uvae* (Schw.) B. & C., of relatively minor importance as a parasite of the berries of the American species of gooseberry (*Grossularia*), has become very destructive when introduced into Europe where the European species of gooseberry are exceedingly susceptible to injury by it. The same thing is true of *Uncinula necator* (Schw.) Burr., which also is a minor enemy of the American species of grape (*Vitis*), but when introduced into Europe proved very harmful to the susceptible *Vitis vinifera* L. *Podosphaera oxyacanthae* (DC.) de Bary and *P. leucotricha* (E. & E.) Salmon are sometimes harmful, especially to nursery stock and young trees, to *Prunus* and *Malus* respectively, in America, Europe, and other parts of the world. *Sphaerotheca humuli* (DC.) Burr. is destructive to the hop (*Humulus lupulus* L.) wherever the host is grown in large plantings. Roses in Europe are frequently seriously injured by *S. pannosa* (Wallr.) Lev., but according to Salmon the similar injury to roses in America is more often due to another species. In America *Microsphaera alni* (Wallr.) Salmon is exceedingly common as the cause of the powdery mildew on the common lilac (*Syringa vulgaris* L.) but it almost never occurs on that host in Europe. *Erysiphe cichoracearum* DC. is probably the most widely distributed species of powdery mildews, attacking hosts in the most varied families. It is not in general very destructive. *E. graminis* DC. on the contrary is often very destructive to various small grains in all parts of the world.

Family Meliolaceae (Perisporiaceae of most authors; see footnote on p. 307). In this family, too, the mycelium is mostly superficial and spreads in a network from the initial point of infection by conidium or ascospore. The hyphae, in contrast to those of the Erysiphaceae, are usually dark in color. They may form a dense crust or be more or less separate. They may send haustoria into the epidermal cells or even into the next layer of cells beneath or may, without producing haustoria, adhere closely to the epidermis whose outer wall becomes more or less corroded with evidence of some injury to the contents of the cell. In *Meliola* and some other genera the creeping hyphae produce short opposite or alternate two-celled branches, the terminal cell of which is enlarged, and rounded or angular. These are the hyphopodia. They serve the double function of anchoring the hypha in place and of producing a haustorium which penetrates the cuticle to enter an epidermal cell of the host. The swollen cell of the pair is homologous to the structure called the appressorium in many fungi. In some genera of the order the hyphopodium may consist of only one cell. In the genus *Pampolysporium* the mycelium and perithecia are subepidermal, in *Alina* and *Lasiobotrys* subcuticular. In *Stomatogene* and *Piline* the mycelium forms a sort of foot which enters through the stoma into the substomatal chamber. Conidia produced singly on the superficial mycelium are reported in one or two genera. In several genera ostiolate pycnidia containing numerous conidia are the characteristic mode of asexual reproduction. It has been shown that most of the reported cases of conidia not in pycnidia are based upon fungi parasitic on the mycelium of Meliolaceae, e.g., *Arthrobotryum* and *Helminthosporium* of each of which Stevens (1918) has described several species.

The perithecia are typically without appendages, mostly black or dark brown, without ostioles in the majority of genera. The asci arise in a single layer or tuft in the base of the perithecial cavity. The ascospores are usually four or eight in number. They are one-celled and hyaline in one genus but in most cases are two- to many-celled or even muriform and either hyaline or brown. In general the ascospores do not represent types that would seem to be primitive. (Figs. 100 B, 102 A.)

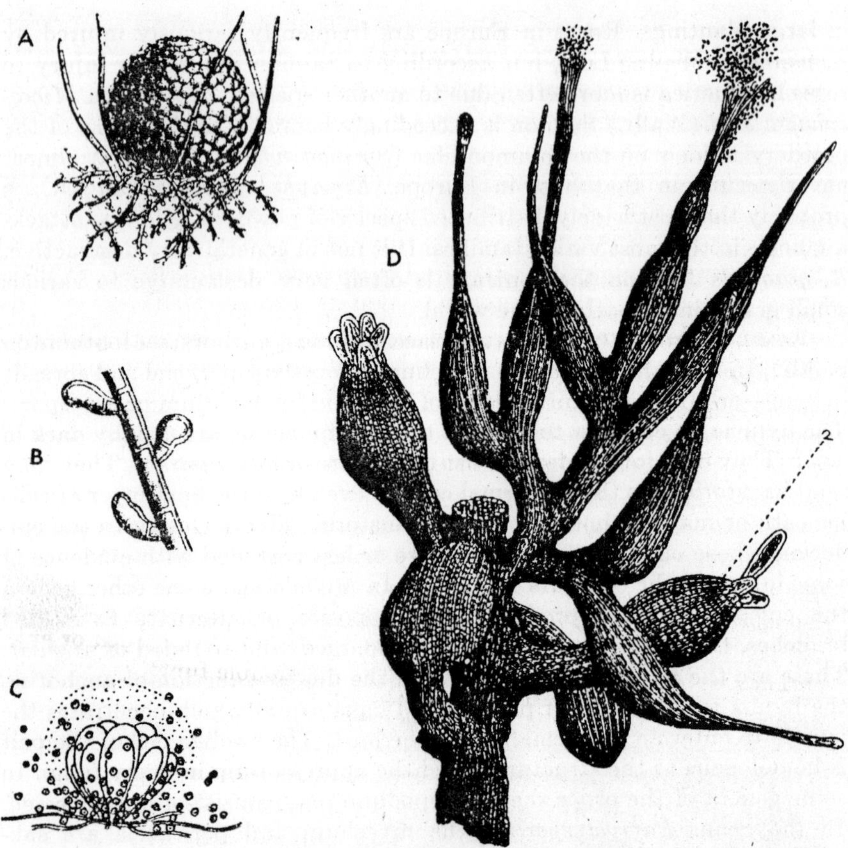

Fig. 102. Erysiphales. (A, B) Family Meliolaceae. (A) *Meliola corallina* Mont. Perithecium. (B) *Irene echinata* (Gaill.) Th. & Syd. Hypha with hyphopodia. (C) Family Englerulaceae. *Englerula effusa* (Cke. & Mass.) Theiss. Perithecium with wall dissolved into individual cells and a mass of slime. (D) Family Capnodiaceae. *Scorias spongiosa* Schw. Perithecia and pycnidia. (A–B, after Engler and Prantl: Die Natürlichen Pflanzenfamilien, Leipzig, W. Engelmann. C, after Theissen and Sydow: *Ann. Mycolog.*, **15**(6):389–491. D, after Ellis and Everhart: The North American Pyrenomycetes.

The process of sexual reproduction has been worked out carefully in *Meliola circinans* Earle by Graff (1932). Close to one another on nearby hyphae there arise an ovoid uninucleate oogone with short stalk cell and a slender somewhat spirally wound uninucleate antherid, also with a short stalk. The two become appressed near their tips and an opening is produced. The antherid nucleus disappears and what appears to be a fusion nucleus is visible in the oogone, although the passage of the male nucleus into the oogone and its fusion with the female nucleus was not observed. Over the united antherid and oogone the surrounding vegetative hyphae grow to produce a dark-colored, shield-like stroma. From the stalk cells of the oogone and antherid hyphae grow out and form a perithecium around these organs, under the stromatic shield. The fertilized oogone elongates and divides into a number of uninucleate cells of which two or three near the apex send out several branched ascogenous hyphae which produce typical hooks which give rise to the asci. A few paraphyses appear among the asci while from the upper part of the perithecium arise periphyses which grow up together and pierce the stromatic shield and spread apart to produce the ostiole. The eight nuclei produced in each ascus in the usual manner are taken up by two's into the four-developing ascospores. Two of these binucleate ascospores are destroyed by the growth of the other two which eventually become five-celled, with one nucleus in each cell except the middle cell which is binucleate. The presence of a true ostiole lined by periphyses and the occurrence of paraphyses would seem to suggest that perhaps this genus is more closely related to the Sphaeriales.

In this family Theissen and Sydow (1917) distinguish 19 genera of which *Meliola* and *Irene* with many hundreds of species each are very abundant in the tropics. A few species occur even in temperate regions. Stevens (1927, 1928) wrote a monograph of the genus *Meliola* which is indispensable for the recognition of the species of this difficult genus. He differs considerably from Theissen and Sydow in his interpretation of the relationships of the genera centered about *Meliola*. In a subfamily Meliolineae he includes *Actinodothis*, placed by those authors in Family Polystomellaceae of the Hemisphaeriales, and *Amazonia*, assigned by them to Family Microthyriaceae, of the same order. Both of these genera have the same type of asci and ascospores. The perithecia of *Amazonia* have been found to be complete, not incomplete below, and the spreading mycelium has hyphopodia like those of *Meliola* and *Irene*. (Fig. 102 B.)

The fact that some genera of Family Capnodiaceae have much this same type of superficial mycelium, with hyphopodia or bristles would also indicate that the classification of the Meliolaceae and Capnodiaceae, as well as of the Englerulaceae (see below) is rather artificial. Arnaud's studies in this field (1925) suggest lines for a better arrangement when more extensive investigations can be completed.

Family Englerulaceae. These are leaf parasites whose perithecial cells dissolve into slime at maturity, exposing the enclosed asci. The 13 to 15 genera and 20 to 30 species are with few exceptions tropical. Petrak (1928) after an extensive study of most of the genera assigned by Theissen to this family decided that it is a collection of heterogeneous forms placed together on account of the one common character, the slimy dissolution of the perithecia. Since this characteristic is known in other orders, e.g., Hemisphaeriales in some genera of Microthyriaceae and of Polystomellaceae; and in some Hypocreales, it does not seem to him to be of sufficient importance to warrant the erection of a separate family. He distributes the genera examined by him among the Microthyriaceae, Polystomellaceae, Mollisiaceae, Myriangiales, etc. The presence of hyphopodia in some genera would suggest relationship to *Meliola* or related genera. Miss Doidge (1942) included the genus *Englerulaster* in the genus *Asterina* in the Microthyriaceae. (Fig. 102 C.)

FAMILY CAPNODIACEAE. These are transferred by Arnaud (1925) to the Sphaeriales but it seems best to follow Theissen and Sydow in retaining them in the Erysiphales. It must be noted that the work of Miss Fraser (1935) on the development of the ascocarp of *Capnodium* shows that this starts as a stroma in which develops an archicarp with ascogenous hyphae and asci. This would indicate that the relationship of this genus lies with the Pseudosphaeriales. The investigations of Graff (1932) on *Meliola* and of various authors on Erysiphaceae show that in those fungi the structure is a true perithecium. Their relationship is probably with the Sphaeriales. When further developmental studies are carried on in other genera of the families tentatively placed in the Erysiphales a better realization as to their true position may be obtained. The dark-colored mycelium of the Capnodiaceae is usually superficial and in many cases saprophytic on "honeydew," the sugary deposit forming on plant parts from the droppings of aphids, scale insects, etc. This mycelium sometimes forms a black papery layer that can be peeled off from the underlying leaf. In a few genera hyphopodia are present. Miss Fraser (1937) has made an extensive study of the physiology of the "sooty mold" fungi, including their food requirements, and relations to temperature, light, and to the presence of other fungi on the same leaf. The conidia are borne in pycnidia of various shapes, sometimes elongated like a long necked bottle. The external perithecial walls are formed of parallel, laterally adhering hyphae, not of polygonal cells as in the Erysiphaceae and Meliolaceae. These component hyphae may be dematioid, i.e., moniliform, or perisporioid, i.e., with parallel sides. In the same genus may be found species with ostiolate perithecia and other species whose perithecia lack ostioles. The perithecia may be sessile or more or less stalked, sometimes elongated like the pycnidia. The ascospores vary from colorless to colored and from two- to many-celled. Theissen and Sydow recognize 25 or more genera and 50 or 60 species, mostly tropical or subtropical, but some found in temperate regions. *Capnodium salicinum* Mont. occurs upon willow (*Salix*) leaves and twigs in Europe. *C. citri* Berk. & Desm. causes sooty mold on oranges, etc. wherever they are cultivated. A few species of *Limacinia* occur on leaves of trees and shrubs in temperate regions, even as far north as Germany and England. (Fig. 102 D.)

FAMILY TRICHOTHYRIACEAE. This family of 5 or more genera and 20 or so species has been tossed about by mycologists since its members were first studied. Mostly these forms have been assigned to the Erysiphales or to the Hemisphaeriales, close to the Microthyriaceae. They are tropical, parasitic on the epiphyllous mycelium of *Meliola* and on other fungi. It was demonstrated by von Höhnel (1917) that the perithecia at maturity lie inversely, with the original point of attachment outermost and the asci radiating downward and outward from this point toward the ontogenetically upper side. The perithecium arises terminally on a stout upright hypha which curves over as the development progresses. Eventually

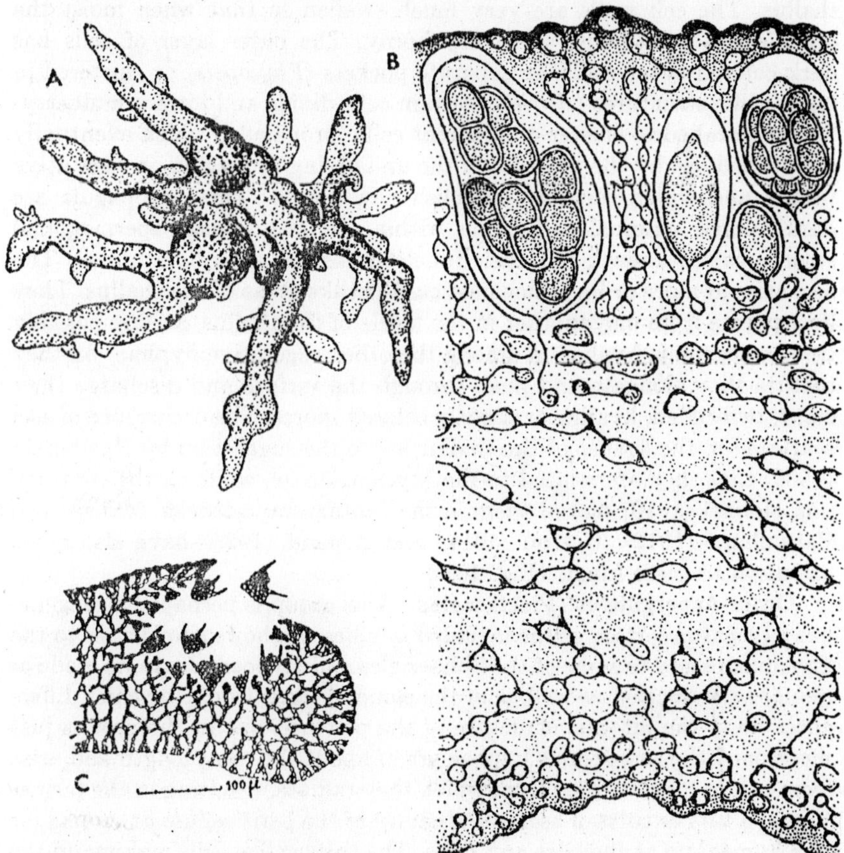

FIG. 103. Erysiphales, Family Atichiaceae. *Atichia millardeti* Racib. (= *Seuratia coffeicola* Pat.). (A) Thallus when wet, showing ascigerous cushions. (B) Section through portion of an ascigerous cushion. (C) Propagula. (A–B, courtesy, Arnaud: *Ann. sci. nat. Botan.*, 7:643–723. C, courtesy, Mangin and Patouillard, *Compt. rend.*, **154**(23):1475–1481.)

on the apparently upper side the tissues break away forming an ostiolar opening. The outer perithecial wall is radial in its structure, resembling many of the Hemisphaeriales in that particular. The ascospores are two- to several-celled and colorless or brown. *Trichothyrium* is the first described and best known genus.

Family Atichiaceae. Because of their epiphyllous habit, apparently entirely external to the leaf tissues and perhaps feeding saprophytically on honeydew or parasitically on hyphae of other fungi this family may be placed here, but with great doubt. There are no separate hyphae but a gelatinous mass of anastomosing threads forming a cushion or a stellate thallus. The cell walls are very much swollen so that when moist the organism is gelatinous, when dry horny. The outer layer of cells has dark-colored outer walls. In separate pockets (*Phycopsis*) or clustered in basket-like structures (*Atichia*) certain cells divide and form a mulberry-like or tetrahedral cluster of adherent cells (propagula) which eventually are pushed out by the pressure of the underlying propagula or vegetative tissues and which serve to establish new plants. These propagula are dark-colored externally, hyaline within. In one species spermogonial structures are known, but their function has not been determined. The asci arise in more or less thickened cushion-like areas of the thallus. They lie separate from one another in the tissue of the thallus, arising possibly from ascogenous hyphae in among the other vegetative hyphae. As they mature they elongate and push through the surface and discharge their eight, two-celled, hyaline or slightly colored spores. The occurrence of asci scattered in the tissues of the thallus led to the suggestion by Raciborski (1909) that they are related to the Myriangiaceae, while on the contrary von Höhnel (1910) placed them in the Saccharomycetaceae. Mangin and Patouillard (1912), Cotton (1914) and Arnaud (1925) have also given attention to these fungi. (Fig. 103.)

Order Aspergillales (Plectascales). This group is perhaps heterogeneous as regards certain of the included families. It shows similarities to the Erysiphales and to some of the Sphaeriales and Hypocreales in its mode of sexual reproduction as well as in the conidial formation. The chief differences lie in the internal structure of the perithecium. In the groups just mentioned the ascogenous hyphae are of about the same length and arise from one or more centers from which they radiate, resulting in the formation of a tuft or tufts of asci in the cavity of the perithecium or stroma, or of a hymenium at the base and sides. The thin-walled cells making up the interior portion of the perithecium or stroma give way before the outgrowing asci and eventually disappear completely or nearly so or the asci and paraphyses grow out into the perithecial cavity. In the Aspergillales, on the contrary, the ascogenous hyphae are of varying lengths so that instead of arising in a tuft the asci are produced scattered throughout the interior of the perithecium. The latter, as in the Erysiphales and Sphaeriales, consists of a firmer exterior cortex (this is sometimes of loose hyphae) and a thin-walled interior portion. Some of these thin-walled cells are pushed aside or destroyed as the branching ascogenous hyphae grow among them so that eventually the asci appear to be imbedded here and there in the internal "nucleus" of the perithecium. Finally most of these interior cells as well as the ascus walls and the remains of the ascogenous hyphae are dissolved, leaving the ascospores loose in the perithecial cavity. True ostioles are lacking in most genera of this order. In the Aspergillaceae they are present in *Microascus* and *Emericella.* If the suggestion of Nannfeldt (1932) is followed and the Ophiostomataceae and Chaetomiaceae are transferred from the Sphaeriales to this order these families will add to the ostiolate forms in the order. The asci are formed at the ends of the ascogenous hyphae, sometimes by the hook method, or in chains by the transformation of successive dicaryon cells of the hypha into asci.

Conidial formation is frequently catenulate, with the apical cell the oldest, as in the Erysiphaceae. Many species occur mostly in the asexual stage, only very rarely producing perithecia. This is especially true of the very numerous species of *Penicillium* and *Aspergillus* which form some of the commonest molds upon organic matter of every kind. The details of sexual reproduction still remain to be studied for the great majority of genera. Even in those cases that have been studied much still remains to be learned, particularly regarding the behavior of the sexual nuclei. In general a straight or coiled ascogonium and coiled antherid are produced. The former may be several-celled, the terminal perhaps corresponding to a trichogyne, or only one-celled. In a few cases the formation of an opening has been observed between the antherid and the tip of the oogone (or of the trichogyne), following which ascogenous hyphae grow out of one or more of the ascogonial cells. The eminent French mycologist P. A. Dangeard (1907) has observed and figured the antherid and ascogonium in numerous species of this order. He even figured the opening from the antherid into the ascogonium or trichogyne in a few cases. He denied, however, that this is a sexual process and considered that the antherid has entirely lost its primary function as a male organ, perhaps functioning now as a nutritive organ, which he therefore calls a trophogone. In the main the mycologists, apart from Dangeard's students and associates, do not agree with him and look upon the antherid as a functional sexual organ, at least in the majority of cases where it is present. The actual observation by Schikorra (1909) and by Young (1931) of the passage of nuclei from antherid to ascogonium in *Monascus* shows that this does occur. On the other hand there are well authenticated cases where a coiled or straight ascogonium is produced and no organ that can be in any way interpreted

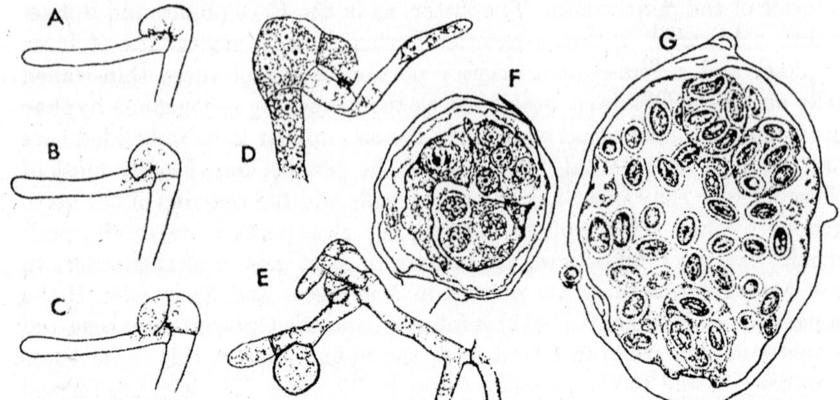

Fig. 104. Aspergillales, Family Aspergillaceae. *Monascus ruber* van Tiegh. Sexual reproduction. (A) Ascogone growing up sympodially at base of elongated antherid. (B) Ascogone separated by septum from hypha. (C) Ascogone divided into a basally located oogone and a trichogyne which has formed an opening to the antherid. (D) Antherid has collapsed and the nuclei have passed through the trichogyne into the oogone. (E) Two investing hyphae are growing up from just below the oogone. A conidium is shown on the antheridial branch. (F) Perithecium wall completed; one of the curved ascogenous hyphae is shown with three terminal asci, successively younger from apex toward base. (G) Section of mature perithecium with ascus walls mostly dissolved. (Courtesy, Young: *Am. J. Botany*, **18**(7):499–517.)

as an antherid. From this ascogonium arise binucleate cells which give rise to ascogenous hyphae. (Fig. 104.)

The formation of the ascus has been reported by Schikorra to take place by the hook method in *Monascus* but this is denied by Young. DeLamater (1937) has demonstrated the formation of croziers in *Arachniotus aureus* (Eidam) Schroet., of Family Gymnoascaceae. In general one or more of the binucleate cells at the end of an ascogenous hypha enlarge, the nuclei fuse and the young ascus is initiated. The perithecium is formed by the growth of hyphae from near the point of attachment of the ascogonium.

In perhaps the majority of species investigated cytologically the mycelial cells and the conidia are plurinucleate. The young ascogonium and young antherid may be plurinucleate or uninucleate. It is difficult to determine which represents the more primitive condition.

Several families are recognized in this order, following Fischer (1896) in the main. Those first to be considered show the closest affinity to the Erysiphales and may well have arisen from or have given rise to that order. The sexual organs and mode of origin of the perithecium are quite similar in some of the Aspergillales to the corresponding structures of the Erysiphaceae. If the short ascogenous hyphae of the latter should become longer and of various lengths, pushing in among the tissues of the central portion of the perithecium we would have the structure as it is found in the Aspergillaceae. If the central and cortical tissues of the perithecium of the latter instead of forming a continuous structure should remain more or less loose and cottony we would have the Gymnoascaceae. If the perithecium should become much enlarged (up to several centimeters) with a firm cortex several layers of cells thick and a more or less permanent mass of loose central tissue traversed by more or less well-developed sterile "veins," in which the asci are scattered, the structure would be that of the Elaphomycetaceae. In the Onygenaceae the ascocarp is differentiated into a basal sterile portion and a somewhat larger head in which the asci are scattered in a central mass of tissue which eventually breaks up into a sort of capillitium. The Trichocomaceae are forms whose development is not well understood and whose relationship to other families of the order is more or less a matter of doubt. The Terfeziaceae, placed by Fischer (1896) in this order were more recently (1938) placed by him in the Tuberales. *Myriangium* and probably several other genera rather similar in structure, forming the Family Myriangiaceae, possibly also belong in this neighborhood but are better placed in a distinct order.

FAMILY ASPERGILLACEAE. The chief genera of this family are the ubiquitous molds, *Penicillium* and *Aspergillus*, of which the commonest species are the blue and green molds found on all sorts of organic matter. Most of their species are saprophytes but a few are animal parasites, causing cases of mycosis. The small orange or yellow perithecia of a species of *Aspergillus* (perhaps more properly called *Eurotium*) are very frequently found on jams and various conserves, as well as on imperfectly dried herbarium specimens where it is called "herbarium mold." Usually when examined these are found to consist of nothing but a thin peridium surrounding a cavity in which are found a large number of ascospores shaped much like a pulley wheel. Rarely a few asci will be found but usually their thin walls have completely dissolved away. In *Penicillium* perithecia are much more rarely formed and are more often sclerotium-like. The two genera are best distinguished by their conidiophores. In *Aspergillus* the conidiophore is swollen at its apex into a head from which radiate numerous short sterigmata bearing at the apex of each a chain of spherical or nearly spherical, smooth or roughened conidia, or sometimes producing secondary sterigmata which bear in their turn the chains of conidia. The species with the secondary sterigmata were formerly set apart in the genus *Sterigmatocystis* but the modern usage is to include these in *Aspergillus*. In *Penicillium* the conidiophore branches several times without enlarging, the terminal portions lying more or less parallel, or at least not widely divergent, and bearing one to three or four sterigmata with chains of conidia. Thom (1914) described in detail the method of formation of the conidia of this genus. From the apex of the sterigma is produced a

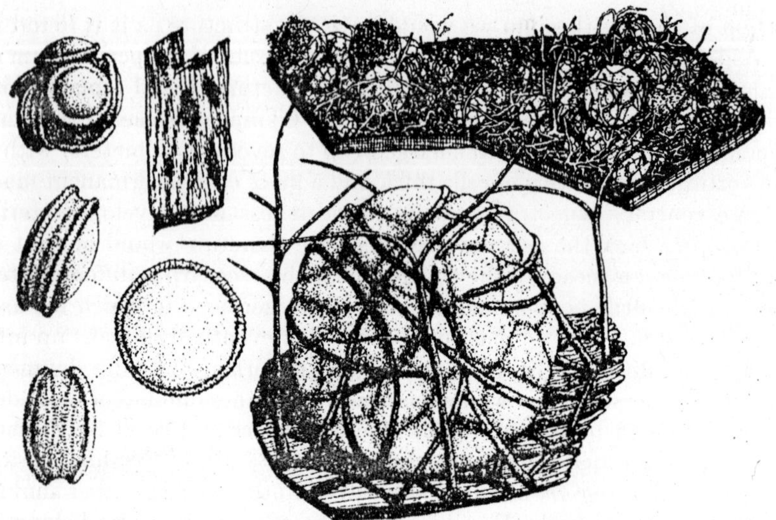

Fig. 105. Aspergillales, Family Aspergillaceae. *Aspergillus* (*Eurotium*) sp. Perithecia, ascus and ascospores. (After Ellis and Everhart: The North American Pyrenomycetes.)

narrow tubular extension near the tip of which a nucleus, formed by the division of a nucleus in the sterigma, takes its position. A delicate cross wall is then formed and the basal portion of the tube elongates and forms a second cell in the same manner. This process continues until a chain of cells is formed, the oldest at the tip. Within the segments of this tubular extension the spores round up and each secretes a thicker wall within and separate from the wall of the tube to which it may adhere or from which it may be free. Thus the conidia are in reality formed internally and are on that account called "endogenous" by Sartory and Sydow (1913). The thin-walled portions of the original tube, lying between the conidia, are often spoken of as "connectives." (Figs. 105, 194.)

In recent years the genus *Penicillium*, especially *P. notatum* Westling and other species of this and related genera have acquired great importance because of the discovery that some of the products of the growth of the mycelium are of great value as antibiotics. These are substances which are able to bring about the destruction of various other organisms, harmful or otherwise, even when introduced into the living bodies of man and other animals infested by these organisms.

Citromyces is much like *Penicillium* but the conidiophores or their branches are somewhat thickened. The chief distinction is a biological one, viz., the production of large quantities of citric acid when grown on a medium containing sugar. *Monascus* is usually distinguished by its red or pink mycelium. The conidia are on short chains. The mature perithecium consists of one or two layers of cortical cells around the cavity enclosing numerous ellipsoid or globose ascospores, the interior perithecial tissues and the walls of the eight-spored asci having dissolved rather early. This was at first interpreted as a single many-spored ascus, hence the name *Monascus* (meaning one ascus). One species is the "pink mold" found in ensilage that was put into the silo in too dry condition. Other species are used in the Orient for the fermentation of rice in the preparation of alcoholic beverages. (Fig. 104.)

The genus *Lilliputia* was established by Boudier and Patouillard (1900) for a fungus shown much later by Dennis and Wakefield (1946) to be a member of this family and in which the conidial stage belongs to the form genus *Gliocladium*. This resembles closely *Penicillium* but the conidia are surrounded by a slimy exudate so that the branched "brush" at the upper end of the main stalk of the conidiophore is involved in a glistening drop filled with hundreds of spores which early lose connection with one another so that they appear to be single or at most only in short chains. The perithecia are relatively large for the family, from 0.5 to 1.0 mm. in diameter, with a thick firm cortical region and round to ovoid asci, each containing eight large spherical, yellowish to brown ascospores which are rough or prickly. The most common species is *L. insignis* (Wint.) Dennis and Wakefield (1946) based on *Eurotium insigne* Wint. This occurs in Europe and in the United States on various types of organic substrata, such as dung of geese and kangaroos, dead seaweeds along the shore, and old stable manure. Brefeld (1908) gave the name *Lysipenicillium* to this genus but it is antedated by *Lilliputia*. It must be noted that the asci and ascospores are very similar to those of *Terfezia* to which Boudier and Patouillard considered them closely related. This casts doubt on the correctness of transferring the Terfeziaceae to Order Tuberales. *Magnusia* produces depressed globose-oblong, dark-colored perithecia with elongated, apically circinate appendages arising from near the base. (Fig. 106 A.)

Sexual reproduction has been studied in several members of this family. Both Schikorra (1909) and Miss Young (1931) have studied this process in *Monascus*. A slender plurinucleate antherid is produced at the end of a hypha. Sympodially from the cell below arises the ascogonial branch which bends so as to lie parallel to or to coil somewhat around the antherid. A basal cell cuts this off from the main hypha and soon an apical trichogyne cell is set off by another septum. The trichogyne and oogone both contain several nuclei. An opening appears between the trichogyne and antherid through which the nuclei of the latter pass, the trichogyne nuclei having previously disappeared. The septum between trichogyne and oogone dissolves out and the male nuclei pass into the latter, the septum then being regenerated. The nuclei pair by twos in the oogone but fusion of these pairs has not been observed. In the meantime hyphal branches arising from just below the oogone form the beginning of the perithecial wall which is one or two layers of cells in thickness. From the base of the fertilized and enlarged oogone grow several ascogenous hyphae, each of several dicaryon cells. The apical and second and third cells may develop into asci, each with eight ascospores. Ascus walls and the remainder of the ascogenous hyphae and of the oogone and antherid dissolve and leave the ascospores free in the perithecial cavity. De Bary (1870) and many years later Miss Dale (1909) reported that in *Aspergillus glaucus* Lk. (= *Eurotium herbariorum* Lk.) a tightly coiled ascogonium is formed, at first one-celled but soon dividing into several multinuclear cells. An antherid rising from lower down or from another branch creeps up the side of the coiled ascogonium. A fusion of antherid and ascogonium was not observed and possibly does not occur. It is not certain that this is really an antherid. Possibly it is one of the investing hyphae that grow up from below the ascogonium to form the perithecial wall. Soon the middle cells of the ascogonium divide into binucleate cells from which arise the branched ascogenous hyphae. From the cells supporting the ascogonium arise the hyphae which form the perithecium with a cortex of one layer of polygonal cells the interior being filled with thin-walled cells. In the species called by Brefeld (1874) *Penicillium crustaceum* Fr., this mycologist reports the formation of short coiled antherid and ascogonium from adjacent cells of the same hypha. These fuse at the apex and give rise to a several-celled structure from which branch out the ascogenous hyphae while around them a dense sclerotium-like perithecium develops. The nuclear behavior has not been followed. Dangeard (1907) reports in *P. vermiculatum* Dang. that there is formed a long straight multinucleate ascogonium around which coils a slender antheridial hypha with an enlarged uninucleate antherid at the apex. This fuses with the ascogonium and the male nucleus may pass into the latter or may remain in the antherid. The ascogonium then divides into binucleate cells from which arise the ascogenous hyphae. Dangeard reports that the male nucleus is not functional. Derx (1925) reported that the ascospores in *P. luteum* Zukal produced mycelia of two sexual phases that are mutually compatible and self-sterile.

Family Gymnoascaceae. This family is characterized by the lack of a firm-walled perithecium, this being represented by a more or less loosely tangled mass of mostly branched hyphae among the bases of which the asci are clustered. When mature they are usually visible through this tangle of hyphae. The chief generic distinctions are based upon the character of these interwoven filaments and the color of the ascospores.

The ascocarps are mostly not over 1–2 mm. in diameter, often smaller, and sessile, globose or depressed globose almost colorless or dark-colored.

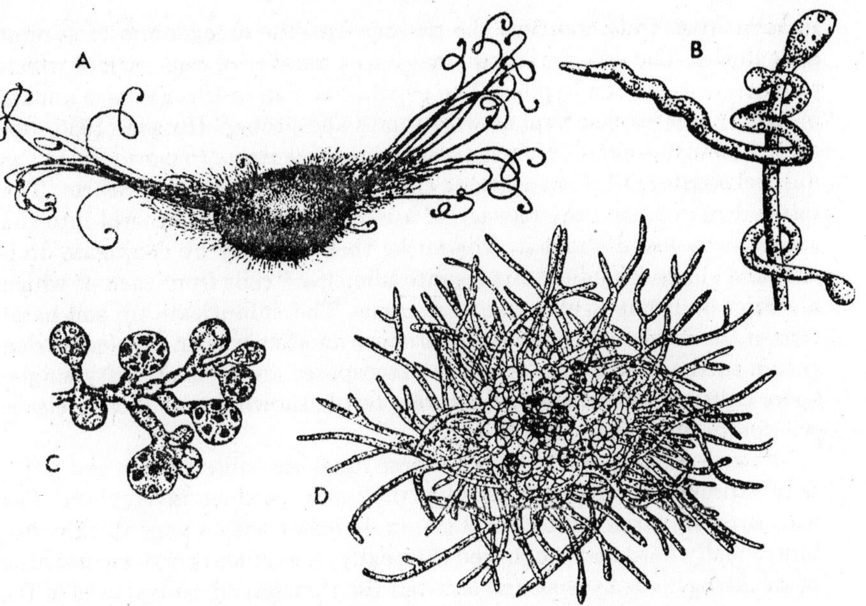

FIG. 106. Aspergillales. (A) Family Aspergillaceae. *Magnusia nitida* Sacc. Perithecium. (B–D) Family Gymnoascaceae. *Arachniotus trisporus* Hotson. (B) Ascogonial branch coiled around the antherid. (C) Ascogenous branch with asci. (D) Mature perithecium with cortex of loosely woven hyphae surrounding the asci. (A, courtesy, Ames: *Mycologia*, **29**(2):222–225. B–D, courtesy, Hotson: *Mycologia*, **28**(6):497–502.)

In some cases the color of the ascospores gives the color to the ascocarp. Conidia are formed in some species, singly on short conidiophores or in chains. Chlamydospores also are known in some species. These fungi are often saprophytic on animal matter such as feathers, dead animal bodies, excrement, etc. *Arachniotus trachyspermus* Shear (1902) was found associated with diseased cranberries and *A. trisporus* Hotson (1936) was isolated from contaminated milk. A number of skin parasites of man and other animals have been credited to this family, but the evidence is not too convincing.

The details of sexual reproduction have been described by several authors. In *Ctenomyces*, Eidam (1880) described a multinucleate ascogonium coiled around the straight multinucleate antherid. Passage of nuclei from the latter to the former was observed. The ascogonium soon divides into numerous short binucleate cells each of which produces a more or less coiled ascogenous hypha. During this time the near-by hyphae have formed a loosely woven perithecial wall around the developing ascogenous hyphae. In *Gymnoascus reessii* Baran. the union of antherid to the tip of the coiled ascogonium was observed by Baranetzky in 1872.

In 1903 Miss Dale observed the passage into the ascogonium of several male nuclei. The ascogonium divides into a number of cells each of which produces ascogenous hyphae. The perithecial wall in this genus is a mass of loosely interwoven hyphae with spines and prongs. Hotson (1936) observed similar reproductive structures in *Arachniotus trisporus* Hots., as did DeLamater (1937) in a species close to *A. aureus* (Eidam) Schröt. The latter demonstrated the passage of a nucleus from the antherid into the ascogonium coiled about it. The nuclei then multiply by conjugate division and the ascogonium divides into binucleate cells from each of which a crozier buds out to give rise to an ascus. The uninucleate tip and basal cells of the crozier often unite and produce another crozier. Nuclear fusion occurs in the young asci and eight ascospores are formed. Fifty single-spore cultures were made and all were fertile showing that this species is self-compatible. (Fig. 106 B–D.)

Family Elaphomycetaceae. These fungi are subterranean and probably saprophytic although possibly they may produce mycorrhiza. The ascocarps are large, up to 2 or 3 cm. in diameter with a very thick peridium, usually hard and roughened externally, the central portion consisting of the ascogenous hyphae and asci and the thin-walled central cells of the ascocarp, traversed radially by the "veins" which probably are conductors of foodstuffs. The cell walls of the central portion dissolve leaving the numerous ascospores free in the center of the ascocarp. Clemencet (1932) reported that in *Elaphomyces* no sexual organs are to be found but that in *Ascoscleroderma* (a segregate of that genus) a stout ascogonial filament coils around a straight antherid which however is sexually functionless. From the ascogonium arise branching ascogenous hyphae which produce at their tips rectascous asci, i.e., the cell elongates and the nuclei fuse without any curving or hook formation. In *Elaphomyces* on the contrary the asci are subterminal and, judging by the illustrations, are formed by means of croziers. C. W. Dodge (1929) recognized 24 species of *Elaphomyces*, mainly from Central and Southern Europe and the United States. One species is reported from Italy, France, and Australia.

Family Onygenaceae. This family contains a few species in the single genus *Onygena*, found mostly in temperate Europe and North America. They grow on old feathers, hair, hoofs, horn, felt, and other animal matter. The ascocarps are a few millimeters up to 1 or 2 cm. tall, consisting of a stalk and a somewhat enlarged head within which the asci are scattered in the manner characteristic of this order. The tissues break up into a sort of capillitium. (Fig. 107.)

Family Trichocomaceae. In this family consisting of the one genus *Trichocoma*, and only a few species, the ascocarp is sessile. When young it is closed with a firm peridium which is thinner above. The ascogenous portion develops centrally and consists of irregularly scattered asci with vertical sterile plates. The latter grow at the base so that at maturity a column of vertical honeycomb structure bursts the upper peridium and pushes upward. In the meshes of the honeycomb there are developing asci at the base while near the top the ascus walls are dissolved and the ascospores lie free. As the upper ends of the column weather away the ascospores are released. Tropical and warm temperate species. Developmental studies of the earlier stages are lacking, in default of which the exact relationship of this family is uncertain.

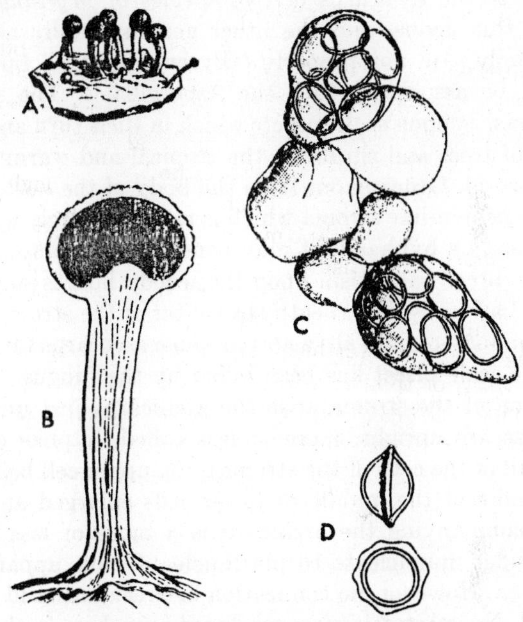

Fig. 107. Aspergillales, Family Onygenaceae. (A–C) *Onygena equina* Pers. ex Fr. (A) Habit sketch. (B) Vertical section. (C) Asci. (D) Ascospores of *Onygena caprina* Fckl. (After Engler and Prantl: Die Natürlichen Pflanzenfamilien, Leipzig, W. Engelmann.)

Family Terfeziaceae. Although the fungi of this family with their large subterranean ascocarps resemble in structure in many ways some of the genera of the order Aspergillales, and especially in ascus and ascospore structure those organs in *Lilliputia*, the author follows Fischer (1938) and Miss Gilkey (1939) in placing them in the Tuberales, but with considerable doubt as to the correctness of this transfer.

Order Myriangiales. This order, in the first edition of this textbook considered to be closely related to the Pseudosphaeriales, appears, from the studies of Julian H. Miller (1938) discussed below, to be more nearly related to the Aspergillales. Some of the earlier mycologists suggested a close relationship to the Pezizales. Whether the order should be considered to consist of but one family or of several depends upon the weight given to the arrangement of the asci in the spore fruit, i.e., in a single layer or scattered at various depths below the outer surface, and to the septation and color of the ascospores. Until life-history studies like those by Miller have been completed on the other genera of this order it seems advisable for the present to include all in one family.

FAMILY MYRIANGIACEAE. Julian Miller (1938) described in detail the development of the ascocarps of two species of *Myriangium* and concluded that this genus, and the other genera usually associated with it in this family, are not properly "Pyrenomycetes" but on the contrary should be associated with the Aspergillales. The species of this genus are parasitic upon scale insects which in their turn are feeding upon many kinds of trees and shrubs in the tropical and warm temperate regions of the world. Growing out from the body of the host the mycelium forms a firm cushion-like stroma which is rather strongly anchored to the bark of the host by hyphae that may penetrate the lenticels. That there may be some direct parasitism upon the woody host is suggested by the fact that the bark cells underneath the center of the stroma die. Furthermore a stroma may remain alive and produce ascocarps for several years, long after the scale insect has been killed by the fungus. At the central thicker portion of the stroma arise the archicarps just under the upper surface. These are upright, more or less coiled, hyphae of uninucleate cells (as are all of the cells of the stroma), the upper cell being attenuated and one or more of the middle or lower cells enlarged and soon multinucleate. Accompanying the archicarp is a more or less loosely coiled hypha of slender uninucleate to plurinucleate cells, apparently the antheridial hypha. However, no connection between antherid and archicarp was observed. No spermatia were produced anywhere in the stroma. The enlarged ascogonial cells divide repeatedly producing a layer of large multinucleate cells. From the apex of each such cell arises an ascogenous hypha of binucleate cells. Thus is produced a flat or concave disk-like thickening on the stroma consisting entirely of parallel, closely packed ascogenous hyphae. These are more or less branched. Terminal and intercalary cells of these hyphae divide longitudinally or diagonally. The two nuclei of one of these daughter cells divide again and two septa are formed, producing a row of three cells, the central one binucleate. These two nuclei unite and the cell enlarges to become the ascus. Since many cells in each ascogenous hypha may produce asci and these hyphae are packed side by side to form the disk of the ascocarp the resultant asci are scattered at various depths below the upper surface but are separated by the tissues consisting almost entirely of ascogenous hyphae. There is no dissolution to form monascous cavities such as is found in the Pseudosphaeriales.

In *M. duriaei* Mont. & Berk. only one archicarp is formed in the young ascocarp, the whole disk being the product of the divisions of its ascogonial cells and of the branching ascogenous hyphae arising from them. In *M. curtisii* Mont. & Berk. 30 to 50 archicarps are produced and the resultant ascogonia and ascogenous hyphae do not form as compact and extensive a structure as in the other species. (Fig. 108 A–C.)

The asci when mature have a two-layered wall. When the ascocarp is wet by rain it becomes soft and the mature asci expand, pushing their

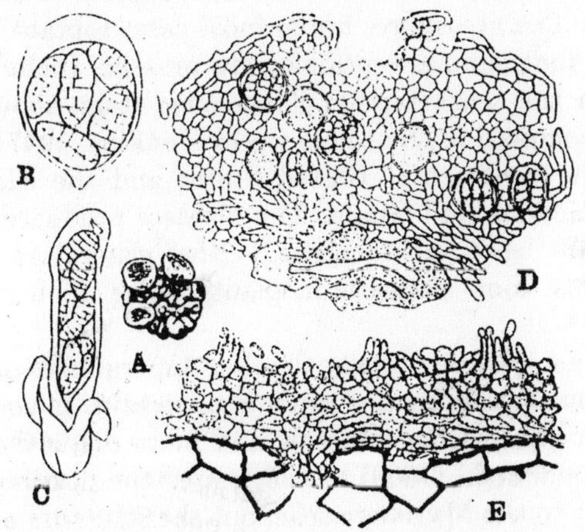

FIG. 108. Myriangiales, Family Myriangiaceae. (A–C) *Myriangium duriaei* Mont. & Berk. (A) Stroma, top view; the cuplike bodies are ascogenous. (B) Ascus. (C) Dehiscing ascus, the outer wall ruptured and the inner wall elongated but still intact. (D, E) *Elsinoe veneta* (Speg.) Jenkins. (D) Vertical section of ascocarp. (E) Acervulus (*Sphaceloma*) stage. (A–C, after Petch: *Brit. Mycol. Soc. Trans.*, **9**:45–80. D–E, after Burkholder: *Cornell Agr. Expt. Sta. Bull.*, **395**:155–183.)

tips through to the surface. The outer ascus wall splits and contracts around the base of the ascus while the thin-walled inner layer pushes out further and discharges its eight greenish yellow muriform ascospores a distance of several centimeters. The lower asci push up through the spaces left by the disappearance of the upper asci and repeat the process. There is no weathering away of the tissues as previously believed. The ascocarps of the other genera usually placed in the same family with *Myriangium* have not been given sufficiently intensive ontogenetic study to make it certain that they have the same type of development although that probably is so. Apparently the asci in these other genera are of the same type as those of *Myriangium*.

Elsinoe consists of a good many species of plant parasites which produce their stromata under or within the epidermis. By the rupture of the latter the stroma containing the scattered asci becomes exposed. Unlike *Myriangium*, in which no conidia have been observed, there are produced acervuli or sporodochia bearing unicellular, hyaline conidia, of the form genus *Sphaceloma*. *E. veneta* (Speg.) Jenkins is the cause of anthracnose of *Rubus*, *E. ampelina* (de Bary) Shear of anthracnose of *Vitis*, *E. piri* (Wor.) Jenkins of anthracnose of pear (*Pirus*) and apple (*Malus*). Other species of *Elsinoe* cause serious diseases of species of *Citrus*, *Canavalia*, *Phaseolus*, etc. The ascospores are in most cases septate transversely but in a few cases longitudinal septa also are present. Including those so far known only in the characteristic *Sphaceloma* stage about 175 species of *Elsinoe* are recognized in the whole world (Jenkins, 1947) of which above 40 occur in the continental United States and the island possessions, Puerto Rico, Hawaii, and Guam. The diseases produced are in the later literature called "spot anthracnoses." Other genera are reported mainly from the tropics, some parasitic in plants, some upon insects. (Fig. 108 D, E.)

The relationship of this family to other Aspergillales or its justification as a separate but closely related order is uncertain. In some particulars it reminds one of the Trichocomaceae, and more distantly of the Elaphomycetaceae. Raciborski (1909) has suggested the possible relationship of the Atichiaceae to the Myriangiaceae, but the structure and development of the ascocarp appears to be too fundamentally different in the two families in so far as the meager information is available on these points in the former family. The possible relationship of the Myriangiaceae to the Pseudosphaeriales is discussed in Chapter 17.

Order Saccharomycetales. This order represents, in the opinion of the author, the ultimate degree of simplification in the Class Ascomyceteae. On the other hand it must be noted that by many eminent mycologists, such as de Bary (1884), Brefeld (1889), Gäumann (1926), Dangeard (1907), Atkinson (1915), and others, the members of this order are looked upon as primitively simple, representing almost the first steps in the evolution of the Ascomyceteae from the Phycomyceteae. To agree with this theory the arrangement of the orders of this class should be the reverse of that used in this book, placing the Saccharomycetales first and the Pezizales and their allies last.

The fungi making up this order are filamentous or unicellular (Yeasts). In the latter case the cells often remain attached until several cell divisions have occurred forming an irregular mass but usually not a cylindrical hypha. The fundamental common character of all the members of the order is the production of single asci, usually as the result of a sexual process but sometimes parthenogenetically, instead of a cluster of asci from branched ascogenous hyphae. Furthermore there is lacking the formation of a protective structure, the perithecium or apothecium.

The mycelial forms of this order mostly take on the unicellular (i.e., yeast) habit when growing in nutritive media of considerable concentration. On the other hand many of the yeasts can be made to develop more or less typical hyphae when grown on special media, usually those not too rich in soluble organic substances. In the main the species that have the yeast habit possess uninuclear cells. That is also true of many of the filamentous forms, although some of the latter may have several nuclei to the cell.

The majority of Saccharomycetales are saprophytes on organic matter, mostly of vegetable origin. However, some are parasites in plants (e.g., species of *Nematospora*) or in Man or other animals. When growing as saprophytes in organic media of plant origin they frequently cause fermentation, the products usually being CO_2 and various alcohols or organic acids. Some species of *Saccharomyces* are of very great importance, especially in the commercial production of alcohol for industrial or beverage purposes, or of CO_2 for leavening bread. Many of these fungi can grow anaerobically if sufficient soluble organic foods are present in the medium.

As in most of the Ascomyceteae in which such studies have been made the cell walls are mostly lacking in cellulose but it is not agreed whether or not chitin is present. In *Schizosaccharomyces octosporus* Beij. and *Sch. versatilis* Wick. & Duprat (1945) the ascus walls give a distinct blue reaction with iodine-containing media, indicating the presence of carbohydrates closely related to dextrin or starch. This should be compared with the similar staining of the apical portions of the asci of some Pezizales and of the basidiospore walls of some Agaricales.

In the filamentous forms asexual reproduction may take place by terminal or lateral budding from the hyphal cells or by the breaking up of the hyphae into short unicellular pieces, often called "oidia." Short conidiophores may be produced in some species which bear conidia singly or in short chains. The unicellular forms may reproduce by fission into equal cells which remain adherent at first and eventually separate from one another. These are the Fission Yeasts. The majority of the yeasts however reproduce by the formation of small buds which enlarge until they equal the mother cell in size. In the meantime other buds may arise on the mother cell and on the buds already in process of development so that a colony results consisting of many cells of various sizes and ages. In the process of budding a small bud is produced and then into it projects an extension of the nucleus of the mother cell. According to Lindegren (1945, 1949) this extension separates by mitosis from the mother nucleus and then the opening between the two cells is closed except for a small central pore which always remains through the separating septum in all the Higher Fungi. This is closed before the cells separate from one another. Fabian and McCullough (1934) reported that by growing some yeasts under certain cultural conditions or by the addition of lithium salts to the culture media the vegetative cells may be induced to break up into minute cells which differ, not only in size and shape but in their ability to cause fermentation, withstand changed environment, etc. On inoculation upon the customary culture media they resume the normal size and characteristics of the original cells. This is of interest as it shows a parallelism in behavior to that demonstrated by Hadley (1927) for some of the true bacteria.

Sexual reproduction consists of the union of two equal or unequal cells and the almost immediate fusion of the two gamete nuclei. This diploid cell or some of its products by division enlarges to become a single ascus within which are produced usually eight or four or sometimes fewer ascospores (even as low as one in the genus *Monospora*). In *Ascoidea* and *Dipodascus* many ascospores are formed. Unlike the case in many of the Ascomyceteae the ascospores in this order are mostly not expelled violently by the bursting of the asci.

In the older classifications of the families within this order the forms normally with mycelium are placed in two or three families and the yeast forms in about as many. The fact that the same type of ascospores may occur in genera assigned to several of these families throws doubt on the validity of these classifications (Zender, 1926). Thus the hat-shaped ascospores occur in the unicellular genus *Hansenula* (*Willia*), in some species of the filamentous genus *Endomyces*, and in the filamentous *Ascoidea rubescens* Bref. These three genera are usually placed in three separate families. This spore type is apparently closely related to that found in those species of the genus *Aspergillus* in which there are two parallel ridges with a furrow between them.

It seems to the author that the yeast forms are to be considered as derived more or less independently from the Endomycetaceae and perhaps other families. In that case the customary yeast family or families are not true phylogenetic units but must eventually be merged with the families from which they are derived. Pending further research by students of these groups the more customary family distinctions are followed here. The most recent extensive work on these organisms with the modern system of classification is that by Miss Stelling-Dekker (1931), Miss Lodder (1934), and Diddens and Lodder (1942).

Asporogenous Yeasts. Closely resembling the yeasts and the filamentous genera of this order are many forms sometimes called the Asporogenous Yeasts. These do not produce asci nor are there any evidences of sexual reproduction. Possibly some of them may be true yeasts that have permanently lost the power of producing asci. Others, however, may represent derivatives of entirely different orders or even classes of fungi which have evolved into yeast-like forms. Under certain conditions of culture yeast-like cells are formed by some Mucorales, Ustilaginales, and many families of Ascomyceteae. Although their true relationships are doubtful they are given consideration here. Apart from these are two families that possess sexual reproduction and a simplicity of structure that implies either great reduction or, on the contrary, a great degree of primitiveness: the Spermophthoraceae and Pericystaceae. Their relationships are very doubtful. Their consideration in this chapter is due to a lack of conviction on the part of the author as to where else they might more properly find their relationship, rather than because of any evident kinship with the Saccharomycetales.

The first two families to be discussed are those that are normally filamentous and ascus-producing.

Family Endomycetaceae (Including Eremascaceae of Some Authors). Typically forming a branched mycelium with the cells uninucleate or containing several nuclei. Asexual reproduction by the breaking up of branches of the mycelium into "oidia" or by the formation of budding branches. Some species are capable of alcoholic fermentation. Sexual reproduction takes place by the union of projections of usually (but not always) adjacent cells to form an ascus perched upon the tips of the united processes. In some cases the gametangia are distinctly a small antherid terminal to a slender antheridial branch and a large oogone. In some species the two processes do not unite or even only one is produced so that the ascus is formed parthenogenetically. Ascospores are eight or four or less, frequently fewer. They are ellipsoidal or hat-shaped.

Eremascus fertilis Stoppel and *E. albus* Eidam produce eight ascospores after the union of the tips of adjacent cells which may be tightly coiled around one another in the latter species. Frequently union fails to occur and the parthenogenetically produced asci contain four or fewer ascospores. The mycelium produces no conidia (either oidia or budding cells). *Endomyces* is distinguished from *Eremascus* by the production of four or fewer ascospores per ascus and by asexual reproduction by means of oidia. *Endomycopsis* differs from both by the production of budding cells as well as occasionally of oidia. In both genera ascus formation may be parthenogenetic or may be the result of the union of two equal or unequal gametangia. Some species of both genera are able to ferment various sugars but some lack this power. ***Endomycopsis albicans*** **(Vuill.)** Dekker has been found in cases of the disease of the mouth known as thrush although this mostly seems to be due to an asporogenous filamentous yeast. The various other species of the two genera are found in soil and on various vegetable products. It must be noted that these two genera are very close to the true yeasts and possibly the distinction be-

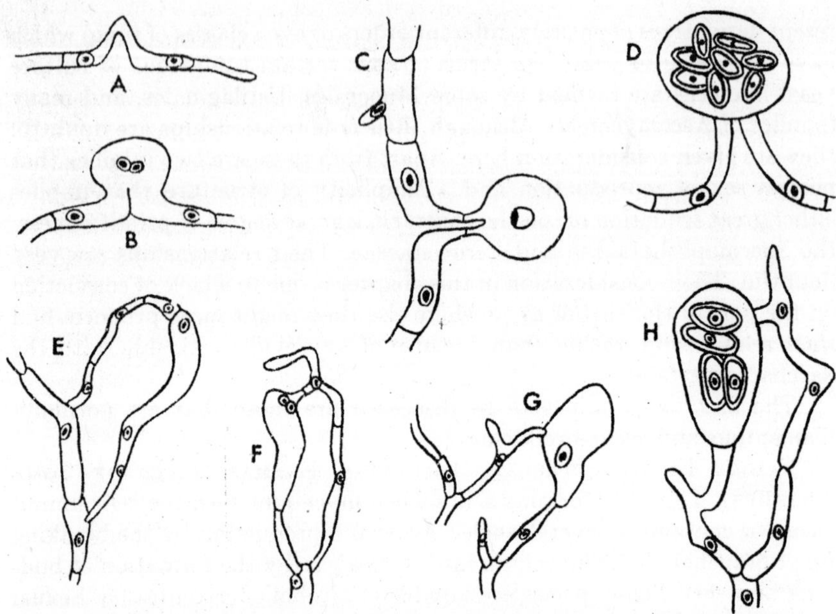

FIG. 109. Saccharomycetales, Family Endomycetaceae. (A–D) *Eremascus fertilis* Stoppel, stages in isogamic sexual reproduction. (E–H) *Endomyces magnusii* Ludwig, stages in anisogamic sexual reproduction. (After Guilliermond: *Ann. fermentations*, 2:129–151, 257–277.)

tween this family and the two families of yeasts will have to be abolished. (Fig. 109.)

Attention should be drawn to two filamentous fungi which agree with this order in simplicity of structure and the formation of asci singly and not in a hymenium. Both are parasitic in the hymenium of species of *Corticium*. In the authors' opinions these are highly derived, simplified forms possibly derived from apothecial genera. They are *Trichomonascus* (Jackson, 1947) and *Myriogonum* (Cain, 1948). Their relationship to the filamentous Saccharomycetales is doubtful, but they may be placed here until they have been studied in culture and their full ontogeny observed.

FAMILY ASCOIDEACEAE. In many respects this family shows relationship to the Endomycetaceae, but the differences seem large enough to warrant their being kept distinct. The two genera *Dipodascus* and *Ascoidea* form well-developed mycelium. In the former asexual reproduction is by the breaking up of the terminal portion of the hypha from the apex downwards to form oidia in one species and is lacking in the other. In *Ascoidea* the conidia are produced singly at the tips of the hyphae which by sympodial development grow by the conidium and produce another one until the hypha appears to have one terminal and many lateral conidia. The asci are elongated and many-spored. They open at the apex and the ascospores escape, sometimes in a worm-like mass.

Dipodascus albidus Lag. was found growing in the gummy exudate of the cut surface of a tropical species of Bromeliaceae. Its septate mycelium consists of multinucleate cells. From two adjacent cells in a hypha arise multinucleate branches which come into contact and fuse at their tips. These branches are then separated from the main hypha by septa. At the tips of each of the gametangia a single "privileged" nucleus enlarges and these two unite in or near the passageway formed by the

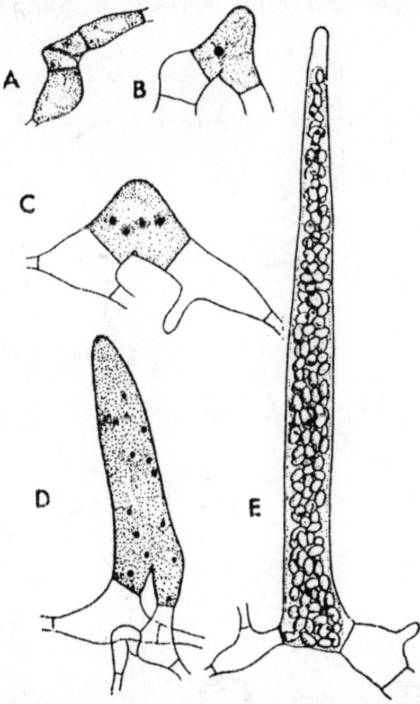

FIG. 110. Saccharomycetales, Family Ascoideaceae. *Dipodascus uninucleatus* Biggs. Stages in sexual reproduction. (A) Early stage in conjugation of adjacent cells. (B) Zygote cell enlarged and nuclei united, thus forming primordium of ascus. (C) Young ascus in 4-nucleate stage. (D) Young ascus with 16 nuclei. (E) Mature ascus. (After Biggs: *Mycologia*, 29(1):34–44.)

dissolution of the intervening walls. From this region grows upward the somewhat tapering ascus into which pass not only the cytoplasm of the gametangia but also their many nuclei and the larger zygote nucleus. The latter divides, apparently meiotically, and probably the four nuclei thus produced continue to divide. It is assumed that it is the nuclei thus formed that serve as the centers about which the ascospores are produced. There are many degenerating nuclei in the ascus which may be the original haploid nuclei introduced from the gametangia. Eventually the apex of the ascus dissolves and the ellipsoidal ascospores escape in a gummy mass. The vegetative hyphae reproduce asexually by the formation of oidia. In *D. uninucleatus* Biggs it was shown by Miss Biggs (1937) that the cells of the mycelium and the gametangia are uninucleate. As a result there are no supernumerary nuclei in the ascus, the many nuclei there produced being the product of the division of the zygote nucleus. Otherwise the ascus resembles that of *D. albidus*. No asexual reproduction has been observed in this species. (Fig. 110.)

Ascoidea, with the single species *A. rubescens* Bref., also grows in plant exudates (such as the beech, *Fagus silvatica* L.) and the mycelium consists of multinucleate cells. No fusion of separate gametangia occurs but the oval asci are produced terminally on upright hyphae. They are from the beginning multinucleate and produce very numerous hat-shaped ascospores like those of *Hansenula* and some species of *Endomyces*.

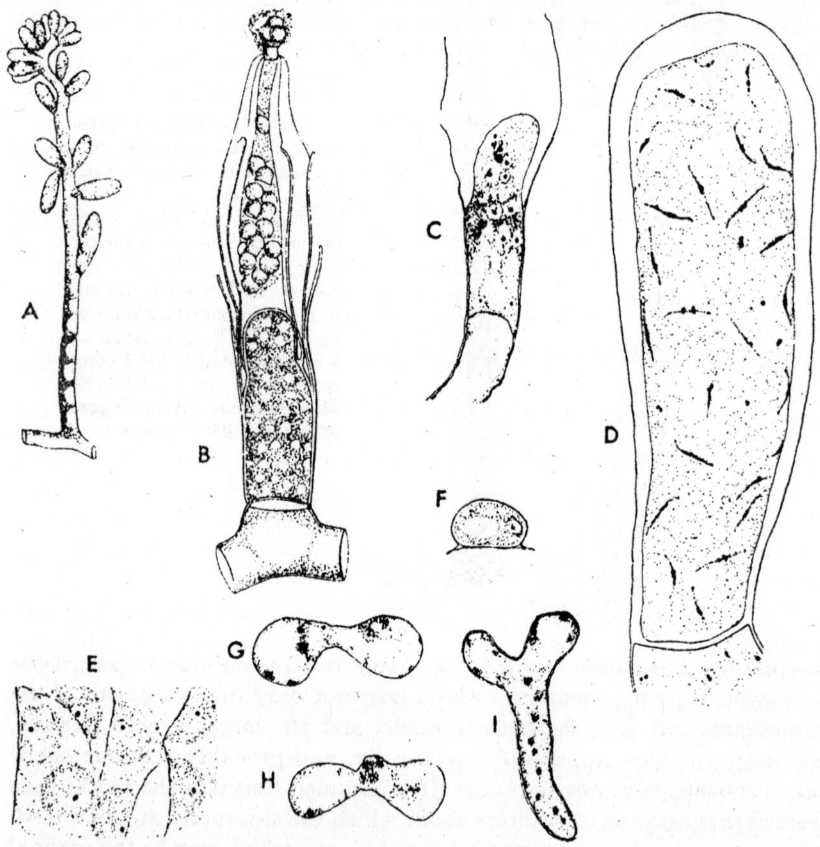

FIG. 111. Saccharomycetales, Family Ascoideaceae. *Ascoidea rubescens* Bref. (A) Sympodial formation of conidia. (B) Proliferated asci. (C) Young multinucleate ascus proliferating in discharged ascus. (D) Ascus approaching spore formation, nuclear divisions simultaneous. (E) Portion of ascus with spores. (F) Mature spore. (G–I) Stages in conjugation and germination of spores. (A–B, After Engler and Prantl: Die Natürlichen Pflanzenfamilien, Leipzig, W. Engelmann. C–I, after Walker: *Mycologia*, **27**(2):102–127.)

According to Varitchak (1931) two of the many nuclei in the young ascus enlarge and then unite. From the division of this zygote nucleus it is supposed that the nuclei of the ascospores are formed. After the ascospores are set free by the dissolution of the apex of the ascus the basal septum arches upwards and a new ascus is produced by proliferation. This may occur many times as in the formation of sporangia in *Saprolegnia.* Miss Walker (1935) was unable to confirm Varitchak's report of the union of two privileged nuclei and believes that the ascus develops parthenogenetically. She noted earlier (1931) that the ascospores frequently fuse by twos after their escape and produce mycelium originating from the conjugation tube connecting the two ascospores. It should be noted that in *Dipodascus uninucleatus* frequently the ascus may be produced parthenogenetically. If that should occur also in *Ascoidea* it may be that both Varitchak and Miss Walker were correct in their reports. (Fig. 111.)

The true yeasts, i.e., those forms of this order that are normally not hyphal, may be divided into the families Schizosaccharomycetaceae in which the cells divide by fission and Saccharomycetaceae in which they divide by budding. Many of them are of great industrial value because of their power of fermenting various sugars and producing alcohol and CO_2, but many are unable to ferment the commoner sugars. Some of the alcohol-producing yeasts grow at the top of the solution to be fermented, the so called "top yeasts," while others are more abundant where the oxygen supply is less, in the lower portion, the "bottom yeasts."

Sexual reproduction is quite varied. Perhaps the more common mode is the union of two cells, more often equal in size but sometimes of unequal size, by means of short tubes. The intervening walls are dissolved and the two nuclei unite, usually in the conjugation tube. The number of ascospores is more often four but in many cases is eight. Probably by failure of some of the nuclei to function in the production of ascospores the number may be less than four, in a few cases only one. The mature ascospores may be distributed evenly or unevenly between the two conjugating cells or may be found in only one. They are more often ellipsoidal but may be hat-shaped or spherical or needle-shaped. A modification of the foregoing type is the conjugation of a mother cell with a bud produced from it. Both cells may produce conjugation tubes or the partition between the larger cell and the bud may dissolve away. This process has been called pedogamy. It is clear that this is a modification of the preceding type. In both types the vegetative cells may be assumed to have haploid nuclei with immediate meiotic division of the zygote (diploid) nucleus to bring about the ascospore production.

In some yeasts the four ascospores unite by twos within the ascus. From the conjugation tube a process pushes through the ascus wall and produces a daughter cell outside. This cell multiplies by budding or fission depending upon the type of asexual reproduction characteristic of the species. Eventually some of these cells enlarge a little and the nucleus divides (probably meiotically) into four and the four ascospores are produced. These in their turn conjugate by twos. Yeasts of this type are apparently diploid during their whole vegetative life, the ascospores alone being haploid. Winge (1935) showed that sometimes a pair of ascospores may be so situated in the ascus that they do not unite. In that case each buds vegetatively. The colonies thus formed consist of haploid cells which are smaller and more rounded than the normal diploid cells of the species. Soon one or more pairs of these haploid cells, of varying degrees of relationship, fuse and the resulting diploid cell multiplies by budding to produce the larger more elongated cells characteristic of the normal vegetative development of the species.

A third category of yeasts has been assumed to produce its asci parthenogenetically. *Saccharomyces cerevisiae* Meyen, the common yeast of bread and beer, does not show conjugation of the ascospores within the ascus nor is such conjugation observed immediately prior to ascus formation. However the Lindegrens (1943, 1949) have shown that the four ascospores of a normal four-spored ascus when germinated separately in the proper culture media produce colonies of haploid cells resembling those mentioned above in the case of failure of a pair of ascospores to conjugate within the ascus. Mostly these remain of the haploid type although occasionally conjugation between two cells may occur to form an "illegitimate" diploid colony. This may, under conditions favorable for ascosporogenesis, produce asci but these usually contain only two ascospores which are mostly incapable of germination. These illegitimate diploid colonies when perpetuated in conditions favorable for asexual multiplication may be of industrial value since these strains are frequently very constant and free from mutations. When the haploid colonies from the four ascospores are mated vigorous diploid colonies develop in certain matings and these under conditions favorable for ascospore formation produce four-spored asci whose ascospores are capable of germination. These matings show that two of the four ascospores in the normal ascus are of one mating type and two of another. The haploid strains from the original four ascospores may develop into lines that under no mating combinations are able to produce ascospores. In other words they may by mutation become asporogenous yeasts to which the name *Torulopsis* (*Torula*) has been given in the past (see below). *Saccharomyces paradoxus* Batsch, as shown by Guilliermond (1936), is variable as to the point where sexual reproduction occurs. Under some conditions the diploid phase vegetative cells become asci and produce ascospores which conjugate in the ascus. From the zygotes thus formed arise by budding the usual vegetative cells of the yeast. Sometimes in the same species under

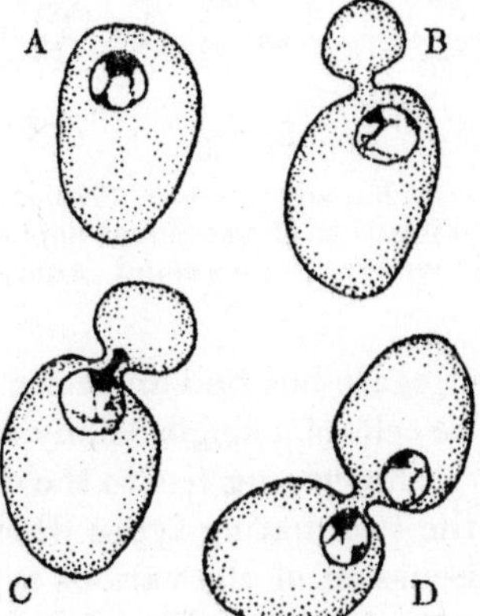

Fig. 112. Saccharomycetales, Family Saccharomycetaceae. *Saccharomyces cerevisiae* Meyen. (A) Cell before beginning of formation of bud. (B) Bud formed but still without nucleus. (C) Nucleus pushing out into bud. (D) Nucleus divided, with one daughter nucleus in each cell. (After Guilliermond: *Ann. Mycolog.*, 2(2):184–189.)

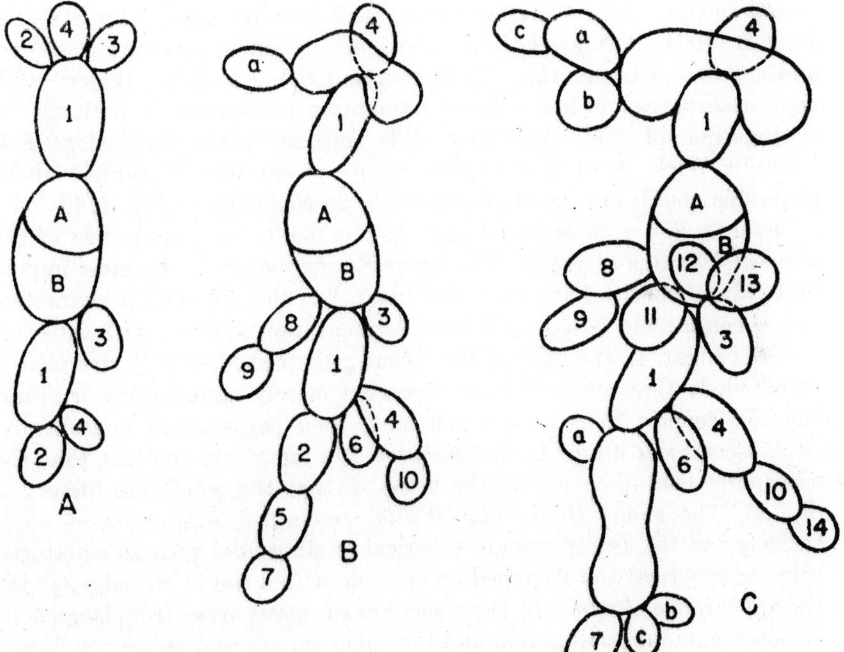

Fig. 113. Saccharomycetales, Family Saccharomycetaceae. *Saccharomyces paradoxus* Batsch. (A) Two ascospores (A and B) have, without conjugation, formed systems of budding cells (1–4) with haploid nuclei. (B) Cells 2 and 3 from ascospore A have conjugated and formed bud (a) with diploid nucleus, and there has occurred further formation of haploid buds in the system from ascospore B. (C) Cells 2 and 5 from ascospore B have conjugated and produced the diploid buds (a), (b), and (c), and further diploid buds (b) and (c) have developed from the previously formed zygote. (Courtesy, Guilliermond: *Ann. fermentations*, 2:129–151, 257–277.)

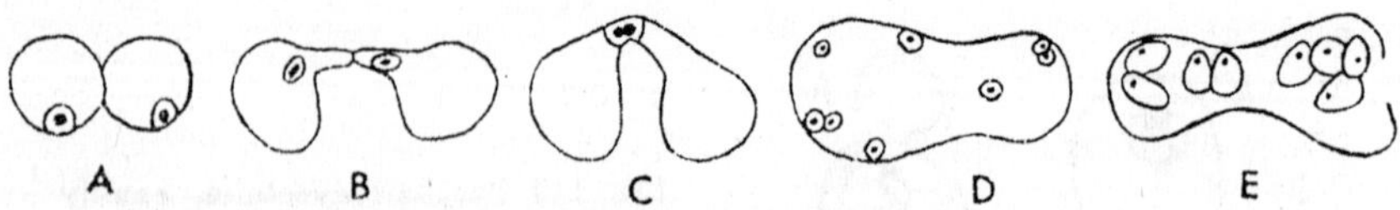

FIG. 114. Saccharomycetales, Family Schizosaccharomycetaceae. *Schizosaccharomyces octosporus* Beijer. (A–E) Stages in the conjugation of two equal haploid cells and the formation of the ascus and ascospores. (Courtesy, Guilliermond: *Ann. fermentations*, **2**:129–151, 257–277.)

other conditions the ascospores do not conjugate but bud and form small colonies of adhering haploid yeast cells. The cells of a single colony arising from a single ascospore conjugate by pairs, thus giving rise to the diploid phase again. This species does not have the two mating types which are possessed by *S. cerevisiae*. A very full discussion of the various types of sexuality in the yeasts is given by Guilliermond (1936). (Figs. 112, 113.)

FAMILY SCHIZOSACCHAROMYCETACEAE. The only genus is *Schizosaccharomyces*. Only two species are definitely recognized by Stelling-Dekker (1931), *Sch. pombe* Lindn., with mostly four (or fewer) ascospores, and capable of fermenting saccharose, and *Sch. octosporus* Beijer., with eight ascospores and incapable of fermenting saccharose. In both species conjugation of two vegetative cells initiates ascus formation. *Sch. versatilis* Wick. & Duprat (1945), by its production of hyphae under anaerobic conditions, shows relationship to *Endomyces*. (Fig. 114.)

FAMILY SACCHAROMYCETACEAE. In this family are found most of the industrial strains of yeasts. The genus *Saccharomyces* is the most important. Many species have been described, but the differential characters have been largely based upon their fermentation abilities with different kinds of sugar. In the light of the recent genetic studies of *S. cerevisiae* it seems likely that many of these represent merely segregations or mutations of genetic characters which have been perpetuated nonsexually. *Saccharomycodes* differs from *Saccharomyces* mainly in the fact that the ascospores conjugate within the ascus so that the whole life history is diploid. The genus *Hansenula* (*Willia*) resembles *Saccharomyces* vegetatively but the ascospores are spherical or ellipsoidal with an equatorial ridge or spherical and flattened on one side with a rim at the edge, giving the appearance of a hat. In *Debaryomyces* the ascus arises from isogamous or heterogamous conjugation and the spherical ascospores are roughened with fine warts. They are usually one (rarely two to three) per ascus. The genus *Nematospora* contains two or three species of yeasts that are parasitic in plant tissues but easily cultivated in various culture media. Asexual reproduction is by budding. In some species short mycelia are produced. The single yeast-like cells or the cells of the mycelium may become transformed into asci within which are produced mostly eight (rarely four or fewer) ascospores. These are usually in two bundles of four spores each, at the two ends of the ascus. The spores are needle-shaped, tapering at one end to a slender flagellum-like thread which is not motile. Sexual union of two cells to produce the ascus has been claimed for *N. lycopersici* Schneid. by Schneider (1916), by Wingard (1925) in *N. phaseoli* Wing., and in *N. coryli* Peglion by Miss Manuel (1938), but is denied by Guilliermond (1928). Probably closely related to this genus is *Ashbya gossypii* (Ashby & Now.) Guill. which causes injury to the bolls of cotton (*Gossypium*). It differs in having a very limited yeast stage, mainly developing a coenocytic, variously branching hypha, one end of which, without a separating septum, functions as the ascus. Guilliermond claims that the nuclei functioning in the ascospore development do not arise from the division of one original ascus nucleus but by the division of several nuclei already present in that portion of the hypha. Both *Nematospora* and *Ashbya* are apparently spread by the punctures of various sucking insects. The genus *Eremothecium* is very similar to *Ashbya*. *E. ashbyii* Guill. (1936a) was, like it, isolated from diseased cotton bolls. It differs chiefly in the slightly different form of the ascospores. In several respects these three genera suggest the gametophytic generation of *Spermophthora gossypii* Ashby & Nowell, which is discussed below (under Family Spermophthoraceae). (Fig. 115.)

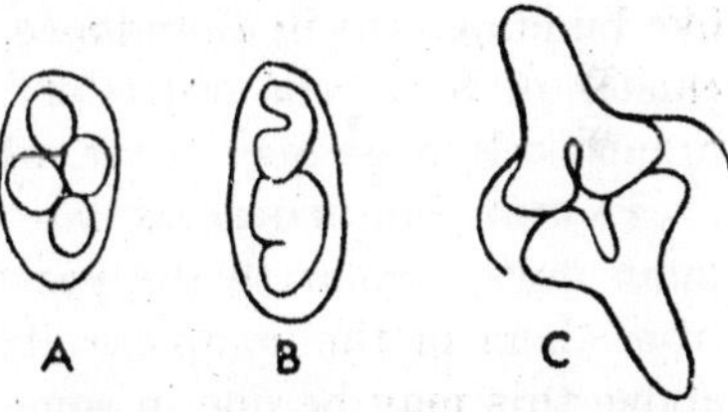

Fig. 115. Saccharomycetales, Family Saccharomycetaceae. *Saccharomycodes ludwigii* Hansen. (A) Ascus with four ascospores. (B) Conjugation of ascospores by twos. (C) Zygotes germinating to form new diploid cells. (Courtesy, Guilliermond: *Ann. fermentations*, 2:129–151, 257–277.)

Monosporella bicuspidata (Metschn.) Keilin, parasitic in *Daphnia* (a Crustacean), and *Coccidiascus legeri* Chatton, parasitic in a species of the fly *Drosophila*, have similar ascospores and possess a budding yeast-like stage. They probably are closely related to the foregoing genera.

The intensive studies by Miss Stelling-Dekker (1931) upon the large collection of yeasts in the Centraalbureau voor Schimmelcultures in Baarn, Netherlands, have led her to the conclusion, already expressed by Van Tieghem in 1906, that the distinction between the three families Endomycetaceae, Saccharomycetaceae, and Schizosaccharomycetaceae cannot be maintained and that they must be united under the older name Endomycetaceae with the four subfamilies Eremascoideae, with the single genus *Eremascus;* Endomycoideae, containing *Endomyces* and *Schizosaccharomyces*; and Saccharomycoideae, with *Endomycopsis* and most of the sporogenous genera of yeast habit; and Nematosporoideae. This scheme of classification was adopted by Guilliermond (1936b) in a contribution in which the various genera are discussed and illustrated.

It seems likely that this tentative scheme may in the future form the best basis for the classification of the sporogenous yeasts and near relatives. Further cytological studies are desirable before this can be considered as entirely settled.

The Asporogenous Yeasts are those in which no formation of asci is known. Since it has been demonstrated by Lindegren and Lindegren (1943) and suggested by others, also, that races of species of the genus *Saccharomyces* may lose their power of ascus production, probably as a result of some gene mutation, the distinction between the two groups is very doubtful as regards some genera.

FAMILY NECTAROMYCETACEAE. This very doubtful family contains the single genus *Nectaromyces* named by the Sydows (1918). This fungus grows in the nectar of flowers and is easily cultivated. It produces yeast-like budding cells in abundance, but under certain conditions produces a cluster of four cells cruciately arranged or "airplane-like." Also long branched hyphae may occur, which bear conidia at their tips.

FAMILY TORULOPSIDACEAE. The members of this family are in the main quite similar to the yeasts of the Saccharomycetaceae but differ from them in the complete absence of ascus formation. As mentioned above this may be due in some cases to gene mutations in sporogenous yeasts. On the other hand it must be borne in mind that yeast-like cells are produced under the proper conditions of environment and nutrition in such widely distinct fungus orders as the Mucorales, Ustilaginales, and some of the Heterobasidiae and Eubasidiae, besides various species of different orders of the Ascomyceteae. Hence it is conceivable that this family may be of heterogeneous origin. Many of the species are capable of fermenting various sugars with production of CO_2 and alcohol or related substances; others can oxidize the sugars in the presence of oxygen but do not cause fermentation. Many of the latter and a few of the former are pathogenic in Man and other animals. Following Lodder (1934) and Diddens and Lodder (1942) the family is divided into two subfamilies: Torulopsoideae, in which no mycelial stages occur and Mycotoruloideae, in which mycelia or pseudomycelia occur, from which at the apex or along the hypha budding takes place. Besides this the fungus may grow as simple budding cells.

In the Torulopsoideae Lodder recognizes seven genera of which *Torulopsis* and *Pityrosporum* may be mentioned. *Torulopsis* is known in medical literature more commonly as *Torula* or as *Cryptococcus* and sometimes as *Blastomyces*. The name *Torula* properly belongs to a genus of filamentous fungi of the Family Dematiaceae of the Fungi Imperfecti. On that account Berlese proposed, according to Lodder, the name *Torulopsis* in 1894 for the yeast-like, asporogenous organisms to which the other names are so often given. The species of this genus resemble those of *Saccharo-*

myces except for the failure to produce asci. The cells are mostly short or long ellipsoidal, propagated by budding and never producing true hyphae. The genus is divided into two subgenera: the species of the first cause the fermentation of various sugars and are almost without exception nonpathogenic to animals, the species of the second subgenus produce no fermentation in sugars although they may oxidize them as a source of energy. A number of these are capable of parasitic life in animals. The genus *Pityrosporum* contains a few species that grow in the skin of Man or other animals, causing scaliness and falling of hairs (pityriasis capitis). Besides the short, oval, budding cells there occur some cells that are flask-shaped, with a larger basal portion and a smaller rounded apical portion. The genus *Mycoderma* of Persoon as amended by Lederle (see Lodder, 1934) is much like *Torulopsis* but under some conditions the budding cells remain attached to form a short branching pseudohypha. The species grow in wine, beer, etc., forming a surface film. They feed upon the sugars and alcohols but produce no true fermentation. The genus *Kloeckera* resembles *Torulopsis* but the cells are in part apiculate, resembling a lemon. Some of the species cause fermentation of various sugars.

In the Mycotoruloideae (established under the name Mycotoruleae by Ciferri and Redaelli in 1929) the cells form definite septate or nonseptate hyphae or pseudohyphae from which sprout crowns or tree-like growths of buds. All the species bud freely and in certain media do not produce mycelium. Many are pathogenic for animals. In medical literature a great many are reported under the generic name *Monilia*. This name is usually reserved by mycologists for saprophytes or parasites on plant tissues for many of which the perfect stage is known to belong to the Discomycetous family Sclerotiniaceae. *Blastodendrion* produces pseudomycelia from the cells of which sprout crown-like or branching clusters of ellipsoidal cells. A number of species growing more or less pathogenically occur in Man in the alimentary canal or on moist skin, nails, etc. *Candida* (synonyms: *Monilia*, *Geotrichoides*, etc.) produces regularly septate hyphae at whose apices chains of yeast-like cells may be produced as well as single such cells or clusters at the joints of the hypha. Many species are pathogenic. The species reported as *C. albicans* (Robin) Berkh. is found as a cause of the disease of the mouth known as "thrush." It is placed by C. W. Dodge (1935) in a closely related genus *Syringospora*. This fungus is often confused with *Endomycopsis albicans* (Vuill.) Dekker which occurs in cases of thrush and is similar to the foregoing in many respects, but produces asci and ascospores. The classification by C. W. Dodge (1935) of the sporogenous and asporogenous yeasts and nearly related filamentous fungi that are pathogenic to animals is very different from that given above which is based mainly upon Ciferri and Redaelli (1929) and Dekker (1931) and Lodder (1934). This monumental work of

Dodge should be consulted by all mycologists interested in medical mycology.

FAMILY RHODOTORULACEAE (THE RED YEASTS). This family was set apart by Lodder from the Torulopsidaceae because of the production in the yeast-like cells of carotinoid pigments, thus giving the colonies a red to orange color. *Rhodotorula* is the one genus recognized. The cells are round to elongated and bud freely. Occasionally the resulting cells remain attached in a row to form a short moniliform pseudohypha. No true fermentation of sugars occurs but dextrose and various other sugars are oxidized in aerobic respiration. In some species growth occurs in ethyl alcohol. Some species are soil inhabitants, probably saprophytes, but some are found in human sputum and others apparently attack hair. The relationship of this genus is uncertain. It may be a true yeast that has lost its power of producing asci or it may be a much reduced, yeast-like fungus of entirely different origin.

FAMILY SPOROBOLOMYCETACEAE. In 1930, Derx recognized two genera, with seven species in *Sporobolomyces* and two in *Bullera*, and another genus, *Tilletiopsis*, which he suggested showed affinities to these. In 1948, the same author described another genus, *Itersonilia*, which together with the other three he now recognizes as belonging to the family. They are all characterized by the production of aerial "ballistospores," a name proposed by Derx for spores usually asymmetrically perched at the apex of

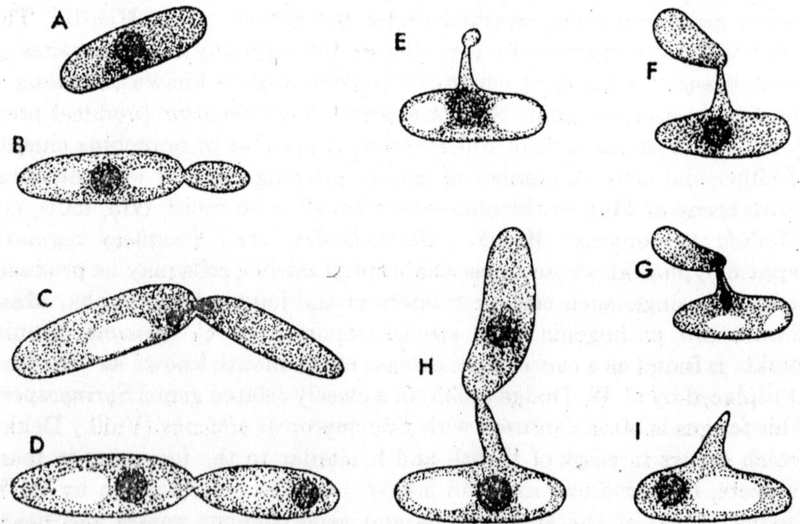

FIG. 116. Saccharomycetales (?), Family Sporobolomycetaceae. *Sporobolomyces roseus* Kl. & van N. (A) Vegetative cell. (B–D) Stages in formation of bud. (E–H) Stages in formation of aerial spore. (I) Mother cell after discharge of spore. (Courtesy, Buller: Researches on Fungi, vol. 5, London, Longmans, Green & Co.)

the sterigmata and discharged violently, often with the production of a droplet of water at the base of the spore just before its discharge. As in the Subclasses Teliosporeae and Heterobasideae these spores are capable of germination by repetition, i.e., by forming sterigmata upon which similar, but somewhat smaller ballistospores are produced. In the first two genera colonies are formed in which the usual mode of reproduction is by budding, as in yeasts, producing considerable mucilage. In the other two genera true hyphae are produced and no budding occurs. Nyland (1949) has described still another genus clearly belonging to this family, to which he has given the name *Sporodiobolus*. Like *Itersonilia* the mycelium has an abundance of clamp connections. The cells are binucleate. An abundance of chlamydospores is produced in *Sporidiobolus*. These are binucleate but soon the nuclei unite and the spore becomes golden-brown. In the younger stages of growth the cells bud, as in yeasts, forming a yeast-like colony similar to that of *Sporobolomyces*. No asci or ascospores occur in any of the four genera. These fungi occur on leaves covered with sooty mold or injured by insects or by fungus parasites, and frequently may be isolated from the soil. The aerial ballistospores are distributed by air currents. (Fig. 116.)

In *Sporobolomyces* a carotinoid pigment is present so that the colonies are red to salmon colored, but in *Bullera* the pigment is lacking and the colonies are pallid to yellowish. Neither of these two genera normally produces hyphae, but it should be noted that sometimes in *Sporobolomyces*, in old cultures, a few short, branching hyphae may arise and on them develop laterally and terminally the characteristic spore-bearing sterigmata. In normal development the first extensive growth is by budding. Later some of the surface cells send out one (or even two or three) sterigma each, which in some cases may become forked. Perched obliquely at the tip of the sterigma a usually asymmetrical spore is formed, like a basidiospore on its sterigma.

The genera *Tilletiopsis* and *Itersonilia* form definite hyphae but not yeast-like colonies of budding cells. From these hyphae arise sterigmata from which the spores are discharged as described above. In *Itersonilia* a clamp connection is formed at every septum while these are lacking in *Tilletiopsis*.

Kluyver and van Niel (1924) suggested that this mode of spore discharge of the aerial spores warrants the idea that these fungi may be very much reduced Basidiomycetes. Buller (1933) investigated this process more fully and accepted the idea. Lohwag (1926) and Guilliermond (1927) did not agree with this suggestion because the cells of the then recognized species are uninucleate and there is no fusion of nuclei prior to spore formation, so that these cells can not be considered as much reduced basidia. Buller, on the other hand, pointed out that in a number of the

Basidiomyceteae the cells are uninucleate, including the basidia, and there is no nuclear fusion in the latter. This can be explained, according to him, by loss of sexuality, as has happened in many of the Saccharomycetales and other fungi, or by the occurrence of + and – strains which in the absence of the reciprocal strain proceed to produce their basidiospores parthenogenetically. With the discovery of *Itersonilia*, with its mycelium made up of dicaryon cells (as evinced by the clamp connections), and of *Sporidiobolus* it seems that probably these are really much reduced forms of Basidiomyceteae from either the Subclass Teliosporeae or Subclass Heterosporeae, in both of which the teliospores often germinate by repetition.

The two families whose discussions follow are of very uncertain relationship. It is possible that the Spermophthoraceae are not far from the Endomycetaceae. The Pericystaceae are of still more doubtful kinship. Both need further investigation in order to confirm the validity of the reported life histories and the conclusions drawn therefrom as to their relationship.

FAMILY SPERMOPHTHORACEAE. The only known species, *Spermophthora gossypii* Ashby & Nowell, was determined to be the cause of the disease called stigmatomycosis affecting the seeds of cotton (*Gossypium*) and fruits of tomato (*Lycopersicon*) in the West Indies. This was studied first by Ashby and Nowell (1926) and further by Guilliermond (1928). The germinating ascospores give rise to a nonseptate, coenocytic, dichotomously branching mycelium with apical growth. The apices of the branches may continue to grow while the older parts die and are cut off by callose plugs as in some Phycomyceteae. No cellulose reaction is shown upon treatment with chloriodide of zinc. Sometimes small cells bud off from this mycelium and these may bud in turn, but it has not been ascertained whether these are functional asexual spores. A short distance back from the hyphal tips the mycelium forms spindle-shaped swellings, the gametangia, which become separated from the usually short tip cells by a cross wall, another septum setting this swelling off from the main portion of the mycelium. The gametangia contain at first from four to ten nuclei. These divide twice simultaneously and the resulting nuclei and most of the cytoplasm form a dense axial strand surrounded by a vacuolate epiplasm. The axial portion divides into fusiform, uninucleate cells which enlarge at the expense of the epiplasm. The rupture of the gametangium wall permits the nonmotile gametes to escape. Whenever two lie in contact they unite by a conjugation tube within which, usually, the two nuclei unite. From this conjugation tube there grows out a rather limited branched septate mycelium of uninucleate cells. The ends of the hyphae enlarge and are cut off by a septum and become spherical asci. The nucleus divides three times and around each nucleus is formed a more or

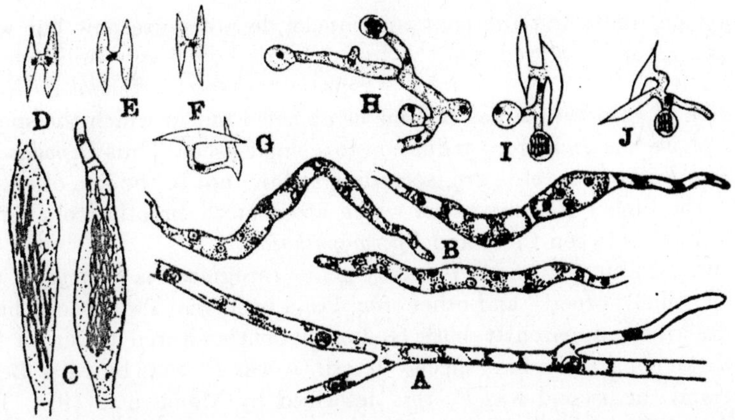

FIG. 117. Saccharomycetales (?), Family Spermophthoraceae. ***Spermophthora gossypii*** Ashby & Nowell. (A) Gametophytic, coenocytic mycelium. (B) Developing gametangia. (C) Mature gametangia containing gametes. (D–G) Stages in union of gametes. (H–J) Sporophytic mycelium bearing asci. (After Guilliermond: ***Rev. gén. botan.***, **40**:328–704.)

less fusiform or lemon-shaped ascospore. As the eight ascospores enlarge they use up the epiplasm. They usually remain in a rather compact bundle even after the ascus wall disappears. Rarely a gamete germinates parthenogenetically to form a septate mycelium bearing apparently normal asci. Although the cytological details could not be made out Guilliermond believed that the first two divisions of the ascus are meiotic so that the fungus shows an alternation of a gametophyte with haploid nuclei and a sporophyte with diploid nuclei. In 1936 he reported that in cultures maintained for ten years the fungus had lost its power to produce the ascogenous mycelium. The gametes from the gametangia no longer underwent conjugation but germinated directly to repeat the gametophytic generation. He suggested that *Ashbya* and *Eremothecium* may represent a similar development in which sexuality has been lost and only an asexual generation is present. (Fig. 117.)

Guilliermond (1928) believed that *Spermophthora* occupies a position intermediate between the Phycomycetes and Ascomycetes. Comparing it with higher Ascomyceteae (e.g., *Pyronema*) he considered its sporophytic stage with uninucleate cells and diploid nuclei to be homologous to the system of ascogenous hyphae with cells containing pairs of haploid nuclei. In *Spermophthora* the single diploid nucleus in the ascus divides to form the nuclei of the ascospores; in *Pyronema* the two haploid nuclei of the ascus unite and then divide to form the nuclei of the ascospores. In both the ascospores with their haploid nuclei give rise to a gametophyte. In *Spermophthora* the gametangia produce individual gametes which are set free and unite; in the higher Ascomyceteae two gametangia (antherid and oogone) unite but the contained nuclei do not unite now but wait until the ascus is formed. On the basis of this homology Guilliermond considered the Saccharomycetales to represent a lateral offshoot in which the union of gametangia has become established but in which the sporophytic phase has undergone reduction to a single ascus. Thus *Dipodascus, Ascoidea, Endomyces,* etc., are, according to him, not in the line of evolution to the higher Ascomyceteae which arose more directly from forms intermediate between them and *Spermophthora.*

FAMILY PERICYSTACEAE. The only genus recognized is *Pericystis,* the cause of "chalk-brood" and other troubles in beehives. Two species have been described, apparently both feeding upon the stored pollen in the cells of the comb. The first species described was *P. alvei* by Miss Betts (1912) and the second was *P. apis* described by Maassen in 1916. The latter besides feeding upon the pollen appears to attack the larvae in the cells of the comb. It was given careful study by Claussen (1921). The systematic position of this genus is in doubt. Fitzpatrick (1930) placed it among the doubtful Phycomyceteae. Varitchak (1933) who studied the life history and nuclear behavior of *P. apis,* concluded that it is a primitive Ascomycete related to *Dipodascus* and *Ascoidea* but far nearer the supposed phycomycetous ancestors. The mycelium, as in *Ascoidea,* contains chitin and not cellulose. As in that fungus it is septate with multinucleate segments, the septa being centrally perforated, thus permitting the flow of cytoplasm from segment to segment. The mycelia are of two sexes, the fungus being heterothallic. When they come into contact multinuclear gametangia are produced, each separated by its septum from the main hypha. They resemble at first the early stages of conjugation in *Mucor.*

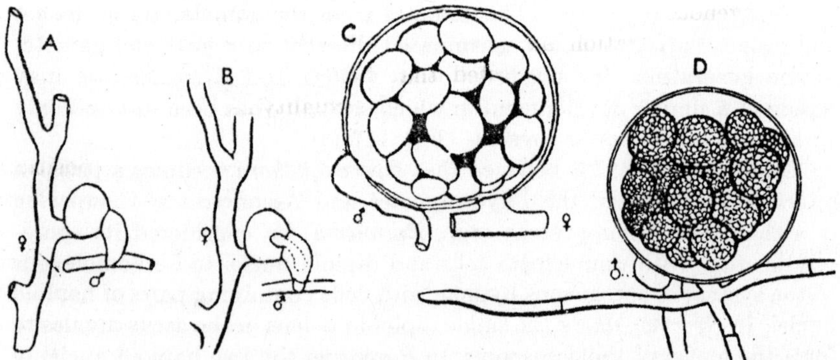

FIG. 118. Fungus of uncertain relationship. Family Pericystaceae. *Pericystis apis* Maassen. (A) Conjugating gametangia. (B) The same, in optical section. (C) Oogone with zygotes. (D) Oogone with zygotes which have produced spores and thereby have become spore balls. (After Claussen: *Arb. biol. Reichsanstalt Land- u. Forstw.,* **10**:467-521.)

The female gametangium enlarges and the male gametangium sends a conjugation tube into it. Numerous male nuclei and some cytoplasm enter the oogone, whose nuclei divide several times as do the introduced male nuclei. Many nuclei of both sexes degenerate but many pairs of uniting nuclei are found. Around each such zygote nucleus is organized a mass of cytoplasm, called by the investigator an "egg." There is no cell wall between this egg and the rest of the protoplasm. The number of such zygote nuclei and eggs varies, usually being large but being reduced to one in rare cases. Each zygote nucleus divides several times and the egg cytoplasm undergoes cleavage until as many spores are formed as there were nuclei produced. Spore walls are formed and within the gametangium are now found as many spore balls as there were uniting pairs of nuclei. Varitchak called each ball of spores an ascus and the whole structure a "synascus." By reduction of the number of uniting nuclei and resultant spore balls a condition would be attained, according to him, similar to that in *Dipodascus* whose ascus he calls a "hemiascus." By reduction of the number of nuclei in the uniting gametangia to one in each as in *Endomyces* a true ascus would be formed. (Fig. 118.)

Key to Families of Order Erysiphales

Aerial mycelium hyaline. Outer colored layer of perithecial wall one cell in thickness, of polygonal cells, brittle. Family Erysiphaceae

Aerial mycelium (when present) dark. Outer peridium layer not brittle.

- Peridium parenchymatous, not slimy; mycelial hyphae cylindrical, not slimy. Family Meliolaceae
- Peridium dissolving into slime; mycelial hyphae moniliform or cylindrical. Family Englerulaceae
- Perithecium walls built of moniliform or cylindrical hyphae, more or less slimy. Mycelium moniliform or cylindrical, united in strands. Family Capnodiaceae
- Mycelium dark-colored, parasitic on epiphyllous Meliolaceae and similar fungi. Perithecium inverted at maturity, with the morphological base upward. Family Trichothyriaceae
- Forming rounded or stellate cushions on leaves, gelatinous when wet, horny when dry, with no free mycelium. Asci scattered in thickened area of the thallus. Family Atichiaceae

Key to the Genera of Family Erysiphaceae

Ascospores one-celled.

- Perithecia normally containing only one ascus.
  - Appendages hypha-like, usually unbranched. *Sphaerotheca*
  - Appendages stiff, dichotomously forked at tip. *Podosphaera*
- Perithecia normally containing several to many asci.
  - Appendages hypha-like, mycelium external, conidia in chains. *Erysiphe*
  - Appendages hypha-like, mycelium internal as well as somewhat external, conidia falling off singly, sometimes producing short chains. *Leveillula*

Appendages more or less stiff, dichotomously forked at tip. *Microsphaera*
Appendages hooked or coiled at tip.
Mycelium entirely external, conidiophores straight. *Uncinula*
Mycelium internal as well as external, conidiophores with spirally twisted base. *Uncinulopsis*
Appendages with swollen base, straight and pointed, mycelium external and in substomatal chambers. *Phyllactinia*
Appendages wanting, surface of perithecium without gelatinous cells. *Brasiliomyces* (Viegas, 1944)
True equatorial appendages wanting, on upper surface of perithecium numerous gelatinous penicillate cells. *Typhulochaete*
Ascospores two-celled.
Appendages hypha-like. *Chilemyces*
Appendages dichotomously forked at tip. *Schistodes*
Appendages lacking, mycelium and spores cinnamon-yellow. *Astomella*
Ascospores four-celled, appendages hypha-like. *Leucoconis*

Key to the Commoner Genera of Family Meliolaceae

Perithecia or mycelium intramatrical.
Perithecia subepidermal, asci many-spored, spores hyaline, two-celled. *Pampolysporium*
Perithecia extramatrical, hyaline, mycelium in the epidermis but breaking out as a superficial dark mass. Spores two-celled, brown, perithecia scattered. *Alina*
Perithecia and dark mycelium external but emerging in strands through the stomata.
Perithecia on peg-like strands from the stomata. *Stomatogene*
Mycelium superficial but penetrating the stomata in narrow bundles. Perithecia on short mycelial branches. *Piline*
Perithecia and mycelium entirely external.
Mycelium with hyphopodia but no bristles. *Irene*[3]
Mycelium with hyphopodia and bristles. *Meliola*
Mycelium without hyphopodia.
With bristles, spores brown, two-celled. *Phaeodimeriella*
With bristles, spores brown, four- to five-celled. *Meliolina*
Without bristles, spores hyaline, two-celled *Dimerina*
Without bristles, spores brown, two-celled, perithecium rust-colored. *Parodiopsis*
Without bristles, spores brown, two-celled, perithecium black. *Dimerium*

Key to the Commoner Genera of Family Englerulaceae

(*Based on Theissen and Sydow, 1917, but see Petrak, 1928*)

Perithecial walls of rounded, soft cells, deliquescing into slime in which these cells are scattered.

[3] Stevens (1927) segregates *Irenopsis* and *Irenina* from the genus *Irene* on some minor characters, and also includes in this family (Meliolaceae) *Actinodothis* and *Amazonia* which are placed by Theissen and Sydow in Order Hemisphaeriales.

Perithecia sessile, with many asci.
  Mycelium with hyphopodia.
    Spores two-celled, colorless. *Schiffnerula*
    Spores two-celled, brown. *Phaeoschiffnerula*
  Mycelium without hyphopodia, spores two-celled, brown *Englerula*
Perithecia stalked, with one ascus, spores two-celled, brown *Thrauste*
Perithecial walls of parallel hyphae which separate at maturity. Mycelium extensive, with hyphopodia, ascospores two-celled, brown. *Parenglerula*

Key to the Commoner Genera of Family Capnodiaceae

Perithecia stalked or elongated vertically, wall of parallel hyphae.
  Mycelium with parallel walls, forming a thick spongy mass, perithecium long-stalked, round, spores four-celled. *Scorias*
  Mycelium of moniliform hyphae, perithecia vertically elongated, sessile or stalked, spores muriform. *Capnodium*
Perithecia sessile or on a very short stalk, spherical, walls mostly of monilioid hyphae.
  Bristles on mycelium or perithecia.
    Perithecia without bristles, spores two-celled, hyaline. Tangled bristles on the mycelium which has no hyphopodia. *Dimerosporina*
    Perithecia without bristles, spores two-celled, brown, ascus single; short, dark bristles on the mycelium which has hyphopodia. *Balladyna*
    Perithecia with several asci, spores four- to more-celled, hyaline, bristles on perithecia or mycelium or both. *Chaetothyrium*
  Bristles lacking.
    Spores several-celled, colorless. *Limacinia*
Various other genera based on form and color of spores, etc.

Key to Some Genera of Family Trichothyriaceae

Forming ribbon-like strands of hyphae covering the mycelium of the host fungus.
  Spores colorless, two-celled. *Trichothyrium*
  Spores colored, two-celled. *Trichothyriella*
  Spores colorless, three- to four-celled. *Trichothyriopsis*
Mycelium disappearing early, on stromata of fungi, spores colorless, two-celled. *Loranthomyces*

Key to the Genera of Family Atichiaceae

Propagula clustered in basket-like structures. *Atichia*
Propagula in separate pockets. *Phycopsis*

Key to the Families of Order Aspergillales

Ascocarps subterranean, 1 cm. or more in diameter, cortex of many cell layers, surrounding a "gleba" of irregularly arranged asci, with a central sterile portion. Conidia not known. Family Elaphomycetaceae
Ascocarps not subterranean, external or in some cases buried in the substratum, mostly not over 2–4 mm. in diameter.

Ascocarps with peridium consisting of loosely interwoven hyphae, conidia formed in chains. Family Gymnoascaceae

Ascocarps at maturity with thin cortex or more stroma-like with a thick cortex surrounding one or more glebal masses of asci. Conidia mostly catenulate. Family Aspergillaceae

Ascocarps stromatic at base, the mass of asci and ascospores pushing out of the top as a columnar structure. Conidia not known. Family Trichocomaceae

Ascocarps stalked, with a spherical head which dehisces variously at maturity to release the spores. Conidia not known. Family Onygenaceae

(If Nannfeldt (1932) is followed the two families below must be added to this key.)

Perithecia thin-walled, with long slender ostiolate necks, from whose apex the hyaline ascospores are exuded in a slimy drop. Family Ophiostomataceae

Perithecia thin-walled, without neck, covered with numerous long, dark-colored, variously branched or coiled hairs. Ascospores dark and extruded in a slimy drop. Family Chaetomiaceae

Key to the Commoner Genera of the Family Aspergillaceae

Perithecia without neck or ostiole.

Perithecia without distinct hair coating or appendages, at least at maturity.

Perithecia small, usually bright-colored, at maturity reduced to a thin cortex, containing the ascospores freed by the dissolution of the internal tissues and the eight-spored asci. Ascospores biconvex, with a grooved edge, often with two ridges. Conidial stage *Aspergillus*. *Eurotium* or *Aspergillus*

Perithecia small, bright- or dark-colored, of two types: sclerotial with thick, firm cortex with the central portion occupied by the eight-spored asci and with ascospores as in the preceding, or with loose, hyphal cortex and the ascospores without marginal groove. Conidial stage *Penicillium*. *Penicillium*

Perithecia small, at maturity with thin cortex, at first containing many eight-spored asci, but these and the surrounding tissues dissolve so that the spherical or ellipsoidal hyaline or red ascospores lie free. Mycelium red or purple in mass. Conidia single or in short chains. *Monascus*

Perithecia 0.5–1 mm. in diameter, with a thick, firm cortex containing numerous ovoid asci, each with eight spherical, roughened, yellow to brown ascospores. Conidial stage *Gliocladium*. *Lilliputia*

Perithecia stalked, 2–3 mm. in diameter, with firm cortex, and containing several irregular, gleba-like masses of sporogenous tissue. Asci two- to eight-spored, ascospores with several ridges running in different directions. Conidial stage resembling *Penicillium* or *Sterigmatocystis*. Tropical. *Penicilliopsis*

Perithecia with distinct hair coating or coiled appendages.

Spherical or depressed, with simple or club-shaped hairs. Cortex firm and thick. Ascospores small, ellipsoidal or lenticular, colored. *Cephalotheca*

Depressed, often bluntly angled, with long appendages at each corner, coiled at the tip. *Magnusia*

Perithecia with neck or ostiole. Doubtfully belonging to this order. Here could be placed *Microascus* and *Emericella* if they are transferred from the Order Sphaeriales.

Key to Some Genera of Family Myriangiaceae[4]

Parasitic upon scale insects, but possibly eventually entering and parasitic upon the tissues of the plant host. Basal stroma well developed, bearing one or more disk-like ascigerous portions (ascomata) throughout whose tissues the asci are scattered in no definite layers. On stems and leaves of various plants, largely tropical.
- Ascomata usually several; ascospores strongly muriform, hyaline or only pale-colored. *Myriangium*
- Ascoma usually single, with a narrower base; ascospores several times transversely septate, occasionally one or two cells longitudinally divided, hyaline or only pale-colored. *Kusanoa*

Parasitic within the tissues of leaves, stems, and fruits, or on their hairs, or in some cases possibly parasitic upon fungi within the plant hosts.
- Attacking only the hairs of the host plants, forming little ascomata, within which only a few asci are produced; ascospores mostly muriform.
  - Ascoma globular, with homogeneous tissue. *Mölleriella*
  - Ascoma globular, strongly gelatinized except a few traversing brown hyphae. *Nostocotheca*
  - Ascoma flattened. *Saccardinula*
- Attacking the synnemata of *Helostroma*, and forming a disk-like or pulvinate ascoma, attached by a narrow base. Ascospores strongly muriform and dark-colored when mature. *Cookella*
- Parasitic within the tissues of the host plant.
  - Ascigerous portion strongly developed outside the host tissue, spreading from a narrower base. *Uleomyces*[5]
  - Ascigerous portion not strongly distinct from the internal stromatic tissue, sometimes barely rupturing the epidermis, or forming a well-developed external cushion. Asexual reproduction by acervuli characteristic of the form-genus *Sphaceloma*. *Elsinoe* (*Plectodiscella*)

Key to the Genera of Family Gymnoascaceae

Hyphae of the ascocarp without appendages.
- Ascospores hyaline, yellowish or red. *Arachniotus*
- Ascospores brown to brown violet. *Amaurascus*

Hyphae of the ascocarp with appendages.
- Appendages consisting of spines or prongs. *Gymnoascus*
- Appendages circinate. *Myxotrichum*
- Appendages comb-like. *Ctenomyces*

Key to the Genera of Family Elaphomycetaceae

Ascocarps with sterile base. *Ascoscleroderma*
Ascocarps without sterile base. *Elaphomyces*

---

[4] This tentative key is modified from Theissen and Sydow (1917) and Arnaud (1925). Further studies are needed to reveal whether the many genera ascribed to this family are valid.

[5] The genera *Myriangina*, *Myrianginella*, and *Kusanoopsis* are probably closely related to *Uleomyces* and *Elsinoe* and perhaps all five should be united in the one genus *Uleomyces*

Key to the Families of Order Saccharomycetales and Appended Fungi

Asci formed, at least under special conditions.
  Under normal conditions producing a mycelium of uninucleate or multinucleate cells.
    Asci containing from one to eight ascospores. Family Endomycetaceae
    Asci containing very numerous ascospores. Family Ascoideaceae
    An alternation of haploid and diploid generations. Family Spermophthoraceae
  Under normal conditions producing a yeast type of growth.
    Cells dividing by fission. Family Schizosaccharomycetaceae
    Cells multiplying by budding. Family Saccharomycetaceae

(The following key is based upon Lodder, 1934.)

Ascus formation unknown (Asporogenous Yeasts).
  No violently expelled aerial spores formed (ballistospores)
    Without carotinoid pigments
      Cells reproducing by budding, but also producing aerial conidia on short conidiophores or sterigmata Family Nectaromycetaceae
      Cells reproducing by budding, but aerial conidia lacking. In some cases short chains of buds formed or even hyphae. Many parasitic in Man and other animals. Family Torulopsidaceae
    Cells with carotinoid pigments, reproduction mostly by budding, no true conidia or hyphae Family Rhodotorulaceae
  Aerial conidia formed which are violently expelled, usually also budding present Family Sporobolomycetaceae

Keys to the Commoner Genera of Saccharomycetales and Appended Fungi

Key to the Commoner Genera of Sporogenous Yeasts Included in Families Endomycetaceae, Schizosaccharomycetaceae, and Saccharomycetaceae

(Based upon Stelling-Dekker, 1931)

Spores spindle-shaped.
  Only one spore per ascus. Parasitic in Arthropods. *Monosporella*
  At least four spores per ascus.
    Spores with a nonmotile, flagellum-like extension at one end. Parasitic in plant tissues.
      Mostly growing in yeast-like colonies. *Nematospora*
      Mostly growing as hyphae; ascospores formed in the terminal portion of a hyphal branch, not separated by a septum from the remainder of the hyphae. *Ashbya*
    Spores without flagellum-like extension.
      Parasitic in plant tissues, forming hyphae. *Eremothecium*
      Parasitic in insects, hyphal formation lacking. *Coccidiascus*
Spores not spindle-shaped.
  Producing true hyphae.
    No production of oidia or budding cells. Asci mostly eight-spored. *Eremascus*
    Oidia produced, asci mostly with four or fewer spores. *Endomyces*
    Asexual reproduction by budding, rarely by oidia. *Endomycopsis*

Normally not producing hyphae or only meagerly.
Asexual reproduction by fission. *Schizosaccharomyces*
Asexual reproduction by budding.
Spores hat-shaped (see also species of *Endomyces*).
Buds produced from all sides of the cells. *Hansenula*
Budding bipolar, i.e., from opposite ends of the cells. *Hanseniaspora*
Spores round with weakly or strongly verrucose walls, one, rarely two per ascus.
Buds produced from all sides of the cells. *Debaryomyces*
Budding bipolar (mostly), buds with a narrow base. Spores with an equatorial ridge. No completed conjugation, though rudimentary conjugation tubes are produced. *Schwanniomyces*
Budding bipolar, but with a broad base. Spores without equatorial ridge. Zygotes produced by union of a mother and a daughter cell. *Nadsonia*
Spores round to oval, smooth, one to four per ascus. Conjugation taking place between pairs of ascospores in the ascus. Vegetative cells large, with bipolar buds with broad base. *Saccharomycodes*
Spores round to oval, smooth, one to four per ascus. Conjugation occurring immediately before ascus formation or between vegetative cells in colonies arising from germination of the spores. Not forming a thick surface layer on beerwort. Fermentation of glucose marked. *Saccharomyces*
Spores round to angular, one to four per ascus. Conjugation before ascus formation or unknown. Forming a thick surface layer almost immediately on beerwort. Fermentation feeble or lacking. *Pichia*

Key to the Commoner Genera of Family Torulopsidaceae

(Based upon Lodder, 1934; Diddens and Lodder, 1942)

Not forming true mycelium. Cells mostly single or in irregular groups. Subfamily Torulopsoideae
Cells prevailingly limoniform, with bipolar budding. *Kloeckera*
Cells prevailingly triangular, budding at the three angles. *Trigonopsis*
Cells prevailingly flask-shaped, budding mainly at the broader basal part. *Pityrosporum*
Cells round, ovoid or cylindrical.
Producing a thick surface growth on beerwort. No fermentation of glucose.
Buds often separating from mother cells by fission. *Schizoblastosporion*
Buds separating from mother cells in the usual manner. *Mycoderma*
Not producing a thick surface growth on beerwort. Fermentation of glucose in some species, no fermentation in others. *Torulopsis*
Forming pseudomycelium or true mycelium. Many species pathogenic to animals. Subfamily Mycotoruloideae
Pseudomycelium or true mycelium, reproduction by budding.
Cells often abruptly pointed, under aerobic conditions strong acid production. *Brettanomyces*

Cells not pointed, under aerobic conditions not strongly acid producing.
Chains of yeast-like buds produced at the ends of the hyphae and single cells or clusters at the joints of the hyphae. Many species are pathogenic to animals. *Candida*
Crown-like or branching clusters of yeast-like cells produced from the cells of the pseudomycelium. Mostly pathogenic to animals. *Blastodendrion*
In addition to budding cells, arthrospores (conidia) are also produced. Mostly pathogenic to animals. *Trichosporon*

Keys to Other Families Possibly Related to Yeasts

Family Nectaromycetaceae
Only genus. *Nectaromyces*
Family Rhodotorulaceae
Only genus. *Rhodotorula*
Family Spermophthoraceae
Only genus. *Spermophthora*
Family Pericystaceae
Only genus. *Pericystis*
Family Sporobolomycetaceae
Forming yeast-like colonies reproducing mainly by budding, rarely few short hyphae.
Cells with red or salmon-colored pigment. *Sporobolomyces*
Cells pallid to yellowish. *Bullera*
Forming hyphae, yeast-like colonies, reproduction by budding usually wanting.
Mycelium without clamp connections, spores falcate. *Tilletiopsis*
Mycelium with clamp connections, spores not falcate, reniform or ellipsoid or ovoid. *Itersonilia*
Forming hyphae with clamp connections, yeast-like budding present in the younger colonies. *Sporidiobolus*

Literature Cited

ALLEN, RUTH F.: A cytological study of Erysiphe polygoni on Delphinium, *J. Agr. Research,* **53**(11):801–818. *Pls.* 1–8. 1936.

AMES, L. M.: A new species of Magnusia, *Mycologia,* **29**(2):222–225. *Figs.* 1–14. 1937.

ARNAUD, G.: Les Astérinées: III. Étude sur les champignons parasites, *Ann. épiphyt.,* **7**:1–115. *Pls.* 1–10. *Figs.* 1–25. 1921.

———: Les Astérinées: IV. Études sur la systématique des champignons Pyrénomycètes, *Ann. sci. nat. Botan.,* Xme sér., **7**:643–723. *Pls.* 1–16. *Figs.* 1–25. 1925.

ASHBY, S. F., and W. NOWELL: The fungi of stigmatomycosis, *Ann. Botany,* **40**(157):69–84. *Pls.* 4–5. 1926.

ATKINSON, G. F.: Phylogeny and relationships of the Ascomycetes, *Ann. Missouri Botan. Garden,* **2**(1–2):315–376. *Figs.* 1–10. 1915.

BARANETZKY, J.: Entwickelungsgeschichte des Gymnoascus Reessii, *Botan. Ztg.,* **30**(10):145–160. *Pl.* 3*A*. 1872.

DEBARY, ANTON: Eurotium, Erysiphe, Cicinnobolus, nebst Bemerkungen über die Geschlechtsorgane der Ascomyceten, *Abhandlungen herausgegeben von der Senckenbergischen Naturforschenden Gesellschaft,* **7**(3–4):361–455. *Pls.* 7–12. 1870.

———: Vergleichende Morphologie und Biologie der Pilze, Mycetozoen und Bacterien, xvi + 558 pp. 198 *figs.* Leipzig, Wilhelm Engelmann, 1884.

Bergman, Bengt: Zytologische Studien über die Befruchtung und Askokarpbildung bei Sphaerotheca castagnei Lev., *Svensk Botan. Tid.*, **35**(2):194–210. *Figs.* 1–3. 1941.

Berlese, Augusto Napoleone: Studi citologici sui funghi: I. Divisione del nucleo e formazione dei conidi in Oidium, *Riv. patol. vegetale*, **6**:66–75. *Pls.* 1–2. 1898.

Betts, Annie D.: A beehive fungus, Pericystis alvei gen. et spec. nov., *Ann. Botany*, **26**(103):795–799. *Pls.* 75–76. 1912.

Biggs, Rosemary: Dipodascus uninucleatus, *Mycologia*, **29**(1):34–44. *Figs.* 1–50. 1937.

Blumer, S.: Die Erysiphaceen Mitteleuropas mit besonderer Berücksichtigung der Schweiz, *Beiträge zur Kryptogamenflora der Schweiz*, **7**(1):1–483. *Figs.* 1–167. 1933.

Boudier, Jean Louis Emile, et N. Patouillard: Note sur deux champignons hypogés, *Bull. soc. mycologique de France*, **16**:141–146. *Pl.* 5. 1900.

Brefeld, Oscar: Die Entwicklungsgeschichte von Penicillium, in Botanische Untersuchungen über Schimmelpilze, Heft. 2, pp. 1–98. *Pls.* 1–8. Leipzig, Arthur Felix, 1874.

———: Basidiomyceten: III. Autobasidiomyceten und die Begründung des natürlichen Systems der Pilze, in Untersuchungen aus dem Gesammtgebiete der Mykologie, Heft. 8, pp. 1–305. *Pls.* 1–12. Leipzig, Arthur Felix, 1889.

———: Lysipenicillium insigne Bref., in *ibid.*, Heft. 14, pp. 209–210. *Pl.* 7. *Figs.* 1–7. Münster i. W., Heinrich Schöningh, 1908.

Brodie, Harold J.: Further observations on the mechanism of germination of the conidia of various species of powdery mildews at low humidity, *Can. J. Research, C*, **23**(4):198–211. *Figs.* 1–7. 1945.

Buller, A. H. Reginald: Sporobolomyces, a Basidiomycetous yeast genus, in Researches on Fungi, vol. 5, pp. 171–206. *Figs.* 84–99. London, Longmans, Green and Company, 1933.

Burkholder, Walter H.: The anthracnose disease of the raspberry and related plants, *Cornell Univ. Agr. Expt. Sta. Bull.*, **395**: 155–183. *Figs.* 12–21. 1917.

Cain, Roy F.: Myriogonum, a new genus among simplified Ascomycetes, *Mycologia*, **40**(2):158–167. *Figs.* 1–56. 1948.

Ciferri, R., and T. Redaelli: Studies on the Torulopsidaceae, *Ann. Mycolog.*, **27**(3–4):243–295. *Pls.* 4–6. 1929.

Claussen, P.: Entwicklungsgeschichtliche Untersuchungen über den Erreger der als "Kalkbrut" bezeichneten Krankheit der Bienen, *Arb. biol. Reichsanstalt Land- u. Forstw.*, **10**:467–521. *Pls.* 3–5. *Figs.* 1–24. 1921.

Clemencet, Marion: Contribution à l'étude du développement et de l'anatomie des Ascomycètes hypogés, *Le Botaniste*, **24**(1–2):2–104. *Pls.* 1–12. *Figs.* 1–7. 1932.

Cotton, A. D.: The genus Atichia, *Roy. Botan. Gardens, Kew, Bull. Misc. Inform.*, **1914**(2):54–63. 5 *figs.* 1914.

Dale, Elizabeth: Observations on Gymnoascaceae, *Ann. Botany*, **17**(67):571–596. *Pls.* 27–28. 1903.

———: On the morphology and cytology of Aspergillus repens de Bary, *Ann. Mycolog.*, **7**(3):215–225. *Pls.* 2–3. 1909.

Dangeard, P. A.: L'origine du périthèce chez les Ascomycètes, *Le Botaniste*, **10**:1–385. *Pls.* 1–91. 1907.

DeLamater, Edward D.: Crozier formation in the Gymnoascaceae: A preliminary note, *Mycologia*, **29**(2):187–198. *Figs*. 1–33. 1937.

Dennis, R. W. G., and E. M. Wakefield: New and interesting British Fungi, *Brit. Mycolog. Soc. Trans.*, **29**(3):141–169. 1946. (Pp. 145–147 and Fig. 5 are devoted to Lilliputia insignis (Winter) n. comb.)

Derx, H. G.: Hétérothallie dans le genre Penicillium (Note préliminaire), *Bull. soc. mycologique de France*, **41**:375–381. 1925.

———: Étude sur les Sporobolomycètes, *Ann. Mycolog.*, **28**(1–2):1–23. *Pl*. 1. 1930.

———: Itersonilia, nouveau genre de Sporobolomycètes à mycélium bouclé, *Bull. Bot. Gard. Buitenzorg Sér*. III, **17**(4):465–472. *Figs*. 1–3. 1948.

Diddens, H. A., und J. Lodder: Die Hefesammlung des "Centraal-Bureau voor Schimmelcultures." Beiträge zu einer Monographie der Hefearten: Teil II. Die anaskosporogenen Hefen, Zweite Hälfte, xii + 511 pp. 99 *figs*., Amsterdam, N. V. Nord-Hollandsche Uitgevers Maatschappij, 1942.

Dodge, Carroll W.: The higher Plectascales, *Ann. Mycolog.*, **27**(3–4):145–184. *Pls*. 1–2. *Figs*. 1–2. 1929.

———: Medical Mycology, 900 pp. 142 *figs*. St. Louis, C. V. Mosby Co., 1935.

Doidge, Ethel M.: A revision of the South African Microthyriaceae, *Bothalia*, **4**(2):273–420. *Pls*. 1–76. 1942.

Eidam, E.: Beitrag zur Kenntnis der Gymnoasceen, *Cohn's Beiträge zur Biologie der Pflanzen*, **3**(2): 267–305. *Illust*. 1880.

Ellis, J. B., and B. M. Everhart: The North American Pyrenomycetes, iii + 793 pp. 41 *pls*. Newfield, N. J., published by the authors, 1892.

Fabian, F. W., and N. B. McCullough: Dissociation in yeasts, *J. Bacteriol.*, **27**(6):583–622. *Pls*. 1–2. 1934.

Fischer, E.: Plectascineae, in A. Engler, und K. Prantl: Die Natürlichen Pflanzenfamilien, Erster Teil, Abt. 1:290–320. *Figs*. 210–227. Leipzig, Wilhelm Engelmann, 1896.

———: Tuberineae, *ibid*., Zweite Auflage, 5b(VIII). Leipzig, Wilhelm Engelmann, 1938.

Fitzpatrick, Harry Morton: The Lower Fungi—Phycomycetes, xi + 331 pp. 112 *figs*. New York, McGraw-Hill Book Co., 1930.

Foëx, Étienne: De la présence de deux sortes de conidiophores chez Oidiopsis taurica, *Compt. rend.*, **154**(4):225–226. 1912.

Fraser, Lilian: An investigation of the "sooty moulds" of New South Wales, I–V, *Proc. Linnean Soc. N. S. Wales*, **58**:375–395. 1933; **59**:123–142. *Figs*. 1–59. 1934; **60**:97–118. *Figs*. 1–65. 1935; 159–178. *Figs*. 1–91. 280–290. *Figs*. 1–39. 1935.

———: Notes on the occurrence of the Trichopeltaceae and Atichiaceae in New South Wales, and on their mode of nutrition, with a description of a new species of *Atichia*, *ibid*., **61**:277–284. *Pls*. 13–14. *Figs*. 1–10. 1936.

———: The distribution of sooty-mould fungi and its relation to certain aspects of their physiology, *ibid*., **62**:25–56. *Pl*. 3. *Figs*. 1–12. 1937.

Gäumann, Ernst: Vergleichende Morphologie der Pilze, 626 pp. *Many figs*. Jena, Gustav Fischer, 1926.

Gilkey, Helen M.: Tuberales of North America, *Oregon State Monographs. Studies in Botany*, **1**:1–63. *Pls*. 1–5. 1939.

Graff, Paul Weidemeyer: The morphological and cytological development of Meliola circinans, *Bull. Torrey Botan. Club*, **59**(5):241–266. *Pls*. 16–17. 1932.

Guilliermond, Alexandre: Sur le noyau de la levure, *Ann. Mycolog.*, **2**(2):184–189. 1 *fig*. 1904.

———: Étude cytologique et taxinomique sur les levures du genre Sporobolomyces, *Bull. soc. mycologique de France*, **43**:245–258. *Pl.* 10. *Figs.* 1–6. 1927.

———: Recherches sur quelques Ascomycètes inférieurs isolés de la stigmatomycose des graines de cotonnier. Essai sur la phylogénie des Ascomycètes, *Rev. gén. botan.*, **40**:328–342, 397–414, 474–485, 555–574, 606–624, 690–704. *Pls.* 22–23. *Figs.* 1–46. 1928.

———: La sexualité, le cycle de développement et la phylogénie des levures d'après les travaux récents, *Ann. fermentations*, **2**:129–151, 257–277. *Figs.* 1–28. 1936a.

———: La classification des levures, *ibid.*, **2**:474–491, 540–551. *Figs.* 1–23. 1936b.

HADLEY, PHILIP: Microbic dissociation. The instability of bacterial species with special reference to active dissociation and transmissible autolysis, *J. Infectious Diseases*, **40**(1):1–312. 1927.

HARPER, ROBERT A.: Die Entwicklung des Peritheciums bei Sphaerotheca Castagnei, *Ber. deut. botan. Ges.*, **13**(10):475–481. *Pl.* 39. 1895.

———: Ueber das Verhalten der Kerne bei der Fruchtentwickelung einiger Ascomyceten, *Jahrb. wiss. Botan.*, **29**(4):655–685. *Pls.* 11–12. 1896.

———: Sexual reproduction and the organization of the nucleus in certain mildews, *Carnegie Inst. Wash. Pub. no. 37*, pp. 1–104. *Pls.* 1–7. 1905.

HASHIOKA, YOSHIO: Specialization in Sphaerotheca fuliginea (Schlecht.) Poll., *Ann. Phytopathological Soc. Japan*, **8**:113–123. 1 *fig.* 1938.

HEIN, ILLO: Studies on morphogenesis and development of the ascocarp of Sphaerotheca castagnei, *Bull. Torrey Botan. Club*, **54**(5):383–417. *Pls.* 28–29. *Figs.* 1–5. 1927.

VON HÖHNEL, FRANZ: Atichia treubii v. Höhnel (Saccharomycetes), *Ann. Jardin Botan. de Buitenzorg, Supplement*, **3**(1):19–28. 1910.

———: Ueber die Trichothyriaceen, *Ber. deut. botan. Ges.*, **35**(5):411–416. 2 *figs.* 1917.

HOMMA, YASU: Homothallism in Sphaerotheca fuliginea (Schlecht.) Pollacci, *Proc. Imp. Acad. (Tokyo)*, **9**:186–187. 1 *fig.* 1933.

HOTSON, J. W.: A new species of Arachniotus, *Mycologia*, **28**(6):497–502. *Figs.* 1–13. 1936.

JACKSON, H. S.: Trichomonascus, a new genus among simple Ascomycetes, *Mycologia*, **39**(6):709–715. *Figs.* 1–20. 1947.

JENKINS, ANNA E.: Spot anthracnoses, *Yearbook of Agriculture for 1943–1947*, pp. 451–454. 1947.

KLUYVER, A. J., UND C. B. VAN NIEL: Über Spiegelbilder erzeugende Hefearten und die neue Hefengattung Sporobolomyces, *Centr. Bakt. Parasitenk., Zweite Abt.*, **63**(1–8):1–20. *Pls.* 1–2. 1924.

LINDAU, G.: Perisporiales, in A. ENGLER und K. PRANTL: Die Natürlichen Pflanzenfamilien, Erster Teil, Abt. 1:325–343. *Figs.* 228–234. Leipzig, Wilhelm Engelmann, 1897.

LINDEGREN, CARL C.: An analysis of the method of budding in yeasts and some observations on the structure of the yeast cell, *Mycologia*, **37**(6):767–780. *Figs.* 1–21. 1945.

———: The yeast cell, its genetics and cytology, xviii + 366 pp. *Illustrated.* St. Louis, Educational Publishers, 1949.

———, AND GERTRUDE LINDEGREN: Segregation, mutation and copulation in Saccharomyces cerevisiae, *Ann. Missouri Botan. Garden*, **30**(4):453–468. *Pls.* 15–17. 2 *figs.* 1943.

LINDER, DAVID H.: A new species of Phyllactinia, *Mycologia*, **35**(4):465–468. *Figs.* 1–5. 1943.

LODDER, JACOMINA: Die Hefesammlung des "Centraal-Bureau voor Schimmelcultures." Beiträge zu einer Monographie der Hefearten. Teil II. Die anaskosporogenen Hefen, Erste Hälfte, Verhandelingen der Koninklijke Akademie van Wetenschappen te Amsterdam. Afdeeling Natuurkunde. (Tweede Sectie)Deel 32, ix + 256 pp. 114 *figs.* 1934.

LOHWAG, HEINRICH: Sporobolomyces—Kein Basidiomyzet, *Ann. Mycolog.*, **24**(3–4):194–202. 1926.

MAASSEN, ALBERT: Pericystis apis, *Mitt. Kaiserlichen Biologischen Anstalt Land- und Forstw.*, **16**:51–58. 1916.

MANGIN, L., ET N. PATOUILLARD: Les Atichiales, groupe aberrant d'Ascomycètes inférieurs, *Compt. rend.*, **154**(23):1475–1481. *Figs.* 1–2. 1912.

MANUEL, JANE: Sur la formation de l'asque de Nematospora Coryli après un phénomène sexuel, *ibid.*, **207**(24):1241–1243. 11 *figs.* 1938.

MILLER, JULIAN H.: Studies in the development of two Myriangium species and the systematic position of the order Myriangiales, *Mycologia*, **30**(2):158–181. *Figs.* 1–4. 1938.

NANNFELDT, J. A.: Studien über die Morphologie und Systematik der nichtlichenisierten inoperculaten Discomyceten, *Nova Acta Regiae Soc. Sci. Upsaliensis*, ser. IV., **8**(2):1–368. *Pls.* 1–20. *Figs.* 1–47. 1932.

NYLAND, GEORGE: Studies on some unusual Heterobasidiomycetes from Washington State, *Mycologia*, **41**(6):686–701. *Figs.* 1–6. 1949.

PETCH, T.: Studies in entomogenous fungi: V. Myriangium, *Brit. Mycolog. Soc. Trans.*, **9**:45–80. *Pls.* 2–3. *Fig.* 1. 1924.

PETRAK, F.: Über Englerula und die Englerulaceen, *Ann. Mycolog.*, **26**(5–6): 385–413. 1928.

RACIBORSKI, MARYAN: Parasitische und epiphytische Pilze Javas, *Bull. Acad. Sci. de Cracovie. Classe Sci. Math. et Nat.*, **1909**:346–394. *Ilus.* 1909.

REED, GEORGE M.: The powdery mildews of Avena and Triticum, *Univ. Missouri Agr. Expt. Sta. Research Bull.*, **23**:1–19. 1916.

SALMON, ERNEST S.: A monograph of the Erysiphaceae, *Mem. Torrey Botan. Club*, **9**:1–292. *Pls.* 1–9. 1900.

———: On specialization of parasitism in the Erysiphaceae, I, *Botan. Centr. Beihefte*, **14**(3):261–316. *Pl.* 18. 1903.

———: Critical experiments with "biologic forms" of the Erysiphaceae, *Trans. Roy. Soc. London, B*, **197**:107–122. 1904a.

———: Recent researches on the specialization of parasitism in the Erysiphaceae, *New Phytologist*, **3**(2):55–60. *Figs.* 63–65. 1904b.

———: On specialization of parasitism in the Erysiphaceae, II, *ibid.*, **3**(5):109–121. 1904c.

———: On specialization of parasitism in the Erysiphaceae, III, *Ann. Mycolog.*, **3**(2):172–184. 1905.

———: On the stages of development reached by certain biologic forms of Erysiphe in cases of non-infection, *New Phytologist*, **4**(9):217–222. *Pl.* 5. 1905.

SARTORY, A., ET H. SYDOW: Étude biologique et morphologique d'un Aspergillus nouveau, Aspergillus Sartoryi Syd. n.sp., *Ann. Mycolog.*, **11**(2):156–160. *Pl.* 8. 1913.

SCHIKORRA, W.: Über die Entwicklungsgeschichte von Monascus, *Z. Botan.*, **1**(6):379–410. *Pl.* 2. *Figs.* 1–3. 1909.

SCHNEIDER, ALBERT: A parasitic Saccharomycete of the tomato, *Phytopathology*, **6**(5):395–399. *Figs.* 1–4. 1916.

SCHRÖTER, J.: Hemiascineae, in A. ENGLER und K. PRANTL: Die Natürlichen Pflanzenfamilien, Erster Teil, Abt. 1:143–149. *Figs.* 128–132. Leipzig, Wilhelm Engelmann, 1894.

SHEAR, C. L.: Arachniotus trachyspermus, a new species of the Gymnoascaceae, *Science*, N.S., **16**:138. 1902.

STELLING-DEKKER, N. M.: Die Hefesammlung des "Centraal-Bureau voor Schimmelcultures." Beiträge zu einer Monographie der Hefearten. Teil I. Die sporogenen Hefen, Verhandelingen der Koninklijke Akademie van Wetenschappen te Amsterdam. Afdeeling Natuurkunde. (Tweede Sectie) Deel 28(11), vii + 547 pp. *Illus.* 1931.

STEVENS, F. L.: Some meliolicolous parasites and commensals from Porto Rico, *Botan. Gaz.*, **65**(3):227–249. *Pls.* 5–6. *Figs.* 1–5. 1918.

———: The Meliolineae, *Ann. Mycolog.*, **25**(5–6):405–469. *Pls.* 1–2. 1927; **26**(3–4):165–383. *Pls.* 2–6. 1928.

SYDOW, H. UND P.: Mykologische Mitteilungen, *Ann. Mycolog.*, **16**(3–6):240–248. 1918.

THEISSEN, F., UND H. SYDOW: Synoptische Tafeln, *Ann. Mycolog.*, **15**(6):389–491. *Figs.* 1–38. 1917.

THIRUMALACHAR, M. J.: *Astomella*, a new genus of the Perisporiales, *New Phytologist*, **46**(2):268–273. *Pl.* 6. *Figs.* 1–7. 1947.

THOM, CHARLES: Conidium production in Penicillium, *Mycologia*, **6**(4):211–215. *Fig.* 1. 1914.

TULASNE, LOUIS RENÉ ET CHARLES: Selecta fungorum carpologia, vol. 1, pp. i–xxviii, 1–242. *Pls.* 1–5. Paris, Typographie Impériale, 1861.

VAN TIEGHEM, PHILIPPE: Elements de Botanique, IVme éd., Paris, 1906.

VARITCHAK, BOGDAN: Contribution à l'étude du développement des Ascomycètes, *Le Botaniste*, **23**:1–183. *Pls.* 1–20. *Figs.* 1–20. 1931.

———: Deuxième contribution à l'étude du développement des Ascomycètes: Évolution nucléaire dans le sac sporifère de Pericystis apis Maassen et sa signification pour la phylogénie des Ascomycètes, *ibid.*, **25**(5–6):343–390. *Pls.* 17–23. *Figs.* 1–4. 1933.

VIEGAS, J. H.: Alguns fungos do Brazil: II. Ascomicetos, *Bragantia*, **4**(1–6):5–392. *Illus.* 1944. (Pp. 17–20 and Pl. XVIII and text-figure 2 treat of *Brasiliomyces*.)

WALKER, LEVA B.: Studies on Ascoidea rubescens: I. History and development, *Mycologia*, **23**(1):51–76. *Figs.* 1–5. 1931; II. Cytological observations, *ibid.*, **27**(2):102–127. *Figs.* 1–78. 1935.

WICKERHAM, LYNFERD J., AND ENRIQUE DUPRAT: A remarkable fission yeast, Schizosaccharomyces versatilis n.sp., *J. Bacteriol.*, **50**(5):597–607. 2 *figs.* 1945.

WINGARD, S. A.: Studies on the pathogenicity, morphology, and cytology of Nematospora Phaseoli, *Bull. Torrey Botan. Club*, **52**(6):249–290. *Pls.* 7–9. 1925.

WINGE, Ö.: On haplophase and diplophase in some Saccharomycetes, *Compt. rend. trav. lab. Carlsberg, sér. physiol.*, **21**(4):77–111. *Pls.* 1–3. *Figs.* 1–16. 1935.

YARWOOD, C. E.: Heterothallism of sunflower powdery mildew, *Science*, N.S., **82**(2131):417–418. 1935.

YOUNG, ELAINE M.: The morphology and cytology of Monascus ruber, *Am. J. Botany*, **18**(7):499–517. *Pls.* 39–41. 1931.

ZENDER, JUSTIN: Sur la classification des Endomycétacées, *Bull. soc. botan. Genève*, 2me sér., **17**:272–302. 1925 (1926).

12

CLASS BASIDIOMYCETEAE: SUBCLASS TELIOSPOREAE

Introduction

In contrast to the condition in the Class Ascomyceteae the ultimate reproductive spores in the Basidiomyceteae are produced externally upon usually terminal cells within which caryogamy and subsequent meiosis produce successively a diploid and then four haploid nuclei. In the Ascomyceteae the haploid nuclei formed as a result of the meiotic divisions of the diploid nucleus become the centers of ascospores formed within the ascus. The difference is apparently that in the one class the spores are produced internally but in the other externally. In reality even in the latter case distinct spore walls are sometimes produced around the spores within the pocket-like outgrowths from the main cell so that the basidiospore is an internally produced spore surrounded by and usually adhering to the wall of the projecting pocket from the basidium. (Fig. 119.)

The basidiospores or sporidia throughout the class are universally unicellular when first formed and usually uninucleate (except in the more unusual cases where two nuclei enter the spore from the basidium). In many species the nucleus divides within the spore before the latter is set free. In *Dacrymyces* and some other genera the basidiospore becomes transversely septate into several cells at or shortly after maturity. In color the spores in this class vary from hyaline (the majority of species) to pink or to yellow-brown, brown, purple-brown, and black.

Typically in the Ascomyceteae there are various means by which the ascospores are expelled from the ascus, thus providing for their distribution. Similarly in most of the Basidiomyceteae where the spores are produced in the open air there is usually a special provision for the discharge of these spores from their points of attachment. The spores are usually discharged with considerable violence, all four spores being discharged successively at intervals of several seconds to several minutes,

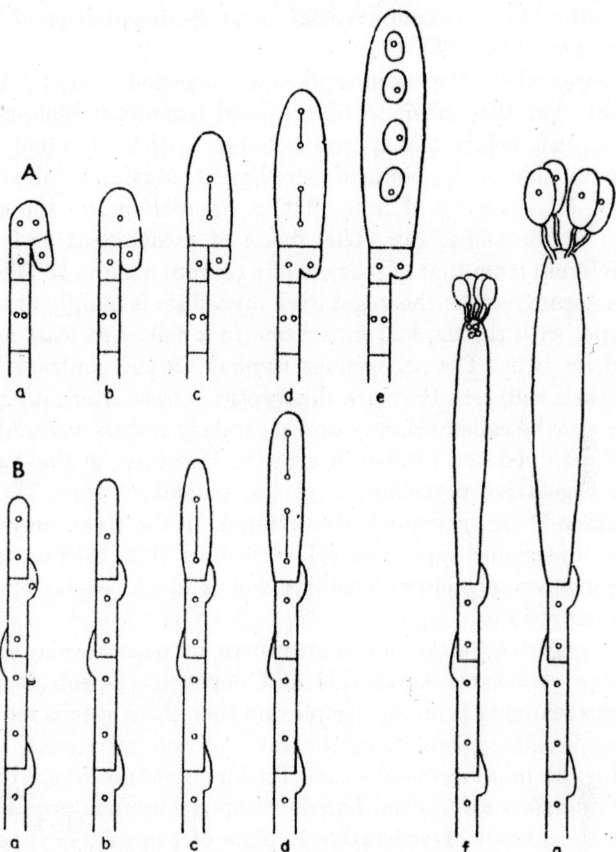

FIG. 119. Class Basidiomyceteae. Diagrammatic sketches comparing the nuclear behavior in the formation of the ascus and the basidium and the further steps leading to the production of the ascospores and basidiospores, respectively. (A) Development of the ascus and ascospores. (B) Development of the basidium and basidiospores. (a) Dicaryon terminal cell of a hypha. (b) union of the two nuclei to form a diploid nucleus; (c, d) as the cell enlarges the first and second meiotic divisions occur, the daughter nuclei in each division being indicated by a connecting line; (e) ascospores are formed around each nucleus in the ascus; (f) external pockets are formed near the apex of the basidium; (g) a wall is formed surrounding the cytoplasm and nucleus and separating them from the sterigma. Spores are now mature.

according to the observations of Buller (1909). The distance to which the spores are discharged may be ten to twenty times the length of the spore. By affixing a clean glass coverslip a short distance above the surface of the mature hymenium, e.g., a gill of a mushroom, the basidiospores will sometimes be found sticking to the glass in groups of four. In this manner it is possible to obtain and make cultures from the four spores arising from a single basidium. Buller has made very extensive studies of this discharge of basidiospores and its relation to the distribution of the spores by air currents. (Figs. 120, 121.)

In all cases where the basidiospores are expelled violently Buller has pointed out that they seem to be attached somewhat obliquely to the sterigmata. Just before the spore discharge a drop of liquid begins to appear at one side of the point of attachment, attaining full size in 5 to 20 seconds. It is worthy of note that in *Sporobolomyces* the spores are discharged in the same way. This mode of attachment and release is entirely different from that of true conidia (except conidia of Tilletiaceae).

In the Ascomyceteae the vegetative mycelium is mainly of the monocaryon type, with the haploid nuclei one to a cell or at least not closely associated by twos. The ascogenous hyphae on the contrary have two nuclei to each cell, i.e., they are dicaryotic. These alternating types of mycelium may be called primary and secondary respectively. Mainly the latter is short-lived and limited in growth. However, in the Taphrinales the main vegetative mycelium is of the secondary type, the primary mycelial stage being very much abbreviated. In the Basidiomyceteae the secondary "dicaryon" type of mycelium is more often the preponderating stage, the primary mycelium usually being of much shorter life than the secondary mycelium.

In very many of the Ascomyceteae there are definite male and female reproductive structures (antherids and oogones, respectively). Yet in some groups sexuality is on the decline and then these special reproductive organs may be substituted for by the union of vegetative cells. This latter condition is the more frequent in the Basidiomyceteae where only in the Rusts (Uredinales) are special female receptive hyphae produced. Even in this group contact of vegetative hyphae of compatible types is sufficient for sexual reproduction. In the Rusts special male gametes (sperm cells or spermatia) are formed in definite organs called spermogonia, but only on monocaryon mycelium. In many other members of the Basidiomyceteae there are produced small hyaline, one-celled and uninucleate spores called oidia. They arise on oidiophores and as they are set free they cling together in a drop of sticky liquid, forming little shining balls. The oidiophores may be simple or branched. The branches or the apical portions of the unbranched oidiophores break up into oidia successively from the tip toward the base, leaving eventually only a short stub. Brodie (1931), Vandendries (1924), and others have shown that mostly the oidia arise only from primary mycelium. Long ago Dangeard (1894–1895) described the formation of uninucleate oidium-like cells from the dicaryon mycelium of *Dacrymyces* and more recently Vandendries and Martens (1932) have demonstrated their occurrence on the dicaryon mycelium of *Pholiota aurivella* (Batsch ex Fr.) Quel. In this species from the same secondary mycelium may be produced thick-walled binucleate

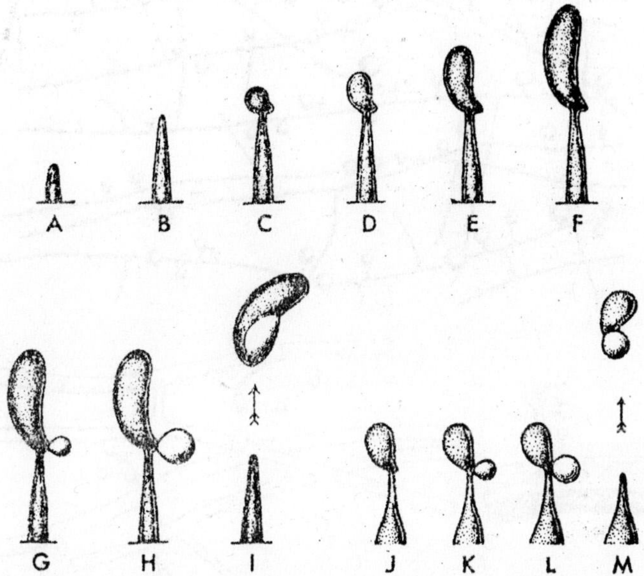

Fig. 120. Class Basidiomyceteae. Steps in the discharge of a basidiospore or of a sporidium from its sterigma. (A–I) Subclass Heterobasidiae, Order Dacrymycetales, Family Dacrymycetaceae, *Calocera cornea* Fr. (J–M) Subclass Teliosporeae, Order Uredinales, Family Pucciniaceae, *Endophyllum euphorbiae-sylvaticae* (DC.) Wint. (Courtesy, Buller: Researches on Fungi, London, Longmans, Green and Co. A–I, vol. 2, p. 7; J–M, vol. 3, p. 54.)

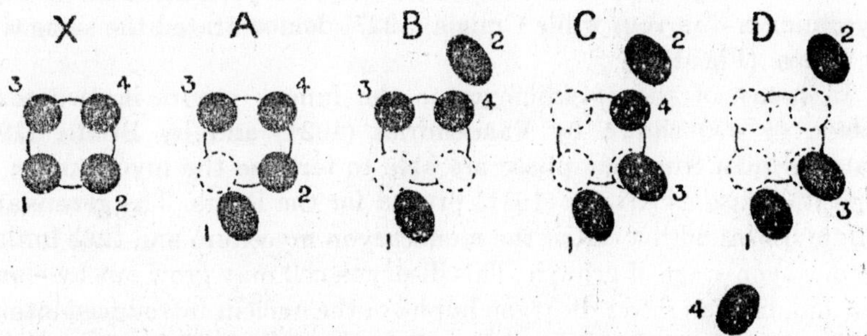

Fig. 121. Class Basidiomyceteae. The successive discharge of four basidiospores from one basidium of *Agaricus campestris* Fr. (Courtesy, Buller: Researches on Fungi, London, Longmans, Green and Co., vol. 1, p. 52.)

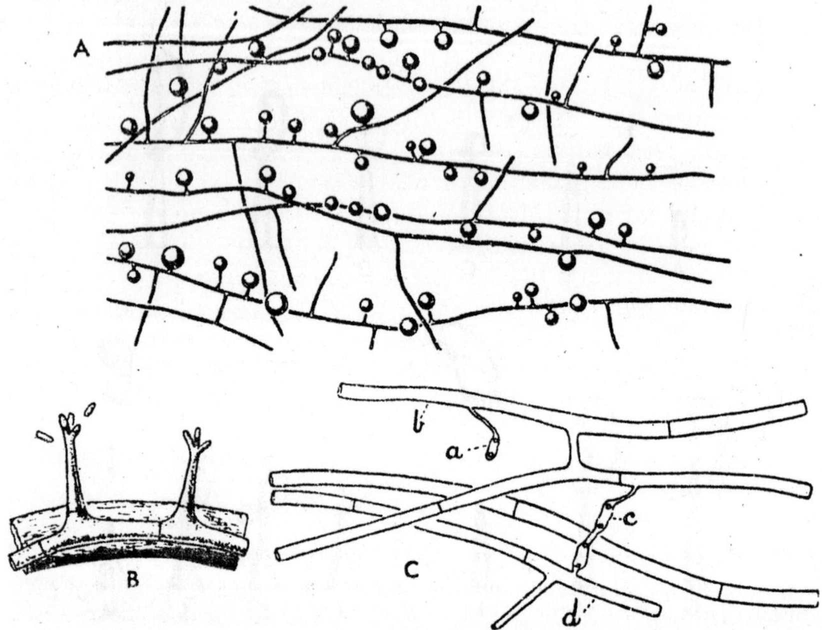

FIG. 122. Class Basidiomyceteae, *Coprinus lagopus* Fr. (A) Monocaryon mycelium bearing heads of oidia in mucilaginous drops. (B) Detail of oidiophore. (C) Oidia (*a*, *c*) of one sexual phase fusing with monocaryon hyphae (*b*, *d*) of the opposite (compatible) sexual phase. (Courtesy, Brodie: *Ann. Botany*, **45**(178):315–344.)

"gemmae" and thin-walled binucleate "oidia," both of which give rise to dicaryon mycelium, and also uninucleate oidia. The latter begin as binucleate, spindle-shaped cells which become divided by a septum into two uninucleate cells which fall apart to form the oidia. These latter upon germination give rise to monocaryon mycelium. The sticky droplets containing the oidia attract insects. Brodie (1931) has demonstrated that small flies carry these oidia from monocaryon mycelium to monocaryon mycelium in *Coprinus* while Craigie (1927) demonstrated the same thing for Rusts. (Fig. 122.)

In many of the Basidiomycetae the fungus occurs in two sexual phases. It was shown by Vandendries (1924) and by Brodie (1931) that the oidia from one phase are able to fertilize the mycelium of the opposite phase as Craigie (1931) proved for the Rusts. The germinating oidium unites with a cell of the monocaryon mycelium and thus initiates the dicaryon stage of growth. This dicaryon cell may grow out by elongation and division into a dicaryon hypha or the nucleus introduced into this cell by the oidium divides and one of the daughter nuclei passes through the wall into the next cell and so on until the whole hypha is "diploidized" as demonstrated by Lehfeldt (1923) and by Buller (1930). The latter made

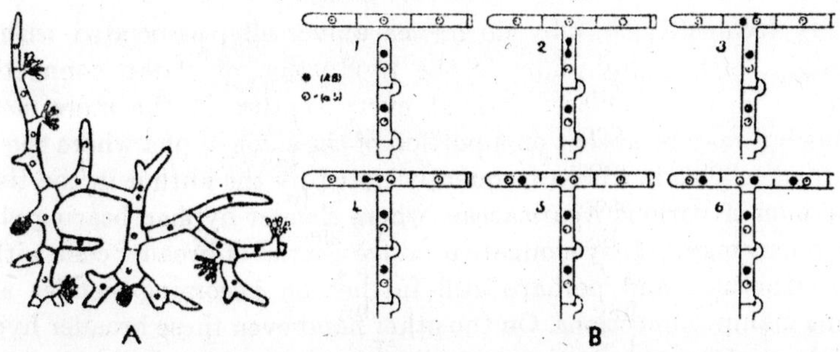

Fig. 123. Basidiomyceteae. Diploidization of monocaryon mycelium. (A) Two monocaryon mycelia, mutually compatible, have united and diploidization has been initiated in one cell. (B) Diagrammatic representation of the steps (1–6) in the diploidization of a monocaryon mycelium by a dicaryon one. (Courtesy, Buller: *Nature*, **126**(3183):686–689.)

experiments which showed that the rate of progression of the diploidizing nuclei through the mycelium was 1.2 to 1.5 mm. per hour in *Coprinus lagopus* Fr. When the introduced nucleus or one of its descendants arrives in the terminal cell of the hyphal branch the two nuclei divide by conjugate division and thenceforth a typical dicaryon mycelium is produced from this hypha. The initiation of the dicaryon phase does not depend alone upon the oidia. Two monocaryon hyphae of opposite sexual phase may unite and the result is the same as when an oidium unites with a hyphal cell. Brodie demonstrated that the oidia of *Coprinus lagopus* are capable of germinating and forming a mycelium made of very slender monocaryon hyphae. From these arise very numerous oidiophores and heads of oidia. When two such oidial mycelia of opposite sexual phase meet diploidization occurs. Also the oidia from one mycelium may fertilize the oidial mycelium of the opposite sexual phase. Possibly the oidia and the oidial mycelium may represent a residual male sexual structure while any cell of the normal monocaryon vegetative mycelium possesses a female tendency, this not being restricted merely to an oogone as in many of the Ascomyceteae. Vandendries and Brodie (1933) find however that in some Basidiomyceteae the oidia are also capable of functioning as conidia, producing a typical monocaryon mycelium which is capable of diploidizing and being diploidized by a monocaryon mycelium of appropriate sexual phase. (Fig. 123.)

The disappearance of definite sexual organs has not done away with the three fundamental phenomena of sexual reproduction, cytogamy, caryogamy, and meiosis, nor with the modification into various types of compatibility and incompatibility of sexual strains such as occur in many Ascomyceteae. Indeed, it seems possible that this development of sexually compatible and incompatible strains has become greatly increased in the Basidiomyceteae.

Very frequently, but by no means universally, associated with the secondary type of mycelium is the production of clamp connections. These are produced more often at every septum in the more slender hyphae but may be absent on a portion of the same hypha where the cells are broader. This has been frequently noted by the author in the tissues of the pilei of various Agaricaceae, where slender hyphae bearing clamp connections may as they elongate produce a series of broader cells without these structures and perhaps still further on becoming slender again bearing clamp connections. On the other hand even these broader hyphae may show clamp connections, as illustrated in some of the figures by Kühner (1926) though other figures show the broader hyphae without

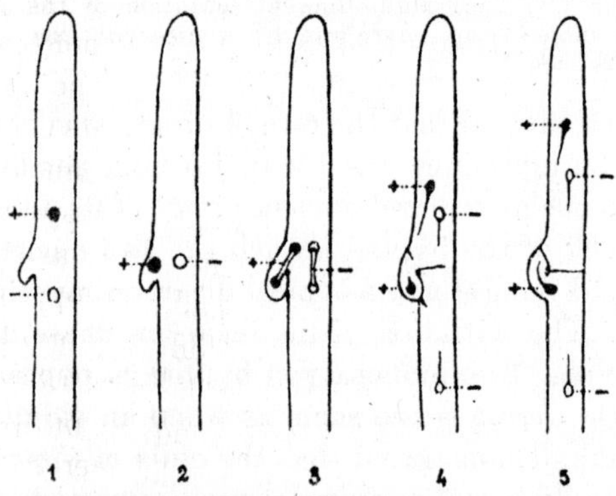

Fig. 124. Class Basidiomyceteae. Diagrammatic representation of the steps in the formation of clamp connections. (Courtesy, Bensaude: Recherches sur le cycle évolutif et la sexualité chez les Basidiomycètes, Nemours, published by author.)

such structures. Whenever clamp connections are present the mycelium is thereby indicated as being dicaryotic, i.e., of the secondary type, but from the foregoing it is evident that the absence of these structures does not in all cases indicate the primary monocaryon nature of the hyphae.

The mode of development of clamp connections was reported independently by Kniep (1915) and Mlle. Bensaude (1918). The details of the formation are as follows: The two nuclei of the terminal cell of a dicaryon hypha lie a short distance apart in the longitudinal axis of the cell. Between them a lateral pocket is formed in the wall. The two nuclei now divide simultaneously (conjugate division) and the lower daughter nucleus of the upper pair passes into the pocket. This is now cut off from the main cell by the formation of a septum. At the same time the upper daughter nucleus of the lower pair, lying at about the level of the upper end of the pocket becomes separated from its sister nucleus by a transverse wall. At this stage the terminal cell now has two nuclei, the cell below it has one nucleus, while the fourth nucleus lies in the lateral pocket. The latter curves around until it is in contact with the lateral wall of the upper end of the penultimate cell and the intervening walls are dissolved and the nucleus passes into the latter cell. The lateral pocket has acted as a by-pass through which a nucleus has been transferred from the terminal to the penultimate cell in such a manner as to provide each cell with a daughter nucleus of each of the two nuclei originally in the terminal cell. This by-pass is known as a clamp connection. Buller (1933) has followed the formation of these clamp connections in living mycelium and found that the process requires only a short time. In *Coprinus lagopus* Fr. the time elapsed from the first appearance of the projecting lateral pocket until the passage of the nucleus out of the pocket into the penultimate cell was 23 minutes, while in *C. sterquilinus* Fr. it was 40 to 45 minutes. The conjugate division of the nuclei was completed in the first species in from 12 to 14 minutes. (Fig. 124.)

The phenomena involved in the formation of the clamp connections are generally considered (e.g., by Kniep, Bensaude, and most students since then) as being homologous to those occurring in an ascogenous hypha when an ascus is forming by the hook or crozier method. The Moreaus (1928) have shown that the formation of the hook does not necessarily lead immediately to the formation of an ascus, for the terminal binucleate cell may elongate and again form a hook while the tip of the original hook unites with the cell below. This may continue several times until a series of dicaryon cells is produced, each connected to the cell below by a clamp connection. De Ferry de la Bellone (1886) and Mattirolo (1887) described and figured typical clamp connections on the mycelium of *Tuber lapideum* Matt., and other species. Unless they mistook some intermingled strands of Basidiomycetous mycelium for that of the fungi they were studying the occurrence of this structure in the Ascomyceteae as well as in the Basidiomyceteae must be considered substantiated.

Buller (1933) did not believe that the formation of clamp connections is at all homologous to the processes occurring in the ascogenous hyphae. The clamp connections play, he believed, an important part in the transfer of food through the mycelium. An actual flow of protoplasm was observed by him through the clamp connection and its centrally perforated upper septum. Possibly the occurrence of whorls of clamp connections in some fungi would support Buller's view as to their function in nutrition. Kemper (1937) studied the development and cytology of the clamp connections which may occur in whorls instead of singly at the septa of *Coniophora cerebella* Pers. (now called *C. puteana* (Schum. ex Fr.) Karst.). In this species the uninucleate young basidiospore becomes binucleate by the division of the nucleus before discharge. It gives rise to a coenocytic mycelium with cross walls at rather long intervals. The nuclei divide apparently independently of one another and no conjugate divisions can be noted. In the younger mycelium clamp connections arise singly but as the mycelium becomes older they may arise by twos and in still older mycelium in whorls. Their formation has no apparent connection with nuclear divisions. From one to several nuclei may pass into each pocket which fuses with the cell below the septum and discharges all the contained nuclei into that cell, or a lateral branch may be formed into which the nuclei pass. Sometimes the terminal cell of the hypha is left without any nuclei. The author believes that these whorls of clamp connections serve to distribute various types of nuclei to the new branches. (Fig. 125.)

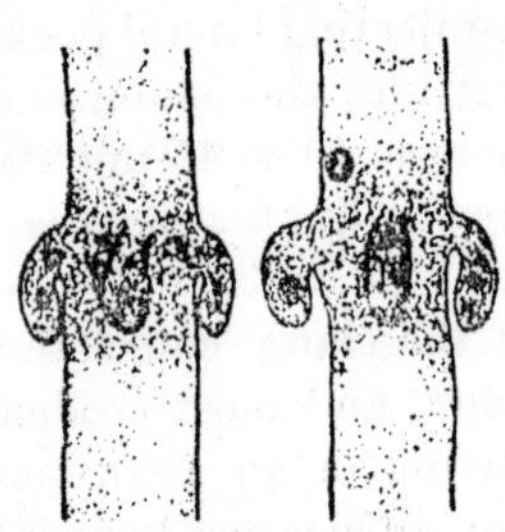

FIG. 125. Class Basidiomyceteae. Whorls of clamp connections in *Coniophora puteana* (Schum. ex Fr.) Karst. (After Kemper: *Zentr. Bakt., Parasitenk. und Infektionskr.*, Zweite Abt., 97(4–8):100–124.)

The greater prevalence of clamp connections in the Basidiomyceteae is probably due in the first place to the fact that in most members of this class the dicaryon mycelium represents a much greater portion of the life history of the plant than do the rather transitory or entirely wanting ascogenous hyphae of the Ascomyceteae. Furthermore, in the latter group these hyphae are broader in general, so that at conjugate division the two dividing nuclei may lie side by side instead of some distance apart in the longitudinal axis of a narrow hypha. Only in the latter case is a by-pass really necessary. Some whole genera of the Basidiomyceteae lack clamp connections entirely.

The prevalence of clamp connections varies greatly in different parts of the same mycelium and is, furthermore, modified greatly by the environment. Thus mycelium submerged in liquid media may have but few or even no clamp connections while the aerial portions may produce them in abundance. They may be present on the slenderer hyphae in the pileus of the mushroom and absent in the broader extensions or branches of these same hyphae. In some species they are found only at great intervals while in others they occur at every septum. In some species they are only found in the subhymenial tissues of the spore fruit but not elsewhere, even when all the tissues consist of dicaryon mycelium. Hirmer (1920) observed conjugate divisions in the mycelium of *Agaricus campestris* Fr., although clamp connections were completely absent. Within the genus *Coprinus* Brunswik (1924) has found some species which lack the clamp connections entirely and other closely related species in whose mycelium they are abundant. He interprets this as the gradual loss of a structure inherited from the Ascomycetous ancestors but whose function is no longer indispensable.

Mlle. Bensaude and Kniep were among the first to show the existence of different sexual phases (or as they called it "heterothallism") in the Basidiomyceteae. Miss Mounce (1922) and Miss Newton (1926) and various other investigators have shown that many of this class are "homothallic," i.e., will produce the dicaryon mycelium in culture from a single basidiospore while other species are always "heterothallic." Rarely a monocaryon mycelium of a heterothallic species after a considerable time begins to produce dicaryon hyphae in a manner not yet satisfactorily explained. In the homothallic species in which the basidiospore is uninucleate it must be assumed that the genetic factors for incompatibility are absent or are both present and mutually cancelling in the same chromosome. Sass (1929) studied the behavior of the nuclei in certain homothallic forms in which the nuclei are two in number in the basidiospore. Thus in *Coprinus ephemerus* Fr. there exist forms in which four uninucleate basidiospores are produced on each basidium; these forms are heterothallic. In *C. ephemerus* forma *bisporus* only two basidiospores are produced each with two nuclei. Mostly these give rise to homothallic mycelia but sometimes they show heterothallism. Sass found that from ninety per cent of such binucleate spores there is produced a coenocytic mycelium which as it grows begins to become septate until the apical portions of the hyphae consist of uninucleate cells. Some of these hyphal branches on coming into contact with other uninucleate hyphae from the same mycelium fuse with them to form typical dicaryon mycelium (with clamp connections) and from this mycelium arise the normal fruiting bodies. Evidently the two nuclei of the basidiospore in this case represented opposite sexual phases. In about ten per cent of the basidia the two nuclei of the spore are clearly of the same sexual phase. Such spores produce a mycelium consisting from the first of uninucleate cells. Only when two such mycelia of opposite sexual phases come into contact is the secondary, fruiting mycelium produced. The situations described by Sass are very similar to those described by Ames (1932) in *Schizothecium* (see Chapter 10).

The four basidiospores may represent two or sometimes four sexual phases which are mutually fertile by twos. Because of the presence of male sexual cells (oidia) on the mycelium of both uniting mycelia and of the fact that each mycelium or its oidia can diploidize the other mycelium it is manifest that we can not look upon these as representing opposite sexes. As in homologous cases in the Ascomyceteae it is a question of self-incompatibility, comparable in a way to that occuring within a given horticultural variety of pear which may be sterile to its own pollen but fertile to pollen of another (but not of every other) variety. Since the term heterothallism as originally applied refers to mycelia representing different sexes it seems inadvisable to use the terms heterothallic and homothallic in these higher fungi, at least without qualification.

Apparently the factors governing the compatibility are two allelomorphic factors borne on different chromosome pairs. Kniep (1928) and others have pointed out that only those unions of cells lead to diploidization that bring about a combination that is heterozygous for both these sets of factors. This need not involve a union of the nuclei into a diploid nucleus, a process that occurs only in the basidium, but concerns merely the bringing together, without union, of two haploid nuclei in the same cell, so that so far as the cell is concerned diploidization has taken place even though the two haploid nuclei remain separate for the present. Thus a monocaryon mycelium with the compatibility factors A and B (these being on separate chromosomes) could unite with another monocaryon mycelium of the formula ab; similarly a mycelium with the formula aB could unite with one of the formula Ab. In either case the formula for the resultant cells would be AaBb, a condition heterozygous for both sets of characters and fulfilling the requirements for the mating of the mycelia. In the basidium the diploid nucleus undergoes two meiotic divisions to consummate the reduction process. If both pairs of chromosomes bearing the factors for incompatibility or compatibility undergo disjunction in the first division it will be a matter of chance whether the resultant daughter nuclei will be Ab and aB or AB and ab. Since each nucleus divides again, this time by splitting the chromosome, there will be four nuclei (one for each spore), two each of the formula Ab and aB or AB and ab respectively. If the disjunction does not occur in either chromosome pair until the second meiotic division the first division may represent merely the splitting of the chromosome so that the two daughter nuclei will be like the parent nucleus with the formula AaBb. When these divide by the disjunction division it will give this time Ab and aB or AB and ab, for each of the two nuclei. If the separation of the chromosomes or the chromosome pairs occurs in the same direction in both nuclei there will result four nuclei, alike two by two, but this separation may be in opposite directions in the two dividing nuclei for one of the chromosome pairs so that the four nuclei may all be different. If disjunction of one chromosome pair occurs at the first division and of the other pair at the second division the four resultant nuclei will also be all different, viz., Ab, aB, AB, ab. Miss Newton (1926) showed that in *Coprinus lagopus* all of the foregoing arrangements may be found in the various basidia of the same hymenium. Apparently disjunction occurs more often at the first division for both sets of chromosomes, for in the majority of cases two of the basidiospores of a given basidium will be of one sexual phase and the other two of the opposite phase. In some species only two sexual phases appear to occur.

These are spoken of as possessing bipolar sexuality in contrast to those species with quadripolar sexuality such as described above. In a bipolar species only one allelomorphic pair of incompatibility factors need be assumed.

Not infrequently a monocaryon mycelium may produce spore fruits but in this case the basidia are less numerous and are either sterile or give rise to but two, uninucleate basidiospores, both of the same sexual phase as the parent mycelium. In *Coprinus fimetarius* Fr., Oort (1930) has shown that two monocaryon mycelia may intermingle and together build spore fruits when they represent phases alike genetically for one factor and heterozygous for the other, e.g., Ab and AB. Such fruits are not normal but may produce two kinds of two-spored basidia which are about equally divided between the component phases, in this case Ab and AB. It is apparent that no true sexual union has occured in such a case. That this is more than a simple intermingling of separate monocaryon mycelia is demonstrated by the fact that clamp connections may be formed, varying from incomplete or abnormal structures to those of perfectly normal appearance. They vary from only occasional to frequent. The external conditions appear to have considerable effect upon the number of clamp connections produced. These unions do not arise in all incompatible combinations. Vandendries and Brodie (1933) showed that in *Hypholoma candolleanum* (Fr.) Quelet, a quadripolar species, the mycelia to which they ascribe the formula ab′ when mated with a′b′, or ab when mated with a′b cause mutual partial inhibition of growth so that when growing close together they are much smaller than when grown apart or in the combination a′b × a′b′. It should be noted that the formulae ab, a′b, ab′, and a′b′ correspond to AB, aB, Ab, and ab in the preceding portion of the paragraph. The combination studied by Oort which produced a spore fruit corresponds to the formula a′b × a′b′. Oort also found that such spore fruits were not produced in certain other combinations. Quintanilha (1935) made genetic and cytological studies of normal and "illegitimate" crosses in *Coprinus fimetarius* and found that nuclear fusions may occur in the basidia even in the latter type of cross.

Vandendries and Brodie (1933) and Vandendries (1934) and Brodie (1934, 1936) described what they termed "barrage sexuel." This had been noted before by Oort (1930) and by Brunswik (1924) but not studied intensively. This amounts to a mutual repulsion of the hyphae of some of the incompatible matings. When two such cultures are established in the surface of an agar medium the mycelia as they grow leave a gap between the two colonies. This is especially marked in the aerial mycelium whose hyphae show abrupt curvatures away when they approach the other mycelium at a distance of 3 to 5 mm. These authors showed further that the interposition of thin low plates of glass did not prevent this repulsion nor did sheets of mica, very thin sheets of silver, lead, and other substances. Apparently the repulsion is due to some sort of emanation from the mycelium, perhaps gaseous. Experiments showed that certain nuclear combinations give these mycelial repulsions but not others. Thus the combinations ab × ab′ and a′b × a′b′ show "barrage" but not ab × a′b or ab′ × a′b′. Clearly the repulsion is between the b and b′ mycelia. The repulsion was also demonstrated between dicaryon mycelium and monocaryon mycelium and between dicaryon mycelia where the two sets of nuclei were different, e.g., (ab + a′b′) × (a′b + ab′). This phenomenon of barrage as well as the inhibition of growth does not occur in all species. Brodie (1935) showed that barrage or "aversion" occurs also in some strains of fungi growing naturally in wood.

A peculiar phenomenon in connection with the occurrence of bipolar and quadripolar sexual phases is that of "geographic races." This has been studied in *Coprinus* by Hanna (1925) and by Vandendries (1924) and extensively by Brunswik (1924) and in *Ustilago* by Bauch (1930, 1931). In some species of *Coprinus* the two sets of incompatibility factors may be alike in fungi growing in the same locality but one or both sets may be different in fungi (of the same species) growing in different localities. In the latter case all sexual phases of one fungus would be compatible with all sexual phases of the other fungus, while if only one pair is different in the two fungi certain combinations will be incompatible. It is evident that these are explained best on the hypothesis of multiple allelomorphs, familiar to geneticists in both animals and plants. When either of the factors produces an allelomorph by mutation a legitimate cross becomes possible. Since these mutated factors are frequently discovered in fungi from different geographic locations the term "geographic races" is often used (see Kniep, 1928). In *Ustilago longissima* (Sow.) Tul., Bauch recognized over 15 allelomorphic strains of the A-a factors and about 8 of the B-b factors. The resulting interfertility of all sexual phases of one collection of species with all sexual phases of another collection has been found to be rather general. Thus Mounce and Macrae (1936) found complete compatibility between collections made in different localities or on different hosts in *Gloeophyllum saepiarium* (Wulf.) Karst. (*Lenzites saepiaria* (Wulf.) Fr.) and in *Trametes americana* Overh., and Vandendries (1936) showed its occurrence in *Leptoporus adustus* (Willd. ex Fr.) Quelet. Barnett (1937) made similar observations for some of the Heterobasidiae.

The Class Basidiomyceteae may be divided into three subclasses: Teliosporeae (as a class in the earlier work of the author), Heterobasidiae, and Eubasidiae. They may be distinguished as follows:

Teliosporeae: parasitic in the leaves, stems, fruits, and sometimes in the roots, of Pteridophyta, Strobilophyta (Coniferae), and Anthophyta (Angiospermae). The septate parasitic mycelium is mostly intercellular in the host and consists of primary (monocaryon) and secondary (dicaryon) phases. In the latter certain cells enlarge and their two nuclei unite to form a diploid nucleus. This cell is a teliospore. It usually has a thick, colored wall and serves as a resting spore. With the proper environmental conditions this teliospore germinates by sending out a thin-walled straight or curved, four-celled or one-celled hypha, the promycelium, within which the zygote nucleus divides meiotically to form four (or, by further mitotic division, more) nuclei. Usually four sporidia are formed on the promycelium and into each of these a single nucleus enters. These sporidia are shot off with violence or (in the Order Ustilaginales) fall off, and give rise to the primary type of mycelium or by union of two sporidia of suitable sexual compatibility the secondary type of mycelium arises. The teliospores are formed within or upon the host tissues in so-called sori. Various secondary types of (asexual) reproduction are produced.

Heterobasidiae: mostly saprophytic, but in many cases parasitic upon various types of Bryophyta or Vascular Plants. Mycelium septate, mostly falling into two phases, the monocaryotic and dicaryotic. From the latter are produced the spore fruits characteristic of the subclass. The basidia are usually produced in a hymenium, i.e., they stand side by side like a palisade, but may be more scattered. They arise as enlarged binucleate cells whose nuclei fuse. Only rarely does the basidium become dark-colored and thick-walled. Usually the formation of the basidiospores takes place without much delay. The basidia may be one-celled and forking (Dacrymycetales), transversely four-celled (Auriculariales), longitudinally four-celled (Tremellales), or one-celled with four enlarged epibasidia bearing the basidiospores (Tulasnellales). Spore fruits are more often gelatinous when wet, or waxy. The basidiospores often germinate by budding at many points in water. Possibly the Sporobolomycetaceae were derived from this subclass.

Eubasidiae: Mostly saprophytic but some species parasitic. Very often living in the wood of living or dead trees. Basidia one-celled, without epibasidia but bearing usually four long or short sterigmata, each bearing a single basidiospore which usually germinates by a germ tube, even when fallen into water. Basidia produced with some exceptions in definite hymenial layers which may be exposed to the air before the spores become mature (the "Hymenomycetes"), or enclosed in the spore fruit until after the spores are completely mature (the "Gasteromycetes"). In this subclass are found the largest spore fruits of the whole Phylum Carpomyceteae (Higher Fungi).

Subclass Teliosporeae

The subclass Teliosporeae corresponds to Dietel's limits of subclass Hemibasidii in the second edition of Engler and Prantl's Die Natürlichen Pflanzenfamilien (1928). These fungi have been sometimes called the Brand Fungi. Usually the two orders Uredinales and Ustilaginales have been considered to be more or less closely related. They were placed close together by Fries (1832) as Ordo IV. Hypodermii,[1] and by Plowright (1889) and by most of the later botanists. Bennett and Murray (1899)

[1] It should be noted that Fries did not believe that these were true fungi but that the spores arose by the transformation of the tissues of the host plant and were not used for the propagation of the rust or smut.

separated the orders, placing the Ustilaginales in their Class Zygomycetes while the Class Uredineae was placed between the two Classes Ascomycetes and Basidiomycetes, the affinities being considered to lie with the former. Because of the fancied resemblance of the teliospores of some Rusts to asci, Charles E. Bessey (1894) was inclined to include the Uredinales and Ustilaginales in the Class Ascomyceteae, a position from which he receded when the cytological phenomena of these groups became better known. That they are rather closely related to the other Basidiomyceteae the studies of Brefeld (1881 and later) and the cytological investigations of Sappin-Trouffy (1896), Harper (1898, 1902), and others leave little doubt. That they stand apart from the majority of families of that class is equally certain. The author believes that the differences are sufficiently great to warrant placing them in a separate subclass, a position that does not deny their relationship within the Basidiomyceteae but leaves each subclass much more homogeneous.

The fungi of this subclass are parasitic. In the Order Ustilaginales some of the species are capable of saprophytic growth in media rich in food; in the Uredinales growth is strictly parasitic and the fungi have never been cultured except on the living tissues of the host. The mycelium is long, slender, and branching, growing intercellularly within the host. In the majority of cases studied occasional or frequent haustoria are produced. The cells of the mycelium are mostly uninucleate in one stage of development (monocaryon stage) and binucleate in the remainder of the life cycle (dicaryon stage). As the cells of the latter type of mycelium divide the two nuclei divide simultaneously so that one daughter nucleus of each of the original nuclei is found in each of the two daughter cells. This is the type of nuclear division to which the term "conjugate division" was given. Clamp connections occur in many of the Ustilaginales but have been rarely reported in the Uredinales. Finally, on the dicaryon mycelium are produced certain larger cells, usually terminal to a hypha or its branches, which become thicker walled. Within these cells the two nuclei unite to form the only diploid nuclei in the life cycle of the fungus. These cells are the teliospores. In the Ustilaginales (Smuts) these are often spoken of as "chlamydospores," a misuse of this name which is rightly applied only to vegetative cells which become filled with food and develop thick walls to permit survival over winter or through other unfavorable environments. In true chlamydospores there is no nuclear fusion and their germination is in the manner usual for asexual spores.

Plowright (1889) recognized the essential homology of the teliospores in the two orders. They have a typical manner of development. The diploid nucleus divides by meiotic divisions into four nuclei, in the teliospore, or more often the exospore ruptures and a thin-walled hypha (promycelium) grows out into which the diploid nucleus passes. The meiotic divisions then take place in the promycelium instead of the body of the teliospore. In some cases the nuclei divide once or twice more resulting in the production of 8 or 16 or even more nuclei. The promycelium may remain nonseptate but more often becomes transversely septate into four cells. From each of these cells is produced a sessile or stalked sporidium or several such sporidia. From the nonseptate promycelium 4 to 16 or more sporidia bud out at the apex. If the teliospore is deeply buried in the tissues of the host or among other teliospores the promycelium at first is a slender hypha, its emergent portion developing the typical structure.

The sporidia are borne in the Uredinales and the conidia in Family Tilletiaceae at the tips of pointed sterigmata from which they are flung violently at maturity. Their position is slightly asymmetrical at the tip as is the case in the Heterobasidiae and in the Hymenomyceteae.

Many writers (e.g., Gäumann, 1926, Arthur, 1929, etc.) apply the names basidium and basidiospore, respectively, to the structures here called by their older names promycelium and sporidium. The latter is doubtless homologous to the basidiospore but the term basidium must include both the teliospore, in which the nuclear fusion occurs, and the outgrowth from it, the promycelium, in which the meiotic nuclear division usually occurs and upon which the sporidia are borne. More correctly, as will be seen further on in the next chapter, the teliospore is probably homologous to the hypobasidium and the promycelium to the epibasidium of some of the Auriculariales.

Asexual reproduction is known in both the Uredinales and the Ustilaginales. In the latter it occurs by means of colorless, often spindle-shaped or sickle-shaped conidia which are uninucleate or binucleate. The latter arise only from dicaryon mycelium but the uninucleate conidia may arise from either monocaryon or dicaryon mycelium. They arise on very short conidiophores, or rather sterigmata, from the sides of the mycelial cells. They are usually in the true sense of the words "repeating spores" for they produce the same type of mycelium as that from which they arose, except in the case of uninucleate conidia from dicaryon mycelium. These give rise to monocaryon mycelium. These conidia are mostly produced on the saprophytic mycelium, but in some species are developed on sterigmata which emerge through the epidermis of the living host. In the Uredinales the urediospores (uredospores) are dicaryon repeating spores and are, therefore, true conidia. There are no monocaryon repeating spores in the Rusts. The aeciospores (aecidiospores) are the result of a sexual fusion involving cells but not the nuclei. They bridge over the step from the monocaryon phase to the dicaryon phase. They are not strictly homologous to ordinary conidia but yet show great resemblance to the urediospores.

Sexual reproduction occurs in two usually separated steps: the union of two monocaryon cells to initiate the dicaryon mycelial phase and eventually the union of the nuclei in the teliospores to form diploid nuclei. The points at which the dicaryon phase may be initiated are quite variable in the Smuts but much more definite in the Rusts. The details will be taken up in the discussion of the respective orders.

Order Uredinales (The Rusts). These form a group which manifests a very high degree of evolutionary development in many directions. The 7,000 or more species are all strict parasites of Ferns (Pteridophyta), Conifers (Strobilophyta), and Flowering Plants (Anthophyta). Not only are the Rusts strict parasites but in many cases they are highly specialized into biologic races which are confined to certain species of a host genus or even to special agricultural varieties of a host species much as in the case of *Erysiphe graminis.* In Europe Eriksson (1894, 1902) was the first to make extensive studies into physiologic races confined to related species. He was followed by numerous other investigators on that continent. In the United States Carleton (1899, 1904) was the first to follow up Eriksson's work, followed by Freeman and Johnson (1911), and by Stakman (1914), and others. The study of the physiologic forms on special agricultural varieties of these species has been carried out in this country very extensively by Stakman and various collaborators (1922). For example in *Puccinia graminis tritici* Erikss. & Henn., the physiologic race that attacks common wheat (*Triticum aestivum* L.) and durum wheat (*T. durum* Desf.), over 150 physiologic forms have been distinguished in North America and many additional forms, mostly distinct from these, in Europe. These distinctions were made by means of a study of their power of infecting 10 or more differential varieties of wheat. Similar physiologic forms have been found in leaf rust of wheat (*P. rubigo-vera tritici* (Erikss. & Henn.) Carl.) by Mains and Jackson (1926) in the United States, Scheibe (1929) in Germany, and Sibelia (1936) in Italy. Stakman and others (1928) have also found them in maize rust (*P. sorghi* Schw.). Physiologic forms have also been distinguished in flax rust (*Melampsora lini* (Pers.) Lev.) by Flor (1935), in *Puccinia helianthi* Schw. by Brown (1936), in *P. iridis* (DC.) Wallr., by Mains (1938), in *Uromyces phaseoli typica* Arthur, by Harter and Zaumeyer (1941), etc.

With one or two exceptions the teliospores are produced within the tissues of the host and remain internal or break out through the epidermis separate and free or united together in a waxy mass or attached by stalks. The sporidia are capable of wind distribution but are short-lived and delicate so that they can be carried only short distances in a living condition. Two other spore forms are usually produced, the aeciospores and the urediospores, both of which are relatively thick-walled and capable of remaining alive while being carried many miles by the wind (hundreds

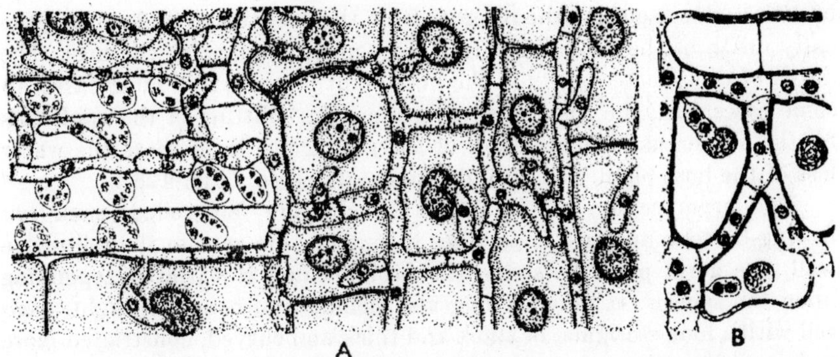

FIG. 126. Monocaryon and dicaryon mycelium and haustoria in Subclass Teliosporeae, Order Uredinales. (A) *Cronartium ribicola* Fischer, monocaryon phase in tissues of *Pinus strobus* L.; intercellular hyphal cells and haustoria uninucleate. (B) *Phragmidium rubi* (Pers.) Wint., dicaryon phase in leaf of *Rubus* sp.; intercellular hyphal cells and haustoria binucleate. (A, courtesy, Colley: *J. Agr. Research*, **15**(12): 619–660. B, after Sappin-Trouffy: *Le Botaniste*, **5**:59–244.)

of miles in some cases). In general in this order the teliospores remain attached in the sorus until after the formation and discharge of the sporidia. However, Pady (1948) has shown that in *Puccinia tumidipes* Peck the very much swollen pedicel will absorb water and rupture, throwing the teliospore some distance (2 to 4 mm.). It is probable that the same phenomenon occurs in other species with enlarged pedicels, such as some species of *Phragmidium* and *Gymnosporangium*.

The mycelium occurs in two well-marked alternating phases, the monocaryon and the dicaryon phase. Clamp connections have been reported by Voss (1903) on the mycelium of the aecial phase but are apparently rare, or at least not readily demonstrated in thin sections. Their presence was confirmed by Wang and Martens (1939) in close proximity to the aecium of *Puccinia coronata* Cda. on *Rhamnus frangula* L. They were demonstrated by maceration of the tissues, not by sections. Haustoria are frequent, sometimes small and spherical, sometimes large and branched, and often partially surrounding the nucleus of the host cell. They are uninucleate or binucleate according to the type of mycelium producing them. Colley (1918) showed for the White Pine Blister Rust (*Cronartium ribicola* Fischer) that the haustorium does not truly penetrate the protoplast of the host cell but causes invagination. Miss Rice (1934) demonstrated this also for *Uromyces caladii* (Schw.) Farl. Apparently the host and parasite stand in a very perfect balance for a very considerable time, especially in the case of rusts that are well adapted to the host. The host cells are not killed outright in such cases. In some of the rusts that are not well adapted to their host Stakman (1914) demonstrated that the host cells are killed immediately around the site of infection, thus isolating the parasite from the host tissues and causing the early death of the fungus before it does much damage to the host. The death of the affected parts of the host plant is gradual in most cases. In some cases, e.g., the rusts of small grains, the splitting of the epidermis by the numerous elongated sori of the rust seems to increase the water loss of the host plant to a very detrimental degree. (Fig. 126.)

The promycelium of the Rusts is normally four-celled, each cell having but one nucleus. It emerges through a thin spot in the teliospore wall, the germ pore. In shape the promycelium is variable depending upon the species. It may be long and slender and nearly straight, each cell with a long sterigma, or short and thick and curved, constricted more or less at the septa. On such a curved promycelium the sporidia always arise on the convex side, often on rather short sterigmata. In a number of genera the promycelium does not entirely emerge from the teliospore. Thus in *Zaghouania* the swelling of the teliospore bursts its thick wall allowing the emergence of a thin-walled, four-celled promycelium whose basal portion still remains enclosed within the old cell wall. In *Coleosporium* and *Gallowaya* the teliospore divides by cross walls into four cells without emerging from the cell wall, thus producing a promycelium that is entirely internal. From each cell a long slender sterigma grows up through the gelatinous stratum that covers the tops of the layer of laterally adhering teliospores. Weir (1912) reported that occasionally in this genus the four cells produced by the division of the teliospore are cruciately arranged, resembling the condition in *Tremella*. In *Chrysopsora* each teliospore of the two forming the stalked compound teliospore divides transversely as in *Coleosporium*. In *Goplana dioscoreae* (B. & Br.) Cummins (1935b) the teliospores arise from the base of a gelatinous matrix and push up into it but not through to the exterior. Each teliospore, as in *Coleosporium*, divides by transverse septa into four cells from each of which a slender sterigma pushes out to the surface where the sporidia are formed. This is very similar to the case in the gelatinous species of Auriculariales.

The uninucleate sporidia may be long, ellipsoid, pointed or rounded at one or both ends, or may approach a spherical shape. The cell contents are usually somewhat yellow. The sporidium is shot off with more or less violence from the tip of the sterigma. Usually the whole contents of the promycelial cell pass into the single sporidium arising from that cell but occasionally a case is met with where the cell nucleus divides and one nucleus and part of the cytoplasm pass into the sporidium, so that a second sporidium may be produced by the same cell, as occurs more frequently in the Ustilaginales. A sporidium that fails to fall upon a suitable host is capable, under proper conditions of moisture, of producing a secondary sporidium at the tip of the sterigma and this in turn a tertiary or even quaternary sporidium, each being shot off from its sterigma. The successive sporidia are smaller and smaller. Under conditions of extreme humidity the author has observed the primary sporidia and the succeeding ones of *Kunkelia nitens* (Schw.) Arthur remaining attached in a short chain of four or five successively smaller cells.

When a sporidium falls upon the epidermis of a suitable host it germinates in a drop of rain or dew or film of water, forming a slender germ tube which, Dr. Ruth F. Allen (1930) showed, penetrates the cuticle and cell wall into the epidermal cell. The actual pore of entry is very small, the hypha on either side being several times as thick. Within the epidermal cell the hypha elongates and becomes divided into several uninucleate cells. From each of these a branch grows through the interior wall of the host cell either into an underlying cell, where it acts just as did the infection hypha in the epidermal cell, or into an intercellular space where growth becomes much more rapid and the hypha becomes larger and more vigorous, sending haustoria into the cells between which it passes. A few cases have been reported where entry took place through a stoma but these seem to be not usual. Pady (1935) reports that in the uninucleate race of *Kunkelia nitens* the sporidia upon germination infect intracellularly the epidermal cells and from these continue to progress as intracellular mycelium for ten days or so, producing then in the phloem tissues of the host the intercellular mycelium which forms haustoria that are coiled more or less. More often after the first two or three cells are formed the production of septa leads to the formation of uninucleate (monocaryon) cells. In *Melampsora lini* (Pers). Lev. the cells remain one to three or four nucleate up to the regions where the definite reproductive cells are formed when monocaryon cells predominate (Allen, 1934a). The infected area of the leaf often becomes thickened, in part at least through rapid formation of large amounts of mycelium which forces the host cells apart and in some cases crushes them. The presence of the rust may cause marked changes in the manner of growth of the host. In normally prostrate species of *Chamaesyce* (*Euphorbia*) the infected shoots become upright, a phenomenon also observed in other families of host plants. Abnormal growth of axillary buds in infected shoots gives rise to

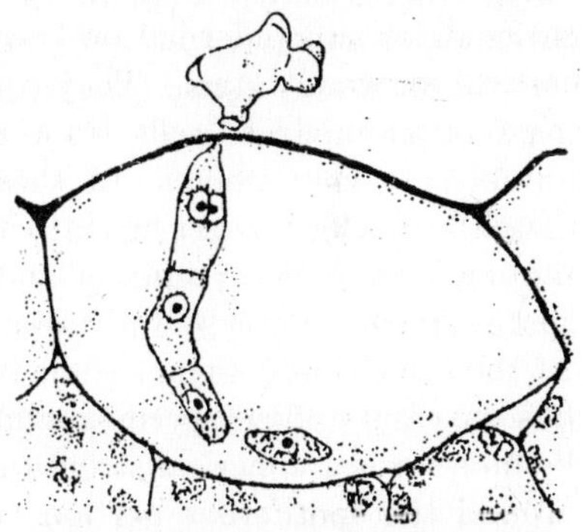

Fig. 127. Subclass Teliosporeae, Order Uredinales. Infection of leaf of *Berberis vulgaris* L. by sporidium of *Puccinia graminis* Pers. Infection hyphae are within epidermal cell and the attached sporidium is empty. (Courtesy, Allen: *J. Agr. Research*, 40(7):585–614.)

the witches' brooms so characteristic of some rust infections on Conifers. Dodge (1923) showed that in leaves of *Rubus* infected by the orange rusts (*Kunkelia nitens* and *Gymnoconia peckiana* (Howe) Trotter) the production of stomata which is usually confined to the lower epidermis is also brought about in the upper epidermis. (Fig. 127.)

The monocaryon mycelium is of rather short duration in the leaves and stems of herbaceous plants. In woody plants it may persist for years, e.g., *Cronartium ribicola* Fischer, in the tissues of the white pine (*Pinus strobus* L.) or *Gymnoconia peckiana* and *Kunkelia nitens* in *Rubus* where the mycelium penetrates to the roots and infects the new shoots next year. From this mycelium arise the spermogonia (pycnia). These may be subcuticular or subepidermal on leaves, green stems or fruits, or even subcortical on woody stems. They consist of a basal pseudoparenchymatous mass of uninucleate cells from which arise numerous parallel slender uninucleate sporophores. In these the nucleus divides and the upper daughter nucleus passes out into a terminal sperm cell (pycniospore) containing very little cytoplasm and surrounded by a thin cell wall. This spore is pushed loose by the formation of a second sperm cell below it, etc. At the same time a sugary slime is secreted which partially or completely fills the cavity of the spermogonium. The subepidermal spermogonia may be more or less spherical structures with a marginal series of paraphyses around the sporiferous portion. These push up through the epidermis, rupturing it and producing an ostiole through which the sugary slime containing the sperm cells exudes as a shining drop which is sweet in taste. Sometimes the mass of spermogonia is fragrant. The more diffuse subcuticular and subcortical spermogonia also rupture the overlying cuticle or cortex respectively, exposing the sugary exudate. Various insects, particularly flies, attracted by the sweet liquid and accompanying fragrance visit the spermogonia and feed on the exudate and go from leaf to leaf and plant to plant. In this way the sperm cells adhering to their feet or mouth parts become scattered widely over various plants or other parts of the same plant. Rain also doubtless helps in the dissemination of the sperms. (Fig. 128.)

From the same mycelium that has given rise to the spermogonia and often near to these or on the opposite side of the leaf the hyphae begin to mass themselves, frequently at first in a substomatal chamber or other large intercellular space. Eventually this becomes a pseudoparenchymatous mass of cells, those nearer the surface being larger and less filled with food, the underlying cells being smaller and better supplied with food, the whole mass being more or less surrounded by several layers of hyphae. All the cells are normally uninucleate. This structure is called an aecial primordium. Buller (1938) suggested for this structure the term proto-aecidium. Massee (1888) reported the presence of an oogone with

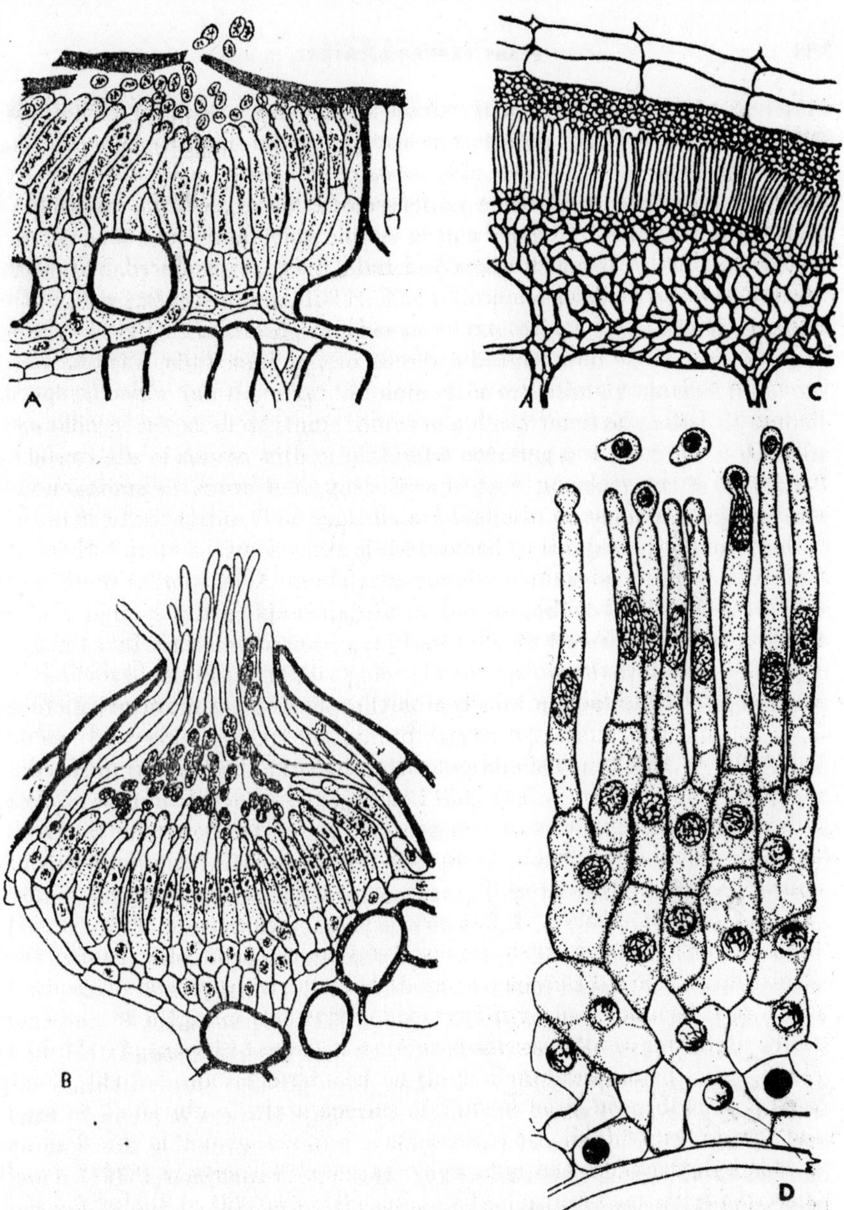

FIG. 128. Subclass Teliosporeae, Order Uredinales. Various types of spermogonia. (A) Subcuticular spermogonium in *Phragmidium violaceum* (Schulz.) Wint. (B) Subepidermal spermogonium in *Gymnosporangium clavariaeforme* (Jacq. ex Pers.) DC. (C, D) Subcortical spermogonia. (C) In *Cronartium comptoniae* Arthur, section of twig of *Pinus* sp. (D) Portion of spermogonium of *Cronartium ribicola* Fischer, showing development of sperm cells. (A–B, courtesy, Blackman: *Ann. Botany*, **18**(71):323–373. C, after Adams: *Penn. State Coll. Agr. Exp. Sta. Bull.*, **160**:31–77. D, courtesy, Colley: *J. Agr. Research*, **15**(12):619–660.)

attached antherid in this primordium. The more recent investigations seem to show that this report is incorrect. In the dense portion of the pseudoparenchymatous tissue there begin to appear cells with two or more nuclei, whose origin will be discussed in the succeeding paragraphs. These cells elongate and may branch. At the apex binucleate aeciospores are budded off. After each spore is formed there is produced a smaller "disjunctor" cell, likewise binucleate, then another aeciospore, etc., until a chain of alternate spores and disjunctor cells results. Between the original binucleate or plurinucleate cells others push their way in from the basal side of the aecial primordium and these also give rise to spore chains. In the meantime the large almost empty cells begin to collapse and digest, forming a space into which the chains of spores are pushed. Finally all of this tissue is crushed or destroyed. Around the sporogenous area a peridium may be produced or it may be wanting. The mass of spores eventually ruptures the overlying host tissues and the mass of spore chains and loose spores is exposed to the air. The spores are usually light to dark yellow or orange and somewhat roughened.

Craigie demonstrated in 1927 that for *Puccinia graminis* Pers., growing on barberry (*Berberis vulgaris* L.) and *P. helianthi* Schw., growing on sunflower (*Helianthus*) an infection with a single sporidium produces a sorus within which are produced numerous spermogonia and aecial primordia. If this is protected from the visits of insects the primordia never develop to spore production. However, if insects are allowed free access to this sorus and other sori as well, the aecia will develop within a few days. Furthermore, by transferring the sperm cells from one sorus to another in about half of the cases aecia will develop. Craigie finally demonstrated that the four sporidia of a promycelium were of two sexual phases, two sporidia of each phase, and that sperm cells from the sorus of one sexual phase, would fertilize the sorus of the other sexual phase, and vice versa. In later papers (1928, 1931, 1933) he has added *P. coronata* Corda, *P. pringsheimiana* Klebh. (*P. caricis grossulariata* Arth.) and *Gymnosporangium* sp. to the list. Miss Allen (1932a) showed this to be true for *P. triticina* Erikss. (*P. rubigo-vera tritici* (Erikss. & Henn.) Carl.) and several other species of rusts. Indeed it seems probable that almost all rusts producing spermogonia have these two sexual phases. If two sori from sporidia of opposite sexual phases lie close together aecia will develop where their mycelia come into contact, without the help of the sperm cells. Brown (1932) showed that a dicaryon mycelium (e.g., uredial sorus) can also induce aecial formation in a sorus from a sporidial infection. Hanna (1929a) found that 48 hours after a mixture of sperm cells of opposite sexual phases was applied to a sorus of *Puccinia graminis* Pers., on the leaf of barberry, the cells at the base of the aecial primordium became binucleate. Using mixtures of sperm cells from different physiologic races of *Puccinia graminis* on wheat, Waterhouse (1929), in Australia, obtained two hitherto unknown forms which are probably to be interpreted as crosses. Similar experiments by Miss Newton (1930) and others in Canada and by Stakman (1930) and collaborators in Minnesota resulted not only in interracial crosses but in crosses between *P. graminis tritici* and *P. graminis secalis* and between the former and *P. graminis agrostidis*. In the course of the various crosses several new physiological races have been produced. The study of the results reveals that many of these races must be heterozygous. This was further demonstrated by the production of selfed strains. Thus physiologic race #17 when selfed (i.e., fertilized by sperm cells of the same race but of opposite sexual phase) produced aeciospores which gave rise to #17 and seven others besides, while race #53 gave besides itself seventeen other races. On the other hand race #9 proved to be homozygous.

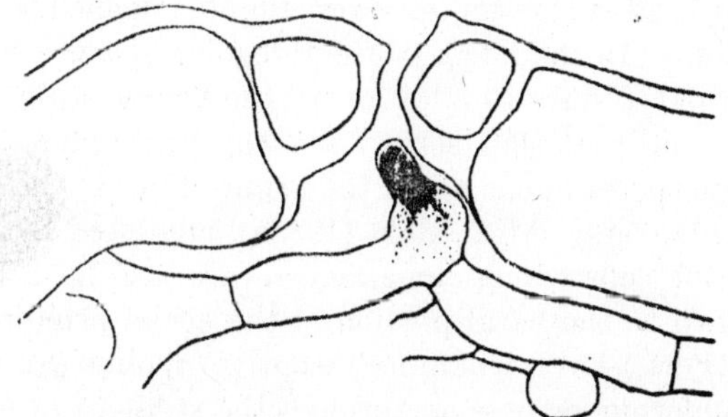

FIG. 129. Subclass Teliosporeae, Order Uredinales. Receptive hypha with attached sperm, just below mouth of stoma. (Courtesy, Andrus: *J. Wash. Acad. Sci.*, **23**(12):544–557.)

Andrus (1931, 1933) demonstrated for the rusts *Uromyces phaseoli typica* Arth. and *U. phaseoli vignae* (Barcl.) Arth. that there are certain elongated filaments, which he called "trichogynes," whose tips emerge from the stomata or from between epidermal cells and to which the sperm cells adhere. The nucleus of the sperm enters the trichogyne and passes down through it. Only then do the multinucleate cells appear in the aecial primordium. Miss Allen (1932a) demonstrated the occurrence of similar receptive hyphae in *Puccinia triticina* and other species of rusts and J. L. Forsberg (in an unpublished thesis, 1932) has shown their presence in *Kunkelia nitens* (Schw.) Arth. and *Gymnoconia peckiana* (Howe) Trotter, but not in the monocaryon race of the former in which functional spermogonia are not produced. Miss Rice (1933) showed that in a number of other rusts similar structures are formed. According to Miss Allen (1932a) the receptive cells, into which the one or more (up to ten or so) sperm nuclei pass, elongate and branch and form a mycelium which penetrates into the basal portion of the aecial primordium where dicaryon or plurinucleate cells give rise to the chains of spores. In a later paper Miss Allen (1933c) showed that in *Puccinia sorghi* Schw., whose aecial stage is produced in species of *Oxalis*, the paraphyses extending from the ostiole of the spermogonium are receptive hyphae as was demonstrated for *P. helianthi* Schw. by Craigie. The union of sperms to these paraphyses and the actual passage of the sperm nuclei into them was illustrated. In her work on *P. triticina* and more particularly on *P. graminis* she reported that receptive hyphae may emerge from stomatal openings or from between epidermal cells in addition to the spermogonial receptive hyphae. Pierson (1933) and Buller (1938) demonstrated the presence of slender flexuous nonseptate receptive hyphae growing up from the base of the spermogonium and out into the drop of spermogonial nectar in *Cronartium ribicola* Fischer and *Puccinia graminis* Pers., respectively. When a drop of mixed nectar is added sperm cells were found attached to these hyphae within a few hours. In the latter of the foregoing species there are in the spermogonium paraphyses in addition to the flexuous hyphae but in no case did Buller find that they served as receptive hyphae. (Fig. 129.)

Just how the sperm nuclei reach the primordium of the aecium is still a matter of controversy. Miss Allen (1933a) suggested that they passed into the receptive cells which elongated inward and formed a mycelium which grew down to the basal portion of the aecial primordium. In *Melampsora lini* (Pers.) Lev., where no receptive hyphae are present either piercing the epidermis or passing through the stomata or emerging from the spermogonia she reported (1934a) that the sperm cell dissolves a hole through the outer epidermal wall of the host (*Linum usitatissimum* L.) through which it gains access to the lumen of the cell where it germinates or through which a slender germ tube passes. In either case a slender branched mycelium arises which grows toward the aecial primordium where abundant pairing of hyphae occurs, resulting in the formation of the basal cells from which the chains of aeciospores are produced. The suggestion has been made that the sperm nuclei possibly initiate a general diploidization of the mycelium of the sorus so that all of the mycelium between the point of union of sperm and receptive hypha to the aecial primordium is diploidized, as Lehfeldt (1923) and Buller (1930) showed to be the case in some of the Eubasidiae. Savile (1939) showed that in several rusts investigated by him the cells of this intervening mycelium are not diploidized, suggesting that the sperm nuclei travel from cell to cell, without accomplishing diploidization, until the aecial primordium is reached. He also demonstrated that sori close together in the leaf tissue can fertilize one another without intervention of the sperm cells, probably by contact of the mycelial cells and mutual diploidization. Andrus (1933) obtained differential staining of the sperm nuclei and of those of the vegetative hyphae in the sorus and was able to confirm Savile's suggestion that the former pass through the hyphae without bringing about their diploidization. Apparently there is lacking a standard method of diploidizing the fertile cells of the aecial primordium. Miss Allen (1932b) has demonstrated that in *Puccinia coronata* Corda the sperm cells unite abundantly with the receptive hyphae and that mycelium (carrying the sperm nuclei) grows from these down to the primordium. In the latter the union of basal cells, to be described in the next paragraph, rarely occurs. In the case of *Puccinia sorghi* Schw., whose spermogonia and aecia occur on species of *Oxalis,* Miss Allen (1934b) reported that functional receptive hyphae occur in the spermogonium, and also project out of stomata. Sometimes sperm cells germinate on the outside of the leaf and the slender hyphae enter the stomata. Within 24 hours after the "spermatization" of a sorus with compatible sperms 60 per cent of the mycelial cells of the mycelium are found to possess more than one nucleus, through rapid division and migration of the sperm nuclei. Six days after spermatization the first aecia set free their aeciospores.

Prior to the discovery of the active participation of the sperm cell Blackman (1904), Christman (1905, 1907), Mme. Moreau (1913), Colley (1918), and many others described the manner by which the aeciosporic chains are originated. A layer of uninucleate cells in the basal portion of the aecial primordium either shows fusion by twos, the walls between the upper portions of these cells dissolving out, resulting in a binucleate two-legged cell, or nuclei pass from adjacent cells through small pores to form a dicaryon cell. From this by conjugate division of the nuclei and the formation of successive cells comes a chain of aeciospores and disjunctor or intercalary cells. Just how the two sets of uniting cells originated was not made clear by these authors. It is now apparent that the two uniting cells contain respectively a nucleus originating from the sperm and one from the mycelium developed from the sporidium. Wang and Martens (1939) do not believe that "Christman" conjugations at the base of the aecium are the normal mode of diploidization but that it occurs earlier, perhaps as far back as spermatial fusions with one another or with receptive cells of the rust. (Fig. 130.)

From the foregoing accounts it is clear that in some and probably all rusts which produce sperm cells they are functional. Furthermore, the same haploid mycelium produces both sperm cells and receptive hyphae so that both male and female structures are present, yet self-fertilization does not occur. Like the condition in *Schizothecium* (*Pleurage*) reported by Ames (1932), in *Neurospora* as demonstrated by Shear and Dodge (1927), and in *Stromatinia* (*Sclerotinia*) *gladioli* (Drayton) Whetzel studied by Drayton (1934), two sexual phases are present, each hermaphroditic but incapable of self-fertilization.

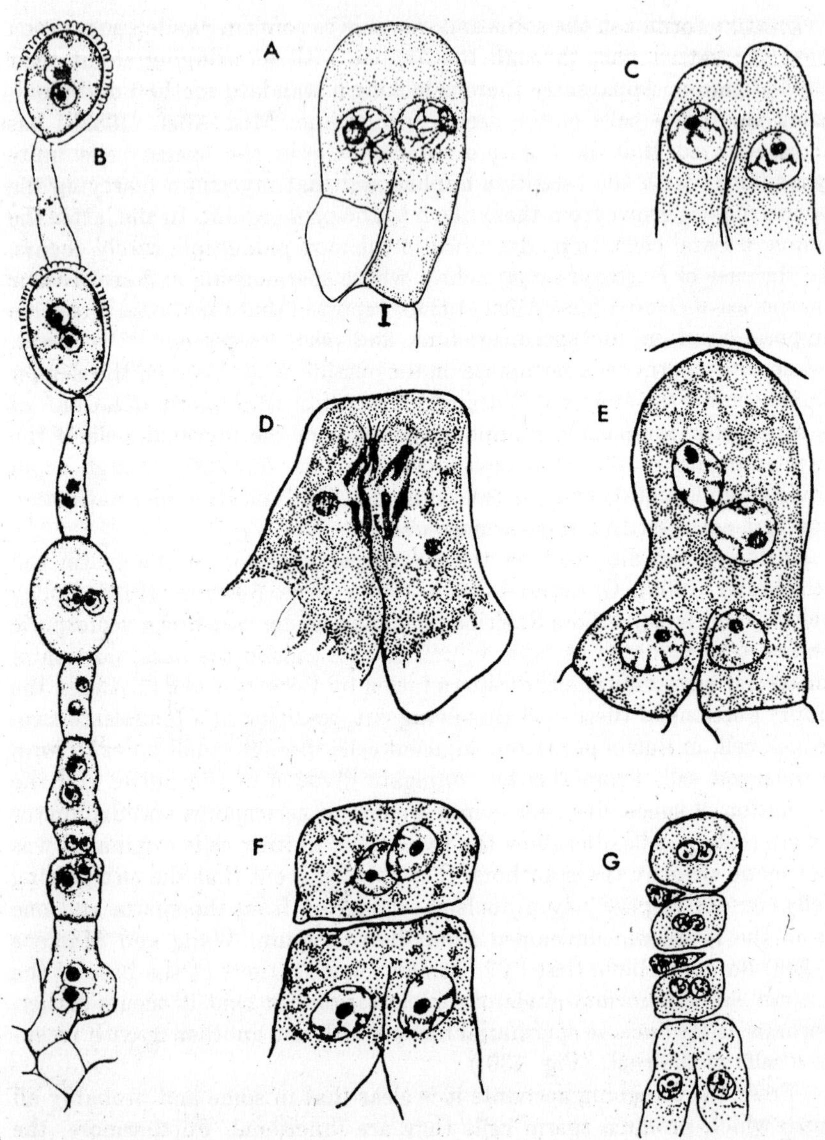

Fig. 130. Subclass Teliosporeae, Order Uredinales. Origin of chains of aeciospores from union of basal cells in the aecium; *Cronartium ribicola* Fischer. (A) Basal cell of future aeciospore chain. (B) Chain of aeciospores and intercalary cells. (C–G) Steps in the formation of a chain of aeciospores and intercalary cells in *Phragmidium speciosum* (Fr.) Cke. (A–B, courtesy, Colley: *J. Agr. Research*, **15**(12):619–660. C–G, courtesy, Christman: *Botan. Gaz.*, **39**(4):267–275, Univ. Chicago Press.)

The binucleate aeciospores are capable of remaining viable for a long time and can be carried great distances by the wind and still cause infection. Falling upon a suitable host plant the spore germinates in a drop of water (dew, rain, etc.) forming a stout germ tube which usually seeks out and enters a stoma. Pady (1935) showed that in *Gymnoconia peckiana* the germ tube penetrates the epidermis directly as do the germ tubes from the sporidia, but this seems to be an exception. Just within the stomatal opening the hypha enlarges to form a "substomatal vesicle" containing many nuclei from which arise hyphae of dicaryon cells which penetrate the host tissue in various directions, growing intercellularly and sending haustoria into some of the cells between which they pass. Since the normal aeciospore is binucleate this mycelium arising from it is dicaryotic and has two nuclei in each cell, one descended from that of the sperm and one from that of the parental monocaryon mycelium. Usually the size of the infected area from one aeciosporic infection is limited, i.e., the fungus in this phase is not capable of indefinite growth. When its full extent of development is nearly attained a subepidermal mass of hyphae is formed making a pseudoparenchymatous layer of dicaryon cells from which grow short upright two-celled branches. These raise and eventually rupture the epidermis, thus producing a uredial sorus or uredium. The outermost of the two cells enlarges to form a binucleate urediospore while the other elongates to form its long stalk. The urediospores are colorless or more usually yellow to orange red in color, mostly finely verrucose or echinulate. There are several germ pores for each urediospore and their number and position on the spore are of great assistance in the identification of the various species. They break loose from the stalk and like the aeciospores may be carried long distances by the wind. They germinate and infect the host through the stomata in exactly the same manner as do the aeciospores. As long as the host plant has not passed a certain stage of development these new infections produce other crops of urediospores. As the host plant becomes more mature teliospores begin to appear, often at first intermingled with the urediospores in the same sorus, but eventually in sori containing only teliospores. They arise from the same type of pseudoparenchymatous subepidermal mycelium as do the urediospores. From this basal layer there grow upward series of dicaryon cells. Some of these may differentiate into stalk cells and teliospores or all may become teliospores. Malençon (1936) observed the production rarely of teliospores in the aecial sori of *Puccinia atropae* Mont. In some of the most primitive rusts no basal layer is formed and the teliospores are produced singly or by twos or threes as enlarged cells of the mycelium in the interior of the leaf as occurs in *Uredinopsis*. The young teliospores, whatever their shape or location, are at first binucleate. The two nuclei unite to form a diploid nucleus. Usually the teliospore becomes thick-walled (mostly with one or more germ pores) and more or less colored, from light yellow-brown to almost black. In some of the rusts on ferns they are colorless and thin-walled.

From the mature uninucleate teliospore the promycelium emerges at once (various species of *Cronartium*, *Puccinia malvacearum*, etc.) or only after over-wintering (*P. graminis* Pers.) or after certain unfavorable environmental conditions have passed. It emerges through one of the germ pores and into it may pass the undivided nucleus of the teliospore or the first of the two divisions of the nucleus may occur before the nuclei enter the promycelium. The final division usually takes place in the promycelium whereupon septa are formed dividing it into four cells. Sometimes a further septum cuts off the teliospore from the basal cell of the promycelium. From the promycelial cells sterigmata grow bearing at their tips the sporidia which are shot off violently at maturity. In the case of a compound teliospore, such as is characteristic of *Puccinia*, *Phragmidium*, *Ravenelia*, etc., each of the component cells produces its own mycelium, thus demonstrating that each cell is to be considered a teliospore, not the whole compound structure.

Rusts may be either autoecious or heteroecious. In the former the sporidial and aeciosporic infections take place on the same or closely related species of host while in a heteroecious species the host infected by the sporidia belongs to a family not at all closely related to that containing the host for the aeciosporic infection. Heteroecism was first proved by actual inoculation experiments by de Bary in 1865 for *Puccinia graminis* and by Oersted for *Gymnosporangium sabinae* (Dicks.) Wint. in the same year. The former showed that the aeciospores produced on the common barberry (*Berberis vulgaris* L.) would not infect that species but would infect the small grains such as wheat, barley, etc. The urediospores would infect the same and often also closely related species of grains. The sporidia from the teliospores on the overwintered straw or stubble would, on the contrary, infect only the barberry. Thus was brought the scientific explanation of a phenomenon known for a hundred years or more that the presence of barberry plants was detrimental to small grains. This much had been definitely proved by careful observation and experiment long before, but the actual connection of the rust on barberry with that on the grain was left for de Bary to prove. Plowright (1889) gives an excellent account of these early observations and beliefs as to the harmful effect of barberry on grain. Since that time the heteroecism of hundreds of species has been demonstrated. It is worthy of note that Waterhouse (1929) discovered one physiologic race of wheat rust that is incapable of infecting barberry.

In an autoecious species the aeciospores infect the same or closely related host species as do the sporidia. Thus *Puccinia helianthi* Schw., the rust of the sunflower (*Helianthus annuus* L. and other species) has its sporidial infection on the sunflower with the production of a monocaryon mycelium which produces the spermogonia and the aecial primordia. Infection by aeciospores produces in the same host species or even in the same plant the dicaryon mycelium from which arise the urediospores and teliospores. It may happen that both types of infection may occur on the same leaf.

Of the rusts of more or less economic importance the hosts are indicated below for the different stages of a few species, showing that both heteroecious and autoecious rusts may be enemies of cultivated plants. The list contains only a very few of the many rusts that attack important crop plants. The customary symbols are used, viz., 0, spermogonial development, I, aecial stage, II, uredial stage and III, telial stage.

Heteroecious Species

Puccinia graminis Pers., Black stem rust of small grains: 0 and I on *Berberis vulgaris* L., barberry; II and III on small grains (wheat, rye, barley, oats) and various other grasses.

P. rubigo-vera tritici (Erikss. & Henn.) Carl., leaf rust of wheat: 0 and I on species of *Thalictrum;* II and III on wheat (*Triticum*).

P. coronata Corda, crown rust of oats: 0 and I on various species of *Rhamnus;* II and III on oats (*Avena sativa* L.).

P. sorghi Schw.: 0 and I on *Oxalis stricta* L.; II and III on maize (*Zea mays* L.).

Gymnosporangium juniperi-virginianae Schw.: 0 and I on apple (*Malus sylvestris* Mill.); III on red cedar (*Juniperus virginiana* L.).

Uromyces dianthi Niessl. (*U. caryophyllinus* Wint.): 0 and I on *Tithymalus* sp.; II and III on carnation (*Dianthus caryophyllus* L.).

Cronartium ribicola Fischer, white pine blister rust; 0 and I on white pine (*Pinus strobus* L.); II and III on various species of currant (*Ribes*) and gooseberry (*Grossularia*).

Uredinopsis spp.: 0 and I on species of fir (*Abies*); II and III on various ferns.

Autoecious Species

Gymnoconia peckiana (Howe) Trotter: 0, I, and III on blackberry, dewberry, and black raspberry (*Rubus* spp.).

Uromyces phaseoli typica Arth.: 0, I, II, and III on American beans (*Phaseolus vulgaris* L.).

Puccinia asparagi DC.: 0, I, II, and III on asparagus (*Asparagus officinalis* L.).

P. helianthi Schw.: 0, I, II, and III on sunflower (*Helianthus annuus* L. and some other species).

Phragmidium spp.: 0, I, II, and III on various species of rose (*Rosa* sp.).

In the case of heteroecious rusts efficient control can be obtained by the elimination of the alternate host in case the rust is unable to overwinter on the host that is of economic importance. Thus in the northern portions of the United States and of Europe the extermination of the barberry has greatly reduced the ravages of the black stem rust because this rust cannot survive the winter except as teliospores on the overwintering wheat or rye or other grains. In the southern portions of the United States and Europe the rust is not killed out by the cold of winter and so perpetuates itself by its urediospores. Under such conditions the eradication of the barberry has little effect. The apple rust has been found to be susceptible to control by removal of its alternate host, the red cedar, over an area to a distance of a mile or more from the orchard that is to be protected. White pine blister rust requires the destruction of all currants and gooseberries to a distance of half a mile (one mile for *Ribes nigrum* L.) from the trees it is desired to protect. *Puccinia rubigo-vera tritici*, although heteroecious, cannot be controlled in this manner for the rust that infects the wheat in the fall survives the winter on this host and produces urediospores in the spring from which the disease is spread. In fact this rust is exceedingly abundant in many parts of the United States where the host for the aecial stage does not occur.

The rust life cycle described above is the typical one. Rusts possessing such a cycle are called macrocyclic or long-cycle rusts. Arthur and his collaborators (1929), Jackson (1931), and others believe that these represent forms with the more primitive life cycle so far as present rusts are known. Many species of rusts have shortened their life cycle by the omission of one or more stages. Dietel (1928), Olive (1908), and Grove (1913), on the contrary, consider the forms with the short life cycle to be more primitive. As examples of the omission of certain stages the following may be mentioned. Most species of *Gymnosporangium* produce no binucleate repeating spores (urediospores) from the dicaryon stage of growth but they do occur in *G. nootkatense* (Trel.) Arth. Some rusts produce only spermogonia and telia while still others omit the spermogonia also (e.g., *Puccinia malvacearum* Bert.). These last two are properly speaking microcyclic rusts. There are also rusts in which no true telia are produced but whose aeciospores germinate in the manner of teliospores by the formation of a promycelium. These rusts also are microcyclic. *Kunkelia nitens* (Schw.) Arth., on *Rubus* spp. is of this type as are the various species of *Endophyllum*. The microcyclic species are of especial interest as regards the origin of the binucleate condition of the young teliospore. In *Puccinia arenariae* (Schum.) Wint., Lindfors (1924) described the formation of a two-celled promycelium, each cell giving rise to a binucleate sporidium. This produces a dicaryon mycelium and no monocaryon mycelium occurs.

Another anomaly in the life cycle of a short-cycle rust is described by Thirumalachar (1946) for *Uromyces aloës* (Cke.) Magn. In this species the spermogonia appear to be normal and exude drops of sweet liquid filled with sperm cells. However, there are no flexuous hyphae. The appearance of the spermogonia is followed immediately by the development of telia. The monocaryon mycelium within the sorus shows here and there contacts between two adjacent hyphae where the intervening walls are dissolved and a nucleus passes from one cell into the other. From this cell arise dicaryon hyphae which occupy that portion of the sorus while the monocaryon hyphae degenerate. These dicaryon hyphae give rise to binucleate teliospores on slender binucleate stalks. The nuclei unite in the teliospore and the pedicel breaks, so that the teliospores are distributed by air currents. The zygote nucleus passes into the apical promycelium where the two normal divisions of meiosis occur. Usually but one septum is formed dividing the promycelium into a basal uninucleate cell which soon degenerates, and a large terminal cell with three nuclei. This sends out a slender germ tube into which the nuclei enter and infect the host. After the epidermis is penetrated the nuclei divide further and septa are formed, dividing the resultant mycelium into uninucleate cells. In this species the terminal cell of the promycelium takes the place of a sporidium in infecting the host while the distribution of the teliospore by air currents makes distribution by means of sporidia unnecessary. Apparently the spermogonia and sperm cells do not function.

In other microcyclic forms there are other ways in which the life cycle is completed. Miss Allen (1933b) found that in *Puccinia malvacearum* the teliospores arise from dicaryon mycelium. The two nuclei unite and then divide in the promycelium in the usual way. The nucleus then undergoes division in the sporidium but Miss Ashworth (1931) finds that it gives rise to a monocaryon mycelium. According to her investigations certain cells in the telial sorus show nuclear migrations producing the dicaryon phase. Miss Allen (1935) reported that occasional conidia are formed at the tips of hyphae emerging from the stomata and suggested that possibly these may have a part in the diploidization of the mycelium that forms the telial sorus. However, in the main this appears to result from the intermingling and fusion of hyphae from mycelia produced from two separate but adjacent sporidial infections. In *Puccinia prostii* Duby, according to I. M. Lamb (1934), no receptive hyphae occur in the spermogonia which are produced. The mycelium within the host (*Tulipa* sp.) remains monocaryotic. The teliospores arise from the lateral fusion of two adjacent hyphal tips or from the passage of a nucleus through a small opening in the septum from a basally placed cell to the cell above. Although normal teliospores are produced, apparently infection rarely if ever occurs by means of sporidia, the mycelium being carried over from year to year through the bulbs.

In most of the short-cycled rusts studied in which normal spermogonia are produced (spermogonia and telia, or spermogonia and aecia whose spores function as teliospores) it has been shown that the mycelium is of monocaryon type until the telium or aecium is formed, when dicaryon cells appear. It is probable that this dicaryon phase arises in the same way as is described above for macrocyclic rusts.

The transfer of the production of the mycelium back from the teliospore to the aeciospore has apparently taken place independently several times so that the microcyclic rusts of the formula 0, I are not necessarily closely related but have probably developed in separate lines from 0, I, II, III forms. It is clear that *Kunkelia nitens*, the microcyclic orange rust of *Rubus*, is derived from *Gymnoconia peckiana*, a long-cycle form (0, I, III) on the same hosts. The genus *Endophyllum* represents a similar series of cases. Some of the species of this genus correspond in aecial host and

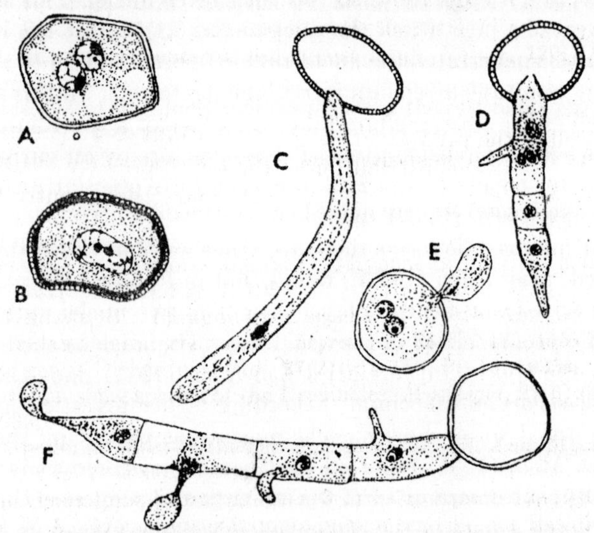

FIG. 131. Subclass Teliosporeae, Order Uredinales, Family Pucciniaceae. A–D, *Endophyllum sempervivi* (A. & S.) de By. (A) Binucleate aeciospore. (B) Aeciospore with the nuclei united. (C) Aeciospore germinating. (D) Promycelium formed. (E, F) *Endophyllum euphorbiae-sylvaticae* (DC.) Wint. (E) Aeciospore germinating without union of nuclei. (F) Promycelium and sporidia. (After Moreau and Moreau: *Bull. Soc. Botan. France*, **66**:14–44.)

structure to the aecial stage of known macrocyclic species. These abbreviated forms may even lack the fusions of the two nuclei within the spores before the formation of the promycelium. Dodge and Gaiser (1926) showed that in *Kunkelia nitens* the two nuclei pass out into the promycelium where each divides again to form the four nuclei which enter the sporidia. The Moreaus (1919) have found the same to be true for *Endophyllum euphorbiae-silvaticae* Lev., but in other species of the genus they demonstrated nuclear fusion in the aeciospore before the promycelium began to be formed. Dodge (1924) reported that in one form of *Kunkelia nitens* no dicaryon mycelium was produced at all and the aeciospores remain uninucleate and produce a two-celled promycelium. The spermogonia in this race are never more than rudimentary and no receptive hyphae are formed. (Fig. 131.)

A third type of life cycle is the one usually designated as 0, II, III, i.e., a cycle in which the typical aecial structure is lacking. The spermogonia are succeeded by sori containing spores exactly resembling typical urediospores. In the same way as are produced the dicaryon basal cells of the chains of aeciospores Christman (1907) showed that there are produced dicaryon cells giving rise to urediospore-like structures. The latter give rise to dicaryon mycelium from which may arise later another series of urediospores. The usual interpretation of this phenomenon is that this is really a macrocyclic rust of the formula 0, I, II, III in which the aeciospores are not produced in chains but singly on stalks, like urediospores. The primary (first produced) urediospores are therefore modified aeciospores while the secondary ones are true urediospores.

The aecium may be cup-shaped (cupulate) with a well-developed peridium or it may be very tall so as to make a horn-like (cornute) structure. The peridium may be lacking so that the aecium is diffuse. Other forms are known, the most curious of which is the "hyphoid" aecium of *Dasyspora foveolata* B. & C. (*D. gregaria* (Kze.) Henn.) in which a branching dicaryon mycelium emerges through various stomatal openings, forming a colorless mass of hyphae. These are terminated by single, not catenulate, aeciospores which drop off while the hypha elongates sympodially and produces another spore and so on. The cytology of this type of aecium needs careful investigation to determine where the diploidization occurs. Sydow (1925) interprets these as urediospores but Arthur and co-authors (1929) consider them to be aeciospores. (Fig. 132.)

The uredium may be merely a cluster of stalked urediospores bursting through the epidermis of the host or it may be surrounded by paraphyses. In *Cronartium* and *Pucciniastrum* and closely related genera the uredium possesses a true peridium. In some genera, e.g., *Coleosporium* the urediospores, like the aeciospores, are produced in chains but they arise from a dicaryon mycelium and so differ from the latter. In a few rusts the uredium is cupulate and resembles the aecium but spermogonia are lacking. They are sometimes called secondary aecia. Cummins (1937) reported that the uredium of *Prospodium* is a salver-shaped structure whose slender stalk about three cells thick emerges through a stoma and spreads out as a flat plate with upright fringe-like marginal paraphyses. On the flat surface of the salver arise the stalked urediospores. In some species of the genus the telial sorus is quite similar. (Fig. 133.)

The telium is the most variable structure in the order. Properly speaking a teliospore is a single cell, binucleate at first but becoming uninucleate by the fusion of the two nuclei and giving rise immediately or after a delay

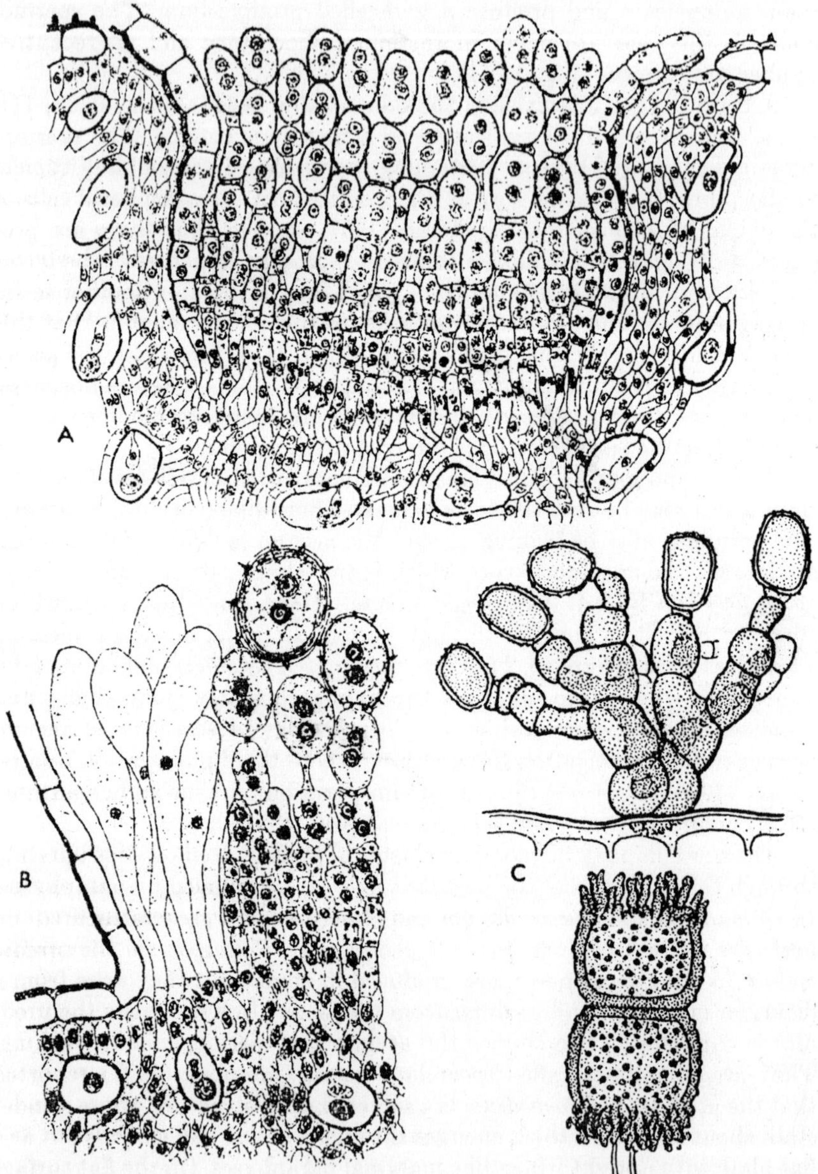

FIG. 132. Subclass Teliosporeae, Order Uredinales. Various types of aecia. (A) Cupulate aecium of *Uromyces erythronii* (DC.) Pass. (B) Margin of caeomoid aecium of *Phragmidium rubi* (Pers.) Wint., showing absence of peridium but presence of paraphysate hyphae. (C) Hyphoid aecium of *Dasyspora foveolata* B & C. and a compound teliospore. (A–B, after Sappin-Trouffy: *Le Botaniste*, 5:59–244. C, after Sydow: *Mycologia*, 17(6):255–262.)

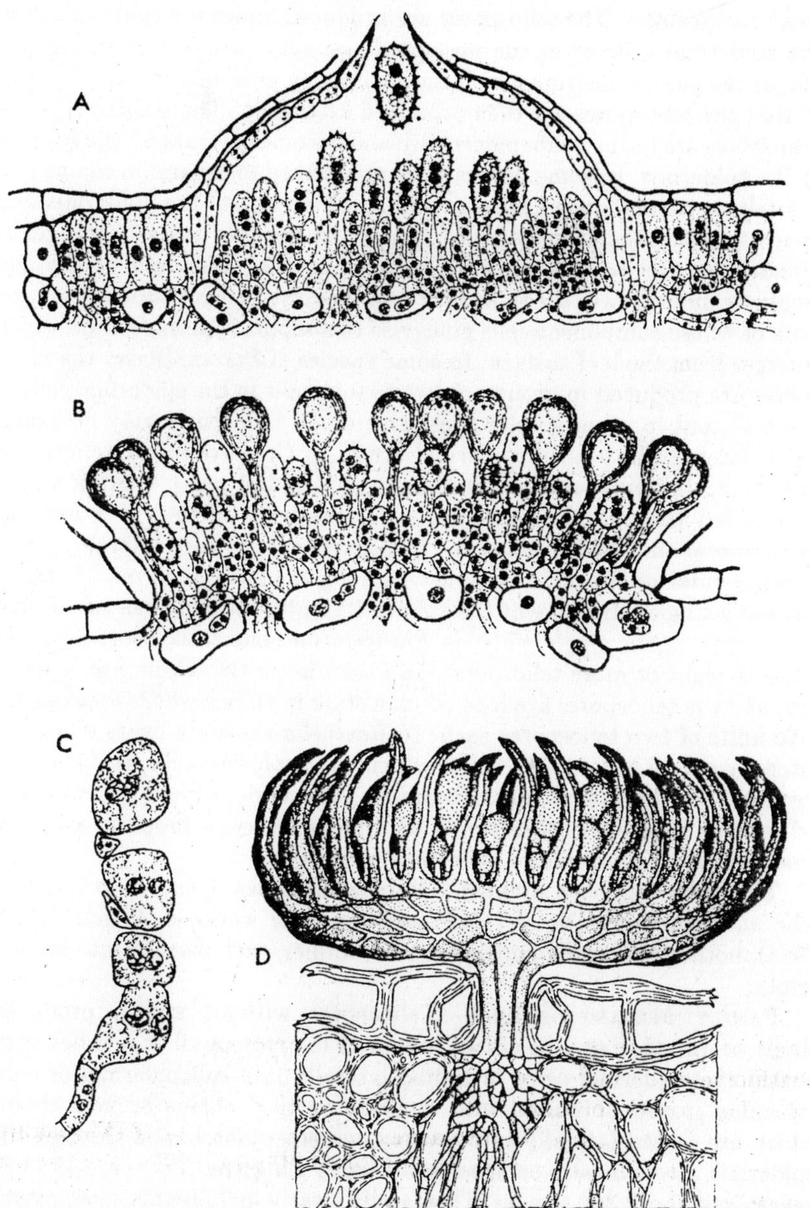

FIG. 133. Subclass Teliosporeae, Order Uredinales. Types of uredia. (A) Uredium with peridium in *Melampsoridium betulinum* (Pers.) Kleb.; on either side a portion of a telial sorus. (B) Uredium without peridium but with numerous capitate paraphyses among the urediospores, in *Melampsora helioscopiae* (Pers.) Cast. (C) Urediospores of *Coleosporium solidaginis* (Schw.) Thüm., in a chain. (D) Extrastomatal uredium of *Prospodium plagiopus* (Mont.) Arth. (A–B, after Sappin-Trouffy: *Le Botaniste*, **5**:59–244. C, after Christman: *Botan Gaz.*, **44**(2):81–101. D, after Cummins: *Ann. Mycol.*, **35**(1):15–21.)

to a promycelium. The teliospores are produced under the epidermis or in the epidermal cells or in the mesophyll, or very rarely (*Cystospora*, *Goplana*) the basal cells from which they arise may push out through a stoma so that the teliospores are then produced externally. The stalked types of teliospores are formed subepidermally and become external by the rupture of the epidermis. In some genera they are formed in separable chains and the telium is surrounded by a peridium which bursts the epidermis and opens to allow the teliospores to escape. In the fern rusts of the genus *Uredinopsis* the teliospores are produced in the mesophyll of the leaf singly or united, two, three, or four together, into a compound teliospore, each of whose component cells gives rise to a separate promycelium which emerges from the leaf surface. In some species of *Pucciniastrum* the teliospores are produced in groups of two to four cells in the epidermal cells of the host, but in other species these clusters of teliospores may be aggregated laterally into a subepidermal crust. In *Cronartium* the teliospores are joined laterally and longitudinally into a tall waxy column which pushes out through the epidermis to a length of up to 6 or 8 mm. In *Melampsora* and *Coleosporium* the teliospores are crowded laterally into a subepidermal or subcuticular crust. Among the stalked forms *Uromyces* has but a single teliospore at the apex of its stalk, *Puccinia* has two united teliospores on the single stalk, in *Phragmidium* one stalk bears a row of three to eight or more teliospores. In *Pucciniosira* the telium has a peridium and the teliospores are formed on a stalk in chains which break apart into units of two teliospores each. In *Ravenelia* the stalk bears a head of laterally united teliospores, below which hang colorless cells, the so-called "cysts." Usually in systematic literature these various types of compound teliospores are spoken of as single teliospores although properly each cell from which a promycelium arises is a teliospore.

The order has been divided into many families or into two families. The author follows Dietel (1928) and the later works of Arthur (1929, 1934), both of whom recognize but two families, each divided into several tribes.

Family Melampsoraceae. Teliospores without stalks, produced singly or united in groups of two to four in the mesophyll or just below or within the epidermal cells, or united laterally into subepidermal or subcuticular crusts, or united into separate vertical chains or into chains which are united laterally into a waxy column which bursts through the epidermis. Aecia mostly on species of Family Pinaceae. There are 15 to 20 genera and about 300 species. This family clearly includes the most primitive living representatives of the order. The genus *Uredinopsis* with its colorless thin-walled teliospores, single or united by twos, threes, or fours, in the mesophyll of the leaves of ferns, its two kinds of colorless urediospores, thick-walled and thin-walled, the sori surrounded by peridia, and

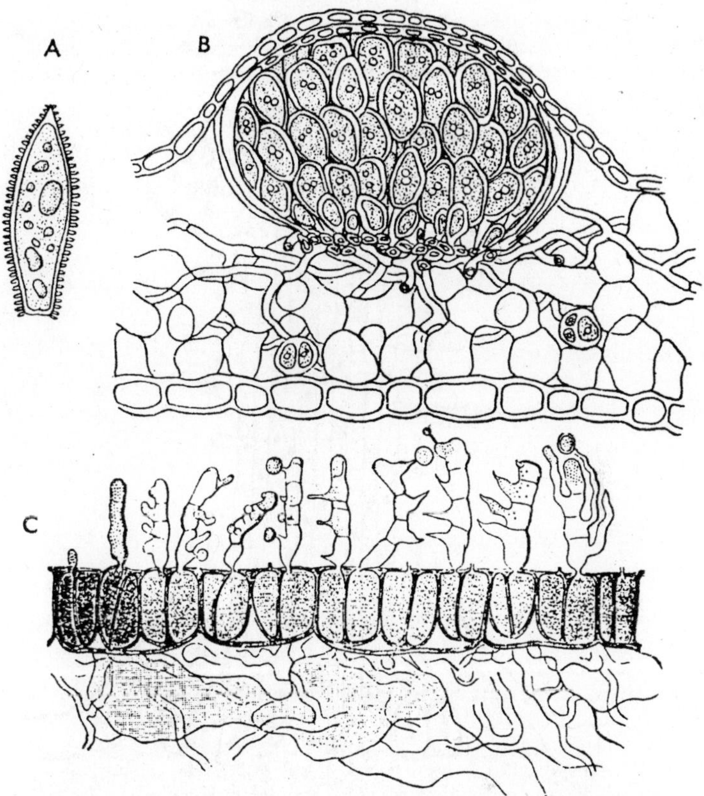

Fig. 134. Subclass Teliosporeae, Order Uredinales, Family Melampsoraceae. (A, B) *Uredinopsis struthiopteridis* Störmer. (A) Urediospore. (B) Section through leaf of host showing uredial sorus with peridium, and teliospores scattered through the mesophyll. (C) *Pucciniastrum goeppertianum* (Kühn) Klebahn; telial stage in epidermis of *Vaccinium* sp. (A–B, after Dietel: *Ber. deut. botan. Ges.*, **13**(7):326–332. C, after Hartig: "Lehrbuch der Baumkrankheiten," Berlin, J. Springer, 1889.)

with its aecia and spermogonia on the needles of species of *Abies* probably combines the greatest number of primitive characters of any rust, viz., telial hosts in the ancient group Pteridophyta, colorless urediospores and teliospores, the latter scattered in the mesophyll, uredia surrounded by peridium etc. Two other genera (*Hyalopsora* and *Milesia*) with colorless teliospores formed in the epidermal cells are also found in the ferns. *Pucciniastrum goeppertianum* (Kühn) Klebahn is of interest because it is one of the few rusts in which the sporophytic mycelium is perennial. The aecia occur on the leaves of the fir (*Abies*) as is true of the other genera mentioned above. The telial stage (no uredia are known) occurs in *Vaccinium*, its presence causing the development of a sort of witches' broom with an upright thickened stem and small distant leaves. The mycelium grows out

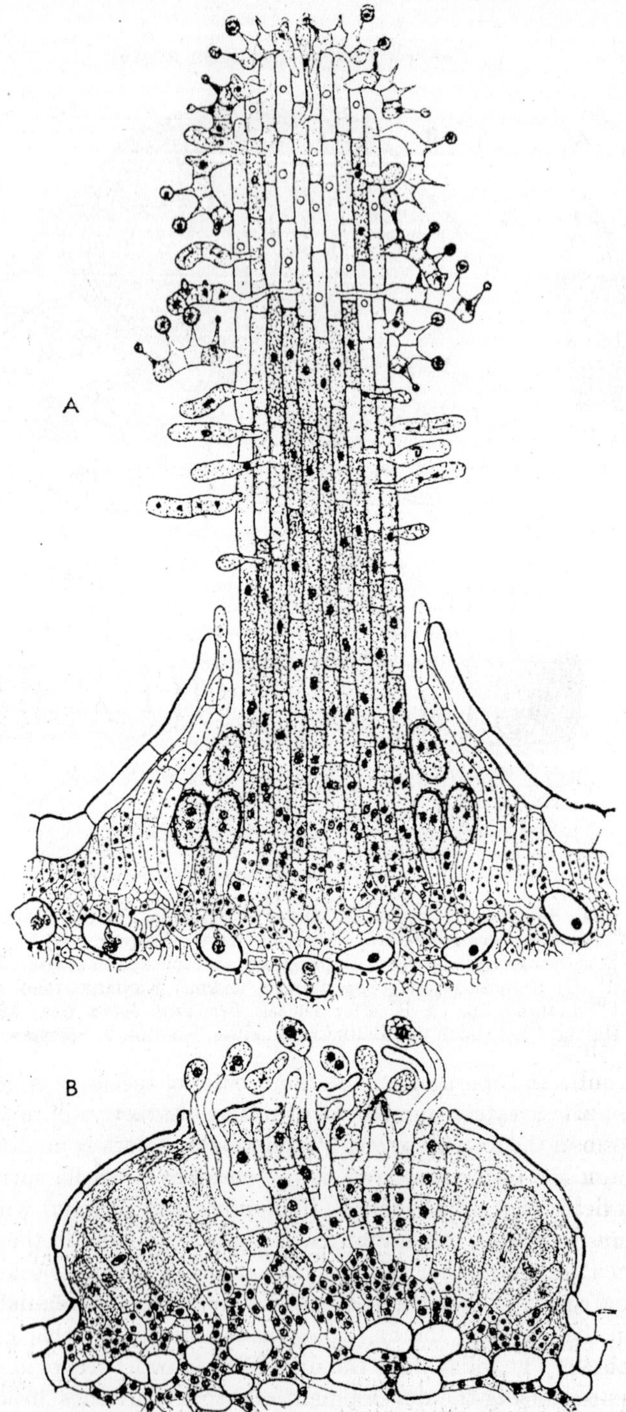

FIG. 135. (See legend on facing page.)

into the epidermal cells of the stem, there forming clusters of two to four closely united teliospores which send forth their promycelium almost immediately, through the epidermal cell walls. In *Cronartium* the gametophytic (monocaryon) mycelium is perennial in the twigs and cortex of older limbs of pine (*Pinus*), often causing the formation of galls which may attain great size and age, up to the size of a human head in *C. quercuum* (Berk.) Miyabe. The uredia are small, with a peridium. The telia are rows of teliospores united laterally into a waxy column 6 to 8 rows in thickness and bursting through the epidermis and projecting several millimeters. These columns continue to grow at the base for some time. Each of the hundreds of teliospores of the column germinates immediately by a curved promycelium producing almost spherical sporidia. The dicaryon stage is found in various Flowering Plants (Anthophyta), apparently only in Dicotyledoneae. In *Chrysomyxa* the teliospores are in separate chains of three or more spores each which arise at the base of the sorus. The urediospores are also in chains. The sporophytic phase occurs in the Family Ericaceae in the wider sense and in a few other families. *Coleosporium* and *Melampsora* both produce subcuticular or subepidermal crusts one cell thick of laterally united teliospores. In the former the urediospores are produced in short chains and the promycelium is "internal." In the latter the urediospores are single and the promycelium is of typical structure. *Coleosporium solidaginis* (Schw.) Thüm., on *Aster* spp. and *Solidago* spp. is abundant even in those parts of the United States where the aecial hosts (species of *Pinus*) are not found. It is apparently able to maintain itself by overwintering mycelium or urediospores. *Melampsora medusae* Thüm. causes the spotting of leaves of various species of poplars (*Populus*) with the dark-colored telial sori, these being preceded by the small powdery yellow uredial sori. The aecial host is the larch (*Larix* sp.) on whose young needles the almost white aecia appear. Most of the species of this genus have their aecia on Pinaceae, but *M. lini* (Pers.) Lev., on flax (*Linum usitatissimum* L.) is autoecious as is *M. euphorbiae* (Schub.) Cast. In *M. ribesii-purpureae* Kleb., the telial stage is on *Salix* but the aecial stage occurs on species of *Ribes* and *Grossularia*, while three other species with *Salix* as their telial host have as their aecial hosts respectively Saxifragaceae, and *Larix* and *Abies* in the Pinaceae. One microcyclic species *M. farlowii* (Arth.) J. J. Davis, occurs in the United States on *Tsuga canadensis* (L.) Carr. of the Pinaceae. (Figs. 134, 135.)

Family Pucciniaceae. Teliospores usually stalked, simple or compound, sometimes without stalks and produced successively as simple or

---

Fig. 135. Subclass Teliosporeae, Order Uredinales, Family Melampsoraceae. (A) *Cronartium flaccidum* (A. & S.) Wint.; teliospore column growing up through uredium. (B) *Coleosporium sonchi-arvensis* (Pers.) Lév.; vertical section through telium. (After Sappin-Trouffy: *Le Botaniste*, **5**:59–244.)

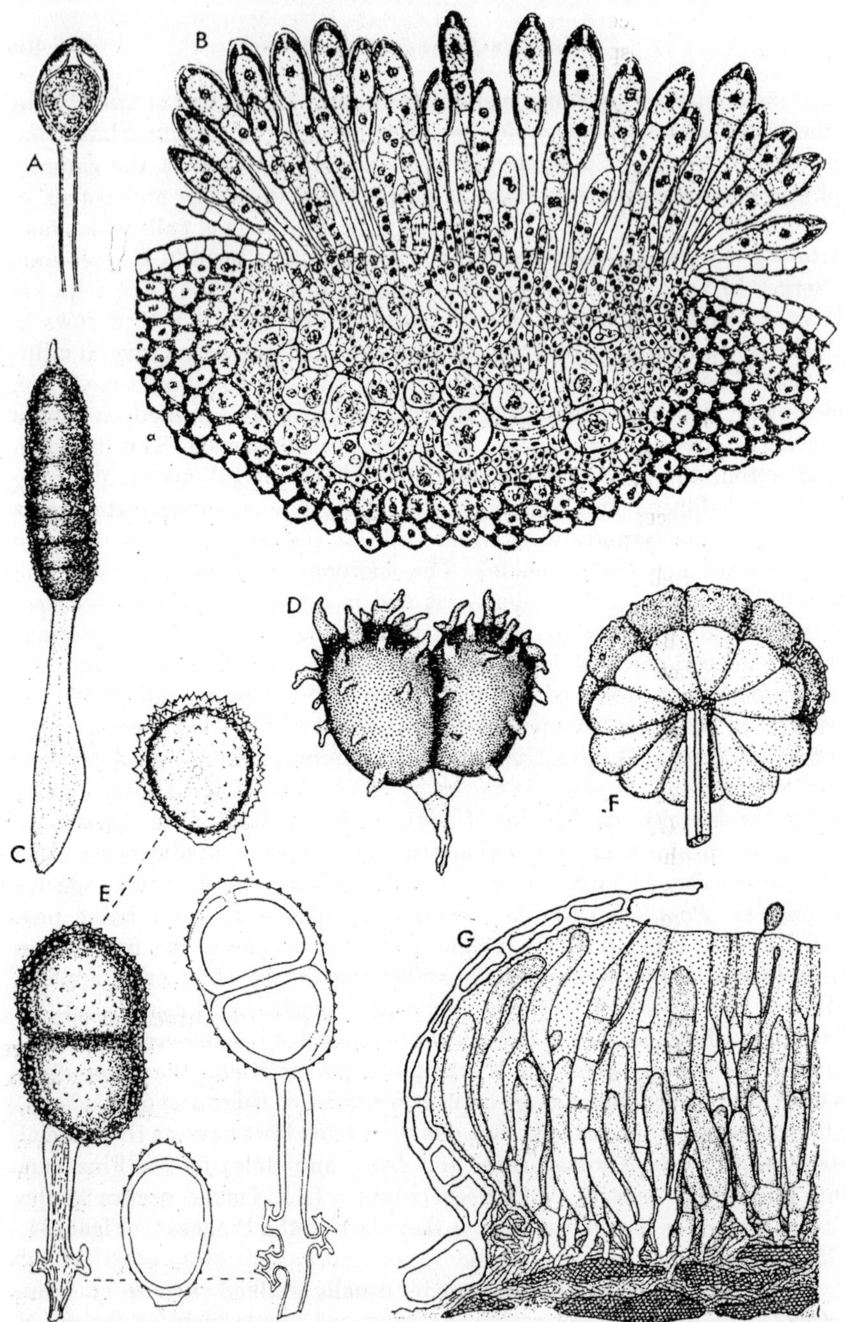

FIG. 136. Subclass Teliosporeae, Order Uredinales, Family Pucciniaceae. Various types of teliospores, all compound except *Uromyces* and *Goplana*. (A) *Uromyces fabae* (Pers.) de By. (B) *Puccinia graminis* Pers. (C) *Phragmidium rubi-idaei* (DC.) Karst. (*Continued on facing page.*)

compound teliospores which escape from the sorus dry or embedded in slime. Aecia only very exceptionally on Pinaceae. The stalked forms of this family are easily distinguished from the Melampsoraceae but the forms with teliospores produced in loose, quickly fragmenting chains are a sort of connecting group difficult to segregate definitely from one or the other family. Dietel (1928) recognized 83 genera and about 3000 species, while some authors add at least 20 genera by the segregation of such larger genera as *Uromyces* and *Puccinia*. These two are among the more important genera of the family with, respectively, about 600 and over 1800 species. The location of the spermogonia is of importance taxonomically. They are subcuticular and rather flattened or subepidermal and then more spherical. The uredia may have paraphyses or these may be lacking. The aecia may be cupulate, or cornute, or hyphoid, or caeomoid, or uredioid. The teliospores may be single on the pedicel (*Pileolaria, Trachyspora, Mainsia, Uromyces*) or several on one pedicel (compound teliospores). In *Puccinia* there are two; in *Phragmidium*, *Xenodochus*, etc., three to many in a single row on the simple pedicel. In *Ravenelia* the teliospores are numerous in a head on a usually compound pedicel, being subtended by hyaline "cysts." As in the Melampsoraceae many species are autoecious and long-cycled, others are heteroecious, and many lack one or more spore forms. In the case of *Puccinia* and *Uromyces* the relationship is very close. The spermogonia are subepidermal, the aecia are cupulate and the uredia are without peridium with urediospores single on long stalks, in both genera. The teliospores are brown and stalked, emerging from a ruptured epidermis. In *Uromyces* they are simple, in *Puccinia* compound, formed of two teliospores closely united in a row with a one-celled stalk. A number of cases are known where the aecia of heteroecious species of the two genera are borne on the same host and the urediospores and teliospores on the same alternate host, the aeciospores and urediospores of the respective species of *Uromyces* and *Puccinia* being practically indistinguishable. The only essential difference is that in the one genus the pedicel is topped by a single teliospore and in the other by two. Arthur (1934) believes that the frequency of such cases indicates a very close relationship of the two genera. In some species of *Puccinia*, intermingled in the same sorus may be found a varying number of *Uromyces*-

---

FIG. 136—(*Continued*)

(D) *Dicheirinia binata* (B. & C.) Arth. (E) *Prospodium plagiopus* (Mont.) Arth. (F) *Ravenelia acaciae-micranthae* Diet. (G) Section through portion of telium of *Goplana dioscoreae* (B. & Br.) Cummins. (A, after Engler and Prantl: Die Natürlichen Pflanzenfamilien, Leipzig, W. Engelmann. B, after Sappin-Trouffy: *Le Botaniste*, **5**:59–244. C, courtesy, Cummins: *Mycologia*, **23**(6):433–445. D, courtesy, Cummins: *Mycologia*, **27**(2):151–159. E, courtesy, Cummins: *Lloydia*, **3**(1):1–78. F, after Dietel: *Botan. Centr. Beihefte, Zweite Abt.*, **20**:343–413. G, courtesy, Cummins: *Mycologia*, **27**(6):605–614.)

like teliospores ("mesospores") among the typical compound teliospores. In *Phragmidium* the row of teliospores is longer, three to eight or more in a row, and the stalk is long and enlarged toward the base. In *Gymnosporangium* (with 40 or more species) the stalks are very long and their walls as well as the outer walls of the usually two-celled compound teliospores swell when wet so that the masses of teliospores are extruded from the telial galls as gelatinous tongues sometimes 2 to 3 cm. in length. The aecial hosts of this genus with a few exceptions belong to the Malaceae and the telial hosts are species of *Juniperus* or closely allied genera. This genus forms rather an exception in that it is the dicaryon stage that is perennial, the galls on the host persisting sometimes for several years. In *Dicheirinia* two, rarely three, verrucous teliospores are borne side by side at the apex of a common pedicel whose upper cell is divided into two or three short cells from which the teliospores arise. *Prospodium*, like *Puccinia*, has two teliospores in a single row but they differ in minor points. *Goplana* is sometimes placed in this family because of its stalked spores. These grow up into a gelatinous matrix and, in the manner of *Coleosporium* produce an internal promycelium. *Ravenelia* is a genus whose teliospores form a head of one layer of fertile spores subtended by colorless cysts which possibly represent sterile teliospores. The head is supported by a centrally attached stalk usually several cells in thickness. The species of this genus are mostly tropical and subtropical and are autoecious so far as known. *Gymnoconia* resembles *Puccinia* in its two-celled compound teliospores but the aecia are diffuse (i.e., caeomoid) without peridium, and the spermogonia are subcuticular. Urediospores are lacking. (Fig. 136.)

In addition to the genera and species assigned to these two families there are over 1000 species of which the telial stage is unknown or its connection with the other stages not determined. *Peridermium*, *Aecidium* (600 species) and *Caeoma* represent different types of aecia. *Uredo* (450 species) consists of species of which the uredial stage only is known. In most of these cases there are probably other stages as yet unknown or whose connection with these has not yet been demonstrated, but it is possible that in some of these species the other stages have been omitted during the course of evolution.

The Uredinales are of great economic interest on account of their harmful effects on many important crops. The following may be mentioned since they frequently cause great damage on small grains: *Puccinia graminis* Pers., several varieties on wheat, rye, oats, barley; *P. rubigo-vera* (DC.) Wint., several varieties on wheat, barley, rye; *P. coronata* Corda, on oats; *P. glumarum* (Schm.) Erikss. & Henn., on wheat, rye, barley; *P. sorghi* Schw., on corn (maize); *Gymnosporangium juniperi-virginianae* Schw., on apple; *Tranzschelia pruni-spinosae* (Pers.) Diet., on peach; *Gymnoconia peckiana* (Howe) Trotter and *Kunkelia nitens* (Schw.) Arth., on *Rubus* spp.; *P. asparagi* DC., on asparagus; *Uromyces phaseoli typica*

Arth., on *Phaseolus vulgaris* L.; *Melampsora lini* (Pers.) Lev., on flax; *Cronartium ribicola* Fischer, on white pine (*Pinus strobus* L.). Several species of rusts are harmful to ornamental plants.

Order Ustilaginales (The Smuts). The Smuts are parasitic, but capable of growth as saprophytes on substrata rich in organic material, e.g., well-manured fields. Flerov (1923) and Sartoris (1924) grew several species of *Ustilago* through to teliospore production on artificial culture media apart from their hosts. Recently Leach and Ryan (1946) grew *Ustilago striiformis* (West.) Niessl. from teliospore to teliospore. Kniep (1911) accomplished this with *Urocystis anemones* (Pers.) Schroet. and Wernham (1938) with *U. gladioli* (Req.) Sm. In the host plant the mycelium is at first intracellular (Kolk, 1930) and later intercellular, with or without haustoria, and actually growing in and keeping pace in its growth with that of the meristematic regions of the host, usually dying in the portions that have passed that stage. A few species parasitic on perennial hosts live over winter in the crown of the plant so that the new growth becomes diseased. Seyfert (1927), investigating several smuts in Europe and Stakman and Christensen (1927) studying *Ustilago zeae* (Beckm.) Unger, and Hanna (1929b), investigating the same species seem to have demonstrated clearly that clamp connections are often present in the dicaryon stage of growth. Sleumer (1932) however, claims that these are not true clamp connections but abortive branches. The nuclear behavior as reported by Seyfert would support the idea that they are typical clamp

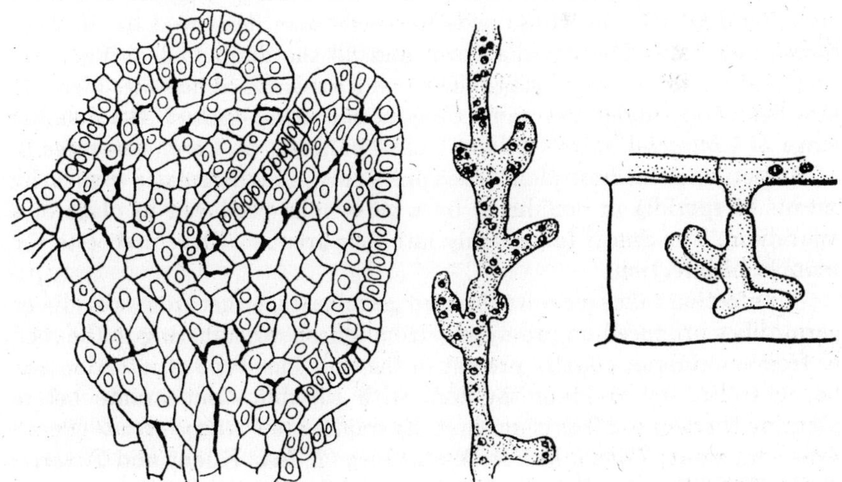

FIG. 137. Subclass Teliosporeae, Order Ustilaginales. (*Left* and *center*) Family Ustilaginaceae, *Ustilago levis* (Kellerm. & Swingle) Magn. (*Left*) Mycelium in meristem of growing tip of seedling of *Avena sativa* L. (*Center*) Piece of intercellular mycelium. (*Right*) Family Tilletiaceae, *Urocystis anemones* (Pers.) Wint.; haustorium. (Courtesy, Lutman: *Trans. Wisconsin Acad. Sci.*, **16**(2):1191–1244.)

connections, especially since Stempell (1934–1935) grew various species of *Entyloma* in culture and obtained typical clamp connections there. The teliospores are produced largely in the flowers, fruits, and inflorescences, but in many cases are produced in the leaves or stems. More often the masses of teliospores are dusty at maturity, and with the rupture of the host tissues are set free for distribution by the wind or other means. The conidia are shot from the mycelium with considerable force in the Family Tilletiaceae accompanied by the formation of a droplet of water at the apex of the sterigma a moment before the discharge of the spore. Conidia are mostly produced on the saprophytic mycelium but in *Tuburcinia* and *Entyloma* they are produced abundantly on the surface of the living leaves of the host plant on long conidiophores which emerge from the epidermis and give a whitish appearance to the affected leaf. When a conidium arises from a cell of monocaryon mycelium the nucleus of the cell divides and one of the daughter nuclei passes out into the conidium which grows out from one side of the hyphal cell. If the cells are binucleate both nuclei divide simultaneously and one daughter nucleus of each pair passes out into the developing conidium, or but one nucleus enters the conidium (Paravicini, 1917). Thus it is possible for a conidium from a monocaryon mycelium to produce a new mycelium only of the same sexual phase, but that from a dicaryon mycelium may produce a dicaryon mycelium or monocaryon mycelia of one or the other sexual phase depending upon which nucleus entered the conidium. (Fig. 137.)

Liro (1935–1938) described conidia in the anthers of plants of *Cardamine bellidifolia* L., in whose pods the seeds were destroyed by *Ustilago cardamines* Liro. The conidia cover and fill the discolored anthers. He proposed for this conidial stage the name *Rhombiella cardamines* Liro. He also described, under the name *Crotalia cintractiae-fischeri*, the conidial stage of *Cintractia fischeri* (Karst.) Liro, parasitic on *Carex canescens* L.

Infection of the host plant takes place only in meristematic tissues, by means of sporidia or conidia or by a germ tube produced in place of a sporidium. The Smuts fall roughly into four groups with reference to the manner of infection:

1. Infection takes place as the seed germinates, either from sporidia or germ tubes produced on promycelia from teliospores adhering to the seed or from sporidia or conidia present in the soil. This type of infection can be controlled by treating the seed with suitable disinfectants before planting them in soil free from smut. Examples are *Ustilago avenae* (Pers.) Jens., oat smut; *Tilletia foetida* (Wall.) Liro (*T. levis* Kühn) and *T. caries* (DC.) Tul. (*T. tritici* (Bjerk.) Wint.), both of which cause stinking smut or bunt of wheat; *Urocystis occulta* (Wall.) Rab., causing stem smut of rye.

2. Any actively growing meristem may be infected by sporidia or by conidia. In the case of maize smut, *Ustilago zeae*, the infection may occur on young roots, at any joint of the stem (i.e., the meristem at the base of

each internode), on young, not yet unrolled leaves, on the male inflorescence (tassel) before it emerges, on the young ear or certain grains of the ear and even on the elongated styles ("silks") on which N. F. Petersen has shown that it may cause small galls.[2] Treatment of the grain with disinfectants is of little value in controlling this type of infection. Planting must be done in soil free from the fungus.

3. Infection of the flower takes place at times of blooming. This was worked out by Brefeld and Falck (1905). It occurs in the loose smuts of wheat and barley, *Ustilago tritici* (Pers.) Rostr. and *U. nuda* (Jens.) Kellerm. & Swingle, respectively. When the host plant has headed out the flowers are normally self-pollinated before opening. When they open the teliospores from nearby diseased plants germinate and produce their sporidia or germ tubes on the stigmas and infect them, the germ tubes growing down into the ovary and entering the developing embryo, in the growing point of whose stem the mycelium becomes dormant until the grain is planted. Then it grows rapidly in the apical meristem, apparently causing little injury to the host plant until the head is being produced. Within this developing head the mycelium grows very vigorously and reduces it to a skeleton surrounded by the powdery masses of spores which are set free at just the time the healthy plants are coming into flower. Control is possible by soaking the infected grain and then dipping it into hot water at a temperature and for a length of time that will kill the contained mycelium without killing the grain.

4. Another type of infection has been shown by Mundkur (1943) to occur in the case of *Neovossia indica* (Mitra) Mundkur. In this smut the spores or the smutted grains fall to the ground and under favorable conditions the long promycelia produce up to 150 sporidia which are wind borne and infect the kernels in the dough stage or a little earlier. These kernels then become smutted that season.

Smuts may cause large galls consisting in part of host tissues and in part of fungus tissues. The galls of maize smut (*Ustilago zeae*) may attain a large size and are edible when young. Various leaf smuts cause the production of galls, e.g., *Doassansia* on the leaves of *Sagittaria*. Many smuts on the other hand do not show their presence until their teliospores are formed. *Ustilago violacea* (Pers.) Fuckel, attacks several species of Dianthaceae and Alsinaceae. Its teliospores are formed only in the anthers. When the female plant of a dioecious species of *Lychnis* is infected the presence of the fungus causes the flower to produce stamens, within which the fungus produces its spores although the normal female flower lacks stamens.

The order is divided into two families which are almost certainly closely related and to which a third family, Graphiolaceae, is probably related.

[2] In a letter to the author.

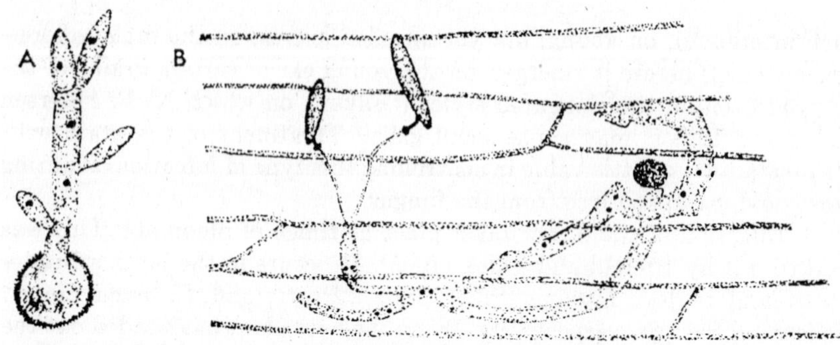

FIG. 138. Subclass Teliosporeae, Order Ustilaginales, Family Ustilaginaceae, *Ustilago zeae* (Beckm.) Unger. (A) Germination of teliospore to form promycelium and sporidia. (B) Two sporidia of opposite sexual phase, infecting epidermal cell of host (*Zea mays* L.), the slender germ tubes uniting to form a stout multinucleate hypha. (Courtesy, Hanna: *Phytopathology*, **19**(5):415–442.)

FAMILY USTILAGINACEAE. Promycelium transversely septate into several, mostly four, cells. The teliospores arise from transformed dicaryon cells of the mycelium. Sometimes the cell walls of the usually tangled hyphae swell and undergo gelatinization, thus separating the protoplasts to some distance. Around these then, within the gelatinized walls, new walls are laid down and the cells enlarge while the gelatinized walls disappear. At about the same time the two nuclei unite and the teliospores come to maturity. The teliospores are produced singly or united into spore balls. They are usually dark-colored, with the wall more or less thickened and smooth or rough. From each cell of the promycelium usually several sporidia are produced, for, unlike the majority of Uredinales, the nucleus of the cell divides several times and one of the daughter nuclei enters each sporidium. In some cases each promycelial cell grows out into a slender germ tube instead of producing sporidia. This difference in behavior is connected for some species with differences in temperature, moisture, etc. In *Ustilago nuda* (Jens.) Kellerm. & Swingle and *U. tritici* (Pers.) Rostr. sporidia are apparently never found. (Fig. 138A.)

Sexual reproduction in this family is accomplished in various ways. In several species it has been determined that two of the promycelial cells represent one sexual phase and the remaining two the other phase. The distribution of the two sexual phases in the four cells of the promycelium may occur apparently in every possible order. In *Ustilago zeae* (Beckm.) Ung., Hanna (1929b) showed that there occur four sexual phases. Probably there are two allelomorphic pairs of genes on separate chromosome pairs that control this sexual behavior. If both chromosome pairs undergo their reduction (disjunction) division in the first or in the second of the two divisions occurring in the production of the promycelium two of the promycelial cells will be of one sexual phase and two of the opposite phase. This is the common case. But if one chromosome pair undergoes disjunction in the first nuclear division and the other chromosome pair waits until the second division before undergoing disjunction the result will be four nuclei with four different combinations of the sexual factors, i.e., four sexual phases in the same promycelium. This doubtless occurs in the other smut species also. The sexuality of smuts has been studied also by Paravicini (1917), Bauch (1922 and later publications), Dickinson (1927, 1928), Kniep (1926), Sleumer (1932), and others. Liro (1924) described a peculiar method of sexual reproduction in *Ustilago vuijckii* Oud. & Beijer., on *Luzula multiflora* (Hoffm.) Lej. This should be reinvestigated

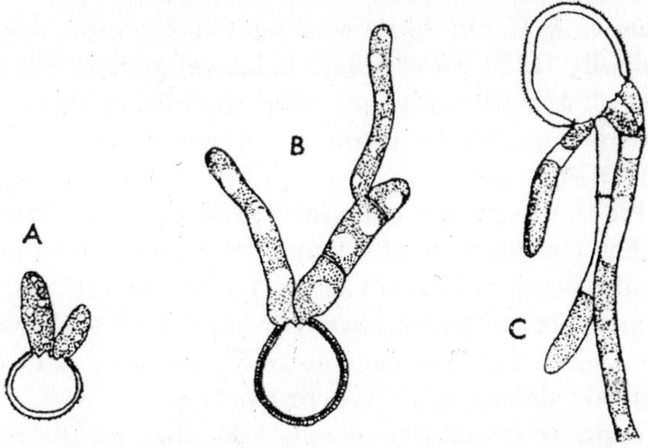

FIG. 139. Subclass Teliosporeae, Order Ustilaginales, Family Ustilaginaceae. Germination of teliospores by two or more promycelia. (A) *Sphacelotheca columellifera* (Tul.) Yen. (B) *Sphacelotheca schweinfurthiana* (Thüm.) Sacc. (C) *Sorosporium consanguineum* E. & E. (After Yen: *Rev. mycol.*, **2**(2):76–84.)

and similar studies undertaken on other species for it is very different from the method usually accepted for this family. The sporogenous hyphae in the ovaries of the host are slender and more or less dichotomously branched. The cells are binucleate. From the base toward the apex of these hyphae the cells swell successively, usually leaving a slender portion containing the septum between each cell and the one next above it. At the upper part of a cell a branch is produced, containing one nucleus. This branch is separated from the cell below by a septum. It pierces the cell above by a very fine tube and the cytoplasm and nucleus pass into it, making it 3-nucleate. In the meantime a similar process from this cell takes one of the two original nuclei and transfers it to the next overlying cell, and so on. Then the two nuclei unite and a thick wall is formed, within and free from the original wall. Sometimes an "antherid," as Liro calls these small branches, from another hypha fertilizes a cell and in that case the antherid from the cell below does not fertilize it but degenerates, unless it finds an adjacent hypha one of whose cells it can fertilize. Liro calls the receptive cells "oogones," and suggests the similarity to the sexual process in the Peronosporaceae. Oudemans (1895) saw these bridging branches and called them clamp connections, but that can hardly be correct since they grow upward from the base toward the apex, and connect two dicaryon cells.

Occasionally more than one promycelium may grow from a single teliospore. This has been observed by Brefeld (1895) and by Yen (1937), both in *Sphacelotheca schweinfurthiana* (Thüm.) Sacc. Yen observed a similar development of two promycelia from a teliospore of *Sorosporium consanguineum* E. & E. On dilute beer malt the sporidia are produced in clusters terminally (as in *Tilletia*) and in lateral groups, not one at a time from each cell of the promycelium. The sporidia in liquid media often bud in the manner of yeasts, forming a monilioid chain or a cluster of yeast-like cells. (Fig. 139.)

Hüttig (1933) has shown that in *Ustilago avenae* (Pers.) Jens. the temperature has considerable effect upon the proportions of disjunction in the first and second meiotic divisions. Disjunction in the first division (called by him pre-reduction) occurred in 14 per cent of the cases at 9° C., 17.4 per cent at 19°, 31.5 per cent at 25.5°, and 18.7 per cent at 29.5°. Various chemicals also modify these proportions.

Union of opposite sexual strains may take place by the conjugation of two sporidia, the nucleus of one passing into the other sporidium. When this germinates it gives rise to a dicaryon mycelium. Both Dickinson (1927, 1928) and Boss (1927) showed that the dicaryon phase initiated by the union of two sporidia is often only transitory in cultures of the fungus, the further growth of the mycelium consisting of monocaryon hyphae, some of one, some of the other sexual phase. For such species it seems probable that only when the union occurs in the tissues of the host is the dicaryon phase permanent. The binucleate sporidium may produce a binucleate conidium. Rawitscher (1912) and others have shown that instead of producing sporidia the promycelial cells of the opposite sexual phase may conjugate by short conjugation tubes, through which the nucleus of one of the cells passes into the other. This latter then produces binucleate sporidia. It may happen that uninucleate sporidia germinate and produce monocaryon mycelium. When two hyphae of such monocaryon mycelia of opposite sexual phases meet they unite and a dicaryon mycelium results. A uninucleate conidium from one mycelium may unite with a hypha of a mycelium of opposite sexual phase, etc. Bauch (1922) showed that in *Ustilago violacea* (Pers.) Fuckel, the two sexual phases of mycelium differ in their reaction to various nutrients present, as evinced by the degrees of growth and distribution of the mycelia in various culture

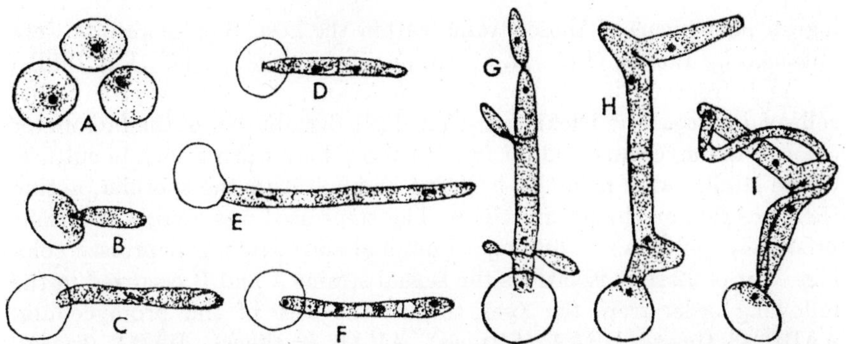

FIG. 140. Subclass Teliosporeae, Order Ustilaginales, Family Ustilaginaceae, *Ustilago hordei* (Pers.) Kellerm. & Swingle. Normal and abnormal germination of teliospores. (A–G) Stages in the normal development. (H, I) Conjugation between promycelial cells of opposite sexual phase. (After Hüttig: *Z. Botan.*, 24(6):529–577.)

media. Both of these differ from the mode of growth of the dicaryon mycelium produced when the two cultures are allowed to grow together. Dickinson (1927, 1928) showed for *Ustilago levis* (Kellerm. & Swingle) Magn., causing smut on oats, that infection does not take place unless mycelia of two opposite sexual phases are present. Monocaryotic structures do not infect the host plants. It must be noted that Flerov (1923) studied a strain of *U. avenae* (Pers.) Jens., also causing a smut on oats, in which a monocaryon sporidium brought about infection by a monocaryon mycelium which eventually produced uninucleate teliospores in which no nuclear fusion occurred. From such teliospores arose a two-celled promycelium. Boss (1927) found the same to be true in *Sphacelotheca ischaemi* (Fuckel) Clinton (*Ustilago ischaemi*). These are similar to the monocaryon strain of *Kunkelia nitens* studied by Dodge (1924). (Fig. 140.)

Hanna (1929b) demonstrated for *U. zeae* on maize that single sporidia or conidia from cultures from single sporidia are able to infect the tender meristem with a very slender germ tube composed of monocaryon cells. This infection is usually of very limited extent and except in rare strains no smut galls or teliospores are produced. When two sporidia of opposite sexual phase infect the tissue in rather close proximity the slender monocaryon hyphae approach one another and unite and thenceforth develop as a stout mycelium of dicaryon cells which penetrates the meristem in all directions. These hyphae show numerous clamp connections. Eventually a smut gall is produced, filled with numerous teliospores. Similar conditions exist according to this author in *Sorosporium reilianum* (Kühn) McAlpine, another species producing smut galls on maize. Christensen and Stakman (1926) studied various mutations of corn smut. Christensen (1929) reported three strains in which infection and production of large galls occurred with monosporidial cultures. He did not follow the cytological phenomena of these strains within the host. Similar results were obtained by Eddins (1929) in his studies of the same species. (Fig. 138B.)

Aside from the distribution of the two sexual phases in the promycelial cells of *Ustilago levis* Dickinson studied the distribution of the factors for color of the mycelium and for form of the colony when grown in culture. These studies were made by isolating and culturing the sporidia, noting their position on the promycelium. The smut used was a cross of strains producing yellow and white mycelium and corrugated or depressed colonies. Out of 22 such isolations the sexual strains A and B occurred in the following order from the apex toward the base of the promycelium: AABB (3 times), BBAA (5 times), ABAB (6 times), BABA (twice), ABBA (3 times), and BAAB (3 times). The distribution of the other two sets of allelomorphic factors was apparently entirely independent of the distribution of the sexual strains and of each other.

Hybridization has been brought about in cultures and also appears to occur in nature. E. and J. Hirschhorn (1935) have shown that in Argentina there may be present in the same smut gall on maize the three species *Ustilago zeae* (Beckm.) Unger, *U. fischeri* Pass., and *Sorosporium reilianum* (Kühn) McAlpine, and that there occur crosses between all three, so that all types of intermediate forms may occur together. Only very rarely are *U. zeae* and *S. reilianum* found alone in the pure state in Argentina. Fischer and Holton (1941) report the result of crosses between *U. avenae* (Pers.) Jens. on oats (*Avena sativa* L.) and *U. perennans* Rostr. on *Arrhenatherum elatius* (L.) Mert. & Koch. The race of the former that was used had naked sori with indurated spore masses while the latter had covered sori (i.e., the glumes were not entirely destroyed by the fungus) and powdery spores. The F_1 grew on *A. fatua* L. but not on *Arrhenatherum*. The F_2 generations grew on *A. sativa* but still not on the tall oatgrass. It was demonstrated that the covered character is recessive to naked and the indurate recessive to powdery. Other hybrids between species of Ustilaginaceae have been reported by Dickinson (1927–1928), Hanna and Popp (1931), Kniep (1926), and others.

The occurrence of geographic races has been studied intensively in *Ustilago longissima* (Sow.) Tul., by Bauch (1930, 1931). He showed that multiple allelomorphy occurs in both the A and B factors.

That the Ustilaginaceae are degenerate forms seems to be indicated by the fact that their types of sexual reproduction are not at all standardized. In the majority of cases studied the teliospore upon germination produces a four-celled promycelium, from each cell of which one or more sporidia are produced, but this is not universal. In some species germ tubes take the place of the sporidia. In *Ustilago striiformis* (West.) Niessl. several races are recognized on different hosts (Davis, 1935, Fischer, 1940). Fischer has shown that the teliospores of forma *Hordei* Fisch. germinate by the production of promycelia and sporidia, and that the latter pair by twos and then give rise to actively growing hyphae. On the contrary in forma *Poae-pratensis* Davis it was shown by Leach and Ryan (1946) that the teliospores germinate by the production of branched germ tubes of indeterminate length. The earlier divisions of the teliospore nucleus are regarded as meiotic so that this branching mycelium is considered to be haploid. Here and there in this mycelium the nuclei begin to assort in pairs and to undergo fusion. These hyphae with diploid nuclei grow and the cells divide but soon the cells separate and form teliospores by thickening and becoming fusiform or limoniform with thickening and darkening of the spore wall. These phenomena can be followed in artificial cultures on agar.

Another indication of the degeneration of this family is the fact that in some species of *Ustilago* no promycelium is formed but from the teliospore there grow out cells that form short branching hyphae of yeast-like cells. Bauch (1923) found that in *Ustilago longissima* and its variety *macrospora* Davis the meiotic divisions of the diploid nucleus of the mature teliospore take place in the teliospore itself and that then successively sporidia are budded off from the latter. Into the first such sporidium two nuclei enter while into those subsequently produced one nucleus may enter or two, the latter being made possible by division of the nuclei remaining in the teliospore. More often the two nuclei in the first sporidium are of opposite sexual phase. In that case the sporidia become two-celled by the formation of a septum and then these two cells unite through conjugation tubes and grow out as slender dicaryon hyphae. The other sporidia if uninucleate mostly remain one-celled but conjugate with sporidia of opposite sexual phase and then they too form the slender dicaryon "Suchfäden." Only under exceptional circumstances is a typical four-celled promycelium formed, from the cells of which are budded off the sporidia.

The teliospores of the Ustilaginaceae vary in their longevity. Fischer (1936) reports viable spores in herbarium specimens up to 3–10 years in many cases and as high as 23 years for *Ustilago hordei* (Pers.) Kellerm. & Swingle.

Mutant forms have been observed frequently in smuts. Johnson et al. (1940) report the occurrence of a mutant race of *Sorosporium syntherismae* (Pk.) Farl., in which the teliospores are almost colorless but still retain their ability to infect the host.

When grown in liquid culture media with an abundance of soluble carbohydrates the mycelium of smuts may break up into separate cells resembling and multiplying like the asporogenous yeasts. Teliospores sown in such media will frequently produce hyphae which become yeast-like, instead of typical promycelia.

The family includes over 450 species in about 12 genera. About 61 per cent of the North American species occur in members of the grass family (Poaceae) and about 14 per cent on sedges (Cyperaceae) with about 19 per cent in Dicotyledons, particularly in the Polygonaceae and the Asteraceae. In the majority of species the spores are produced in the ovaries or other parts of the flower, or the whole inflorescence may be involved. Some species produce their sori in the leaves or stems of their hosts.

The largest genus is *Ustilago* with over 300 species. Its spores are produced singly in a powdery mass. Many serious enemies of cultivated plants are found in this genus, e.g., *U. avenae* (Pers.) Jens. and *U. levis* (Kellerm. & Swingle) Magn., destroying the spikelets and inflorescence of oats (*Avena sativa* L.), *U. zeae* (Beckm.) Unger, producing smut galls on maize (*Zea mays* L.), *U. hordei* (Pers.) Kellerm. & Swingle and *U. nuda* (Jens.) Kellerm. & Swingle on barley (*Hordeum sativum* L.), *U. tritici* (Pers.) Rostr. on wheat (*Triticum* spp.), *U. striiformis* (West.) Niessl on *Poa pratensis* L., *U. violacea* (Pers.) Fuckel on Dianthaceae, etc. *Sphacelotheca* differs from *Ustilago* in having the powdery mass of teliospores surrounded by a pseudoparenchymatous layer of fungus tissue. *S. sorghi* (Link) Clinton is injurious to the ovaries of *Sorghum vulgare* Pers. and related species. In *Schizonella* the teliospores are in twos, otherwise much as in *Ustilago*. *S. melanogramma* (DC.) Schroet. produces long black sori

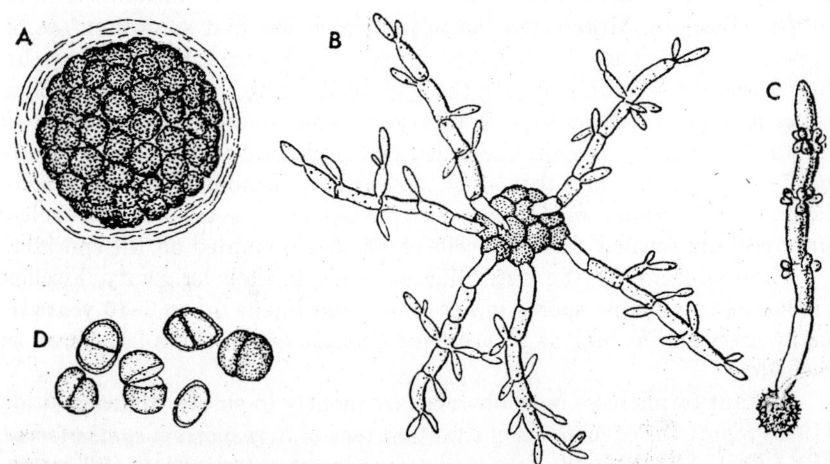

FIG. 141. Subclass Teliosporeae, Order Ustilaginales, Family Ustilaginaceae. (A) Spore ball of *Sorosporium saponariae* Rud. (B) Spore ball of *Tolyposporium junci* (Schroet.) Woron., some of the spores germinating to form promycelia. (C) *Ustilago kuehneana* Wolff, teliospore germinating with whorls of sporidia at the septa of the promycelium. (D) *Schizonella melanogramma* (DC.) Schroet. Teliospores joined in twos. (A, after Dietel, in Engler und Prantl: Die Natürlichen Pflanzenfamilien, Zweite Auflage, vol. 6, pp. 1–98, Leipzig, W. Engelmann. B–C, after Brefeld: Untersuchungen aus dem Gesammtgebiete der Mykologie, vol. 12, pp. 99–236. D, after Clinton: *Connecticut State Geological and Natural History Survey Bull.*, **5**:1–44.)

in the leaves of various species of *Carex*. In *Sorosporium*, *Tolyposporium* and other genera the teliospores are united into more or less firm balls. They are largely parasitic on grasses. (Fig. 141.)

FAMILY TILLETIACEAE. Promycelium nonseptate, the sporidia being formed at its apex in a dense cluster or in a whorl. The number of sporidia varies usually from 4 to 30–50, but in *Neovossia indica* (Mitra) Mundkur may reach 150 (Mundkur, 1943). The teliospores arise as terminal cells of hyphae or of short lateral branches or as intercalary cells. The two nuclei unite and the cells round up and secrete a heavier wall which may be dark or light in color and is smooth or more often reticulately marked or spiny. The teliospores occur as single spores in a dusty mass or united into small or large balls of spores with or without a covering or a core of sterile cells. These spores or spore balls may escape as a powdery mass or remain within the host tissue. The sporidia are fusiform or sickle-shaped and are inclined to unite by twos while still attached to the promycelium or after they have fallen off. A binucleate conidium may be shot off violently from a pair of united sporidia. As in the preceding family the sporidia appear to be of at least two sexual phases. From one of the sporidia of the united pair a germ tube of dicaryon cells may grow out and infect the host plant. Infection may take place from a dicaryon conidium set free from the united sporidia or arising on a dicaryon mycelium. The sexual cycle which is initiated by the fusion of the sporidia is completed by the union of the nuclei in the teliospore. (Fig. 142A, B.)

About 13 genera and over 250 species are recognized in the family. *Tilletia* (about 40 species) corresponds to *Ustilago* in producing its teliospores as single cells in a powdery mass. *T. caries* (DC.) Tul. (*T. tritici* (Bjerk.) Wint.), with rough teliospores and *T. foetida* (Wallr.) Liro (*T. levis* Kühn), with nearly smooth teliospores, cause stinking smut or bunt of wheat. Flor (1932) has crossed these two species by picking off single sporidia and allowing the two monocaryon mycelia produced from them to unite. These hybrid mycelia were used successfully to inoculate wheat plants. The teliospores produced on these plants resembled most closely those of *T. foetida*. Certain species of *Tilletia* have been described from the capsules of *Sphagnum* and *Anthoceros*, both in the Bryophyta, but Bauch (1938) studying *T. sphagni* Nawaschin has demonstrated that it does not belong to this order, the supposed teliospores actually representing the conidial stage of *Helotium schimperi* Naw., one of the Pezizales.

Entyloma produces its teliospores singly and in the tissues of the host from which they do not escape as a powdery mass, but germinate within the host sending their elongated promycelia out through the epidermis and forming the sporidia externally. Conidia are also formed on conidiophores which emerge through the stomata. Kaiser (1936) observed clamp connections on the mycelium of *E. calendulae* (Oud.) de Bary. Frequently

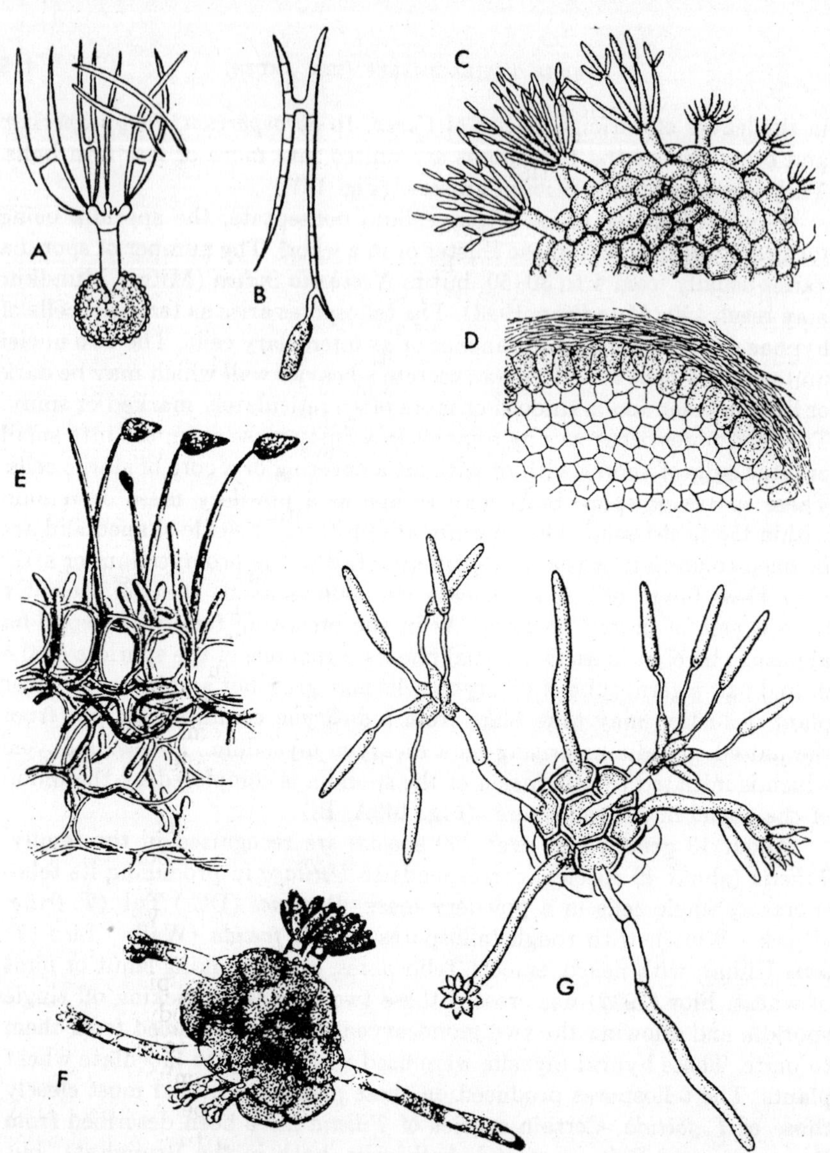

FIG. 142. Subclass Teliosporeae, Order Ustilaginales, Family Tilletiaceae. (A, B) *Tilletia caries* (DC.) Tul. (A) Teliospore with sporidia united in H form. (B) The conjugated sporidia have produced a dicaryon conidium. (C) *Doassansia sagittariae* (West.) Fisch., with some of the teliospores germinating. (D) Section through a portion of a spore ball of *Doassansiopsis martianoffiana* (Thüm.) Diet. (E, F) *Tuburcinia trientalis* (B. & Br.) Wor. (E) Section of portion of infected leaf with external conidia. (F) Germinating spore ball. (G) Germinating spore ball of *Urocystis violae* (Sow.) Fisch. de Waldh. (A–B, after Plowright: A Monograph of British Uredineae and Ustilagineae, London, Kegan, Paul, Trench, & Co. C, G, after Brefeld: Untersuchungen Gesammtgebiete der Mykologie, Heft 12, pp. 99–236; D, after Dietel, in Engler und Prantl: Die Natürlichen Pflanzenfamilien, Zweite Auflage, vol. 6, pp. 1–98, Leipzig, W. Engelmann. E–F, after Woronin: *Abhandl. Senckenberg. Natur. Gesell.*, **12**:559–591.)

the teliospores arise as outgrowths of these structures. Stempell (1935) obtained cultures of *Entyloma* on agar media from conidia on infected leaves. The mycelia were of two types, monocaryon and dicaryon, but clamp connections were produced only on the latter. He reported that sickle-shaped, uninucleate conidia were formed on the first type of mycelium and lunate, binucleate conidia on the dicaryon type. Both types of conidia are discharged violently with the formation of a droplet of liquid at the apex of the sterigma, just as occurs in the discharge of the sporidia from the promycelium. Hanna (1938) found that in several species of *Entyloma* two types of conidia were produced; sickle-shaped, uninucleate conidia which are discharged violently and slender filiform or needle-shaped conidia which are not shot off. In the species studied by Hanna the sickle-shaped spores were always uninucleate and lunate binucleate spores were not observed. The 100 or more species are found on Grasses (Poaceae), Ranunculaceae and other families, but particularly on the Asteraceae. *Urocystis* (about 60 species) produces its teliospores in balls of from two (rarely one) to four or five, the ball being partly or completely surrounded by a layer of small sterile cells. The mass of spore-balls is powdery and they escape upon rupture of the host tissue. *U. occulta* (Wallr.) Rab. causes longitudinal, lead-colored, slightly raised sori on the stems and leaf sheaths of rye (*Secale cereale* L.), the head being killed by the presence of the fungus in the stalk below, it being rarely entered by the fungus. The very similar *U. tritici* Körn, parasitic on wheat (*Triticum aestivum* L.) has been shown by Yu and his associates (1936) to occur in several physiologic races in China. *U. violae* (Sow.) Fisch. de Waldh. and *U. anemones* (Pers.) Wint. form their sori in the leaves, respectively, of *Viola* and of various species of the Ranunculaceae. In *Tuburcinia* all the cells of the spore ball are fertile spores. It must be noted that Liro (1922) and others combine the genera *Tuburcinia* and *Urocystis* under the former name which is the earlier. About 20 species of *Doassansia* produce their sori in the leaves of various Alismataceae and related aquatic plants. The large spore balls have very numerous teliospores and an external layer of hyaline sterile cells. *Doassansiopsis,* also on Alismataceae, produces large spore balls made up of a central core of hyaline pseudoparenchymatous cells surrounded by a single layer of larger dark teliospores, these in their turn being surrounded by a filamentous sheath. (Fig. 142C–G.)

As in the preceding family, the teliospores in the Tilletiaceae may live a long while. Fischer (1936) obtained germination from specimens preserved in the herbarium for 10 years in the case of *Entyloma dahliae* Syd. and for 25 years for *Tilletia foetida* (Wallr.) Liro.

Family Graphiolaceae. Parasitic in the leaves of palms. Sori formed under the epidermis and immediately underlying tissues and tearing these so as to permit the emergence of the spores. They consist of a thick, cup-like, dark outer peridium surrounding the sporogenous central portion which may project some distance above the rim of the opened peridium. A thin hyaline inner peridium may surround this projecting mass of sporogenous hyphae. The spores are formed in parallel chains and bud laterally to form two to four (or more) sporidia which become more or less colored, with somewhat thickened walls. Bundles of sterile hyphae scattered throughout the chains of spores probably serve the same function as the capillitium in Mycetozoa, Lycoperdaceae, etc. They are lacking in one genus. (Fig. 143.)

Graphiola has been carefully studied as to its morphology by Eduard Fischer (1883, 1920, 1922) and by Killian (1924) who showed the cyto-

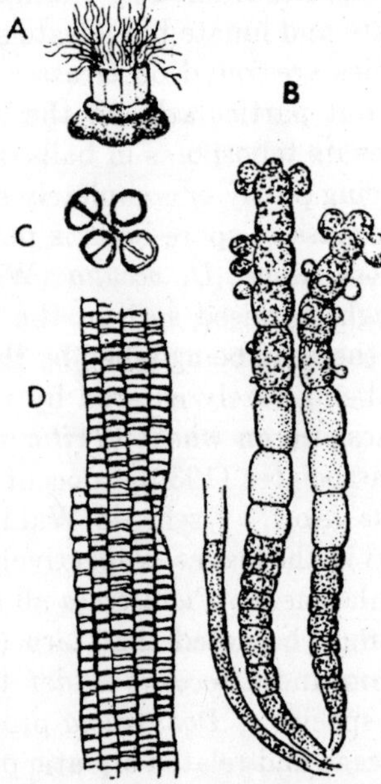

Fig. 143. Subclass Teliosporeae, Order Ustilaginales (?), Family Graphiolaceae. (A, B) *Graphiola phoenicis* (Moug.) Poit. (A) Sorus on leaf of palm. (B) Sporogenous hyphae, the cells near the top producing sporidia. (C, D) *Graphiola thaxteri* Fisch. (C) Group of four sporidia from one spore, three of them once septate. (D) Bundle of sterile hyphae in sorus. (A–B, after Fischer: *Botan. Ztg.*, **41** (45):745–756. C–D, after Fischer: *Ann. Mycolog.*, **20**(3–4):228–237.)

logical features. The sporidia are uninucleate and produce a monocaryon mycelium in the host leaf. Eventually a mass of monocaryon hyphae is formed beneath the epidermis and immediately underlying layers of cells and from the marginal portion of the cushion arise vertically closely packed, branched, thick-walled monocaryon hyphae that form the outer peridium. This arches over the whole structure at first. From the central portion of the hyphal cushion arise the sporogenous hyphae and, if present, the sterile hyphal bundles. The cells are all at first monocaryon, and this remains true of the cells of the hyphal bundles. The cells of the closely packed sporogenous hyphae elongate rapidly and become multinucleate, but very soon cross walls divide these hyphae into chains of dicaryon cells. These soon show nuclear fusion. At this stage the sporogenous hyphae show at their base monocaryon cells, a little further up cells with several nuclei, followed by a series of dicaryon cells, while the upper portion consists of cells each with a single diploid nucleus. These, according to Killian, are the cells which correspond to the teliospores of the Ustilaginaceae. In them the nucleus undergoes two divisions and the four nuclei pass out into the sporidia which bud out of the teliospores. The latter may remain attached, so that the upper portion of the sporogenous hypha shows teliospores with sporidia in various stages of development, or the teliospores may break apart before the sporidial development is completed. The sporidia may remain one-celled or may become two-celled by the formation of a septum. In their germination they may bud like yeasts or form germ tubes. The true relationship of this family is not a matter of agreement among mycologists. Fischer and Killian incline to the idea of kinship with the Ustilaginaceae in some of which (e.g., *Sphacelotheca*) a peridium of hyphal tissue surrounds the sporogenous part of the sorus. The budding of the sporidia from the teliospores instead of the formation of typical promycelia caused Killian to refer to the somewhat similar case in *Ustilago longissima* referred to previously in this chapter. Fischer (1922) recognized two genera: *Graphiola,* with bundles of sterile hyphae among the sporogenous hyphae, the latter separating into their individual cells at maturity, and *Stylina* which lacks the sterile hyphae and whose sporogenous hyphae do not separate into individual cells. He described the four species of *Graphiola* studied by him and the one species of *Stylina*.

Relationships Within the Subclass Teliosporeae

It is very apparent that there is not a very close relationship between the more highly developed Uredinales and Ustilaginales. In vegetative structures the similarities are greater, the mycelium in both orders being intercellular with nucleated haustoria. Both types of mycelium are present, monocaryon and dicaryon, although the former may have a very brief life, especially in the parasitic life of the Ustilaginales. Eventually in both orders the dicaryon mycelium produces special cells, the teliospores, within which the nuclei unite to form a diploid nucleus. This nucleus undergoes two meiotic divisions to form four haploid nuclei which apparently in both Rusts and Smuts are two of one sexual phase and two of the other. The meiotic divisions may occur in the teliospore but most often occur in the promycelium. In Ustilaginales the dicaryon phase of the mycelium frequently bears clamp connections but these have been demonstrated rarely in the Uredinales. In the sexual reproduction the Ustilaginales produce no definite male gametes. Any two cells of opposite sexual phase may unite to initiate the dicaryon phase, be it nearby promycelial cells, sporidia, conidia, or mycelia. In the Uredinales two mycelia of opposite sexual phase may diploidize one another when they come in contact within the host but the usual mode of reproduction is by the union of sperm cells with special receptive hyphae of the opposite sexual phase. The Rusts have reached a much higher stage of specialization of spore forms and of adaptation to alternate hosts, while the Smuts are parasitic but able to grow saprophytically, thus eliminating the necessity for two growth phases in the same or different species of hosts. Because of the formation of sperm cells in typical spermogonia and the production of receptive hyphae the Rusts in this regard hark back to more primitive ancestral forms among the Ascomyceteae where such structures are present. Clearly the Ustilaginales must have branched off from the earlier Rusts, with loss of these special organs, but at an early stage when the inherited tendency to produce clamp connections had not been lost. Jackson (1931) points out the similarity in life cycles of many Rusts and Florideae. This is worthy of further careful consideration. The relationships of the Teliosporeae to the Subclass Heterobasidiae are discussed in the next chapter.

Key to the Orders and Families of Subclass Teliosporeae

Obligate parasites in Pteridophyta, Strobilophyta (Coniferae) and Anthophyta (Angiospermae). Teliospores single or united into crusts or columns or several together in compound spores, remaining within the host tissue or bursting through the epidermis or cuticle. Spermogonia normally produced, the sperm cells diploidizing special receptive hyphae. Typically three types of spores are produced, aeciospores, the product of the diploidization of monocaryon mycelium which arises from the sporidia; urediospores (repeating spores) and teliospores from which arise the promycelia and sporidia. Sporidia always expelled violently. Order Uredinales

Teliospores without stalks, produced singly or in groups of two to four in the mesophyll or just below or within the epidermal cells or united laterally into subepidermal or subcuticular crusts or united into separate vertical chains or into chains that are joined laterally into a waxy column which emerges through the epidermis. Aecia mostly on species of Pinaceae. Family Melampsoraceae

Teliospores usually stalked, simple or compound, sometimes without stalks and produced successively as simple or compound teliospores which escape from the sorus dry or embedded in slime. Aecia only very exceptionally produced on Pinaceae. Family Pucciniaceae

Obligate parasites in Anthophyta or in many cases facultative saprophytes. Teliospores single or united in columns or balls, remaining within or bursting out of the host tissue, mostly distributed by air currents. No spermogonia or special receptive hyphae. Diploidization by means of union of compatible spores, hyphae, etc. Typically only teliospores and often hyaline thin-walled conidia are produced. Sporidia expelled violently in one family, not so in the two others. Order Ustilaginales

Promycelium transversely septate into several, mostly four, cells. Teliospores arising in the tissues of the host from transformed hyphal cells, and mostly distributed by air currents. Sporidia not expelled from the promycelium. Family Ustilaginaceae

Promycelium not septate, the four to many sporidia at its blunt apex. Teliospores mostly arising as lateral outgrowths from hyphal cells or intercalarly. Conidia expelled from the conidiophores. Family Tilletiaceae

Promycelium lacking, but the teliospores bud directly to form four sporidia which form thick, dark walls. Teliospores in vertical rows in compact sori. Parasites in leaves of palms. Sporidia not expelled violently. Family Graphiolaceae

Key to the Commoner North American Genera of Family Melampsoraceae

(*Based upon Arthur*, 1934)

Teliospores single or united laterally into groups of two or more.
- Telia in ferns (Polypodiaceae and Osmundaceae), aecia in *Abies*.
  - Teliospores subepidermal. *Uredinopsis*
  - Teliospores in the epidermal cells.
    - Aeciospores and urediospores colorless. *Milesia*
    - Aeciospores and urediospores with yellow contents. *Hyalopsora*
- Telia in Anthophyta (Angiospermae).
  - Teliospores with brown walls. *Pucciniastrum*
  - Teliospores with colorless walls.
    - Peridium of uredium delicate, with a central pore. *Melampsorella*
    - Peridium of uredium firm, with long-pointed ostiolar cells. *Melampsoridium*

Teliospores united laterally into subcuticular or subepidermal crusts.
- Promycelium external.
  - Teliospore walls colored. *Melampsora*
  - Teliospore walls colorless. *Aplospora*
- Promycelium within the teliospore. *Coleosporium*

Teliospores in chains which are united laterally into long columns emerging through the epidermis. Urediospores borne singly on stalks. *Cronartium*

Teliospores in chains which are not united laterally and not emerging as a column. Urediospores in chains. *Chrysomyxa*

Teliospores united laterally into crusts of two or more layers, mainly tropical. Life histories not well known. *Bubakia, Physopella, Cerotelium, etc.*

Key to the Commoner North American Genera of Family Pucciniaceae

Teliospores borne singly.
- Teliospores colorless; aecia peridermioid; autoecious. *Mainsia*
- Teliospores colored.
  - Teliospores flattened vertically; aecia uredioid; autoecious. *Pileolaria*
  - Teliospores spherical or short ellipsoidal, aecia uredioid, autoecious; spermogonia subcuticular. *Trachyspora*
  - Teliospores spherical or short ellipsoidal, with thickened apex; aecia cupulate (rarely uredioid); autoecious or heteroecious; spermogonia subepidermal. *Uromyces*

Teliospores compound, two to a pedicel (see also *Earlea*).
  Teliospores easily separable, not surrounded by a conspicuous common membrane, pedicels fascicled at base; spermogonia subcuticular. *Tranzschelia*
  Teliospores not separable, surrounded by a distinct common membrane.
    Outer membrane gelatinizing when wet; spermogonia subcuticular. *Uropyxis*
    Outer membrane not gelatinizing when wet (except in the case of *Gymnosporangium*)
      Spermogonia subcuticular; aecia caeomoid; no uredia. *Gymnoconia*
      Spermogonia subepidermal.
        Pedicel and teliospore wall gelatinizing when wet; aecia cupulate or cornute; uredia wanting except in one species. *Gymnosporangium*
        Pedicels not gelatinous when wet; aecia cupulate or (more rarely) uredioid; apex of terminal teliospore thickened or papillate.
          Teliospores pedicellate. *Puccinia*
          Teliospores sessile, usually in a cluster of 2 or more.
Teliospores compound, several in a row (one species of *Earlea* has simple teliospores and several only two teliospores to a pedicel); spermogonia subcuticular.
  Teliospores with two to three lateral pores, smooth or more often verrucose; basal portion of pedicel inflated in water; uredia present; aecia caeomoid. *Phragmidium*
  Like the preceding but basal portion of pedicels not inflated in water; uredia absent except in one species; aecia caeomoid (often included in *Phragmidium*). *Earlea*
  Terminal teliospores with one pore, the others with two, near the top; mostly smooth, pedicel short, not swelling in water; uredia wanting; aecia caeomoid. *Xenodochus*
  Teliospores with one pore in each cell, firmly united, smooth; pedicel elongated, not swelling in water, aecia uredioid. *Frommea*
  Teliospores with one pore in each cell, smooth, easily separable, pedicel very short; aecia uredioid. *Kuhneola*
Teliospores compound, consisting of a basal teliospore attached to a long pedicel and two horizontally lying teliospores at its top; aecia uredioid, spermogonia subcuticular.
  Teliospores verrucose; uredia with paraphyses. *Triphragmium*
  Teliospores spiny; uredia wanting. *Nyssopsora*
Teliospores numerous in a head, subtended by hyaline cysts. *Ravenelia*

Key to the Commoner Mostly North American Genera of Family Ustilaginaceae

(*Based on Clinton, 1906*)

Teliospores separate.
  Sori dusty at maturity.
    Without definite false membrane. *Ustilago*
    With false membrane of fungus cells. *Sphacelotheca*
  Spores more or less agglutinated at maturity.
    Spores firmly agglutinated into tubercular nodules. *Melanopsichium*
    Spores developed around a central columella (rarely dusty). *Cintractia*

Teliospores mostly adhering in pairs.
  Agglutinated, in leaves. *Schizonella*
  Dusty, in inflorescences. *Mycosyrinx*
Teliospores in balls.
  Sori dusty or granular.
    Spore balls often evanescent; olive-brown or black-brown. *Sorosporium*
    Spore balls rather permanent; yellowish or reddish. *Thecaphora*
    Spore balls rather permanent; spores adhering by folds or thickenings of the outer coat. *Tolyposporium*
  Sori agglutinated.
    Spore balls of thick-walled spores. *Tolyposporella*
    Spore balls of central sterile cells surrounded by the peripheral functional teliospores. *Testicularia*

(It must be noted that *Ustilago* and *Sphacelotheca* are in some cases distinguished with difficulty. Furthermore, in some species of *Sorosporium* the teliospores do not remain adherent in the spore balls, so that they are not easily distinguishable from the other genera. In some genera the germination of the teliospores has not been studied, or perhaps in only a few species of the genus.)

Key to the Commoner Genera of Family Tilletiaceae

Teliospores single.
  Spores dusty and escaping at maturity. *Tilletia*
  Spores remaining embedded in the host tissue. *Entyloma*
Teliospores in balls.
  Sori dusty; spore balls more or less surrounded by an adhering layer of sterile cells; escaping from the ruptured sorus. *Urocystis*
  Spore balls more or less permanently remaining in the host tissue.
    Spore balls lacking a cortex of sterile hyphae.
      Spore balls dark, no sterile core. *Tuburcinia*
      Spore balls light-colored, with or without a core of sterile pseudoparenchymatous cells. *Burrillia*
      Spore balls light-colored, with a core of septate hyphae. *Tracya*
    Spore balls with filamentous cortex.
      No central core of sterile cells. *Doassansia*
      A single layer of spores surrounding a large core of pseudoparenchymatous cells. *Doassansiopsis*

Key to the Genera of Family Graphiolaceae

Bundles of sterile hyphae among the sporogenous hyphae; the latter separating at maturity into their individual teliospores. *Graphiola*
No bundles of sterile hyphae present; the sporogenous hyphae do not separate at maturity into their individual teliospores. *Stylina*

Literature Cited

Adams, J. F.: Sexual fusions and development of the sexual organs in the Peridermiums, *Penn. State Coll. Agr. Exp. Sta. Bull.*, **160**:31–77. *Pls.* 1–5. *Text Figs.* 1–8. 1919.

Allen, Ruth F.: A cytological study of Puccinia glumarum on Bromus marginatus and Triticum vulgare, *J. Agr. Research*, **36**(6):487–513. *Pls.* 1–12. 1928.

———: A cytological study of heterothallism in Puccinia graminis, *ibid.*. **40**(7): 585–614. *Pls.* 1–17. 1930.

Allen, Ruth F.: A cytological study of heterothallism in Puccinia triticina, *ibid.*, **44**(10):734–754. *Pls.* 1–11. 1932a.

———: A cytological study of heterothallism in Puccinia coronata, *ibid.*, **45**(9): 513–541. *Pls.* 1–16. 1932b.

———: Further cytological studies of heterothallism in Puccinia graminis, *ibid.*, **47**(1):1–16. *Pls.* 1–6. 1933a.

———: A cytological study of the teliospores, promycelia and sporidia in Puccinia malvacearum, *Phytopathology*, **23**(7):572–586. *Figs.* 1–4. 1933b.

———: The spermatia of corn rust, Puccinia Sorghi, *ibid.*, **23**(11):923–925. *Fig.* 1. 1933c.

———: A cytological study of heterothallism in flax rust, *J. Agr. Research*, **49**(9): 765–791. *Pls.* 1–13. 1934a.

———: A cytological study of heterothallism in Puccinia sorghi, *ibid.*, **49**(12): 1047–1068. *Pls.* 1–7. 1934b.

———: A cytological study of Puccinia malvacearum from the sporidium to the teliospore, *ibid.*, **51**(9):801–818. *Pls.* 1–9. 1935.

Ames, Lawrence M.: An hermaphroditic self-sterile but cross-fertile condition in Pleurage anserina, *Bull. Torrey Botan. Club*, **59**(6):341–345. *Fig.* 1. 1932.

Andrus, C. F.: The mechanism of sex in Uromyces appendiculatus and U. vignae, *J. Agr. Research*, **42**(9):559–587. *Figs.* 1–11. 1931.

———: Sex and accessory fusions in the Uredineae, *J. Wash. Acad. Sci.*, **23**(12): 544–557. 3 *figs.* 1933.

Arthur, Joseph C.: Manual of the rusts in the United States and Canada, xv + 438 pp. 487 *figs.* Lafayette, Ind., Purdue Research Foundation, 1934.

———, in collaboration with F. D. Kern, C. R. Orton, and Others: The Plant Rusts (Uredinales), v + 446 pp. 186 *figs.* New York, John Wiley and Sons, 1929.

Ashworth, Dorothy: Puccinia malvacearum in monosporidial culture, *Brit. Myco. Soc. Trans.*, **16**:177–202. *Pls.* 8–9. *Figs.* 1–7. 1931.

Barnett, Horace L.: Studies in the sexuality of the Heterobasidiae, *Mycologia*, **29**(5):626–649. *Figs.* 1–3. 1937.

de Bary, Anton: Neue Untersuchungen über die Uredineen, insbesondere die Entwicklung der Puccinia graminis und den Zusammenhang derselben mit Aecidium Berberidis, *Monatsber. kgl. preuss. Acad. Wiss. Berlin*, 1865:15–49. 1 *pl.* 1865.

Bauch, Robert: Kopulationsbedingungen und sekundäre Geschlechtsmerkmale bei Ustilago violacea, *Biol. Centr.*, **42**(1):9–38. 1922.

———: Untersuchungen über die Entwicklungsgeschichte und Sexualphysiologie der Ustilago grandis, *Z. Botan.*, **17**(3):129–177. *Figs.* 1–4. 1925.

———: Über multipolare Sexualität bei Ustilago longissima, *Arch. Protistenk.*, **70**(2):417–466. *Pl.* 15. *Figs.* 1–2. 1930.

———: Geographische Verteilung und funktionelle Differenzierung der Faktoren bei der multipolaren Sexualität von Ustilago longissima, *ibid.*, **75**(1):101–132. 2 *figs.* 1931.

———: Die Sexualität von Ustilago Scorzonerae und Ustilago Zeae, *Phytopath. Z.*, **5**(3):315–321. *Figs.* 1–4. 1933.

———: Über die systematische Stellung von Tilletia Sphagni Nawaschin, *Ber. deut. botan. Ges.*, **56**(2):73–85. *Figs.* 1–7. 1938.

Bennett, Alfred W., and George Murray: A Handbook of Cryptogamic Botany, viii + 473 pp. *Figs.* 1–382. London, Longmans, Green and Co., 1899.

Bensaude, Mathilde: Recherches sur le cycle évolutif et la sexualité chez les Basidiomycètes, 156 pp. *Pls.* 1–13. *Figs.* 1–30. Nemours, H. Bouloy, 1918.

Bessey, Charles E.: The homologies of the Uredineae (The Rusts), *Am. Naturalist*, **28**(336):989–996. *Pl.* 32. 1894.

Blackman, Vernon H.: On the fertilization, alternation of generations and general cytology of the Uredineae, *Ann. Botany*, **18**(71):323–373. *Pls.* 2–24. 1904.

Boss, Georg: Beiträge zur Zytologie der Ustilagineen, *Planta*, **3**(4):597–627. *Figs.* 1–20. 1927.

Brefeld, Oscar: Basidiomyceten I. in Botanische Untersuchungen über Schimmelpilze, Heft. 3, pp. 1–266. *Pls.* 1–11. Leipzig, Arthur Felix, 1877.

———: Basidiomyceten II. Protobasidiomyceten, in Untersuchungen aus dem Gesammtgebiete der Mykologie, Heft. 7, pp. 1–178. *Pls.* 1–11. Leipzig, Arthur Felix, 1888.

———: Die Brandpilze I. in Botanische Untersuchungen über Hefenpilze. Fortsetzung der Schimmelpilze, Heft. 4, pp. 1–191. *Pls.* 1–10. Leipzig, Arthur Felix, 1881.

———: Die Brandpilze II. in Untersuchungen aus dem Gesammtgebiete der Mykologie, Heft. 11, pp. 1–98. *Pls.* 1–5. Münster i. W., Heinrich Schöningh, 1895a.

———: Die Brandpilze III. Hemibasidii, *ibid.*, Heft. 12, pp. 99–236. *Pls.* 6–12. Münster i. W., Heinrich Schöningh, 1895b.

———: Die Brandpilze IV. Hemibasidii, *ibid.*, Heft. 13, pp. 1–75. *Pls.* 1–2. Münster i. W., Heinrich Schöningh, 1905. (R. Falck, co-author.)

———: Die Brandpilze V. *ibid.*, Heft. 15, pp. 1–151. *Pls.* 1–7. Münster i. W., Heinrich Schöningh, 1912.

Brodie, Harold J.: The oidia of Coprinus lagopus and their relation with insects, *Ann. Botany*, **45**(178):315–344. *Pl.* 10. *Figs.* 1–24. 1931.

———: The occurrence in Nature of mutual aversion between mycelia of Hymenomycetous fungi, *Can. J. Research, C*, **13**:187–189. *Pl.* 1. 1935.

———: The barrage phenomenon in Lenzites betulina, *Genetica*, **18**:61–73. 1936.

Brown, A. M.: Diploidisation of haploid by diploid mycelium of Puccinia helianthi, *Nature*, **130**(3290):777. *Fig.* 1. 1932.

———: Studies on the interfertility of four strains of Puccinia helianthi Schw., *Can. J. Research, C*, **14**:361–367. *Pl.* 1. 1936.

Brunswik, H. Untersuchungen über die Geschlechts- und Kernverhältnisse der Hymenomycetengattung Coprinus, *Botan. Abhand. herausgegeben von Dr. K. Goebel*, **1**(5):1–152. *Figs.* 1–3. 1924.

Buller, A. H. Reginald: Researches on Fungi, 7 vols. London, Longmans, Green and Co., 1909–1934. Vol. 1, xi + 287 pp., 5 *pls.*, 83 *figs.*, 1909; vol. 2, xii + 492 pp., 157 *figs.*, 1922; vol. 3, xii + 611 pp., 227 *figs.*, 1924; vol. 4, xiii + 329 pp., 4 *pls.*, 149 *figs.*, 1931; vol. 5, xiii + 416 pp., 174 *figs.*, 1933; vol. 6, xii + 513 pp., 231 *figs.*, 1934; vol. 7, xx + 458 pp., 124 *figs.*, 1950.

———: The biological significance of conjugate nuclei in Coprinus lagopus and other Hymenomycetes, *Nature*, **126**(3183):686–689. *Figs.* 1–7. 1930.

———: Fusions between flexuous hyphae and pycnidiospores in Puccinia graminis, *ibid.*, **141**(3557):33–34. *Figs.* 1–3. 1938.

Carleton, Mark A.: Cereal Rusts of the United States, *U. S. Dept. Agr., Div. Vegetable Physiol. Pathol. Bull.* **16**:1–73. *Pls.* 1–4. *Fig.* 1. 1899.

———: Investigations of rusts, *U. S. Dept. Agr., Bureau of Plant Industry, Bull.* **63**:1–29. *Pls.* 1–2. 1904.

Christensen, J. J.: Mutation and hybridization in Ustilago zeae: II. Hybridization, *Univ. Minnesota Agr. Expt. Sta. Tech. Bull.* **65**:89–108. *Pls.* 1–3. 1929.

CHRISTENSEN, J. J., AND E. C. STAKMAN: Physiologic specialization and mutation in Ustilago zeae, *Phytopathology*, **16**(12):979–999. *Figs.* 1–11. 1926.

CHRISTMAN, A. H.: Sexual reproduction in the rusts, *Botan. Gaz.*, **39**(4):267–275. *Pl.* 8. 1905.

———: Alternation of generations and the morphology of the spore in the rusts, *ibid.*, **44**(2):81–101. *Pl.* 7. 1907.

CLINTON, GEORGE PERKINS: The Ustilagineae, or smuts, of Connecticut, *Connecticut State Geological and Natural History Survey Bull.*, **5**:1–44. *Figs.* 1–55. 1905.

———: Ustilaginales, *N. Am. Flora,* **7**(1):1–82. 1906.

COLLEY, R. H.: Parasitism, morphology, and cytology of Cronartium ribicola, *J. Agr. Research*, **15**(12):619–660. *Pls.* 48–59. 1918.

CRAIGIE, J. H.: Experiments on sex in rust fungi, *Nature*, **120**(3012):116–117. *Fig.* 1. 1927a.

———: Discovery of the function of the pycnia of the rust fungi, *ibid.*, **120**(3030): 765–767. *Figs.* 1–2. 1927b.

———: On the occurrence of pycnia and aecia in certain rust fungi, *Phytopathology*, **18**(12):1005–1015. *Figs.* 1–3. 1928.

———: An experimental investigation of sex in the rust fungi, *ibid.*, **21**(11):1001–1040. *Figs.* 1–14. 1931.

———: Union of pycniospores and haploid hyphae in Puccinia helianthi Schw., *Nature*, **131**(3297):25. *Fig.* 1. 1933.

CUMMINS, GEORGE B.: Phragmidium species of North America: Differential teliospore and aecial characters, *Mycologia*, **23**(6): 433–445. *Pl.* 32. 1931.

———: The genus Dicheirinia, *ibid.*, **27**(2):151–159. *Pl.* 16. 1935a.

———: Notes on some species of the Uredinales, *Mycologia*, **27**(6):605–614. *Figs.* 1–5. 1935b.

———: Prospodium; notes on the morphology of the sori, *Ann. Mycol.*, **35**(1):15–21. *Pl.* 1. *Fig.* 1. 1937.

———: The genus Prospodium, *Lloydia*, **3**(1):1–78. *Figs.* 1–12. 1940.

DANGEARD, P. A.: Mémoire sur la reproduction sexuelle des Basidiomycètes, *Le Botaniste*, **4**:119–181. *Figs.* 1–24. 1894–1895.

DAVIS, W. H.: Summary of investigations with Ustilago striaeformis parasitizing some common grasses, *Phytopathology*, **25**(8):810–817. 1935.

DE FERRY DE LA BELLONE, C.: Du mycélium des champignons hypogés en général et de celui des Tubéracées en particulier, *Assoc. franç. avance. sci., Comp. rend.* 14*e session, Grenoble, 1885, Seconde partie*: 437–443. *Pl.* 11. 1886.

DICKINSON, SYDNEY: Experiments on the physiology and the genetics of the smut fungi, *Proc. Roy. Soc. London, B*, **101**(708):126–136. *Pl.* 10. *Figs.* 1–3. 1927; **102**(715):174–176. 1927; **103**(726):547–555. *Pl.* 1. 1928.

DIETEL, P.: Zur Kenntnis der Gattung Uredinopsis Magnus, *Ber. deut. botan. Ges.*, **13**(7):326–332. *Pl.* 26. 1895.

———: Über den Generationswechsel der Rostpilze, *Natur und Schule*, **1**:205–216. 1902.

———: Monographie der Gattung Ravenelia Berk, *Botan. Centr. Beihefte*, Zweite Abt., **20**:343–413. *Pls.* 5–6. 1906.

———: Unterklasse Hemibasidii (Ustilaginales und Uredinales), in A. ENGLER und K. PRANTL: Die Natürlichen Pflanzenfamilien, Zweite Auflage, **6**:1–98. *Figs.* 1–80. 1928.

DODGE, B. O.: Effect of the orange rusts of Rubus on the development and distribution of stomata, *J. Agr. Research*, **25**(12):495–500. *Pl.* 1. *Fig.* 1. 1923.

———: Uninucleated aecidiospores in Caeoma nitens and associated phenomena, *ibid.*, **28**(10):1045–1058. *Pls.* 1–5. 1924.

———, AND L. O. GAISER: The question of nuclear fusions in the blackberry rust, Caeoma nitens, *ibid.*, **32**(11):1003–1024. *Pls.* 1–4. 1926.

DRAYTON, F. I.: The sexual mechanism of Sclerotinia Gladioli, *Mycologia*, **26**(1): 46–72. *Pls.* 5–7. *Figs.* 1–4. 1934.

EDDINS, A. H.: Pathogenicity and cultural behavior of Ustilago zeae (Beckm.) Ung. from different localities, *Phytopathology*, **19**(10):885–916. *Figs.* 1–7. 1929.

ERIKSSON, JAKOB: Ueber die Specialisierung des Parasitismus bei den Getreiderostpilzen, *Ber. deut. botan. Ges.*, **12**(9):292–331. 1894.

———: Ueber die Specialisierung des Getreideschwarzrostes in Schweden und in anderen Ländern, *Centr. Bakt., Parasitenk.*, Zweite Abt., **9**(16):590–607; (17–18):654–658. 1902.

FISCHER, E.: Beitrag zur Kenntnis der Gattung Graphiola, *Botan. Ztg.*, **41**(45): 745–756; (46):761–773; (47):777–788; (48):791–801. *Pl.* 6. 1883.

———: Zur Kenntnis von Graphiola und Farysia, *Ann. Mycol.*, **18**(4–6):188–197. *Figs.* 1–7. 1920.

———: Weitere Beiträge zur Kenntnis der Gattung Graphiola, *ibid.*, **20**(3–4): 228–237. *Figs.* 1–4. 1922.

FISCHER, GEORGE W.: The longevity of smut spores in herbarium specimens, *Phytopathology*, **26**(12):1118–1127. 1936.

———: Fundamental studies of the stripe smut of grasses (Ustilago striaeformis) in the Pacific Northwest, *ibid.*, **30**(2):93–118. *Figs.* 1–4. 1940.

———, AND C. S. HOLTON: Inheritance of sorus characters in hybrids between Ustilago avenae and U. perennans, *Mycologia*, **33**(5):555–567. *Figs.* 1–3. 1941.

FLEROV, B. K.: Sur la cytologie de l'Ustilago Avenae Pers. d'après des cultures in vitro, in Travaux de la Section de Mycologie et de Phytopathologie de la Société Botanique de Russe: Tome I. Travaux de la Division de Moscou, pp. 23–36. *Pl.* 3. 1923. (Russian, with French summary.)

FLOR, H. H.: Heterothallism and hybridization in Tilletia tritici and T. levis, *J. Agr. Research*, **44**(1):49–58. 1932.

———: Physiologic specialization of Melampsora lini on Linum usitatissimum, *ibid.*, **51**(9):819–837. *Pl.* 1. 1935.

FREEMAN, E. M., AND E. C. JOHNSON: The rusts of grains in the United States, *U. S. Dept. Agr. Bureau of Plant Industry Bull.* **216**:1–87. *Pl.* 1. 1911.

FRIES, ELIAS: Ordo IV. Hypodermii, Systema Mycologicum, sistens fungorum ordines, genera et species hucusque cognitas, vol. 3, pp. 504–519. Greifswald, Ernest Mauritius, 1832.

GÄUMANN, ERNST ALBERT: Comparative Morphology of Fungi. Translated and revised by Carroll William Dodge, xiv + 701 pp. 406 *figs.* New York, McGraw-Hill Book Company, 1928.

GROVE, W. B.: The evolution of the higher Uredineae, *New Phytologist*, **12**(3):89–106. *Figs.* 1–2. 1913.

HANNA, W. F.: The problem of sex in Coprinus lagopus, *Ann. Botany*, **39**(154): 431–457. 1925.

———: Nuclear association in the aecium of Puccinia graminis, *Nature*, **124** (3120):267. 1929a.

———: Studies in the physiology and cytology of Ustilago zeae and Sorosporium reilianum, *Phytopathology*, **19**(5):415–442. *Pl.* 7. *Figs.* 1–3. 1929b.

———: The discharge of conidia in species of Entyloma, *Mycologia*, **30**(5):526–536. *Fig.* 1. 1938.

HANNA, W. F., AND W. POPP: Relationship of the oat smuts, *Phytopathology*, **21** (1):109. 1931. (Abstract.)

HARPER, R. A.: Nuclear phenomena in certain stages in the development of the smuts, *Trans. Wisconsin Acad. Sci.*, **12**:475–498. *Pls.* 8–9. 1898.

———: Binucleate cells in certain Hymenomycetes, *Botan. Gaz.*, **33**(1):1–25. *Pl.* 1. 1902.

HARTER, L. L., AND W. J. ZAUMEYER: Differentiation of physiologic races of Uromyces phaseoli typica, *J. Agr. Research*, **62**(12):717–731. *Pls.* 1–3. 1941.

HARTIG, ROBERT: Lehrbuch der Baumkrankheiten, Zweite verbesserte und vermehrte Auflage, ix + 291 pp. 1 *colored plate. Figs.* 1–137. Berlin, Julius Springer, 1889.

HIRMER, MAX: Zur Kenntnis der Vielkernigkeit der Autobasidiomyzeten, I, *Z. Botan.*, **12**:657–674. *Pl.* 5. *Figs.* 1–10. 1920.

HIRSCHHORN, ELISA Y JULIO: Los carbonas del maiz en Argentina, *Rev. facultad agron. Univ. nacl.* (*La Plata*), **20**(2):108–139. *Pls.* 1–5. 1935.

HÜTTIG, WERNER: Über den Einfluss der Temperatur auf die Keimung und Geschlechterverteilung bei Brandpilzen, *Z. Botan.*, **24**:529–577. *Figs.* 1–26. 1931.

———: Über physikalische und chemische Beeinflussungen des Zeitpunkts der Chromosomenreduktion bei Brandpilzen, *ibid.*, **26**(1):1–26. *Figs.* 1–9. 1933.

JACKSON, HERBERT S.: Present evolutionary tendencies and the origin of life cycles in the Uredinales, *Mem. Torrey Botan. Club*, **18**(1):1–108. 1 *fig.* 1931.

JOHNSON, H. W.; H. A. RODENHISER; AND C. L. LEFEBVRE: Two types of fall Panicum smut, *J. Agr. Research*, **61**(12):865–875. *Figs.* 1–7. 1940.

KAISER, WALTER: Zur Biologie und Entwicklungsgeschichte einiger Entyloma-Arten, *Angew. Botan.*, **18**(2):81–131. *Figs.* 1–4. 1936.

KEMPER, WILHELM: Zur Morphologie und Zytologie der Gattung Coniophora, insbesondere des sogenannten Kellerschwamms, *Zentr. Bakt.*, *Parasitenk.*, Zweite Abt., **97**(4–8):100–124. *Figs.* 1–20. 1937.

KILLIAN, CHARLES: Le développement du Graphiola phoenicis Poit., et ses affinités, *Rev. gen. botan.*, **36**:385–394, 451–460. *Pls.* 7–10. 1924.

KNIEP, HANS: Beiträge zur Kenntnis der Hymenomyceten III, *Z. Botan.*, **7**(6): 369–398. *Pl.* 2. *Figs.* 1–20. 1915.

———: Urocystis anemones (Pers.) Winter, *ibid.*, **13**:282–311. 1921.

———: Über Artkreuzungen bei Brandpilzen, *Z. Pilzk.*, **5**(14–15):217–247. *Figs.* 1–35. 1926.

———: Die Sexualität der niederen Pflanzen, 544 pp. 221 *figs.* Jena, Gustav Fischer, 1928.

KOLK, LAURA ALMA: Relation of host and pathogen in the oat smut, Ustilago Avenae, *Bull. Torrey Botan. Club*, **57**(7):443–507. *Pls.* 27–30. *Figs.* 1–21. 1930.

KÜHNER, R. E.: Contribution à l'étude des Hyménomycètes et spécialement des Agaricacées, *Le Botaniste*, **17**:5–218. 4 *pls.* 37 *figs.* 1926.

LAMB, IVAN M.: On the morphology and cytology of Puccinia Prostii Moug., a micro-form with pycnidia, *Trans. Roy. Soc. Edinburgh*, **58**(1):143–162. *Pls.* 1–2. 1934.

LEACH, J. G., AND MARY ALICE RYAN: The cytology of Ustilago striiformis forma Poae-pratensis in artificial culture, *Phytopathology*, **36**(10):876-886. *Figs.* 1–6. 1946.

LEHFELDT, W.: Über die Entstehung des Paarkernmycels bei heterothallischen Basidiomyceten, *Hedwigia*, **64**:30–51. *Pl.* 1. *Figs.* 1–3. 1923.

LINDFORS, T.: Studien über den Entwicklungsverlauf bei einigen Rostpilzen aus zytologischen und anatomischen Gesichtspunkten, *Svensk Botans. Tid.*, **18**:1–84. *Illustrated.* 1924.

LIRO, J. IVAR: Die Ustilagineen Finnlands, I, *Ann. Acad. Sci. Fennicae, A*, **17**(1): 1–636. *Figs.* 1–9. 1924.

———: Die Ustilagineen Finnlands, II, *ibid.*, **42**:i–xiii, 1–720. *Figs.* 1–8. 1 *map.* 1935–1938.

LUTMAN, B. F.: Some contributions to the life history and cytology of the smuts, *Trans. Wisconsin Acad. Sci.*, **16**(2):1191–1244. *Pls.* 87–95. 1910.

MAINS, E. B.: Additional studies concerning the rust of Iris, *Phytopathology*, **28**(1):67–71. 1938.

———, AND H. S. JACKSON: Physiologic specialization in the leaf rust of wheat, Puccinia triticina Erikss., *Phytopathology*, **16**(2):89–119. *Pls.* 6–9. 1926.

MALENÇON, G.: Notulae mycologicae Maroccanae, *Rev. mycol.*, N.S., **1**(2):43–74. *Pls.* 2–7. 1936.

MASSEE, GEORGE: On the presence of sexual organs in Aecidium, *Ann. Botany*, **2**(5):47–51. *Pl.* 4*A*. 1888.

MATTIROLO, O.: Sul parassitismo dei tartufi e sulla questione delle mycorhizae, *Malpighia*, **1**:359–369. *Pl.* 10. 1887.

MOREAU, MME. FERNAND: Les phénomènes de la sexualité chez les Urédinées, *Le Botaniste*, **13**:145–284. *Pls.* 15–28. 2 *figs.* 1913.

MOREAU, M. ET MME. FERNAND: L'évolution nucléaire chez l'Endophyllum Sempervivi Lév., *Bull. soc. mycol. France*, **33**:70–72. *Figs.* 1–6. 1917.

——— ET ———: L'ecidiospore de l'Endophyllum Euphorbiae-silvaticae (D.C.) Winter est-elle le siège d'une karyogamie?, *ibid.*, **33**:97–99. *Figs.* 1–5. 1917.

——— ET ———: Les Urédinées du groupe Endophyllum, *Bull. Soc. Botan. France*, **66**:14–44. *Figs.* 1–15. 1919.

——— ET ———: Les phénomènes cytologiques de la reproduction chez les champignons des lichens, *Le Botaniste*, **20**(1–2):1–67. *Figs.* 1–35. 1928.

MOUNCE, IRENE: Homothallism and heterothallism in the genus Coprinus, *Brit. Mycol. Soc. Trans.*, **8**:256–269. 1922.

———, AND RUTH MACRAE: The behavior of paired monosporous mycelia of Lenzites saepiaria (Wulf.) Fr., L. trabea (Pers.) Fr., L. thermophila Falck, and Trametes americana Overh., *Can. J. Research, C*, **14**:215–221. *Pl.* 1. 1936.

MUNDKUR, B. B.: Karnal bunt, an air-borne disease, *Current Science* (*India*), **12**(8):230–231. *Fig.* 1. 1943.

NEWTON, DOROTHY E.: Bisexuality of individual strains of Coprinus rostrupianus, *Ann. Botany*, **40**(157):105–128. *Pl.* 6. *Figs.* 1–17. 1926.

———: The distribution of spores of diverse sex on the hymenium of Coprinus lagopus, *ibid.*, **40**(160):891–917. 1926.

NEWTON, MARGARET; T. JOHNSON; AND A. M. BROWN: A preliminary study in the hybridization of physiologic forms of Puccinia graminis tritici, *Sci. Agr.*, **10**(11):721–731. *Figs.* 1–4. 1930.

———; ———; ———: A study of the inheritance of spore color and pathogenicity in crosses between physiologic forms of Puccinia graminis tritici, *ibid.*, **10**(12):775–798. *Figs.* 1–8. 1930.

OERSTED, A. S.: Forelöbig beretning om iagttagelser, hvored det godtgjores at Podisoma Sabinae (Dicks.) og Roestelia cancellata (Jacq.) ere vexlende generationer of samme svampeart, *Botan. Notiser*, **1865**:105–107. 1865.

OLIVE, EDGAR W.: Sexual cell fusions and vegetative nuclear divisions in the rusts, *Ann. Botany*, 22(87):331–360. *Pl.* 22. 1908.

OORT, A. J. P.: Die Sexualität von Coprinus fimetarius, *Rec. trav. botan. néerland.*, **27**:85–148. *Pls.* 4–6. *Figs.* 1–5. 1930.

ORTON, C. R.: A working hypothesis on the origin of rusts, with special reference to the phenomenon of heteroecism, *Botan. Gaz.*, **84**(2):113–138. 1927.

OUDEMANS, C. A. J. A.: Over twee nog onbekende fungi: Septoria Dictyotae en Ustilago Vuijckii, *Verslagen. Zittingen Wis- Natuurk. Afdeel. Koninkl. Akad. Wetenschap. Amsterdam*, **3**:54–57. 1895.

PADY, S. M.: The rôle of intracellular mycelium in systemic infections of Rubus with the orange-rust, *Mycologia*, **27**(6):618–637. *Figs.* 1–42. 1935.

———: Teliospore discharge in Puccinia tumidipes Peck, *ibid.*, **40**(1):21–33. *Figs.* 1–9. 1948.

PARAVICINI, EUGEN: Untersuchungen über das Verhalten der Zellkerne bei der Fortpflanzung der Brandpilze, *Ann. Mycol.*, **15**(1–2):57–96. *Pls.* 1–6. *Figs.* 1–5. 1917.

PIERSON, ROYALE K.: Fusion of pycniospores with filamentous hyphae in the pycnium of the white pine blister rust, *Nature*, **131**(3316):728–729. 1933.

PLOWRIGHT, CHARLES B.: A monograph of British Uredineae and Ustilagineae, vii + 347 pp. *Pls.* 1–8. *Figs.* 1–13. London, Kegan, Paul, Trench, & Co., 1889.

QUINTANILHA, A.: Cytologie et génétique de le sexualité chez les Hyménomycètes, *Bol. soc. Broteriana*, IIme sér., **10**:289–336. 4 *pls.* 1935.

RAWITSCHER, FELIX: Beiträge zur Kenntnis der Ustilagineen, *Z. Botan.*, **4**:673–706. *Pl.* 8. *Figs.* 1–20. 1912.

RICE, MABEL A.: Reproduction in the rusts, *Bull. Torrey Botan. Club*, **60**(1): 23–54. *Pls.* 4–6. 1933.

———: The relation of Uromyces Caladii and other rusts to their hosts, *ibid.*, **61**(3):155–162. *Pls.* 6–8. 1934.

SAPPIN-TROUFFY, P.: Sur la signification de la fécondation chez les Urédinées, *Le Botaniste*, **5**:32–37. *Figs.* 1–2. 1896a.

———: Recherches mycologiques, *ibid.*, **5**:44–58. *Figs.* 1–6. 1896b.

———: Recherches histologiques sur le famille des Urédinées, *ibid.*, **5**:59–244. *Figs.* 1–70. 1896c.

SARTORIS, GEORGE B.: Studies in the life history and physiology of certain smuts, *Am. J. Botany*, **11**(10):617–647. *Pls.* 39–41. 1924.

SASS, JOHN E.: The cytological basis for homothallism and heterothallism in Agaricaceae, *Am. J. Botany*, **16**(9):663–701. *Pls.* 64–67. *Figs.* 1–6. 1929.

SAVILE, D. B. O.: Nuclear structure and behavior in species of the Uredinales, *Am. J. Botany*, **26**(8):585–609. *Figs.* 1–106. 1939.

SCHEIBE, A.: Studien zum Weizenbraunrost, Puccinia triticina Eriksson: I. Methoden und Ergebnisse bei der Bestimmung seiner physiologischen Formen (Biotypen), *Arb. biol. Reichsanstalt Land- u. Forstw.*, **16**(4):575–608. 2 *colored plates.* 6 *figs.* 1929.

SEYFERT, RICHARD: Über Schnallenbildung im Paarkernmyzel der Brandpilze, *Z. Botan.*, **19**:577–601. *Figs.* 1–22. 1927.

SHEAR, C. L., AND B. O. DODGE: Life histories and heterothallism of the red bread-mold fungi of the Monilia sitophila group, *J. Agr. Research*, **34**(11):1019–1042. *Pls.* 1–4. 1927.

SIBELIA, C.: Ricerche sulle ruggini dei cereali: V. Ulteriori ricerche sulla specializzazione delle "Puccinia triticina" Erikss. in Italia, *Boll. staz. patol. vegetale, Roma*, N.S., **16**(2):69–75. 1 *map*. 1936.

SLEUMER, HERMANN OTTO: Über Sexualität und Zytologie von Ustilago Zeae (Beckm.) Unger, *Z. Botan.*, **25**(3):209–263. *Pl.* 1. *Figs.* 1–33. 1932.

STAKMAN, E. C.: A study in cereal rusts, physiological races, *Univ. Minnesota Agr. Expt. Sta. Bull.* **138**:1–56. *Pls.* 1–9. 1914.

———; R. C. CASSELL; AND M. B. MOORE: The cytology of Urocystis occulta, *Phytopathology*, **24**(8):874–889. *Pls.* 1–3. 1934.

———, AND J. J. CHRISTENSEN: Heterothallism in Ustilago zeae, *Phytopathology*, **17**(12):827–834. 1927.

———; ———; AND H. E. BREWBAKER: Physiologic specialization in Puccinia sorghi, *Phytopathology*, **18**(4):345–354. *Figs.* 1–2. 1928.

———; ———; C. J. EIDE; AND BJORN PETURSON: Mutation and hybridization in Ustilago zeae: I. Mutation, *Univ. Minnesota Agr. Expt. Sta. Tech. Bull.* **65**: 1–87. *Pls.* 1–10. 1929.

———, AND M. N. LEVINE: The determination of biologic forms of Puccinia graminis in Triticum spp., *Univ. Minnesota Agr. Expt. Sta. Tech. Bull.* **8**: 1–10. *Fig.* 1. 1922.

———; ———; AND R. V. COTTER: Origin of physiologic forms of Puccinia graminis through hybridization and mutation, *Sci. Agr.*, **10**(11):707–720. 1930.

———, AND F. J. PIEMEISEL: Biologic forms of Puccinia graminis on cereals and grasses, *J. Agr. Research*, **10**(9):429–496. *Pls.* 53–59. 1917.

STEMPELL, K. L.: Studien über die Entwicklungsgeschichte einiger Entyloma-Arten und über die systematische Stellung der Familie der Sporobolomycetes, *Z. Botan.*, **28**(4):225–259. *Pl.* 1. *Figs.* 1–9. 1935.

SYDOW, H.: Rusts of British Guiana and Trinidad, *Mycologia*, **17**(6):255–262. *Fig.* 1. 1925.

THIRUMALACHAR, M. J.: A cytological study of Uromyces aloës, *Botan. Gaz.*, **108**(2):245–254. *Figs.* 1–25. 1946.

VANDENDRIES, RENÉ: Contribution nouvelle à l'étude de la sexualité des Basidiomycètes, *La Cellule*, **35**:129–157. *Pls.* 1–2. *Figs.* 1–4. 1924.

———: Les barrages sexuels chez Lenzites betulina (L.) Fr., *Compt. rend.*, **198**(2): 193–195. 1934.

———: Les tendences sexuelles chez les Polyporés, *Rev. mycol.*, N.S., **1**(2):85–92. (4):181–190, (6):294–302. 10 *figs.* 1936.

———, ET H. J. BRODIE: Nouvelles investigations dans la domaine de la sexualité des Basidiomycètes et étude expérimentale des barrages sexuels, *La Cellule*, **42**(2):165–209. 1 *pl.* 27 *figs.* 1933.

———, ET P. MARTENS: Oïdies haploïdes et diploïdes sur mycélium diploïde chez "Pholiota aurivella" Batsch, *Acad. roy. Belg., Bull. classe sci.*, 5e. sér., **18**: 468–472. 1 *fig.* 1932.

VOSS, W.: Über Schnallen und Fusionen bei den Uredineen, *Ber. deut. botan. Ges.*, **21**(7):366–371. *Pl.* 19. 1903.

WANG, YUN-CHANG, ET P. MARTENS: Sur l'origine de la dicaryophase chez quelques Urédinées, *La Cellule*, **48**:213–245. *Figs.* 1–81. 1939.

WATERHOUSE, W. L.: A preliminary account of the origin of two new Australian physiologic forms of Puccinia graminis tritici, *Proc. Linnean Soc. New South Wales*, **54**(2):96–106. *Pl.* 4. 1929.

WEIR, JAMES R.: A short review of the general characteristics and cytological phenomena of the Uredineae, with notes on a variation in the promycelium of Coleosporium pulsatillae (Str.), *New Phytologist*, **11**(4):129–239. 1912.

WERNHAM, C. C.: Chlamydospore production on artificial media by Urocystis gladioli, *Phytopathology*, **28**(8):598–600. *Fig.* 1. 1938.

WORONIN, MICHAEL: Beitrag zur Kenntniss der Ustilagineen, in A. DE BARY und M. WORONIN: Beiträge zur Morphologie und Physiologie der Pilze, V. Reihe, *Abhandlungen der Senckenbergischen Naturforschenden Gesellschaft*, **12**:559–591. *Illustrated.* 1882.

YEN, WEN-YU: Note sur quelques Ustilaginées marocaines, *Rev. mycol.*, N.S., **2**(2):76–84. 2 *pls. Figs.* 1–5. 1937.

YU, T. F.; L. HWANG; AND C. T. TSIANG: Varietal resistance and susceptibility of wheats to flag smut (Urocystis tritici Koern.): III. Physiologic specialization in Urocystis tritici Koern., *Bull. Chinese Botan. Soc.*, **2**(2):111–113. 1936.

13

CLASS BASIDIOMYCETEAE: SUBCLASS HETEROBASIDIAE

IN THE basidial structure of one of its orders, the Auriculariales, this subclass shows many points of similarity with the more typical forms of Subclass Teliosporeae. By some authors (e.g., Patouillard, 1900) the latter are included in this order. Taking all their characteristics into consideration it seems to the author to be a more satisfactory arrangement to keep them apart in separate, but more or less closely related, subclasses. Thus limited the Heterobasidiae consist of fungi most often living as saprophytes upon wood or other plant materials but in some cases parasitic upon living plant tissues or even upon insects (e.g., *Septobasidium*). They form their spore fruits as thin or thick layers or projecting cushions or clubs or shelves of basidium-forming tissue. The basidia are formed mostly in a more or less recognizable layer, the hymenium, often crowded side by side in a palisade-like arrangement, and shoot their basidiospores off into the air with the exception of one or two genera. The basidiospores frequently become septate as they mature or after they are discharged and as a general rule (to which there are exceptions), germinate by the formation of numerous small buds when they fall into water, though capable of producing germ tubes under favorable conditions. In many cases a basidiospore will form a small sterigma and a secondary spore will be shot off, just as occurs frequently in the formation of secondary sporidia in the Uredinales. The spore fruits are more often waxy, or gelatinous when wet, and frequently dry down to rather inconspicuous horny masses when dry. This is made possible by the fact that the outer portion of the wall of each hypha swells greatly upon absorbing water.

The basidial characters that further distinguish this subclass from Subclass Eubasidiae are the following: In the Eubasidiae the basidium remains as one cell usually bearing the basidiospores at its upper end on short or long sterigmata. *Tulostoma* and one or two related genera form an exception in that the sterigmata may be formed laterally on the basidium. In the Heterobasidiae the basidium may be (1) elongated and transversely divided by three septa into four cells from each of which a sterigma arises and produces a single basidiospore, or (2) it may be rounded and divided into four cells by vertical septa, each cell producing a sterigma and a basidiospore, or (3) elongated and forked into two prongs, at the apex of each of which a sterigma bears one basidiospore, or (4) the rounded basidium produces (usually) four rounded cells at its upper end, these usually being separated by a septum from the collapsed basidium and frequently falling free from it. Each of these cells produces a sterigma and a basidiospore. In all of these four types, and in the Eubasidiae, the basidia arise as enlarged binucleate cells on mycelium of the secondary (dicaryon) type. Mostly these cells are terminal to the hyphae bearing them but sometimes (*Sirobasidium*) several cells at the terminal portion of the hypha become transformed into a chain of basidia. In a few cases a single intercalary cell may become a basidium.

In this subclass the author recognizes four orders, based upon the four types of basidia. It must be noted that Martin (1944) considers these to be more closely related and unites them into one order, Tremellales. The orders recognized by the author are:

Auriculariales: with elongated basidium divided into a row of four cells by cross septa. Nuclear divisions stichobasidial.

Dacrymycetales: with nonseptate forked basidium ("tuning fork" type) bearing two basidiospores. Nuclear divisions stichobasidial.

Tremellales: with rounded basidium divided cruciately into four cells by vertical septa. Nuclear divisions chiastobasidial.

Tulasnellales: with rounded basidium bearing large rounded epibasidia, usually separated from the hypobasidium by septa and often falling free. Nuclear divisions chiastobasidial.

In these four orders there is a marked parallelism of evolution of spore fruit structure from effused, sometimes almost film-like, structures to cushions, shelves, stipitate pilei, etc.

The arrangement followed below is not to be considered as representing a single progressing line of evolution. On the contrary the groups have been produced by a more or less parallel evolution from primitive forms that have been lost. The orders Auriculariales and Tremellales are sometimes set apart from the other orders as the Phragmobasidiae, since the basidium is divided into four cells. The Dacrymycetales and Tulasnellales with undivided basidia are then united with the remaining orders of the class under the name Holobasidiae. Gilbert (1921) has studied the nuclear behavior in the Heterobasidiae and finds that in their essentials they are identical with those occurring in the Eubasidiae.

Within this subclass the basidium in many genera has been considered as being made up of two parts, hypobasidium and epibasidium or epibasidia (Juel, 1898; Neuhoff, 1924; Rogers, 1934). These are then homologized with the teliospore and promycelium (as hypobasidium and epibasidium respectively) of the Uredinales and Ustilaginales. Ontogenetically, where this distinction can be made, the hypobasidium is a more or less spherical dicaryon cell within which the fusion of the nuclei occurs. It may remain thin-walled and proceed immediately with the formation of the epibasidium or may develop a thicker wall and become a resting cell with diploid nucleus, from which, under proper conditions the epibasidium then grows.

The difficulty with the foregoing interpretation is that in some of these fungi there is no such distinction in some species of a genus while it is pronounced in other species. Furthermore, different structures have been designated as hypobasidium and epibasidium. In the Auriculariales, in the genus *Auricularia*, the somewhat elongated dicaryon basidium primordium elongates after the nuclei have united and then meiosis occurs and the four nuclei become separated by septa. There is no hypobasidium nor epibasidium that can be distinguished. In some species of *Septobasidium* the basidium primordium becomes a somewhat thick-walled resting cell, and when favorable conditions come on this sends out a short stout hypha like the whole basidium of *Auricularia*, within which meiosis occurs and septa are formed. In other species no hypobasidium is formed at all and within the basidial primordium occur the union of the two nuclei, and their meiotic division, followed by septation to form the four celled basidium. In still other species of this genus whether the resting stage, usually interpreted as a hypobasidium, develops or not depends upon the environment. Thus it is apparent that this is perhaps not such a fundamental character as has been supposed. In other words in the development of the basidium in this group the presence or absence of a hypobasidium and consequently of an epibasidium depends upon whether the ontogeny of the basidium is interrupted or not.

In the type of basidium found in the Dacrymycetales the basal portion is interpreted as the hypobasidium and the two arms as epibasidia. But here the meiotic divisions occur in the "hypobasidium," not in the "epibasidium" as in *Septobasidium*. In *Tremella* the "hypobasidium" is the rounded basal portion within which meiosis occurs before the vertical walls divide it cruciately into four cells. The four arms at whose tips the sterigmata produce the basidiospores are interpreted as "epibasidia." In *Tulasnella* the same structures are found, except that the "hypobasidium" is not septate but is separated from the four epibasidia by septa.

Order Auriculariales. The fungi included in this order are some of them cosmopolitan but many are confined to the Tropics. They are largely saprophytes on wood. A few are true parasites, such as some species of *Eocronartium* and *Jola*, occurring on mosses; *Septobasidium* and *Ured-*

inella, parasitizing upon scale insects; *Herpobasidium filicinum* (Rostr.) Lind and *Platycarpa* upon ferns; etc. According to Boedijn and Steinmann (1931) the East Indian species of *Helicobasidium* are parasitic upon roots, while H. *purpureum* (Tul.) Pat. is parasitic on various plant roots and crowns in Europe.

Clamp connections have been recognized in *Auricularia, Phleogena, Helicobasidium, Jola, Helicogloea*, etc. H. L. Barnett (1937) reported that the mycelium of *Auricularia* derived from single spores, and therefore monocaryotic, lacks clamp connections but that when two such mycelia of opposite sexual phase come into contact the resulting dicaryotic mycelium may be recognized immediately by the presence of these structures.

The spore fruits are external and vary from a more or less felt-like film to a thin crust or to firm shelf-like structures standing out from the substrata. In *Phleogena* they are upright and stipitate with an enlarged head. In size they vary from a few millimeters to several centimeters. *Septobasidium* forms a more or less felted layer over the twigs and larger branches of the host with various tunnels and chambers in which the scale insects are protected or through which the larvae may travel. The basidia are formed on the outer surface of the felted mycelium. In some genera the spore fruits are gelatinous when wet, drying down to a horny crust or cushion. Other genera on the contrary, such as *Phleogena* and *Septobasidium*, do not become gelatinized with moisture.

In the gelatinous types of spore fruits, e.g., *Auricularia*, the basidia may be situated below the surface or at least with only the upper end reaching the air. This necessitates the formation of tube-like extensions from the four cells of the basidium to the surface where each develops a sterigma and bears a spore. This is very similar to the structure of the germinating teliospore of *Coleosporium*. Where the structure is not highly gelatinized the basidia are superficial and the sterigmata are very short, without any elongated supporting tubes, e.g., *Phleogena*.

The basidia are usually the terminal cells of the hyphae but in some species of *Helicogloea* (*Saccoblastia*) the basidial primordia may be intercalary according to Miss Baker (1936).

No distinction of hypobasidium and epibasidium is apparent in *Auricularia, Phleogena*, and *Herpobasidium*. Within the basidial primordium the two nuclei unite and then undergo the two meiotic divisions, the spindles of the dividing nuclei being parallel to the long axis of the enlarging cell, thus producing a stichobasidial structure. In *Helicogloea* (*Saccoblastia*) the primordial cell of the basidium produces a lateral sack-like "hypobasidium" in which the union of the nuclei occurs. Then sometimes at one end of this sack, but more often from the apex of the primordium the "epibasidium" grows, and in it the meiotic divisions occur. Since this genus includes species with greater or less gelatinization of the vegetative hyphae there arise from the four cells of the epibasidium shorter or longer arms reaching to the surface and there bearing the sterigmata and spores. The basidial primordium may be an intercalary cell of the hypha although more often it is terminal. In the genus *Septobasidium* it was pointed out by Couch (1938) that the basidia among the more than 160 species represent a number of different types. In the majority there is a distinct rather thin-walled and hyaline hypobasidium which may send out the epibasidium immediately or the hypobasidium may have a thick, colored wall and serve as a resting spore until favorable conditions are present. The epibasidium is more often four-celled but in some species it is three-celled, in others two-celled, and in two species it is one-celled. The basidiospores become segmented into two to several cells and usually germinate by sprout cells, when placed in water.

Perhaps three families should be recognized in this order, although Gäumann and Dodge (1928) recognized four.

Auriculariaceae: parasitic on plants or saprophytic on dead plant material, usually wood. Spore fruits often gelatinous but not so in some genera. Basidia with no distinction into hypobasidium and epibasidium or these well distinguished. Clamp connections present in some genera, absent in others.

Phleogenaceae: saprophytic on wood, bark, etc. No distinction into hypobasidium and epibasidium. Spore fruit stalked with a head of radiating, more or less coiled hyphae, among which the curved basidia are formed, bearing the four spores without visible sterigmata. Clamp connections observed.

Septobasidiaceae (Order Septobasidiales, according to Couch, 1938): parasitic upon scale insects with some of which they live in symbiotic relation. Basidia usually with well-developed thin-walled or thick-walled hypobasidia and one- to four-celled (mostly the latter) epibasidia, but hypobasidia sometimes lacking. Basidiospores produced on distinct sterigmata. Conidia often produced. Clamp connections apparently lacking.

Family Auriculariaceae. In the Auriculariaceae the following genera should be mentioned as illustrating the various types of structure. *Eocronartium*, with a single species *E. muscicola* (Fr.) Fitzp. is a perennial fungus parasitic in the gametophytes of many mosses (see Fitzpatrick, 1918a, b). The cells of the mycelium are always binucleate and there are no clamp connections. The mycelium is intracellular, passing from cell to cell of the host but apparently doing little harm except that the production of the sporophyte appears to be suppressed in the infected plants. At the apex of the stem the hyphae pass out into the spaces between the leaves and grow upward and parallel and form a gelatinous, club-shaped sporophore on the outer surface of which the basidia are produced in great abundance. The longitudinal hyphae of the sporophore turn outward to the surface and there give rise to the basidia which as they elongate bend almost at right angles so as to lie nearly parallel to the surface. The basidial primordium is at first cylindrical or clavate, the major portion (epibasidium) becoming separated by a septum from the short cylindrical thin-walled hypobasidium which is not enlarged as in some genera. The epibasidia are transversely three septate and from each cell is produced a long tube at whose apex a sterigmatic structure is formed on which arises the one-celled uninucleate basidiospore. This fungus is found on many species of mosses in Europe and America. (Fig. 144 A.)

Jola is also parasitic upon mosses but mostly in the sporophytes. It seems to be almost exclusively tropical. It forms its small spherical or elongated spore fruit at the apex of the sporophyte. It appears to be more or less gelatinous. The binucleate hyphal cells at the surface enlarge at the ends and in these terminal cells (hypobasidia) the nuclei unite. From the apex of each emerges the epibasidium within which the meiotic division of the nucleus takes place. After the three transverse septa have been formed short or long tube-like hyphal growths reach the surface of the hymenium and bear the uninucleate spores at their tips. Below the terminal basidium the next cell grows out sympodially and produces another basidium and this process is repeated until a dozen or more basidia are produced (see Gäumann, 1922). Closely related, but growing on fungi instead of mosses is the likewise tropical genus *Cystobasidium*. (Fig. 144 B.)

Herpobasidium filicinum (Rostr.) Lind was studied by Jackson (1935). It is parasitic upon the leaves of ferns in which the mycelium is intercellular, producing massive coiled haustoria in the cells of the host. The internal mycelium emerges through the stomata, forming small white patches. On the surface the basidia arise as terminal cells of upright hyphae. They are usually slightly bent. No distinction of hypobasidium and epibasidium is observable. The fusion nucleus divides once, by the first step of meiosis, so that the two nuclei now have the haploid number of chromosomes. A single septum is formed and one basidiospore develops on the tip of a sterigma on each cell. The spore is uninucleate and no nuclear division has been observed. It may germinate by repetition. The origin of the dicaryon phase is unknown as is also the case in the two foregoing genera. No clamp connections have been observed. This species occurs in Europe and North America.

Another genus, *Platycarpa* (Couch, 1949), is parasitic upon tropical ferns. The fruit body is resupinate, very small, dry to subcartilaginous, separable from the host at maturity. The vegetative mycelium forms coiled haustoria within the epidermal and mesophyll cells or in the sporogenous cells of the host. No clamp connections have been observed. Externally a more or less felty mass of hyphae occurs. These are more or less wavy or loosely coiled near the surface and produce terminally on short branches the ovoid probasidia. These are hyaline-walled and germinate by the production of a straight or, more often, curved and mostly four-

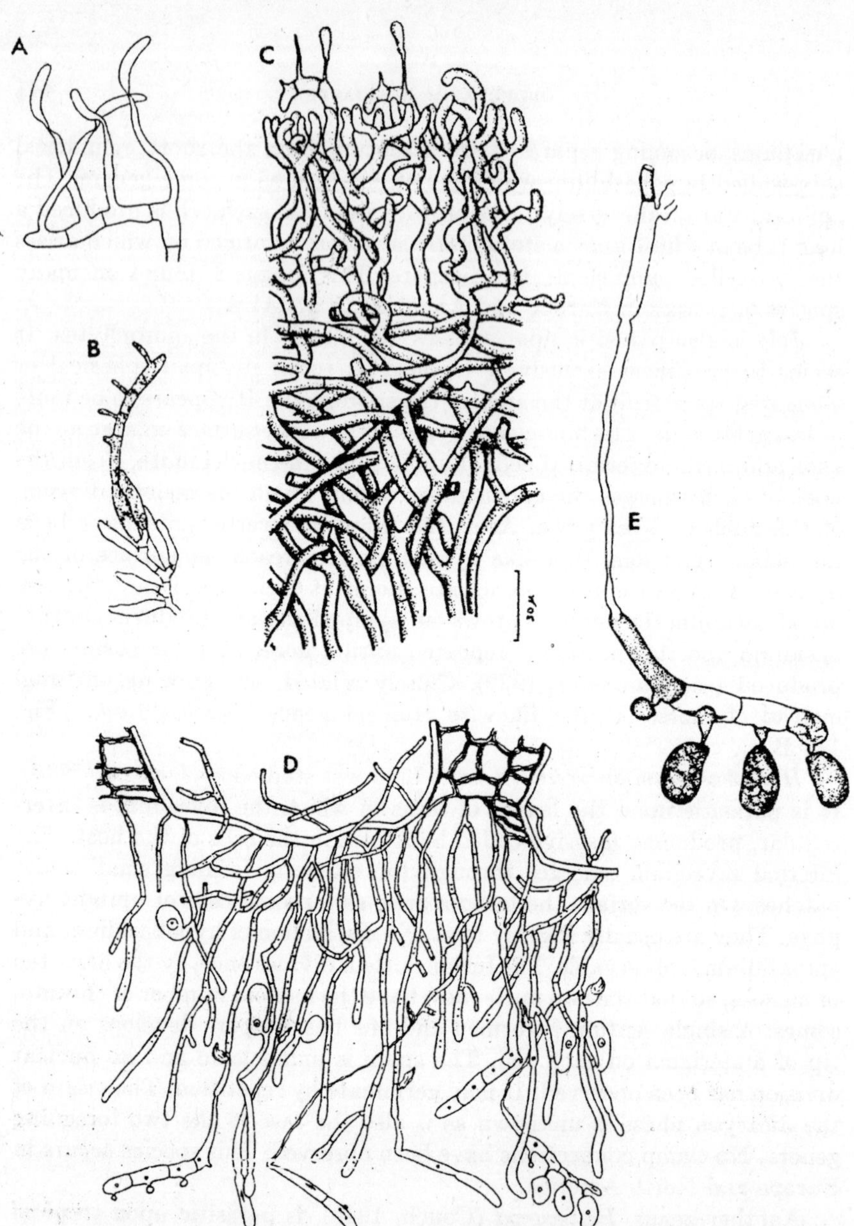

FIG. 144. Order Auriculariales, Family Auriculariaceae. (A) *Eocronartium muscicola* (Fr.) Fitzp. Basidium with tubular extensions terminated by sterigmata bearing spores. (B) *Jola javensis* Pat., basidiophoric hypha showing sympodial mode of growth, with almost mature basidium at the tip. (C) *Helicobasidium compactum* Boedijn, section through the hymenium. (D, E) *Helicogloea lagerheimi* Pat. (*Saccoblastia*). (D) Fructification on wood showing basidia in various stages of development. (E) Mature basidium shedding spores, the hypobasidium collapsed. (A, courtesy, Fitzpatrick: *Phytopathology*, **8**(5):197–218. B, after Gäumann: *Ann. Mycolog.*, **20**(5–6):272–289. C, after Boedijn and Steinmann: *Bull. Jardin Botan. Buitenzorg*, Série III, **11**(2):165–219. D–E, courtesy, Baker: *Ann. Missouri Botan. Garden*, **23**(1):69–128.)

celled epibasidium. The somewhat allantoid basidiospores are produced on distinct sterigmata. In structure of the sporiferous portion of the fruit body the fungi of this genus show considerable similarity to *Septobasidium*, to which Couch suggests that they may be transitional forms.

Helicobasidium is parasitic on the roots or crowns of trees and shrubs or perennial herbaceous plants, and also may grow as a saprophyte. *H. purpureum* (Tul.) Pat. has been studied by Buddin and Wakefield (1927). On the surface of the substratum it forms a nongelatinized, effused, felted layer. The basidial primordia are often more or less coiled. They become hypobasidia from whose apex grows out the terminal hooked epibasidium which produces sometimes rather long sterigmata. In *H. candidum* Martin (1940) the hypobasidium is small, forming practically only a short slender stalk for the stout curved basidium. Clamp connections are present. *Syzygospora*, with a single species *S. alba* was described by Martin (1937a). It forms a white gelatinous spore fruit up to 10 cm. long and 2.5 cm. thick, the surface covered by the hymenium. The basidia occur in sympodial fascicles. They have but a single septum with a basal cell more or less elongated and clavate and the terminal cell spherical. A single spherical basidiospore is produced from each cell, near the base of the apical cell of the basidium and near the top of the basal cell. Thus the spores come into contact and unite to form an ellipsoid spore. Martin compares this with the union of two sporidia while still attached to the promycelium of some smuts. The hyphae are abundantly provided with clamp connections. (Fig. 144 C.)

Platygloea is waxy or gelatinous, resupinate, and effused, mostly growing on wood. Usually there is no distinction into hypobasidium and epibasidium. *Helicogloea* (*Saccoblastia*) has many of the same characteristics but is more gelatinous. It also grows on dead wood. A distinct, usually lateral and hanging pyriform hypobasidium is conspicuous. The basidial primordia are usually terminal but may be intercalary. (Fig. 144 D, E.)

Auricularia forms the largest and most conspicuous spore fruits of this family. They are borne on branches or trunks of trees and are especially abundant in the Tropics. The common species of the Temperate Zones is *A. auricularis* (S. F. Gray) Martin (*A. auricula-judae* or *Hirneola auricula-judae* of most authors). The spore fruits are gelatinous and more or less ear-shaped when moist, and of a translucent brown color, but dry down to small dark horny masses. The basidia form a hymenium on the lower surface, standing parallel in a sort of palisade immersed in the gelatinous matrix. The elongated basidial primordia become divided by three transverse septa and from each of the cells grows a tube-like extension to the surface of the matrix where a sterigma is formed bearing a single basidiospore. The tropical genus *Tjibodasia* is waxy, and more or less pezizoid in

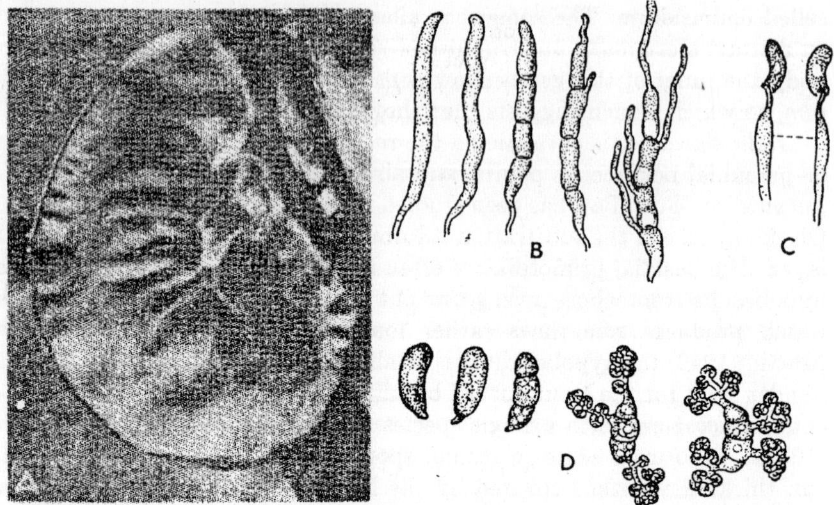

Fig. 145. Order Auriculariales, Family Auriculariaceae. *Auricularia auricularis* (S. F. Gray) Martin. (A) Expanded, moist spore fruit. (B) Stages in the development of the basidium. (C) Tubular extensions from basidium, each with its sterigma and basidiospore. (D) Stages in germination of basidiospore. (A, after Buller: Researches on Fungi, vol. 2, p. 162, London, Longmans, Green and Co. B–D, after Brefeld: Untersuchungen aus dem Gesammtgebiete der Mykologie, Heft 7, pp. 1–178.)

appearance. No sharp distinctions into hypobasidium and epibasidium are apparent. (Fig. 145.)

Family Phleogenaceae. In this family the fleshy or gelatinous spore fruit is a stalked structure of ascending hyphae. At the top the hyphae flare outward to form a head. The outer hyphae form a sort of loose peridium. Within this the straight or curved basidia arise. They are transversely one to three septate and the basidiospores are practically sessile. There is no distinction between the hypobasidium and epibasidium. Clamp connections are abundant. The only common genus in the temperate regions is *Phleogena* (*Pilacre* of some authors), with a single species *P. decorticata* (Schw.) Mart. (*P. faginea* (Fr.) Link). This grows on dead stumps, logs, etc., and forms colonies of stalked structures 5 to 7 mm. tall, including the head which is 1 to 3 mm. in diameter. When young they are fleshy but at maturity dry. The spores are yellow-brown. Superficially they resemble the Ascomycetous genus *Onygena*. Shear and Dodge (1925) described the life history and cytology of this species very fully. Other genera more or less probably belonging in this family are *Pilacrella*, fleshy, with a disk-shaped head and hyaline spores; *Hoehnelomyces*, tropical, slimy cartilaginous, or even waxy, with hyaline spores and with a round head with loose wavy hairs; and perhaps *Stilbum*, in which the stalk and head are fleshy but the surrounding peridial hyphae are lacking. The ba-

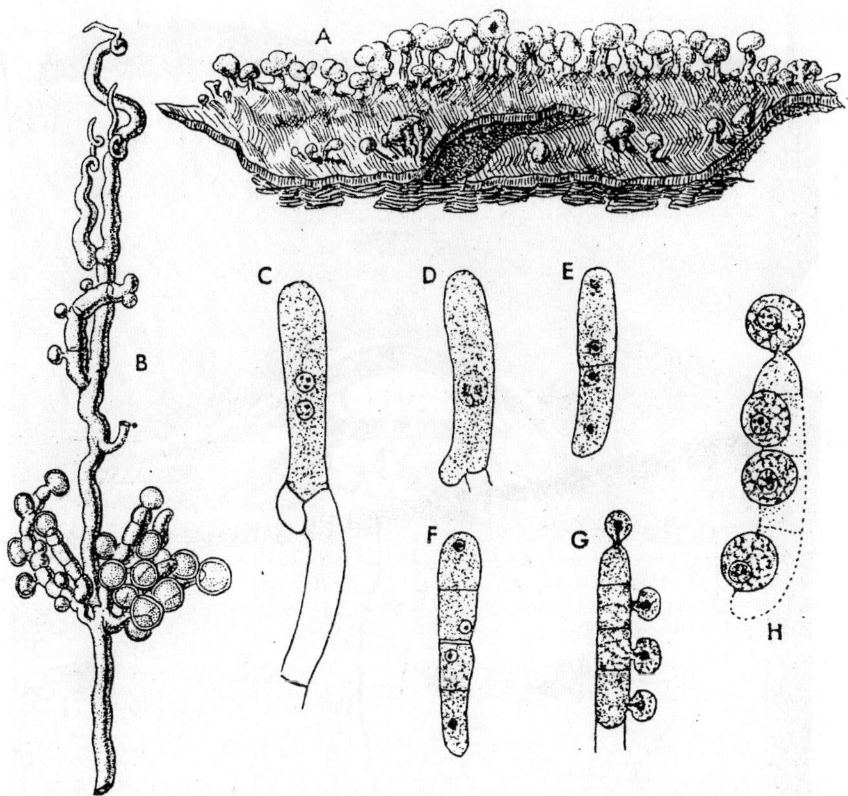

FIG. 146. Order Auriculariales, Family Phleogenaceae. *Phleogena decorticata* (Schw.) Mart. (A) Habit study of the fungus. (B) Basidiophoric hypha with apex extended to form part of the so-called peridium. (C–H) Stages in the development of the basidium and basidiospores. (A–B, after Brefeld: Untersuchungen aus dem Gesammtgebiete der Mykologie, Heft 7, pp. 1–178. C–H, courtesy, Shear and Dodge: *J. Agr. Research,* **30**(5):407–417.)

sidia are two-celled and the spores hyaline. There is one species in Europe and North America and three others in the tropics. There is uncertainty as to whether the genus really belongs here. (Fig. 146.)

FAMILY SEPTOBASIDIACEAE. This is considered by Couch (1937 and 1938) to be worthy of ordinal rank. The two genera included in the family are parasitic upon scale insects, with which they live in a sort of symbiotic relationship much as the lichen fungus does with the imprisoned algae. For the majority of these insects the fungus provides a home and shelter where they feed upon the woody host plant and produce their young. Some of the insects, however, are parasitized and continue to feed upon the host plant but give up their food to the fungus which penetrates their bodies with numerous coiled or knotted haustoria. These insects may out-

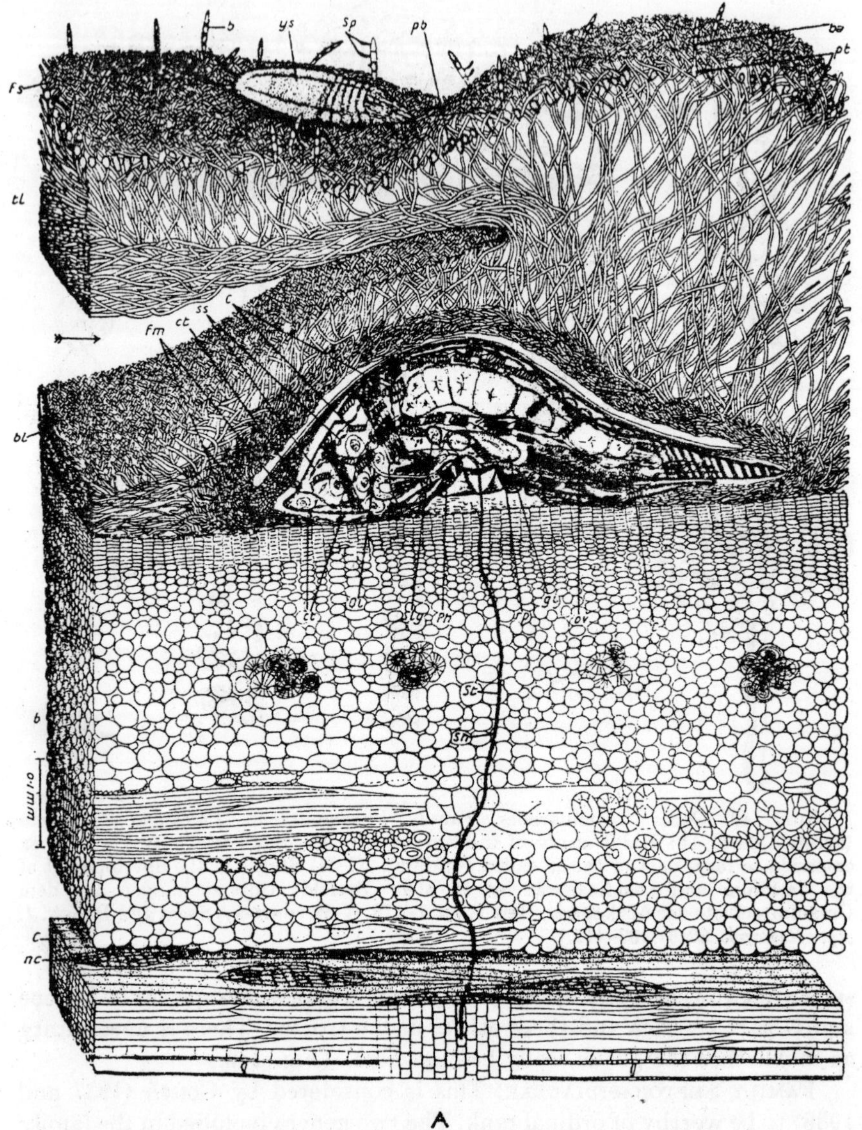

Fig. 147. Order Auriculariales, Family Septobasidiaceae. *Septobasidium burtii* Lloyd. (A) Vertical section through a portion of the fungus showing a parasitized living scale insect (*Aspidiotus*) and the hymenium on the upper surface. (*In upper portion of figure*) (*b*, *ba*) Epibasidium; (*pb*) hypobasidium; (*Sp*) spore; (*ys*) young scale insect; (*Fs*) fruiting surface of fungus; (*tl*) top layer of fungus; (*bl*) bottom layer of fungus; (*arrow*) tunnel in fungus. (*Referring to host plant*) (*b*) Bark; (*c*) cambium; (*nc*) medullary ray. (*Referring to parasitized insect and surrounding fungus*) (*Fm*) Enveloping fungus mat; (*ct*) fungus thread connecting fungus mat with insect; (*C*) coiled haustorium within insect; (*ss*) spindle-shaped threads within insect; (*St*) stylet; (*sh*) sheath secreted around stylet. (*Ol*, *Stg*, *Ph*, *rp*, *gl*, *ov*) Various organs of the insect.

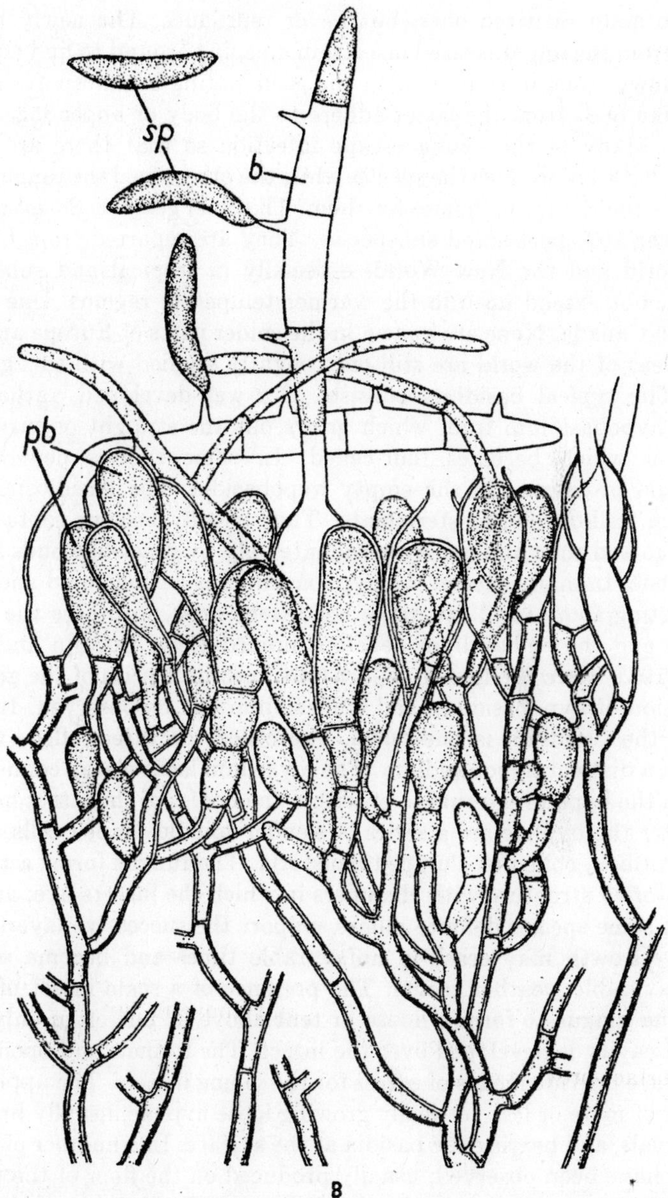

FIG. 147—(*Continued*). (B) Section of the hymenium showing hypobasidia and fully developed basidium with basidiospores. (Courtesy, Couch: The Genus Septobasidium, Chapel Hill, Univ. North Carolina Press.)

live the nonparasitized ones, but never reproduce. The newly hatched young from the unparasitized insects, in creeping around to find a place to settle down come in contact with the germinating basidiospores and the yeast-like buds from the latter adhere to the body or appendages of the insects. Many of the young escape infection so that there are always enough left to reproduce the species while the others feed the fungus which provides the sheltering homes for them. The chief genus is *Septobasidium*, with over 160 species and subspecies. They are reported from both the Old World and the New World, especially in tropical and subtropical regions, but extend up into the warmer temperate regions. One species grows in Canada. None are known in the colder parts of Europe and Asia. Vast areas of the world are still only slightly studied with this genus in view. The typical basidium consists of a well-developed, rather firm-walled hypobasidium from which grows out the straight or curved epibasidium which becomes four-celled. In some species new basidial primordia proliferate in the empty hypobasidia. The basidiospores are borne on well-developed sterigmata. They usually become septate after discharge and then apparently germinate only by yeast-like buds and not by hyphae. In many species the hypobasidium is dark-colored and serves as a resting spore until favorable conditions arise, in others the wall is hyaline and the epibasidium develops immediately. Boedijn and Steinmann (1931) were the first to report that in some species of the genus no distinction of hypobasidium and epibasidium can be observed. In many species the basidium is one-celled, two-celled, or three-celled, with or without a distinct hypobasidium. The mycelium lacks clamp connections, both on the hyphae external to or within the bodies of the insect hosts. In the latter the hyphae are a series of slender spindle-shaped cells with or without thick, coiled or clumped haustoria. The fungus forms a two-, or more, storied structure with chambers in which the insects live, and tunnels. In some species definite pillars support the successive layers of the fungus. Growth may cease in unfavorable times and become renewed when favorable weather recurs. The presence of a scale insect often induces the fungus to form a house or tent above it, but often only a low vaulted cavity is developed over the insect. The tunnels and openings to the surface provide means of egress for the young insects. The upper layer consists of more or less vertically growing loose hyphae, usually branched at intervals, and bearing the basidia at the surface. In a number of species conidia have been observed, usually produced on the floor of the tunnels or chambers or on short branches from the ascending hyphae of the upper layer. (Fig. 147.)

The *Septobasidium* colony may be but a few millimeters in diameter or up to 20 or 30 cm. or more. It may be a fraction of a millimeter in thickness or up to 1 cm., depending upon the species. It often resembles a lichen and the color varies from very light gray to dark brown, purple or even almost black. The infected trees or branches are injured, perhaps not by direct action of the fungus itself but by the protection it offers to the many scale insects it harbors.

The genus *Uredinella* is perhaps best placed in this family. Like *Septobasidium* it is parasitic upon scale insects and forms small circular dark-colored spots on the bark. The fungus is annual, not perennial as is *Septobasidium*. The top layer is a hymenium of ovoid to club-shaped, brown basidial primordia (called "teleutospores" by Couch, 1937). These are binucleate when young and nuclear fusion occurs within them. The mature cells have two or three thick layers of wall especially at the apex, and a distinct germ pore. When the fungus is wet with rain an epibasidium emerges through the germ pore and forms a straight four-celled structure on which four basidiospores are produced on distinct sterigmata. The epibasidium may break off from the hypobasidium and apparently can float around in a film of rain water. The insects are infected through the mouth region and the dicaryon mycelium within the host produces coiled haustoria much like those of *Septobasidium*. Besides the basidia somewhat similar "uredo-mother cells" are produced and from their apices grow out binucleate, elongated ellipsoid, slightly bent spores, called by Couch "uredospores." It is suggested by Couch that this genus may well represent a stage intermediate between the Uredinales and *Septobasidium*.

Order Dacrymycetales. As in the preceding order the spore fruits are mostly gelatinous or waxy, drying down to a thin sheet or horny mass. They vary from thin, broadly effused sheets to cushion-like, cupulate or pileate structures, or cornute, coralloid or spatulate upright forms. No species are known to be parasitic. They are almost exclusively confined to dead wood, with or without bark. With very few exceptions the spore fruits are colored some shade of yellow or orange to deep brown and the basidiospores are mostly yellowish in mass. Conidia are frequently produced. Whether these correspond to the "oidia" of some of the other orders is uncertain, for cultural studies and attempts to match and diploidize different strains are as yet much to be desired. The basidiospores are one-celled when ready to be discharged but in most species septa are formed immediately after discharge, dividing the spore into 2, 4, or even up to 12 cells. In a specimen of *Dacrymyces* studied by the author the basidiospores became once septate before they were discharged from the sterigmata. They germinate by the formation of hyphae or of yeast-like buds from the various cells of the spore.

The basidia are formed in a close hymenium or intermingled with sterile hyphae, on the outer surface of resupinate forms or on definite surfaces in various other forms. They start as binucleate terminal cells soon thicker than the rest of the hypha. They are at first long cylindrical or somewhat clavate. The two nuclei unite and then divide once stichobasidially, and then a second time in the same direction. The broadened apex of the basidium becomes lobed to produce two "epibasidia" of the same diameter and often of almost the same length as the "hypobasidium." The upper two nuclei migrate into the two epibasidia and then through the sterigma at the tip of each into the basidiospores. The two nuclei remaining in the hypobasidium degenerate. The mature basidium often resembles a tuning fork. Clamp connections are frequent in the spore fruits but in the primary mycelium produced by the germination of the spores they are lacking. Just how and when diploidization occurs has not been demonstrated. Eight to ten genera are recognized, in temperate as well as in tropical regions.

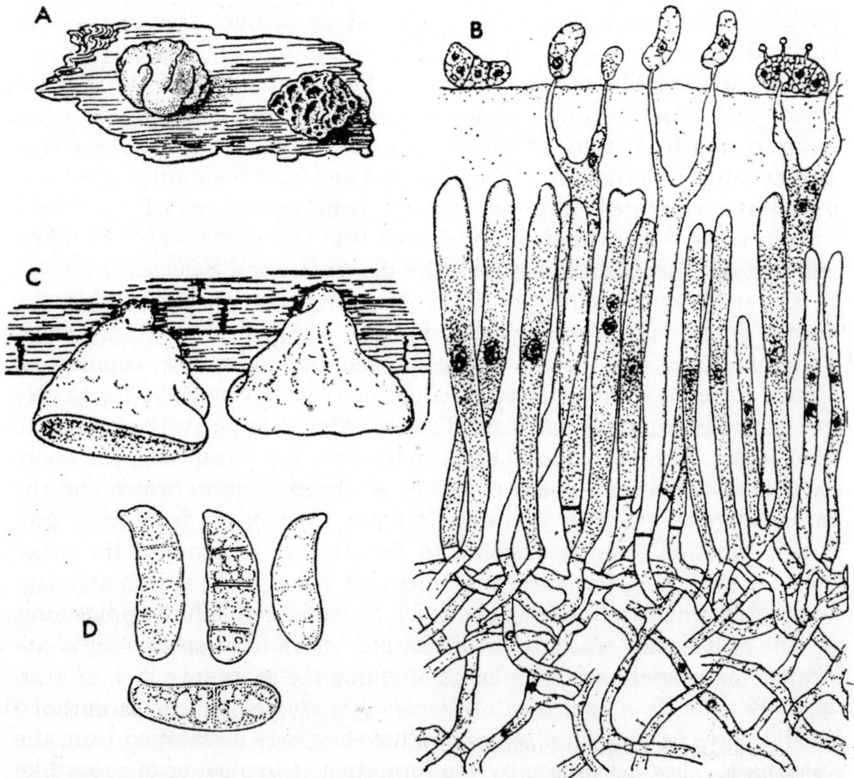

FIG. 148. Order Dacrymycetales, Family Dacrymycetaceae. (A) *Dacrymyces lutescens* Bref., habit sketch. (B) *Dacrymyces deliquescens* Duby, various stages in the development of basidia and basidiospores. (C) *Guepiniopsis* sp., habit sketch. (D) Basidiospores of *Guepiniopsis*. (A, after Brefeld: Untersuchungen aus dem Gesammtgebiete der Mykologie, Heft 7, pp. 1–178. B, after Dangeard: *Le Botaniste*, **4**:119–181. C–D, courtesy, Martin: *Mycologia*, **24**(2):215–220.)

Among the effused forms is *Cerinomyces* Martin, 1949 (*Ceracea* of authors), forming a thin waxy to fleshy layer, without definite mycelial roots. The basidiospores may remain nonseptate or in some species may become transversely septate. The genus *Arrhytidia* forms tough waxy, at first discoid, then broadly effused spore fruits with a centrally rooting base. The spore fruits of *Dacrymyces* are sessile, attached by a point or rhizoids, or substipitate. They are gelatinous to waxy, pulvinate, discoid, or cerebriform, or even cupulate. The entire exposed area is covered by the hymenium which may become wrinkled or folded when older. Martin (1944) recognizes seven species in the United States and Canada. *Guepiniopsis* (*Heterotextus* of earlier works) produces substipitate, cupulate spore fruits, with the cup often turned downward so that the hymenium is inferior. The cortex consists of swollen, thick-walled cells. The interior hyphae are strongly gelatinized. *Femsjonia* is also discoid or cupulate, sometimes substipitate, but is white-villous or tomentose externally. *Dacryopinax* (Martin, 1948, a name substituted for the more familiar but preoccupied name *Guepinia*), also may be cupulate and erect when young, becoming spatulate, fan-shaped or petaloid, with hymenium on the lower side. The spore fruits are definitely stipitate and tough or cartilaginous. *Dacryomitra* also is stipitate, with a distinct pileus more or less morchelloid in appearance and gelatinous. The hymenium covers all sides of the head. *Calocera* is cornute to coralloid, with hymenium on all sides. It resembles *Clavaria* but differs in the basidial structure—gelatinous to tough. (Fig. 148.)

Order Tremellales. Spore fruits varying from adhering, waxy or gelatinous sheets to foliose or cushion-like or pileate structures, sometimes upright and branched, forming more or less leaf-like lobes, or funnel-shaped. In the genus *Hyaloria* the spore fruit is soft with a rounded head. The consistency of the fruit body in this order may be very soft-gelatinous to almost leathery or waxy and may become horny when dry. The colors range from white to yellow, brown or almost black.

The basidial primordium is terminal and rounded but in *Sirobasidium* the basidia are produced basipetally from the apex in the same hypha. The young basidium is binucleate. After the fusion of the nuclei in the typical cases the diploid nucleus undergoes its two meiotic divisions at right angles to the axis of the cell (chiastobasidially) and then a vertical septum is formed, followed almost immediately by another vertical septum at right angles to the first in each of the two cells. At the top of each of the four cells so formed a sterigma may be produced bearing one basidiospore. More often, since the majority of the species are gelatinous with the basidia embedded a short distance below the surface, a tube-like extension of the basidial cell grows upward to the surface, there producing the sterigma and basidiospore. These tubular outgrowths are homologous

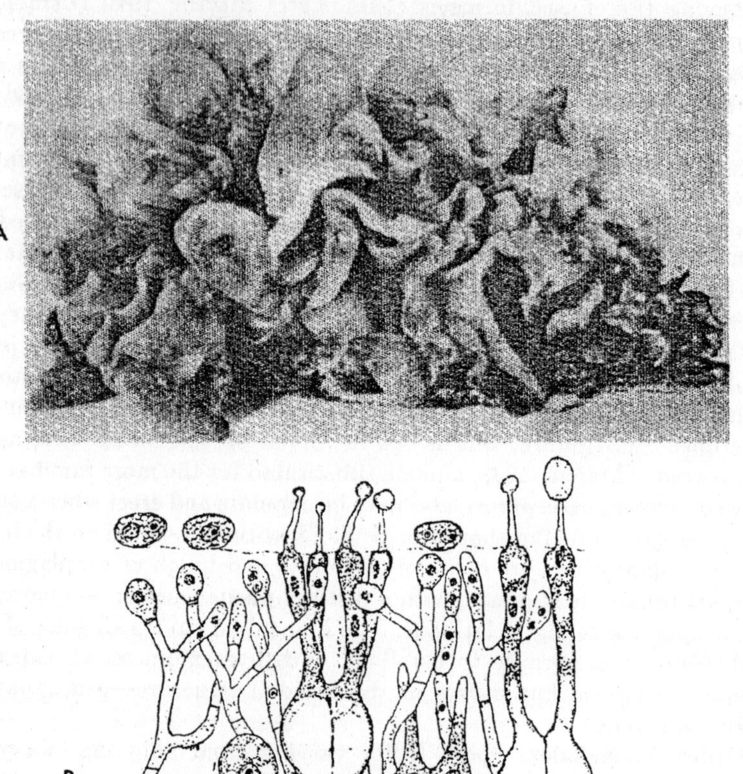

FIG. 149. Order Tremellales, Family Tremellaceae; (A) *Tremella foliacea* (S. F. Gray) Pers. (B) *Tremella mesenterica* (S. F. Gray) Pers. Stages in the development of basidia and basidiospores. (A, courtesy, Atkinson: Studies of American Fungi, Ithaca, N. Y., Andrus and Church. B, after Dangeard: *Le Botaniste*, **4**:119–181.)

to the similar structures growing out of the cells of the basidia of *Auricularia* and cannot rightly be called epibasidia, for the epibasidia of Auriculariales and the promycelium of Uredinales are probably homologous, but are entirely different from these tubular extensions to bring the spore production to the surface. The basidiospores are hyaline or only slightly colored. They are without septa when set free. They may then become once septate, but more often not. Frequently they produce a sterigma and a secondary spore, a process several times repeated. In abundance of moisture the spores bud out innumerable "oidia" on their surface. They may, if on a proper substratum, germinate by hyphae. The primary mycelium produces an abundance of oidia and lacks clamp connections. Eventually clamp connections appear (secondary mycelium) and oidial production usually ceases. Barnett (1937) showed for several species of *Exidia* that the spores from a given spore fruit exist in two sexual phases. Multiple allelomorphy of the compatibility factors was demonstrated also, as occurs in *Auricularia* and in various Ustilaginales.

Over 100 species in 17 or more genera are recognized in this order, of which about half are found in the North Central United States and Canada (Martin, 1944).

Three families are usually distinguished:

Family Tremellaceae. Basidia are single and terminal on the supporting hyphae, with a more or less elongated extension from each cell, terminated by a sterigma. They are gelatinous, waxy or somewhat dry. Basidia normally are cruciately four-celled but sometimes three-celled or even two-celled. The primary septum is vertical or oblique. Hymenial surface is exposed. Spore fruits may be flat, cushion-shaped, lobed, or pileate. Clamp connections present in secondary mycelium of many species. (Fig. 149.)

Family Sirobasidiaceae. Gelatinous, cushion-shaped, hymenium external. Basidia are formed in chains by successive transformation of the cells of the basidiogenous hyphae into basidia, beginning at the apex. Basidia are four- or two-celled, the septum in the latter case oblique. Basidiospores are sessile and therefore probably not discharged violently. Mostly tropical but known in the United States from North Carolina. Contains a single genus, *Sirobasidium*. (Fig. 150.)

Family Hyaloriaceae. Spore fruits are stalked, with a head; stalk and head somewhat gelatinous externally or sessile, forming a filmy layer on the substratum. Basidia are two- to four-celled with tubular extensions which may taper to a long fine thread, the basidiospore being borne symmetrically at the apex and breaking off with part of the supporting thread attached, not discharged from a sterigma. Clamp connections are present. There are two genera: *Hyaloria*, with one tropical species, *H. pilacre* A. Möll. (see Martin, 1937b), and one European one, *H. europaea* Killer-

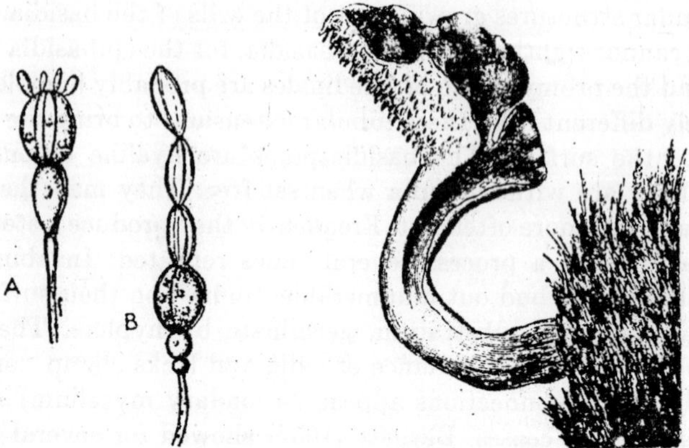

FIG. 150 (*Left*). Order Tremellales, Family Sirobasidiaceae. *Sirobasidium albidum* Lagerh. & Pat. (A) Chain of two basidia, the terminal one almost mature. (B) Chain of several basidia, the upper three collapsed, the fourth approaching maturity. (After Lagerheim and Patouillard: *J. Botan.*, 6(24):465–469.)

FIG. 151 (*Right*). Order Tremellales, Family Tremellaceae. *Protohydnum gelatinosum* (Fr.) Karst. (*Tremellodon* of most authors). Habit sketch. (After A. Möller, from Killermann, in Engler und Prantl: Die Natürlichen Pflanzenfamilien, Zweite Auflage, vol. 6, pp. 99–290, Leipzig, W. Engelmann.)

mann (1936) and *Xenolachne,* growing as a parasite on a minute Discomycete in Oregon (Rogers, 1947). *Hyaloria* is stalked and externally gelatinous with the basidia on the head, much overtopped by long hairs (or cystidia). *Xenolachne* forms a thin film on the apothecium of the host and lacks a gelatinous coat or cystidia but the extremely long extensions of the two-celled basidia, with the basidiospores at the apex give a felty appearance to the fungus. The type of basidium in these two genera resembles that characteristic of most of the Gasteromycetes and for this structure Rogers adopts the name apobasidium proposed by Gilbert (1928).

In the Tremellaceae the genus *Stypella* produces a small felty mass of tangled, more or less gelatinous hyphae in whose upper layer the basidia arise. These spore fruits are clustered, separate or anastomosing on a dry floccose subiculum, the whole patch sometimes reaching a diameter of several centimeters. In *Sebacina* there is a more or less waxy or gelatinous crust with the hyphae of the upper portion directed perpendicularly to the surface, the terminal cells of these rather closely packed hyphae being the basidia. Cushion-like or lobed, gelatinous spore fruits are characteristic of *Exidia* and *Tremella.* Gland-like dots occur in the spore fruit of the former and are absent in the latter. The basidiospores of the former are mostly allantoid, those of the latter straight and ellipsoidal to nearly spherical. When these spores germinate the sprout-conidia of *Exidia* are mostly curved, those of *Tremella* yeast-like.

Phlogiotis (*Gyrocephalus*) has funnel-shaped spore fruits with the hymenium on the outer surface only. *Tremella* is found in all parts of the world. *T. reticulata* (Berk.) Farl. forms large white masses of gelatinous leaf-like lobes, the whole mass sometimes being 10 to 12 cm. in diameter. It is considered edible. Other species are usually smaller and some are bright-colored. Other genera less common in the temperate zones or confined to the tropical or subtropical regions are among those described below. *Patouillardina* has the basidia spindle-shaped. The first septum is oblique and in each cell thus formed another septum is produced at right angles to the first one. Because of the shape of the basidium these two septa do not intersect the first septum opposite one another. In *Protomerulius* the soft fleshy or waxy spore fruit is resupinate and rather thin. Its hymenium is poroid. *Protodontia* and *Protohydnum* are waxy or gelatinous and resupinate or stalked, but the hymenium instead of being poroid is borne on downward directed teeth. The earlier name *Protohydnum* must, according to Martin (1948), be used instead of the more familiar but later *Tremellodon*. In *Protohydnum* the stalk of the basidium is separated from it by a septum but not so in *Protodontia*. *Heterochaete* is somewhat similar, but the blunt teeth are peg-like and pierce the hymenium, not being covered with basidia themselves. *Tremellodendron* is erect, more often branched, resembling *Clavaria* or some species of *Tremella*. *Eichleriella* (*Hirneolina*) is cupulate or broadly attached. (Fig. 151.)

The Tremellales must be considered as a group which has developed with more or less parallelism to the Auriculariales. The low, felty or gelatinous waxy crust, bearing basidia on the upper surface, seems to be the most primitive in each order, and from this simple structure have developed the more complex forms of spore fruit. It must be emphasized again that until the life histories have been more fully worked out the true relationships are only a matter of conjecture. The rather frequent occurrence of conidial production in these orders would hint at relationship to the Ascomyceteae in which conidia are produced abundantly.

Order Tulasnellales. The fungi making up this order are mostly found on dead wood or on old fungi on which they are saprophytic. The spore fruits are resupinate, gelatinous or dry, usually thin, sometimes being only a slight film-like coating on the substratum. Clamp connections are found on the hyphae of most species but are lacking in some. The characteristic feature of the single family, Tulasnellaceae, is the structure and development of the basidia. These are typical in the genera *Tulasnella* and *Gloeotulasnella*. The genus *Ceratobasidium* (Rogers, 1935) was tentatively placed by Martin (1944) in this family but later (1948) segregated by him in a distinct family Ceratobasidiaceae, assigned to a position close to the Tulasnellaceae. The basidium in all three genera is a holobasidium, i.e., is not divided by vertical septa as in the Tremellales nor by transverse septa as in the Auriculariales. The basidial primordium is subglobose, pyriform or broadly clavate. From the upper portion of this cell arise usually four stout cells which are narrowed at the tip to form a sterigma upon which a single basidiospore is formed. These four cells are considered by Rogers, Martin, and others to be homologous to the stout arms growing out of the four cells of the basidium of the Tremellaceae, and which they call "epibasidia."

In *Tulasnella* and *Gloeotulasnella* the "epibasidia" are separated from the "hypobasidium" by a septum at the base of each. They may remain attached or may fall off. At the apical end they elongate and bear a sterigma on which is borne the nonseptate basidiospore. This spore when set free germinates "by repetition," i.e., it produces a lateral or terminal sterigma and a secondary spore into which the whole cytoplasm and nucleus pass, this spore then being discharged as in the case of the parent spore. These two genera are distinguished as follows: *Tulasnella*, "arid-pruinose to waxy, basidia not imbedded in a gelatinous matrix; probasidia globose to obovate, sessile or with a short scarcely differentiated stalk; epibasidia with subulate tips merging into the sterigmata; gloeocystidia never present" (Martin, 1944). A dozen or more species. *Gloeotulasnella*, "waxy gelatinous to mucous gelatinous, basidia imbedded in a gelatinous matrix; probasidia clavate capitate, with a more or less cylindrical stalk and a swollen head; epibasidia extended into cylindrical tubular filaments sharply constricted at the base of the sterigmata; gloeocystidia present or absent" (Martin). Ten or more species. The

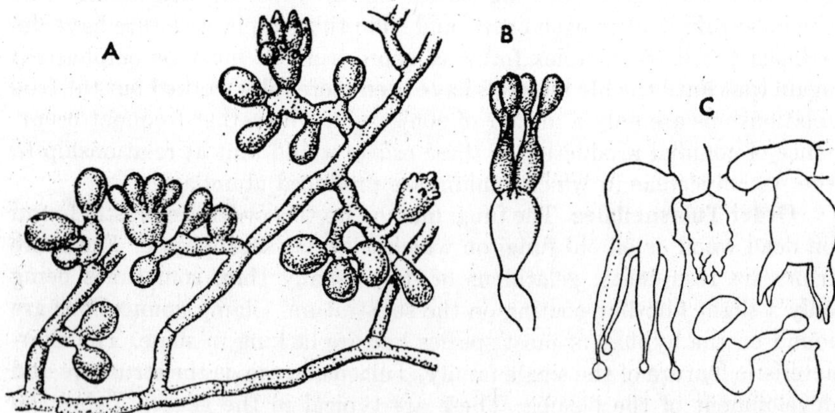

FIG. 152. Order Tulasnellales, Family Tulasnellaceae. (A, B) *Tulasnella violea* (Quel.) Bourd. & Galz. (*Pachysterigma fugax* Johan-Olsen). (A) Mycelium with several clusters of basidia. (B) Mature basidium. (C) *Ceratobasidium cornigerum* (Bourd.) Rogers. Basidia and basidiospores in various stages of development. (A–B, after Brefeld: Untersuchungen aus dem Gesammtgebiete der Mykologie, Heft 8, pp. 1–305. C, courtesy, Martin: *Univ. Iowa Studies in Natural History*, **18**(3):1–88.)

cytology and taxonomy of these genera were studied by Rogers (1932, 1933). (Fig. 152 A, B.)

From the two foregoing genera *Ceratobasidium* is distinguished by the stout "epibasidia," elongate cornute or flexuous, continuous with the "hypobasidium" (rarely a cross septum). The spore fruit is arid or waxy. Six or more species. By the usual absence of septa cutting off the "epibasidia" and their cornute shape they approach on the one hand *Dacrymyces* in *C. sterigmaticum* (Bourd.) Rogers, in which only two such horns are produced, while the remaining species with their four "epibasidia" approach *Pellicularia* in the Thelephoraceae (Eubasidiae). From the latter they differ by the germination by repetition, of the basidiospores. (Fig. 152 C.)

Summary of Subclasses Teliosporeae and Heterobasidiae

A comparison of these subclasses demonstrates more or less basic similarities throughout, in basidium production. Between the Auriculariales, Uredinales, and Ustilaginales there are such basidial similarities that they are often placed together in one group. On the other hand the tendency for the septation of the basidium in the Tremellales to be oblique instead of vertical is taken by some mycologists who have studied these groups intensively to indicate gradation from one to the other. Rogers (1934) suggested that the septa at the base of the "epibasidia" of *Tulasnella* may be accounted for by displacement upward of the vertical septa of the basidium of *Tremella*. The peculiar tuning-fork type of basidium of *Dacrymyces* could be considered a derivation from the *Tulasnella* type by the loss of their cross septa entirely and the reduction of the "epibasidia" to two (as actually does occur in *Ceratobasidium sterigmaticum*).

The question of the phylogeny of these groups is treated more fully in Chapter 17, but the following suggestions may well be made here:

Studies by Juel (1898), Neuhoff (1924), Martin (1931), and Rogers (1934) have led the latter to an interpretation of the relationships in the class somewhat different from that of the author. He holds in common with some of the others mentioned, that the primitive basidium consisted of two parts, the basal hypobasidium, binucleate at first, within which the nuclear union occurs, and one or more outgrowths, the epibasidia. Into the latter the nuclei pass from the hypobasidium. Meiotic division may occur either in the latter or in the epibasidium, if there be but one. Each epibasidium produces a true sterigma which bears a basidiospore. The genus *Tulasnella* is considered by Rogers to represent a primitive form. In the ovoid or pyriform hypobasidium of this genus the fusion nucleus divides into usually four nuclei. One of these passes out into each of the four (sometimes fewer) stout epibasidia which usually become separated from the now almost empty hypobasidium by a basal septum. Each epibasidium produces a terminal sterigma and basidiospore. The nucleus may divide within the epibasidium and both nuclei pass into the basidiospore. By producing but two epibasidia, not separated by septa from the hypobasidium the typical tuning-fork basidium of the Dacrymycetales can be evolved. By crowding the basal septa down into the hypobasidium so as to divide that longitudinally into four cells the basidium of the Tremellales may be derived. By reduction of the size of the epibasidia until only the sterigmata are left, is developed the basidium characteristic of the Hymenomycetes. Rogers attempts to explain the derivation of the hypobasidium and single, transversely septate epibasidium of *Septobasidium* and of the corresponding teliospore and promycelium of the Uredinales and Ustilaginales as being due to the delay in the meiotic division of the fusion nucleus until it passes out into an epibasidium which naturally would be single for a single nucleus. He believes that the hypobasidium of *Tulasnella* represents an ascus, perhaps of some form resembling *Ascocorticium* (Order Taphrinales), in which the ascospores have pushed out into pockets, germinating there to form secondary spores. These pockets have become the epibasidia and the secondary spores have become the basidiospores. Martin (1938) discusses the morphology of the basidium in connection with Heterobasidiae and Eubasidiae.

Linder (1940) would, on the other hand, derive the Uredinales from the Ascomyceteae, in the vicinity of the Sphaeriales or Dothideales. From the Uredinales he would derive the Auriculariales, Tremellales and Dacrymycetales. The Corticiae (Family Thelephoraceae), and thence the other Eubasidial families he would derive from the Tremellales.

On the other hand some students of these fungi consider the primitive basidium to have been of the holobasidium type from which the forked and septate types have been derived. The distinction of hypobasidium and epibasidium are, in this viewpoint, specializations to meet the need of holding the basidium over until a more favorable period. The thick-walled hypobasidium (or "probasidium") such as is found in the Teliosporeae or in *Septobasidium* cannot, because of the thickened wall, develop in the manner normal to basidia, and so a thin-walled new structure, the epibasidium or promycelium, was developed. The thickened apical branches of the basidia of the Tremellales and Dacrymycetales and of *Tulasnella* are not considered to be epibasidia but merely modified sterigmata.

Key to the Families and More Important Genera of Order Auriculariales

(*Modified from Martin, 1944*)

Parasitic on plants or saprophytic on dead plant material, usually wood. Basidia with distinct sterigmata. Division into hypobasidium and epibasidium present or absent. Basidia forming a more or less distinct exposed, loose or compact hymenium. Family Auriculariaceae

Parasitic on the gametophytes of mosses at whose apex a gelatinous, club-shaped spore fruit is produced. Hypobasidium short, not enlarged, very soon collapsing, and not conspicuous. No clamp connections. *Eocronartium*

Parasitic on the sporophytes of mosses forming a felty or gelatinous, more or less spherical spore fruit. Hypobasidium and epibasidium distinct. No clamp connections. Tropical. *Iola*

Parasitic on the leaves of ferns; effused; haustoria coiled; no distinct difference of hypobasidium and epibasidium. Conidial stage sometimes present. No clamp connections observed. *Herpobasidium*

Parasitic on the leaves of ferns. Definite coiled haustoria formed. No clamp connections present. Definite ovoid hypobasidia give rise to distinct, usually four-celled epibasidia. *Platycarpa*

Parasitic on the roots or crowns of vascular plants or saprophytic on dead plant tissues. Dry, floccose, effused. Hypobasidia cylindrical or not obvious, epibasidia curved at the top. *Helicobasidium*

Saprophytic on plant tissues or parasitic in spore fruits of fungi.

Resupinate; gelatinous, forming a large cushion; basidia two-celled, the basal cell clavate; the terminal one spherical; the two spores uniting to form an ellipsoid spore. Numerous clamp connections. Tropical. *Syzygospora*

Resupinate; soft gelatinous, hypobasidia in the form of lateral, reflexed sacks. Clamp connections sometimes present. *Helicogloea* (*Saccoblastia*)

Resupinate; firmly gelatinous or waxy, parasitic, with diffusely scattered basidia: hypobasidia lacking in some species, clamp connections sometimes present. *Platygloea*

Pileate or ear-shaped, tough gelatinous; clamp connections obvious; no distinction into hypobasidium and epibasidium. *Auricularia*

Saprophytic on wood, bark, etc. No distinction into hypobasidium and epibasidium. Spore fruit stalked with a head of radiating, more or less coiled hyphae among which the curved basidia are found, bearing their two to four spores without visible sterigmata. Clamp connections present in some species. Family Phleogenaceae[1]

Stalked with the hyphae flaring at the top to form a head, the outer ends forming a loose peridium-like structure. Basidia not borne on sterigmata. Forming colonies of stalked spore fruits on dead wood, fleshy when young, then becoming dry. *Phleogena* (*Pilacre*)

Parasitic upon scale insects, with some of which they live in symbiotic relation. Basidia usually with well-developed thin-walled or thick-walled hypobasidia and one- to four-celled (mostly the latter) epibasidia; but hypobasidium sometimes lacking. Basidiospores produced on distinct sterigmata. Conidia often produced. Clamp connections not observed. Family Septobasidiaceae

---

[1] Other genera sometimes assigned to this family are *Pilacrella*, fleshy, with disk-shaped head; *Hoehnelomyces*, slimy-cartilaginous or waxy, with round head with loose wavy hairs; and perhaps *Stilbum* with fleshy stalk and head but no surrounding peridial hyphae.

Fungus perennial, forming a two- or three-storied structure with chambers and tunnels within which the scale insects live. Basidia formed on the outside, more or less felty, layer. Often lichen-like. *Septobasidium*

Fungus annual, not forming distinct "houses" for the parasitized scale insects. Hypobasidia thick-walled with an apical germ-pore through which a hypha grows to form a straight, four-celled epibasidium with definite sterigmata. In addition conidia are produced. *Uredinella*

Key to the More Important Genera of Family Dacrymycetaceae

(*Based on Martin, 1944*)

Fructifications broadly effused.

Broadly effused from the first, without root-like bases; arid to waxy-gelatinous. *Cerinomyces* (*Ceracea* of authors)

At first discoid or pustulate, soon becoming effused, attached to radicating bases, tough waxy or waxy-gelatinous. *Arrhytidia*

Fructifications remaining distinct even when anastomosis occurs.

Sessile and attached by a point or on a constricted root-like base.

Pulvinate or discoid or rarely pezizoid, often cerebriform; hymenium opposite substratum, usually inferior. *Dacrymyces*

Definitely pezizoid; hymenium concave, at least until very late.

Cortex concolorous; spores finally three to seven septate. *Guepinionsis*

Cortex conspicuously white-tomentose; spores tardily multiseptate. *Femsionia*

Distinctly stipitate and pileate.

Cornute to coralloid, *Clavaria*-like; hymenium amphigenous. *Calocera*

Pileate, pileus much broader than stalk.

Tough or cartilaginous, spatulate or cupulate; hymenium unilateral, inferior. *Dacryopinax* (*Guepinia* of authors)

Gelatinous, pileus conical, subglobose, flattened or morchelloid; hymenium amphigenous. *Dacryomitra*

Key to the Commoner Genera of Family Tremellaceae

(*Based on Martin, 1944*)

Fructification of thickly clustered, more or less anastomosing papillae, borne on a thin floccose subiculum. *Stypella*

Fructification continuous, at least from an early stage, frequently enlarged by anastomosis.

Resupinate, broadly effused, with indeterminate margins, probasidia globose to ovate to pyriform, the first septum mainly longitudinal.

Hymenium smooth or nearly so; arid or tough to waxy or gelatinous. *Sebacina*

Hymenium with spines or spine-like structures.

Spines sterile, piercing the hymenium; texture coriaceous to waxy or tough gelatinous. *Heterochaete*

Spines fertile, texture soft to tough gelatinous.
Soft gelatinous; subiculum delicate; probasidia without stalk becoming separated as a stalk cell. *Protodontia*
Tough gelatinous; subiculum thick; probasidia with stalk becoming separated as a stalk cell. *Protohydnum*
Erumpent or pileate, or, if appearing effused, with determinate margins.
Tough or coriaceous to somewhat waxy when moist.
Cupulate to broadly attached with a free margin; aspect of *Stereum*. *Eichleriella*
Erect, branched or rarely simple; aspect of *Clavaria* or *Thelephora*. *Tremellodendron*
Gelatinous; horny when dry.
Erect-cerebriform to lobate.
Spores subglobose or ovate. *Tremella*
Spores allantoid; gloeocystidia lacking. *Exidia*
Spores allantoid; gloeocystidia present. *Seismosarca*
Pileate and stipitate or substipitate.
Stipitate or dimidiate; hymenium on teeth. *Protohydnum* (*Tremellodon* of authors)
Infundibuliform; hymenium inferior, smooth or somewhat wrinkled. *Phlogiotis* (*Gyrocephalus*)

Key to the Genera of Family Tulasnellaceae

(*Based on Martin, 1944*)

"Epibasidia" at first bluntly cylindrical, at length fusiform, not separated by septa from the "hypobasidium." *Ceratobasidium*
"Epibasidia" at first globose, becoming ovate, pyriform or ventricose-cylindrical, separated by septa from the "hypobasidium."
Arid-pruinose to waxy; basidia short-stalked, not embedded in mucus; gloeocystidia never present. *Tulasnella*
More or less gelatinous; basidia long-stalked, embedded in mucus; gloeocystidia present or absent. *Gloeotulasnella*

Literature Cited

Atkinson, George F.: Studies of American Fungi. Mushrooms edible, poisonous, etc., vi + 275 pp. 6 *colored plates*. 223 *figs*. Ithaca, N. Y., Andrus and Church, 1900.

Baker, Gladys E.: A study of the genus Helicogloea, *Ann. Missouri Botan. Garden*, **23**(1):69–128. *Pls*. 7–14. 1936.

Barnett, Horace L.: Studies in the sexuality of the Heterobasidiae, *Mycologia*, **29**(5):626–649. *Figs*. 1–3. 1937.

Boedijn, K. B., et A. Steinmann: Les espèces des genres Helicobasidium et Septobasidium des Indes Néerlandaises, *Bull. Jardin Botan. Buitenzorg*, sér. III, **11**(2):165–219. *Pls*. 14–18. *Figs*. 1–31. 1931.

Brefeld, Oscar: Basidiomyceten II. Protobasidiomyceten, in Untersuchungen aus dem Gesammtgebiete der Mykologie, Heft. 7, pp. i–ix, 1–178. *Pls*. 1–11. Leipzig, Arthur Felix, 1888.

———: Basidiomyceten III. Autobasidiomyceten und die Begründung des natürlichen Systems der Pilze, *ibid*., Heft. 8, pp. i–iv, 1–305. *Pls*. 1–12. Leipzig, Arthur Felix, 1889.

BUDDIN, W., AND E. M. WAKEFIELD: Studies in Rhizoctonia crocorum (Pers.) DC. and Helicobasidium purpureum (Tul.) Pat., *Brit. Mycolog. Soc. Trans.*, **12**:116–140. *Pls.* 11–14. 1927.

COUCH, JOHN N.: A new fungus intermediate between the Rusts and Septobasidium, *Mycologia,* **29**(6):665–673. *Figs.* 1–30. 1937.

———: The genus Septobasidium, ix + 480 pp. *Frontispiece and* 114 *plates.* 60 *text figs.* Chapel Hill, N. C., Univ. North Carolina Press. 1938.

———: The taxonomy of Septobasidium polypodii and S. album, *Mycologia,* **41**(4):427–441. *Figs.* 1–25. 1949.

DANGEARD, P. A.: Mémoire sur la reproduction sexuelle des Basidiomycètes, *Le Botaniste,* **4**:119–181. *Figs.* 1–24. 1895.

FITZPATRICK, HARRY M.: The life history and parasitism of Eocronartium muscicola, *Phytopathology,* **8**(5):197–218. *Pl.* 1. 4 *text figs.* 1918a.

———: The cytology of Eocronartium muscicola, *Am. J. Botany,* **5**(8):397–419. *Pls.* 30–32. 1918b.

GÄUMANN, ERNST: Über die Entwicklungsgeschichte von Iola javensis, *Ann. Mycolog.,* **20**(5–6):272–289. *Pl.* 3. *Figs.* 1–36. 1922.

———: Comparative Morphology of Fungi. Translated by Carroll William Dodge, xiv + 701 pp. 406 *figs.* 43 *diagrams.* New York, McGraw-Hill Book Co., 1928.

GILBERT, E.: Bribes mycologiques: VI. Conjectures sur la classification et la filiation des espèces, *Bull. soc. mycologique de France,* **44**(3):225–227. 1928.

GILBERT, E. N.: Cytological studies of the lower Basidiomycetes, *Trans. Wisconsin Acad. Sci.,* **20**:387–397. *Pl.* 29. *Fig.* 1. 1921.

JACKSON, H. S.: The nuclear cycle in Herpobasidium filicinum with a discussion of the significance of homothallism in Basidiomycetes, *Mycologia,* **27**(6):553–572. *Portrait of author. Figs.* 1–4. 1935.

JUEL, H. O.: Die Kerntheilungen in den Basidien und die Phylogenie der Basidiomyceten, *Jahrb. wiss. Botan.,* **32**:361–388. *Pl.* 4. 1898.

KILLERMANN, S.: Unterklasse Eubasidii: Reihe Hymenomyceteae (Unterreihen Tremellineae und Hymenomycetineae), in A. ENGLER und K. PRANTL: Die Natürlichen Pflanzenfamilien, Zweite Auflage, vol. 6, pp. 99–290. 5 *pls. Figs.* 81–157. 1928.

———: Eine europäische Hyaloria Art, *Ber. deut. botan. Ges.,* **54**(2):165–167. *Pl.* 25. 1936.

DE LAGERHEIM, G., ET N. PATOUILLARD: Sirobasidium, nouveau genre d'Hyménomycètes hétérobasidiés, *J. Botan.,* **6**(24):465–469. *Figs.* 1–2. 1892.

LEACH, J. G.: Insect transmission of plant diseases, xviii + 615 pp. *Frontispiece. Figs.* 1–238, New York, McGraw-Hill Book Co., 1940.

LINDER, DAVID H.: Evolution of the Basidiomycetes and its relation to the terminology of the basidium, *Mycologia,* **32**(4):419–447. *Figs.* 1–6. 1940.

MARTIN, G. W.: Notes on Iowa Fungi, 1929–30. The genus Tulasnella in Iowa, *Univ. Iowa Studies in Natural History,* **13**(5):4–10. 1931.

———: On certain species of Heterotextus, *Mycologia,* **24**(2):215–220. *Pl.* 5. 1932.

———: The application of the generic name Guepinia, *Am. J. Botany,* **23**(9):627–629. 1936.

———: A new type of heterobasidiomycete, *J. Wash. Acad. Sci.,* **27**(3):112–114. *Fig.* 1. 1937a.

———: New or noteworthy fungi from Panama and Colombia, I, *Mycologia,* **29**(5):618–625. *Figs.* 1–29. 1937b.

———: The morphology of the basidium, *Am. J. Botany,* **25**(9):682–685. 1938.

———: Some Heterobasidiomycetes from Eastern Canada, *Mycologia*, **32**(6):683–695. *Figs.* 1–9. 1940.

———: The Tremellales of the North Central United States and adjacent Canada, *Univ. Iowa Studies in Natural History*, **18**(3):1–88. *Pls.* 1–5. 1944.

———: New or noteworthy tropical fungi IV, *Lloydia*, **11**(2):111–122. *Figs.* 1–5. 1948.

———: The genus Ceracea Cragin, *Mycologia*, **41**(1):77–86. *Figs.* 1–13. 1949.

Möller, Albert: Protobasidiomyceten, *Botan. Mitt. Tropen*, **8**:1–179. *Pls.* 1–6. 1895.

Neuhoff, W.: Zytologie und systematische Stellung der Auriculariaceen und Tremellaceen, *Botan. Arch.*, **8**(3–4):250–297. *Pls.* 1–4. *Figs.* 1–7. 1 *diagram.* 1924.

Patouillard, N.: Essai taxonomique sur les familles et les genres des Hyménomycètes. Thèse pour l'obtention du diplôme de Docteur de l'Université de Paris, École Supérieure de Pharmacie, année 1900–1901, No. 2. 184 pp. 74 *figs.* Lons-le-Saunier, 1900.

Rogers, Donald P.: A cytological study of Tulasnella, *Botan. Gaz.*, **94**(1):86–105. *Figs.* 1–79. 1932.

———: A taxonomic review of the Tulasnellaceae, *Ann. Mycolog.*, **31**(3):181–203. *Pls.* 6–7. 1933.

———: The basidium, *Univ. Iowa Studies in Natural History*, **16**:160–183. *Pl.* 7. 1934.

———: Notes on the lower Basidiomycetes, *ibid.*, **17**(1):1–43. *Pls.* 1–3. 1935.

———: A new gymnocarpous Heterobasidiomycete with gasteromycetous basidia, *Mycologia*, **39**(5):556–564. 1 *fig.* 1947.

Shear, C. L., and B. O. Dodge: The life history of Pilacre faginea (Fr.) B. & Br., *J. Agr. Research*, **30**(5):407–417. *Pls.* 1–2. 1925.

14

CLASS BASIDIOMYCETEAE: SUBCLASS EUBASIDIAE, "HYMENOMYCETEAE"

Subclass Eubasidiae

IN CONTRAST with the Heterobasidiae the Eubasidiae possess basidia which are one-celled and which mostly do not show a sharp distinction into hypobasidium and epibasidium. The nuclear divisions occur in the body of the basidium and after the spores have arisen at the ends of true sterigmata the nuclei pass through the latter into the spores. The nuclear spindles may be parallel to the longitudinal axis of the basidium (stichobasidial) or at right angles to this axis (chiastobasidial). In a few species both types of basidia may be present (*Exobasidium* and some Boletaceae).

In the genera included in the "Hymenomyceteae" the basidia occur in a hymenium (in a few species they are scattered so that no continuous hymenium is present), which becomes exposed to the air *before* the spores are shot off from the sterigmata to whose tips they are obliquely attached. These characters are in contrast to the situation in the group of orders to which collectively the name "Gasteromyceteae" has been applied (see next chapter). In these the spores are mostly attached symmetrically to the tips of the sterigmata and are not thrown off. They are set free by the opening of the basidiocarps to the air in various manners *after* the spores are mature.

The fungi included as Hymenomycetes have been considered to form one order, Agaricales; or two orders, Polyporales (or Aphyllophorales) and Agaricales, or several orders (Heim, 1934). In the main two series can be distinguished; those in which the basidia so far as studied are stichobasidial (most of the Polyporales) and those in which the nuclear division where investigated has been shown to be chiastic (Agaricales). It must be noted that of the thousands of species in this subclass only a few in each group have been studied cytologically.

It is not by any means settled whether the families considered to belong to this subclass really form a monophyletic series or whether some of them have arisen from one or more of the gasteromycetal orders. In the arrangement here set forth the majority are believed to form one series with a few forms of more doubtful origin.

Apparently the most primitive forms produced resupinate spore fruits with the basidia scattered or packed close together, and without a very definite limit of growth to the hymenium. By the formation of folds, ridges, teeth, etc., the hymenial surface became increased in the various families. At the same time a tendency appeared toward confining the hymenium to the under side of laterally attached or centrally stalked spore fruits. These spore fruits vary from felty to fleshy or leathery or corky or woody in consistence and may function for only a few hours or days, in some of the fleshy sorts, to many years in some of the corky or woody species. The size may vary from a few millimeters to over a meter in diameter and from a few layers of cells in thickness to 30 or 40 cm. A few species are obligate parasites in the stems, leaves, fruits, and flowers of Anthophyta (Angiosperms), e.g., *Exobasidium*. Some are parasitic or saprophytic depending upon the environment or opportunity, e.g., *Pellicularia filamentosa* (Pat.) Rogers (*Corticium vagum* var. *solani* Burt ex Rolfs). Many are parasitic upon the roots of plants, killing them and some extend up into the stem, killing the bark. Some are saprophytic on humus, decaying leaves, stems, etc. The nonliving woodcells of living trees may be attacked and the wood rotted without actually any parasitic action upon the living cells of the trunk. The weakening of the stem by the decaying of its woody elements may cause it to break, thus leading to its death. *Schizophyllum commune* Fr. may attack some trees as parasites but may be saprophytic on others. Many of the more woody or corky or leathery species are capable of culture in the laboratory but many of the fleshy forms have resisted all such attempts as yet.

The hymenium may consist entirely of basidia all of the same age or with younger basidia pushing up between the older ones. The newer basidia may push out beyond those first formed so as to increase the thickness of the hymenium which may then show definite or indefinite layers. In a great number of forms where the development of the spore fruit is not limited by the formation of a definite border of different structure the centrally produced basidia are the oldest and around them additional basidia arise successively further and further from the center. As a result in such species all ages of basidia may be found on making a radial section, from the oldest ones near the center to those near the margin which are just beginning to develop.

The basidia vary greatly in shape in the Hymenomycetes. They are usually round in cross section but when crowded laterally may become somewhat angular. They may be cylindrical, tapering at the very base and rounded at the apex, or urn-shaped or clavate or almost globose.

The sterigmata may be long in proportion to the length of the basidium or short, stout or slender, straight or curved. In some species they are hardly different from those of *Ceratobasidium*, tentatively placed in the Tulasnellales (see Chapter 13). Indeed these may represent transition forms.

In a great many forms some of the potential basidia do not develop far enough to produce sterigmata and spores. When they otherwise resemble not much modified basidia they are often spoken of as paraphyses. It must be noted however that they arise from the same type of hyphae that give rise to the basidia, so that in normal cases the paraphyses as well as the young basidia are binucleate. In the Ascomyceteae, on the contrary, the asci normally arise from dicaryotic hyphae and the paraphyses from monocaryotic hyphae. Thus the term paraphysis in the Hymenomycetes is based on morphology and location of the structure, not upon its phylogenetic implications (Kühner, 1925a).

Besides the paraphyses some of the hyphae underlying the basidial layer may insert themselves between the basidia in the form of much modified terminal cells. Except for the fact that in some cases they come from hyphae more deeply located than those from which the basidia arise they differ from the paraphyses mainly by their greater differentiation. They are called cystidia. These are given special names depending upon location, shape, contents, or function. They may be simple or branched, colorless or colored, thin-walled or with thick walls, obtuse or pointed, barely projecting from the hymenium or far exserted. In some species of *Coprinus* they serve to hold the gills apart (trabecular cystidia) while in *Hymenochaete* the stiff, sharp-pointed, bristle-like cystidia probably protect the hymenium from snails, slugs, or other soft bodied animals that otherwise might destroy the basidia. The term gloeocystidium is given to cystidia containing mucilaginous or oleaginous contents, usually at the ends of conducting hyphae underneath the hymenium. In a few Hymenomycetes there are produced in the trama or in the hymenium stellately branched, thick-walled cells which may be considered as being specially modified cystidia. By their location cystidia may be called cheilocystidia, when they develop at the edges of the pores or lamellae, or pleurocystidia when they occur in the hymenium that lines the pores or the surfaces of the lamellae. Sometimes cystidium-like structures that develop on the upper side of the pileus are called pileocystidia and similar structures on the stipe caulocystidia. It must be recognized that these last two categories although resembling cystidia are perhaps better considered special types of pubescence, confining the use of the term cystidium to structures in the hymenium.

The spore fruits show a very great variability of size and complexity of structure. It seems probable that the simple forms in many cases represent more primitive organisms phylogenetically, although as elsewhere among the fungi there is ample evidence that retrogression from more complex to simpler structures has occurred frequently. Assuming, as appears to the author to be most likely, that in some cases at least the simpler forms of the Hymenomycetes represent more primitive forms, the original type of spore fruit may have been a thin structure, not many layers of hyphae in thickness, adhering at all points to the outer surface of the substratum within which the vegetative mycelium was actively growing and accumulating food reserves. This young spore fruit spreads from a central spot more or less radially and its size is limited by external obstacles or by the exhaustion of the supply of food. Thus the sporocarp is theoretically to be considered as of unlimited ability to extend itself. From the external layer of hyphae there turn outward short branches which produce the basidia as their terminal cells. These upright hyphae may branch sympodially as the basidia approach maturity so that not only are new basidia appearing on the new radial growth of the spore fruit but also are arising among the older basidia. At the same time that the basidia are developing, the paraphyses, cystidia, etc., make their appearance but probably not to any great extent in the most primitive forms. In the species in which the mycelial growth outside of the substratum is more rapid and widespread than the production of the basidia the hymenium may consist of interrupted groups of sympodially produced basidia, as occurs in some species of the genera *Pellicularia* (*Botryobasidium*) and *Tomentella*. The spore fruit may then be more or less cottony with scattered clusters of basidia. Such fungi are indiscriminately lumped in the older works, in the genus *Hypochnus*.

From the simple type of fruit body illustrated by the foregoing, evolutionary development progressed in various directions. The subhymenial structure became more complicated, often with two or three distinct layers, each several or many cells in thickness (*Stereum*, etc.). The sporocarps more and more developed on the substratum in such a position that the hymenium faces downward, especially in those forms in which the edges pull away from the substratum to form a kind of shelf, the effuse-reflexed forms. This tendency persists until we find that in many genera the spore fruits are shelf-like, with little or no resupinate portion. These shelf-like structures may be narrowed toward the point of attachment and often a distinct stipe occurs. This stipe may be attached at the edge of the pileus or show a tendency to be attached excentrically or even centrally to the under side of the pileus (many Polyporaceae, Boletaceae, most Agaricaceae, and other related families).

In consistency the sporocarp may be cottony, papery, leathery, corky, woody, or fleshy. It may be short lived or may persist for many years.

The colors may be white or shades of gray or bright-colored or almost black.

The mycelium in perhaps the majority of the Hymenomycetes shows clamp connections. These may be found generally in the vegetative mycelium as well as in the spore fruit or may be lacking in the latter or in both. In abnormal forms whose mycelium is of the monocaryon type only, clamp connections are lacking, but their absence in any given species does not necessarily indicate such an abnormal type. Cytological study is required to determine whether hyphae without clamp connections are monocaryotic or dicaryotic.

Tischler (1927) and others who have made cytological studies in the Higher Fungi report that for the Hymenomycetes as well as most other Basidiomycetes the haploid number of chromosomes is mostly two, although in a number of species it may be four, six or eight.

The basidiospores are various in shape; globose to ellipsoidal to ovoid and in some genera angular or knobbed (*Rhodophyllus*). They are rarely symmetrical in more than one plane, that which passes through the spore, the sterigma and the center of the apex of the basidium. Apparently almost without exception they are perched in a slightly oblique manner on the tips of the sterigmata from which they are expelled with violence. They are nonseptate but in *Exobasidium* may become transversely septate before germinating or even before being discharged. Normally they germinate by a germ tube which may arise from any point on the spore wall or only from a specially located germ pore. They vary in color from hyaline, pink, red, yellow, ochre, ferruginous, to purple and black.

Conidia are produced in this group but in a rather limited number of species scattered throughout the two orders. They occur on various types of conidiophores. When produced internally in the spore fruit in a cushion-shaped or spherical structure they are usually placed in the "form genus" *Ceriomyces* but when formed externally may be called *Paramyces*. Chlamydospores are produced abundantly in *Nyctalis asterophora* Fr. and elsewhere. Oehm (1937) concludes that there are no true conidia in the Hymenomycetes but that they are all to be considered as various forms of chlamydospores.

Besides the foregoing the monocaryon stage of the mycelium of very many species produces oidia which appear to be capable of functioning as sexual cells (see Chapter 12), but which in some cases serve as conidia. More rarely they are produced as two-celled oidia on dicaryon mycelium but usually the two cells then fall apart and function like the uninucleate oidia from the monocaryon mycelium. This was reported for *Pholiota aurivella* (Fr.) Quélet by Vandendries and Martens (1932).

In the author's earlier book all of the Hymenomycetes were included in one order, the Agaricales, but the modern tendency is toward further division, even to the recognition of four or five orders. The author conservatively recognizes two orders, Polyporales (Aphyllophorales) and Agaricales. It must be confessed that the distinction between these two groups is not sharp at some points, either from convergence in structure of two otherwise very distinct orders or because one grades into the other, representing a phylogenetic relationship.

The great Swedish mycologist Elias Fries made extensive studies upon the Hymenomycetes as well as upon other fungi for about sixty years. His original classification was based upon the studies of Persoon with whose later work his earlier publications were contemporaneous. Persoon's chief works were his Synopsis Methodica Fungorum, 1801, and Mycologia Europaea, 1822–1828. The work of Fries, which is used as the standard upon which the nomenclature of many groups of fungi (including the Hymenomycetes) is based, is his Systema Mycologicum, in four volumes, 1821–1832. His latest important work was Hymenomycetes Europaei, 1874. The majority of the students of this group of fungi have followed the Friesian system with minor modifications. He recognized five families within the limits of the Hymenomycetes as follows:

Agaricaceae: hymenium on radiating gills or lamellae.
Polyporaceae: hymenium lining the surfaces of small pores or tubes.
Hydnaceae: hymenium spread over spines or protuberances.
Thelephoraceae: hymenium unilateral, spread over a firm, smooth or corrugated, under or upper surface. Mostly membranous, leathery, etc.
Clavariaceae: hymenium spread over the surface of smooth, simple or branched clubs. Mostly fleshy.

Several families have been segregated from those above. In the author's first book in addition to the foregoing there were recognized the Exobasidiaceae, segregated from the Thelephoraceae, and the Boletaceae and Fistulinaceae, separated from the Polyporaceae.

As mentioned in the preceding chapter the studies by Juel (1896, 1916) and of Maire (1900, 1902) showed that in the Basidiomyceteae a distinction could be made as to the position of the spindle in the meiotic division of the diploid nucleus of the basidium. This appeared to be correlated more or less with the relationship of the groups. In some the basidium is more slender and the diploid nucleus occupies an approximately central position. The spindle of the first meiotic division is parallel to the longitudinal axis of the basidium as are the two spindles of the subsequent divisions of the two daughter nuclei. The nuclei then mostly migrate to the upper part of the basidium where the sterigmata are formed. Such basidia are stichobasidial. In other Basidiomyceteae the basidium is broader above and the diploid nucleus is located in the somewhat widened upper portion. The first nuclear spindle is more or less transverse and the next two spindles also transverse, usually at right angles to the axis of the first spindle. Such basidia are called chiastobasidial. It was shown that many of the Thelephoraceae and Clavariaceae and some of the genera hitherto placed in the Agaricaccae were stichobasidial while most of the Polyporaceae and Agaricaceae and Boletaceae were chiastobasidial.

As more and more species and genera of the Hymenomycetes became known it was necessary to find other characters for their classification than the external morphological ones which largely formed the bases for the studies by Fries. The form and color and surface characters of the spores showed their importance, but soon the anatomy of the spore fruit and especially of the trama, the tissue upon which the hymenium is produced, proved to be of great value, as well as the mode of development of the spore fruit. The old families had to be broken up and recombined in order that a more logical systematic treatment could be expressed. Patouillard (1900) recognized two main groups: "Aphyllophoracées," with the hymenium naked from the first and capable of continued expansion, and "Agaricacées" with the hymenium more or less lamellar and hemiangiocarpic, i.e., at first enclosed by a more or less fugacious veil which is variously ruptured at maturity. Later authors have called these Polyporales and Agaricales respectively. The old families Boletaceae and Agaricaceae made up the latter order, the Polyporales containing the other Friesian families and one or two stichobasidial genera from the old family Agaricaceae (e.g., *Cantharellus*). In the main the Agaricales as so delimited were chiastobasidial, but the Polyporales had both types of basidia. This led Gäumann (1926) and others to divide the included families into two series, stichobasidial and chiastobasidial. The fact that in the same hymenium of *Exobasidium* both types may occur throws doubt upon the validity of this as a fundamental character, although it is apparently correlated sufficiently with other characters to make it important.

Another character has been emphasized recently as of perhaps great importance: the blue or violet coloration of the basidiospore exospore or of the warts or network of lines on the spores upon treatment with a solution containing free iodine. In some cases even certain of the hyphae of the spore fruit give the same reaction. Such spores and hyphae are said to be amyloid (i.e., starch-like in their reaction). Just how far this can be used in revising the arrangement of genera and families is uncertain for within certain genera (e.g., *Mycena*) occur some species with amyloid and some with non-amyloid spores.

Although the presence of clamp connections on the mycelium is well known in a great many of the Hymenomycetes yet it has been found that they may be absent in some genera on the hyphae in the interior of the spore fruit. This has been used as a supplementary generic character but cannot be considered as of fundamental importance. In the genus *Coprinus* there occur species that are quite similar, some of which possess clamp connections and some lacking them.

In the sequence of families followed below it must be considered that a logical arrangement should be based upon the supposed phylogenetic relationships of the group. In the absence of decisive fossil remains we have to fall back upon a comparison of species and genera now existent. The surmised origins of the different families are quite various. Thus Singer (1936) and others have been led to believe that the Agaricales have descended from the Gasteromyceteae while others (e.g., Heim, 1937) believe the reverse to be the case. In the following discussion the writer follows somewhat the latter's interpretation although admitting that the evidence for Singer's view is quite strong.

Order Polyporales (Aphyllophorales). Hymenium always gymnocarpic, i.e., from its inception exposed to the air and not enclosed by a veil, with the possible exception of *Cryptoporus volvatus* (Pk.) Hulb. Growth of the hymenium is more often without definite morphological limits. In the supposedly more primitive forms (at least forms of simpler structure) no highly developed morphological or anatomical differentiation occurs beyond the resupinate layers of hyphae upon which arise the more or less separate clusters of basidia or a continuous hymenial layer which may continue to enlarge in all directions at the margin. In the higher forms the subhymenial portion of the spore fruit becomes distinguished into several layers differing more or less morphologically and anatomically and the tendency to develop upright or reflexed (shelf-like) portions becomes more strongly marked. With evolution progressing in various directions there appear clavate or dendroid structures or pileate forms which are sessile or laterally, excentrically or even centrally stipitate. Along with this external differentiation the incipient hymenium may be increased in surface area by being thrown into folds or by the production of emergent tubercles or teeth or by reticulate outward growth leaving shallow or deep pits (pores) which are lined by hymenium. With some of the simpler types of structure is correlated the stichobasidial type of basidia, but the more complex structures mostly have the chiastobasidial type. The chief distinction in the following order, the Agaricales, is the fact that in the latter the poroid or more often lamelloid hymenial portion, though in its younger stages sometimes gymnocarpic, often becomes secondarily enclosed (pseudoangiocarpic) or may arise from the beginning as an internal development (angiocarpic). In both the latter cases prior to the maturation of the basidia the spore fruits become opened in a regular manner to permit the distribution of the spores.

In both of these orders the basidiospores are perched obliquely at the tips of the sterigmata and are shot off violently so that spore distribution is effected by air currents. This distinguishes the Hymenomycetes from the Gasteromycetes in which the basidiospores are perched symmetrically on the tips of the sterigmata and are not shot off, the spores reaching their maturity before the spore fruit opens.

Key to Families of Polyporales

Parasitic in the living tissues of leaves, growing stems, or fruits of Anthophyta (Angiosperms), producing the clavate basidia externally in a continuous or interrupted layer. Basidiospores becoming septate upon germination, each cell giving rise to a few spindle-formed conidia. Family Exobasidiaceae

Saprophytic in plant tissues or on plant debris, less often parasitic. Basidia oval to clavate, basidiospores germinating directly.

Hymenium interrupted or continuous, smooth (rarely slightly roughened by small wart-like emergences), entirely resupinate or reflexed or partially stipitate, the hymenial surface being on the under side. Spore fruit a thin weft of hyphae or a more definite structure, more often papery, leathery, or even corky or slightly woody. Both stichobasidial and chiastobasidial types of basidia present. Family Thelephoraceae

Spore fruits with short stipes, funnel-shaped to trumpet-shaped or almost clavate, but somewhat broader and truncate at the top; fleshy; the hymenium on the outer surface, smooth or reticulate or with low, broad, rounded longitudinal ridges. Basidia stichobasidial. Basidiospores white or light-colored. Family Cantharellaceae

Hymenium continuous on all sides of upright filiform or clavate or much ramose spore fruits which are fleshy or leathery. Both stichobasidial and chiastobasidial types present. Family Clavariaceae

Spore fruit resupinate or reflexed or stipitate, both laterally or centrally. Hymenium smooth with a number of projecting pegs or teeth or the latter very numerous, being directed downward in all but a few of the resupinate forms with few emergences. Spore fruits leathery, corky, woody, or fleshy. The teeth are round in cross section or flattened. Basidia usually chiastic but some stichic. Family Hydnaceae

Spore fruit resupinate or reflexed, or laterally or centrally stipitate. The hymenial surface is increased by the development of shallow or deep pores which may be round or elongated radially with greater emphasis on the radial ridges so as to form lamellae with secondary cross connections. Papery, leathery, corky, woody, or fleshy. Basidia always chiastic. Family Polyporaceae

Spore fruits resupinate or reflexed or laterally stipitate, fleshy, the lower fruiting surface growing out to form numerous, elongated, separate tubes which are lined internally by the hymenium. Basidia chiastic. Family Fistulinaceae

Spore fruits resupinate or partially reflexed, the hymenial surface at first smooth and then thrown into thick shallow ridges which often anastomose to form low-walled, broad, shallow pits. Hymenium continuous over the sides and edges of the ridges as well as the floor of the pits. Fruit body often more or less gelatinous, with the surface in almost any direction. Many species are very destructive to wood both in trees and in structural timbers. Family Meruliaceae

The division of families in this order is not always as above. Patouillard (1900) makes an entirely different arrangement. In the older works the Boletaceae and Fistulinaceae were united with the Polyporaceae. Maire (1937) and Singer (1936) removed the Cantharellaceae from the Agaricales to the Polyporales, but Heim (1934) retains this family in the Agaricales. He recognizes two more orders intermediate between the Polyporales and Agaricales: the Boletales and the Asterosporales (including the Russulaceae).

FAMILY THELEPHORACEAE. Hymenial surface smooth or at most only slightly warty or folded. Spore fruits membranous, leathery, or in two or more genera fleshy; closely appressed to the substratum or forming a shelf or funnel or simple or divided pileus with hymenium on one surface only. Twenty or more genera are recognized and probably about 1000 species. The most complete study of the North American species of this family is that by E. A. Burt (1914–1926). Rogers and Jackson (1943) have made a thorough nomenclatorial study of many of the resupinate species of this family.

The genus *Corticium* forms a thin spore fruit growing closely appressed to the substratum and not distinguishable into several layers. The hymenium arises directly from the mycelium and consists of a layer of closely packed basidia. The margin of the spore fruits may be definite or indefinite. When dry the hymenium is often cracked. There are no true cystidia among the basidia but gloeocystidia may be present in some species. Most of the species of the genus are saprophytic on wood or bark, a few are destructive to wood.

Miss Nobles (1937) demonstrated, by mating monocaryon cultures of *Corticium incrustans* von Höhn. & Litsch., that this species falls into two sexual phases, i.e., is of the bipolar type of sexuality. Aerial hyphae of monocaryon mycelium give rise to allantoid hyaline uninucleate conidia, often many to a hyphal cell, which upon germination produce again the monocaryon phase. The dicaryon aerial hyphae, which have a clamp connection at every septum, produce a single binucleate conidium on each hyphal cell, leaving two nuclei behind in that cell. These conidia upon germination give rise immediately to dicaryon hyphae.

The old genus *Corticium* has been divided into several genera, the distinctions being based chiefly upon the structure of the basidia and the nature of the hymenium whether loose or compact. One of these genera, *Ceratobasidium*, with two to six sterigmata so long and so much thickened as to be called "epibasidia" shows close relationship to the Tulasnellaceae and has been considered in the discussion of that family in the preceding chapter. It represents about a halfway step between the Thelephoraceae in the Eubasidiae and the Tulasnellaceae and Dacrymycetaceae in the Heterobasidiae. Another genus, *Pellicularia* (*Botryobasidium*) has been segregated by Rogers (1943) for those fungi formerly included in *Corticium* which have a thin film of short, broad-celled mycelium on the substratum

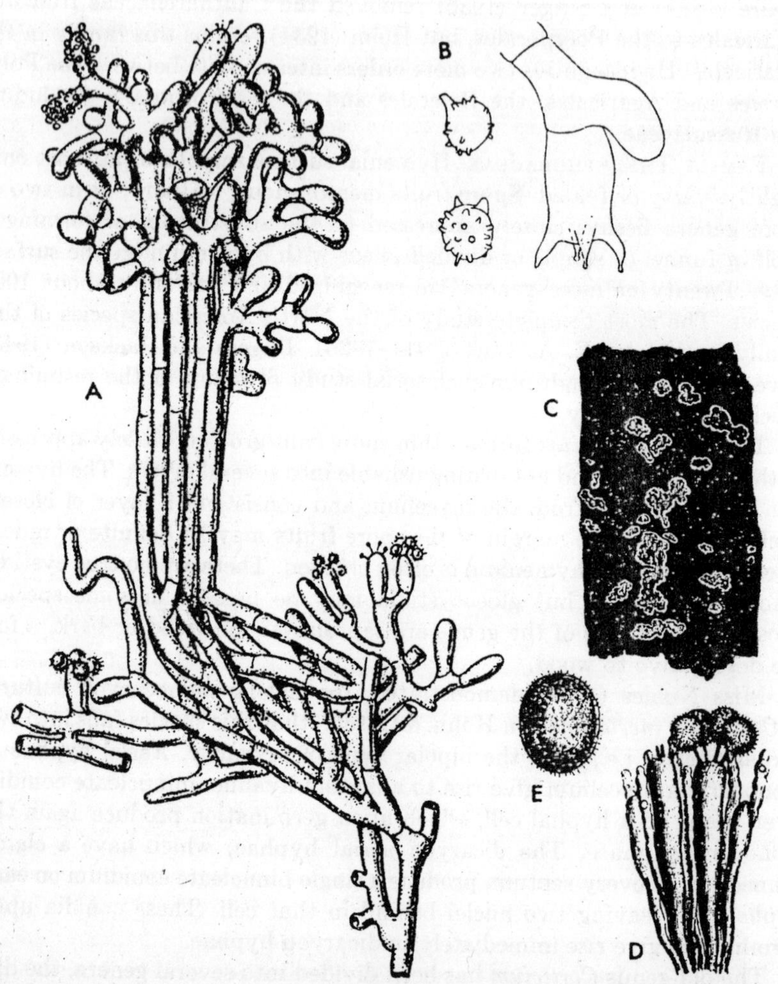

FIG. 153. Polyporales, Family Thelephoraceae. (A, B) *Pellicularia isabellina* (Fr.) Rogers (*Tomentella flava* Bref.) (A) Loose weft of mycelium bearing below scattered basidia and above conidiophores. (B) Basidium and spores. (C–E) *Aleurodiscus amorphus* (Pers.) Rabenh. (C) Habit sketch. (D) Section of hymenium with basidia and "paraphyses." (E) Basidiospore. (A, after Brefeld: Untersuchungen aus dem Gesammtgebiete der Mykologie, Heft 8, pp. 1–305. B, courtesy, Rogers: *Univ. Iowa Studies in Natural History*, **17**(1):1–43. C–E, after Killermann, in Engler und Prantl: Die Natürlichen Pflanzenfamilien, Zweite Auflage, vol. 6, pp. 124–288, Leipzig, W. Engelmann.)

and cymose tufts of short, broad basidia with four, less often six to eight, sterigmata. The commonest parasitic species of this genus, *P. filamentosa* (Pat.) Rogers (*Corticium vagum* var. *solani* Burt, *Hypochnus solani* Prill. & Del., etc.) occurs as a parasite upon the stems and roots of potato (*Solanum tuberosum* L.), bean (*Phaseolus vulgaris* L.), and very many other plants of economic value. It produces cankers at or below the surface of the soil which kill or seriously injure the parts affected. Small sclerotia are formed which enable the fungus to overwinter. On the stem of the host plant the mycelium creeps up as a thin gray or white hyphal layer on which the oval basidia are produced in groups. The sclerotial stage is known under the name of *Rhizoctonia*. (Fig. 153A, B.)

The genus *Tomentella* (*Hypochnus*, as interpreted by Burt, 1916) produces its basidia more or less scattered or in tufts on a loose cottony mycelium. The basidiospores are nearly spherical and spiny. They are mostly saprophytic.

The species of *Corticium* in which gloeocystidia occur, are placed by many authors in a separate genus, *Gloeocystidium*, but Burt (1926) and Rogers and Jackson (1943) do not approve of this segregation. *Peniophora* is practically a *Corticium* with true, fusiform, pointed cystidia (not gloeocystidia). *Asterostroma* is similar but has stellately branched, thick-walled bristles or cystidia in the hymenium. *Epithele* has the hymenium interrupted here and there by sterile projections or pegs consisting of bundles of hyphae. These differ from the teeth of the Hydnaceae which are covered with basidia. *Coniophora* is practically a *Corticium* with ferruginous spores. Some species have cystidia (*Coniophorella* Karst.) and others not. Donk (1933) and Singer (1944) place the Family Meruliaceae close to these genera (see p. 484).

In *Aleurodiscus* the spore fruit instead of remaining flat against the bark on which it develops curls up a little at the edge to form a flat saucer or shallow cup. The basidia and spores are rather large and there are present various types of structures called by some cystidia, by others paraphyses. Much similar is *Vararia* (*Asterostromella*) but the modified cells in the hymenium are many times dichotomously branched. (Fig. 153, C–E.)

All of the foregoing genera have a rather thin and not much differentiated spore fruit below the hymenium. The following genera have a subhymenial structure much thicker and often in several distinct layers. At the edges the spore fruit bends away from the substratum to form a sort of shelf, with the hymenium on the smooth lower surface. Even a sort of central stipe may be developed in some species, the spore fruit being more or less funnel-shaped in that case. In *Stereum* conspicuous cystidia are lacking, but in *Hymenochaete* the hymenial layer has numerous long stiff, usually brown, pointed setae, whose probable function is

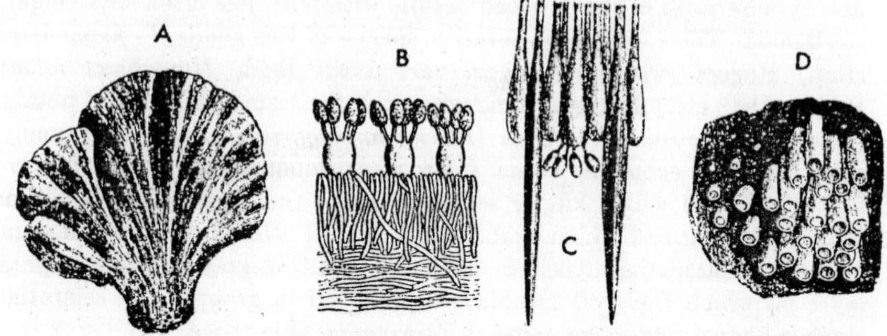

Fig. 154. Polyporales, Family Thelephoraceae. (A, B) *Thelephora terrestris* (Ehrh.) Fr. (A) Lower view of lobe. (B) Section through hymenium. (C) *Hymenochaete cacao* Berk. Portion of hymenium showing basidia and setae. (D) *Solenia candida* (Hoffm.) Fr., habit sketch. (A–D, after Killermann, in Engler und Prantl: Die Natürlichen Pflanzenfamilien, Zweite Auflage, vol. 6, pp. 124–288, Leipzig, W. Engelmann.)

the protection of the hymenium from snails and other harmful animals. In *Thelephora* (called by the name *Phylacteria* Pat. by Patouillard, 1887, 1900; Rea, 1922, and others) the more or less leathery fruit body is upright, stalked, pileate or fan-shaped or much lobed or in an overlapping series, the hymenium being on the under side and smooth or slightly warty. In *Cyphella* and *Solenia* (united into one genus by some authors) the spore fruit is cup-like or tubular, with the opening usually directed downward, and lined internally by the hymenium. In some species of *Solenia* many spore fruits arise close together, but separate from one another, from a common mycelial mass, the subiculum. Some mycologists place these two genera in Family Fistulinaceae (see p. 484). (Fig. 154, A, D.)

Sparassis (often included in the Clavariaceae, but see Cotton, 1912) is fleshy, much branched, with the terminal portions flattened and bearing the hymenium on the under side only. The edible *S. ramosa* (Schaef.) Schroet. occurs in Europe and North America, often attaining the size and shape of a small, loose-leafed cabbage.

Family Cantharellaceae. Perhaps intermediate between the Thelephoraceae and Clavariaceae is the genus *Craterellus*. This is fleshy, club- or funnel- or trumpet-shaped with the hymenium on the outer side which is smooth or more or less longitudinally ribbed or reticulate. Several species are edible. The genus is stichobasidial as is the genus *Cantharellus* which is undoubtedly closely related but differs in having the longitudinal ribs further developed so as to be low, thick lamellae. Perhaps, as Singer suggests, the correct name is *Gomphus*. *Cantharellus* was formerly included in the Agaricaceae but probably belongs here. Perhaps these two genera should be united into a distinct family, the Cantharellaceae (as was done

FIG. 155. Polyporales, Family Cantharellaceae. *Cantharellus floccosus* Schw. (*Gomphus floccosus* (Schw.) Singer). (Courtesy, M. B. Walters.)

by Maire 1900; Rea, 1922). This family shows close affinities with the following and may be really more closely related to it than to the Thelephoraceae. (Fig. 155.)

FAMILY CLAVARIACEAE. Spore fruits fleshy or waxy or even gelatinous, rarely leathery; upright, clavate or branched in a coralloid manner; round or flattened; usually covered by the hymenium on all sides over the whole spore fruit or over special more terminal and often enlarged portions. A dozen or more genera and probably over 500 species. Almost all are saprophytic and possibly some form mycorrhizae on tree roots. A few species are parasitic upon plants. Many of the larger fleshy forms are edible. Some of the genera seem to have their origins in the Thelephoraceae and some are perhaps nearer to the Cantharellaceae, and perhaps some are related to *Irpex* in the Polyporaceae or to *Hericium* in the Hydnaceae. In other words, the Clavariaceae probably do not represent a phylogenetic unit. Some are stichobasidial and some chiastobasidial. How great weight this should have upon the division into genera must await further studies in an attempt to determine whether the structural differences within the group show definite correlations with the stichic or chiastic position of the nuclear spindle in the meiotic divisions of the diploid nucleus of the basidium.

Pistillaria is a small fungus usually not over 2 to 5 mm. tall, with no sharp distinction between stalk and hymenium-bearing portion, which is clavate. No sclerotium is present. Some of the species have two-spored basidia, others have four spores to each basidium. Growing on soil or dead stems, leaves, etc. Also small, but much taller than the foregoing, is the genus *Typhula* whose spore fruits grow from sclerotia and form very slender filiform stalks, sometimes branched, with the terminal portions thickened to form spindle-formed hymenophores. A number of species are parasitic upon grasses, sugar beets, potatoes, etc. Miss Remsberg (1940) studied the two genera and determined the presence and absence of sclerotia to be the best distinguishing character.

Physalacria is also small, up to 2 cm. tall, consisting of a slender stalk and a downturned, hollow, fleshy head on whose lower surface the hymenium is most abundantly developed. For this reason McGuire (1939) suggested that the genus should be placed in the Thelephoraceae as was

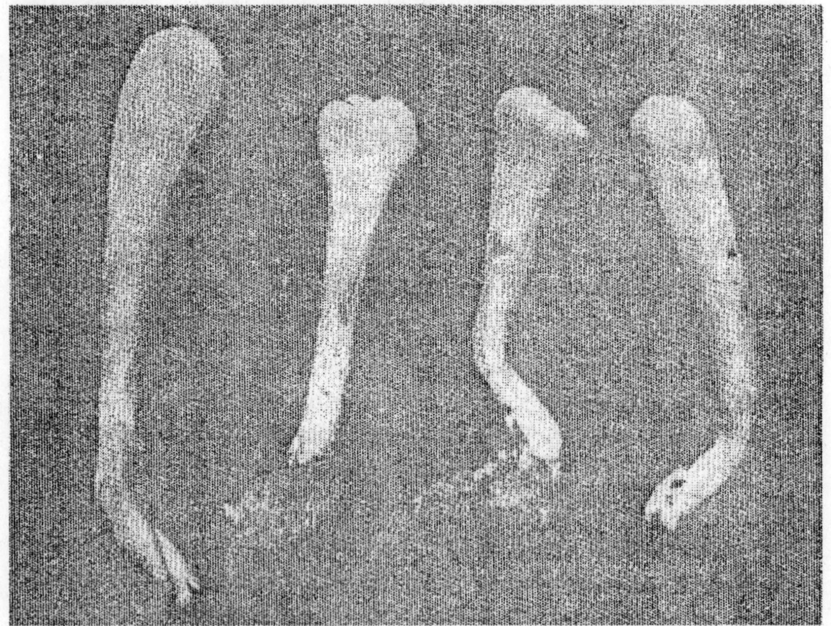

FIG. 156. Polyporales, Family Clavariaceae. *Clavariadelphus pistillaris* (Fr.) Donk. (Courtesy, Coker: The Clavarias of the United States and Canada, Chapel Hill, Univ. North Carolina Press.)

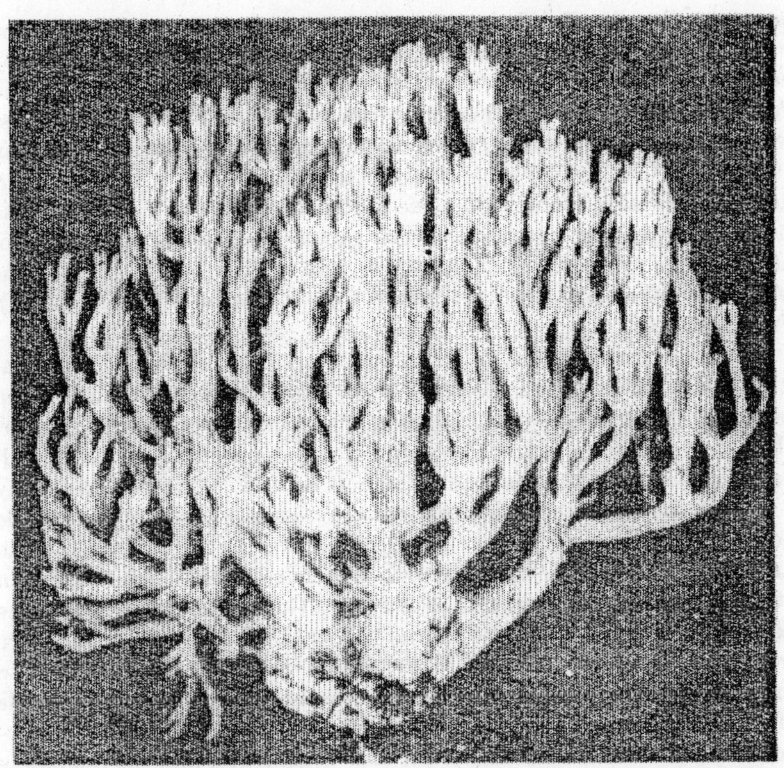

FIG. 157. Polyporales, Family Clavariaceae. *Clavariella subbotrytis* (*Ramaria subbotrytis* (Coker) Corner). (Courtesy, Coker: The Clavarias of the United States and Canada, Chapel Hill, Univ. North Carolina Press.)

done for *Sparassis* by Cotton (1912) because the ultimate flattened branches of the much ramose spore fruit of the latter bear their hymenium on the lower surface only. Killermann (1928) retains both these genera in the Clavariaceae.

The larger forms of this family are clavate and unbranched or only slightly branched or very much branched in a more or less coralloid manner. Among the clavate forms is the very large *Clavariadelphus pistillaris* (Fr.) Donk (*Clavaria pistillaris*), often up to 10 to 15 cm. tall with a thickness of 2 to 3 cm. It is edible. In many ways it resembles some of the members of Family Cantharellaceae but differs in possessing chiastic instead of stichic basidia. Somewhat smaller and usually brighter-colored are the unbranched species of *Clavaria* which are not enlarged much upwards and have smaller spores. Some species are much branched and fleshy. Where the branching consists of very slender dry cartilaginous, cylindrical and tapering branches, forming a bush-like structure we have the genus *Pterula*. *Eriocladus* (*Lachnocladium*) is larger, with flattened or rounded branches which are hairy. Some of the much branched fleshy species of *Clavaria* have been set off as separate genera: *Clavulina* and *Clavariella* (*Ramaria*). The majority (but not all) of the species of these two genera are respectively stichobasidial and chiastobasidial. Gäumann (1926) and Donk (1933) accordingly place *Clavulina* in a distinct order, the Cantharellales, along with some segregates from the Thelephoraceae and Hydnaceae and Agaricaceae. The genus *Clavicorona*, with apices of the branches truncate or cup-shaped, and with gloeocystidia in the hymenium layer has been set apart from *Clavaria* by Doty (1947). (Figs. 156, 157.)

Family Exobasidiaceae. The fungi composing this family are parasitic in the leaves, green stems or even fruits of higher plants, often distorting the affected parts or causing the formation of galls. The intercellular mycelium apparently bears no clamp connections. It sends rod-shaped or branched haustoria into the host cells. The basidia arise singly or in tufts between the epidermal cells, eventually piercing the cuticle and forming their usually four to eight spores externally. There are no paraphyses or cystidia. Eftimiu and Kharbush (1927) have made an extensive study of a species of *Exobasidium*. The mycelial cells within the host are elongated but become shorter nearer the epidermis and there are binucleate. The fusion nucleus in the basidium divides meiotically into four nuclei, one passing into each basidiospore. The nuclei may divide before the spores are formed and then the eight nuclei enter eight basidiospores. Sometimes one nucleus of the four degenerates and the remaining three divide so that six spores are formed. Sometimes only two nuclei are found in the basidium and then only two basidiospores. These spores germinate by budding like yeasts often becoming once septate first. In *E. rhododendri* Cramer, the spores divide by a septum and send out germ tubes from each of the two cells thus formed. In this species the nuclear divisions within the basidium may be either stichobasidial or chiastobasidial. Gadd and Loos (1948) report that in *E. vexans* the basidiospores become once septate before they are discharged. These two-celled spores may develop thick walls after becoming free. *E. vaccinii* (Fuckel) Wor. is frequent upon the cranberry (*Oxycoccos macrocarpus* (Ait.) Pursh) and related plants. The affected shoots become upright and take on a pink color and the leaves are increased in size and the stem thickened. Sometimes only a small portion of a leaf may be infected or a spot on one side of a fruit. Such spots are thickened and reddened. The basidia are club-shaped with four basidiospores which become septate before germination and send out short branching sterigmata bearing spindle-shaped spores. About 30 species are known in this genus. The genus *Kordyana* probably also belongs in this family. It is tropical. The basidia are two-spored, rarely four-spored and stichobasidial and emerge from the stomata of the host, sometimes intermingled with long, slender, hyaline hyphae or "paraphyses." This was described rather fully by Gäumann (1922). (Fig. 158.)

The following genera have been assigned to this family but later study has shown that they belong elsewhere: *Microstroma*, *Protocoronospora*, and *Urobasidium*. Wolf (1927, 1929) showed that the first is not a Basidiomycete at all because its supposed basidia are multinuclear, the mycelium lacks conjugate nuclei, and the spores are not borne upon the sterigmata in the manner typical of the class. Wolf (1920) and Karakulin (1923) showed that the second, like the former, belongs to the Fungi Imperfecti, with the name properly *Kabatiella*. *Urobasidium* described by Giesenhagen (1892) was shown by Mason (1941) to be still another Imperfect Fungus. Its proper name is *Zygosporium*. (Fig. 201B, C.)

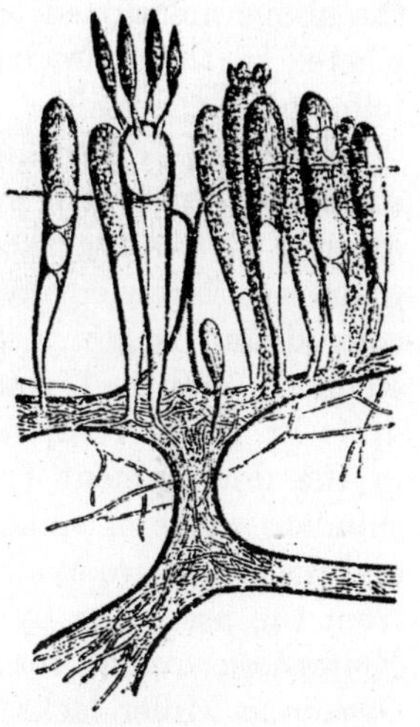

FIG. 158. Polyporales, Family Exobasidiaceae. *Exobasidium vaccinii* (Fuckel) Wor. Basidia in various stages of maturity emerging through the epidermis of the host leaf. (After Killermann, in Engler und Prantl: Die Natürlichen Pflanzenfamilien, Zweite Auflage, vol. 6, pp. 124–288, Leipzig, W. Engelmann.)

The relationship of this family is not certain. Perhaps it represents a line derived from the simplest Thelephoraceae but highly modified by its extreme parasitism.

FAMILY HYDNACEAE. These fungi are mostly saprophytic. Some cause serious decay of timber. The spore fruit may be small or large and resupinate or shelf-like or with a pileus borne on a lateral or central stipe. In some forms the pileus is divided into many small pilei. The consistency varies from fleshy to woody and there is a great range of color. The under side of the fruit body is at first smooth but as the hymenium develops it grows out into hymenium-covered spines or teeth. Under the classification of Fries all such fungi were classed in one family. However, Bourdot and Galzin (1927) placed the genera with hyaline or light-colored spores in the group Hydnés (except *Irpex* which they place in the Porés) and those with brown spores in the Phylactériés (i.e., close to *Thelephora* or *Phylacteria*). Donk (1933) places the stichobasidial forms in Tribe Hydneae of the Cantharelloideae and the remaining, chiastobasidial forms in the Phylacteroideae close to *Thelephora*, except some species of *Irpex* placed by him in the Polyporoideae, Tribe Daedaleae. *Irpex* is placed by Singer (1944) in the Polyporaceae, and by Murrill (1907) in the same family under the name *Irpiciporus*. Miller (1933) regards *Hydnochaete*, *Sistotrema*, *Irpex*, and *Echinodontium*, whose teeth are formed by the breaking up or unequal growth of the pore walls, as properly placed in that family. *Hydnum*, as recognized by Fries in 1821, included all of the genera now included in the family as well as some of the above mentioned ones that may eventually have to be placed elsewhere. In the following the generic distinctions of Miller (1933) are followed in the main.

Grammothele Berk. & Curt. is resupinate with its porose-reticulate surface covered with numerous small warts over which the hymenium extends. *Caldesiella* Sacc. is in habit much like *Tomentella* of the Thelephoraceae, but is covered with soft conical spines or teeth. The spores are colored and rough. *Asterodon* Pat. is similar but has dark, simple or stellately branched, setae, and the smooth spores are subhyaline. *Grandinia* Fr. is also resupinate and resembles *Corticium* from which it differs in the development toward maturity of hemispherical, cylindrical or subulate warts or spines, covered with hymenium. Cystidia are lacking and the spores are hyaline and smooth or roughened. *Odontia* Pers. differs from the preceding by the presence of cystidia. It therefore resembles *Peniophora* except for the conical to subulate or cylindrical spines. *Oxydontia* Miller lacks cystidia and has long subulate and conspicuous teeth. It is resupinate or effused-reflexed. This is the genus called *Acia*, Karst., an untenable name because of its prior use for one of the Rosaceae. *Radulum* Fr. (including *Phaeoradulum* Pat.) is resupinate or reflex, with coarse, blunt, irregularly scattered or confluent teeth. Spores hyaline or light-colored. *Phlebia* with hymenium covered with more or less notched wrinkles is sometimes placed in this family. These last four genera suggest how the Hydnaceae may have arisen from the resupinate Thelephoraceae in which the smooth hymenium often shows little warts or even projecting bundles of hyphae. However, these latter are not covered by the hymenium. *Mucronella* Fr. is essentially a cluster of subulate teeth arising from a fugacious mycelium.

The following genera are attached laterally or have stalks. *Steccherinum* S. F. Gray is sessile or substipitate and laterally attached. It has terete or flattened spines and develops cystidia. The spores are white and small. It is wood inhabiting. *Auriscalpium* S. F. Gray is laterally stipitate and has short subulate spines with scarcely differentiated cystidia. Growing on cones of conifers. The following three genera are centrally stipitate with subulate spines, and grow on the ground: *Dentinum* S. F. Gray, fleshy, pale, spores white and smooth, subspherical; *Hydnum* L. emend. S. F. Gray, fleshy, dark-colored, spores subspherical, angular or echinulate, brown; *Calodon* Quél. (*Hydnellum* and *Phellodon* of Karsten), fibrous,

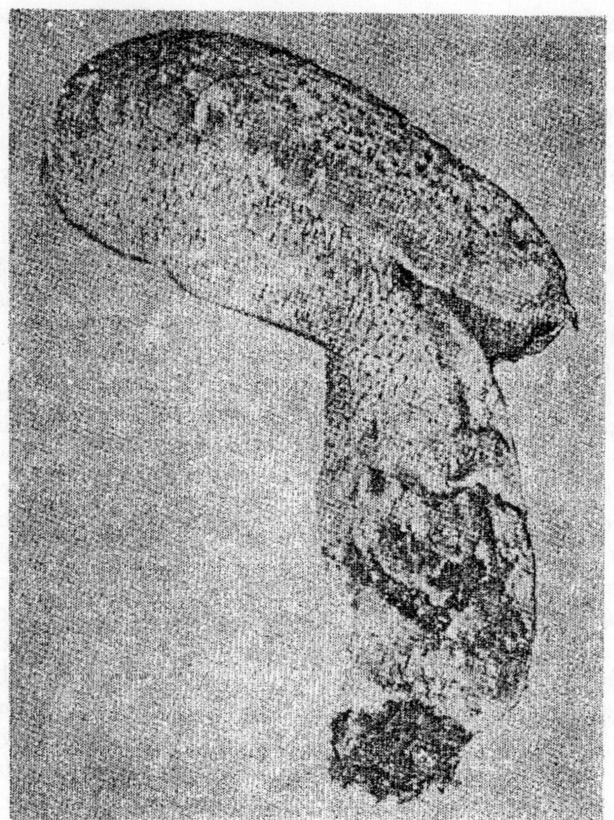

FIG. 159. Polyporales, Family Hydnaceae. *Hydnum imbricatum* L. ex S. F. Gray. (Courtesy, M. B. Walters.)

tough and sometimes woody, dark-colored, spores shaped as in *Hydnum*, brown or subhyaline. (Fig. 159.)

Usually placed in this family is the genus *Hericium* Pers. ex S. F. Gray (*Dryodon* Quél., *Manina* Scop. ex Banker). This is fleshy and unbranched or more often branched, with subulate spines mostly long and pendent. Spores spherical or subspherical and amyloid (i.e., walls staining blue with iodine). In *H. coralloides* Pers. ex S. F. Gray the pileus is but little developed, the fruit body consisting essentially of branching stalks bearing at their tips pendent tufts of pointed teeth. The writer questioned in 1935 whether such a fungus belongs in this family at all and suggested that it might be more closely related to the Clavariaceae. Singer (1936) suggests its origin from much branched species of that family with which they have in common the amyloid type of spore. For *Irpex* and *Echinodontium* see Polyporaceae (p. 494).

FAMILY FISTULINACEAE. This small family of only a few species and two, possibly more, genera is distinguished by its fleshy spore fruits and by the separate though closely crowded hollow tubes which hang down underneath the pileus. The latter may be centrally or laterally stipitate in *Fistulina*, or resupinate in other genera. Each separate tube is formed as an open cup which elongates and becomes lined internally by the hymenium. If one would imagine hundreds of spore fruits of *Solenia* of the Thelephoraceae arising from a common pileus the characteristic structure of this family would be realized. *Fistulina hepatica* Fr., the beefsteak fungus, is found in nearly all temperate regions of the world on oaks and other deciduous trees. The spore fruit is more or less fan-shaped with a short, thick, lateral stipe. It is brown-red, blood-red internally, and the closely crowded but separate tubes are reddish brown. When young it is edible. *Porothelium* perhaps belongs here. It forms a resupinate membranous or crust-like spreading hymenophore on which develop numerous scattered wart-like projections which elongate, leaving a central pore in each. This perhaps represents a connecting link between *Solenia* and *Fistulina*. The relationship of this family to the Polyporaceae is doubtful. Lohwag and Follmer (1936) and Elrod and Blanchard (1939) studied the development of *Fistulina* and showed that its tubes are of essentially the same structure as the cup-like spore fruits of *Solenia* and *Cyphella*. There is no similarity in the development to the Boletaceae to which the easily separable pores of the latter had suggested relationship.

FAMILY MERULIACEAE. Like so many others of the Hymenomycetes the members of this family are wood destroyers. The spore fruits at first develop a smooth hymenium on which basidia develop to spore bearing maturity. As this enlarges the surface produces low, rather thick folds or ridges on the sides and rounded edges of which new basidia continue to develop so that eventually basidia of all ages are to be found on the ridges as well as on the intervening hymenium at the original level. The ridges may anastomose so as to form a net-work, the pores that form the meshes being shallow. Sometimes the ridges are more or less radiate or may be interrupted. The spore fruits and the ridges are fleshy or waxy or gelatinous. The low walls separating the pores do not necessitate their pointing downward since air currents can carry off the spores when discharged from the basidia regardless of the direction in which the pores point. There is no agreement as to what genera should be included in this family. Bourdot and Galzin (1927) include seven genera, three of which have a smooth hymenium (*Coniophora*, *Coniophorella*, and *Jaapia*), the other four having projecting ridges or tubercles. Of these *Phlebia* is often placed in the Hydnaceae and *Plicatura* (*Trogia*) in the Agaricaceae, leaving *Merulius* (with colorless spores) and *Gyrophana* Pat. (with brown spores). According to Singer (1944) this should be called *Serpula* Pers. ex S. F. Gray. Rea (1922) has the same arrangement except that he does not separate *Gyrophana* from *Merulius*. These last two genera may be looked upon as of Thelephoraceous origin, with the added feature of the folds, ridges or tubercles that increase the hymenial surface. The best known species are *M. lacrymans* (Wulf.) Schum. (*Gyrophana lacrymans*

Fig. 160. Polyporales, Family Meruliaceae. *Merulius lacrymans* (Wulf.) Schum. (Courtesy, Falck, in Möller: Hausschwammforschungen in amtlichem Auftrage, vol. 6, pp. 1–405.)

(Wulf.) Pat.) and its close relatives which have been given intensive study by Falck (1912). This author believes that it is distinct enough as a house fungus ("Hausschwamm"), both in habits and details of structure to deserve specific distinction under the name *M. domesticus* Falck. It is a very destructive enemy of floor boards, beams and other wood construction in buildings, causing a red-colored dry rot. It spreads between timbers in sheets and strands and in openings between the woodwork forms great cottony masses of mycelium. It forms resupinate sheets on the floors and walls and even ceilings of rooms. These enlarge, with white margins and brown central portion on which the hymenium develops. This rapidly becomes wrinkled and porose and sometimes in older stages with flattened teeth much like those of *Irpex*. It occurs very extensively over Europe and Asia but is comparatively rare in America. Some species of *Merulius* are laterally attached and form shelf-like spore fruits. (Fig. 160.)

FAMILY POLYPORACEAE. The fungi here included in one family are placed in two or more families by some of the more modern authors (Rea, 1922; Donk, 1933; Singer, 1944). Until further studies show the definite limits of these different families it may be well to take the more conservative stand and retain the one family.

In the sense that the cells invaded by the hyphae of the fungi of this family are mostly no longer living, i.e., wood fibers and tracheary tissue, these fungi are saprophytes. Many of them, however, attack only the sap wood of living trees in which living cells are intermingled with the dead fiber and tracheary cells, bringing about a "sap rot" and death of the tree. Others, though confined to the heart wood, which contains few if any living cells, attack this wood only in living standing trees. Still others attack only the wood of dead trees or of structural timbers. So there are all grades of practical parasitism even though the particular cells invaded are not living. A very few species of this family are directly parasitic upon other fungi. Although most species are wood inhabiting some grow on the ground, obtaining their nourishment from buried pieces of wood or from the vegetable matter in the soil. Such species are true saprophytes.

The spore fruits may be fleshy or fleshy-leathery when young but at maturity are, with few exceptions, papery, leathery, corky, or even woody. They range in size from a few millimeters in width and 1 or 2 mm. in thickness to a width of 75 cm. (specimens of *Ganoderma applanatum* (Pers. ex Fr.) Pat. collected by the author) and 30 to 50 cm. thick (specimens of *Fomitopsis* (*Fomes*) *officinalis* (Vill.) B. & S. seen by the author), and a width of over two meters in *Polyporus squamosus* (Huds.) Fr., according to Clements (1910). They may be evanescent or may live many years; according to Atkinson over 80 years in the case of specimens of *Phellinus* (*Fomes*) *igniarius* (Fr.) Pat.

The spore fruits may be closely appressed to the sides of tree branches, logs, boards, etc., without a free margin, or may grow out laterally like a shelf or bracket, or may be stalked laterally or centrally. The underside is usually smooth when young, as in the Thelephoraceae, but develops unevenly so as to leave numerous pits (pores) of various shapes on whose inner face the hymenium develops. With but few exceptions the pores are directed downward so that as the spores are shot off from the sterigmata of the basidia that line the pore they drop down and out of the pore into the open air, where they are carried off by currents of air. Cystidia of various types may be present in the hymenium and in some species stiff, pointed, brown setae similar to those in *Hymenochaete* of the Thelephoraceae. This has led to the suggestion that the various species with setae now distributed among several genera, but especially in *Phellinus,* in the Polyporaceae, should be placed in the same group as the above mentioned genus. With very few exceptions the hymenium is confined to the sides of the pores, not usually being formed on the edges as in Meruliaceae. Conidiophores are produced on the upper surface of the sporophore in some species. Under certain environmental conditions a large portion of the tissue of the spore fruit may be converted into chlamydospores, a condition upon which were based the genera *Ceriomyces* and *Ptychogaster*.

The vegetative mycelium is slender and branching, the individual cells often being rather long. Some species of Polyporaceae produce large tuber-like subterranean sclerotia the size of a man's head (e.g., *Pachyma cocos* Fr.). From these the spore fruits arise when conditions are favorable.

The processes of sexual reproduction are known in only a few cases in the family. The mycelium of the spore fruits of many species appears mostly to have clamp connections as does to a large degree the vegetative mycelium growing in the wood. The basidia eventually have four or eight nuclei, four of which pass into the four basidiospores whose nuclei in some cases divide so that the spores become binucleate. In such species the germ tubes show by the presence of clamp connections that secondary or dicaryon mycelium is produced immediately upon germination of the spores. Cultures of uninucleate basidiospores produce monocaryon mycelia which lack clamp connections. When compatible mycelia come into contact diploidization occurs and the dicaryon mycelium may develop clamp connections, or in rarer cases they may fail to develop although the mycelium is dicaryotic.

Mounce and Macrae (1936) showed that in *Gloeophyllum saepiarium* (Wulf.) Karst. (*Lenzites saepiaria*), *Coriolopsis trabea* (Pers.) B. & S. (*L. trabea*) and *Trametes americana* Overh. the monocaryon mycelia arising from germination of the basidiospores fall into only two mutually compatible classes, therefore these species are of "bipolar" sexual behavior. However, there is complete compatibility between all monocaryon mycelia derived from sporophores collected in different geographic regions. This indicates that geographical races exist in these fungi as has been demonstrated in the Ustilaginales (see p. 378) and some of the Heterobasideae (see p. 453). Robak (1936) reported that *Hirschioporus abietinus* (Dicks. ex Fr.) Donk (*Polystictus abietinus*) is, on the contrary, quadripolar in its sexual behavior. He also showed that the monocaryon mycelium of *Coriolellus serialis* (Fr.) Murr. (*Trametes serialis*), a bipolar species, is just as capable of causing rot when inoculated into wood blocks as is the dicaryon mycelium. The production of oidia by the monocaryon mycelia of Polyporaceae has been demonstrated by Vandendries (1936) in *Leptoporus adustus* (Fr.) Quél. (*Bjerkandera adusta* (Fr.) Karst.). Such oidia do not occur on the dicaryon mycelium of this species. This species is quadripolar as are *Leucoporus brumalis* (Fr.) Quél. and *L. arcularius* (Batsch) Quél., neither of which produces any oidia.

The number of species in this family is very uncertain. Several thousand have been described but it is probable that very many of these are synonymous. The opinions as to the validity of described species vary greatly. Thus in Gäumann-Dodge, Comparative Morphology of Fungi (1928) the genus *Polystictus* is credited with nearly 1000 species while Killermann (1928) in the second edition of Engler and Prantl, Die Natürlichen Pflanzenfamilien, admits only "some hundreds." The agreement is still less as to generic limits. Killermann recognizes 16 genera in the family limits adopted in this work, but Murrill, in North American Flora (1907–1908) recognizes 78 genera for North America alone. Bondarzew and Singer (1941) have made a very complete and radical revision of the genera of this family. They exclude 11 genera from the old family limits but still retain 53 genera. Their basis for segregation and classification of the genera is largely anatomical, so that the 11 old Friesian genera are broken up into many sharply defined and not so unwieldy ones. Singer (1944) adds one genus and gives further information as to the systematic arrangement within the family. William Bridge Cooke (1940) recognizes 46 genera from North America (most of them also occurring in Europe) and 19 more from the Tropics and the Southern Hemisphere. The anatomical studies by Miss Ames (1913) contributed considerably to the work leading to the further subdivision of the older genera.

In the following discussion of the more important genera of the family the attempt has been made to give the modern names of the genera and species mentioned as well as the names that are to be found in the older standard works.

The genus *Poria* was in its customary limits used for the completely resupinate members of this family, regardless of the color and consistency of the trama, color of the spores, etc. Studies by Baxter (1929–1949) and others have shown that many species closely related to other genera may develop in a resupinate manner and thus be assigned to the genus *Poria*, which as a consequence became a catchall for unrelated forms which agreed only in their resupinate habits. Nevertheless, it seems that there remains a body of species that are more or less closely related and which properly may be given this name. *Poria* produces resupinate spore fruits which adhere to the substratum and consist mainly of a thin layer of underlying mycelium and the pore layer. Some species produce their fruiting bodies only on the under side of branches or logs so that the pores point directly downward, but this is not the case in all species. The pores are mostly rather small, angular or round, and not very deep. The spore fruits vary in size and color as well as in color of the spores. Some species ascribed to *Poria* have the structure of trama and hymenium characteristic of some of the species of *Corticium* and other closely related Thelephoraceae, differing only in their poroid habit. They probably belong in that family. It may be that they are intermediate forms between the two groups. Aside from these forms and those that are closely related to nonresupinate genera the remaining resupinate species have been divided further on the basis of texture, and color of trama and spores (Murrill, 1907; Donk, 1933; Cooke, 1940; Bondarzew and Singer, 1941).

The annual-fruited forms with effused-reflexed, or shelf-like or stipitate structures and with membranous or leathery texture and with pileus and pore trama similar and continuous were formerly included in the old genus *Polystictus* with many hundred species. By the more recent students of this group this genus has been broken up into six or eight or more genera. Perhaps the commonest species of this group is *Coriolus versicolor* (L. ex Fr.) Quél. It is very common on dead stumps, logs, etc., and forms great numbers of overlapping semicircular or kidney-shaped, velvety-haired pilei, which are strongly marked by zones of various colors. The individual pilei are 2 to 5 cm. in diameter and may grow together at the margin to form broad sheets if they are emerging from the cut top of a stump. On the under side of a log they may be resupinate or effused-reflexed. *Hirschioporus abietinus* is grayish white and hairy above and concentrically furrowed, the edge of the pileus and the pore surface being violet colored in fresh specimens. They are formed on branches, logs, etc., being resupinate on the under side of the substratum but forming shelves at the sides. Mostly on coniferous wood. The pores in age break up into flattened teeth. *Coltricia perennis* (L. ex Fr.) Karst. grows on the ground usually in coniferous forests and forms a funnel-shaped, centrally stipitate spore fruit, brown and velvety above when young, glabrate with age, more or less strongly concentrically marked.

The genus *Polyporus* in its older limits differed mainly from *Polystictus* in being as a rule larger and thicker and more fleshy when young, and with the trama of the pileus usually different from that of the pore layer so that the latter sometimes separates from the former in age. At maturity the spore fruits become cheesy or leathery or corky, rarely woody or membranous. Fifteen or more genera have been segregated by some of the modern students of this genus. Donk (1933) and Singer (1944) separated off some of the species forming the genera *Boletopsis* Fayod and *Scutiger* Murrill, and *Grifola* (S. F. Gray) and placed them near the

Thelephoraceae and Clavariaceae because of the similarity of the structure of the spore fruit and spores to those families, in spite of the occurrence of the hymenium in pores. The genus *Polyporus* in the limited sense consists of tough, fleshy, centrally or laterally stipitate forms. *P. tuberaster* (Jacq.) Fr., considered by Donk and Singer to be the type species of the genus, develops, underground, large, hard, sclerotium-like structures of intermingled hyphae and particles of soil, called pietra fungaia in Italy. When placed in a warm, moist situation several sporophores develop from each sclerotium. These are centrally stipitate, with pileus somewhat funnel-shaped at maturity, scaly, yellowish, with rather large pores. This is fleshy when young and is prized for food by the Italians who collect the sclerotia and preserve them for some time and grow from them the edible spore fruits. Closely related is *P. squamosus* (Huds.) Fr. (*P. caudicinus* (Scop.) Murr.), which is very common and destructive to many kinds of deciduous trees. Its stipe is usually lateral or eccentric and the pileus may exceed a diameter of 50 cm. and a thickness of 3.5 cm. It occurs in imbricated masses growing from the trunks of the infected trees. It also is edible when young. Other fungi included in the old genus ***Polyporus*** are the following: ***Laetiporus sulphureus*** (Bull. ex Fr.) Murr.,

Fig. 161. Polyporales, Family Polyporaceae. *Grifola berkeleyi* (Fr.) Murr. (Courtesy, M. B. Walters.)

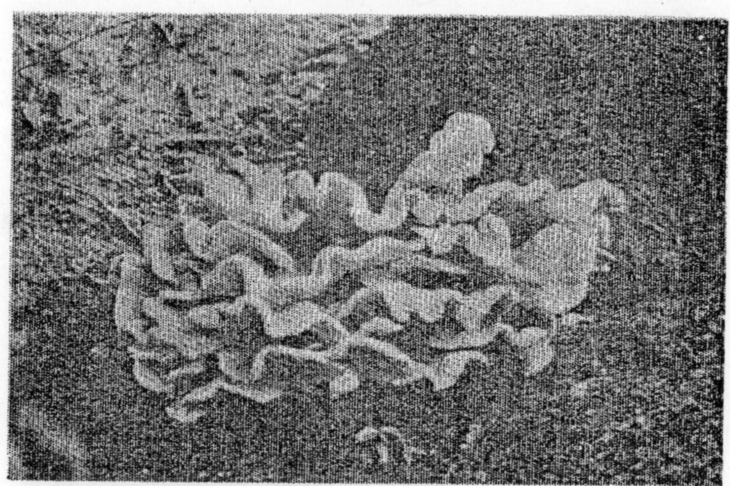

FIG. 162. Polyporales, Family Polyporaceae. *Laetiporus sulphureus* (Bull. ex Fr.) Murr. (Courtesy, F. C. Strong.)

which forms a series of shelves up to 60 cm. broad, bright yellow to orange in color, and rather fleshy and edible at first and dry and cheesy at maturity. The dried sporophores when ground up and soaked in water are edible upon cooking. It occurs at the bases of trunks of deciduous trees, often oak, whose wood it destroys. Occurring very commonly on standing trunks of birch (*Betula*) are the spore fruits of *Piptoporus betulinus* (Bull. ex Fr.) Karst. The more or less hoof-shaped, reniform or globose sporophores are attached by a narrowed, almost stipe-like portion to the side of the trunk. The gray or whitish surface consists of a thin layer which flakes off with age from the white pilear trama. The layer of 3 to 8 mm. long pores separates easily from the thick layer of the pileus. (Figs. 161, 162.)

The old genus *Trametes* was supposed to be characterized by the uninterrupted continuation of the pilear trama into that of the pore layers, with the further character that the pores were of different depths in the same spore fruit. These are true of some species but also occur in some of the genera formerly included in *Polyporus*, *Polystictus*, and *Fomes*. As a result the genus *Trametes* has been much reduced and segregated into six or eight genera. Among these is *Pogonomyces hydnoides* (Schw.) Murr., a very common species of Florida and the Tropics. Its dimidiate, sessile, sometimes imbricate spore fruits may be 5 to 10 cm. broad and up to 1 cm. thick. The upper surface is covered by long, black, stiff, branched fibers which resemble considerably the teeth of some species of Hydnaceae. The pileus trama is dark brown, punky to corky and that of the tubes light brown. *Pycnoporus cinnabarinus* (Jacq. ex Fr.) Karst.

FIG. 163. Polyporales, Family Polyporaceae. *Ganoderma applanatum* (Pers. ex Fr.) Pat. Habit view on dead trunk of maple (*Acer*). (Courtesy, F. C. Strong.)

is also dimidiate, 4 to 10 cm. in diameter and up to 1 cm. thick. It is bright orange to cinnabar-red and the trama also is red, as are the pores. It occurs on dead wood of various deciduous trees in America, Europe, and Asia. At maturity it is corky to punky.

The old genus *Fomes* is now mostly broken up into several genera. Its chief characters were the corky to woody texture of the pileus and the perennial nature of the spore fruit, so that in successive growing seasons new layers of pores are formed below the ones last formed. The genus *Fomes* in the restricted sense has a corky to punky pilear trama which is more or less rusty brown in color. The spore fruits are somewhat hoof-shaped. *F. fomentarius* (L.) Gill. forms its sporophores on standing trees. They are provided with a hard crust, black and shining with age, and are 8 to 10 cm. in width laterally, 7 to 9 cm. from front to back, and 3 to 10 cm. tall, depending upon age. These were formerly used as a source of tinder for kindling fires and are often still called "punks." *Fomitopsis* differs in having its pilear trama whitish or light-colored (not rust brown). *F. officinalis* (Vill.) B. & S. grows on larch (*Larix*) in the Northern Hemisphere. Its chalky white, intensely bitter, friable trama has been used for medicine for many centuries. The large spore fruits may become almost cylindrical, about 15 cm. in diameter and up to

Fig. 164. Polyporales, Family Polyporaceae. *Ganoderma applanatum* (Pers. ex Fr.) Pat. Vertical section through three-year-old sporophore. (Courtesy, Buller: Researches on Fungi, London, Longmans, Green and Co.)

30 or more cm. in height. *Phellinus igniarius* (L. ex Fr.) Quél. resembles somewhat *Fomes fomentarius* but lacks the horny crust and is pubescent when young. The hymenium of the pores has numerous sharp brown spines. *Ganoderma* differs from the preceding genera in possessing spores truncated at one end and two-layered, the brown endospore being spiny, the spines projecting up into the hyaline exospore. The surface of the spore fruit has a hard crust formed by a palisade of thick-walled elongated cells. A stipe may be present in some species. A varnish-like coating may be present over the whole surface or only on the stipe or may be entirely lacking. In the narrow use of this name only those with the varnished layer would properly belong to the genus. *G. lucidum* (Leyss. ex Fr.) Karst. occurs on coniferous and deciduous trees and is annual. It is varnished over the whole top surface as well as the stipe. *G. curtisii* (Berk.) Murr. is perennial and may produce several layers of pores. It loses its laccate covering early. *G. applanatum* (Pers. ex Fr.) Pat. lacks the laccate surface entirely but has a whitish to gray crust. It is one of the commonest species in North America, being found on fallen trees and old stumps of deciduous species almost everywhere. Its spore fruits may attain a diameter of 75 cm. White (1920) estimated that a large spore fruit of this species may liberate 30 billion spores a day for several months attaining a total of 5500 billion spores for the season. Yet of all this vast number of spores carried far and wide by the wind all but a very few must perish. With so many spores in the air it is not to be wondered at that the fungus is very common wherever a deciduous tree has died or its trunk has fallen. This fungus does not attack healthy uninjured trees. (Figs. 163, 164.)

Probably closely related to the genera grouped about th genus

Trametes are those forming the Tribe Daedaleae (of Bondarzew and Singer, 1941). In these the pores are elongated radially or are labyrinthiform. *Daedalea* is distinguished by having the pores elongated or labyrinthiform. Its spore fruits are shelf-like and corky. *D. confragosa* (Bolt.) Fr. is very common in North America and Europe. In *Lenzites* the pores are elongated radially from the point of attachment so as to resemble gills, with occasional cross connections which may disappear with age. The spore fruit is more or less corky. *L. betulinus* (L.) Fr. is common on birch and other trees in North America and northern Eurasia. Its fruiting bodies are 3 to 7 cm. broad, 3 to 10 mm. thick, velvety and zonate above. But for the cross connections in the young specimens this might well be placed in the Family Agaricaceae. Donk (1933) and some others follow Schroeter in dividing the genus *Daedalea* by separating off *D. confragosa* to form the genus *Daedaleopsis*. Karsten (1882) separated off from *Lenzites* the species with rust-brown trama as the genus *Gloeophyllum*. This would include the common *Lenzites saepiaria* (Wulf.) Fr., which causes the decay mainly of coniferous wood. The species of this tribe are very variable as to pore form. In the same species some specimens may have poroid, labyrinthiform, lamelloid, or even irpiciform hymenophores. The distinctions between some species of *Trametes, Daedalea*, and *Lenzites* are therefore rather arbitrary.

Undoubtedly belonging to the Polyporaceae are the species of *Irpex* (*Irpiciporus* of Murrill) and probably the genus *Echinodontium*. *Irpex* may be resupinate, effused-reflexed, or shelf-like. The younger parts of the hymenophore are poroid but with increasing age the walls of the pores grow unevenly so as to produce flattened teeth. Thus the specimen comes to resemble closely some of the species of the old genus *Polystictus*. *Echinodontium* has woody, shelf-like, or unguliform fruiting bodies resembling some of the forms of the genus *Fomes* but with the hymenophore composed of irpiciform plates which have small lateral teeth along their edges. The spore fruits are brightly colored and were formerly used by the Indians of Northwestern United States as a source of a red dye. It causes decay of the hemlock (*Tsuga*) and of Fir (*Abies*) in Alaska and northwestern United States. Both *Irpex* and *Echinodontium* were formerly placed in the Hydnaceae.

The course of evolution in this family is very uncertain. Some of the *Poria*-like forms may be primitive but as very many of the normally pileate or shelf-like genera may become resupinate this latter habit cannot always be considered to be a proof of primitiveness. Some of the *Poria* group may have arisen from Thelephoraceae that were more or less *Corticium*-like. It may be possible that from the Hydnaceae, by union of the teeth into pores some Polyporaceae may have developed. Probably the stratose species like *Fomes* have developed from annual species. The labyrinthiform or lamelloid species like *Daedalea* and *Lenzites* may show relationship to the Agaricaceae but they may represent merely a parallel course of development. Some of the genera close to *Polyporus* have gills radially elongated.

Order Agaricales. Largely fleshy, but some leathery or even corky or woody at maturity. Basidia usually chiastic (after removal of *Cantharellus* to the forgoing order). Hymenium on lamellae (gills) or pores (Boletaceae), gymnocarpic or pseudoangiocarpic or angiocarpic. Spore fruits mostly stipitate, more often centrally, but sometimes attached laterally or even resupinately, without stipe. Hymenium usually not formed on the edges of the pores or lamellae. The distinction between Polyporales and Agaricales is not always sharp and it is possible that these intergrade so completely that the two groups cannot justifiably remain as distinct orders.

FAMILY BOLETACEAE. This family consists of fungi growing on the ground almost exclusively. The spore fruits are fleshy and stipitate, centrally so in most species. The pileus is thick and convex and the layer of pores is mostly easily separable from it. The pores are usually easily broken apart from each other. In the genus *Ixechinus*, described by Heim (1939) from Madagascar, the trama separating the young pores splits so that at maturity they are separate and become divergent with the uprolling of the pileus. Thus they have a superficial resemblance to *Fistulina*, but the origin of the separate pores is entirely different. In some genera a veil covers the layer of pores in the young spore fruits, extending from the edge of the pileus to the stipe. The pores in some genera are elongated somewhat in a radial direction, suggesting a transition to or from the Agaricaceae.

Some species form large spore fruits. Heim (1936) described *Boletus* (*Phlebopus*) *colossus* from Madagascar with a pileus up to 60 cm. broad and 4 to 6 cm. thick, and with a stipe up to 25 cm. tall and 22 cm. thick in the lower swollen basal portion. The whole fungus weighed 6 kg. The flesh of the various species in the family may be mainly white or pink or yellow. In many species it becomes blue or blue-green when bruised, in others remaining unchanged. Treatment with KOH, NH_4OH, and other chemicals brings about various color changes which are of value in the distinction of species and genera. The spores vary from pale to yellowish to purplish or yellow-brown and may be small or above 20μ in length. They may be thin-walled and smooth or may have external ridges, warts or reticulations. In some cases the endospore is covered with prickles which reach into or through the exospore. Germ pores are produced in a few species. Some species are edible and some are known to be poisonous. *Boletus edulis* Bull. ex Fr., according to Mez (in a verbal communication to the author), contains a toxalbumin that when injected into the blood stream is extremely poisonous but whose toxic character is destroyed in the process of digestion when the fungus is eaten.

The spore fruits of various members of the family have been studied as to their ontogeny by Kühner (1927), Elrod and Snell (1940), and others. In some cases they are entirely gymnocarpic, i.e., as they develop they are more or less cylindrical and then at the top the hyphae spread outward to form the pileus. The basidia begin to appear and to bear spores along the stipe, especially the upper portion, and the under side of the pileus, before the pores begin to develop. They may even be formed on the upper surface of the pileus. The enlarging pileus curves downward at the margin as the pores develop but in the truly gymnocarpic forms never curves inward so far as to come into contact with the stipe. In the pseudoangiocarpic species the development is like the foregoing except that the edge of the pileus eventually comes into contact with the stipe, forming a circular enclosed chamber lined above by the developing pores and centrally by the stipe. The marginal tissues of the pileus and the portion of the stipe with which they join may enlarge as the pileus grows in diameter and flattens out, so as to produce an annulus which in *Paragyrodon sphaerosporus* (Pk.) Sing. (*Boletus sphaerosporus* Pk.) spreads from near the base of the stipe to the margin of the mature pileus as a grayish-white sheet up to 6 or more cm. broad, leaving a chamber between it and the pore layer. Whether an annulus is formed or not the enlargement and flattening out of the pileus eventually exposes the pores to the air so that the spores may be carried away by air currents. It is doubtful whether true angiocarpy occurs in this family, i.e., development of the pores in a cavity formed internally in the spore fruit and not by the curving downward and inward of the pileus margin. In the genus *Gastroboletus* the surface layer of the pileus and stipe remain connected and include the young hymenophore like a peridium. Here, as in *Paragyrodon sphaerosporus*, the development is probably pseudoangiocarpic.

The several hundred species here included in one family were divided by Singer (1936) into two families, Boletaceae and Strobilomycetaceae with a total of 21 genera, to which (1945–1947) he added three other families, Gomphidiaceae, Paxillaceae and Jugasporaceae, usually placed in the old family Agaricaceae. These five families then form his Suborder Boletineae. Coker and Beers (1943), on the other hand, recognize only three genera of Boletaceae in the usual sense, in North Carolina. Murrill (1910) recognizes 11 genera in North America.

It has long been recognized that the genus *Paxillus* of the old family Agaricaceae has many points of similarity to the Boletaceae: frequent occurrence of transverse ridges between the lamellae, ease of separation of the lamellae from the pilear trama, gymnocarpic development of the spore fruit and certain cytological and chemical similarities. Whether these justify transferring *Paxillus* to the Boletineae, as Singer does, or indicate the close relationship of the latter group to the Agaricaceae is more a matter of opinion. In this book *Paxillus* will be retained in the Agaricaceae.

The Boletaceae are found in the temperate, subtropical and tropical regions of both hemispheres, but especially where the rainfall is fairly abundant or in the season of the year when considerable rain falls. They do not occur in arid regions. Many, perhaps most, of them occur in connection with mycorrhizal development on roots of mostly woody plants. Some are confined to the roots of conifers and some of this group are limited to certain genera (e.g., *Larix*, *Pinus*, etc.). *Gyrodon merulioides* (Schw.) Sing. (*Boletinus porosus* (Berk.) Pk.) is known only in proximity to trees of ash (*Fraxinus*) and *Paragyrodon sphaerosporus* (Pk.) Sing. (*Boletus sphaerosporus* Pk.) only near species of oak (*Quercus*), etc.

Among the larger and commoner species several may be mentioned. *Boletus edulis* Bull. ex Fr. has a reddish brown pileus, white or yellowish within, the flesh not becoming blue upon wounding. The pores are yellowish and become greenish with age. The stipe is reticulately marked. The pileus is 6 to 20 cm. broad and 2 to 4 cm. thick and the stipe cylindrical or enlarged below and 5 to 10 cm. tall and 3 to 4 cm. thick. The spores are yellowish to ochraceous brown. The flesh has a pleasant nutty taste. This highly prized edible species occurs in frondose woods. The species related to *Boletus luridus* Schaeff. ex Fr. (*Suillellus luridus* (Schaeff.) Murr.) are often about the same size as the foregoing and the pileus has much the same appearance above. The flesh is whitish or yellowish but becomes blue very rapidly when exposed to the air. The pores are yellowish with red mouths. The stipes are 5 to 10 cm. tall and 1 to 2 cm. thick, reddish below, yellow above, reticulated near the top. This species is reputed to be poisonous. *Tylopilus felleus* (Bull. ex Fr.) Karst. (*Boletus felleus*), also may be confused by a beginner with both the foregoing. It differs in its pink spores, its white flesh, which may turn pink on wounding, and its intensely bitter taste. The pores, which are white, become flesh colored as the spores are produced in large numbers. The stipe may be reticulate above or completely reticulate. All three species may be found in the same woods. *Suillus luteus* (L. ex Fr.) S. F. Gray grows in the vicinity of species of pine. The yellowish to reddish brown pileus is very viscid. The flesh is pale yellowish, not changing color when wounded. The stipe is pale yellow to reddish brown and glandular dotted, and has a large persistent annulus. It is edible. *Strobilomyces floccopus* (Vahl ex Fr.) Karst. (*S. strobilaceus* (Scop. ex Fr.) Berk.) has dark spores, completely covered by a network and with a distinct germ pore. The pileus and stipe are gray when young but the numerous shaggy scales quickly become dark. The gray to white pores become reddish or black on wounding or bruising, and are more or less lamellar near the stipe, which has an annulus. This edible species grows in frondose or mixed frondose and coniferous woods. *Gyrodon merulioides* has an eccentric or lateral stem and the pores are formed by radiating lamellae connected by numerous cross veins not quite so high as the main lamellae. The pileus is reddish brown, with yellow flesh slowly turning bluish green when wounded. The hymenial surfaces are yellow becoming slightly blue on wounding. The spores are yellowish brown. The pores are decurrent somewhat on the hollow stipe. This approaches closely some of the species of *Paxillus*. (Fig. 165.)

Fig. 165. Agaricales, Family Boletaceae. *Tylopilus felleus* (Bull. ex Fr.) Karst. (Courtesy, Atkinson: Studies of American Fungi, Ithaca, N. Y., Andrus and Church.)

Family Agaricaceae. This family in its broader and more customary usage included those fungi whose fruit bodies increased the hymenial surface by the production of radiating lamellae which are entirely covered, or all but the edge, by the hymenium. The latter may or may not extend from gill to gill on the interlamellar surface of the pileus. The interior tissue of the lamella (the trama) may continue unchanged up into the pileus or the pilear trama may be distinct in structure, color, etc., from the lamellar trama, paralleling the conditions in the Polyporaceae.

In contrast with the Polyporaceae where the spore fruits are prevailingly rather dry at maturity those of the Agaricaceae are mostly fleshy, although some dry forms occur. In the vast majority of cases they are centrally stipitate, rarely laterally so, occasionally attached laterally without a stipe, or even partially resupinate. In size the pileus may vary from a very few millimeters in diameter in some species of *Marasmius* to 40 cm. in specimens of an exannulate form of *Agaricus arvensis* Schaeff. ex Fr., collected by the author. A specimen of this size must be capable of producing an enormous number of spores since Buller (1909) has shown that a not unusually large specimen of *A. campestris* L. ex Fr. can produce 1800 million spores.

In general the basidia are club-shaped, varying to ovoid or cylindrical. Usually four basidiospores are produced although species or races frequently occur in which the number is two. In the latter case this may result from the development of spore fruits on monocaryon mycelium and the consequent lack of nuclear fusion and meiotic divisions in the basidium, there being only one division and that mitotic. The spore fruit may arise from a dicaryon mycelium with normal nuclear phenomena in the basidium and yet produce only two spores on the basidium with two nuclei passing into each. In some members of the family, as occurs frequently on other Eubasidiae a mitotic division subsequent to the second meiotic division results in the production of eight nuclei in the basidium of which four may enter into the basidiospores and four remain behind, or two proceed into each spore. In a stained specimen of *Coprinus* sp. studied by the author each basidium possessed eight nuclei but after spore discharge four still remained in the basidium. With the exclusion of *Cantharellus* all Agaricaceae are chiastobasidial.

The hymenium may consist entirely of basidia or there may be cystidia of various types. In *Coprinus* the basidia are separated by large sterile cells called paraphyses, not so tall as the basidia but much broader so that the latter are arranged in squares, one at each corner where four paraphyses meet. Such paraphyses are not produced in most of the genera of the family. The cystidia may differ in appearance from the basidia only in the absence of sterigmata or they may be elongated, pointed, or forked, or knobbed so as to resemble a tenpin. The upper portion of the cystidium is often covered with crystals. Some cystidia are the terminal cells of laticiferous hyphae or of hyphae containing mucilaginous substances. The cystidia occurring at the edge of the lamellae are called cheilocystidia and may resemble or differ from those borne on the faces of the gills, the pleurocystidia. Where, as in some species of *Coprinus*, the large, stout pleurocystidia extend across the space between two gills and assist in holding them apart they are sometimes called trabecular cystidia. (Fig. 166.)

Fayod (1889) distinguished several types of gills by their tramal structures. Other mycologists since then have recognized the value of these distinctions in the classification of the Agaricaceae. The outer surface always consists of the vertically standing basidia (and cystidia if present) arising from a subhymenium, a thin or thick layer immediately below the basidia. This may be indistinguishable. In the *mixed* or *irregular* trama the hyphae are arranged without apparent order, being sinuous or interlaced. The *regular* trama consists of elements clearly parallel. In the *bilateral* trama the trama proper is reduced to a thin median plane from which the hyphae diverge obliquely in a curved line toward the strongly developed subhymenium. In the *inverse* trama the young lamella has the "regular" structure with a distinct subhymenium. As maturity approaches hyphae grow from the subhymenium obliquely inward, filling the space formerly occupied by the vanished median portion of the trama. It must be recognized that these four types grade into one another and that at times it is difficult to decide which type is present, especially if the gill that is being studied is a little too young or too old.

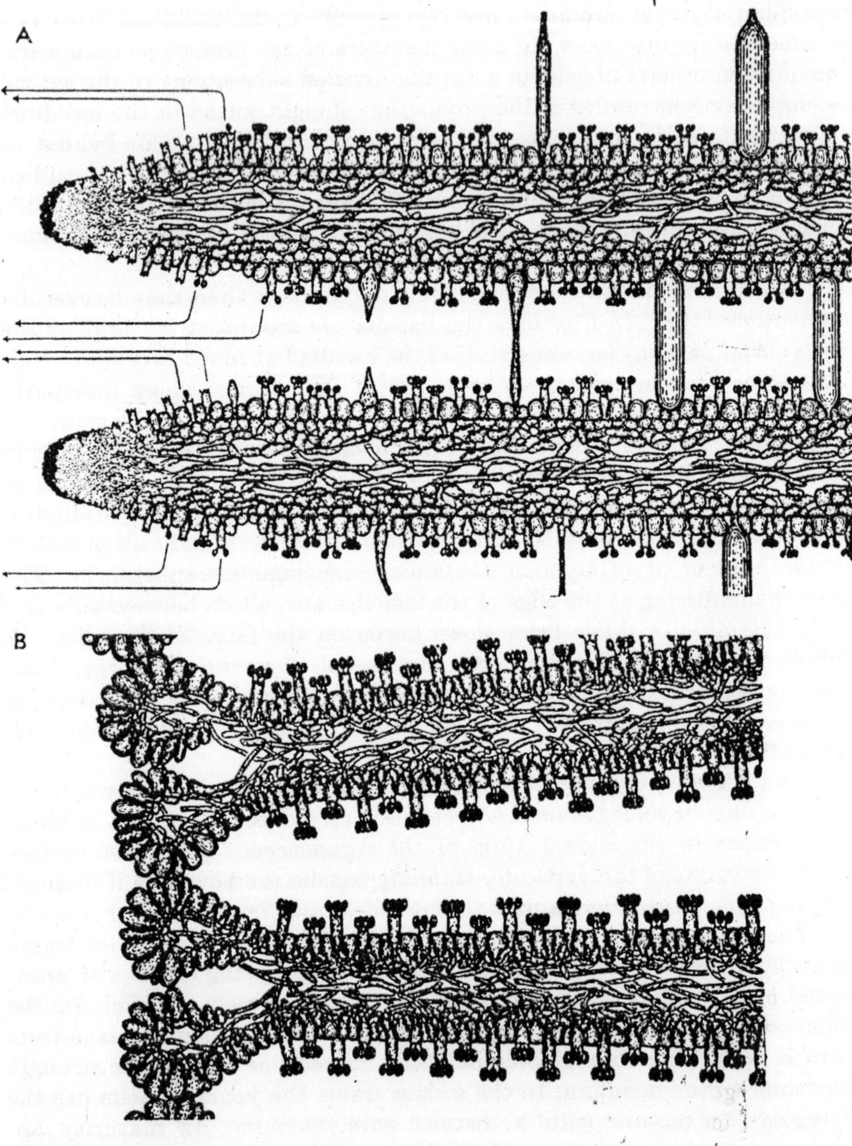

Fig. 16C. Agaricales, Family Agaricaceae. Mechanisms which, by holding the gills apart, permit the falling of the spores. (A) *Coprinus atramentarius* Fr. with gills held apart by trabecular cystidia; autolysis beginning at edge. (B) *C. sterquilinus* Fr. with thickened edges that hold the gills apart. Note the long and short basidia. (Courtesy, Buller: Researches on Fungi, London, Longmans, Green and Co.)

Many Agaricaceae show clamp connections at the septa of almost all the hyphae of the lamellar and pilear trama, the cortex of the pileus, and the tissues of the stipe as well as the mycelium from which the spore fruit arose. In many cases the clamp connections occur only at occasional septa and only in special portions of the pileus or stipe, often only in the cortical regions. In *Russula* and *Lactarius* groups of spherical cells (sphaerocysts) are found in the pileus and even in the lamellae in addition to the usual slender hyphae. The pilear trama of other genera may show enlargements of the hyphae but not so characteristic as those of the genera mentioned. Slender hyphae bearing clamp connections frequently do not develop these structures where cells much larger in diameter are developed as they grow. In the genus *Lactarius* the tissues of the pileus, stipe and gills possess branching hyphal tubes filled with latex which flows out of breaks in the tissue and coagulates. The fresh latex may be white, pink, yellow, green, blue, or even colorless. It may change color on exposure to the air or remain unchanged. Aside from the protection to wounds afforded by the coagulated latex it may be that the laticiferous tubes serve for food transportation. Besides this genus laticiferous vessels are reported by Heim (1936b) in three other genera of Agaricaceae from Madagascar, *Bertrandia*, *Mycena*, and *Rhodophyllus*, as well as in some species of Gasteromycetes.

The species of Agaricaceae may be "homothallic" or "heterothallic" and bipolar or quadripolar. Most genera of the Agaricaceae have not been grown in culture so that it is not known how extensively the appearance of two or more sexual phases is to be found. The phenomena of sexuality have been discussed rather fully in the preceding chapters. Oidia are produced in abundance in monocaryon cultures but are usually absent from dicaryon mycelium. Where they do occur, as Vandendries and Martens (1932) described for *Pholiota aurivella*, they may be binucleate and give rise to dicaryon mycelium or divide into two uninucleate cells which give rise to monocaryon mycelium.

The Agaricaceae are mainly saprophytic, living on the ground, decaying leaves, bark, wood, manure, etc. Some species of *Nyctalis* and *Volvaria* are parasitic on other Agaricaceae. *Armillariella mellea* (Vahl) Karst. (*Armillaria mellea*) attacks the roots of trees and kills their cortical tissue, growing up in the living portions of the bark of the tree trunk and causing the death of the tree. It is especially destructive to apple and cherry trees planted where previously there were oaks or other trees on whose roots this fungus once grew parasitically, continuing to live as a saprophyte until the apple or cherry roots became available. *Agaricus campestris* L. ex Fr. and one or two other species are cultivated for food and many wild species are collected for this purpose. Buller (1922) pointed out that *Marasmius*

FIG. 167. Agaricales, Family Agaricaceae. *Armillariella mellea* (Vahl) Karst. (Courtesy, Dow V. Baxter.)

oreades (Bolt. ex Fr.) Fr. when grown in manure produces large amounts of edible fungi much larger than the wild form. (Fig. 167.)

Among the edible wild species are various kinds of *Lepiota*, e.g., *L. naucina* (Fr.) Quél., *L. procera* (Scop. ex Fr.) Quél. (*Leucocoprinus procerus* (Scop. ex Fr.) Pat.). It must be noted that some species of this genus are poisonous. Most of the larger species of *Agaricus* (*Psalliota* of some authors) are edible, e.g., *A. campestris* L. ex Fr., *A. rodmani* Pk., *A. arvensis* Schaeff. ex Fr., etc. *Pluteus cervinus* (Schaeff. ex Fr.) Quél., *Coprinus comatus* Fr., *C. micaceus* Fr., *C. atramentarius* Fr. (but see under poisonous species below), *Marasmius oreades*, *Lactarius volemus* Fr., *L. deliciosus* (L.) Fr., *Pleurotus ostreatus* (Jacq. ex Fr.) Quél., *P. ulmarius* (Bull. ex Fr.) Quél., *Collybia radicata* (Rehl.) Berk., and various other species of *Collybia*, *Tricholoma personatum* (Fr.) Quél., *Armillariella mellea*, and many others are edible. Of the genus *Amanita*, *A. rubescens* (Pers. ex Fr.) Fr. and *A. caesarea* Schaeff. ex Fr. are highly prized, but see note concerning this genus among the poisonous species below. Aside from the hundreds of edible species of mushrooms there are many that are poisonous, some mildly so, some dangerously. Besides these edible and poisonous sorts the larger number of species are too small or tough or slimy in consistency, or of disagreeable odor or taste even though not poisonous, or occurring only occasionally here and there so as to be too scattered for collection in sufficient numbers for a meal, so that the majority are not usually subjects of concern as possible edible products.

Among the poisonous mushrooms several species of *Amanita* are exceedingly dangerous. *A. phalloides* (Bull.) Fr. and some of its close rela-

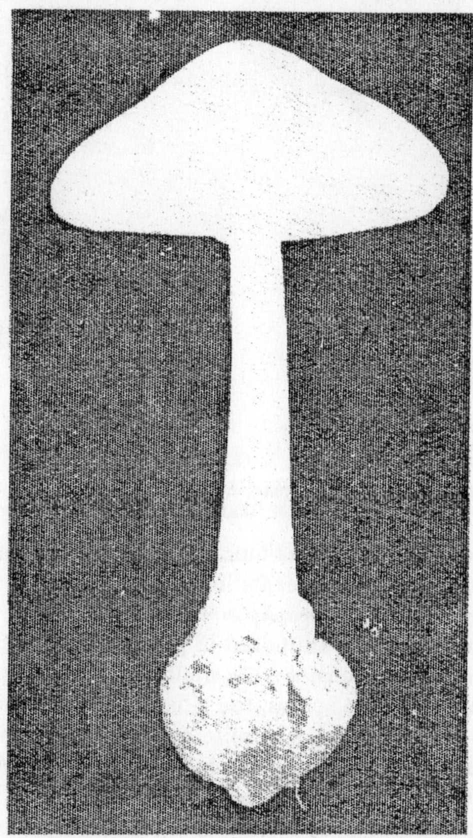

Fig. 168. Agaricales, Family Agaricaceae. *Amanita verna* (Fr.) Quél. (Courtesy, M. B. Walters.)

tives (e.g. *A. verna* (Fr.) Quél.) are so poisonous that a piece one cubic centimeter in size will cause severe illness or even death, there being no known antidote for the poisonous principle. *A. muscaria* (L.) Fr. was formerly used for the purpose of destroying flies as the sticky cuticle is very poisonous. In some parts of Europe and Siberia after careful removal of the cuticle the remainder of the fungus is eaten with apparent safety. Some tribes in Siberia make an intoxicating beverage out of the fungus.

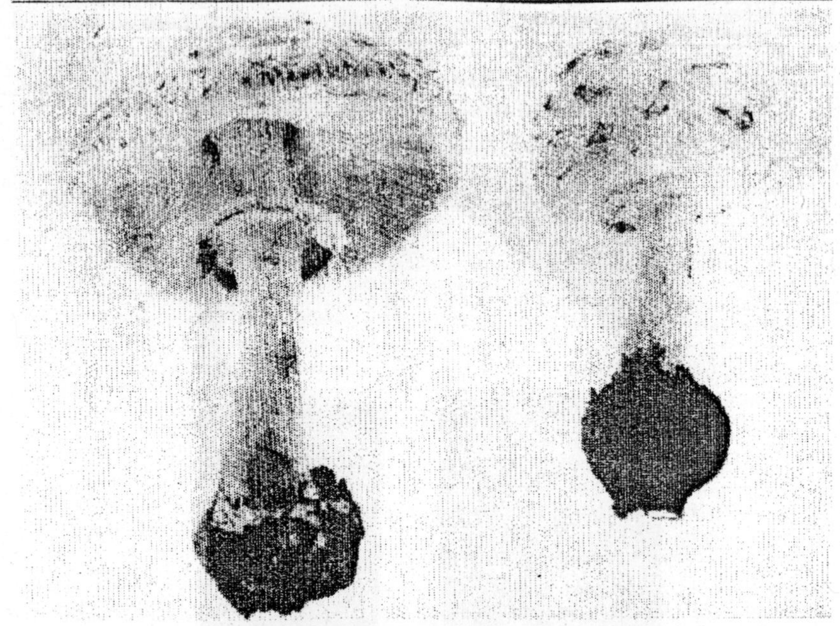

FIG. 169. Agaricales, Family Agaricaceae. *Lepiota rachodes* (Vitt.) Quél. (Courtesy, M. B. Walters.)

It is wiser to avoid this species as numerous cases of poisoning and some of death have been recorded after its ingestion. Atropin is a partial antidote. *Lepiota morgani* Pk. (*Chlorophyllum esculentum* Mass., *Leucocoprinus molybdites* (Meyer) Heim) is abundant in parts of North America as well as in the Tropics. It has considerable similarity to the edible *L. procera* but is stouter and has green spores, the gills taking on a greenish tinge as the spores develop. This species is quite poisonous for some persons but may be eaten with apparent impunity by others. *L. helveola* Bres. is also poisonous, sometimes very dangerously so (Josserand, 1931). Many of the pink-spored fungi (*Entoloma* or *Rhodophyllus* of some authors) are poisonous although the edible *Pluteus cervinus* has pink spores and is safe. Several species of *Tricholoma* are unsafe. Recently a case of slight poisoning occurred with fresh specimens of a species of *Agaricus* where other persons eating specimens of the same lot were not affected. (Figs. 168, 169.)

The general conclusion to be drawn from the foregoing is that because some species of a genus are safe this does not necessarily give a clean bill of health to other species of the same genus. Besides that there are personal susceptibilities to the poisonous characteristics of some fungi. It has been shown (by Suss, 1936, and Hugon, 1938) that *Coprinus comatus* and

C. atramentarius may cause quite severe poisoning if alcoholic beverages (beer or wine, as well as distilled liquors) have been drunk shortly before or within several hours after eating the fungi. The method of preparation may have considerable effect upon the poisonous nature of some species. Parboiling and pouring off the water will often remove the poison but not in all cases. Of course almost all fungi which have reached the state of incipient decay are dangerous. This is the basis of the "silver spoon" test, for the blackening of the silver surface is not due to the poison naturally resident in the fungus but to the products of incipient decay. It is unsafe to eat a fungus unless its identity is certain and then only perfectly fresh specimens or dried unspoiled specimens.

Botanically it is insisted that the words mushroom and toadstool are practically synonymous and may be used indiscriminately for both edible and poisonous species. It must be recognized, however, that originally the word toadstool was derived from a word meaning death's chair (Todesstuhl).

The poisonous principles in the Agaricaceae appear to belong to the chemical groups of "toxalbumins" in some cases and alkaloids in others.

Studies of the development of the spore fruits of centrally stipitate members of this family show that with reference to the hymenial origin they may be gymnocarpic, pseudoangiocarpic or truly angiocarpic. In the first the hymenium is external from the beginning and never enclosed in a cavity. At the upper part of the young columnar spore fruit the tissues spread out laterally, to form the young pileus. On the under side of this and often on the upper part of the stipe the hymenium begins to develop, gradually producing radial folds, the gills. At maturity these bear the mature basidia and basidiospores and are at no time cut off from the outside air by tissue of any kind. (Fig. 170 A–C.) In the pseudoangiocarpic species development is as above at first, but the broadening pileus curves downward at the edge and finally curves back to the stipe with which it comes into loose contact or to which it becomes united by the intermingling of hyphae from the stipe and edge of the pileus. Thus a closed circular cavity develops on whose roof the lamellae are produced and become covered with the hymenium. At the approach of maturity the pileus flattens out and its edge breaks free from the stipe, so that now the hymenium with its ripening spores is exposed to the air. There may be left on the stipe a collar (annulus) or the sheet of connecting tissues may break away from the stipe and remain hanging at the edge of the pileus as a cortina. This may be in broken sheets or like a spider web. (Fig. 170 D–G.)

In the angiocarpic forms there develops within the tissues of the pileus a circular layer of palisade cells, the hymenium primordium, on the lower or inner side of which a circular cavity is formed into which the radiating

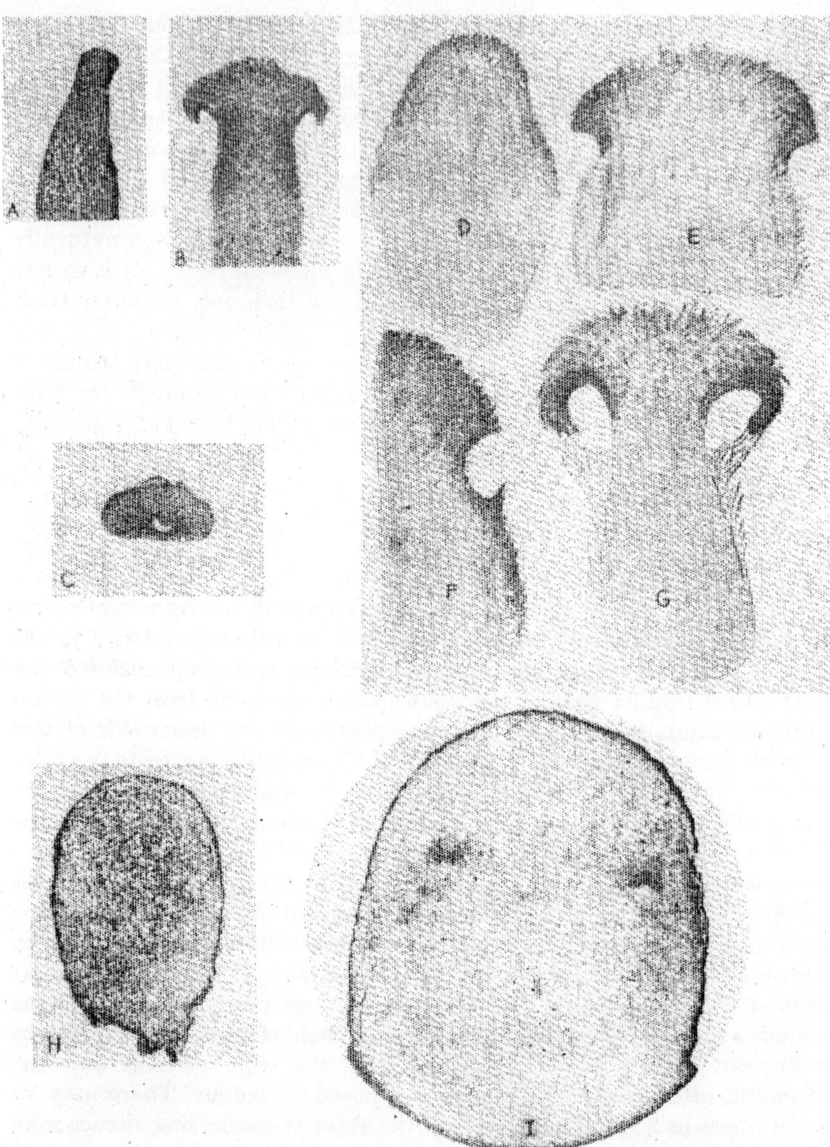

Fig. 170. Family Agaricaceae. Methods of development of the hymenophore. (A–C) Gymnocarpic type, in *Omphalia chrysophylla* Fr. (D–G) Pseudoangiocarpic type, in *Lentinus tigrinus* (Bull.) Fr. (H, I) Angiocarpic type, in *Agaricus campestris* L. ex Fr. (A–C, courtesy, Blizzard: *Am. J. Botany*, **4**(4): 221–240. D–G, courtesy, Kühner: *Compt. rend.*, **181**(3): 137–139; H–I, courtesy, Atkinson: *Botan. Gaz.*, **42**(4): 241–264, Univ. Chicago Press.)

gills protrude as they develop. According to Levine (1922) the early appearance of the annular cavity is an artefact and the openings arise first between the developing gills and only later below their edges if at all. In *Coprinus* the gills remain in contact with the upper part of the stipe until the spore fruit is nearly mature and the pileus begins to expand. Eventually this circular cavity is bounded on the inner side by the portion of the spore fruit that becomes the stipe, and by the flattening of the pileus a circular break next to the stipe permits the pilear expansion so that the gills now become exposed to the air and shed their spores. (Fig. 170 H–I.) In these forms the surfaces destined to bear the hymenium arise within the pileus while in the other two types it was at first exposed to the air and became ultimately enclosed only in the pseudoangiocarpic type of development. For details as to gymnocarpic and pseudoangiocarpic development consult Kühner (1925b, 1926), Blizzard (1917), Douglas (1918), Walker (1919), Reijnders (1933), Heim (1936b, 1937), and others, and for the angiocarpic type Atkinson (1906, 1914, 1915, 1916), Levine (1922), Douglas (1916, 1920), and some of the foregoing list.

The 5000 to 8000 species of the Agaricaceae are divided into 50 to 112 genera, according to the ideas of the various students of the group. Fries (1821) recognized only two genera in the limits of the family as here treated, *Agaricus* and *Schizophyllum*. Killermann (1928) recognized 66 in the second edition of Engler and Prantl. Singer (1936) included about 112 genera. More often these are all included in only one family but more recently Heim, Singer, and others have divided them into 10 or 11 families. It will probably be desirable to make a segregation, but until there is greater agreement as to just what the generic distinctions must be and how the families should be delimited the author will be conservative and include all in one family.

Fries divided his genus *Agaricus* into 38 tribes, most of which were later recognized as genera by him and by other mycologists. The primary basis of the division into two genera was a splitting of the lamellae in *Schizophyllum* and the entire lamellae in *Agaricus*. The latter was divided into series on the basis of spore color: white, rose, ochraceous, rusty, purple-brown, and black. In the "Sylloge Fungorum" (Saccardo, 1887) these primary subdivisions were called respectively Leucosporae, Rhodosporae, Ochrosporae (including both ochraceous and rust-colored spores), Melanosporae (including both purple-brown and black spores). The location of the stipe (central, eccentric, or lateral), the presence or absence of annulus and volva, the shape of the pileus, the character of the stipe, the relation of lamellae to stipe (i.e., free, attached, decurrent, etc.), the color of the gills and of the pileus were all characters used in dividing these spore-color groups into the lesser groups (tribes or genera). All these characters were largely external and did not take into consideration the internal structures such as have been found of so great importance in the Polyporaceae. Fayod (1889) and others (Karsten, Heim, Singer, etc.) have emphasized more and more the necessity of basing the generic distinctions upon these anatomical and chemical characters. The type of gill structure as emphasized by Fayod, the amyloid or nonamyloid character of the spore wall, the structure of the cuticle of the pileus, the presence or absence of clamp connections, all these have proved to be of great importance. Just how far these characters should be used in distinguishing genera or whether they should only be used for lesser units of classification is not yet agreed upon, hence the above-noted great differences in the number of recognized genera or even families.

The characters of size—of spores, cystidia, pilear hyphae, etc., as well as of the whole spore fruit—are not always reliable. Thus Togashi and Oda (1934) showed that the age of the pileus has considerable effect upon the spore size. In *Armillariella mellea* and in *Pholiota adiposa* (Fr.) Quél. the length and width of the spores shed on the fourth day were 12 to 15 per cent less than of the spores shed on the first day from the same sporophores. On the other hand *Collybia velutipes* (Curt.) Fr. (*Myxocollybia*) does not exhibit any such change in spore size with the age of the pileus. The size and even the shape and color of the pileus, and sometimes of the gills, depends considerably upon the temperature, humidity, and illumination, and to a large degree upon the amount of food available in the substratum. The position of the stipe, whether central, eccentric, or lateral, may vary in the different spore fruits growing from the same tree or log, depending upon the position of the point of exit from the host. If this is from the top it may be centrally stipitate and from the side the stipe may be eccentric or lateral.

In general the spores may represent four rather distinct types: (a) With an endospore and epispore, often with a distinct germ pore. Such spores are mostly ellipsoidal or short cylindrical, or oval, and are often somewhat truncate at one end. (b) With only one spore wall layer visible, usually subspherical, oval, ellipsoidal or cylindrical. (c) Very strongly angular or with large knobs. Spores of this type are mostly rose-colored to pink. (d) More or less ellipsoidal or oval or cylindrical with mostly six or eight longitudinal grooves so that in cross section they are not round but angular or lobed. These spores also are more often rose or pink. The outer surface of the spores of all types may be smooth (more often the case) or somewhat roughened or tuberculate or reticulately marked by ridges. These tubercles or ridges are stained blue by iodine reagents in amyloid spores such as occur in species of *Russula* and *Lactarius*.

The angiocarpic spore fruits such as those of *Amanita* show an external universal veil or cortex before they open. Upon completion of the expansion the portion of this veil on the upper surface of the pileus may be visible as patches of tissue, as in *A. muscaria,* where the veil does not keep pace with the remainder of the expanding pileus and so is torn to pieces. On the other hand if it does keep pace in its growth it is detectable as a thin fibrous layer or cuticle of some other type. Around the foot of the stipe the universal veil is left as a loosely or closely adhering cup, the volva. In angiocarpic forms like *Agaricus* there is no volva, as the rupture is so low down that no free basal portion of the veil is distinguishable. In angiocarpic forms with an annulus or secondary veil this arises as a sheet of tissue below the circular hymenial cavity, either separating this cavity from the outside air below or, if attached high on the stipe, sheathing the upper part of it. The annulus may be simple (as in *Agaricus campestris*) or of two distinct layers (as in *A. arvensis*). It may break free from the edge of the pileus and also from the stipe to form a movable ring (as in *Lepiota procera*) or at one or the other place, leaving a ring attached to the stipe or broken sheets or threads attached to the edge of the pileus (a cortina). In pseudoangiocarpic forms the annulus, as in the Boletaceae, may be composed of tissues from the stipe, that grew out and attached themselves to the incurving pileus margin or of tissues from the edge of the pileus that became attached to the surface of the stipe. The truly gymnocarpic forms show neither annulus nor volva nor patches or cuticular structure on the pileus that represent in any manner the remnants of a universal veil.

Reijnders (1933) assembled from literature and from his own investigations a list of 79 species of Agaricaceae of which the mode of development of one is doubtful, 21 gymnocarpic, 2 pseudoangiocarpic and 55 angiocarpic. Heim (1937) added two pseudoangiocarpic species to this list. It must be noted that some writers list the latter type of development as gymnocarpic so that probably there should be in Reijnders' list a higher proportion of species with pseudoangiocarpic type of development. All of the Ochrosporae and Melanosporae among the 81 species concerned are angiocarpic; of the four Rhodosporae listed one is pseudoangiocarpic and three gymnocarpic. Of the 41 Leucosporae 19 are angiocarpic, 19 gymnocarpic, 2 pseudoangiocarpic and 1 doubtful. If the angiocarpic species, following Singer (1936) and others, are the more primitive, arising from the Gasteromycetes, it is apparent that the Agaricaceae with colored spores are the more primitive and those with white or pink spores further from the ancestral forms. If the light-spored forms are considered the more primitive then the angiocarpic genera in general are further advanced in evolution. The peculiar type of hymenium of *Coprinus* and of some species of *Pseudocoprinus*, with broad paraphyses and the more slender basidia arranged in squares at the corners is not found except in these and perhaps a few other black-spored species. Aside from this arrangement of the basidia and paraphyses these two genera have certain other specialized structures that appear far from primitive. In the stout, more fleshy *Coprini* the broad gills are close together but not in contact and the original downward curve of the pileus keeps them in contact with the stipe. Thus the outer portion of the gills furthest from the center of the pileus is situated at the bottom of the gill mass. The maturing of the basidia and discharge of the basidiospores occur first in a narrow band at this outer edge of the gill. The spores have only a short distance to fall to escape into the air. Then this band from which the spores have been dislodged undergoes autodigestion into a dark-colored fluid which either dries up or drops off and the next strip of the hymenium produces its spores. Thus the spores never have to fall far between the crowded gills. The black inky drops are not colored by the spores and are not the means by which the latter are distributed. In the *Coprini* with narrow gills, quite widely spaced, also in *Pseudocoprinus,* the tissue of the pileus is very thin. It splits radially over the median plane of each gill which spreads out in a V-shaped cross section, thus making the fall of the spores more efficient. Buller (1909) gave a thorough discussion of these different types of pilei. In most of the remainder of the Agaricaceae the basidia reach maturity not in a band but here and there all over the surface of the gills. In *Panaeolus,* another dark spored genus, the spores mature in patches on the gills, giving them a more or less variegated appearance. Possibly this indicates a tendency toward the habit in the fleshy *Coprini.*

It is worthy of note that Pietro Antonio Micheli published in 1729 what is probably the first key to the species of this family, which he recognized merely as a single genus *Fungus.* Among the characters used by him in his key were the clustered or separate growth of the spore fruits, their branching or nonbranching, presence or absence of volva, presence or absence of annulus, nature of annulus, i.e., whether free or attached to the stipe, presence or absence of striations on the pileus, locations of striations if present, whether pileus and gills were of the same or different colors, presence or absence of latex, etc. Although his system, too, was largely artificial it served to distinguish the species known to Micheli and in some particulars was no more artificial than the one more recently employed.

In spite of the various systems of classification proposed in the last few decades the relationships within the Hymenomycetes are still very uncertain. The simplest forms appear to be those related to *Corticium* in the Thelephoraceae. In many respects this genus is morphologically very much like *Ascocorticium* of the Taphrinales. It may be that by the extrusion of the spores into external pockets the basidium has arisen from the ascus and thus the gap between the Ascomyceteae and Basidiomyceteae was bridged. On the other hand the basidium of a form like *Auricularia* may represent a four-spored ascus in which the ascospores instead of escaping germinate *in situ* and produce secondary spores. From this structure may have arisen finally the holobasidium of *Corticium*. From the latter as a starting point the greater complication of the hymenium leading to increased spore-bearing surface brings us to such families as Clavariaceae, Hydnaceae, and Polyporaceae. Within the latter there is a tendency toward radially elongated pores or even lamellae. From some of the fleshy *Polypori* with central stems may have arisen forms like the Boletaceae. Some of the latter have pores that are semilamellate. So there is a possibility of the origin of the Agaricaceae from either of these families. On the other hand there is an undeniable close relationship between a number of the Agaricaceae and several genera of the Gasteromycetes. Singer (1936) held that this indicates a descent of the former from the latter. In this case the universal veil would represent a reduced peridium and evolution within the Agaricaceae would have to be considered as progressing from the angiocarpic through the pseudoangiocarpic to the gymnocarpic forms. The chief objection to this theory is that it would involve the transformation of the spore attachment from a symmetrical one to the obliquely perched spore of the Agaricaceae. Inasmuch as the theory further presupposes that the Gasteromycetes were evolved from simple forms with obliquely attached spores this would involve the mutation back again to the latter condition with an ancestral interphase of symmetrical spore attachment. The further phylogenetic considerations of these groups will be given more in detail in Chapter 17.

Key to the Genera of Family Exobasidiaceae

Mycelium intracellular; clusters of basidia, often among elongated paraphyses, emerging from the stomata; basidia with two sterigmata. *Kordyana*

Mycelium intercellular; basidia emerging from between the epidermal cells, usually four- to six-spored; no paraphyses. *Exobasidium*

Key to the More Important Genera of Family Thelephoraceae

Hymenophore resupinate, floccose or continuous, of only one layer.

Hymenophore floccose or felty or pellicular; basidia in scattered clusters, mostly not forming a continuous hymenial layer.

Fructification tenuous with scanty subiculum, the hyphae distinct, loose; basidia subglobose or short-claviform, with two to four stout, elongate, more or less cornute or flexuous "epibasidia"; spores smooth, germinating by repetition. See *Ceratobasidium* in Chapter 13.

Fructification mucedinoid, reticulate-pellicular, finely granulose; hyphae short-celled, branching at right angles, often with formation of cruciform cells; basidia in cymose, often candelabrum-like clusters, short cylindric with four or six to eight sterigmata; spores mostly smooth but in a few species spinulose; in some species spores germinating by repetition. *Pellicularia* (*Botryobasidium*)

Fructification felt-like or "hypochnoid," composed of loosely interwoven hyphae; basidia in scattered clusters or sometimes in a compact, smooth or papillar hymenium; basidiospores usually colored, rough-walled to echinulate. *Tomentella* (*Hypochnus* of some authors)

Hymenophore membranous or coriaceous, not reflexed or saucer-shaped: basidia mostly forming a continuous smooth or papillose hymenium.

Basidiospores white or rarely bright-colored.

Cystidia lacking. *Corticium*

Cystidia present in hymenium or in subhymenial tissues or both. *Peniophora*

Brown stellate organs present in subhymenial tissue. *Asterostroma*

Hymenium interrupted by sterile pegs or projections. *Epithele*

Basidiospores ochraceous, ferruginous, or fuscous, smooth.

Cystidia lacking. *Coniophora*

With cystidia. *Coniophorella*

Hymenophore leathery or corky, saucer-shaped, or at least upturned at the edges; basidiospores pale-colored, usually large; cystidia of various types. *Aleurodiscus*

Resembling the foregoing but with groups of dendrophyses (feather-like cystidia) forming projections above the hymenium. *Dendrothele*

Saucer or cup-shaped, more or less gelatinous-fleshy. *Cytidia*

Hymenophore adhering closely to the substratum; antler-like cystidia forming a felty layer above the basidia. *Vararia* (*Asterostromella*)

Hymenophore of three layers, resupinate or more often reflexed-effuse or attached laterally.

Leathery, without cystidia. *Stereum*

Leathery, sometimes almost woody or corky, variable in shape; with stiff, brown, pointed setae extending from the hymenium. *Hymenochaete*

Hymenophore upright, often stalked, simple or branched or funnel-shaped, or pendent cup-shaped or separate tubes.

Hymenium lining the inner surface of pendent cups or tubes.

Sessile or stalked cups; separate. *Cyphella*

Tube-like arising together, but not united, from a common subiculum. *Solenia*

Hymenium lining the outer or lower surface of funnels or flattened lobes.

Leathery; spores mostly brown and roughened.

Hymenium almost smooth or warty. *Thelephora*

Hymenium with woody ribs; tropical. *Cladoderris*

Other uncommon genera. *Skepperia*, *Hypolyssus*

Fleshy, much branched, forming a round, cabbage-like structure; spores hyaline, smooth. *Sparassis*

Key to the Genera of Family Cantharellaceae

(*After Smith and Morse, 1947*)

Fruit body typically fleshy and centrally stipitate or stipe eccentric, but well-developed (or whole fruit body funnel-shaped and stipe not distinct).

Hymenium smooth or nearly so. *Craterellus*

Hymenium in the form of radiating, sometimes reticulate, ridges or folds or as obtuse, forked lamellae. *Cantharellus* (*Gomphus*)

Fruit body very thin and delicate, stipe small or rudimentary.

Spores hyaline. *Leptotus* (*Dictyolus*)

Spores tawny to incarnate. *Arrhenia*

Key to the Genera of Family Clavariaceae

(*Based on Doty, 1948*)

Spores hyaline (rarely tinted), mostly smooth and thin-walled; basidia two- or four-spored.

Fructification with broad, flattened terminal portions of the branches (probably better in Family Thelephoraceae). *Sparassis*

Fructification not as above.

Spores large (7 μ or over), globose; basidia two-spored; hymenium white or gray; not staining green with Fe_2SO_4 solution. *Clavulina*

Spores distinctly smaller or ellipsoid; basidia typically four-spored if spores spherical; hymenium often colored otherwise or staining green with Fe_2SO_4 solution.

With stout hyphae in the trama producing setae or gloeocystidia in the hymenium or subhymenium, or the apices of the branches truncate or cup-shaped; spores under 8 μ long; not staining green with Fe_2SO_4 solution.

Apices of some branches truncate to cup-shaped, with gloeocystidia in the hymenium layer. *Clavicorona*

Apices of branches acute; with strong setae in the hymenium or subhymenium.

Setae loosely dichotomously branched (dichophyses). *Eriocladus* (*Lachnocladium*)

Setae closely dichotomously branched (asterophyses). *Stelligera*

Without such hyphae or apices; spores various; or staining green with Fe_2SO_4 solution.

Fructifications branched.

Very finely branched (under 1 mm.); toughish. *Pterula*

Branches larger; flesh fragile or putrescent.

Hymenium staining green with Fe_2SO_4 solution. *Clavariella*

Hymenium not staining green with Fe_2SO_4 solution. *Clavaria*

Fructifications simple or rarely branched above.

Fructifications enlarged above or over 1 cm. in diameter above; spores ellipsoid.

Minute (not over 5 mm. tall); or with an abruptly inflated head; not staining green with Fe_2SO_4 solution.

With an inflated, down-turned head. *Physalacria*

With the apex merely enlarged.

Hymenium on the expanded blunt apex. *Pistillina*

Hymenium on the sides of the club. *Pistillaria*

Large (over 2 cm. tall); or turning green with Fe_2SO_4 solution. *Clavariadelphus*

Fructifications not enlarged above (i.e., slenderly clavate to filiform clubs); spores various.

Fructifications over 2 cm. tall; not obviously restricted to specific hosts or substrata; or the fructifications fascicled fleshy forms; often globose spores.

Tramal hyphae with many secondary cross walls; clamp connections rare. *Clavaria*

Tramal hyphae with secondary cross walls only rarely; clamp connections on most cross walls. *Clavulinopsis*

Fructifications smaller, restricted to specific hosts or substrata which may be sclerotia or living plants; spores ellipsoid or flattened on one side.

With a sclerotial base; stipe slender and distinct; mostly over 5 mm. tall. *Typhula*

Without a sclerotial base; stipe not distinct; mostly less than 2 mm. tall. *Pistillaria*

Spores typically ochraceous, mostly roughened or obdurate walls; basidia four-spored.

Toughish to woody; spores echinate or sharply warty; hymenium sometimes unilateral; not becoming green with Fe_2SO_4 solution.[1]

Coarse, leathery to woody fungi; hymenium often unilateral or branches flattened. *Thelephora* (some species)

Delicate toughish to woody fungi; hymenium covering all surfaces of the rounded branches. *Scytinopogon*

Fleshy; spores smooth, verrucose to echinulate; hymenium on all lateral surfaces of the branches; becoming green with Fe_2SO_4 solution.

Simple, unbranched fungi with broadened sterile apices; sometimes "mushroom-like" in form. *Gomphus* (*Cantharellus* in part)

Branched fungi; coralloid in form. *Clavariella*

Key to the More Important Genera of Family Hydnaceae

(*Based on Miller, 1933*)

Fructification with a porose-reticulate hymenial surface covered over with minute warts over which the hymenium continues. *Grammothele*

Fructification with distinct warts or teeth, never poroid.

Trama dark; spores roughened, subhyaline to dark, usually brown.

Resupinate, soft, floccose, growing on wood. *Caldesiella*

Stipitate, fleshy or coriaceous; growing on the ground.

Fleshy. *Hydnum*

Fibrous tough. *Calodon*

Trama pale; spores smooth or sometimes echinulate, hyaline or slightly colored.

Teeth arising directly from the woody substratum. *Mucronella*

Teeth developed on a distinct hymenophore.[2]

---

[1] Some of the following genera may belong in other families but are clavarioid in form.

[2] The genera *Hydnochaete*, *Irpex*, and *Echinodontium*, in which the flattened teeth develop by the breaking up of pores should be sought in Family Polyporaceae.

Resupinate or reflexed, spines borne on tough branching processes which are partially submerged in a brownish tomentum. *Gloiodon*
Resupinate, reflexed or stipitate; teeth or spines not borne on such processes.
  Resupinate, thin, floccose, crustaceous, ceraceous, or subcoriaceous.
    Ceraceous; teeth thick, occasionally slender, obtuse, deformed or irregularly scattered. *Radulum*
    Texture otherwise; teeth varying from short fragile warts to long conspicuous teeth or spines.
      Stellate setae present. *Asterodon*
      Cystidia present. *Odontia*
      Cystidia or setae absent.
        Warts short, hemispheric, cylindrical, or subulate and fragile. *Grandinia*
        Teeth or spines conspicuous, long, slender, usually terete. *Oxyodontia*
  Reflexed to stipitate, rarely resupinate; fleshy to coriaceous.
    Fleshy, growing on the ground.
      With central stipes; spores smooth. *Dentinum*
      Pileus irregular, with deformed stipe, spores minutely echinulate. *Hydnodon*
    Fleshy or coriaceous; growing on a woody substratum.
      Richly branched or pulvinate; soft; fleshy. *Hericium*
      Cap coriaceous with long, laterally attached stipe, spores slightly roughened. *Auriscalpium*
      Reflexed to obscurely laterally stipitate, occasionally resupinate; subfleshy to coriaceous; spores smooth. *Steccherinum*

Key to the Genera of Family Meruliaceae[3]

(Based in Part on Bourdot and Galzin, 1927)

Spores hyaline.
  Hymenium with irregular tubercules or radiating (not anastomosing) folds; waxy, then indurated; mostly resupinate (sometimes placed in the Hydnaceae). *Phlebia*
  Hymenium with lamelliform folds, crisped; membranous; sessile centrally or laterally. *Plicatura (Trogia)*
  Hymenium with folds anastomosed into alveoli or pores; sessile or resupinate; more or less gelatinous. *Merulius*
Spores rust-colored or ochraceous; resupinate, rarely reflexed. *Gyrophana (Serpula)*

Key to the Genera of Family Fistulinaceae[4]

Fleshy, mostly forming a laterally attached pileus, sometimes resupinate. Pores remaining separate although in close contact. *Fistulina*
Membranous or crusty; resupinate; with scattered warts which elongate to become pores. *Porothelium*

---

[3] Some authors (Bourdot and Galzin, 1927; Rea, 1922, etc.) include in this family the following genera with smooth hymenium: *Coniophora, Coniophorella, Jaapia*.

[4] *Cyphella* and *Solenia* possibly belong here also, instead of in the Thelephoraceae.

Key to the More Important Genera of Family Polyporaceae

Always resupinate; annual (except *Fomitoporia*); with thin-walled, not truncate[5] spores; without setae, but sometimes with cystidia (see also some resupinate species or specimens of *Gloeoporus*, *Hapalopilus*, *Tyromyces*, *Coriolus*, *Coriolellus*, etc. which are mainly more or less pileate).
- Context white or light-colored, not becoming brown.
  - Not becoming brighter on being bruised or with age. *Poria*
  - Becoming some bright color on being bruised or with age. *Podoporia*
- Context brown or black.
  - Strongly irpiciform or hydnoid; at the margin with remains of tubes. *Hydnochaete*
  - Definitely porose.
    - Fructification black. *Melanoporia*
    - Fructification brown, spores hyaline; perennial. *Fomitoporia*
    - Fructifications brown, spores brown. *Physiporus*

Usually pileate; stalked or sessile; reflexed-effuse; under some conditions sometimes resupinate.
- Spores truncate at the apical end, two-layered, the epispore smooth and the endospore with spines or other types of projections into the epispore; sessile or stipitate; upper surface with a hard, often laccate, crust. *Ganoderma*
- Spores not as above.
  - Volva-like structure present. *Cryptoporus*
  - Volva-like structure wanting.
    - Hymenophore definitely porose, in a few species the pores breaking up into flattened tooth-like plates; annual.
      - Context white or light-colored, not pronounced brown.
        - Fleshy or tough; stipitate; context homogeneous.
          - Stipe much branched, at the bases or near trees. *Grifola* (*Polypilus*)
          - Stipe simple; pileus fleshy; terrestrial.
            - Spores strongly warted, light brown. *Boletopsis*
            - Spores smooth, hyaline. *Scutiger* (*Albatrellus*)
          - Stipe simple; pileus fleshy to tough; large to medium size; growing on wood or from a sclerotium; stipe central, eccentric or lateral; pores round or radially somewhat elongated. *Polyporus* (including *Melanopus* and *Leucoporus*)
          - Stipe curved, attached at the top of the minute turbinate hymenophore; emerging from the lenticels of dead twigs; pores round, small. *Porodisculus*
          - Stipe lateral, very short, attached at the top of a large hoof-shaped to bell-shaped hymenophore which has a thin, separable pellicle; on dead trunks of *Betula*, pores round, small. *Piptoporus*

---

[5] Based on Cooke, 1940. In addition, Bondarzew and Singer (1941) recognize several other genera of resupinate Polyporaceae, distinguished by entirely different characters than shown here. Murrill's (1907) genera differ also in some ways.

Stipe central or lateral; tubes large, alveolar, often radially elongated; growing on wood. *Favolus*

Context duplex; spongy above, woody below; centrally stipitate; tomentose. *Abortiporus* (*Heteroporus*)

Without stipe; sessile to effused-reflexed (rarely resupinate).

Pileus more or less firm, flexible or rigid.

Context duplex, spongy above, firm below. *Spongipellis*

Context uniform; hymenium at maturity more or less smoke-colored. *Bjerkandera*

Context uniform, hymenium white or pallid.

Fleshy to fleshy-tough; friable when dry. *Tyromyces*

Punky to corky; not friable when dry. *Trametes*

Pileus thin, leathery and more or less flexible; surface usually zonate.

Hymenophore preceded by a cup-shaped sterile body. *Poronidulus*

Hymenophore normally pileate; tubes small and mostly regular; pore layer not violet-colored; true cystidia lacking. *Coriolus*

Hymenophore semi-resupinate; tubes large and irregular; dentate but not irpiciform. *Coriolellus*

Hymenophore pileate; tubes early breaking up into flat, tooth-like plates; margin and pore surface when young violet-colored; true cystidia present. *Hirschioporus*

Effused-reflexed or resupinate; poroid remains visible only at margin. *Irpex* (*Irpiciporus*)

Context bright-colored; yellow to red; not stipitate.

Tubes broad and long; trama orange-red. *Aurantioporellus*

Tubes small; trama orange to red; of moderate size. *Pycnoporus*

Tubes small; trama and tubes yellow; large; imbricate. *Laetiporus*

Context brown; spores hyaline.

Duplex; upper layer forming upright tufts of stiff bristles; tubes small. *Pogonomyces*

Duplex; upper layer spongy; tubes large. *Funalia*

Duplex; upper layer spongy; tubes small, sometimes lamellate. *Coriolopsis*

Context simple; light brown.

At first fleshy, becoming slightly corky. *Ischnoderma*

Tough from the first; surface encrusted. *Antrodia*

Tough from the first; surface not encrusted, glabrous. *Hapalopilus*

Context simple; dark brown; context friable; often stipitate. *Phaeolus*

Context brown; spores brown.

Sessile. *Inonotus*

With central stipe; hymenium with setae. *Polystictus*

With central stipe; hymenium lacking setae. *Coltricia*

Hymenophore large, woody, perennial; mostly sessile but sometimes stipitate.

Pores breaking down to form flat teeth with denticulate margin. *Echinodontium*
Not forming denticulate teeth.
Context wood-colored or pale brown. *Fomes*
Context white or light-colored. *Fomitopsis*
Context brown; surface not encrusted. *Phellinus* (*Pyropolyporus*)
Context brown; surface encrusted; several pilei massed together. *Globifomes*
Walls of pores breaking down to form concentric lamellae. *Cyclomyces*
Pore walls breaking down to form radiating lamellae or labyrinthiform passages.
Context brown. *Gloeophyllum*
Context light-colored.
Hymenophore lamellate; woolly and zonate above. *Lenzites*
Hymenophore typically labyrinthiform, but sometimes broadly lamellate; glabrous but rough above. *Daedalea*
Like *Daedalea* but the labyrinthiform tubes small, with thin walls, becoming lamellate with age.[6] *Daedaleopsis*

In the more conservative classifications the genera included in this Key to the Polyporaceae are left united into a much smaller number of large, polymorphous genera, as follows:

Poria includes also *Podoporia, Hydnochaete, Melanoporia, Fomitoporia, Physiporus.*
Polyporus includes also *Abortiporus, Bjerkandera, Boletopsis, Cryptoporus, Grifola, Hapalopilus, Inonotus, Ischnoderma, Laetiporus, Phaeolus, Piptoporus, Porodisculus, Pycnoporus, Scutiger, Spongipellis.*
Polystictus is sometimes included in *Polyporus;* if maintained as a distinct genus it may include also *Coriolus, Coriolellus, Coriolopsis, Coltricia, Funalia, Hirschioporus, Poronidulus.*
Trametes includes also *Antrodia, Pogonomyces,* and species out of some of the foregoing genera grouped in *Polyporus* and *Polystictus.*
Daedalea includes also *Daedaleopsis.*
Lenzites includes also *Gloeophyllum.*
Irpex and *Echinodontium* are included in Family Hydnaceae.

Key to the Subfamilies and More Important Genera of Family Boletaceae[7]

(*Based partly on Snell, 1941 and 1942, and Singer, 1945–1947*)

Tubes adhering to one another even to maturity although usually easily separable mechanically.
Spores mostly dark and ornamented with warts, spines, reticulations, or longitudinal ridges, sometimes smooth spores present on the same hyme-

[6] See also *Coriolopsis trabea* (Pers.) B. & S., whose pores are often lamellate.
[7] Gilbert (1931) and Singer (1945–47) divide these genera among two families: Strobilomycetaceae, including *Strobilomyces, Boletellus,* and *Porphyrellus;* and Boletaceae, including all the other genera. Coker and Beers (1943) recognize only one family and three genera: *Boletus, Boletinus,* and *Strobilomyces.*

nium, spores with a germ pore and often truncate at one end, often over 20 μ in length; no clamp connections observed.
Subfamily Strobilomyceteae

Spores globose to short ellipsoid; spore print black; tubes white or gray at first, becoming darker; pileus and stipe warty, woolly or spinose-squarrose. *Strobilomyces*

Spores more elongated; spore print reddish-brown; tubes white to cream color, becoming pink vinaceous to sordid gray to porphyry brown; spores smooth or with short warts or spines perforating the exospore. *Porphyrellus*

Spores elongated (or short elliptical in species with reticulate markings); spore print olivaceous brown to blackish but not reddish; spores smooth or with longitudinal wing-like ridges or ribs, or with warts, or reticulately marked. *Boletellus*

Spores pale pink to yellowish or ferruginous or olive brown; germ pores not visible; spores mostly less than 20 μ in length.
Subfamily Boleteae

Tubes short (see also *Phlebopus*), radially elongated, often with the radial walls higher than the cross connections; clamp connections regularly present; stipe sometimes eccentric or lateral.

Hymenophore depressed or subfree around the apex of the stipe.

Spore print yellow. *Gyroporus*

Spore print olive brown. *Phaeogyroporus*

Hymenophore more or less arcuate-decurrent.

Veil present. *Paragyrodon*

Veil absent. *Gyrodon*

Tubes long (see also *Phlebopus*, below); clamp connections entirely lacking or not numerous.

Pores radially elongated or lamellate (see also *Suillus* under Pores round).

Pores distinctly lamellate, but always with low anastomoses; veil lacking; spore print olive brown. *Phylloporus*

Pores radially elongated but not lamellate; veil present; pileus rarely viscid; stipe without glandulae; spore print olive brown. *Boletinus*

Pores radially elongated but not lamellate; pileus always viscid; without veil, or if with veil then pores round; stipe frequently with glandulae; spore print olive brown. *Suillus*

Pores round; spores olive brown (see also *Suillus* under Pores radially elongated).

Pileus always viscid; veil present; stipe frequently with glandulae. *Suillus*

Pileus tomentose; rarely viscid; stipe without glandulae. *Xerocomus*

Surface of pileus and stipe more or less pulverulent; spores small, ellipsoidal to ovoid; tubes long, adnate or depressed around stipe; veil sometimes present; stipes mostly cylindrical or only slightly thickened downward, often viscid. *Pulveroboletus*

Surface of pileus not noticeably pulverulent, sometimes slightly viscid; tubes short or of medium length; hymenophore arcuate, at least when young, and somewhat decurrent; veil absent; spores small to medium size; stipes frequently swollen-ventricose; not viscid. *Phlebopus*

Veil none; stipe not viscid; pileus sometimes sub-tomentose, not pulverulent; cuticle often consisting of a trichodermium (i.e., a palisade of parallel, vertical hyphae); stipe usually swollen below but sometimes cylindrical, often reticulately marked but not provided with glandulae; stipe not with furfuraceous or squamulose scabrosities; pores small; tubes usually long. *Boletus*

Veil none; stipe not viscid, slender and tapering upward, scabrous; spores naviculate; pores very small; tubes long, depressed around stipe. *Leccinum*

Pores round; spore print rusty yellow; spores under microscope golden yellow; veil absent; stipe entirely smooth, equal or ventricose; stipe never reticulate. *Xanthoconium*

Pores round; spore print flesh color or vinaceous; veil absent; tubes more or less flesh-colored; stipe often reticulate. *Tylopilus*

Tubes divergent from one another at maturity as the pileus spreads; only known from Madagascar. *Ixechinus*

Keys to the More Important Genera of Family Agaricaceae[8]

(Modified from Key by A. H. Smith, 1938)

Trama of pileus and usually of the gills composed of nests of sphaerocysts surrounded by connective tissue and with lactifers irregularly dispersed throughout. (By many recent students considered as Family Russulaceae.)

Cut or broken parts of the fruiting body exuding a watery to milk-like or colored latex. *Lactarius*

No latex present; fruiting body often very fragile. *Russula*

Trama of pileus not with sphaerocysts.

Parasitic upon other agarics; flesh of cap breaking down into a mass of chlamydospores. *Asterophora* (*Nyctalis*)

If parasitic on agarics, flesh not breaking down into chlamydospores.

Hymenium typically waxy; spores smooth.

Spore print white. *Hygrophorus* (including *Hygrocybe, Camarophyllus, Limacium*)

Spore print smoky gray to blackish. *Gomphidius*

Hymenium not waxy, or if appearing so, the spores echinulate.

Fruiting body typically rather tough, if fleshy, or membranous and very pliant; reviving when remoistened.

Gills with distinctly dentate to serrate edges. *Lentinus* (including *Panus*)

Gills with edges even, or merely slightly fimbriate.

Stipe eccentric, lateral, or wanting (usually not reviving when remoistened). *Pleurotus*

---

[8] Used in the wider sense of the term. It should be noted that these keys of the Agaricaceae are merely artificial keys by which to determine the genera and do not represent a system of classification based upon phylogenetic considerations. Singer (1936, 1951) has divided the gill fungi (Agaricaceae in the older sense) into many families and split up many more of the genera; the bases of segregation being largely the structure of the gills and type of surface structure (cuticle, etc.) and the structure of the spores, as well as the chemical character of the spore and hyphal walls. His system is, so far as possible, based upon suppesed true phylogenetic relationships.

Stipe central, typically 0.5–7.0 mm. in diameter; spores never amyloid. *Marasmius*

Fruiting body typically woody or semiwoody.

Lamellae splitting along the edges. *Schizophyllum*

Lamellae arranged concentrically around the stipe. *Cyclomyces*

Lamellae more or less poroid. *Daedalea* and *Lenzites*[9]

Fruiting body typically soft, or if membranous, rather fragile; not usually reviving when remoistened.

Stipe eccentric; lateral, or wanting.

Spore print white to tinged lilac or creamy-vinaceous. *Pleurotus*

Spore print pinkish.

Spores longitudinally striate; stipe lacking or rudimentary. *Octojuga* (*Clitopilus* in part)

Spores as above; stipe well developed. *Clitopilus*

Spores angular. *Rhodophyllus*[10]

Spore print yellow to rusty brown; lamellae separating easily from the pilcus; the decurrent lamellae sometimes anastomosing to become poroid near the stipe. *Paxillus*

Spore print as above; lamellae not separating easily from pileus. *Crepidotus*

Stipe typically central.

Spore print white to creamy or pale creamy vinaceous. See below Key to Leucosporae

Spore print pink to flesh color. See below Key to Rhodosporae

Spore print yellow to rusty brown or earth brown. See below Key to Ochrosporae

Spore print cocoa color, chocolate color or purplish to black. See below Key to Melanosporae

Key to the Cenrally Stipitate, Soft, Putrescent Genera of Leucosporae

Stipe slender, if more than 5 mm. thick then with a distinct cartilaginous cortex.

Stipe somewhat horny in consistency; gills decurrent to adnate; fruiting bodies maramoid in appearance but spores amyloid. *Xeromphalina*

Not as above.

Cap margin typically straight, or if incurved then gills not truly decurrent; if margin of pileus incurved and pileus brown to gray to blackish, then pileus with a differentiated hypoderm. *Mycena*

Cap margn strongly incurved or inrolled and gills typically decurrent. *Omphalina* (*Omphalia* of many authors)

Cap margin strongly incurved and gills typically adnate to adnexed; pileus lacking a differentiated hypoderm.

[9] These three genera, *Cyclomyces*, *Daedalea*, and *Lenzites*, are usually included in Family Polyporaceae.

[10] Includs several other genera with similar spores.

Basidia with darkly staining granules with aceto-carmine. *Lyophyllum*

Basidia not as above. *Collybia* (including *Myxocollybia*)

Stipe usually over 3–5 mm. in thickness, typically fleshy.

Stipe readily separable, with a clean break, from the pileus.

Volva present and annulus absent. *Vaginata* (*Amanitopsis*)

Volva and annulus present. *Amanita* (*Venenarius*)

Volva absent; annulus present.

Gill trama divergent; pileus viscid; annulus large to small or obsolete, not movable. *Limacella* (*Lepiota* in part)

Gill trama interwoven to parallel.

Spores thick-walled, with germ pore; annulus movable. *Leucocoprinus*

Spores thin-walled, without germ pore; annulus fixed or movable. *Lepiota*

Stipe and pileus confluent (not readily separable); annulus present or absent.

Universal veil (or cuticle of cap and the outer layers of sheath on stipe) powdery to granulose. *Cystoderma*

Veil if present not granulose on outer surface (hyphal cells not readily separable from each other).

Partial veil present.

Veil typically leaving a membranous annulus.

Lamellae clearly decurrent. *Armillariella*

Lamellae emarginate or sinuate, with a tooth *Armillaria*

Veil fibrillose, merely leaving a fibrillose zone on the stipe. *Tricholoma*

Partial veil absent or rudimentary.

Spores amyloid; fruiting body typically large and fleshy; pileus appressed fibrillose to dull and unpolished, rarely hygrophanous and glabrous; conspicuous mycelium surrounding base of stipe and extending through the surrounding debris (if hygrophanous with rough spores, see *Melanoleuca*); pigment, if present, intracellular. *Leucopaxillus*

Spores not amyloid or if so fruiting body not as above.

Gills broadly emarginate, with a waxy luster and some shade of flesh color; spores typically echinulate and not amyloid *Laccaria*

Gills not waxy-appearing as above; adnexed to emarginate; spores typically roughened but amyloid; harpoon-like cheilocystidia present; pigment, if present, epicellular. *Melanoleuca*

Gills decurrent to adnate; fruiting bodies often blackening when bruised; typically hygrophanous with somber colors; growing in tufts; basidia with dark granules when stained with aceto-carmine. *Lyophyllum*

Gills adnate to sinuate; not hygrophanous; never turning black; spores typically not amyloid. *Tricholoma*

Gills decurrent; spores not amyloid. *Clitocybe*

Key to the Mostly Centrally Stipitate, Soft, Putrescent Genera of Rhodosporae

Spores angular. *Rhodophyllus*[11]
Spores not angular.
 Spores longitudinally striate.
  Stipe eccentric to central; terrestrial. *Clitopilus*
  Stipe lateral or lacking; not terrestrial. *Octojuga*
 Spores smooth, globose to ellipsoid.
  Volva well developed; annulus lacking. *Volvaria*
  Volva typically wanting.
   Annulus present. *Annularia*
   Annulus absent.
    Gills free; cap and stipe readily separable. *Pluteus*
    Gills attached to stipe; pileus and stipe not separable. *Psathyrella*

Key to the Centrally Stipitate, Soft, Putrescent Genera of Ochrosporae

Partial veil cobweb-like; spores typically with a slightly wrinkled to warty exospore and never truncate; typically terrestrial. *Cortinarius*
Not as above.
 Cuticle of pileus in the form of a viscid pellicle, or radially arranged as interwoven filamentous hyphae typically 1.5–5 μ in diameter.
 Stipe typically fleshy, about 4–20 (up to 40) mm. thick.
  Typically terrestrial.
   Annulus membranous; rudimentary universal veil often evident. *Rozites*
   Annulus not membranous, but a fibrillose zone may be present.
    Pileus viscid. *Hebeloma*
    Pileus typically dry and fibrillose. *Inocybe*
  Typically lignicolous.
   Annulus present or stipe conspicuously scaly or both. *Pholiota*
   Annulus lacking; scales if present on the stipe easily obliterated. *Flammula*
 Stipe typically cartilaginous or slender and fragile.
  Spores very pale and thin-walled; gills typically subdecurrent to decurrent and pale to bright cinnamon. *Tubaria*
  Spores typically thick-walled and yellowish brown; gills typically adnate or notched.
   Margin of pileus straight at first. *Galerina* (*Galera* in part)
   Margin of pileus at first incurved or inrolled. *Naucoria*
 Cuticle of pileus formed by a palisade of pyriform cells, or if the cells are irregularly arranged, vesiculose and nearly isodiametric (best determined in young pilei); spores typically truncate.
 Pileus plicate-striate; paraphyses coprinoid; more or less deliquescent. *Bolbitius*

---

[11] This includes the genera known as *Entoloma*, *Nolanea*, *Leptonia*, and *Eccilia*.

Not as above.
Stipe typically fleshy; spore print typically dull earthy brown to deep rusty brown. *Agrocybe* (*Pholiota* in part)
Stipe typically cartilaginous; spore print yellowish brown to rusty brown. *Conocybe* (*Galera* in part)

Key to the More Important Centrally Stipitate, Soft, Putrescent Genera of Melanosporae

Gills deliquescing at maturity; basidia separated by broad paraphyses. *Coprinus*
Gills not deliquescing.
Pileus with cuticle formed of pyriform or vesiculose cells arranged in a palisade, or in a compact layer one or more cells thick; fruiting body typically fragile.
Gills spotted by the maturing spores; spores not discoloring rapidly in concentrated sulfuric acid. *Panaeolus*
Gills not conspicuously spotted by the maturing spores.
Pileus piicate-striate; paraphyses coprinoid. *Pseudocoprinus*
Pileus not plicate-striate, or, if so, paraphyses not coprinoid. *Psathyrella* (including *Hypholoma* in part)
Pileus with cuticle formed of slender filamentous hyphae (cells vesiculose in a few, but then formed by the breaking up of chains, and surface of the cap appears powdery).
Annulus typically present.
Gills free; stipe readily separating from pileus. *Agaricus* (*Psaliota*)
Gills attached; stipe not readily separable. *Stropharia*
Annulus not present; veil if present adhering to cap or along its margin.
Gloeocystidia present in hymenium; typically lignicolous, or on peat or muck (if on dung or humus, see *Stropharia*). *Naematoloma* (*Hypholoma* in part)
Gloeocystidia not present though other types of pleurocystidia may be present and abundant. *Psilocybe*

Literature Cited

AMES, ADELINE: A consideration of structure in relation to genera of the Polyporaceae, *Ann. Mycol.*, **11**(3):211–253. *Pls.* 10–13. 1913.

ATKINSON, GEORGE F.: The development of Agaricus campestris, *Botan. Gaz.*, **42**(4):241–264. *Pls.* 7–12. 1906.

———: The development of Lepiota clypeolaria, *Ann. Mycol.*, **12**(3):346–356. *Pls.* 13–16. 1914a.

———: The development of Amanitopsis vaginata, *ibid.*, **12**(4):369–392. *Pls.* 17–19. 1914b.

———: The development of Armillaria mellea, *Mycolog. Centr.*, **4**(3):113–121. *Pls.* 1–2. 1914c.

———: Homology of the "universal veil" in Agaricus, *ibid.*, **5**(1):13–19. *Pls.* 1–3. 1915.

———: Origin and development of the lamellae in Coprinus, *Botan. Gaz.*, **61**(2):89–130. *Pls.* 5–12 *and* 6 *diagrams*. 1916.

BAXTER, DOW V.: Some resupinate Polypores from the region of the Great Lakes, (some of these under the title, Some Porias from the region of the Lake States), *Papers Mich. Acad. Sci.*, **6**:67–76. *Pls.* 1–6. 1927; **9**:39–46. *Pls.* 26–29. 1929; **15**:191–228. *Pls.* 17–26. 1932; **17**:421–439. *Pls.* 41–50. 1933; **19**:305–332. *Pls.* 58–65. 1934; **20**:273–281. *Pls.* 55–60. 1935; **21**:243–267. *Pls.* 30–39. 1936; **22**:275–295. *Pls.* 31–37. 1937; **23**:285–305. *Pls.* 1–9. 1938; **24**:167–188. *Pls.* 1–7. 1939; **25**:145–170. *Pls.* 1–12. 1940; **26**:107–121. *Pls.* 1–7. 1941; **27**:139–161. *Pls.* 1–11. 1942; **28**:215–233. *Pls.* 1–6. 1943; **29**:85–109. *Pls.* 1–6. 1943 (1944); **30**:175–191. *Pls.* 1–14. 1944(1945); **31**:117–130. *Pls.* 1–5. 1945(1947); **32**:189–211. *Pls.* 1–10. 1946(1948); **33**:9–30. *Pls.* 1–9. 1947(1949).

BLIZZARD, A. W.: The development of some species of Agarics, *Am. J. Botany*, **4**(4):221–240. *Pls.* 6–11. 1917.

BONDARZEW, A., UND R. SINGER: Zur Systematik der Polyporaceen, *Ann. Mycol.*, **39**(1):43–65. 1941.

BOURDOT, H., ET A. GALZIN: Hyménomycètes de France. Hétérobasidiés-Homobasidiés Gymnocarpes. Contribution a la Flore Mycologique de la France, volume 1, iv + 761 pp. 185 *figs.* Sceaux, Marcel Bry, 1927.

BREFELD, OSCAR: Basidiomyceten: III. Autobasidiomyceten und die Begründung des natürlichen Systems der Pilze, in Untersuchungen aus dem Gesammtgebiete der Mykologie, Heft. 8, pp. 1–305. *Pls.* 1–12. Leipzig, Arthur Felix, 1889.

BULLER, A. H. REGINALD: Researches on Fungi, vol. 1, xi + 287 pp., 5 *pls.*, 83 *figs.*, 1909; vol. 2, xii + 492 pp., 157 *figs.*, 1922; London and New York, Longmans, Green and Company.

BURT, EDWARD A.: The Thelephoraceae of North America: I. Thelephora, *Ann. Missouri Botan. Garden*, **1**(2):185–228. *Pls.* 4–5. 1914; II. Craterellus, *ibid.*, **1**(3):327–350. *Pls.* 15–17. 1914; III. Craterellus borealis and Cyphella, *ibid.*, **1**(4):357–382. *Pl.* 19. 1914; VI. Hypochnus, *ibid.*, **3**(2):203–241. *Figs.* 1–30. 1916; VIII. Coniophora, *ibid.*, **4**(3):237–269. *Figs.* 1–19. 1917; IX. Aleurodiscus, *ibid.*, **5**(3):177–203. *Figs.* 1–14. 1918; X. Hymenochaete, *ibid.*, **5**(4):301–372. *Pls.* 16–17. *Figs.* 1–32. 1918; XI. Tulasnella, Veluticeps, Mycobonia, Epithele and Lachnocladium, *ibid.*, **6**(4):253–280. *Pl.* 5. *Figs.* 1–15. 1919; XII. Stereum, *ibid.*, **7**(2–3):81–248. *Pls.* 2–6. *Figs.* 1–48. 1920; XIII. Cladoderris, Hypolyssus, Cymatella, Skepperia, Cytidia, Solenia, Matruchotia, Microstroma, Protocoronospora and Asterostroma, *ibid.*, **11**(1):1–36. *Pl.* 1. 1924; XIV. Peniophora, *ibid.*, **12**(3):213–357. 1925; XV. Corticium and supplement to the whole series, *ibid.*, **13**(3):173–354. 3 *figs.* 1926.

CLEMENTS, FREDERICK E.: Minnesota Mushrooms, Minnesota Plant Studies, No. 4, 169 pp. 124 *figs.* University of Minnesota, 1910.

COKER, WILLIAM CHAMBERS: The Clavarias of the United States and Canada, 209 pp. 92 *pls.* Chapel Hill, Univ. North Carolina Press, 1923.

———, AND ALMA HOLLAND BEERS: The Boletaceae of North Carolina, viii + 96 pp. *Frontispiece, colored, and* 65 *pls.* (5 *colored*), 7 *figs.* Chapel Hill, Univ. North Carolina Press, 1943.

COOKE, WILLIAM BRIDGE: A nomenclatorial survey of the genera of pore fungi, *Lloydia*, **3**(2):81–104. 1940.

COTTON, A. D.: On the structure and systematic position of Sparassis, *Brit. Mycol. Soc. Trans.*, **3**:333–339. 1912.

DONK, M. A.: Revision der Niederländischen Homobasidiomyceteae-Aphyllophoraceae, II, *Mededeelingen van het Botanisch Museum en Herbarium van de Rijks Universiteit te Utrecht*, **9**:1–278. 1933.

DOTY, MAXWELL S.: Clavaria, the species known from Oregon and the Pacific Northwest, *Oregon State Monographs. Studies in Botany*, **7**:1–91. *Pls.* 1–11. *Figs.* 1–9. 1944.

DOTY, MAXWELL S.: Clavicorona, a new genus among the clavarioid fungi, *Lloydia*, **10**(1):38–44. *Figs.* 1–6. 1947.

———: A preliminary key to the genera of Clavarioid fungi, *Chicago Acad. Sci. Bull.* **8**(5):173–178. 1948.

DOUGLAS, GERTRUDE E.: A study of development in the genus Cortinarius, *Am. J. Botany*, **3**(6):319–335. *Pls.* 8–13. 1916.

———: The development of some exogenous species of Agarics, *ibid.*, **5**(1):36–54. *Pls.* 1–7. 1918.

———: Early development of Inocybe, *Botan. Gaz.*, **70**(3):211–220. *Pls.* 18–22. 1920.

EFTIMIU, PANCA, ET SAID KHARBUSH: Recherches histologiques sur les Exobasidiées, *Rev. path. végétale entomol. agr. France*, **14**:62–88. 1 *pl. Figs.* 1–9. 1927.

ELROD, R. P., AND DOROTHY L. BLANCHARD: Histological studies of the Boletaceae and related genera, *Mycologia*, **31**(6):693–708. *Figs.* 1–2. 1939.

———, AND WALTER H. SNELL: Development of the carpophores of certain Boletaceae, *Mycologia*, **32**(4):493–504. *Figs.* 1–2. 1940.

FALCK, RICHARD: Die Merulius-Fäule des Bauholzes, in A. MÖLLER: Hausschwammforschungen in amtlichem Auftrage, vol. 6, pp. i–xvi, 1–405. 17 *pls.* 75 *text figs.* 1912.

FAYOD, V.: Prodrome d'une histoire naturelle des Agaricinés, *Ann. sci. nat. Botan.*, 7me sér., **9**:181–411. *Pls.* 6–7. 1889.

FRIES, ELIAS MAGNUS: Systema mycologicum, sistens fungorum ordines, genera et species huc usque cognitas, vol. 1, lvii + 520 pp., 1821; vol. 2, pt. 1, pp. 1–274, 1822; vol. 2, pt. 2, pp. 275–621, 1823; vol. 3, pt. 1, pp. i–viii, 1–259, 1829; vol. 3, pt. 2, pp. 261–524, index, pp. 1–202, 1832. Greifswald, Ernestus Mauritius.

———: Hymenomycetes Europaei sive epicriseos systematis mycologici, editio altera, pp. 1–756, Uppsala, E. Berling, 1874.

GADD, C. H., AND C. A. LOOS: The basidiospores of Exobasidium vexans, *Brit. Mycol. Soc. Trans.*, **31**(3–4):229–233. *Fig.* 1. 1948.

GÄUMANN, ERNST ALBERT: Über die Gattung Kordyana, *Ann. Mycol.*, **20**(5–6):257–271. *Figs.* 1–7. 1922.

———: Vergleichende Morphologie der Pilze, 626 pp. 398 *figs.* Jena, Gustav Fischer, 1926.

———: Comparative Morphology of Fungi. Translated and revised by Carroll William Dodge, xiv + 701 pp. 406 *figs.* 43 *diagrams.* New York, McGraw-Hill Book Company, Inc., 1928.

GIESENHAGEN, KARL: Hexenbesen an tropischen Farnen, *Flora oder Allgemeine Botan. Ztg.*, **76**:130–156. *Pls.* 12–13. 1892.

GILBERT, E. J.: Les Bolets, Les Livres du Mycologue, vol. 3, 254 pp. 16 *pls.* Paris, Librairie E. le François, 1931.

HEIM, ROGER: Observations sur la flore mycologique malgache: III. Trois bolets gigantesques d'Afrique et de Madagascar, *Rev. mycol.*, N.S., **1**(1):3–18. *Pls.* 1–4. 1936a.

———: Observations sur la flore mycologique malgache: V. Les Lactario-Russulés à anneau: ontogénie et phylogénie, *Rev. mycol.*, N.S., **2**(1):4–17; (2):61–75; (3):109–117. *Figs.* 1–20. 1937.

———: Observations sur la flore mycologique malgache: VII. Les Bolets à tubes libres, *ibid.*, **4**(1):5–20. *Pl.* 1. *Figs.* 1–6. 1939.

———: Sur trois Agarics à latex de la flore malgache, *Compt. rend.*, **202**:1450–1452. 1936b.

———: Les Lactario-Russulés du Domaine Orientale de Madagascar. Essai sur la classification et la phylogénie des Astérosporales, pp. 1–196. 8 *pls.* (4 *colored*). 59 *figs.* 2 *phylogenetic diagrams.* Paris, Laboratoire de Cryptogamie du Muséum National d'Histoire Naturelle, 1937 (1938).

———, avec la collaboration du Dr. P. Font Quer et du Dr. J. Codina: Fungi Iberici. Observations sur la flore mycologique catalane, *Publications de la Junta de Ciències Naturals de Barcelona, 1934. Treballs del Museu de Ciències de Barcelona 15, Série Botánica,* **3**:1–146. *Pls.* 1–4. *Figs.* 1–22. 1934.

Hennings, P.: Hymenomycetineae, in A. Engler und K. Prantl: Die Natürlichen Pflanzenfamilien, Teil 1, Abt.,1**, pp. 105–276. *Figs.* 66–125. Leipzig, Wilhelm Engelmann, 1897.

Hugon, L.: Une intoxication par Coprinus atramentarius, *Rev. mycol. Supplement,* **3**(3):48–49. 1938.

Josserand, M.: Note sur un empoisonment grave causé par une Lepiote du groupe helveola, *Bull. trimestriel de la soc. mycologique de France,* **47**:52–71. *Pl.* 3. 1931.

Juel, H. O.: Die Kerntheilungen in den Basidien und die Phylogenie der Basidiomyceten, *Jahrb. wiss. Botan.,* **32**:361–388. *Pl.* 4. 1898.

———: Cytologische Pilzstudien: I. Die Basidien der Gattungen Cantharellus, Craterellus und Clavaria, *Nova Acta Regiae Soc. Sci. Upsaliensis,* Ser. IV, **4**(6):1–36. *Pls.* 1–3. 1916.

Karakulin, B. P.: On the question of the systematic position of fungi belonging to the type of Exobasidiopsis mihi, *Notulae Systematicae Inst. Crypt. Horti Botanici Petropolitani,* **2**:101–108. 1923. (In Russian.)

Karsten, P. A.: Rysslands, Finlands, och den Skandinavska Halföns Hattsvampar, *Bidrag till kännedom af Finlands Natur och Folk,* **32**:i-xxviii, 1–550. 1879; **37**:v–xix, 1–245. 1882. Helsingfors.

Killermann, S.: Unterklasse Eubasidii. Reihe Hymenomyceteae (Unterreihen Tremellineae und Hymenomycetineae), in A. Engler und K. Prantl: Die Natürlichen Pflanzenfamilien, Zweite Auflage, **6**:99–290. *Pls.* 1–5. *Figs.* 81–157. Leipzig, Wilhelm Engelmann, 1928.

Kühner, Robert: Sur la nature des cystides chez les Basidiomycètes, *Compt. rend.,* **180**(6):454–457. 5 *figs.* 1925a.

———: Le développement du *Lentinus tigrinus* Bull., *ibid.,* **181**(3):137–139. 1 *fig.* 1925b.

———: Contribution à l'étude des Hyménomycètes et spécialement des Agaricacés, *Le Botaniste,* **17**:3–224. *Pls.* 1–4. *Figs.* 1–37. 1926.

———: Le développement du Boletinus cavipes (Opat.) Kalch., *ibid.,* **18**:177–181. *Pls.* 6–7. 1927.

Levine, Michael: The origin and development of lamellae in Agaricus campestris and in certain species of Coprinus, *Am. J. Botany,* **9**(9):509–533. *Pls.* 28–35. *Diagrams* 1–12. 1922.

Lohwag, Heinrich, und Leopoldine Follmer: Die Hymenophore von Fistulina hepatica, *Ann. Mycol.,* **34**(6):456–469. *Figs.* 1–5. 1936.

Maire, René: Sur la cytologie des Hyménomycètes, *Compt. rend.,* **131**(2):121–124. 1900.

———: Sur la cytologie des Gastromycètes, *ibid.,* **131**(26):1246–1248. 1900.

———: Recherches cytologiques et taxonomiques sur les Basidiomycètes, *Bull. soc. mycologique de France,* **18**:1–209. *Pls.* 1–8. 2 *diagrams.* 1902. (Special fascicle with special page and plate numbers.)

———: Fungi Catalaunici. Series altera. Contribution à l'étude de la Flore Mycologique de la Catalogne, *Pubs. institut botan. Barcelona,* **3**(4):1–128. 8 *figs.* 1937.

Mason, E. W.: Zygosporium. Annotated Account of Fungi Received at the

Imperial Mycological Institute, List II, Fascicle 3, Special part, pp. 134–144. *Figs.* 40–44. London, 1941.

McGuire, J. M.: The morphology of Physalacria inflata, *Mycologia*, **31**(4):433–438. *Figs.* 1–14. 1939.

Micheli, Pietro Antonio: Nova plantarum genera juxta Tournefortii methodum disposita, xxii + 234 pp. 108 *pls.* Florence, 1729.

Miller, L. W.: The genera of Hydnaceae, *Mycologia*, **25**(4):286–302. 1933.

Mounce, Irene, and Ruth Macrae: The behavior of paired monosporous mycelia of Lenzites saepiaria (Wulf.) Fr., L. trabea (Pers.) Fr., L. thermophila Falck, and Trametes americana Overh., *Can. J. Research, C*, **14**(6):215–221. *Pl.* 1. 1936.

Murrill, Wm. A.: Polyporaceae, *N. Am. Flora*, **9**:1–131. 1907.

———: Boletaceae, *ibid.*, **9**:133–161. 1910.

Nobles, Mildred K.: Production of conidia by Corticium incrustans, *Mycologia*, **29**(5):557–566. *Figs.* 1–28. 1937.

Oehm, Gustav: Kritische Bemerkungen zur Nebensporenfrage bei höheren Pilzen mit besonderer Berücksichtigung der Gattung Ceriomyces, *Botan. Centr. Beihefte*, **57**(Abt. A):252–263. 1937.

Patouillard, N.: Les Hyménomycètes d'Europe. Anatomie générale et classification des champignons supérieurs, 166 pp. 4 *pls.* Paris, Paul Klincksieck, 1887.

———: Essai taxonomique sur les familles et les genres des Hyménomycètes. Thèse pour l'obtention du diplôme de Docteur de l'Université de Paris, École Supérieure de Pharmacie, année 1900–1901, No. 2, 184 pp. 74 *figs.* Lons le-Saunier, 1900.

Persoon, Christiaan Hendrik: Synopsis methodica fungorum, Pars prima et secunda, xxx + 706 pp. and Index. *Pls.* 1–5. Göttingen, Heinrich Dietrich, 1801.

———: Mycologia Europaea seu completa omnium fungorum in variis Europeae regionibus detectorum enumeratio, methodo naturali disposita, descriptione succincta, synonymia selecta et observationibus criticis additis, 3 sections (volumes), 852 pp. 30 *colored plates.* Erlangen, Johann Jacob Palm, 1822–1828.

Rea, Carleton: British Basidiomyceteae. A Handbook to the Larger British Fungi, xii + 799 pp. Cambridge, Cambridge Univ. Press, 1922.

Reijnders, A. F. M.: De studie van de ontwikkeling van de basidiocarpen der Agaricales, *Mededeelingen van de Nederlandsche Mykologische Vereeniging*, **21**:135–208. *Pls.* 4–7. 2 *text figs.* 1933.

Remsberg, Ruth E.: Studies in the genus Typhula, *Mycologia*, **32**(1):52–98. *Figs.* 1–58. 1940.

Robak, Haakon: Studies in the biology of wood-destroying Hymenomycetes: I. Contribution to the knowledge of homothally and heterothally in some species of Thelephoraceae and Polyporaceae. II. The ability of haploid mycelia to produce rot, *Nytt Magasin for Naturvidenskaberne*, **76**:5–20. 1936.

Rogers, Donald P.: Notes on the Lower Basidiomycetes, *Univ. Iowa Studies in Natural History*, **17**(1):1–43. *Pls.* 1–3. 1935.

———: The genus Pellicularia (Thelephoraceae), *Farlowia*, **1**(1):95–118. *Figs.* 1–11. 1943.

———, and H. S. Jackson: Notes on the synonymy of some North American Thelephoraceae and other resupinates, *ibid.*, **1**(2):263–328. 1943.

Saccardo, P. A.: Sylloge fungorum omnium hucusque cognitorum, vol. 5, pp. 1–1146. Pavia, published by author, 1887.

Singer, Rolf: Das System der Agaricales, *Ann. Mycol.*, **34**(4–5):286–378. 1936.

———: Notes on taxonomy and nomenclature of the Polypores, *Mycologia*, **36**(1):65–69. 1944.

———: The Boletineae of Florida with notes on extralimital species: I. The Strobilomyceteae, *Farlowia*, **2**(1):97–141. *Pl.* 1. 1945; II. The Boletaceae (Gyroporoideae), *ibid.*, **2**(2):223–303. 1945; III. The Boletoideae of Florida, *Am. Midland Naturalist*, **37**(1):1–135. *Pls.* 1–2. 1947; IV. The lamellate families (Gomphidiaceae, Paxillaceae and Jugasporaceae), *Farlowia*, **2**(4):527–567. *Pl.* 1. 1946.

———: The Agaricales. (Mushrooms) in modern taxonomy, *Lilloa Revista de Botanica*, **22**:1–834. *Pls.* 1–29. 1949 (1951).

Smith, Alexander H.: Common edible and poisonous mushrooms of southeastern Michigan, *Cranbrook Inst. Sci. Bull.* **14**:1–71. *Illustrated.* 1938.

———, and Elizabeth E. Morse: The genus Cantharellus in the Western United States, *Mycologia*, **39**(5):497–534. *Figs.* 1–13. 1947.

Snell, Walter H.: The genera of the Boletaceae, *Mycologia*, **33**(4):415–423. *Fig.* 1. 1941.

———: New proposals relating to the genera of the Boletaceae, *ibid.*, **34**(4):403–411. *Fig.* 1. 1942.

Süss, W.: Zur Geniessbarkeit des Faltentintlings (Coprinus atramentarius), *Z. Pilzkunde*, **15**:54–56. 1936.

Tischler, G.: Pflanzliche Chromosomen-Zahlen, *Tabulae Biologicae*, **4**:1–83. 1927.

Togashi, Kogo, and Kojiro Oda: Spore-size variability in subsequent spore prints of some Hymenomycetous fungi, *Trans. Sapporo Natural History Soc.*, **13**(3):121–125. 1034.

Vandendries, René: Les tendances sexuelles chez les Polyporés, *Rev. mycol.*, N.S., **1**(2):85–92; (4):181–190; (6):294–302. 10 *figs.* 1936.

———, et P. Martens: Oïdies haploïdes et diploïdes sur mycélium diploïde chez "Pholiota aurivella" Batsch, *Bull. classe sci., Acad. roy. Belg.*, 5e *sér.*, 18:468–472. 1 *fig.* 1932.

Walker, Leva B.: Development of Pluteus admirabilis and Tubaria furfuracea. *Botan. Gaz.*, **68**(1):1–21. *Pls.* 1–5. *Figs.* 1–8. 1919.

White, J. H.: On the biology of Fomes applanatus, *Trans. Roy. Can. Inst.*, **12**(2):133–174. *Pls.* 2–7. 1920.

Wolf, Frederick A.: A little known vetch disease, *J. Elisha Mitchell Sci. Soc.*, **36**(1–2):72–85. *Pls.* 2–6. 1920.

———: The morphology and systematic position of the fungus Microstroma juglandis (Bereng.) Sacc., *ibid.*, **43**(1–2):97–99. *Pl.* 4. 1927.

———: The relationship of Microstroma juglandis (Bereng.) Sacc., *ibid.*, **45**(1):130–135. *Pl.* 6. 1929.

Woronin, Michael: Exobasidium Vaccinii, *Verhandlungen der Naturforsch. Gesellschaft zu Freiburg i. B.*, **4**(4):397–416. 3 *pls.* 1867.

15

CLASS BASIDIOMYCETEAE: SUBCLASS EUBASIDIAE, "GASTEROMYCETEAE"

Gasteromyceteae

The name Gasteromyceteae (often spelled Gastromyceteae) is given to a group including those orders of Basidiomyceteae which have one-celled, two- to four-spored basidia (rarely with more spores) produced within closed spore fruits. These may have definite means of dehiscence by which the spores reach the exterior or they may depend upon animals which feed upon the spore fruits and thus carry the spores to various places, or the spores escape only upon the decay of the spore fruits. Mostly the basidia are produced in hymenia which line closed cavities, but the latter instead of being exposed to the air at the time of spore maturation, as in the Hymenomyceteae, remain closed until after the spores have been produced. Frequently the spore fruits are produced underground, emerging as they approach maturity or remaining subterranean. Other forms grow on stumps, logs, etc. The mycelium usually shows the presence of clamp connections. Conidia are known in a few forms. Sexual reproduction is practically unknown, except for the union of the two nuclei in the young basidium. Few species of this group have been grown successfully in culture. Even the germination of the spores has been found impossible of accomplishment in many species. The basidia may be clavate or cylindrical, as in the majority of the Hymenomyceteae, but in the main are more inclined to be ovoid or globose with sterigmata varying from short to long or even lacking. The spores are perched symmetrically at the tips of the sterigmata and are not discharged with violence as in the preceding orders. Sometimes the long sterigma or a portion of it remains attached to the spore, like the handle of a drumstick.

The young basidia are binucleate and the two nuclei unite, this union being followed by meiotic divisions leading to the production of four haploid nuclei. The first meiotic spindle is transverse in the majority of cases (chiastobasidial) but is sometimes longitudinal (stichobasidial), e.g., in *Tulostoma*. In some cases subsequent divisions produce eight nuclei or even more. Usually the number of basidiospores is four, but not rarely six to eight and in rare cases up to 12 spores are produced. These are almost always at first uninucleate but in most of the few cases studied they early become binucleate by the division of the original nucleus. Whether such binucleate spores give rise to the dicaryon phase of mycelium is not proved, but seems probable. Lorenz (1933) has shown that in *Sphaerobolus grandis* Lorenz the uninucleate basidiospores give rise to monocaryon mycelia. These mycelia show two sexual phases, i.e., this species is sexually bipolar, and only when the appropriate mycelia are mated does a dicaryon mycelium with clamp connections arise. In this species the basidiospores are uninucleate, although some species of this genus have been described as producing binucleate spores. *Crucibulum leve* (DC.), Kambly (*C. vulgare* Tul.) and *Cyathus striatus* (Willd.) Pers. are both, according to Nils Fries (1936), quadripolar.

In general the structure of the spore fruit in the Gasteromyceteae is as follows: Externally there is a peridium consisting of one or more layers. It may be firm and hard (*Scleroderma*) or soft and papery or may even disappear during the development of the spore fruit, in *Gautieria graveolens* Vittad., being present, according to Fitzpatrick (1913), only in the very young stages. Within the peridium the tissue may consist simply of the gleba or of the gleba traversed by "veins" or by a "columella" or by both. The gleba consists of a more or less fleshy mycelial growth containing usually numerous hymenium-lined cavities (hymenial cavities) but only one in a few cases. The columella and veins are slender or stout strands of hyphae having several functions, food conduction, support, and in some cases the dehiscence of the spore fruit at maturity. The gleba in many genera undergoes autodigestion after the basidiospores have been formed. The tissues involved are the basidia and the fungous tissue lying between the hymenial cavities, the trama. As a result nothing may be left of the gleba except the basidiospores, and in a few cases the basidia also, or also a few stiff threads, the capillitium. These are simple or branching thick-walled hyphae, which develop in the interhymenial tissues of the gleba before maturity of the spore fruit. They are rarely septate. The tangled mass of capillitial hyphae prevents the escape of all the spores at once, permitting them to sift out a few at a time. In some cases hymenial cavities with a definite layer of hymenium are wanting but clusters of basidia appear here and there in the gleba.

The ontogeny of the spore fruit exhibits an early gymnocarpic stage in some species while in others the earliest stages are angiocarpic. The gymnocarpic stage usually becomes closed early so that the development corresponds to the pseudoangiocarpic mode of growth found in various

Boletaceae and Agaricaceae. The gymnocarpic origin or pseudoangiocarpic origin of the hymenial surfaces is found in *Hemigaster* (Juel, 1895), *Chamonixia*, *Hydnangium*, *Arcangeliella*, *Elasmomyces*, and probably other forms. In the majority of genera however, the development is angiocarpic.

The structure of the spore fruits is not at all on a common plan but they are very heterogeneous. It is not certain that the 1200 to 1500 species included in about 120 genera form a group of monophyletic ancestry. Certain genera are beyond doubt closely related to the Agaricaceae and may have arisen from that family or may be, as Lohwag (1925), Singer (1936 and 1951) and others have suggested, in the ancestral line leading to those fungi. If these are excluded from the Gasteromyceteae the remainder form a more coherent group.

The structure of the basidiospores is of several types and probably should be given further study with reference to its importance in the determination of the relationship of the genera. The more or less lemon-shaped spore with dark color and warty surface is very striking. This type is found in *Gasterella* and *Gasterellopsis*, *Hymenogaster*, *Chondrogaster*, *Dendrogaster*, and perhaps others. The spores in the Phallales are smooth, ellipsoidal to cylindrical, and nearly colorless. In many genera the spores are spherical and spiny, in others smooth. In *Podaxis* the spore is plainly two layered, with a terminal germ pore. In *Gautieria* and *Chamonixia* the spores are longitudinally ridged and furrowed, as in *Clitopilus* and *Octojuga* in the Agaricaceae. In *Nigropogon* they are angled, somewhat as in *Rhodophyllus* (*Entoloma*) in that family. Until the ontogeny of the spore fruits from their earliest stages has been studied it is uncertain to what degree these different spore types represent relationships within or without the Gasteromyceteae.

In *Secotium*, *Elasmomyces* and *Gasterella*, and some other forms cystidia are present in the hymenium. This has been considered by some investigators to indicate relationship with the Hymenomyceteae. In *Arcangeliella* latex vessels are present and also in *Lactariopsis* (which may perhaps better be placed in the Agaricaceae close to *Lactarius*). They are also described in *Battarrea* and *Phellorinia*.

Since the youngest stages of development of the spore fruit have not been studied, except in a relatively small number of the subterranean genera, and also are unknown in many of the tropical species whose earlier stages are not subterranean, the relationships of these fungi to other groups and the interrelationships of the genera and families making up the Gasteromyceteae are still far from settled.

In general there appear to be four types of structure, as pointed out by Lohwag (1924b, 1925, 1926) and Eduard Fischer (many publications, especially in Engler and Prantl, 1933). These may be called the lacunar, coralloid, multipileate, and unipileate types. The first stands rather by itself but the other three represent a gradation from coralloid to unipileate types. Briefly they are as follows: (1) *Lacunar type.* Within the fleshy mass of hyphae destined to form the gleba the tissues pull apart at various places to form cavities which become lined by a palisade of basidia. A modification of this type is found in scattered species, genera, and whole families in which the hyphae producing the tufts of basidia instead of forming an even layer around a cavity grow irregularly into it, thus forming nests of basidia which obliterate the hymenial cavities. Such a structure is called "plectobasidial." (2) *Coralloid type.* From the basal portion of the spore fruit there grow upward and outward branching masses of hyphae, all within the enlarging peridium so that a coralloid structure develops. The ends of these tramal branches may unite with the inner surface of the peridium and spread along it to form the inner peridial layer, which is therefore of tramal origin. The spaces between the branches are at first continuous so that there is in fact only a single cavity very much interrupted by the coral-like tramal branches which are covered by a continuous hymenium. Probably the lacunar and coralloid types grade into one another. Lohwag believes that the former is derived from the latter. In *Lycoperdon* the basal part of the spore fruit is built on the lacunar plan and the upper, fertile part develops, according to Rehsteiner (1892) in the coralloid manner. (3) *Multipileate type.* In this type, clearly derived from the coralloid type, a number of the tramal branches are enlarged and reaching the inner surface of the peridium spread out along it, producing abundant tramal development on the under side of each "pileus." (4) *Unipileate type.* The apical pileus of a spore fruit of the multipileate plan becomes large, clothing the inner surface of practically the whole peridium, the other pilei remaining rudimentary and not producing fertile glebal branches. The main central stalk of the unipileate type may exist as a "columella" or may elongate downward so that a stipe is formed. (Fig. 171.)

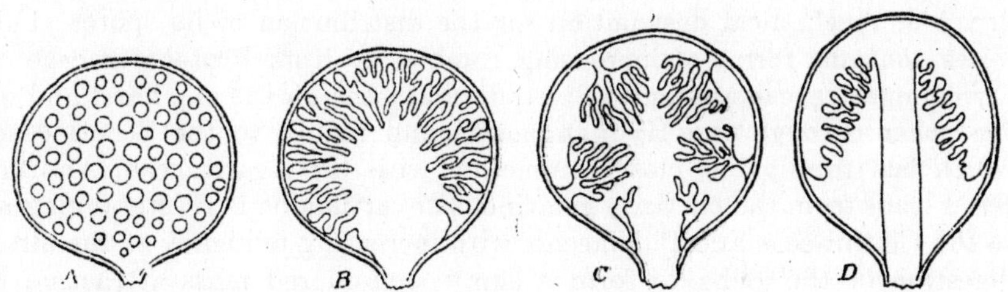

Fig. 171. Schematic representation of the basic structures of the spore fruits of various Gasteromyceteae. (*A*) Lacunar type. (*B*) Coralloid type. (*C*) Multipileate type. (*D*) Unipileate type. (After Fischer, in Engler und Prantl: Die Natürlichen Pflanzenfamilien, Zweite Auflage, vol. 7a, Leipzig, W. Engelmann.)

The logical classification into orders, families, and genera of the Gasteromycetes depends upon the probable course of evolution in the group. The undeniable points of similarity between some of the Agaricaceae and various unipileate Gasteromyceteae, e.g., *Elasmomyces*, *Secotium*, *Montagnea*, etc., have already been mentioned. Heim (1934) in fact includes Family Hydnangiaceae in the Agaricales, and others have placed *Secotium* and *Montagnea* in the same order. The fact that the basidiospore in all Gasteromyceteae is symmetrically perched at the tip of the sterigma and is not shot off from it makes a derivation of the Agaricales from them difficult, while the loss of these characters would not be so difficult to imagine if some Agaricaceae became gasteromycetous in structure and habit. It must be remembered that the genus *Hyaloria* in the Heterobasidiae has the basidia enclosed in a sort of loose peridium and the basidiospores there are symmetrically placed and not shot away. Practically all of the other Heterobasidiae have the usual type of basidium. The occurrence of transitional forms from *Secotium* and the apparent close relationship to some other genera of Gasteromyceteae is an argument in favor of the theory that there is a close connection between both groups. Yet it is difficult to imagine simple forms like *Protogaster* and *Gasterella* as being derived from the very complicated Secotiaceae. In view of the fact that so much investigation still remains to be done on the development of the spore fruit of many species and genera before a completely satisfactory conclusion can be arrived at as to the classification and arrangement within the group the author has decided to follow Eduard Fischer (1933) in the main, with modifications that seem to him to be necessary because of more recent studies, especially on *Protogaster*, *Gasterella* and *Gasterellopsis*, and some of the more secotioid genera, *Gyrophragmium*, *Battarrea*, etc.

The Gasteromyceteae may be arranged in several parallel series, each ranging from simple to complex structures (or possibly the reverse). As outlined below the groups with simple structures are placed first. In one series the spore fruit is mainly fleshy, with well-developed hymenial chambers. It does not undergo partial autodigestion and depends upon decay or mechanical destruction for the distribution of its spores. This series contains forms of increasing complexity from Protogastraceae to Hymenogastraceae and on to Sclerodermataceae on the one hand and on the other through the Hydnangiaceae and finally to the Secotiaceae, which last family has close connection with the Agaricaceae. Another series leads from the partially gelatinous or cartilaginous Hysterangiaceae to the Clathraceae and Phallaceae, with increasing tendency to the autodigestion of the gleba to form a slimy, evil-odored mass attractive to insects which distribute its spores. A third series, arising again in the Hymenogastraceae leads to the Lycoperdaceae and Geastraceae, and

Tulostomataceae and possibly the Podaxaceae. Some genera usually assigned to the Sclerodermatales probably belong in this series. The gleba breaks down by autodigestion and leaves the spores and sometimes also the basidia and the hyphae making up the capillitium, as a dry powdery mass, spore distribution taking place by air currents. A fourth tendency shown by the Sphaerobolaceae, Nidulariaceae and the genus *Pisolithus*, in the Sclerodermataceae, is the formation of firmer walls around definite regions of the gleba so that these are distributed as units, either by mechanical means or currents of water or by violent expulsion from the fruit body. These probably do not form a connected series but represent separate evolutionary changes from both the Hymenogastrales and the Sclerodermatales.

The orders tentatively recognized here are Protogastrales, Hymenogastrales, Sclerodermatales, (possibly better distributed among other orders), Lycoperdales, Nidulariales, Sphaerobolales, Phallales. The late Sanford M. Zeller (1948, 1949) recognized 9 orders and 32 families instead of the 7 orders and considerably fewer families recognized in this work. Until much more intensive study has been given to the ontogeny of the sporocarps of the Gasteromyceteae any arrangement must be considered to be more or less tentative.

Order Protogastrales. In this order the minute spore fruits have but a single hymenial cavity. The hymenium consists of basidia without cystidia, and the basidiospores are light-colored and smooth. *Protogaster* was described by Zeller (1934). It is a minute fungus growing on the roots of plants in Maine and has been found but once and then only in the mature stages. It consists of a nearly spherical spore fruit, less than a millimeter in diameter. It contains a single large cavity lined by basidia bearing

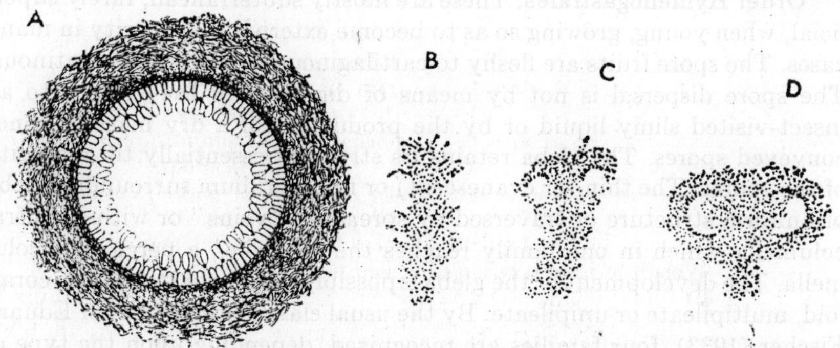

FIG. 172. Protogastrales. (A) Family Protogastraceae. *Protogaster rhizophilus* Thaxt.; diagrammatic median section of spore fruit. (B–D) Family Hemigastraceae. *Hemigaster candidus* Juel; three stages in the pseudoangiocarpic development of the spore fruit. (A, courtesy, Zeller: *Ann. Missouri Botan. Garden*, **21**(2):231–240. B–D, from Comparative Morphology of Fungi by Gäumann and Dodge, New York, McGraw-Hill Book Co., Inc.)

colorless ellipsoid spores. There are no noticeable projections or folds on the surface of the hymenial cavity and no cystidia have been observed. In the absence of younger stages it is impossible to locate this with certainty. Perhaps it is wiser to place this in a distinct family Protogastraceae in the Order Protogastrales. (Fig. 172 A.)

Possibly Juel's genus and family, *Hemigaster* and Hemigastraceae, respectively, should be placed in this same order. The minute spore fruit grows on rabbit excrement. It is 2 to 3 mm. tall and at maturity about as broad. It arises as an upright tuft of parallel hyphae which spread at the top like a sheaf of grain. The spreading hyphae curve downward and inward and eventually join with hyphae growing out from the stipe to form a circular chamber centrally pierced by the percurrent columella which is in reality the upper portion of the stipe. Two layers are visible in the peridium, an outer loosely woven portion and an inner denser subhymenial layer from which arise the basidia which cover the upper and outer side of the circular hymenial cavity, but not the inside formed by the columella. The basidia bear four nearly spherical, pale, flesh-colored basidiospores with smooth surfaces. There are no cystidia or paraphyses in the hymenium. From the columella there grow out into the cavity slender hyphae which bear colorless chlamydospores wound about by slender hyphae. Eventually the basidiospores and chlamydospores fill the cavity. The mode of development of the spore fruit is of the type called pseudoangiocarpic. Since the younger stages of *Protogaster* are not known it is not possible to determine definitely whether these two fungi are related or not. Juel (1895) concluded that *Hemigaster* is related to the Thelephoraceae but it seems to the author that it belongs rather in the Gasteromyceteae. (Fig. 172 B–D.)

Order Hymenogastrales. These are mostly subterranean, rarely superficial, when young, growing so as to become external at maturity in many cases. The spore fruits are fleshy to cartilaginous or somewhat gelatinous. The spore dispersal is not by means of digestion of the gleba into an insect-visited slimy liquid or by the production of a dry mass of wind-conveyed spores. The gleba retains its structure essentially till maturity of the spores. The thin (or evanescent) or firm peridium surrounds a gleba of uniform structure or traversed by spreading "veins" or with a central columella which in one family reaches the apex, i.e., a percurrent columella. The development of the gleba is possibly lacunar or more often coralloid, multipileate or unipileate. By the usual classification, that of Eduard Fischer (1933), four families are recognized, depending upon the type of development. Many spore types are found, colorless or colored, smooth or verrucose, or ribbed, etc. Possibly it may be feasible, when the youngest stages of development have been studied for most of the genera, to correlate development, spore type and mature morphology to produce a more satisfactory system of classification.

Family Hymenogastraceae. The type of this family is the genus *Hymenogaster*. This is subterranean or with the upper surface projecting above the ground at maturity. In very young specimens Rehsteiner (1892) showed that a palisade layer of cells develops as a boundary setting off the thick sterile base from the upper peridium. This layer arches upward with the growth in size of the spore fruit forming a single central cavity into which grow downward from the roof various branches and plates which extend to and apparently in places grow fast to the base. These plates anastomose with one another so that labyrinthiform hymenial cavities are formed. With the increase in size of the spore fruits the relative size of the sterile base becomes much smaller. Eventually the spore fruit shows very numerous irregular cavities, sometimes radiating a little from the sterile base. The peridium is rather firmly attached to the outer side of the gleba. The hymenial cavities are lined with clavate basidia which bear two, rarely four, spores. These vary with the species but are mostly ellipsoidal, ovoid or limoniform, mostly yellow to brown in color, smooth or more often verrucose or wrinkled. No cystidia are described for this genus. Apparently closely related to the foregoing is the unilocular genus *Gasterella*, described by Zeller and Walker (1935) and Miss Walker (1940). It grows on the surface of the soil and reaches the diameter of 300 to 700 μ and even up to over 1200 μ. According to Routien (1939) rhizomorphs are attached to the spore fruits and their hyphae show clamp connections although these are not visible in the basidiocarp. In the specimens that reach maturity in the smaller dimensions the cavity may have smooth walls but in the larger spore fruits the hymenium sends folds and projections into the cavity from above and from the sides. These greatly increase the hymenial surface but do not reach the bottom, so that the spore fruit remains unilocular. The basidiospores arise in twos or more often fours and are dark-colored, verrucose and somewhat apiculate. When detached a piece of the sterigma often remains attached to the spore. Cystidia with black verrucose heads are sometimes found, especially in the smaller specimens that have developed under less favorable conditions. They appear to be, perhaps, aborted basidia. In the very young spore fruits an arching palisade layer of densely staining cells appears in the midst of the loose tuft of hyaline hyphae. As it broadens and arches up further a cavity is formed into which eventually push the plates or folds which partially divide up the single cavity. Miss Walker (1940) suggested that *Gasterella* should be placed in Order Protogastrales but in a separate family Gasterellaceae. In the author's opinion the similarity of *Gasterella* to the young stages of *Hymenogaster rehsteineri* Bucholtz is too great to allow their separation into different families and orders. (Fig. 173 A–E.)

Gasterellopsis (Routien, 1940) begins its development much as in *Gasterella* except that there is a central percurrent columella so that when

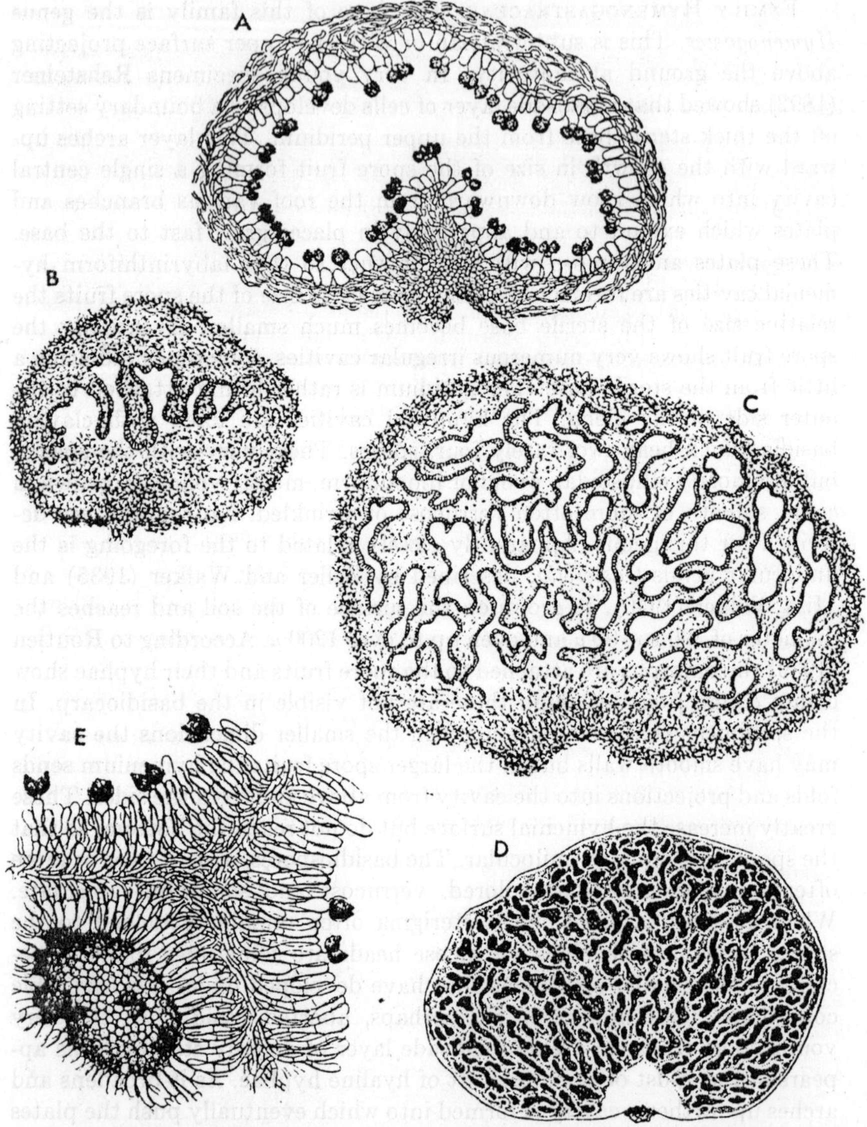

FIG. 173. Hymenogastrales, Family Hymenogastraceae. (A) *Gasterella lutophila* Zeller & Walker; vertical section through a spore fruit showing the tendency toward the formation of folds from the hymenial wall. (B, C) *Hymenogaster rehsteineri* Bucholtz, showing vertical sections of very young and somewhat further developed spore fruits. (D, E) *Hymenogaster tener* Berk. (D) Section through mature spore fruit. (E) Portion of mature gleba. (A, courtesy, Zeller and Walker: *Mycologia*, **27**(6):573–579. B–E, after Tulasne, from Fischer, in Engler und Prantl: Die Natürlichen Pflanzenfamilien, Zweite Auflage, vol. 7a, Leipzig, W. Engelmann.)

the hymenial layer begins to appear it forms the roof of a circular cavity instead of a depressed spherical one. From all sides of this cavity, except the columella, centripetally directed folds begin to form and become covered by the mature hymenium. They may stop short of the columella leaving the circular cavity not completely divided but with numerous radial lobes, or some or all of these folds may grow to the columella in which case separate radial cavities are produced. Eventually the peridial tissues in contact with the basal portion of the columella dissolve, thus forming a basal circumscissile opening and then the remainder of the peridium and finally of the tramal tissues dissolve so that a single cavity filled with spores is left. Its outer wall then is the remains of the subhymenial layer. The spores are like those of *Gasterella* and are produced by twos or fours on the basidia. No cystidia were observed. Routien suggested that this represents a further step in complication from *Gasterella*, and perhaps should be placed in the same family with it, since it starts as a unilocular fungus and often remains so. On the other hand it shows great similarities, although it is much simpler in structure, to the Secotiaceae. *Rhizopogon* is a genus of thirty or more species with subterranean basidiocarps whose surface is covered with numerous loose or adherent branching fibrils which lead into rhizomorphs. Its spores are more or less ellipsoidal and smooth. The young spore fruit has a central portion of loosely branching coralloid structure with the interconnecting open spaces lined with hymenium. The basidia are two- to eight-spored. If a much enlarged *Protogaster* should develop invaginating and branched ridges and lobes it would show many of the characteristics of *Rhizopogon*.

Pilat (1934) discussed the genus *Gastrosporium* and based upon it the family Gastrosporiaceae. In the latter the peridium is double while in the Hymenogastraceae it is, according to him, simple.

FAMILY MELANOGASTRACEAE. This is an assemblage of several more or less related genera that differ from the Hymenogastraceae in having their lacunar hymenial cavities more or less filled or obliterated by a gelatinous mass which in *Leucogaster* appears, according to Zeller and Dodge (1924), to be the product of the gelification of conidia or chlamydospores which were produced prior to basidial development. Into these jelly filled cavities long, slender basidia push their way, partially filling them with a crisscross tangle. The basidiospores are often coated with a gelatinous layer and arise two to eight per basidium. They are almost colorless in *Leucogaster* and dark brown in *Melanogaster*. The spore fruits are subterranean or partially emerging at maturity, and without a stalk. The gleba is traversed by veins or sheets of tramal tissue that divide it into polyhedric or rounded units each of which is a "basidial nest" as some authors call it. Fischer (1933) places *Alpova* tentatively in this family. It has been reported only from the United States so far. It is partially subterranean at maturity and reaches a diameter of 5 to 20 mm. The glebal chambers are at first filled with large spherical cells which then gelatinize. Long hyphae traverse these cavities and on them sit the long, slender basidia which bear 5 to 11 pale brown, almost sessile, ellipsoid, smooth spores. Dodge (1931) who described the genus considered it to belong to the Rhizopogonaceae, a family segregated from the Hymenogastraceae, and including a number of genera placed by Fischer in the Melanogastraceae. Zeller (1939) held that its development suggests closer relationship to the latter family than to *Rhizopogon*.

FAMILY HYDNANGIACEAE. In this family of Hymenogastrales the coralloid development of the gleba has become unipileate. Like *Hemigaster* the spore fruits are pseudoangiocarpic in their development. The upper portion of the stipe becomes the percurrent columella. Unlike *Hemigaster* the palisade layer of the under side of the recurving pileus is thrown into folds which anastomose with one another and with the columella so that a multilocular gleba is produced. The columella may become reduced with age to a slender, scarcely recognizable strand, in some species. The stipe below the pileus is represented by only a small projection, if visible. Cystidia and spiny basidiospores two to four per basidium on long sterigmata are found in *Hydnangium* and *Arcangeliella*. In the latter genus laticiferous tubes are present. In *Chamonixia* the spores are longitudinally ribbed, much as in *Gautieria*.

FAMILY SECOTIACEAE. In this family the general plan is much like that of the preceding one, but the stipe is more pronounced in most forms. Development is pseudoangiocarpic in *Elasmomyces* and angiocarpic in most of the remaining genera. *Elasmomyces* represents probably an intermediate form between *Hydnangium* and *Secotium*. The fruit body is mostly eventually epigeous. Its development is pseudoangiocarpic like that of *Hydnangium*. In the tissue of the stipe are nests of enlarged, bladder-like cells, resembling those of *Russula* in the Agaricaceae. The spores are marked with verrucosities, sometimes connected by ridges, as in *Russula*, and as in that genus they are stained blue with reagents containing free iodine. Bucholtz (1903) considered these two genera, as did Malençon (1931), to be closely related. Heim (1938) also emphasizes the relationship of *Lactarius* and *Russula* to *Elasmomyces*.

The genus *Secotium*, with which *Elasmomyces* is sometimes united, is angiocarpic in its development in the species studied (*S. agaricoides* (Czern.) Hollos, by Conard, 1915, *S. novae-zelandiae* Cunningh. and *S. erythrocephalum* Tul., by Cunningham, 1924 and 1925). (Fig. 174.) In the earlier stages of development it shows great resemblance to that of *Agaricus* (Atkinson, 1906, 1915), but instead of forming radial lamellae separating the annular opening into radial cavities the tramal plates are irregular in the direction of their growth and anastomose to form closed

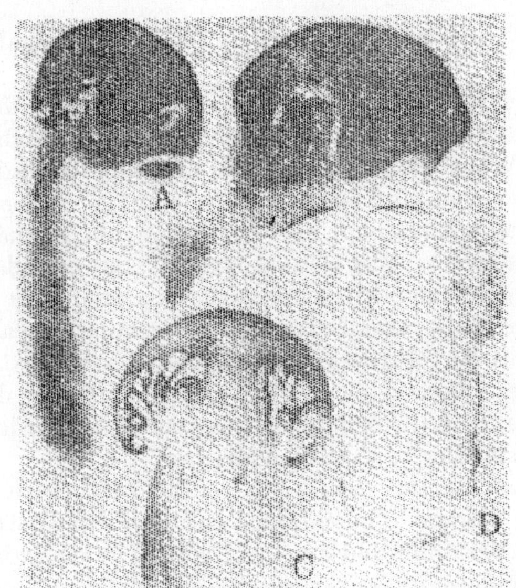

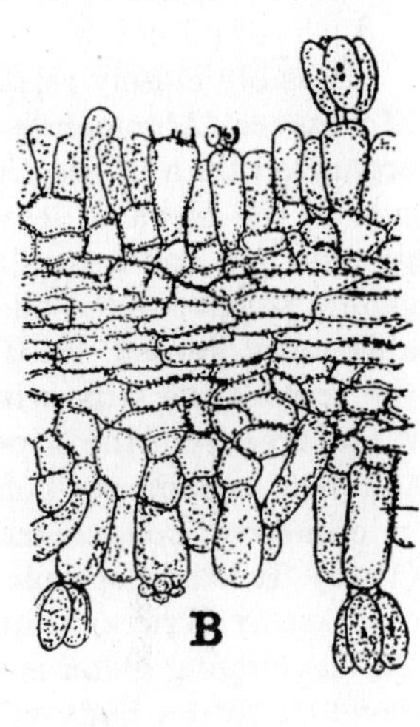

Fig. 174. Hymenogastrales, Family Secotiaceae. *Secotium erythrocephalum* Tul. (A) Mature plants. (B) Section of tramal plate. (C, D) Vertical sections through young and nearly mature plants. (Courtesy, Cunningham: *Brit. Mycol. Soc. Trans.*, 10:216–224.)

cavities. Apparently within the main tissues of the pileus other hymenial cavities develop as well as in the thick tramal plates in the gleba so that finally the latter is made up of very many narrow, hymenium-lined cavities, separated by thin tramal plates. As the stipe elongates the edge of the pileus pulls loose from it and the gleba becomes partly exposed to the air. Since the glebal cavities are not continuous the spores are distributed by the action of insects infesting the spore fruits and also by the decay of the latter. The presence of cystidia in some species of *Secotium* is another point of similarity to the Agaricaceae. There are some very small species, e.g., *S. coprinoides* Routien (1940), about 4 mm. tall with a pileus about 2 mm. broad. Its gleba consists of numerous (about 18) radial hymenial cavities extending from the lateral peridium to the stipe. The hymenium consists of two- or four-spored basidia intermixed with paraphyses. The ellipsoidal spores are smooth, black, and with a short pedicel formed by the apical portion of the sterigma. *S. olbium* Tul. is 4 to 6 mm. tall but the spores are smaller, spherical, and wrinkled. The structure of the gleba of the former is somewhat similar to that of *Gasterellopsis*, but is not sometimes unilocular and the spores are of different types. There is no deliquescence so that this is not a *Coprinus* and the position of the spore on the sterigma is typical of the Gasteromyceteae. At maturity the edge of the pileus breaks loose from the base of the stipe as the latter elongates and leaves no volva.

Possibly closely related to *Secotium* are *Gyrophragmium*, *Longula*, and *Montagnea* (*Montagnites*). In them the mature spore fruit is more highly organized than in *Secotium*. When the pileus expands, exposing the under side of the gleba, it leaves a volva, and sometimes an annulus, in *Gyrophragmium*, and a two-layered annulus, but no volva in *Longula*. In these genera the expanded pileus is convex. The gleba is lamellar, but considerably anastomosed. In *Montagnea* the pileus at maturity is a small disk at the apex of the stipe with the black radially lamellar gleba hung beneath it and free from the stipe as in the foregoing genera. This extends beyond the disk as separate lamellae. There is a volva but no annulus. The development of *Longula texensis* (B. & C.) Zeller has been studied by Barnett (1943). It is angiocarpic and similar to that of *Agaricus*.

FAMILY HYSTERANGIACEAE. In this family the coralloid structure of the developing gleba is very marked. The enlarged end of a rhizomorph develops into a body with a peridium and a central core. The enlarging core begins to form folds and plates under the expanding peridium. These anastomose so that eventually a multilocular gleba is formed with a

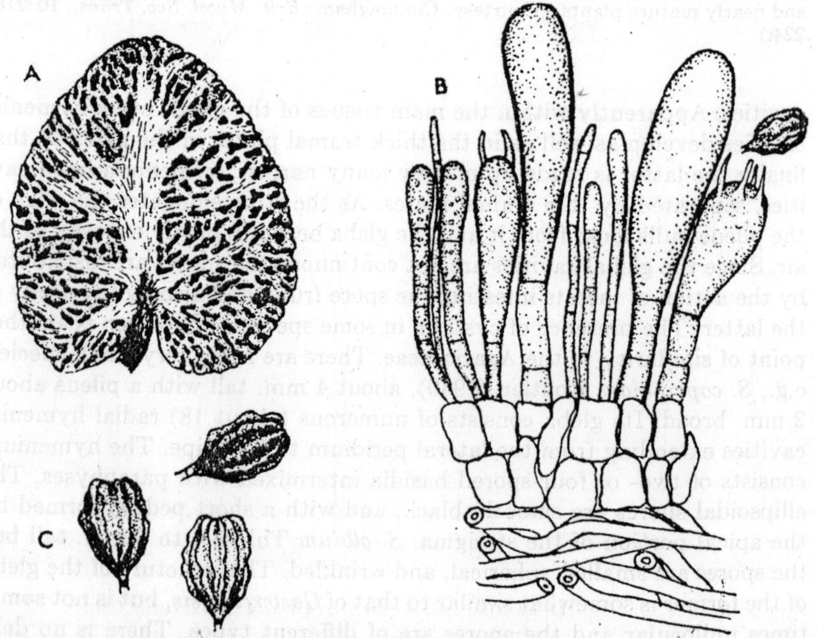

FIG. 175. Hymenogastrales, Family Hysterangiaceae. *Gautieria plumbea* Zell. & Dodge. (A) Vertical section of fruiting body. (B) Portion of hymenium. (C) Basidiospores. (Courtesy, Zeller and Dodge: *Ann. Missouri Botan. Garden*, 5(2):133–142.)

dendroidally branching system of supporting branches arising at the base. The peridium disappears at an early stage of development in *Gautieria* so that the gleba is exposed. The spores in this genus are ribbed longitudinally and resemble those of *Clitopilus* and *Octojuga* in the Agaricaceae and *Chamonixia* in the Hydnangiaceae. In *Hysterangium* the peridium persists and the glebal branches may grow fast to and spread along its inner surface. A gelatinous subperidial layer may develop from the trama where it comes into contact with the peridium. The cartilaginous-gelatinous tramal character and the production of the gelatinous subperidial layer are the main distinctions between this family and the Hymenogastraceae. *Protubera* and *Phallogaster* are genera that show further transitional steps toward the Clathraceae in the Order Phallales. (Fig. 175.)

Order Phallales. These are noteworthy because of the dissolution of their gleba into a usually evil-smelling slimy mass filled with spores. This attracts flies, especially those that feed upon and lay their eggs in carrion. They serve to carry the spores far and wide. It was shown by Cobb (1906) that the spores were not injured in their passage through the alimentary canal of these insects. At first the spore fruits are completely or partially subterranean, more or less spherical, with a firm, somewhat leathery peridium, underneath which is a thick layer of slime resembling the white of a raw egg. This, according to Lohwag (1925), is a modified outer portion of the gleba. The functional portion of the gleba is supported upon the surface of a pileate "receptacle" in the Phallaceae or upon or between a framework of radiating or anastomosing branches in the Clathraceae. In

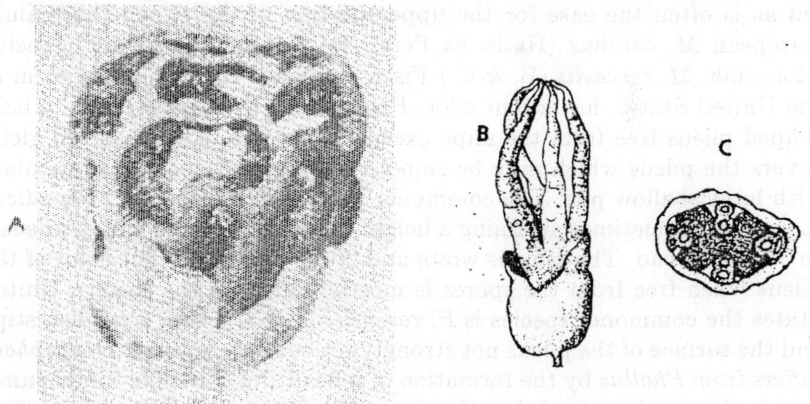

FIG. 176. Phallales, Family Clathraceae. (A) *Clathrus ruber* Mich. ex Pers. (B, C) *Pseudocolus javanicus* (Penz.) Lloyd. (B) Fully expanded spore fruit. (C) Cross section of "egg" just about to open, showing volva and four receptacular arms surrounding the mass of gleba. (A, courtesy, Lloyd: *Mycological Writings*, vol. 3. B–C, after Bernard: *Ann. Jardin Botan. Buitenzorg*, **31**:93–102.)

the former there is a stout, rapidly expanding stipe whose expansion tears open the peridium at the top leaving it around the base of the stipe as a volva. (Fig. 178.) In the Clathraceae the stipe may be present (*Simblum*, *Pseudocolus* (Fig. 176 B–C), *Aseroe*, etc.) bearing the receptacle at its top, or absent, the enlarging of the network of the receptacle rupturing the volva (*Clathrus*). The young gleba is more or less coralloid in its development and forms a complicated system of branching and anastomosing branches and plates covered by the hymenium. These dissolve completely, except the spores, to form a slimy malodorous mass.

The simple Phallales have many points of resemblance to the Hysterangiaceae, in some of the genera of which the gleba eventually dissolves (e.g., *Phallogaster*). The majority of the order are tropical or subtropical, but several genera are common in the temperate regions. Among these are, in the Clathraceae, *Clathrus ruber* Mich. ex Pers. (Fig. 176 A) which forms a pyriform coarse net with thick receptacular branches, arising from the ruptured volva. The dissolved gleba lines the inner surface of the hollow receptacle. The fungus is 6 to 8 cm. tall and red, rarely yellow, in color. *Lysurus australiensis* Cke. & Mass. (*Anthurus borealis* Burt) has a white to pink stipe with several connate, outwardly furrowed receptacular arms at its top, the whole reaching a height of 10 cm. In the Phallaceae the commonest genera of the temperate regions are *Mutinus*, *Phallus*, and *Dictyophora*. In the first the receptacle is a closely appressed cap on the upper portion of the stipe which, as in all the stalked members of the order, stands in the ruptured volva. The pileus is usually some shade of red as is often the case for the upper portion of the stipe. The mainly European *M. caninus* (Huds. ex Pers.) Fr. is usually without offensive odor while *M. ravenelii* (B. & C.) Fisch., which is the commoner form in the United States, has a foul odor. *Phallus* and *Dictyophora* have a bell-shaped pileus free from the stipe except at the top. The dissolved gleba covers the pileus which may be smooth but in most species is reticulate with large shallow pits. The commoner species in Europe is *P. impudicus* L. ex Pers., sometimes attaining a height of 15 cm. or more with a pileus 3 to 3.5 cm. broad. The stipe is white and 2 to 3 cm. thick. The color of the pileus when free from the spores is mostly white. In the eastern United States the commoner species is *P. ravenelii* B. & C., with a reddish stipe and the surface of the pileus not strongly reticulately marked. *Dictyophora* differs from *Phallus* by the formation of a beautiful skirt-like "indusium" attached near the top of the stipe beneath and free from the pileus. This is white in color and reticulate with large meshes. The tropical *D. indusiata* (Pers.) Fisch. has a larger indusium than *D. duplicata* (Bosc ex Fr.) Fisch. which is frequent in the eastern United States and occasional in Europe. (Figs. 177, 178.) The tropical genus *Itajahya*, first described from Brazil, has recently been discovered in Arizona and New Mexico by Long

Fig. 177. Phallales, Family Phallaceae. *Dictyophora duplicata* (Bosc ex Fr.) Fisch. (Courtesy, Walters: *Mycologia*, **35**(1).)

and Stouffer (1943). In it the pileus has numerous overlapping trama plates between which the gleba is formed. When the latter is washed away the pileus resembles a wig perched at the apex of the stipe.

The genus *Claustula,* whose ovoid receptacle remains enclosed in the volva until the gleba is completely mature, has been placed by Cunningham (1931) in a separate family, the Claustulaceae, considered by him to be more primitive than the remainder of the order. Fischer (1933) includes this genus in the Clathraceae with a total of 15 genera and recognizes 10 genera in the Phallaceae.

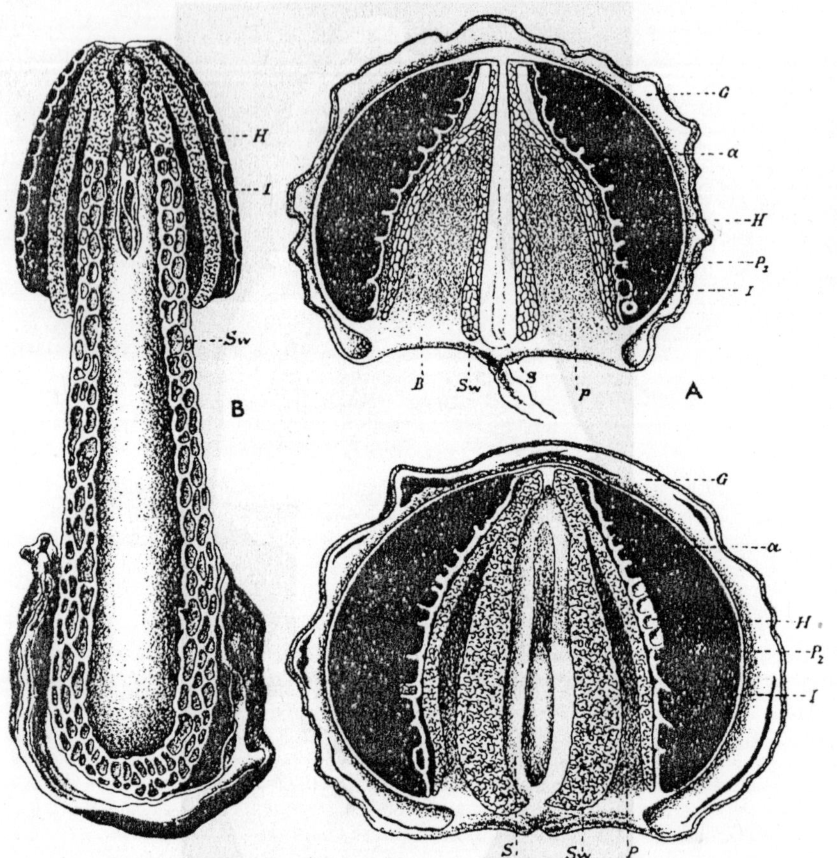

FIG. 178. Phallales, Family Phallaceae. *Dictyophora indusiata* (Pers.) Fisch. (A) Vertical sections through two unexpanded eggs, the lower one further advanced. (B) Vertical section through an expanded spore fruit whose indusium has not yet expanded. (*G*) Gelatinous layer of volva. (*H*) Pileus. (*I*) Indusium. (*Sw*) Wall of stalk. (*S*) Stalk axis becoming the hollow at maturity. (*P*) Primordial tissue between stalk and indusium. (P_2) Primordial tissue between pileus and indusium. (α) Gleba. (After Fischer: *Ann. Jardin Botan. Buitenzorg*, Série I, **6**:1–51.)

Order Sclerodermatales (Plectobasidiales). These were set apart from the Hymenogastrales by Schroeter (1897) and recognized by Fischer (1933, 1936). The distinction is based upon the structure of the gleba which mostly does not exhibit sharply defined hymenial cavities lined by an even layer of basidia as is typical of most Gasteromyceteae. Even in this order a tendency toward chamber formation is apparent in the younger basidiocarps but the hyphae whose terminal cells are destined to become basidia grow out into these incipient cavities to different lengths so that they are more or less completely obliterated, being represented by nests of basidia. According to Fischer these chambers arise in the lacunar manner. There is great need of developmental studies before the relationships within this order can be determined as well as to the Hymenogastrales, from which they undoubtedly have arisen. At maturity the gleba mostly becomes a powdery mass of spores with more or less capillitium. The peridium may be thin but is often several-layered and thick and firm, hence the name of the principal genus, *Scleroderma*. The spores are from four to six or more on the basidium, sessile or nearly so. They are usually dark-colored. Only one family, Sclerodermataceae, appears to the author to belong in this order. It is really doubtful whether the obliteration of the hymenial cavities is of sufficient importance to warrant the removal of this family from the Hymenogastrales, especially in view of the fact that *Melanogaster*, *Leucogaster*, *Alpova*, and others have a tendency toward this structure. (Fig. 179.)

Fischer (1933) included 10 genera in the family, of which he indicates five to be in doubt. Of the typical members of the family *Scleroderma* and *Pisolithus* may be noted. The former is subterranean or growing on the surface of the soil. It forms rounded spore fruits, in some species reaching a diameter of 10 cm. If superficial, there is a rooting mass of mycelial strands, the base being sometimes slightly stipe-like. The peridium is thick and at maturity more or less leathery. The surface may be smooth or roughened or in some cases forming large overlapping scales. The mature gleba shows numerous dark basidium-producing areas, separated by sterile veins or sheets. There is no definite hymenial layer in the hymenial cavities. The pyriform basidia bear two to five nearly sessile, rounded or ellipsoid, smooth or sculptured spores. The basidia disappear

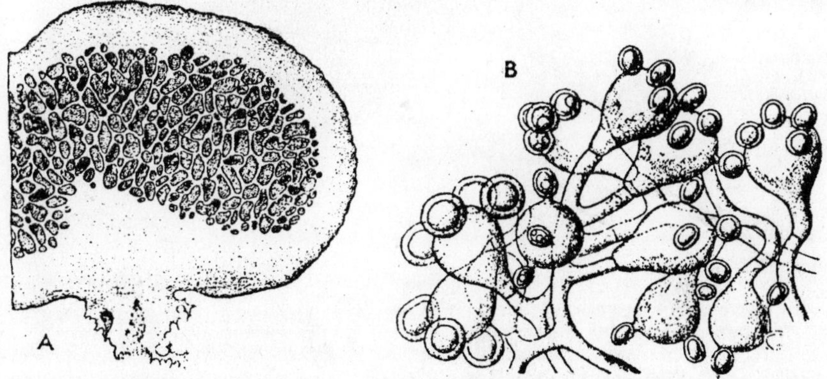

FIG. 179. Sclerodermatales, Family Sclerodermataceae. *Scleroderma aurantiacum* Pers. (A) Vertical section through an almost mature spore fruit. (B) Basidia, with sessile spores, with completely obliterated hymenial cavity. (After Tulasne, from Fischer, in Engler und Prantl: Die Natürlichen Pflanzenfamilien, Zweite Auflage, vol. 7a, Leipzig, W. Engelmann.)

early and the spores are in some species enveloped in a sheath of nurse hyphae till maturity when everything disappears except the spores and a few capillitial threads. The spore fruit may crack open stellately or may be opened by attacks of insects or rodents. *Pisolithus* does not have so thick a peridium and the tramal sheets split in such a manner that the basidium-producing areas form numerous ellipsoid or round "peridioles" retained within the peridium. Finally these fall apart into a powdery spore mass. The capillitial threads are few. The basidia bear two to six almost sessile, rounded spores. There is no special means of dehiscence provided in this genus.

Order Nidulariales. In this order the spore fruits are external, not subterranean. They have a thin or thick peridium and the gleba contains several to innumerable hymenial cavities, apparently formed in the lacunar manner, each lined internally by a layer of basidia. The tramal tissue in Family Nidulariaceae forms a firm several-layered wall around each cavity with its contained spores and these hard-walled bodies, called "peridioles" lie in the bottom of the cup formed by the dissolution of the top peridium of the spore fruit and of the tissues surrounding the peridioles. In Family Arachniaceae placed in this order by Fischer, the peridioles are innumerable and the tramal wall of each is thin and fragile.

FAMILY NIDULARIACEAE (BIRD'S NEST FUNGI). Several peridioles, over 20 in *Nidularia*, are formed in each spore fruit. At first connected in a continuous gleba the peridioles early become separated from each other and lie free in the cavity of the spore fruit or are connected to the peridium by long slender strands, the "funiculi." The principal genera are *Crucibulum*, *Cyathus*, and *Nidularia*. The spore fruits are several millimeters

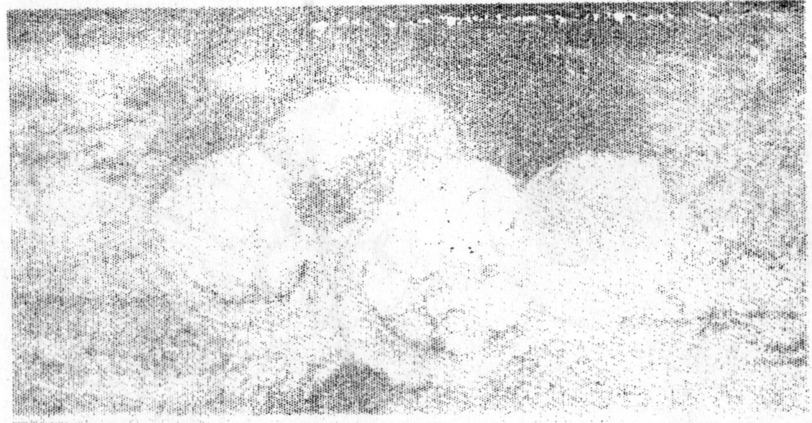

FIG. 180. Nidulariales, Family Nidulariaceae. *Crucibulum leve* (DC.) Kambly (*C. vulgare* Tul.) Three opened fruit bodies showing peridioles and one unopened spore fruit. (Courtesy, F. C. Strong.)

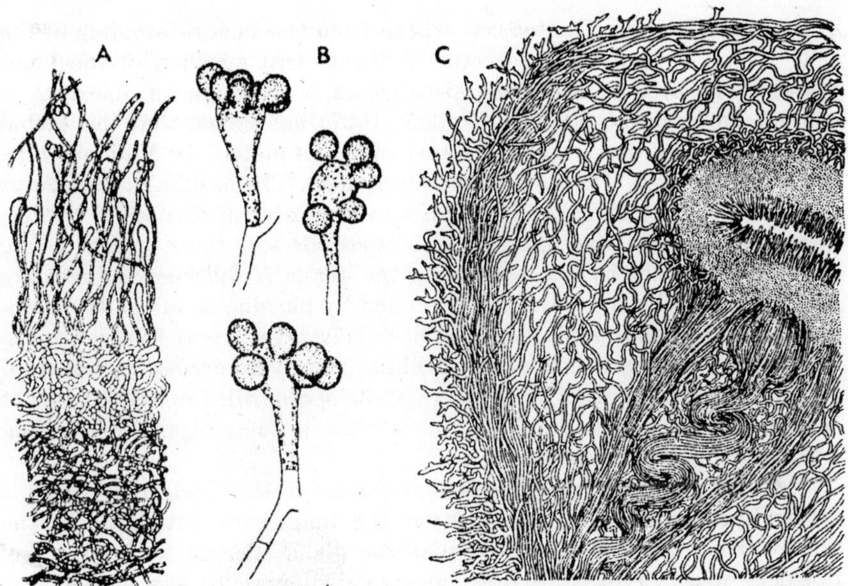

FIG. 181. Nidulariales, Family Nidulariaceae. (A, B) *Cyathus stercoreus* (Schw.) DeToni. (A) Section through wall of mature peridiole. (B) Basidia with their sessile basidiospores. (C) *Crucibulum leve* (DC.) Kambly; vertical section through a portion of immature spore fruit. (A, B, courtesy, Coker and Couch: The Gasteromycetes of the Eastern United States and Canada, Chapel Hill, Univ. North Carolina Press. C, after Sachs: *Botan. Ztg.*, **13**(48):833–845; (49):849–861.)

up to a centimeter in height and funnel-formed or almost spherical, with a flattened top. The peridium on this flattened upper portion ruptures and exposes the peridioles lying like eggs in a nest, whence the common name of the fungi. B. O. Dodge (1941) reports that they are discharged from the spore fruit at maturity, in some cases to a height of 3 or 4 meters. In forms with a funiculus (e.g., *Cyathus*) the latter remains attached to the peridiole when it is discharged and, being sticky, attaches it to various objects with which it may come in contact. Germination of the basidiospores occurs within the peridiole from whose outer surface numerous germ tubes emerge in all directions. According to Martin (1927), in the development of the peridiole the basidia collapse while the spores are not yet mature or fully grown. The spores are then nourished by a weft of hyphae surrounding each spore, much as occurs in some species of *Scleroderma*. (Figs. 180 and 181.)

Order Sphaerobolales. This order, included in the preceding one by many mycologists, has but one family, the Sphaerobolaceae. In the two genera *Sphaerobolus* and *Nidulariopsis*, the peridium has three or more rather thick layers, the middle one of which is lacking in the apical region of the second genus. The tension arising from the osmotic swelling of the cells of one of these layers results in the violent eversion of the inner peridial wall so that the whole glebal mass, 1 to 2 mm. in diameter, is ejected. In *Sphaerobolus* Miss Walker (1927) has shown that this glebal ball may be shot upward to a distance of over 4 meters. In *Nidulariopsis*, according to Greis (1935), the distance is short. The basidia in *S. stellatus* Tode ex Pers. are arranged irregularly in a number of clusters of hyphae intermingled with the basidia, as in *Scleroderma*, these groups being separated by thin hyphal sheets. In the genus *Nidulariopsis* there are numerous definite hymenial cavities lined by basidia, as in the Hymenogastraceae and Lycoperdaceae. The basidia bear from four to nine spores. The ejected gleba germinates as in Nidulariales by numerous hyphae from all sides. In the course of the rupture of the spore fruit the outer layers of the peridium are split into five or more lobes, in some regards resembling a partially opened *Geastrum*.

Order Lycoperdales. The remaining order of the Gasteromyceteae is Order Lycoperdales. This carries on the tendencies noticeable in the Sclerodermataceae of destruction of the glebal tissues to form a dry powdery mass of basidiospores intermingled more or less with sterile hyphae which form the capillitium. The function of the capillitial hyphae in the young spore fruits may well be that of water and food conduction as suggested by Fischer (1936) and Zeller (1939). At maturity these hyphae, now empty and more or less thick-walled, serve to keep the spores loosened up so that they do not escape all at once but gradually, over a longer period. The spore distribution therefore has become entirely dependent upon air currents. The fully grown gleba undergoes autodigestion which involves the hyphae of the trama (except those that become the capillitium) and mostly the basidia also. At this stage the contents of the spore fruit form a soggy, water-soaked mass. The water is quickly evaporated or perhaps also returned to the mycelium so that soon the gleba is dry, colored brown to purple, depending upon the color of the spores and of the capillitial threads.

Fischer (1933) recognized only two families in this order, Lycoperdaceae and Geastraceae, sometimes united into one family. In spite of the difference in glebal structure in some genera it seems to the author that the Tulostomataceae and Podaxaceae perhaps should find their position here, as representing in this order the same evolutionary trend toward the stipitate pilear structure as is shown in the series Hydnangiaceae to Secotiaceae in the Hymenogastrales. The knowledge of the early developmental stages is lacking in many of the genera but in those that have been studied it seems that the gleba may be lacunar or coralloid in its development, often the former in the basal, first developed portion of the gleba and the latter in the upper, fertile portion. Whether *Podaxis* represents a

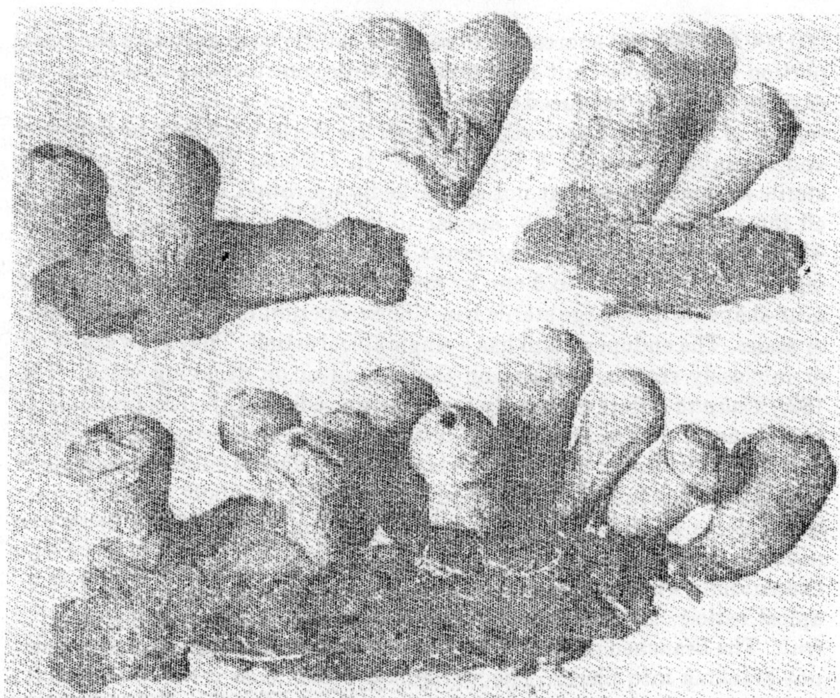

FIG. 182. Lycoperdales, Family Lycoperdaceae. *Lycoperdon pyriforme* Pers. (Courtesy, F. C. Strong.)

unipileate modification of the coralloid type of development needs careful study of the very early stages.

FAMILY LYCOPERDACEAE. The spore fruits are external from the first or may be shallowly subterranean when young, becoming external at maturity. They consist of a flexible peridium of two or three well-marked layers enclosing the gleba. The basidia are ovoid with short or long sterigmata and four to eight basidiospores. After the spores are mature the basidia and the tramal tissues dissolve, except for the brown, thick-walled capillitial threads. As the spore fruit enlarges the outer peridium ruptures in various ways, scaling off in granules or larger pieces. The inner peridium may also break up in pieces or more often forms one or more ostioles, usually in the apical region. As wind or firmer objects strike the spore fruit the spores are puffed out through these ostioles. The spore fruits vary from a few millimeters in diameter up to 1.6 m. in length, 1.35 m. in width and 24 cm. in height, in the case of a specimen of *Calvatia gigantea* (Batsch ex Pers.) Lloyd, collected in New York State many years ago and reported by C. E. Bessey (1884). Such a puffball would produce approximately

160,000,000,000,000 spores. As in some of the Hymenogastrales the basal portion of the spore fruit may remain sterile, not forming basidia in the cavities which are produced in the sterile base. This is particularly characteristic of the genus *Lycoperdon* in which the sterile base may be narrower and resemble somewhat a broad stipe. In *Calvatia* the basal portion is sterile in some species but not so markedly narrowed. The spore fruits of all the species of this family appear to be edible when young, while still white and rather brittle. The genus *Lycoperdon* has many species varying in size from 1 cm. to 5 cm. or more. The spore fruits are more or less pear-shaped with a large sterile base. The exoperidium scales off as granules or scales. The endoperidium has a single apical ostiole. They grow scattered or in closely crowded masses on the ground or on decaying wood. (Fig. 182.) *Calvatia* differs externally from the preceding in the less pronounced narrowing of the sterile base and in the absence of an ostiole in the endoperidium. The latter breaks off in large pieces. The capillitium consists of long, tangled, somewhat branched threads which usually break up into short pieces at maturity. *C. gigantea*, occurring in the Fall in fields and pastures, is collected for food while still firm and white. *Bovista* has a thin exoperidium which sloughs off and a slightly thicker endoperidium with an apical ostiole. There is no sterile base, the gleba filling the whole spore fruit. The branched capillitial hyphae are slender and smooth and usually not breaking into pieces. The sterigmata break loose from the basidium and remain attached to the basidiospores. *B. plumbea* Pers. is 3 to 5 cm. in diameter, nearly round and with a lead-gray endoperidium. *Mycenastrum* has a rather thick endoperidium which cracks open in a more or less stellate manner. The capillitium is composed of thick, branched, spiny hyphae, tapering from the middle to the acute tips. The sterigmata are very short so that the spores are almost sessile. *Disciseda* (*Catastoma*) has a firm exoperidium which splits equatorially. The endoperidium pulls free from the basal half of the exoperidium while remaining attached to the upper half. The result is that the spore fruit escapes and blows around, leaving the basal portion of the exoperidium still attached to the ground. The ostiole pierces the endoperidium at the center of the exposed portion, which is morphologically its base. *Disciseda candida* (Schw.) Lloyd (*C. circumscissum*) is common in grassy places in the prairie regions of the United States and in similar regions in Eastern Europe. (Fig. 183.) *Lanopila* has a very thin peridium which separates in irregular pieces from the gleba. This consists at maturity of a capillitium of much entangled, slender hyphae with the intermingled spores. The generic name was given because of its resemblance to a ball of wool. Swoboda (1937) studied the structure of the not quite mature spore fruits of *L. bicolor* (Lev.) Pat. and found that the spores are borne on one side of the basidium on short

Fig. 183. Order Lycoperdales, Family Lycoperdaceae, *Disciseda candida* (Schw.) Lloyd (*Catastoma circumscissum* (B. & C.) Morgan). Spore fruits in various stages of dehiscence. (After Morgan, from Fischer, in Engler und Prantl: Die Natürlichen Pflanzenfamilien, Zweite Auflage, vol. 7a, Leipzig, W. Engelmann.)

sterigmata. The basidia are produced sympodially on branched hyphae of a finely divided coralloid gleba, not on a hymenium clothing definite cavities. He concluded that the genus should form the type of a family to be named Lanopilaceae in the Sclerodermatales. Its abundant capillitium and the position of the spores on the basidium would seem to indicate possible relationship to *Tulostoma*.

Two genera, *Broomeia* and *Diplocystis*, from the warmer parts of the world, are characterized by the production of their spore fruits crowded side by side on a stroma, which is thick and often with a stout stalk in the former and thin and saucer-shaped in the latter. A few other genera are recognized. In the mountains in the western part of the United States occurs *Calbovista*, described by Miss Morse (1935). It resembles *Calvatia sculpta* (Hark.) Lloyd, but has a capillitium resembling that of *Bovista*.

Zeller (1944) segregated four genera from the Lycoperdaceae to form the Mesophelliaceae. These are *Radiigera*, so far only found in the United States, *Abstoma*, from New Zealand, Australia, and California, *Mesophellia*, from Europe and Australia, and *Castoreum*, from Australia. The distinction is based on a usually three-layered peridium which is indehiscent or rupturing irregularly at the apex. In this group the hymenial cavities are not well defined. The basidia are borne in clusters on short branches of radial hyphae. This seems to indicate a tendency toward the basidial arrangement found in *Phellorinia* and *Podaxis*.

Family Geastraceae. In most mycological works the genera included in this family are placed in the Lycoperdaceae, but the author follows Fischer (1933) in making the segregation. The chief distinction is that in the latter the outer layer of the peridium lacks a fibrous layer and disintegrates at maturity, while in the Earthstars, as the Geastraceae are called, the outer peridium does possess such a layer and splits stellately from the top toward the base, spreading out in a star-like manner. The

FIG. 184. Lycoperdales, Family Geastraceae. *Geastrum rufescens* Pers. (Courtesy, Coker and Couch: The Gasteromycetes of the Eastern United States and Canada, Chapel Hill, Univ. North Carolina Press.)

chief genus is *Geastrum.*[1] In this genus the inner peridium remains intact when the outer peridium splits open. It is sessile or on a short stalk and has a single apical ostiole. (Fig. 184.) In *Myriostoma* the inner peridium stands upon several slender stalks and there are several to numerous ostioles.

In the two foregoing genera the gleba is like that of the Lycoperdaceae, made up of tramal tissue with numerous closed, basidium-lined hymenial cavities. There is usually but not always a columella. Formerly included in this family was *Astraeus hygrometricus* (Pers.) Morg., in which the chief difference is the partial obliteration of the hymenial cavities by ingrowing tufts of basidia. The outer peridium is exceedingly hygrometric, opening out when moist, closing to almost the original position when dry. Because of the difference in glebal structure Fischer (1899, 1933) placed this genus in Family Calostomataceae in the Order Sclerodermatales. In view of the fact that in the family Sphaerobolaceae both types of gleba are present without even leading to the division of the family it appears best to retain *Astraeus* in the Geastraceae. (Fig. 185 A.)

The two remaining genera of the family were called by Fischer *Geasteropsis* and *Trichaster*, but Long (1945) pointed out that the former name is not available nor is his own earlier name (*Geasteroides*, 1917) so that he applied a new name *Terrostella.* This differs from *Geastrum* in the pos-

[1] In most publications and in the author's earlier book this genus is called *Geaster.* Since, however, the international rules of botanical nomenclature designate Persoon's Synopsis Methodica Fungorum (1801) as the basis for the nomenclature of the Gasteromyceteae his name *Geastrum* must be used, not the name preferred by later mycologists.

session of a prominent sterile base, and a central ostiole on the endoperidium, whose whole upper portion soon is more or less caducous. In *Trichaster*, apart from a subligneous columella there is no sterile base, and the endoperidium has no ostiole and either falls off in pieces or adheres to the exoperidium when the latter dehisces.

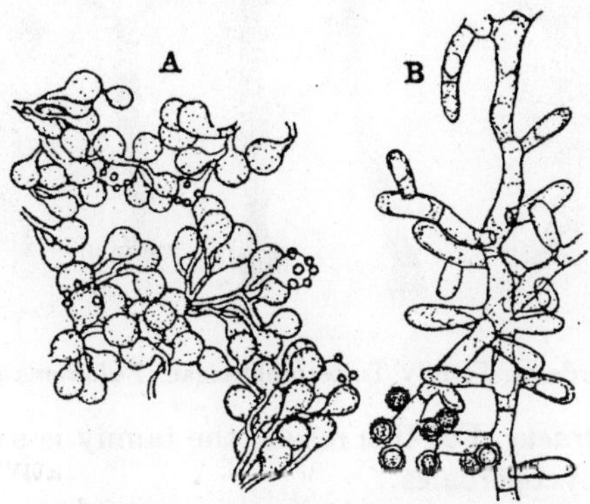

FIG. 185. Lycoperdales. (A) Family Geastraceae. *Astraeus hygrometricus* (Pers.) Morg.; portion of gleba showing partial obliteration of hymenial cavities. (B) Family Tulostomataceae. *Tulostoma simulans* Lloyd; portion of gleba showing basidia with laterally produced spores. (Courtesy, Coker and Couch: The Gasteromycetes of the Eastern United States and Canada, Chapel Hill, Univ. North Carolina Press.)

FAMILY TULOSTOMATACEAE. This family is based upon the genus *Tulostoma*.[2] These fungi are popularly called the stalked puffballs. The spore fruits of this family originate hypogeously but by the elongation of a basal stalk become epigeous. The commonest genus is *Tulostoma* with a spherical spore fruit 1 to 3 or 4 cm. in diameter and a slender stalk sometimes 5 or more cm. in length and 2 to 4 mm. thick, often with a small volva remaining at its base. (Fig. 186.) The endoperidium opens by an apical ostiole which may be irregular or which may have a projecting striate margin. The 40 or more species are mostly found in drier regions. In some species the basidia are described as producing four basidiospores laterally instead of apically. Whether this is true for the whole genus or for all other genera of the family is not known. (Fig. 185 B.) The hymenial cavities in this genus and possibly other genera become obliterated by the growth into them of hyphae bearing the irregularly arranged basidia as in *Scleroderma* and *Sphaerobolus stellatus* (not as in *Nidulari-*

[2] Persoon (1801) used this spelling instead of the more usual *Tylostoma* and for this reason his spelling must be followed.

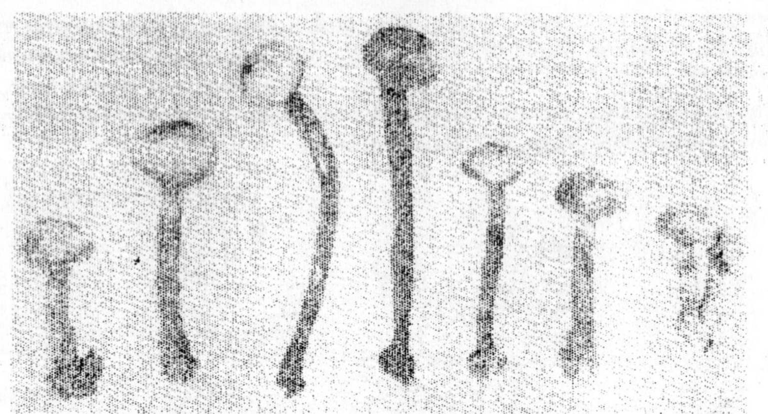

Fig. 186. Lycoperdales, Family Tulostomataceae. *Tulostoma campestre* Morgan.

opsis) and in *Astraeus*. For this reason the family is sometimes removed from the Order Lycoperdales.

The stipe may be a rather broad and relatively short one or tall and slender. In *Tulostoma* the rounded main body of the spore fruit reaches its full size before a small mass of tissue at its base, and within the exoperidium, begins its rapid elongation to become the stipe. The exoperidium may remain as a fragmentary cup at its base and shred off from the main portion of the body. The capillitium is abundant and grows fast to the peridium. The basidia bear the spores, four in number, on short sterigmata laterally. This led von Tavel (1892) to suggest the origin of this genus from *Phleogena* in the Auriculariales, with the loss of the cross septa in the basidia of the latter.

Queletia mirabilis Fr., the only species of this genus, has been found in France, England, and the United States, but appears to be rare. Like *Tulostoma* its spore fruits develop underground. Only as the gleba approaches the spore-forming stage does the base start to elongate, breaking through the peridium and forming a stout stipe 8 to 15 cm. tall and 3 to 4 cm. thick, bearing a rounded sporocarp 3 to 7 cm. in diameter. There is a sharp line of distinction between the concave base of the latter and the rounded apex of the stipe. The gleba is similar to that of *Tulostoma*, with no trace of hymenial cavities at maturity. The 1 to 4 (mostly 3) basidiospores are terminal and lateral as in that genus. Normally the head breaks away from the stipe and is blown about in the wind, scattering its spores, but if the attachment is too firm the stipe shreds away and thus exposes the under side of the powdery gleba which consists only of spores and capillitium. The genus *Calostoma* was placed by Fischer (1933) in the Family Calostomataceae along with *Astraeus*, with which it has little in agreement except the plectobasidial type of gleba. In *Calostoma* the stipe

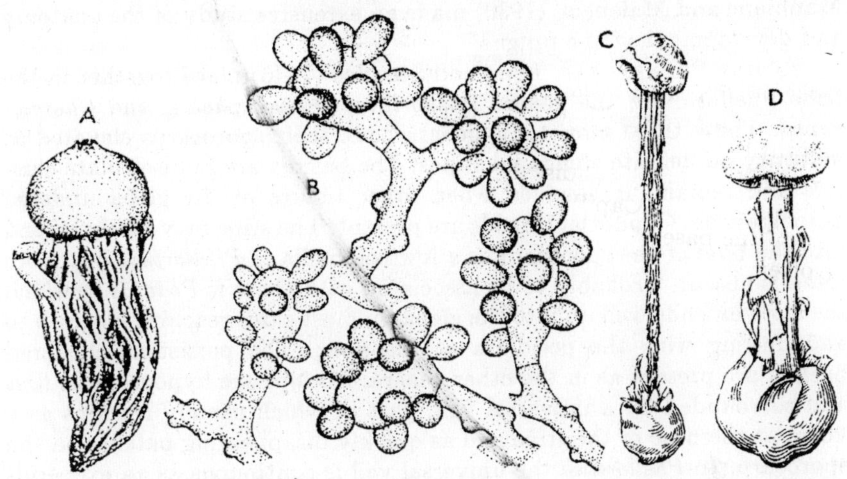

FIG. 187. Lycoperdales, Family Tulostomataceae. (A, B) *Calostoma cinnabarinum* Desv. (A) Spore fruit from which most of the volva has disappeared, through deliquescence. (B) Several basidia showing the sessile spores on all sides. (C, D) *Battarrea phalloides* (Dicks.) Pers.; two specimens which grew in different environments. (A–B, courtesy, Burnap: *Botan. Gaz.*, **23**(3):180–192, Univ. Chicago Press. C–D, courtesy, Rea: *Mycologia*, **34**(5):563–574.)

is mostly subterranean. The basidia have 5 to 12 sessile spores borne on the apex and along the sides, in which they give some suggestion of their relationship to *Tulostoma*. The capillitium is sometimes marked with annular or spiral thickenings, resembling the elaters of some fungi. In the author's opinion *Calostoma* probably belongs in the Tulostomataceae. In this he follows Burnap (1897). (Fig. 187 A–B.)

Battarrea has a stout scaly stalk that may reach over 30 cm. in height and with a large volva at its base. (Fig. 187 C–D.) The endoperidium of *B. phalloides* (Dicks.) Pers. splits circumscissilely, rolling upwards a little at a time as the spores and capillitium escape and in *B. digueti* Pat. & Har. is perforated by many pores. The gleba in this genus contains typical hymenial cavities. It is found in the sandy or gravelly soil of the foothill regions of the southwestern United States and in Europe, South America, Australia, etc. Rea (1942) recognizes two species in the United States with one or the other of which perhaps all the other described species are synonymous. There is some capillitium, evidently the remains of the tramal tissues. In addition there are numerous elaters with annular or spiral thickenings, reminding one of those of the Trichiaceae among the Mycetozoa or in the Hepaticae. In the younger, not fully developed specimens single cells or several in succession in a hypha enlarge and become "laticiferous." These do not become transformed into capillitial hyphae as in *Phellorinia*. Clamp connections are numerous throughout the spore fruit.

Maublanc and Malençon (1930) made an extensive study of the anatomy and development of the fungus.

FAMILY PODAXACEAE. Long and Stouffer (1946) place together in the tribe Phellorinieae the genera *Phellorinia*, *Dictyocephalos*, and *Chlamydopus*. These three monotypic genera have their sporocarps elevated at maturity on definite elongated stipes. The basidia are in fasciculate clusters and remain undissolved when other tissues of the gleba undergo autodigestion. Capillitial threads are present. The stipe may rarely extend into the base of the sporocarp as a low columella in *Phellorinia* but more often is absent. Probably to be associated with these is *Podaxis* in which the stipe extends well up into the gleba, in most cases reaching clear up to and uniting with the peridium at the apex. The persistent clustered basidia are present as in the other genera. All four are hypogeous at first and surrounded by a universal veil, part of which usually remains as a volva at the base of the stipe and as quickly disappearing patches on the sporocarp. In *Phellorinia* the universal veil is continuous as an exoperidium and the outer layer of the stipe, rarely breaking away well up on the stipe to show indications of a tightly adhering volva. The endoperidium is an extension of the stipe surrounding the gleba for two-thirds or more of its height, with a thin inner layer extending over its top. This falls away at maturity. Malençon (1935) showed that in the immature gleba there are numerous sinuously curved and branching hymenial cavities lined by the basidial primordial cells. The latter branch sympodially and form clusters of basidia on hyphae of various lengths which eventually obliterate these cavities. This led earlier investigators who studied only the practically mature fungi to classify them with plectobasidial forms such as Sclerodermatales. At maturity the major part of the hyphae making up the trama between the cavities dissolves, leaving spores and basidia and the hyphae upon which these arose undissolved, as well as a few of the thicker tramal hyphae which in the earlier stages of development served as laticiferous tubes. These hyphae and those bearing the clusters of basidia make up the so-called capillitium. This genus occurs in North and South America, Europe, Asia, and Australasia.

Dictyocephalos (see Long and Plunkett, 1940) occurs in the southwestern part of the United States and also northern Africa. There is but one species, *D. attenuatus* (Pk.) Long and Plunk. (*D. curvatus* Underw.). It originates 4 to 20 cm. beneath the surface of the soil and may attain a height of 7 to 56 cm. with a head 5 to 13 cm. broad. The universal veil breaks as the stipe elongates, leaving a volva at the base and a fleshy or gelatinous exoperidium over the sporocarp. This at maturity becomes a series of hairy scales which break off and expose the tough endoperidium. The latter breaks away irregularly exposing the gleba. When young the latter is cellular. *Chlamydopus*, with the single very variable species *C.*

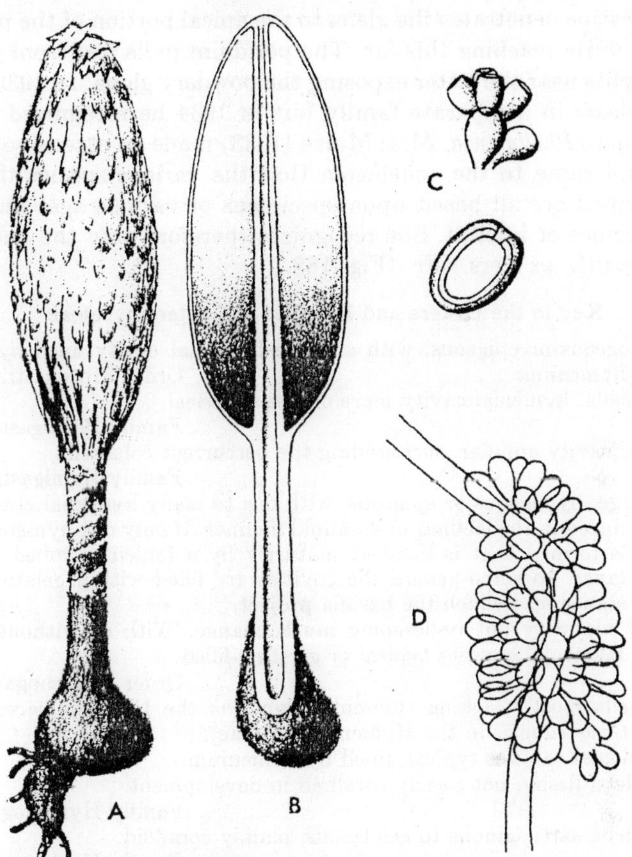

FIG. 188. Lycoperdales, Family Podaxaceae. *Podaxis pistillaris* (L. ex Pers.) Fr. (A) Spore fruit, external view. (B) Somewhat diagrammatic longitudinal section of spore fruit. (C) Basidium with sessile spores and single spore more highly magnified. (D) Cluster of basidia on piece of capillitial thread. (After Fischer, in Engler und Prantl: Die Natürlichen Pflanzenfamilien, Zweite Auflage, vol. 7a, Leipzig, W. Engelmann.)

meyenianus (Klotzsch) Lloyd, has been found in North and South America and Australia (Long and Stouffer, 1946). In this genus the portion of the universal veil which forms the exoperidium is verrucose and fragile and brittle, usually breaking very early and exposing the tough endoperidium which at maturity produces a single apical mouth. The large volva is conspicuous. The stout clavate stipe is 4 to 15 cm. tall, tapering toward the base and becomes woody to corky. The sporocarp is depressed globose up to 30 mm. wide. The gleba shows no signs of having been cellular. *Podaxis* has a slender woody stipe, arising from a narrow volva and bearing a more or less pyriform or rounded head. The central part of the stipe penetrates the gleba to the apical portion of the peridium, rarely not quite reaching this far. The peridium pulls free from the stipe below or splits near the latter exposing the powdery gleba. In 1933 Fischer placed *Podaxis* in a separate family but in 1934 he recognized its close relationship to *Phellorinia*. Miss Morse (1933) made an extensive study of *Podaxis* and came to the conclusion that the various species that have been described are all based upon specimens of various ages and grown under extremes of habitat. She recognizes therefore only the one species *P. pistillaris* (L. ex Pers.) Fr. (Fig. 188.)

Key to the Orders and Families of Gasteromyceteae

Small, hypogeous or epigeous, with a single hymenial cavity lined by an even hymenium. Order Protogastrales

- No columella, hymenial cavity more or less spherical. Family Protogastraceae
- Hymenial cavity annular, surrounding the percurrent columella. Family Hemigastraceae

Small to large, hypogeous or epigeous, with one to many hymenial cavities produced in the lacunar or coralloid manner. If only one hymenial cavity is formed that is lined at maturity by a folded or lobed hymenial layer. In some genera the cavities are filled with a gelatinous substance into which the basidia project.

- Gleba at maturity not undergoing much change. With or without a stipe. Hymenial cavities typical or gelatine-filled. Order Hymenogastrales
  - Columella mostly lacking (present in some of the Hysterangiaceae and in *Gasterellopsis* in the Hymenogastraceae).
    - Hymenial cavities typical, lined by hymenium.
      - Gleba fleshy, not clearly coralloid in development. Family Hymenogastraceae
      - Gleba cartilaginous to gelatinous, plainly coralloid. Family Hysterangiaceae
    - Hymenial cavities filled with gelatinous substance or with irregular masses of basidia. Family Melanogastraceae
  - Columella reaching the apex and spreading to form the pileus, no marked stipe at maturity. Development pseudoangiocarpic. Family Hydnangiaceae
  - Columella produced downward below the pileus to form a distinct stipe. Gleba free from the stipe at maturity, at least below. Family Secotiaceae
- Gleba at maturity deliquescing to a slimy, usually evil-smelling mass, which covers or is supported by a definite framework (the receptacle). Order Phallales
  - Receptacle lattice-like or irregularly branched or lobed, with or without a stipe. Family Clathraceae
  - Receptacle occupying the upper portion of a stout, hollow stipe, either grown fast to it or forming a bell-shaped structure (pileus) attached at the top of the stipe. Family Phallaceae

Medium to large, mostly epigeous at maturity and usually with a thick peridium. Hymenial cavities lacunar in origin, obliterated or replaced by nests or clusters of basidia. At maturity the whole gleba or portions of it become a powdery mass of spores with only rudimentary or lacking capillitium. Order Sclerodermatales

Only family. Family Sclerodermataceae

Medium-sized to small, not hypogeous at maturity or not at all. Hymenial cavities with definite lining of basidia. Toward maturity the tramal tissue surrounding each cavity encloses it in a thin or thick wall, producing separate structures called peridioles. Order Nidulariales

Peridioles very numerous, with thin walls, escaping by irregular breaking of the thin peridium. Family Arachniaceae

Peridioles few, with thick hard walls. Peridium beaker-like, opening by rupture of the diaphragm-like top, leaving the peridioles like eggs in a nest. Family Nidulariaceae

Small (up to 5 mm.), mostly growing on decayed wood or on manure, the whole gleba being expelled as a single ball by the eversion of the inner layer of the thick peridium which splits stellately at the top. Gleba with many distinct hymenial cavities or these obliterated by the ingrowing basidia. Order Sphaerobolales

Only family. Family Sphaerobolaceae

Medium-sized to large, epigeous, at least at maturity (a very few exceptions). Gleba mostly with definite hymenial cavities, or these sometimes obliterated. At maturity the gleba becomes a dry powdery mass of spores and capillitium, in one family the basidia also remaining intact. Order Lycoperdales

Without stipe. Outer peridium shedding in patches or granules, inner peridium mostly thin (thick in *Mycenastrum*), opening by a mouth (ostiole) or by breaking away in pieces. Columella sometimes present. Family Lycoperdaceae

Without stipe. Outer layers of the peridium splitting and turning back stellately. Inner peridium opening by an ostiole or falling away in pieces. Columella mostly present. Family Geastraceae

With stipe. Basidia lining definite cavities or these more or less completely obliterated. At maturity the whole gleba except capillitium and spores dissolving into a powdery mass. Family Tulostomataceae

With stipe. Basidia occurring in clusters on glebal hyphae. At maturity the basidia and their supporting hyphae are not destroyed but make up, with the spores and capillitium, the powdery contents of the sporocarp. Family Podaxaceae

Key to the More Important Genera of Gasteromyceteae

Family Protogastraceae.

Single genus, United States. *Protogaster*

Family Hemigastraceae.

Single genus, Europe. *Hemigaster*

Family Hymenogastraceae.

No percurrent columella.

At maturity a single hymenial cavity with folded or lobed hymenial lining. United States. *Gasterella*

At maturity many hymenial cavities.

Peridium almost absent at maturity. Spores colorless or pale brown, spiny or verrucose. North America, Australia. *Gymnomyces*

Peridium well developed at maturity.
Sporocarp without root-like mycelial strands.
Spores ellipsoid, ovoid or fusiform.
Trama plates irregular or radiating somewhat from a small sterile base. Old and New World. *Hymenogaster*
Trama plates arising from a dendroidally branching axial strand (possibly better assigned to the Hysterangiaceae). Europe. *Dendrogaster*
Spores spherical, spiny.
No sterile base, but glebal chambers converging toward the center of the base. Europe. *Martellia*
Glebal chambers converging toward a sterile base. Europe, North America, Australia. *Octaviania*
Sporocarp with root-like mycelial strands.
Spores spherical, verrucose, cystidia present. Europe and North America. *Sclerogaster*
Spores ellipsoid, smooth, no cystidia. Europe, North and South America, Asia, Australia. *Rhizopogon*
Spores angular, no cystidia. North America. *Nigropogon*
Percurrent columella. Annular glebal cavity divided by centripetally developing plates into several lobes or separate cavities. North America. *Gasterellopsis*
Genera of doubtful relationship, fusiform to pyriform, sometimes shortly stalked: *Gymnoglossum, Protoglossum, LeRatia, Clavogaster*.
Family Hysterangiaceae.
Spores spiny or ribbed.
Peridium wanting at maturity, spores oblong, ribbed. Europe, North Africa, and North America. *Gautieria*
Peridium present at maturity, spores spherical.
No gelatinous layer beneath exoperidium.
Basidia two- to four-spored. Europe. *Maccagnia*
Basidia five- to six-spored. Java. *Hoehneliogaster*
A gelatinous layer present beneath the thin exoperidium. Basidia two-spored. East India. *Clathrogaster*
Spores smooth, ellipsoid or rod-shaped.
Columella a basal cushion or a mycelial strand, often branched, not percurrent to the apex.
Peridium with large gelatinous outgrowths. Australia. *Phallobata*
Peridium more or less uniformly thick.
Gleba at maturity collapsing into a thin layer on the inner surface of the peridium. Australia. *Gallacea*
Gleba not collapsing.
Branches of the columella not dividing the gleba into sharply bounded portions.
Sporocarp sessile. Europe, Africa, North and South America, Australia. *Hysterangium*
Sporocarp stalked. Europe. *Jaczewskia*
Branches of the columella dividing the gleba into sharply bounded portions.
Sporocarp tuberoid, sessile. South America, Ceylon. *Protubera*
Sporocarp pyriform, stalked. North America. *Phallogatser*

Columella unbranched, percurrent to the apex. North America. *Rhopalogaster*

Family Hydnangiaceae.

Spores ovoid to fusiform, with longitudinal furrows. Europe. *Chamonixia*

Spores spherical, spiny, laticiferous vessels wanting. Europe. *Hydnangium*

Spores spherical or oblong, spiny or verrucose, laticiferous vessels present. Europe, North America, South America. *Arcangeliella*

Family Melanogastraceae.

Gleba not clearly chambered, basidia in nests between sterile veins. Africa. *Corditubera*

Gleba with numerous cavities which become filled with hyphae or gelatinous substance.

Without stipe.

Chambers loosely filled with hyphae among which the basidia are placed, spores roughened. North Africa. *Chondrogaster*

Chambers formed by enlarged cells which then gelatinize. On hyphae traversing the chambers clusters of basidia arise. Spores smooth. North America. *Alpova*

Basidia forming an irregular hymenium surrounding the hymenial cavities filled with loose hyphae or gelatine.

Spores ellipsoidal, brown. Europe and North America. *Melanogaster*

Spores spherical, almost colorless. Europe and North America. *Leucogaster*

With stipe at maturity, the exoperidium remaining as a basal volva and as shreds on the sporocarp which is hemispherical, and concave below. Relationship uncertain. *Torrendia*

Family Secotiaceae.

Fleshy. Development pseudoangiocarpic. Spores hyaline or light-colored, spiny or reticulate.

Numerous groups of pseudoparenchymatous cells in the tissues. Cystidia present. Europe and North America. *Elasmomyces*

Groups of pseudoparenchymatous cells wanting. South Africa. *Macowanites*

Firm, stipe almost woody. Development angiocarpic. Gleba chambers more or less radially elongated, sometimes lamelloid. Spores colored at maturity.

Lower edge of the peridium pulling loose from the stipe, leaving no volva (at least not conspicuous), cystidia often present. World-wide in its distribution. *Secotium*

Lower edge of peridium pulling loose, leaving a distinct annulus but no volva. Western United States. *Longula*

Lower edge of slender conical peridium pulling loose and leaving a cortina, but no annulus or volva. Spores resembling those of *Galerula* and *Bolbitius*. Europe, Asia, United States. *Galeropsis*

Peridium opening circumscissilely, leaving a large volva. Europe, Africa, North and South America, Asia. *Gyrophragmium* (including *Polyplocium*)

Peridium leaving a large volva, persisting on the pileus only as a small central patch beyond which the slender radial glebal lamellae project. No distinct annulus. Europe, Africa, North America, Australasia. *Montagnea* (*Montagnites*)

Family Clathraceae.

Receptacle lattice-like or of meridionally curved branches united at their tips.

Without stipe, not strongly narrowed below.

Branches of the lattice-like receptacle heavy and thick. Europe, United States, Ceylon. *Clathrus*

Branches of the lattice-like receptacle slender, meshes large. Tropics of Australia and Asia. *Ileodictyon*

Meridionally curved branches heavy. Tropical regions of Old and New World. *Colonnaria*

Meridionally curved branches slender. West Indies. *Laternea*

With stipe or strongly narrowed below.

No true stipe, lower meshes like the upper ones or slightly elongated. Tropics of Old and New Worlds. *Clathrella*

Stipe short. Lower series of meshes much elongated, upper ones isodiametric. Mediterranean region. *Colus*

Stipe tall, stout. Lattice-like receptacle with isodiametric meshes. Temperate and tropical regions of Old and New Worlds. *Simblum*

Stipe distinct. Several arms bowed out and united at the tip. Warmer regions of Old and New Worlds. *Pseudocolus*

Receptacle stipitate, of arms not united at their tips at maturity.

Arms spreading horizontally from the margin of a disk-like widening of the upper end of the stipe. Tropical regions of Old and New Worlds. *Aseroë*

Arms vertically parallel, spreading at their tips.

Arms surrounded by gleba laterally, and wholly or partially dorsally. Australia, Asia, America, Europe. *Lysurus*

Gleba only on inner side of arms. Africa, Australia, Europe. *Anthurus*

Arms projecting in all directions, knobby. Africa. *Kalchbrennera*

Receptacle stipitate and surrounding the gleba, splitting stellately into five lobes. New Zealand. *Claustula*

The tropical genera *Blumenavia* and *Mycopharus* closely resemble *Colonnaria* and *Pseudocolus*, respectively.

Family Phallaceae.

Receptacle closely clothing the upper part of the stipe, not on a campanulate pileus.

No pseudoparenchymatous ridges or reticulations on the gleba at maturity.

Gleba covering the tip as well as the upper portion of the stipe. South America. *Xylophagus*

Gleba forming a belt some distance below apex of stipe. South America. *Staheliomyces*

Gleba closely investing the upper portion of the stipe but not overrunning the tip. Old and New Worlds. *Mutinus*

Like *Mutinus*, but with pseudoparenchymatous projections or ridges from the gleba. East Indies and Australasia. *Jansia*

Like *Mutinus*, gleba invested with a loose net. Africa. *Floccomutinus*

Receptacle forming a campanulate pileus, attached centrally at the upper end of the stout stipe.

Indusium growing from between pileus and stipe.

Pileus perforated like lattice-work. Volva mostly spiny. East Indies. *Echinophallus*

Pileus not perforated like lattice-work. Volva not spiny. Tropics of Old and New World, one species in temperate North America. *Dictyophora*

No indusium between pileus and upper portion of stipe.

Pileus and gleba continuous over the apex of the stipe, leaving no apical perforation. Brazil. *Aporophallus*

Apex of pileus perforate, rarely covered temporarily by a fragment of the volva. Most regions of the world. *Phallus*

Apex of the pileus covered by a sort of cap, the receptacle, with many lobes or branches so that at maturity, after the gleba has disappeared, it resembles a wig. Southwestern United States and tropical and temperate South America. *Itajahya*

Family Sclerodermataceae.

Gleba marked by veins into many distinct regions. Capillitium rudimentary.

Gleba breaking up into a powdery mass, the veins remaining or disappearing. Peridium firm, thick.

Peridium covered externally by conical spines. The spiny spores borne on long sterigmata. East Indies. *Caloderma*

Peridium not spiny, thick.

Spores before maturity surrounded by a coat of hyphae. All over the world. *Scleroderma*

Spores without hyphal sheath. Europe. *Pompholyx*

Gleba breaking up into numerous separate, thin-walled peridioles. Peridium thin. falling to pieces at maturity. Europe, America, Asia, Australasia. *Pisolithus*

Gleba not marked by veins, capillitium well developed. Peridium simple, thin, opening by an apical pore. Spores borne laterally on the basidia. Europe. *Glischroderma*[3]

Family Arachniaceae.

Single genus. Americas, Africa, Australasia. *Arachnion*

Family Nidulariaceae.

Peridioles without funiculus.

Spore fruit roundish, without typical epiphragma. All continents. *Nidularia*

Spore fruit beaker-formed, with epiphragma. North and South America, Asia, Australasia. *Nidula*

Peridioles with funiculus.

Spore fruit cup-shaped, peridium of one layer, peridioles with a thick white tunica. All continents. *Crucibulum*

Spore fruit bell- or goblet-shaped, peridium of three layers, tunica of peridioles thin, hence their color, black or gray. All continents. *Cyathus*

Family Sphaerobolaceae.

Peridium of three layers, including the apical region. Widely distributed. *Sphaerobolus*

Middle layer of peridium lacking in apical region. Europe and North America. *Nidulariopsis*

Family Lycoperdaceae (including Mesophelliaceae). The little known tropical or African genera, *Lycoperdopsis*, *Lasiosphaera*, *Hippoperdon*, and *Bovistoides*, are not included in the key.

---

[3] This genus is put in a separate family, Glischrodermataceae, by Rea (1922) and Fischer (1933).

Sporocarps single or gregarious, not on a stroma.
Inner peridium thin, opening variously.
Capillitium of more or less uniform hyphae, branched or simple.
Outer peridium separating in granules or flakes from the inner peridium.
Inner peridium breaking up into flakes or fragments.
Capillitial hyphae interwoven into a woolly ball. North and South America, Africa. *Lanopila*
Capillitial threads not tightly interwoven, breaking into short pieces at maturity. In most continents. *Calvatia*
Inner peridium opening by an apical pore. The world over. *Lycoperdon*
Outer peridium firm, remaining attached to the upper half of the inner peridium and splitting circumscissilely. Inner peridium breaking free from the basal portion of the outer peridium, opening by a pore in the originally basal portion. Europe, America, Australasia. *Disciseda* (*Catastoma*)
Outer peridium firm, remaining attached to the inner peridium. At first hypogeous in most cases. The whole peridium rupturing irregularly. (This is Family Mesophelliaceae of Zeller.)
Spores spherical, echinulate, reticulated or verrucose.
Gleba without a sterile base. Australia and Western United States. *Abstoma*
Gleba with a sterile base. Western United States. *Radiigera*
Spores ellipsoidal, smooth or irregularly roughened.
Gleba with a central core. Australasia and Europe. *Mesophellia*
Gleba without core. Australasia. *Castoreum*
Capillitium hyphae much branched, consisting of a thicker main stem and tapering branches.
Inner peridium opening by apical pore.
At maturity attached to the ground. Eurasia and Australasia. *Bovistella*
At maturity breaking loose and blown by the wind. Europe, North America, and Australasia. *Bovista*
Inner peridium breaking up in flakes. Western United States. *Calbovista*
Inner peridium thick and corky, opening irregularly or somewhat stellately.
Capillitium threads thick, short, spiny, with numerous thorn-like processes. The world over. *Mycenastrum*
Sporocarps perched close together on a stroma.
Stroma thick, often columnar. South Africa. *Broomeia*
Stroma shallow, thin, shell-like. South Africa and West Indies. *Diplocystis*

Family Geastraceae.
Columella wanting. Hymenial chambers lacking (plectobasidial). Europe and North America. *Astraeus*
Columella present. Mostly with typical hymenial chambers.
A prominent sterile base in addition to the columella. Endoperidium with an apical pore but soon the whole upper part more or less caducous. America and Africa. *Terrostella* (Syn., ***Geasteropsis*** **and** ***Geasteroides***)

No sterile base in addition to columella.

Endoperidium sessile or on a single short stalk, opening by a single apical pore. All parts of the world. *Geastrum*

Endoperidium on several slender stalks, opening by several pores. Europe, South Africa, North and South America. *Myriostoma*

Endoperidium with no pore, breaking away in pieces or adhering in pieces to the dehisced exoperidium. Europe and South Africa. *Trichaster*

Family Tulostomataceae.

Stipe (largely subterranean) stout, composed of parallel or interwoven cartilaginous strands growing from the base of the endoperidium. Basidia bearing 5 to 12 sessile spores. Ostiole stellately bordered by several, usually colored, lobes. North and South America, Asia, East Indies, Australia. *Calostoma*

Stipe stout or slender, of parallel hyphae, sometimes scaly with age.

With elaters marked with rings or spirals. Gleba with hymenial cavities. At maturity stipe with a large volva and covered with overlapping scales.

Sporocarp low bell-shaped (convex above and concave below). Widely distributed. *Battarrea*

Sporocarp spherical (by some considered a species of *Battarrea*). Africa. *Sphaericeps*

Elaters lacking, but with typical capillitium. Hymenial cavities obliterated.

Inner peridium with apical mouth. Stipe slender. Widely distributed. *Tulostoma*

Inner peridium opening by stellate lobes. Stipe slender. Africa. *Schizostoma*

Inner peridium opening irregularly. Stipe stout. Europe and United States. *Queletia*

Family Podaxaceae.

Columella wanting or very low.

With volva.

Dehiscing by an apical pore. North America, Africa, Australasia. *Chlamydopus*

Dehiscing by irregular rupture of upper portion of endoperidium. Western North America, and Africa. *Dictyocephalos*

Volva wanting; urceolate at maturity. North and South America, Africa, Asia, Australasia. *Phellorinia*

Columella continuing as an extension of the stipe to or nearly to the top of the sporocarp.

Peridium pulling loose at its lower edge. Spores sessile. North America, Africa, Asia, Australia. *Podaxis*

Peridium remaining attached below, splitting laterally. Spores on sterigmata. Australia. (Sometimes considered a form of *Podaxis*.) *Chainoderma*

(For a more detailed key to the genera of the Gasteromyceteae the student is referred to the latest publication by the late Dr. S. M. Zeller (1949). His arrangement is not entirely in agreement with the ideas of the author.)

Literature Cited

ATKINSON, GEORGE F.: The development of Agaricus campestris, *Botan. Gaz.*, **42**(4):241–264. *Pls.* 7–12. 1906.

ATKINSON, GEORGE F.: The homology of the "universal veil" in Agaricus, *Mycolog. Centr.*, **4**(3):113–121. *Pls.* 1–2. 1915.

BARNETT, HORACE L.: The development and structure of Longia texensis, *Mycologia*, **35**(4):399–408. *Figs.* 1–3. 1943.

BERNARD, C.: Une très rare Phalloidée, Pseudocolus javanicus (Penzig) Lloyd, *Ann. Jardin Botan. Buitenzorg*, **31**:93–102. *Pls.* 15–18. 1921.

BESSEY, CHARLES E.: An enormous puff-ball, *Am. Naturalist*, **18**(5):530. 1884.

BUCHOLTZ, F.: Zur Morphologie und Systematik der Fungi hypogaei, *Ann. Mycol.*, **1**(2):152–174. *Pls.* 4–5. 1903.

BURNAP, CHARLES EDWARD: Notes on the genus Calostoma. Contributions from the Cryptogamic Laboratory of Harvard University XXXVIII, *Botan. Gaz.*, **23**(3):180–192. *Pl.* 19. 1897.

COBB, N. A.: Fungus maladies of the sugar cane, *Hawaiian Sugar Planters' Assoc. Exp. Sta., Div. Pathol. Physiol. Bull.*, **5**:1–254. *Pls.* 1–7. *Figs.* 1–102. 1906.

CONARD, HENRY S.: The structure and development of Secotium agaricoides, *Mycologia*, **7**(2):94–104. *Pl.* 157. *Fig.* 1. 1915.

CUNNINGHAM, G. H.: A critical revision of the Australian and New Zealand species of the genus Secotium, *Proc. Linnean Soc. New South Wales*, **49**(2):97–119. 4 *pls.* 1924.

———: The structure and development of two New Zealand species of Secotium, *Brit. Mycol. Soc. Trans.*, **10**:216–224. *Pls.* 11–12. 1925.

———: The Gasteromycetes of Australasia: X and XI. The Phallales, *Proc. Linnean Soc. New South Wales*, **56**(2):1–15. *Pls.* 1–2; (3):182–200. *Pls.* 8–10. 1931.

DODGE, B. O.: Discharge of the sporangioles of bird's nest fungi, *Mycologia*, **33**(6):650–654. *Figs.* 1–2. 1941.

DODGE, CARROLL W.: Alpova, a new genus of Rhizopogonaceae, with further notes on Leucogaster and Arcangeliella, *Ann. Missouri Botan. Garden*, **18**(3):457–464. *Pl.* 40. 1931.

FISCHER, EDUARD: Zur Entwicklungsgeschichte der Fruchtkörper einiger Phalloideen, *Ann. Jardin Botan. Buitenzorg*, **6**:1–51. *Pls.* 1–5. 1887.

———: Plectobasidiineae (Sclerodermineae), in A. ENGLER und K. PRANTL: Die Natürlichen Pflanzenfamilien, Teil 1, Abt. 1**, pp. 329–346. *Figs.* 171–182. Leipzig, Wilhelm Engelmann, 1899.

———: Unterklasse Eubasidii. Reihe Gastromyceteae, *ibid.*, Zweite Auflage, **7a**:i–iv, 1–122. *Figs.* 1–91. Leipzig, Wilhelm Engelmann, 1933.

———: Zur Kenntnis der Fruchtkörperentwicklung von Podaxis, *Ber. schweiz. botan. Ges.*, **43**(1):11–18. 5 *figs.* 1934.

———: Neue Beiträge zur Kenntnis der Verwandtschaftsverhältnisse der Gastromyceten. Eine kritische Untersuchung, *Ber. schweiz. botan. Ges.*, **45**:231–247. *Figs.* 1–3. 1936.

FITZPATRICK, H. M.: A comparative study of the development of the fruit body in Phallogaster, Hysterangium and Gautieria, *Ann. Mycol.*, **11**(2):119–149. *Pls.* 4–7. 1913.

FRIES, NILS: Crucibulum vulgare Tul. und Cyathus striatus Pers., zwei Gasteromyceteae mit tetrapolar Geschlechtsverteilung, *Botan. Notiser*, **1936**:567–594. *Figs.* 1–2. 1936.

GÄUMANN, ERNST ALBERT: Comparative Morphology of Fungi. Translated and revised by Carroll William Dodge, xiv + 701 pp. 406 *figs.* 43 *diagrams.* New York, McGraw-Hill Book Co., Inc., 1928.

GREIS, H.: Nidulariopsis melanocarpa Greis nov. gen. nov. spec. und eine neue

Form von Sphaerobolus iowensis, *Hedwigia*, **75**(4):255–266. *Pls.* 3–4. *Fig.* 1. 1935.

HEIM, ROGER: Les Lactario-Russulés du Domaine Orientale de Madagascar. Essai sur la classification et la phylogénie des Astérosporales, pp. 1–196. 8 *pls.* (4 *colored*). 59 *figs.* 2 *phylogenetic diagrams.* Paris, Laboratoire de Cryptogamie du Muséum National d'Histoire Naturelle, 1937 (1938).

———: avec la collaboration du DR. P. FONT QUER et du DR. J. CODINA: Fungi Iberici. Observations sur la flore mycologique catalane, *Publications de la Junta de Ciències Naturals de Barcelona*, 1934. *Treballs del Museu de Ciències Naturals de Barcelona* 15, *Série Botánica*, **3**:1–146. *Pls.* 1–4. *Figs.* 1–22. 1934.

JUEL, O.: Hemigaster; ein neuer Typus unter den Basidiomyceten, *Bihang till Kgl. Svenska Vetenskapsakad. Handl.*, **21.** Afd. 3, No. 4: 5–22. *Pls.* 1–2. 1895.

LLOYD, C. G.: Synopsis of the known Phalloids, *Mycological Writings*, **3**: (1909–1912). Separate pagination 1–96. 107 *figs.* 1909.

LOHWAG, HEINRICH: Entwicklungsgeschichte und systematische Stellung von Secotium agaricoides (Czern.) Holl., *Oesterreichische Botan. Z.*, **73**(7–9):161–174. *Pl.* 2. 1924a.

———: Zur Stellung und Systematik der Gastromyceten, *Verhandlungen der Zoolog. Botan. Ges. Wien*, **74–75**:38–55. 1924b.

———: Zur Entwicklungsgeschichte und Morphologie der Gastromyceten. Ein Beitrag zur Systematik der Basidiomyceten, *Botan. Centr. Beihefte*, Zweite Abt., **42**(2–3):177–334. *Pls.* 1–2. *Figs.* 1–49. 1925.

———: Die Homologien im Fruchtkörperbau der höheren Pilze, *Biologia Generalis*, **2**(1–2):148–182; (6):575–608. *Pls.* 18 *and* 30. 13 *figs.* 1926.

———: Mykologische Studien: IX. Über die Fruchtkörperentwicklung der Geastraceen, *Botan. Centr. Beihefte*, Abt. A, **52**(2):269–289. *Pl.* 15. *Figs.* 1–6. 1934.

LONG, W. H.: Notes on new or rare species of Gasteromycetes, *Mycologia*, **9**(5):271–274. 1917.

———: Studies in the Gasteromycetes: XI. The genera Trichaster and Terrostella, *ibid.*, **37**(5):601–608. *Figs.* 1–4. 1945.

———, AND O. A. PLUNKETT: Studies in the Gasteromycetes: I. The genus Dictyocephalos, *ibid.*, **32**(6):696–709. *Figs.* 1–13. 1940.

———, AND DAVID J. STOUFFER: Studies in the Gasteromycetes: IX. The genus Itajahya in North America, *ibid.*, **35**(6):620–628. *Figs.* 1–10. 1943.

———, AND DAVID J. STOUFFER: Studies in the Gasteromycetes: XIV. The genus Chlamydopus, *ibid.*, **38**(6):619–629. *Figs.* 1–7. 1946.

LORENZ, FRANZ: Beiträge zur Entwicklungsgeschichte von Sphaerobolus, *Arch. Protistenk.*, **81**(2):361–398. *Figs.* 1–16. 1933.

MALENÇON, GEORGES: La série des Astérosporés. Travaux Cryptogamiques dédiés à Louis Mangin, pp. 377–396. 1 *pl.* 1 *fig.* Paris, Laboratoire de Cryptogamie du Muséum National d'Histoire Naturelle, 1931.

———: Considérations sur les spores des Russules et des Lactaires, *Bull. trimestriel de la soc. mycologique de France*, **47**(1):72–86. *Pl.* 4. *Figs.* 1–3. 1931.

———: Études sur les Phellorinés: I. Le Phellorinia Delestrei (Dur. et Mtgn.) E. Fischer; II. Le Dictyocephalus curvatus Underwood, *Ann. cryptogam. exotique*, **8**(1–2):5–48. *Pls.* 1–4. *Figs.* 1–8. 1935; (3–4):101–133. *Pls.* 5–6. *Figs.* 1–6. 1936.

MARTIN, G. W.: Basidia and spores of the Nidulariaceae, *Mycologia*, **19**(5):239–247. *Pls.* 22–23. 1927.

MAUBLANC, A., ET G. MALENÇON: Recherches sur le Battarrea Guicciardiniana

Ces., *Bull. trimestriel de la soc. mycologique de France*, **46**:43–73. *Pls.* 2–5. *Figs.* 1–6. 1930.

MORGAN, A. P.: North American Fungi, V, *J. Cincinnati Soc. Natural History*, **14**:141–148. *Illustrated.* 1892.

MORSE, ELIZABETH EATON: A study of the genus Podaxis, *Mycologia*, **25**(1):1–33. *Pls.* 1–12. 1933.

———: A new puffball, *ibid.*, **27**(2):96–101. *Pls.* 12–15. 1935.

PERSOON, CHRISTIAAN HENDRIK: Synopsis methodica fungorum. Pars prima et secunda, xxx + 706 pp. and Index. *Pls.* 1–5. Göttingen, Heinrich Dieterich, 1801.

PILAT, ALBERT: Sur le genre Gastrosporium Mattirolo (Gastéromycètes), *Bull. trimestriel de la soc. mycologique de France*, **50**(1):37–49. *Pls.* 1–3. 1934.

REA, CARLETON: British Basidiomyceteae. A Handbook to the Larger British Fungi, xii + 799 pp. Cambridge, Cambridge Univ. Press, 1922.

REA, PAUL MARSHALL: Fungi of Southern California, I, *Mycologia*, **34**(5):563–574. *Figs.* 1–3. 1942.

REHSTEINER, H.: Beiträge zur Entwicklungsgeschichte der Fruchtkörper einiger Gastromyceten. *Botan. Ztg.*, **50**(47):761–771; (48):777–792; (49):801–814; (50):823–839; (51):843–863; (52):865–878. *Pls.* 10–11. 3 *text figs.* 1892.

ROUTIEN, JOHN B.: Observations on Gasterella lutophila, *Mycologia*, **31**(4):416–417. 1939.

———: Two new Gasteromycetes, *ibid.*, **32**(2):159–169. *Figs.* 1–23. 1940.

SACHS, JULIUS: Morphologie des Crucibulum vulgare Tulasne, *Botan. Ztg.*, **13**(48):833–845; (49):849–861. *Pls.* 13–14. 1855.

SCHROETER, J.: Fungi (Pilze), in A. ENGLER und K. PRANTL: Die Natürlichen Pflanzenfamilien. Teil I, Abt. 1**, pp. 42–64. Leipzig, Wilhelm Engelmann, 1897.

SINGER, ROLF: Das System der Agaricales, *Ann. Mycol.*, **34**(4–5):286–378. 1936.

———: The Agaricales (Mushrooms) in modern taxonomy, *Lilloa Revista de Botanica*, **22**:1–834. *Pls.* 1–29. 1949 (1951).

SWOBODA, FRANZ: Über den Fruchtkörperbau und die systematische Stellung von Lanopila Fries, *Ann. Mycol.*, **35**(1):1–14. *Figs.* 1–11. 1937.

VON TAVEL, FRANZ: Vergleichende Morphologie der Pilze, 208 pp. 90 *figs.* Jena, Gustav Fischer, 1892.

TULASNE, LOUIS RENE: Fungi hypogaei. Histoire et monographie des champignons hypogés, xix + 222 pp. 21 *pls.* (9 *colored*). Paris, Fr. Klincksieck, 1851. (Second edition in 1863.)

WALKER, LEVA B.: Development and mechanism of discharge in Sphaerobolus iowensis n. sp. and S. stellatus Tode, *J. Elisha Mitchell Sci. Soc.*, **42**(3–4):151–178. *Pls.* 16–25. 1927.

———: Development of Gasterella lutophila, *Mycologia*, **32**(1):31–42. *Figs.* 1–45. 1940.

WHITE, V. S.: The Tylostomaceae of North America, *Bull. Torrey Botan. Club*, **28**(8):421–444. *Pls.* 31–40. 1901.

ZELLER, S. M.: Protogaster, representing a new order of the Gasteromycetes, *Ann. Missouri Botan. Garden*, **21**(2):231–240. 2 *pls.* 1934.

———: Developmental morphology of Alpova, *Oregon State Monographs. Studies in Botany*, **2**:1–19. *Pls.* 1–4. 1939.

———: Representatives of the Mesophelliaceae in North America, *Mycologia*. **36**(6):627–637. *Figs.* 1–6. 1944.

———: A new name, *Mycologia*, **37**(5):636. 1945.

———: Notes on certain Gasteromycetes, including two new orders, *ibid.*, **40**(6):639–668. 1948.

———: Keys to the orders, families, and genera of the Gasteromycetes, *ibid.*, **41**(1):36–58. 1949.

———, AND CARROLL W. DODGE: Gautieria in North America, *Ann. Missouri Botan. Garden*, **5**(2):133–142. *Pl.* 9. 1918.

———, AND ———: Leucogaster and Leucophlebs in North America, *ibid.*, **11**(4):389–410. *Pl.* 11. 1924.

———, AND LEVA B. WALKER: Gasterella, a new uniloculate Gasteromycete, *Mycologia*, **27**(6):573–579. *Figs.* 1–13. 1935.

16

FUNGI IMPERFECTI: THE IMPERFECT FUNGI

THERE are a great many species of fungi of which the perfect stage is not known and which therefore cannot find a place in the classes already discussed. By the term "perfect stage," as here used, is meant that stage in which the ultimate sexual structures are formed, e.g., zygospores, oospores, asci, basidia, and teliospores. Most of the Phycomyceteae are so characteristic in their mycelial structure as well as in their modes of asexual reproduction that ordinarily the genus and often even the species can be determined from the asexual stage alone. Thus the Imperfect Fungi are practically confined to those Higher Fungi in which the stage is lacking in which the asci, basidia, or teliospores are produced. Since most of the Uredinales have very characteristic asexual stages the imperfect forms of this order are readily assigned to that group and are placed in one of the imperfect genera there, e.g., *Aecidium*, *Uredo*, *Caeoma*, etc., if their host requirements and other features make it impossible to assign them to described species in recognized perfect genera of that order. Thus it comes about that the Fungi Imperfecti, as ordinarily considered, include those fungi not otherwise referable to their natural relationship (e.g., Phycomyceteae or Uredinales) whose true relationship cannot be determined in the absence of the perfect stage. Judging by the rather exceptional presence of clamp connections, as well as by the similarity of the conidial stages to those in the Class Ascomyceteae, it is probable that a great majority of species of imperfect fungi really belong to that class, and the perfect stage is not present in the specimens examined. It is possible that some fungi have lost entirely the power to produce a perfect stage and so are truly imperfect fungi. A few assigned to this class are doubtless imperfect stages of Ustilaginales or other groups of Basidiomyceteae. In the first edition of Engler and Prantl, "Die Natürlichen Pflanzenfamilien," Lindau (1899, 1900) recognized about 600 genera and 15,000 to 20,000 species. In a more recent, as yet unpublished, work on this group Dr. Harold B. Bender (1931) recognized as valid 1331 genera.

Since for a great many fungi the asexual and sexual stages of reproduction may be separated in time and substratum, many fungi have been described under different names according to whether one or the other stage was studied. As time goes on the connection between the two stages is recognized in many cases. Theoretically, therefore, the asexual stage should cease to be known by its name among the Imperfect Fungi and it should no longer be included in that group. Practically, however, it is desirable to retain this name among the Fungi Imperfecti since it would be sought there in attempts to identify it, unless the perfect stage were found along with it. Thus we still seek for such genera as *Aspergillus*, *Penicillium*, *Sphaceloma*, *Ramularia*, *Cercospora*, etc. in the manuals describing the Imperfect Fungi although the perfect stages of many species of these genera are known and provided with names.

The distribution of the many thousand species of Fungi Imperfecti into genera, families, and orders must necessarily be based upon vegetative and asexual reproductive structures instead of upon the perfect reproductive stages. Inasmuch as it has been demonstrated that fungi whose perfect stages show them to be of very different families may possess rather similar types of asexual reproduction it follows that genera based upon the asexual reproductive forms are not necessarily assemblages of related species. As an example attention may be drawn to the genus *Gloeosporium*. In this genus the one-celled, hyaline, ellipsoidal, straight or slightly curved conidia are produced, usually embedded in a gummy substance, from short conidiophores packed in a palisade underneath the host epidermis which is ruptured by the developing mass of conidia. This acervulus type of asexual reproduction is found in some species of *Gnomonia* and *Glomerella*, both being genera in the Gnomoniaceae of the Sphaeriales. The fungus commonly known as *Pseudopeziza ribis* Kleb., of Family Mollisiaceae, Order Pezizales, also has a similar type of asexual reproduction. A number of similar cases are known. Since, then, the genera based upon asexual structures do not necessarily indicate true relationships of the included species the term "form genus" was suggested for such groups by Schroeter. It is in this sense that the term genus is used in this class. On the other hand, as has been pointed out by several authors including Petrak and Sydow (1926–1927) there is frequently a similarity in asexual structures among fungi considered to be closely related as judged by their perfect stage. Careful study has revealed that in many cases some of the form genera of the Imperfect Fungi can be subdivided into groups of species correlated with the perfect stages. Thus has come about on the part of some mycologists the breaking up of the larger genera into smaller more compact ones, on characters that would otherwise be considered of rather minor importance except for their correlation with groups of perfect fungi.

The fact that many Imperfect Fungi possess several different spore

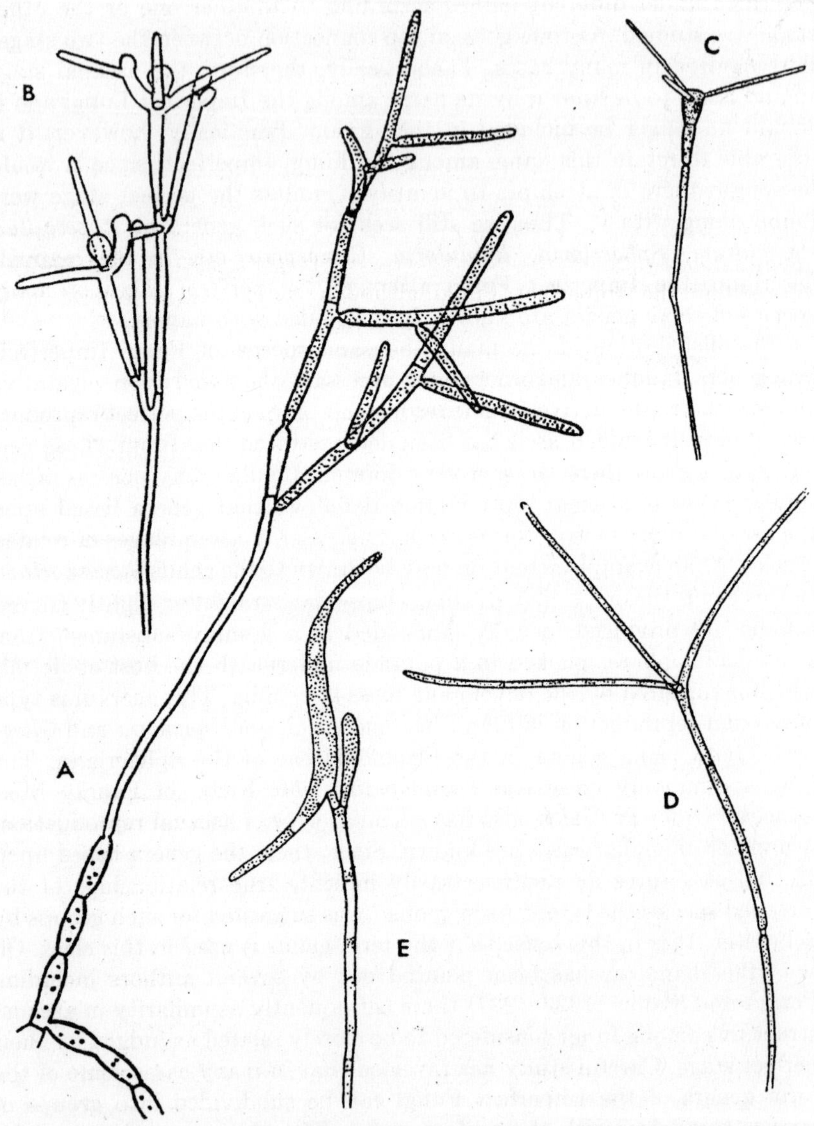

FIG. 189. Moniliales, Family Moniliaceae. Various types of adaptation for the production of submerged spores. (A) *Varicosporium elodeae* Kegel. (B) *Tetracladium marchalianum* De Wild. (C) *Clavariopsis aquatica* De Wild. (D) *Tetrachaetum elegans* Ingold. (E) *Lunulospora curvula* Ingold. (Courtesy, Ingold: *Brit. Mycol. Soc. Trans.*, 25(4):339–417.)

forms makes their recognition as definite species difficult, even where the perfect stages are not discovered. Furthermore the adaptation of the type of spore or sporophore to special habitats may be responsible for similarities among these that perhaps do not reflect real relationships. Thus the fungi that live in wet habitats and frequently produce their spores on hyphae that are completely submerged show many points of likeness; as in the genera *Varicosporium*, *Tetracladium*, *Heliscus*, *Lemonniera*, *Tricladium*, *Tetrachaetum*, *Lunulospora*, etc. The spores are usually slender and with long branches which give them a great ability to float. In some genera these are aleuriospores, in the sense of Vuillemin (1910, 1911) and in others are phialospores and radulaspores, in the sense of Mason (1933, 1937). To these Ingold (1942) would add the type "aquatic spore." A phialospore is borne at the apex of a phialide, a "fusiform truncate, fusiform beaked or acuminate terminal portion of a hypha, from the apex of which, or within which, thin-walled conidia are abstricted" (Mason). The septum separating the spore from the phialide is not produced until the spore is fully grown. An aleuriospore is a terminal portion of a hypha that is early separated by a septum from the parent hypha. A radulaspore is according to Mason a type "in which each spore is borne on a little sterigma, without any reference to the growing-point of a hypha," as in *Botrytis cinerea*. There is no direct evidence that these various aquatic genera are closely related except in habitat. Ingold (1942, 1943, 1944) has studied many of these and described several new genera. (Fig. 189.)

As the life histories of various fungi are studied by pure culture methods or by means of carefully controlled inoculations, from time to time an imperfect fungus is connected up with its perfect stage. This may perhaps be a species already known or may have been unknown theretofore. Klebahn (1918), the German mycologist, has made many such connections. It often happens that the same species has several types of asexual reproduction so that it may appear in several different form genera.

The many species and genera are usually divided into four form orders as follows:

Sphaeropsidales: conidia produced within pycnidia or modifications of such structures. A pycnidium is a perithecium-like structure and may be complete, like the perithecium of the Sphaeriales and Hypocreales, or only the top half may be present as in the perithecium-like spore fruit of the Hemisphaeriales, or it may open by a longitudinal slit as in the apothecia of the Hysteriales or may be closed at first and finally open into a cup or saucer-shaped structure, much like a miniature apothecium.

Melanconiales: conidia produced singly or in chains, often surrounded by a gummy mass, from conidiophores packed closely in a usually subepidermal or subcortical layer, the acervulus.

Moniliales: conidia formed on conidiophores which are separate, at least at their apical portions, or the vegetative mycelium breaking up into conidia. The conidiophores may be simple or branched, short or long, similar to the vegetative mycelium or very distinct from it, but are never enclosed within a pycnidium nor packed laterally into a subepidermal or subcortical acervulus. They are almost always external at time of conidium production.

Mycelia Sterilia: imperfect fungi which lack all conidial formation, and which produce sclerotia, rhizomorphs, and various other forms of mycelium without spores.

The usual extremely artificial classification of the Fungi Imperfecti separates many genera which perhaps more logically should be placed nearer each other. An example is the series of fungi in which the conidia ("endoconidia") are produced in the interior of the conidiophore and pushed out successively from an opening at the apex. These genera are found in several different "form families" and have conidia that are colored or colorless; one-celled to several-celled; pushed out in chains or singly; surrounded by slime or not, etc. In some of these genera other imperfect spore-forms also occur while in some the endoconidia are the only ones known. Some are believed to be imperfect forms of Ascomyceteae while it is suspected that the perfect stage of others may be Basidiomycetous. The following genera include most of the endosporous Fungi Imperfecti: *Cadophora*, *Thielaviopsis*, *Hymenellc*, *Chalara*, *Sporoschisma*, *Sporendonema*, *Endoconidium*, *Chalaropsis*, and probably *Catenularia* and *Phialophora*. (Fig. 190.) In the last two the conidiophore approaches the type of phialide (or sterigma) found in *Cephalosporium*, *Gliocladium*, and *Penicillium*, in which the conidia appear to be almost endogenous in origin.

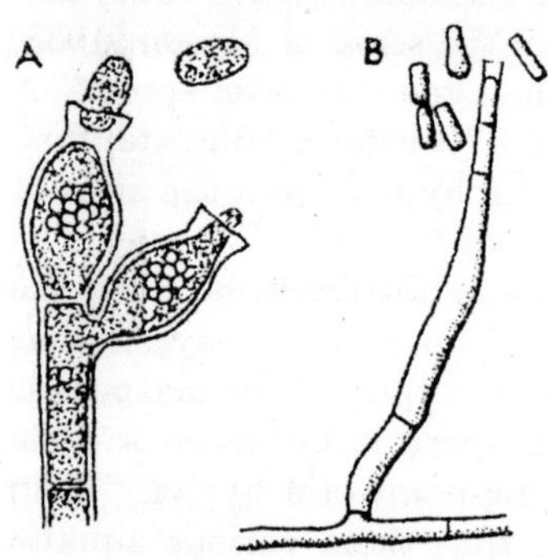

FIG. 190. Moniliales. Endogenous production of conidia (endospores). (A) *Cadophora obscura* Nannfeldt. (B) *Thielaviopsis paradoxa* (de Seynes) von Höhnel (conidial stage of *Ophiostoma* (*Ceratostomella*) *paradoxum*). (A, after Melin and Nannfeldt, *Svenska Skogsvårdsföreningens Tidskrift*, Häfte III–IV, pp. 397–616. B, courtesy, Dade: *Brit. Mycol. Soc. Trans.*, **13**:184–194.)

Order Sphaeropsidales. The 568[1] genera (with over 2300 species in North America alone) ascribed to this order are divided into four form families.

FAMILY SPHAEROPSIDACEAE (SPHAERIOIDACEAE OF SOME AUTHORS). Pycnidia resembling typical perithecia or forming pycnidial cavities in a

[1] The figures for this order are taken from H. B. Bender's pamphlet (1934) on the Sphaeropsidales.

stroma; tough leathery to brittle, and dark-colored. The spores often exude from the ostiole in damp weather in a worm-like mass, or cirrhus, consisting of gum and embedded spores. This family contains 359 genera so that it has become necessary to devise some means of subdividing it in a practical and easily applied manner. The scheme most often used was suggested by the great Italian mycologist P. A. Saccardo (1899) and is based on the structure and the color of the spores as follows:

Amerosporae: spores one-celled, spherical, ovoid or somewhat elongated.
- Hyalosporae: spores hyaline.
- Phaeosporae: spores colored some shade of light brown to black.

Didymosporae: spores similar to the foregoing, but two-celled.
- Hyalodidymae: spores hyaline.
- Phaeodidymae: spores colored.

Phragmosporae: spores three- or more-celled by transverse septa.
- Hyalophragmiae: spores hyaline.
- Phaeophragmiae: spores colored.

Dictyosporae: spores divided by both transverse and longitudinal septa.
- Hyalodictyae: spores hyaline.
- Phaeodictyae: spores colored.

Scolecosporae: spores very slender, thread- or worm-like, one-celled to several-celled, hyaline or colored.

Helicosporae: spores cylindrical and more or less spirally coiled, one-celled to several-celled, hyaline or colored.

Staurosporae: spores radiately lobed or star- or cross-shaped, one-celled to several-celled, hyaline or colored.

This same scheme is used for other orders and families of Imperfect Fungi in which conidia are produced, with the omission of such subdivisions as are not represented. Where only a few forms are present in the major subdivisions the minor ones based on spore color are often omitted. In the discussion of representative genera of this class the group name, based on the foregoing scheme, follows the generic name in parenthesis.

The following genera should be mentioned: *Phyllosticta* (Fig. 191A–C), *Phoma, Dendrophoma* (Fig. 191D–E) and *Macrophoma* (all Hyalosporae). Their 2500 or more species are parasitic on leaves and stems of plants. A possible exception is the fungus named by L. R. Fitzgerald (1943) *Phoma stenobothri* (Holl. & Mor.) Fitzg., which is parasitic upon grasshoppers. The first genus produces leaf spots with definite borders, the others produce less definite spots and occur on other parts of the host as well. *Macrophoma*, as the genus is usually defined, but incorrectly, according to Petrak and Sydow (1926–1927), has large conidia, over 15 μ long, but otherwise is like the others. In these genera the pycnidium is buried in the host tissue but the short ostiole emerges at maturity. The distinctions between these genera are clearly entirely artificial. A leaf spot of the beet (*Beta vulgaris* L.) is caused by *Phyllosticta tabifica* Prill. The same fungus

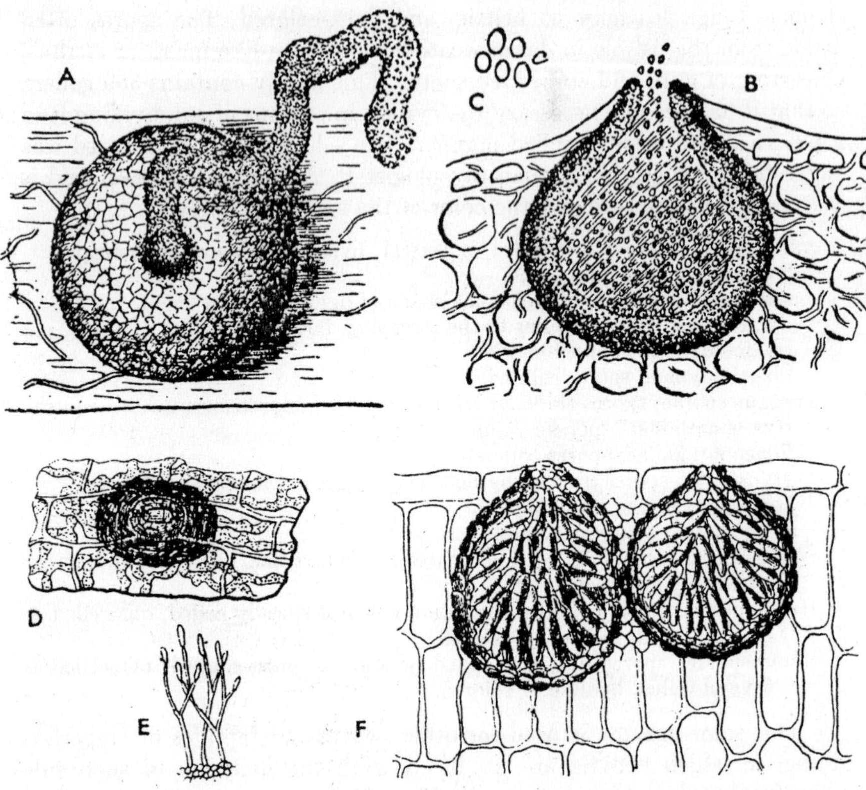

FIG. 191. Sphaeropsidales, Family Sphaeropsidaceae. (A–C) *Phyllosticta tabifica* Prill. (A) Pycnidium viewed from above, showing cirrhus of spores. (B) Vertical section through pycnidium. (C) Spores. (D, E) *Dendrophoma convallariae* Cav. (D) Pycnidium from above. (E) Conidiophores. (F) *Ascochyta dianthi* (A. & S.) Berk.; section through pycnidium. (A–C, after Prillieux: *Bull. soc. mycol. France*, 7(1):15–19. D–E, after Briosi and Cavara: Fascicle IV, No. 89. F, *ibid.*, Fascicle XIV, No. 342.)

on the roots and inflorescence has been called *Phoma betae* Fr. Various species of these genera have been shown to have as their perfect stage species of *Guignardia*, *Mycosphaerella*, etc., in the Sphaeriales or Pseudosphaeriales. The specific distinctions have to a considerable extent been based on the host species attacked. Until very extensive inoculation experiments can be carried out, this basis for segregation of species will probably continue to be used. *Cytospora* and *Ceuthospora* (Hyalosporae) produce their pycnidial cavities in subepidermal or subcortical stromata. The spores emerge from the separate or united ostioles in an amber-colored cirrhus. Many of the species are the imperfect stages of *Valsa* and related genera in the Sphaeriales. (Fig. 192A.) *Sphaeropsis* (Phaeosporae) corresponds to *Macrophoma* except that the conidia are dark-colored. *S.*

malorum Pk. causes twig cankers on apple and quince and the black rot of the fruits. Its perfect stage is *Physalospora.* The imperfect form genus is broken up by the more recent authors into several form genera. The name *Sphaeropsis* having been first applied to the perfect stage of an Ascomycete should, according to Petrak and Sydow, be replaced by *Haplosporella.* Some of the species usually included here they transfer to *Botryodiplodia*, including among those so transferred *S. malorum. Coniothyrium* has smaller pycnidia and very much smaller conidia which emerge from the ostiole in a black cirrhus. *C. fuckelii* Sacc., the cause of the cane blight of various species of *Rubus*, has as its perfect stage *Leptosphaeria coniothyrium* (Fckl.) Sacc., in the Sphaeriales. *Ascochyta* (Hyalodidymae)

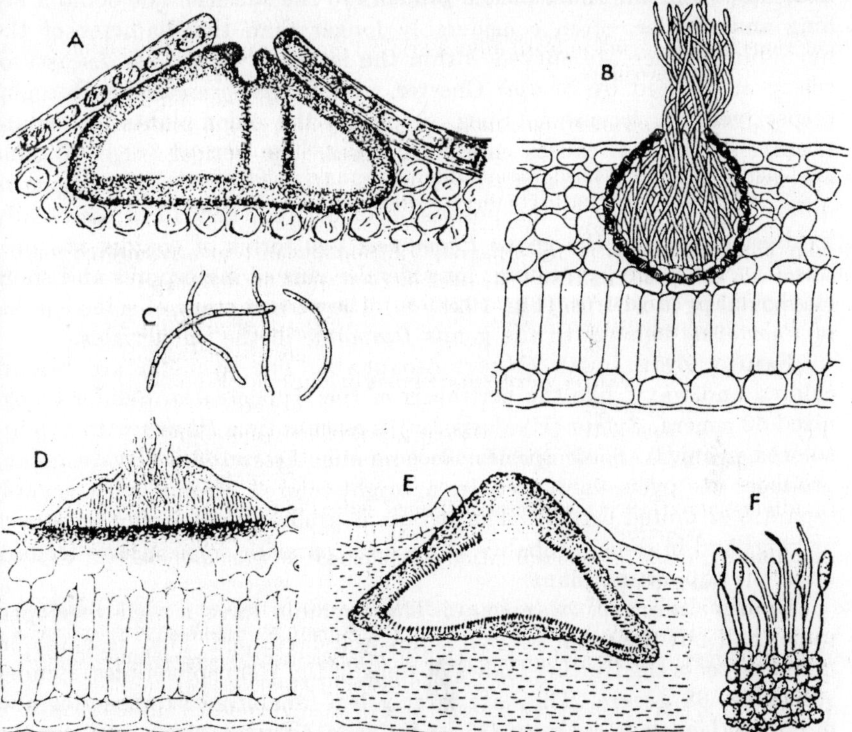

FIG. 192. Sphaeropsidales, Family Sphaeropsidaceae. (A) *Ceuthospora abietina* Delacr.; section through stroma with several pycnidial cavities with one common ostiole. (B, C) *Septoria aesculi* (Lib.) West. (B) Section through pycnidium. (C) Spores. (E–F) *Phomopsis citri* Fawcett. (E) Section through pycnidium. (F) Portion of wall of pycnidium showing pycnospores and the long, slender, curved stylospores. (D) Family Leptostromataceae. *Leptothyrium acerinum* (Kunze) Corda; section through pycnidium. (A, after Delacroix: *Bull. soc. mycol. France*, 6(4):181–184. B–C, after Briosi and Cavara: Fascicle V, No. 120. E–F, after Fawcett: *Phytopathology*, 2(3):109–113. D, after Briosi and Cavara: Fascicle II, No. 40.)

is essentially a *Phoma* with two-celled conidia. (Fig. 191F.) *A. pisi* Lib. is very destructive to cultivated peas. Its perfect stage is *Mycosphaerella pinodes* (B. & Bl.) Stone. *Diplodia* (Phaeodidymae) is like *Macrophoma* with colored two-celled spores. In fact some species of *Diplodia* have been described as *Macrophoma*, *Sphaeropsis*, and *Diplodia*, depending upon the age of the conidia. Such species are placed by some authors in *Botryodiplodia*. All three types of spores may be present in the same pycnidium and all are viable. The 500 or more described species occur on leaves, stems, etc., and are often the cause of serious diseases of their host plant. *Septoria* (Scolecosporae) is a genus of over 1000 species, practically all parasites. (Fig. 192B–C.) Like *Phoma* the pycnidium is immersed in the host tissue and the short ostiole projects to the surface. The conidia are long and slender, often considerably longer than the diameter of the pycnidium, hence are curved within the latter. Two serious diseases of celery are caused by *S. apii* Chester, and *S. apii-graveolentis* Dorogin, respectively. *S. lycopersici* Speg. defoliates the older plants of tomato (*Lycopersicom*) and causes enormous losses. The perfect stage of some species of *Septoria* belongs to the genus *Mycosphaerella*. *Phaeoseptoria* differs from *Septoria* in the possession of colored, instead of essentially hyaline, spores. In the genus *Phomopsis* two forms of conidia are produced in the same pycnidium; long slender curved stylospores and short ellipsoidal pycnospores. (Fig. 192E–F.) The perfect stage of some species of *Phomopsis* belongs to the genus *Diaporthe* in the Sphaeriales.

Family Zythiaceae (Nectrioidaceae). The pycnidia are bright-colored and waxy, like the perithecia of the Hypocreales. Bender recognized 62 genera. *Zythia* (Hyalosporae) is essentially a *Phoma* with bright-colored pycnidia. Some species are parasitic. *Aschersonia* (Hyalosporae) produces its pycnidia buried in a bright-colored stroma with several separate or united ostioles. The stroma is produced externally on leaves or twigs and in some, probably all, cases is parasitic upon insects feeding upon the supporting plant.

Family Leptostromataceae. The pycnidia have a well-developed upper half but the basal portion is not well-developed. They resemble in many respects the fruiting bodies of some of the Hemisphaeriales. Bender recognizes 88 genera. They are largely leaf-inhabiting saprophytes and parasites. *Leptothyrium pomi* (M. & F.) Sacc. (Hyalosporae) produces the so-called "fly-specks" of apple fruits. These are the minute flattened, round pycnidia. (Fig. 192D.) *Leptostroma* (Hyalosporae) produces elongated pycnidia with slit-like ostioles, on leaves, stems, etc. *Entomosporium maculatum* Lev. (Hyalophragmiae) forms four-celled conidia, the cells forming a sort of square, each with a bristle. It causes leaf and fruit spots of the pear. Its perfect stage is *Diplocarpon soraueri* (Kleb.) Nannf. (*Fabraea maculata* (Lev.) Atk.), Family Mollisiaceae, Order Pezizales.

Family Excipulaceae. The pycnidia open out early to form a more or less deep, cup or saucer-shaped structure, tough or hard and black, either arising subepidermally or subcortically and breaking through to the outside, or in some cases external from the first. Largely saprophytic on twigs, stems, etc., less often on leaves. It is sometimes difficult to distinguish certain subcortical species of this family from the following order (Melanconiales). Bender recognizes 59 genera. *Excipula* and *Discula* (Hyalosporae) and *Discella* (Hyalodidymae) are among the genera with the greatest number of species.

Tehon (1940) separated from among the genera usually included in the Leptostromataceae two families, Pycnothyriaceae and Rhizothyriaceae, which he combines in a distinct order Pycnothyriales. These differ from the Leptostromataceae in bearing their spores on the under side of the pycnidial cover instead of basally.

Order Melanconiales. This order of 92 genera[2] and over 600 North American species consists of but one family, Melanconiaceae. To a large extent the species are parasitic, causing the type of plant disease known as anthracnose. The three form genera, *Gloeosporium* (Fig. 193A), *Colletotrichum* (Fig. 194A), and *Myxosporium* (Hyalosporae), contain some of the most destructive parasites of cultivated plants. They differ by arbitrary characters. *Colletotrichum* produces stiff colorless or colored bristles (setae) around the acervulus while these are lacking in the other two. *Gloeosporium* occurs on herbaceous host structures while *Myxosporium* occurs on woody stems. How artificial these distinctions are may be seen in the case of the fungus called *Colletotrichum gloeosporioides* Penz., a common parasite of the leaves, young twigs, and fruits of species of *Citrus* and many other genera. When inoculated upon the mango (*Mangifera indica* L.) setae are produced on some of the acervuli on the twigs

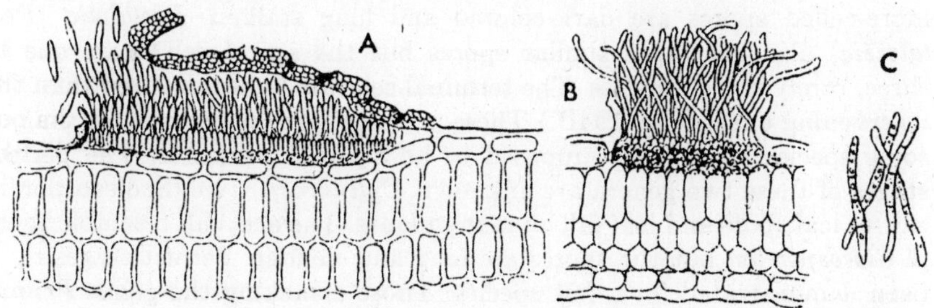

Fig. 193. Melanconiales, Family Melanconiaceae. (A) *Gloesporium populi-albae* Desm.; section through acervulus. (B, C) *Septogloeum mori* (Lév.) Briosi & Cav. (B) Section through acervulus. (C) Spores. (A, after Briosi and Cavara: Fascicle VI, No. 147. B–C, *ibid.*, Fascicle I, No. 21.)

[2] The figures for this and the following orders and families are taken from H. B. Bender (thesis, 1931).

and leaves, but on the fruits the setae are lacking. George L. Fawcett, then (1906--1908) a colleague of the author, grew this fungus on over 50 different hosts in Miami, Florida. These had been described in literature under 25 or more species names in the two genera *Gloeosporium* and *Colletotrichum*. On many of these hosts the perfect stage was produced and was found to be *Glomerella cingulata* (St.) Sp. & von S., of the Gnomoniaceae. Since it grows upon twigs as well as upon the leaves and fruits the distinction between *Myxosporium* and *Gloeosporium* breaks down. The acervuli begin at first as tangled subepidermal masses of hyphae from which arise numerous closely packed conidiophores which partially raise the epidermis. From the apex of each conidiophore are developed one or more conidia, embedded in gum. When moisture is abundant this gum swells and the epidermis is burst open and the spores are exuded in a sticky mass. Insects or other objects coming into contact with these spores distribute them, as does the rain, whose drops striking such a spore mass are broken up into smaller wind-borne droplets, each carrying its burden of spores. *Gloeosporium ribis* (Lib.) Mont. & Desm. is the imperfect form of *Pseudopeziza ribis* Kleb., in Family Mollisiaceae, Order Pezizales. Some species of *Myxosporium* are imperfect stages of *Diaporthe* in the Sphaeriales. *Marssonina panattoniana* (Berl.) Magn. (Hyalodidymae) causes injury resulting in the formation of holes in leaves of lettuce (*Lactuca sativa* L.) and is sometimes very destructive in greenhouses. *Septogloeum* (Hyalophragmiae) is mostly parasitic on leaves. It resembles *Gloeosporium* but the somewhat elongated spores are several times transversely septate. (Fig. 193B–C.) *Coryneum* (Phaeophragmiae) has numerous species. *C. beijerinckii* Oud. occurs in Europe and various parts of the United States as the cause of a serious disease of the peach (*Amygdalus persica* L.). It kills twigs or may attack the buds and merely kill these and the surrounding tissues, or may cause injury to the fruit. The four- or more-celled spores are dark-colored and long stalked. *Pestalotia* (*Pestalozzia*) also has quite similar spores but the apical cell bears one to three, rarely more, bristles. The terminal cells are lighter colored than the intervening ones. (Fig. 194B.) These are saprophytes on many hosts but some species have been supposed to be harmful parasites. The perfect stages of these two genera are unknown. *Cylindrosporium* (Scolecosporae) causes leaf spots and leaf fall on many plants. The acervuli resemble those of *Gloeosporium* but the spores are long and slender. Perfect stages have been demonstrated for a few species. Those attacking the genus *Prunus* have as their perfect stage apothecial fungi belonging, according to Higgins (1914), to the genus *Coccomyces*, of the Family Phacidiaceae; but according to Nannfeldt (1932) they do not belong to that genus and family but to the Mollisiaceae, and a genus to which he gave the name *Higginsia*. This name, however, is preoccupied and must be replaced if the genus is held distinct from *Coccomyces*.

Sphaceloma (Hyalosporae) was long confused with *Gloeosporium*. Like the latter its conidia are ellipsoid and produced on short conidiophores. The acervulus differs considerably in forming a rather firm fungus cushion, in some cases almost intermediate between the sporodochium of Family Tuberculariaceae and a typical acervulus. On leaves it may cause considerable malformation. It also attacks herbaceous stems, fruits, etc.

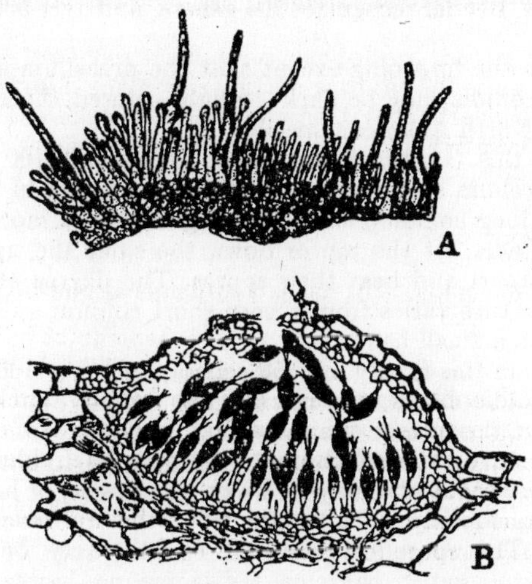

FIG. 194. Melanconiales, Family Melanconiaceae. (A) *Colletotrichum malvarum* (A. Br. & Casp.) Southw. (B) *Pestalotia versicolor* Speg. (A, after Southworth: *J. Mycology*, **6**(2):45–50. B, after Klebahn: *Mycol. Cent.*, **3**(3):97–115.)

The perfect stage of many species has been demonstrated to belong to the genus *Elsinoë* of Order Myriangiales. Many species are destructive to economic plants.

Order Moniliales (Hyphomyceteae). This order contains 651 genera and toward 10,000 species divided into four families. The first two are distinguished by the color of the mycelium and conidiophores. The older distinction placed in the first family those genera with light-colored or bright-colored mycelium and conidiophores *and conidia*. Forms with dark-colored mycelium and conidiophores were placed in the second family, whether the conidia were light or dark, as well as forms with light-colored conidiophores but colored conidia. The author follows Bender in making the distinction as follows:

Moniliaceae: mycelium and conidiophores hyaline or bright-colored (not brown, smoky, or black), conidia hyaline or colored. These are formed on the ends of short conidiophores not distinguishable from the other branches of the mycelium or are terminal or lateral on distinct, unbranched or branched, separate conidiophores. In a few cases the mycelium breaks up into more or less rounded conidia. Some species of *Aspergillus* with ochre-colored to black conidia have conidiophores that are distinctly brown in their upper portion. Because in their structure they correspond to the species of this form genus that have colorless conidiophores they are not transferred to the following family. Members of this family are saprophytes or parasites, including some of the most ubiquitous molds and some very serious enemies of economic plants. Bender recognizes 204 genera, and over 500 North American species.

Dematiaceae: like the foregoing except that the mycelium and conidiophores are dark. The conidia may be dark- or light-colored. There are 206 genera, and over 1000 North American species.

Stilbellaceae[3]: in this family of 89 genera, and about 100 North American species the mycelium spreads through the substratum in the usual manner but the rather long conidiophores arise together in a more or less compact column or synnema. At the top or down the sides the tips of the conidiophores spread apart and bear their spores. The degree of union is various so that the structure varies from a very short column and bushy head to a tall column with a small head.

Tuberculariaceae: in this family are 152 genera, and over 400 North American species. The conidiophores arise more or less radially, packed close together or separate, from the surface of a somewhat rounded mass of hyphae forming a sort of cushion, or sporodochium. This may be hyphal in structure or pseudoparenchymatous and may be waxy, gelatinous, or horny. The conidia are produced terminally or laterally from the unbranched or branched conidiophores. The sporodochium and conidia may be light-colored or dark-colored.

Family Moniliaceae. In the Moniliaceae the classification is based upon the number of cells in the conidia, whether these are borne upon special conidiophores or are merely modified portions of the vegetative mycelium, the number of conidia at the apex of the conidiophore and their arrangement there, whether in a chain or a head of separate spores caught in a mucilaginous drop, etc. The genus *Oospora*, as mostly interpreted by mycologists, consists of a slender, branched or unbranched mycelium which breaks up into ellipsoidal or spherical, hyaline or light-colored conidia (often called "oidia"). There is no sharp line of distinction between the vegetative hyphae and those that are breaking up into conidia. *O. lactis* (Fres.) Sacc. forms a thick wrinkled skin on the surface of sour milk and of other liquids containing considerable organic matter. It is quite similar to some of the asporogenous yeasts, many of which are the causes of disease in Man and other animals, and which are discussed in

---

[3] Since the type species of the genus *Stilbum*, formerly assigned to this family, has been determined to belong to Order Auriculariales (see p. 444) this name is not available for a genus in the Fungi Imperfecti and accordingly some mycologists have proposed the name *Stilbella* for the imperfect forms, wherefore the family name becomes Stilbellaceae instead of Stilbaceae.

Chapter 11 under the Saccharomycetales. Many of the asporogenous yeasts are truly Fungi Imperfecti, in that the ascogenous stage is unknown. The plant-inhabiting species of *Oospora* such as *O. nicotianae* Splend. & Sacc. are probably not at all closely related to *O. lactis*, but since we are dealing with form genera the matter is not so important unless good characters can be discovered that will permit the rather unwieldy genus to be broken up into several smaller genera.

The genus *Actinomyces*, in the older, wider sense of the name, represents beyond doubt a series of several closely related genera whose relationships within the fungi are not agreed upon by various investigators. The mycelium consists of branching, rarely septate hyphae of great slenderness, approximately 1 μ in thickness or more or less. These may be aerobic or almost anaerobic in their mode of growth. The hyphae submerged in the substratum eventually break up, by the formation of numerous septa, into short, cylindrical spores. In the genus *Streptomyces* there is formed a mat of external, branched, aerial conidiophores which become septate basipetally into conidia which eventually break apart and are oval or cylindrical or ellipsoidal. These germinate by 1 to 4 germ tubes. The conidiophores are more often spirally wound. The presence of true nuclei (Drechsler, 1919 and Newcomer and KenKnight, 1939) and this mode of germination by germ tubes seem to justify Drechsler and some other students in considering these genera as true fungi. Until more is known as to any perfect stage they may well be placed in the Moniliaceae near *Oospora*. Perhaps Drechsler (1935a) is right in drawing attention to the close similarity of the vegetative structures and asexual mode of reproduction between *Streptomyces* and some of the Zoopagaceae in the Phycomycetes.

Waksman and Henrici (1943) place the genera related to *Actinomyces* in two families, which they unite with the Mycobacteriaceae in a separate order the Actinomycetales, locating it in a position intermediate between the Bacteria and the Fungi Imperfecti. These two families are the Actinomycetaceae, with no aerial conidia, and with two genera *Actinomyces*, anaerobic, and *Nocardia* (syn. *Proactinomyces*), aerobic, and Streptomycetaceae, with aerial conidia, in chains in *Streptomyces* and singly at the tips of upright conidiophores in *Micromonospora*. The chief species of *Actinomyces* is *A. bovis* Harz, the organism causing the disease of animals (especially cattle) known as "lumpy jaw." The genus *Streptomyces* contains many soil inhabitants feeding upon dead vegetable matter. *S. scabies* (Thaxt.) Waks. & Henrici is one of the organisms causing the scab of potato tubers and other subterranean organs of living plants. For the use of *Streptomyces sp.* in the production of antibiotics, see p. 603.

In *Cephalosporium* (Hyalosporae) the conidiophores produce the spherical or ellipsoidal conidia successively at the apex, each new conidium pushing aside the one last formed and all of these adhering by means of a small amount of slime and forming a head at the apex of the conidiophore. Miss Pinkerton (1936) has shown that these conidia are produced endogenously, each one pushing out the previously formed one into the drop of slime extruded at the tip of the conidiophore. Some species are parasitic on fruiting bodies of *Polyporus*. The microspores of some species of *Fusarium* are of this type. *Haplotrichum* is essentially a *Cephalosporium* with larger, upright, unbranched conidiophores. Other genera in which conidia are produced on heads but not in chains are *Oedocephalum*, *Rhopalomyces*, *Sigmoideomyces*, in all of which the conidia arise on the swollen tips of the conidiophores. Some of these are probably in reality Mucorales, but until the sexual reproductive stages are discovered it is unwise to say that of all. B. O. Dodge (1937) definitely proved that one species of *Oedocephalum* is the conidial stage of *Peziza pustulata* Pers. (*Aleuria umbrina* Gill.).

There is a large series of genera in which the conidia are produced in basigenous chains. In many of these the ultimate, spore-bearing element of the conidiophore is more or less flask- or bottle-shaped, producing the spores successively at the neck of the bottle. These ultimate segments are called phialides, or sterigmata. In *Aspergillus* the perfect stage is known for a number of species, and it is probable that many more where this stage is not now known will be found to possess it. Since it is only rarely or never found in most of the species the genus is to be sought not only in the Order Aspergillales of the Ascomyceteae but also in the Fungi Imperfecti. The conidiophores are upright with a swollen head from which arise at all sides or only on the upper portion numerous phialides or sterigmata. At the tips of these are produced successively the conidia which often form very long chains, the terminal conidium being the oldest. Although classed in the Family Moniliaceae because of similarity of structure, the conidia of the various species may be hyaline, yellow, green, ochre, or even black, and the upper portion of the conidiophore may be dark. In some species when grown with an abundance of nutrients the primary sterigmata bear at their apices groups of secondary sterigmata which produce the conidial chains. Such species have been segregated as the genus *Sterigmatocystis*, but since the distinguishing character is largely dependent upon the nutrition and other environmental conditions, it does not seem to be of great validity. Closely related to *Aspergillus* is the genus *Penicillium*. Here the conidiophore produces two to several branches which grow more or less parallel to it and at about the same level produce at their tips several sterigmata, each bearing a chain of conidia. The branches may arise in a whorl and these themselves bear whorled branches or some of the branches may arise singly at the level below a whorl, thus producing an asymmetrical penicillus or brush of sterigmata and supporting branches. These substerigmatal branches are the metulae. As in *Aspergillus* the conidia vary in color among the different species but none are as dark as those of *Aspergillus niger* van Tiegh. Both of these genera are ubiquitous molds, mostly saprophytes, although one or two species of *Aspergillus* have been described as parasitic in the human ear. Animals eating moldy hay heavily infested with these molds sometimes inhale so many spores that fatal mycosis of the lungs ensues. *Penicillium notatum* Westling and *P. chrysogenum* Thom are of great economic importance because of their production of antibiotic substances (see closing paragraphs of this chapter). Resembling *Penicillium* in the general plan of its conidiophore is the genus *Gliocladium* which differs in the secretion of a mucilaginous drop at the top of the penicillus in which the successively formed conidia are held. The perfect stage of *G. penicillioides* Corda is *Lilliputia*, which belongs to the Order Aspergillales. Matruchot (1895) considered that it belonged to the Perisporiales (Erysiphales) and Winter (1873) named it *Eurotium insigne* and placed it in the Aspergillales. (Fig. 195A–D.)

Another genus with catenulate conidia borne on phialides is *Spicaria*. In this the conidiophore bears definite whorls of phialides and does not

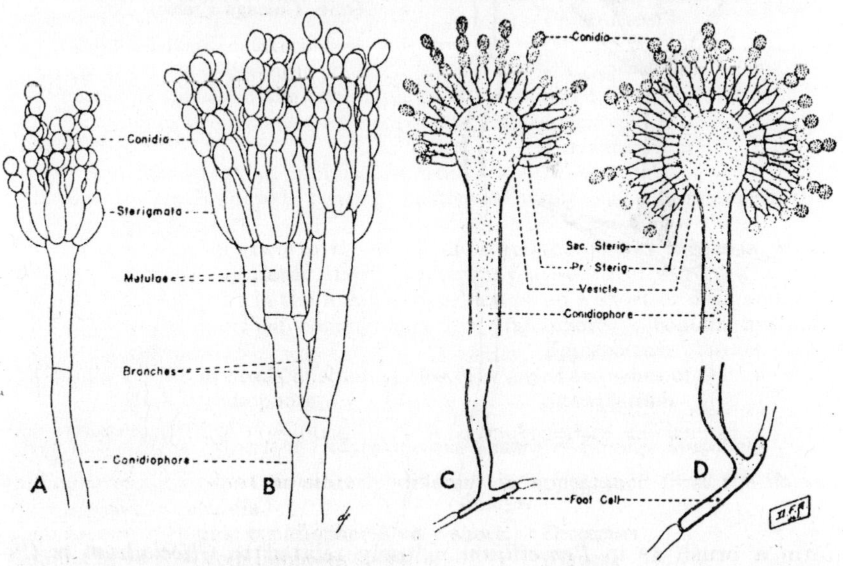

FIG. 195. Moniliales, Family Moniliaceae. (A) *Penicillium frequentans* Westling, illustrating simple type of conidiophore. (B) *Penicillium expansum* Link, illustrating branched type of conidiophore. (C) *Aspergillus niveo-glaucus* Thom and Raper, illustrating conidiophore with one series of sterigmata. (D) *A. versicolor* (Vuill.) Tirab., showing conidiophore with two series of sterigmata. (A–B, courtesy, Raper and Thom: A Manual of the Penicillia, Baltimore, Williams & Wilkins Co. C–D, courtesy, Thom and Raper: A Manual of the Aspergilli, Baltimore, Williams & Wilkins Co.)

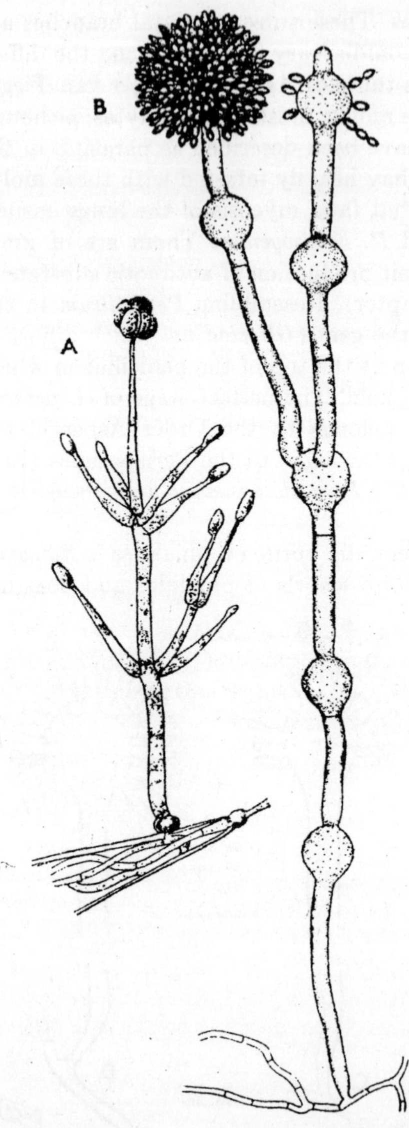

FIG. 196. Moniliales, Family Moniliaceae. (A) *Verticillium albo-atrum* Reinke & Berthold. (B) *Gonatorrhodiella parasitica* Thaxt. (A, after Lindau, in Engler und Prantl: Die Natürlichen Pflanzenfamilien, Leipzig, W. Engelmann. B, after Thaxter: *Botan. Gaz.*, 16(6):201–215, Univ. Chicago Press.)

form a brush as in *Penicillium*. Closely related to *Gliocladium* is *Clonostachys* in which the successive whorls of branches of first and second rank are more symmetrical and the conidia are in elongated mucilaginous masses.

Among the Moniliales there are several genera in which the conidia arise from intercalary enlarged cells in the upright conidiophores. In some cases well-developed phialides are present but in other cases the conidia appear to be more properly "radulaspores." They may be single or in chains, one-celled or septate, hyaline, or brown. Probably the fungi with this type of conidiophore are not closely related. Among such fungi are *Gonatorrhodiella, Arthrobotrys, Gonatobotryum, Gonatorrhodum, Gonytrichum.* (Fig. 196B.)

The genus *Verticillium* is named because of the whorled arrangement of the branches of the conidiophore. The conidia arise singly at the tips of phialide-like branchlets. *V. alboatrum* Reinke & Berth. is the cause of a disease called "hadromycosis," producing death in very many plants, woody and herbaceous, e.g., species of *Acer, Solanum tuberosum* L., *Dahlia*, etc. No perfect stage is known. (Fig. 196A.) *Botrytis* has branched conidiophores, but these are not verticillate, nor swollen at the ends. Various serious diseases of plants are caused by species of this genus. Some species form sclerotia in the host tissues. Some of the Sclerotiniaceae (Ascomyceteae) have a *Botrytis* stage, but for the majority of the species of *Botrytis* no perfect stage is known. The genus *Phymatotrichum* differs from the foregoing in that the terminal spore-bearing portions of the branches of the conidiophore are somewhat swollen. *P. omnivorum* (Shear) Duggar is the conidial stage of the fungus causing the root rot of cotton and many other plants in Texas and some adjacent states. Closely related is the genus *Oidium*, more often known as *Rhinotrichum*. (Fig. 197A.) All of the foregoing genera are hyalosporous, i.e., have one-celled conidia which are colorless or light-colored (except as noted for *Aspergillus*). Among the Hyalodidymae, *Trichothecium*, and *Cephalothecium* both produce somewhat pear-shaped hyaline or pale pink spores that are two-celled, the basal cell being smaller than the terminal one. They are mostly saprophytic although *C. roseum* Corda enters the fruit of the apple through the lesions caused by the scab organism (*Venturia inaequalis* (Cke.) Winter) and causes rot. In *Cephalothecium* the conidia are produced in heads while in *Trichothecium* they are single at the tip of the conidiophore. *Arthrobotrys* bears similar conidia but they are clustered at various swollen intercalary cells of the unbranched conidiophore. Some species are saprophytic, but Drechsler (1937) described six species parasitic upon terricolous nematodes which are caught in net-like structures of the mycelium. *Mycogone perniciosa* Magn. is a dangerous pest in the commercial culture of the common mushroom (*Agaricus campestris* L. ex Fr.). Its short conidiophores, lateral to the hyphae of the mycelium, bear terminally single conidia, two-celled and constricted between the upper (larger) and lower cells. These conidia may be slightly roughened. They are sometimes considered to be a form of chlamydospore since on slender upright hyphae there sometimes occur verticillate branches bearing small one-celled spores. Other Hyalodidymae growing upon fungi are *Diplo-*

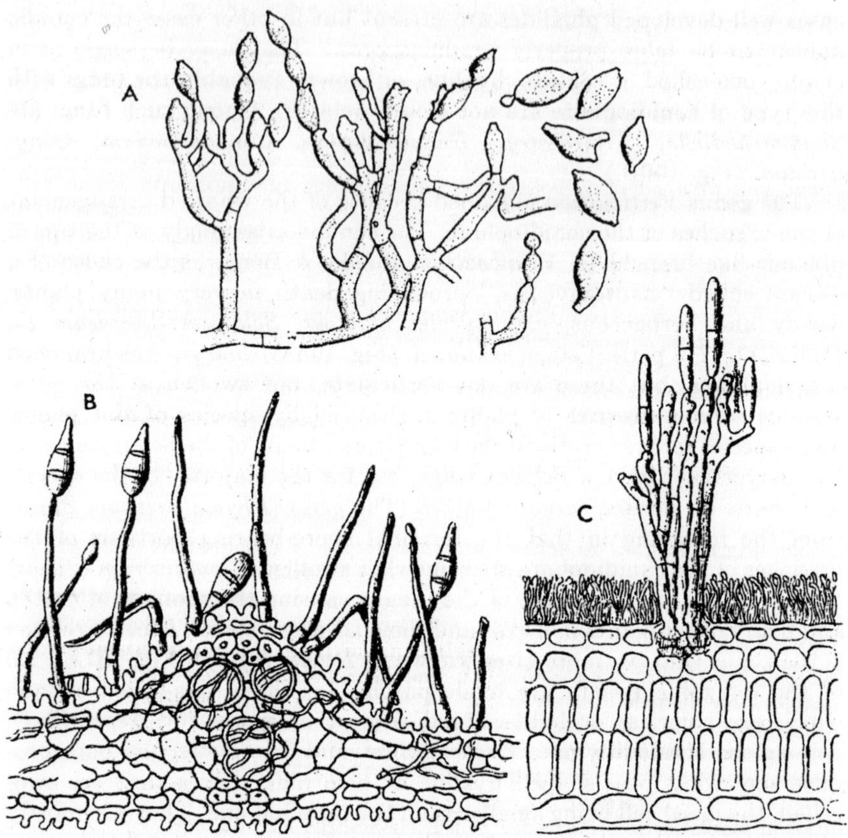

FIG. 197. Moniliales, Family Moniliaceae. (A) *Oidium* (*Rhinotrichum*) *aureum* Corda; conidiophores and conidia. (B) *Piricularia oryzae* Briosi & Cavara; conidiophores. (C) *Ramularia rosea* (Fckl.) Sacc.; conidiophores with conidia. (A, courtesy, Linder: *Lloydia*, 5(3):165–207. B, after Briosi and Cavara: Fascicle VIII, No. 188. C. *ibid*., Fascicle IV, No. 77.)

cladium and *Didymocladium*, both with creeping vegetative mycelium and upright branched conidiophores, the former with single or a few not catenulate conidia at each branch tip, and the latter with more strongly verticillate branching and chains of conidia at the tips.

Among the Hyalophragmiae there are about a dozen unimportant saprophytic genera and two that are serious parasites of cultivated plants. These are *Ramularia* and *Piricularia*. In addition some species of *Septocylindrium* are more or less parasitic. In *Ramularia* the conidiophores emerge through the stomata of the host in bundles of two or three to half a dozen. At first each conidiophore bears a single terminal conidium which is ellipsoid and one-celled. As it enlarges it becomes elongated, ellipsoid and three- or more-celled. (Fig. 197C.) In some species at the distal end of the conidium another conidium develops, and so on until a short, acrogenously produced chain of conidia is formed. The conidiophore produces sympodially a lateral projection on which another conidium arises, and this may be repeated until the upper part of the conidiophore is a little zigzag or shows several small teeth, each representing the position where a conidium was borne. The perfect stage of a few species has been shown to belong to the genus *Mycosphaerella* in the Sphaeriales. *Ramularia armoraciae* Fckl. is the cause of the very abundant leaf spot of horse radish (*Radicula armoracia* (L.) B. L. Rob.). The perfect stage of the strawberry leaf spot (*Mycosphaerella fragariae* (Tul.) Lind.) develops only on the dead overwintered leaves, but the conidial stage, *Ramularia tulasnei* Sacc., is sometimes very harmful to the growing leaves of the host. *Piricularia* causes severe damage to rice (*Oryza sativa* L.) and produces leaf spots on many other grasses. Its conidiophores are simple, emerging from the host's surface and bearing narrowly pyriform, three- to several-celled hyaline conidia. (Fig. 197B.) In *Septocylindrium* from short conidiophores arise long, sometimes branched, chains of ellipsoidal, three- to several-celled conidia. Perfect stage unknown.

The genus *Dactylella* has unbranched conidiophores with the conidia single at the apex while *Dactylaria* has the conidia in clusters at the apex. Many of the species have been described as saprophytic but Drechsler (1935b, 1937) described some species parasitic on terricolous Amoebae and nematodes. (Fig. 198C–H.)

The genus *Helicomyces* is saprophytic. It must be mentioned as one of the series of probably related genera which have been distributed, because of the spore color or arrangement of the vegetative hyphae and conidiophores among several groups of the four form families of the Moniliales. In all of these the conidia are two- to many-celled and spirally rolled. In some the spiral is in one plane, in others it is drawn out somewhat like a screw. The color of the conidiophores is hyaline in *Helicomyces*, *Hobsonia*, and some others, but in *Helicoma* and *Helicoceras* they are dark-colored. For a fuller understanding of this group the reader should consult Linder (1929, 1931a and b). (Fig. 199.)

In the Dematiaceae many genera parallel closely those in the preceding family, differing in the dark color of the mycelium and conidiophores. More often the conidia are dark, also. *Haplographium* (Phaeosporae), except for its dark-colored conidiophores and conidia, closely resembles *Penicillium.* Its species are probably all saprophytic. The genus *Coniosporium* corresponds closely to *Chromosporium* in the Moniliaceae. The round or ellipsoid dark conidia arise on very short stalks from the scanty, dark, mostly saprophytic mycelium. Corresponding to *Oospora* in the Moniliaceae are *Torula* and *Hormiscium* in which portions of or the

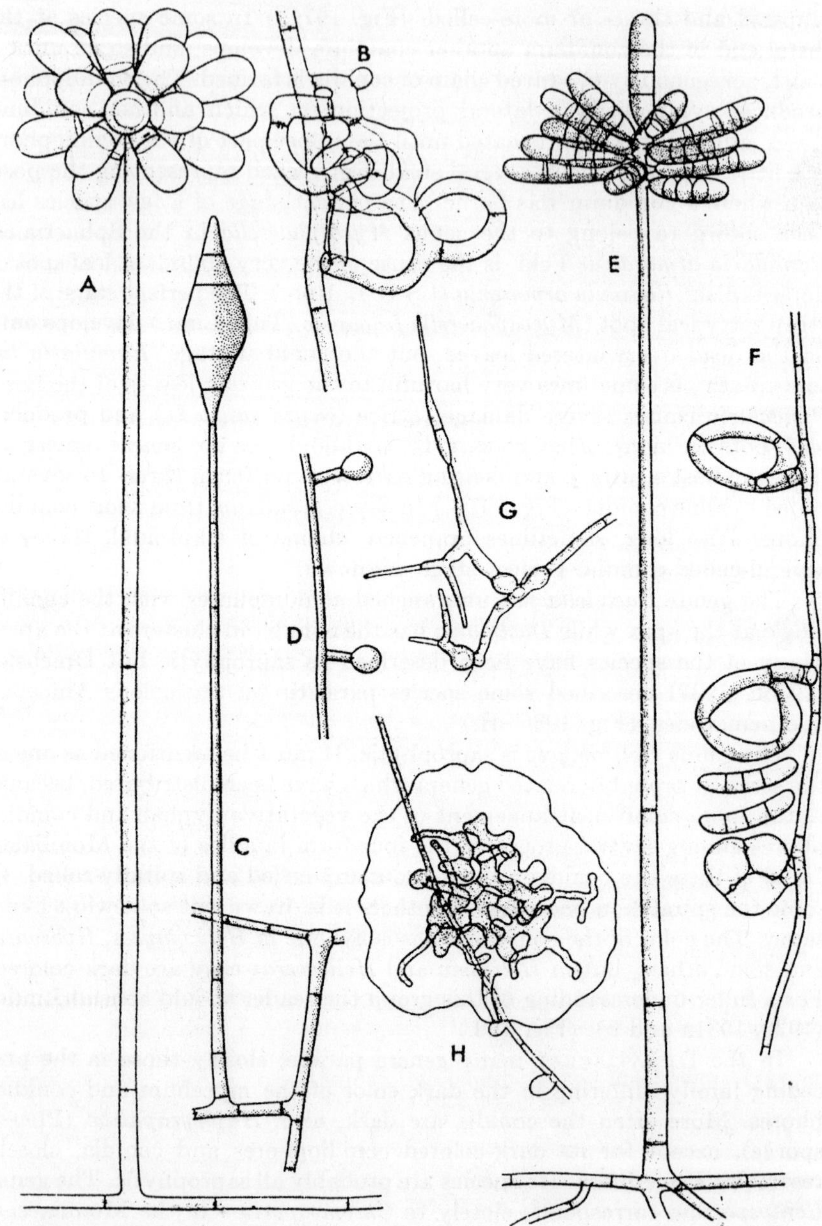

FIG. 198. Moniliales, Family Moniliaceae. Forms adapted to capturing and destroying microscopic animals. (A, B) *Arthrobotrys conoides* Drechsler. (A) Creeping hypha and conidiophore. (B) Net in whose meshes nematodes become entangled. (C, D) *Dactylella ellipsospora* Grove. (C) Conidiophore. (D) Portion of hypha with

(*Continued on facing page.*)

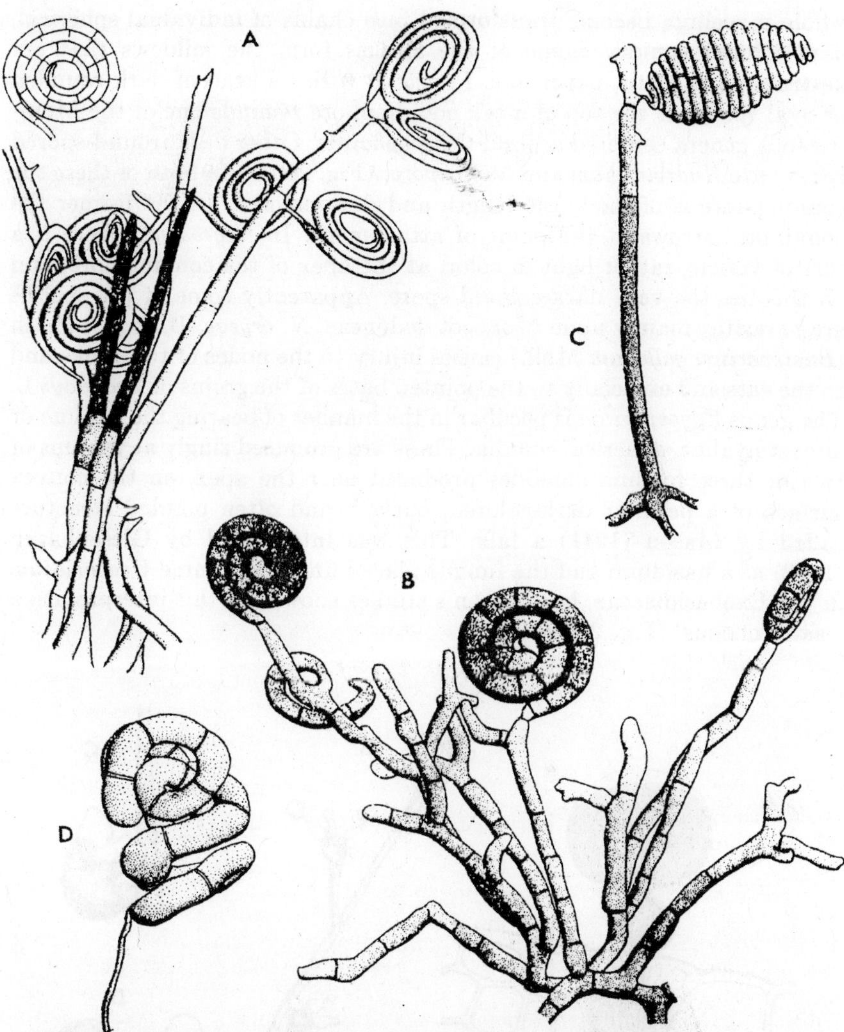

FIG. 199. Moniliales, Families Moniliaceae and Dematiaceae. Various helicosporous members of these families. (A) *Helicomyces scandens* Morgan. (B) *Helicoma perelegans* Thaxt. (C) *Helicoon auratum* (Ellis) Morgan. (D) *Hobsonia mirabilis* (Peck) Linder. (Courtesy, Linder: *Ann. Missouri Botan. Garden*, **16**(3):227–388.)

FIG. 198—(*Continued*)

sticky knobs on short lateral branches. (E, F) *Dactylaria brochophaga* Drechsler. (E) Conidiophore. (F) Nematode-capturing rings, the two upper ones not constricted, the lower one constricted by contact with a nematode which has escaped. (G, H) *Dactylella tylopaga* Drechsler, a fungus parasitic upon *Amoeba* sp. (G) External mycelium with conidiophores. (H) An infected animal, with an external hypha bearing three sticky knobs. (A–F, courtesy, Drechsler: *Mycologia*, **29**(4):447–552. G–H, *ibid.*, **27**(2):216–223.)

whole mycelium become transformed into chains of individual spherical, dark-colored conidia. Some of the species form the mildews that are destructive to cloth, paper, etc. *Periconia* with its head of dark rounded or oval spores at the top of a tall conidiophore reminds one of the Moniliaceous genera centered around *Oedocephalum.* Other dark, round-spored genera are *Hadrotrichum* and *Nigrospora.* (Fig. 200A.) In both of these the conidiophore is of moderate length and dark-colored. In the former the conidium narrows at the point of attachment. In *Nigrospora* there is a sort of vesicle, rather light in color, at the apex of the conidiophore, and on this lies the very dark-colored spore. Apparently some of the species are parasitic, mainly upon Monocotyledoneae. *N. oryzae* (B. & Br.) Petch (*Basisporium gallarum* Moll.) causes injury to the nodes of the stalks and to the ears and especially to the pointed bases of the grains of *Zea mays* L. The genus *Zygosporium* is peculiar in the manner of bearing the hyaline or almost hyaline, spherical conidia. These are produced singly at the tips of two or three hyaline phialides produced near the apex, on the convex surface of a peculiar dark-colored, curved, and often pointed structure called by Mason (1941) a falx. This was interpreted by Giesenhagen (1892) as a basidium and the fungus placed under the name *Urobasidium* in the Exobasidiaceae, but Mason's studies show that this interpretation was erroneous. (Fig. 200B–C.)

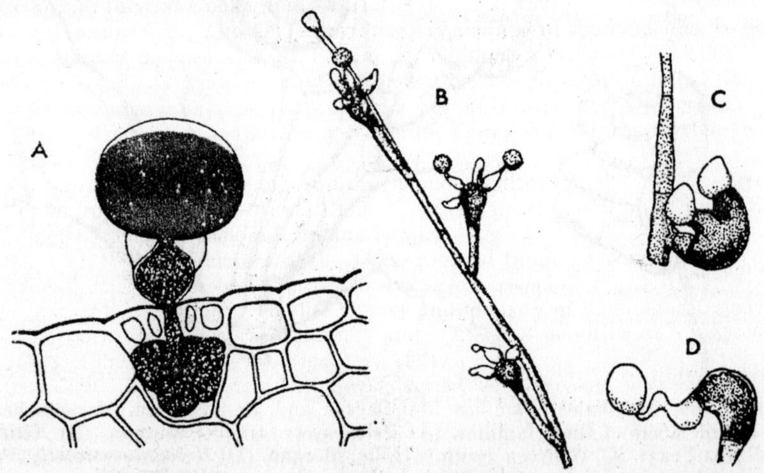

FIG. 200. Moniliales. Some peculiar fungi. Family Dematiaceae. (A) *Nigrospora panici* Zimmermann (*Basisporium* of some authors). (B) *Zygosporium echinosporum* Bunting & Mason, showing the characteristic falces each bearing three phialides with spores. (C) *Zygosporium oscheoides* Mont. with two phialides on each falx. (D) Lateral view of a falx showing one phialide with spore. (A, after Zimmermann: *Centr. Bakt. Parasitenk.*, Zweite Abt., 8(7):216–221. B–D, courtesy, Mason: Annotated account of fungi received at the Imperial Mycological Institute, List II, Fascicle 3 (Special Part), pp. 134–144.)

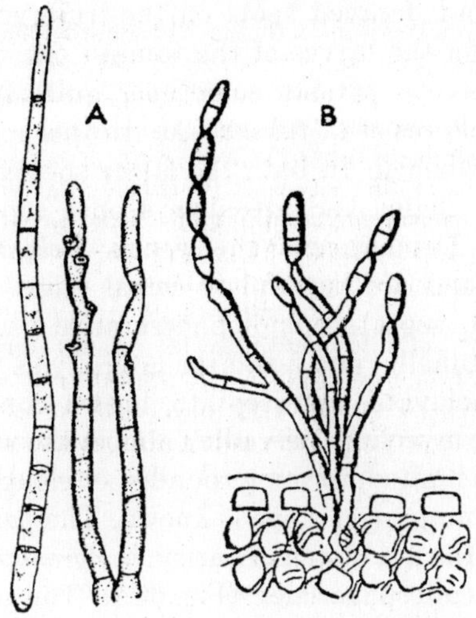

FIG. 201. Moniliales, Family Dematiaceae. (A) *Cercospora zeae-maydis* Tehon & Daniels. (B) *Cladosporium fulvum* Cke. (A, after Tehon and Daniels: *Mycologia*, 17(6):240–249. B, after Prillieux and Delacroix, *Bull. soc. mycol. France*, 7(1):19–21.)

Polythrincium (Phaeodidymae) produces its dark-colored wavy conidiophores in tufts emerging through the epidermis of the host leaf. They bear apically the colored two-celled conidia. *P. trifolii* Schm. & Kze, is parasitic on the leaves of clover (*Trifolium*). Wolf (1935) has proved that the perfect stage of this fungus is *Cymadothea trifolii* (Pers.) Wolf, so that really this species should no longer be given consideration under the Fungi Imperfecti but under Order Dothideales in the Ascomyceteae. The genera *Hormodendron* and *Cladosporium* although maintained separately in the reference books are scarcely worthy of distinction. In both of them the conidiophores are colored, septate, and variously branched. Near the tip the branches are more numerous and bear acrogenously produced chains, often branching also, of conidia. In *Hormodendron* these are one-celled, in *Cladosporium* the younger spores are one-celled and those further down in the chain may become two-celled and sometimes three- to four-celled. (Fig. 201B.) Some species of the former represent the stage in culture on artificial media of certain of the pathogens of Man that cause a serious disease of the skin known as dermatitis verrucosa (see Emmons and Carrión, 1937). The genus *Cladosporium* with over 160 described species is an assemblage of species representing many different and probably not properly congeneric types. *C. herbarum* Link ex Fr. is found the world over on dead organic material and apparently occurs sometimes as a plant parasite. It probably does not represent a single species but a host of closely similar and very variable species. The perfect stage of one form has been described as *Mycosphaerella tulasnei* Jancz., parasitic on various cereals. Another very dissimilar parasitic species is *C. cucumerinum* Ell. & Arth., with rather short, unbranched conidiophores, causing spots on leaves and decayed spots on the fruits of the cucumber, *Cucumis sativus* L. Upon the leaves of the tomato one often finds brown- or violet-colored, velvety patches sometimes quite large. These are the conidia and conidiophores of *C. fulvum* Cke. In this species the conidiophores are but little branched. In both the latter species the conidia rarely if at all appear in chains and are always once septate.

Among the phragmosporous Dematiaceae the genus *Helminthosporium* includes several species that are harmful to cereal crops. The upright, unbranched or branched, septate conidiophores arise usually several from a stoma and bear terminally, and less often laterally as well, the more or less cylindrical or obclavate, multiseptate, brown conidia. The infection may be systemic, the mycelium pervading almost the whole plant, or the infected areas may be limited, forming rounded or elongated leaf spots. For most of the species no perfect stage is known. The perfect stage of *H. teres* Sacc., the cause of the net blotch of barley, is *Pyrenophora teres* (Died.) Drechs., one of the Pseudosphaeriales. (Fig. 96A.) The genus *Heterosporium* has shorter conidiophores and ellipsoid, roughened spores, usually single at the apex of the conidiophore but sometimes in short chains. The perfect stage of *H. gracile* (Wallr.) Sacc., cause of leaf spots and killing of leaves of *Iris*, is *Didymellina iridis* (Desm.) v. Höhn., also of the same order as the foregoing. The perfect stages of the remaining 35 or 40 species of *Heterosporium* are unknown. There are many other genera of the phragmosporous Dematiaceae, e.g., *Dendryphium*, with dendroidally branched conidiophores and the conidia single or in short chains, *Napicladium*, with short, weak conidiophores and large terminal conidia, some species parasitic and others not.

The dictyosporous genera of this family are largely saprophytes. Their conidia are septate transversely and to a certain degree longitudinally as well, and occur singly or in chains on short or long, unbranched or branched, stiff or weak conidiophores. In *Alternaria* the conidiophore is rarely branched and the conidia are produced in acrogenously developing chains or singly. The apical portion of each conidium is narrowed and often much elongated, bearing at its tip the next narrowly or broadly ovoid, tapering conidium. In the older literature a distinction was made between *Alternaria* and *Macrosporium* with conidia single at the apex of the conidiophore. It was shown, however, by Wiltshire (1933, 1938), Groves and Skolko (1944a and b) and by Neergaard (1945) that the latter name was first used for species with catenulate spores and only later for those with spores produced singly. Therefore they proposed that the name be dropped and *Stemphylium* be used for some species called in literature by the name *Macrosporium*. Those forms with spores tapering at the upper end should be called *Alternaria*, even when they may usually fail to produce chains. *Stemphylium* has its spores rounded at both ends

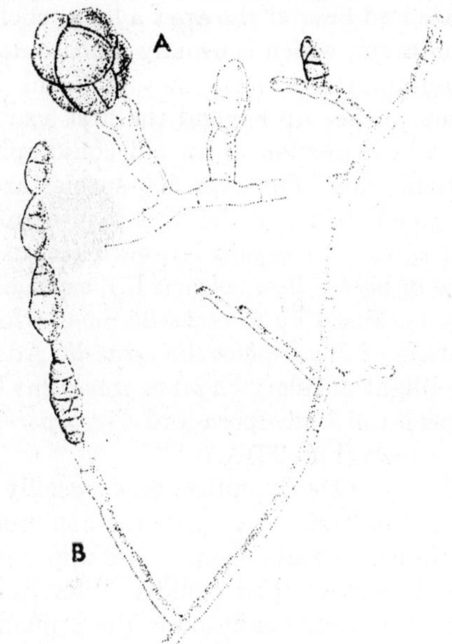

FIG. 202. Moniliales, Family Dematiaceae. (A) *Stemphylium sarcinaeforme* (Cav.) Wilts. (B) *Alternaria tenuis* Auct. (sensu Wiltshire 1933). (A, courtesy, Groves and Skolko: *Can. J. Research*, **22**(4):190–199. B, *ibid.*, **22**(5):217–234.)

and in some species has a constriction at the median transverse septum. The conidiophore may be slightly swollen at its apex. The perfect stage of *S. botryosum* Wallr. is *Pleospora herbarum* (Pers.) Rabenh., a fungus belonging to the Sphaeriales or Pseudosphaeriales and widespread the world over. It occurs, apparently, in many strains on all sorts of hosts, probably developing mostly only in the saprophytic stage of growth. The perfect stages of the other species of *Stemphylium* are not known. *S. sarcinaeforme* (Cav.) Wilts. is a frequent cause of sometimes serious leaf injury of various species of *Trifolium.* Its spores are smooth whereas those of *S. botryosum* are roughened. The genus *Sporodesmium,* with often roughened conidia occurring singly on rather short conidiophores is of interest because it sometimes is the imperfect stage of species of Hysteriales. (Fig. 202.)

Two genera of great economic importance with long, slender spores (Scolecosporae) are *Cercospora* and *Cercosporella.* These differ only in the presence of a brown pigment in the conidiophores and sometimes the spores of the former and its absence in the latter. Since, however, this pigmentation varies in degree with the age of the fungus and external conditions there is no real basis for their separation except convenience. *Cercospora* is placed in the form family Dematiaceae and *Cercosporella* in the Moniliaceae. The conidiophores emerge through the stomatal openings in tufts of varying number, from a pseudoparenchymatous mass of cells that lies beneath the stoma or sometimes projects through it a little.

They are stiff, and usually septate and bear at the apex a long, obclavate, sometimes slenderly tapering conidium, which is usually multiseptate. As in *Ramularia* as soon as the conidium is formed another forms closely below it on a short branch which pushes up beyond the first scar of conidial attachment, so that the apical portion of an old conidiophore is crooked and marked by numerous scars. The over 500 species are with few exceptions parasitic in the green tissues of the host plants on which they produce characteristic leaf spots. *Cercospora beticola* Sacc. causes a very destructive leaf spot disease of beets (*Beta vulgaris* L.), especially the sugar beet. A leaf spot of cherry is caused by *C. cerasella* Sacc. which has been shown to be the conidial stage of *Mycosphaerella cerasella* Aderh. *C. apii* Fr. is the cause of the early blight of celery (*Apium graveolens* L.). Of the many hundred described species of *Cercospora* and *Cercosporella* the perfect stage is known for only a few. (Fig. 201A.)

FAMILY STILBELLACEAE. In many of the Moniliaceae, especially in the genus *Penicillium*, under certain conditions of growth the conidiophores will be massed together into columns, coremia, from whose upper portion the conidiophores spread out and produce their conidia. What is here a response to special conditions is the normal condition in the Stilbellaceae. The coremium or synnema may be relatively short, the upper half or more being covered by the spreading tips of the conidiophores, or these may appear only at the top, the closely united hyphae producing no spore-bearing branches except at their upper ends. In *Isaria* the upright, simple or branched, colorless coremia are covered from near the base to the apex by slender hyphae bearing terminally the single, small, spherical or ellipsoidal, hyaline, one-celled conidia. Many of the species grow saprophytically on plant tissues but a few grow on insects, probably in many cases parasitically. These latter may be the conidial stages of species of *Cordyceps* of the Hypocreales. In *Graphium* the coremium is dark-colored and the spore-bearing head is only at its top. The conidia are hyaline or almost so, ovoid or ellipsoid, not in chains, the whole head being enclosed in a drop of slime. (Fig. 203C.) Such fungi are mostly insect distributed. Some species are the conidial stages of *Ophiostoma* (*Ceratostomella*) of the Ascomyceteae. Most destructive is *G. ulmi* Buis. whose perfect stage is *O. ulmi* (Buis.) Nannf., the cause of the so-called Dutch elm disease, so destructive to American elm (*Ulmus americana* L.) in America. For the majority of species the perfect stage is unknown. In the genus *Stysanus* the coremial stalk is colored and as in *Graphium* the conidia are light-colored or hyaline. They are borne in chains, covering the upper half or so of the fruiting body. Most of them are saprophytic but some are suspected of being weak plant parasites. Perfect stages are unknown. (Fig. 203B.)

FAMILY TUBERCULARIACEAE. The genera customarily assigned to this family are almost certainly not closely related. The conidia are borne on

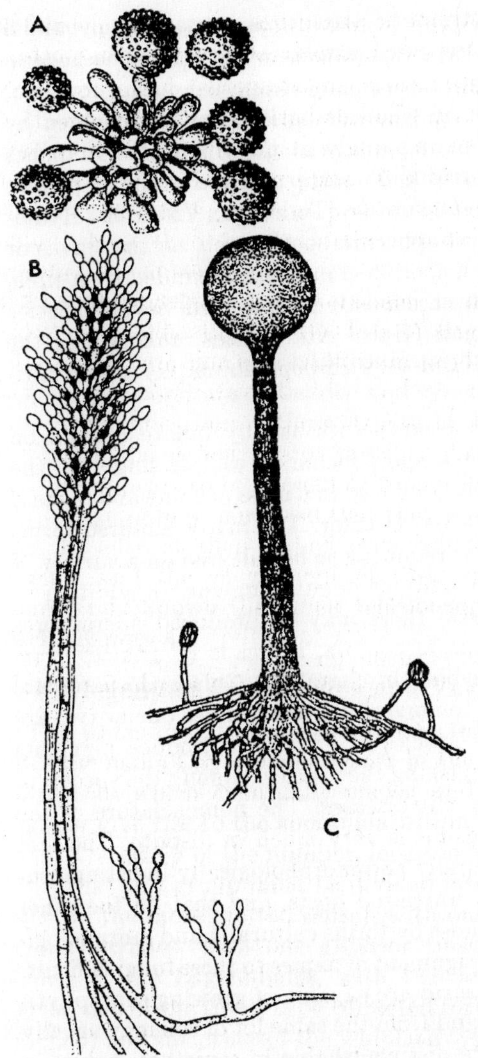

Fig. 203. Moniliales. (A) Family Tuberculariaceae. *Epicoccum nigrum* Link. (B, C) Family Stilbellaceae. (B) *Stysanus stemonites* (Pers.) Corda. (C) *Graphium rigidum* (Pers.) Sacc. (A, after Lindau, in Engler and Prantl: Die Natürlichen Pflanzenfamilien, Leipzig, W. Engelmann. B, after Hasselbring: *Botan. Gaz.*, **29**(5):312–322, Univ. Chicago Press. C, after Hedgcock: *Missouri Botan. Garden Ann. Rep.*, **17**:59–114.)

short or rarely long conidiophores arising from a cushion of fungal tissue (sporodochium). This varies in color and consistency and the conidia and conidiophores also show parallelism to the Moniliaceae and Dematiaceae. The artificiality of the group is seen in the fact that the production or nonproduction of the sporodochium often depends upon the cultural conditions. Thus the author has grown *Colletotrichum gloeosporioides* Penz. so that it produced its normal acervuli, but in older drier cultures true sporodochia appeared. Later the latter enlarged and became thrown into folds and eventually became stromatic structures containing pycnidial cavities. Montemartini (1899) reported similar observations in several groups of Imperfect Fungi. *Tubercularia* (Hyalosporae) forms rounded, bright-colored cushions, mostly on wood or bark. They are covered by fine, branching conidiophores bearing singly at the tips of the branches the small, ellipsoidal hyaline conidia. *T. vulgaris* Tode is the imperfect stage of *Nectria* (*Creonectria*) *cinnabarina* (Tode) Fr. *Volutella* (Hyalosporae) produces its small, almost spherical sporodochia on the leaves or stems of herbaceous plants. Each sporodochium is surrounded by a circle of long bristles. The conidiophores are mostly unbranched. Several species cause diseases of plants. *V. dianthi* (Hals.) Atk. causes cankers near the base of the stem of carnation (*Dianthus caryophyllinus* L.) which ultimately kill the plant.

Fusarium (Hyalophragmiae) produces its usually lunate conidia on rather broad, indefinitely bordered sporodochia as well as singly on the mycelium. The conidia are usually produced in a mass of slime and in mass may be white, yellow, orange, or red in color. The many hundred forms are distinguishable with difficulty, requiring to be cultured on a variety of culture media, under standard conditions of environment. In addition to this type of conidia (macrospores) there may be produced microspores which are rounded or short ellipsoid conidia, in heads (*Cephalosporium* type), as well as spores intermediate in character. Chlamydospores are often produced in abundance in the mycelium. Sometimes one or two are formed in a macrospore. Many species of *Fusarium* produce pigments which are of assistance in distinguishing the species. Among the species of this genus are many that cause wilt diseases. The nomenclature of the different wilt-producers in this genus is very much in dispute. The fact that on the one hand forms culturally and morphologically indistinguishable may be limited to distinct, unrelated hosts, and that on the other hand wilt may be caused in one host by forms culturally and morphologically quite different makes the assignment of names to these fungi difficult. All wilt-producing species of *Fusaria* are capable of growing saprophytically in the soil for many years, and from the same lot of soil may usually be isolated several forms that are distinguishable in structure, habits in culture, and pathogenicity. In some forms sporodochia appear only very rarely if at all in culture. It is only a very artificial classification to place the genus in the Tuberculariaceae. Wilt diseases are produced by various strains of *Fusarium* in tomato (*Lycopersicon esculentum* Mill.), flax (*Linum usitatissimum* L.), cotton (various species of *Gossypium*), watermelon (*Citrullus vulgaris* Schrad.), cowpea (*Vigna sinensis* Endl.), potato (*Solanum tuberosum* L.), celery (*Apium graveolens* L.), and many other important crops. *Fusarium*-like fungi are known as the imperfect stages of *Gibberella zeae* (Schw.) Petch, and of various other Hypocreales.

Among the dark-spored members of this form family is the genus

Epicoccum, which forms little black dots on leaves and stems and other plant parts. These consist of short, stout conidiophores radiating from small black sporodochia. The dark conidia are spherical, and slightly spiny or reticulately marked. In the latter case the spore is apparently many-celled (Goidanich, 1938), the reticulate lines marking the edges of the walls of the separate cells. *E. oryzae* Ito & Iwadara is reported by Iwadara (1934) to be the cause of injury to rice grains, producing pink or red lesions. Some of the other species are also parasitic but some seem to be saprophytes. (Fig. 203A.) *Exosporium* (Phaeophragmiae) forms similar sporodochia and very short conidiophores from which arise the obclavate or ellipsoidal conidia each with several transverse septa. Some species appear to be parasitic and some merely saprophytic *Spegazzinia* (Phaeodictyae or perhaps more properly Staurosporae) occurs on leaves, stems, and other plant debris and most species are probably saprophytic. The author (1907) showed that two kinds of conidia are produced on the small black sporodochia. The commoner kind is very long-stalked, the conidium consisting of four somewhat rounded spiny cells united at a common central point to which the conidiophore is attached. The second kind has very short conidiophores and the conidia are square, except for rounded corners, being divided into four cells by diagonal septa. They are smooth The conidiophore is attached to the edge of one of the triangular cells making up the conidium and arises directly from the sporodochium and not, as described by some authors, from the long-stalked conidia. The spiny conidia of the long-stalked type germinate by bladder-like outgrowths from which radiate many germ tubes. The smaller type of conidium gives rise to a single germ tube from each cell of the conidium

Another fungus sometimes ascribed to this family and sometimes to the Melanconiaceae is *Haincsia lythri* (Desm.) von Höhn. This is parasitic upon strawberry leaves, fruits, and roots (*Fragaria* sp.) and many other plants. It has a perfect stage, *Pezizella oenotherae* (C. & E.) Sacc This is rarely observed. Much more abundant are the two fruiting stages *Hainesia*, usually assigned to this family, and *Sclerotiopsis concava* (Dum.) Shear & Dodge, belonging to Family Sphaeropsidaceae of the Imperfect Fungi. According to Shear and Dodge (1921) the former has been described under seven generic names in four families while the pycnidial stage has been described under four genera in two families Part of this multiple naming has been due to incorrect identification but in part this is the result of the different appearances of the fungus in various media and stages of development.

Order Mycelia Sterilia. This order is madé up of 20 genera and 400 or more species. This exceedingly heterogeneous group does not at all consist of closely related species. Any nonsporiferous mycelial structure, whether sclerotium, rhizomorph, dense or loose mass of mycelium, etc is placed here. The presence of clamp connections in some species shwso that those forms belong to Class Basidiomyceteae, their absence in very many cases may or may not indicate that they belong elsewhere. The form genus *Sclerotium* includes species producing sclerotia with more or less definite form, usually light-colored internally and with a brown or black rind. For many forms only the sclerotia and associated mycelium are known, but many similar sclerotia produce apothecia (*Sclerotinia* in the Pezizales) or basidiomycetous spore fruits (*Typhula* in Family Clavariaceae). *Pachyma* is an enormous subterranean sclerotium, more or less fibrous-fleshy internally and rough or irregular externally. *P. cocos* Fr., the tuckahoe, sometimes larger than a man's head, was formerly used for food by the Indians inhabiting the regions now comprised by Virginia and adjacent states. Its perfect stage is a species of *Polyporus* which grows out of the sclerotium. *Rhizoctonia* consists of superficial, irregular, scale-like, black sclerotia growing on the surface of the subterranean portions of the host plant, these being preceded and accompanied by a superficial dark-colored, short-celled, abundantly branching, rather stout mycelium. This mycelium is entirely different in appearance from the slender, hyaline mycelium growing within the host tissues. The perfect stage of *Rhizoctonia solani* Kühn is *Pellicularia filamentosa* (Pat.) Rogers, more commonly referred to as *Corticium vagum* var. *solani* Burt ex Rolfs, or *C. solani* (Prill. & Del.) Bourd. & Galz., Family Thelephoraceae. It is a very destructive enemy of many species of cultivated plants as well as of a large number of wild plants. *Rhizomorpha* is the name given to strands of mycelium, dark externally and usually white internally. *R. subcorticalis* Pers. consists of the rhizomorphs of *Armillariella mellea* (Dahl) Karst., Family Agaricaceae. *Ozonium* consists of loose masses of usually bright-colored mycelium, frequently united into strands and again spreading out. Some species are parasitic, others saprophytic. *O. omnivorum* Shear, destructive to cotton and many other cultivated as well as wild plants in Texas and adjacent states, spreads through the soil and attacks the roots of the host plants. It sometimes produces a conidial stage, *Phymatotrichum omnivorum* (Shear) Duggar, mentioned earlier in this chapter. Once a species of *Hydnum* was found associated with it (Shear, 1925), but its connection with this fungus has not been demonstrated beyond question.

Penicillin, Streptomycin, and Other Antibiotics

Bacteriologists and students of fungi have observed for years that frequently in cultures of organisms certain contaminating species of *Penicillium*, *Aspergillus*, and other molds would destroy the adjacent portions of the colonies of bacteria or fungi. Dr. Alexander Fleming had such an experience in 1928, but instead of discarding his culture of *Staphylococcus* as spoiled and useless, he began a study of the contaminant and of the antibiotic substance that it produced. It proved to be a species of

Penicillium, later determined to be *P. notatum* Westl., and in 1929 he gave the name penicillin to the active substance. His experiments demonstrated that it would destroy many species of bacteria in cultures to which it was added, while many other kinds of bacteria were unharmed. Dr. Florey, of Oxford, about ten years later, his interest having been aroused by Dr. Fleming, tested the effect of the injection of penicillin into the blood stream of laboratory animals which had been inoculated with pathogenic organisms which had been found by Fleming to be destroyed by penicillin in culture. The result was marvelous and the animals recovered. Then it was tried on Man with promising results. The result is the widespread commercial production of penicillin on a wholesale basis, using improved strains of *P. notatum* as well as of *P. chrysogenum* Thom, in media and under conditions that produce many times more of the penicillin than in the original experiment. *Aspergillus* has been observed to produce antibiotic substances but up to the present the products have shown harmful effects when injected into the blood stream of animals. Probably some strain of this genus may be discovered eventually that will lack the harmful factor. Noting that some of the forms related to *Actinomyces*, e.g., *Streptomyces* as delimited by Waksman and Henrici (1943), also seemed promising in this regard various soil inhabiting species of this group were studied and from one of them was selected a strain which produced the substance to which the name streptomycin has been given. This also is of great value in medicine since it destroys in the body many organisms which are not affected by penicillin (see Fleming, 1946; Duemling et al., 1946; Herrell, 1945).

What the role of these antibiotic substances is in the fungi that produce them is not yet clear. That these are in the nature of waste products has been suggested. They reach their maximum production in relatively young (4 to 7 days), actively growing cultures, well supplied with oxygen. The production is greatly increased by the addition of certain organic substances which possibly stimulate the growth of the fungi. This is a matter that must be left for the researches by mycologic physiologists

Key to the Orders and Families of Fungi Imperfecti[4]

Asexual spores regularly produced.
- Spores produced in pycnidia; other types of spores may sometimes be present. Order Sphaeropsidales
  - Pycnidia complete, with or without an ostiole, sometimes representing merely pycnidial cavities in a stroma.
    - Pycnidia (or stroma) dark-colored, usually rather hard. Family Sphaeropsidaceae
    - Pycnidia (or stroma) bright-colored, usually fleshy or leathery. Family Zythiaceae

[4] Note that these groups are with few exceptions artificial and do not represent phylogenetic relationships.

Pycnidia with well-developed roof but basal portions poorly differentiated.
Family Leptostromataceae
Pycnidia with well-developed base, soon opening wide to form a cup-like or saucer-like structure.
Excipulaceae
Spores produced in an acervulus; often other spore types present.
Order Melanconiales
Single family.
Family Melanconiaceae
Spores produced neither in pycnidia nor in acervuli; but on or within free sporophores free from one another or grouped in various ways.
Order Moniliales
Conidiophores or spore-bearing hyphae more or less separate, not united into synnemata nor on sporodochia (cushion-like structures).
Conidia and conidiophores hyaline or bright-colored, not brown or black.
Family Moniliaceae
Conidiophores, and usually but not always the conidia, brown or black.
Family Dematiaceae
Conidiophores united into synnemata, the conidia arising on the spreading tips of the conidiophores in a head at the top or all up and down the sides.
Family Stilbellaceae
Conidiophores usually rather short, arising more or less radially from cushion-like sporodochia.
Family Tuberculariaceae
No asexual spore forms known.
Order Mycelia Sterilia

Keys to the More Important Genera of Fungi Imperfecti

Key to the More Important Genera of Family Sphaeropsidaceae

(Based largely upon Lindau in Engler and Prantl, 1899–1900)

Conidia one-celled, spherical, ellipsoid, or oval. Amerosporae
Conidia hyaline. Hyalosporae
Stroma lacking.
Pycnidia smooth, without appendages.
Conidia not produced in chains.
Conidia without appendages.
Pycnidia not surrounded by a subiculum; free or sunk in the substratum.
Pycnidia with regular ostiole or papillate.
Conidiophores simple or but little branched.
Pycnidia at first covered by the epidermis, later becoming free, at least around the ostiole.
Conidia less than 15 μ long.
Forming definitely margined spots on leaves.
Phyllosticta
On any part of the host, but if on leaves not forming definitely delimited spots.
Phoma
(see also *Phomopsis*)
Conidia over 15 μ long. *Macrophoma*
Pycnidia free from the first, free on wood or bark.
Aposphaeria
Conidiophores dendroidally or verticillately branched.
Dendrophoma

Pycnidia without ostiole or with irregular opening.
Conidia oblong, pointed at each end. *Sclerotiopsis*
Conidia oblong, ends rounded. *Plenodomus*
Pycnidia with long beak. *Sphaeronema*[5]
Pycnidia in a radiating subiculum. *Asteroma*
Conidia with several early disappearing apical appendages. *Neottiospora*
Conidia produced in chains. *Sirococcus*
Pycnidia with appendages or hairs.
Appendages unbranched.
Bristles all over the pycnidium; spores ellipsoid, curved. *Vermicularia*
Bristles mostly around the ostiole; spores straight. *Pyrenochaeta*
Appendages stellate or stellately branched at the tip. *Staurochaeta*
Pycnidia (rarely) single or several in a stroma or frequently represented only by pycnidial cavities in the stroma.
Pycnidium mostly single, often without definite, distinct stroma. Conidia of two types, oval on long conidiophores and long slender stylospores, usually curved like a hook at one end, on short conidiophores. *Phomopsis*
Pycnidial cavities several in a stroma, with separate ostioles.
Conidia large, more or less fusiform, straight. *Fusicoccum*
Conidia cylindrical or ovoid, straight. *Dothiorella*
Conidia small, allantoid. *Cytospora*
Pycnidial cavities often with a common ostiole, conidia straight. *Ceuthospora*
Conidia colored. Phaeosporae
No stroma, rarely with a subiculum.
Pycnidia glabrous, with regular ostiole.
Conidia oval or elongated, strikingly large. *Sphaeropsis*
Conidia spherical or ellipsoidal, very small. *Coniothyrium*
Pycnidia glabrous, with pronounced beak. *Naemosphaeria*
Pycnidia glabrous, with lobed, irregular mouth. *Harknessia*
Pycnidia with external hairs or bristles. *Chaetomella*
Stroma present; pycnidia superficial or at first in the stroma. *Haplosporella*
Conidia two-celled. Didymosporae
Conidia hyaline, pale greenish in some species. Hyalodidymae
Pycnidia free, without subiculum or stroma.
Pycnidia glabrous.
Pycnidia without beak.
Pycnidia in definite spots on leaves or stems.
Spores without appendages. *Ascochyta*

---

[5] Species of *Ophiostoma* with dissolved asci may be confused with this.

Spores with apical bristles. *Robillardia*

Pycnidia not in definite spots.

Spores without appendages; parasites or saprophytes on higher plants. *Diplodina*

Spores with slimy or hair-like appendages at each end; parasitic upon Uredinales. *Darluca*

Pycnidia with a beak. *Rhynchophoma*

Pycnidia with hairs or bristles. *Vermiculariella*

Pycnidia embedded in a radiate subiculum. *Actinonema*

Conidia colored. Phaeodidymae

Pycnidia free from one another, without stroma.

Pycnidia subepidermal or subcortical, then emerging.

Pycnidia glabrous.

Spores without outer slime layer. *Diplodia*[6]

Spores with external slime coat. *Macrodiplodia*[6]

Pycnidia with hairs or bristles. *Chaetodiplodia*[6]

Pycnidia from the beginning superficial, on wood. *Diplodiella*[6]

Pycnidia crowded, with stroma. *Botryodiplodia*[6]

Conidia 3- or more-celled by transverse septa only. Phragmosporae

Conidia hyaline. Hyalophragmiae

Conidia without appendages.

Pycnidia more or less globose or depressed globose, mostly remaining covered, except the ostiolar area (practically a *Hendersonia* with hyaline spores). *Stagonospora*

Pycnidia vertically elongated, emerging in groups. *Mastomyces*

Conidia with a bristle-like appendage at the apex. *Kellermania*

Conidia colored. Phaeophragmiae

Without stroma (sometimes a stromatic crust in *Dilophospora*).

Conidia without appendages, the terminal cells of the conidia hyaline in some species. *Hendersonia*

Conidia with appendages.

Appendages a single fine thread at each end of the spore; no stroma. *Cryptostictis*

Appendages a tuft of several, fine, branched threads at each end; pycnidia sometimes with a stromatic crust; parasitic on grass leaves. *Dilophospora*

Pycnidia embedded in a stroma; conidia not appendaged. *Hendersonula*

Conidia several-celled by transverse and some longitudinal septa; colored. Phaeodictyae

Without stroma; breaking out through the bark. *Camarosporium*

Without stroma; superficial on wood. *Cytosporium*

With pulvinate stroma. *Dichomera*

---

[6] These distinctions are very artificial, for under different cultural conditions any of these may develop into almost any other form. Also when younger they may represent *Diplodina* or *Sphaeropsis* stages.

Conidia slender, many times as long as thick, more often hyaline or only slightly colored, with or without cross septa. Scolecosporae

Without stroma.

Pycnidia tapering upward to a point; carbonaceous.

Conidia filiform, one-celled. *Sphaerographium*

Conidia several-celled, constricted at each septum. *Cornularia*

Pycnidia globose to depressed globose, ostiole at most with a small papilla; carbonaceous.

Pycnidia opening by a rather small round ostiole.

Pycnidia more or less sunken in the host tissue, emergent somewhat at maturity.

Pycnidia glabrous; parasitic in leaves or herbaceous stems.

Conidia hyaline or at most light-colored. *Septoria*

Conidia distinctly brown. *Phaeoseptoria*

Pycnidia glabrous; growing on wood or bark. *Rhabdospora*

Pycnidia hairy. *Trichoseptoria*

Pycnidia superficial; conidia straight, cylindrical, or narrowly fusoid. *Collonema*

Pycnidia opening by a wide mouth; not completely developed. *Phleospora*

Pycnidia opening by a narrow slit; not completely developed. *Phlyctaena*

Pycnidia globose-conical; tough or leathery.

Pycnidia separate. *Micula*

Pycnidia crowded in heaps. *Micropera*

With stroma.

Conidia hyaline, filiform, curved, one-celled. *Cytosporina*

Conidia dark, several-celled. *Septosporiella*

Key to the Important Genera of Family Zythiaceae

Stroma lacking; pycnidia globose, with small ostiole, conidia ovoid or oblong, resembling the perithecia of *Nectria*. *Zythia*

With a cushion-like stroma containing several pycnidial cavities; superficial. parasitic in some cases on leaf-sucking insects. *Aschersonia*

Key to the More Important Genera of Family Leptostromataceae

(Including the Pycnothyriaceae and Rhizothyriaceae, Tehon, 1940)

Conidia arising from the roof of the external pycnidium. (Order Pycnothyriales, Tehon, 1940)

Pycnidia ("pycnothyria") radial in structure, often connected with an external mycelium or subiculum. (Family Pycnothyriaceae, von Höhnel, 1910)[7]

---

[7] Tehon recognizes 15 genera, with probably not over 40 described species, mostly tropical or subtropical. The following are included in Leptostromataceae by Lindau, in Engler and Prantl: (Hyalosporae) ***Eriothyrium***, ***Trichopeltulum;*** (Phaeosporae)

Pycnidia ("pycnothyria") radial in structure, mounted on columellae whose bases are connected with the internal mycelium.
(Family Rhizothyriaceae, Tehon, 1940)[8]

Conidia arising from the floor of the dimidiate subcuticular pycnidium.
(Family Leptostromataceae)

Conidia one-celled, hyaline.
 Stroma lacking.
  Pycnidia shield-shaped, opening by a pore or short slit. *Leptothyrium*
  Pycnidia irregular in shape. *Piggotia*
  Pycnidia mostly oblong; opening by a long slit. *Leptostroma*
 Stroma present (probably the conidial stage of *Rhytisma*). *Melasmia*
Conidia phragmosporous, hyaline, with a hair at each end. *Discosia*
Conidia hyaline, cruciform, with two large cells and usually two lateral cells, all but the basal cell with a slender hair (conidial stage of *Diplocarpon*). *Entomosporium*
Conidia muriform, hyaline. *Dictyosporium*
Conidia filiform, hyaline or nearly so. *Leptostromella*

Key to the More Important Genera of Family Excipulaceae[9]

Conidia globose, ellipsoid, oblong subcylindrical or fusoid, hyaline or only faintly colored.
 Pycnidia without hairs or bristles.
  Pycnidia eventually discoid; wall composed, at least in part, of modified host tissue. *Discula*
  Pycnidia eventually cup-shaped; walls of typical fungus tissue.
   Pycnidia long buried, opening by laciniae. *Sporonema*
   Pycnidia erumpent, opening by round pore. *Excipula*
   Pycnidia erumpent, opening by a wide torn margin. *Dothichiza*
 Pycnidia with setae.
  Conidia without appendages. *Amerosporium*
  Conidia with a delicate bristle at each end. *Dinemasporium*
Conidia once septate; pycnidia opening by laciniae. *Discella*
Conidia oblong-fusoid, pluriseptate.
 Conidia hyaline; not prolonged into a subulate beak, pycnidia without bristles. *Excipulina*
 Conidia colored; pycnidia with bristles. *Excipularia*
 Conidia hyaline; prolonged into a subulate beak; pycnidia without bristles. *Heteropatella*

Asterostromella; (Hyalodidymae) *Leptothyriella;* (Phaeodidymae) *Diplopeltis.* Tehon recognizes ten other genera.

[8] Tehon recognizes 6 genera of which Lindau, in Engler and Prantl includes in the Leptostromataceae the following: (Hyalosporae) *Actinothecium;* (Phaeosporae) *Pirostoma;* (Scolecosporae) *Actinothyrium.* Besides these Tehon includes *Pirostomella, Rhizothyrium,* and *Cylindrothyrium.*

[9] In some cases an old acervulus of Family Melanconiaceae may be mistaken for the saucer-shaped pycnidium of the Excipulaceae.

Key to the More Important Genera of Family Melanconiaceae

Conidia nonseptate, hyaline, globose to ellipsoid to cylindrical, sometimes fusoid or allantoid.

Acervulus immersed, but early erumpent, bright-colored, somewhat gelatinous; some species, at least, parasitic upon Uredinales. *Hainesia*

Acervulus immersed but remaining covered by the host epidermis for a considerable time; pale or eventually dark-colored; spores surrounded by slime; mostly parasitic or saprophytic on herbaceous parts of higher plants, not on Uredinales. Setae wanting. *Gloeosporium*

Much like *Gloeosporium* but the very short conidiophores arising from a firm base, almost stromatic, or even resembling a small sporodochium. Perfect stage *Elsinoë*. *Sphaceloma*

Similar to *Gloeosporium* but with pale or colored setae around or in the acervulus, their presence often depending upon the substratum. *Colletotrichum*[10]

Similar to *Gloeosporium* but growing only on branches of woody plants.

Conidia straight or almost so. *Myxosporium*

Conidia allantoid. *Naemospora*

Conidia nonseptate, globose to ovoid, etc., dark-colored.

Conidia more or less globose; mostly saprophytic on twigs of woody plants. *Melanconium*

Conidia fusiform, often curved; mostly on herbaceous plants. *Cryptomela*

Conidia hyaline, with one transverse septum.

Parasitic on leaves; conidia more or less oblong. *Marssonina*

Usually on twigs (rarely on leaves), conidia oblong to fusoid. *Septomyxa*

(*Rhynchosporium*, with conidia strongly beaked, and growing mostly on grasses is sometimes placed in this group.)

Conidia colored, with one transverse septum. Mostly saprophytic on twigs. *Didymosporium*

Conidia hyaline, with two or more transverse septa; on leaves or twigs. *Septogloeum*

Conidia colored, with two or more transverse septa.

Conidia not beaked, emerging in a black drop or cirrhus. *Stilbospora*

Conidia not beaked, not emerging in a drop or cirrhus.

Basal cell of conidium usually colorless; the elongated conidiophore persistent on the conidium; walls of the cells thin and collapsing slightly on drying. *Coryneopsis*

Basal cell of conidium colored; walls of the cells of conidium usually much thickened; conidiophores usually shorter than the spore. *Coryneum*

Conidia with colorless upper cell forming a sort of beak.

Conidia three to five septate, arcuate, only the two middle cells colored. *Toxosporium*

---

[10] Some species may be confused with *Vermicularia* or with *Volutella*.

Conidia five to eleven septate, straight or only slightly curved, all the cells colored except the basal cell and the one- to three-celled beak. *Scolecosporium*

Conidia with filiform appendages.

Appendages only from the apical cell.

Appendage single. *Monochaetia*

Appendages several. *Pestalotia*[11]

Appendages from both apical and basal cells.

Appendages single. *Hyaloceras*

Appendages two. *Diploceras*

Conidia colored, with transverse and some longitudinal septa. *Steganosporium*

Conidia colorless, filiform or rod-shaped, many times as long as thick.

Parasitic in leaves or fruits. *Cylindrosporium*

Saprophytic (or sometimes parasitic?) on twigs.

Conidia filiform, narrow, more or less curved; spore mass yellow or reddish *Libertella*

Conidia cylindric or fusoid, broader; spore mass whitish. *Cryptosporium*

Keys to Special Groups of Order Moniliales; Endosporeae[12]

Endoconidiophores forming a layer on the surface of a sporodochium; endoconidia globose, colored, embedded in a slimy layer (Tuberculariaceae). *Hymenella*

Endoconidiophores not borne on a sporodochium, endoconidia one-celled.

Endoconidia one-celled, hyaline, pushed out in a chain from the spreading opened apex of the conidiophore; "macroconidia" in chains, colored. Parasites in roots. *Thielaviopsis*

Endoconidia as in the foregoing, but the apex of the conidiophore not flaring: "macroconidia" single, sessile or short-stalked. Parasites in roots and upper portions of plants. *Chalaropsis*

Endoconidia hyaline, produced in chains in the more or less flask-shaped dark-walled conidiophores which do not flare at the top. Parasites and saprophytes. *Chalara*

Endoconidia mostly becoming brown-walled with age; pushed out successively into the flaring mouth of the conidiophore and gathering there into a slimy ball. Saprophytic in woody material, or parasitic on leaves, one species pathogenic to Man, causing chromoblastomycosis. The saprophytic species are mostly called *Cadophora*. *Phialophora*

Endoconidiophores with flaring collars, producing a short chain of colored dry conidia, and then proliferating so that a chain of conidiophores is formed, each with its flaring collar. *Catenularia*

---

[11] More often written by the later name *Pestalozzia*.

[12] In this group of probably not all closely related fungi some of the conidia are produced in tubular conidiophores from whose apical end the spores are pushed out successively. These special conidiophores are perhaps to be considered as especially modified phialides. In addition to the endoconidia other usually larger conidia of the more usual types ("macroconidia") may occur. By the system of classification ordinarily used for Order Moniliales these genera would be distributed among three different families: Moniliaceae, Dematiaceae, Tuberculariaceae, and perhaps, under certain conditions of growth, Stilbellaceae.

Endoconidiophores not on a sporodochium; endoconidia colored, cylindrical, several-celled, pushed out of the not flaring opening. *Sporoschisma*

The following genera with endosporous conidia have only one or two species each and are not well understood: *Sporendonema, Malbranchea, Glycophila.* The spores produced on the phialides of *Cephalosporium, Gliocladium,* and some other genera have been shown by Miss Pinkerton (1936) to have an endogenous origin.

Key to the Helicosporous Genera of Order Moniliales[13]

(Based upon Linder, 1929, 1931a, 1931b)

Conidiophores forming a loose arachnoid, cottony or velvety colony or else apparently obsolete; not forming a compact fruiting body (Moniliaceae and Dematiaceae).

Conidia coiled in three planes to form a cylindrical or barrel-shaped spore body.

Conidia in chains. *Helicodendron*

Conidia not in chains. *Helicoön*

Conidia coiled in two planes (sometimes but slightly inclined to coiling) or if in three planes not as above.

Parasitic on vascular plants; conidiophores obsolete or as swellings of the vegetative hyphae; spores toruloid, often nearly straight. *Helicoceras* (*Gyroceras*)

Saprophytic or some doubtfully parasitic on other fungi; conidiophores present, in some not conspicuous, but then the spores not toruloid.

Conidial filaments thick in proportion to their length, not hygroscopic.

Conidia in chains. *Helicodendron*

Conidia not in chains.

Conidia longitudinally and transversely septate. *Xenosporella*

Conidia only transversely septate. *Helicoma*

Conidial filaments thin in proportion to their length, hygroscopic.

Conidiophores and conidia hyaline, the conidiophores as teeth on, or short erect branches from, the creeping vegetative mycelium. *Helicomyces*

Conidiophores or conidia, or both, fuscous in shade; conidiophores conspicuous. *Helicosporium*

Conidiophores aggregated to form a stele upon which the spores are borne acrogenously (Stilbellaceae). *Helicostilbe*

Conidiophores aggregated to form a flattened pulvinate or irregularly globose sporodochium (Tuberculariaceae).

Sporodochia effuse-pulvinate, at first covered by the epidermis of the host, then erumpent; conidia once coiled, with thick hyaline walls. *Drepanoconis*

Sporodochia pulvinate to irregularly globose, dry, horny, or gelatinous; conidia without a conspicuously thickened wall.

Conidia coiled in three planes to form a conical or oblong-ellipsoidal spore body. *Troposporium*

---

[13] The genera here assembled in one key represent members that are usually distributed in the form families Moniliaceae, Dematiaceae, Tuberculariaceae, and Stilbellaceae.

Conidia not coiled in three planes, or if so then the filaments irregularly twisted and contorted.

Conidia once coiled, one to three septate; fructifications gelatinous. *Delortia*

Conidia not one to three septate; or if so, then sporodochia not gelatinous.

Conidial filaments 7 μ or more in width; conidia coiled in three planes, twisted and contorted. *Hobsonia*

Conidial filaments less than 7 μ in width; conidia not coiled in three planes.

Conidiophores slender, even; fructifications horny when dry. *Everhartia*

Conidiophores moniliform; fructifications not as above. *Troposporella*

Key to the More Important Amerosporous Genera of Family Moniliaceae

(Based chiefly on Lindau, in Engler and Prantl, 1899–1900)

No conspicuous conidiophores, the globose conidia forming a pulverulent layer on the surface of the scanty mycelium; saprophytes. *Chromosporium*

Conidia produced in chains by the basipetally progressing segmentation of the vegetative hyphae or conidiophores.

Mycelium very slender ($< 2\ \mu$), mostly nonseptate, producing aerial, more often coiled conidiophores breaking up into chains of cylindrical or ellipsoidal conidia. Sometimes still included in the genus *Actinomyces*. Soil organisms, parasitic or saprophytic on plant material. *Streptomyces*

Mycelium larger than the foregoing, septate.

Conidia ovoid or globose.

Mycelium forming spreading or cushion-like masses on the substratum. On plant or animal debris, occasionally parasitic. *Oospora*

Mycelium extensive in the substratum, producing tufts of external hyphae which break up into chains of moderately large to large conidia. Parasitic or saprophytic on plant material. Some species are the imperfect stage of Sclerotiniaceae. *Monilia*

Mycelium mostly external, with haustoria penetrating the epidermal cells of the host, and forming upright chains of basigenous conidia. Asexual stage of various Erysiphaceae. *Oidium*[14]

Conidia fusiform. *Fusidium*

Conidiophores simple or only slightly branched; conidia not catenulate, grouped in heads.

Conidiophores mostly stout, swollen at the top into a distinct sphere or disk, with radiating globose or fusiform conidia. Some, perhaps all, of the genera falling into this category really belong to the Mucorales.

---

[14] Commonly, but erroneously so called (see Linder, 1942).

Conidiophores straight, unbranched.
Conidiophores ending in a globose head.
Surface of the head not, or only slightly, areolate. *Oedocephalum*
Surface of the head plainly hexagonally areolate. *Rhopalomyces*
Conidiophores ending in a lobed disk. *Kickxella* (sometimes called *Coronella*)
Conidiophores repeatedly curved, in S-form; conidial heads lateral. *Sigmoideomyces*
Conidiophores slender, not, or only slightly, enlarged at the apex.
Conidia issuing one at a time from the apex of the conidiophore and remaining embedded in a slime drop. Conidiophores short and straight from the mycelium. Saprophytic on plant remains, and some species found in skin lesions in Man. *Cephalosporium*
Conidiophores branched; mycelium forming cushions on the surface of the substratum; conidia becoming green. Saprophytic on plant material. *Trichoderma*
The genus *Coemansiella* placed here by Lindau belongs in *Kickxella* in the Mucorales.
Conidia in basigenous chains, grouped in heads.
Conidiophores enlarged at the apex, covered by the sterigmata (phialides) which bear at the apex the chains of conidia: sometimes primary and secondary sterigmata present.[15] *Aspergillus*
The genus *Dispira* (including *Dimargaris*) of the Mucorales might be keyed here.
Conidiophore not markedly enlarged at the apex, but branching in regular or irregular whorls of branches which terminate in two or more sterigmata.
Conidia and upper portion of the "penicillus" embedded in a drop of slime. *Gliocladium*
Slime drop none or small; conidia thin-walled, without ring or collar at base.
Sterigmata irregularly produced, partly variously arranged on fertile branches, partly in verticils; mostly tapering to long slender points which are commonly curved or bent from the main axis of the sterigmata; conidial areas never green. *Paecilomyces*
Sterigmata in characteristic verticils with the conidium-producing points or tubes straight; conidial areas commonly some shade of green, blue-green or gray-green during their growing period; white, yellow, or reddish forms occasional, but few. *Penicillium*[16]
Slime drop lacking, conidia rather thick-walled, with a thickened ring or collar at the truncate base. *Scopulariopsis*
Conidiophores unbranched, bearing a tuft of conidial chains without sterigmata. *Briarea*

---

[15] The species with two series of sterigmata have been called *Sterigmatocystis*.

[16] The genus *Citromyces* represents only citric acid producing species of this genus and is not morphologically distinguishable from *Penicillium*.

Conidiophores more or less richly branched, rarely simple.
Conidia arising from especially differentiated intercalary cells of conidiophore.
Conidia produced singly.
Fertile cells of the conidiophore as well as the intervening sterile cells cylindrical. *Gonatobotrys*
Fertile cells spherical, intervening cells bone-shaped. *Nematogonium*
Conidia produced in chains on the globose intercalary or terminal fertile cells. *Gonatorrhodiella*
Conidia not arising from especially differentiated intercalary cells.
Branching of the conidiophore very various but never purely verticillate.
Conidia typically pleurogenous on the conidiophore, never terminal.
Conidiophore forked two to several times. *Haplaria*
Conidiophores unbranched; conidia globose or ellipsoid. *Acladium*
Conidia pleurogenous and acrogenous.
Conidiophores typically unbranched.
Parasitic within leaf tissues, clusters of conidiophores emerging from the stomata, bearing at their tips several (usually 6) conidia. *Microstroma*
Mycelium creeping, producing numerous upright unbranched conidiophores each with a single hyaline or brightly colored conidium. *Acremonium*
Conidiophores with short teeth along the enlarged or not enlarged upper end; conidia single or in chains from these teeth, large in proportion to the diameter of the conidiophore, globose to ellipsoid. Mainly saprophytic (including *Olpitrichum* and *Physospora*). *Oidium* (syn., *Rhinotrichum*)
Resembling the foregoing but parasitic on leaves; conidia hyaline, globose or oval, single, rarely in short chains. *Ovularia*
Conidiophores almost always branched.
Mycelium creeping, conidiophores not upright. *Sporotrichum*
Conidiophores always upright.
Conidia single at the tips of the branches.
Conidiophores dendroidally branched; conidia single at the tips of the branches, hyaline or bright-colored, oval or globose. *Monosporium*
Conidia clustered near the ends of the usually branched conidiophores.
Branches of the conidiophores slender, almost uniform. Conidia forming loose groups at the tips. *Botrytis*
Mostly thicker than the foregoing; the conidia on small sterigmata; sclerotium formation frequent. *Polyactis*[17]
The tips of the rather slender branches somewhat enlarged and bearing the conidia on distinct sterigmata. *Phymatotrichum*[17]

[17] Frequently included in *Botrytis*.

Conidia borne along the upper or the lower side of lateral, somewhat curved, several-celled branches. *Martensella* and *Coemansia* (Mucorales)

Branching of the conidiophores verticillate.

Conidia in chains. *Spicaria*

Conidia not in chains.

Conidiophores verticillately branched, more often in threes and twos; conidia hyaline or light-colored, globose or oval, soon falling off.

Ultimate branches straight, conidia mostly terminal. *Verticillium*

Ultimate branches subulate, zigzag toward the tip, bearing a conidium at each angle, this portion with the adhering conidia resembling the rachis of a head of wheat with attached spikelets. *Tritirachium* (Limber, 1940)

Like *Verticillium*, but the conidia cylindric. *Acrocylindrium*

Key to the More Important Genera of the Hyalodidymous Moniliaceae

Conidia not produced in chains.

Conidia smooth.

Conidiophores unbranched; conidia oval or pyriform, the upper cell often larger than the basal cell; hyaline or bright-colored; mostly saprophytes.

Conidium single, at the apex of the conidiophore. *Trichothecium*

Conidia forming a head at the apex of the conidiophore. *Cephalothecium*

Conidia clustered on somewhat enlarged, intercalary swellings on the conidiophore. *Arthrobotrys*

Conidiophores always branched; mostly saprophytes.

Conidiophores verticillately branched; growing saprophytically on fungi. *Diplocladium*

Branching of conidiophores irregular, conidia single at the branch tips. *Diplosporium*

Conidiophores mostly unbranched, parasitic on leaves.

Conidiophores straight. *Didymaria*

Conidiophores spirally curved. *Bostrichonema*

Conidia warty, parasitic on fleshy fungi. Upper cell of conidium larger than the lower one. *Mycogone*

Conidia in chains.

Conidiophore branches verticillate. On decaying fungi. *Didymocladium*

Key to the More Important Genera of the Phragmosporous Moniliaceae

Conidiophores not much developed; conidia borne singly, curved, fusoid; parasites. *Fusoma*

Conidiophores well developed.

Conidiophores unbranched; not parasitic on vascular plants.

Conidia single on conidiophores. *Dactylella*

Conidia several in a head on the conidiophore. *Dactylaria*

Conidiophores unbranched, parasitic on vascular plants.

Conidia cylindric-oval, single or in short chains, growing out of the stomata of the host plants. *Ramularia*

Conidia pyriform; single; mostly parasitic on grasses, sometimes saprophytic and aquatic. *Piricularia*

Conidiophores simple or slightly branched, bearing the conidia in chains. Saprophytic or some species parasitic on vascular plants. *Septocylindrium*

Key to the Scolecosporous Moniliaceae

Parasitic on vascular plants, mostly producing leaf spots.

Only genus. *Cercosporella*

Key to the More Important Staurosporous Moniliales

Conidia hyaline, borne successively on definite hyaline phialides.

Conidia when mature consisting of four slender, ultimately septate, widely diverging arms, attached to the phialide at their point of divergence; conidiophore long and slender, with two to eight phialides near the apex; aquatic. *Lemonniera*

Conidia consisting of a main curved axis attached at one end to the phialide and with two, nearly opposite, lateral, nonseptate branches produced at the high point of the curve; conidiophores long, with one to four phialides. Aquatic. *Alatospora*

Conidia consisting of a clavate, main axis which may be one to three septate or nonseptate, bearing at the broader, upper end three slender, nonseptate arms or only short divergent processes (the spore then resembling a clove); conidiophore long, unbranched or branched, bearing at the apex one to four phialides. Mostly aquatic; the divergent arms sometimes lacking on conidia produced in the air; sporodochia sometimes produced. *Heliscus*

Conidia (aleuriospores or radulaspores) not borne on phialides.

Conidia and conidiophores hyaline.

Aquatic saprophytes, conidia produced submersed.[18]

Conidia with a main cylindrical, or more often clavate to pyriform, axis at or near whose upper end usually three divergent branches arise successively.

Main axis of conidium narrowly pyriform or broadly clavate, once septate; on a long slender conidiophore; the three branches from the upper cell diverging at angles of about 120° and at about right angles to the main axis of the spore. In cultures pycnidia sometimes formed. *Clavariopsis*

Main axis of conidium narrowly clavate, eventually once or twice septate, giving rise to three unequal, divergent tapering branches with a knob or thick finger-like process on the upper side, near the base, of the first two branches produced. *Tetracladium*

Main axis of the conidium and the three somewhat longer branches about equal in thickness, nearly cylindrical, one to three septate at maturity. *Articulospora*

Conidia with a straight or curved, septate, main axis from which arise laterally, near together or from separate cells, the strongly divergent septate branches which may, in their turn, bear lateral branches.

[18] For details of these aquatic genera, see Ingold (1942, 1943, 1944).

Main axis of conidium slender, with two lateral branches diverging at the same level from the convex side of the curved axis, the two branches and the lower and upper portion of the axis forming a four-armed spore; a short "separating cell" present at the upper end of the conidiophore. *Tetrachaetum*

Main axis of the conidium stout, septate, curved, the usually two eventually septate branches arising from adjacent cells of the axis or at opposite ends of the same long cell. *Tricladium*

Main axis of the conidium septate, curved, bearing from separate cells on the convex side several branches approximately in the same plane; these may also produce lateral branches, the whole spore resembling a character of Chinese writing; conidiophore slender, bearing one to several conidia on its distal quarter. *Varicosporium*

Main axis of conidium long and slender, much septate, straight, bearing from the cells of the lower half one or more septate, lateral branches, which in similar manner bear shorter lateral branches; main axis of spore a continuation of the long, slender, septate conidiophore. *Dendrospora*

Not all aquatic, producing the conidia in the air; saprophytes or parasites.

Conidiophore slender, nonseptate, bearing a single conidium with three cylindrical, septate arms, one of which is attached by its lower end to the conidiophore. *Trinacrium*

Conidiophore slender, nonseptate; conidia four- to five-armed, nonseptate, all but one or two tapering into slender bristles; parasitic on fungi. *Titaea*

Conidiophore long and slender, hyaline, septate, sometimes branched; the single hyaline conidium inversely pyramidal, the lower part septate and tapering to the conidiophore, the broad upper portion with four one- to two-celled conical arms. Parasitic on terricolous nematodes. *Triposporina*

Conidia colored.

Conidia practically sessile, dictyosporous, at the apex with four long, septate, cylindrical, spreading extensions. Saprophytic on herbaceous stems. *Tetraploa*

Conidia sessile, three to several radiate, the arms obclavate and septate, saprophytic on woody substrata. *Ceratosporium*

Conidia sessile, horseshoe-shaped, therefore two-armed, multiseptate; parasites. *Hirudinaria*

(The systematic placing of the three foregoing genera is very uncertain.)

Conidiophores elongated, upright, septate; the single terminal conidium with a short axial cell from whose top radiate horizontally three septate arms tapering to a blunt apex. Saprophytic on woody or herbaceous stems or parasitic on leaves. *Triposporium*

Conidia of two kinds, usually arising from a small dark-colored sporodochium: (1) long-stalked conidiophore bearing at its apex four spiny cells, the attachment at the center where the cells meet; (2) four-celled, smooth spores, attached by the edge of one cell to the short conidiophore. The latter kind of spore and the sporodochium are apparently lacking in some species. Saprophytic on dead herbaceous leaves or stems, one species possibly parasitic on leaf fungi. *Spegazzinia*

Key to the More Important Amerosporous Genera of Family Dematiaceae

Conidia sessile or very short-stalked, vegetative mycelium not extensive.
- Conidia round, oval, or discoid. *Coniosporium*
- Conidia elongated, almost fusiform. *Fusella*

Whole mycelium or the more or less elongated branches developing into chains of dark-colored conidia.[19]
- Conidia falling apart easily; spherical or elongate. *Torula*
- Conidia remaining attached and separating with difficulty. *Hormiscium*

Clusters of spreading ovoid or limoniform conidia at the tips of the short conidiophores. *Echinobotryum*

Conidiophores well developed; vegetative hyphae abundant.
- Conidia dark-colored (except some species of *Acrotheca* and *Trichosporium*).
  - Conidia not in chains.
    - Conidia terminal in heads.
      - Conidia curved or unsymmetrical on unbranched conidiophores which are hyaline but with black rings at each septum; conidia dark. *Camptoum*
      - Conidia spherical or ovoid, dark.
        - With sterigmata. *Stachybotrys*
        - Practically sessile. *Periconia*
      - Conidia fusiform, dark to hyaline, practically sessile. *Acrotheca*
    - Conidia lateral, mostly in whorls.
      - Conidia smooth, not angular.
        - Conidiophores dark, unbranched or dichotomously branched. *Gonatobotryum*
        - Conidiophores hyaline, with thick dark septa, mostly unbranched *Arthrinium*
      - Conidia angular, opposite or in whorls. *Goniosporium*
    - Conidia terminal on branched or inflated conidiophores.
      - Conidiophores distinct from the mycelium, simple or forked, straight. *Virgaria*
      - Conidiophores arising on short lateral upright twigs from the creeping mycelium.
        - Conidia sessile. *Trichosporium*
        - Conidia on sterigmata. *Rhinocladium*
    - Conidiophores branched, the branches curved or hooked; conidia crowded near the ends of the twigs, colored or almost hyaline. *Campsotrichum*
    - Conidia terminal and sometimes also lateral on unbranched conidiophores (in the sense of Saccardo). *Monotospora*
    - Conidiophores short, unbranched, in a crowded cluster with a single dark conidium at the apex, narrowed at point of attachment. *Hadrotrichum*
    - Conidiophores unbranched or branched, the black conidium sessile on the vesicle-like enlargement at the apex of the conidiophoric branch. *Nigrospora*

---

[19] Forms with bent or somewhat spirally wound rows of spores should be sought in the Helicosporae under the genus *Helicoceras* (*Gyroceras*).

Conidia formed in chains, mostly colored.
Conidiophore branched at the summit in a manner resembling *Penicillium*. *Haplographium*
Conidiophore bearing a few lateral branches with chains of conidia. *Dematium*
Conidiophore dendroidally much branched, the ends of the branches bearing acrogenously produced chains of conidia, resembling some species of *Cladosporium* but conidia not septate. *Hormodendron*
Conidiophore nodulose, with clusters of phialides at the nodes. *Gonatorrhodium*
Conidia hyaline or almost so; conidiophores dark, or hyaline and then dark sterile hyphae present.
Conidiophores unbranched, straight, usually hyaline, upright, close together, among tall usually dark sterile hyphae, sometimes growing from the foot of such hyphae.
Conidia ellipsoid or allantoid.
Sterile hyphae wavy, unbranched. *Sarcopodium*
Sterile hyphae circinate, unbranched. *Helicotrichum*
Sterile hyphae repeatedly forked, with circinately curved branches. *Circinotrichum*
Conidia fusiform, curved. *Ellisiella*
Conidiophores bearing the conidia on lateral branches or sessile on the sides.
Conidiophores unbranched, conidia sessile near the upper end. *Chloridium*
Conidia arising on lateral branches from the middle portion of the conidiophore, cylindrical. *Chaetopsis*
Fertile branches of the conidiophore not confined to the middle portion.
Branches verticillately borne; conidia globose or ovoid. *Verticicladium*
Branches irregular, conidia fusiform. *Menispora*
Conidiophores branched, bearing clusters of numerous sterigmata at the joints. *Gonytrichum*
Conidia arising on (two to three) phialides on the convex side of a dark, curved, pyriform "falx," which may arise directly from a brown "foot cell" in the hyaline mycelium or on a short or elongated, or even branched conidiophore (or "falciphore"); conidia hyaline, spherical. *Zygosporium*
Conidia arising in heads of slime at tips of whorled branches of the upright, dark conidiophore. *Stachylidium*

Key to the More Important Didymosporous Genera of Family Dematiaceae

Conidiophores very short or scarcely different in appearance from the dark-colored conidia.
Conidia not in chains; conidiophores very short. *Dicoccum*
Conidia in chains; conidiophores short. *Bispora*
Conidiophores distinctly different from the mycelium, mostly erect.
Conidia not in chains.
Conidia exclusively apical on conidiophores.
Conidiophores not twisted or swollen.
Conidiophores quite long, with several septa. *Passalora*

Conidiophores short, with only one or two septa. *Fusicladium*

Conidiophores regularly twisted or swollen. *Polythrincium*

Conidiophores short, mostly clustered; conidia terminal and lateral, conidiophore not branched. *Scolecotrichum*

Conidiophores longer, branched. *Cladosporium* (some species)

Conidia in mostly acrogenously developed chains.

Conidiophores showing no swelling.

Conidia in short, branched clusters, at first nonseptate but mostly some of the older conidia once septate (or even twice septate). *Cladosporium*

Conidial chains clustered nearer the apex of the long upright conidiophore; conidia once septate. *Diplococcium*

Conidiophores branched, septate, the individual cells swollen above; conidia in acrogenously produced chains, early becoming once septate. *Cladotrichum*

Key to the More Important Phragmosporous Genera of Family Dematiaceae

Conidiophores very short or scarcely different in appearance from the dark-colored conidia.

Conidia single, not in chains.

Conidia not drawn out into a long tail.

Conidiophores here and there on the creeping mycelium. *Clasterosporium*

Conidiophores standing close together. *Stigmina*

Conidia drawn out into a long, pale, often curved tail. *Ceratophorum*

Conidia produced in chains. *Septonema*

Conidiophores well developed.

Conidia not in chains, nor in whorls, nor clustered in a head.

Conidia smooth.

Conidiophores firm.

Conidia elongate. *Helminthosporium*[20]

Conidia ovoid. *Brachysporium*[20]

Conidiophores weak. *Napicladium*[20]

Conidia rough. *Heterosporium*

Conidia formed in whorls, laterally, on the conidiophores. *Spondylocladium*

Conidia clustered in a head at the apex of the unbranched conidiophore. *Acrothecium*

Conidia in chains at the tips of the much-branched conidiophore.[21] *Dendryphium*

Key to the More Important Dictyosporous Genera of Family Dematiaceae

Conidiophores very short or lacking; conidia mostly sitting directly on the mycelium.

---

[20] These three genera grade into one another and have no sharp distinguishing characters.

[21] In some species the spore chains are short or lacking.

Conidia not in chains.
Cells of the conidia not arranged in regular vertical rows.
Conidia of irregular shape with irregular divisions; mostly rather short sessile. *Coniothecium*
Conidia elongated, sessile or nearly so, on scanty mycelium, rather uniform in size and in manner of septation. *Sporodesmium*
Conidia more or less ovoid, muriform, in little epiphyllous dense heaps. *Stigmella*
Cells of the conidium arranged in regular vertical rows which may separate and spread at maturity. *Speira*
Conidia in chains. *Sirodesmium*
Conidiophores distinct, mostly erect.
Spores obclavate, or attenuate at the distal end, borne singly or under favorable circumstances forming acrogenous chains. *Alternaria*[22]
Spores rounded at both ends, often sarciniform in appearance. *Stemphylium*[22]

Key to the Scolecosporous Dematiaceae

Conidiophores colored; conidia lightly colored to almost hyaline; never occurring in chains; parasitic. *Cercospora* (closely related to *Cercosporella*) Single genus.

Key to the More Important Genera of Family Stilbellaceae[23]

Conidia and conidiophores hyaline.
Conidia one-celled, not in chains; synnema with a capitate sporogenous portion, conidia embedded in a slime drop. *Stilbella* (*Stilbum* of many authors)
Conidia one-celled, not in chains; sporogenous portion of synnema cylindrical or clavate; conidia not embedded in a slime drop. *Isaria*
Conidia one-celled; not in chains; synnema covered by numerous lateral conidial heads; parasitic on insects or spiders (probably not properly placed in this form family). *Gibellula*
Conidia one-celled, produced in chains on more or less verticillate sterigmata (phialides); in many cases merely cultural forms of *Penicillium* whose conidiophores often are united into synnemata. *Coremium*
Conidia phragmosporous, straight, not in chains. *Arthrosporium*
Conidia phragmosporous, slender falcate, not in chains. *Atractium*
Conidia or conidiophores, or both, dark-colored.
Conidia one-celled.
Conidia not in chains, globose, ellipsoid, or oval.

---

22 In mycological and pathological literature the name *Alternaria* is often reserved for species occurring in chains and the genus *Macrosporium* for species never producing chains. Sometimes the name *Stemphylium* is used in a manner synonymous with *Macrosporium*.

23 For helicosporous Stilbellaceae, see key on p. 611 for helicosporous Moniliales. Many Moniliaceae and Dematiaceae under special cultural conditions will produce coremia, e.g., *Aspergillus*, *Penicillium*, etc.

Conidia hyaline or pale, mostly embedded in slime. *Graphium*

Conidia dark-colored. *Sporocybe*

Conidia not in chains, more or less falcate, hyaline. *Harpographium*

Conidia hyaline, in chains; spore-bearing portion of synnema globose or elongated, without hairs. *Stysanus*

Conidia in chains; but spore-bearing portion of synnema with long black hairs. *Trichurus*

Conidia phragmosporous.

Spore-bearing portion covering the whole synnema; conidia smoke-colored. *Podosporium*

Spore-bearing portion only at the upper end of the synnema.

Synnema of loose hyphae; spore-bearing head loose; conidia hyaline or pale. *Isariopsis*

Synnema of stiff, firmly united hyphae; spore-bearing head distinct conidia pale or smoky. *Arthrobotryum*

Key to the More Important Genera of Family Tuberculariaceae

(After Lindau in Engler and Prantl, 1900)

Conidia and conidiophores and sporodochium hyaline or the last bright-colored (not dark).

Conidia amerosporous.

Sporodochia without hairs or bristles.

Conidia single, not occurring in chains.

Conidia not embedded in slime.

Sporodochia of various shapes but never cup-shaped with definite margin.

Conidiophores unbranched or more rarely a little branched.

Sporodochia almost spherical, superficial. *Aegerita*

Sporodochia cushion-shaped or tuberculate or indefinite in form.

Conidia globose; sporodochia very small, later becoming hard. *Tuberculina*

Conidia ovoid; sporodochia disk-like, larger. *Hymenula*

Conidiophores always branched, mostly considerably.

Conidia ovoid or elongate.

Conidiophores not verticillately branched. *Tubercularia*

Conidiophores verticillately branched. *Dendrodochium*

Conidia fusiform to falcate or cylindrical. *Fusicolla*

Sporodochia cup-shaped, with definite margin. *Patellina*

Conidia embedded in slime; sporodochia disk-shaped, gelatinous-waxy *Illosporium*

Conidia occurring in chains.

Sporodochia more or less globose, sometimes stalked. *Sphaeridium*

Sporodochia more or less disk-shaped, not stalked. *Cylindrocolla*

Sporodochia sessile or short-stalked; at the margin with bristles. *Volutella*[24]

Conidia phragmosporous.

Conidia straight, at the upper end somewhat thickened and angular. *Heliscus*

Conidia straight, elongate-cylindrical, not thickened at upper end. *Bactridium*

Conidia falcate-fusiform.

Sporodochia conical or cushion-like, delicate; often parasitic upon insects. *Microcera*

Sporodochia extended, or large and cushion-formed, more often lacking, the spores then arising from the mycelium. *Fusarium*

Conidia brown, rarely almost hyaline; sporodochia and usually the conidiophores brown.

Conidia one-celled.

Sporodochia without hairs or bristles.

Sporodochia cushion-like or tuberculate or almost spherical, black.

Conidiophores short; conidia globose; parasites, or more often, saprophytes. *Epicoccum*

Conidiophores longer, mostly branched; conidia ovoid to elongate, sometimes curved. Saprophytes. *Strumella*

Conidiophores slender or clavate at the apex; conidia ovoid, elongate, or pyriform, borne singly or in chains. *Epidochium*

Sporodochia flat, not convex; conidiophores cylindrical, conidia ovoid, elongated, or rod-shaped. *Hymenopsis*

Sporodochia with hairs or bristles.

Hairs black, marginal. *Chaetostroma*

Hairs hyaline, marginal. *Myrothecium*

Conidia phragmosporous.

Conidia single on conidiophores. *Exosporium*

Conidia in chains on conidiophores. *Trimmatostroma*

Conidia spherical, reticulately marked, each areola representing a distinct cell. *Epicoccum* (some species)

Key to a Few Genera of Order Mycelia Sterilia

Sclerotium-like bodies produced.

Sclerotia flattened, usually on surface of host, connected by brown fibrillose hyphae. Imperfect forms of Basidiomyceteae. *Rhizoctonia*

Sclerotia round, ellipsoid or elongated, often more or less flattened on lower surface, often several growing together into a compound structure; internally firm and usually white, with a definite, black, brown, or light brown cortical layer. Many species are imperfect forms of Ascomyceteae, others of Basidiomyceteae. *Sclerotium*

[24] Parasitic species when young may be confused with *Colletotrichum*.

Sclerotium subterranean, consisting of a large mass of tangled hyphae enclosing particles of soil, plant debris, etc., the outer surface not forming a hard cortical layer; giving rise to spore fruits of Polyporaceae.
Pachyma

Root-like, branching, growing through the soil or between wood and bark or in timber; hard, white within, with black cortex. Usually representing a stage of some Basidiomycete, most frequently *Armillariella mellea.*
Rhizomorpha

Coarsely interwoven hyphae; mostly parasitic on roots of vascular plants.
Ozonium[25]

Literature Cited

BENDER, HAROLD B.: The genera of Fungi Imperfecti: North American Species and Hosts, with Particular Reference to Connecticut, 2000 pp. (Unpublished thesis: Yale University, New Haven, Conn., 1931.)

———: The Fungi Imperfecti: Order Sphaeropsidales. With Keys and References for the Genera, 52 pp. North Woodbury, Conn., published by the author, 1934.

BESSEY, ERNST A.: Spore forms of Spegazzinia ornata Sacc., *J. Mycology*, **13**(2):43–45. *Pl.* 101. 1907.

BRIOSI, GIOVANNI, E FRIDANO CAVARA: I funghi parassiti delle piante coltivate od utile, essicati, delineati e descritti, Fascicles I–XVII, containing Nos. 1–425. 1888–1909.

DADE, H. A.: Ceratostomella paradoxa, the perfect stage of Thielaviopsis paradoxa (de Seynes) von Höhnel, *Brit. Mycol. Soc. Trans.*, **13**:184–194. *Pls.* 10–12. 1928.

DELACROIX, G.: Quelques espèces nouvelles de champignons inférieurs, *Bull. soc. mycol. France*, **6**(4):181–184. *Pl.* 20. 1890.

DODGE, B. O.: The conidial stage of Peziza pustulata, *Mycologia*, **29**(6):651–655. *Fig.* 1. 1937.

DRECHSLER, CHARLES: Morphology of the genus Actinomyces, *Botan. Gaz.*, **67**(1):65–83; (2):147–168. *Pls.* 2–9. 1919.

———: Some conidial Phycomycetes destructive to terricolous Amoebae, *Mycologia*, **27**(1):6–40. *Pls.* 1–7. 1935a.

———: A new mucedinaceous fungus capturing and consuming Amoeba verrucosa, *ibid.*, **27**(2):216–223. *Fig.* 1. 1935b.

———: Some Hyphomycetes that prey on free-living terricolous nematodes, *ibid.*, **29**(4):447–552. *Figs.* 1–18. 1937.

DUEMLING, WERNER WILLIAM; GARFIELD G. DUNCAN; WILLIAM H. FELDMAN; ET AL.: Antibiotics: I. Microbiological; II. Pharmacological, *Ann. New York Acad. Sci.*, **48**(2):31–218. 1946.

EMMONS, C. W., AND A. L. CARRIÓN: Sporulation of the Phialophora type in Hormodendron, *Mycologia*, **29**(3):327–333. *Figs.* 1–6. 1937.

FAWCETT, H. S.: The cause of stem-end rot of citrus fruits (Phomopsis citri n. sp.), *Phytopathology*, **2**(3):109–113. *Pls.* 8–9. 1912.

FITZGERALD, L. R.: Phoma stenobothri, a fungus parasite of the grasshopper. *Am. Midland Naturalist*, **29**(3):761–767. 2 *pls.* 1943.

FLEMING, ALEXANDER: Penicillin—Its Practical Application, x + 380 pp. *Illustrated.* Philadelphia, The Blakiston Company, 1946.

[25] Besides Ozonium many names have been given to various hyphal forms.

GIESENHAGEN, KARL: Hexenbesen an tropischen Farnen, *Flora oder Allgemeine Botan. Ztg.*, **76**:130–156. *Pls.* 12–13. 1892.

GOIDANICH, G.: La microflora della pasta-legno: Funghi, In GOIDANICH ET AL.: Ricerche sulle alterazioni e sulla conservazione della pasta di legno destinata alla fabbricazione della carta. Capitolo II, pp. 43–192. *Figs.* 12–80. Ente Nazionale per la Cellulosa e per la Carta. Reale Stazione di Patologia Vegetale di Roma, 1938.

GROVES, J. W., AND A. J. SKOLKO: Notes on seed-borne fungi: I. Stemphylium, *Can. J. Research, C*, **22**(4):190–199. *Pls.* 1–5. 1944a; II. Alternaria, *ibid.*, 22(5):217–234. *Pls.* 1–10. 1944b.

HASSELBRING, H.: Comparative study of the development of Trichurus spiralis and Stysanus stemonites, *Botan. Gaz.*, **29**(5):312–322. *Pls.* 23–24. 1900.

HEDGCOCK, G. G.: Studies upon some chromogenic fungi which discolor wood, *Missouri Botan. Garden Ann. Rep.*, **17**:59–114. *Pls.* 3–12. 3 *text-figures*. 1906.

HERRELL, WALLACE EDGAR: Penicillin and other antibiotic agents, xv + 348 pp. *Illustrated.* Philadelphia and London, W. B. Saunders Co., 1945.

HIGGINS, BASCOMBE BRITT: Contribution to the life history and physiology of Cylindrosporium on stone fruits, *Am. J. Botany*, **1**(4):145–173. *Pls.* 13–16. 1914.

VON HÖHNEL, FRANZ: Fragmente zur Mykologie: X. Mitteilung Nr. 468 bis 526, *Sitz. ber. Kaiserlichen Akad. Wiss. (Wien) Math. naturw. Klasse*, **119**:393–473. 1910.

INGOLD, C. T.: Aquatic Hyphomycetes of decaying alder leaves, *Brit. Mycol. Soc. Trans.*, **25**(4):339–417. *Pls.* 12–17. *Figs.* 1–48. 1942.

———: Further observations on aquatic Hyphomycetes of decaying leaves, *ibid.*, **26**(3–4):104–115. *Pls.* 5–6. *Figs.* 1–6. 1943.

———: Some new aquatic Hyphomycetes, *ibid.*, **27**(1–2):35–47. *Pls.* 3–4. *Figs.* 1–8. 1944.

IWADARA, SATORU: Studies on Epicoccum oryzae Ito et Iwara n. sp., *Trans. Sapporo Natural History Soc.*, **13**(3):210–217. *Figs.* 1–2. 1934.

KLEBAHN, H.: Haupt- und Nebenfruchtformen der Askomyzeten: Erster Teil. Eigene Untersuchungen, xi + 395 pp. 274 *figs.* Leipzig, Gebrüder Borntraeger, 1918.

LIMBER, DONALD P.: A new form genus of the Moniliaceae, *Mycologia*, **32**(1):23–30. *Figs.* 1–2. 1940.

LINDAU, G.: Fungi Imperfecti, In A. ENGLER und K. PRANTL: Die Natürlichen Pflanzenfamilien. Teil 1, Abt. 1**, pp. 347–523. *Figs.* 183–263. Leipzig, Wilhelm Engelmann, 1899–1900.

LINDER, DAVID H.: A monograph of the helicosporous Fungi Imperfecti, *Ann. Missouri Botan. Garden*, **16**(3):227–388. *Pls.* 12–31. *Figs.* 1–17. 1929.

———: The genus Helicoceras, *ibid.*, **18**(1):1–8. *Pl.* 1. 1931a.

———: Brief notes on the Helicosporeae with descriptions of four new species, *ibid.*, **18**(1):9–16. *Pl.* 2. 1931b.

———: A contribution towards a monograph of the genus Oidium (Fungi Imperfecti), *Lloydia*, **5**(3):165–207. *Pls.* 1–7. 1942.

MASON, E. W.: Annotated account of fungi received at the Imperial Mycological Institute, List II, Fascicle 2, 1933; List II, Fascicle 3, General Part, 1937.

———: Zygosporium, *ibid.*, List II, Fascicle 3 (Special Part), pp. 134–144. *Figs.* 40–44. 1941.

MATRUCHOT, L.: Structure, développement et forme parfaite du Gliocladium, *Rev. gén. botan.*, **7**:321–331. *Pl.* 16. 1895.

Melin, Elias, and J. A. Nannfeldt: Researches into the blueing of ground woodpulp, *Svenska Skogsvårdsföreningens Tidskrift 1934*, Häfte III–IV, pp. 397–616. *Figs.* 1–60. 1934.

Montemartini, Luigi: Ricerche sopra la struttura delle Melanconiee ed i lori rapporti cogli Ifomiceti e colle Sferossidee, *Atti ist. botan. reale univ. Pavia*, N.S., vol. 6, 45 pp. *Pls.* 11–12. 1899.

Nannfeldt, J. A.: Studien über die Morphologie und Systematik der nichtlichenisierten inoperculaten Discomyceten, *Nova Acta Regiae Soc. Sci. Upsaliensis*, Ser. IV, **8**(2):1–368. *Pls.* 1–19. *Figs.* 1–47. 1932.

Neergaard, Paul: Danish species of Alternaria and Stemphylium. Taxonomy, parasitism, economical importance, Communication from the Phytopathological Laboratory of J. E. Ohlsens Enke, Copenhagen, pp. 1–560. *Figs.* 1–159. Copenhagen, Einar Munksgaard, Publisher, 1945.

Newcomer, Earl H., and Glenn KenKnight: Nuclei in Actinomyces, *Papers Mich. Acad. Sci.*, **25**:85–87. 1939 (1940).

Petrak, F., und H. Sydow: Die Gattungen der Pyrenomyzeten, Sphaeropsideen und Melanconien: Erster Teil. Die phaeosporen Sphaeropsideen und die Gattung Macrophoma, *Repertorium Specierum Novarum Regni Vegetabilis, Beihefte*, **42**(1–3):1–551. 1926–1927.

Pinkerton, M. Elizabeth: A comparative study of conidial formation in Cephalosporium and some related Hyphomycetes, *Ann. Missouri Botan. Garden*, **23**(1):1–68. *Pls.* 1–6. 1936.

Prillieux, G.: La pourriture du coeur de la betterave, *Bull. soc. mycologique de France*, **7**(1):15–19. *Pl.* 3. 1891.

———, et G. Delacroix: Sur une maladie des tomates produite par le Cladosporium fulvum Cooke, *ibid.*, **7**(1):19–21. *Pl.* 3. 1891.

Raper, Kenneth B., and Charles Thom: A manual of the Penicillia, ix + 875 pp. 10 *colored plates*. 172 *figs*. Baltimore, Williams & Wilkins Co. 1949.

Reinke, J., und G. Berthold: Die Zersetzung der Kartoffel durch Pilze, Untersuchungen aus dem Botanischen Laboratorium der Universität Göttingen, Heft 1, pp. 1–100. *Pls.* 1–8. Berlin, Verlag von Wiegandt, Hempel und Parey, 1879.

Saccardo, P. A.: Sylloge fungorum hucusque cognitorum, vol. 14, pp. 1–1316. Pavia, published by author, 1899. (Contains an explanation of an arrangement of genera by spore form and color.)

Shear, C. L.: The life-history of the Texas root rot fungus, Ozonium omnivorum Shear, *J. Agr. Research*, **30**(5):475–477. *Pl.* 1 (*colored*). *Fig.* 1. 1925.

———, and B. O. Dodge: The life history and identity of "Patellina fragariae," "Leptothyrium macrothecium," and "Peziza oenotherae," *Mycologia*, **13**(3):135–170. *Pls.* 8–10. *Figs.* 1–5. 1921.

Southworth, E. A.: A new hollyhock disease, *J. Mycology*, **6**(2):45–50. *Pl.* 3. 1890.

Tehon, L. R.: The pycnothyrium in the taxonomic system of the Fungi Imperfecti, *Trans. Illinois State Acad. Sci.*, **33**(2):63–65. *Figs.* 1–3. 1940.

———, and Eve Daniels: Notes on the parasitic fungi of Illinois, II, *Mycologia*, **17**(6):240–249. *Pl.* 25. 1925.

Thaxter, Roland: On certain new or peculiar North American Hyphomycetes: II. Helicocephalum, Gonatorrhodiella, Desmidiospora nov. genera and Everhartia lignatilis n. sp., *Botan. Gaz.*, **16**(6):201–205. *Pls.* 19–20. 1891.

Thom, Charles: The Penicillia, xiii + 644 pp. 99 *figs*. Baltimore, The Williams and Wilkins Company, 1930.

———, AND KENNETH B. RAPER: A manual of the Aspergilli, ix + 373 pp. 7 *colored plates*. 76 *figs*. Baltimore, The Williams and Wilkins Company, 1945.

VUILLEMIN, P.: Matériaux pour une classification rationnelle des Fungi Imperfecti, *Compt. rend.*, **150**(14):882–884. 1910.

———: Les Aleurisporés, *Bull. soc. sci. Nancy*, pp. 151–175. 1911.

WAKSMAN, S. A., AND A. T. HENRICI: The nomenclature and classification of the Actinomycetes, *J. Bacteriol.*, **46**(4):337–341. 1943.

WILTSHIRE, S. P.: The foundation species of Alternaria and Macrosporium, *Brit. Mycol. Soc. Trans.*, **18**:135–160. *Pls.* 6–8. *Figs.* 1–6. 1933.

———: The original and modern conceptions of Stemphylium, *ibid.*, **21**:211–239. *Pls.* 14–15. *Figs.* 1–17. 1938.

WINTER, GEORG: Mycologische Notizen, *Hedwigia*, **12**:145–147. 1873.

WOLF, FREDERICK A.: Morphology of Polythrincium, causing sooty blotch of clover, *Mycologia*, **27**(1):58–73. *Figs.* 1–5. 1935.

ZIMMERMANN, A.: Ueber einige an tropischen Kulturpflanzen beobachtete Pilze. II, *Centr. Bakt. Parasitenk.*, Zweite Abt., **8**(7):216–221. *Figs.* 5–8. 1902.

17

THE PHYLOGENY OF THE FUNGI

It is a commonly accepted axiom that evolution in both plants and animals generally proceeds from the simpler to the more complex structures, although it is recognized that in many cases a retrogression from complex to simpler structures may occur. Therefore it is natural that from the time that the ideas of evolution began to be given consideration in the attempts to develop taxonomy along supposed phylogenetic lines search was made for simple organisms as the probable primitive types from which the complex phylogenetic trees of plants and animals had developed.

Since animals lack the power of synthesizing organic food stuffs out of inorganic materials it has usually been assumed that the predecessors of the animal kingdom must have been organisms that did possess that power. The same assumption is necessary for the establishment of a logical phylogenetic arrangement of the fungi and other more or less plant-like organisms that are not autotrophic. There are a number of groups of microscopic organisms ordinarily classed among the Bacteria in which the energy necessary for the synthesis of organic materials is obtained by various types of chemical reaction independent of light. Thus the sulphur bacteria by the oxidation of sulphur or of H_2S are able to gain the necessary energy for their life processes. The same is true of those bacteria which oxidize iron. There are several other types of energy-obtaining processes. It is, however, only in those plants which develop chlorophyll-like substances able to react with light to obtain the energy necessary for synthesis that a comparatively large amount of energy became available and that production of large amounts of food occurred. Probably not until then did the evolution of larger, more complex organisms begin and not till then was the available food supply great enough to permit the extended evolution of the various orders of animals and of fungi.

From the fact that most of the simpler animals, green plants, and fungi for the major portion of their life cycle consist of single, uninucleate cells with flagella or cilia or produce such cells for the purpose of multiplication it seems logical to conclude that the ancestors of these groups were unicellular, motile organisms. They must have been already highly developed in their cell structure for we find in the green algae, lower fungi, and animals essentially the same basic nuclear structure, sexual reproduction whose fundamental phenomena are the union of two haploid nuclei intc a diploid nucleus followed sooner or later by meiosis, and similar laws of heredity. Therefore we must believe that the forms ancestral to these three lines had already reached the height of development attained by the unicellular, motile or zoospore-producing algae. The sulphur or iron bacteria mentioned above lack such definite nuclei and are much too simple in structure to have been the immediate ancestors sought.

It appears to the author that probably the lines of evolution after the invention of chlorophyll progressed rapidly in various directions. The known Myxophyceae have not reached the type of cell structure that could have given rise immediately to the unicellular, flagellate green algae or to the lower fungi or animals. It is to the organisms more like the simple chlorophyll-bearing flagellate algae that we must turn in our search. In such algae we have a constant production of organic material, by means of photosynthesis, and thus chlorophyll-lacking organisms could develop, depending upon them either directly or indirectly for their food. We find that the motile organs of such simple alga-like organisms as we now know are usually one to four in number, connected to little organelles, the blepharoplasts, at or just below the plasma membrane, these usually being then connected to the nucleus. Evolution led to the production of two main types of flagella. In one type there is a firmer outer tubular envelope surrounding a softer more flexible portion which extends beyond this outer portion and is capable of very active lashing motion. These are the "whiplash" flagella. Another type consists of a much more flexible axis from whose sides arise at right angles numerous very fine cilioles which are capable of movement. This is the "tinsel type" of flagellum. Some of the single-celled plants possess one or more of the one type or of the other type, but in one group of algae one flagellum of each type occurs. This is the group called Heterocontae or Xanthophyceae. Here the flagella may be equal in size but usually with the tinsel flagellum directed anteriorly and the other turned back. The latter is frequently shorter and in some cases appears to be lacking.

Another characteristic of the unicellular algae is the fact that the motile cells do not normally produce a true cell wall, the outer layer being simply a plasma membrane. This becomes encysted with a true cell wall which serves as a protection. In the further evolution of the various groups of algae the encysted stage becomes the normal vegetative condition, the naked, swimming cells being produced for multiplication We find this same trend of evolution in the fungi that we speak of as the Lower Fungi In the line of evolution of animals the contrary feature is emphasized The motile stage becomes more and more the prevailing one, the encystment occurring only at the reproductive stage and finally dis appearing altogether. In some of the simple animals the plasma membrane has become less firm and amoeboid motion is possible, together with the amoeboid engulfment of food particles, yet even here encysted stages may occur.

We may perhaps assume that some of the chlorophyll-containing, one-celled, naked, swimming organisms, whether we call them plants or not, and which had possibly not yet reached the level of the present simplest green algae, lost their power of producing chlorophyll and so had to become parasitic in the tissues of animals or plants or saprophytic, feeding upon the organic products of plant-like organisms. Among the group commonly called the Flagellata we find numerous examples of this type in which the lack of chlorophyll is the chief distinguishing character. Also such examples of lost chlorophyll occur among the true green algae, as for example in *Rhodochytrium*, in which the carotinoid pigments are still produced although the chlorophyll has disappeared. This plant lives parasitically in the tissues of higher plants. Intermediate steps are found in certain unicellular algae that still retain their chlorophyll but which are obligate endophytes in aquatic higher plants. Examples of these are *Chlorochytrium* and *Eremosphaera*.

Phylogeny of Mycetozoa

The Mycetozoa are characterized by their naked plasmodial vegetative stage and by the ultimate rounding up and encystment of the cells as "spores," the only stage (with minor exceptions, such as the stalk cells of Acrasiales) in which cell walls are found. With the exception of the few described Acrasiales and Labyrinthulales, the encysted spores germinate by production of anteriorly biflagellate swarm cells or sometimes one flagellum seems to be lacking although two blepharoplasts are present. These two flagella are both of the whiplash type. The presence of flagellate, amoeboid planospores and inclination toward the formation of plasmodia or pseudoplasmodia is found also in some of the Protozoa included in the Rhizopoda or Sarcodina and therefore it seems to the author only reasonable to conclude that the Mycetozoa (in the broader sense of this term) are derived from those organisms. Where they came from remains to be determined but possibly the amoeboid Protozoa are further developments of chlorophyll-less Flagellata. The Plasmodiophorales are classified by Sparrow (1943) and some others in the Biflagellatae, a group which includes also the Saprolegniales, Leptomitales, Lagenidiales, and Peronosporales. Since, however, Ellison (1945) has shown that the two flagella of *Plasmodiophora* are both of the whiplash type while those of the other groups assigned to the Biflagellatae have one each of the tinsel and whiplash type, the relationship of *Plasmodiophora* appears more likely to be with the biflagellate slime molds whose flagella are both of the whiplash type and whose plasmodial structure is more nearly similar. Whether the other genera at present assigned to the Plasmodiophorales really belong there needs further study, especially of their flagellar structure.

The nonflagellate order Acrasiales is possibly closely related to the true slime molds. Besides the loss of flagellum production there is no definite peridium nor is a capillitium produced unless possibly the slime in which the spores are embedded and which helps to make up the growing stalk are modifications of these structures. Furthermore the production of true plasmodia is lost except possibly just before spore formation, but sexuality appears to be present (Skupienski, 1920). As for the Labyrinthulales it can only be surmised that they possibly are related to the Mycetozoa because of their naked cells, which form a sort of pseudoplasmodium, and the encystment of their spores. Their life history is so incompletely known that their relationship cannot be determined more closely at present.

Phylogeny of Chytridiales and Vegetatively Similar Forms

The whole group of chiefly aquatic fungi formerly grouped together under the name Chytridiales agrees in the production of flagellate naked swarm cells which eventually encyst and enlarge and become the sporangia or give rise to the sporangia within which the swarm cells are produced. The study of the number, location and structure of the flagella has shown that this group must be divided into at least three orders: Chytridiales (in the more limited sense), with one posterior whiplash flagellum; Hyphochytriales, with one anterior tinsel type flagellum; and certain families of the Lagenidiales, with two anteriorly or laterally attached flagella, one of the tinsel type (directed forward if the flagella are laterally attached) and the other of the whiplash type (directed posteriorly in most cases). Correlated with these flagellar characters are chemical differences in the composition of the cell wall. In the Lagenidiales this gives the cellulose test readily with chloriodide of zinc, in the Hyphochytriales this reaction may or may not appear and may depend upon the age of the organism, while in the Chytridiales only rarely is cellulose revealed by this test and in some cases fungus chitin occurs.

In 1942, the author suggested that these three orders which vegetatively and in their mode of reproduction are so closely parallel, might have had a common origin in some of the green, heterocont, one-celled, motile, anteriorly biflagellate algae (Heterocontae or Xanthophyceae). It is necessary to make the assumption that the loss of chlorophyll and the adoption of a saprophytic or parasitic mode of life occurred early in their evolution. The simpler Lagenidiales represent forms in which both flagella are retained, as well as the cellulose composition of the cell walls. For the Hyphochytriales we must assume a gradual modification of the heterocontal habit until finally (as sometimes occurs in the algae) the shorter, posteriorly inclined flagellum entirely disappeared, leaving only the single anterior tinsel type flagellum. At the same time the cellulose became in part replaced by or concealed by an admixture of fungus chitin. For the Chytridiales (in the narrower use of the term) the anterior (tinsel type) flagellum has disappeared leaving the single, posteriorly directed whiplash flagellum. Concurrently the cellulose reaction of the cell walls has become less and less marked, appearing only in a minority of the species so far studied. (Fig. 204.)

These three groups have undergone parallel evolutionary develop-

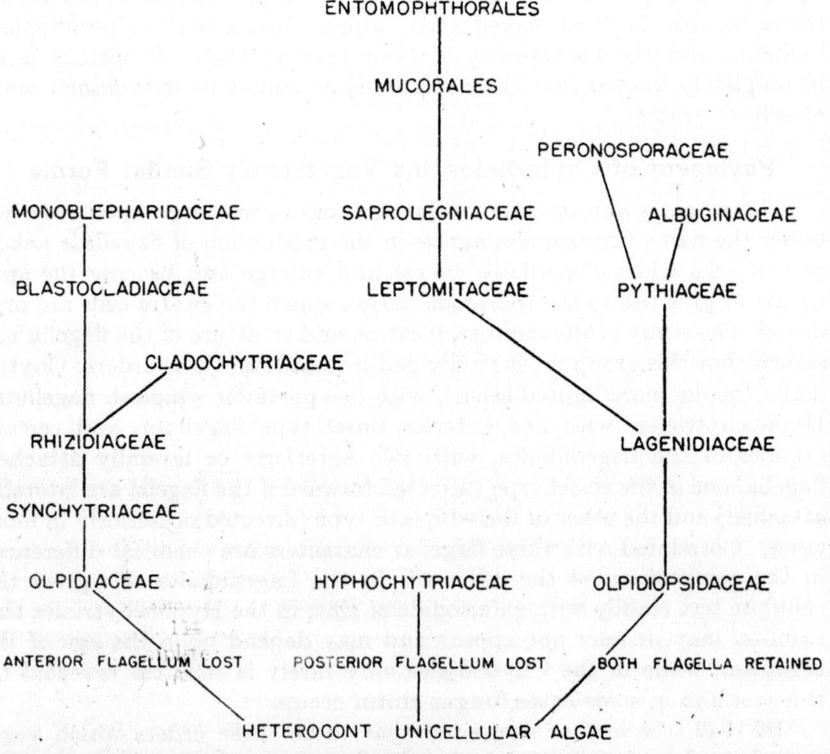

FIG. 204. Suggested lines of evolution of the Phycomyceteae, based upon the idea of their origin from unicellular algae. (After Bessey: *Mycologia*, **34**(4):355–379.)

ments from holocarpic to eucarpic monocentric forms and finally to eucarpic polycentric forms. So far as known the Hyphochytriales have not given rise to further advanced, hyphal forms of a truly mycelial nature. The Chytridiales, on the contrary, are certainly closely related to the Blastocladiales and Monoblepharidales, culminating in the latter in an oosporic mode of sexual reproduction. This line seems to end blindly. From the Lagenidiales we can draw lines of ascent to the Saprolegniales and Peronosporales.

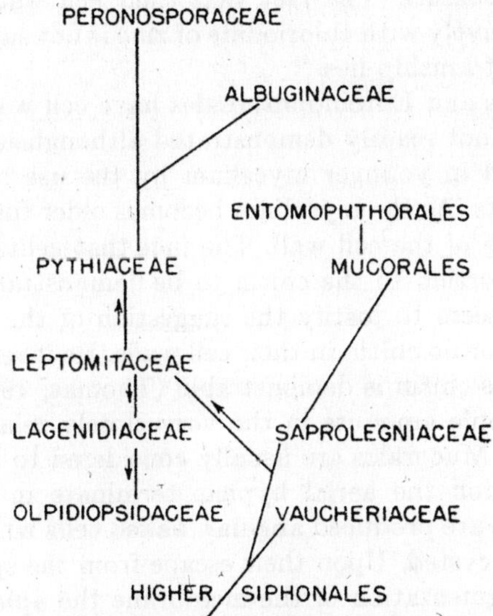

FIG. 205. Suggested lines of evolution of the biflagellate Phycomyceteae, based upon Sachs's and Mez's idea of the origin of the Saprolegniales from the Siphonales. (After Bessey: *Mycologia*, **34**(4):355–379.)

On the contrary Mez (1929), de Bary (1884), and many others have suggested that the evolution may have been regressive from the Saprolegniales or from the Pythiaceae, to the Lagenidiales, by a process of simplification From the simple, holocarpic Olpidiopsidaceae, by the loss of the posterior or anterior flagellum could have arisen the Hyphochytriales and Chytridiales respectively. No forms are known from which the Monoblepharidales and Blastocladiales might have arisen and then by regression led to the production of the Chytridiales. The Saprolegniales are assumed by Mez and by Sachs (1874) under this hypothesis to have arisen by loss of chlorophyll from some alga similar to *Vaucheria* in the Siphonales, at a point in the evolution of this alga prior to the substitution of a single compound zoospore for the many separate biflagellate zoospores characteristic of most Siphonales and of the Saprolegniales. The close similarity of the serum reactions by *Saprolegnia* and *Vaucheria* as determined by Mez would seem to add force to this idea. (Fig. 205.)

Phylogeny of Higher Phycomyceteae

The phylogeny of the Mucorales and Entomophthorales is probably properly tied up with that of the Zoopagales. Of the Eccrinales so much still remains to be learned that their origin and their relationships to other groups are very obscure. The fact that their cell walls respond to the cellulose test positively with chloriodide of zinc is not sufficient to indicate where their closest kinship lies.

The Mucorales and Entomophthorales have cell walls at maturity in which cellulose is not readily demonstrated although in some Mucorales it can be detected in younger mycelium by the use of suitable iodine-containing reagents. As the mycelium becomes older fungus chitin makes up more and more of the cell wall. The fact that cellulose is sufficiently abundant in proportion to the chitin to be demonstrable in the younger mycelium would seem to justify the suggestion of the possibility of ancestors with little or no chitin in their cell walls. Be it remembered that in *Pythium* no fungus chitin is demonstrable (Thomas, 1942, 1943) while it occurs in measurable amounts in the very closely related *Phytophthora*. The more typical Mucorales are usually considered to be those in whose asexual reproduction the aerial hyphae terminate in sporangia within which by cleavage are produced angular, naked cells which quickly round up and become encysted. Upon their escape from the sporangium by the dissolution or fragmentation of the membrane the aplanospores may be distributed by water or by air currents or even by insects. Except for the nonflagellate condition of the spores this type of asexual reproduction is found in the Blastocladiales, Monoblepharidales, Saprolegniales, and some of the terrestrial and aquatic Peronosporales. The fact that the mycelium of the Mucorales is mostly stout and that large numbers of species are soil inhabitants would seem to exclude many of the Pythiaceae and Blastocladiales and Monoblepharidales from consideration. In the Saprolegniaceae we find the genus *Aplanes* in which the spores produced in the sporangium become aplanospores, without undergoing the swimming stage. Thus, so far as the asexual mode of reproduction is concerned, there is no serious barrier to the belief that the Mucorales may have evolved from some Saprolegniaceous soil fungus with aplanospores. In general, however, the conjugating gametangia in the Mucorales are almost equal in size while in the Saprolegniales they consist of a small antherid and a large oogone, the former usually producing a conjugation tube which penetrates through the oogone wall and opens at its tip when nearly or quite in contact with the egg. Yet in *Brevilegnia diclina* no such conjugation tube is formed, there being developed simply an opening through the antherid and oogone walls which permits the entry of the male nucleus. However, the supposed isogamy in the Mucorales is more apparent than real. Even where the gemetangia are equal in size the nuclei and part of the cytoplasm of one pass through an opening in the walls into the other so that we actually have a functioning antherid although it is equal in size with the oogone. In *Dicranophora* one gametangium is very large and the other very small and the "zygospore" wall includes the oogone alone or rarely part of the antherid also. So we can perhaps with justification look to some soil-inhabiting member of the Saprolegniales for the ancestral form of the Mucorales. It must be noted that Jaczewski (1929–30) suggested that these latter fungi arose directly from some of the Chytridiales (in the wider sense). Others have suggested the Monoblepharidaceae or the Cladochytriaceae as the ancestral stock of the Mucorales.

Within the Mucorales the evolution has apparently been in various directions of modification of the sporangium, terminating in small indehiscent sporangioles or portions of elongated sporangia. The Entomophthorales appear to be terminal lines of evolution in which the mycelium is much reduced, and the "conidia" represent usually violently discharged, mostly indehiscent, sporangioles (but in *Basidiobolus* these produce internal aplanospores). The sexual reproduction of the various genera represents modifications of some of the different types found in the Mucorales.

In the Zoopagales the sporangia are reduced to indehiscent sporangioles (or "conidia") and the sexual reproduction is sometimes isogamous and reminiscent of some of the Mucorales or more often heterogamous. Not enough is known of their cytology, the chemical composition of the cell walls etc., to permit more definite suggestions as to their phylogeny.

Origin of Ascomyceteae

There are two main schools of thought regarding the phylogenetic origin of the Higher Fungi (Ascomyceteae, Basidiomyceteae, etc.). In the one it is held that the fungi as a whole form a monophyletic series and, consequently, it is believed that the Ascomyceteae and other higher fungi arose from the Phycomyceteae. There is no general agreement as to the definite paths along which such derivation occurred, nor whether the higher fungi are monophyletic or polyphyletic in their origin from the lower fungi. The other school holds that the fungi are not necessarily monophyletic and that some, if not all, of the higher fungi arose from algae that had some of the characteristics of the simpler Florideae. In both theories the points of connection between the Ascomyceteae and the Rusts and Smuts and the more characteristic Basidiomyceteae are not in agreement, regardless of the ideas as to the ancestry of the Ascomyceteae. The author follows the school of Sachs in the belief that the evidence is stronger in favor of the Floridean ancestry of the Ascomyceteae, but he will attempt to indicate the viewpoints of this hypothesis which are most unsatisfactory and to exhibit the evidence that convinced de Bary and many of his successors that the Phycomyceteae actually gave rise to the higher fungi.

Among the points against the Floridean ancestry, the composition of the cell wall has been emphasized. In the higher fungi as well as the lower fungi the basic compounds making up the wall are various types of carbohydrates, among which may be found cellulose, callose, various pectin compounds, etc. Admixed with these, especially as the mycelium becomes older, are various substances which, when in sufficiently large proportion, prevent the reaction of the cell wall to the various cellulose tests, such as chloriodide of zinc and certain stains and solvents. These admixtures are sometimes fats, deposited in the surface layers of the wall, as is the case in *Monoblepharis*, *Blastocladia*, etc. In these, treatment with substances that will saponify the fat (such as KOH solution, warm but not hot, followed by thorough washing) will then permit the cellulose reaction to appear. Far more often the hindering agent is a substance which, although it is usually called chitin, has been shown to be sufficiently different from animal chitin to warrant its designation as fungus chitin (see Chapter 1, p. 3). This is found in some of the Chytridiales (e.g., *Synchytrium*), to a small extent in *Phytophthora*, but not enough to prevent the cellulose reaction entirely, and in the Mucorales and Entomophthorales. Even in the former it may not block the cellulose reaction in the younger hyphae but usually does so in the more mature mycelium. The author's investigations on *Pilobolus* (1948) have shown that in *P. kleinii* van Tiegh. the cellulose coloration by chloriodide of zinc is limited to the slightly thickened lip surrounding the opening at the top of the subsporangial vesicle after the sporangium has been discharged, while in *P. longipes* van Tiegh. the lower two-thirds or three-quarters of the sporangiophore and the trophocyst, the basal swelling, and much of the vegetative mycelium show this reaction. In the Entomophthorales the chitin appears to make up a large portion of the cell wall but cellulose prevails in *Basidiobolus* (Couch, 1939). In the Ascomyceteae and other higher fungi, fungus chitin usually blocks all cellulose tests although the presence of the latter may be shown by chemical analysis. In the ascogenous hyphae of some lichens the cellulose reaction shows up well. Some of the yeasts do not indicate the presence of fungus chitin though cellulose seems to be replaced by other carbohydrates. From the foregoing it seems that there is a growing trend as we pass from the lower fungi to the higher fungi for an increased amount of fungus chitin in proportion to the cellulose but in general the higher fungi are much like the lower fungi in cell wall composition. In the Florideae cellulose is present usually accompanied by a considerable amount of pectic substances but there is no evidence of fungus chitin

The hypothesis of Floridean ancestry of the higher fungi requires that the phylogenetic evolution of the latter must be assumed to progress from complex forms (Pezizales, Sphaeriales, etc.) to simpler forms such as *Taphrina*, *Endomyces*, *Saccharomyces*, etc., contrary to the usual direction of evolution accepted for most groups of fungi. On the contrary from the relatively simply-built Phycomyceteae the rather simple Endomycetaceae may be assumed to have arisen without too great difficulty.

The chief arguments for a Floridean ancestry lie in the similarity between the sexual reproduction in the simpler Florideae and in many of the Ascomyceteae, including:

1. Nonmotile, naked or at most very thin-walled, spermatia, with a relatively large nucleus, mostly produced one or more at a time endogenously in scattered or crowded antherids. These depend for distribution in the Florideae upon water currents and in the Ascomyceteae upon streaming surface layers of water or upon insects.

2. The production of a receptive filament (trichogyne) projecting from the oogone.

3. The multiplication within the oogone of the zygote nucleus (or of the paired but not united gamete nuclei) and their passage through outgrowths from the oogone to terminal cells which become carpospores or asci in the Florideae or Ascomyceteae respectively.

4. Usually, but not always, the formation of enclosing envelopes of various types around the oogone, sometimes before but more often after fertilization.

Aside from these reproductive similarities the vegetative structure shows similarities.

5. The filaments (hyphae) of both groups of organisms mostly grow in length by the elongation and division of the terminal cells, which however occurs also in the Phycomyceteae.

6. The septa which arise by circular shelf-like growth from the lateral wall of the cell do not entirely close in the Ascomyceteae as they do in the Phycomyceteae (according to Buller, 1933) but leave a central pore through which the protoplasmic continuity of adjacent cells is maintained, a condition which also occurs in the Florideae.

In the Phycomyceteae in general each sexual act results in the production of a single zygote: zygospore or oospore (e.g., *Monoblepharis*, *Saprolegnia*, *Peronospora*, *Mucor*, *Entomophthora*, *Zoopage*, etc.) but in the Ascomyceteae in which sexual reproduction occurs by the union of nonmotile sperm cells with a trychogyne the one sexual act leads to the production of many asci borne on ascogenous hyphae and usually surrounded and protected by a structure made up of vegetative hyphae (apothecium, perithecium, etc.). The same is true of the Florideae where the union of one sperm cell with the trichogyne will bring about the formation of a spore fruit (as Sachs called it) with many carpospores, either surrounded by a protective envelop or not. It should be noted that in the very simple Endomycetaceae and Saccharomycetaceae, considered by the author to be extreme simplifications from very complex ancestors, a single ascus is produced by each sexual act. The proponents of the hypothesis that the Ascomyceteae have arisen directly from the Phycomyceteae naturally consider the foregoing fungi to represent connecting links between the two groups.

A further argument against the phycomycetous origin of the Ascomyceteae is the type of mycelium, which is prevailingly coenocytic in the former and cellular in the latter. To be sure cross walls occur frequently in some Phycomyceteae but they are mainly (1) to set off reproductive organs from the main mycelium, (2) to fence off emptied portions of the mycelium from those regions in which living protoplasm is still present, and (3) to wall off injured regions. Furthermore the production of the septa in the coenocyte is entirely independent of any immediately preceding nuclear division. In the higher fungi the mycelium is usually cellular and the cell walls separate uninucleate or binucleate cells. Cell division and septum formation are the immediate consequences of the division of the nucleus or of the two nuclei of the cell. It cannot be denied that the mycelium of many Ascomyceteae is composed of coenocytic segments separated by cross walls, but these forms are relatively few in number. On the other hand the somewhat anomalous *Basidiobolus* in the Entomophthoraceae has its mycelium made up of uninucleate cells.

If we accept Sachs's suggestion as to the possible origin of Ascomyceteae from the Florideae we must prepare a phylogenetic tree that will include as more primitive those forms whose characters include the greater number of those found in the Florideae. Therefore we must give first place to those families or orders in which receptive hyphae (trichogynes) are fertilized by nonmotile, naked or thin-walled, usually endogenously produced sperms and from whose oogones arise several or many asci or ascogenous hyphae bearing the asci. Thus we should place first in the Ascomyceteae the Laboulbeniales, Lecanorales, Sphaeriales, and some Pezizales, granting that many of the last two orders accomplish their sexual reproduction by direct contact of the antherids with trichogynes or oogones. All of these forms are rather complex in their structure, far more so than the very simple yeasts and Taphrinales, which are placed at the base of the class by so many mycologists. On the same grounds of reasoning the Uredinales also, must be considered to proceed from near the orders believed by the author to be primitive in the Asco myceteae. H. S. Jackson (1944) gives very convincing arguments for the hypothesis that the closely parallel life histories of the Uredinales and Florideae indicate something more than a mere convergence, probably true relationship.

From the Pezizales it is usually considered that the Tuberales have arisen and possibly the Hysteriales. The fructifications of the Ascocorticiaceae may be looked upon as representing much reduced apothecia, thin and without a definite margin. They still retain the characteristic hook or crozier mode of formation of the asci. The family Taphrinaceae might be looked upon as producing a much further simplified apothecium in which the hook or crozier are lacking and where the whole vegetative mycelium is dicaryotic, perhaps representing the ascogenous hyphae of the usual types of Pezizales.

The Sphaeriales show in the genus *Mycosphaerella* numerous cases of the production of rounded oogones with unbranched, nonseptate, apical, usually uninucleate trichogynes which are fertilized by sperm cells produced internally (sometimes in fours) in the spermogonial structure. In this order as in the Pezizales the union of oogone and antherid may take the place of spermatization of a trichogyne. The Erysiphales represent mostly Sphaeriaceous forms with usually superficial, nonostiolate perithecia. Probably the Pseudosphaeriaceae arose from typical Sphaeriaceae as also did the Aspergillales and Myriangiales. The Gymnoascaceae represent a development in the Aspergillales toward the loosening up of the perithecial wall which is entirely lacking in the Endomycetaceae as are the ascogenous hyphae, so that a single sexual act produces but one ascus. (Fig. 206.)

B. O. Dodge (1914) gave a very scholarly exposition of the similarities in the sexual reproductive structures of the Ascomyceteae and Florideae with special reference to the probable origin of the former from the latter. This should be read in connection with the excellent explanation (by Atkinson, 1914) of the origin of the simpler Ascomyceteae from the Phycomyceteae.

For those who look to the Phycomyceteae as the possible source of the higher fungi several hypotheses have been proposed. For many years that of Brefeld (1889) was the prevalent one. He started from the assumption that sexuality among the fungi was confined to the Phycomyceteae and was entirely lacking in the Ascomyceteae and Basidiomyceteae. He assumed that the ascus is a sporangium homologous to that of the Mucorales. Accordingly he considered the polysporous asci to represent a more primitive stage and the fungi with the standard eight-spored asci as more advanced. The basidium he considered to be a conidiophore bearing a not standardized number of conidia in the more primitive forms (the Ustilaginales), the number finally becoming fixed at four per conidiophore

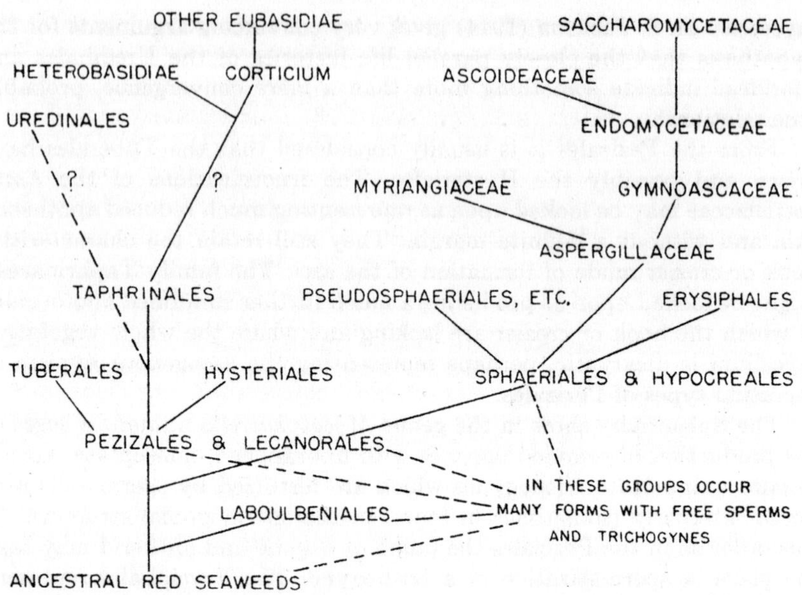

Fig. 206. Suggested lines of evolution within the Ascomyceteae and origin of the Basidiomyceteae, based upon Sachs's idea of their derivation from the Florideae. (After Bessey: *Mycologia*, **34**(4):355–379.)

(i.e., the typical four-spored basidium). The "conidial" Phycomyceteae (some Mucorales, the Entomophthorales, etc.) were looked upon as the origin of the Basidiomycetous line. Although this Brefeldian hypothesis has become discredited with the demonstration of the actual occurrence of sexuality in the higher fungi and by the general acceptance of the homologies of ascus and basidium and of the crozier and clamp connections his theory still exerts a strong influence in most systems of classification of these fungi.

With the acceptance of true sexuality in the higher fungi various mycologists have turned to the morphologically simple families of the Ascomyceteae as probably representing the intermediate stages between the Phycomyceteae and the Ascomyceteae. The discovery by Lagerheim (1892) of *Dipodascus albidus* and the careful working out by Juel (1902) of the cytology of its sexual reproduction have revealed a form that lends itself well to consideration as such a link. In this fungus the mycelium is septate, each segment being coenocytic. Usually adjacent segments produce upright, multinucleate gametangia which unite near their apices. A "privileged nucleus" of each gametangium unites with the corresponding one from the other to form a large zygote nucleus. From the junction of the gametangia or from one of the latter the long ascus develops, tapering toward its apex. In it the zygote nucleus divides repeatedly and finally around each nucleus is formed an ascospore. Numerous nuclei degenerate without acting as centers for spore formation. It is assumed that these were the supernumerary nuclei of the gametangia which did not become the privileged nuclei. The interpretation by Dangeard (1907), Atkinson, Gäumann (1926), and others is that the gametangia are homologous to those of the Mucorales and that the ascus represents the zygospore of the latter and the sporangium which it usually produces upon germination after a period of rest. In *Dipodascus*, however, there is no resting period and the zygote develops immediately to become the ascus. In *D. uninucleatus* Biggs, the mycelium is cellular, consisting of uninucleate cells. The gametangia are also uninucleate and there are no supernumerary nuclei. The ascus and ascospores develop as in *D. albidus*, the number of ascospores being typically large but often small in poorly developed asci. This species forms a close connection to *Endomyces* and *Eremascus* in the Endomycetaceae in which single asci, of eight ascospores or less, arise as the result of the sexual union of the two uninucleate gametangia. By increase of the tendency toward the budding habit of vegetative growth we can derive the Saccharomycetaceae. It is assumed that in some forms similar to *Dipodascus* the elongated multinucleate ascus gave rise to a series of branching hyphae containing the as yet unfused gamete nuclei and their products, and producing terminal cells where the pairs of nuclei unite and undergo meiotic division. Thus is interpreted the origin of the dicaryon ascogenous hyphae and asci. By the growth of loose protective vegetative hyphae between and around the ascogenous hyphae it is assumed that such a loose perithecium is developed as we find in the Gymnoascaceae. By the development of a firmer cortex the Aspergillaceae were derived. From these organisms one can imagine the development of Erysiphales, Myriangiales, Pseudosphaeriales, Sphaeriales, Pezizales, etc., a complete reversal of the phylogenetic tree based upon the idea of the origin of the Ascomyceteae from Florideae. (Fig. 207.)

Whichever of the two phylogenetic series mentioned above is considered to be nearer the truth there are certain observed facts or interpretations of facts that disagree. In both hypotheses the ascus is postulated as being derived from a sporangium: for the Floridean ancestry it must come from a tetrasporangium and for the Phycomycetous derivation from a sporangium more or less on the plan of that of the Mucorales. In either case the spores arise by the complete segmentation of the protoplasm, with no enucleate epiplasm left over. Yet, in the ascus, part of the cytoplasm gathers around the nuclei and becomes encysted, leaving a portion without nuclei, the epiplasm, surrounding the ascospores. Whether or not this difference in manner of spore formation is so fundamental as many have considered, it must be settled by further studies upon the various types of sporangia in the Phycomyceteae and different groups of algae.

In the Phycomycetous hypothesis of the origin of the Ascomyceteae there is presented the difficulty of accounting for the production of the nonmotile sperms and of the trichogynes. Apparently very early in the development of the Phycomyceteae sexual reproduction took place by the

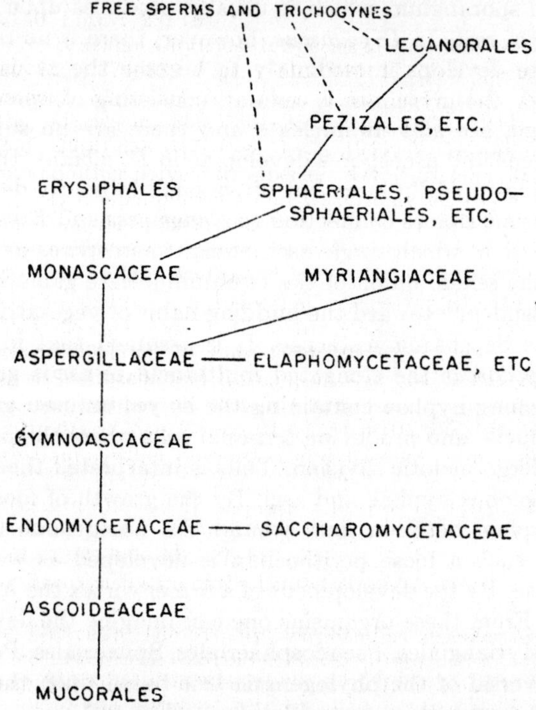

FIG. 207. Suggested lines of evolution within the Ascomyceteae, following in the main the ideas of Dangeard, Atkinson, Gäumann, Mez, et al. (After Bessey: *Mycologia*, **34**(4):355–379.)

union of two equal-sized motile gametes, as in some of the Chytridiales and Blastocladiales, by the union of two motile gametes of unequal size, as in some species of *Allomyces*, or by the union of a flagellate sperm with a nonflagellate egg, as in *Monoblepharis*. In all the other Phycomyceteae in which sexual reproduction is known, this occurs by the union of two nonmotile cells (*Olpidiopsis*) or by the passage of nuclei and cytoplasm from one gametangium to another one through a simple opening or through a conjugation tube. In *Dipodascus*, *Endomyces*, and most of the forms considered to be the primitive Ascomyceteae, if their Phycomycetous origin is accepted, we have this same general method of sexual reproduction, but in the forms considered under that hypothesis to be the furthest advanced we find separate nonmotile sperms uniting with special receptive organs (trichogynes). Whence have these been derived? They represent an entirely new development. Attempts have been made to consider the spermatia to be specialized conidia that have assumed the sexual function. That might possibly be so but it is far easier to surmise that they are the reproductive structures inherited from the Floridean ancestors where that is the standard mode of sexual reproduction.

The case of *Liagora tetrasporifera* Börgesen (1927) indicates how a short-cycled red seaweed has developed a life history very closely parallel to that of such an Ascomycete as some species of *Mycosphaerella*. In this representative of the Order Nemalionales, after spermatization of the trichogyne there soon grow out from the oogone a number of gonimoblasts whose terminal cells instead of becoming carpospores, as is the usual case in red seaweeds, become tetrasporangia which produce four tetraspores each. Although in this particular case the cytological behavior of the nuclei in the oogone after fertilization and in the tetrasporangium has not been studied it may well be assumed, from the analogy of other Nemalionales and other orders of the Florideae, that the gamete nuclei united normally in the oogone but did not undergo immediate meiosis, the result being that the nuclei in the gonimoblasts remained diploid. The terminal cells instead of becoming carpospores became tetrasporangia in which meiosis occurred and tetraspores with haploid nuclei were produced. Compare this with *Mycosphaerella*. A uninucleate sperm cell spermatizes a one-celled unbranched trichogyne and passes down into the oogone where it associates itself with, but does not unite with, the egg nucleus. The two nuclei then divide simultaneously by "conjugate division" until many pairs of nuclei are produced which pass out into the ascogenous hyphae. At or near the tip a pair of nuclei passes into a terminal or subterminal cell where they unite. Then this diploid nucleus undergoes meiosis to produce four haploid nuclei and, after the fashion of the majority of the Ascomyceteae, a further mitotic division occurs, so that eight ascospores arise in each ascus. It should be remembered that in many Ascomyceteae where typical sexuality is known only four nuclei and therefore not over four ascospores are produced in each ascus. Although one cannot assume that *Liagora* is a direct ancestor of the Ascomyceteae this shows that in the red seaweeds the conditions exist now that might have given rise ages ago to this group of fungi.

Possibly the acquisition of the land habit might have led to the production of ascospores in an ascus where the epiplasm is of importance in the scattering of the spores, whereas the tetraspores which depend upon the water currents for their distribution slip out from the tetrasporangium as naked cells.

Origin of Rusts and Smuts and Heterobasidiae

As mentioned above, Jackson (1944) has emphasized the parallelism of the life history types in the Uredinales and the Florideae. He has demonstrated beyond chance of contradiction that the various types of life history, such as alternation of haploid and diploid generations and shortened life cycles occur in both groups of organisms, together with a number of structural similarities (spermatia, receptive hyphae, etc.).

In view of the extreme, obligate type of parasitism of the rusts and the usual nonparasitic mode of life of the red seaweeds, and of the type of teliospores with promycelium—characters not indicated in the latter group—it appears evident that the relationship is not necessarily very close. Indeed the structure and function of these organs seem to indicate relationship with the Basidiomyceteae under which they are more often now classified. Bound up with the Uredinales is the question as to the relationship of the Ustilaginales with which they have very often been associated in classification. There is no denying the fact that in very many regards the latter order is far less specialized in structure and life history than the former. Yet there are a number of characters common to both: (1) a high degree of intercellular parasitism confined to hosts among the vascular plants; (2) a more or less well-marked distinction of two types of mycelium, sometimes associated with special types of spores and hosts, the monocaryon and dicaryon types of mycelium; (3) the production on the dicaryon mycelium of binucleate cells, usually with colored and thickened walls, in which the two nuclei unite; (4) the outgrowth from this teliospore of a thin-walled hypha (promycelium) of limited growth within which meiosis occurs and upon which are then borne the sporidia which are of two or four sexual phases; (5) the production by the latter of the monocaryon phase of mycelial growth; (6) the possession of various methods by which this mycelium becomes diploidized to form the dicaryon mycelial phase which produces the teliospores. Both orders give rise to repeating spores, i.e., asexual spores which repeat the mycelial phase upon which they are borne. In the Uredinales these are known only upon the dicaryon mycelium (being known as urediospores) but in some of the Ustilaginales they occur on both monocaryon and dicaryon types of mycelium.

In some regards the Ustilaginales appear the more primitive in that their dicaryon mycelium often has typical clamp connections, while these have been definitely discovered in only a few cases in the Uredinales. In the latter order there are present typically definite spermogonia which produce spermatia capable of spermatizing receptive hyphae (trichogynes) and thus diploidizing special cells from which are produced typical dicaryon spores (aeciospores). In the Ustilaginales the diploidization may occur in various ways but does not include spermatia and receptive hyphae.

If, then, the two orders Ustilaginales and Uredinales are considered to be related we must conclude that they have arisen probably (1) from a group of fungi in which clamp connections occurred or their homologues, the croziers, characteristic of ascus formation; (2) from fungi in which spermatia were produced in well-organized spermogonia; (3) from fungi in which sexual reproduction was initiated by the spermatization of receptive hyphae; and (4) from fungi in which the monocaryotic and dicaryotic mycelial phases were both present. Clamp connections are found in most groups of the Basidiomyceteae and in some lichens while their homologues, the croziers, occur in Lecanorales, Pezizales, Tuberales, Sphaeriales, Aspergillales, and in the Ascocorticiaceae and a few other forms. They are absent in the Laboulbeniales, Erysiphales, Saccharomycetales, and a number of other groups. Spermogonia are found in Laboulbeniales, Lecanorales, Sphaeriales, Pyrenulales, Dothideales, and some Pezizales. Special receptive hyphae (trichogynes) are found in Lecanorales, Pezizales, Laboulbeniales, Sphaeriales, Hypocreales, Pyrenulales, Dothideales, and a few other groups. The two mycelial phases are absent in the Saccharomycetales and Taphrinaceae. We must therefore consider that the Rusts and Smuts may have been derived from some Ascomyceteae possibly before the Pezizales, Sphaeriales, and Dothideales had become strongly distinguished from one another. Possibly these early ancestral Ascomyceteae if they were derived from Florideae that were diplobiontic might have had their two mycelial phases much more strongly marked than in the forms now known, in which the dicaryon phase has been reduced to or has not yet proceeded beyond the formation of the ascogenous hyphae.

Linder (1940), in a very thought-provoking discussion, has sought to derive the Uredinales from forms near the Dothideales. He assumed that the disappearance of clamp connections or croziers may have been connected with the parasitism entirely within the host tissues. He believed that the Ustilaginales (which still retain clamp connections) were derived from the Uredinales with reduced life cycles in which loss of spermatia and receptive hyphae had been recompensed by the ability of any two cells of opposite sexual phase to fuse and thus initiate the dicaryotic type of mycelium. It must be remembered that even in the Uredinales there appears to develop mutual diploidization of mycelia of opposite sexual phase when they meet in the tissues of the host, even though spermatization of receptive hyphae is prevented.

Linder's view of the origin of the teliospore and promycelium from the

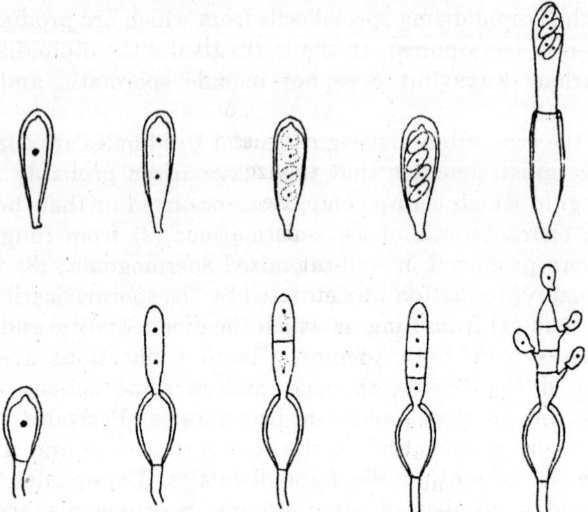

FIG. 208. Diagram showing Linder's suggestion as to the origin of the teliospore and promycelium of Uredinales from an ascus. (Courtesy, Linder: *Mycologia*, **32**(4): 419–447.)

ascus assumes that the type of ascus was that with a two-layered wall, such as occurs in *Pleospora*, *Leptosphaeria*, etc. The outer layer is firm and the inner layer thin and elastic. Upon the softening of the apex of the ascus where there is a germ pore the inner layer expands so that the endoascus and contained ascospores project some distance beyond the outer layer and eventually ruptures to set the spores free. He assumes that such an ascus became more thick-walled so as to become a resting spore but with the oncoming of favorable conditions the inner portion pushed out as described above, the meiotic division occurring usually within this projection. The contents of this extruded portion (promycelium or basidium) divide to form four spores by simple cross walls instead of by free spore formation as is usual in the Ascomyceteae. These four spores remaining attached to each other permit their contents to escape by forming external sterigmata from which secondary spores (sporidia or basidiospores) are set free. (Fig. 208.)

Accompanying the gradual loss of obligate parasitism and simplification of the life cycle the Uredinales according to Linder led to the Auriculariales, probably to those forms in which a "probasidium" is present such as *Septobasidium* and *Helicobasidium*. With the complete loss of the formation of a distinct probasidium, forms such as the entirely saprophytic *Auricularia* could be reached. In other more or less similar ways he postulates the origin of the Tremellales and from them the Autobasidiomyceteae (Eubasidiae).

Another view as to the origin of the Heterobasidiae (excluding the Uredinales and Ustilaginales) is that they may have arisen from some Ascomycetous form somewhat like *Ascocorticium*. In this fungus the asci are developed with the formation of croziers. There are no conspicuous paraphyses and between the wood or bark on which the fructification is produced there are only a few layers of hyphae running parallel to the surface. From the outer hyphae arise the asci. There is no definite limiting margin to this structure which may be looked upon as a much simplified apothecium. It is not known whether the subhymenial hyphae are dicaryotic or monocaryotic or a mixture of the two or whether the mycelium within the substratum is one or the other. Also, it is not known whether spermatia and receptive hyphae are present or how the dicaryophase arises. Therefore we do not have much on which to base a suggestion that a fungus of this structure is a possible forerunner of the *Corticium*-like Basidiomyceteae. If we do make this assumption, we must account for the origin of the basidium (holobasidium in this case) from the ascus. It is known that in all fungi with holobasidia the meiotic division of the diploid nucleus of the young basidium occurs within the latter until four nuclei, or very often eight through a subsequent mitotic division, are produced. Then there arise slender outgrowths from the upper portion of the basidium and these sterigmata enlarge terminally. At this point the nuclei, usually one to each, squeeze through the slender sterigmata and round up in their enlarged ends. Then, and not until then, is another wall laid down within this "spore." This wall is always visible separating the opening of the sterigma from the spore and frequently can be traced all over the inner surface of the outer spore wall, sometimes not tightly appressed to it. In other words, after the nucleus enters the terminal swelling of the sterigma it proceeds to induce the formation of a wall around the nucleus and accompanying cytoplasm. We may then interpret such a basidium as an ascus on which arise evaginations or external pockets within which the ascospores are produced. The so-called basidiospores therefore may be looked upon as ascospores enclosed within the evaginated ascus pockets. This is somewhat different from the suggestion of Linder by whom the septate promycelium of the teliospore was looked upon as a row of four ascospores, separated by simple septa, not rounded up as occurs in a normal ascus. From each such ascospore a short stalk (sterigma) produces a basidiospore which must then be looked upon as a secondary spore, budded off from the ascospore. Rogers (1932, 1934) looks upon *Tulasnella* as perhaps derived from an *Ascocorticium*-like ancestor. He interprets the four large pockets at the top of the basidium as ascus pockets, each containing the homologue of an ascospore. The spore produced on the slender sterigma from such a pocket is then, as in Linder's hypothesis, not homologous to an ascospore but, as Linder intimated, a secondary spore budded off from the sterigma sent out by the ascospore.

The fact that the asymmetrical manner of attachment of the basidiospore to the sterigma is found throughout the Heterobasidiae (except for a few "angiocarpous" forms) and in the Polyporales and Agaricales probably indicates that they all have a common origin, for it is unlikely that so intricate a mechanism for spore discharge has been polyphyletic in origin. Since the same structure occurs in the Uredinales and in the Tilletiaceae these too must be descended from the same ancestors. There is considerable evidence for the surmise that the Basidiomyceteae with symmetrically attached basidiospores which are not shot off from the tips of the sterigmata are secondary modifications from the asymmetrical types, associated with the angiocarpous development of the spore fruit (but see below for discussion of the Agaricales-Gasteromyceteae relationships).

If then the Uredinales (and Ustilaginales) are of common phylogenetic origin with the Heterobasideae and the Eubasidial Hymenomyceteae we must decide whether the heterobasidium or eubasidium is the more primitive. Linder's suggestion has been mentioned above that from the Uredinales arose all other types of Basidiomycetous fungi. Rogers, on the contrary, would derive the Uredinales from *Tulasnella*, *Tremella*, and Auriculariales, while deriving the Eubasidiae in another direction from the same primitive genera. If either of these hypotheses is correct then the objects that we call basidiospores (or sporidia) are not homologous with ascospores but are secondary spores produced from them to provide for aerial distribution, the homologues of the ascospores being the individual cells of the transversely or longitudinally septate basidium. If Rogers is right in postulating the change in position of the four septa in *Tulasnella* so as to divide the "probasidium" or "hypobasidium" vertically as in *Tremella* and that then by a further shifting of the septa (intermediate forms seem to occur) they become transverse, as in *Auricularia*, we still must insist upon the homology of the external basidiospores in all these cases. If the Uredinales arose from the Auriculariales by the emphasis on the development of the "probasidium" into a thick-walled resting spore, as occurs in many species of *Septobasidium*, the sporidia of these too are homologous to the foregoing basidiospores, i.e., are secondary spores. The close relationship of the Auriculariales, Tremellales, Dacrymycetales, Tulasnellales, Uredinales, and Ustilaginales would seem to be emphasized by the wide-spread occurrence of the production of secondary spores from the basidiospores (sporidia) by budding. The germination of these spores directly by single germ tubes may occur but in water or damp air they may give rise to yeast-like or to allantoid buds, often accompanied by transverse divisions of the basidiospore. Even in the Uredinales (e.g., as observed by the author in *Kunkelia nitens*) the sporidia in damp air while still attached to the promycelium may produce successively secondary, tertiary, and further sporidia, all obliquely perched on short sterigmata from the sporidia below. In addition there is a great tendency toward the production of spore fruits whose cell walls swell and become gelatinous when wet, although this is not universal. In the Uredinales we find this in *Gymnosporangium*, *Coleosporium*, *Uropyxis*, etc., chiefly in the teliospores and their pedicels. In the Auriculariales, Tremellales, Dacrymycetales, and some Tulasnellales so many genera have this character throughout the whole spore fruit that they are often grouped under the common name "Jelly Fungi." Both of the foregoing characters are rare in the Eubasidiae. However in *Exobasidium* the basidiospores may divide transversely and produce external buds but this is scarcely known in any other Eubasidial genera. Scattered here and there are occasional species with gelatinous cell walls but these are marked exceptions.

If the Thelephoraceae are to be considered as derived from near *Tulasnella* it must be by the elimination of the cross walls between the main body of the basidium and the four large spore-bearing pockets, accompanied by a reduction in size of these pockets until finally in most cases these have entirely disappeared leaving merely their apical sterigmata. In that case the homologues of the ascospores have, as it were, drawn back into the body of the basidium and entirely disappeared. If this is the way that evolution occurred we must still regard the basidiospores as secondary spores from hypothetical ascospores, as in the Tulasnellales, etc.

Let us consider the case if the holobasidium is the more primitive type, as it is beyond controversy the most frequent in the whole class. We can still derive it theoretically from one of the Pezizales somewhat similar to *Ascocorticium*, although probably spermatia were still produced, perhaps even in spermogonia. We can imagine that the basidiospores, as suggested by the author on page 647, are ascospores, in some cases provided with true ascospore walls, inside the everted pockets. The sterigmatal apparatus for their discharge may have to be looked upon as a new development to permit their being discharged violently, although perhaps this is a modification of the discharge mechanism found in most asci in which four or more spores contained in the ascus are discharged simultaneously or in quick succession through the single opening at the apex of the ascus. This is brought about by the increase of the osmotically produced tension of the ascus wall until it gives way at the apex and the epiplasm and spores are violently ejected. In the new modification as found in the basidium the discharge affects only one spore at a time at the apex of each sterigma. This new habit is clearly not confined to the primary spore for in the case of *Kunkelia* mentioned above the secondary and tertiary spores are similarly discharged and in the Tilletiaceae this seems to extend to secondary sporidia and even to conidia in some species. So we might imagine that this might have been a habit acquired for asexual spores that has been extended to the sporidium and basidiospore as well.

If the holobasidium was a primary step in the evolution from the *Ascocorticium*-like Ascomyceteae we can assume that in some cases the normally slender sterigmata as in *Corticium* became stouter, bearing at their tips the points on which the basidiospores perch and finally attaining the separable condition found in *Tulasnella*. The basidium itself may be considered to have formed vertical walls as in *Tremella* and finally transversely septate walls would bring about a structure like the basidium of *Auricularia*. From *Ceratobasidium* which has stout sterigmata, only two in number in some species, it is an easy step to *Dacrymyces*.

From the simple, resupinate forms like *Corticium*, *Tomentella*, etc. it is easy to make surmises as to the modifications which increase the hymenial surface so as to derive the Hydnaceae, Clavariaceae, Polyporaceae, and eventually Agaricaceae.

In the immediately preceding paragraphs we have left the Uredinales hanging without a connection. They cannot be derived from any Auriculariales such as we now know, for well organized spermogonia and receptive hyphae are lacking in the latter. We can only guess that the transformation from ascus to holobasidium occurred deeper down in the Ascomyceteae than the present *Ascocorticium* and that the development of holobasidium to phragmobasidium took place perhaps at several levels or when once accomplished proceeded in various directions. The earlier Uredinales must have come from the newly developed holobasidiomycete while the latter still retained the primitive characters of its ancestors, including spermogonia, receptive cells, diplobiontic life cycle, croziers or clamp connections, and probably well pronounced parasitism. We must assume that at an early stage of development along this line thick-walled resting probasidia began to develop. This would be the beginning of teliospore formation. On resuming their growth these probasidia would grow out as thin-walled promycelia except in such cases as *Coleosporium* where the teliospore wall was thin enough to permit of stretching as a whole, thus producing a so-called "internal promycelium" instead of the usual external one characteristic of the order. We need not homologize the separate cells of the promycelium whether external or internal with ascospores but consider these cells as parts of a phragmobasidium. Under this assumption the sporidia may be considered as ascospores in external pockets not as secondary spores. It must be admitted that this whole suggestion stands on shaky foundations and the whole matter needs further study. (Fig. 209.)

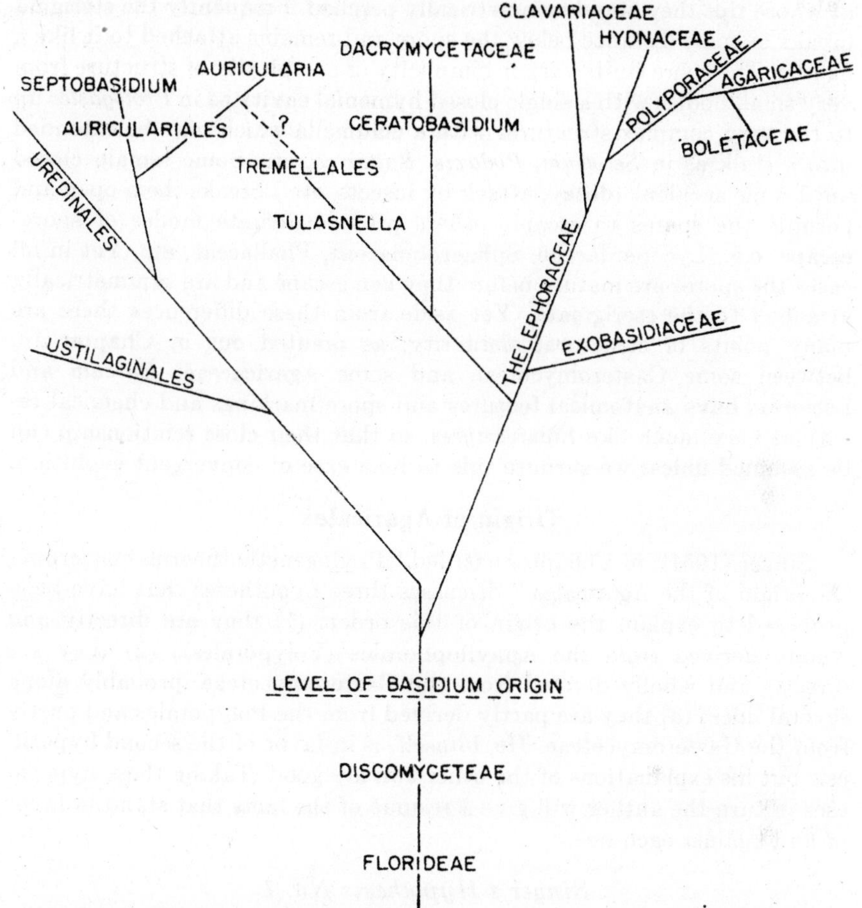

FIG. 209. Diagram showing the author's ideas as to the possible origin of and relationships between the chief families of Subclasses Teliosporeae, Heterobasidiae, and Hymenomycetous Eubasidiae.

Phylogeny of Hymenomyceteae and Gasteromyceteae

Another very uncertain phylogenetic tree is that concerned with the relationship of the Gasteromyceteae with the Hymenomyceteae. The former, be it remembered, differ from the latter by the fact that the spore fruits do not open and permit the escape of the spores until after the basidia and basidiospores have reached maturity, while in the latter the hymenium becomes exposed to the air before the basidia have produced their spores. Associated with this Gasteromycetous structure of the spore fruit is the fact that the basidiospores are not shot off from the sterigmata at whose tips they stand symmetrically perched. Frequently the sterigma breaks at some distance below the spore and remains attached to it like a handle. The spore fruits vary in simplicity or complexity of structure from very small bodies with a single closed hymenial cavity as in *Protogaster* up to the very complex structures with a columella which may be extended into a stalk as in *Secotium*, *Podaxis*, *Battarrea*, etc. Some remain closed until some accident (decay, attack by insects, etc.) breaks them open and permits the spores to escape, others provide definite modes of spore-escape, e.g., Lycoperdaceae, Sphaerobolaceae, Phallaceae, etc. Yet in all cases the spores are mature before they can escape and are symmetrically attached to the sterigmata. Yet aside from these differences there are many points of structural similarity, as pointed out in Chapter 15, between some Gasteromyceteae and some Agaricaceae. *Russula* and *Lactarius* have anatomical features and spore markings and chemical reactions very much like *Elasmomyces*, so that their close relationship can be assumed unless we surmise this to be a case of convergent evolution.

Origin of Agaricales

Singer (1951), in a chapter entitled "Phylogenetic theories concerning the origin of the Agaricales," discusses three hypotheses that have been proposed to explain the origin of this order: (1) they are directly and wholly derived from the Aphyllophorales (Polyporales); (2) they are directly and wholly derived from the Gasteromyceteae, probably along several lines; (3) they are partly derived from the Polyporales and partly from the Gasteromyceteae. He, himself, is in favor of the second hypothesis but his explanations of the other two are good. Taking these hypotheses in turn the author will give a résumé of the facts that stand in favor of and against each one.

Singer's Hypothesis No. 1

Agaricales Derived Wholly from Polyporales. This is perhaps the oldest theory since the time that the idea of evolution began to affect the classification of fungi. Briefly sketched, it assumes that from a simple resupinate spore fruit with a smooth, or possibly rather loose, cottony, hymenium, evolution progressed toward more compact structures some of which assumed a lateral position or even became stipitate. These forms with flat hymenium made up the old family Thelephoraceae. The amount of hymenial surface in proportion to the size of the spore fruit was limited, so long as the hymenium was a plane surface. This surface became increased by the formation of hymenium-covered pegs or teeth projecting from the surface, leading to the Hydnaceae. In a slightly different manner the surface became increased by a reticulate pattern of outgrowth so that eventually there were very numerous pits or pores of various depths, lined by hymenium, the Polyporaceae. By the radial elongation of these pores so that they extended from the point of attachment (more often a central stipe) to the edge of the spore fruit a system of gills or lamellae was developed, the Agaricaceae. The lateral branches producing the Exobasidiaceae and Clavariaceae were not in the main line of upward evolution and need not be given further consideration in this hasty review. Up to the Agaricaceae and the closely related Boletaceae practically all the foregoing fungi were gymnocarpous in their ontogeny. The hymenium was developed externally over the whole surface of the spore fruit or over part of the surface beginning as a plane which in the more advanced forms became thrown into teeth, pores, or gills. Naturally, therefore, those Agaricaceae were considered the most primitive in which this gymnocarpous ontogeny continued. Upon study of the development of the fruiting body from very early stages of growth it has been shown that in many Agaricaceae and in the Boletaceae the primordium of the hymenium is laid down gymnocarpously, but that by the outward spreading and downward and inward curving of the edge of the pileus this hymenial surface is finally enclosed in a circular tunnel surrounding the top of the stipe and sealed from the exterior by the inturned edge of the pileus which has grown fast to the stipe. Into this circular space the gills grow from the pileus and the basidia begin to form on them but before they have matured sufficiently to produce their spores the edge of the pileus breaks loose from the stipe and pulls away from it so that the gills at their maturity are exposed to the exterior. This is called the pseudoangiocarpous mode of development. Beyond this type of development we find many Agaricaceae in which the hymenial primordium originates in a circular layer entirely closed from its first inception within the tissues of the spore fruit. Not until the gills have reached almost their full development does the growth of the fruiting body bring about a circular rupture when the expanding pileus pulls away from the stipe, so that finally the gills are exposed to the air and the basidia shed their spores. This angiocarpous mode of development is found especially in those genera with a volva or with a well-developed annulus or cortina or a universal veil, even though this may be rather thin. Since the Gasteromyceteae have a peridium that in some forms, at least, seems homologous to the universal veil and since, as mentioned above, other Agaric-like structures are present in some members of this group it has been suggested that these represent further stages of evolution from the Agaricaceae. Thus Heim (1937, 1948) suggested that the Asterosporales are a closely related group, containing the genera *Russula* and *Lactarius* of the Agaricales and *Arcangeliella, Elasmomyces, Maccagnia,* and some others usually included in the Gasteromyceteae. He indicates his belief that these last mentioned genera have been derived from *Russula* and *Lactarius.*

The chief merit of this suggested system is that the Agaricales appear to be an extension, as it were, of the Polyporales with a gradual increase in complexity of ontogeny from strict gymnocarpy to pseudoangiocarpy and finally to complete angiocarpy. From the latter it would seem logical to extend this development to the Gasteromyceteae. Here, where the spores are set free from the basidia before the spore fruit opens we find that they are symmetrically attached to the sterigmata or even practically sessile on the surface of the basidium.

There are two obstacles to this system. In the first place the type of the trama and character of the basidia and cystidia are in the main quite different in the Agaricales and Polyporales. So that even where, externally, similarities and apparent transitions seem to occur careful study of the anatomy and chemical reaction of the hymenium and trama show that these are perhaps better explained as morphological convergences but not true relationships. Furthermore it would seem more reasonable to expect that the Agaricaceae which possess a well-developed universal veil, as revealed by the production of volva, annulus, etc., and which have an angiocarpic mode of development, would be far more likely to have arisen from the mostly strictly angiocarpic Gasteromyceteae and that the reduction and final disappearance of a universal veil and the appearance of pseudoangiocarpy and finally of gymnocarpy would indicate a gradual degeneration from the complex to simplified forms, now that the firm universal veil (peridium) had become no longer necessary. A third and very strong objection is that the Gasteromyceteae that show the closest similarity to the Agaricaceae lead by gradually simplified structures to the forms assumed by almost all mycologists to be the simplest ones, the forms that well may have developed from organisms related to *Aleurodiscus*, *Corticium*, *Tomentella*, etc.

Singer's Hypothesis No. 2

Agaricales Derived Wholly from Gasteromyceteae. This is the hypothesis, favored by Singer and at least in part by Bucholtz (1903), that the Polyporales do not give rise to the Agaricales; the latter arise from the Gasteromyceteae and at some points, mostly by convergent evolution, assume forms externally similar to some of those of the Polyporales. Its chief virtue is that it explains the presence of the universal veil and of angiocarpy in some of the Agaricales and allows a logical explanation of their disappearance as the rather definite affinities to the Gasteromyceteae become more distant. The undeniable close similarities of some Agaricales to some Polyporales is difficult to explain away. To the author the most outstanding objection to this hypothesis is the necessity of assuming the evolution anew of the sterigmatal function of spore discharge at several points, for if this hypothesis is adopted the Agaricaceae probably arose from the Gasteromyceteae at several different places. In this viewpoint the simplest organisms of this group are supposed to have originated from simple Thelephoraceae, which already had their spores perched asymmetrically on the tips of the sterigmata from which they are discharged. Several cases are known where near relatives of the Tremellales have adopted the angiocarpous structure and with it have lost this peculiarity of sterigmatal discharge. It is difficult however to understand how so complex a structure could have been lost at the beginning of the Gasteromyceteae and regained several times independently as some Agaricaceae developed from them.

Singer's Hypothesis No. 3

Agaricales Derived Both from Polyporales and from Gasteromyceteae. This is really a sort of compromise. Its chief objection is that we have to find a line of separation within the Agaricaceae between those descended from the Polyporales and from the Gasteromyceteae. This is perhaps even more difficult than the first or second hypothesis.

The relationships within the Gasteromyceteae as well as their origin are very uncertain. To be sure Eduard Fischer (1933 and earlier) proposed a system that is fairly logical, progressing from simpler to more complex structures, and this has been the basis for much of the work in this group for the last fifty years. The characters that have been considered as the more important are the modes of development of the gleba, the progressive development of the columella (and its downward extension, the stipe) and the tendency to progress from a coralloid to a multipilar and eventually unipilar structure. Other important characters are the development of the capillitium, the formation of definite hymenial cavities or their obliteration by ingrowing hyphae (plectobasidial structure), the autodigestion of the gleba to an evil smelling mass attractive to insects, angiocarpous or pseudoangiocarpous development, etc.

One very important point, in the author's opinion, that has been subordinated to a secondary position is the character of the spores. It appears to him that there are several types of spores that cut across the family and ordinal boundaries as customarily recognized. These, perhaps, are of much greater importance and it may be that they indicate closer relationships than have been recognized. Besides spore structure a much more extensive study of the very early stages of the formation of the spore fruits is absolutely essential before a well-grounded system can be established.

In the genera *Gasterella*, *Hymenogaster*, and *Gasterellopsis* we have, respectively, very small, unicameral spore fruits, a moderate sized organism with very numerous hymenial chambers with a more or less coralloid arrangement, and a small, at first unicameral, structure with a percurrent columella and gill-like outgrowths from the top of the chamber which may divide it into several radial cavities. The spores in all three are ovoid or limoniform, dark-colored and symmetrically perched on the sterigmata. Many of the spores when set free bear at their base a piece of the upper end of the sterigma. In their spore character they are practically identical. The first two could find a place in the same family, Hymenogastraceae, but the third with its percurrent columella and radial hymenial chambers and basally circumscissile dehiscence shows some characters that might be considered as belonging to the Secotiaceae. This is merely an example to show how, possibly, the current classifications are faulty.

Phylogenetic System of Gasteromyceteae

In default of further studies on the anatomy and ontogeny of the very young spore fruits and of intensive comparison of spore types throughout the whole group the following may be suggested as a tentative phylogenetic system of the Gasteromyceteae. The primitive (or the simplest) forms are minute, with one closed hymenial chamber. This might be supposed to have arisen from a small *Aleurodiscus*-like fungus in which the upward curvature at the edges continued until a closed cavity was produced, lined by the hymenium. As seems to be the case when angiocarpy develops the basidia come to bear sterigmata with symmetrically attached spores. The genus *Protogaster* represents one of these small, unicameral fungi. Much like this, but with an entirely different type of spores, is *Gasterella*. In the spore fruit of this a somewhat arched hymenial primordium arises angiocarpously and gradually a basidial layer appears with the basidia directed downwards to line a shallow cavity which soon becomes larger and nearly spherical. In the more vigorous specimens the roof of this cavity may be thrown into folds and convolutions which do not reach to the base so that although the hymenial surface is increased the cavity is not divided. At this stage it resembles closely the early developmental stages of *Hymenogaster* as described by Rehsteiner (see Chapter 15). Probably the next step beyond the unicameral condition arose by the increase in the folds and convolutions to form many cavities lined by hymenium instead of just the one. In the Hysterangiaceae the coralloid development is much more marked along with the production of a more or less pronounced columella. The cartilaginous and gelatinous character of the gleba is possibly a modification to permit the dissolution of the spore fruit at maturity. In the Lycoperdaceae the glebal tissues enclosing the numerous hymenial cavities dissolve after the spores have become mature, leaving one large cavity filled with the spores entangled in the filamentous capillitium. By the flaking off of the outer layer of the peridium and the formation of an ostiole in the inner peridium or some other mode of dehiscence the spores are enabled to escape and be scattered by air currents.

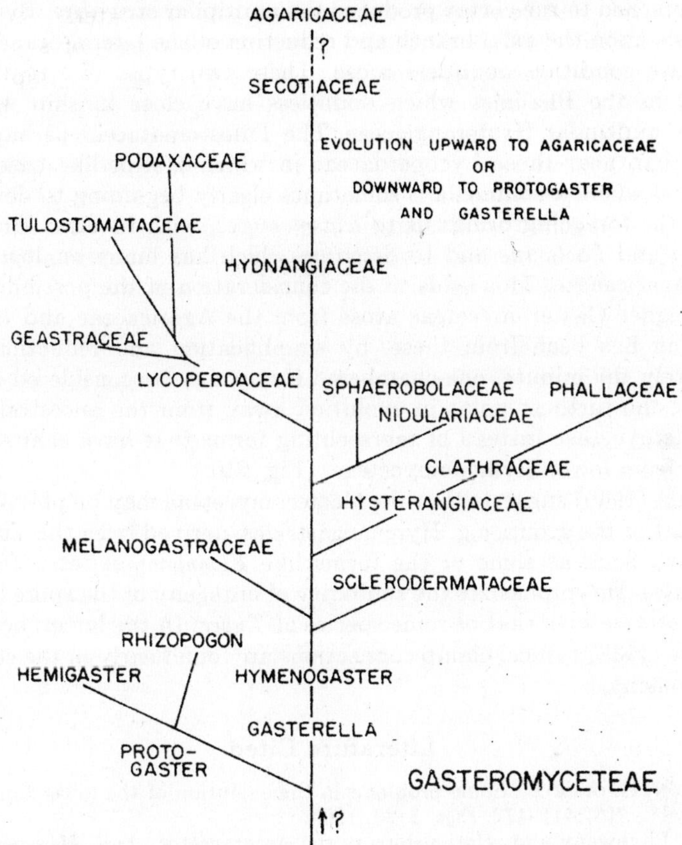

FIG. 210. Diagram showing the possible relationships within the Gasteromyceteae, based largely upon Fischer (1933), with some additions.

The plectobasidial structure of some Gasteromyceteae appears, in the author's opinion, to have arisen at many points in evolution by the failure of the hymenium to form a distinct layer bounding each hymenial cavity when produced. In its place the basidiogenous hyphae grow unequally and fill these cavities with a loose mass of hyphae bearing scattered basidia. Since this appears in one or two species of Sphaerobolaceae while other species have normal hymenial cavities, in *Astraeus* while the rest of the Geastraceae have typical glebal structure, and in some genera of Tulostomataceae while some genera are not plectobasidial, it seems more reasonable to distribute such plectobasidial forms among genera with normal hymenial cavities with which they show the closest resemblance instead of putting them all together in one series.

The coralloid structure by emphasis on a number of main branches which reached to the cortex produced the multipilar structure. By further emphasis upon the axial branch and reduction of the lateral branches the unipileate condition doubtless arose. These two types of structure are present in the Phallales which doubtless have close kinship with the simpler, multipilar Hysterangiaceae. The Tulostomataceae perhaps have their origin near those Lycoperdaceae in which a stipe-like base shows itself and where a columella is sometimes clearly beginning to develop.

All the foregoing brings us to *Elasmomyces*, with its close kinship to *Russula* and *Lactarius* and to *Secotium* which has many analogies with other Agaricaceae. This leads to the consideration of the possibility that these higher Gasteromyceteae arose from the Agaricaceae and that the evolution has been from these, by simplification and reduction, until ultimately the minute, one-chambered forms may be considered to have reached the furthest point in evolution away from the ancestral higher Hymenomyceteae instead of representing forms that have changed only a little from lower Hymenomyceteae. (Fig. 210.)

Holm (1949) suggests that the Gasteromyceteae may be polyphyletic, with part of the group, e.g. Hymenogastrales, derived from the Tuberales and part, such as some of the forms like *Elasmomyces*, etc., from the Agaricales. He emphasizes the similarity of ontogeny of the spore fruits of *Hymenogaster* with that of some species of *Tuber*. In the latter, according to Greis (1938) typical clamp connections are found early in the course of development.

Literature Cited

ATKINSON, GEORGE F.: Some problems in the evolution of the lower fungi, *Ann Mycol.*, **7**(5):441–472. *Figs.* 1–20. 1909.

———: Phylogeny and relationships in the Ascomycetes, *Ann. Missouri Botan. Garden*, **2**(1–2):315–376. *Figs.* 1–10. 1914.

DE BARY, ANTON: Vergleichende Morphologie und Biologie der Pilze, Mycetezoen und Bacterien, xvi + 588 pp. 198 *figs.* Leipzig, Wilhelm Engelmann, 1884.

BESSEY, ERNST A.: Some problems in fungus phylogeny, *Mycologia*, **34**(4):355–379. *Figs.* 1–5. 1942.

———: Studies on Pilobolus: *P. kleinii* and *P. longipes*, *Papers Mich. Acad. Sci.*, **32**:15–25. *Pls.* 1–3. 1946 (1948).

BÖRGESEN, F.: Marine algae from the Canary Islands: III. Rhodophyceae, 1. Bangiales and Nemalionales, *Kgl. Danske Videnskab. Selskab Biol. Medd. VI*, **6**:1–97. Köbenhavn. 1927.

BREFELD, OSCAR: Basidiomyceten: III. Autobasidiomyceten und die Begründung des natürlichen Systems der Pilze, Untersuchungen aus dem Gesammtgebiete der Mykologie, Heft. 8, pp. 1–305. *Pls.* 1–12. Leipzig, Arthur Felix, 1889.

BUCHOLTZ, F.: Zur Morphologie und Systematik der Fungi hypogaei, *Ann. Mycol.*, **1**(2):152–174. *Pls.* 4–5. 1903.

BULLER, A. H. REGINALD: Researches on Fungi, vol. 5, pp. i–xiii, 1–416. *Figs.* 1–174. London, Longmans, Green and Co., 1933. (Especially p. 154, *Fig.* 78.)

COUCH, JOHN N.: A new Conidiobolus with sexual reproduction, *Am. J. Bot.*, **26**(3):119–130. *Figs.* 1–53. 1939.

DANGEARD, P. A.: L'origine du périthèce chez les Ascomycètes, *Le Botaniste*, **10**:1–385. *Pls.* 1–91. 1907.

DODGE, B. O.: The morphological relationships of the Florideae and the Ascomycetes, *Bull. Torrey Botan. Club*, **41**(3):157–202. *Figs.* 1–13. 1914.

ELLISON, BERNARD R.: Flagellar studies on zoospores of some members of the Mycetozoa, Plasmodiophorales and Chytridiales, *Mycologia*, **37**(4):444–459. *Figs.* 1–4. 1945.

FISCHER, EDUARD: Gastromyceteae, in A. ENGLER und K. PRANTL: Die Natürlichen Pflanzenfamilien, Zweite Auflage, vol. 7a, pp. 1–122. *Figs.* 1–91. Leipzig, Wilhelm Engelmann, 1933.

GÄUMANN, ERNST ALBERT: Vergleichende Morphologie der Pilze, pp. 1–626. *Figs.* 1–398. Jena, Gustav Fischer, 1926.

GREIS, HANS: Die Sexualvorgänge bei Tuber aestivum und Tuber brumale, *Biol. Zentralb.*, **58**(11–12):617–631. *Figs.* 1–3. 1938.

HEIM, ROGER: Les Lactario-Russulés du Domaine Orientale de Madagascar. Essai sur la classification et la phylogénie des Astérosporales, pp. 1–196. 8 *pls.* (*4 colored*). 59 *figs.* 2 *phylogenetic diagrams.* Paris, Laboratoire de Cryptogamie du Muséum National d'Histoire Naturelle, 1937 (1938).

———: Phylogeny and natural classification of macro-fungi, *Brit. Mycol. Soc. Trans.*, **30**:161–178. *Figs.* 1–19. 1948.

HOLM, LENNART: Some aspects on the origin of the Gastromycetes, *Svensk Botanisk Tidskrift*, **43**(1):65–71. 1949.

JACKSON, H. S.: Life cycles and phylogeny in the higher fungi. Presidential address, *Trans. Roy. Soc. Can.*, Ser. 3, *V*, **38**:1–32. *Figs.* 1–5. 1944.

VON JACZEWSKI, A. A.: Zur Phylogenie der Pilze, *Phytopath. Z.*, **1**(2):117–150. 1929–30.

JUEL, H. O.: Über Zellinhalt, Befruchtung und Sporenbildung bei Dipodascus, *Flora oder Allgemeine Botan. Ztg.*, **91**:47–55. *Pls.* 7–8. 1902.

DE LAGERHEIM, G.: Dipodascus albidus, eine neue geschlechtliche Hemiascee, *Jahrb. wiss. Botan.*, **24**:549–565. *Pls.* 24–26. 1892.

LINDER, DAVID H.: Evolution of the Basidiomycetes and its relation to the terminology of the basidium, *Mycologia*, **32**(4):419–447. *Figs.* 1–6. 1940.

MEZ, CARL: Versuch einer Stammesgeschichte des Pilzreiches, *Schriften königsberg. gelehrten Ges. Naturw. Klasse*, **6**:1–58. 1 *fig.* 1929.

ROGERS, DONALD P.: A cytological study of Tulasnella, *Botan. Gaz.*, **94**(1):86–105. *Figs.* 1–79. 1932.

———: The basidium, *Univ. Iowa Studies in Natural History*, **16**:160–183. *Pl.* 7. 1934.

SACHS, JULIUS: Lehrbuch der Botanik. Vierte, umgearbeitete Auflage, xvi + 928 pp. 492 *figs.* Leipzig, Wilhelm Engelmann, 1874.

SINGER, ROLF: Phylogenetic theories concerning the origin of the Agaricales, a chapter in The Agaricales (Mushrooms) in modern taxonomy, *Lilloa Revista de Botanica*, **22**:1–834. *Pls.* 1–29. 1949 (1951).

SKUPIENSKI, F. X.: Recherches sur le cycle évolutif de certains Myxomycètes, 83 pp. 2 *pls.* 2 *figs.* Paris, Imprimerie M. Flinikowski, 1920.

SPARROW JR., FREDERICK K.: Aquatic Phycomycetes Exclusive of the Saprolegniaceae and Pythium, xix + 785 pp. 69 *figs.* Ann Arbor, Univ. Mich. Press, 1943.

THOMAS, R. C.: Composition of fungus hyphae: III. The Pythiaceae, *Ohio J. Sci.*, **42**:60–62, 1942; IV. Phytophthora, *ibid.*, **43**:135–138. 1943.

18

GUIDE TO THE LITERATURE FOR THE IDENTIFICATION OF FUNGI

The vast number of fungi makes a knowledge of even a considerable part of the species beyond the ability of all but a few specialists. To be able to identify a given specimen requires that the necessary literature be available. Before 1880 the great Italian mycologist, P. A. Saccardo, began to work on a compilation of all species of fungi described up to that time. These descriptions were brought together in a compendious work entitled "Sylloge Fungorum," the first volume of which appeared in 1882. The descriptions, in Latin, were arranged in accordance with the systematic classification of fungi then recognized by the author. He completed the work with Volume 8 in 1889, but in the meantime, such was the stimulus afforded by this great work bringing together in one place descriptions of all known species of fungi, that thousands of additional species had been recognized and described. Thus it became necessary to publish supplementary volumes, the last of which, Volume 25, appeared in 1931. For many years species of fungi have been described at the rate of 1500 to 2500 species a year. Thus it is inevitable that such a work must be from two to five years behindhand. A further difficulty is the language of the description, Latin, a knowledge of which is unfortunately all too meager among the later generation of botanists. Furthermore the lack of illustrations and the necessary scattering of the descriptions among the original volumes and the many supplements make the work difficult to use. Besides this, the necessary high cost precludes its purchase by most botanists so that they must depend upon copies owned by libraries.

To obviate these difficulties local fungus floras have been issued in various countries or even subdivisions of countries. These publications are usually in the language of the country and are furthermore smaller, inasmuch as only the species occurring in the limited areas concerned are included.

Still another type of publication is the monograph, or intensive study of some smaller group of fungi, such as a family or a genus or even a section of a genus. By virture of the limitation of effort to this relatively small number of species it is possible for such a monograph to devote larger space to the description of the individual species and to make their identification easier.

In the following pages the more general works are first listed; then under different headings are included the publications that are confined more to special groups. These are arranged systematically in accordance with the classification of the fungi. It is the aim of the author to list the more recent publications of this nature from all parts of the world, but it is certain that there are many omissions, especially for parts of the world whose mycological literature is not so widely available in the United States as is the literature of this country and of the larger European countries. Even for the latter the large number of publications of somewhat limited scope from botanical or mycological societies makes a complete list difficult to obtain as in no one library will even a majority of such works be found. The disruption of communication and the destruction due to World War II and the subsequent disturbed political and economic conditions have made it increasingly difficult to keep in touch with the mycological work of other regions of the world.

The student is advised to turn first to the portion of the list where these papers of monographic nature are to be found. In case no such paper is listed for the fungus he has, he must turn to the more general lists.

Because of the various systems of classification used by different authors the arrangements of the items in the following lists do not follow any one system in all its details. Cross references are necessary in the cases where one work includes groups now segregated although formerly united.

It must be remembered that certainty of identification depends not only upon the availability of the necessary literature but also upon the fullness of the knowledge of the structure, development, etc., of the fungus in question. The literature should not be consulted until the main morphologic and anatomic details have been ascertained, including measurements of spores, sporophores, etc. The fuller the knowledge of details as to the substratum on which the fungus grows, its habitat, color, appearance when fresh, appearance at different stages of development, etc., the more easily will the identification be effected.

List 1. General Works Covering the Whole Field of Systematic Mycology

SACCARDO, P. A.: Sylloge fungorum omnium hucusque cognitorum, 25 vols. issued up to 1931. Pavia, Italy, published by the author.

1:1–768. 1882. Pyrenomyceteae.

2:1–959. 1883. Pyrenomyceteae (continued).

3:1–860. 1884. Sphaeropsideae and Melanconieae.

4:1–807. 1886. Hyphomyceteae.

5:1–1146. 1887. Hymenomyceteae: I. Agaricineae.

6:1–928. 1888. Hymenomyceteae: II. Polyporeae, Hydneae, Thelephoreae, Clavarieae, Tremellineae.

7:1–941. 1888. Gasteromyceteae, Phycomyceteae, Myxomyceteae, Ustilagineae, Uredineae.

8:1–1143. 1889. Discomyceteae, Phymatosphaeriaceae, Tuberaceae, Elaphomycetaceae, Onygenaceae, Laboulbeniaceae, Saccharomycetaceae, Schizomycetaceae.

9:1–1141. 1891. Supplement 1: Hymenomyceteae, Gasteromyceteae, Hypodermeae (Ustilaginaceae and Uredinaceae), Phycomyceteae, Pyrenomyceteae, Laboulbeniaceae.

10:1–964. 1892. Supplement 2: Discomyceteae, Onygenaceae, Tuberoideae, Myxomyceteae, Sphaeropsideae, Melanconieae, Hyphomyceteae, Fossil Fungi.

11:1–753. 1895. Supplement 3: All groups of Fungi. Generic index to all volumes.

12:1–1053. 1897. Index to vols. 1–11.

13:1–1340. 1898. Host index.

14:1–1316. 1899. Supplement 4: All groups of Fungi. Sterile Mycelia.

15:1–455. 1901. Supplement 5: Synonyms.

16:1–1291. 1902. Supplement 6: All groups of Fungi. Generic index to all volumes.

17:1–991. 1905. Supplement 7: Hymenomyceteae, Gasteromyceteae, Uredinaceae, Ustilaginaceae, Phycomyceteae, Pyrenomyceteae, Laboulbeniomyceteae.

18:1–838. 1906. Supplement 8: Discomyceteae (including Saccharomycetaceae, Exoascaceae, Gymnoascaceae, Tuberaceae, etc.), Myxomyceteae, Deuteromyceteae (Fungi Imperfecti). Generic index to all volumes.

19:1–1158. 1910. Index of illustrations of Fungi, A–L.

20:1–1310. 1911. Index of illustrations of Fungi, M–Z.

21:1–928. 1912. Supplement 9: Hymenomyceteae, Gasteromyceteae, Ustilaginaceae, Uredinaceae, Phycomyceteae.

22:1–612. 1913. Supplement 10: Ascomyceteae, Deuteromyceteae, Sterile Mycelia.

23:1–1026. 1925. Supplement 10 (continued): Hymenomyceteae, Ustilaginales, Uredinales.

24(Section I):1–703. 1926. Supplement 10 (continued): Phycomyceteae, Laboulbeniales, Pyrenomyceteae, in part.

24(Section II):704–1438. 1928. Supplement 10 (continued): Remainder of Pyrenomyceteae, Discomyceteae; Appendix, consisting of additions to vols. 23 and 24.

25:1–1093. 1931. Supplement 10 (continued): Myxomyceteae, Myxobacteriaceae, Deuteromyceteae, Mycelia Sterilia.

(Volumes 1, 10, and 17 contain bibliographies; Vol. 14 contains an explanation of the arrangement of the genera by the spore form and color scheme.)

ENGLER, A., UND K. PRANTL: Die natürlichen Pflanzenfamilien, Leipzig, Wilhelm Engelmann. The parts devoted to fungi (including the Lichens) are the following:

Teil I, Abteilung 1:1–513. *Figs.* 1–293. 1897. Myxomyceteae, Phycomyceteae, Ascomyceteae.

Teil I, Abteilung 1*:1–249. *Figs.* 1–125. 1907. Lichens.

Teil I, Abteilung 1**:1–570. *Figs.* 1–263. 1900. Basidiomyceteae, including Hemibasidii (Ustilaginales) and Uredinales. Fungi Imperfecti.

(This work will enable one to determine the genus of almost any fungus but not the species. It is very helpful because of the illustrations.)

———, UND ———: Die natürlichen Pflanzenfamilien, Zweite Auflage. Leipzig, Wilhelm Engelmann.

2:304–339. *Figs.* 425–447. 1928. Mycetozoa.

5b:1–42. 22 *figs.* 1938. Tuberineae.

6:1–290. *Pls.* 1–5. *Figs.* 1–157. 1928. Ustilaginales, Uredinales, Hymenomyceteae.

7a:1–122. *Figs.* 1–91. 1933. Gastromyceteae.

8:1–270. *Figs.* 1–127. 1926. Lichens.

CLEMENTS, FREDERICK E., AND CORNELIUS L. SHEAR: The Genera of Fungi, iv + 496 pp. 58 *pls.* New York, H. W. Wilson Company, 1931.

MARTIN, G. W.: Outline of the fungi, *Univ. Iowa Studies in Natural History*, **18**(supplement):1–64. *Figs.* 1–118. 1941.

RABENHORST, L.: Kryptogamen-Flora von Deutschland, Oesterreich und der Schweiz, Zweite Auflage. Leipzig, Verlag von Eduard Kummer.

Band 1. WINTER, GEORG: Die Pilze. This "Volume" on Fungi is so extensive that it is issued as ten separately bound "Abteilungen," as follows:

1:1–924. 1 *pl. and numerous text figs.* 1884. Schizomyceten, Saccharomyceten, und Basidiomyceten. By A. DE BARY, H. REHM, and GEORG WINTER.

2:1–928 and Index, 1–112. *Numerous text figs.* 1887. Ascomyceten: Gymnoasceen und Pyrenomyceten. By A. DE BARY, H. REHM, and GEORG WINTER.

3:1–1275 and Index, 115–169. *Numerous text figs.* 1896. Ascomyceten: Hysteriaceen und Discomyceten. By H. REHM.

4:1–505. *Figs.* 1–74. 1892. Phycomyceten. By ALFRED FISCHER.

5:1–131. *Numerous text figs.* 1897. Ascomyceten: Tuberaceen und Hemiasceen. By EDUARD FISCHER.

6:1–1016. *Numerous text figs.* 1901. Fungi Imperfecti: Hyalin-sporige Sphaerioideen. By ANDREAS ALLESCHER.

7:1–993 and Index to Abteilungen 6 and 7, pp. 995–1072. *Numerous figs.* 1903. Fungi Imperfecti: Gefärbt-sporige Sphaerioideen, sowie Nectrioideen, Leptostromaceen, Excipulaceen und Melanconieen. By ANDREAS ALLESCHER.

8:1–852. *Numerous text figs.* 1907. Fungi Imperfecti: Hyphomyceten: Mucedinaceen und Dematiaceen (Phaeosporae und Phaeodidymae). By G. LINDAU.

9:1–983. *Numerous text figs.* 1910. Fungi Imperfecti: Dematiaceen (Phaeophragmiae bis Phaeostaurosporae), Stilbaceen und Tuberculariaceen. By G. LINDAU.

10:1–474. 182 *figs.* 1920. Myxogasteres (Myxomycetes, Mycetozoa). By HANS SCHINZ.

Band 8. KEISSLER, KARL VON: Die Flechtenparasiten, xi + 712 pp. 135 *figs.* 1930.

Band 9. ZAHLBRUCKNER, ALEXANDER: Die Flechten. This is divided into Abteilungen, Teile und Lieferungen of which, so far as the records are available, the following have appeared to date:

Abteilung I, Teil 1. VON KEISSLER, KARL: Moriolaceae, pp. 1–43. *Figs.* 1–15. 1933. ZSCHACKE, HERMANN: Epigloeaceae, Verrucariaceae und Dermatocarpaceae, pp. 44–695. *Figs.* 16–344. 1934. Teil 2. VON KEISSLER, KARL: Pyrenulaceae, Trypetheliaceae, Pyrenidiaceae, Xanthopyreniaceae, Mycoporaceae und Coniocarpineae, pp. 1–846. *Figs.* 1–210. 1937–1938.

Abteilung II, Teil 1. REDINGER, K.: Arthoniaceae, Graphidaceae, Chiodectonaceae, Dirinaceae, Rocellaceae, Lecanatidiaceae, Thelotremaceae, Diploschistaceae, Gyalectaceae, Coenogoniaceae, pp. 1–404. 2 *pls. Figs.* 1–107. 1937–1938.

Abteilung III. MAGNUSSON, A. K.: Lecideaceae. (The numbers of Lieferungen and pages and dates were not available.)

Abteilung IV, Erste Hälfte. FREY, EDUARD: Cladoniaceae (unter Ausschluss der Gattung Cladonia), Umbilicariaceae, pp. 1–426. *Pls.* 1–8. *Figs.* 1–64. 1933. Zweite Hälfte. SANSTEDE, HEINRICH: Die Gattung Cladonia, pp. 1–531. *Pls.* 1–34. *Figs.* 1–8. 1931.

Abteilung V, Teil 1. MAGNUSSON, A. K.: Acarosporaceae und Thelocarpaceae, pp. 1–320. *Figs.* 1–64. 1935. ERICKSEN, C. F. E.: Pertusariaceae, pp. 321–728. *Figs.* 1–74. 1935–36. Teil 3, Lieferungen 1–2. HILLMANN, JOHANNES: Parmeliaceae, pp. 1–309. Register 1–10. *Figs.* 1–16. 1936.

Abteilung VI, Lieferung 1. HILLMANN, JOHANNES: Teloschistaceae, pp. 1–36. *Figs.* 1–4. 1935. LYNGE, BERNT: Physciaceae, pp. 37–188. *Pls.* 1–12. *Figs.* 1–10. 1935.

Kryptogamenflora der Mark Brandenburg, Leipzig, Gebrüder Borntraeger. This work appears in nine or more volumes of which the following concern the fungi:

5. 630 pp. 151 *figs.* 1915. Schizomycetes, by R. KOLKWITZ; Myxobacteriales, by E. JAHN; Chytridiineae, Ancylistineae, Monoblepharidineae, Saprolegniineae, by M. VON MINDEN.

5a. 946 pp. 380 *figs.* 1914. Uredineen, by H. KLEBAHN; Ustilagineen, Auriculariineen, Tremellineen, by G. LINDAU.

6. Not complete. Issued so far only:

Heft 1:1–92. *Many illustrations.* 1910. By W. HERTER. Autobasidiomycetes: Dacryomycetaceae, Exobasidiaceae, Tulasnellaceae, Corticiaceae, Thelephoraceae, Cyphellaceae, Craterellaceae, Clavariaceae, Sparassiaceae, Hydnaceae (incomplete).

6a. 264 pp. 114 *figs.* 1935. Mucorineae. By H. ZYCHA.

7. Not complete. Issued so far only:

Heft 1:1–160. *Many illustrations.* 1905. Hemiasci, by G. LINDAU; Saccharomycetineae, by P. LINDNER; Protoascineae, by G. LINDAU; Exoascaceae, by F. NEGER; Ascocorticiaceae and Gymnoascaceae, by G. LINDAU; Aspergillaceae, by F. NEGER; Onygenaceae, Elaphomycetaceae and Terfeziaceae, by P. HENNINGS; Erysiphaceae and Perisporiaceae, by F. NEGER; Tuberaceae (incomplete) by P. HENNINGS.

Heft 2:161–304. *Many illustrations.* 1911. Tuberaceae (completed), by

P. HENNINGS; various families of the Sphaeriales, by W. KIRCHSTEIN.

Heft 3:305–448, 1938. Ascomycetes (continued), by W. KIRCHSTEIN.

8. No parts yet issued.

9. 962 pp. 339 *figs.* 1915. Sphaeropsideen, Melanconien, by H. DIEDICKE.

OUDEMANS, C. A. J. A.: Révision des Champignons tant supérieurs qu'inférieurs jusqu'a ce jour trouvés dans les Pays-Bas.

1: 1–638. Hymenomycetes, Gasteromycetes, Hypodermeae. Amsterdam, J. Müller, 1893.

2: 1–491. *Pls.* 1–14. Phycomycetes and Pyrenomycetes. Separate reprint from *Verh. Kon. Akad. Wet. Amsterdam* 2. Ser II. 1897.

SCHROETER, J.: Die Pilze Schlesiens, in FERDINAND COHN: Kryptogamen-Flora von Schlesien, vol. 3. Breslau, J. V. Kern's Verlag.

Erste Hälfte: pp. 1–814. 1889. Myxomycetes, Schizomycetes, Chytridiei, Zygomycetes, Oomycetes, Protomycetes, Ustilaginei, Uredinei, Auriculariei, Basidiomycetes. Host index for this half volume.

Zweite Hälfte: pp. 1–597. 1908. Ascomycetes and a small part of the Fungi Imperfecti. Host index for second half volume.

MIGULA, W.: Krytogamen-Flora von Deutschland, Deutsch-Österreich und der Schweiz, in D. W. Thomé: Flora von Deutschland etc., Zweite Auflage. Gera, Friedrich von Zezschwitz.

Band 3, Teil 1. iv + 510 pp. 92 *pls.* 1910. Myxomycetes, Phycomycetes, Basidiomycetes (Ordnungen Ustilagineae und Uredineae).

Teil 2. iv + 814 pp. 304 *pls.* 1912. Basidiomycetes (completed).

Teil 3, Abteilung 1. iv + 1–684. *Pls.* 1–100. 1913. Hemiasci, Saccharomycetineae, Protodiscineae, Plectascineae, Pyrenomycetes (Perisporiales und Sphaeriales).

Teil 3, Abteilung 2. iv + 685–1404. *Pls.* 101–200. 1913. Dothideales, Hypocreales, Hysteriales, Discomycetes, Laboulbeniaceae.

Teil 4, Abteilung 1. iv + 614 pp. 90 *pls.* 1921. Fungi Imperfecti.

Band 4, Teil 1. viii + 527 pp. 82 *pls.* 1929. Flechten.

Teil 2. iv + 868 pp. 143 *pls.* 1931. Flechten (conclusion).

JACZEWSKI, A. A.: Identification of Fungi, vol. 1, 1913; vol. 2, 1917. (In Russian.)

———, AND P. A. JACZEWSKI: Identification of Fungi. Perfect Forms (Diploid Stages), Tom I. Phycomycetes, ed. 3, 294 pp. 329 *figs.* Leningrad and Moscow, 1931. (In Russian.)

COOKE, M. C.: Handbook of British Fungi with Full Descriptions of all the Species and Illustrations of the Genera, 2 vols. ii + 981 pp. 7 *pls.* 408 *figs.* London, Macmillan and Co., 1871; ed. 2, 1883.

———: Handbook of Australian Fungi, xxxii + 457 pp. 36 *pls.* London, Williams and Norgate, 1892.

MASSEE, GEORGE: British Fungus Flora, a Classified Text-book of Mycology, London, George Bell and Sons.

Vol. 1. xii + 432 pp. *Illustrated.* 1892. Gastromycetes; Tremellineae; Clavarieae; Thelephoreae; Hydneae; Polyporeae; Agaricineae: Melanosporeae and Porphyrosporeae.

Vol. 2. vii + 460 pp. *Illustrated.* 1893. Agaricineae: Ochrosporeae, Rhodosporeae, Leucosporeae.

Vol. 3. viii + 512. *Illustrated.* 1893. Agaricineae: Leucosporeae; Hyphomycetes.

Vol. 4. viii + 522 pp. *Illustrated.* 1895. Ascomycetes.

———: British Fungi with a Chapter on Lichens, 551 pp. *Colored pls.* 1–40. *Pls.* A–B. 19 *unnumbered figs.* London, George Routledge and Sons. Undated (about 1911).

MASSEE, GEORGE, AND IVY MASSEE: Mildew, Rusts and Smuts: A Synopsis of the Families Peronosporaceae, Erysiphaceae, Uredinaceae and Ustilaginaceae, 229 pp. *Pls.* 1–5. London, Dulau and Co., 1913.

STEVENS, F. L.: The Fungi which Cause Plant Disease, ix + 754 pp. 449 *figs.* New York, Macmillan, 1913.

———: Hawaiian Fungi, *Bernice P. Bishop Museum Bulletin* **19**:i–ii, 1–189. *Pls.* 1–10. *Figs.* 1–35. Honolulu, 1925.

North American Flora, published by the New York Botanical Garden, New York. Of this work various parts describing fungi have been issued, as follows:
- Vol. 1, Pt. 1. 1949. Mycetozoa.
- Vol. 2, Pt. 1. 1937. Blastocladiales, Monoblepharidales and Saprolegniales.
- Vol. 3, Pt. 1. 1910. Hypocreales and Fimetariales.
- Vol. 6, Pt. 1. 1922. Phyllostictales (part).
- Vol. 7, Pts. 1–13. 1906–1931. Ustilaginales and Uredinales.
- Vol. 9, 1–542. 1907–1916. Polyporaceae (part), Boletaceae, Agaricaceae (part).
- Vol. 10, Pts. 1–5. 1917–1932. Agaricaceae (continued).

LARSEN, P.: Fungi of Iceland, vol. 2, pt. 3, in L. KOLDERUP ROSENVINGE and E. WARMING: The Botany of Iceland. Copenhagen and London, Oxford Univ. Press, 1932.

HEIM, ROGER, (ED.): Flore mycologique de Madagascar et Dépendances, Paris, Laboratoire de Cryptogamie du Muséum National d'Histoire Naturelle,
- I. HEIM, ROGER: Les Lactario-Russulés. 196 pp. 8 *pls.* 60 *figs.* 1938.
- II. ROMAGNESI, H.: Les Rhodophylles. 146 pp. 46 *figs.* 1941.
- III. METROD, GEORGES: Les Mycènes. 144 pp. 88 *figs.* 1949.
- IV. LEGAL, MARCELLE: Les Discomycètes Operculés. (To appear 1950.)
- V. BUCHET, SAMUEL: Les Myxomycètes. (To appear 1951.)
- VI. HEIM, ROGER, ET RAYMOND DECARY: Les Phalloïdées. (In preparation.)
- VII. BOURIQUET, GILBERT: Les Rouilles. (In preparation.)

Flora Italica Cryptogama. Firenze, published under the auspices of the Società Botanica Italiana. Pars I. Fungi.

TRAVERSO, G. B.: Elenco bibliografico della mycologia Italiana, Fasc. 1:1–118, Supplemento I:119–135. 1905. Supplemento II, Fasc. 9:1–151. 1912.

———: Pyrenomyceteae: Xylariaceae, Valsaceae, Ceratostomataceae, Fasc. 2:1–352. *Figs.* 1–68. 1906; Sphaeriaceae allantosporae, hyalosporae, phaeosporae. Fasc. 3:353–492. *Figs.* 69–97. 1907; Sphaeriaceae hyalodidymae, Fasc. 11:493–700. *Figs.* 98–116. 1913.

TROTTER, ALEX: Uredinales, Fasc. 4:1–519. *Figs.* 1–110. 1908.

PETRI, L.: Gasterales, Fasc. 5:1–139. *Figs.* 1–83. 1909.

FERRARIS, T.: Hyphales: Tuberculariaceae, Stilbaceae, Fasc. 6:1–198. *Figs.* 1–53. 1910; Dematiaceae, Fasc. 8:199–534. 1912; Mucedinaceae, Fasc. 10:535–846. 1913; Indice generale, Fasc. 13:847–979. 1914.

SACCARDO, P. A., ADIUVANTE HIER. DALLA CORDA: Hymeniales: Leucosporae et Rhodosporae, Fasc. 14:1–576. *Pls.* 1–6. *Figs.* 1–7. 1915; Hymeniales: ceterae Agaricaceae, Polyporaceae, Hydnaceae, Thelephoraceae, Tremellaceae, Fasc. 15:577–1386. *Pls.* 7–11. 1916.

COLLA, S.: Laboulbeniales, Fasc. 16:1–157. *Figs.* 1–108. 1934.

CIFERRI, RAPHAEL: Ustilaginales, Fasc. 17:1–443. *Figs.* 1–23. 1938.

(Other fascicles have probably appeared but their citations are not available.)

Pars. III. Lichenes.

JATTA, A.: Lichenes, Fasc. 1: i–xxii, 1–958. *Figs.* 1–80. 1909.

BARGHOORN, E. S., AND D. H. LINDER: Marine fungi: their taxonomy and biology, *Farlowia*, **1**(3):395–467. *Pls.* 1–7. *Figs.* 1–3. 1944.

VAN OVEREEM, C., UND J. WEESE: Icones fungorum Malayensium Abbildungen und Beschreibungen der Malayischen Pilze, Hefte 1–16. 16 *colored pls.* 1 *pl. in black and white.* Weesp, Holland, Mycol. Museum, 1923–1926.

CORDA, A. C. I.: Icones fungorum hucusque cognitorum, Prag, J. G. Calve (vols. 1–4), Fr. Ehrlich (vols. 5–6).

1:1–32. *Pls.* 1–7. 1837. Mostly Fungi Imperfecti; a few Uredinales, Mycetozoa and miscellaneous fungi.

2:1–43. *Pls.* 8–15. 1838. Fungi Imperfecti, Mucorales, Mycetozoa, a few Uredinales, Ustilaginales, Pezizales and miscellaneous fungi.

3:1–55. *Pls.* 1–9. 1839. Uredinales, Fungi Imperfecti, Agaricales, Miscellaneous.

4:1–53. *Pls.* 1–10. 1840. Erineum galls, Fungi Imperfecti, Uredinales, Mycetozoa, Agaricales and various Ascomyceteae.

5:1–92. *Pls.* 1–10. 1842. Erineum galls, Fungi Imperfecti, Mucorales, Mycetozoa, Gasteromycetes, Tuberales, Agaricales and Miscellaneous.

6:i–xix, 1–91. *Pls.* 1–20. 1854. (Plates by A. C. I. CORDA, text by J. B. ZOBEL). Uredinales, Fungi Imperfecti, Mucorales, Mycetozoa, Gasteromycetes, Tuberales, and Miscellaneous.

JUILLARD-HARTMANN, G.: Iconographie des champignons supérieurs, 5 vols. 250 *colored pls.* Epinal, Juillard et Fils. vol. 1, 1919, others not dated. Illustrations in color of approximately 2400 species of fungi. Vols. 1–3 and part of vol. 4 represent Agaricaceae, the remainder of vol. 4 illustrates Polyporaceae, Boletaceae, Fistulinaceae; vol. 5 illustrates Hydnaceae, Clavariaceae, Thelephoraceae, Exobasidiaceae, Gastromyceteae, Dacryomycetales, Tremellales, Auriculariales, Helvellaceae, and a few subterranean Ascomyceteae. No descriptions accompany the plates.

KONRAD, P., ET A. MAUBLANC: Icones selectae fungorum. 5 vols. of plates totaling 500; 1 vol. text, over 500 pp. Paris, Paul Lechevalier, 1924–1937.

BRESADOLA, J.: Fungi Tridentini novi vel nondum delineati, descripti et iconibus illustrati, Trieste. Published by author.

1:1–114. *Pls.* 1–105. 1881.

2:118 pp. *Pls.* 106–217. 1892.

———: Iconographia mycologica, edited by J. TRAVERSO, L. FENAROLI, G. CATONI, AND J. B. TRAVERSO. 24 vols. 1200 *pls.* Milan, Società Bot. Italica, Seg. Lombard. 1927–1932. Vols. 1–18, Agaricaceae; vols. 19–21, Polyporaceae and part of Hydnaceae; vol. 22, remainder of Hydnaceae, Thelephoraceae, part of Clavariaceae; vol. 23, Clavaria, Auriculariaceae, Tremellaceae, Dacryomycetaceae, Gastromyceteae; vol. 24, Helvellaceae, Leotiaceae, Pezizaceae.

COUPIN, HENRI: Album general des Cryptogames. Fungi (Champignons). Les Champignons du Globe, 5 vols. 473 *pls.* Paris, E. Orlhac, about 1920–1925. (Gives illustrations of nearly all the recognized genera of fungi except the lichens.)

SCHWARZE, CARL A.: The parasitic fungi of New Jersey, *New Jersey Agr. Expt. Sta. Bull.* **313**:1–226. *Figs.* 1–1056. 1917. (Contains beautiful illustrations of very many genera and species of parasitic fungi.)

List 2. Host Indexes, Local Fungus Lists with Host Indexes, Lists of Fungi on Special Hosts or Substrata, Bibliographies, Fungi of Man and Other Animals

SACCARDO, P. A. Sylloge fungorum omnium hucusque cognitorum, 13:1–1340. 1898. Also there is a generic host index (Repertorium) at the close of vols. 14, 16, 17, 18, 21, 25, covering the fungi included in the respective volumes.

OUDEMANS, C. A. J. A.: Enumeratio systematica fungorum, The Hague, Martin Nijhoff.

1. cxxvi + 1230 pp. 1919. Host Index of Algae, Fungi, Bryophyta, Pteridophyta, Gymnosperms and Monocotyledons.
2. xix + 1069 pp. 1920. Dicotyledons: Salicaceae—Basellaceae.
3. xvi + 1313 pp. 1921. Dicotyledons: Caryophyllaceae—Vitaceae.
4. xiii + 1231 pp. 1923. Dicotyledons: Elaeocarpaceae—Compositae. Supplement.
5. vii + 999 pp. 1924. Index to species of hosts and fungi in vols. 1–4. (This is a host index of all parasitic fungi reported in any part of the world on plants native to Europe or introduced into Europe.)

SEYMOUR, ARTHUR BLISS: Host Index of the Fungi of North America, xiii + 732 pp. Cambridge, Harvard Univ. Press, 1929.

ANDERSON, PAUL J.; ROYAL J. HASKELL; WALTER C. MUENSCHER; CLARA J. WELD; JESSIE I. WOOD; and G. HAMILTON MARTIN: Check list of diseases of economic plants in the United States, *U.S. Dept. Agr. Dept. Bull.* **1366**:1–111. *Figs.* 1–4. 1926.

WEISS, FREEMAN: Check list revision, *Plant Disease Reptr.*, **24–33**, various numbers, 1940–1949. (A revision of the foregoing.)

ALSTATT, G. E.: Diseases of plants reported in Texas since 1933, *Plant Disease Reptr.*, supplement **135**:37–50. 1946.

WATERSTON, J. M.: The fungi of Bermuda, *Dept. Agr. Bermuda, Bulletin* **23**: i–iii, 1–305. *Figs.* 1–38. 1947.

BISBY, G. R., with the collaboration of A. H. R. BULLER, JOHN DEARNESS, W. P. FRASER, R. C. RUSSELL, and with a preface by H. T. GÜSSOW: The Fungi of Manitoba and Saskatchewan, 189 pp. 1 *map.* 49 *figs.* Ottawa, National Research Council of Canada, 1938. (Lists over 2700 species of fungi, with host index.)

BUTLER, E. J., AND G. R. BISBY: Fungi of India, Imperial Council of Agricultural Research of India. *Science Monograph* **1**:i–xviii, 1–237. 1931.

MUNDKUR, B. B., AND M. J. THIRUMALACHAR: Revisions of and additions to Indian Fungi, I, *Mycological Papers. Commonwealth Mycological Inst.*, **16**:1–27. 19 *figs.* 1946.

———, AND SULTAN AHMAD: Revisions of and additions to Indian Fungi, II, *ibid.*, **18**:1–11. 8 *figs.* 1946.

BROWN, CHARLES C.: Contributions toward a host index to plant diseases in Oklahoma, *Oklahoma Agr. Expt. Sta. Circ.* 33. 1939. Revised edition 1941.

———: Supplement No. 1. *ibid., Mimeographed Circular M.* **104**:1–32. 1943.

PRESTON, D. A.: Host Index of Oklahoma plant diseases, supplement, 1948, *Plant Disease Reptr.*, **32**(9):398–401. Sept. 15, 1948.

MELCHERS, L. E.: A check list of plant diseases and fungi occurring in Egypt, *Trans. Kansas Acad. Sci.*, **34**:41–106. 1931.

COOKE, WM. BRIDGE: Preliminary host index to fungi of Mt. Shasta, California, *Plant Disease Reptr.*, supplement **123**:125–133. 1940.

COONS, G. H.: A preliminary host index of the fungi of Michigan, exclusive of the Basidiomycetes, and of the plant diseases of bacterial and physiological origin, *Mich. Acad. Sci. Rept.*, **14**:232–276. 1912.

DAVIS, J. J.: A provisional list of the parasitic fungi of Wisconsin, *Trans. Wisconsin Acad. Sci.*, **17**(2):846–984. 1914.

———: Parasitic Fungi of Wisconsin, 157 pp., Madison, Wis., published for the author posthumously. 1942.

PARRIS, G. K.: A check list of fungi, bacteria, nematodes and viruses occurring in Hawaii, and their hosts, *Plant Disease Reptr.*, supplement **121**:1–91. 1940.

NOBLE, R. J.; H. J. HYNES; F. C. MCCLEERY; AND W. A. BIRMINGHAM: Plant diseases recorded in New South Wales, *Dept. Agr. of New South Wales Sci. Bull.* **46**:1–47, 1934; Supplement to the foregoing, *ibid.*, supplement **1**:1–7, 1937.

MCALPINE, D.: Systematic arrangement of Australian fungi together with host-index and list of works on the subject, Department of Agriculture, Victoria, vii + 236 pp. Melbourne, Government Printer, 1895.

SCHADE, ARTHUR L.: A preliminary list of the parasitic fungi of Idaho, *Plant Disease Reptr.*, supplement **95**:77–113. 1936.

STEVENSON, JOHN A.: A check list of Porto Rican fungi and a host index, *J. Dept. Agr. Porto Rico*, **2**:125–264. 1918.

KAWAMURA, S.: The Japanese Fungi, ed. 3, Tokyo, 1930. (In Japanese.)

MANEVAL, WILLIS E.: A list of Missouri fungi with special reference to plant pathogens and wood destroying species, *Univ. Missouri Studies*, **12**(3):1–150. 1937. (A list of the reported species of fungi from Missouri and a host index.)

GILMAN, JOSEPH C.: First supplementary list of parasitic fungi from Iowa, *Iowa State Coll. J. Sci.*, **6**:357–365. 1931.

———: Second supplementary list of parasitic fungi from Iowa, *ibid.*, **23**(3):261–272. 1949.

———, AND W. ANDREW ARCHER: The fungi of Iowa parasitic on plants, *ibid.*, **3**(4):299–507. 2 *figs.* 1929.

SPRAGUE, RODERICK: A revised check list of the parasitic fungi on cereals and other grasses in Oregon, *Plant Disease Reptr.*, supplement **134**:1–36. 1942.

BOYCE, J. S.: Host relationships and distribution of Conifer rusts in the United States and Canada, *Trans. Connecticut Acad. Arts Sci.*, **35**:329–482. 1943.

THIRUMALACHAR, M. J.: Some fungal diseases of Bryophytes in Mysore, *Brit. Mycol. Soc. Trans.*, **31**(1–2):7–12. *Figs.* 1–8. 1947.

CHARDON, CARLOS E., AND RAFAEL A. TORO: Mycological exploration of Venezuela, *Monographs of the Univ. Puerto Rico, Physical and Biological Sciences, B*, **2**:1–353. *Pls.* 1–33 (1 *colored*). 1 *map*. 1934. (Keys to various families. Host index.)

MOLLER, F. H.: Fungi of the Faeroes: I. Basidiomycetes, 295 pp. *Map.* 3 *pls.* (*colored*). 134 *figs.* Copenhagen, Ejnar Munksgaard Forlag, 1945.

DARKER, G. D.: A brief host index of some plant pathogens and virus diseases in Eastern Asia, *Plant Disease Reptr.*, supplement **122**:93–123. 1940.

GARCIA RADA, GERMAN, Y J. A. STEVENSON: La flora fungosa Peruana. Lista preliminar de hongos que atacan a las plantas en el Peru, 112 pp. Lima, Estación Experimental Agricola de la Molina, 1942. (List of Peruvian fungi and host index.)

OSTERO, JOSÉ I., AND MELVILLE T. COOK: A bibliography of mycology and phytopathology of Central and South America, Mexico and the West Indies, *J. Agr. Univ. Puerto Rico*, **2**(3):249–486. 1937.

REINKING, OTTO A.: Higher Basidiomycetes from the Philippines and their hosts, I, *Philippine J. Sci.*, **15**(5):479–490. 1919; II, *ibid.*, **16**(2):167–179. 1920; III, *ibid.*, **16**(5):527–537. 1920.

———: Host index of diseases of economic plants in the Philippines, *Philippine Agr.*, **8**(1–2):38–54. 1919.

TEODORO, NICANOR G.: An enumeration of Philippine fungi, *Commonwealth of the Philippines, Department of Agriculture and Commerce, Technical Bull.* **4**: 1–585. 1937. (Contains a host index.)

RAABE, ACHILLES: Parasitische Pilze der Umgebung von Tübingen. Ein Beitrag zur Kryptogamenflora Südwestdeutschlands, *Hedwigia*, **78**(1–2):1–106. *Pl.* 1. *Figs.* 1–7. 1938.

SĂVULESCU, TRIAN, ET C. SANDU-VILLE: Contribution à la connaissance des Micromycètes de Roumanie, *Bull. trimestr. soc. mycol. France*, **46**(3–4):177–192. 1930. (See the three further contributions following below.)

———, UND ———: Beiträge zur Kenntnis der Micromyceten Rumäniens, *Hedwigia*, **73**:71–132. 1933.

———, UND ———: Beiträge zur Kenntnis der Micromyceten Rumäniens, *ibid.*, **75**:159–233. 1935.

———, ET ———: Quatrième contribution à la connaissance des Micromycètes de Roumanie, *Acad. Română Mem. Sect. Ştiinţifice (Bucharest)*, **15**:397–502. 15 *pls.* 1939–40 (1941).

———, UND ———: Die Erysiphaceen Rumäniens, *Annales Scientifiques de l'Académie de Hautes Études Agronomiques de Bucarest*, **1**:47–123. *Pls.* 1–24. 1929. (Keys and host index.)

———: Contributions à la connaissance des Ustilaginées de Roumanie, *Ann. inst. recherches agron. Roumanie*, **7**:1–86. *Pls.* 1–35. 1935. (Distribution list and host index.)

———, ET T. RAYSS: Contribution à l'étude de la Mycoflore de Palestine, *Ann. cryptogam. exotique*, **8**(1–2):49–87. 1935.

SIEMASZKO, WINCENTY: Badania mykologiczne w górach Kaukazu (Recherches mycologiques dans les montagnes du Caucase), *Archiwum Nauk Biologicznych Towarzystwa Naukowego Warszawskiego (Disciplinarum Biologicarum Archivum Societatis Scientiarum Varsaviensis)*, **1**(14):1–57. *Fig.* 1. 1923.

BREMER, HANS; H. ISMEN; G. KAREL; H. OZKAN; UND M. OZKAN: Beiträge zur Kenntnis der parasitischen Pilze der Türkei, III, *Istanbul Universitesi Fen Fakultesi Mecmuasi (Rev. Fac. Sci. Univ. Istanbul), Ser. B., Sci. Nat.*, **13**(1):1–53. 1948.

CUMMINS, GEORGE B.: Annotated check list and host index of the rusts of Guatemala, *Plant Disease Reptr.*, supplement **142**:79–131. 1943.

FAULL, JOSEPH HORACE: Tropical fern hosts of rust fungi, *J. Arnold Arboretum Harvard Univ.*, **28**(3):309–319. 1947.

LIENEMAN, CATHERINE: Host index of the North American species of the genus Cercospora, *Ann. Missouri Botan. Garden*, **16**(1):1–52. 1929.

MAKJU, NAZEER AHMED: Contribution to our knowledge of Indian coprophilous fungi, *J. Indian Botan. Soc.*, **12**(2):153–164. *Pls.* 1–2. 1933.

GINAI, MOHAMMED ASGHER: Further contributions to our knowledge of Indian coprophilous fungi, *ibid.*, **15**(5):269–284. *Pls.* 20–22. 1936.

NIETHAMMER, A.: Die mikroskopischen Boden-Pilze, 1–193. *Pls.* 1–6. *Figs.* 1–57. The Hague, W. Junk, 1937.

GILMAN, J. C.: A manual of soil fungi, 1–392. *Figs.* 1–135. Ames, Iowa State College Press, 1945.

JENSEN, C. N.: Fungous flora of the soil, *Cornell Univ. Agr. Expt. Sta. Bull.* **315**: 415–501. *Illustrated.* 1912.

SABET, YOUNIS S.: On some fungi isolated from soil in Egypt, *Fouad I Univ. Bull. Fac. Sci.*, **19**:61–112. *Figs.* 1–45. 1939.

BISBY, G. R.: An Introduction to the Taxonomy and Nomenclature of Fungi, 117 pp. Kew, England, Imperial Mycological Institute, 1945.

LINDAU, G., ET P. SYDOW: Thesaurus litteraturae mycologicae et lichenologicae, Leipzig, Gebrüder Borntraeger.

1:1–903. 1908. Authors A to L, up to 1906, incl.

2:1–808. 1909. Authors M to Z, up to 1906, incl.

3:1–766. 1913. Corrections and additions up to 1910, incl.

4:1–609. 1915. Subject lists. Applied mycology, geographical distribution, pathology.

5:1–526. 1917. Subject lists (continued). Systematically arranged.

(A nearly complete bibliography of all mycological literature up to the close of 1910, arranged alphabetically by the authors. Vols. 4 and 5 are arranged by subjects, the plants diseased under their hosts.)

GUBA, E. F., AND P. A. YOUNG: Check list of important references dealing with the taxonomy of fungi, *Trans. Am. Microscop. Soc.*, **43**:17–67. 1924.

CHARLES, VERA K.: A preliminary check list of the entomogenous fungi of North America, *Insect Pest Survey Bull.* **21**:707–785. 1941. (Supplement to No. 9.)

AINSWORTH, G. C., AND G. R. BISBY: A Dictionary of the Fungi, ed. 2, viii + 431 pp. 138 *figs.* Kew, Surrey, The Imperial Mycological Institute, 1945.

THAXTER, ROLAND: On certain peculiar fungus-parasites of living insects, *Botan. Gaz.*, **58**(3):235–253. *Pls.* 16–19. 1914. (See also Lists 11, 12, and 14, under Entomophthorales, Zoopagales, Eccrinales, and Laboulbeniales.)

PETCH, T.: A list of the entomogenous fungi of Great Britain, *Brit. Mycol. Soc. Trans.* **17**(3):170–178. 1932. (A list with hosts and localities of all entomogenous fungi known to occur in Great Britain, with the exception of the Laboulbeniales.)

———: A revised list of British entomogenous fungi, *ibid.*, **31**(3–4):286–304. 1948.

———: Studies in entomogenous fungi, *ibid.*, **7–12,** about 1922–1927. (Scattered papers.)

———: Notes on entomogenous fungi, *ibid.*, **16–27,** 1931–1944. (Scattered papers.)

WATSON, W.: List of British fungi parasitic on lichens or which have been included as lichens (or vice versa), with some notes on their characters and distinctions, *Brit. Mycol. Soc. Trans.*, **31**(3–4):305–339. 1948.

DODGE, CARROLL W.: Medical Mycology. Fungous Diseases of Men and Other Mammals, 900 pp. 142 *figs.* St. Louis, C. V. Mosby Co., 1935.

OTA, MASAO, ET MAURICE LANGERON: Nouvelle classification des Dermatophytes, *Ann. parasitol.*, **1**(4):305–336. *Figs.* 1–8. 1923.

SARTORY, A.: Champignons parasites de l'homme et des animaux, 895 + 47 pp. 50 *pls.* 91 *figs.* Paris, Lefrançois, 1920.

———: Champignons parasites de l'homme et des animaux, 1er supplément, pp. 1–78. *Pls.* 1–2. *Figs.* 1–11. Paris, Lefrançois, 1923.

———, ET J. BAILLY: Champignons parasites de l'homme et des animaux, 2me supplément, pp. 1–95. Paris, Editions Clinique et Laboratoire, 1927.

———; A. GODEAU; R. SARTORY; L. BAILLY; ET J. MEYER: Champignons parasites de l'homme et des animaux, 3me supplément, pp. 1–159. Paris, Editions Clinique et Laboratoire, 1933.

VUILLEMIN, PAUL: Les champignons parasites et les mycoses de l'homme, Encyclopédie Mycologique, vol. 2, pp. 1–290. *Figs.* 1–140. Paris, Paul Lechevalier et Fils, 1931.

CONANT, N. F.; D. S. MARTIN; D. T. SMITH; R. D. BAKER; AND J. L. CALLAWAY: Manual of Clinical Mycology, 348 pp. *Illustrated.* Philadelphia, W. B. Saunders Co., 1945.

List 3. Mycetozoa, Including Myxogastrales, Plasmodiophorales, Acrasiales, Labyrinthulales

LISTER, ARTHUR: A monograph of the Mycetozoa, ed. 3, revised by Gulielma Lister, xxxii + 296 pp. 222 *pls.* 60 *figs.* London, Trustees of the British Museum, 1925.

MACBRIDE, THOMAS H., AND G. W. MARTIN: The Myxomycetes, a descriptive list of the known species with special reference to those occurring in North America, 339 pp. 20 *pls.* New York, Macmillan Co., 1934.

HAGELSTEIN, ROBERT: The Mycetozoa of North America, based upon the specimens in the New York Botanical Garden, pp. 1–306. *Pls.* 1–16. Mineola, N. Y., published by the author, 1944.

MARTIN, G. W.: Fungi, Myxomycetes. Ceratiomyxales, Liceales, Trichiales, Stemonitales, Physarales, *North American Flora,* **1**(1):1–151, with Bibliography, pp. 153–178, by Harold W. Rickett and Index, pp. 179–190, by Gussie M. Miller. 1949.

TORREND, C.: Flore des Myxomycètes, 271 pp. 9 *pls.* 1908. (Reprinted from *Broteria.*)

JACZEWSKI, A. A.: Mycological Flora of European and Asiatic Russia: II. Myxomycetes, *Materialien zur Kenntnis der Fauna und Flora des Russichen Reiches, Botanischer Teil.* Heft **6**:1–140. 84 *figs.* 1907. (In Russian.)

BROOKS, TRAVIS E.: Myxomycetes of Kansas, I, *Trans. Kansas Acad. Sci.,* **44**:130–157. 1941.

GREENE, H. C.: Wisconsin Myxomycetes, *Trans. Wisconsin Acad. Sci.,* **27**:141–181. 6 *pls.* 1932.

EMOTO, YOSHIKADZU: Die Myxomyceten Japans, *Botanical Magazine* (*Tokyo*), **48,** various numbers, 97 *figs.,* 1934; **49,** various numbers, 168 *figs.,* 1935; **50,** various numbers, *numerous figs.,* 1936. (Text in Japanese.)

HATTORI, H.: Myxomycetes of Nasu District (Japan), pp. 1–280. 23 *colored pls.* 320 *figs.* 1935. (Text in Japanese.)

DENNISON, MARY LOUISE: The genus Lamproderma and its relationships, I, *Mycologia,* **37**(1):80–108. *Figs.* 1–22. 1945; II, *ibid.,* **37**(2):197–204. *Fig.* 1. 1945. (Part I contains a key and descriptions of all recognized species of *Lamproderma.* Part II includes a discussion of the Family Stemonitaceae and key to the 13 genera recognized by the author.)

KARLING, JOHN S.: The Plasmodiophorales, ix + 144 pp. 17 *pls.* 17 *figs.* New York, published by the author, 1942.

COOK, W. R. IVIMEY: A monograph of the Plasmodiophorales, *Arch. Protistenk.,* **80**(2):179–254. *Pls.* 5–11. *Figs.* 1–14. 1933.

OLIVE, EDGAR W.: A preliminary enumeration of the Sorophorae, *Proc. Am. Acad. Arts Sci.*, **37**:333–344. 1901.

———: Monograph of the Acrasiae, *Proc. Boston Soc. Natural History*, **30**:451–513. *Pls.* 5–8. 1902.

YOUNG III, EDWARD LORRAINE: Studies on Labyrinthula. The etiologic agent of the wasting disease of eel-grass, *Am. J. Botany*, **30**(8):586–593. *Figs.* 1–2. 1943. (Includes a brief monograph of the genus.)

List 4. General Works on Phycomyceteae

SPARROW JR., FREDERICK K.: Aquatic Phycomycetes Exclusive of the Saprolegniaceae and Pythium, xx + 785 pp. 634 *figs.* Ann Arbor, Univ. Mich. Press, 1943.

———: A contribution to our knowledge of the aquatic Phycomycetes of Great Britain, *J. Linnean Soc. London*, **50**(334):417–478. *Pls.* 14–20. *Figs.* 1–7. 1936.

FITZPATRICK, HARRY M.: The lower fungi. Phycomycetes, xi + 331 pp. *Figs.* 1–112. New York, McGraw-Hill Book Co., 1930.

ITO, S.: Mycological flora of Japan: I. Phycomycetes, 340 pp. 125 *figs.* Tokyo, 1936.

JACZEWSKI, A. A.: Opredelitel gribov: I. Fikomitsety. (Determination of Fungi: I. Phycomycetes), 294 pp. 329 *figs.* Moscow and Leningrad, 1931. (In Russian.)

VON MINDEN, M.: Chytridiineae, Ancylistineae, Monoblepharidineae, Saprolegniineae, in Krytogamenflora der Mark Brandenburg, vol. 5, pt. 2, pp. 193–352, 1911; pt. 3, pp. 353–496, 1911; pt. 4, pp. 497–608, 1912; pt. 5, pp. 609–630, 1915. (*Illustrated.*)

PETERSEN, H. E.: An account of Danish freshwater Phycomycetes, with biological and systematical remarks, *Ann. Mycol.*, **8**(5):494–560. *Figs.* 1–27. 1910.

TIESENKAUSEN, MANFRED BARON: Beiträge zur Kenntnis der Wasserpilze der Schweiz, *Arch. Hydrobiol. Planktonkunde*, **7**:261–308. *Figs.* 1–24. 1912.

VIÉGAS, A. P., E A. R. TEIXEIRA: Alguns fungos do Brasil (Phycomycetos), *Bragantia*, **3**(8):223–245. 22 *pls.* 4 *figs.* 1943.

List 5. Chytridiales

(NOTE: In some of these references are included fungi now segregated in the Hyphochytriales and in the nonfilamentous Lagenidiales.)

PETERSEN, H. E.: Contributions à la connaissance des Phycomycètes marins (Chytridineae Fischer), *Oversigt over Kongelige Danske Videnskabernes Selskab Forhandlinger*, **1905**:439–488. *Illustrated.* 1905.

SCHERFFEL, A.: Beiträge zur Kenntnis der Chytridineen: II. Einiges über neue oder ungenügend bekannte Chytridineen, *Arch. Protistenk.*, **54**(2):167–260. 3 *pls.* 1926.

———: Beiträge zur Kenntnis der Chytridineen, III, *ibid.*, **54**(3):510–528. 1926. (Includes descriptions of all known species of *Harpochytrium.*)

ATKINSON, GEORGE F.: The genus Harpochytrium in the United States, *Ann. Mycol.*, **1**(6):479–502. *Pl.* 10. *Fig.* 1. 1903.

———: Notes on the genus Harpochytrium, *J. Mycology*, **10**(1):3–8. *Pl.* 72. *Text figs.* 24–33. 1904.

JANE, FRANK W.: A revision of the genus Harpochytrium, *J. Linnean Soc. London*, **53**(348):28–40. *Figs.* 1–28. 1946.

KARLING, JOHN S.: Brazilian Chytrids: I. Species of Nowakowskiella, *Bull. Torrey Botan. Club*, **71**(4):374–389. *Figs.* 1–69. 1944; II. New species of Rhizidium, *Am. J. Botany*, **31**(5):254–261. *Figs.* 1–72. 1944; III. Nephrochytrium amazonensis, *Mycologia*, **36**(4):351–357. *Figs.* 1–28. 1944; IV. Species of Rozella, *ibid.*, **36**(6):638–647. *Figs.* 1–28. 1944; V. Nowakowskiella macrospora n. sp. and other polycentric species, *Am. J. Botany*, **32**(1):29–35. *Figs.* 1–51. 1945; VI. Rhopalophlyctis and Chytriomyces, two new chitinophilic operculate genera, *ibid.*, **32**(7):362–369. *Figs.* 1–61. 1945; VII. Observations relative to sexuality in two new species of Siphonaria, *ibid.*, **32**(9):580–587. *Figs.* 1–53. 1945; VIII. Additional parasites of rotifers and nematodes, *Lloydia*, **9**(1):1–12. *Pls.* 1–2. 1946; IX. Species of Rhizophydium, *Am. J. Botany*, **33**(5):328–334. *Figs.* 1–37. 1946; X. New species with sunken opercula, *Mycologia*, **39**(1):56–70. *Figs.* 1–56. 1947.

———: A synopsis of Rozella and Rozellopsis, *ibid.*, **34**(2):193–208. 1942.

CANTER, HILDA M.: Studies on British Chytrids: II. Some new monocentric Chytrids, *Brit. Mycol. Soc. Trans.*, **31**(1–2):94–105. *Pls.* 1–10. *Figs.* 1–8. 1947; III. Zygorhizidium willei Löwenthal and Rhizophidium columnaris n. sp., *ibid.*, **31**(1–2):128–135. *Pl.* 11. *Figs.* 1–4. 1947.

TOKUNAGA, YOSIO: Studies on the aquatic Chytrids of Japan: II. Olpidiaceae, *Trans. Sapporo Natural History Soc.*, **13**(2):78–84. *Pl.* 5. 1933; III. Rhizidiaceae, *ibid.*, **13**(4):388–393. *Pl.* 11. 1934.

FARLOW, W. G.: The Synchytria of the United States, *Botan. Gaz.*, **10**(3):235–240. *Pl.* 4. 1885.

TOBLER-WOLFF, GERTRUD: Die Synchytrien. Studien zu einer Monographie der Gattung, *Arch. Protistenk.*, **28**:143–238. *Pls.* 10–13. 1913.

COOK, MELVILLE T.: Species of Synchytrium in Louisiana: I. Descriptions of species found in the vicinity of Baton Rouge, *Mycologia*, **37**(3):284–294. *Figs.* 1–4. 1945; II. Species of Louisiana Synchytrium, *ibid.*, **37**(5):571–575. 1 *fig.* 1945; III. The development and structure of the galls, *ibid.*, **37**(6):715–740. *Figs.* 1–12. 1945; IV. Two new species of Synchytrium, *ibid.*, **39**(3):351–357. *Figs.* 1–4. 1947; V. A new species on Sambucus canadensis, *ibid.*, **41**(1):24–27. *Figs.* 1–9. 1949.

———: Synchytrium decipiens and Synchytrium chrysosplenii, *ibid.*, **38**(3):300–305. *Figs.* 1–3. 1946.

QUINTANILHA, A.: Contribuição ao estudo dos Synchytrium, *Boletim da Sociedade Broteriana II*, serie 3, 110 pp. 1 *fig.* 1926.

MHATRE, J. R., AND B. B. MUNDKUR: The Synchytria of India, *Lloydia*, **8**(2):131–138. 1945.

CEJP, KAREL: Some remarks to the knowledge of the parasitic Phycomycetes of Conjugates in Bohemia, *Académie Tchèque des Sciences Mathématiques,*

Naturelles et de la Médecine, **33**:17–23. *Pls.* 1–2. 1932. (Includes descriptions of the three known species of *Micromycopsis*.)

COUCH, JOHN N.: Rhizophidium, Phlyctochytrium and Phlyctidium in the United States, *J. Elisha Mitchell Sci. Soc.*, **47**(2):245–260. *Pls.* 14–17. 1932.

———: Notes on the genus Micromyces, *Mycologia*, 29(5):592–596. *Figs.* 1–14. 1937.

CANTER, HILDA M.: Studies on British Chytrids: VI. Aquatic Synchytriaceae, *Brit. Mycol. Soc. Trans.*, **32**(1):69–94. *Pls.* 7–11. *Figs.* 1–13. 1948.

KARLING, JOHN S.: The genus Physoderma (Chytridiales), *Lloydia*, **13**(1):29–71. 1950. (The genus *Physoderma* including *Urophlyctis*.)

List 6. Hyphochytriales

KARLING, JOHN S.: The life history of Anisolpidium ectocarpii gen. nov. et sp nov., and a synopsis and classification of other fungi with anteriorly uniflagellate zoospores, *Am. J. Botany*, **30**(8):637–648. *Figs.* 1–21. 1944.

List 7. Blastocladiales and Monoblepharidales

(See also SPARROW, 1943, in List 4.)

COKER, W. C., AND VELMA D. MATTHEWS: Blastocladiales, Monoblepharidales and Saprolegniales, *North American Flora*, **2**(1):1–76. 1937.

SPARROW JR., FREDERICK K.: The Monoblepharidales, *Ann. Botany*, **47**(187):517–542. *Pl.* 20. *Figs.* 1–2. 1933.

KANOUSE, BESSIE B.: A monographic study of special groups of water molds: I. Blastocladiaceae, *Am. J. Botany*, **14**(6):287–306. *Pls.* 32–34. 1927.

WOLF, FREDERICK TAYLOR: The aquatic Oomycetes of Wisconsin, Pt. I, 64 pp. *Pls.* 1–6. Madison, Univ. Wisconsin Press, 1944.

EMERSON, RALPH: An experimental study of the life cycle and taxonomy of Allomyces, *Lloydia*, **4**(2):77–144. *Figs.* 1–16. 1941.

COUCH, JOHN N., AND ALMA J. WHIFFEN: Observations on the genus Blastocladiella, *Am. J. Botany*, **29**(7):582–591. *Figs.* 1–66. 1942.

———: Observations on the genus Catenaria, *Mycologia*, **37**(2):163–192. *Figs.* 1–78. 1945.

CEJP, KAREL: Sur les affinités des Blastocladiaceae. Révision du genre Gonapodya, sa position systématique, *Bull. trimestr. soc. mycol. France*, **62**(3–4):246–257. 1946 (1947).

COUCH, JOHN N.: Revision of the genus Coelomomyces, parasitic in insect larvae, *J. Elisha Mitchell Sci. Soc.*, **61**(1–2):124–136. *Pls.* 1–2. 1945.

———, AND H. R. DODGE: Further observations on Coelomomyces, parasitic on mosquito larvae, *ibid.*, **63**(1):69–79. *Pls.* 15–20. 1947.

SPRINGER, MARTHA E.: Two new species of Monoblepharella, *Mycologia*, **37**(2):208–216. *Figs.* 1–51. 1945.

INDOH, HIROHARO: Studies on Japanese aquatic fungi: II. The Blastocladiaceae, *Science Repts. Tokyo Bunrika Daigaku, B*, **4**:237–384. 34 *figs*. 1940.

List 8. Lagenidiales

(See also SPARROW, 1943, and others in List 4.)

KARLING, JOHN S.: The simple holocarpic biflagellate Phycomycetes. Including a complete host index and bibliography, x + 123 pp. *Pls*. 1–25. 3 *diagrams*. New York, published by the author, 1942.

TOKUNAGA, YOSIO: Studies on the aquatic Chytrids of Japan: I. Woroninaceae, *Trans. Sapporo Natural History Soc.*, **13**(1):20–28. *Pl*. 2. 1933.

———: Notes on the Lagenidiaceae of Japan, *ibid.*, **13**(3):227–232. 3 *figs*. 1934.

List 9. Saprolegniales (Including Leptomitales)

(See also SPARROW, 1943, and others in List 4.)

HUMPHREY, J. E.: The Saprolegniaceae of the United States with notes on other species, *Trans. Am. Phil. Soc.*, **17**:64–148. *Pls*. 14–20. 1892.

COKER, W. C.: The Saprolegniaceae with notes on other water molds, 201 pp. *Pls*. 1–63. Chapel Hill, Univ. North Carolina Press, 1923.

———, AND VELMA D. MATTHEWS: Saprolegniales, *North American Flora*, **2**(1):15–67. 1937.

NAGAI, MASAJI: Studies on the Japanese Saprolegniaceae, *J. Faculty Agr. Hokkaido Imp. Univ.*, **32**(1):1–43. *Pls*. 1–7. 1931.

———: Additional notes on the Japanese Saprolegniaceae, *Botanical Magazine (Tokyo)*, **47**(554):136–137. *Fig*. 1. 1933.

MAURIZIO, A.: Zur Entwicklungsgeschichte und Systematik der Saprolegniaceen, *Flora*, **79**:109–158. *Pls*. 3–5. 1894.

WOLF, FREDERICK TAYLOR: The aquatic Oomycetes of Wisconsin, Pt. I, pp. 1–64. *Pls*. 1–6. Madison, Univ. Wisconsin Press, 1944.

APINIS, ARV.: Untersuchungen über die in Lettland gefundenen Saprolegniaceen nebst Bemerkungen über einige andere Wasserpilze, *Acta Horti Botan. Univ. Latviensis*, **4**:201–246. *Figs*. 1–4. 1929.

CHAUDHURI, H., AND P. L. KOCHHAR: Indian water molds, I, *Proc. Indian Acad. Sci., B*, **2**:137–154. *Pls*. 5–12. 1935.

———, AND S. S. LOTUS: Indian water molds, II, *ibid.*, **3**:328–333. 1936.

CROOKS, KATHLEEN M.: Studies on Australian aquatic Phycomycetes, *Proc. Roy. Soc. Victoria*, **49**(2):206–232. *Pl*. 10. *Figs*. 1–11. 1937.

KANOUSE, BESSIE B.: A monographic study of special groups of water molds: II. Leptomitaceae and Pythiomorphaceae, *Am. J. Botany*, **14**(7):335–357. *Pl*. 48. 1927.

INDOH, HIROHARO: Studies on the Japanese aquatic fungi: I. On *Apodachlyella completa* sp. nov., with revision of the Leptomitaceae, *Science Repts. Tokyo Bunrika Daigaku, B,* **4**:43–50. *Pl.* 7. *Figs.* 1–11. 1939.

COKER, W. C., AND JOHN N. COUCH: Revision of the genus Thraustotheca with a description of a new species, *J. Elisha Mitchell Sci. Soc.,* **40**(3–4):197–202. *Pls.* 38–40. 1924.

CUTTER JR., VICTOR M.: Observations on certain species of Aphanomyces, *Mycologia,* **33**(2):220–240. *Figs.* 1–15. 1941.

List 10. Peronosporales, Also Protomycetales

MATTHEWS, VELMA DARE: Studies on the genus Pythium, 136 pp. 29 *pls.* Chapel Hill, Univ. North Carolina Press, 1931.

SIDERIS, C. P.: Taxonomic studies in the Family Pythiaceae: I. Nematosporangium, *Mycologia,* **23**(4):252–295. *Figs.* 1–12. 1931; II. Pythium, *ibid.,* **24**(1):14–61. *Figs.* 1–21. 1932.

DRECHSLER, CHARLES: Some new species of Pythium, *J. Wash. Acad. Sci.,* **20**(16):398–418. 1930.

SPARROW JR., FREDERICK K.: Two new species of Pythium parasitic on algae, *Ann. Botany,* **45**(178):255–277. 1 *pl.* 2 *figs.* 1931.

BUTLER, E. J.: An account of the genus Pythium and some Chytridiaceae, *Mem. Dept. Agri. India, Botan. ser.,* **1**(5):1–161. *Pls.* 1–10. Feb. 1907.

FITZPATRICK, H. M.: Generic concepts in the Pythiaceae and Blastocladiaceae, *Mycologia,* **15**(4):166–173. 1923.

MIDDLETON, JOHN T.: The taxonomy, host range and geographic distribution of the genus Pythium, *Mem. Torrey Botan. Club,* **20**(1):1–171. *Figs.* 1–17. 1943.

TUCKER, C. M.: Taxonomy of the genus Phytophthora de Bary, *Univ. Missouri Agr. Expt. Sta. Research Bull.* **153**:1–208. *Figs.* 1–30. 1931.

ROSENBAUM, J.: Studies of the genus Phytophthora, *J. Agr. Research,* **8**(7):233–276. 7 *pls.* 1917.

BAKER, R. E. D.: Notes on Trinidad fungi: I. Phytophthora, *Trop. Agr. Trinidad,* **13**(12):330–332. *Figs.* 1–3. 1936. (The species of Phytophthora on cacao in Trinidad.)

FOISTER, C. E.: The white tip disease of leeks and its causal fungus, Phytophthora Porri n. sp., *Trans. Proc. Botan. Soc. Edinburgh,* **30**(4):257–281. *Pl.* 18. *Figs.* 1–3. 1931. (Describes a new species and gives for comparison the description of 10 other rather similar species.)

WILSON, G. W.: Studies in North American Peronosporales: I. The genus Albugo, *Bull. Torrey Botan. Club,* **34**(2):61–84. 1907; II. Phytophthoreae and Rhysotheceae, *ibid.,* **34**(8):387–416. 1907; III. New or noteworthy species (Species of Albugo and Peronospora), *ibid.,* **35**(7):361–365. 1908; IV. Host Index, *ibid.,* **35**(11):543–554. 1908; V. A Review of the genus Phytophthora, *Mycologia,* **6**(2):54–83. *Pl.* 119. 1914; VI. Notes on Miscellaneous Species, *ibid.,* **6**(4):192–210. *Pls.* 135–136. 1914; VII. New and Noteworthy Species, *ibid.,* **10**(3):168–169. 1918.

FARLOW, W. G.: Enumeration of the Peronosporaceae of the United States *Botan. Gaz.,* **8**(10):305–315, (11):327–337, 1883.

FARLOW, W. G.: Additions to the Peronosporaceae of the United States, *ibid.*, **9**(1):37–40. 1884. (These two papers contain descriptions of every species of this order known from the United States up to the close of 1883, with full notes as to known distribution, hosts, etc.)

SWINGLE, W. T.: Some Peronosporaceae in the Herbarium of the Division of Vegetable Pathology, *J. Mycology*, **7**(2):109–130. 1892.

BERLESE, A. N.: Saggio di una monografia delle Peronosporaceae, *Riv. patol. vegetale*, **6**(1):78–101. (2):237–268. *Pls.* 7–10. 1897; **7**:19–37. 1898; **9**:1–126. *Figs.* 1–21. 1900; **10**:185–298. *Figs.* 22–69. 1902.

———: Icones fungorum ad usum Sylloges Saccardianae accomodatae. Phycomycetes: Fasc. I. Peronosporaceae, pp. 1–40. *Pls.* 1–67. Padua, 1898, published by the author. (Beautifully executed illustrations of most of the species of this family.)

WAKEFIELD, E. M.: The genus Cystopus in South Africa, *Bothalia*, **2**(Ib):242–246. 1927.

JACZEWSKI, A. A.: Mycological Flora of European and Asiatic Russia: I. Peronosporaceae, *Materialien zur Kenntnis Fauna und Flora des Russischen Reiches, Botanischer Teil*, Heft 4. 1907. (In Russian.)

THIND, KARTAR SINGH: The genus Peronospora in the Punjab, *J. Indian Botan. Soc.*, **21**(3–4):197–215. 1942.

GÄUMANN, ERNST: Über die Specialisation der Peronospora auf einigen Scrophulariaceen, *Ann. Mycol.*, **16**(1–2):189–199. *Figs.* 1–6. 1918.

———: Ueber die Formen der Peronospora parasitica (Pers.) Fries. Ein Beitrag zur Speziesfrage bei den parasitischen Pilzen, *Beihefte Botan. Centr., Erste Abteilung*, **35**:1–143, 305–533. *Figs.* 1–47. 1918.

———: Zur Kenntnis der Chenopodiaceen bewohnenden Peronospora-Arten, *Mitteilungen der Naturforschenden Gesellschaft in Bern aus dem Jahre* **1918**: 45–66. *Figs.* 1–15. 1919.

———: Beiträge zu einer Monographie der Gattung Peronospora Corda, *Beiträge zur Kryptogamenflora der Schweiz*, **5**(4):1–360. *Figs.* 1–166 1923.

WARTENWEILER, ALFRED: Beiträge zur Systematik und Biologie einiger Plasmopara-Arten, *Ann. Mycol.*, **16**(3–6):249–299. *Pls.* 1–3. *Figs.* 1–12. 1918.

SĂVULESCU, TRIAN: Les espèces de Peronospora Corda de Roumanie, *Sydowia, Ann. Mycol.*, **2**(1–6):255–307. 1948.

VON BÜREN, GÜNTHER: Die schweizerischen Protomycetaceen mit besonderer Berücksichtigung ihrer Entwicklungsgeschichte und Biologie, *Beiträge zur Kryptogamenflora der Schweiz*, **5**(1):1–95. *Pls.* 1–7. *Figs.* 1–28. 1915.

———: Weitere Untersuchungen über die Entwicklungsgeschichte und Biologie der Protomycetaceen, *ibid.*, **5**(3):1–94. *Pls.* 1–2. *Figs.* 1–27. 1922.

List 11. Mucorales, Entomophthorales

ZYCHA, H.: Mucorineae, Kryptogamenflora der Mark Brandenburg, vol. 6a, pp. 1–264. 114 *figs.* Leipzig, Gebrüder Borntraeger, 1935.

NAUMOV, N. A.: Opredelitel Mukorovych (Determination of Mucorales), ed. 2, pp. 1–136. *Figs.* 1–49. 1 *Diagram.* Moscow and Leningrad, Botanical Institute of the Academy of Sciences of U.S.S.R., 1935. (In Russian.)

———: Clés des Mucorinées (Mucorales). Translated by S. Buchet and I. Mouraviev, in Encyclopédie Mycologique, ed. 2, vol. 9, pp. 1–137, with Appendix i–xxxvi. 83 *figs*. Paris, Paul Lechevalier, 1939.

LENDNER, ALF.: Les Mucorinées de la Suisse, *Beiträge zur Kryptogamenflora der Schweiz*, **3**(1):1–180. *Pls*. 1–3. *Figs*. 1–59. 1908.

HAGEM, O.: Untersuchungen über norwegische Mucorineen, *Videnskapselskapets-Skrifter. Mat. naturv. Klasse*, **1907**(7):1–50. *Figs*. 1–22. 1908; **1910**(4):1–152. 1910.

———: Neue Untersuchungen über norwegische Mucorineen, *Ann. Mycol.*, **8**(3):265–286. *Figs*. 1–11. 1910.

LINNEMANN, G.: Beitrag zu einer Flora der Mucorineae Marburgs, *Flora*, **130**(N.S. **30**):176–217. *Figs*. 1–25. 1936.

SUMSTINE, D. R.: The North American Mucorales, I, *Mycologia*, **2**(3):125–154. 1910.

CHRISTENBERRY, GEORGE A.: A taxonomic study of the Mucorales in the Southeastern United States, *J. Elisha Mitchell Sci. Soc.*, **56**(2):333–366. *Pls*. 13–19. 1946.

OU, SHIH-KUANG: Phycomycetes of China, I, *Sinensia*, **11**(1–2):33–57. 18 *figs*. 1940. (Descriptions and keys to the Mucorales found in Szechuan.)

CAMPBELL, MARIE E.: An investigation of the Mucorales in the soil, *Trans. Roy. Soc. Edinburgh*, **59**(2):411–436. 3 *pls*. 15 *figs*. 1938.

POVAH, A. H. W.: A critical study of certain species of Mucor, *Bull. Torrey Botan. Club*, **44**(5):241–259. (6):287–313. *Pls*. 17–20. 1917.

ZACH, FRANZ: Zur Kenntnis der Formenkreis von Mucor plumbeus Bonorden, *Oesterreichische Botanische Zeitschrift*, **84**(2):117–122. *Fig*. 1. 1935.

———: Beitrag zum Formenkreis von Mucor plumbeus Bonorden, *ibid.*, **85**(2):151–153. 1936.

GROVE, W. B.: A systematic account and arrangement of the Pilobolidae, in A. H. REGINALD BULLER: Researches on Fungi, vol. 6, chap. 4, pp. 190–224. *Figs*. 97–111. 1934.

PALLA, E.: Zur Kenntnis der Pilobolus-Arten, *Oesterreichische Botanische Zeitschrift*, **50**(10):349–370, (11):397–401, *Pl*. 10. 10 *text figs*. 1900.

MORINI, FAUSTO: Materiali per una monografia delle Pilobolee, *Mem. reale accad. sci. inst. Bologna, ser*. 6, **3**:111–129. 1906.

HANZAWA, J.: Studien über einige Rhizopus-Arten, *Mycolog. Centr.*, **5**(5):230–246. (6):257–281. *Figs*. 1–12. 1915. (A morphological, systematic, and physiological study of 12 species.)

YAMAMOTO, YOSHIHIKO: Ein Beitrag zur Kenntnis der Gattung Rhizopus: I. Morphologisches, *J. Faculty Agr. Hokkaido Imp. Univ.*, **28**(1):1–101. *Pls*. 1–4. 1930. (Divides Rhizopus into 2 sections. Gives diagnoses for 15 species.)

YAMAZAKI, MOMOJI: On the classification of Rhizopus species, *Bull. Utsunomiya Agr. Coll.*, **5**:1–16. 1934. (Japanese, with English summary.)

DAUPHIN, J.: Contribution à l'étude des Mortierellées, *Ann. sci. nat. Botan.*, 9me sér., **8**:1–112. *Figs*. 1–45. 1908.

VUILLEMIN, PAUL: Sur les Mortierelia des groupes polycephala et nigrescens, *Bull. soc. mycol. France*, **34**:41–46. *Figs*. 1–3. 1918.

ALCORN, GORDON D., AND CHARLES C. YEAGER: A monograph of the genus Cunninghamella with additional descriptions of several common species, *Mycologia*, **30**(6):653–658. *Figs*. 1–2. 1938.

CUTTER JR., VICTOR M.: The genus Cunninghamella (Mucorales), *Farlowia*, **2**(3):321–343. *Pls*. 1–2. 1946.

LINDER, D. H.: The genera Kickxella, Martensella, and Coemansia, *Farlowia*, **1**(1):49–77. *Pls.* 1–4. 1943.

BUCHOLTZ, FEDOR: Beiträge zur Kenntnis der Gattung Endogone Link, *Beihefte Botan. Centr., Zweite Abt.*, **29**:147–225. *Pls.* 3–10. 1912. (Besides morphological and cytological data this paper also contains descriptions of most of the known species.)

BOEDIJN, K. B.: The genera Endogone and Sclerocystis in the Netherlands Indies, *Bull. Jardin Botan. de Buitenzorg, III*, **13**(3):503–508. *Illustrated.* 1935.

THAXTER, ROLAND: A revision of the Endogoneae, *Proc. Am. Acad. Arts Sci.*, **57**(12):289–350. *Pls.* 1–4. 1922.

———: The Entomophthorae of the United States, *Mem. Boston Soc. Natural History*, **4**:133–201. *Pls.* 14–21. 1888.

BERDAN, HELEN: Revision of the genus Ancylistes, *Mycologia*, **30**(4):396–415. *Figs.* 1–22. 1938.

List 12. Zoopagales, Eccrinales

(NOTE: For both these orders no publication brings together in one place a discussion of the described species. Therefore the many papers bearing on these fungi must be referred to.)

DRECHSLER, CHARLES: Some conidial Phycomycetes destructive to terricolous Amoebae, *Mycologia*, **27**(1):6–40. *Pls.* 1–7. 1935.

———: Some non-catenulate conidial Phycomycetes preying on terricolous Amoebae, *ibid.*, **27**(2):176–205. *Figs.* 1–5. 1935.

———: A new species of conidial Phycomycete preying on nematodes, *ibid.*, **27**(2):206–215. *Fig.* 1. 1935.

———: A new species of Stylopage preying on nematodes, *ibid.*, **28**(3):241–246. *Fig.* 1. 1936.

———: New conidial Phycomycetes destructive to terricolous Amoebae, *ibid.*, **28**(4):363–389. *Figs.* 1–7. 1936.

———: New Zoopagaceae capturing and consuming soil Amoebae, *ibid.*, **30**(2):137–157. *Figs.* 1–4. 1938.

———: A few new Zoopagaceae destructive to large soil rhizopods, *ibid.*, **31**(2):128–153. *Figs.* 1–7. 1939.

———: Five new Zoopagaceae destructive to rhizopods and nematodes, *ibid.*, **31**(4):388–415. *Figs.* 1–5. 1939.

———: Four Phycomycetes destructive to nematodes and rhizopods, *ibid.*, **33**(3):248–269. *Figs.* 1–5. 1941.

———: New species of Acaulopage and Cochlonema destructive to soil Amoebae, *ibid.*, **34**(3):274–297. *Figs.* 1–6. 1942.

———: Several additional Phycomycetes subsisting on nematodes and Amoebae, *ibid.*, **37**(1):1–31. *Figs.* 1–3. 1945.

———: Three new Zoopagaceae subsisting on soil Amoebae, *ibid.*, **38**(2):120–143. *Figs.* 1–6. 1946.

———: Three zoopagaceous fungi that capture and consume soil-inhabiting rhizopods, *ibid.*, **39**(3):253–281. *Figs.* 1–7. 1947.

———: Three new species of Zoopage predaceous on terricolous rhizopods, *ibid.*, **39**(4):379–408. *Figs.* 1–7. 1947.

———: Three Zoopagaceae that subsist by capturing soil Amoebae, *ibid.*, **40**(1):85–105. *Figs.* 1–4. 1948.

(The following five papers concern Family Harpellaceae, including Genistellaceae).

LEGER, LOUIS, ET O. DUBOSCQ: Harpella melusinae n. g. et n. sp. Entophyte eccriniforme parasite des larves de Simulie, *Compt. rend.* **188**(14):951–954. *Figs.* 1–6. 1929.

———, ET MARCELLE GAUTHIER: Endomycètes nouveaux des larves aquatiqúes d'insectes, *ibid.*, **194**(26):2262–2265. *Figs.* 1–3. 1932.

———, ET ———: La spore des Harpellacées (Leger & Duboscq), champignons parasites des insectes, *ibid.*, **200**(17):1458–1460. 1935.

———, ET ———: Graminella bulbosa nouveau genre d'entophyte parasite des larves d'Ephémérides du genre Baetis, *ibid.*, **204**(1):27–29. *Figs.* 1–5. 1937.

GAUTHIER, MARCELLE: Sur un nouvel entophyte du groupe des Harpellacées Leg. & Dub., parasite des larves d'Ephémérides, *ibid.*, **202**(12):1096–1098. *Figs.* 1–4. 1936.

(The following papers treat of the Eccrinales.)

LEGER, LOUIS, ET O. DUBOSCQ: Eccrinoides Henneguyi n. g. et n. sp. et la systématique des Eccrinides, *Archives d'Anatomie Microscopique*, **25**:309–324. *Figs.* 1–6. 1929. (The order is classified into families with lists of the contained genera, but no generic distinctions are given. Therefore the following papers will have to be consulted.)

DUBOSCQ, O., L. LEGER, ET O. TUZET: Contribution à la connaissance des Eccrinides: les Trichomycètes, *Arch. Zool. exptl. et gén.*, **86**(2):29–144. *Pls.* 1–5. *Figs.* 1–42. 1948.

LEIDY, JOSEPH: Enterobrus, a new genus of Confervaceae, *Proc. Acad. Nat. Sci. Phila.*, **4**:225–227. 1849.

———: A flora and fauna within living animals, *Smithsonian Inst. Pubs. Contribs. to Knowledge*, **5**(2):1–67. *Pls.* 1–10. 1851 (1853).

LEGER, LOUIS, ET O. DUBOSCQ: Les Eccrinides, nouveau groupe de Protophytes parasites, *Compt. rend.*, **141**(9):425–427. 1905.

———, ET ———: Sur les Eccrinides des Hydrophilides, *Arch. zool. exptl. et gén.* **56**(2):21–31. *Figs.* 1–4. 1916.

———, ET ———: L'évolution du Paramoebidium, nouveau genre d'Eccrinides, parasite des larves aquatiques d'insectes, *Compt. rend.*, **189**(2):75–77. *Figs.* 1–15. 1929.

LICHTENSTEIN, JEAN L.: Sur un Amoebidium à commensalisme interne du rectum des larves d'Anax imperator Leach: Amoebidium fasciculatum n. sp., *Arch. zool. exptl. et gén.*, **56**(3):49–62. *Figs.* 1–7. 1917.

———: Sur un mode nouveau de multiplication chez les Amoebidiacées, *ibid.*, **56**(4):95–99. *Fig.* 1. 1917.

POISSON, RAYMOND: Recherches sur quelques Eccrinides parasites de Crustacés amphipodes et isopodes, *Arch. zool. exptl. et gén.*, **69**(3):179–216. *Figs.* 1–23 1929.

———: Recherches sur les Eccrinides. Deuxième contribution, *ibid.*, **74**(4):53–68. *Figs.* 1–7. 1931. (Volume jubilaire.)

POISSON, RAYMOND: Asellaria Caulleryi N. G., N. Sp., type nouveau d'entophyte parasite intestinal des Aselles (Crustacés isopodes). Description des stades connus et d'une partie de son cycle évolutif, *Bull. biol. France Belg.*, **66**:232–254. *Pl.* 3. *Figs.* 1–9. 1932.

List 13. Ascomyceteae, Miscellaneous

TULASNE, L. R., ET C.: Selecta fungorum carpologia. Paris, Typographie Impériale. Vol. 1. Erysiphei, pp. xxviii + 1–242, *Pls.* 1–5, 1861; vol. 2. Xylariei, Valsei, Sphaeriei, pp. xx + 1–319, *Pls.* 1–34, 1863; vol. 3. Nectriei, Phacidiei, Pezizei, pp. xvi + 1–221, *Pls.* 1–22, 1865.

GILMAN, J. C.: A partial list of the parasitic Ascomycetes of Iowa, *Proc. Iowa Acad. Sci.*, **32**:235–264. 1925.

FINK, BRUCE, AND SYLVIA C. FUSON: An arrangement of the Ascomycetes of Indiana, *Proc. Indiana Acad. Sci.*, **1919**:113–133. 1921.

MILLER, JULIAN H.: The Ascomycetes of Georgia, *Plant Disease Reptr.*, supplement **131**:31–93. 1941.

VIÉGAS, A. P.: Alguns fungos do Brasil: II. Ascomicetos, *Bragantia*, **4**(1–6):5–392. 220 *pls.* 34 *figs.* 1944.

List 14. Laboulbeniales

THAXTER, ROLAND: Contribution toward a monograph of the Laboulbeniaceae, I, *Mem. Am. Acad. Arts Sci.*, **12**:195–429. *Pls.* 1–26. 1895; II, *ibid.*, **13**:219–469. *Pls.* 28–71. 1908; III, *ibid.*, **14**:309–426. *Pls.* 1–12. 1924; IV, *ibid.*, **15**(4):427–580. *Pls.* 1–24. 1926; V, *ibid.*, **16**(1):1–435. *Pls.* 1–60. 1 *fig.* 1931.

———: Laboulbeniales parasitic on Chrysomelidae, *Proc. Am. Acad. Arts Sci.*, **50**(2):17–50. 1914.

———: New Laboulbeniales from Chile and New Zealand, *ibid.*, **54**:207–232. 1918.

———: New Dimorphomyceteae, *ibid.*, **55**:211–282. 1920.

PAOLI, G.: Nuovi Laboulbeniomiceti parassiti di Acari, *Redia*, **7**:283–295. 1 *pl.* 1911.

SPEGAZZINI, CARLOS: Contribucion al estudio de las Laboulbeniomicetas Argentinas, *Anales del Museo Nacional de Historia Natural de Buenos Aires*, **23**:167–244. 71 *figs.* 1912.

———: Laboulbeniali ritrovate nelle collezioni de alcuni musei italiani, *ibid.*, **26**:451–511. *Illustrated.* 1915.

———: Revision de las Laboulbeniales Argentinas, *ibid.*, **29**:445–688. *Figs.* 1–213. 1917.

SIEMASZKO, JANINA I WINCENTY: Owadorosty polskie i palearktyczne (Laboulbeniales polonici et palaearctici), I–III, *Polskie Pismo Entomologiczne* (*Bul-*

letin Entomologique de la Pologne), **6**(3–4):188–211. *Pl.* 7. 1928; **10**(3–4): 149–188. *Pls.* 7–10. *Figs.* 1–2. 1931; **12**(1–4):115–138. *Pls.* 9–10. 1933.

COLLA, S.: Laboulbeniales: Peyritschiellaceae, Dimorphomycetaceae, Laboulbeniaceae Heterothallicae, Laboulbeniaceae Homothallicae, Ceratomycetaceae, *Flora Italica Cryptogama, Fasc.* **16**:1–157. 108 *figs.* 1934.

MIDDELHOEK, A.: Laboulbeniaceae in Nederland, *Nederlandsch Kruidkundig Archief*, **53**:86–115. 12 *pls.* 1943; **56**:249–260. 10 *figs.* 1949.

LEPESME, P.: Revision des Rhachomyces paléarctique (Laboulbeniaceae), *Bull. trimestr. soc. mycol. France*, **58**:57–80. *Pls.* 2–6. 1942.

List 15. Lecanorales and Pyrenulales

(See also in List 1, ZAHLBRUCKNER, Band 8 in RABENHORST: Kryptogamen-Flora, etc., and MIGULA in THOMÉ: Flora von Deutschland, etc.)

FINK, BRUCE: The Lichen Flora of the United States. Completed for publication by Joyce Hedrick, x + 426 pp. 47 *pls.* 4 *figs.* Ann Arbor, Univ. Mich. Press, 1935.

TUCKERMAN, EDWARD: A synopsis of the North American lichens: Pt. I. Parmeliacei, Cladoniei and Coenogoniei, xx + 262 pp. Boston, S. E Cassino, 1882. Pt. II. Lecideacei and (in part) Graphidacei. 177 pp. New Bedford, Mass., E. Anthony and Sons, 1888.

SCHNEIDER, ALBERT: A Text-book of General Lichenology, xvii + 230 pp. 76 *pls.* Binghamton, N. Y., Williard N. Clute and Co., 1897. (Contains a systematic account with keys to families and genera but not to species.)

SMITH, ANNIE LORRAIN: Lichens, xxviii + 464 pp. 135 *figs.* Cambridge, Cambridge Univ. Press, 1921. (Pp. 313–342 give keys to families and genera.)

———: A monograph of the British lichens, ed. 2, London, published by the Trustees of the British Museum. Pt. I. xxiii + 519 pp., *Pls.* 1–71, 1918; Pt. II. ix + 447 pp., *Pls.* 1–63, 1926.

LEIGHTON, W. A.: The lichen-flora of Great Britain, Ireland and the Channel Islands, ed. 2, pp. i–xlvii, 1–502. Shrewsbury, published by the author, 1872.

MAAS GEESTERANUS, R. A.: Revision of the lichens of the Netherlands. I. Parmeliaceae, *Blumea* **6**(1): 1–199. 18 *figs.* 1947–1948.

ZAHLBRUCKNER, ALEXANDER: Catalogus lichenum universalis, 9 vols. Leipzig, Gebrüder Borntraeger, 1922–1940. (Vols. 1–9: A systematically arranged catalogue of all of the families, genera and species of lichens with complete citation of places of publication, illustrations, etc., but without keys or descriptions. Vol. 10: Index.)

FRIES, THEODOR MAGNUS: Lichenographia Scandinavica sive dispositio lichenum in Dania, Suecia, Norvegia, Fennia, Lapponia, Rossica hactenus collectorum, lv + 639 pp. *Illustrated.* Uppsala, E. Berling, 1873–1874.

GALLØE, OLAF: Natural History of the Danish lichens. Copenhagen, H. Aschehoug, Dansk Forlag. Pt. 1. 93 pp., 160 *pls.*, 1927; Pt. 2. 84 pp., 129 *pls.*, 1929; Pt. 3. 114 pp., 127 *pls.*, 1930; Pt. 4. 81 pp., 133 *pls.*, 1932; Pt. 5. 117 pp., 140 *pls.*, 1936; Pt. 6. 103 pp. 88 *pls.*, 1939; Pt. 7. 72 pp., 101 *pls.*, 1948; Pt. 8. 81 pp., 122 *pls.*, 1950.

MOREAU, F.: Les lichens: Morphologie, biologie, systématique, in Encyclopédie Biologique, vol. 3, pp. 1–148. 2 *pls.* 65 *figs.* Paris, Paul Lechevalier, 1929.

Vainio, E. A.: Lichenographia fennica: I. Pyrenolichenes iisque proximi Pyrenomycetes et Lichenes Imperfecti, *Acta Societatis pro Fauna et Flora Fennica*, **49**(2):1–274, 1921; II. Baeomyceae et Lecideales, *ibid.*, **53**(1):1–340, *Map*. 1922; III. Coniocarpeae, *ibid.*, **57**(1):1–138, 1927; IV. Lecideales, pt. 2 (edited by B. Lynge), *ibid.*, **57**(2):1–531, *Pls.* 1–4, 1934.

———: Étude sur la classification naturelle et la morphologie des lichens du Brésil, *ibid.*, **7**(1):1–247, (2):1–256, 1890.

Dodge, Carroll W.: The foliose and fruticose lichens of Costa Rica, I, *Ann. Missouri Botan. Garden*, **20**(3):373–467. 1933.

Magnusson, A. H., and A. Zahlbruckner: Hawaiian lichens: I. The families Verrucariaceae to Peltigeraceae, *Arkiv för Botanik*, **31A**(1):1–96. 1944; II. The families Lecideaceae to Parmeliaceae, *ibid.*, **31A**(6):1–109. 1944; III. The families Usneaceae to Physciaceae, *ibid.*, **32A**(6):1–89. *Pls.* 1–10. 1945.

van der Byl, P. A.: Korsmosse van die Unie van Suid-Afrika: I. Familie Rocellaceae, II. Familie Teloschistaceae, *Ann. Univ. Stellenbosch*, **11A**(6):1–18. 10 *figs.* 1933; III. Familie Cladoniaceae, *ibid.*, **11A**(4):1–13. 5 *figs.* 1933; IV. Die geslag Ramalina, *ibid.*, **13A**(1):1–13. 12 *figs.* 1935; V. Familie Collemaceae, *ibid.*, **13A**(4):1–11. 4 *figs.* 1935.

Hasse, Hermann E.: The lichen flora of Southern California, *Contribs. U.S. National Herbarium*, **17**:1–132. 1913.

Herre, Albert W. C. T.: The lichen flora of the Santa Cruz Peninsula, California, *Proc. Wash. Acad. Sci.*, **12**:27–269. 1910.

Nearing, G. G.: The Lichen Book, 624 pp. 690 *drawings*. Ridgewood, N.J., published by the author, 1947.

Dodge, Carroll W., and Gladys E. Baker: The second Byrd Antarctic expedition—Botany: II. Lichens and lichen parasites, *Ann. Missouri Botan. Garden*, **25**(2):515–718. *Pls.* 38–65. 1938.

Magnusson, A. H.: Lichens from Central Asia, Report of the Scientific Expedition in the North-Western Provinces of China under Sven Hedin, The Sino-Swedish Expedition Publication No. 13, pp. 1–168, 12 *pls.*, 1940; No. 22, pp. 1–71, 8 *pls.*, 1944.

Fink, Bruce: The lichens of Minnesota, *Contribs. United States National Herbarium*, **14**:1–269. *Pls.* 1–51. *Figs.* 1–18. 1910.

Räsänen, Veli: Das System der Flechten. Übersicht mit Bestimmungstabellen der natürlichen Flechtenfamilien, ihrer Gattungen, Untergattungen, Sektionen und Untersektionen, *Acta Botanica Fennica*, **33**:1–82. 1943

Zahlbruckner, A.: Additamenta ad Lichenographiam Japoniae, *Botanical Magazine* (*Tokyo*), **41**:313–364. *Pls.* 11–12. 1927.

Vainio, E. A.: Monographia Cladoniarum universalis, *Acta Societatis pro Fauna et Flora Fennica*, **4**:1–509, 1887; **10**:1–499, 1894; **14**:1–268, 1897.

Sandstede, H.: Ergänzungen zu Wainio's Monographia Cladoniarum universalis unter besonderer Berücksichtigung des Verhaltens der Cladonien zu Asahina's Diaminprobe, *Repertorium Specierum Novarum Regni Vegetabilis*, Beiheft 103, 103 pp. 16 *pls.* 1938.

Evans, Alexander W.: The Cladoniae of Connecticut, *Trans. Connecticut Acad. Arts Sci.*, **30**:357–510. 1930.

———: Notes on the Cladoniae of Connecticut, *Rhodora*, **34**(403):121–142, (404):153–164, 1932; **40**(469):4–26, 1938; **52**(617):77–123. *Pl.* 1161. 1950.

———: The Cladoniae of New Jersey, *Torreya*, **35**(4):81–109. 1935.

———: The Cladoniae of New Jersey. Supplement, *ibid.*, **38**(6):137–149. 1938.

———: A study of certain North American Cladoniae, *Bryologist*, **50**(1):14–31. *Pls.* 1–5. *Figs.* 1–7. 1947.

———: Supplementary report on the Cladoniae of Connecticut, *Trans. Connecticut Acad. Arts Sci.*, **35**:519–626. 1944.

DES ABBAYES, H.: Revision monographique des Cladonia du sous-genre Cladina (Lichens), *Bull. soc. sci. Bretagne*, **16**(2):51–156. 1939.

SAVICZ, V. P.: Die Cladonien Kamtschatkas, *Repertorium Specierum Novarum Regni Vegetabilis*, **19**:337–372. 1924.

HANSEN, H. MOLHOLM, OG MOGENS LUND: De Danske arter af slaegten Cladonia med angivelse af deres udbredelse og forekomst, *Botanisk Tidsskrift*, **41**(1):1–80. 4 *pls*. 37 *figs*. 1929.

TORREY, RAYMOND H.: Cladoniae in the range of the Torrey Botanical Club, *Torreya*, **33**(5):109–129. *Pls*. 1–4. 1933.

———: Cladoniae of the North Woods, *ibid.*, **34**(3):57–74. *Pls*. 1–3. 1934.

HOWE JR., R. HEBER: Preliminary notes on the genus Usnea, as represented in New England, *Bull. Torrey Botan. Club*, **36**(6):309-327. *Pls*. 21–23. *Figs*. *A–C*. 1909.

———: A manual of the genus Usnea as represented in North and Middle America, north of the 15th parallel, *ibid.*, **37**(1):1–18. *Pls*. 1–7. 1910.

———: Classification de la famille des Usneaceae dans l'Amérique du Nord, Théses presentées à la Faculté des Sciences de l'Université de Paris, série 60, ordre 71, pp. 1–32. 1912.

———: A monograph of the North American Usneaceae, *Rept. Missouri Botan. Garden*, **23**:133–146. *Pl*. 7. 1912. (Part 1, general, apparently all that was issued.)

———: The genus Evernia as represented in North and Middle America, *Botan. Gaz.*, **51**(6):431–442. *Pls*. 24–25. 1911.

———: Oropogon loxensis and its North American distribution, *Mycologia*, **4**(3):152–156. *Figs*. 1–2. 1912.

———: North American species of the genus Ramalina, *The Bryologist*, **16**:65–74, *Pls*. 5–7. 81–89, 2 *pls*. 1913; **17**:1–7, 2 *pls*. 17–27, 33–40, 2 *pls*. 3 *figs*. 49–52, 65–69, 81–87, 2 *pls*. 1914.

———: The genus Teloschistes in North America, *Bull. Torrey Botan. Club*, **42**(10):579–583. *Figs*. 1–2. 1915.

———: American species of Alectoria occurring north of the fifteenth parallel, *Mycologia*, **3**(3):106–150. *Pls*. 41–47. 1911.

DU RIETZ, G. EINAR: Vorarbeiten zu einer "Synopsis Lichenum": I. Die Gattungen Alectoria, Oropogon und Cornicularia, *Arkiv för Botanik*, **20A**(4):1–43. 2 *pls*. 2 *figs*. 1926.

———: Morfologi och systematik hos släket Ramalina, särskillt dess skandinavisks arter, *Svensk Botan. Tid.*, **20**:295–298. 1926. (Keys to sections and to the Scandinavian species of Ramalina.)

———: Bestämningskema öfver Skandinaviens Stereocaulonarter, *ibid.*, **20**(1):95–96. 1926.

DODGE, CARROLL W.: A synopsis of Stereocaulon with notes on some exotic species, *Ann. cryptogam. exotique*, **2**(2):93–153. 1929.

LAMB, I. MACKENZIE: A monograph of the lichen genus Placopsis Nyl., *Lilloa*, **13**:151–288. *Pls*. 1–16. *Figs*. 1–7. 1947.

LOWE, JOSIAH L.: The genus Lecidea in the Adirondack Mountains of New York, *Lloydia*, **2**(4):225–304. *Pls*. 1–4. *Fig*. 1. 1939.

FREY, EDUARD: Beiträge zur Biologie, Morphologie und Systematik der Umbilicariaceen, *Hedwigia*, **69**(5):219–252. *Figs*. 1–9. 1929.

———: Weitere Beiträge zur Kenntnis der Umbilicariaceen, *ibid.*, **71**(1–2):94–119. *Figs*. 1–8. 1931

Scholander, P. F.: On the apothecia in the lichen family Umbilicariaceae, *Nytt Magasin för Naturvidenskapene,* **75**:1–31. *Pls.* 1–5. *Figs.* 1–12. 1936.

Berry, Edward Cain: A monograph of the genus Parmelia in North America, north of Mexico, *Ann. Missouri Botan. Garden,* **28**(1):31–146. 1941.

Tavares, C. das Neves: Contribuicão para o estudo das Parmeliaceas portuguesas, *Portugaliae Acta Biologica, ser. B,* **1**(1–2):1–233. 1945.

Herre, Albert W.: The Parmelias of California, *Contribs. Dudley Herbarium of Stanford Univ.,* **3**(10):313–350. 1946.

———: The Gyrophoraceae of California, *Contribs. U.S. National Herbarium,* **13**:313–321. *Pls.* 68–73. 1911.

du Rietz, G. Einar: Die europäischen Arten der Gyrophora "anthracina" Gruppe, *Arkiv för Botanik,* **19**(12):1–14. 1925.

Gyelnik, V.: Über einige Arten der Gattung Parmeliopsis (Stizenb.) Nyl., *Ann. Mycol.,* **30**(5–6):456–459. 1932. (Descriptions of some species and key to all known species of this genus of lichens.)

———: Lichenologische Substratstudien (Squamaria radiosa-Gruppe), *Hedwigia,* **71**(1–2):120–132. 1931.

———: Nephroma-Studien, *ibid.,* **72**(1):1–30. *Figs.* 1–2. 1932.

Darbishire, Otto V.: Monographia Rocellarum: Ein Beitrag zur Flechtensystematik, *Bibliotheca Botanica,* **45**:1–102. *Pls.* 1–30. 1898.

Hillmann, Johannes: Studien über die Flechtengattung Teloschistes Norm, *Hedwigia,* **69**:303–343. *Figs.* 1–2. 1930.

———: Übersicht über die Arten der Flechtengattung Xanthoria (Th. Fr.) Arn., *ibid.,* **63**(3–4):198–208. 1922.

———: Bemerkungen über einige Arten der Flechtengattung Parmelia, I, *ibid.,* **78**(5–6):249–267. 1939.

Asahina, Yasuhiko: Leptogium (Section Mallotium) aus Japan. *J. Japanese Botany,* **11**:544–566. 20 *figs.* 1935. (In Japanese.)

Schede, Alwin: Die Sächsischen Arten der Flechtengattung Rhizocarpon (Ram.) Th. Fr., *Beihefte Botan. Centr., Abt. B,* **54**(1–2):75–107. *Figs.* 1–18. 1935.

Lynge, B. A.: A monograph of Norwegian Physciaceae, *Videnskapselskapets-Skrifter, I, Mat.-naturv. Klasse,* No. 8, 110 pp. 3 *pls.* 1916.

Räsänen, Veli: Bestimmungsschlüssel für die gelben Rhizocarpon-Arten, -Varietäten und Formen, *Repertorium Spec. Nov. Reg. Veg.,* **52**:127–136. 1943.

Grespi, L.: Notas liquenalogicas. El género Rhizocarpon en España, *Boletin Sociedad Española de Historia Natural,* **30**(5):261–269. 1930.

Bachmann, E.: Die Moriolaceen, *Nyt Magazin för Naturvidenskaberne,* **64**:170–328. 3 *pls.* 13 *figs.* 1926.

———: A revision of the genus Rhizocarpon (Ram.) Th. Fr. in Greenland, Norges Svalbard og Ishavs-undersökelser, *Skriften om Svalbard og Ishavet,* **47**:1–30. 1932.

Thompson Jr., John W.: The Wisconsin species of Peltigera, *Trans. Wisconsin Acad. Sci.,* **38**:249–271. *Pls.* 1–6. 1946 (1947).

Zahlbruckner, A.: Die Gattung Lecanora. *Report of the scientific results of the Norwegian expedition to Novaya Zemlya 1921, Botany,* **44**:1–32. 4 *pls.* 1928.

Lettau, Georg: Monographische Bearbeitung einiger Flechtenfamilien, *Repertorium Specierum Novarum Regni Vegetabilis,* Beiheft **69** (1, Lief. 1):1–96. 3 *pls.* 1932. (Contains key to species of Lecanactidaceae.)

Magnusson, A. H.: A monograph of the genus Acarospora, *Kgl. Svenska Vetenskapsakad. Handl.,* tredje ser., **7**(4):1–400. 18 *maps.* Stockholm. 1929.

———: The yellow species of Acarospora in North America, *Mycologia,* **21**(5):249–260. 1929.

———Supplement to the monograph of the genus Acarospora, *Ann. cryptogam. exotique*, 6(1):13–48. 1933.

———: The lichen genus Acarospora in New Mexico, *Meddelanden från Göteborgs Botaniska Trädgård*, **5**:55–72. 1930.

———: New or otherwise interesting Lecanora species, *ibid.*, **6**:1–20. 1930.

———: Beiträge zur Systematik der Flechtengruppe Lecanora fusca, *ibid.*, **7**:65–87. 1931.

———: Studien über einige Arten der Lecidea armeniaca und elata Gruppen, *ibid.*, **6**:93–144. 1930.

———: On saxicolous species of the genus Lecidea proper to North America, *ibid.*, **10**:1–53. 1935.

———: Contribution to the taxonomy of the Lecidea goniophila group, *ibid.*, **16**:125–134. 1944.

———: A monograph of the lichen genus Ionaspis, *ibid.*, **8**:1–46. 1933.

———: Die Flechtengattung Maronea Mass., *ibid.*, **9**:41–66. 1934.

———: On the species of Biatorella and Sarcogyne in America, *Ann. cryptogam. exotique*, **7**(3–4):115–146. 1934.

———: On North American, nonsaxicolous species of the genus Rinodina, *Botan. Notiser*, **1947**(1):32–54. 1947.

Ericksen, C. F. E.: Lichenologische Beiträge, *Hedwigia*, **70**:216–233. *Fig.* 1. 1930. (Contains a key to the bark-inhabiting crustose lichens of the lowlands of North Germany, mainly sterile or nearly sterile specimens.)

Olivier, H.: Les lichens pyrénocarpés de la flore d'Europe, *Bull. de Géographie Botanique*, **28**:146–152, 168–183. 1918; **29**:6–16, 35–48, 97–110. 1919.

Zschacke, H.: Die mitteleuropäischen Verrucariaceen, *Hedwigia*, **60**:1–9, 1918; **62**:90–154, 1921; **65**:46–64, 1924; **67**:45–85, 1927.

Malme, Gust. O. A.-N.: De sydsvenska formerna af Rinodina sophodes (Ach.) Th. Fr. och Rinodina exigua (Ach.) Th. Fr., *Bihang till Kgl. Svenska Vetenskapsakad. Handl.*, Band 21, Afd. III, No. **11**:1–40. *Pls.* 1–2. 1895.

(The following series of papers throws light upon the lichen flora of Paraguay and Southern Brazil, being based almost entirely upon the collections of Regnell.)

Malme, Gust. O. A.-N.: Die Pannariazeen des Regnellschen Herbars, *Arkiv för Botanik*, **20A**(3):1–23. 1926.

———: Lichenes blasteniospori herbarii Regnelliani, *ibid.*, **20A**(9):1–51. 1926. (Theloschistaceae.)

———: Die im Regnellschen Herbar aufbewahrten Arten der Flechtengattung Lecanactis (Eschw.) Vainio, *ibid.*, **20B**(2):1–6. 1926.

———: Buelliae itineris Regnelliani primi, *ibid.*, **21A**(4):1–42. 1 *fig.* 1928.

———: Lichenes pyrenocarpi aliquot in herbario Regnelliano asservati, *ibid.*, **22A**(6):1–11. 1928.

———: Pyrenulae et Anthracothesia herbarii Regnelliani, *ibid.*, **22A**(11):1–40. 3 *figs.* 1929.

———: Porinae et Phylloporinae in itinere Regnelliani collectae, *ibid.*, **23A**(1, paper 1):1–37. 1930.

———: Die Ramalinen der ersten Regnell'schen Expedition, *ibid.*, **26A**(12):1–9. 2 *pls.* 1934.

———: Die Gyalectazeen der ersten Regnell'schen Expedition, *ibid.*, **26A**(13):1–10. 1934.

———: Die Stictazeen der ersten Regnell'schen Expedition, *ibid.*, **26A**(14):1–18. 3 *pls.* 2 *figs.* 1934.

MALME, GUST. O. A.-N.: Bacidiae itineris Regnelliani primi, *ibid.*, **27A**(5):1–40. 1935.

———: Lecideae expeditionis Regnellianae primae, *ibid.*, **28A**(7):1–53. 1936.

———: Thelotremaceae Brasilienseae imprimis ex herbario Regnelliano cognitae praeteraque in herbariis Krempelhaberi, Müller Arg., Nylanderi, Wainionis et Zahlbruckneri asservatae, *ibid.*, **28A**(8):1–122. *Figs.* 1–75. 1936.

———: Pertusariae expeditionis Regnellianae primae, *ibid.*, **28A**(9):1–27. 1936.

REDINGER, KARL M.: Die Graphidineen der ersten Regnell'schen Expedition nach Brasilien 1892–1894: I. Glyphis, Medusulina, und Sarcographa, *Arkiv. för Botanik.*, **25A**(13):1–20. 1 *pl.* 7 *figs.* 1933; II. Graphina und Phaeographina, *ibid.*, **26A**(1):1–103. 7 *pls.* 1934; III. Graphis und Phaeographis, nebst einem Nachtrage zu Graphina, *ibid.*, **27A**(3):1–103. 7 *pls.* 1935.

List 16. Pezizales Operculati and Inoperculati, Including "Phacidiales"

(See BIGEARD ET GUILLEMIN, J. B. CLELAND, J. RAMSBOTTOM, V. O. GRAHAM in List 34; MASSEE, vol. 4, in List 1.)

BOUDIER, ÉMILE: Histoire et classification des Discomycètes d'Europe, vii + 223 pp. Paris, Librairie des Sciences Naturelles, Paul Klincksieck, 1907.

———: Icones mycologicae ou iconographie des champignons de France, principalement Discomycètes, 4 vols. 600 *colored plates.* Paris, Paul Klincksieck, 1905–1910.

SEAVER, FRED JAY: The North American Cup-Fungi (Operculates), 284 pp. *Frontis.* (*col.*) 45 *pls.* 15 *figs.* New York, published by the author, 1928.

———: Supplement to North American Cup-Fungi, pp. i–viii + 285–377. *Pls.* 46–74. *Figs.* 16–23. New York, published by the author, 1942.

———: The North American Cup-Fungi (Inoperculates), 429 pp. *Pls.* 76 (colored) to 150. New York, published by the author, 1951.

———: Photographs and descriptions of Cup-Fungi: I. Peziza, *Mycologia*, **7**(2):90–93. *Pls.* 155–156. 1915; II. Sepultaria, *ibid.*, **7**(4):197–199. *Pl.* 161. 1915; III. Peziza domiciliana and Peziza repanda, *ibid.*, **8**(4):195–198. *Pls.* 188–189. 1916; IV. Peziza clypeata, *ibid.*, **8**(5):235–238. *Pl.* 191. 1916; V. Peziza proteana and Peziza violacea, *ibid.*, **9**(1):1–3. *Pl.* 1. 1917; VI. Discina venosa, *ibid.*, **9**(2):53–54. *Pl.* 5. 1917; VII. The genus Underwoodia, *ibid.*, **10**(1):1–3. *Fig.* 1. 1918; VIII. Elvela infula and Gyromitra esculenta, *ibid.*, **12**(1):1–5. *Pl.* 1. 1920; IX. North American species of Discina, *ibid.*, **13**(2):67–71. *Pl.* 4. 1921; X. Ascotremella, *ibid.*, **22**(2):51–54. *Pls.* 11–12. 1930; XI. Solenopezia, *ibid.*, **22**(3):122–124. *Pl.* 16. 1930; XII. Elvelaceae, *ibid.*, **22**(4):163–164. *Pls.* 17–19. 1930; XIII. Subhypogeous forms, *ibid.*, **22**(5):215–218. *Pls.* 22–23. 1930; XIV. A new genus, *ibid.*, **23**(4):247–251. *Pls.* 23–24. 1931 (the genus Chloroscypha with key and description of the 4 known species); XV. The giant Elvela, *ibid.*, **23**(6):409–410. *Pl.* 29. 1931; XVI. Stamnaria, *ibid.*, **24**(1):1–3. *Pl.* 1. 1932; XVII. A new species of Godronia, *ibid.*, **24**(4):353–354. *Pl.* 9. 1932; XVIII. Rare species of Godronia, *ibid.*, **25**(1):55–57. *Pl.* 15. 1933; XIX. The cabbage-head fungus, *ibid.*, **25**(3):157–159. *Pls.* 24–25. 1933; XX. A new Lamprospora, *ibid.*, **26**(1):102–103. *Pl.* 14. 1934; XXI. The genus Calycina, *ibid.*, **26**(4):344–347. *Pl.* 40. 1934; XXII. Dasyscypha. *ibid.*, **28**(1):1–6. *Fig.* 1. 1936; XXIII. Stamnaria, *ibid.*, **28**(2):186–187. *Fig.* 1. 1936;

XXIV. Chlorociboria, *ibid.*, **28**(4):390–394. *Figs.* 1–2. 1936; XXV. Urnula geaster, *ibid.*, **29**(1):60–65. *Figs.* 1–3. 1937; XXVI. The genus Diplocarpa, *ibid.*, **29**(2):174–177. *Fig.* 1. 1937; XXVII. Pezicula on Cornus, *ibid.*, **29**(3):334–337. *Figs.* 1–2. 1937; XXVIII. A proposed genus, *ibid.*, **29**(6):678–680. *Fig.* 1. 1937 (*Wolfina*, nov. gen.); XXVIII (bis). A new Helotium, *ibid.*, **30**(1):79–81. *Fig.* 1. 1938; XXIX. Chloroscypha, *ibid.*, **30**(5):594–596. *Fig.* 1. 1938; XXX. Arachnopeziza, *ibid.*, **30**(6):659–663. *Fig.* 1. 1938; XXXI. Mollisiella, *ibid.*, **31**(1):93–95. *Fig.* 1. 1939; XXXII. Podophacidium, *ibid.*, **31**(3):350–353. *Fig.* 1. 1939; XXXIII. A new Boudiera, *ibid.*, **31**(4):499–501. *Fig.* 1. 1939; XXXIV. A new Humarina, *ibid.*, **31**(5):533–536. *Fig.* 1. 1939; XXXV. A new species of Patella, *ibid.*, **32**(4):567–569. *Fig.* 1. 1940; XXXVI. A new species and genus, *ibid.*, **34**(3):298–301. *Fig.* 1. 1942 (Pestalopezia nov. gen.); XXXVII. Pezicula purpurascens, *ibid.*, **34**(4):412–415. *Fig.* 1. 1942; XXXVIII. The genus Kriegeria, *ibid.*, **35**(4):492–493. 1943; XXXIX. A new Helotium, *ibid.*, **37**(2):267–269. *Fig.* 1. 1945; XXXIX (bis). The genus Godronia and its allies, *ibid.*, **37**(3):333–359. 3 *figs.* 1945; XLI. Catinella nigro-olivacea, *ibid.*, **38**(4):473–476. *Fig.* 1. 1946; XLII. Gorgoniceps, *ibid.*, **38**(5):549–553. 2 *figs.* 1946; XLIII. Seaverinia, *ibid.*, **39**(1):113–119. *Fig.* 1. 1947.

NANNFELDT, J. A.: Studien über die Morphologie und Systematik der nicht lichenisierten inoperculaten Discomyceten, *Nova Acta Regiae Soc. Sci. Upsaliensis*, Ser. IV, **8**(2):1–368. *Pls.* 1–19. *Figs.* 1–47. 1932.

PHILLIPS, WILLIAM: A manual of British Discomycetes, in International Science Series, vol. 61, 462 pp. 12 *pls.* London, Kegan Paul, Trench, Trübner and Co., 1887; ed. 2, 1893.

GRELET, L. J.: Les Discomycètes de France d'après la classification de Boudier, Fascicules 1–8, *Bull. soc. botanique du Centre-Ouest*, 1932–1940; Fasc. 9–19, *Rev. mycol.*, N. S., vols. **7–14**, 1942–1949 (to be continued.)

LEGAL, MARCELLE: Florule mycologique des Bois de la Grange et de l'Étoile (Seine-et-Oise): Discomycètes (Operculés), *Rev. mycol.*, N.S., **2**:150–162, 197–222. *Figs.* 1–27. 1937; Discomycètes (Inoperculés), *ibid.*, **3**:129–147. *Figs.* 1–9. 1938; **4**:25–63. *Figs.* 10–29. 1939.

———: Quelques Galactinia de la flore Française, *ibid.*, **4**:169–186. *Figs.* 1–8. 1939; **5**:102–112. *Figs.* 1–3. 1940; **10**:90–95. *Figs.* 1–3. 1945.

———: Les Aleuria et les Galactinia, *ibid.*, **6**(supplément 3):56–82. *Figs.* 1–4. 1941.

LAGARDE, J.: Contribution à l'étude des Discomycètes charnus, *Ann. Mycol.*, **4**(2):125–201, (3):203–256. *Pls.* 3–4. *Figs.* 1–58. 1906.

———: Discomycètes de France: I. Les morilles, pp. 1–35, *Pls.* 1–5, 1923; II. Les Helvelles, pp. 39–82, *Pls.* 6–12, 1924. Paris, La Pensée Française.

RAMSBOTTOM, J.: A list of the British species of Discomycetes arranged according to Boudier's system with a key to the genera, *Brit. Mycol. Soc. Trans.*, **4**(2):343–381. 1914.

KANOUSE, BESSIE B.: A survey of the Discomycete flora of the Olympic National Park and adjacent areas, *Mycologia*, **39**(6):635–689. *Figs.* 1–35. 1947.

———: The genus Plectania and its segregates in North America, *ibid.*, **40**(4):482–497. *Figs.* 1–12. 1948.

———: Studies in the genus Otidea, *Mycologia*, **41**(6):660–677. *Figs.* 1–21. 1949.

VELENOVSKY, J.: Monographia Discomycetum Bohemiae, vol. 1:1–436; vol. 2: *pls.* 1–31. Prague, 1934.

KILLERMANN, S.: Bayerische Becherpilze: I. Eupezizaceen. Mit kritischen Bemerkungen, *Kryptogamische Forschungen herausgegeben von der Krypto-*

gamenkommission der Bayerischen Botanischen Gesellschaft zur Erforschung der Heimischen Flora, **2**(1):27–47. *Pls.* 1–3. 1929; II. Pezizeae, *ibid.*, **2**(3):255–296. *Pls.* 4–5. 1935.

RICK, J.: Monographia Pezizinearum Riograndensium, *Broteria, Sér. Botan.*, **25**(2):77–122. 1931; continued in *Broteria, Série Trimestral, Ciencias Naturais*, **1**(1):35–46, (2):89–96, (3):97–105. 1932.

MCLENNAN, ETHEL, AND ISABEL COOKSON: Addition to the Australian Ascomycetes, I, *Proc. Roy. Soc. Victoria*, N.S., **35**(2):153–158. *Pls.* 8–10. *Fig.* 1. 1923; II, *ibid.*, **38**:69–76. *Pls.* 4–6. 1926.

———, AND FRANCIS HALSEY: Addition to the Australian Ascomycetes, III, *ibid.*, **49**:51–62. *Pl.* 2. 1936.

RODWAY, L.: Tasmanian Discomycetes, *Papers and Proc. Roy. Soc. Tasmania*, **1924**:90–122. 1925.

KILLERMANN, S.: Morcheln und andere Helvellaceen aus Bayern, *Kryptogamische Forschungen der Bayerischen Botanischen Gesellschaft*, **3**:148–154. 1 *fig.* 1918.

IMAI, SANSHI: Contributions to the knowledge of the classification of the Helvellaceae, *Botanical Magazine (Tokyo)*, **46**(544):172–175. 1932.

HEIM, ROGER: Tableaux pratiques de détermination des principales morilles, in "Le culture des Morilles," *Rev. Mycol.*, **1**(supplement to 2):22–25. 2 *pls.* 1936.

MOSER, MEINHARD: Über das Massenauftreten von Formen der Gattung Morchella auf Waldbrandflächen, *Sydowia, Ann. Mycol.*, **3**(1–6):176–195. *Figs.* 1–4. 1949. Includes critical descriptions of the several species occurring in the burned area.

OVERHOLTS, L. O.: The morels of Pennsylvania, *Proc. Penna. Acad. Sci.*, **8**:108–114. *Figs.* 1–8. 1934.

RICK, J.: Monographia Helvellinearum Riograndensium, *Broteria, Sér. Botan.*, **25**(2):72–76. 1931.

KLIKA, JAROMIR: Přispěvek k poznáni hub chrapáčovitých v Ceskoslovensku (Czechoslovakian Helvellaceae), *Věstnik Královski České Společnosti Nauk Tř. II*, **1924**:1–54. 8 *figs.* 1925.

BOUDIER, É.: Révision analytique des morilles de France, *Bull. soc. mycol. France*, **13**:129–153. 1897.

HONE, DAISY S.: Minnesota Helvellineae, *Minnesota Botanical Studies*, **3**:309–321. *Pls.* 48–52. 1904.

ANDERSON, P. J., AND MARGUERITE G. ICKIS: Massachusetts species of Helvella, *Mycologia*, **13**(4–5):201–229. *Pls.* 11–12. 1921.

MASSEE, GEORGE: A monograph of the Geoglossaceae, *Ann. Botany*, **11**(42):225–306. *Pls.* 12–13. 1897.

DURAND, E. J.: The Geoglossaceae of North America, *Ann. Mycol.*, **6**(5):387–477. *Pls.* 5–22. 1908. (A very valuable work with full descriptions of all species and illustrations of many of them.)

———: New or noteworthy Geoglossaceae, *Mycologia*, **13**(3):184–187. 1921. (Supplementary to the preceding.)

SINDEN, J. W., AND H. M. FITZPATRICK: A new Trichoglossum, *Mycologia*, **22**(2):55–61. *Pl.* 13. 1930. (A new species and corrections to Durand's paper.)

LLOYD, C. G.: The Geoglossaceae, *Mycological Writings* 5, pp. 1–24 (separate pagination). *Figs.* 782–807. 1916. (Based largely on Durand's paper but with some modification of generic names and limits.)

VAN LUIJK, A.: Fungi van Nederland: I. Geoglossaceae van Nederland, *Nederlandsch Kruidkundig Archief*, **1918**:111–144. 12 *figs.* 1919.

IMAI, SANSHI: Studies on the Geoglossaceae of Japan: I. *Trans. Sapporo Natural History Soc.*, **13**(3):179–184. *Pl.* 7. 1934; II. The genus Leotia, *Botanical*

———, AND ———: New discoveries of Myriangiales in the Americas, *Proc. Eighth Am. Sci. Congr., Biol. Sci.: Botany, Washington, 1940,* **3**:149–172, 1942.

HANSFORD, C. G.: Contributions toward the fungus flora of Uganda: III. Some Uganda Ascomycetes, *Proc. Linnean Soc. London,* **153**:4–52. 1940–41. (Has descriptions of seven new species of Sphaceloma.)

JENKINS, ANNA E.; A. A. BITANCOURT; AND FLORA G. POLLACK: Spot anthracnoses in the Pacific Coast States, *J. Wash. Acad. Sci.,* **36**(12):416–421. *Figs.* 1–2. 1946.

———, AND ———: Spot anthracnoses in the United States and some island possessions, *Plant Disease Reptr.,* **31**(3):114–117. 1 *map.* 1947. (This has no descriptions but gives the distribution of 23 spot anthracnoses in the United States, Puerto Rico, Guam, and Hawaii as well as references to their published descriptions.)

THIRUMALACHAR, M. J.: Some new Sphaceloma diseases of economic plants in Mysore, *Brit. Mycol. Soc. Trans.,* **31**(1–2):1–6. *Figs.* 1–9. 1947.

(In addition to the foregoing consult *Mycologia* for various articles describing separate species of Elsinoe (Sphaceloma) in the last eight years.)

List 29. Aspergillales (Plectascales)

(For extensive systematic studies of Aspergillus and Penicillium and closely related genera see List 49.)

DODGE, CARROLL W.: The higher Plectascales, *Ann. Mycol.,* **27**(3–4):145–184. *Pls.* 1–2. *Figs.* 1–2. 1929. (This treats of the Trichocomaceae and Elaphomycetaceae.)

IMAI, SANSHI: Fourth note on Elaphomyces in Japan, *Proc. Imp. Acad. (Japan),* **15**:146–147. 1939.

———: Elaphomyces Japoniae et Cordyceps fungieola, *Acta Phytotex. et Geobat.,* **13**:75–83. 1942. (In Japanese.)

GOIDANICH, GABRIELE: Il genere di Ascomiceti "Grosmannia" G. Goid., *Boll. staz. patol. vegetale,* N.S., **16**(1):26–60. *Pl.* 1. *Figs.* 1–19. 1936.

List 30. Saccharomycetales and Asporogenous Yeasts

STELLING-DEKKER, N. M.: Die Hefesammlung des "Centraalbureau voor Schimmelcultures." Beiträge zu einer Monographie der Hefesorten. Erster Teil: Die sporogenen Hefen, *Verhandelingen der Koninklijke Akademie van Wetenschappen te Amsterdam. Afdeeling Natuurkunde (Tweede Sectie),* Deel 28, No. 1, vii + 547 pp. *Illustrated.* 1931.

LODDER, J.: Die Hefesammlung des "Centraalbureau voor Schimmelcultures." Beiträge zu einer Monographie der Hefearten: Teil II. Die anaskosporogenen Hefe. Erste Hälfte, *ibid.,* Deel 32, ix + 256 pp. 114 *figs.* 1934.

DIDDENS, H A., UND J. LODDER: Die Hefesammlung des "Centraalbureau voor Schimmelcultures." Beiträge zu einer Monographie der Hefearten: Teil II. Die anaskosporogenen Hefe. Zweite Hälfte, xii + 511 pp. 99 *figs*. Amsterdam, N.V. Nord-Hollandsche Uitgevers Maatschappij, 1942.

GUILLIERMOND, ALEXANDRE: Recherches cytologiques et taxonomiques sur les Endomycètées, *Rev. gén. botan.*, **21**:353–391, 401–419. *Pls*. 12–19. *Figs*. 1–33. 1909.

———: Les levures, xii + 565 pp. 163 *figs*. Paris, O. Doin et fils, 1912.

———: The Yeasts. Translation of the foregoing by F. W. Tanner, xix + 424 pp. 163 *figs*. New York, Wiley and Sons, 1920.

———: Clef dichotomique pour la detérmination des levures, 124 pp. *Illustrated*. Paris, Librairie le François, 1928.

———: La classification des levures, *Ann. fermentations*, **2**:474–491, 540–551. *Figs*. 1–23. 1936.

ZENDER, JUSTIN: Sur la classification des Endomycètacées, *Bull. soc. botan. Genève*, **17**:272–302. 1925.

BEDFORD, C. L.: A taxonomic study of the genus Hansenula, *Mycologia*, **34**(6):628–649. 1942.

NICKERSON, WALTER J.: Studies in the genus Zygosaccharomyces: I. Transfer of pellicle-forming yeasts to Zygopichia, *Farlowia*, **1**(3):469–481. 1944. (With key to and description of the species of Zygopichia.)

KONOKOTINA, A. G., AND N. A. KRASIL'NIKOV: Yeasts of the genus Debaryomyces Klock. and their distribution in nature, *J. Microbiologie de l'Institut Bactériologique Pasteur (Leningrad)*, **9**(1):93–107. 1 *pl*. 1929. (In Russian.)

GUILLIERMOND, ALEXANDRE: Étude cytologique et taxinomique sur les levures du genre Sporobolomyces, *Bull. trimestr. soc. mycol. France*, **43**:245–258. *Pl*. 10. *Figs*. 1–6. 1927.

DERX, H. G.: Étude sur les Sporobolomycètes, *Ann. Mycol.*, **28**(1–2):1–23. *Pl*. 1. 1930.

VERONA, O., E R. CIFERRI: Revisione dei lieviti asporogeni del genero Sporobolomyces Kluyver et van Niel, *Atti ist. botan. "Giovanni Briosi" e lab. crittogam. Italiano univ. Pavia*, ser. IV, **10**:241–255. 1938.

CIFERRI, RAFFAEL, E PIERO REDAELLI: Caratteri e posizione sistematica dell' agente della "malattia di Darling," Histoplasma capsulatum Darling e note sugli H. farciminosum, H. pyriforme e H. Muris, *ibid.*, ser. IV, **6**:247–309. *Pl*. 1. *Figs*. 1–27. 1935. (Gives keys to the families Nectaromycetaceae, Histoplasmaceae, and Torulopsidaceae.)

———, E ———: Monografia delle Torulopsidaceae a pigmento rosso, *Atti ist. botan. regia univ. Pavia*, ser. III, **2**:147–303. 8 *pls*. 1925.

———, E ———: Studies on the Torulopsidaceae, *Ann. Mycol.*, **27**(3–4):243–295. *Pls*. 4–6. 1929.

———, E ———: Contribuzione alla sistematica delle Torulopsidaceae, XV–XXXIII, *Arch. Mikrobiol.*, **6**:9–72. 1935.

———: Studi sulle Torulopsidaceae–Sui nomi generici di Torula, Eutorula, Torulopsis, Cryptococcus e sul nome di gruppo Torulaceae, *Atti ist. botan. regia univ. Pavia*, ser. III, **2**:129–142. 1925.

———: Morphological relations of the genera of asporogenous yeasts, *Ann. Mycol.*, **28**(5–6):372–376. 1930.

———, E O. VERONA: Chiavi analitiche dei lieviti segnalati nell' uve, nei mosti e nei vini, *Mycopathologia*, **4**(3):243–248. 1948.

HARRISON, F. C.: Cheese Torulae, *Trans. Roy. Soc. Can.*, **21**(Sect. 5, part 2):341–

380. 5 *pls*. 1927. (Divides the genus *Torula* into three genera and describes the species.)

———: A systematic study of some Torulae, *ibid.*, **22**:(Sect. 5, part 2):187–225. 5 *pls*. 1928.

CASTELLANI, ALDO: I miceti della blastomicosi Nord-Americana, *Ann. med. navale e coloniale*, 2(516):239–257. 2 *pls*. 1929. (All species of *Blastomycoides* are described.)

NEGRONI, P., Y C. A. N. DAGLIO: Sobre el género Nectaromyces, *Anales soc. cient. argentina*, **144**:484–491. *Figs*. 1–2. 1947.

BERKHOUT, CHRISTINE MARIE: De schimmelgeslachten Monilia, Oidium, Oospora en Torula, Doctor's Thesis, Univ. Utrecht, pp. 1–77. *Pls*. 1–14. Scheveningen, Edauw and Johannissen, 1923.

CONANT, NORMAN F.: Studies in the genus Microsporum: I. Cultural studies, *Arch. Dermatol. and Syphilol.*, **33**:665–683. 1936 (reprint with additions and change of pagination); II. Biometric studies, *ibid.*, **34**:79–89. 1936 (reprint with additions and change of pagination).

MARTIN, DONALD S.; CLAUDIUS P. JONES; K. F. YAO; AND L. E. LEE, JR.: A practical classification of the Monilias, *J. Bact.*, **34**(1):99–128. *Pls*. 1–3. 1937. (Deals with the species of Monilia (Candida) parasitic upon Man.)

List 31. Ustilaginales (Including Graphiolaceae)

(See also PLOWRIGHT in List 32.)

CLINTON, GEORGE P.: North American Ustilagineae, *Proc. Boston Soc. Natural History*, **31**:329–529. 1904.

———: The Ustilagineae, or smuts, of Connecticut, *Connecticut State Geological and Natural History Survey Bull.* **5**:1–45. *Figs*. 1–55. 1905.

———: Ustilaginales, *North American Flora*, **7**(1):1–82. 1906.

ZUNDEL, GEORGE LORENZO INGRAM: Additions and corrections to Ustilaginales, *North American Flora*, **7**(14):971–1045. 1939.

CUNNINGHAM, G. H.: The Ustilagineae, or smuts of New Zealand, *Trans. Proc. N. Zealand Inst.*, **55**:307–433. 7 *pls*. 1924.

———: Third supplement to the New Zealand Uredinales and Ustilaginaceae, *ibid.*, **56**:74–80. 1926.

———: Fourth supplement to the Uredinales and Ustilaginales of New Zealand, *ibid.*, **57**:186. 1926.

———: Fifth supplement to the Uredinales and Ustilaginales of New Zealand, *ibid.*, **58**:47–50. 1927.

———: Sixth supplement to the Uredinales and Ustilaginales of New Zealand, *ibid.*, **59**:491–505. 1928.

———: Seventh supplement to the Uredinales and Ustilaginales of New Zealand, *ibid.*, **61**:402–418. 1930.

MCALPINE, D.: The Smuts of Australia, vii + 288 pp. *Pls*. 1–56. *Figs*. 1–15. Melbourne, Department of Agriculture, Victoria, 1910.

SCHELLENBERG, H. C.: Die Brandpilze der Schweiz, *Beiträge zur Kryptogamenflora der Schweiz*, **3**(2): i–xlvi, 1–180. *Figs*. 1–79. 1911.

CIFERRI, R.: Prima contribuzione allo studio degli "Ustilaginales," *Boll. soc. botan. ital.*, **2–3**:46–59. 1924. (Discusses 22 species of Tolyposporium, Entyloma, and Melanotaenium.)

———: Seconda contribuzione allo studio degli Ustilaginales, *Atti ist. botan. regia univ. Pavia*, ser. III, **1**(2):77–97. 1924. (Describes various species of Tuburcinia and Entyloma.)

———: Terza contribuzione allo studio degli Ustilaginales. Alcuni micromiceti della flora Spagnola e Svizzera, *ibid.*, **2**:7–14. 1925. (Describes several species of Entyloma and of other genera.)

———: Quarta contribuzione allo studio degli Ustilaginales, *Ann. Mycol.*, **26**(1–2):1–68. *Pl.* 1. 1928. (Includes a synopsis of the known species of Entyloma.)

———: A few interesting North American smuts: I. Revision of the smuts on Bouteloua spp., *Brit. Mycol. Soc. Trans.*, **18**(4):257–262. 1934.

BUBÁK, F.: Die Pilze Böhmens: Teil II. Brandpilze (Hemibasidii), *Archiv der Naturwissenschaftlichen Landesdurchforschung von Böhmen*, **15**(3):1–81. *Figs.* 1–24. 1916.

VIÉGAS, A. P.: Alguns fungos do Brasil: III. Ustilaginales, *Bragantia*, **4**(12):739–762. *Pls.* 1–10. *Figs.* 1–4. 1944.

NAGORNY, P. I.: Caucasian species of the genus Ustilago Pers., *Zapiski Nautchno-prikladnikh Otdyelov Tiflisskovo Botanitcheskovo Sada*, **5**:109–128. *Pls.* 1–2. 1926. (In Russian.)

———: Caucasian species of the genus Urocystis Rabenhorst, *ibid.*, **6**:104–108. 1929. (In Russian.)

———: Caucasian representatives of the genus Tilletia Tulasne, *Vyestnik Tiflisskovo Botanitcheskovo Sada*, **1926–27**(3–4):89–96. 1927. (In Russian.)

———: Caucasian species of the genus Entyloma DeB., *Stravopolskaya Stantsia Zashtchity Rastyenii*, **1926**:49–52. 1926. (In Russian.)

———: Caucasian species of the genus Doassansia Cornu, *Izvyestia Terskoi Okruzhnoi Stantsii Zashtchity Rastyenii*, **1–2**:84–85. 1927. (In Russian.)

GUTNER, L. S.: Golovnevye griby (po materialam A. A. Jachevskovo) (The smut fungi of the U.S.S.R. after the materials of the late A. A. Jaczewski), 383 pp. 137 *figs.* Moscow and Leningrad, Lenin Academy of Agricultural Science. Institute of Plant Protection, 1941. (In Russian.)

KOCHMAN, JOZEF: Grzby glowniowe polski. Ustilaginales Poloniae, *Planta Polonica Materjaly do Flory Polskiej. Wydawane przez Towarzystwo Naukowa Warszawskie*, **4**:1–161. 12 *pls.* 1 *fig.* 1936. (In Polish.)

BEELI, M.: Notes Mycologiques: II. Relevé des Ustilaginées récoltées dans le bassin du Congo, *Bull. Jardin Botanique de l'État. Bruxelles*, **8**(1):12–15. 1922.

———: Notes Mycologiques: III. Relevé des Ustilaginées d'Afrique et de leurs hôtes, *ibid.*, **8**(1):16–22. 1922.

VERWOERD, LEN: n'Bydrae tot ons kennis van die Suid-Afrikaanse Ustilaginales of Brandswamme, *Ann. Univ. Stellenbosch*, **4A**(2):1–34. *Figs.* 1–6. 1926.

MUNDKUR, B. B.: A contribution towards a knowledge of Indian Ustilaginales, *Brit. Mycol. Soc. Trans.*, **23**(1):86–121. 2 *figs.* 1939.

———: A second contribution towards a knowledge of Indian Ustilaginales, *ibid.*, **24**(3–4):312–336. 1940.

———: Some rare and new smuts from India, *Indian J. Agr. Sci.*, **14**(1):49–52. *Figs.* 1–2. 1944.

———: Fungi of the Northwestern Himalayas. Ustilaginales, *Mycologia* **36**(3):286–292. 1944.

YEN, WEN-YU: Première note sur quelques Ustilaginées de Chine, *Ann. Crypto. Exotique,* **7**:11–18. *Pls.* 1–2. 1934.

———: Deuxième note sur quelques Ustilaginées de Chine, *ibid.,* **7**:85–95. *Pls.* 3–4. *Figs.* 1–4. 1934.

———: Note sur les Ustilaginées de Chine, *Contrib. Inst. Bot. Nat. Acad. Peiping,* **1**:165–175. 1934; **3**:5–15, 41–58. 1935.

———: Recherches systématiques, biologiques et cytologiques sur les Ustilaginées de Chine, Thèses présentées à la Faculté des Sciences de l'Université de Paris. Sér. A, No. 341. No. d'ordre 365:157–310. *Pls.* 8–25. *Figs.* 1–52. Paris, 1937.

LING, LEE: Taxonomic notes on Asiatic smuts, I., *Sydowia, Ann. Mycol.,* **3**(1–6):123–134. 1949; II. *ibid.,* **4**:74–81. *Fig.* 1. 1950.

WHETZEL, H. H., AND F. D. KERN: The smuts of Porto Rico and the Virgin Islands, *Mycologia,* **18**(3):114–124. *Pl.* 16. 1926.

LIRO, J. IVAR: Uber die Gattung Tuburcinia Fries, *Ann. univ. Fennicae Aboensis,* Ser. A, **1**:1–153. 1922.

———: Die Ustilagineen Finnlands, I, *Ann. Acad. Sci. Fennicae,* Serie A, **17**:1–636. *Figs.* 1–9. 1924; II, *ibid.,* **42**:i–xiii, 1–720. *Figs.* 1–8. 1 *map.* 1935–1938.

ZUNDEL, GEORGE LORENZO INGRAM: The Ustilaginales of South Africa, *Bothalia,* **3**(3):283–320. 1938.

———: Monographic studies on the Ustilaginales attacking Andropogon, *Mycologia,* **22**(3):125–158. 1930.

SAVILE, D. B. O.: A study of the species of Entyloma on North American composites, *Can. J. Research, C,* **25**(3):105–120. 1 *pl.* 1947.

GARRETT, A. O.: The Ustilaginales or Smuts of Utah, *Bull. Univ. Utah,* Biological Series 4(2), **29**(9):1–23. *Pls.* 1–4. 1939. (A list of all smuts known in Utah, with hosts and distribution, and a host index. A key to the genera and illustrations of every genus and of many species.)

FISCHER, GEORGE W.: The stem smuts of Stipa and Oryzopsis in North America, *Butler Univ. Botan. Studies,* **7**:25–39. 6 *figs.* 1945.

———, AND ELISA HIRSCHHORN: A critical study of some species of Ustilago causing stem smut on various grasses, *Mycologia,* **37**(2):236–266. *Figs.* 1–6. 1945.

———, AND ———: Observations on certain species of Ustilago on Hilaria, Stenotaphrum, and Muhlenbergia, *ibid.,* **37**(3):318–325. 2 *figs.* 1945.

DAVIS, W. H.: Summary of investigations with Ustilago striaeformis parasitizing some common grasses, *Phytopathology,* **25**(8):810–817. 1935.

FISCHER, GEORGE W.: Fundamental studies of the stripe smut of grasses (Ustilago striaeformis) in the Pacific Coast, *Phytopathology,* **30**(2):93–118. *Figs.* 1–4. 1940.

HIRSCHHORN, ELISA, Y JULIO HIRSCHHORN: Los carbones del maiz en Argentina, *Rev. facultad agron. Univ. nacl. La Plata,* **20**(2):108–139. *Pls.* 1–5. 1935.

JACKSON, H. S.: The Ustilaginales of Indiana, *Proc. Indiana Acad. Sci.,* **1917**:119–132. 1918; **1920**:157–164. *Fig.* 1. 1921. (Distribution lists and host index.)

SĂVULESCU, TRIAN: Contributions à la connaissance des Ustilaginées de Roumanie, *Ann. inst. recherches agron. Roumanie,* **7**:1–86. *Pls.* 1–35. 1936.

THIRUMALACHAR, M. J.: Species of the genera Doassansia, Doassansiopsis, and Burrillia in India, *Mycologia,* **39**(5):602–611. *Figs.* 1–9. 1947.

FISCHER, EDWARD: Weitere Beiträge zur Kenntnis der Gattung Graphiola, *Ann. Mycol.,* **20**(3–4):228–237. *Figs.* 1–4. 1922. (Distinguishes all the recognized species of the genus.)

List 32. Uredinales

GENERAL WORKS

SYDOW, P. ET H.: Monographia Uredinearum seu specierum omnium ad hunc usque diem descriptio et adumbratio systematica, Leipzig, Gebrüder Borntraeger. Vol. 1, pp. i–xxxv, 1–972. 45 *pls.* 1904 (The genus Puccinia); vol. 2, pp. i–xix, 1–396. 14 *pls.* 1910 (The genus Uromyces); vol. 3, pp. 1–728. 32 *pls.* 1915 (The remainder of the Pucciniaceae and families Melampsoraceae, Zaghouaniaceae and Coleosporiaceae); vol. 4, pp. i–iv, 1–670. 1924 (Uredineae Imperfecti: Peridermium, Aecidium, Monosporidium, Roestelia, Caeoma, Uredo, Mapea).

KLEBAHN, H.: Die wirtswechselnden Rostpilze, xxxvii + 447 pp. *Pls.* 1–6. Berlin, Gebrüder Borntraeger, 1904.

HARIOT, Paul: Les Urédinées (Rouilles des plantes), xv + 392 pp. 47 *figs.* Paris, Octave Doin, 1908. (This is the first volume issued of the Bibliothèque de Botanique Cryptogamique directed by L. Mangin, one of the series of Toulouse, Encyclopédie Scientifique.)

GUYOT, A. L.: Les Urédinées (ou rouilles des végétaux). Étude morphologique et biologique des champignons de ce groupe, qui vivent en Europe, Asie Occidentale, Afrique Septentrionale et révision des espèces connues dans les autres parties du monde: Tome I. Genre Uromyces (a) Espèces parasites des plantes appartenant aux familles des Graminées, Cyperacées, Juncacées, Renonculacées, Polygonacées, Ombellifères, Campanulacées, in Encyclopédie Mycologique, **8**:1–439. *Figs.* 1–83. Paris, Paul Lechevalier, 1938.

———, ET AL.: Uredineana: Tome 1, pp. 1–205, *Figs.* 1–11, Encyclopédie Mycologique, 8 (suppl.), 1939; Tome 2, pp. 1–228, *Figs.* 1–5, *ibid.* vol. 13, 1946.

ARTHUR, JOSEPH CHARLES: Manual of the rusts in the United States and Canada. Illustrations by George B. Cummins, xv + 438 pp. 487 *figs.* 1 *map.* Lafayette, Ind., Purdue Research Foundation, 1934.

PLOWRIGHT, CHARLES B.: A monograph of the British Uredineae and Ustilagineae, vii + 347 pp. *Pls.* 1–8. *Figs.* 1–13. London, Kegan Paul, Trench and Co., 1889.

GROVE, W. B.: The British rust fungi (Uredinales): their biology and classification, xi + 412 pp. 290 *figs.* Cambridge, Cambridge Univ. Press, 1913.

VIÉGAS, A. P.: Alguns fungos do Brasil: IV. Uredinales, *Bragantia,* **5**(1):1–144. *Pls.* 1–48. *Figs.* 1–89. 1945.

FISCHER, E.: Die Uredineen der Schweiz, *Beiträge zur Kryptogamenflora der Schweiz,* **2**(2):i–xciv, 1–591. *Figs.* 1–342. 1909.

CUNNINGHAM, G. H.: The rust fungi of New Zealand together with the biology, cytology and therapeutics of the Uredinales, xx + 261 pp. 177 *figs.* Dunedin, N. Z., McIndoc, 1931.

———: The Uredinales, or rust fungi, of New Zealand: I. Pucciniaceae, tribe Puccineae, *Trans. Proc. N. Zealand Inst.,* **54**:619–704. 1 *pl.* 1923.

———: The Uredinales, or rust fungi, of New Zealand: Supplement to Part I; and Part II, *ibid.,* **55**:1–58. 1924.

———: Second supplement to the Uredinales of New Zealand, *ibid.,* **55**:392–396. 1924. (For the third to seventh supplement see List 31 under CUNNINGHAM: The Ustilagineae of New Zealand.)

MCALPINE, D.: The Rusts of Australia, vii + 349 pp. 55 *pls.* 28 *figs.* Melbourne, Department of Agriculture, Victoria, 1906.

Bubák, Franz: Die Pilze Böhmens: Erster Teil. Rostpilze (Uredinales), *Archiv der Naturwissenschaftlichen Landesdurchforschung von Böhmen*, **13**(5):1–234. *Figs.* 1–59. 1908.

Săvulescu, Trian et Olga: Matériaux pour la flore des Urédinées de Roumanie, *Academie Romăna Secţiuneă Stiinţifică Memoriile*, **17**:114–261. 18 *figs.* 1941–42 (1943).

———: Matériaux pour la flore des Uredinées de Roumanie. Supplément, *Académie Roumaine Bull., Sect. Sci.*, **26**(5):308–332. 2 *figs.* 1944.

Doidge, Ethel M.: A preliminary study of the South African rust fungi, *Bothalia*, **2**(1a):1–228. 6 *col. pls.* 221 *figs.* 1926.

Fragoso, Romualdo Gonzáles: Flora Iberica. Uredales, Museo Nacional de Ciencias Naturales, Madrid, **1**:i–lxxi, 1–416. *Figs.* 1–208. 1924; **2**:i–viii, 1–421. *Figs.* 1–174. 1925. (The first volume contains the genus Puccinia, alone; the second volume the other genera of rusts.)

Fraser, W. P.: The rusts of Nova Scotia, *Proc. Trans. Nova Scotian Inst. Sci. Halifax*, **12**(4):313–443. 1913.

Burrill, T. J.: Parasitic fungi of Illinois: I. Uredinales, *Bull. Illinois State Laboratory of Natural History*, **2**:141–255. 1885.

Ramsbottom, J.: Some notes on the history of the classification of the Uredinales, *Brit. Mycol. Soc. Trans.*, **4**(1):77–105. 1913. (Contains keys to the families and genera and lists of the species occurring in Great Britain.)

Yoshinaga, Torama, and Naohide Hiratsuka: A list of Uredinales collected in the Province of Tosa, *Botanical Magazine* (*Tokyo*), **44**(528):627–667. 1930.

Jørstad, Ivar: A study on Kamtchatka Uredinales, *Skrifter Norske Videnskaps-Akad. Oslo.* I. *Mat.-naturv. Klasse*, **1933**(9):1–183. *Figs.* 1–22. 1934. (Keys to the genera and species and a host index.)

———: Uredinales of Northern Norway, *ibid.*, **1940**(6):1–145. 1940. (A list of the rusts of the districts Nordland, Troms, and Finmark. Host index.)

Camara, Emmanuele de Sousa da; Antonis Lopes Branquinho de Oliveira; et Carlos Gomes da Luz: Uredales aliquot Lusitaniae, III, *Agronomia Lusitana*, **5**(4):317–347. 1943.

Arthur, J. C., and George B. Cummins: Phillippine rusts in the Clemens collection, 1923–1926, I, *Philippine J. Sci.*, **59**(3):437–449. *Pls.* 1–3. 1936.

Gobi, Christian J., and W. Tranzschel: On the rust fungi of St. Petersburg and some adjacent portions of Estland, Viborg and Novgorod Governments. *Scripta Botanica Horti Universitatis Petropolitanae*, **3**(2):65–123. 1891. (In Russian.)

Kern, F. D.; H. W. Thurston, Jr.; and H. H. Whetzel: Annotated index of the rusts of Colombia, *Mycologia*, **25**(6):448–503. 1933.

———, R. Ciferri, and H. W. Thurston, Jr.: The rust-flora of the Dominican Republic, *Ann. Mycol.*, **31**(1–2):1–40. 1933.

Sydow, H.: Fungi Venezuelani, *Ann. Mycol.*, **28**(1–2):29–224. 1930. (Uredineae in pp. 37–52.)

Kern, F. D.; H. W. Thurston, Jr., and H. H. Whetzel: Uredinales, in Mycological exploration in Venezuela, *Monograph of the Univ. Puerto Rico, B*, **2**:262–303. 1934.

———: Additions to the Uredinales of Venezuela, I, *Mycologia*, **30**(5):537–552. 1938.

———, and H. W. Thurston, Jr.: Additions to the Uredinales of Venezuela, II–III, *ibid.*, **35**(4):434–445. 1943; **36**(1):54–64. 1944.

Hiratsuka, Naohide: Zweiter Beitrag zur Uredineen-flora von Sudsachalin, *Trans. Tottori Soc. Agr. Sci.*, **2**(3):233–246. 1931.

HASHIOKA, YOSHIO: Matériaux pour la flore des Urédinées de l'Ile de Saghaline septentrionale, *J. Japanese Botany*, **12**:882–886. 1936. (A list and host names of 29 species from the Russian portion of Sakhalin.)

JACKSON, H. S.: The Uredinales of Indiana, *Proc. Indiana Acad. Sci.*, **1915**:429–475. 1916.

———: The Uredinales of Delaware, *ibid.*, **1917**:311–385. 1918.

———: The Uredinales of Oregon, *Brooklyn Botanic Garden Mem.*, **1**:198–287 1918.

———: The rusts of South America based on the Holway collection, *Mycologia*, **18**(4):139–162. *Pl.* 19. 1926; **19**(2):51–65. 1927; **23**(2):96–116. *Pl.* 11. *Figs.* 1–5. (5):332–364, (6):463–503. 1931; **24**(1):62–186. 1932. (Keys to the genus Mainsia and to the rusts on various groups of hosts.)

GARRETT, A. O.: The Uredinales or rusts of Utah, *Bull. Univ. of Utah*, **28**(7):1–81. *Pl.* 1–8. 1937.

HOTSON, JOHN WILLIAM: Key to the rusts of the Pacific Northwest, *Univ. Washington Pub. Biology*, **3**:193. *Illustrated.* 1934.

BARCLAY, A.: Descriptive list of the Uredineae occurring in the neighborhood of Simla (Western Himalaya), *J. Asiatic Soc. Bengal*, **56**:350–375. 4 *pls.* 1887; **58**:232–251. 3 *pls.* 1889; **59**:75–112. 4 *pls.* 1890.

———: Additional Uredineae from the Neighborhood of Simla, *ibid.*, **60**:211–230. 2 *pls.* 1891.

CUMMINS, GEORGE B.: Uredinales of New Guinea, *Mycologia*, **32**(3):359–373. *Figs.* 1–14. 1940; **33**(1):64–68. 1 *fig.* (2):143–154. *Figs.* 1–7. (4):380–389. *Figs.* 1–14. 1941.

TAI, F. L.: Uredinales of Western China, *Farlowia*, **3**(1):95–139. 27 *figs.* 1947.

Melampsoraceae

MAINS, E. B.: Species of Melampsora occurring upon Euphorbia in North America, *Phytopathology*, **7**(2):101–105. 1917.

THIRUMALACHAR, M. J., AND FRANK D. KERN: Notes on some species of Phakopsora and Angiopsora, *Mycologia*, **41**(3):283–290. *Figs.* 1–3. 1949. (Contains a key for distinguishing the seven nearly related genera of rusts, Phakopsora, Angiopsora, Bubakia, Baeodromus, Dasturella, Arthuria, and Cerotelium.)

HIRATSUKA, NAOHIDE: Studies on the Melampsoraceae of Japan, *J. Faculty Agr. Hokkaido Imp. Univ.*, **21**(1):1–42. 2 *figs.* 1927.

———: Notes on the Melampsoraceae of Japan: II. Chrysomyxa of Japan, *Botanical Magazine* (*Tokyo*), **43**(513):466–478. 1929.

———: Notes on the Melampsoraceae of Japan: III. Pucciniastrum of Japan, *ibid.*, **44**(521):261–284. 1930.

———: Beiträge zu einer Monographie der Gattung Pucciniastrum Otth., *J. Faculty Agr. Hokkaido Imp. Univ.*, **21**(3):63–119. 1 *pl.* 1927.

———: A contribution to the knowledge of the Melampsoraceae of Hokkaido, *Japanese J. Botany*, **3**(4):289–322. 1927.

———: Additional notes on the Melampsoraceae of Hokkaido, *Botanical Magazine* (*Tokyo*), **42**(503):503–504. 1928.

———: Additional notes on the Melampsoraceae of Hokkaido, II, *Trans. Tottori Soc. Agr. Sci.*, **4**(2):111–115. 1932.

———: Additional notes on the Melampsoraceae of Saghalien, *Trans. Sapporo Natural History Soc.*, **10**:119–121. 1929.

———, AND Y. UEMURA: On Japanese species of Hyalopsora, *Trans. Tottori Soc. Agr. Sci.*, **4**(1):11–27. 2 *figs.* 1932. (In Japanese.)

———: Thekopsora of Japan, *Botanical Magazine* (*Tokyo*), **43**(505):12–22. 1929.

———: Phakopsora of Japan, *ibid.*, **49**(587):781–788, (588):853–860. 1935; **50**(589):2–8. 1936.

———: A monograph of the Pucciniastreae, ix + 374 pp. *Pls.* 1–11. Tottori, Japan, 1936. (Reprinted from *Mem. Tottori Agr. Coll.*, 4.)

FAULL, JOSEPH HORACE: Taxonomy and geographical distribution of the genus Milesia, *Contribs. Arnold Arboretum Harvard Univ.*, **2**:1–138. *Pls.* 1–9. 1932.

———: Taxonomy and geographical distribution of the genus Uredinopsis, *ibid.*, **11**:1–120. *Pls.* 1–6. 1938.

———: Tropical fern hosts of rust fungi, *J. Arnold Arboretum Harvard Univ.*, **28**(3):309–319. 1947.

HIRATSUKA, NAOHIDE, AND Y. YOSHIDA: Two species of Milesina on some Japanese species of Polystichum, *Trans. Tottori Soc. Agr. Sci.*, **4**(1):7–10. 1932.

———: On some new species of Milesina, *Botanical Magazine* (*Tokyo*), **48**(565):39–47. *Figs.* 1–6. 1934.

KAMEI, SENJI: On new species of heteroecious fern rusts, *Trans. Sapporo Natural History Soc.*, **12**(3):161–174. 1932.

BELL, H. P.: Fern rusts of Abies, *Botan. Gaz.*, **77**(1):1–30. *Pls.* 1–5. 1924.

HUNTER, LILIAN M.: Comparative studies of spermogonia of rusts of Abies, *Botan. Gaz.*, **83**(1):1–23. *Pls.* 1–4. *Figs.* 1–2. 1927.

WEIR, JAMES R., AND ERNEST E. HUBERT: Observations on forest tree rusts, *Am. J. Botany*, **4**(6):327–335. *Figs.* 1–2. 1917.

HEDGCOCK, GEORGE G., AND N. REX HUNT: Notes on some species of Coleosporium, I, *Mycologia*, **14**(5):244–257. *Pls.* 20–21. 1922. (Distinguishes several species of Coleosporium inhabiting Composites.)

WEIR, JAMES R.: The genus Coleosporium in the northwestern United States, *Mycologia*, **17**(6):225–239. *Pls.* 22–24. *Fig.* 1. 1925.

HEDGCOCK, George G.: A key to the known aecial forms of Coleosporium occurring in the United States and a list of the host species, *Mycologia*, **20**(2):97–100. 1928.

ARTHUR, J. C., AND F. D. KERN: North American species of Peridermium, *Bull. Torrey Botan. Club*, **33**(8):403–438. 1906.

———: North American species of Peridermium in pine, *Mycologia*, **6**(3):109–138. 1914.

LUDWIG, C. A.: Notes on some North American rusts with Caeoma-like sori, *Phytopathology*, **5**(5):273–281. 1915. (Gives keys for the determination of such rusts.)

RHOADS, A. S.; G. G. HEDGCOCK; E. BETHEL; AND C. HARTLEY: Host relationships of the North American rusts other than Gymnosporangium which attack Conifers, *Phytopathology*, **8**(7):309–352. 1918.

MAINS, E. B.: Angiopsora, a new genus of rusts on grasses, *Mycologia*, **26**(2):122–132. *Pls.* 17–20. 1934.

WEIR, JAMES R.: The genus Chrysomyxa, *Mycologia*, **15**(4):183–187. *Pl.* 17. 1923.

Pucciniaceae

DIETEL, P.: Monographie der Gattung Ravenelia, *Beihefte Botan. Centr.* Zweite Abt., **20**:343–413. *Pls.* 5–6. 1906.

MASSEE, GEORGE: Revision of the genus Hemileia, *Roy. Botan. Garden, Kew, Bull. Misc. Inform.*, **1906**:35–42. 1 *pl.* 1906.

MILESI, M., E G. B. TRAVERSO: Saggio di una monografia del genere Triphragmium, *Ann. Mycol.*, **2**(2):143–156. *Pl.* 5. 1904.

Dietel, P.: Zur Umgrenzung der Gattung Pileolaria Cast, *Ann. Mycol.*, **19**(5–6):300–303. 1921.

Olive, E. W., and H. H. Whetzel: Endophyllum-like rusts of Porto Rico, *Am. J. Botany*, **4**(1):44–52. *Pls.* 1–3. 1917.

Moreau, M. et Mme. F.: Les Urédinées du groupe Endophyllum, *Bull. soc. botan. France*, **66**:13–44. 1919.

Cummins, George B.: The genus Prospodium (Uredinales), *Lloydia* **3**(1):1–78. *Figs.* 1–12. 1940.

Gymnosporangium

Kern, Frank D.: A biologic and taxonomic study of the genus Gymnosporangium, *Bull. New York Botanical Garden*, **7**:391–483. *Pls.* 151–161. 1911.

Eriksson, Jakob: Die schwedischen Gymnosporangien, ihr Wirtswechsel und ihre Spezialisierung, *Kgl. Svenska Vetenskapsakad. Handl.*, **59**(6):1–82. *Pls.* 1–4 (*two colored*). *Figs.* 1–13. 1918.

Tanaka, Tyôzaburo: New Japanese fungi. Notes and translations, XII, *Mycologia*, **14**(5):282–295. 1922. (Discusses the Japanese species of Gymnosporangium, with a key.)

Prince, Alton Ernest, and Ferdinand Henry Steinmetz: Gymnosporangium rusts in Maine and their host relationships, *Maine Bull.*, **42**(12):1–46. 1 *map.* 1940.

Hiratsuka, Naohide: Gymnosporangium of Japan, *Botanical Magazine* (*Tokyo*), **50**(597):481–488, (598):549–555, (599):593–599, (600):661–668. *Pls.* 8–9. 1936; **51**(601):1–8. 1937.

———: Japanese species of Gymnosporangium. *Botany and Zoology*, **7**:748–749. 1939. (In Japanese.) (Eleven species are enumerated and a key provided for their identification.)

Phragmidium and Kuehneola

Arthur, J. C.: Relationship of the genus Kuehneola, *Bull. Torrey Botan. Club*, **44**(11):501–511. 1912.

Hiratsuka, Naohide: Kuehneola of Japan, *J. Japanese Botany*, **12**:809–815. *Illustrated.* 1936. (List of species and key to their determination.)

Dietel, P.: Ueber die Arten der Gattung Phragmidium, *Hedwigia*, **44**:112–132, 330–346. *Pl.* 4. 1905.

Arthur, J. C.: North American rose rusts, *Torreya*, **9**(2):21–28. *Figs.* 1–3. 1909.

Cummins, George B.: Phragmidium species of North America: differential teliospore and aecial characters, *Mycologia*, **23**(6):433–445. *Pl.* 32. 1931.

Hiratsuka, Naohide: Phragmidium of Japan, *Japanese J. Botany*, **7**(3–4):227–299. *Pls.* 3–4. 1935.

———: On some new species of Phragmidium, *Trans. Sapporo Natural History Soc.*, **13**(3):134–138. 1934.

Uromyces and Puccinia and Their Segregates

Holway, E. W. D.: North American Uredineae, Parts I–V. *Pls.* 1–54. Minneapolis, Published by the author, 1905–1924. (Text and illustrations from microphotographs of species of Puccinia on various families of host plants.)

Aref'yef, L. A.: Species of the genus Puccinia in the Baltic Province, 85 pp. St. Petersburg, 1916. (In Russian.)

———: Species of Uromyces in the Baltic Province, *Izvyestia i Trudy Sel'skokhozyaistovo otdyelya Rizhskovo Politekhnitchevo Instituta*, **3**:117–156. 1916. (In Russian.)

POLE-EVANS, I. B.: The South African rust fungi: I. The species of Puccinia on Compositae. *Trans. Royal Soc. South Africa*, **5**:637–646. 5 *pls*. 1916.

BISBY, G. R.: Short cycle Uromyces in North America, *Botan. Gaz.*, **69**(3):193–217. *Pl*. 10. 1920.

ITO, SEIYA: On the Uredineae parasitic on the Japanese Gramineae, *J. Coll. Agr. Tohuku Imp. Univ., Sapporo, Japan*, **3**(2):179–362. *Pls*. 10–12. 1909.

———: Uromyces of Japan, *J. Coll. Agr. Hokkaido Imp. Univ., Sapporo, Japan*, **11**(4):211–287. *Pls*. 7–9. 1922. (Describes 56 species of *Uromyces* and 3 of *Pileolaria* from Japan; also an index of the rust species and a host index.)

———: Additional notes on Uromyces in Japan, *Botanical Magazine* (*Tokyo*), **40**(473):276–280. *Fig*. 1. 1926.

HIRATSUKA, NAOHIDE: Studies on Uromyces fabae and its related species, *Japanese J. Botany*, **6**(3):329–379. *Pls*. 16–17. 1933.

JACKSON, H. S.: Carduaceous species of Puccinia: I. Species occurring on the tribe Vernoniae, *Botan. Gaz.*, **65**(4):289–312. 1918.

———: New or noteworthy rusts on Carduaceae, *Mycologia*, **14**(3):104–120. 1922.

PETRAK, F.: Beiträge zur Kenntnis der auf Achillea vorkommenden Arten der Gattung Puccinia, *Sydowia, Ann. Mycol.*, **1**(1–3):44–48. 1947.

SCHILLING, MAX: Die Spezialisierung des Puccinia taraxaci Plow., *Sydowia, Ann. Mycol.*, **3**(1–6):201–233. *Figs*. 1–20. 1949.

GÄUMANN, ERNST A.: Zur Kenntnis einiger Umbelliferen-Puccinien, *Ber. schweiz. botan. Ges.*, **51**:143–164. 8 *figs*. 1941.

———, UND O. JAAG: Über Kleinarten aus dem Formenkreis der Puccinia campanulae, *Hedwigia*, **75**(3):121–129. *Figs*. 1–3. 1935.

POEVERLEIN, HERMANN: Die Saxifraga-Roste Süddeutschlands, *Ann. Mycol.*, **35**(1):53–58. 1937. (Contains a key to the species of Puccinia on Saxifraga in South Germany.)

SPEGAZZINI, CARLOS: Breve nota sobre Uredinales berberidicolas sudamericanas, *Revista Chilena de Historia Natural, Pura y Aplicada*, **25**:263–279. 2 *pls*. 1921.

CUMMINS, GEORGE B.: Revisionary studies of the tropical American rusts of Panicum, Paspalum and Setaria, *Mycologia*, **34**(6):669–695. *Figs*. 1–24. 1942.

———: The full-cycle Puccinias on Onagraceae in North America, *Am. J. Botany*, **19**(4):334–339. *Figs*. 1–4. 1932.

———: New species of Puccinia on Lauraceae from China, *Bull. Torrey Botan. Club*, **76**(1):31–38. *Figs*. 1–12. 1949.

BISBY, G. R.: The Uredinales found upon the Onagraceae, *Am. J. Botany*, **3**(10):527–561. 1916.

KERN, FRANK D.: North American species of Puccinia on Carex, *Mycologia*, **9**(4):205–238. 1917.

———: North American rusts on Cyperus and Eleocharis, *ibid.*, **11**(3):134–147. 1919.

———: The microcyclic species of Puccinia on Solanum, *ibid.*, **25**(6):435–441. *Pl*. 48. 1933.

ARTHUR, J. C.: The Uredineae occurring upon Phragmites, Spartina, and Arundinaria in America, *Botan. Gaz.*, **34**(1):1–20. *Figs*. 1–4. 1902.

REES, C. C.: The rusts occurring on the genus Fritillaria, *Am. J. Botany*, **4**(6):368–373. *Figs*. 1–3. 1917.

ORTON, C. R.: North American species of Allodus, *Mem. New York Botanical Garden*, **6**:173–208. 1916.

———: Notes on some Polemoniaceous rusts, *Mycologia*, **11**(4):168–180. 1919.

DIETEL, P.: Über Leptopuccinien auf Artemisia-Arten, *Ann. Mycol.*, **39**(2–3):150–154. 1941.

FAHRENDORFF, E.: Ueber die Brachypuccinia der Artemisia-Arten, *Ann. Mycol.*, **39**(2–3):158–203. *Illustrated.* 1941.

LINDROTH, J. J.: Die Umbelliferen-Uredineen, *Acta Societatis pro Fauna et Flora Fennica*, **22**:(1). 1922.

HOLWAY, E. W. D.: North American Salvia rusts, *J. Mycology*, **11**(4):156–158. 1905.

LINDQUIST, JUAN C.: Las Puccinias parásitas de Geranium en la República Argentina, *Notas Mus. La Plata*, **13**:63–71. 1948.

———: Uredineas parásitas de Amarantáceas, en la República Argentina, *ibid.*, **13**:243–251. *Pls.* 1–2. 1948.

HIRATSUKA, NAOHIDE: On the microcyclic species of the Pucciniaceae collected in some mountains of Japan, *Trans. Tottori Soc. Agr. Sci.*, **3**:211–253. 1 *pl.* 1 *fig.* 1931. (In Japanese.)

VON TAVEL, C.: Zur Speziesfrage bei einigen Allium-bewohnenden Uredineen, *Ber. schweiz. botan. Ges.*, **4**(1):123–169. *Pls.* 1–2. 1932.

SAVILE, D. B. O. and I. L. CONNERS: The rusts of Armeria and Limonium in North America, *Mycologia*, **43**(2):185–195. 1951.

List 33. Heterobasidiae

(See also List 34, BOURDOT ET GALZIN, C. REA, RAMSBOTTOM, DONK; List 35, BURT.)

MARTIN, G. W.: The Tremellales of the North Central United States and adjacent Canada, *Univ. Iowa Studies in Natural History*, **18**(3):1–88. *Pls.* 1–5. 1944. (Includes Tremellales, Auriculariales and Dacrymycetales.)

COKER, WM. C.: Notes on the lower Basidiomycetes of North Carolina, *J. Elisha Mitchell Sci. Soc.*, **35**(3–4):113–182. *Pl.* 23 (*colored*) *and* 30–67. 1920.

BURT, EDWARD A.: Some North American Tremellaceae, Dacryomycetaceae and Auriculariaceae, *Ann. Missouri Botan. Garden*, **8**(4):361–396. *Pl.* 3. *Figs.* 1–6. 1921.

BOURDOT, H., ET A. GALZIN: Hyménomycètes de France: I. Hétérobasidiés, *Bull. soc. mycol. France*, **25**:15–36. 1909.

———: Heterobasidiae nondum descriptae, *ibid.*, **39**:261–266. 1924. (Descriptions of various Tremellaceae and of 10 species of Tulasnella and one of Gloeotulasnella.)

NEUHOFF, WALTHER: Die Gallertpilze (Tremellineae), *Die Pilze Mitteleuropas*, **2**(1a–4a; 7a):1–56. *Pls.* 1–9 (*colored*), 1–4 (*not colored*). Leipzig, Werner Klinkhart, 1935–1938. Uncompleted (?).

———: Die Gallertpilze Schwedens (Tremellaceae, Dacrymycetaceae, Tulasnel. laceae, Auriculariaceae), *Arkiv för Botanik*, **28A**(1):1–57. *Pls.* 1–8. *Fig.* 1- 1936. (Includes descriptive keys to all European species of Exidia and Dacrymyces.)

TEIXEIRA, A. RIBEIRO: Himenomicetos Brasileiros: Auriculariales e Dacrymycetales, *Bragantia*, **5**(2):153–186. *Pls.* 1–14. 1945.

VIÉGAS, A. P.: Alguns Fungos do Brasil: V. Basidiomycetos-Auriculariales, *Bragantia*, **5**(3):197–212. *Pls.* 1–4. 2 *figs.* 1945. (Septobasidium.)

KOBAYASI, YOSIO: On the genus Tremella and its allies from Japan, *Tokyo Bunrika Daigaku Science Repts.*, B, **4**(64–65):1–26. 6 *pls.* 15 *figs.* 1938.

———: On the genus Holtermannia of Tremellaceae, *ibid.*, **3**(50):75–81. 1 *pl.* 2 *figs.* 1937.

———: Fungorum ordinis Tremellalium Studia Monographica: III. On the Dacrymyces-group, *ibid.*, **4**(70):105–128. 3 *pls.* 4 *figs.* 1939.

———: Fungorum ordinis Tremellalium Studia Monographica: IV. On the genera Femsjonia, Guepinia and Calocera from Japan, *ibid.*, **4**(74):215–227. 2 *pls.* 5 *figs.* 1939.

Couch, John N.: Septobasidium in the United States, *J. Elisha Mitchell Sci. Soc.*, **51**(1):1–77. *Pls.* 1–44. 1935.

———: The genus Septobasidium, ix + 480 pp. *Frontispiece and* 114 *pls.* 60 *figs.* Chapel Hill, Univ. North Carolina Press, 1938.

Rogers, Donald P.: A taxonomic review of the Tulasnellaceae, *Ann. Mycol.*, **31**(3):181–203. *Pls.* 6–7. 1933.

———: Some noteworthy fungi from Iowa, *Univ. Iowa Studies in Natural History*, **15**(3):9–29. *Pls.* 1–3. 1933. (Contains keys to the Iowa species of Sebacina (incl. Bourdotia) and Heterochaetella.)

———: Notes on the lower Basidiomycetes, *ibid.*, **17**(1):1–43. *Pls.* 1–3. 1935. (Includes keys to Ceratobasidium, Botryobasidium, and Sebacina, subgenus Bourdotia.)

van der Byl, P. A.: Suid-Afrikaanse Dacryomycetaceae, Tremellaceae en Auriculariaceae, *Ann. Univ. Stellenbosch*, **1A**(3):1–14. *Figs.* 1–8. 1923.

McGuire, J. M.: The species of Sebacina (Tremellales) of Temperate North America, *Lloydia*, **4**(1):1–43. *Pls.* 1–5. 1941.

Rick, J.: Dacryomycetaceae Riograndenses, *Broteria. Serie Trimestral. Ciências Naturais*, **5**:74–79. 1936.

Brasfield, T. W.: The Dacrymycetaceae of Temperate North America, *Am. Midland Naturalist*, **20**(1):211–235. *Pls.* 1–4. 1938.

Bodman, Sister Mary Cecilia: The genus Tremellodendron, *ibid.*, **27**(1):203–216. *Pls.* 1–3. 1942.

Baker, Gladys E.: A study of the genus Helicogloea, *Ann. Missouri Botan. Garden*, **23**(1):69–128. *Pls.* 7–14. 1936.

———: Addenda to the genera Helicogloea and Physalacria, *Mycologia*, **38**(6):630–638. *Figs.* 1–25. 1946.

Bjørnekaer, K.: Floristiske Undersøgelser over danske Baevresvampe (Tremellaceae), *Friesia*, **3**(2):1–34. 3 *figs.* 1944.

Boedjin, K. B., et A. Steinmann: Les espèces des genres Helicobasidium et Septobasidium des Indes Néerlandaises, *Bull. Jardin Botanique de Buitenzorg, Sér. III*, **11**(2):165–219. *Pls.* 14–18. *Figs.* 1–31. 1931.

Lloyd, C. G.: The genus Naematelia, *Mycological Notes*, **7**:1149–1150. *Figs.* 2223–2226. 1922.

Olive, Lindsay S.: Notes on the Tremellales of Georgia, *Mycologia*, **39**(1):90–108. *Figs.* 1–16. 1947.

———: Taxonomic notes on Louisiana Fungi, I–II, *ibid.*, **40**(1):6–20. *Figs.* 1–3; (5):586–604. *Figs.* 1–3. 1948. (Tremellales.)

Martin, G. W., and Edna E. Huber: Notes on the Tremellales of Iowa, with keys, *Univ. Iowa Studies in Natural History*, **12**(4):91–104. 1 *pl.* 1927.

List 34. "Hymenomyceteae": General Works

SMITH, W. G.: Synopsis of the British Basidiomycetes, A descriptive catalogue of the drawings and specimens in the Department of Botany, British Museum, 531 pp. *Pls.* 1–5. *Figs.* 1–145. London, 1908.

COOKE, M. C.: Illustrations of British Fungi (Hymenomycetes), Vols. 1–2, 1881–1883; 3–4, 1884–1886; 5–6, 1886–1888; 7, 1889–1890; 8, 1889–1891. 1198 *col. pls. in all.* London, Williams and Norgate.

REA, CARLETON: British Basidiomyceteae. A Handbook to the Larger British Fungi, xii + 799 pp. Cambridge, Cambridge Univ. Press, 1922.

———: Appendix to British Basidiomyceteae. Additions and corrections, *Brit. Mycol. Soc. Trans.*, **12**:205–230. 1927.

———: Appendix II to British Basidiomyceteae, *ibid.*, **17**:35–50. *Pl.* 17. 1932.

PEARSON, A. A., AND R. W. G. DENNIS: Revised list of British Agarics and Boleti, *Brit. Mycol. Soc. Trans.*, **31**(3–4):145–190. 1948.

BOURDOT, H., ET A. GALZIN: Hyménomycètes de France. Hétérobasidiés-Homobasidiés Gymnocarpes. Contribution à la Flore Mycologique de la France, vol. 1, iv + 761 pp. 185 *figs.* Sceaux, France, Marcel Bry, 1927.

HEIM, ROGER: Les champignons. Tableaux d'un monde étrange, 144 pp. 230 *pls.* (*from photos*). 6 *colored pls.* Paris, Editions Alpina, 1948.

BUCHOLTZ, FEDOR V.: Illustrated guide to the fungi of Central Russia: I. Hymenomycetes; II. Agaricaceae. Riga, 1909. (In Russian.)

LINDAU, GUSTAV: Die höheren Pilze: Basidiomyceten mit Ausschluss der Brand- und Rostpilze, in Kryptogamenflora für Anfänger, ed. 3, revised by Eberhard Ulbrich, vii + 497 pp. 14 *pls.* Berlin, Julius Springer, 1928.

RAMSBOTTOM, JOHN: A Handbook of the Larger British Fungi, 222 pp. 141 *figs.* London, Trustees of the British Museum, 1923. (Includes all of the genera and the more important species of British Basidiomyceteae and fleshy Ascomyceteae.)

DONK, M. A.: Revisie van de Nederlandse Heterobasidiomyceteae (uitgez. Uredinales en Ustilaginales) en Homobasidiomyceteae-Aphyllophoraceae, I, *Mededeel. Nederland. Mycol. Ver.*, **18–20**:67–200. 1931.

———: Revision der Niederländischen Homobasidiomyceteae-Aphyllophoraceae, II, *Mededeel. Botanisch Museum en Herbarium van de Rijks Univ. Utrecht*, **9**:1–278. 1933.

KARSTEN, PETER ADOLPH: Kritisk öfversigt af Finlands Basidsvampar (Basidiomycetes; Gastero- und Hymenomycetes), 482 pp. Helsingfors, Finska Vetenskaps-Societaten, 1889. (There are three supplements, as follows: Tillagg I: 179–230, 1892; II: 157–186, 1893; III: 3–36, 1898. The whole work is part of a set of several volumes entitled "Bidrag till Kännedom af Finlands Natur och Folk.")

MÖLLER, F. H.: Fungi of the Faeroës: Part I. Basidiomycetes, 295 pp. 3 *colored pls.* 134 *figs.* 1 *colored map.* Copenhagen, Einar Munksgaard, 1945.

MOFFATT, WILL SAYER: The higher fungi of the Chicago region: I. The Hymenomycetes, *Chicago Acad. Sci. Natural History Survey Bull.*, **7**(1):1–156. *Pls.* 1–24. 1909.

GRAHAM, VERNE OVID: Mushrooms of the Great Lakes region, *Chicago Academy of Science Special Bulletin*, **5**:i–vii, 1–390. *Pls.* 1–49. 1944. (Hymenomycetes, and also Gasteromycetes and fleshy Ascomycetes.)

WHITE, EDWARD A.: A preliminary report on the Hymeniales of Connecticut, *Connecticut State Geological and Natural History Survey Bull.* **3**:1–81. *Pls.*

1–40. 1905. (Keys to the genera of Agaricaceae, Polyporaceae, Thelephoraceae, and Clavariaceae.)

———: Second report on the Hymeniales of Connecticut, *ibid.*, **15**:1–70. *Pls.* 1–28. 1910. (Keys to the Connecticut species of Agaricaceae.)

CLELAND, JOHN BURTON: Toadstools and Mushrooms and Other Larger Fungi of South Australia: Part I. Introduction and the Toadstools and Mushrooms, pp. 1–178, *col. pls.* 1–6, *Figs.* 1–35. 1934; Part II. Polypores, coral fungi and the remaining Hymenomycetes and the puff-balls, jelly-like fungi, the larger Ascomycetes and the Myxomycetes, pp. 179–362. *Col. pls.* 7–10, *Figs.* 6–79, 1935. Adelaide, British Science Guild (South Australian Branch), Government Printer.

DE LAPLANCHE, MAURICE C.: Dictionnaire iconographique de Champignons supérieurs (Hyménomycètes) qui croissent en Europe, Algérie et Tunisie suivi des tableaux de concordance de Barrellier, Batsch, Battarra, etc., 544 pp. Paris, Paul Klincksieck, 1894.

GILLET, C. C.: Les champignons (Fungi Hyménomycètes) qui croissent en France. Description et iconographie, propriétés utiles ou vénéneuses, 828 pp. 738 *pls.* Paris, J. B. Baillière et fils, 1878–1890.

BIGEARD, RENÉ, ET HENRI GUILLEMIN: Flore des champignons supérieurs de France les plus importants à connaître (comestibles et vénéneux), 600 pp. 56 *pls.* Chalon-sur-Saône, E. Bertrand, 1909. (Describes 1607 species, mainly Hymenomycetes and Gasteromycetes, but also Tuberales and Pezizales.)

———, ET ———: Flore des champignons supérieurs de France. Complément, ou Tome II, xx + 791 pp. 44 *pls.* Paris, Paul Klincksieck, 1913. (2200 species not included in the foregoing work, but covering the same groups of fungi.)

QUÉLET, LUCIEN: Flore mycologique de la France et des pays limitrophes, xviii + 492 pp. Paris, Octave Doin, 1888. (Includes Basidiomyceteae and some Ascomyceteae.)

VELENOVSKY, J.: Ceske houby, 950 pp. 179 *figs.* Praz. 1920. (In Czech language—Bohemian.) (All Basidiomycetes and Discomycetes and Tuberales.)

SINGER, ROLF: The Laschia-complex (Basidiomycetes), *Lloydia* **8**(3):170–230. 3 *pls.* 1 *fig.* 1945.

List 35. Thelephoraceae and Exobasidiaceae (Including Hymenolichens)

(See also List 34)

BURT, EDWARD A.: The Thelephoraceae of North America: I. Thelephora, *Ann. Missouri Botan. Garden*, **1**(2):185–228. *Pls.* 4–5. 1914; II. Craterellus, *ibid.*, **1**(3):327–350. *Pls.* 15–17. 1914; III. Craterellus borealis and Cyphella, *ibid.*, **1**(4):357–382. *Pl.* 19. 1914; IV. Exobasidium, *ibid.*, **2**(3):627–658. *Pl.* 21. 1915; V. Tremellodendron, Eichleriella and Sebacina, *ibid.*, **2**(4):731–770. *Pls.* 26–27. *Figs.* 1–7. 1915; VI. Hypochnus, *ibid.*, **3**(2):203–241. *Figs.* 1–30. 1916; VII. Septobasidium, *ibid.*, **3**(3):319–343. *Figs.* 1–14. 1916; VIII. Coniophora, *ibid.*, **4**(3):237–269. *Figs.* 1–19. 1917; IX. Aleurodiscus, *ibid.*, **5**(3):177–203. *Figs.* 1–14. 1918; X. Hymenochaete, *ibid.*, **5**(4):301–372. *Pls.* 16–17. *Figs.* 1–32. 1918; XI. Tulasnella, Veluticeps, Mycobonia, Epithele and Lachnocladium, *ibid.*, **6**(4):253–280. *Pl.* 5. *Figs.* 1–15. 1919; XII. Stereum, *ibid.*, **7**(2–3):81–248. *Pls.* 2–6. *Figs.* 1–48. 1920; XIII. Cladoderris, Hypolyssus, Cymatella, Skepperia, Cytidia, Solenia, Matruchotia, Microstroma, Protocoronospora, and Asterostroma, *ibid.*, **11**(1):1–36. *Pl.* 1. 1924; XIV. Peniophora, *ibid.*, **12**(3):213–357. 1925; XV. Corticium and supplement to the whole series, *ibid.*, **13**(3):173–354. 3 *figs.* 1926.

———: Corticiums causing Pellicularia disease of the coffee plant, hypochnose of Pomaceous fruits, and Rhizoctonia disease, *ibid.*, **5**(2):119–132. *Figs.* 1–3. 1918.

Rogers, Donald P., and H. S. Jackson: Notes on the synonymy of some North American Thelephoraceae and other resupinates, *Farlowia*, **1**(2):263–328. 1943.

Lloyd, C. G.: Synopsis of the genus Cladoderris, *Mycological Writings*, **4**, *Figs.* 520–530. 1913. (Separate pagination, pp. 1–11.)

———: Synopsis of the stipitate Stereums, *ibid.*, **4**, *Figs.* 531–564. 1913. (Separate pagination, pp. 14–44.)

Coker, W. C.: Notes on the Thelephoraceae of North Carolina, *J. Elisha Mitchell Sci. Soc.*, **36**(3–4):146–196. *Pls.* 1 *and* 14–35. 1921.

Emmons, C. W.: The Thelephoraceae of Iowa, *Univ. Iowa Studies in Natural History*, **12**(4):49–90. 2 *pls.* 1927.

Lentz, Paul L.: The genus Thelephora in Iowa, *Proc. Iowa Acad. Sci.*, **49**:175–184. *Figs.* 1–11. 1942 (1943).

van der Byl, P. A.: Die Suid-Afrikaanse Thelephoraceae, *Ann. Univ. Stellenbosch*, **7A**(3):1–52. *Pls.* 1–3. 1929.

Massee, G.: A monograph of the Thelephoraceae, *J. Linnean Soc. London, Botany*, **25**(170):107–155. 3 *pls.* 1890; **27**(181–182):95–205. 3 *pls.* 1891.

Litschauer, V.: Ueber einige Tomentella-Arten aus Schweden und Macedonien, *Ann. Mycol.*, **39**(4–6):360–378. 7 *figs.* 1941.

von Höhnel, Franz, und Viktor Litschauer: Beiträge zur Kenntnis der Corticieen, *Sitzber. Kaiserlichen Akad. Wiss. Wien Math.-naturw. Klasse*, **115**:1549–1620. *Figs.* 1–10. 1906; **116**:739–852. *Pls.* 1–4. *Figs.* 1–20. 1907; **117**:1081–1124. *Figs.* 1–10. 1908.

Sartory, A., et L. Maire: Synopsis du genre Fistulina Bull., Paris, 1924.

Lentz, Paul Lewis: Some species of Cyphella, Solenia and Porothelium, *Proc. Iowa Acad. Sci.*, **54**:141–152. *Pls.* 1–2. 1947 (1948).

Pilat, Albert: Beiträge zur Kenntnis der Thelephoraceen: I. Die Cyphellaceen Böhmens, *Ann. Mycol.*, **22**(1–2):204–218. *Pl.* 1. 1924. (Cyphella and Solenia.)

———: Zweiter Beitrag zur Kenntnis der tschechoslowakischen Cyphellaceen, *ibid.*, **23**(1–2):144–173. *Figs.* 1–23. 1925.

———: Monographia Cyphellacearum Cechosloveniae, I–II, *Publications de la Faculté des Sciences de l'Université Charles à Prague*, Nos. 28–29. 1925.

———: Zwei neue Arten der Gattung Cyphella aus der Tschechoslowakei, *Hedwigia*, **66**:261–264. *Fig.* 1. 1926.

———: Ein kleiner Beitrag zur Kenntnis der Gattung Cyphella Fr. in Tschechoslowakei, *ibid.*, **67**:113–118. *Pl.* 1. 1927.

———: Československé dřevni houby: I. Stereum Pers., *Československá Akademie Zemědělská Sbornik*, **5**(3):361–420. 3 *pls.* 2 *figs.* 1930. (Detailed descriptions (in Czech) of the 20 species recognized in Czechoslovakia up to 1930.)

———: Monographie der Mitteleuropäischen Aleurodiscineen, *Ann. Mycol.*, **24**(3–4):203–230. *Pl.* 16. 1926.

———: Monographie der europäischen Stereaceen, *Hedwigia*, **70**:10–132. *Pls.* 1–3. *Fig.* 1. 1930.

RICK, J.: Monographia Thelephoracearum resupinatarum riograndensium, *Broteria. Série Trimestral. Ciências Naturais*, **3**:31–48, 66–80, 151–173. 1934.

OVERHOLTS, L. O.: Mycological notes for 1933, *Mycologia*, **26**(6):502–515. *Pls*. 54–55. *Fig*. 1. 1934. (Keys and descriptions of species of Corticium, Section Botryodea.)

———: The genus Stereum in Pennsylvania, *Bull. Torrey Botan. Club*, **66**(8):515–537. *Pls*. 14–18. 1939.

ITO, TOKUTARO: Symboles ad Mycologiam Japonicam: I. Aleurodiscus; II. Peniophora; III. Corticium, Gloeocystidium et Asterostroma; IV. Asterostromella et Hymenochaete; V. Hymenochaete, *Botanical Magazine* (*Tokyo*), **43**(513):460–466, (514):515–524, (516):633–643, 1929; **44**(518):89–93, (519):151–157, 1930.

LITSCHAUER, VIKTOR: Beitrag zur Kenntnis der Gattung Aleurodiscus (mit besonderer Berücksichtigung schwedischer Arten), *Ann. Mycol.*, **42**(1–2):1–23. 1944.

ROGERS, DONALD P.: Notes on the Lower Basidiomycetes, *Univ. Iowa Studies in Natural History*, **17**(1):1–43. *Pls*. 1–3. 1935. (Ceratobasidium, Botryobasidium, and other segregates of Corticium.)

———: The genus Pellicularia (Thelephoraceae), *Farlowia*, **1**(1):95–118. *Figs*. 1–11. 1943.

———: The genera Trechispora and Galzinia (Thelephoraceae), *Mycologia*, **36**(1):70–103. *Figs*. 1–14. 1944.

TOMASELLI, R.: Appunti sulla sistematica e distribuzione geografica dei Basidiolicheni, *Archivio Botanico*, **26**, terza serie **10**(2):100–116. *Figs*. 1–5. 1950.

List 36. Clavariaceae

COKER, W. C.: The Clavarias of the United States and Canada, 209 pp. 92 *pls*. Chapel Hill, Univ. North Carolina Press, 1923.

———: Further notes on Clavarias, with several new species. *J. Elisha Mitchell Sci. Soc.*, **63**(1):43–69. *Pls*. 1–14. 1947.

BURT, EDWARD A.: The North American species of Clavaria with illustrations of the type specimens. *Ann. Missouri Botan. Garden*, **9**(1):1–78. *Pls*. 1–11. 1922.

DOTY, MAXWELL S.: Clavaria, the species known from Oregon and the Pacific Northwest, *Oregon State Monographs. Studies in Botany*, **7**:1–91 *Pls*. 1–11. *Figs*. 1–9. 1944.

———: A preliminary key to the genera of Clavarioid fungi, *Bull. Chicago Acad. Sci*. **8**(5):173–178. 1948.

———: Proposals and notes on some genera of Clavarioid fungi and their types, *Lloydia*, **11**(2):123–138. 1948.

COTTON, A. D., AND E. M. WAKEFIELD: A review of the British Clavariae, *Brit. Mycol. Soc. Trans.*, **6**:164–198. 1918.

FAWCETT, STELLA G. M.: Studies in the Australian Clavariaceae, *Proc. Roy. Soc. Victoria*, N.S., **51**:1–20. *Pls*. 1–5. 1938; 265–280. *Pls*. 17–23. 1939.

COOL, C.: Overzicht van de in Nederland groeiende Clavaria-soorten, *Mededeel. Nederland. Mycol. Ver.*, **16–17**:96–159. *Pls*. 1–5. 1928.

IMAI, SANSHI: On the Clavariaceae of Japan, I, *Trans. Sapporo Natural History Soc.*, **11**(1):38–45. 1929. (English with Japanese summary.)

———: On the Clavariaceae of Japan, II, *ibid.*, **11**(2):70–77. 1930.

IMAI, SANSHI: On the Clavariaceae of Japan, III. The species of Clavaria found in Hokkaido and Southern Saghalien, *ibid.*, **12**(1):9–12. 1931.

———: On the Clavariaceae of Japan, IV, *ibid.*, **13**(4):377–384. 1934.

———: On the Clavariaceae of Japan, V. The species found in the Northern Honshu, *ibid.*, **16**(4):209–217. 1941.

———: On the Clavariaceae of Japan, VI. The species found in the Central Honshu, *ibid.*, **16**(4):244–250. 1941.

VAN DER BYL, PAUL A.: Oor enige Suid-Afrikaanse Clavaria-soorte of knotsswamme, *S. African J. Sci.*, **29**:317–323. *Pl.* 3. 1932.

HARPER, EDWARD T.: The Clavaria fistulosa group, *Mycologia*, **10**(2):53–57. *Pls.* 3–5. 1918.

KILLERMANN, SEB.: Die Gattungen Typhula und Pistillaria. Kritische Darstellung und neue Arten, *Z. für Pilzkunde*, N.F., **13**(4):98–108. 1 *fig.* 1934.

REMSBERG, RUTH E.: Studies in the genus Typhula, *Mycologia*, **32**(1):52–96. *Figs.* 1–58. 1940. (Key and descriptions of 16 species.)

VANG, J.: Typhula species on agricultural plants in Denmark, *K. Vet.- og Landbohojskole Aarskr.* (*Copenhagen*), **1945**:1–46. 18 *figs.* 1945.

PETCH, T.: British species of Hirsutella, *Naturalist*, **1932**:45–49. 1932.

LINDER, DAVID H.: The genus Myxomycidium, *Mycologia*, **26**(4):332–343. *Pl.* 39. *Figs.* 1–6. 1934.

BAKER, GLADYS E.: Studies in the genus Physalacria, *Bull. Torrey Botan. Club*, **68**(5):265–288. *Figs.* 1–105. 1941. (Includes key and descriptions of 13 species.)

———: Addenda to the genera Helicogloea and Physalacria, *Mycologia*, **38**(6):630–638. *Figs.* 1–25. 1946.

CORNER, E. J. H.: A monograph of Clavaria and allied genera. xv + 740 pp. 16 *colored pls.* 298 *figs. Annals of Botany Memoirs No.* 1. London, Oxford Univ. Press. 1950.

List 37. Hydnaceae

(See also List 39 for some of the dentate forms among the Polyporaceae. Some species ascribed to the genus Irpex probably are more correctly placed in that family.)

COKER, W. C.: The Hydnums of North Carolina, *J. Elisha Mitchell Sci. Soc.* **34**(4):163–197. *Pls.* 1–29. (1 *and* 19 *colored*). 1919.

———: Further notes on Hydnums, *ibid.*, **41**(3–4):270–286. *Pls.* 51–65. 1926.

———: New or noteworthy Basidiomycetes, *ibid.*, **55**(2):373–387. *Pls.* 34–44. 1939. (Contains a key to fleshy stipitate species of Hydnum of the Eastern United States.)

BEARDSLEE, H. C.: Notes on the scaly species of the Hydnaceae, *Mycologia*, **16**(6):255–258. 1924.

BANKER, H. J.: A preliminary contribution to a knowledge of the Hydnaceae, *Bull. Torrey Botan. Club*, **28**(4):199–222. 1901.

———: A contribution to a revision of the North American Hydnaceae, *Mem. Torrey Botan. Club*, **12**:99–194. 1906.

———: Type studies in the Hydnaceae: I. The genus Manina, *Mycologia*, **4**(5):271–278. 1912; II. The genus Steccherinum, *ibid.*, **4**(6):309–318. 1912; III. The genus Sarcodon, *ibid.*, **5**(1):12–17. 1913; IV. The genus Phellodon, *ibid.*, **5**(2):62–66. 1913; V. The genus Hydnellum, *ibid.*, **5**(4):194–205. 1913; VI. The genera Creolophus, Echinodontium, Gloiodon and Hydnodon, *ibid.*, **5**(6):293–298. 1913; VII. The genera Asterodon and Hydnochaete, *ibid.*, **6**(5):231–234. 1914.

MILLER, L. W.: The genera of the Hydnaceae, *Mycologia*, **25**(4):286–302. 1933. (Gives a key to the recognized genera of this family.)

———: The Hydnaceae of Iowa: I. The genera Grandinia and Oxydontia, *ibid.*, **25**(5):356–368. *Pl.* 43. 1933; II. The genus Odontia, *ibid.*, **26**(1):13–32. *Pls.* 2–3. 1934; III. The genera Radulum, Mucronella, Caldesiella and Gloiodon, *ibid.*, **26**(3):212–219. *Pl.* 27. 1934; IV. The genera Steccherinum, Auriscalpium, Hericium, Dentinum and Calodon, *ibid.*, **27**(4):357–373. *Pl.* 33. 1935.

HENRY, LEROY K.: A review of the Hydnaceae (Fungi) of Western Pennsylvania, *Ann. Carnegie Museum*, **31**:19–32. *Pls.* 1–2. 1948. (Article 3.)

LLOYD, C. G.: The genus Radulum, *Mycological Writings*, **4**, *Figs.* 961–984. 1917. (Separate pagination, pp. 1–12.)

BOURDOT, H., ET A. GALZIN: Hyménomycètes de France: V. Hydnées, *Bull. soc. mycol. France*, **30**:243–280. 1914.

CEJP, K.: Monografie Hydnacei Republiky Československé. Praze, 1928.

VAN DER BYL, P. A.: Die Suid-Afrikaanse Hydnaceae of Stekelswamme, *Ann. Univ. Stellenbosch*, **12A**(1):1–9. *Figs.* 1–11. 1934.

PILAT, ALBERT: Revision der zentraleuropäischen resupinaten Arten der Gattung Irpex Fr., *Ann. Mycol.*, **23**(3–6):302–307. 1925.

JØRSTAD, IVAR: Norske resupinate Hydnaceer, *Friesia*, **1**(1):2–20. 1932.

WAKEFIELD, E. M.: Australian resupinate Hydnaceae, *Trans. Proc. Roy. Soc. Australia*, **54**:155–158. 1930. (A key to the genera Acia, Grandinia, and Odontia.)

BATAILLE, F.: Flore analytique-descriptive des Hydnes terrestres d'Europe, *Bull. trimestr. soc. mycol. France*, **39**:201–216. 1924.

List 38. Meruliaceae

(See also some of the more general works on Polyporaceae in List 39 and in List 34, in some of which the Meruliaceae are not regarded as a separate family.)

BURT, EDWARD A.: Merulius in North America, *Ann. Missouri Botan. Garden*, **4**(4):305–362. *Pls.* 20–22. *Figs.* 1–39. 1917.

———: Merulius in North America. Supplementary Notes, *ibid.*, **6**(2):143–145. 1919.

BOURDOT, H., ET A. GALZIN: Hyménomycètes de France: IX. Meruliés, *Bull. soc. mycol. France*, **39**:96–118. 1923.

BUCHWALD, N. FABRITIUS: De danske arter af slaegten Merulius (Hall.) Fr. med en saerlig omtale af gruppen Coniophori Fr., *Dansk Botanisk Arkiv*, **5**(6):1–46. 1 *pl.* 1928.

NIKOLAIEVA, T. L.: The genus Merulius in U.S.S.R. *Sovietskaia Botanika*, **1933**(5):96–111. 1933. (In Russian.)

List 39. Polyporaceae

(The first three references will serve to distinguish the genera in accordance with the more recent, but not as yet, fully accepted segregations and arrangements. The distinction of species will be found in the references further on in this list.)

BONDARZEW, A., UND R. SINGER: Zur Systematik der Polyporaceen, *Ann. Mycol.*, **39**:43–65. 1941.

COOKE, W. BRIDGE: A nomenclatorial survey of the genera of pore fungi, *Lloydia*, **3**(2):81–104. 1940.

SINGER, ROLF: Notes on taxonomy and nomenclature of the Polypores, *Mycologia*, **36**(1):65–69. 1944.

MURRILL, W. A.: Polyporaceae, *North American Flora*, **9**:1–131. 1907–1908.

———: Northern Polypores, 64 pp. New York, published by the author, 1914. (Covers the area of Northeastern United States and Canada included in the limits for Britton and Brown's Illustrated Flora.)

———: Western Polypores, 36 pp. New York, published by the author, 1915. (Covers the area of Alaska, British Columbia and the Pacific Coast States.)

———: Southern Polypores, 66 pp. New York, published by the author, 1915.

———: Tropical Polypores, 113 pp. New York, published by the author, 1915.

———: A key to the white and bright-colored sessile Polyporeae of temperate North America, *Torreya*, **8**(1):14–16, (2):28–29, (6):130–132. 1908.

———: Corrections and additions to the Polypores of temperate North America, *Mycologia*, **12**(1):6–24. 1920.

———: Light-colored resupinate Polypores, *ibid.*, **12**(2):77–92, (6):299–308. 1920; **13**(2):83–100, (3):171–178. 1921.

———: Florida resupinate Polypores, *ibid.*, **34**(5):595–596. 1942. (A key to the resupinate genera found in Florida.)

LOWE, JOSIAH L.: The Polyporaceae of New York State (except Poria), revised ed., *N. Y. State Coll. Forestry Syracuse Univ. Tech. Pub.* **60**:1–128. *Illustrations* 1–2. 1942.

———: The Polyporaceae of New York State (The genus Poria), *ibid.*, **65**: 1–91. *Figs.* 1–20. 1946.

———: Studies in the genus Poria: II. White and brightly-colored type material, *Lloydia*, **10**(1):45–59. 1947.

———: Studies in the genus Poria: IV. Brown type material, *ibid.*, **11**(3):162–170. *Figs.* 1–11. 1948.

OVERHOLTS, L. O., AND J. L. LOWE: New species of Poria, *Mycologia*, **38**(2):202–212. 2 *figs.* 1946.

———: The Polyporaceae of the Middle-Western United States, *Wash. Univ. Studies*, **3** (Pt. I, No. 1):1–98. *Pls.* 1–8. 1915.

———: The Polyporaceae of Ohio, *Ann. Missouri Botan. Garden*, **1**(1):81–155. 1914.

———: The species of Poria described by Peck, *New York State Museum Bull.* **205–206**:67–166. *Pls.* 1–23. 1918.

———: The species of Poria described by Schweinitz, *Mycologia*, **15**(5):207–232. *Pls.* 21–24. *Figs.* 1–30. 1923.

———: Diagnoses of American Porias, I, *ibid.*, **14**(1):1–11. *Pl.* 1. *Figs.* 1–6. 1922; II, *Bull. Torrey Botan. Club*, **50**(7):245–253. *Pls.* 13–14. *Figs.* 1–3. 1923.

———: Diagnoses of American Porias: III. Some additional brown species with a key to the common brown species of the United States and Canada, *Mycologia*, **23**(2):117–129. *Pls.* 12–14. 1931.

———: The Polyporaceae of Pennsylvania: I. The genus Polyporus, *Penna. Agr. Exp. Sta. Tech. Bull.* **298**:3–28. *Pls.* 1–2. 1933; II. The genera Cyclomyces, Daedalea, Favolus, Fomes, Lenzites and Trametes, *ibid.*, **316**:3–16. *Figs.* 1–12. 1935; III. The genus Poria, *ibid.*, **418**:3–64. 1942.

HENRY, LEROY K.: Pore fungi of Western Pennsylvania: I. The more common small members of the genus Polyporus, *Carnegie Museum Botany Pamphlet* **2**:1–16. 41 *figs.* 1939.

———: Pore fungi of Western Pennsylvania: II. The more common larger members of the genus Polyporus and some common members of other genera, *ibid.* **3**:1–15. 37 *figs.* 1942.

———: A review of the pileate Polypores of Western Pennsylvania, *Ann. Carnegie Museum*, **28**:221–272. *Pls.* 26–29. 1941. (Article XIII.)

———: New and noteworthy polypores from Western Pennsylvania, *Proc. Pennsylvania Acad. Sci.*, **22**:87–93. 1948.

NEUMAN, J. J.: The Polyporaceae of Wisconsin, *Wisconsin Geological and Natural History Survey Bull.* **33**, Scientific Series (10):1–206. *Pls.* 1–25. 1914.

SHOPE, PAUL F.: The Polyporaceae of Colorado, *Ann. Missouri Botan. Garden*, **18**:287–456. *Pls.* 16–39. 1931.

BAXTER, DOW V.: Some Porias from the region of the Lake States, *Papers Mich. Acad. Sci.*, **6**:67–76. *Pls.* 1–6. 1927; **9**:39–46. *Pls.* 26–29. 1929.

———: Some resupinate Polypores from the region of the Great Lakes, *ibid.*, **15**:191–228. *Pls.* 17–26. 1932; **17**:421–439. *Pls.* 41–50. 1933; **19**:305–332. *Pls.* 58–65. 1934; **20**:273–281. *Pls.* 55–60. 1935; **21**:243–267. *Pls.* 30–39. 1936; **22**:275–295. *Pls.* 31–37. 1937; **23**:285–305. *Pls.* 1–9. 1938; **24**:167–188. *Pls.* 1–7. 1939; **25**:145–170. *Pls.* 1–12. 1940; **26**:107–121. *Pls.* 1–7. 1941; **27**:139–161. *Pls.* 1–11. 1942; **28**:215–233. *Pls.* 1–6. 1943; **29**:85–109. *Pls.* 1–6. 1943 (1944); **30**:175–191. *Pls.* 1–14. 1944 (1945); **31**:117–130. *Pls.* 1–5. 1945 (1947); **32**:189–211. *Pls.* 1–10. 1946 (1948); **33**:9–30. *Pls.* 1–9. 1947 (1949); **34**:41–56. *Pls.* 1–6. 1948 (1950).

ERIKSSON, JOHN: The Swedish species of the "Poria vulgaris" group, *Svensk Botanisk Tidskrift*, **43**(1):1–25. *Pls.* 1–5. 1949.

COKER, W. C.: The United States species of Coltrichia, *J. Elisha Mitchell Sci. Soc.*, **62**(1):95–107. *Pls.* 17–22. 1946.

———: Notes on Carolina fungi, *ibid.*, **64**(2):287–303. *Pls.* 37–54. 1948. (Keys to, and descriptions of, the fleshy stipitate Polypores of North Carolina.)

NOBLES, MILDRED K.: Studies in forest pathology: VI. Identification of cultures of wood-rotting fungi, *Can. J. Research, C*, **26**:281–431. *Pls.* 1–18. *Figs.* 1–3. 1948.

PILAT, ALBERT: Polyporaceae, Monographie des espèces européennes en vue des espèces de l'Asie du Nord, in Atlas des Champignons d'Europe, vol. 3, pt. 1–2, pp. 1–624. 374 *pls.* Prag, Kavina et Pilat, 1936–1942.

VAN DER BYL, PAUL A.: A contribution to our knowledge of the Polyporaceae of South Africa, *S. African J. Sci.*, **18**:246–293. 1921.

———: Descriptions of additional South African Polyporeae, *ibid.*, **21**:308–313. 1924.

TORREND, C.: Les Polyporacées du Brésil, *Broteria. Revista de Sciencias Naturaes, Série Botánica*, **18**:23–43, 121–143. *Pls.* 1–8. 1920; **20**:107–112. 1922; **21**:12–42. *Pls.* 1–4. 1924.

———: Les Polyporacées stipitées du Brésil, *ibid.*, **22**:5–19. *Pl.* 1. 1926.

———: Les Polyporacées du Brésil. Le genre Hexagonia, *Broteria. Série Trimestral: Ciências Naturais*, **4**:108–120. 1935.

BOSE, S. R.: Descriptions of fungi in Bengal, I, *Proc. Indian Assoc. Cultivation Sci.*, **4**:109–114. *Pls.* 1–11. 1918; II. *ibid.*, **1918**:136–143. *Pls.* 1–13. 1920.

Bose, S. R.: Polyporaceae of Bengal, III, *Carmichael Medical College, Belgachia, Bull.*, **1**:1–5. 12 *pls.* 1920.

———: Polyporaceae of Bengal, IV. Descriptions of some polypores new to Bengal, *ibid.* 2, pp. 1–5. 15 *pls.* 1921.

———: Polyporaceae of Bengal, V, *ibid.* 3, pp. 20–25. *Pls.* 1–9. 1927.

———: Polyporaceae of Bengal, VI, *Proc. Indian Assoc. Cultivation Sci.*, **1919**:55–62. *Pls.* 1–6. 1922.

———: Polyporaceae of Bengal, VII, *ibid.*, **1920–21**:27–36. *Illustrated.* 1923.

———: Polyporaceae of Bengal, VIII, *J. Dept. Sci., Calcutta Univ.*, **9**:27–31. *Illustrated.* 1928.

———: Polyporaceae of Bengal, IX, *ibid.*, **9**:35–44. *Illustrated.* 1928.

———: Polyporaceae of Bengal, X, *ibid.*, **11**:1–18. *Pls.* 1–5. 1934.

Lloyd, C. G.: Synopsis of the genus Fomes, *Mycological Writings*, **4**:211–288. *Figs.* 570–610. 1915.

———: Synopsis of the genus Hexagona, *ibid.*, **3**:1–46. *Figs.* 279–330. 1910.

———: Synopsis of the sections Microporus, Tabacinus and Funales of the genus Polystictus, *ibid.*, **3**:49–70. *Figs.* 336–356. 1910.

———: Synopsis of the section Ovinus of Polyporus, *ibid.*, **3**:73–94. *Figs.* 496–509. 1911.

———: Synopsis of the stipitate Polyporoids, *ibid.*, **3**:95–208. *Figs.* 395–500. 1912.

———: Synopsis of the section Apus of the genus Polyporus, *ibid.*, **4**:291–392. *Figs.* 631–706. 1915.

———: The tessellate Favoli, *Mycological Notes*, **7**:1156–1157. *Figs.* 2281–2284. 1922.

Espinosa Bustos, Marcial R.: Sobre las especies chilenas del genero Fomes, *Revista Chilena de Historia Natural, Pura y Aplicada*, **25**:321–343. 7 *pls.* 3 *figs.* 1921.

Rick, J.: Polypori riograndenses: I. Mesopodes aut Pleuropodes, *Broteria. Série Trimestral. Ciências Naturais*, **3**(4):180–189. 1934; II. Merisma, *ibid.*, **4**:17–27. 1935; III. Apodes, *ibid.*, **4**:84–94. 1935.

———: Polysticti riograndenses, *ibid.*, **4**:121–138. 1936; **5**:171–178. 1936.

Patouillard, N.: Le genre Ganoderma, *Bull. soc. mycol. France*, **5**(3):64–80. *Pls.* 10–11. 1889.

Haddow, W. R.: Studies in Ganoderma, *J. Arnold Arboretum Harvard Univ.*, **12**(1):25–46. 2 *pls.* 1 *fig.* 1931.

Humphrey, C. J.: A partial revision of the Ganoderma applanatum group with particular reference to its oriental variants, *Philippine J. Sci.*, **45**(4):483–589. *Pls.* 1–36. *Fig.* 1. 1931.

———, and Simeona Leus: Studies and illustrations in the Polyporaceae, III, *ibid.*, **49**(2):159–184. *Pls.* 1–12. 1932. (A continuation of the previous paper on Ganoderma.)

Cunningham, G. H.: The Polyporaceae of New Zealand, *Trans. New Zealand Inst.*, **58**(3):202–250. 11 *pls.* 1927.

———: New Zealand Polyporaceae: I. The genus Poria, *Bull. Dept. Sci. Industr. Res. New Zealand* 72, 43 pp., 7 *pls.*, 37 *figs.*, 1947; II. The genus Fuscoporia, *ibid.* 73, 14 pp., 1 *pl.*, 10 *figs.*, 1948; III. The genus Polyporus, *ibid.* 74, 39 pp., 7 *pls.*, 1948; IV. The genus Coriolus, *ibid.* 75, 10 pp., 2 *pls.*, 1948; V. The genus Fomitopsis, *ibid.* 76, 8 pp., 3 *pls.*, 1948; VI. The genus Coltrichia, *ibid.* 77, 10 pp., 3 *pls.*, 1948. VII. The genus Inonotus, *ibid.* 78; 5 pp., 2 *pls.*, 1948; VIII. The genus Fomes, *ibid.* 79, 24 pp., 6 *pls.*, 1948. IX. Trametes, Lenzites and Daedalea, *ibid.* 80, 10 pp., 4 *pls.*, 1948.

CLELAND, J. BURTON, AND EDWIN CHEEL: Australian fungi: Notes and descriptions: II. The sclerotia-forming polypores of Australia, *Trans. Proc. Roy. Soc. South Australia*, **43**:11–22. *Pls.* 1–5. 1919.

———, AND ———: Notes on Australian Fungi: IV. Polyporus, Fomes and Hexagona, *J. Proc. Roy. Soc. New South Wales*, **51**:473–557. 1918.

———, AND L. RODWAY: Notes on the genus Poria. I. *Roy. Soc. Tasmania: Papers and Proc.*, **1928**:31–43. 1928; II. The Porias and Poria-like fungi with definitely coloured hyphae, *ibid.*, **1928**:73–86. 1929; III. The Australian Porias and Poria-like fungi with hyphae not deeply coloured, *ibid.*, **1929**:7–24. 1929. (A continuation of Pt. II.)

IMAZEKI, ROKUYA: Studies in Ganoderma of Nippon, *Bull. Tokyo Science Museum*, **1**:29–52. 15 *figs.* 1939. (In Japanese with English summary.)

GRAFF, PAUL W.: North American Polypores: I. Polyporus squamosus and its varieties, *Mycologia*, **28**(2):154–170. 1936; II. Polyporus biennis and its varieties, *ibid.*, **31**(4):466–484. 1939.

COOKE, WILLIAM BRIDGE: Oxyporus nobilissimus and the genus Oxyporus in North America, *Mycologia*, **41**(4):442–455. *Figs.* 1–10. 1949.

List 40. Boletaceae

(See also some of the references in List 39, Polyporaceae, and R. SINGER: Phylogenie und Taxonomie der Agaricales in List 41.)

MURRILL, WM. A.: Boletaceae, *North American Flora*, **9**:133–161. 1910.

PECK, CHARLES H.: Boletinus, *New York State Museum Bull.* **8**:74–80. 1889.

———: Boletus, *ibid.*, **8**:80–150. 1889.

———: Strobilomyces, *ibid.*, **8**:158–159. 1889.

GILBERT, E. J.: Les Bolets, in Les Livres du Mycologue, vol. 3, 254 pp. 16 *pls.* Paris, Librairie E. le François, 1931.

———Notules sur les Bolets, I. *Bull. trimestr. soc. mycol. France*, **52** (2):249–260. *Figs.* 1–5. 1936.

———: Notules sur les Bolets, II, *ibid.*, **56**(2):120–124. 1940.

SARTORY, A., ET L. MAIRE: Les Bolets. Monographie du genre Boletus Dill, Fascicules 1–2, 512 pp. 3 *pls.* Nancy, V. Idoux et Cie, 1931.

KALLENBACH, FRANZ: Die Röhrlinge (Boletaceae), in Die Pilze Mitteleuropas, vol. 1, pts. 1–21, pp. 1–54. *Pls.* 1–55 (*in greater part colored*). Leipzig, Werner Klinkhart, 1926–1941 (?). Uncompleted (?).

SNELL, WALTER H.: Notes on Boletes, *Mycologia*, **24**(3):334–341. *Fig.* 1. 1932; **25**(3):221–232. 1933; **26**(4):348–359. 1934; **28**(1):13–23. (5):463–475. 1936.

———: Tentative keys to the Boletaceae of the United States and Canada, *Rhode Island Botanical Club Publication*, **1**:1–25. 1935.

———: The genera of the Boletaceae, *Mycologia*, **33**(4):415–423. 1941.

———: New proposals relating to the genera of the Boletaceae, *ibid.*, **34**(4):403–411. *Fig.* 1. 1942.

SLIPP, ALBERT W., AND WALTER H. SNELL: Taxonomic-ecologic studies of the Boletaceae in Northern Idaho and adjacent Washington, *Lloydia*, **7**(1):1–66. *Pls.* 1–8. 1944.

STRAUSER, M. C.: The Boletaceae of Pennsylvania, *Proc. Penna. Acad. Sci.*, **4**:17–24. *Pls.* 1–2. 1930. (Contains a key to the Pennsylvania species of Boletaceae.)

HENRY, LEROY K.: A review of the Boletes (fungi) of Western Pennsylvania, *Ann. Carnegie Museum*, **30**:213–240. *Pls.* 1–3. 1946. (Article XIII.)

CHIU, WEI-FAN: The Boletes of Yunnan, *Mycologia*, **40**(2):199–231. 1948.

KILLERMANN, S.: Ueber den Hexenpilz (Boletus luridus Schäff.) und Verwandte, *Kryptogamische Forschungen der Bayerischen Botanischen Gesellschaft*, **4**:336–343. 3 *figs*. 1919.

LOHWAG, HEINRICH: Kritische Bemerkungen zur Luridus-Gruppe, *Hedwigia*, **63**:323–328. 1922.

SMOTLACHA, FRANTIŠEK: Monografie českych hub hřibovitych (Boletinei), *Sitzber. kgl. Böhmischen Ges. Wiss. Math.-naturw. Klasse*, **1911**(8):1–73. 1912.

BOHUS, GABOR: Von der Gruppe scaber, *Acta Mycologica Hungarica*, **1**(1):28, (2):62–65, (3–4):111–118. 1944.

ANONYMOUS: A tinorugombák (Boletus-Boletinus-Gyrodon-Strobilomyces) határozó kulcsa., *ibid.*, **2**(1–2):50–56. 1945. (Hungarian.) (Key to these 4 genera and to 47 species of Boletaceae.)

SINGER, ROLF: Sur les genres Ixocomus, Boletinus, Phylloporus, Gyrodon et Gomphidius, *Rev. mycol.*, N.S., **3**(2):17–53, (3):157–177. *Pl.* 4 (*colored*). 1938.

———: Notes sur quelques Basidiomycètes: IVe Séries. 1. Le genre Krombholzia Karst. *ibid.*, **3**(4):187–191. 1938.

———: The Boletineae of Florida with notes on extralimital species: I. The Strobilomycetaceae, *Farlowia*, **2**(1):97–141. *Pl.* 1. 1945; II. The Boletaceae (Gyroporoideae), *ibid.*, **2**(2):223–303. 1945; III. The Boletoideae of Florida, *Am. Midland Naturalist*, **37**(1):1–135. *Pls.* 1–2. 1947; IV. The lamellate families (Gomphidiaceae, Paxillaceae and Jugasporaceae), *Farlowia*, **2**(4):527–567. *Pl.* 1. 1946.

MURRILL, WILLIAM A.: Florida Boletes, *Lloydia*, **11**(1):21–35. 1948.

List 41. Agaricaceae (in the Broader Sense)

(For the lamellate Boletoid genera see List 40.)

Works (of More or Less Popular Nature) on Edible and Poisonous Mushrooms.

ATKINSON, GEORGE F.: Studies of American Fungi. Mushrooms, Edible, Poisonous, etc., 275 pp. *Pls.* 1–76. *Figs.* 1–223. Ithaca, N. Y., Andrus and Church, 1900.

HARD, M. E.: The Mushroom, Edible and Otherwise, Its Habitat and Time of Growth, 609 pp. *Pls.* 1–66. *Figs.* 1–505. Columbus, Ohio, Ohio Library Co., 1908.

GIBSON, W. H.: Our Edible Toadstools and Mushrooms and How to Distinguish Them, x + 337 pp. 30 *col. pls.* 57 *figs.* New York, Harper and Brothers, 1895.

SMITH, ALEXANDER H.: Common edible and poisonous mushrooms of southeastern Michigan, *Cranbrook Institute of Science Bull.* **14**:1–71. *Illustrated.* 1938.

ANONYMOUS: Edible and poisonous fungi, ed. 6, *Ministry of Agriculture and Fisheries Bull.* 23, 35 pp. 27 *col. pls.* London, 1945.

RICHON, CHARLES, ET ERNEST ROZE: Atlas des champignons comestibles et vénéneux de la France et des pays circonvoisins, Text vol., pp. i–xcviii, 1–265. Vol. of pls. pp. i–xii. *Pls.* 1–72. Paris, Octave Doin, 1888.

BRESADOLA, G.: Funghi mangerecci e velenosi, ed. 3, 2 vols., 647 pp. 224 *pls.* 60 *figs.* Trento Societá Botanica Italiana. 1932, 1934.

SARTORY, A., ET L. MAIRE: Les champignons vénéneux, 251 pp. 10 *col. pls.* Paris, 1922.

BAVENDAMM, W.: Wie unterscheide ich die Speisepilze von den Gift- und Bitterpilze? *Merkbl. Reichsinst. Forst- und Holzw.* 7, 59 pp. 38 *figs.* 1948.

MAUBLANC, A.: Les champignons comestibles et vénéneux, ed. 2, vol. 2, pp. 108–144. 96 *col. pls.* Paris, Paul Lechevalier, 1927.

DUJARRIC DE LA RIVIÈRE, R., ET ROGER HEIM: Les champignons toxiques. Caractères et détermination. Toxines, intoxications, thérapeutique. Aquarelles de A. Bessin, 1 vol. in quarto, 60 pp. 8 *col. pls., Figs.* 1–5. Paris, Encyclopédie Medico-chirurgicale Editeur, 1938.

SCHMIDEG, ARMAND: Mes expériences d'un quart de siècle et mes expériments personnels avec des champignons concernant leur comestibilité, *Acta Mycologica Hungarica,* **3**(1–4):33–47, 1946; **4**(1–2):33–52, 1947.

GÜSSOW, H. T., AND W. S. ODELL: Mushrooms and Toadstools. An account of the more common edible and poisonous fungi of Canada, 274 pp. 128 *pls.* (2 *colored*). Ottawa, Division of Botany, Dominion Experiment Farms of Canada, 1927.

Works on the Taxonomy of the Agaricaceae.

SINGER, ROLF. Phylogenie und Taxonomie der Agaricales, *Schweizerische Z. für Pilzkunde,* **17**(2):23–28, (3):35–39, (4):52–57, (5):71–73, (6):84–87, (7):97–101. 1 *fig.* 1939. (Gives keys to the genera of Boletaceae and Agaricaceae (in their old sense), based upon the new anatomical and chemical studies.)

———: Das System der Agaricales, *Ann. Mycol.,* **34**(4–5):286–378. 1936. (Keys to the families and genera of the Agaricales (narrow sense).)

———: The Agaricales (Mushrooms) in modern taxonomy, *Lilloa Revista de Botanica,* **22**:1–834. *Pls.* 1–29. 1949 (1951).

MURRILL, W. A.: Agaricaceae, *North American Flora,* **9**:162–426. 1910–1916 (keys and descriptions of genera and species of Tribes Chanterellae and Lactarieae and part of the white-spored Agariceae); *ibid.,* **10**:1–348. 1914–1932 (vol. not completed) (remainder of white-spored forms and the pink- and brown-spored forms, including Inocybe and Cortinarius by C. H. Kauffman, and Pholiota and Hypodendron by L. O. Overholts).

———: The Agaricaceae of the Pacific Coast, I, *Mycologia,* **4**(4):205–217. 1912 (white-spored genera); II, *ibid.,* **4**(5):231–262. 1912 (white- and ochre-spored genera); III, *ibid.,* **4**(6):294–308. *Pl.* 77. 1912 (brown- and black-spored genera).

———: The Agaricaceae of Tropical North America, I, *Mycologia,* **3**(1):23–36. 1911 (white-spored genera); II, *ibid.,* **3**(2):79–91. 1911 (white-spored genera); III, *ibid.,* **3**(4):189–199. 1911 (white-spored genera); IV, *ibid.,* **3**(6):271–282. 1911 (genera with rose-colored spores); V, *ibid.,* **4**(2):72–83. 1912 (ochre-spored genera); VI, *ibid.,* **5**(1):18–36. 1913 (ochre-spored genera); VII, *ibid.,* **10**(1):15–33. 1918 (purple-brown- to black-spored genera); VIII, *ibid.,* **10**(2):62–85. 1918 (purple-brown- to black-spored genera).

MURRILL, W. A.: Dark-spored Agarics: I. Drosophila, Hypholoma and Pilosace, *Mycologia*, **14**(2):61–76. 1922; II. Gomphidius and Stropharia, *ibid.*, **14**(3):121–142. 1922; III. Agaricus, *ibid.*, **14**(4):200–221. 1922; IV. Deconica, Atylospora and Psathyrella, *ibid.*, **14**(5):258–278. 1922; V. Psilocybe, *ibid.*, **15**(1):1–22. 1923.

———: The rosy-spored Agarics of North America, *Brooklyn Botanic Garden Mem.*, **1**:334–336. 1918. (Contains a key to the genera of Subtribe Pluteanae.)

FAYOD, V.: Prodrome d'une histoire naturelle des Agaricinées, *Ann. sci. nat. Botan.*, VII Sér., **9**:181–411. *Pls.* 6–7. 1889.

LANGE, JAKOB E.: Flora Agaricina Danica, 5 vols. Copenhagen, published under the auspices of the Society for the Advancement of Mycology and the Danish Botanical Society.

Vol. 1, pp. 1–90. *Pls.* 1–40. 1935. Amanita, Limacella, Lepiota, Armillaria, Tricholoma, Clitocybe.

Vol. 2, pp. 1–105. *Pls.* 41–80. 1936. Collybia, Marasmius, Mycena, Omphalia, Pleurotus, Panus, Volvaria, Pluteus.

Vol. 3, pp. 1–96. *Pls.* 81–120. 1938. Cortinarius, Pholiota, Inocybe, Hebeloma.

Vol. 4, pp. 1–119. *Pls.* 121–160. 1939. Flammula, Naucoria, Tubaria, Galera, Bolbitius, Pluteolus, Crepidotus, Paxillopsis, Paxillus, Psalliota, Stropharia, Lacrymaria, Hypholoma, Psilocybe, Panaeolus, Psathyra, Pseudocoprinus, Coprinus.

Vol. 5, pp. 1–108. *Pls.* 161–200. Gomphidius, Melanomphalia, Nyctalis, Limacium, Camarophyllus, Hygrocybe, Lactarius, Russula, Cantharellus, Lentinus, Schizophyllum. (Supplementary notes. Also title pages and i–xxiv and index for the five volumes.)

CERUTI, ARTURO: Fungi analytice delineati iconibus pictis illustrati, vol. 1, pp. 1–276. *Pls.* 1–35 (*in part colored*). Torino, 1948. (Latin, English and French.) (This first volume of the series includes several species of Amanita and of Russula.)

KÜHNER, R.: Contribution à l'étude des Hyménomycètes et specialement des Agaricacées, *Le Botaniste*, **17**(1–4):1–224. 4 *pls.* 37 *figs.* 1926.

ROMAGNESI, HENRI: Clé pratique pour la détermination générique des Agarics d'Europe, *Rev. mycol.*, N.S., **2**(supplément 1):11–18. 1937.

COSTANTIN, J., ET L. DUFOUR: Nouvelle flore des champignons pour la détermination facile de toutes les espèces de France, ed. 2, 291 pp. 4166 *figs.* 1 *pl. col.* Paris, Paul Dupont, 1895.

HENRY, LEROY K.: Twenty-eight common gilled mushrooms of western Pennsylvania, *Trillia*, **10**:82–100. *Pls.* 12–14. 1939.

HEIM, ROGER: Les Agarics tropicaux à hyménium tubulé (Madagascar, Côte d'Ivoire, Antilles, Insulinde), *Rev. mycol.*, N.S., **10**(1–4):3–61. 4 *pls.* 32 *figs.* 1945.

CLELAND, J. BURTON, AND EDWIN CHEEL: The Hymenomycetes of New South Wales, *Agr. Gaz. New South Wales*, **25**:507–515, 885–888, 1045–1049. 2 *pls.* 1 *fig.* 1914; **26**:325–333. 2 *pls.* 1915; **27**:97–106. 2 *pls.* 1916. (Apparently all that appeared. Descriptions are given of the New South Wales species of Amanita, Amanitopsis, Lepiota, and Armillaria.)

———: Australian Fungi. Notes and Descriptions, I, *Trans. Roy. Soc. S. Australia*, **42**:88–138. *Pls.* 9–12 (*col.*). 1918 (brown- purple- and black-spored Agarics); III, *ibid.*, **43**:262–315. *Pls.* 28–29 (*col.*). 1919 (white-spored Agarics, Polyporaceae, Hydnaceae, Thelephoraceae, Gasteromyceteae, Ascomyceteae); IV, *ibid.*, **47**:58–78. *Pls.* 1–2 (*col.*). 1923 (miscellaneous Agaricaceae, Polyporaceae, Thelephoraceae, Clavariaceae, Tremellaceae, and Gasteromyceteae);

V, *ibid.*, **48**:236–252. 1924 (miscellaneous Agaricaceae, Polyporaceae, Thelephoraceae, Gasteromyceteae, and Ascomyceteae); VI, *ibid.*, **51**:298–306. 1927 (miscellaneous Agaricaceae); VII, *ibid.*, **53**:217–222. 1928 (South Australian species of Cortinarius); VIII, *ibid.*, **55**:152–160. 1931 (miscellaneous Agaricaceae (mainly Amanita and Clitocybe) and new species of Clavaria).

KRIEGER, LOUIS C. C.: Field Key to the Genera of the Gill Mushrooms, pp. 1–8. 1 *chart*. Baltimore, The Norman-Remington Co., 1920.

KAUFFMAN, C. H.: The Agaricaceae of Michigan, *Michigan Geological and Biological Survey Publication* 26, *Biological ser.* 5, Vol. 1, pp. 1–924, *Figs.* 1–4, 1918; Vol. 2, pp. 1–10, *Pls.* 1–172, 1918. (Very full descriptions of all species of Agaricaceae known to occur in Michigan, and in many genera of all species recognized in Northeastern United States. Illustrated by excellent photographs.)

STOVER, WILMER G.: The Agaricaceae of Ohio, *Proc. Ohio State Acad. Sci.*, **5**:462–577. 1912.

SMITH, HURON H.: Mushrooms of the Milwaukee region, *Milwaukee Public Museum Field Guide No.* 1, *Botan. ser.*, 87 pp. 184 *figs. Colored table.* 1931.

GRAHAM, V. O.: Mushrooms of the Chicago region, *Program of the Activities Chicago Acad. Sci.*, **4**(3):42–62. *Figs.* 1–45. 1933.

IMAI, SANSHI: Studies on the Agaricaceae of Hokkaido, *J. Faculty Agr. Hokkaido Imp. Univ.*, Sapporo, **43**:1–378. *Pls.* 1–5. 1938.

STEVENSON, JOHN: Hymenomycetes Brittanici. British Fungi (Hymenomycetes). Edinburgh, William Blackwood and Sons. Vol. 1, pp. 1–372. *Figs.* 1–39. 1886 (Agaricus to Bolbitius); vol. 2, pp. 1–336. *Figs.* 40–103. 1886 (Cortinarius to Dacrymyces).

MASSEE, GEORGE: British Fungus-Flora. A Classified Text-Book of Mycology. London, George Bell and Sons. Vol. 1 pp. 1–432, *numerous* (*unnumbered*) *figs.*, 1892 (Basidiomycetes through the purple-spored Agaricaceae); vol. 2, pp. 1–460, *numerous* (*unnumbered*) *figs.*, 1893 (Agaricaceae: Ochrosporae, Rhodosporae, and part of Leucosporae); vol. 3, pp. 1–512, *numerous* (*unnumbered*) *figs.*, 1893 (Remainder of Agaricaceae-Leucosporae: Hyphomycetes); vol. 4, pp 1–522, *numerous* (*unnumbered*) *figs.*, 1895. (Ascomycetes.)

RICKEN, ADALBERT: Die Blätterpilze (Agaricaceae) Deutschlands und der angrenzenden Länder, besonders Oesterreichs und der Schweiz, xxiv + 480 pp. *Pls.* 1–112. Leipzig, Theodor Oswald Weigel, 1915.

———: Vademecum für Pilzfreunde. Zweite vermehrte und verbesserte Auflage, xxiv + 352 pp. Leipzig, Quelle und Meyer, 1920.

KARSTEN, P. A.: Rysslands, Finlands och den Skandinaviska halföns Hattsvampar, *Bidrag till Kännedom af Finlands Natur och Folk*, **32**:1–571, 1879; **37**:1–257, 1882.

MICHAEL, EDMUND: Führer für Pilzfreunde, systematisch geordnet und gänzlich neu bearbeitet von Roman Schulz, 3 Bände, 144 pp. introductory text. 386 *colored plates* (*with descriptions*). Leipzig, Quelle und Meyer, 1927.

PECK, CHARLES H.: Reports of the State Botanist of New York. (From 1871 up to about 1913. Descriptions of numerous species of fungi, chiefly Agarics, as well as monographs of many genera. Many colored illustrations.) Among the more important are the following:

New York species of Amanita, **33**:38–49. 1880.
New York species of Lepiota, **35**:150–164. 1882.
New York species of Psalliota, **36**:41–49. 1883.
New York species of Lactarius, **38**:111–133. 1884.
New York species of Pluteus, **38**:133–138. 1884.

New York species of Pleurotus, Claudopus, and Crepidotus, **39**:58–73. 1885.
New York species of Clitopilus, **42**:39–46. 1889.
New York species of Armillaria, **43**:40–45. 1890.
New York species of Tricholoma, **44**:38–64. 1891.
New York species of Pluteolus, **46**:58–61. 1893.
New York species of Galera, **46**:61–69. 1893.
New York species of Collybia, **49**:32–55. 1896.
New York species of Flammula, **50**:133–142. 1897.
New York species of Hygrophorus, **60**:47–67. 1907.
New York species of Russula, **60**:67–98. 1907.
New York species of Pholiota, **61**:141–158. 1908.
New York species of Lentinus, **62**:42–47. 1909.
New York species of Entoloma, **62**:47–58. 1909.
New York species of Inocybe, **63**:48–67. 1910.
New York species of Hebeloma, **63**:67–77. 1910.
New York species of Hypholoma, **64**:77–84. 1911.
New York species of Psathyra, **64**:84–86. 1911.
New York species of Clitocybe, **65**:59–89. 1912.
New York species of Laccaria, **65**:90–93. 1912.
New York species of Psilocybe, **65**:94–105. 1912.

———: Cantharellus, *New York State Museum Bull.* **2**:34–43. 1887.

———: Craterellus, *ibid.* **2**:44–48. 1887.

Gilman, J. C.: Illustrations of the fleshy fungi of Iowa, I. The purple-brown spored Agarics, *Proc. Iowa Acad. Sci.*, **47**:83–90. *Figs.* 1–6. 1940; II. The white-spored Agarics, *ibid.*, **48**:99–115. *Figs.* 1–14. 1941; III. The black-spored Agarics, *ibid.*, **49**:153–158. *Figs.* 1–5. 1942; V. The pink-spored Agarics, *ibid.*, **50**:159–163. *Figs.* 1–7. 1943.

Earle, F. S.: The genera of the North American gill fungi, *Bull. New York Botanical Garden*, **5**(18):373–451. 1909.

———: A key to the North American species of Hypholoma, *Torreya*, **2**(2):22–23. 1902.

———: Keys to the North American species of Coprineae, *ibid.*, **2**(3):37–40. 1902.

———: A key to the North American genera and species of the Hygrophoreae, *ibid.*, **2**(4):53–54, (5):73–74. 1902.

———: A key to the North American species of Russula, *ibid.*, **2**(7):101–103, (8):117–119. 1902.

———: A key to the North American species of Lactarius, *ibid.*, **2**(9):139–141, (10):152–154. 1902.

———: A key to the North American species of Cortinarius, *ibid.*, **2**(11):169–172, (12):180–183. 1902.

———: A key to the North American species of Stropharia, *ibid.*, **3**(2):24–25. 1903.

———: A key to the North American species of Lentinus, *ibid.*, **3**(3):35–38, (4):58–60. 1903.

———: A key to the North American species of Panus, *ibid.*, **3**(6):86–87. 1903.

———: A key to the North American species of Pluteolus, *ibid.*, **3**(8):124–125. 1903.

———: A key to the North American species of Galera, *ibid.*, **3**(9):134–136. 1903.

———: A key to the North American species of Inocybe, *ibid.*, **3**(11):168–170, (12):183–184. 1903.

Gams, H.: Schlüssel für die europäischen Familien, Gattungen und wichtigen Untergattungen der Agaricales, *Veröffentlichungen der Österreichischen Mykologischen Gesellschaft*, **2**:24 pp. 1948.

White-Spored Genera.

DONK, M.: De geslachten Cantharellus, Craterellus en Dictyolus in Nederland, *Mededeel. Nederland. Mycol. Ver.*, **16–17**:163–183. 1928.

COKER, W. C.: Craterellus, Cantharellus and related genera in North Carolina; with a key to the genera of gill fungi, *J. Elisha Mitchell Sci. Soc.*, **35**(1–2):24–28. *Pls.* 1 (*col.*)-17. 1919.

SMITH, ALEXANDER H., AND ELIZABETH E. MORSE: The genus Cantharellus in the Western United States, *Mycologia*, **39**(5):497–534. *Figs.* 1–13. 1947.

BATAILLE, FRÉDÉRIC: Flore monographique des Hygrophores, *Mém. soc. d'Émulation du Doubs*, 8me sér., **4**:131–191. 1909.

KAUFMANN, F.: Die in Westpreussen gefundenen Pilze der Gattungen Dermocybe, Myxacium, Hygrophorus und Nyctalis, *Bericht des Westpreussischen Botanisch-Zoologischen Vereins*, **34**:199–233. 1912.

GRAHAM, V. O.: The genus Hygrophorus in the Chicago region, *Trans. Illinois State Acad. Sci.*, **23**:160–168. 1932.

SMITH, ALEXANDER H., AND L. R. HESLER: Studies in North American species of Hygrophorus: I. The subgenus Limacium, *Lloydia*, **2**(1):1–62. *Pls.* 1–24. 1939; II. *ibid.*, **5**(1):1–94. *Pls.* 1–18. 1942.

COKER, W. C.: The smaller species of Pleurotus in North Carolina, *J. Elisha Mitchell Sci. Soc.*, **60**(1):71–95. *Pls.* 40–52. 1944.

PILAT, ALBERT: Pleurotus Fries, in Atlas des Champignons de l'Europe, Vol. II. 193 pp. 80 *pls.* 114 *text figs.* Prag, Kavina et Pilat, 1928.

FORSTER, EDWARD J.: Agarics of the United States. Genus Panus, *J. Mycology*, **4**(2–3):21–26. 1888.

HARPER, EDWARD T.: The species of Lentinus in the region of the Great Lakes, *Trans. Wisconsin Acad. Sci.*, **20**:365–385. *Pls.* 14–28. 1921.

VAN DER BYL, P. A.: Suid-Afrikaanse Lentinus-Soorte, *Ann. Univ. Stellenbosch*, **2A**(1):1–11. *Figs.* 1–7. 1924.

MALKOVSKY, KAREL M.: Über die europäischen Arten der Gattung Panus, *Ann. Mycol.*, **30**(1–2):10–80. *Pls.* 1–2. *Figs.* 1–26. 1932.

PILAT, ALBERT: Revision der tropischen Lentinus-Arten aus dem Herbar des Botanischen Museums in Berlin-Dahlem, *Ann. Mycol.*, **34**(1–2):108–140. 1936.

———: Monographie des espèces européennes du genre Lentinus Fr., in Atlas des Champignons de l'Europe, vol. 5, pp. 1–46. 31 *pls.* Prag, Kavina et Pilat, 1946.

SINGER, ROLF: Studien zur Systematik der Basidiomyceten: I (1). Über Panus Fr. und verwandte Gattungen, *Beihefte Botan. Centr., Abt. B*, **56**(1–2):137–347. 1936. (Gives a key to all the genera of white-spored Agaricaceae with eccentric or lateral stipe or with stipe lacking.)

KAUFFMAN, CALVIN H.: The genus Armillaria in the United States and its relationships, *Papers Mich. Acad. Sci.*, **2**:53–67. *Pls.* 5–9. 1923.

HOTSON, H. H.: The genus Armillaria in western Washington, *Mycologia*, **32**(6):776–790. *Figs.* 1–3. 1940.

KAUFFMAN, CALVIN H.: The genus Clitocybe in the United States with a critical study of all north temperate species, *Papers Mich. Acad. Sci.*, **8**:153–204. 7 *pls.* 1926.

NÜESCH, EMIL: Die Trichterlinge. Monographie der Gattung Clitocybe, 279 pp. St. Gallen, Switzerland, F. Schwald, 1926.

MURRILL, W. A.: The genus Clitocybe in North America, *Mycologia*, **7**(5):256–283. *Pls.* 164–166. 1915.

SINGER, ROLF, AND ALEXANDER H. SMITH: A monograph of the genus Leucopaxillus Boursier, *Papers Mich. Acad. Sci.*, **28**:85–132. *Pls.* 1–8. 1942 (1943).

METROD, G.: Les Tricholomes, *Rev. mycol.*, N.S., **7**(supplément 2):22–50. *Figs.* 1–86. 1942. (Key to the five genera into which Tricholoma is segregated, Rhodocybe, Rhodopaxillus, Melanoleuca, Tricholomopsis, and Tricholoma, and to their species.)

———: Les Clitocybes, *ibid.*, **14**(supplément):4–37. *Figs.* 1–18. 1949.

MURRILL, WILLIAM A.: Florida Tricholomas, *Lloydia*, **12**(1):62–69. 1949.

ATKINSON, GEORGE F.: Collybia campanella Peck, and its near relatives in the eastern United States, *New York State Museum Bull.*, **205–206**:61–65. 1918.

COKER, W. C., AND H. C. BEARDSLEE: The Collybias of North Carolina, *J. Elisha Mitchell Sci. Soc.*, **37**(1–2):83–107. *Pl.* 1 (*col.*). *Pls.* 4–23. 1921.

———, AND ———: The Laccarias and Clitocybes of North Carolina, *ibid.*, **38**(1–2):98–126. 1922.

BOURSIER, M.: Note sur le genre Mucidula Pat., *Bull. trimestr. soc. mycol. France*, **40**:332–333. 1926.

MORGAN, A. P.: North American species of Marasmius, *J. Mycology*, **11**(5):201–212, (6):233–247. 1905; **12**(1):1–9. 1906.

———: Descriptive synopsis of Morgan's North American species of Marasmius, *ibid.*, **12**(4):159–162. 1916. (Species key to the foregoing paper.)

PENNINGTON, L. H.: New York species of Marasmius, *New York State Museum Bull.*, **179**:52–79. 1915.

KÜHNER, R.: Études sur le genre Marasmius, *Le Botaniste*, **25**(1–2):57–114. *Illus.* 1933.

SINGER, ROLF: Studien zur Systematik der Basidiomyceten: I (6). Abgrenzung zwischen Collybia und Marasmius, *Beihefte Botan. Centr.*, *Abt. B*, **56**(1–2):157–163. 1936. (Contains key to the subfamily Marasmioideae and to the sections of the genus Marasmius.)

PETCH, T.: A revision of Ceylon Marasmii, *Brit. Mycol. Soc. Trans.*, **31**(1–2):19–44. *Pls.* 2–4 (*col.*). 1947.

BATAILLE, F.: Flore monographique des Marasmes d'Europe, 37 pp. Besançon, 1919

LINDER, DAVID H.: The genus Schizophyllum: I. Species of the Western Hemisphere, *Am. J. Botany*, **20**(8):552–564. *Pls.* 33–36. *Fig.* 1. 1933.

SMITH, ALEXANDER H.: North American species of Mycena, pp. i–xviii, 1–521. *Pls.* 1–99. *Figs.* 1–56. Ann Arbor, Univ. Michigan Press, 1947.

BEARDSLEE, H. C., AND W. C. COKER: The Mycenas of North Carolina, *J. Elisha Mitchell Sci. Soc.*, **49**(1–2):49–91. *Pls.* 6–30. 1924.

OORT, A. J. P.: De Nederlandsche Mycenas, *Mededeel. Nederland. Mycol. Ver.*, **16–17**:163–183. 1928.

CEJP, KAREL: Omphalia (Fr.) Quél., in Atlas des Champignons de l'Europe, Vol. 4. 157 pp. 58 *pls.* Prag. Kavina et Pilat.

ROMAGNESI, H.: Quelques points de taxonomie: I. Sur un groupe particulier d'Omphalia; II. Sur le genre Fulvidula, *Bull. trimestr. soc. mycol. France*, **58**:81–89. 1942.

SARTORY, A., ET L. MAIRE: Compendium Hymenomycetum, II. Lepiota, 512 pp. *Illustrated.* Paris, Librairie le François. 1925–1927.

KAUFFMAN, CALVIN H.: The genus Lepiota in the United States, *Papers Mich. Acad. Sci.*, **4**:311–344. *Pls.* 15–18. 1925.

MURRILL, WILLIAM A.: Florida Lepiotas, *Lloydia*, **12**(1):56–61. 1949.

MORGAN, A. P.: North American species of Lepiota, *J. Mycology*, **12**(4):154–159 (5):195–203, (6):242–248. 1906; **13**(1):1–18. 1907.

BEELI, M.: Fungi Goassensiansi: IX. Genre Lepiota, *Bull. soc. roy. botan. Belg.* 64, 2me sér., **14**(2):206–219. *Pls.* 25–27. 1912.

KALMÁR, ZOLTÁN: Nouvelle clef pour la détermination des Lepiotes de la Hongrie, *Acta Mycologica Hungarica*, **4**(1–2):15–30. 1947.

KÜHNER, R.: Recherches sur le genre Lepiota, *Bull. trimestr. soc. mycol. France*, **52**(2):177–238. *Figs.* 1–9. 1936.

HUYSMAN, H. S. C.: Observations sur le "genre" Lepiota, *Mededeel. Nederland. Mycol. Ver. Leyden*, **28**:3–54. 12 *figs.* 1943.

———: Un faux Lepiota. Le genre Coolia nov. gen., *ibid.*, **28**:54–60. 1 *fig.* 1943.

SMITH, ALEXANDER H., AND ROLF SINGER: A monograph of the genus Cystoderma, *Papers Mich. Acad. Sci.*, **30**:71–124. *Pls.* 1–5. *Figs.* 1–12. 1944 (1945).

SMITH, HELEN V.: The genus Limacella in North America, *Papers Mich. Acad. Sci.*, **30**:125–147. *Pl.* 1. 1944 (1945).

SARTORY, A., ET L. MAIRE: Compendium Hymenomycetum: I. Amanita, 447 pp. 26 *pls. Numerous text figs.* Paris, Librairie le François, 1922–1923.

ATKINSON, GEORGE F.: Six misunderstood species of Amanita, *Mem. Torrey Botan. Club*, **17**:246–252. 1918.

MARTIN, G. W.: Some Amanitas from eastern Iowa, *Proc. Iowa Acad. Sci.*, **32**:205–218. 3 *pls.* 1925.

GILBERT, JEAN-EDUARD: Le genre Amanita Persoon (Amanita s.st.—Amanitopsis R.—Limacella E.). Étude morphologique des espèces et variétés. Revision critique de la systématique, 188 pp. Lons-le-Saunier, Lucien Declume, 1918.

———: Notules sur les Amanites: Supplément: commentaires et conjectures sur quelques Amanites mal connues, 23 pp. 1 *pl.* Paris, published by the author, 1941.

———: Notules sur les Amanites: XXX. Amanites d'Europe, 3 pp. Paris, published by the author, 1941.

———: Amanitaceae, *Iconographia Mycologica*, **27**(Supplément 1):8–427. 73 *col. pls.* 1940–41. (Besides the foregoing the author has published various other "Notules" on Amanita in *Bull. trimestr. soc. mycol. France* between 1924 and 1930.)

VESELÝ, RUDOLF: Revisio critica Amanitarum europaearum, *Ann. Mycol.*, **31**(4):209–304. *Pls.* 8–25. *Figs.* 1–7. 1933.

———: Amanita, Atlas des Champignons de l'Europe, Série A, Tome 1, Fascicules 1–5, 80 pp. 40 *pls.* Prag, Charles Kavina and Albert Pilat, 1934.

HUYSMAN, H. S. C.: Opmerkingen en problemen betreffende de taxonomie van het geslacht Amanita, *Mededeel. Nederland. Mycol. Ver. Leyden*, **26**:1–27. 7 *figs.* 1942.

IMAI, SANSHI: Studies on the Agaricaceae of Japan: I. Volvate Agarics in Hokkaido, *Botanical Magazine (Tokyo)*, **47**(558):423–432. 1933. (List with no descriptions except some species described as new.) II. Lactarius in Hokkaido, *ibid.*, **49**(585):603–610. 1935.

———: Studia Agaricacearum Japonicarum, I. *ibid.*, **53**(633):392–399. 1939; II. *ibid.*, **55**(658):444–451; III. *ibid.*, **55**(659):514–520. 1941. (This is a continuation of Studies on the Agaricaceae of Japan.)

KALMÁR, ZOLTÁN: A galócagombák (Amanita) határozó táblázata. (Hungarian), *Acta Mycologica Hungarica*, **1**(3–4):101–110. 1944.

GÁLLFY, ZOLTÁN: Legyölö—Amanita muscaria (L.) Pers. (Hungarian), *Acta Mycologica Hungarica*, **3**(1–4):26–28. 1946. (Description of Amanita muscaria and six varieties.)

MURRILL, W. A.: The Amanitas of Eastern United States, *Mycologia*, **5**(2):72–86. *Pls.* 85–86. 1913.

MURRILL, W. A.: Florida Amanitas, *Lloydia*, **11**(2):99–110. 1948. (Contains a key to all known species of Venenarius (Amanita) occurring in Florida.)

SEYOT, PIERRE: Les Amanites et la tribu des Amanitées. 120 pp. 59 *figs*. Nancy, Edit. des Arts Graphiques Modernes, 1930.

BEARDSLEE, H. C.: Notes on the Amanitas of the Southern Appalachians: I. Subgenus Amanitopsis, in LLOYD, C. G.: Mycological Writings, vol. 1., 7 pp. 3 *pls*. 1902.

HOTSON, J. W.: The Amanitae of western Washington, *Mycologia*, **28**(1):63–76. *Figs*. 1–4. 1936.

BURLINGHAM, GERTRUDE S.: A study of the Lactariae of the United States, *Mem. Torrey Botan. Club*, **14**:1–109. *Figs*. 1–15. 1908.

———: New or noteworthy species of Russula and Lactaria, *Mycologia*, **28**(3):253–267. *Figs*. 1–8. 1936.

BEARDSLEE, H. C., AND GERTRUDE S. BURLINGHAM: Interesting species of Lactariae from Florida, *Mycologia*, **32**(5):575–586. *Figs*. 1–4. 1940.

MURRILL, W. A.: Species of Florida Lactarius Fr., *Lloydia*, **11**(2):86–98. 1948. (Contains a key to all of the species of the genus known to occur in Florida.)

COKER, W. C.: The Lactarias of North Carolina, *J. Elisha Mitchell Sci. Soc.*, **34**(1):1–62. *Pls*. 1 (*col*.)-40. 1918.

NÜESCH, E.: Die Milchlinge (Pilzgattung Lactarius). Bestimmungsschlüssel und Beschreibung der Milchlinge Mitteleuropas, 50 pp. St. Gallen, Switzerland, published by author, 1921.

ROMAGNESI, H.: Les lactaires. Clé pratique de détermination des espèces d'Europe. *Rev. mycol.*, N.S., **4**(supplément 1):8–21. 1939.

———: Recherches sur les Lactaires de la section des Fuliginosi Konrad, *ibid.*, **14**(2):103–112. *Figs*. 1–3. 1949.

KNAUTH, B., UND W. NEUHOFF: Die Milchlinge (Lactarius), *Die Pilze Mitteleuropas*, **2**(5b, 6b, 9b–12b):1–64(?). *Colored pls*. 1–14(?). 1 *uncolored pl*. Leipzig, Werner Klinkhart, 1937–1943. Uncompleted.

KONRAD, P.: Les Lactaires. Notes critiques et essai de classification, *Bull. trimestr. soc. mycol. France*, **41**(1):160–191. 1935.

HEIM, ROGER: Les Lactario-Russulés du Domaine Orientale de Madagascar. Essai sur la classification et la phylogénie des Astérosporales, pp. 1–196. 8 *pls*. (4 *col*.). 59 *figs*. 2 *phylogenetic diagrams*. Paris, Laboratoire de Cryptogamie du Muséum National d'Histoire Naturelle, 1937 (1938).

SINGER, ROLF: Monographie der Gattung Russula, *Hedwigia*, **66**:163–260. 1 *pl*. 1926.

———: Wie bestimmt man frische Täublinge? *Z. für Pilzkunde*, **6**(11):169–176. 1927. (Key to the common German species of Russula.)

———: Neue Mitteilungen über die Gattung Russula, *Hedwigia*, **68**(3–4):191–202. 1928.

———: Monographie der Gattung Russula, *Beihefte Botan. Centr.*, Zweite Abt., **49**:205–280. 1932.

———: Sur la classification des Russules, *Bull. trimestr. soc. mycol. France*, **41**(2):281–304. 1935.

———: Supplemente zu meiner Monographie der Gattung Russula, *Ann. Mycol.*, **33**(5–6):297–352. 1935.

———: Notes sur quelques Basidiomycètes, *Rev. mycol.*, N.S., **1**(2):75–84. *Pl*. 6. 1936. (Contains keys to the more or less green species of Russula in France and to the species of Russula associated with the birch in France and bordering countries, also key to distinguish the genera Phyllotopsis, Dochmiopsis, Rhodotus, and Octojuga.)

———: Notes sur quelques Basidiomycètes, 2 me. sér., *ibid.*, **1**(4):279–293. 1936. (Contains a key to the yellow species of Russula in Europe.)

———: New and interesting species of Basidiomycetes, II, *Papers Mich. Acad. Sci.*, **32**:103–150. *Pl.* 1. 1946 (1948).

SCHAEFFER, JULIUS: Russula-Monographie, *Ann. Mycol.*, **31**(5–6):305–516. *Pls.* 26–27. 1933; **32**(3–4):141–243. *Pls.* 1–4. 1934.

———: Bestimmungstabelle für die europäischen Täublinge, *Z. für Pilzkunde*, **12**(2):48–53, (3):83–91. 1933.

———: Le Système naturel des Russules, *Bull. trimestr. soc. mycol. France*, **41**(2):263–276. 1935.

———: Die Täublinge (Russulae), *Die Pilze Mitteleuropas*, **3**(1–3). pp. (?). 1942–1943. Uncompleted.

———, W. NEUHOFF, UND W. G. HEITER: Die Russulaceen. Bestimmungstabelle für die mitteleuropäischen Russula- und Lactarius-Arten, *Sydowia, Ann. Mycol.*, **3**(1–6):150–173. 1949.

MAIRE, RENÉ: Les bases de la classification dans le genre Russula, *Bull. soc. mycol. France*, **26**:49–125. *Figs.* 1–6. 1910.

ROMAGNESI, H.: Contribution à l'étude des Russules de la flore française, *Bull. trimestr. soc. mycol. France*, **59**:61–71. *Figs.* 1–4. 1943.

MELZER, V., ET JAR. ZVARA: České holubinky (Russulae Bohemiae), *Archiv přirod. výzk. Čech.* **17**:1–126. 25 *figs.* Prag, 1927.

———, ET ———: České holubinky (Russulae Bohemiae). Flore monographique des Russules de Bohême. Avec un tableau analytique des espèces. Résumé, *Bull. trimestr. soc. mycol. France*, **44**:135–146. 1928. (A French summary of the preceding book.)

ZVARA, JAR.: Russula atropurpurea Kromb. et ses variétés, *ibid.*, **47**:44–51. *Pls.* 1–2. 1931.

BOHUS, GÁBOR: Wie steht es mit der Russula sardonia? *Acta Mycologica Hungarica*, **1**(3–4):123–127. 1944. (Discussion and key for the distinction of the Russula sardonia-sanguinea-emetica group.)

———: A Russulák-galambgombák határozó kulcsa. *ibid.*, **3**(1–4):55–71. 1946. (In Hungarian.) (Contains key with brief descriptions of 92 species and 2 subspecies or varieties of Russula.)

CRAWSHAY, RICHARD: The spore ornamentation of the Russulas, 188 pp. 48 *pls.* London, Baillière, Tindall and Cox, 1930.

CHIU, WEI-FAN: The Russulaceae of Yunnan, *Lloydia*, **8**(1):31–59. 48 *figs.* 1945.

WINTERS, GRACE: The Iowa species of Russula, *Univ. Iowa Studies in Natural History*, **11**:5–30. 1 *pl.* 1926. (Keys to 39 species with descriptions and spore measurements.)

BATAILLE, FRÉDÉRIC: Flore monographique des Asterosporés, Lactaires et Russules, *Mém. soc. d'Émulation du Doubs*, 8me sér., **2**:163–260. 1907.

BURLINGHAM, GERTRUDE S.: Spore ornamentation of some American Russulas and a new species of Lactaria, *Mycologia*, **34**(1):8–12. *Figs.* 1–8. 1942.

KAUFFMAN, CALVIN H.: Unreported Michigan fungi for 1908 with a monograph of the Russulas of the state, *Mich. Acad. Sci. Rept.*, **11**:55–91. *Figs.* 1–3. 1909.

BEARDSLEE, H. C.: The Russulas of North Carolina, *J. Elisha Mitchell Sci. Soc.*, **33**(4):147–199. *Pls.* 70–111. 1918.

Red- or Pink-spored Genera.

MURRILL W. A.: The genus Lepista, *Mycologia*, **7**(2):105–107. 1915.

BOHUS, GÁBOR: Volvaria. Bocskorósgombák határozó kulcsa. *Acta Mycologica Hungarica*, **2**(3–4):95–96. 1945. (In Hungarian.) (Key to 10 species of Volvaria.)

ROMAGNESI, H.: Quelques observations sur les Rhodophyllus, *Bull. trimestr. Soc. mycol. France*, **48**(3–4):306–323. *Figs.* 1–11. 1932.

———: Les Rhodophylles de Madagascar, in Prodrome à une flore mycologique de Madagascar publié sous la direction de R. Heim, Tome II. 175 pp. 45 *figs.* Paris, Laboratoire de Cryptogamie du Muséum Nationale d'Histoire Naturelle, 1941.

COKER, W. C.: North Carolina species of Volvaria, *J. Elisha Mitchell Sci. Soc.*, 63(2):220–230. 5 *pls.* 1947.

Ochre- or Rust-spored Genera.

(See also above under White-spored genera for yellow-spored species of Lactarius and Russula.)

SINGER, ROLF: Contributions toward a monograph of the genus Crepidotus, *Lilloa* **13**:59–95. *Fig.* 1. 1947.

BATAILLE, FRÉDÉRIC: Flore analytique des Inocybes d'Europe, *Bull. soc. d'Histoire Naturelle Doubs* 18, 27 pp. 1910.

KAUFFMAN, CALVIN H.: Studies in the genus Inocybe, *New York State Museum Bull.* **223–224**:43–60. 1921.

SARTORY, A., ET L. MAIRE: Synopsis du genre Inocybe, 246 *pp.* 2 *pls.* Paris, E. le François, 1923.

HEIM, ROGER: Le genre Inocybe, in Encyclopédie Mycologique, vol. 1, pp. 1–429. *Pls.* 1–35. *Figs.* 1–219. Paris, Paul Lechevalier et Fils, 1931.

BOURSIER, J., ET R. KÜHNER: Notes sur le genre Inocybe, *Bull. trimestr. soc. mycol. France*, **44**(2):170–189. *Figs.* 1–9. 1928.

KÜHNER, R., ET J. BOURSIER: Notes sur le genre Inocybe: I. Les Inocybe goniosporées, *Bull. trimestr. soc. mycol. France*, **48**(2):118–161. *Figs.* 1–31. 1932.

———: Notes sur le genre Inocybe: les Inocybe goniosporées (fin), *ibid.*, **49**(1):81–121. *Figs.* 32–53. 1933.

STUNTZ, D. E.: Studies in the genus Inocybe: I. New and noteworthy species from Washington, *Mycologia*, **39**(1):21–55. *Figs.* 1–50. 1947.

KÜHNER, R.: Le genre Galera, in Encyclopédie Mycologique, vol. 7, 240 pp. 75 *figs.* Paris, Paul Lechevalier, 1935.

METROD, GEORGES: Description de Galera, *Bull. trimestr. soc. mycol. France*, **56**(1):46–55. *Figs.* 1–4. 1940.

ATKINSON, GEORGE F.: The genus Galerula in North America, *Proc. Am. Phil. Soc.*, **57**:357–374. 1918.

KAUFFMAN, CALVIN H.: The genera Flammula and Paxillus and the status of the American species, *Am. J. Botany*, **13**(1):11–32. 1926.

ROMAGNESI, H.: Sur quelques groupements naturels d'Agarics ochrosporés, *Rev. mycol.*, N.S., **1**(4):207–213. 1936. (Includes a key to the genera segregated from Flammula and Pholiota.)

———: Essai sur le genre Tubaria W. Sm., *ibid.*, **5**(1):29–43. 1940.

———: Études complémentaires sur le genre Tubaria et sur deux Naucoria tubarioides, *ibid.*, **8**(3–4):26–35. 8 *figs.* 1943.

———: Description de quelques espèces d'Agarics ochrosporés, *Bull. trimestr. soc. mycol. France*, **58**(3–4):121–149. *Figs.* 1–15. 1942. (Includes discussion of Alnicola, Naucoria, Agrocybe, Flammula, Galerina, and Conocybe.)

OVERHOLTS, L. O.: A monograph of the genus Pholiota in the United States, *Ann. Missouri Botan. Garden*, **14**(2):87–210. *Pls.* 8–24. *Figs.* 1–171. 1927.

HARPER, EDWARD T.: Species of Pholiota in the region of the Great Lakes, *Trans. Wisconsin Acad. Sci.*, **17**:470–502. *Pls.* 26–55. 1913.

———: Species of Pholiota and Stropharia in the region of the Great Lakes, *ibid.*, **17**:1011–1026. *Pls.* 69–77. 1914.

———: Additional species of Pholiota, Stropharia and Hypholoma in the region of the Great Lakes, *ibid.*, **18**:392–431. *Pls.* 11–24. 1916.

SINGER, ROLF: Studien zur Systematik der Basidiomyceten: I (8). Die Gattung Pholiota ist künstlich, *Beihefte Botan. Centr., Abt. B*, **56**(1–2):165–174. 1936. (Contains a key to the European species of Agrocybe and to the modern segregates of the Friesian genus Pholiota.)

BATAILLE, FRÉDÉRIC: Flore monographique des Cortinarius d'Europe, *Extr. from Bull. soc. d'Histoire Naturelle Doubs*, 1 vol., 162 pp. 1912.

KAUFFMAN, CALVIN H.: The genus Cortinarius with key to the species, *J. Mycology*, **13**(1):32–39. *Pls.* 93–100. 1907.

SMITH, ALEXANDER H.: Studies in the genus Cortinarius I, *Contributions from the Univ. Michigan Herbarium*, **2**:1–42. *Pls.* 1–12. Ann Arbor, Univ. Michigan Press, 1939.

———: New and unusual Cortinarii from Michigan, with a key to the North American species of subgenus Bulbopodium, *Bull. Torrey Botan. Club*, **69**(3):44–64. *Figs.* 1–7. 1942.

MURRILL, W. A.: Some Florida gill fungi, *J. Elisha Mitchell Sci. Soc.*, **55**:361–372. 1939. (Contains a key to the Alachua County species of Cortinarius.)

HENRY, R.: Étude de quelques Cortinaires du groupe des Scauri. Deux espèces nouvelles, *Bull. trimestr. soc. mycol. France*, **51**(1):34–101. *Pls.* 1–2. *Figs.* 1–10. 1935.

———: Étude de quelques Cortinaires, *ibid.*, **51**(2):205–241. 2 *figs.* 1935.

———: Révision de quelques Cortinaires, *ibid.*, **53**(1):49–80. 1937.

———: Description de quelques dermocybes du groupe "Anomaliae" Fr., *ibid.*, **53**(2):143–164. 1937.

———: Suite à l'étude du genre Hydrocybe, *ibid.*, **56**:85–119. *Figs.* 1–6. 1940.

———: Quelques Cortinaires "Hinnuloïdes" (Telamonias, Hydro-telamonias et Hydrocybes hinnuloides), *ibid.*, **57**:17–35. 1941 (1942).

———: Cortinaires nouveaux ou rares de la flore française, *ibid.*, **59**:52–60. 1943.

———: Essai d'une clé dichotomique analytique provisoire destinée à faciliter l'étude des Cortinaires du groupe des Scauri, *Rev. mycol.*, N.S., **8**(supplément à 2):1–56. 1943.

———: Essai d'une clé dichotomique provisoire destinée à faciliter l'étude des Cortinaires de groupe des Phlegmacia (Cliduchi et Elastici), *ibid.*, **10**(supplément à 2):44–82. 1945 (1947).

Purple-spored Genera.

HARPER, EDWARD T.: Species of Hypholoma in the region of the Great Lakes, *Trans. Wisconsin Acad. Sci.*, **17**:1142–1164. *Pls.* 72–84. 1914.

PARKER, CHARLES S.: A taxonomic study of the genus Hypholoma in North America, *Mycologia*, **25**(3):160–212. *Pls.* 26–31. *Figs.* 1–2. 1933.

SMITH, ALEXANDER H.: Studies of North American Agarics, I, *Contributions from the Univ. Michigan Herbarium*, **5**:1–73. *Pls.* 1–32. 1941. (Includes a discussion of Psathyrella.)

———: Studies in the genus Agaricus, *Papers Mich. Acad. Sci.*, **25**:107–138. *Pls.* 1–10. *Figs.* 1–4. 1939 (1940).

HOTSON, J. W., AND D. E. STUNTZ: The genus Agaricus in western Washington, *Mycologia*, **30**(2):204–234. *Figs.* 1–10. 1938.

MENDOZA, J. M., AND SIMEONA LEUS-PALO: A revision of the genus Psalliota in the Philippines, *Philippine J. Sci.*, **72**:337–345. 8 *pls.* 1940.

COKER, W. C.: The Chapel Hill species of the genus Psalliota, *J. Elisha Mitchell Sci. Soc.*, **43**(3–4):243–256. *Frontis. Pls.* 38–46, 48. 1928.

KALMÁR, ZOLTÁN: A csiperke (Psalliota) nemzetség europai fajai, *Acta Mycologica Hungarica*, **3**(1–4):10–22. 1946. (In Hungarian.) (Key and descriptions of 16 species of Psalliota (Agaricus).)

Black-spored Genera.

(See also Purple-spored genera.)

MASSEE, GEORGE: A revision of the genus Coprinus, *Ann. Botany*, **10**(38):123–184. *Pls.* 10–11. 1896.

KÜHNER, R., ET M. JOSSERAND: Description de quelques espèces du groupe de Coprinus plicatilis (Curt.) Fr., *Bull. trimestr. soc. mycol. France*, **50**(1):53–63. *Figs.* 1–5. 1934.

ROMAGNESI, H.: Les Coprins, *Rev. mycol.*, N.S., **6**(supplément à 1):20–35. 1 *pl.* 1941.

———: Étude de quelques Coprins (2me série), *ibid.*, **10**(5–6):73–89. 8 *figs.* 1945 (1947).

MORGAN, A. P.: North American species of Agaricaceae, Melanosporae, *J. Mycology*, **13**(2):53–62, (4):143–153, (6):246–255. 1907; **14**(1):27–32, (2):64–75. 1908.

KAUFFMAN, CALVIN H.: The genus Gomphidius in the United States, *Mycologia*, **17**(3):113–126. *Pls.* 12–14. 1925.

SINGER, ROLF: The genus Gomphidius Fries in North America, *Mycologia*, **41**(4):462–489. *Figs.* 1–3. 1949.

SARTORY, A., ET L. MAIRE: Synopsis du genre Gomphidius, 23 pp. Paris, 1922(?).

———, ET ———: Synopsis du genre Chitonia. Paris, 1923.

NÜESCH, E.: Die schwarzsporigen Blätterpilze der Kantone St. Gallen und Appenzell. Bestimmungsschlüssel, Artenbeschreibung mit kritischen Bemerkungen und Fundverzeichnis, *Jahrb. d. St. Gallischen Naturwiss. Ges.* **57**(2): 141–169. 1920–1921.

List 42. Gasteromyceteae—General Works

(See also MATTIROLO in List 17 and GRAHAM in List 34.)

COKER, W. C., AND JOHN N. COUCH: The Gasteromycetes of the Eastern United States and Canada, ix + 201 pp. 123 *pls.* Chapel Hill, Univ. North Carolina Press, 1928.

———: The Gasteromycetes of North Carolina, *J. Elisha Mitchell Sci. Soc.*, **38**(3–4):231–243. *Pls.* 71–83. 1923.

MORGAN, A. P.: North American Fungi. The Gasteromycetes, *J. Cincinnati Soc., Natural History*, **11**:141–149. *Pls.* 2–3. 1883; **12**:8–22, 163–172. *Pls.* 1–2. 1890; **14**:5–21, 141–148. *Pls.* 1, 2, 5. 1891–1892.

JOHNSON, MINNIE MAY: The Gasteromyceteae of Ohio: Puffballs, bird's-nest fungi and stinkhorns, *Ohio Biological Survey Bull.*, **4**(7):273–352. 1929.

KAMBLY, PAUL E., AND ROBERT E. LEE: The Gasteromycetes of Iowa, *Univ. Iowa Studies in Natural History*, **17**(4):121–185. *Pls.* 9–11. 1936.

HENRY, LEROY K.: A review of the Gasteromycetes (fungi) of Western Pennsylvania, *Ann. Carnegie Museum*, **30**:339–362. *Pls.* 1–4. 1947. (Article XX.)

KAUFFMAN, C. H.: Unreported Michigan fungi for 1907, with an outline of the Gasteromycetes of the state, *Mich. Acad. Sci. Rept.*, **10**:63–84. 1908. (Includes a key to the genera and species.)

MOFFAT, WILL S.: The higher fungi of the Chicago region: II. The Gasteromycetes, *Chicago Acad. Sci. Natural History Bull.*, **7**(2):1–24. *Pls.* 1–26. 1923.

GILMAN, JOSEPH C.: Illustrations of the fleshy fungi of Iowa: VII. Some common puff balls, *Proc. Iowa Acad. Sci.*, **52**:113–119. *Figs.* 1–6. 1945; VIII. The stinkhorns, *ibid.*, **53**:147–151. *Figs.* 1–5. 1946; IX. Further Gasteromycetes, *ibid.*, **54**:131–137. *Figs.* 1–7. 1947 (1948).

MASSEE, GEORGE: A monograph of the British Gasteromycetes, *Ann. Botany*, **4**(13):1–103. *Pls.* 1–4. 1891.

FRIES, THORE C. E.: Sveriges Gasteromyceter, *Arkiv för Botanik*, **17**(9):1–63. *Figs.* 1–43. 1921. (Keys to orders, families, and genera, and descriptions of species of all Gasteromycetes known to occur in Sweden. Every genus and most species illustrated.)

LLOYD, C. G.: The genera of the Gasteromycetes, *Mycological Writings*, **1** (1898–1905). (Separate pagination, pp. 1–24.) *Pls.* 1–11. Jan. 1902. (Key to all the genera (American) and good illustrations.)

CUNNINGHAM, G. H.: The Gasteromycetes of Australia and New Zealand, xv + 236 pp. *Pls.* 1–37. 1 *text fig.* Dunedin, New Zealand, published by the author, 1944.

KILLERMANN, S.: Bayerische Gasteromyceten, *Kryptogamische Forschungen. Bayerische Botanische Gesellschaft. Erforschungen der Heimischen Flora*, **7**:498–512. 2 *pls.* 1926.

VIÉGAS, A. P.: Algunos fungos do Brasil: X. Gastromicetos, *Bragantia*, **5**(9):583–595. 1945.

LANGE, MORTEN: Macromycetes. I. The Gasteromycetes of Greenland, *Meddalelser om Grønland*, **147**(4):1–32. *Illustrated.* 1948.

List 43. Hymenogastrales, Sclerodermatales, etc.

HESSE, R.: Die Hypogäen Deutschlands. Natur, und Entwicklungsgeschichte sowie Anatomie und Morphologie der in Deutschland vorkommenden Trüffeln und der diesen verwandten Organismen nebst praktischen Anleitungen bezüglich deren Gewinnung und Verwendung, vol. 1, pp. 1–149. *Pls.* 1–13. Halle a. S., Ludw. Hofstetter, 1891. (Vol. 2 contains the Tuberales.)

BATAILLE, F.: Flore analytique et descriptive des Hyménogastracées d'Europe, *Bull. soc. mycol. France*, **39**:157–196. 1923.

SOEHNER, ERT: Die Formenkreis von Hymenogaster tener Berk. et Br., *Hedwigia*, **64**:192–202. *Figs.* 1–15. 1913. (Key and description of six species and varieties of Hymenogaster of the H. tener series.)

———: Hymenogasterstudien. Das Formenkreis um Hymenogaster verrucosum Buch., *Hedwigia*, **81**(3–4):162–192. *Pl.* 4. 1943.

LLOYD, C. G.: The Hymenogastraceae. The Octaviana group. The genus Arcangeliella, *Mycological Notes*, **7**:1138–1143. *Figs.* 2152–2176. 1922.

VELENOVSKÝ, J.: Les espèces tchéques du genre Rhizopogon Fr., *Mykologia*, **8**:89–94. 1 *fig.* 1931.

Zeller, Sanford M., and Carroll W. Dodge: Rhizopogon in North America. *Ann. Missouri Botan. Garden*, **5**(1):1–36. *Pls.* 1–3. 1918.

———, and ———: Gautieria in North America, *ibid.*, **5**(2):133–142. *Pl.* 9. 1918.

———, and ———: Arcangeliella, Gymnomyces and Macowanites in North America, *ibid.*, **6**(1):49–59. *Figs.* 1–3. 1919.

———, and ———: Leucogaster and Leucophlebs in North America, *ibid.*, **11**(4):389–410. *Pl.* 11. 1924.

———, and ———: Hysterangium in North America, *ibid.*, **16**(1):83–228. *Pls.* 1–3. 1929.

———, and ———: New species of Hydnangiaceae, *ibid.*, **22**:365–373. 1935.

———, and ———: Elasmomyces, Arcangeliella, and Macowanites, *ibid.*, **23**(4):599–638. 1936.

———, and ———: Melanogaster, *ibid.*, **23**(4):639–655. 1936.

Dodge, Carroll W.: Alpova, a new genus of Rhizopogonaceae, with further notes on Leucogaster and Arcangeliella, *Ann. Missouri Botan. Garden*, **18**:457–464. *Pl.* 40. 1931.

———, and Sanford M. Zeller: Hydnangium and related genera, *ibid.*, **23**(4):565–598. 1936.

Zeller, Sanford M.: New and noteworthy Gasteromycetes, *Mycologia*, **31**(1):1–32. *Figs.* 1–54. 1939. (Includes a key to the families of the Hysterangiales.)

———: Further notes on fungi, *ibid.*, **33**(2):186–214. *Figs.* 1–17. 1941. (Mostly Hymenogastrales.)

———: North American species of Galeropsis, Gyrophragmium, Longia and Montagnea, *ibid.*, **35**(4):409–421. 1 *fig.* 1943.

List 44. Lycoperdales, including Tulostomataceae and Podaxaceae

Verwoerd, Len: Suid-Afrikaanse Lycoperdaceae en Nidulariaceae, *Ann. Univ Stellenbosch*, **3A**(1):1–45. *Figs.* 1–14. 1925.

Lloyd, C. G.: The Lycoperdons of the United States, *Mycological Writings*, **2** (1905–1908); *Mycological Notes*, **20**:221–238. *Pls.* 41–67. 1905.

Massee, George: A monograph of the genus Lycoperdon, *J. Roy. Microscop. Soc.*, **1887**(5):701–727. *Pls.* 12–13. 1887.

Peck, Charles H.: New York species of Lycoperdon, *Report of the State Botanist (New York)*, **32**:58–72. 1879.

Lohman, M. L.: The Iowa species of Lycoperdon, *Univ. Iowa Studies in Natural History*, **12**(4):5–28. 2 *pls.* 1927.

Massee, George A.: A revision of the genus Bovista, *J. Botany*, **26**(4):129–137. *Pl.* 282. 1888.

Lloyd, C. G.: Key to American species of Catastoma, *Mycological Notes*, **7**:1167–1168. 1922.

———: The genus Bovistella, *Mycological Writings*, **2** (1905–1908); *Mycological Notes*, **23**:277–287. *Pls.* 86–89. 1905.

Long, W. H., and David J. Stouffer: Studies in the Gasteromycetes: II. Bovistina, a new genus, *Mycologia*, **33**(3):270–273. 1 *fig.* 1941.

ZELLER, SANFORD M.: Representatives of the Mesophelliaceae in North America, *Mycologia*, **36**(6):627–637. *Figs.* 1–6. 1944. (Keys to the Suborder Lycoperdineae, separating the three families Lycoperdaceae, Mesophelliaceae, and Geastraceae. Also a key to the 4 genera of Mesophelliaceae, Abstoma, Radiigera, Mesophellia, and Castoreum, and a monograph of the genus Radiigera.)

POLE-EVANS, I. B., AND A. M. BOTTOMLEY: On the genera Diplocystis and Broomeia, *Trans. Roy. Soc. S. Africa*, **7**:189–192. *Pls.* 19–22. 1919.

LLOYD, C. G.: The Geasters, *Mycological Writings*, **1**(1898–1905). *Figs.* 1–80. 1902. (Separate pagination, pp. 1–44.)

MORGAN, A. P.: The North American Geasters, *Am. Naturalist*, **18**(10):963–970. *Figs.* 1–12. 1884.

DETONI, J. B.: Revisio monographica generis Geastris Mich. e tribu Gasteromycetum, *Rev. mycol.*, **9**(34):61–77, (35):125–133. *Pls.* 62–63. 1887.

DESTRÉE, CAROLINE E.: Revision des Geaster observés dans les Pays-Bas, *Nederlandsch Kruidkundig Archief*, ser. 2, deel 6, stuk. 3. 488 pp. *Illustrated.* Nijmegen, 1894.

LONGNECKER, WILLIAM M.: The Geasters of Iowa, *Univ. Iowa Studies in Natural History*, **12**(4):29–47. 2 *pls.* 1927.

COKER, W. C.: The Geasters of the United States and Canada, *J. Elisha Mitchell Sci. Soc.*, **39**(3–4):170–221. *Pls.* 18–36. 1924.

PONCE DE LÉON, PATRICIO: Contribución al estudio de los Gasteromicetos Cubanos: I. El género Geastrum en Cuba, *Rev. soc. cubana Botan.*, **3**(3):63–70. 4 *figs.* 1946.

REISNER, OTAKÉR: Les espèces du genre Geaster Mich. en Bohême, *Travaux Mycologiques Tchécoslaviques*, **1**:1–9. 1 *pl.* 1924. (Reprinted from *Mykologia*, **1**, 1924.)

WILLIS, J. H.: The Geastrae or "earth-stars" of Victoria, *Victorian Naturalist*, **51**(5):115–124. *Pl.* 24. 8 *figs.* 1934.

LONG, W. H.: Studies in the Gasteromycetes: XI. The genera Trichaster and Terrostella, *Mycologia*, **37**(5):601–608. *Figs.* 1–4. 1945.

———, AND DAVID J. STOUFFER: Studies in the Gasteromycetes: XVI. The Geastraceae of the Southwestern United States, *ibid.*, **40**(5):547–585. *Figs.* 1–52. 1948.

LLOYD, C. G.: The Tylostomeae, *Mycological Writings* **2**:(1905–1908). *Pls.* 10, 11, 20, 28, 74–85. 1906. (Separate pagination, pp. 1–28.)

WHITE, V. S.: The Tylostomaceae of North America, *Bull. Torrey Botan. Club*, **28**(8):421–444. *Pls.* 31–40. 1901.

VAN DER BYL, PAUL A.: The genus Tulostoma in South Africa, *Trans. Roy. Soc. S. Africa*, **9**(2):185–186. *Pl.* 9. 1921.

FRIES, THORE C. E.: Sveriges Tulostoma-arter, *Botan. Notiser*, **1921**(1):33–36. *Fig.* 1. 1921.

PETRI, L.: Sul valore diagnostico del capillizio nel genere "Tylostoma" Pers., *Ann. Mycol.*, **2**(5):412–438. *Pl.* 6 (*col.*). 25 *text figs.* 1904.

MASSEE, GEORGE: A monograph of the genus Calostoma, *Ann. Botany*, **2**(5):25–45. *Pl.* 3. 1888.

BURNAP, CHARLES EDWARD: Notes on the genus Calostoma, *Botan. Gaz.*, **23**(3):180–192. *Pl.* 19. 1897.

LLOYD, C. G.: The genus Mitremyces, *Mycological Writings*, **2**(1905–1908). *Mycological Notes*, **30**:238–243. *Pls.* 8, 9, 68, 69. *Fig.* 87. 1905.

LONG, W. H., AND O. A. PLUNKETT: Studies in the Gasteromycetes: I. The genus Dictyocephalos, *Mycologia*, **32**(6):696–709. *Figs.* 1–13. 1940. (Contains also a key to the genera of Family Tulostomataceae.)

LONG, W. H., AND DAVID J. STOUFFER: Studies in the Gasteromycetes: VII. The genus Schizostoma, *ibid.*, **35**(1):21–32. *Figs.* 1–8. 1943.

———, AND SULTAN AHMAD: The genus Tylostoma in India, *Farlowia,* **3**(2):225–267. *Figs.* 1–25. 1947.

———: Studies in the Gasteromycetes: XV. Notes on new or rare species of Tylostoma, *Lloydia,* **10**(2):115–135. 19 *figs.* 1947.

———: Studies in the Gasteromycetes: XVII. Two interesting species from Argentina, *Lloydia,* **11**(1):57–59. *Figs.* 1–4. 1948. (Schizostoma argentinense and Broomeia congregata.)

REA, PAUL MARSHALL: Fungi of Southern California, I, *Mycologia,* **34**(5):563–574. 3 *figs.* 1942. (A study of Battarrea with a key to the two groups of species.)

LONG, W. H.: Studies in the Gasteromycetes: VIII. Battarrea laciniata, *Mycologia,* **35**(5):546–556. *Figs.* 1–6. 1943.

List 45. Nidulariales, including Arachniaceae

WHITE, V. S.: The Nidulariaceae of North America, *Bull. Torrey Botan. Club* **29**(5):251–280. *Pls.* 14–18. 1902.

LLOYD, C. G.: The Nidulariaceae, *Mycological Writings,* **2**:(1905–1908). *Figs.* 1–20. 1906. (Separate pagination, pp. 1–32.)

KILLERMANN, S.: Die Nidularia Fr.-Gruppe, *Kryptogamische Forschungen herausgegeben von der Kryptogamenkommission der Bayerischen Botanischen Gesellschaft zur Erforschung der heimischen Flora,* **2**(2):194–198. *Pl.* 5. 1931.

LONG, W. H.: Studies in the Gasteromycetes: III. The Family Arachniaceae, *Mycologia,* **33**(4):350–355. *Figs.* 1–7. 1941. (Includes descriptions of a new genus, Araneosa, and a key to the seven species of Arachnion recognized by the author.)

List 46. Phallales

FISCHER, E.: Untersuchungen zur vergleichenden Entwicklungsgeschichte und Systematik der Phalloiden, I, *Denkschrift der Schweizerischen Naturforschenden Gesellschaft,* **32**(1):1–103. *Pls.* 1–6. 1890; II, *ibid.*, **33**(1):1–51. 3 *pls.* 5 *figs.* 1893; III, *ibid.*, **36**(2):1–84. 6 *pls.* 4 *figs.* 1900.

LLOYD, C. G.: Synopsis of the known Phalloids, *Mycological Writings,* **3**:(1909–1912). 107 *figs.* 1909. (Separate pagination, pp. 1–96.)

PETCH, T.: The Phalloideae of Ceylon, *Ann. Royal Botanical Garden Peradeniya,* **4**(4):139–184. *Pls.* 7–17. 1908.

MARTIN, G. W.: Notes on Iowa fungi. 1928, *Proc. Iowa Acad. Sci.*, **36**:127–130. 1 *pl.* 1929. (Descriptions of the Iowa species of Mutinus.)

BOEDIJN, K. B.: The Phallineae of the Netherlands East Indies, *Bull. Jardin Botanique de Buitenzorg, sér. III,* **12**:71–103. 12 *figs.* 1932.

LONG, W. H.: The Phalloideae of Texas, *J. Mycology*, **13**(3):102–114. *Pls.* 102–106. 1907.

———, AND DAVID J. STOUFFER: Studies in the Gasteromycetes: XVIII. The Phalloids of the Southwestern United States, *Lloydia*, **11**(1):60–76. *Figs.* 1–21. 1948.

KOBAYASI, Y.: Revisions of several species of Clathraceae, *J. Japanese Botany*, **13**(4):262–270. *Illustrated.* 1937. (Japanese, with Latin diagnoses of species.)

LOPES, J. PINTO: Lysurus sulcatus (Cke. et Massee) Cunn. e a ordem das Phallales em Portugal, *Bol. soc. Broteriana*, **19**(2a ser., Pt. I):125–162. 8 *pls.* 1944.

List 47. Sphaeropsidales

BENDER, HAROLD B.: The genera of Fungi Imperfecti: North American species and hosts with particular reference to Connecticut, 3 parts, 2000 pages. Unpublished thesis, Yale University, (June) 1931.

———: The Fungi Imperfecti: Order Sphaeropsidales. With keys and references for the genera, 52 pp. North Woodbury, Conn., published by the author, 1934.

GROVE, W. B.: British stem- and leaf-fungi (Coelomycetes). A contribution to our knowledge of the Fungi Imperfecti belonging to the Sphaeropsidales and Melanconiales: I. Sphaeropsidales. To the end of the Sphaerioideae which have colourless or nearly colourless spores, xx + 488 pp., 31 *figs.*, 1935; II. Remainder of Sphaeropsidales and the Melanconiales, xii + 407 pp., 102 *figs.*, 1937; Cambridge, Cambridge Univ. Press.

VON HÖHNEL, FRANZ: Zur Systematik der Sphaeropsideen und Melanconieen, *Ann. Mycol.*, **9**(3):258–265. 1911.

———: System der Fungi Imperfecti Fuckel: I. Histiomyceten; II. Synnematomyceten, *Mykologische Untersuchungen und Berichte*, **1**(3):301–369. 1923.

UNAMUNO, P. LUIS M.: Enumeración y distribución geografica de los Esferopsidales conocidos de la peninsula Ibérica y de las Islas Baleares. Familia Esferioidaceos, *Memorias de la Academia de Ciencias Exactas, Fisicas y Naturales de Madrid. Serie de Ciencias Naturales*, **4**:1–457. 1933. (A list of all species known from the area of the work with no descriptions. Also a host index.)

MARTIN, GEORGE: The Phyllostictas of North America, *J. Mycology*, **2**(2):13–20, (3):25–27. 1886.

ELLIS, J. B., AND B. M. EVERHART: The North American Phyllostictas with descriptions of the species published up to August, 1900, 79 pp. 1900.

ANDERSON, P. J.: Index to American species of Phyllosticta, *Mycologia*, **11**(2):66–79. 1919. (Additions to the species of Phyllosticta that have been recorded for North America since the publication of the preceding article together with a complete host index for the now known North American species.)

SEAVER, FRED. J.: Phyllostictales. Phyllostictaceae (pars), *North American Flora*, **6**:1–84. 1922. (The genus Phyllosticta.)

DACOSTA, G. C., AND B. B. MUNDKUR: A revision of the Genus Phyllosticta in India, *Proc. Nat. Inst. Sci. India*, **14**(2):55–63. 1948.

CIFERRI, R.: Notae mycologicae et phytopathologicae, ser. I, No. 1–11, *Ann. Mycol.*, **20**(1–2):34–53. *Pl.* 1. 1922. (Contains a conspectus of the species of Phyllosticta on Acer and on Cydonia.)

TEHON, L. R., AND E. Y. DANIELS: Notes on the parasitic fungi of Illinois: III, *Mycologia*, **19**(3):110–129. *Pl.* 11. 1927. (Contains, aside from descriptions of new species, keys to the five known species of Phyllosticta parasitic on Syringa vulgaris, to the four known species of Phyllosticta parasitic on Solidago, to the four known species parasitic on Nymphaea advena, to the three known species on Plantago, to the two species known on Sassafras, to the five species known on Chenopodium, to the four species of Septoria known on Festuca, and to the three species of Cercospora known on Setaria.)

———, AND G. L. STOUT: Notes on the parasitic fungi of Illinois: IV, *ibid.*, **21**(4):180–196. *Pl.* 13. 1929. (Contains a key to distinguish five genera of the Family Stigmateaceae, Order Hemisphaeriales; also keys to the American species of Phyllosticta on Rubus, to the three known species attacking Asparagus officinalis, to the two species of Cercospora on Sambucus, to eight species of Diplodia occurring on Acer.)

———: Notes on the parasitic fungi of Illinois: V, *ibid.*, **25**(4):237–257. *Pl.* 34. 1933. (Contains keys to the 15 species of Phyllosticta known in America on Quercus, to the 6 species of Stagonospora reported on Scirpus, and to the 10 species of Marssonina reported on Salix.)

———: Notes on the parasitic fungi of Illinois: VI, *ibid.*, **29**(4):434–446. *Figs.* 1–9. 1937. (Contains keys to 4 species of Phyllosticta on Plantago, 10 species of Macrophoma on grasses, and 5 species of Septoria on Quercus.)

GROVE, W. B.: The British species of Phomopsis, *Roy. Botan. Gardens, Kew. Bull. Misc. Inform.*, **1917**(2):49–73. *Pls.* 1–2. 1917.

———: Species placed by Saccardo in the genus Phoma, *ibid.*, **1919**(4):177–201. *Figs.* 1–23 (10):425–445. *Figs.* 1–6. 1919; **1921**(4):136–157. *Figs.* 1–8. 1921. (Discusses and gives descriptions of many species placed by Saccardo in Phoma but which must be transferred to other genera, e.g., Phomopsis, Dendrophoma, Dothiorella, Cytospora, Diplodia, Camerosporium, Rhabdospora, Gloeosporium, Colletotrichum, etc.)

———: The British species of Cytospora, *ibid.*, **1923**(1):1–30. 1923.

———: The British species of Ceuthospora and Cytosporina, *ibid.*, **1923**(10):353–359. 1923.

GUTNER, L. S.: Materialien zu einer Monographie der Gattung Cytospora, *Acta Instituti Botanici Academiae Scientiarum U.S.S.R.*, ser. II, Fasc. 2, pp. 411–484. *Figs.* 1–66. 1934. (Russian, with German summary.)

VON HÖHNEL, FRANZ: Fragmente zur Mykologie: 973. Über Myxosporella populi Jaap, *Sitzber. kaiserlichen Akad. Wiss. Wien Math. naturw. Klasse*, **125**:76–80. 1916. (Gives key to the genera of the family Sclerophomaceae.)

SYDOW, H. UND P.: Scleropycnis, ein neurer Gattungstypus unter den hyalosporen Sphaeropsideen, *Ann. Mycol.*, **9**(3):277–278. *Figs.* 1–4. 1911.

JACZEWSKI, A. L.: Monographie du genre Sphaeronema, *Nouvelle Mémoire de la Société Impériale des Naturalistes de Moscou* 15, 112 pp. *Illustrated.* 1898.

PETRAK, F.: Mykologische Notizen 225: Über einige Pleurostromella Nebenfruchtformen von Cucurbitariaceen, *Ann. Mycol.*, **21**(3–4):215–224. 1923.

DIEDICKE, H.: Die Gattung Phomopsis, *Ann. Mycol.*, **9**(1):8–35. *Pls.* 1–3. 1911. (Besides a discussion of the known species of this genus and of the structural characters, the author gives a key distinguishing Phomopsis, Plenodomus, Dothiopsis, Sclerophoma, and Sclerotiopsis.)

———: Die Gattung Plenodomus Preuss, *ibid.*, **9**(2):137–141. *Pl.* 8. 1911.

———: Dothiopsis, Sclerophoma und Sclerotiopsis, *ibid.*, **9**(3):279–285. *Pl.* 15. 1911.

———: Die Gattung Asteroma, *ibid.*, **9**(5):534–548. *Pl.* 18. 1911.

———: Myxofusicoccum, nov. gen. Sphaeropsidearum, *ibid.*, **10**(1):68–72. *Figs.* 1–5. 1912.

———: Die Abteilung Hyalodidymae der Sphaerioideen, *ibid.*, **10**(2):135–152. 1912.

DAVIS, J. J.: North American Ascochytae, *Trans. Wisconsin Acad. Sci.*, **19**(2):655–670. 1919.

SWIFT, MARJORIE E.: A new species of Chaetomella on rose, *Mycologia*, **22**(4):165–168. *Fig.* 1. 1930. (Includes a key to the described species of Chaetomella with spore measurements and hosts.)

DIEDICKE, H.: Die braunsporigen Sphaeropsideen, *Ann. Mycol.*, **11**(1):44–53. 1913. (A discussion of a few genera of this group.)

PETRAK, F.: Mycologische Notizen, V, 187. Coniothyrinula n.g., *Ann. Mycol.*, **21**(1–2):2–8. 1923. (Descriptions of genera segregated from Coniothyrium, and key.)

PETRAK, F., AND H. SYDOW: Die Gattungen der Pyrenomyzeten, Sphaeropsideen und Melanconieen: Teil I. Die phaeosporen Sphaeropsideen und die Gattung Macrophoma, *Repertorium Specierum Novarum Regni Vegetabilis Beihefte*, **42**(1–3):1–551. 1926–1927.

DIEDICKE, H.: Die Gattung Septoria, *Ann. Mycol.*, **10**(5):478–487. 1912.

MARTIN, GEORGE: Enumeration and description of the Septoriae of North America, *J. Mycology*, **3**(4):37–41, (5):49–53, (6):61–69, (7):73–82, (8):85–94. 1887.

UPPAL, B. N.: A provisional list of the species of Septoria from Iowa, *Proc. Iowa Acad. Sci.*, **32**:189–199. 1925. (List with descriptions and spore measurements, also host index.)

GARMAN, PHILIP, AND F. L. STEVENS: The genus Septoria presented in tabulation with discussion, *Trans. Illinois Acad. Sci.*, **13**:176–219. 1920.

SPRAGUE, RODERICK: The genus Phaeoseptoria on grasses in the Western Hemisphere, *Mycologia*, **35**(4):483–491. *Figs.* 1–2. 1943.

———: The status of Septoria alopecuri and some related species, *ibid.*, **35**(3):259–263. *Fig.* 1. 1943.

LINDER, DAVID H.: New species of Sphaeropsidales and Melanconiales, *Mycologia*, **35**(5):495–502. 1 *fig.* 1943.

PETCH, T.: British Nectrioideae and allied genera, *Brit. Mycol. Soc. Trans.*, **26**(1–2):53–70. 1943.

DIEDICKE, H.: Die Leptostromaceen, *Ann. Mycol.*, **11**(2):172–184. *Figs.* 1–10. 1913. (A discussion of some of the genera and species of this family and of the Pycnothyriaceae segregated from it.)

ARNAUD, G.: Les Astérinées, *Ann. école nat. agr. Montpellier*, N.S., **16**:1–288. *Pls.* 1–53. *Figs.* 1–22. 3 *maps.* 1918. (Pages 205 to 220 and *pls.* 46–50 are devoted to Asterinoid pycnidioid fungi. A key is given to the genera and these are described.)

VIÉGAS, A. P.: Algunos fungos do Brasil: XI. Fungi imperfecti, *Bragantia.* **5**(12):715–780. 1945. (Sphaeropsidales.)

List 48. Melanconiales

ELLIS, J. B., AND B. M. EVERHART: The North American species of Gloeosporium, *J. Mycology*, **1**(9):109–119. 1885.

———: North American species of Cylindrosporium, *ibid.*, **1**(10):126–128. 1885.

———: Additions to Cercospora, Gloeosporium and Cylindrosporium, *ibid.*, **3**(2):13–22. 1887.

GROVE, W. B.: The British species of Melanconium, *Roy. Botan. Gardens, Kew. Bull. Misc. Inform.*, **1918**:161–178. 1 *plate*. 1918.

KLEBAHN, H.: Beiträge zur Kenntnis der Fungi Imperfecti: III. Zur Kritik einiger Pestalozzia-Arten, *Mycolog. Centr.*, **4**(1):1–19. *Figs*. 34–38. 1914.

GUBA, E. F.: Monograph of the genus Pestalotia de Notaris, I, *Phytopathology*, **19**(3):91–232. *Pl*. 4. *Figs*. 1–7. 1929; II, *Mycologia*, **24**(4):355–397. *Figs*. 1–4. 1932. (Part II contains a key to all the species whose descriptions are distributed in Parts I and II.)

STEYAERT, R. L.: Contribution à l'étude des Pestalotia du Congo belge, *Bull. Jard. Bot. Brux.*, **19**(2): 173–186. *Pls*. 3–6. 1948.

SHEN, C. I.: Species of Pestalozzia and Monochaetia in China, *Contributions of the Biological Laboratory of the Scientific Society of China, Botanical Series*, **7**(5):131–141. 2 *figs*. 1932.

DUKE, MAUD M.: The genera Vermicularia Fr. and Colletotrichum Cda., *Brit. Mycol. Soc. Trans.*, **13**(3–4):156–184. 1 *pl*. 11 *figs*. 1928.

EDGERTON, C. W.: The Melanconiales, *Trans. Am. Microscopical Soc.*, **31**:243–265. *Figs*. 1–9. 1912.

VIÉGAS, A. P.: Algunos fungos do Brasil: XII. Fungi imperfecti—Melanconiales, *Bragantia*, **6**(1):1–37. 11 *pls*. 2 *figs*. 1946.

List 49. Moniliales: Moniliaceae

(For the asporogenous (i.e., imperfect) yeasts see List 30.)

Actinomycetes.

(This group of organisms is often considered to be intermediate between the Bacteria and the Fungi. Other students consider them to be Imperfect Fungi, and in that case they should be placed in or near to the Moniliaceae. They are accordingly so placed in this list.)

BREED, ROBERT S.; E. G. D. MURRAY; AND A. PARKER HITCHENS: Bergey's Manual of Determinative Bacteriology, ed. 6., xvi + 1529 pp. Baltimore, The Williams and Wilkins Co., 1948. (Pages 875–980 are devoted to Actinomycetales, divided into three families: Mycobacteriaceae, Actinomycetaceae, and Streptomycetaceae. Aside from the foregoing, most recent and extensive work the following papers may prove helpful.)

DRESCHLER, CHARLES: Morphology of the genus Actinomyces, *Botan. Gaz.*, **67**(1):65–83, (2):147–168. *Pls*. 2–9. 1919.

WAKSMAN, S. A., AND R. E. CURTIS: The Actinomyces of the soil, *Soil Sci.*, **1**:99–134. *Pls*. 1–3. *Fig*. 1. 1916.

———: Cultural studies of species of Actinomyces, *ibid.*, **8**:71–207. *Pls.* 1–4. 1919.

DUCHÉ, JACQUES: Les Actinomyces du groupe albus, in Encyclopédie Mycologique, vol. 6, pp. 1–375. *Pls.* 1–4. *Figs.* 1–32. Paris, Paul Lechevalier et Fils, 1934.

BREED, R. S., AND H. J. CONN: The nomenclature of the Actinomyceteae, *J. Bact.*, **4**:583–602. 1919.

ERIKSON, D.: Morphology, cytology and taxonomy of the Actinomycetes, *Ann. Rev. Microbiol.*, **3.** 1949.

JENSEN, H. L.: Contributions to our knowledge of the Actinomycetales, I, *Proc. Linnean Soc. New South Wales*, **56**:79–98. 1931; II, *ibid.*, **56**:345–370. *Pls.* 19–20. 1931 (recognizes two families: Proactinomycetaceae and Actinomycetaceae with 3 and 2 genera, respectively. Gives morphological and cultural descriptions of about 20 species of Actinomyces and 6 of Proactinomyces); III. Further observations on the genus Micromonospora, *ibid.*, **57**:173–180. *Illustrated.* 1932; IV. The identity of certain species of Mycobacterium and Proactinomyces, *ibid.*, **57**(5–6):364–376. *Illustrated.* 1932.

DE MELLO, F., ET J. F. ST. ANTONIO FERNANDES: Révision des champignons appartenants au genre Nocardia, *Mem. Asiatic Soc. Bengal*, **7**:103–138. 1919.

BALDACCI, E.: Revisione di alcune specie del genere Actinomyces, *Mycopathologia*, **1**(1):68–76. 1938.

———: Introduzione allo studio degli Attinomiceti, *ibid.*, **2**:84–106. *Pls.* 13–15. 1939.

———: Contributo alla sistematica degli Attinomiceti, IV. Sull' Actinomyces melanosporus Kr., *Atti. ist. botan. "Giovanni Briosi" e lab. crittogam. univ. Pavia*, ser. **10**:321–329. 3 *figs.* 1938.

———: Die Systematik der Actinomyceten, *Mycopathologia*, **4**(1):60–84. 1947.

Moniliaceae.

CONSTANTIN, J.: Les Mucédinées simples. Matériaux pour l'histoire des champignons, vol. 2, viii + 210 pp. *Figs.* 1–190. Paris, Librairie Paul Klincksieck, 1888. (Keys to the families and genera of the Moniliales (not to species).)

FRAGOSO, ROMUALDO GONZÁLES: Estudio sistemático de los Hifales de la flora española, *Mem. real acad. cienc. Madrid. Ser. cienc. nat.*, **6**:1–377. 85 *figs.* 1927.

LINDER, DAVID H.: A contribution towards a monograph of the genus Oidium, *Lloydia*, **5**(3):165–207. 7 *pls.* 1942.

BERKHOUT, CHRISTINE MARIE: De schimmelgeslachten Monilia, Oidium, Oospora en Torula, pp. 1–77. *Pls.* 1–4. Doctor's Thesis, University of Utrecht. Scheveningen, Edauw and Johannissen, 1923.

PINKERTON, M. ELIZABETH: A comparative study of conidial formation in Cephalosporium and some related Hyphomycetes, *Ann. Missouri Botan. Garden*, **23**(1):1–68. *Pls.* 1–6. 1936. (Keys to the human-parasitic species of Cephalosporium and to related saprophytic genera, Clonostachys, Gliocladium, etc.)

BALDACCI, E.; R. CIFERRI; E E. VACCARI: Revisione sistematica del genere Malbranchea, *Atti ist. botan. "Giovanni Briosi" e lab. crittogam. univ. Pavia*, ser. IV, **11**:75–103. 15 *figs.* 1938.

SIMOES BARBOSA, FREDERICO A.: Subsidios para o estudo parasitologico do genero Hyalopus Corda, 1938, 62 pp. 6 *pls.* Recife, Imprensa Industrial, 1941.

THOM, CHARLES: The Penicillia, xiii + 644 pp. 98 *figs.* Baltimore, Williams and Wilkins Co., 1930.

RAPER, KENNETH B., AND CHARLES THOM: A manual of the Penicillia, i–ix, 1–875 pp. 10 *Colored plates.* 172 *Figs.* Baltimore, Williams & Wilkins Co. 1949.

———, AND DOROTHY I. FENNELL: New species of Penicillium, *Mycologia,* **40**(5):507–546. *Figs.* 1–11. 1948.

BIOURGE, P.: Les moisissures du groupe Penicillium Link, *La Cellule,* **33**:1–331. *Col. pls.* 1–13. *Pls.* 1–23. 1923. (An attempt at a monograph of this difficult genus.)

SOPP, OLAV JOHAN-OLSEN: Monographie der Pilzgruppe Penicillium. Mit besonderer Berücksichtigung der in Norwegen gefundenen Arten, *Videnskapselskapets-Skrifter I. Mat.-naturv. Klasse,* **1912**(11):1–208. *Pls.* 1–23. *Fig.* 1. 1912.

WESTLING, R.: Über die grünen Spezies der Gattung Penicillium. Versuch einer Monographie, *Arkiv för Botanik,* **11**(1):1–156. *Figs.* 1–81. 1911.

WÖLTJE, WILHELM: Unterscheidung einiger Penicillium-species nach physiologischen Merkmalen, *Centr. Bakt. Parasitenk.,* Zweite Abt., **48**:97–130. 1918.

SHIH, YOU-KUANG: The Penicillium from Wuchang, Central China, *Trans. Sapporo Natural History Soc.,* **14**(4):286–296. *Pl.* 12. 1936.

NIETHAMMER, ANNELIESE: Zur Systematik der Pilzgruppe Penicillium Link: I. Mitteilung: Die Untersektion Radiata (in der Sektion Velutina der Asymmetrica), *Arch. Mikrobiol.,* **14**(1):46–62. 6 *figs.* 1948.

PETCH, T.: Gliocladium, *Brit. Mycol. Soc. Trans.,* **22**:257–263. *Figs.* 1–2. 1938–39.

THOM, CHARLES, AND KENNETH B. RAPER: A Manual of the Aspergilli, ix + 373 pp. 7 *col. pls.* 76 *figs.* Baltimore, Williams and Wilkins Co., 1945.

BLOCHWITZ, ADALBERT: Die Gattung Aspergillus: I. Neue Spezies, Diagnosen, Synonyme, *Ann. Mycol.,* **27**:205–240. *Pl.* 3. 1929. (Includes a key to the recognized species of the genus.)

———: Die Gattung Aspergillus: II. Neue Spezies, Synonyme und Nachträge, *ibid.,* **31**(1–2):73–83. 1933.

———: Die Gattung Aspergillus: III. Neue Spezies, Varianten und Mutanten der Konidienfarbe, Synonyme und interessante Standorte, *ibid.,* **32**(1–2):83–89. 1934.

———: Die Gattung Aspergillus: IV. Neue Arten, Synonyme, Varianten und Mutationen, *ibid.,* **33**(3–4):238–250. 1935.

NEILL, J. C.: The mould fungi of New Zealand: I. The genus Penicillium, *Trans. Proc. Roy. Soc. New Zealand,* **67**:101–112. *Pls.* 22–24. 1937; II. The genus Aspergillus, *ibid.,* **69**:237–264. 1939. (Keys and descriptions and illustrations of all species of Penicillium recognized in New Zealand and keys and descriptions of all 18 species of Aspergillus recognized as valid, of which 12 are known in New Zealand. Sixteen doubtful species are also described.)

SHIH, Y. K.: A taxonomic study of the genus Aspergillus around Wuchang, Central China (Hyphomycetes), *Lingnan Sci. J.,* **15**(3):365–378. *Pl.* 16. (4):607–612. 1936.

MOSSERAY, RAOUL: Les Aspergillus de la section "Niger" Thom and Church, *La Cellule,* **43**(2):203–286. 4 *pls.* 1934.

ELLIS, J. B., AND B. M. EVERHART: North American species of Ramularia with descriptions of the species, *J. Mycology,* **1**(6):73–83. 1885.

———, AND ———: Supplementary notes on Ramularia, *ibid.,* **1**(8):102. 1885.

———, AND ———: Additions to Ramularia and Cercospora, *ibid.,* **4**(1):1–7. 1888.

SUMSTINE, DAVID ROSS: Studies in North American Hyphomycetes: I. The genera Rhinotrichum and Olpitrichum, *Mycologia,* **3**(2):45–56. *Pls.* 37–39. 1911; II. The tribe Oosporeae, *ibid.,* 5(2):45–61. *Pls.* 82–84. 1913.

HANSFORD, C. G.: The genus Eriomycopsis, *Bothalia,* **4**(2):464–472. 17 *figs.* 1942. ("Hyphomycetea Mucedinea, Macronemea.")

WOLLENWEBER, H. W.: Ramularia, Mycosphaerella, Nectria, Calonectria. Eine morphologisch-pathologische Studie zur Abgrenzung von Pilzgruppen mit cylindrischen und sichelförmigen Konidienformen, *Phytopathology*, **3**(4):197–242. *Pls*. 20–22. 1913.

List 50. Moniliales: Dematiaceae

LINDER, DAVID H.: New Venezuela Fungi Imperfecti, *Mycologia*, **29**(6):656–664. 6 *figs*. 1937. (Includes a key for the differentiation of all the 11 known species of Periconia that possess rough globose spores.)

MASON, E. W.: On species of the genus Nigrospora Zimmermann recorded on Monocotyledons, *Brit. Mycol. Soc. Trans.*, **12**:152–165. *Pl*. 15. 1927.

———: Annotated account of fungi received at the Imperial Mycological Institute, List II, Fascicle 3(special part):101–144. *Figs*. 31–44. 1941. (A careful study of the genera Monotospora Sacc. and Monotospora Corda, Torula Pers., Gliomastix Guegen, Catenularia Grove, Sporocybe Fr., sensu Bonorden, Zygosporium.)

GOIDANICH, GABRIELE: Il genere di Ascomiceti Grosmannia G. Goid., *Boll. staz. patol. vegetale*, N.S., **16**(1):26–60. *Pl*. 1. *Figs*. 1–19. 1936. (Description of the genus Scopularia and its species connected with the ascomycetous genus Grosmannia.)

VAN BEYMA THOE KINGMA, F. H.: Beschreibung der im Centraalbureau voor Schimmelcultures vorhandenen Arten der Gattungen Phialophora Thaxter und Margarinomyces Laxa, nebst Schüssel zu ihrer Bestimmung, *Antonie van Leeuwenhoek J. Microbiol. Serol.*, **9**(1–2):51–76. *Figs*. 1–15. 1943.

BALDACCI, ELIO: Un nuovo genere di micete parassito del pioppa Pollaccia radiosa (Lib.) Baldacci e Ciferri. Revisione dei g. Stigmella e Stigmina, *Atti ist. botan. "Giovanni Briosi" e lab. crittogam. univ. Pavia*, ser. IV, **10**:55–72. 5 *figs*. 1938.

NEERGAARD, PAUL: Danish species of Alternaria and Stemphylium. Taxonomy. Parasitism. Economical significance, Communications from the Phytopathological Laboratory of J. E. Ohlsens Enke, Copenhagen, pp. 1–559. *Figs*. 1–157. Copenhagen, Einar Munksgaard, Publisher, 1945.

GROVES, J. W., AND A. J. SKOLKO: Notes on seed-borne fungi: I. Stemphylium, *Can. J. Research, C*, **22**:190–199. *Illustrated*. 1944; II. Alternaria, *ibid.*, **22**:217–234. *Illustrated*. 1944; III. Curvularia, *ibid.*, **23**:94–104. *Pls*. 1–7. 1945.

ELLIOTT, JOHN A.: Taxonomic characters of the genera Alternaria and Macrosporium, *Am. J. Botany*, **4**(8):439–476. *Pls*. 19–20. 9 *graphs*. 1917.

YOUNG, P. A.: Tabulation of Alternaria and Macrosporium, *Mycologia*, **21**(3):155–166. 1929.

DRECHSLER, CHARLES: Some graminicolous species of Helminthosporium, *J. Agr. Research*, **24**:641–740. *Pls*. 1–33. 1923.

STEVENS, F. L.: Some meliolicolous parasites from Porto Rico, *Botan. Gaz.*, **65**(3):227–249. *Pls*. 5–6. *Figs*. 1–5. 1918. (Contains keys to the Porto Rican species of Arthrobotryum and Helminthosporium that occur on Meliola.)

HUGHES, S. J.: Studies on microfungi. II. The genus Sporoschisma Berkeley & Broome and a redescription of Helminthosporium rousselianum Montagne,

Commonwealth Mycological Institute Mycological Papers, **31**:1–33. *Pl.* 1. *Figs.* 1–9. 1949.

Jacques, J. Émile: Studies in the genus Heterosporium, *Contribs. inst. botan. univ. Montreal No.* **39**:1–46. *Pls.* 1–6. 1941.

Linder, David H.: A monograph of the Helicosporous Fungi Imperfecti, *Ann. Missouri Botan. Garden,* **16**(3):227–388. *Pls.* 12–31. *Figs.* 1–17. 1929.

———: The genus Helicoceras, *ibid.,* **18**(1):1–7. *Pl.* 1. 1931.

———: Brief notes on the Helicosporeae with descriptions of four new species, *ibid.,* **18**(1):9–16. *Pl.* 2. 1931.

Ellis, J. B., and B. M. Everhart: Enumeration of the North American Cercosporae, *J. Mycology,* **1**(1):17–24, (2):33–40, (4):49–56, (5):61–65. 1885.

———, and ———: Additions to Cercospora, Gloeosporium and Cylindrosporium, *ibid.,* **3**(2):13–22. 1887.

———, and ———: Additions to Ramularia and Cercospora, *ibid.,* **4**(1):1–7. 1888.

Overholts, L. A.: Species of Cercospora on Smilax in the United States, *Ann. Missouri Botan. Garden,* **14**(4):425–432. *Pls.* 40–41. 1927.

Lieneman, Catherine: A host index to the North American species of the genus Cercospora, *Ann. Missouri Botan. Garden,* **16**(1):1–52. 1929.

Solheim, William G.: Morphological studies of the genus Cercospora, *Illinois Biological Monographs,* **12**(1):1–84. *Pls.* 1–4. 1929. (Divides the genus into 21 sections listing the species under each and describing in detail the more important species.)

———, and F. L. Stevens: Cercospora Studies: II. Some tropical Cercosporae, *Mycologia,* **23**(5):363–405. *Figs.* 1–12. 1931.

Davis, R. H.: The Cercospora leaf spot of rose caused by Mycosphaerella rosicola, *ibid.,* **30**(3):282–298. *Figs.* 1–7. 1938. (Contains a key to the four species of Cercospora known to grow on Rosa.)

Săvulescu, Trian, et T. Rayss: Les Cercospora parasites des feuilles de vigne en Palestine, *Rev. path. végétale entomol. agr. France,* **22**(3):1–19. *Pls.* 1–6. 1935.

Ray, W. Winfield: Notes on Oklahoma Cercosporae, I, *Mycologia,* **33**(2):174–177. 1941; II, *ibid.,* **34**(5):555–562. 1942.

Müller, Albert S., y Charles Chupp: Las Cercospora de Venezuela, *Bol. soc. venezolana cienc. nat.,* **8**(52):33–59. 1942. (List and host index of 176 species of Cercospora including descriptions of 29 species new to science.)

———, y ———: Cercosporae de Minas Geraes, *Arquiv. inst. biol. vegetal,* **1**(3):213–220. 1935.

———, y ———: Una segunda contribuiçao á 's Cercosporae de Minas Geraes, *ibid.,* **3**(1):91–98. 1936.

Chupp, Charles: Cercosporae. Reprinted from C. E. Chardon and R. A. Toro: Mycological Exploration of Venezuela, pp. 241–255. *Monographs of the Univ. Puerto Rico, Series B, No.* 2. 1934.

———: Cercospora species and their host genera, *Dept. Plant Pathol., New York (Cornell) Agr. Expt. Sta. Mimeographed Circ.,* 23 pp. 1937.

———, and David H. Linder: Notes on Chinese Cercosporae, *Mycologia,* **29**(1):26–33. 1 *fig.* 1937.

Tai, T. L.: Cercosporae of China, I, *Bull. Chinese Botan. Soc.,* **2**:45–66. 5 *pls.* 1936; II, *Lloydia,* **11**(1):36–56. *Figs.* 1–15. 1948.

Viégas, A. P.: Alguns fungos do Brasil: Cercospora, *Bol. soc. brasil. agron. (Rio de Janeiro),* **8**(1):1–160. 86 *pls.* 18 *figs.* 1945.

Yamamoto, Wataro: Cercospora from Formosa, I, *Trans. Sapporo Natural History Soc.,* **13**:139–143. *Figs.* 1–3. 1934.

———: Cercospora-Arten aus Taiwan (Formosa), II, *J. Tropical Agr.*, **6**:599–608. *Figs.* 1–4. 1934.

List 51. Moniliales: Tuberculariaceae, Stilbellaceae and Mycelia Sterilia

Fusarium.

APPEL, O., UND H. W. WOLLENWEBER: Grundlagen einer Monographie der Gattung Fusarium (Link), *Arbeiten aus der Kaiserlichen Anstalt für Land- und Forstwirtschaft*, **8**:1–207. *Pls.* 1–3. *Figs.* 1–10. 1910. (This work is the basis for most of the subsequent systematic work on this genus.)

WOLLENWEBER, H. W.: Fusaria autographice delineata, *Ann. Mycol.*, **15**(1–2):1–56. 1917. Supplementary to this, 659 separate sheets of drawings.

———: Fusarium-Monographie. Fungi parasitici et saprophytici, I, *Z. Parasitenk.*, **3**(3):269–516. *Figs.* 1–71. 1931; II, *Centr. Bakt. Parasitenk.* Abt. II, **106**(5–7):104–135, (8–10):171–202. 71 *figs.* 1943.

———, UND O. A. REINKING: Die Fusarien, ihre Beschreibung, Schadwirkung und Bekämpfung, 355 pp. 95 *figs.* Berlin, Paul Parey, 1935.

SHERBAKOFF, C. D.: Fusaria of potatoes, *Cornell Univ. Agr. Expt. Sta. Mem.*, **6**:89–270. *Col. pls.* 1–7. *Figs.* 1–51. 1915.

REINKING, O. A., AND H. W. WOLLENWEBER: Tropical Fusaria, *Philippine J. Sci.*, **32**(2):103–252. 6 *pls.* (2 *colored*). 47 *figs.* 1927.

DOIDGE, ETHEL M.: Some South African Fusaria, *Bothalia*, **3**(3):331–483. *Pls.* 1–4. *Figs.* 1–48. 1938.

BUGNICOURT, FRANCIS: Les Fusarium et Cylindrocarpon de l'Indochine, Thèse de doctorat ès sciences, Paris. 208 pp. 10 *pls.* 36 *figs.* Paris, Paul Lechevalier, 1939.

JAMALAINEN, E. A.: Ueber die Fusarien Finnlands, I, *Staatl. Landwirtsch. Versuchstätigk. Veröffentlich.*, 122, 22 pp. 5 *figs.* Helsinki, 1943.

SNYDER, W. C.: Notes on Fusaria of the section Martiella, *Zentr. Bakt. Parasitenk.*, Abt. II, **91**(8–10):163–184. *Illustrated.* 1934.

———, AND H. N. HANSEN: The species concept in Fusarium with reference to section Martiella, *Am. J. Botany*, **28**(9):738–742. 1941.

———, AND ———: The species concept in Fusarium with reference to Discolor and other sections, *ibid.*, **32**(10):657–666. 1945.

MILLER, J. J.: Cultural and taxonomic studies on certain Fusaria: II. The taxonomic problem in Fusarium with particular reference to section Elegans, *Can. J. Research, C*, **24**:213–223. 1946. (Keys to 13 species based on the features expressed on potato dextrose agar.)

Other Tuberculariaceae.

HUGHES, S. J.: Studies in microfungi: I. The genus Fusariella Saccardo, *Mycological Papers. The Commonwealth Mycological Institute*, **28**:1–11. *Figs.* 1–5. 1949.

PRESTON, N. C.: Observations on the genus Myrothecium Tode: I. The three classic species, *Brit. Mycol. Soc. Trans.*, **26**(3–4):158–168. 2 *pls.* *Text figs.* 1943.

STEVENS, F. L., AND NORA DALBY: New or noteworthy Porto Rican fungi, *Mycologia*, **11**(1):4–9. *Pls.* 2–3. 1919. (Gives a generic key to the Tuberculariaceae—Scolecosporae.)

Stilbellaceae.

VAN ZINDEREN-BAKKER, E. M.: Stephanoma tetracoccum, *Ann. Mycol.*, **32**(1–2):101–104. 1 *fig.* 1934. (Descriptions of all known species of the genus.)

PETCH, T.: The genus Endocalyx Berkeley and Broome, *Ann. Botany*, **22**(87):389–400. *Pl.* 24. 1908.

———: Gibellula, *Ann. Mycol.*, **30**:386–393. *Fig.* 1. 1932.

Mycelia Sterilia.

HOTSON, J. W.: Notes on bulbiferous fungi with a key to the described species, *Botan. Gaz.*, **64**(4):265–284. *Pls.* 21–23. *Figs.* 1–6. 1917. (Chiefly Papulospora.)

SHAW, F. J. F., AND S. L. AJREKAR: The genus Rhizoctonia in India, *Mem. Dept. Agr. India. Botanical Series*, **7**:177–194. *Pls.* 1–6. 1915.

MATZ, J.: The Rhizoctonias of Porto Rico; *J. Dept. Agr. Porto Rico*, **5**:1–31. *Pls.* 1–28. 1921.

INDEX

Boldface folios indicate main discussion of topic.

A

Abe, 24
Abies, 395, 403, 405, 494
Absidia, 165
  coerulea, 160
  glauca, 160, 163, 164
  septata, 164
  spinosa, 160
Abstoma, 553
Acaulopage, 179
Acer, 235, 589
  rubrum, 235
Acervulus, 231, 263, 333, 334
Achlya, 106, 107, 108, 109, 110, **112**, 115, 130, 142
  flagellata, 110
  glomerata, 34
  oblongata, var. *globosa*, 104
  polyandra, 104
  racemosa, 111
Achlyogeton, 95
Achlyogetonaceae, 50
Acia, 482
Acrasiales, 23, **29**, 30, 35, 36, 631
  key to genera, 38
Acrasis, 30
Actinodothis, 319, 354
Actinomyces, 179, **585**, 603
  bovis, 585
Actinomycetaceae, 585
Actinomycetales, 585
Actinonema, 231
Adams, 387
Aecial primordium, 386, 390
Aecidiospore, 381
Aecidium, **408**, 572
  berberidis, 18
Aeciospore, 381, 388
  origin of chains, 392
Aecium, 399
  caeomoid, 400, 407, 408
  cornute, 399, 407
Aecium—(*Continued*)
  cupulate, 399, 400, 407
  diffuse, 399, 408
  hyphoid, 399, 400, 407
  secondary, 399
  types, 400
  uredioid, 407
Aethalium, 28
Agaricaceae, 9, 10, 11, 12, 193, 372, 467, 469, 470, 480, 484, 494, 495, 496, 497, **498**, 500, 501, 502, 503, 504, 505, 506, 507, 509, 510, 511, 532, 534, 540, 541, 543
  keys to genera, 520
Agaricacées, 470
Agaricales, 335, 464, 469, 471, 473, **495**, 498, 534
  origin, 652
Agaricum, 11
  fungus, 10
Agaricus, 10, 11, 502, 504, 507, **509**, 540, 542
  arvensis, 196, 498, 502, **509**
  campestris, 196, 369, 374, 498, 501, 502, 506, **509**, 589
  rodmani, 502
Ajello, 63
Albuginaceae, 8, 127, **135**, 136, 138, 139
  key to genera, 145
Albugo, **135**, 139, 142
  bliti, 136, 137
  candida, **135**, 137, 139
  ipomoeae-panduranae 138
  portulacae, 136, 137, 138
  tragopogonis, 137
Aleuria umbrina, 586
Aleuriospore, definition of, 575
Aleurodiscus, 475
  amorphus, 474
Algae, blue-green, 217
  green, 217
Alina, 317
Alisma, 65

Alismataceae, 421
Alkaloids, 505
Allantosphaeriaceae, 279, 281, **282**, 284
Allen, Miss, 14, 166, 312, 385, 388, 389, 390, 391, 397
Allomyces, 4, 42, 78, 79, 81, 82, **84**, 86, 91, 107
arbuscula, 85
cystogenus, 81, **86**
javanicus, 85
Alnus, 243
Alpova, **539**, 547
Alternaria, 194, **596**
tenuis, 597
Alternation of generations, 79, 85, 86
Amanita, 10, 502, 503, **508**
caesarea, 9, **502**
muscaria, **503**, 509
phalloides, 503
rubescens, 502
verna, 503
Amaranthus, 137
Amazonia, 309, 319, 354
Ames, L. M., 264, 268, 269, 277, 329, 375, 391
Ames, Miss, 14, 488
Ammiaceae, 143
Amoeba, 24, 178, 179, 182, 591, 593
Amoebidiaceae, 182
Amoebidium, 181, **182**
Amorphomyces falagriae, 214
Amphiloma, 218
Amphisphaeriaceae, 278
Amygdalus persica, 243, 582
Ancylistes, 100, **175**
Ancylistidaceae, 100, 175
Andrus, 389, 390
Angiocarpy, 496
Angiospermae, 2, 126, 127
Anisogametes, 6
Anisogamy, 143
Anisolpidiaceae, 70, **71**
Anisolpidium, 71
ectocarpii, 70, 71
Anisomyxa, 35
Annulus, 505, 509
simple, 509
Anteriorly uniflagellate fungi, key to order, 43
Antherid, 6, 106
amphigynous, 132, 133
diclinous, 129
hypogynous, 129
monoclinous, 129
paragynous, 132
Anthoceros, 419
Anthophyta, 2, 126, 127, 309, 382
Anthostoma, **282**, 284
Anthurus borealis, 544
Antibiotics, 9, 15, 602
Aphanodictyon, 115
papillatum, 115
Aphanomyces, 108, 110, **112**, 114, 115
acinetophagus, 113
exoparasiticus, 113
phycophilus, 113
Aphanomycopsis, 102
Aphragmium, 130
Aphyllophoracées, 470
Aphyllophorales, 464, 469, **471**
Apinis, 115, 116
Apium graveolens, 598, 600
Aplanes, 110, 634
Aplanogametes, 6
Aplanospore, 5, 151
Apobasidium, 454
Apodachlya, **116**, 119
brachynema, 107
Apodachyella, 116, **117**
Apogamous development, 213
Apophysis, 57
Apostemidium, 230
Apothecium, 193, 196, 202, 210, 220, 262
angiocarpic, 204
gymnocarpic, 204
Appendage, 314, 315
Apple, bitter rot, 280
scab, 8
Appressoria, 310, 317
Arachniotus aureus, 324, **330**
trachyspermus, 329
trisporus, 329, 330
Araiospora, 106, 117, **119**
pulchra, 118
Arcangeliella, 532, **540**
Archicarp, 205, 215
Archilegnia, 116
Arcyria, 28
Armillariella mellea, 4, **501**, 502, 508, 602
Arnaud, 271, 308, 309, 314, 319, 320, 321, 322, 357
Aronescu, 269
Arrhenatherum elatius, 416
Arrhytidia, 451
Arthrobotrys, 589
conoides, 592
Arthrobotryum, 317
Arthronia, 220
Arthropoda, 3, 180
Arthur, 381, 396, 399, 402, 407, 425
Arundinella, 182
Arundinula capitata, **181**, 182
Arundinulaceae, 182
Aschersonia, 580
Ascobolus, 226, **227**
carbonarius, 224, **225**, **226**

Ascobolus—(*Continued*)
furfuraceus, 226
immersus, 227
magnificus, 225, 227
stercorarius, 204
Ascocarp, 236, 237
Ascochyta, 579
dianthi, 578
pisi, 298, **580**
Ascocorticiaceae, 241, **243**, 639
Ascocorticium, **243**, 458, 510, 647
Ascogenous hyphae, 208, 209, 210, 211, 220, 222, 313
Ascogonium, 203, 210, 220, 264
Ascohymeniales, 205, 271
Ascoidea, 336, 338, **340**, 352, 353
rubescens, 336, **340**
Ascoideaceae, **338**, 339, 340
Ascoloculares, 205, 271
Ascomycetales, 278
Ascomyceteae, 3, 6, 12, 14, 143, 172, 193, 200, 366, 368, 371, 380, 424, 455, 456, 458, 466, 510, 572, 576
Floridean ancestry, 636, 637
origin, 635
Ascophanus, 227
carneus, 195
granulatus, 211
granuliformis, 204
Ascoscleroderma, 330
Ascospore, discharge, 226
hat-shaped, 336, 337, 340, 341
Ascosporogenesis, 342
Ascus, 145
bilabiate, 225
dehiscence, 202
bilabiate, 202
inoperculate, 202
operculate, 202
development, 367
inoperculate, 219, 227
operculate, 227
Aseroe, 544
Ashby and Nowell, 350
Ashbya, **345**, 351
gossypii, 345
Ashworth, Miss, 397
Asparagus officinalis, 395
Aspergillaceae, 194, 323, 324, **325**, 326, 329
key to genera, 356
Aspergillales, 239, 272, 274, 275, 277, 292, 299, **322**, 331, 334, 586, 587
key to families, 355
Aspergillus, 11, 168, 323, **325**, 326, 336, 573, **586**, 587, 589, 602, 603
glaucus, 328
niger, 587
Aspergillus—(*Continued*)
niveo-glaucus, 587
versicolor, 587
Aspidotus, 446
Aster, 405
Asteraceae, 140, 143, 421
Asterina, **294**, 320
camelliae, 294
Asterodon, 482
Asterophlyctis, 55
Asterosporales, 473
Asterostroma, 475
Asterostromella, 475
Astomella, 311, **314**
Astraeus, 556
hygrometricus, **554**, 555
Astrothelium, 285
Atanasoff, 202
Atichia, 322
millardeti, 321
Atichiaceae, 308, 309, 321, **322**, 334
key to genera, 355
Atkinson, 170, 212, 334, 452, 486, 498, 506, 507, 540, 639, 641
Auricularia, 10, 11, 438, 439, **443**, 453, 510
auricula-judae, 443
auricularis, **443**, 444
Auriculariaceae, 9, **440**, 442, 444
Auriculariales, 197, 379, 381, 384, 436, 437, **438**, **444**, 446, 453, 455, 456, 556, 584
key to families and genera, 458
Auriscalpium, 482
Avena fatua, 416
sativa, 395, 409, 416, 418
Avocado, anthracnose, 280
Ayers, 151
Azygospore, 170, 174, 175, 176, 177

B

Bachmann, Miss, 220, **221**, 224
Backus, 231
Bacteria, 2, 585
Baker, Miss, 439, 442
Baker, Mrak and Smith, 145
Ballistospore, 348, 349
Baranetzky, 329
Barnett, H. L., 378, 439, 453, 542
Barrage sexuel, 160, 377
Barrett, 98, 99
Bartlett, 66
de Bary, 1, 13, 22, 23, 133, 139, 237, 244, 311, 328, 334, 394, 633
Basidial nest, 539
primordium, 440
Basidiobolaceae, 175
Basidiobolus, 172, **174**, 175, 178, 635, 636
ranarum, 174, 176

Basidiomyceteae, 3, 6, 14, 160, 193, 195, 212, 237, 242, 313, 349, 350, 366, 436, 572, 602
Basidiophora, 140
  entospora, 138
Basidiospore, 350, 366, 381
  color, 468
  discharge, 369
  shape, 468
  structure, 532
Basidium, 366, 381
  chiastic, 470, 479, 495, 531
  development, 367
  shape, 465
  stichic, 469, 479, 531
  tuning fork type, 458
Basisporium gallarum, 594
Battarrea, 532, 534, **557**
  digueti, 557
  phalloides, 557
Bauch, 14, 378, 413, 414, 416, 417, 419
Bauhin, 9, 10
Baur, 220, 221
Baxter, 488
Bdellospora, 179
Bell morel, 228
Bender, 572, 576, 580, 581, 583, 584
Benjamin, 168
Bennett and Murray, 379
Bensaude, Mlle., 14, 372, 373, 375
Berberis vulgaris, 18, 385, 388, 394, 395
Berdan, Miss, 64, 100
Bergman, 312
Berlese, 136, 138, 277, 279, 311, 346
Bernard, 543
Bertrandia, 501
Bessey, C. E., 196, 197, 380, 551
Bessey, E. A., 284
Beta vulgaris, 577, 598
Betts, 226, 352
Betula, 491
Biflagellatae, 35, 95, 630
Biflagellate fungi, key to orders, 43
  zygotes, 52
Biggs, Miss, 201, 339
Biologic forms, 8
Bipolarity, 377, 501
Bird's nest fungi, 548
Bisby, 15
Bishop, 106
Bjerkandera adusta, 488
Black scurf, 5
Blackman, 286, 387, 391
Blackwell, Miss, 84, 131
Blackwell, Waterhouse and Thompson, 135
Blakeslea, 155, **167**
  trispora, 158, 160
Blakeslee, 14, 157, 160
Blastocladia, 78, 79, 82, **84**
  globosa, 84
  pringsheimii, 83, **84**
Blastocladiaceae, 79, **82**, 83, 85
Blastocladiales, 49, 64, 78, **79**, 81, 86, 119, 143, 633, 634
  key to families and genera, 91
  sexual reproduction, 79
Blastocladiella, 78, 79, **82**, 91
  asperosperma, 83
  cystogena, 81, 83, **84**, 86
  simplex, 82
  stomophila, 82
  stübenii, **82**, 84
  variabilis, 82
Blastodendrion, 347
Blastomyces, 346
Blepharoplasts, 24
Blizzard, 506, 507
Blumer, 316
Boedijn, 236
Boedijn and Steinmann, 439, 442
Boletaceae, 9, 10, 11, 196, 464, 467, 469, 470, 473, 484, **495**, 496, 497, 498, 507, 509, 511, 532
  key to subfamilies and genera, 518
Boletales, 473
Boletineae, 496
Boletinus porosus, 497
Boletopsis, 489
Boletus, 10, 11, 12
  colossus, 495
  edulis, 495, **497**
  felleus, 497
  luridus, 497
  sphaerosporus, 496, 497
Bondarzew and Singer, 488, 489, 494, **516**
Bonnier, 217
Boss, 414, 415
Botryobasidium, 467, 473
Botryodiplodia, 579, 580
Botryosphaeriaceae, 299
Botryotinia, 232
  fuckeliana, 211, 212, **232**
Botrytis, 8, 11, 232, **589**
  cinerea, 233, 575
Boudier, 227, 327
Bourdot and Galzin, 481, 484, 515
Bovista, 10, **552**, 553
  plumbea, 552
Brand Fungi, 379
Brassica, 34, 47
Brassicaceae, 31, 141
Brefeld, 14, 152, 154, 159, 232, 328, 334, 380, 414, 418, 420, 444, 445, 450, 456, 474, 639
Brefeld and Falck, 411

Bremia, 141
  lactucae, 138, 139, **141**
Brevilegnia diclina, 107, 634
Briosi and Cavara, 578, 579, 581, 590
Brodie, 195, 311, 368, 370, 371, 377
Bromus, 316
Broomeia, 553
Brown, A. M., 382, 388
Brown, H. B., 268
Brown, W. H., 210
Brunswik, 374, 378
Brussels sprouts, 34
Bryopsis, 100
Bucholtz, 170, 171, 172, 237, 540
Buddin and Wakefield, 443
Buellia punctiformis, 220
Buisman, 277
Buller, 155, 166, 195, 196, 226, 270, 348, 349, 367, 369, 370, 371, 373, 386, 390, 444, 493, 498, 500, 501, 510
Bullera, 348, **349**
von Büren, 143, 144
Burgeff, 160
Burkholder, 333
Burmanniaceae, 2
Burnap, 557
Burt, 473, 475
Butler, 130, 132

C

Cabbage, 47
Cadophora, 576
  obscura, 576
Caeoma, **408**, 572
Cain, 338
Calbovista, 553
Caldesiella, 482
Calicium parietinum, 220
Callitriche, 34
Callose, 3, 181, 193, 350
Calocera, 451
  cornea, 369
Calodon, 482
Calostoma, 556
  cinnabarinum, 557
Calostomataceae, 554, 556
Calvatia, 552
  gigantea, 7, 196, 551, **552**
  sculpta, 553
Camp, 25
Canavalia, 334
Candida, 347
  albicans, 347
Cannabis, 141
Canter, Miss, 54
Cantharellaceae, 473, **476**, 477, 479
  key to genera, 512
Cantharellales, 480
Cantharellus, 470, **476**, 499
  floccosus, 477
Capillitium, 25, 325, 422, 531, 550
Capnodiaceae, 308, 309, 318, 319, **320**
  key to genera, 355
Capnodium, 320
  citri, 320
  salicinum, 320
Cardamine bellidifolia, 410
Carex, 419
  canescens, 410
Carleton, 382
Carpogenic cell, 213
Carpomyceteae, **192**, 379
  coloring, 197
  fruiting structures, 196
  nomenclature, 197
  structure, 192
Caryogamy, 371
Cassytha, 2
Castanea dentata, 282
Castoreum, 553
Catastoma, 552
  circumscissum, 552, 553
Catenaria, 64, 78, **81**
  allomycis, 81
  anguillulae, 81
Catenariaceae, **79**, 81
Catenarioideae, 64
Catenariopsis, 71
Catenomyces, 45, 63, **64**
  persicinus, 64
Catenularia, 576
Caulocystidium, 466
Cayley, Miss, 24, 25
Cellular hypha, 3
Cellulin, 78, 87, 104, 116
Cellulose, 3, 26, 45, 134, 151, 181, 182, 193
  reaction, 94, 98, 104, 127, 142, 143
  wall, 100, 145
Celtis, 141
  occidentalis, 310
Centrosome, 208
Cephalosporium, 277, 576, **585**, 586, 600
Cephalothecium, 589
  roseum, 589
Ceracea, 451
Ceramium, 100
Ceratiomyxa, 26
  fruticulosa, 27
Ceratobasidiaceae, 455
Ceratobasidium, 455, **457**, 466, 473
  cornigerum, 456
  sterigmaticum, 457
Ceratostomataceae, 277
Ceratostomella, 267, **277**, 278, 598

Ceratostomella—(*Continued*)
coerulea, 267
pluriannulata, 267
Cercosphaerella, 298
Cercospora, 573, **597**
apii, 598
beticola, 598
cerasella, 298, **598**
liriodendri, 265
zeae-maydis, 595
Cercosporella, **597**, 598
Cerinomyces, 451
Ceriomyces, 468, 487
Cetraria islandica, 223
Ceuthospora, 578
abietina, 579
Chaetocladium, 155, **167**
Chaetomiaceae, 274, **275**, 276, 323
Chaetomium, 275
aterrimum, 276
bostrychoides, 268
kunzeanum, 268
Chalara, 277, 576
Chalaropsis, 576
Chalk-brood, 352
Chamaesyce, 385
Chamonixia, 532, **540**, 543
Chara, **34**
Chaze, 237
Cheilocystidium, 466, 499
Chiastobasidial development, 437, 451, 464, 477, 480, 499
Chilemyces, 314
Chitin, 3, 26, 31, 35, 45, 50, 134, 142, 143, 150, 151, 193
Chlamydopus, 558
meyenianus, 558
Chlamydospore, 5, 110, **143**, 170, 171, 172, 179, 195, 329, 349, 380, 468, 487
Chlorochytrium, 630
Chlorociboria aeruginosa, 197, **232**
Chlorococcum, 217
Chlorophyceae, 215, 217
Chlorophyllum esculentum, 504
Choanephora, 155, **167**
conjuncta, 158
Choanephoraceae, 158, **167**, 168
key to genera, 185
Chondrogaster, 532
Christenberry, 164
Christensen, 409, 415
Christman, 391, 392, 399, 401
Chromosporium, 591
Chroococcus, 217
Chrysomyxa, 405
Chrysopsora, 384
Chrysothrix nolitangere, 193
Chupp, 33
Chytridiaceae, 54
Chytridiales, 6, **44**, 49, 69, 78, 81, 95, 97, 116, 119, 143, 145, 197, 632, 633, 635
composition of cell wall, 45
eucarpic, monocentric, 54, 61
haploid chromosome number, 46
key to families and genera, 71
life history, 45
phylogeny, 631
polycentric, 61
eucarpic, 60
Chytridiochloris, 59
Chytridium, 59
Ciboria, 232
Ciborinia, 232
Ciccinobolus cesatii, 311
Cichoriaceae, 143
Ciferri and Redaelli, 347
Cintractia fischeri, 410
Circinella, 165
minor, 164
umbellata, 164
Citromyces, 326
Citrullus vulgaris, 600
Citrus, 280, 334, 581
Cladochytriaceae, **61**, 64, 69, 81, 635
Cladochytrium, 61
replicatum, 61
tenue, 63
Cladonia, 217
Cladophora, 100
Cladosporium, 194, **595**
cucumerinum, 595
fulvum, 595, **596**
herbarum, 595
Clamp connections, 195, 237, 349, 350, 372, 373, 380, 383, 409, 410, 415, 419, 421, 439, 448, 450, 453, 455, 468, 470, 487, 501, 508, 530, 557
development, 372
prevalence, 374
whorls, 373, 374
Clathraceae, 10, 534, 543, 544, 545
Clathrus, 10, 11, **544**
ruber, 543, **544**
Claussen, 207, 209, 210, 352
Claustula, 545
Claustulaceae, 545
Clavaria, 10, 11, 12, 451, 455, **479**, 480
pistillaris, 479
Clavariaceae, 9, 11, 469, 470, **477**, 478, 479, 483, 490, 511, 602
key to genera, 513
Clavariadelphus pistillaris, 478, **479**
Clavariella, 480
subbotrytis, 479

Clavariopsis aquatica, 574
Claviceps, 5, **289**
  purpurea, 286, **289**, 290
Clavicipitaceae, 285, **287**, 289, 290
Clavicorona, 480
Clavochytrium, 82
Clavulina, 480
Clemencet, 330
Clements, 196, 486
Clinton, 133, 418, 426
Clitopilus, 532, 543
Clonostachys, 588
Closterium, 175
Clusius, 9
Clypeus, 280
Cobb, 543
Coccidiascus legeri, 345
Coccidioides immitis, 145
Coccidioidomycosis, 145
Coccomyces, 231, 582
Cochlonema, 179
Coelomomyces, 78, 79
  lativittatus, 81
Coelomomycetaceae, 79, **81**
Coemansia, 156, **169**
  aciculifera, 161
  erecta, 161
Coenocyte, 3, 151
  tubular, 3, 192
Coenocytic hypha, 3, 127
Coker, 108, 111, 478, 479
Coker and Beers, 496, 518
Coker and Braxton, 115
Coker and Couch, 109, 115, 549, 554, 555
Coker and Matthews, 108, 110
Cokeromyces, 168
  recurvatus, 168
Coleoptera, 215
Coleosporium, 384, 399, 402, **405**, 408, 439
  solidaginis, 401, **405**
  sonchi-arvensis, 405
Collema, 215, 216, **219**, 222, 264
  crispum, 221
  pulposum, 220, 221
Collemodes, 219, 222, 224
  bachmannianum, 220, 221
Colletotrichum, 280, **581**, 582
  gloeosporioides, **581**, 599
  malvarum, 583
Colley, 383, 387, 391, 392
Collybia radicata, 502
  velutipes 508
Coltricia perennis, 489
Columella, 152, 153, 154, 531
  percurrent, 537, 540
Compatibility, 371
  factors, 453
  factors governing, 376
Compatibility—(*Continued*)
  mutual, 269
Completoria, 175
Compositae, 143
Conard, 540
Conidiobolus, 172, **174**
  brefeldianus, 173
Conidiophore, 5
  monopodial, 142
  sympodial, 142
Conidium, 5, 126, 127, 129, 130, 137, 142, 155, 167, 169, 170, 172, 175, 193
  acrogenous development, 194
  alpha, 283
  basigenous development, 194
  beta, 283
  catenulate, 142
  chain, 194
  endogenous, 326, 576
  oïdial mode of formation, 5
Coniophora, **475**, 484
  cerebella, 373
  puteana, 373, 374
Coniophorella, **475**, 484
Coniosporium, 591
Coniothyrium, 579
  fuckelii, 579
Conjugate division, 208, 372, 373, 380
Conjugation tube, 95, 100, 101, 104, 107, 108, 112, 117, 126, 129, 132, 144
Connectives, 326
Cook, M. T., 50
Cook, W. R. I., 34
Cook and Nicholson, 97
Cook and Schwartz, 32, 34
Cooke, W. B., 488, 489, 516
Cooper, 107
Cooper and Porter, 132
Coprinus, 370, 374, 378, 466, 470, 499, 500, 507, **509**, 510, 541
  atramentarius, 500, 502, 505
  comatus, 502, 504
  ephemerus, 375
  ephemerus forma *bisporus*, 375
  fimetarius, 377
  lagopus, 370, 371, 373, 376
  micaceus, 502
  sterquilinus, 373, 500
Copromyxa, 30
Corallofungus, 10
Coralloid, structure, 542
  type, 533
Coralloides, 10, 11
Corda, 12
Cordyceps, **289**, 598
  militaris, 289
  ophioglossoides, 289
Coremium, 194, 598

Coriolellus serialis, 487
Coriolopsis trabea, 487
Coriolus versicolor, 489
Corner, 203, 204
Cornu, 88, 99, 129
Cortex, 216, 312, 313
  primary, 239
  secondary, 239
Corticiae, 458
Corticium, 338, **473**, 475, 482, 489, 494, 510, 511
  incrustans, 473
  solani, 602
  vagum var. *solani*, 465, 475, 602
Cortina, 505
Coryneum, 582
  beijerinckii, 582
Cotner, 78, 88
Cotton, 322, 476, 479
Couch, 53, 56, 57, 64, 69, 78, 79, 81, 82, 98, 100, 102, 107, 113, 114, 115, 158, 173, 175, 440, 441, 443, 445, 447, 449
Couch and Dodge, 81, 82
Couch and Whiffen, 82, 83
Craigie, 14, 370, 388, 390
Cranberry, 53
Craterellus, 476
Cronartium, 392, 394, 399, 402, **405**
  comptoniae, 387
  flaccidum, 405
  quercuum, 405
  ribicola, 383, 386, 387, 390, 395, 409
Crosses, illegitimate, 377
Crotalia cintractiae-fischeri, 410
Crozier, 243, 373
Crucibulum, 548
  leve, 531, 548, 549
  vulgare, 531, 548, 549
Cruciferae, 31
Cryptococcus, 346
Cryptogamia, 11
  algae, 11
  fungi, 11
Cryptoporus volvatus, 471
Ctenomyces, 329
Cucumis melo, 141
  sativus, 141, 596
Cucurbitariaceae, 278, **299**
Cucurbitaria, 299
Cudonia, 234
Cummins, 399, 401, 407, 540, 541, 545
Cunninghamella, 155, **167**, 168
Curtis, Miss, 51, 53
Cuscuta, 2
Cutter, 160, 162
Cyathus, 548
  stercoreus, 549
  striatus, 531
Cylindrosporium, 231, **582**
Cymadothea trifolii, 595
Cyphella, **476**, 484
Cystidium, 473, 487, 499, 537, 541
  trabecular, 466, 499, 500
  types of, 466
Cystobasidium, 441
Cystogenes, 86
Cystopage, 179
Cystopus, 135
Cystosori, 96, 97
Cystospora, 35, 402
Cytogamy, 210, 268, 371
Cytospora, 282, **578**
Cyttaria, 225, **236**
  gunnii, 236
Cyttariaceae, 236

D

Dacrymyces, 366, 368, 449, **451**, 457
  deliquescens, 450
  lutescens, 450
Dacrymycetaceae, 369, 450, 473
  key to genera, 460
Dacrymycetales, 379, 437, 438, **449**, 450, 458
Dacryomitra, 451
Dacryopinax, 451
Dactylaria, 591
  brochophaga, 593
Dactylella, 591
  ellipsospora, 592
  tylopaga, 593
Dade, 278, 576
Daedalea, **494**, 495
  confragosa, 494
Daedaleopsis, 494
Dahlia, 589
Daldinia, 284
  concentrica, 284
Dale, Miss, 328, 330
Daleomyces phillipsii, 225
Dangeard, 14, 30, 101, 160, 207, 210, 212, 312, 323, 328, 334, 368, 450, 452, 641
Daphnia, 345
Darwin, 227
Dasyscypha, 232
Dasyspora foveolata, 399, 400
  gregaria, 399
Davis, 416
Debaryomyces, 344
De Ferry de la Bellone, 237, 373
Delacroix, 579
De Lamater, 324, 330
Dematiaceae, 346, 584, **591**, 593, 594, 595, 597, 599

Dematiaceae—(*Continued*)
key to amerosporous genera, 618
key to dictyosporous genera, 620
key to didymosporous genera, 619
key to phragmosporous genera, 620
scolecosporous key, 621
Dendrogaster, 532
Dendrophoma, 577
convallariae, 578
Dendryphium, 596
Dennis, 327
Dentinum, 482
Dermateaceae, 230
Dermatitis verrucosa, 595
Derx, 328, 348
Desmidiaceae, 172, 175
Desmids, 54
Development, angiocarpic, 506, 507, 511, 531, 532
gymnocarpic, 506, 507, 511, 531, 532
pseudoangiocarpic, 506, 507, 511, 531, 532, 535, 540
unipileate, 540
Dianthaceae, 418
Dianthus caryophyllus, 395, 600
Diaporthaceae, 281, **282**, 283, 286
Diaporthe, 282
arctii, 283, 580, 582
Diasporangium, 135
Diatrypaceae, 280
Diatrype, **282**, 285
virescens, 279
Diatrypella, 282
Diblepharis, 88
Dicaryon, 208, 237
Dicaryon cells, 350
hypha, 241
phase, 4, 6, 383
stage, 380
Dicheirinia, 408
binata, 407
Dickinson, 413, 414, 415, 416
Dicotyledoneae, 2
Dicranophora, 153, 163, **166**, 635
fulva, 157
Dictydium cancellatum, 28
Dictyocephalos, 558
attenuatus, 558
curvatus, 558
Dictyophora, 544
duplicata, **544**, 545
indusiata, **544**, 546
Dictyostelium, 30
discoideum, 29
mucoroides, 29
Dictyuchus, 106, 110, **114**
missouriensis, 115
polysporus, 115
Diddens, 346
Diddens and Lodder, 336, 359
Didymellina iridis, 596
Didymium difforme, 24, 27
squamulosum, 25, 26
Didymocladium, 590
Dietel, 379, 396, 402, 403, 407, 418, 420
Dillenius, 10
Dimorphism, 106, 109
Diplanetism, 104, 106
Diplocarpon, 230
earlianum, 231
soraueri, 580
Diplocladium, 589
Diplocystis, 553
Diplodia, 580
Diploid colony, illegitimate, 342
nucleus, 161
Diploidization, 7, 370, 487
Diplophlyctis, 60
intestina, 60
Dipodascus, 336, **338**, 352, 353
albidus, **339**, 340
uninucleatus, 201, **339**, 341, 641
Diptera, 215
Discella, 581
Disciseda, 552
candida, **552**, 553
Discomycetes, 16, 200, 215, 227, 273, 292
key to orders, 244
Discula, 280, **581**
Disjunctor, 137, 193, 388
Disk lichens, 215
Dispira, 168
americana, 159
cornuta, 151
Dissophora, 170
decumbens, 169
Doassansia, 411, **421**
sagittariae, 420
Doassansiopsis, 421
martianoffiana, 420
Dobbs, 157, 166
Dodge, B. O., 14, 224, 225, 268, 269, 270, 386, 391, 398, 415, 549, 586, 639
Dodge, B. O., and Gaiser, 398
Dodge, C. W., 9, 330, 347, **348**, 540
Doidge, Miss, 320
Donk, 475, 480, 481, 486, 489, 490, 494
Dothideaceae, 290, 291, 309
Dothideales, 271, **290**, 295, 458, 595
key to families, 302
Dothiora, 296
Dothioraceae, 295
Doty, 480, 513
Douglas, 507
Dous and Ziegenspeck, 3
Dowding, Miss, 268, 270

Drayton, 14, 211, 224, 226, 233, 391
Drechsler, 129, 172, 178, 179, 180, 182, 297, 585, 589, 591, 593
Drepanopeziza ribis, 230
Drosophila, 15, 345
Dryodon, 483
Dudresnaya, 268
Duemling et al., 603
Durand, 227, 233, 234
Dutch elm disease, 278

E

Earthstars, 553
Eccrina, 181
Eccrinaceae, 181, 182
Eccrinales, 150, **180**, 182, 634
  key to families, 187
Echinodontium, 482, 484, **494**
Ectocarpus, 70
Ectostroma, 281
Ectrogella, 104
Ectrogellaceae, 94, 104
Eddins, 416
Edgerton, 268
Edson, 128, 130
Eel grass, 30
Eftimiu, 480
Eichleriella, 455
Eidam, 176, 329
Elaphomyces, 330
Elaphomycetaceae, 325, **330**, 334
  key to genera, 357
Elasmomyces, 532, 534, **540**
Elaters, 557
Elfving, 217
Elliott, E. W., 24
Elliott, J. A., 266, 267
Ellis and Everhart, 277, 283, 318, 326
Ellison, 24, 32, 631
Elrod and Blanchard, 484
Elrod and Snell, 496
Elsinoe, **334**, 583
  ampelina, 334
  piri, 334
  veneta, 333, **334**
Elvela, 12
Emerson, 85, 86
Emmons and Carrión, 595
Empusa, 17, 175
Endochytrium, 60
Endocochlus, **178**, 179
  asteroides, 180
Endoconidia, 182, 576
Endoconidium, 576
Endodesmidium, 54
Endogenous oospore, 89
Endogonaceae, 165, **170**, 171, 172
Endogonaceae—(*Continued*)
  key to genera, 186
Endogone, 163, 171, 172
  fasciculata, 170
  lactiflua, **170**, 171, 172
  malleola, **170**, 171
  occidentalis, 170, **171**
  pisiformis, 171
  reniformis, 171
  sphagnophila, **170**, 171, 172
Endomyces, 336, **337**, 340, 344, 345, 352, 353, 637
  magnusii, 338
Endomycetaceae, 336, **337**, 338, 345, 350
Endomycoideae, 345
Endomycopsis, **337**, 345
  albicans, **337**, 347
Endo-operculate exit tubes, 64
Endophyllum, 396, **398**
  euphorbiae-sylvaticae, 369, 398
  sempervivi, 398
Endospore, 137, 201, 508
Endosporeae, 26, 27
Endothia, 282
  parasitica, 282
Englerula effusa, 318
Englerulaceae, 308, 318, **319**
  key to genera, 354
Englerulaster, 320
Enterobryus, 181
  elegans, 181
Entoloma, 504, 532
Entomophthora, 172, **175**, 177
  fresenii, 176
  grylli, 177
  muscae, 174, 176
  sepulchralis, 176
Entomophthoraceae, **174**, **176**
Entomophthorales, 17, 100, **143**, 150, **172**, 173, 178, 180, 192, 193, 634, 635, 636
  key to genera, 186
Entomosporium, 231
  maculatum, 580
Entophlyctaceae, 54, **59**, 60, 61
Entophlyctis, **59**, 60
  vaucheriae, 60
Entostroma, 281
Entyloma, 410, **419**
  calendulae, 419
  dahliae, 421
Eocronartium, 438, **440**
  muscicola, **440**, 442
Epibasidium, 379, 381, 437, 438, 439, 440, 441, 443, 448, 453, 456, 457, 464
Epichloe, 287
Epicoccum, 601

Epicoccum—(*Continued*)
nigrum, 599
oryzae, 601
Epiplasm, 200, 208
Epispore, 137, 193, 200, 508
Epithecium, 238, 239, 273, 291
Epithele, 475
Eremascaceae, 337
Eremascoideae, 345
Eremascus, 345
albus, 337
fertilis, **337**, 338
Eremosphaera, 630
Eremothecium, **345**, 351
ashbyii, 345
Ericaceae, 405
Eriksson, 141, 382
Erinaceus, 11, 12
Eriocladus, 480
Erodium cicutarium, 50
Erysiphaceae, 12, 194, 308, **309**, 311, 313, 314, 315, 316, 317, 320, 324
key to genera, 353
Erysiphales, 16, 272, **307**, 320, 321, 322, 323, 324, 587
key to families, 353
Erysiphe, 307, 314
cichoracearum, 316, 317
graminis, 8, 309, 310, 311, 316, 317, 382
polygoni, 311, 312
Euallomyces, 85, 86, 91
Eubasidiae, 346, 378, 379, 390, 436, 437, 457, 458, 464, 473
Eucarpic, 42
genera, 46
Euglena, 50, 57, 99
Eumyceteae, 197
Euphorbia, 385
Eurotium, 325, 326
herbariorum, 328
insigne, 327, 587
Eurychasma, 104
Eurychasmidium, 104
Eutypella, 282
Evolutionary tendencies, 142
Exciple, 219
proper, 219
thalloid, 219
Excipula, 581
Excipulaceae, 581
key to genera, 608
Excipulum, 203, 210, 219
Exidia, 453, **454**
Exoascales, 241
Exoascus, 211, 243
Exobasidiaceae, 469, **480**, 481, 594
key to genera, 511
Exobasidium, 464, 465, 468, 470, **480**
Exobasidium—(*Continued*)
rhododendri, 480
vaccinii, **480**, 481
vexans, 480
Exogenous oospore, 89, 90
sperm, 214
Exospore, 144
Exosporeae, suborder, 26, 27
Exosporium, 601

F

Fabian, 336
Fabraea maculata, 580
Fagus silvatica, **340**
Fairchild, 176
Falck, 202, 226, 278, 485, 594
Famintzin and Woronin, 27
Fawcett, G. L., 582
Fawcett, H. S., 579
Fayod, 499, 508
Fedorintschik, 32, 34
Femsjonia, 451
Fenner, Miss, 167
Fermentation, 335
alcoholic, 337
Fimetaria, 275
fimicola, 195, 266
Fimetariaceae, 264, 269, 274, **275**
Fink, 245
Fischer, Alfred, 99, 130
Fischer, Eduard, 237, 238, 239, 324, 325, 331, 422, 423, 532, 533, 534, 536, 539, 540, 545, 546, 547, 550, 553, 554, 559, 560, 565, 657
Fischer, G. W., 416, 417, 421
Fischer, G. W., and Holton, 416
Fistulina, 11, **484**, 495
hepatica, 484
Fistulinaceae, 469, 473, 476, **484**
key to genera, 515
Fitzgerald, 577
Fitzpatrick, H. M., 46, 127, 134, 143, 163, 175, 352, 440, 442, 531
Fitzpatrick, R. E., 241, 242
Flagellata, 630
Flagellum, tinsel type, 24, 46, 69, 71, 94, 105
types of, 629
whiplash type, 24, 45, 46, 69, 78, 94, 97, 98, 105
Fleming, 602, 603
Flerov, 409, 415
Flint, 166
Flor, 382, 419
Florey, 603
Florideae, 2, 100, 195, 203, 212, 215, 220, 222, 262, 268, 424, 639

Flowering plants, 126
Foëx, 310
Fomes, 10, 11, 491, **492**, 494
  fomentarius, **492**, 493
  officinalis, 196
  pinicola, 16
Fomitopsis, 492
  officinalis, 486, **492**
Form families, 576
  genus, 573
Forsberg, 389
Fragaria, 298, 601
Fraser, Miss, 224, 320
Fraxinus, 240, 497
Freeman and Johnson, 382
Frey, 266, 298
Fries, Elias, 12, 16, 227, 379, 469, 481, 482, 507
Fries, Nils, 531
Fuligo septica, 23, 26, 28
Fungi, definition, 2
  genetics of, 9
  holocarpic endobiotic, monocentric, 94
  key to major groups, 18
  number of, 1
  phylogeny, 628
  pigments, 197
  saprophytic, 7
Fungi Imperfecti, 6, 12, 16, 263, 346, **572**
  key to form orders, 575
  key to genera, 604
  key to orders and families, 603
Fungi lamellati, 11
  porosi, 11
  pulverentes, 11
  ramosi, 11
Fungoides, 10
Fungus, 9, 10, 11, 510
Fungus chitin, 634
Funiculus, 549
Fusarium, 7, 197, 287, 586, **600**
Fusicladium dendriticum, 298
Fusicoccum, 280

G

Gadd and Loos, 480
Gallowaya, 384
Gametangia, 32
Gamete, 6
  male, 6
Ganoderma, 493
  applanatum, 196, 486, 492, **493**
  curtisii, 493
  lucidum, 493
Gasterella, 532, 534, 537, 539
  lutophila, 538
Gasterellaceae, 537
Gasterellopsis, 532, 534, **537**, 541
Gasteromyceteae, 16, 379, 454, 464, 471, 472, 501, 509, 511, 530
  key to genera, 561
  key to orders and families, 560
  phylogenetic system, 656
  phylogeny, 651
Gastroboletus, 496
Gastrosporiaceae, 539
Gastrosporium, 539
Gaümann, 112, 141, 167, 212, 291, 292, 296, 299, 334, 381, 441, 442, 470, 480, 481, 641
Gäumann and Dodge, 157, 440, 488
Gauthier, Miss, 178
Gautieria, 532, 540, **543**
  graveolens, 531
  plumbea, 542
Geaster, 11, 554
Geasteroides, 554
Geasteropsis, 554
Geastraceae, 534, 550, **553**, 554, 555
Geastrum, 550, **554**
  rufescens, 554
Geitler, 216
Gelasinospora, 269, **275**
  tetrasperma, 270
Gemmae, 111, 126, 370
Genea, 238
  cubispora, 17
  hispidula, 238
Genistella, 178
  ramosa, 177
Genistellaceae, 177, **178**, 180
Geoglossaceae, 233
Geoglossum, 203, 234
  glabrum, 233
Geographic races, 416, 487
Geolegnia, 110
Geopyxis, 228
  cacabus, 196, 225
Geotrichoides, 347
Germ pore, 508
Germination by repetition, 349, 350, 457
Gibberella, 287
  saubinetii, 287
  zeae, **287**, 600
Giesenhagen, 48, 594
Gilbert, E., 454
Gilbert, E. J., 518
Gilbert, E. N., 437
Gilbert, Frank A., 24
Gilbert, H C., 26
Gilkey, Miss, 17, 201, **225**, 238, 239, 255, 33
Glaziella, 172
Gleba, 531
  coralloid, 536

Gleba—(*Continued*)
lacunar, 536
multipileate, 536
unipileate, 536
Gliocladium, 327, 576, **587**, 588
penicillioides, 587
Glischrodermataceae, 565
Gloeocystidium, 466, 473
Gloeocystidium, 475
Gloeoglossum, 234
Gloeophyllum, 494
saepiarium, 378, 487
Gloeosporium, 280, 573, **581**, 582, 583
populi-albae, 581
ribis, 230, **582**
Gloeotulasnella, 455, **456**
Glomerella, **280**, 573
cingulata, 268, 279, **280**, 582
Glotzia centroptili, 178
Gnomonia, 573
veneta, 280
Gnomoniaceae, 279, **280**, 573, 582
Goidanich, 601
Goldring, 167
Gomphidiaceae, 496
Gomphus, 476
floccosus, 477
Gonapodya, 78, **86**
siliquaeformis, 87
Gonatobotryum, 589
Gonatorrhodiella, 589
parasitica, 588
Gonatorrhodum, 589
Gonidia, 216
Gonimochaete horridula, 182
Gonytrichum, 589
Goplana, 402, **408**
dioscoreae, 384, 407
Gossypium, 345, 350, 600
Graff, 318, 320
Grammothele, 482
Grandinia, 482
Graphiola, **422**, 423
phoenicis, 422
thaxteri, 422
Graphiolaceae, 411, **421**, 422
key to genera, 427
Graphium, 278, **598**
rigidum, 599
ulmi, **278, 598**
Greathouse, 277
Green, Miss, 226
Gregory, 139, 140, 166
Greis, 211, 237, 266, 268, 550
Griffiths, 264
Grifola, 489
berkeleyi, 490
Grossularia, 310, 316, 395, 405
Grove, 166, 396
Groves and Skolko, 596, 597
Guepinia, 451
Guepiniopsis, 450, **451**
Guignardia, **296**, 598
bidwellii, 296
Guilliermond, 338, 342, 343, 344, 345, 349, 350, 351, 352
von Gutenburg, 50
Guttulina, 30
Gwynne-Vaughan, 207, 210, 211, 307
Gymnoascaceae, 324, 325, **328**, 329
key to genera, 357
Gymnoascus reessii, 329
Gymnoconia, 408
peckiana, 386, 389, 393, 395, 398, **408**
Gymnosporangium, 11, 383, 388, 396, **408**
clavariaeforme, 387
juniperi-virginianae, 395, 408
nootkatense, 396
sabinae, 394
Gyrocephalus, 455
Gyrodon merulioides, 497, **498**
Gyromitra, 228
esculenta, 229
Gyrophana, 485
lacrymans, 485
Gyrophora muhlenbergii, 218
Gyrophragmium, 534, **542**

H

Hadley, 336
Hadromycosis, 589
Hadrotrichum, 594
Haerangiomycetes, 278
Haerangium, 278
Hainesia lythri, 601
Hanna, 14, 378, 388, 409, 412, 415, 421
Hanna and Popp, 416
Hansenula, 336, 340, **344**
Hanson, 56, 60, 64
Haplographium, 591
Haploid nucleus, 161
Haplosporangium, 170
bisporale, 170
Haplosporella, 579
Haplotrichum, 586
Harder, 45, 79
Harpella, 178
melusinae, 177
Harpellaceae, 177, **178**, 180
Harper, 14, 207, 209, 210, 311, 312, 313, 380
Harpochytrium, 59
viride, 59
Harter and Zaumeyer, 382
Hartig, 403

Harvey, 110
Harveyella mirabilis, 2
Hashioka, 316
Hasselbring, 599
Hatch, 85
Hausschwamm, 485
Haustorium, 7, 132, 139, 196, 212, 217, 307
  binucleate, 383
  dicaryon, 383
  monocaryon, 383
  uninucleate, 383
Hedgcock, 599
Hedrick, 245
Heim, 196, 312, 313, 464, 471, 473, 495, 501, 507, 508, 509, 534, 540
Helianthus, 388
  annuus, 395
Helicobasidium, 439, **443**
  candidum, 443
  compactum, 442
  purpureum, 439, **443**
Helicocephalum, 172
Helicoceras, 591
Helicogloea, 439, **443**
  lagerheimi, 442
Helicoma, 591
  perelegans, 593
Helicomyces, 591
  scandens, 593
Helicoon auratum, 593
Helicostylum, 166
Heliscus, 575
Helminthosporium, 317, **596**
  teres, 296, 297, **596**
Helotiaceae, **231**, 234
Helotiales, 230
Helotium, 71, **231**
  schimperi, 419
Helvella, 228
Helvellaceae, **228**, 233
Helvellales, 233
Hemiascus, 353
Hemibasidii, 379
Hemigaster, 532, **536**, 540
  candidus, 535
Hemigastraceae, 535, **536**
Hemisphaeriaceae, 295
Hemisphaeriales, 271, 272, 291, **292**, 307, 319, 321, 322, 354, 580
  key to families, 303
Hepaticae, 557
Hericium, 477, **483**
  coralloides, 483
Herpobasidium filicinum, 439, **441**
Herrell, 603
Heterobasidiae, 16, 346, 349, 378, 379, 381, **436**, 458, 464, 473, 487, 504
Heterobasidiae—(*Continued*)
  and Teliosporeae, summary, 457
  origin, 644
Heterochaete, 455
Heterocontae, 629, 632
Heteroecism, 394
Heterosporeae, 350
Heterosporium, 596
  gracile, 596
Heterotextus, 451
Heterothallic species, 106, 157, 162, 316, 352, 375, 376, 501
Heterothallism, 157, 226, 227, 375, 376
Higgins, 205, 207, 231, 265, 582
Higginsia, 582
Higher Fungi, 4, 143, **192**
  key to classes, 197
  reproduction, 194, 196
Hirmer, 374
Hirneola auricula-judae, 443
Hirneolina, 455
Hirschhorn, 416
Hirschioporus abietinus, 487, **489**
Hobsonia, 591
  mirabilis, 593
Hoehnelomyces, 444
Hoerner, 141
von Höhnel, 205, 256, 271, 274, 292, 321, 322
Höhnk, 135
Holcus lanatus, 34
Holobasidiae, 437
Holobasidium, 455
Holocarpic, 42
Homma, 314, 316
Homothallic species, 106, 157, 160, 162, 163, 375, 376, 501
Honey, 193
Honeydew, 307, 308, 320
Honig, 31
Hook, 208, 209, 211, 220, 243, 373
Hopkins, 151, 193
Hordeum sativum, 418
Hormiscium, 591
Hormodendron, 595
Hormones, 107
Horn, 104, 112
Hotson, 329, 330
Howard, 23, 24, 26
Howard and Currie, 25
Hugon, 504
Humaria, 227
  granulata, 211
Humarina, 227
Humulus, 141
  lupulus, 316
Huneycutt, Miss, 115
Hüttig, 414, 415

Hyalopsora, 403
Hyaloria, 451, 453, **454**, 534
  europaea, 453
  pilacre, 453
Hyaloriaceae, 453
Hybrids in fungi, 15
Hydatinophagus, 113
Hydnaceae, 11, 469, 475, 477, 480, **481**, 483, 484, 494, 511
  key to genera, 514
Hydnangiaceae, 534, **540**, 543, 550
Hydnangium, 532, **540**
Hydnellum, 482
Hydnés, 481
Hydnochaete, 482
Hydnocystis, 238
Hydnotria, 17
  cubispora, 17
Hydnum, 12, **482**, 602
  imbricatum, 483
Hydrodictyon, 71
Hydrophora, 165
Hymenella, 576
Hymenial cavities, 531, 537
Hymenium, 203, 379, 436, 464, 465, 499
  angiocarpic, 471, 495, 505
  gymnocarpic, 471, 495, 505
  monascous, 295
  primordium, 505
  pseudoangiocarpic, 471, 495, 505
  structure, 465
Hymenochaete, 466, **475**, 487
  cacao, 476
Hymenogaster, 532, **537**
  rehsteineri, 537, 538
  tener, 538
Hymenogastraceae, 534, **537**, 538, 539, 540, 550
Hymenogastrales, 535, **536**, 538, 546, 550, 552
Hymenomyceteae, 12, 25, 379, 381, 458, **464**
  phylogeny, 651
Hymenoptera, 215
Hypha, 3
  acrogenous, 243, 313
  dematioid, 320
  flexuous, 390
  laticiferous, 499
  moniliform, 320
  perisporioid, 320
  receptive, 391
Hyphal bodies, 172, 176
Hyphochytriaceae, 70, **71**
Hyphochytriales, 46, **69**, 95, 632, 633
  key to families and genera, 74
Hyphochytrium catenoides, 70, **71**
Hyphochytrium—(*Continued*)
  hydrodictii, 71
  infestans, 71
Hyphomyceteae, 583
Hyphopodium, 317, 318, 319, 320
Hypobasidium, 381, 437, 438, 439, 440, 441, 443, 448, 456, 457, 464
Hypochnus, 467, 475
  solani, 475
Hypocopra, 275
Hypocreaceae, 272, 285, 287
Hypocreales, 203, 271, 272, 275, **285**, 299, 309, 320, 322, 580, 599, 600
  key to families, 301
Hypoderma, 234
Hypodermataceae, 234, 240
Hypodermii, 16, 379
Hypomyces, 287
Hypostroma, 294
Hypothecium, 203, 219, 241
Hypoxylon, 268, **284**, 287
  marginatum, 283
Hysterangiaceae, 534, **542**, 544
Hysterangium, 543
Hysteriaceae, 240
Hysteriales, 215, **239**, 278, 294, 597, 639
  key to genera, 256
Hysterium insidens, 240
Hysterographium fraxini, 240
  minutum, 240

I

Iceland moss, 223
Imperfect fungi, 572
Incompatibility, 269, 371
  genetic factors, 375
Indusium, 544
Ingold, 574, 575, 616
Inoperculatae, 46, 225, **230**
Inoperculate series, 45
International Botanical Congresses, 15
Introduction, 1
Ipomoea batatas, 138, 151
Irene, 319
  echinata, 318
Irenina, 354
Irenopsis, 354
Iris, 596
Irpex, 477, 481, 482, 484, 486, **494**
Irpiciporus, 482, 494
Isaria, 598
Isoachlya, 110
Isogametes, 6
Isogamy, 143
Itajahya, 544
Itersonilia, 348, **349**, 350
Iwadara, 601
Ixechinus, 495

J

Jaapia, 484
Jackson, 338, 396, 424, 441, 639, 644
Jacobsonia, 236
Jaczewski, 635
Jahn, 26
Jane, 59
Jenkins, 334
Johnson, 417
Jola, 438, 439, **441**
javensis, 442
Jolivette, Miss, 166
Jones, P. M., 34
Jones, S. G., **224**, **235**
Jones and Drechsler, 66, 67
Jones and Torrie, 139
Josserand, 504
Juel, 242, 438, 457, 469, 532, 536, 640
Jugasporaceae, 496
Juniperus, 408
virginiana, 395

K

Kabatiella, 481
Kaiser, 419
Kanouse, Miss, 108, 170, 171
Karakulin, 481
Karling, 34, 35, 39, 58, 61, 63, 69, 70, 74, 95, 96, 99, 100, 102
Karlingia, 45
rosea, 57
Karsten, 494, 508
Karyogamy, 210
Keene, Miss, 160
Keilin, 82
Keitt, 270
Kemper, 373, 374
Kevorkian, 107
Kharbush, 211, 480
Kickxella, 156, **169**
Kickxellaceae, 156, 161, **169**
key to genera, 186
Killermann, 454, 474, 476, 479, 481, 488, 507
Killian, 266, 288, 293, 298, 422, 423
Klebahn, 132, 298, 575, 583
Kloeckera, 347
Kluyver, 349
Kniep, 14, 85, 372, 373, 375, 376, 378, 409, 413, 416
Kny, 223
Kolk, 409
Kordyana, 480
Krafczyk, 155
Kudo, 35
Kühner, 193, 372, 466, 496, 506, 507
Kunkelia nitens, 385, 386, 389, 396, **398**, 408, 415, 648
Kusano, 47, 48, 52, 53, 55
Kusanoopsis, 357

L

Laboulbeniales, 8, 193, 195, 202, 210, **212**, 222, 262, 264
key to families, 244
Labyrinthomyxa, 31, 35
Labyrinthula, **30**, 31, 35
macrocystis, 30, 31
Labyrinthulales, 23, 29, **30**, 35, 36, 631
key to genera, 39
Lachnea scutellata, 227
Lachnocladium, 480
Lactariopsis, 532
Lactarius, 193, **501**, 508, 532, 540
deliciosus, 502
volemus, 502
Lactuca sativa, 582
Lacunar type of sporocarp, 533
Laetiporus sulphureus, **490**, 491
Lafar, 9
Lagena, 100, **102**
radicicola, 103
Lagenidiaceae, 96, **100**, 101, 103, 145, 175
key to genera, 120
Lagenidiales, 31, 35, **94**, 107, 108, 119, 126, 143, 630, 632, 633
key to families and genera, 119
Lagenidium, 100, **102**, 175
giganteum, 102
rabenhorstii, 101
Lagerheim, 88, 640
Lagerheim and Patouillard, 454
Laibach, 89
Lamb, 397
Lambertella, 232
Lamellae, 495
Langford, 270
Lanopila, 552
bicolor, 552
Lanopilaceae, 553
Larix, 232, 405, 492, 497
Lasiobotrys, 317
Latex, 501
Laticiferous hyphae, 501
tubes, 540
Latrostium, 71
Leach and Ryan, 409, 417
Leaf curl, 243
Lecanorales, 200, 210, **215**, 225, 230, 264, 271, 272, 273, 291
Lecidea platycarpa, 218
Ledingham, 32, 33, 34, 35
Leger and Duboscq, 177, 178, 181, 182

Leger and Gauthier, 177, 178
Lehfeldt, 370, 390
Leidy, 181
Lembosia, 294
Lemna, 71
Lemonniera, 575
Lendner, 160, 161, 164
Lentinus tigrinus, 506
Lenzites, **494**, 495
  betulinus, 494
  saepiaria, 378, 487, **494**
  trabea, 487
Leontodon taraxacum, 17
Leotia, 234
  chlorocephala, 233
Lepiota, 502
  helveola, 504
  morgani, 504
  naucina, 502
  procera, 16, 502, 504, 509
  rachodes, 504
Leptolegnia, 110, **114**
  caudata, 107, 114
Leptomitaceae, 95, 108, 111, **116**, 119
  key to genera, 122
Leptomitales, 35, 108, 630
Leptomitus, 116
  lacteus, 111, **116**
Leptoporus adustus, 378, 488
Leptosphaeria, **292**, 296
  coniothyrium, 579
Leptostroma, 580
Leptostromataceae, 234, **580**
  key to genera, 607
Leptothyrium acerinum, 579
  pomi, 580
Leucoconis, 314
Leucocoprinus molybdites, 504
  procerus, 502
Leucogaster, **539**, 547
Leucoporus arcularius, 488
  brumalis, 488
Leucosporae, 507
  key to centrally stipitate genera, 521
Leucostoma, 282
Leveillula, **310**, 314
  taurica, 309
Levine, 507
Levisohn, 174
Liagora tetrasporifera, 643
Lichen, 11
Lichen acids, 218
  aquatic, 218
  crustose, 218
  foliose, 218
  fruticose, 218
  pendent, 218, 219
Lichens, 9, 16, 194
Lichens—(*Continued*)
  apothecial, key to families and genera, 245
Lichtenstein, 181, 182
Lichtheimia, 184
Lieneman, Miss, 265
Ligniera, 34
Lilliputia, **327**, 331, 587
  insignis, 327
Limacinia, 320
Lindau, 227, 236, 273, 280, 307, 572, 588, 599, 604, 612, 622
Lindegren, 14, 269, 271, 335, 342, 346
Linder, 156, 161, 169, 315, 458, 590, 591, 593, 611, 612, 645, 646, 648
Lindfors, 396
Lindsay, 222
Ling-Young, 157, 160, 162, 164
Linnaeus, 10, 11, 15
Linum usitatissimum, 390, 405, 600
Liriodendron tulipifera, 265
Liro, 217, 410, 413, 414, 421
Lister, 23, 28
Literature cited, 18, 39, 75, 92, 123, 146, 187, 198, 257, 303, 360, 427, 461, 524, 567, 624, 658
Literature for identification of fungi, 660
  Agaricaceae, Black-spored genera, 742
    General, 730
    Ochre- or Rust-spored genera, 740
    Purple-spored genera, 741
    Red- or Pink-spored genera, 739
    White-spored genera, 735
  Ascomyceteae, miscellaneous, 682
  Aspergillales (Plectascales), 707
  Asporogenous yeasts, 707
  Atichiaceae, 705
  Blastocladiales and Monoblepharidales, 675
  Boletaceae, 729
  Capnodiaceae, Englerulaceae, Trichothyriaceae, Atichiaceae, 705
  Chytridiales, 673
  Clavariaceae, 723
  Dothideales, 700
  Eccrinales, 680
  Englerulaceae, 705
  Entomophthorales, 678
  Erysiphaceae, 703
  Exobasidiaceae, 721
  Gasteromyceteae, General Works, 742
  General Works, 662
  Hemisphaeriales, 702
  Heterobasidiae, 718
  Host indexes, bibliographies, etc., 668
  Hydnaceae, 724
  Hymenogastrales, Sclerodermatales, etc., 743

Literature for identification of fungi—(*Continued*)
"Hymenomyceteae" (General Works), 720
Hyphochytriales, 675
Hypocreales, 699
Hysteriales, 695
Laboulbeniales, 682
Lagenidiales, 676
Lecanorales and Pyrenulales, 683
Lycoperdales (incl. Tulostomataceae and Podaxaceae), 744
Melanconiales, 750
Meliolaceae (Perisporiaceae), 704
Meruliaceae, 725
Moniliales: Dematiaceae, 753
Moniliaceae, 751
Tuberculariaceae, Stilbellaceae, and Mycelia Sterilia, 755
Monoblepharidales, 675
Mucorales, Entomophthorales, 678
Mycetozoa, 672
Myriangiaceae, 706
Nidulariales (incl. Arachniaceae), 746
Peronosporales, Protomycetales, 677
Pezizales Operculatae and Inoperculatae (incl. "Phacidiales"), 688
Phallales, 746
Phycomyceteae, general works, 673
Polyporaceae, 726
Protomycetales, 677
Pseudosphaeriales, 701
Pyrenulales, 683
Saccharomycetales and Asporogenous yeasts, 707
Saprolegniales (incl. Leptomitales), 676
Sclerodermatales, 743
Sphaeriales, 695
Sphaeropsidales, 747
Taphrinales, 694
Thelephoraceae and Exobasidiaceae, 721
Trichothyriaceae, 705
Tuberales, 693
Uredinales, 712
Ustilaginales (incl. Graphiolaceae), 709
Zoopagales, Eccrinales, 680
Litmus, 197, 219, 223
Ljungh, 196
Lloyd, 543
Lodder, Miss, 336, 346, 347, 348, 359
Lohman, 240
Lohwag, 242, 349, 532, 533, 543
Lohwag and Follmer, 484
Long, 554
Long and Plunkett, 558
Long and Stouffer, 544, 558
Longula, 542
texensis, 542
Lophiostoma, 278
Lophiostomataceae, **278**, 279
Lophium mytilinum, 240
Lophodermium, 235
Lorenz, 531
Lower Fungi, 192, 196
Lunulospora, 575
curvula, 574
Lupo, Miss, 268
Lutman, 409
Luttrell, 293
Luzula multiflora, 413
Lychnis, 411
Lycoperdaceae, 9, 11, 12, 534, 550, **551**, 553
Lycoperdales, 535, **550**
Lycoperdon, 10, 11, 12, 533, **552**
giganteum, 17
poculiforme, 18
pyriforme, 551
Lycopersicon, 350, 580
esculentum, 133, 600
Lysipenicillium, 327
Lysurus australiensis, 544

M

Maassen, 352
Macbride and Martin, 23, 27, 36
Macrochytrium, 60
Macrophoma, **577**, 578, 580
Macrospore, 181, 182
Macrosporium, 596
Magnusia, 327
nitida, 329
Magnusiella, 243
Mains and Jackson, 382
Mainsia, 407
Maire, 469, 473, 477
Malaceae, 408
Malençon, 193, 393, 540, 558
Malus, 280, 316, 334
sylvestris, 395
Mangifera, 280
indica, 581
Mangin, 151, 321, 322
Mango, anthracnose, 280
Manina, 483
Manuel, Miss, 345
Marasmius, 498
oreades, 501, 502
Marssonina, 231
panattoniana, 582
Martensella, 156, **169**
Martin, 1, 437, 443, 450, 451, 453, 455, 456, 457, 458, 460, 461, 549
Mason, 481, 575, 594
Massee, 386

Massospora, 172, **177**
Matruchot, 587
Matthews, Miss, 128, 130
Mattirolo, 237, 273
Maublanc and Malençon, 558
McCranie, 81, 85, 86
McCullough, 336
McDonough, 140
McGuire, 478
McWhorter, 217
Medicago sativa, 230
Medulla, 216
Megachytrium, 63, **64**
Meiosis, 107, 137, 144, 160, 161, 162, 371
Meiotic division, 22
Melampsora, 402, **405**
  euphorbiae, 405
  farlowii, 405
  helioscopiae, 401
  lini, 382, 385, 390, **405**, 409
  medusae, 405
  ribesii-purpureae, 405
Melampsoraceae, **402**, 403, 405, 407
  key to genera, 425
Melampsoridium betulinum, 401
Melanconiaceae, **581**, 583
  key to genera, 609
Melanconiales, 575, **581**, 583
Melanconidaceae, 280
Melanin, 197
Melanogaster, **539**, 547
Melanogastraceae, **539**, 540
Melanospora, 278, 285
  chionea, 276
Melanosporaceae, **275**, 276
Melanosporae, 507
  key to centrally stipitate genera, 524
Melhus, 135, 139
Melin and Nannfeldt, 576
Meliola, 292, 308, 309, **317**, 319, 320, 321
  circinans, 318
  corallina, 318
Meliolaceae, 308 314, **317**, 318, 319
  key to genera, 354
Meliolineae, 319
Mellor, Miss, 218
Melogrammataceae, 280
Meruliaceae, 475, **484**, 485, 487
  key to genera, 515
Merulius, 485, 486
  domesticus, 485
  lacrymans, 485
Mesophellia, 553
Mesophelliaceae, 553
Mesospore, 408
Mestas, 237
Metaphyses, 273, 292
Metula, 587
Mez, 633, 634
Micheli, 10, 11, 510
Microconidia, 233, 264, 297
Microglossum, 234
Micromonospora, 585
Micromyces, **53**, 54
  longispinosus, 53
  zygogonii, 53
Micromycopsis, **53**, 54
  cristata, 45, 54
Micropeltis, 295
Microsphaera, 314
  alni, 311, 317
  berberidis, 315
  quercina, 314
Microspore, 181, 270
Microstroma, 481
Microthyriaceae, 292, 293, **294**, 307, 319, 320, 321
Microthyrium, 294
Middleton, 127, 130
Milbrath, 138, 141
Mildew, powdery, 8, 309
Milesia, 403
Miller, Julian, 203, 272, 274, 277, 285, 292, 331, 332
Miller, L. W., 482, 514
von Minden, 49, 86, 99, 134, 135
Mindeniella, 108, 117, **119**
Mitrula, 234
Mittman, Miss, 267
Mix, 244
Modicella, 171
Molds, blue, 325
  green, 325
  herbarium, 325
  pink, 327
  sooty, 320, 349
Möller, 216, 220, 454
Mollisia, 230
Mollisiaceae, **230**, 320, 573, 580, 582
Monascus, 323, 326, 327
  ruber, 324
Monilia, 347
Moniliaceae, 574, 583, **584**, 586, 587, 588, 590, 591, 592, 593, 597, 598, 599
  key to amerosporous genera, 612
    to hyalodidymous genera, 615
    to phragmosporous genera, 615
    to scolecosporous, 616
Moniliales, 240, 574, 575, 576, **583**, 588
  key to endosporeae, 610
    to form families, 583
    to helicosporous genera, 611
    to staurosporous, 616
Monilinia, 193, 194, **232**
  fructicola, 232
Monoblepharella, 86, **89**, 91

Monoblepharella—(*Continued*)
elongata, 90
mexicana, 90
taylori, 90
Monoblepharidaceae, **86**, 87, 88, 89, 90, 635
Monoblepharidales, 49, 78, 79, **86**, 95, 119, 143, 633, 634
key to family and genera, 92
Monoblepharis, 3, 78, 79, 86, **88**, 90, 91
insignis, 88, **89**
macrandra, 88, 89
polymorpha, 89
Monocaryon phase, 4, 6, 383
stage, 380
Monocentric Chytridiales, 42, 46
Monocotyledoneae, 2
Monomorphic zoospore, 106
Monoplanetic zoospores, 106
Monospora, 336
Monosporella bicuspidata, 345
Monotropa, 2
Monsma, 104
Montagnea, 534, **542**
Montagnellaceae, 291
Montemartini, 600
Morchella, 10, 11, 12, 203, 211, **228**
conica, 229
Moreau, 160, 212, 222, 373, 391, 398
Morels, 228
Morenoella quercina, 293
Morgan, 553
Morse, Miss, 553, 560
Mortierella, 152, 163, **170**, 172
rostafinskii, 152
Mortierellaceae, 152, **169**, 172
key to genera, 185
Mosquitoes, 78, 81
Mougeotia, 53
Mounce, Miss, 14, 375
Mounce and Macrae, 378, 487
Mucor, 11, 12, 151, 152, 153, 163, **165**, 166, 167, 168, 170, 352
hiemalis, 160, 162
mucedo, 154, 161, 162
proliferus, 153
ramannianus, 171
rouxianus, 151
type, sexual process, 162
Mucoraceae, 5, 154, 157, 164, **165**, 170
key to genera, 183
Mucorales, 12, 69, 143, **150**, 151, 152, 156, 157, 172, 175, 178, 179, 180, 192, 193, 226, 337, 346, 586, 634, 635
key to families and genera, 183
Mucronella, 482
Multipileate type, 533
Mundkur, 132, 411, 419
Murphy, 133
Murrill, 481, 488, 489, 496, 519
Musca domestica, 176
Musci, 218
Muscus Saxatilis vel Lichen, 9
Mushroom, 505
Mutinus, 544
caninus, 544
ravenelii, 544
Mycelia, self-sterile, compatible, 268
Mycelia Sterilia, 576, **601**
key to genera, 623
Mycelium, 3
coenocytic, 42, 104, 150
color, 4
dicaryon, 383
diploidization, 371
intercellular, 385
intracellular, 385
monocaryon, 368, 370, 383
primary, 4, 368, 453
secondary, 4, 368, 453
Mycena, 193, 470, 501
Mycenastrum, 552
Mycetozoa, 1, 2, 10, 12, 14, 16, 22, 23, 45, 94, 192, 197, 557
ancestry, 36
key to orders, 36
phylogeny, 630
Mycobacteriaceae, 585
Mycoderma, 347
Mycogone perniciosa, 589
Mycology, 1
history, 9
medical, 9
technical, 9
Mycosphaerella, 210, 296, **298**, 578, 591, 639
cerasella, **298**, 598
fragariae, 16, **298**, 591
pinodes, **298**, 580
sentina, 298
tabifica, 298
tulasnei, 595
tulipiferae, 205, 206, 265
Mycosphaerellaceae, 280, **296**
Mycotoruleae, 347
Mycotoruloideae, 346, **347**
Mycotypha, 155, **167**
Mykes, 9
Myriangiaceae, 322, 325, **332**, 333, 334
key to genera, 357
Myriangiales, 272, 292, 299, 309, 320, **331**, 583
Myriangina, 357
Myrianginella, 357
Myriangium, 325, **332**, 334
curtisii, 333

Myriangium—(*Continued*)
duriaei, 202, **333**
Myrioblepharis, 86
paradoxa, 86
Myriodiscus, 236
Myriogonum, 338
Myriostoma, 554
Myxamoeba, 22, 24
Myxocollybia, 508
Myxogastrales, 23, 30, 35, 36
key to families and genera, 36
Myxogastres, 23
Myxomycetes, 1, 23
Myxophyceae, 215, 217
Myxosporium, **581**, 582
Myzocytium, 100, 102
proliferum, 101
vermicola, 101

N

Nabel, 45, 79, 151
Nannfeldt, 203, 205, 212, 223, 227, 230, 231, 234, 240, 251, 254, 271, 272, 274, 275, 277, 291, 323, 582
Napicladium, 596
Narasinhan, 132
Naumov, 163, 166
Nauss, Miss, 35
Nectaromyces, 346
Nectaromycetaceae, 346
Nectria, 287
cinnabarina, **287**, 289, 600
Nectriaceae, 287, 289
Nectrioidaceae, 580
Neergaard, 596
Nematodes, 78, 79, 589, 591
Nematospora, 335, **344**
coryli, 345
lycopersici, 345
phaseoli, 345
Nematosporangium, 130
Nematosporoideae, 345
Neovossia indica, 411, 419
Nephrochytrium, 60
Nero, 9
Net-plasmodium, 30, 31
Neuhoff, 438, 457
Neurospora, 15, 269, **275**, 391
crassa, 271
sitophila, 269, 270
tetrasperma, 268, 270
Newcomer and KenKnight, 585
Newton, 375, 376, 389
Nidularia, 548
Nidulariaceae, 535, **548**, 549
Nidulariales, 535, **548**
Nidulariopsis, **549**, 550, 555
van Niel, 349
Nienburg, 286
Nigropogon, 532
Nigrospora, 594
oryzae, 594
panici, 594
Nitzschia, 2
Nobles, Miss, 473
Nocardia, 585
Nomenclature, rules for, 15
Nostoc, 11, 216, 217, 219
Nothofagus, 226, 236
Nowakowskiella, 63
macrospora, 63
Nuclear cap, 79, 81
Nucleus, diploid, 6
privileged, 341
Nurse cells, 312, 313
Nyctalis, 501
asterophora, 468
Nyland, 439

O

Obelidium, 57
mucronatum, 57
Ochrosporae, 507
key to centrally stipitate genera, 523
Octojuga, 532, 543
Octomyxa, 34, 35
Octophore, 278
Odontia, 482
Oedocephalum, **586**, 594
Oedogonium, 49
Oehm, 468
Oenothera, 53
Oersted, 394
Oidiophore, 368, 370, 371
Oidium, 194, 335, 337, 368, 370, 371 453, 468, 488, 501
binucleate, 370
uninucleate, 370
Oidium, 589
aureum, 590
Olive, 26, 29, 30, 396
Olpidiaceae, **46**, 55, 97, 99, 100
Olpidiopsidaceae, 32, 46, 49, 50, 96, **97**, 98, 102, 633
key to genera, 120
Olpidiopsis, 2, **98**, 100
luxurians, 98
schenkiana, 49, 99
varians, 98
vexans, 98
Olpidium, **47**, 49, 71, 95
brassicae, 47
radicale, 45, 49
trifolii, 47

Olpidium—(*Continued*)
trifolii, sexual reproduction, 47
viciae, 47, 48
sexual reproduction, 47
Omphalia chrysophylla, 506
Onygena, **330**, 444
caprina, 331
equina, 331
Omygenaceae, 325, **330**, 331
Oogone, 106, 213
aplerotic, 129
plerotic, 129
Oomyceteae, 143
Oort, 377
Oosphere, 112, 136
Oospora, 194, **584**, 585
lactis, **584**, 585
nicotianae, 585
Oospore, 6, 95, 126, 136, 163
exogenous, 90
parthenogenetic, 133
Opegrapha atra, 220
subsiderella, 220
Operculatae, forms, 46, 224, 230
Operculate series, 45
Operculum, 45, 225, 226
Ophiostoma, **277**, 278, 598
adiposum, 267
coeruleum, 267
fimbriatum, 267
paradoxum, **278**, 576
piceae, 267
pluriannulatum, 267
ulmi, 277, **278**, 598
Ophiostomataceae, **277**, 323
Orange, 280
Orcein, 197, 219, 223
Orchidaceae, 2
Orobanche, 2
Orton, 284
Oryza sativa, 591
Osborn, 33
Ostiole, 203, 225, 262, 551
Ostropa, 230
Ostropaceae, **230**, 254
Ostropales, 230
Oudemans, 414
van Overeem, 174
Oxalis, 390, 391
stricta, 395
Oxycoccus macrocarpus, 50, 53, 480
Oxydontia, 482
Ozonium, 602
omnivorum, 602

P

Pachyma, 602
cocos, 487, **602**
Pachysterigma fugax, 456
Padus, 243
Pady, 383, 385, 393
Palm, 137
Palm and Burk, 33, 34
Palmella, 285
Palmeter, 270
Pampolysporium, 317
Panaeolus, 510
Panicum, 55
Papulospora, 5
mytilina, 240
Paragyrodon sphaerosporus, 496, 497
Paramyces, 468
Paraphysis, 203, 208, 210, 220, 273, 319, 466, 499
Paraphysoid, 273
Parasite, balanced, 8
destructive, 8, 132
facultative, 7
obligatory, 8
Parasitism, 7, 193
balanced, 194
destructive, 193
Paravicini, 410, 413
Parenchyma, true, 195, 196
Parmularia, 240, 294
Parodiopsis stevensii, 314
Parr, Miss, 166
Parthenogenesis, 6, 107, 108, 112, 129
Patella, 227
scutellata, 227
Patouillard, 321, 322, 327, 436, 470, 472, 476
Patouillardina, 455
Patterson, 128
Paxillaceae, 496
Paxillus, 496, 497, 498
Peck, 12
Pectose, 3, 151
Pedogamy, 341
Pellicularia, 5, 457, 467, **473**
filamentosa, 465, **475**, 602
isabellina, 474
Peltomyces, 35
Penicillin, 602, **603**
Penicillium, 194, 323, **325**, 326, 327, 573, 576, **586**, 588, 591, 598, 602
chrysogenum, 587, 603
crustaceum, 328
expansum, 587
frequentans, 587
luteum, 328
notatum, 326, 587, 603
vermiculatum, 328
Peniophora, **475**, 482
Perfect stage, 18, 572
Perforations in septa, 195

Periconia, 594
Pericystaceae, 337, 350, **352**
Pericystis, 352
alvei, 352
apis, 352
Peridermium, 408
Peridiole, 548
Peridium, 531
Periphysis, 203, 207, 274, 319
Periplasm, 100, 101, 106, 114, 116, 117, 119, 126, 129, 132, 136, 137, 142
Perisporiaceae, 307, **317**
Perisporiales, 292, 295, **307**, 587
Perisporium, 307
gramineum, 307
Perithecium, 196, 202, 203, 262
typical, 272
Peronoplasmopara, 141
Peronospora, 139, **141**, 142
ficariae, 138
parasitica, 141
spinaciae, 141
Peronosporaceae, 18, 127, **138**, 139, 140, 141, 142, 172
key to genera, 145
Peronosporales, 35, 49, 95, 108, 119, **126**, 135, 142, 143, 151, 193, 631, 633, 634
keys to families and genera, 145
Persea, 280
Persoon, 12, 16, 17, 18, 227, 469, 554, 555
Pestalotia, 582
versicolor, 583
Pestalozzia, 582
Petch, 202, 333
Petersen, H. E., 55, 135
Petersen, N. F., 411
Petersenia, 100
Pethybridge, 132
Petrak, 273, 284, 291, 292, 299, 319, 354
Petrak and Sydow, 573, 577, 579
Peziza, 10, 12, **228**
badia, 228
cacabus, 196
pustulata, 586
repanda, 228
vesiculosa, 228
Pezizaceae, 10, 202, 203, **227**, 234
Pezizales, 197, 200, 210, 215, 223, **224**, 237, 238, 243, 271, 295, 332, 334, 335, 573, 580, 582, 602, 639
key to families and genera of inoperculatae, 251
of operculatae, 249
Pezizella oenotherae, 601
Phacidiaceae, **234**, 239, 240, 582
Phacidium, 234
Phaeoradulum, 482
Phaeoseptoria, 580
Phallaceae, 9, 534, **543**, 544, 545, 546
Phallales, 532, 535, **543**
Phallogaster, 543, 544
Phallus, 10, 11, 12, **544**
impudicus, 544
ravenelii, 544
Phaseolus, 334
limensis, 133
vulgaris, 395, 409, 475
Phellinus, 487
igniarius, 486, **493**
Phellodon, 482
Phellorinia, 532, 553, 557, **558**, 560
Phellorinieae, 558
Phialide, 5, 576, 586
definition, 575
Phialophora, 576
Phlebia, **482**, 484
Phleogena, 439, **444**, 556
decorticata, **444**, 445
faginea, 444
Phleogenaceae, 440, **444**, 445
Phlogiotis, 455
Phlyctidiaceae, 55
Phlyctochytrium, **55**, 71
hallii, 56
Phlyctorhiza, 60
endogena, 60
Pholiota adiposa, 508
aurivella, 368, 468, 501
Phoma, 297, **577**, 580
betae, 298, **578**
stenobothri, 577
Phomopsis, 282
citri, 579
Phragmidium, 383, 394, 395, 402, 407, **408**
rubi, 383, 400
rubi-idaei, 406
speciosum, 392
violaceum, 387
Phragmobasidiae, 437
Phycomyces, 160, 163
blakesleanus, 161
microsporus, 161, 164
nitens, 160, 161, 163
Phycomyceteae, 3, 4, 5, 7, **143**, 182, 192, 193, 195, 334, 350, 572, 585
higher, phylogeny, 634
holocarpic biflagellate, 95
nonflagellate, key to orders, 44
Phycopsis, 322
Phylacteria, **476**, 481
Phylactériés, 481
Phyllachora, 284
graminis, 284
Phyllachoraceae, 282, **284**, 290
Phyllachorineae, 284

Phyllactinia, 310, 314
eleagni, 315
guttata, 309
Phyllosticta, 577
tabifica, 577, 578
Phymatotrichum, 589
omnivorum, **589**, 602
Physalacria, 478
Physalospora, **298**, 579
cydoniae, 298
Physarella oblonga, 23
Physarum nutans, 28
polycephalum, 23, 24, 25
Physcia, 222, 223, 264
Physiologic forms, 8, 382
race, 389
Physocladia, 61
obscura, 61
Physoderma, 64, 66
maculare, **65**, 66
zeae-maydis, 65, **66**
Physodermataceae, 61, **64**, 65, 67
Phytophthora, 127, **130**, 133, 134, 135, 142, 634
arecae, 132
cactorum, 131, **133**
himalayensis, 132
infestans, 133
phaseoli, 133
stellata, 133
Pierce, 242
Pierson, 390
Piersonia, 239
Pietra fungaia, 490
Pigment, 197
Pilacre, 444
Pilacrella, 444
Pilaira, **165**, 166
Pilat, 539
Pileocystidium, 466
Pileolaria, 407
Piline, 317
Pilobolaceae, 155, 156, **165**
key to genera, 184
Pilobolus, 49, 153, 156, 163, **165**, 166, 169, 170, 175, 636
crystallinus, 155
kleinii, 155, 636
longipes, 636
Pinkerton, Miss, 586
Pinus, 387, 405, 497
strobus, 383, 386, 395, 409
Piptocephalidaceae, 156, 159, **168**, 178
key to genera, 185
Piptocephalis, 163, **168**, 170
cruciata, 159
freseniana, 159
Piptoporus betulinus, 491
Piricularia, 590, **591**
oryzae, 590
Pirus, 334
Pisolithus, 535, 547, **548**
Pistillaria, 478
Pityriasis capitis, 347
Pityrosporum, **346**, 347
Placenta, 291
Planocyte, 22, 23, 42
Planogametes, 6
Planont, 79, 82
Planospore, 5, 66
Plasmatoönkosis, 130
Plasmodiophora, **31**, 34, 36, 94, 631
brassicae, **31**, 32, 33, 35
Plasmodiophoraceae, 94
Plasmodiophorales, 23, **31**, 35, 630
key to genera, 39
Plasmodium, 22, 24, 35
Plasmopara, 139, **140**, 141
Platanus, 280
Platycarpa, 439, **441**
Platygloea, **443**
Plectania, 227
coccinea, 228
Plectascales, 205, 272, 275, **322**
Plectobasidial structure, 533
Plectobasidiales, 546
Plectospira, 114
Pleospora, 296
herbarum, 597
Pleosporaceae, 266, 280, 296, **298**
Pleotrachelus, **49**, 100
fulgens, 49
Pleurage, 200, **275**, 391
anserina, 264
Pleurocystidium, 466, 499
Pleurotus ostreatus, 502
ulmarius, 502
Plicatura, 484
Plowright, 379, 380, 394, 420
Plum pockets, 243
Pluteus cervinus, 502, 504
Poa, 316
pratensis, 418
Poaceae, 140, 316, 421
Podaxaceae, 535, 550, **558**, 559
Podaxis, 532, 550, 553, **559**, 560
pistillaris, 559, **560**
Podetium, 218
Podosphaera, 312, 314
leucotricha, 316
oxyacanthae, 316
Podospora, 275
Pogonomyces hydnoides, 491
Poisson, 181
Poitras, 129, 168
Polycentric series, **42**, **46**

Polychytrium, 63
Polymyxa, 35
Polyphagus, **57**, 59
  euglenae, 46, 57
Polyporaceae, 8, 9, 10, 11, 467, 469, 470, 473, 477, 484, **486**, 490, 491, 492, 493, 498, 508, 511
  key to genera, 516
Polyporales, 464, 469, **471**, 473, 474, 481, 485, 490, 491, 492, 493, 495
  key to families, 472
Polyporus, 5, 11, **489**, 490, 491, 586, 602
  caudicinus, 490
  squamosus, 196, 486, **490**
  tuberaster, 490
Polysphondylium, 30
  violaceum, 29
Polystictus, 488, **489**, 491
  abietinus, 487
Polystigma, 287
  rubrum, 286
Polystomella, 294
Polystomellaceae, 290, **294**, 319, 320
Polythrincium, 595
  trifolii, 595
Pontisma, 100
Poole, 243
Populus, 405
Pore, 495
  fungi, 2
Porés, 481
Poria, **488**, 489, 494
Porothelium, 484
Portulaca oleracea, 137
Posteriorly uniflagellate fungi, key to orders, 43
Prillieux, 578
Prillieux and Delacroix, 595
Primordium, basidial, 451
Primula, 227
Pringsheim, 130
Proabsidia, 184
Proactinomyces, 585
Probasidium, 458, 646
Proliferation, 109, 130, 153, 341
Promycelium, 379, 380, 381, 384, 394, 438, 453
  internal, 384, 405
Propagula, 308, 321, 322
Prosenchyma, 196
Prosorus, 52
Prospodium, 399, **408**
  plagiopus, 401, 407
Prosporangium, 54
Proteomyxa, 36
Protoabsidia, 184
Protoachlya, 115
Proto-aecidium, 386
Protococcus, 217, 285
Protocoronospora, 481
Protodontia, 455
Protogaster, 334, **535**, 536, 539
  rhizophilus, 535
Protogastraceae, 534, 535, **536**
Protogastrales, **535**, 536, 537
Protohydnum, 455
  gelatinosum, 454
Protomerulius, 455
Protomyces, **143**, 144
  inundatus, 143
  macrosporus, 144
Protomycetaceae, **143**, 144, 172
Protomycetales, 126, **143**, 145
Protomycopsis, 143
  leucanthemi, 144
Protozoa, 22, 35, 116, 630
Protubera, 543
Prunus, 231, 243, 316, 582
  avium, 243
  cerasus, 243
  domestica, 286
Psalliota, 502
Pseudocolus, 544
  javanicus, 543
Pseudocoprinus, 509, 510
Pseudohypha, 348
Pseudolpidiopsis, **49**, 99
Pseudolpidium, 99
Pseudoparenchyma, 195, 196
Pseudoperonospora, **141**, 142
  celtidis, 141
  cubensis, 141
  humuli, 141
Pseudopeziza, 224, **230**
  medicaginis, 230
  ribis, 230, 573, 582
  trifolii, 224
Pseudoplasmodium, 29, 30
Pseudosepta, 4, 42, 78, 84, 86
Pseudosphaeriaceae, **296**, 297
Pseudosphaeriales, 223, 240, 271, 272, 273, 274, 278, 280, 284, 291, **295**, 331, 332, 334, 578, 596, 597
  key to families, 302
Pseudosphaerita, 50, **100**
  euglenae, 99
Pseudotrichogyne, 266
Pteridophyta, 382
Pterula, 480
Ptychogaster, 487
Puccinia, 11, 17, 394, 402, **407**, 408
  arenariae, 396
  asparagi, 395, 408
  atropae, 393
  caricis grossulariata, **388**
  coronata, 383, 388, 391, 395, 408

Puccinia—(*Continued*)
glumarum, 408
graminis, 14, 18, 385, 388, 389, 390, 394, 395, 406, 408
agrostidis, 389
secalis, 389
tritici, 8, 382, 389
helianthi, 382, 388, 390, 394, 395
iridis, 382
malvacearum, 394, 396, 397
pringsheimiana, 388
prostii, 397
rubigo-vera, 408
tritici, 382, 388, 395, 396
sorghi, 17, 382, 390, 391, 395, 408
triticina, 388, 389, 390
tumidipes, 383
Pucciniaceae, 369, 398, **405**, 406
key to genera, 425
Pucciniastrum, 399, 402
goeppertianum, 403
Puff ball, 2, 7
Pycnidium, 194, 219, 263
definition, 575
Pycniospore, 386
Pycnium, 386
Pycnoporus cinnabarinus, 491
Pycnothyriaceae, 581
Pycnothyriales, 581
Pyrenomycetes, 16, 262, 332
ostiolate, key to orders, 299
Pyrenophora, 202, **296**
teres, **296**, 297, 596
trichostoma, 296
Pyrenula, 285
Pyrenulales, 223, 271, 272, **285**
key to families, 302
Pyronema, 193, 203, 210, 211, 220, 224, 225, **227**, 243, 293, 351
omphalodes, **207**, 209
omphalodes var. *inigneum*, 210
Pyrrhosorus, 97
Pythiaceae, 95, 97, 98, 126, **127**, 128, 131, 133, 134, 138, 633
key to genera, 145
Pythiella, **100**, 126
Pythiogeton, 135
transversum, 134
Pythiomorpha, 135
Pythiopsis, 105, 106, 109, **110**
cymosa, 110
Pythium, 8, 95, 100, 102, 114, **127**, 130, 132, 133, 134, 135, 142, 634
aphanidermatum, 130
debaryanum, 130
dictyosporum, 128
monospermum, 130
polysporum, 129
Pythium—(*Continued*)
proliferum, **127**, 128
torulosum, 128

Q

Quadripolar sexual phases, 377, 501, 531
Queletia mirabilis, 556
Quercus, 239, 293, 497
Quintanilha, 377

R

Rabenhorst, 13
Races, biologic, 382
geographic, 378
Raciborski, 322, 334
Radicula armoracia, 591
Radiigera, 553
Radulaspore, 589
definition of, 575
Radulum, 482
Rafflesia, 2
Ramaria, 480
Ramlow, 201, 224
Ramularia, 573, **590**, 598
armoraciae, 591
rosea, 590
tulasnei, 298, **591**
Ramularisphaerella, 298
Ranunculaceae, 421
Raper, 29, 107, 156
Raper and Thom, 587
Ravenelia, 394, 402, **408**
acaciae-micranthae, 407
Rawitscher, 414
Rea, Carleton, 476, 477, 485, 486, 515, 557, 565
Rea, P. M., 557
Receptacle, 543
Receptive hypha, 389
Red Seaweeds, 2, 97
Reduction division, 112
Reed, 316
Reesia, 71
amoeboides, 71
Rehsteiner, 533, 537
Reijnders, 507, 509
Reinke, 218
Remsberg, Miss, 478
Renn, 30
Reproduction, asexual, 5
sexual, 5
Respiratory hyphae, 224
Resting sporangia, 65, 66, 67, 79, 83, 84, 85, 86, 143, 144
Resting spore, 79, 95, 96, 99, 102, 181, 182

Reticularia, 24, 26
Reticulomyxa, 35
Rhamnus, 395
  frangula, 383
Rheosporangium, 130
Rhinotrichum, 589
Rhipidiaceae, 108, **117**, 118, 119, 126, 127, 142
  key to genera, 122
Rhipidium, 117
  americanum, 118
Rhizidiaceae, 45, 54, **55**, 56, 58, 59, 61, 78, 91
Rhizidiomyces, 71
  apophysatus, **69**, 70
Rhizidiomycetaceae, 70, **71**
Rhizoclosmatium, 55
Rhizoctonia, 5, 475, **602**
  solani, 602
Rhizoid, 7, 81
Rhizomorph, 4, 195
Rhizomorpha, 602
  subcorticalis, 602
Rhizomycelium, 61, 66, 67, 71
Rhizophlyctis, 57
  petersenii, 57
  rosea, 57
Rhizophydium, **55**, 57, 65, 71, 91, 96, 104
  coronum, 56
  couchii, 57
  graminis, 55
  ovatum, 57
Rhizoplast, 24
Rhizopoda, 179, 630
Rhizopogon, **539**, 540
Rhizopogonaceae, 540
Rhizopus, 162, **165**
  nigricans, 11, 57, 160
Rhizothyriaceae, 581
Rhodochytrium, 630
Rhodophyllus, 468, 501, 504, 532
Rhodosporae, 507
  key to centrally stipitate genera, 523
Rhodotorula, 348
Rhodotorulaceae, 348
Rhombiella cardamines, 410
Rhopalomyces, 586
Rhysotheca, 139, **140**, 141
  australis, 138
  viticola, 139, 140, **141**
Rhytisma, 235
  acerinum, 235
Ribes, 230, 395, 405
  nigrum, 396
Riccia, 98
Rice, Miss, 383, 389
Robak, 487
Roberts, J. M., 63
Roberts, Miss Catherine, 244
Rochlin, 31
Rogers, 438, 454, 455, 457, **458**, 473, 474, 648
Rogers and Jackson, 473, 475
Rolland, 193
Rosa, 395
Rosellinia, **275**, 284
  aquila, 276
  necatrix, 275
Rostowzew, 141
Rotifers, 113
Routien, 537, 539, 541
Rozella, 45, 47, **49**
Rozellopsis, 97
Rubus, 334, 383, 386, 395, 396, 398, 408, 579
Rumex, 34
Russula, 193, 501, 508, 540
Russulaceae, 473
Rusts, 8, 14, 194, 197, **382**, **424**
  autoecious, 394, 395
  heteroecious, 394, 395
    control, 395
  macrocyclic, 396
  microcyclic, 396
  origin, 644
  stem, of wheat, 8
Rutstroemia, 232

S

Saccardo, 13, 507, 577
Saccharomyces, 335, **344**, 346, 637
  cerevisiae, 342, 343, 344
  paradoxus, 342, 343
Saccharomycetaceae, 244, 322, 341, 343, **344**, 345, 346
Saccharomycetales, 192, 200, 202, 210, 211, 212, 272, **334**, 350, 352, 585
  key to families, 358
  keys to genera, 358
Saccharomycodes, 344
  ludwigii, 345
Saccharomycoideae, 345
Saccharum officinarum, 140
Saccoblastia, 439, 442, **443**
Saccobolus, 226
Sachs, 549, 633, 636, 638
Sagittaria, 411
Salix, 320, 405
Salmon, 310, 316, 317
Salvin, 110
Sammelzelle, 65, 66, 67
Sappin-Trouffy, 143, 380, 383, 400, 401, 405, 407
Saprolegnia, 100, 108, **110**, 112, 114, 116, 130, 142, 153, 341, 634

Saprolegnia—(*Continued*)
ferax, 98
monoica, 112
var. *glomerata*, 111
Saprolegniaceae, 71, 95, 97, 98, 99, **108**, 109, 111, 113, 114, 115, 116, 119, 127, 142
key to genera, 121
Saprolegniales, 3, 35, 49, 94, 95, **104**, 105, 107, 108, 126, 128, 130, 135, 142, 143, 150, 151, 630, 633, 634, 635
key to families and genera, 121
Sapromyces, 117
Saprophyte, 7, 8
Sarcodina, 630
Sartcris, 267, 409
Sartory, 326
Sass, 375
Satina and Blakeslee, 160
Savile, 390, 391
Sawyer, 177
Saxifragaceae, 405
Scheibe, 382
Scherffel, 45, 53, 99
Schikorra, 327
Schistodes, 314
Schizonella, 418
melanogramma, 418
Schizophyllum, 507
commune, 465
Schizosaccharomyces, **344**, 345
octosporus, 335, **344**
pombe, 344
versatilis, 335, **344**
Schizosaccharomycetaceae, 341, **344**, 345
Schizostoma montellicum, 279
Schizothecium, 200, **275**, 375, 391
anserinum, 264, 266, 268, 269, 270
Schneepia, 294
Schneider, 218, 221, 345
Schostakowitsch, 153
Schroeter, 130, 135, 227, 494, 546, 573
Schultz, 141
Schwartz and Cook, 45
von Schweinitz, 12
Schweizer, 224
Schwendener, 216
Scirrhiineae, 284
Sclerenchymatous cells, 196
Sclerocystis, 171
Scleroderma, 531, **547**, 549, 550, 555
aurantiacum, 547
Sclerodermataceae, 534, 535, 547, 550
Sclerodermatales, 535, **546**, 553, 554, 558
Sclerospora, 140
graminicola, 140
macrospora, 140
Sclerotia, 4, 5, 8, 195, 196, 203, 224
Sclerotia—(*Continued*)
subterranean, 487
Sclerotinia, **232**, 602
fructicola, 232
sclerotiorum, 194, **232**
Sclerotiniaceae, **232**, 347, 589
Sclerotiopsis concava, 601
Sclerotium, 602
rolfsii, 4
Scorias spongiosa, 318
Scott, 66
Scutellinia stercorea, 224
Scutiger, 489
Seaver, 202, 225, 227, 249
Sebacina, 454
Secale cereale, 421
Secotiaceae, 534, 539, **540**, 541, 550
Secotium, 532, 534, **540**, 542
agaricoides, 540
coprinoides, 541
erythrocephalum, **540**, 541
novae-zelandiae, 540
clbium, 541
Self-compatibility, 330
Self-incompatibility, 269, 375
Self-sterility, 226, 227
Senft, 193
Septobasidiaceae, 440, **445**, 446
Septobasidiales, 440
Septobasidium, 436, 438, 439, 440, 443, **448**, 449
burtii, 446
Septochytrium, 63
variabile, 64
Septocylindrium, 590, **591**
Septogloeum, 582
mori, 581
Septolpidium, 50
Septonema spilomeum, 240
Septoria, 265, **580**
aesculi, 579
apii, 580
apii-graveolentis, 580
lycopersici, 580
piricola, 298
Septorisphaerella, 298
Septotinia, 232
podophyllina, 201
Septum formation, 4
Serpula, 485
Sesamum indicum, 7
orientale, 197
Setae, 487
Setaria, 140
Seuratia coffeicola, 321
Sexual phases, 241, 242, 370, 375, 388, 412, 453, 473, 501, 531
bipolar, 378

Sexual phases—(*Continued*)
quadripolar, 378
reproduction, fundamental phenomena, 371
strains, 226, 267, 278
tendencies, 212
Sexuality, bipolar, 377, 487
quadripolar, 377, 487, 488
Seyfert, 409
Shanor, 90, 98, 132, 133, 168
Shear, 269, 391, 602
Shear and Dodge, 444, 445, 601
Shikorra, 323, 324
Sibelia, 382
Sideris, 130
Sigmoideomyces, **167**, 586
Silver spoon test, 505
Simblum, 544
Singer, 471, 473, 475, 481, 483, 485, 486, 488, 489, 490, 496, 507, 508, 509, 511, 518, 532, 652
Sinoto and Yuasa, 23, 24
Siphonales, 107, 634
Siphonaria, 55, **57**
petersenii, 58
Sirobasidiaceae, **453**, 454
Sirobasidium, 437, 451, 453
albidum, 454
Sirolpidiaceae, 96, **100**
key to genera, 120
Sirolpidium, 100
Sistotrema, 482
Sjöwall, 162
Skupienski, 24, 27, 29, 631
Sleumer, 409, 413
Slime molds, 1, 23
Smith, A. H., 520
Smith, A. H., and Miss Morse, 512
Smith, Miss A. L., 219, 223
Smut galls, 418
Smuts, 8, 15, 192, 194, 197, **409**
hybridization, 416
manner of infection, 410
mutant forms, 417
origin, 644
Snell, 518
Solanum tuberosum, 34, 50, 133, 475, 589, 600
Solenia, **476**, 484
candida, 476
Solidago, 405
Sommerstorffia, 113
Sordaria, 275
fimicola, 266
Sordariaceae, 274
Soredium, 219
Sorghum vulgare, 418
Sorodiscus, 34
Sorokin, 67, 69
Sorolpidium, 35
Sorosphaera, 34
radicalis, 34
veronicae, 33, **34**
Sorosporium, 419
consanguineum, **413**, 414
reilianum, 415, 416
saponariae, 418
syntherismae, 417
Sorus, 52, 135
uredial, 393
Southern beech, 226
Southworth, 583
Sparassis, **476, 479**
ramosa, 476
Sparrow, 31, 35, 46, 49, 55, 57, 59, 60, 64, 65, 66, 70, 79, 89, 94, 95, 96, 100, 102, 104, 108, 113, 128, 630
Spathularia, 234
Speare, 177
Species, bipolar, 531
quadripolar, 531
Spegazzini, 13
Spegazzinia, 601
Sperm, 6, 386
Spermatia, 220, 222, 368
Spermatization, 391
Spermogonial nectar, 390
Spermogonium, 205, 206, 207, 216, 220, 222, 223, 368, 386
subcortical, 386, 387
subcuticular, 386, 387, 407
subepidermal, 386, 387, 407
Spermophthora, 352
gossypii, 345, **350**, 351
Spermophthoraceae, 337, 345, **350**, 351
Sphacelia, 289
Sphaceloma, 333, 334, 573, **583**
Sphacelotheca, **418**, 423
columellifera, 413
ischaemi, 415
schweinfurthiana, 413, 414
sorghi, 418
Sphaeria, 273
Sphaeriaceae, 273, **275**, 276, 278, 284
Sphaeriales, 12, 210, 215, 271, 272, **273**, 285, 286, 290, 291, 292, 295, 296, 299, 309, 319, 320, 322, 323, 458, 573, 578, 579, 580, 582, 591
key to families, 300
Sphaerioidaceae, 576
Sphaerita, **50**, 100
dangeardii, 50, 99
Sphaerobolaceae, 535, **549**, 554
Sphaerobolales, 535, **549**
Sphaerobolus, **549**, 550
grandis, 531

Sphaerobolus—(*Continued*)
stellatus, 550, 555
Sphaerocladia, 82
Sphaerocysts, 501
Sphaeropsidaceae, **576**, 578, 579, 601
key to genera, 604
Sphaeropsidales, 16, 240, 575, **576**, 578
Sphaeropsis, **578**, 579
malorum, 298, **578**
Sphaerosporangium, 130
Sphaerostilbe, 287
gracilipes, 289
Sphaerotheca, 314
castagnei, 312, 313
fuliginea, 314
humuli. 315, 316
mors-uvae, 310, 316
pannosa, 317
var. *rosae*, 311
phytoptophila, 310
Sphagnum, 419
Spicaria, 587
Spinacia oleracea, 141
Spinalia, 168
Spirogyra, 53, 99
Sponge mushrooms, 228
Spongospora, 34
subterranea, 33, 34
Sporangia, 5, 32
sympodially produced, 110
Sporangiole, 153, 156, 166, 170, 172
monosporous, 155
Spore, amyloid, 470, 484
aquatic, 575
forms, classification, 577
fruit, 7, 192, 197, 638
gymnocarpic, 496
ontogeny, 496
pseudoangiocarpic, 496
resupinate, 465
non-amyloid, 470
repeating, 381
secondary, 436
types, 508
wall, amyloid, 508
non-amyloid, 508
Sporendonema, 576
Sporidiobolus, **349**, 350
Sporidium, 366, 379
discharge, 369, 384
infection by, 385
quaternary, 385
secondary, 384
tertiary, 385
Sporobolomyces, 348, **349**, 368, 379
roseus, 348
Sporobolomycetaceae, 348
Sporocarp, 170, 171
Sporocarp—(*Continued*)
consistency, 467
Sporocladia, 156, 161, 169
Sporodesmium, 240, **597**
Sporodinia, 165
grandis, 157, 160, 161, 164
Sporodochium, 334, 599
Sporomyxa, 35
Sporonema, 280
Sporoschisma, 576
Spot anthracnose, 334
Stachylina, 178
Stahl, 220
Stakman, 14, 382, 383, 389, 409, 415
Staphylococcus, 602
Starbäck, 196
Steccherinum, 482
Stelling-Dekker, Miss, 336, 345, 358
Stemonitis, 26, 28
fusca, 24
Stempell, 410, 421
Stemphylium, 596
botryosum, 597
sarcinaeforme, 597
Stereocaulon coralloides, 218
Stereum, 11, 467, **475**
Sterigma, 168, 576, 586, 587
primary, 586
secondary, 586
Sterigmatocystis, 325, **586**
Stevens, 136, 137, 317, 319, 354
Stichobasidial development, 437, 464, 477, 480
Stictis, 230
Stigeoclonium, 57
Stigmatea, 293
robertiani, 293
Stigmateaceae, 293
Stigmatomyces baeri, 214
Stigmatomycosis, 350
Stilbaceae, 584
Stilbella, 286, 287, 584
Stilbellaceae, 584, **598**, 599
key to genera, 621
Stilbum, **444**, 584
Stomatogene, 317
Stoneman, 279
Storage organs, 195
Streptomyces, **585**, 603
scabies, 585
Streptomycetaceae, 585
Streptomycin, 602, **603**
Strobilomyces floccopus, 497
strobilaceus, 497
Strobilomycetaceae, 496
Strobilophyta, 382
Stromatinia gladioli, 211, 224, 226, **233**, 391

Strong, 228, 229, 235, 290
Structure, types of, 532
Stylina, 423
Stylopage, 179
Stylospore, 170
Stypella, 454
Stysanus, 598
  stemonites, 599
Subhymenium, 499
Subiculum, 476
Suborder Endosporeae, 25
Suchfäden, 417
Suillellus luridus, 497
Suillus, 11
  luteus, 497
Suss, 504
Swarm cells, anteriorly or laterally biflagellate, 94
  posteriorly uniflagellate, 94
Swarm spore, 23
Swingle, 151
Swoboda, 552
Sydow, 273, 284, 290, 292, 293, 294, 296, 299, 303, 308, 316, 318, 319, 320, 326, 346, 354, 357, 399, 400
Sympodial development, 130
Synascus, 145, 353
Syncephalastrum, 168
Syncephalideae, 180
Syncephalis, 168
  cornu, 159
Synchytriaceae, **50**, 51, 52, 55
  composition of cell wall, 54
Synchytrium, 45, **50**, 53
  decipiens, 52
  endobioticum, **50**, 51, 53
  fulgens, 52, 53
  mercurialis, 45
  papillatum, 50
  vaccinii, 50, 53
Synnema, 278, 598
Syringa vulgaris, 317
Syringospora, 347
Syzigites, 165
Syzygospora, 443
  alba, 443
Szymanek, 132

T

Tabor and Bunting, 135
Taeniellaceae, 182
Taeniellopsis orchestiae, 182
Taphria, 243
Taphridium, **143**, 144
Taphrina, 211, 212, **243**, 637
  cerasi, 243
  communis, 243
Taphrina—(*Continued*)
  deformans, 241, 242, **243**, 244
  epiphylla, 241
  klebahni, 241
  potentillae, 243
  pruni, 243
  robinsoniana, 243
Taphrinaceae, 241, **243**, 639
Taphrinales, 215, **241**, 368, 458, 510
  key to genera, 256
Taraxacum, 17
Targioni-Tozzetti, 11
von Tavel, 556
Tear stain, 280
Tehon, 581, 607, 608
Tehon and Daniels, 595
Teleutospore, 449
Teliospore, 5, 379, 380, **399**, 438
  compound, 394
Teliosporeae, 349, 350, 366, 378, **379**, 383, 436
  and Heterobasidiae, summary, 457
  key to orders and families, 424
  relationships, 423
  sexual reproduction, 382
Telium, 399
Terfezia, 327
Terfeziaceae, 239, 325, 327, **331**
Ternetz, 195
Terrostella, 554
Tetrachaetum, 575
  elegans, 574
Tetrachytrium, 67, **69**
  triceps, 69
Tetracladium, 575
  marchalianum, 574
Tetramyxa, 34
Thallus, crustose, 217
  foliose, 217
  fruticose, 217
  pendent, 217
Thamnidiaceae, 157, **166**
  key to genera, 184
Thamnidium, 153, **166**, 167
  elegans, 157
Thamnocephalis, 167
Thaxter, 83, 86, 87, 88, 118, 159, 169, 174, 176, 214, 215, 244, 588
Theissen, 205, 271, 273, 284, 290, 292, 293, 294, 296, 299, 303, 308, 316, 318, 319, 320, 354, 357
Thelebolus stercoreus, 200, 201
Thelephora, **476**, 481
  terrestris, 476
Thelephoraceae, 457, 458, 469, 470, **473**, 474, 476, 477, 480, 481, 482, 484, 486, 487, 489, 490, 494, 510, 536, 602

Thelephoraceae—(Continued)
key to genera, 511
Thielaviopsis, 277, 278, 576
paradoxa, 576
Thirumalachar, 311, 396
Thom, 325
Thom and Raper, 587
Thomas, 134, 193, 634
Thraustochytriaceae, 94, 96, **104**
key to family, 121
Thraustochytrium, 104
Thraustotheca, 109, 110, 115
clavata, 110
primoachlya, 109
Thrush, 337, 347
Tibicina septendecim, 177
van Tieghem, 159, 345
Tieghemella coerulea, 160
glauca, 160
Tilletia, 414, **419**
caries, 410, **419**
foetida, 410, **419**, 421
levis, 410, 419
sphagni, 419
tritici, 410, 419, 420
Tilletiaceae, 368, 381, 409, 410, **419**, 420
key to genera, 427
Tilletiopsis, 348, **349**
Tinsel type flagellum, 629
Tischler, 45, 468
Tisdale, 66
Tithymalus, 395
Tjibodasia, 443
Toadstool, 505
Tobler, 217, 218
Togashi and Oda, 508
Tolyposporium, 419
junci, 418
Tomentella, 467, **475**, 482
flava, 474
Torula, 342, 346, **591**
Torulopsidaceae, **346**, 348
key to genera, 359
Torulopsis, 342, **346**, 347
pulcherrima, 244
Torulopsoideae, 346
Tournefort, 10
Toxalbumins, 505
Trabutiineae, 284
Trachysphaera, 135
Trachyspora, 407
Tragopogon porrifolius, 137
Trama, 498, 531
bilateral, 499
inverse, 499
irregular, 499
mixed, 499
regular, 499
Trametes, 11, **491**, 494
americana, 378, 487
serialis, 487
Tranzschelia pruni-spinosae, 408
Trematophlyctis, 35
Tremella, 11, 384, 438, **454**, 455, 457
mesenterica, 452
reticulata, 452, **455**
Tremellaceae, 452, **453**, 454, 456
key to genera, 460
Tremellales, 197, 379, 437, **451**, 452, 454, 456, 458
Tremellodendron, 455
Tremellodon, 454, **455**
Trentepohlia, 217, 285
Trichaster, 554
Trichia, 28
varia, 24
Trichiaceae, 557
Trichocoma, 330
Trichocomaceae, 325, **330**, 334
Trichoglossum, 234
Trichogyne, 210, 211, 213, 214, 215, 220, 222, 389
Tricholoma, 504
personatum, 502
Trichomonascus, 338
Trichopeltaceae, 295
Trichophore, 213
Trichoscyphella, 232
wilkommii, 232
Trichothecium, 589
Trichothyriaceae, 308, **321**
key to genera, 355
Trichothyrium, 322
Tricladium, 575
Trifolium, 595, 597
pratense, 224
Triticum, 55, 316, 395, 418
aestivum, 102, 382, 421
durum, 382
Trogia, 484
Trophic hyphae, 42
Trophogone, 208, 268, 323
Trow, 128
Truffle, 10, 239
Trypethelium, 285
Tsuga, 494
canadensis, 405
Tuber, 10, 11, 237, **239**
aestivum, 237, 238, **239**
brumale, 237
candidum, 201
lapideum, 237, 273
magnatum, 238
melanosporum, 237, **239**
mesentericum, 237
panniferum, 237

Tuber—(*Continued*)
  rufum, 238
Tubera, 10
Tuberales, 203, 215, **236**, 273, 291, 325, 327, 639
  key to genera, 255
Tubercularia, 287, **600**
  vulgaris, 600
Tuberculariaceae, 584, **598**, 599
  key to genera, 622
Tuburcinia, 410, **421**
  trientalis, 420
Tucker, 132
Tulasne, 12, 13, 238, 289, 315, 538, 547
Tulasnella, **456**, 457
  violea, 456
Tulasnellaceae, 455, 456, 473
  key to genera, 461
Tulasnellales, 379, 437, **455**, 466
Tulipa, 397
Tulostoma, 436, 531, 553, **555**, 557
  campestre, 556
  simulans, 555
Tulostomataceae, 535, 550, **555**, 556, 557
Turbinate cell, 66
  organs, 61
Tylopilus felleus, **497**, 498
Type family, 16
  genus, 16
  species, 16
  specimen, 16
Typha, 167
Typhella, 178
Typhula, **478**, 602
Typhulochaeta, 314

U

Ulmus americana, 598
Umbelliferae, 143
Uncinula, 314
  necàtor, 310, 316
  salicis, 315
Uncinulopsis, 314
Unipileate type of sporocarp, 533
Universal veil, 508
Uredinales, 5, 8, 14, 16, 368, 379, 380, 381, **382**, 385, 392, 398, 400, 401, 403, 405, 406, 423, 436, 438, 449, 453, 458, 572, 639
Uredineae, 380
Uredinella, 438, **449**
Uredinopsis, 393, 395, **402**
  struthiopteridis, 403
Urediospore, 381
  primary, 399
  secondary, 399
Uredium, 393, **399**

Uredium—(*Continued*)
  cupulate, 399
  extrastomatal, 401
  type, 401
  with peridium, 401
Uredo, **408**, 572
  linearis, 18
  mother cells, 449
  sorghi, 17
Uredospore, 381, 449
Urobasidium, 481, 594
Urocystis, 421
  anemones, 409, **421**
  gladioli, 409
  occulta, 410, **421**
  tritici, 421
  violae, 420, **421**
Uromyces, 402, **407**
  aloës, 396
  caladii, 383
  caryophyllinus, 395
  dianthi, 395
  erythronii, 400
  fabae, 406
  phaseoli typica, 382, 389, 395, 408
  phaseoli vignae, 389
Urophlyctis, 64, **66**
  alfalfae, 66, 67
Urtica, 141
Usnea barbata, 218
Ustilaginaceae, 6, 172, 379, 409, **412**, 413, 415, 418, 423
  key to genera, 426
  teliospore longevity, 417
Ustilaginales, 8, 14, 16, 194, 337, 346, 379, 380, 381, 384, **409**, 412, 413, 415, 418, 420, 422, 423, 424, 438, 453, 487, 572
Ustilago, 378, 409, **418**, 419
  avenae, 410, 414, 415, 416, 418
  cardamines, 410
  fischeri, 416
  hordei, 415, 417, 418
  ischaemi, 415
  kuehneana, 418
  levis, 409, 415, 416, 418
  longissima, 378, 416, 417, 423
  longissima var. *macrospora*, 417
  nuda, 411, 412, 418
  perennans, 416
  striiformis, 409, 416, 418
  striiformis forma *Hordei*, 416
  striiformis forma *Poae-pratensis*, **417**
  tritici, 411, 412, 418
  violacea, 411, 414, 418
  vuijckii, 413
  zeae, 409, 410, 411, 412, 415, 416, **418**

V

Vaccinium, 403
Vaillant, 10
Valid publication, 16
Valkanov, 30
Valsa, **282**, 578
Valsaceae, 280
Valsella, 282
Vandendries, 14, 368, 370, 371, 377, 378, 488
  and Brodie, 160
  and Martens, 368, 468, 501
Vanterpool and Ledingham, 102, 103
Vararia, 475
Varicosporium, 575
  elodeae, 574
Varitchak, 267, 341, 352, 353
Vaucheria, 71, 97. 107, 633, 634
Veins, 239, 330, 531
Venturia inaequalis, 8, 265, 266, 270, **298**, 589
Veronica, 34
Verpa, **228**, 232
Verrucaria, 285
Verticillium, 589
  albo-atrum, 588, **589**
Vesicle, 95, 100, 102, 103, 117, 128, **129**, 130, 134, 135
  subsporangial, 153, 165, 166
  substomatal, 393
Vetch, 47
Vibrissea, 230
Vicia, 47
Vigna sinensis, 600
Vincens, 274, 277
Viola, 421
Vitis, 141, 275, 310, 316, 334
  vinifera, 297, 316
Volkartia, 143
Volutella, 600
  dianthi, 600
Volva, 509
Volvaria, 501
Voss, 383
Vuillemin, 168, 575

W

Wager, 57
Waite, 243
Wakefield, 327
Waksman and Henrici, 585, 603
Walker, Miss, 170, 171, 340, 341, 507, 537, 550
Wang and Martens, 383, 391
Wartenweiler, 141
Water molds, 105
Waterhouse, 389, 394
Webb, 34
Weber and Wolf, 158
Wehmeyer, 274, 277, 280, 281, 283
Weir, 384
Welsford, 286
Wernham, 409
Weston, 140
von Wettstein, 3, 26, 50, 79, 134, 151, 193
Whetzel, 201, 232
Whiffen, Miss, 46, 54, 59
Whiplash type flagellum, 629
White, 196, 493
White rust, 8, 137
Wieben, Miss, 211, 241, 242
Wilcox, 269
Willia, 336, **344**
Williamson, 207, 210, 211
Wilson, M. and Cadman, 24
Wilson, O. T., 66
Wiltshire, 596, 597
Wingard, 345
Winge, 342
Winter, 587
van Wisselingh, 26, 31
Witches' broom, 241, 386
Withertip, 280
Wolf, F. A., 231, 265, 481, 495
Wolf, F. T., 85
Woronin, 14, 420
Woronina, 31, 35, **97**
  polycystis, 97
Woroninaceae, **96**, 97
  key to genera, 119

X

Xanthophyceae, 629, 632
Xenodochus, 407
Xenolachne, 454
Xylaria, 268, **284**
  hypoxylon, 16, **284**
  polymorpha, 284
  subterranea, 283
Xylariaceae, 277, 280, 282, 283, **284**, 287

Y

Yarwood, 316
Yeast-like growth, 151
Yeasts, 6, 200, 334, 335
  asporogenous, **336**, 342, 346, 585
  bottom, 341
  fission, 335
  key to related families, 360
  red, 348
  sporogenous, key to genera, 358
  top, 341

Yeasts—(*Continued*)
  true, 341
Yen, 413, 414
Young, E. L., 30, 31
Young, Miss, 323, 324, 327
Yu, 421
Yuasa, 24

Z

Zaghouania, 384
Zahlbruckner, 223
Zea mays, 15, 66, 140, 395, 418, 594
Zeller, 225, 535, 540, 550, 553
Zeller and Dodge, 539, 542
Zeller and Walker, 537, 538
Zender, 336
Ziegenspeck, 202
Ziegler, 109
Zimmermann, 594
Zodiomyces vorticellarius, **213**, 214
Zoopagaceae, **178**, 179, 180, 585
Zoopagales, 150, **177**, 634, 635
  key to families and genera, 186
Zoopage, 179
  phanera, 180
Zoosporangium, 5
  filamentous, 129
  formation by proliferation, 105
  proliferated, 111, 112, 128, 142
  resting, 119
  smooth-walled, 119
Zoosporangium—(*Continued*)
  spiny, 118, 119
  sympodial formation, 105
  toruloid, 128, 130
Zoospore, 5, 193
  biflagellate, 78
  dimorphic, 116
  encysted, 113
  primary, 100, 102, 104, 105, 109, 110, 112, 115, 116
  secondary, 95, 100, 102, 105, 110, 112, 115, 116, 117, 126, 132
Zopf, 49, 99, 101, 155, 218
Zostera marina, 30
Zycha, 163
Zygnemataceae, 49
Zygochytrium, 67, **69**
  aurantiacum, 69
Zygomyceteae, 143, 380
Zygorhynchus, 163, **165**
  macrosporus, 157
Zygosporangium, 179
Zygospore, 6, 150, 152, 156, 163, 176, 179
  anisogamous formation, 157
Zygosporium, 481, **594**
  echinosporum, 594
  oschioides, 594
Zygote, 6, 7, 24
  biflagellate, 82, 83, 85
Zythia, 580
Zythiaceae, 580
  key to genera, 607